Student Friendly Quantum Field Theory

Volume 2

The Standard Model

Robert D. Klauber

Maharishi International University

Student Friendly Quantum Field Theory, Volume 2, The Standard Model

Published by

Sandtrove Press
Fairfield, Iowa
sandtrovepress@gmail.com

Cover by Aalto Design
Beach photo by Jan Dineen
Galaxy photo courtesy of NASA

November 2021 (2nd revision with corrections and pedagogic improvements July 2025)

Library of Congress Control Number: 2021949566

ISBN: Hard cover 978-0-9845139-7-0
Soft cover 978-0-9845139-8-7

Printed in the United States of America

To

Mom, Dad, and Susan,

without whose unwavering support and devotion,
I would have accomplished nothing of measure.

"Study hard what interests you the most in the most undisciplined, irreverent and original manner possible."

Richard Feynman

Table of Contents

Table of Wholeness Charts viii
Preface x
Prerequisites xiii
Acknowledgements xiii

Book Overview

1. **Bird's Eye View: The Standard Model..... 1**
   - 1.0 Purpose of this Chapter................................. 1
   - 1.1 This Book's Approach and Content 1
   - 1.2 Structure of this Book.................................. 3
   - 1.3 Units, Notation, and Field Dimensions.......... 4
   - 1.4 Chapter Summary: Our Goal in this Book 5
   - 1.5 Suggestions?.. 5
   - 1.6 Problem 5

Part One: Mathematical Preliminaries

2. **Group Theory... 8**
   - 2.0 Introduction 8
   - 2.1 Overview of Group Theory 8
   - 2.2 Lie Groups... 11
   - 2.3 Lie Algebras 23
   - 2.4 "Rotations" in Complex Space and Associated Symmetries 38
   - 2.5 Singlets and Multiplets 43
   - 2.6 Matrix Operators on States vs Fields........... 45
   - 2.7 Cartan Subalgebras and Eigenvectors 47
   - 2.8 Group Theory Odds and Ends 50
   - 2.9 Chapter Summary..................................... 58b
   - 2.10 Appendix: Proof of Particular Determinant Relationship 61
   - 2.11 Problems.. 61
3. **Relevant Mathematical Topics 64**
   - 3.0 Preliminaries... 64
   - 3.1 Green Functions.. 65
   - 3.2 Grassmann Fields 87
   - 3.3 The Generating Functional.......................... 92
   - 3.4 Odds and Ends.. 100
   - 3.5 Chapter Summary..................................... 101
   - 3.6 Appendices 103
   - 3.7 Problems.. 107

Part Two: QED Revisited: Path Integrals

4. **The Many Paths Approach to QED...... 110**
   - 4.0 Preliminaries.. 110
   - 4.1 Overview of the Path Integral Approach ... 111
   - 4.2 A Key Relation for the PI Approach.......... 116
   - 4.3 Green Functions in the PI Approach.......... 118
   - 4.4 Generating Functional in the PI Approach 120
   - 4.5 Chapter Summary: What Have We Proven and Why? 125
   - 4.6 Appendix: Evaluating the Fermionic Path Integral 127
   - 4.7 Problems 130

Part Three: Electroweak Interactions

5. **Electroweak Toolkit................................ 132**
   - 5.0 Preliminaries... 132
   - 5.1 Spinor Representations 133
   - 5.2 Chirality 148
   - 5.3 Spinors and Boosts.................................... 153
   - 5.4 Massive Spin 1 (Vector) Fields................. 154
   - 5.5 Chapter Summary 157
   - 5.6 Appendices.. 158
   - 5.7 Problems 160
6. **The High Energy Symmetric Universe. 162**
   - 6.0 Preliminary Remarks 162
   - 6.1 Background for Electroweak Symmetry and Its Breaking 162
   - 6.2 The Scalar Field Potential at High Energy. 168
   - 6.3 The High Energy Lagrangian..................... 171
   - 6.4 Summary of Postulates for High Energy Electroweak Theory 182
   - 6.5 A Return to Charges: Weak Hypercharge, Weak Isospin, Electric........................... 182
   - 6.6 Typical High Energy Interactions............. 184
   - 6.7 Quarks and Electroweak Theory................ 188
   - 6.8 Other Things to Note 190
   - 6.9 Chapter Summary 194
   - 6.10 Appendices.. 196
   - 6.11 Problems 205
   - 6.12 Potentials in QFT (Added July 2025)...... 206

7. **The Low Energy Universe: Broken Symmetry** **207**
- 7.0 Preliminaries 207
- 7.1 Goldstone Model 208
- 7.2 Higgs Model 216
- 7.3 Glashow/Salam/Weinberg Model 220
- 7.4 Other Things to Note 241
- 7.5 Chapter Summary 249
- 7.6 Appendix. Transforming the Lagrangian to the True Vacuum 253
- 7.7 Problems 264

8. **Experiment: Scattering and Decay** **265**
- 8.0 Preliminaries 265
- 8.1 Feynman Rules for Electroweak Interactions 266
- 8.2 Particle Decay 266
- 8.3 Electroweak Interaction Scattering 278
- 8.4 Bhabha Scattering 287
- 8.5 Things to be Aware of 288
- 8.6 Chapter Summary 288
- 8.7 Appendices 293
- 8.8 Problems 297
- 8.9 Addendum to Appendix A 297

9. **Electroweak Symmetries: Continuous & Discrete** **298**
- 9.0 Chapter Overview 298
- 9.1 Noether's Theorem and Electroweak Theory 298
- 9.2 Lorentz Transformation Revisited 311
- 9.3 Charge, Parity, and Time Transformations 315
- 9.4 Other Conserved Quantities 339
- 9.5 Chapter Summary 340
- 9.6 Appendices 344
- 9.7 Problems 345

10. **Neutrinos** **346**
- 10.0 Preliminaries 346
- 10.1 Neutrino Oscillations 347
- 10.2 Neutrinos as Possible Majorana Particles 352
- 10.3 The Seesaw Mechanism 353
- 10.4 Neutrino Odds and Ends 359
- 10.5 Appendices 359
- 10.6 Problems 361

11. **Additional Topics in Electroweak Theory** **362**
- 11.0 Chapter Overview 362
- 11.1 Specific Topics 362
- 11.2 General Topics Pervading Overall Theory 368
- 11.3 Summary of Topics in this Chapter 378
- 11.4 Appendix. Resolving the Spin 1 Quantization Issue 380
- 11.5 Problems 382

Part Four: Strong Interactions

12. **Quantum Chromodynamics: The Basics** **384**
- 12.0 Preliminaries 384
- 12.1 The "Free" Quark and "Free" Gluon Lagrangian 385
- 12.2 The QCD Interaction Lagrangian 387
- 12.3 Typical QCD Interactions Simplified 391
- 12.4 Chapter Summary 397
- 12.5 Problems 397

13. **QCD Symmetries: Continuous and Discrete** **399**
- 13.0 Chapter Overview 399
- 13.1 Symmetry of Any $SU(n)$ Quantum Field Theory 399
- 13.2 Noether's Theorem and QCD 404
- 13.3 Noether's Theorem and Quark Flavor Symmetry 407
- 13.4 The Strong CP Problem 410
- 13.5 Chapter Summary 414
- 13.6 Problems 416

14. **Hadron Composition** **418**
- 14.0 Chapter Overview 418
- 14.1 More Group Theory 418
- 14.2 Example from NRQM: Spin of Composites 423
- 14.3 Generalizing Direct Products and Direct Sums 429
- 14.4 Hadrons: Combining Flavors 430
- 14.5 Things to be Aware of 438
- 14.6 Chapter Summary 439
- 14.7 Problems 440

15. **QCD Interaction Theory........................ 441**

15.0 Preliminaries... 441
15.1 The Faddeev-Popov Method 443
15.2 QCD Perturbation Theory 452
15.3 Feynman Rules for Strong Interactions... 458
15.4 Chapter Summary................................... 461
15.5 Problems.. 462

16. **QCD Renormalization and Coupling Constant.. 463**

16.0 Preliminaries... 463
16.1 Primitive Divergences 464
16.2 Renormalization in Different Theories.... 473
16.3 Other Interaction Terms.......................... 473
16.4 The Renormalization Procedure 476
16.5 $\overline{\text{MS}}$ Renormalization of QED 478
16.6 $\overline{\text{MS}}$ Renormalization of QCD................. 485
16.7 Chapter Summary................................... 490
16.8 Appendix: Deriving (16-74).................... 492
16.9 Problems.. 492

17. **QCD Experiments.................................. 493**

17.0 Preliminaries... 493
17.1 QCD Coupling vs Energy........................ 494
17.2 High Energy Scattering off Nucleons...... 494
17.3 Precise Makeup of the Proton................. 499
17.4 Hadron Jets from Electron-Positron Collisions 500
17.5 Chapter Summary................................... 502
17.6 Problem 504

Part Five: Summary of the Standard Model and Beyond

18. **Looking Backward and Looking Forward: QFT Summary and What's Next 506**

18.0 Preliminaries... 506
18.1 Big Picture Overview of QFT 507
18.2 What's Next.. 513

Index.. 515

Table of Wholeness Charts

Book Overview

1-1. The Standard Model and Gravity 1
1-2. The Fundamental Interactions 2
1-3. Both Quantization Methods Yield Same Transition Amplitude 2
1-4. The Major Parts of this Book 3
1-5. Book Topic vs Volume, Part, and Chapter... 3
1-6. Dimensions of Quantum Fields for Natural Units... 4

Part One: Mathematical Preliminaries

2-1. Synopsis of Groups, Fields, Vector Spaces, and Algebras 9
2-2. Types of Operations Involving Groups and Vector Spaces 21
2-3. Overview of Types of Groups.................... 21
2-4. Lie Group Representation Parametrizations and Characterizations............................. 34
2-5. Exponentiation of Lie Algebra Generators 37
2.6. Symmetries in Various Spaces 42
2-7a. *SU*(2) Cartan Subalgebra Generator Eigenvalues for 1st Generation Fermions 48
2-7b. *SU*(3) Cartan Subalgebra Generator Eigenvalues for Fermions 49
2-8. Lie Groups and Their Lie Algebra Generators 51
2-9. Various Terms in Group, Matrix, and Tensor Theory.. 53
2-10. Imposition of Unitarity and Special-ness on a General Matrix.................................... 57
2-11. Schematic Overview of Chapter 2............ 59
2-12. Table Overview of Chapter 2 60
2-13. Quantum Field Theory Vector Spaces ... 58b
3-1. Finding Transition Amplitude from the Generating Functional............................ 65
3-2. Comparing Connected, Disconnected, and Vacuum Bubble Green Function Diagrams 81
3-3. Summary of QED Green Functions for Canonical Approach 86
3-4. Summary of Properties: Grassmann Numbers and Fields 91
3-5. A Different Form for *Z* for Comparing Later On with Path Integral *Z* 96
3-6. Deducing the Whole Theory from the Free Generating Functionals........................ 101
3-7. Summary of QED Green Functions and Generating Functional.......................... 102
3-8. Fermion Sign Switching in Transition Amplitudes and Green Functions.......... 103
3-9. Transition Amplitudes and Green Functions in Three Pictures 105

Part Two: QED Revisited: Path Integrals

4-1. Some Differences between the CA and PIA 114
4-2. Going from Green Functions to Transition Amplitudes for CA and PIA................. 119
4-3. Showing Our Guess for Path Integral Green Function is Correct............................... 120
4-4. Comparison of Two Approaches to QED Green Functions and Generating Functional 126

Part Three: Electroweak Interactions

5-1. Overview of Weyl Representation........... 141
5-2. Advantages of Different Reps 148
5-3. Summary of Chiral Field Effects on States 151
5-4. Development of Chiral Operators and Fields 152
6-1. Differences Between Goldstone, Higgs, and Glashow/Salam/Weinberg Models........ 169
6-2. Lagrangians We Will Examine................. 171
6-3. Finite *SU*(2) and *U*(1) Local Transformations in Electroweak Theory 179
6-4. Infinitesimal *SU*(2) and *U*(1) Local Transformations in Electroweak Theory180
6-5. Infinitesimal *SU*(2)X*U*(1) Local Transformations in Electroweak Theory181
6-6. Charges: Electric, Weak Isospin, and Hypercharge 183
6-7. Non-Abelian Theories Have Non-Linear Field Equations..................................... 192
6-8. Overview of the Development of High Energy GSW Theory............................ 195
6-9. Mass Terms in High Energy Electroweak $\mathcal{L}$ and Symmetry 196
6-10. Various Charges for Elementary Particles 196
6-11. Potential Fiels in QED 206

7-1. Goldstone Model 215
7-2. Difference Between Goldstone and Higgs Models 216
7-3. Higgs Model 219
7-4. Difference Between Goldstone, Higgs, and Glashow/Salam/Weinberg Models 221
7-5. Comparison of Two Possible Lepton Basis Choices 232
7-6. Comparison of Two Possible Quark Basis Choices 237
7-7. SM Particle Masses Numerically 239
7-8. Glashow/Salam/Weinberg Model 240
7-9. Various Charges for Elementary Particles 243
7-10. Electroweak Symmetry Breaking: Overview of Three Models 250
7-11. Mass vs Flavor Eigenstate Bases 252a
7-12. The Standard Model in a Nutshell 252b
8-1. Vector Fields Polarization Vectors and Propagators 295
9-1. Electroweak Four-Currents and Charges . 341
9-2. Lorentz Transformation Properties 342
9-3. C, P, and T Transformations 342
9-4. Comparison of Helicity and Chirality 343
10-1. Properties of Fields in Alternative Notation 353
10-2. Weak Charge and Lepton Number Conservation 355
11-1. Fierz Identities Coefficients 366

Part Four: Strong Interactions

13-1. General Principles for $SU(n)$ Symmetry 414
13-2. QCD Four-Currents and Charges 415
13-3. Quark Flavor Symmetries 416
14-1. Our Use So Far of Operators and Eigenvalues for Fields and States 423
14-2. Summary of Matrix Representations of $SU(2) \otimes SU(2)$ 429
14-3. Some $SU(n)$ Direct Product, Direct Sum Relations 430
14-4. Quantum Numbers of Pions 432
14-5. Quantum Numbers for *u, d,* and *s* Quarks 434
15-1. Steps of QCD Perturbation Theory 453
16-1. Examples of Proper and Improper QED Feynman Diagrams 465
16-2. Primitively Divergent vs Non-primitively Divergent QED Feynman Diagrams 465
16-3. Superficial Divergences for the Fermion In/Fermion Out Proper Diagrams 467
16-4. All Possible Primitively Divergent Graphs for QED 469
16-5. All Possible Primitively Divergent Graph Types for QCD 472
16-6. Comparison of On-Shell and MS Renormalization Schemes 477

Part Five: Summary of the Standard Model and Beyond

18-1. Milestones in the Canonical Development of Quantum Theories 507
18-2. Milestones in the Path Integral Development of Quantum Theories 508
18-3. The Interaction Threads of the SM Tapestry 508
18-4. Counting the Particles in the SM 509
18-5. Counting the Parameters in the SM 509

Preface

"Only the student knows."
Edwin Taylor[1]

With the resounding popularity of what is now Volume 1 of *Student Friendly Quantum Field Theory*, subtitled *Basic Principles and QED*, came a flurry of fervent requests from students to write a second volume on the weak and strong interactions. This book is the response to those requests.

As with Vol. 1, I do things in this book that no other texts I am aware of do. For but one example among many from Vol. 1, Chap. 3 therein shows the explicit, far from trivial, derivation from the quantization postulates of the coefficient commutation relations, which form the very foundation of the entire theory of quantum fields. Yet, such a derivation does not seem to be presented elsewhere, and students are typically informed that "it can be shown", asked to do the derivation themselves in their spare time, or told it is simply a postulate of the theory.

In this Vol. 2, there are more such things – things that seem to have gone missing from many presentations of the theory. Chap. 5 herein, for example, presents the explicit transformations between the three most common gamma matrix representations, which otherwise seem to be very hard to find in the literature. Chap. 6 delineates, step-by-step, how every term in the electroweak Lagrangian transforms to make that Lagrangian symmetric under $SU(2)\text{X}U(1)$. In Chap. 7, every term in the low energy Lagrangian is derived from the terms in the high energy Lagrangian. Chap. 9 deduces the four electroweak four-currents and their corresponding charge operators, then explains why two of those operators are taken as zero. Chap. 9 also takes over twenty pages to explain and develop charge, parity, and time transformations in depth, and provides a one page wholeness chart summary (pg. 342) of all relevant terms and their transformations under C,P, and T. And there is more, much more, you will not find in other texts.

In the Preface of Vol. 1, I listed some of the pedagogic principles I employ in that and this book. In this preface, I add a few more below, which should have been included in that list. After that, I discuss the effort, frustration, and time consumption involved in extirpating errata. And finally, I close with a bit of personal philosophical musings on the teaching of science, engineering, and mathematics.

Pedagogic Principles Employed

As a brief review, in Vol. 1, I enumerated the following of my teaching philosophy basic tenets: avoidance of brevity, holistic previews, wholeness chart overviews, reviews of background material, sticking to basic concepts without peripheral tangents, minimal time/maximal learning problems, small-step by small-step derivations, liberal use of concrete examples, margin overview notes, key equations highlighted in boxes, fundamental definitions underlined, and shunning of debilitating terms like "trivial", "obvious", etc. In that volume and this, I also use several more, as delineated below.

Bottom line summaries

After detailed treatment of a subtopic within a chapter, students, even those who followed every step, all too often wonder "Now, what exactly did we just do? What is the meaning and final result of all this?". I answer this question for them with clearly marked "Bottom Line" (underlined and/or bold face) paragraphs at the end of many presentations within chapters.

Under and over brackets in equations

In going through line after line of equations, where former equations are plugged in and many manipulations occur, I use brackets above and/or below parts of those equations indicating what those parts are equivalent to. See, for example, Chap. 5, equations (5-33) and (5-35). I have found this can save students a lot of time and eliminate confusion.

Numbering terms in equations with many terms

Sometimes, evaluations of relations entail a large number of terms in a given line, followed by re-arranging and combining of such terms in subsequent lines, where keeping track of which terms go where can be challenging. To help in keeping things straight in such cases, I put a number or letter, enclosed in a box above or below each term, and keep that boxed number/letter with the term in subsequent lines, as the derivation proceeds. See Chap. 6, (6-157) to (6-160), as one such example. This simple procedure helps one keep tabs on which terms go where, and can save considerable time, as well as frustration.

Simple, sometimes imprecise, preliminary expositions

On occasion, in introducing new, mentally challenging concepts, I present things in a manner that may not be exactly correct, but is easier to understand. This allows the new learner to get a feeling for what is going on, without becoming bewildered by a fully precise, and thus far more complex, rendition of the topic. By seeing it for the first time in this

[1] E. F. Taylor, Guest comment: Only the student knows, *Am. J. Phys.* **60**(3), March 1992, 201-202.

manner, the student gains a sense of where we are headed, what the ultimate objective is, its importance, and how it fits into the overall structure of knowledge we are in the process of exploring. We then generally avoid the common "what in blazes is going on?" student syndrome.

After we cover enough ground for students to become more comfortable with the theory and concepts involved, I either then point out the imprecisions in our earlier introduction, or I assign a problem asking them to determine those imprecisions themselves. The final result is close alignment of student understanding with the precise essence of the theory.

One example of this is the introduction to transition amplitudes in Chap. 1 of Vol. 1. The presentation is simple, but not (as noted clearly there) completely correct. Had it been, virtually no students would have understood it. In Chap. 8, 252 pages later, when students have covered the necessary ground to truly understand transition amplitudes, they get a problem asking them to detail the imperfections of the original introduction.

I realize many lecturers will take issue with this approach and consider it unsuitable, even though it is little different from using analogies, which are inherently imperfect and thus, incorrect, but whose use, nonetheless, is widely recognized as a valuable teaching aid.

In practice, I have found this method greatly helps students reach the goal line faster, and with deeper comprehension. Feedback from students has been ample in this regard.

Errata

Errata are the bane of every author, each of whom learns, with her/his first book, how insidious and ubiquitous they are, and how excruciatingly difficult and time consuming they are to eliminate.

This book, for instance, has on the order of two million characters in it. If I were 99.9% accurate, that would still leave two thousand or so errors.

Editors, reviewers, and readers in substantial number can proof read a book and detect many errata, but later readers always seem to come up with more. One seasoned editor told me that she was continually amazed by how many typos slip through the cracks of repeated edits, by skilled editors.

The late and truly great theorist Joe Polchinski noted, in his memoirs, that when planning his now famous string theory text, he vowed to have zero errata. He ended up with over 400. For Vol. 1 of *Student Friendly Quantum Field Theory*, I also had hundreds, and estimate I spent several hundred hours correcting them (plus several hundred more revising portions to make them clearer and easier to understand).

So, if you the reader feel frustrated with the number of corrections posted on the website for this book or those in other books, please think kindly of us authors. Our frustration much exceeds yours. We and our teams of editors/reviewers are doing our best. Our very, very best, in spite of how it might seem to the uninitiated.

Problems with "The System" of Present-Day Science and Technology Education

Problems I see inherent in physics/math/engineering education, as presently structured, are these.

1. Professors and teaching assistants spend 50% of their time teaching, yet zero percent of their training was on how to do so. Very few have taken even an afternoon seminar on teaching itself, let alone an actual course in it. Doesn't this seem strange? Doesn't it seem conducive to poor quality teaching?

2. Communication skills of technically trained people tend, on average, to be inferior to those of their liberal arts leaning brethren, and teaching, if it involves anything, involves communication.

3. Those who become professors, those who teach and write the textbooks, were typically the very brightest students, who had far less trouble with the extant texts and teaching skill level of their instructors. So, they are less likely to understand problems the more typical student encounters, and more likely to simply parrot the modes of teaching they themselves encountered as students.

4. After years of working in a subject area, one becomes so familiar with it, and it seems so obviously true, that it can be difficult to understand what exactly it is that new learners could have a difficult time with.

5. Material is almost invariably presented at such a fast pace that few, if any, students completely understand all of it. And since later subject matter is structured upon the foundation of prior subject matter, this leads to holes in that foundation. I call this the "Swiss cheese" model of education. I think almost all of us have had the experience, hours, days, months, or even years later, of thinking "Oh, now I see what that meant." Surely, it would be better if that hole had been plugged the first time around. And just as surely, an educational system that left fewer such holes would be better. Should we educators not be asking ourselves if there is any way we can move in the direction of such a system?

Suggestions for Improving "The System"

The solution I propose is simple.

1) Require university instructors to take courses in teaching, particularly of technical subjects,

2) have them solicit, and listen open mindedly to, student feedback on how to best present material, and

3) have them study and emulate the teaching methods of others whose methods students have responded to enthusiastically.

Granted, this would require a fair degree of instructor humility. But, I submit, the rewards, such as higher grades, student expressions of gratitude, and fewer conceptual holes in each student's knowledge base, would provide a sense of fulfillment for teachers that would be hard to beat.

Pedagogic Philosophy

My high school football coach was fond of saying "defense is 90% desire". So, I suggest, is good teaching. Desiring to help students means putting in the time to learn what we, as educators, can do to help them learn better and more efficiently. A major part of that is soliciting feedback from them. Have them tell us how we should present the material.

In doing this myself, I've learned quite a number of things. The two most important, I believe, are these.

First, show them, at the start, an overview of what we are doing, where we are headed, and how it fits into the overall framework of their studies. And return to this theme after they have covered a given section of material. Don't let them flounder in the "what is going on?" wasteland. Begin with a bird's eye roadmap view, then present the details, then finish with the bird's eye view again, in somewhat greater depth. Tell them what they are going to cover, then cover it, then tell them again what they just covered.

Second, put every little step in derivations and analyses. I, personally, have, cumulatively, wasted untold hours trying to see how an author/teacher got from one line to the next. Many times, in formally working things out myself, I found the missing steps were numerous, at times extending even to two or three pages. This, I submit, is not how the learning process should go.

Those who study education as a life's work have long known that learning comes fastest in the smallest bites. Yet, this lesson seems to have made it into few STEM educational circles. I sincerely hope it starts making greater inroads soon.

So, in general, as I noted in the preface of Vol. 1, I'd rather have readers spending their study hours learning new material than squandering it trying to figure out where they got the wrong sign or incorrect factor of two in some routine, but time devouring, algebraic marathon. That is why, in these volumes, I do pedagogically fruitless things like this for the reader, step-by-step. I do this because it saves time, and I consider efficiency in learning the name of the game.

Yet, I've been criticized by field theory experts for being too simple. But, as I said in Vol. 1, the proof is in the pudding. In 150 or so messages from students sent to me (see the book website) or posted on Amazon, all praised the clarity of Vol. 1, and not one felt it was beneath them. Can we not let the new students judge pedagogy, and let the critics form judgements based on what they have to say. "The student knows", as Edwin Taylor has advised us.

He also said "Why not ask, then learn ourselves – from the only person who can teach us." The students can make our texts and classroom teaching better, if we will only listen. Since I began writing Vol. 1, I have actively solicited help from students on how best to present material to them. And I have listened. My books are virtually co-written by numerous student contributors. Vol. 1 has been revised four times in eight years, all such revisions (plus the original) incorporating extensive input from new learners.

And is it not a general practice among successful companies to solicit advice from their customers on how to improve their product or service? And does not that solicitation lead to more success for the company and more satisfaction for the customer? Why should we not do the same?

It is very, very hard for seasoned practitioners in a field to present material in a way that is too simple and too easy for students. It is very, very easy to make it too difficult. As educators, we should strive to err on the side of simplicity. Such an error costs very little in time or energy, as such material can be covered quickly. Errors in the other direction, on the other hand, can cost new learners a great deal of time and energy, with the resulting lack of progress generating much frustration, exasperation, and self-doubt.

There are many things I have learned from students. Most are summarized in the preface to Vol. 1 and above herein. We teach them the physics, but they can teach us how to teach that physics. Let's listen to them.

And Finally, My Thanks to Many of You

I, and those who learn from the books later, are indebted to the many, many readers who, by sending me feedback, have improved the books' quality and clarity immeasurably. On the behalf of the learners yet to come, and myself, I thank you all, profoundly.

Robert D. Klauber
October 2021

Things to Note for the Second Revision of July 2025

In addition to typo corrections, multiple other changes have been made to improve pedagogy in this second revision of this book. These encompass the re-wording of several sections and the addition of a few new pages spread throughout the text, all of which should improve the learning experience for many.

The effect on page and equation numbers is the same as noted in the Preface to the Second Edition in Vol. 1 of *Student Friendly Quantum Field Theory*. For example, two pages have been inserted after page 58 and are numbered 58a and 58b. After equation (4-31), three equations have been added and numbered (4-31)+1, (4-31)+2 and (4-31)+3. Hence, all page and equation numbers in prior versions match the same material in this revision, and confusion in communication between users of different versions should be minimized.

Prerequisite

Before beginning this book, one should have a reasonably good background in path integrals for non-relativistic quantum mechanics plus quantum electrodynamics via the canonical quantization approach. *Student Friendly Quantum Field Theory Volume 1*, by the present author, covers that material, as do a number of other texts. As *SFQFT Vol. 1* ends where this book begins, and as it is the foundation upon which this book is constructed, it should come as no surprise to anyone that it is my first recommendation. However, provided one has the requisite background noted above, it is not necessary to have read *Vol. 1* before beginning this *Vol. 2*. Be forewarned, however, that, as background for developing particular concepts, relevant sections of *Vol. 1* are commonly referenced herein.

Acknowledgements

"Service to others is the rent you pay for your room here on Earth."

Muhammad Ali

Though my name is on the title page, the writing of this book was a group effort. I would write the first draft of each chapter and submit it to my team of brilliant, hard-working, devoted reviewers. Each of them would, in turn, read every word, understand every concept, digest every equation, correct innumerable errata of every stripe, re-write certain parts, and offer a bounty of suggestions for improving pedagogy.

Some of these folks were already well grounded in quantum field theory. Others were learning it for the first time. The former were best at detecting conceptual errors; the latter, at finding typos and making the material clearer, more transparent, and easier for newcomers to assimilate.

These people gave immeasurably to me, the author, and to you, the reader, with no thought of material compensation. They were motivated by a simple desire to help others. So, how can I express my thanks to them? I cannot, adequately, but will try, nonetheless.

Thank you, profusely, my dear friends and colleagues, Luc Longtin, Ezequiel Lozano, Jimmy Snyder, Holger Teutsch, and Mike Worsell. This book, if it had even come into existence without you, would be so much less. My feelings of gratitude far exceed my ability to express them.

Others helped, in lesser, yet still very meaningful ways. They include Nadeem Afana, Alexandra Barnett, Lou Biegelesien, Chinmaya Bhargava, Albert Clement, Ali Dehghani, Jim Dimech, Mikhail Denisov, Jack Freinhar, Blaise Gassend, Vasudev Godbole, Park Hensley, David Lear, Doug McKenzie, Morgan Orcutt, Papia Panda, Glen Rowe, David Scharf and his QFT class, Anthony Soong, Robin Ticciati, Jon Wesick, and Mengliang Yao. Deep thanks to each of you, as well.

Yet, just as I said for Vol. 1, any errors or insufficiencies that still remain herein are my responsibility, and mine alone.

Special Acknowledgement

After the first version of this volume was printed, Luc Longtin read the entire book a second time and came up with over 200 additional corrections/improvements, which are included in the revision you now hold in your hands. The majority of these changes are minor typos, but a substantial number comprise marvelous suggestions that made presentations of material more accurate, clearer, and easier to understand.

Not only I, but all readers of this text, owe Luc an enormous debt of gratitude for his efforts on our behalf. For each of us, I say this.

Thank you profoundly, Luc. Thank you, over and over. What a truly amazing job!

The website for this book is

www.quantumfieldtheory.info

It contains presentations of advanced topics, topics in other fields of physics, and a list of corrections and improvements to this version of the text. Please use the site to report any errors you might find and to suggest ways to make material presented in this book easier for students to understand.

Problem solutions booklet

Solutions to the problems in this book are provided in the booklet *Solutions to Problems for Student Friendly Quantum Field Theory Volume 2*. The website above contains a link to where the booklet can be obtained.

Chapter 1

Bird's Eye View: The Standard Model

In teaching, start with wholeness and end with wholeness. In between, do the parts.

1.0 Purpose of this Chapter

In *Student Friendly Quantum Field Theory: Basic Principles and Quantum Electrodynamics* (hereinafter "Vol. 1"), Chapter 1, we started with a grand, but much oversimplified, schematic vision of the entire scope of quantum field theory (QFT) and the subset within it contained in that book. Here we do the same thing.

1.1 This Book's Approach and Content

This book is structured in two ways, which are, figuratively speaking, orthogonal. These are via

1) interaction type (quantum electrodynamic, weak, or strong), and
2) quantization type (canonical vs path integral).

Two-way structure of our path:
1) interaction type
2) quantization type

By "orthogonal", we mean that our development of theory at any time, like a node point in a 2D coordinate grid, can have any "component" (quantum electrodynamic, weak, or strong) of 1) along with any independent "component" (canonical approach vs path integral approach) of 2).

For example, we develop quantum electrodynamics (QED) herein via the path integral approach, whereas, in Vol. 1, we did QED via the canonical quantization approach. We will also fully develop weak interaction theory via canonical quantization, and after that, briefly look at how it can be developed using path integrals. Strong interaction theory is most commonly developed using path integrals, and that is what we will do here, though as with electroweak theory, we will discuss how it can alternatively be done with canonical quantization.

In other words, any one of 1) above can be done via either approach of 2). We will focus on the easiest-to-learn approach for a given interaction type, at least as an introduction to the concomitant theory (which is what we did in Vol. 1 for QED.)

1.1.1 Interaction Types: The Standard Model

If you are reading this book, you are well aware of the three interactions (electrodynamic, weak, and strong) found in nature that are encompassed by contemporary QFT. The field of theoretical physics encompassing these three interactions is known as the standard model (SM). You are also no doubt aware that a fourth interaction, gravity, is not, as of this writing, a part of the standard model. There is no adequate theory of quantum gravity, though that is not due to lack of trying. Most researchers in the field believe that a viable theory of quantum gravity, if one exists, lies beyond the structure of the SM, as we know it.

Wholeness Chart 1-1. The Standard Model and Gravity

Standard Model			Odd Man Out
↙	↓	↘	↓
QED	Weak	Strong	Gravity

Nature's 4 forces, 3 of which are handled quantum mechanically by the SM

As we will see, weak interaction theory is actually closely meshed with electromagnetic interaction theory, and one commonly finds use of the term "electroweak" interactions. Weak interactions are responsible for particle decays, and thus, radioactivity. Weak interactions affect both leptons (electrons, muons, taus, three types of neutrinos, and antiparticles for all of them) and quarks. Strong interactions hold quarks together inside hadrons such as protons or neutrons, and mesons such as the pi meson. As a residual effect, strong forces associated with baryons hold nuclei together. Wholeness Chart 1-2 summarizes the various interactions, their associated charge types, virtual bosons that mediate them, and the type of fundamental fermions affected by each. (For a summary of bosons vs fermions, see Vol. 1, pg. 65, Wholeness Chart 3-1.)

Charges, bosons, and fermions associated with the 4 interactions

Wholeness Chart 1-2. The Fundamental Interactions

Interaction	Charge Type	Mediating Boson(s)	Fundamental Fermions Affected
Electromagnetic	electric	photon	all quarks and charged leptons (all fermions except neutrinos)
Weak	weak	3 intermediate vector bosons (IVBs) W^+, W^-, Z	all quarks and leptons
Strong	color	8 gluons	all quarks
Gravity	mass-energy	graviton (conjectured)	all

Note that QFT can, in fact, handle the effect of weak (as in feeble, not as in weak force) gravitational fields on fundamental particles (quarks, leptons, photons, other bosons) by modeling the usual quantum fields as behaving in a classical spacetime background. That is, gravity is treated in the classical general relativistic way such that it bends the spacetime within which the quantum fields interact[1]. This is not quantum gravity as gravity in such a theory is classical, not quantized. In quantum gravity, the gravitational force should be mediated by a boson, such as the conjectured graviton, rather than effected by curved classical spacetime.

No quantum theory of gravity, but can model 3 SM forces in weak gravity limit

1.1.2 Canonical vs Path Integral Approaches

In Vol. 1, we developed QFT primarily via the canonical quantization approach, which as mentioned therein, I and many others consider the easier and more intuitive way to be introduced to the subject. In Chap. 18 of that volume, we investigated the path integral (also known as functional quantization, many paths, or sum over histories) approach to non-relativistic quantum mechanics (NRQM) and relativistic quantum mechanics (RQM), and from there, extrapolated to the first few steps of its use in QFT.

Two approaches to QFT: canonical and path integral

In Part Two of this volume, we will start path integration (PI) from the point where we left off in Vol. 1, and use it to re-derive QED. Later in the book, we will note its application in electroweak theory and then use it in vigor to develop strong interaction theory.

The two approaches differ primarily in the manner in which one arrives at transition amplitudes. From there, calculations of cross sections (see Vol. 1, Chap. 17) and other experimental results (Vol. 1, Chap. 16) remain the same. PI does, however, make deducing the partition function of statistical mechanics easier (which we will not study herein). All this is illustrated in Wholeness Chart 1-3.

Wholeness Chart 1-3. Both Quantization Methods Yield Same Transition Amplitude

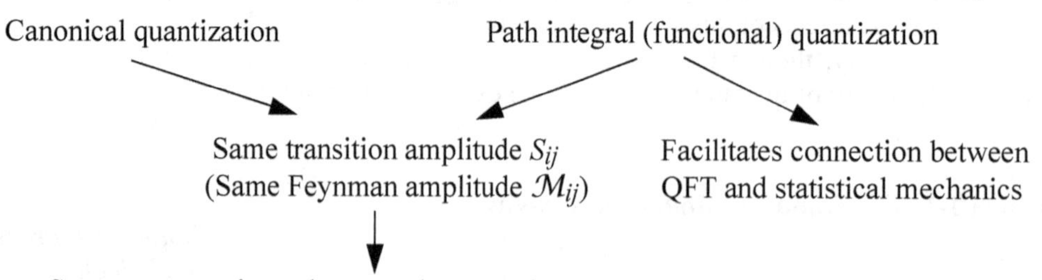

Both approaches yield the same transition amplitude

PI also very helpful connecting QFT and thermodynamics

[1] When you are ready for it (probably not now), see Mukhanov, V. F., and Winitzki, S., *Introduction to Quantum Effects in Gravity* (Cambridge, 2007) as well as the relevant notes on the website for the present text (URL on opposite page from pg. 1 herein.)

1.2 Structure of this Book

1.2.1 Major Parts

After this first chapter, this book has five major parts, as shown in Wholeness Chart 1-4.

Wholeness Chart 1-4. The Major Parts of this Book

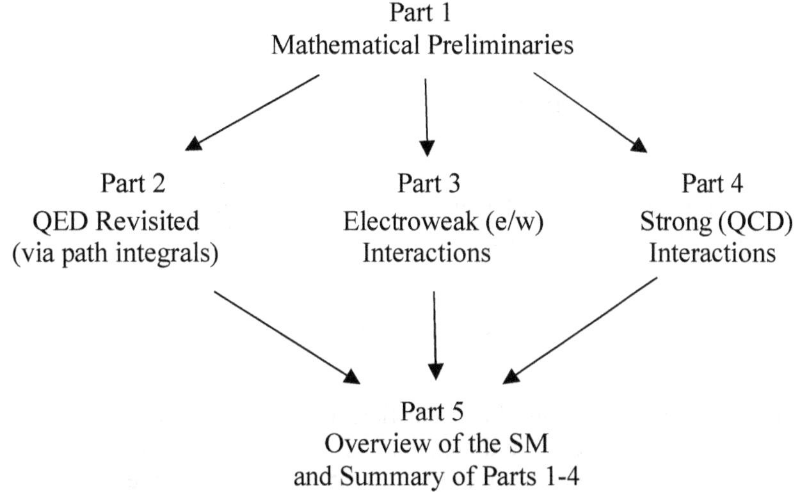

The five major parts of this book

Part 1 includes introductions to group theory, the mathematical structure underlying the SM, and other areas of mathematics that play key roles in advanced QFT.

Part 2 rederives QED (which we originally derived in Vol. 1 using canonical quantization) via the path integral quantization formulation for fields.

Part 3 develops electroweak theory (sometimes abbreviated as e/w theory) starting from basic principles and covers symmetry breaking, the Higgs particle, weak interactions including decay, neutrino behavior, and more.

Part 4 derives strong interaction theory, more formally, quantum chromodynamics (QCD), and investigates the manner in which quarks form composite particles (hadrons).

Part 5 puts all the pieces together into the wholeness of the SM and presents a final summary of the prior four parts. It is an overview of the whole book.

1.2.2 Chapter and Part Content

Wholeness Chart 1-5 outlines the part and chapter content of Vol. 1 (shaded) and this Vol. 2 (not shaded).

Wholeness Chart 1-5. Book Topic vs Volume, Part, and Chapter

Where particular content is located in Vol. 1 and this book (Vol.2)

Topic	**Canonical Quantization**	**Functional Quantization**
Basic principles	**Vol. 1** Prep, Parts 1 & 2 Chaps. 1-7	**Vol. 1** Addenda Chap. 18
QED	**Vol. 1** Parts 2, 3 & 4 Chaps. 8-17	**Vol. 2** Part 2 Chap. 4
Electroweak	**Vol. 2** Parts 1 & 3 Chaps. 3, 5-11	**Vol. 2** Part 3 Chap. 11*
QCD	**Vol. 2** Parts 1 & 4 Chap. 15*	**Vol. 2** Part 4 Chap. 15
Whole SM	**Vol. 2** Part 5 Chap. 18	

* We will only briefly discuss this alternative approach. The other approach is more common in the literature.
Vol. 2 chapter numbers not shown cover material involved in both quantization methods.

1.3 Units, Notation, and Field Dimensions

1.3.1 Units

As we did in Vol. 1, and as is common practice, we will use natural units where $c = 1$ and $\hbar = 1$, both dimensionless. For a complete explanation and description of natural units, see Chap. 2 of Vol. 1, Sect. 2.1, pgs. 11-15.

Natural units in this book

1.3.2 Notation

We will follow the same notation as used in Vol. 1, which is also common (though not universal) practice. See that volume, Sect. 2.2, pgs. 15-16, plus Appendix A therein, pgs. 32-36.

Notation and dimensions as in Vol. 1

1.3.3 Dimensions

Dimensions for various quantities in physics in general, and QFT in particular, in the natural units system are equal to the value of the symbol M (M=[mass]=[energy]=[length]$^{-1}$=[time]$^{-1}$) in Vol. 1, Chap. 2, Wholeness Chart 2-1, pg. 14.

Natural unit dimensions for different field types

1.3.4 Dimensions of Quantum Fields

Spin 0 Field Dimension

In Chap. 2 of Vol. 1, problem 3, we deduced the dimension of a scalar field ϕ as equal to one. I should have had similar problems in Chaps. 4 and 5 to deduce the dimensions of spinor fields ψ and vector fields such as the photon field A^{μ}. These are things we should know for the future, so we examine them here.

Spin ½ Field Dimension

The Lagrangian L (not density) has units of energy, which in natural units has dimension $M = 1$. (See Vol. 1, pg. 14.) Length has dimension $M = -1$, so volume has dimension –3. Energy density (energy per unit volume) thus has dimension 4, and so, the Lagrangian density $\mathcal{L}$ has a dimension M = 4.

The free spinor Lagrangian (density) has form (where each term has dimension 4)

$$\mathcal{L}_0^{1/2} = i\bar{\psi}\partial\!\!\!/\psi - m\bar{\psi}\psi \tag{1-1}$$

Picking the second of these terms on the RHS, we see that m has dimension 1 and both of $\bar{\psi}$ and ψ have the same dimension. For the term to have total dimension 4, ψ must have dimension 3/2.

Spin 1 Field Dimension

The free photon Lagrangian (density) (Vol. 1, pg. 294, equation (11-36)) is

$$\mathcal{L}_0^1 = -\tfrac{1}{2}\left(\partial_\beta A_\nu \partial^\beta A^\nu - \partial_\nu A_\beta \partial^\beta A^\nu\right). \tag{1-2}$$

Each term in (1-2) has two derivatives, each such derivative having length in the denominator (ct is spatial), with dimension $M = -1$, i.e., the equivalent of $M = 1$ in the numerator. Each photon factor in the term must then have the dimension 1 in order for the entire term to be dimension 4.

Do **Problem** 1 for a little practice in determining field dimensions.

Wholeness Chart 1-6 summarizes the results of this section.

Wholeness Chart 1-6. Dimensions of Quantum Fields for Natural Units

Summary of dimensions for different field types

Field Type	Examples	Symbols	Dimension M
Scalar	Higgs	ϕ, H	1
Spinor	leptons, quarks	ψ	3/2
Vector	photon, IVBs, gluons	A^{μ}, W^{μ}, Z^{μ}, $A^{\mu}{}_a$ (gluons)	1

1.4 Chapter Summary: Our Goal in this Book

With our foundational basis in basic principles and QED from Vol. 1 (or other sources), we seek, in this book, to extend our understanding of QFT to the entire range of the SM, i.e., to weak and strong interactions and to the path integral approach. The way in which this endeavor will be structured is shown in Wholeness Charts 1-3 (pg. 2), 1-4 (pg. 3), and 1-5 (pg. 3).

1.5 Suggestions?

If you have suggestions to make the material anywhere in this book easier to learn, or if you find any errors, please let me know via the web site address for this book posted herein opposite pg.1. Thank you.

1.6 Problem

1. Using the first term after the equal sign in (1-1), show, as we did with the second term, that the dimension of ψ is 3/2. Then, show that a massive vector boson field (spin 1, by definition), symbolized by W^{μ} and having a term in the Lagrangian of form $m_W^2 W_\mu^\dagger W^\mu$, where m_W is its mass, has dimension $M = 1$. After that, note that since the fine structure constant has dimension zero, as we showed in Prob. 2, Chap. 2 of Vol. 1, the constant e, the absolute value of the charge on the electron, has dimension zero. With that knowledge, show that the interaction term in the QED Lagrangian $e\bar{\psi}\gamma^\nu\psi A_\nu$ (Vol. 1, pg. 294, equation (11-36)) has dimension 4, and thus is consistent dimension-wise with the free Lagrangian terms.

This page intentionally left blank.

Part One
Mathematical Preliminaries

"Mathematics is the language in which God has written the universe."
Galileo Galilei

Chapter 2 ***Group Theory***

Chapter 3 ***Relevant Mathematical Topics***

Chapter 2

Group Theory

"I believe that ideas such as absolute certitude, absolute exactness, final truth, etc. are figments of the imagination which should not be admissible in any field of science... This loosening of thinking seems to me to be the greatest blessing which modern science has given to us."
Max Born

2.0 Introduction

In Part One of this book, we cover topics that could be ignored for the canonical approach to QED (and were, in fact, ignored in Vol. 1), but play major roles elsewhere in QFT. These are

- group theory (this chapter), and
- other relevant math topics (Chapter 3).

The former is used throughout the theories of electroweak and strong interactions. The latter play essential roles in the path integral approach to all standard model (SM) interactions and include Green functions, Grassmann variables, and the generating functional, all of which are probably only names for most readers at this point.

Hopefully, for many, at least part of the group theory presentation will be a review of course work already taken. Also, hopefully, it will be sufficient for understanding, as treated in latter parts of this book, the structural underpinning that group theory provides for the SM of QFT.

As always, we will attempt to simplify, in the extreme, the presentations of these topics, without sacrificing accuracy. And we will only present the essential parts of group theory needed for QFT. For additional applications of the theory in physics (such as angular momentum addition in QM), presented in a pedagogic manner, I suggest Jeevanjee[1], McKenzie[2] and Schwichtenberg[3].

2.1 Overview of Group Theory

Areas of study: groups, fields, vector spaces, and algebras

Group theory is, in one sense, the simplest of the theories about mathematical structures known as groups, fields, vector spaces, and algebras, but in another sense includes all of these, as the latter three can be considered groups endowed with additional operations and axioms. Wholeness Chart 2-1 provides the basic defining characteristics of each of these types of structures and provides a few simple examples. The "In Common" column lists characteristics all structures share. "Particular" lists specific ones for that particular structure. Hopefully, there is not too much new in the chart for most readers, but even if there is, it, though compact, should be a relatively intelligible introduction.

Note that for algebras, the first operation has all the characteristics of a vector space. The second operation, on the other hand, does not necessarily have to be associative, have an identity element or inverse elements in the set, or be commutative.

Algebras may or may not be associative, unital, or commutative

An algebra with (without) the associative property for the second operation is called an associative (non-associative) algebra. An algebra with (without) an identity element for the second operation is called a unital (non-unital) algebra or sometimes a unitary (non-unitary) algebra. We will avoid the second term as it uses the same word (unitary) we reserve for probability conserving operations. A unital algebra is considered to possess an inverse element under the 2nd operation (for the first operation, it already has one, by definition), for every element in the set.

[1] Jeevanjee, N., *An Introduction to Tensors and Group Theory for Physicists*, 2nd ed., (Birkhäuser/Springer 2015).
[2] McKenzie, D., An Elementary Introduction to Lie Algebras for Physicists, https://www.liealgebrasintro.com/
[3] Schwichtenberg, J., *Physics from Symmetry*, 2nd ed., (Springer 2018).

Wholeness Chart 2-1. Synopsis of Groups, Fields, Vector Spaces, and Algebras

Type of Structure	Elements	Main Characteristics	Examples (A, B, C, D = elements in set)	Other Characteristics In Common	Other Characteristics Particular
Group	Set of elements	1 (binary) operation "binary" = between two elements in the set	#1: Real numbers under addition. **A** + **B** = **C**, e.g., 2+3 = 5 #2: 2D rotations (can be matrices) under multiplication. **AB** = **C**	Closure; associative; identity; inverse;	May or may not be commutative
Field	As above	2 (binary) operations Commonly, addition and multiplication	#1: Real numbers under addition & multiplication. 2•3 + 4 = 10 #2: Complex numbers under addition & multiplication. $(1+2i)(1-2i) + (2+4i) = 7+4i$	1st operation: As in block above 2nd operation: All but inverse as in block above	Both operations commutative: 2nd distrib. over 1st 2nd operation: Inverse not required
Vector Space	Set of (vector) elements & 2nd set of scalars	1 (binary) operation & 1 scalar multiplication "scalar" = element of a field	#1: 3D vectors under vec addition & scalar multip. 3**A** + 2**B** = **C** #2: Elements are matrices with matrix addition & scalar multip. 3**A** + 2**B** = **C** #3: Hilbert space in QM	As in top block above	Commutative; distributive for scalar multip with vector operation; may have inner product, i.e., **A**•**B** = scalar
Algebra	As above	2 (binary) operations & 1 scalar multiplication	#1: 3D vectors under vec addition, vec cross product, scalar multip 3**A** X **B** + 2**D** = **C** #2: Matrices under matrix addition, matrix multip, scalar multip. 3**AB** + 2**D** = **C** #3: Matrices under matrix addition, matrix commut, scalar multip. 3[**A**, **B**+**D**] = **C**	1st operation (often addition): As in top block above 2nd operation: Closure	2nd operation distrib over 1st 1st operation: Commutative 2nd operation: Not required to be associative, have identity, have inverses, be commutative

Definitions (**A**, **B**, **C** represent any & all elements in set. ∘ denotes a binary operation)		**Examples**
Closure	All operations on set elements yield an element in the set. **A**∘**B** = **C**	All **C** in Examples column above are in original set of elements.
Associative	**A**∘(**B**∘**C**) = (**A**∘**B**)∘**C**	Real numbers, rotations, matrices, vectors, all under addition or multiplication.
Identity	There is an element **I** of the set with the property **A**∘**I** = **I**∘**A** = **A**	Real number addition, **I** = 0. Matrix multiplication, **I** = identity matrix.
Inverse	For each **A** in the set, there is a unique element $\mathbf{A}^{-1}$ of the set with the property $\mathbf{A}\circ\mathbf{A}^{-1} = \mathbf{A}^{-1}\circ\mathbf{A} = \mathbf{I}$	Real number addition, $\mathbf{A}^{-1} = -\mathbf{A}$. Matrix multiplication, $\mathbf{A}^{-1}$ = matrix inverse of **A**
Commutative	**A**∘**B** = **B**∘**A**	Real number addition and multiplication. Vector addition. Non-commutative examples: 3D rotation, creation & destruction operators in QFT under multip.
Distributive	For two binary operations (✟ = another binary operation) **A**∘(**B** ✟ **C**) = (**A**∘**B**) ✟ (**A**∘**C**) and (**B** ✟ **C**)∘**A** = (**B**∘**A**) ✟ (**C**∘**A**)	Real numbers: ✟ as addition, ∘ as multip. Matrices: ✟ as addition, ∘ as multip.

Doing **Problems 1 and 2** may help in understanding groups, fields, vector spaces, and algebras.

In this chapter, we will focus on a particular kind of group, called a Lie group, which will be defined shortly. We will then show how a Lie group can be generated from the elements of an associated algebra, called, appropriately, a Lie algebra. After that, we see how Lie groups and Lie algebras play a key role in the SM of QFT. For an overview chart of where we are going, check out Wholeness Chart 2-11 in the chapter summary on page 59. Don't worry too much about some of what is now undefined terminology in that chart. You will understand that terminology soon enough.

SM underlying structure → Lie groups and Lie algebras

2.1.1 A Set of Transformations as a Group

Consider the set of rotation transformations in 3D (in particular, it will be easier to think in terms of active transformations [see Vol. 1, pg. 164]), a typical element of which is symbolized by **A** herein. Such transformations can act as operations on a 3D vector, i.e., they rotate the 3D vector, which we designate by the symbol **v**. In the transformation, the vector **v** is rotated to a new position, designated **v′**. The transformations **A** comprise an abstract expression of rotation in physical space, i.e., **A** signifies rotation independent of any particular coordinate system. Any element **A** does the same thing to a given vector regardless of what coordinate system we choose to view the rotation operation from.

Rotation transformations = a set of elements

Now, if we select a given coordinate system, we can represent elements **A**, in one manner, as matrices, whose components depend on the coordinate system chosen. For practical applications, and for aid in learning, we almost always have to express rotations as matrices.

One way to represent rotations is via matrices

$$\underbrace{\mathbf{A}\mathbf{v}=\mathbf{v}'}_{\text{Abstract form}} \xrightarrow[\text{and column vectors}]{\text{expressed as matrix}} \begin{bmatrix} a_{11} & a_{12} & a_{13} \\ a_{21} & a_{22} & a_{23} \\ a_{31} & a_{32} & a_{33} \end{bmatrix}\begin{bmatrix} v_1 \\ v_2 \\ v_3 \end{bmatrix}=\begin{bmatrix} v_1' \\ v_2' \\ v_3' \end{bmatrix}. \qquad (2\text{-}1)$$

Note that **A** has the characteristics delineated in Wholeness Chart 2.1 for a group. In particular, if $\mathbf{A}_1$ and $\mathbf{A}_2$ are two members of this set of transformations, then rotating the vector first via $\mathbf{A}_1$ and then via $\mathbf{A}_2$ is symbolized by

$$\mathbf{A}_2 \circ \mathbf{A}_1\mathbf{v}=\mathbf{v}''=\mathbf{C}\mathbf{v} \qquad \text{where } \mathbf{C}=\mathbf{A}_2 \circ \mathbf{A}_1 \text{ is a member of the set of 3D rotations}. \qquad (2\text{-}2)$$

Therefore, the rotation transformation set has closure and the binary operation, when the transformation is expressed in matrix form, is matrix multiplication. Further, the operation on set members (we are talking operation between matrices here [in the matrix representation], not the operation of matrices on vectors) is associative, and inverses ($\mathbf{A}^{-1}$ for each **A**) and an identity (**I**) exist. Further, there is no other operation, such as addition, involved for the members of the set. (In the matrix representation, two successive transformations involve matrix multiplication, not matrix addition.) Hence, the transformations **A** form a group.

Set of rotation transformations satisfy criteria to be a group

Note, this rotation example is a non-commutative group, since

$$\text{in general, } \mathbf{A}_2 \circ \mathbf{A}_1 \neq \mathbf{A}_1 \circ \mathbf{A}_2. \qquad (2\text{-}3)$$

You can prove this to yourself by rotating a book 90° ccw (counterclockwise) along its binder axis first, then 90° ccw along its lower edge second; and then starting from the same original book position and reversing the order of the rotation operations. The book ends up in different final positions.

A non-commutative group is denoted a non-Abelian group. Note that some pairs of elements in a non-Abelian group can still commute, just *not any and all* pairs. A group in which all elements commute is an Abelian group.

Set of 3D rotation transformations is non-Abelian (non-commutative)

2.1.2 Groups in QFT

As insight into where we are going with this, recall from Vol. 1 (see pg. 196, first row of eq. (7-49)) that the S operator in QFT transforms an initial (eigen) state $|i\rangle$ into a final (general) state $|F\rangle$ (that's what happens during an interaction).

$$S\,|\,i\rangle=|\,F\rangle. \qquad (2\text{-}4)$$

But that state could be further transformed (via another transformation) into another state $|\,F'\rangle$. So, for two such transformations S_1 and S_2, we would have

In a similar way, the set of S operator transformations on QFT states forms a group

$$S_2 S_1 | i \rangle = S_3 | i \rangle = | F' \rangle . \tag{2-5}$$

Recall also from Vol. 1 (pg. 195, eq. (7-43)) that the S operator could be represented by a matrix (S matrix) and the initial and final states by column vectors. The parallels between (2-1) and (2-4), and between (2-2) and (2-5), should allow us to surmise directly that the set of all transformations (interactions) in QFT form a group. And so, the mathematics of groups should help us (and it does help us as we will eventually see) in doing QFT.

2.1.3 Quick Summary

1st Bottom line: A set of transformations on column vectors (or on QM states) can form a group. We can apply group theory to them (with or without considering the column vectors [or QM states]).

2nd Bottom line: The column vectors (or QM states) can form a vector space.

Do **Problem 3** to show this.

So, the group elements act as operators on the vectors (or QM states). Discern between *operations* (which are transformations) by group members *on vector space members* from the *group operation between group members* (matrix multiplication in our sample representation.)

Group operation between group elements differs from what some groups have as operation of a group element on a vector

2.1.4 Notation

We will generally use bold capital letters, such as **A**, for abstract group elements (which could characterize some operation in physical space, such as rotation); and non-bold capital letters, such as A, for matrix representations of abstract group elements. We will generally use bold lower-case letters, such as **v**, for abstract vector elements in a vector space; and non-bold lower-case letters, such as v, for column matrix (or row matrix) representations of those vectors. The binary operation on abstract elements $\mathbf{A} \circ \mathbf{B}$, for matrix multiplication in the matrix representation, will be expressed simply as AB.

***Bold** = abstract elements of group or vector space*
Non-bold = matrix representation of elements

2.1.5 Our Focus: Matrix Representations of Groups

Group theory as the study of abstract groups is extensive, deep, and far from trivial. When restricted to representations of groups as matrices, it becomes easier. Since practically, for our purposes and in much of the rest of physics, matrix group representation theory covers the bases one wants to cover, we will focus on that. So, when we use terms such as "group", "group theory', "group element", etc. from here on out, they will generally mean "matrix group", "matrix group theory", "matrix group element", etc. unless otherwise stated or obvious.

Our focus: groups with elements represented by matrices

2.2 Lie Groups

A Lie group is a group whose elements are continuous, smooth functions of one or more variables (parameters) which vary continuously and smoothly. (See examples below.)

This definition is a bit heuristic, but will suffice for our work with matrix groups. There are fancier, more mathematically precise definitions, particularly for abstract groups.

Lie groups are named after the Norwegian mathematician Sophus Lie, who was among the first to develop them in the late 1800s.

Lie group elements vary continuously and smoothly with a continuous, smooth parameter

2.2.1 Representations of Lie Groups

A representation of an abstract Lie group, for our purposes, is a matrix that acts on a vector space and depends on one or more continuous, smoothly varying independent parameters, and that itself varies continuously and smoothly. (More formally, it is an action of a Lie group on a vector space.) The vector space is called representation space. The term representation is also used to mean *both* the vector space and the matrix operators (matrix group elements) together.

A representation of a Lie group, for us, is a set of matrices

2.2.2 A One Parameter Lie Group

A simple example of a Lie group is rotation in 2D, which can be represented by a matrix $\hat{M}$ (with a "hat" because we will use the symbol M for something else later) that operates on a 2D vector, i.e., it rotates the vector counter clockwise,

$$\hat{M}(\theta)=\begin{bmatrix} \cos\theta & -\sin\theta \\ \sin\theta & \cos\theta \end{bmatrix}. \qquad (2\text{-}6)$$

A simple example

This group is characterized by a single parameter, the angle of rotation θ. As (2-6) is just a special case of the 3D rotations of (2-1), it forms a group. And because all of its elements can be generated continuously and smoothly by a smooth, continuous variation of a parameter, it is a Lie group.

Note that for $\theta = 0$, $\hat{M}(0) = I$, and this is a property all of our Lie groups will share, i.e., when all parameters equal zero, the matrix equals the identity. Some authors include this as part of the definition of Lie groups, as virtually all groups that prove useful in physical theory have this property.

When $\theta = 0$, $\hat{M}(0) = I$ is key for representing real world

An example of a non-Lie group would be the set of 2D rotations through increments of 90°. The set of matrix elements would be (take $\theta = 0°, 90°, 180°, 270°$ in (2-6))

$$\hat{M}_1=\begin{bmatrix} 1 & 0 \\ 0 & 1 \end{bmatrix} \quad \hat{M}_2=\begin{bmatrix} 0 & -1 \\ 1 & 0 \end{bmatrix} \quad \hat{M}_3=\begin{bmatrix} -1 & 0 \\ 0 & -1 \end{bmatrix} \quad \hat{M}_4=\begin{bmatrix} 0 & 1 \\ -1 & 0 \end{bmatrix} \qquad (2\text{-}7)$$

Example of a non-Lie group: discrete elements

Obviously $\hat{M}_1$ is the identity. $\hat{M}_4$ is the inverse of $\hat{M}_2$, and $\hat{M}_1$ and $\hat{M}_3$ are each their own inverses. Additionally, (you can check if you like) any two of (2-7) multiplied together yield one of the other elements (closure). Further, matrix multiplication is always associative. So, (2-7) form a group under matrix multiplication. But since the elements are discrete and do not convert one into the other via a continuous variable parameter, it is not a Lie group.

A property of the particular Lie group (2-6) (for two values of θ such as α and ϕ) is

$$\hat{M}(\alpha)\hat{M}(\phi)=\hat{M}(\alpha+\phi). \qquad (2\text{-}8)$$

Property of this particular Lie group

This should be evident from our general knowledge of 2D rotations. Rotating first through 30°, then second through 13° degrees is the same as rotating through 43° degrees. The parameter θ varies continuously and smoothly and so does $\hat{M}(\theta)$.

As an aside, the 2D rotation group is commutative (Abelian), as rotating first by 13°, then second by 30° is the same as rotating first by 30° and second by 13°.

2.2.3 Orthogonal vs Special Orthogonal Groups

Both groups (2-6) and (2-7) are what are termed <u>special orthogonal</u> groups. "<u>Orthogonal</u>" means the elements of the group (represented by the matrices) are real and the transpose of the matrix is the same as its inverse, i.e., $\hat{M}^T = \hat{M}^{-1}$. Recall from linear algebra that the magnitude of a vector remains unchanged under an orthogonal transformation (as in rotation). "<u>Special</u>" means the determinant of each matrix in the group is unity. det $\hat{M} = 1$.

Orthogonal, $O(n)$, $\hat{M}$ a real matrix $\hat{M}^T = \hat{M}^{-1}$ (vector magnitude invariant under $\hat{M}$)

Do **Problem 4** to investigate a 2D orthogonal Lie group matrix that is *not* special orthogonal and to help understand the significance of special orthogonal transformations.

Special Orthogonal, $SO(n)$, Det $\hat{M} = 1$

In the above problem solution, an orthogonal matrix acting on a vector maintains the norm (magnitude) of the vector unchanged. This, as noted above, is a general rule. The solution also suggests that non-special orthogonal Lie matrices do not produce continuous rotation of a vector.

Note further, that for the matrix of Problem 4, when $\theta = 0$, we do not have the identity matrix, so its action on a vector would change the direction of the vector. It is more advantageous in representing real world phenomena if there is no change in a vector when the continuous parameter(s) on which it depends is (are) zero, as in (2-6).

Hopefully, from these examples, we can begin to see some of the advantages of working with special orthogonal matrix Lie groups which equal the identity when the independent parameter(s) is (are) zero.

The shorthand notation for our special orthogonal example groups (2-6) and (2-7) is $SO(2)$. If the rotations were in 3D instead of 2D, as in (2-1), we denote it an $SO(3)$ group. For n dimensional space rotations, we would have <u>$SO(n)$</u>. If the determinant were not constrained to be equal to positive unity, the group would be simply an <u>orthogonal group</u>, symbolized by <u>$O(n)$</u>.

The number <u>n</u> in $SO(n)$ and $O(n)$ is known as the <u>degree of the group</u>. In our examples above, it is equal to the dimension of the vector space upon which our matrices act. However, there are subtleties

n in $O(n)$ and $SO(n)$ is the degree of the group

involved regarding the actions of groups on particular dimension spaces, which we will address later in the chapter.

Note that orthogonal and special orthogonal groups do not have to be Lie groups. For example, (2-7) is special orthogonal, but not a Lie group. We define these more general terms in this Section 2.2, which is specifically on Lie groups, because it is easiest to understand them in the context of the examples presented herein.

O(n) and SO(n) groups can be Lie or non-Lie groups

2.2.4 Different Parametrizations of the Same Group

Note that we can represent the *SO*(2) rotation group in a different way as

$$\hat{M}(x) = \begin{bmatrix} \sqrt{1-x^2} & -x \\ x & \sqrt{1-x^2} \end{bmatrix} \tag{2-9}$$

Same group can have different parametrizations

where $x = \sin\theta$. We say that (2-9) is another parametrization of the same *SO*(2) group of (2-6). A third parametrization is the transpose of (2-6),

$$\hat{M}(\theta') = \begin{bmatrix} \cos\theta' & \sin\theta' \\ -\sin\theta' & \cos\theta' \end{bmatrix} \tag{2-10}$$

where $\theta' = -\theta$.

Bottom line: A particular group is an abstract structure (which can, as in the above example, characterize 2D rotations; in the example of (2-1), 3D rotations) that can be represented explicitly via matrices and by using different parameters in different ways, called parametrizations.

2.2.5 A Lie Group with More than One Parameter: SO(3)

Of course, there are many Lie groups with more than one parameter. As one example, let us express the *SO*(3) group of 3D rotations (2-1) as a function of certain angles (successive ccw rotations about different axes) θ_1, θ_2, and θ_3. Typically, for a solid object with three orthogonal axes visualized as attached to the object, the first rotation is about the x_3 axis; the second, about the x_2 axis; the third, about the x_1 axis. In this perspective, it is an active transformation about each axis.

Consider **A** in (2-1) as an abstract group element (characterized simply in that it performs rotations), and *A* as the particular parametrized matrix representation under consideration.

Example of Lie group with more than one parameter → 3D rotation

$$A(\theta_1,\theta_2,\theta_3)v = v' \quad \rightarrow \quad \begin{bmatrix} a_{11} & a_{12} & a_{13} \\ a_{21} & a_{22} & a_{23} \\ a_{31} & a_{32} & a_{33} \end{bmatrix}\begin{bmatrix} v_1 \\ v_2 \\ v_3 \end{bmatrix} = \begin{bmatrix} v_1' \\ v_2' \\ v_3' \end{bmatrix} \qquad a_{ij} = a_{ij}(\theta_1,\theta_2,\theta_3) \tag{2-11}$$

Any *A* can be expressed via different parametrization choices, and with an eye to the future we will choose to build *A* from the following particular building block parametrizations.

$$A_1(\theta_1) = \begin{bmatrix} 1 & 0 & 0 \\ 0 & \cos\theta_1 & -\sin\theta_1 \\ 0 & \sin\theta_1 & \cos\theta_1 \end{bmatrix} \quad A_2(\theta_2) = \begin{bmatrix} \cos\theta_2 & 0 & \sin\theta_2 \\ 0 & 1 & 0 \\ -\sin\theta_2 & 0 & \cos\theta_2 \end{bmatrix} \quad A_3(\theta_3) = \begin{bmatrix} \cos\theta_3 & -\sin\theta_3 & 0 \\ \sin\theta_3 & \cos\theta_3 & 0 \\ 0 & 0 & 1 \end{bmatrix}. \tag{2-12}$$

There are a number of ways the matrices (2-12) can be combined to embody different (equivalent) forms of *SO*(3), but to be consistent with our above noted order of rotations, we will use (2-13). Note the operations proceed from the right side to the left side in (2-13). (I suggest you save yourself the time by just accepting the algebra involved, though you can prove it to yourself, if you wish.)

$$\begin{aligned} A(\theta_1,\theta_2,\theta_3) &= A_1(\theta_1)A_2(\theta_2)A_3(\theta_3) \\ &= \begin{bmatrix} \cos\theta_2\cos\theta_3 & -\cos\theta_2\sin\theta_3 & \sin\theta_2 \\ \cos\theta_1\sin\theta_3 + \sin\theta_1\sin\theta_2\cos\theta_3 & \cos\theta_1\cos\theta_3 - \sin\theta_1\sin\theta_2\sin\theta_3 & -\sin\theta_1\cos\theta_2 \\ \sin\theta_1\sin\theta_3 - \cos\theta_1\sin\theta_2\cos\theta_3 & \sin\theta_1\cos\theta_3 + \cos\theta_1\sin\theta_2\sin\theta_3 & \cos\theta_1\cos\theta_2 \end{bmatrix}. \end{aligned} \tag{2-13}$$

One of many ways to embody that group using building blocks

The matrix *A* varies continuously and smoothly with continuous, smooth variation in the three parameters θ_1, θ_2, and θ_3. We could, of course, define the order of operation on the RHS of (2-13) differently and have a different embodiment of the same *SO*(3) rotation group. Similarly, we could define our building blocks with different parameters (similar to the x in (2-9) for $SO(2)$ rotations) and have yet other, different parametrizations. Again, we see that a group itself is an abstract structure (which can characterize 3D rotations here) that can be expressed mathematically in different ways.

Key point:

Note the A_i of (2-12) are not bases in the vector space sense of spanning the space of all possible 3X3 matrices. 'Basis' refers to addition and scalar multiplication but the matrices we are dealing with involve multiplication. However, all possible 3D rotations, expressed as matrices, can be obtained from the three A_i, so they are the foundation of the group.

Don't confuse building blocks of groups with bases of vector spaces. They are generally different.

The 3D rotation group matrices form a subset of all 3X3 real matrices, and the reader should be able to verify this by doing the problem referenced below. Further, any group element can be formed, in this representation, by matrix multiplication of three group building blocks. But in a typical vector space, any element can be expressed as a linear combination (adding, not multiplying) of basis vectors (which are matrices here).

Similarly, in 2D rotations, we only had one matrix, such as (2-6), which is a function of one parameter (θ in the referenced parametrization). For a basis for matrices in 2D, we would need four independent matrices. Don't confuse the building blocks of a Lie group with the basis vectors of a linear vector space.

Do **Problem 5** to help illustrate the difference between matrices as vector space elements and matrices as group elements.

Small Values of the Parameters

For future reference, we note that for small (essentially infinitesimal) parameters θ_i, (2-13) becomes

$$A(\theta_1,\theta_2,\theta_3) \approx \begin{bmatrix} 1 & -\theta_3 & \theta_2 \\ \theta_3 & 1 & -\theta_1 \\ -\theta_2 & \theta_1 & 1 \end{bmatrix} \qquad |\theta_i| \ll 1 . \qquad (2\text{-}14)$$

2.2.6 Lorentz Transformations Form a Lie Group

The Lorentz transformation, with $c = 1$ and boost velocity v in the direction of the x^1 coordinate, is (where we use the Einstein summation convention, as in Vol. 1)

$$\Lambda^{\alpha}{}_{\beta}(v)dx^{\beta} = \begin{bmatrix} \frac{1}{\sqrt{1-v^2}} & \frac{-v}{\sqrt{1-v^2}} & & \\ \frac{-v}{\sqrt{1-v^2}} & \frac{1}{\sqrt{1-v^2}} & & \\ & & 1 & \\ & & & 1 \end{bmatrix} \begin{bmatrix} dx^0 \\ dx^1 \\ dx^2 \\ dx^3 \end{bmatrix} = \begin{bmatrix} dx'^0 \\ dx'^1 \\ dx'^2 \\ dx'^3 \end{bmatrix} = dx'^{\alpha} . \qquad (2\text{-}15)$$

The set of transformations Λ^{α}_{β} satisfies our criteria for a group. The elements Λ^{α}_{β} are subject to a single operation (matrix multiplication) under which the set has an identity member (when $v = 0$), obeys closure, is associative, and possesses an inverse for every member. The Lorentz transformations constitute a group.

Lorentz transformations form a group

Due to the Minkowski metric in special relativity's 4D spacetime, things get a little tricky comparing the Lorentz transformation to matrices in Euclidean space. So, even though the magnitude (vector "length" as in (2-47), pg. 32, of Vol. 1, sometimes called the "Minkowski norm") of the four-vector $|dx^{\beta}|$ remains invariant under the transformation (see Section 2.2.3 above), the inverse of $\Lambda^{\alpha}_{\beta}(v)$ is not its transpose, but $\Lambda^{\alpha}_{\beta}(-v)$. (You can check this or save time by just taking my word for it, as this material is a bit peripheral.) In a special relativity sense, therefore, the Lorentz transformation is considered orthogonal. As the determinant of Λ^{α}_{β} is unity (you can check using (2-15) and the rules for calculating 4D matrix determinants), it is special. To discern the special relativistic nature of this particular kind of orthogonality, the Lorentz group is denoted by $SO(3,1)$ [for 3 dimensions of space, and one of time.]

Note that the 4D transformation matrix is a continuous function of v, the velocity between frames, whose possible values vary continuously. So, the Lorentz group is a Lie group. The addition property (2-8) holds, but for relativistic velocity addition, i.e.,

$$\Lambda^{\alpha}{}_{\beta}(v)\Lambda^{\beta}{}_{\delta}(v') = \Lambda^{\alpha}{}_{\delta}(v'') \quad \left(v'' = \text{relativistic velocity addition of } v \text{ and } v'\right). \qquad \text{(2-16)}$$

Extending the form of (2-15) to include the more general cases of 3D coordinate axes rotation plus boosts in any direction leads to the same conclusions. Lorentz transformations, i.e., boosts plus all possible 3D rotations, comprise a Lie group $SO(3,1)$, the so-called Lorentz group.

In particular, a special orthogonal Lie group, SO (3,1)

2.2.7 Complex Groups: Unitary vs Special Unitary

So far, we have looked exclusively at groups represented by real matrices. But since QM is replete with complex numbers, we need to expand our treatment to include representations of groups using complex matrices. See Vol.1, Box 2-3, pg. 27, to review some differences and similarities between real, orthogonal transformations and complex, unitary transformations.

Unitary, U(n), M complex matrix $M^{\dagger} = M^{-1}$ *(vector magnitude invariant under M)*

Unitary groups (symbol $U(n)$ for degree n) are effectively the complex number incarnation of (real number) orthogonal groups. A matrix group is unitary if for every element M in it, $M^{-1} = M^{\dagger}$. (Compare to orthogonal matrices, which are real, and for which $\hat{M}^{-1} = \hat{M}^{T}$.)

Special unitary groups (symbol $SU(n)$) are the complex number incarnation of special orthogonal groups. A matrix group is special if for every element M in it, Det $M = 1$.

Special Unitary, SU(n) Det M = 1

The Simplest Unitary Lie Group $U(1)$

As a simple case of a representation of a unitary group of degree 1 (n = 1), consider the set of "matrices" (of dimension 1) for continuous, real θ,

$$U(\theta) = e^{i\theta} \quad \left(\text{a representation of } U(1)\right). \qquad \text{(2-17)}$$

Do **Problem 6** to show that $U(1)$ of (2-17) forms a group.

U in (2-17) is unitary because $U^{\dagger}U = I$. It is not special because Det U does not equal 1 for all θ. The only 1X1 matrix for which the determinant equals one corresponds to $\theta = 0$, i.e., the trivial case where the only member of the (special unitary) group is $U = 1$. Hence, we won't have much use for $SU(1)$. We will, however, make good use of $U(1)$.

Also, similar to an orthogonal transformation on a real vector, a unitary transformation on a complex vector leaves the magnitude of the vector unchanged.

Do **Problem 7** to prove the last statement.

If the vector happens to be a normalized QM state, this means that the total probability (to find the quantum system in *some* quantum state) remains unity under the action of the transformation U.

Unitary operator transformations on quantum state vectors leave probability unchanged

Additionally, since the set elements of (2-17) vary continuously and smoothly with the continuous, smooth variation of the parameter θ, (2-17) comprises a Lie group.

Given that QFT teems with complex numbers (of which operators and states are composed) and the theory is inherently unitary (conservation of total probability = 1 under transformations), we can expect to be focusing herein on unitary groups. And, as we will see, they will also be special.

Problem 8 can help in understanding how some physical world phenomena can be described by different kinds of groups.

The $SU(2)$ Lie Group

The $SU(2)$ Lie group is a very important group in QFT, as it is intimately involved in theoretical descriptions of the weak force. We will examine the group as represented by matrices that are 2X2, have complex components, and may operate on two component vectors in a complex vector space.

The SU(2) Lie group, important in physics

A general two-dimensional complex matrix M has form

$$M = \begin{bmatrix} m_{11} & m_{12} \\ m_{21} & m_{22} \end{bmatrix} \qquad m_{ij} \text{ complex, in general}. \tag{2-18}$$

2X2 complex matrices

But since the group we will examine is special unitary, it must satisfy

$$M^{\dagger}M = I \qquad Det\, M = 1. \tag{2-19}$$

satisfying special unitary group requirements

I submit (we will show it) that (2-20) below satisfies (2-19), where a and b are complex.

$$M = \begin{bmatrix} a & b \\ -b^* & a^* \end{bmatrix} = \begin{bmatrix} a_{Re} + ia_{Im} & b_{Re} + ib_{Im} \\ -b_{Re} + ib_{Im} & a_{Re} - ia_{Im} \end{bmatrix} \quad \text{with } aa^* + bb^* = 1 \tag{2-20}$$

One parametrization that does satisfy them

The RHS (last part) of (2-19) is obvious from the constraint imposed at the end of (2-20). To save you the time and tedium, I show the LHS (first part) of (2-19) below.

$$M^{\dagger}M = \begin{bmatrix} a^* & -b \\ b^* & a \end{bmatrix}\begin{bmatrix} a & b \\ -b^* & a^* \end{bmatrix} = \begin{bmatrix} a^*a + b^*b & a^*b - a^*b \\ ab^* - ab^* & b^*b + a^*a \end{bmatrix} = \begin{bmatrix} 1 & 0 \\ 0 & 1 \end{bmatrix} \tag{2-21}$$

Now consider M as a matrix representation of a Lie group, where the a and b are continuous and smoothly varying. Since a and b are complex numbers, there are four real number variables $a_{Re}, a_{Im}, b_{Re}, b_{Im}$, which vary continuously and smoothly. From the constraint at the end of (2-20), only three of these are independent. They are related by

The $Det\, M = 1$ constraint means one parameter dependent on 3 independent ones

$$a_{Re}^2 + a_{Im}^2 + b_{Re}^2 + b_{Im}^2 = 1, \tag{2-22}$$

and we choose a_{Im}, b_{Re}, b_{Im} to be independent, and $a_{Re} = a_{Re}\left(a_{Im}, b_{Re}, b_{Im}\right)$. For future reference, we find the partial derivative of a_{Re} with respect to each of the independent variables via (2-22).

$$\frac{\partial a_{Re}}{\partial a_{Im}} = \frac{\partial}{\partial a_{Im}}\left(1 - a_{Im}^2 - b_{Re}^2 - b_{Im}^2\right)^{\frac{1}{2}} = \tfrac{1}{2}\left(1 - a_{Im}^2 - b_{Re}^2 - b_{Im}^2\right)^{-\frac{1}{2}}\left(-2a_{Im}\right)$$
$$= -\frac{a_{Im}}{\sqrt{1 - a_{Im}^2 - b_{Re}^2 - b_{Im}^2}} = -\frac{a_{Im}}{a_{Re}} \quad \text{and} \quad \frac{\partial a_{Re}}{\partial b_{Re}} = -\frac{b_{Re}}{a_{Re}} \quad \frac{\partial a_{Re}}{\partial b_{Im}} = -\frac{b_{Im}}{a_{Re}} \tag{2-23}$$

Evaluating derivatives of dependent parameter

Note that when $a_{Im} = b_{Re} = b_{Im} = 0$, $a_{Re} = 1$ (where we assume the positive value for the square root in (2-22)), and the partial derivatives in the last line of (2-23) all equal zero.

For the Lie group, when the continuously variable independent parameters are all zero, nothing has changed, so we must have the identity element, and with (2-22), this is true for (2-20), i.e.,

$$M\left(a_{Im} = b_{Re} = b_{Im} = 0\right) = I. \tag{2-24}$$

For all independent parameters $= 0$, $M = I$

Thus, we have shown that M of (2-20) represents the $SU(2)$ Lie group of three real parameters.

Do **Problem 9** to show that M obeys the group closure property.

Using the *Re* and *Im* subscripts helped us keep track of which real parameters went where in the 2X2 matrix, but it will help us in the future if we change symbols, such that $a_{Re} = \alpha_0$, $b_{Im} = \alpha_1$, $b_{Re} = \alpha_2$, and $a_{Im} = \alpha_3$. Then,

$$M = \begin{bmatrix} a_{Re} + ia_{Im} & b_{Re} + ib_{Im} \\ -b_{Re} + ib_{Im} & a_{Re} - ia_{Im} \end{bmatrix} = \begin{bmatrix} \alpha_0 + i\alpha_3 & \alpha_2 + i\alpha_1 \\ -\alpha_2 + i\alpha_1 & \alpha_0 - i\alpha_3 \end{bmatrix} = i\begin{bmatrix} -i\alpha_0 + \alpha_3 & \alpha_1 - i\alpha_2 \\ \alpha_1 + i\alpha_2 & -i\alpha_0 - \alpha_3 \end{bmatrix}. \tag{2-25}$$

The *SU*(3) Lie Group

The $SU(3)$ Lie group is also an important group in QFT, as it is intimately involved in theoretical descriptions of the strong force. We will examine the group as represented by matrices, which are 3X3, have complex components, and may operate on three-component vectors in a vector space. As you may be surmising, in strong interaction theory, the three components of the vectors will represent three quark eigenstates, each with a different color charge (eigenvalue). More on that later in the book.

The $SU(3)$ Lie group relevant to the strong interaction

A general three-dimensional complex matrix N has form

$$N=\begin{bmatrix} n_{11} & n_{12} & n_{13} \\ n_{21} & n_{23} & n_{23} \\ n_{31} & n_{32} & n_{33} \end{bmatrix} \qquad n_{ij} \text{ complex, in general} \tag{2-26}$$

Unitary & special properties a 3X3 matrix must have to represent SU(3)

Since the group we will examine is special unitary, it must satisfy

$$N^{\dagger}N=I \qquad Det\,N=1 \tag{2-27}$$

Using (2-27) with (2-26) would lead us, in similar fashion to what we did in $SU(2)$, to one dependent real variable and eight independent ones upon which the dependent one depends. But doing so requires an enormous amount of complicated, extremely tedious algebra, and there are many possible choices we could make for independent variables. Further, with eight such variables, the final result would be far more complicated than (2-13) for $SO(3)$ or (2-25) for $SU(2)$, both of which have only three (where α_0 depends on the other α_i in (2-25)).

SU(3) Lie group relations parallel those of SO(3)

So, we will instead jump to a result obtained by Murray Gell-Mann in the mid-20th century. That result uses a particular set of independent variables that end up working best for us in QFT. Additionally, it is a much-simplified version of N that applies only in cases where the independent variables are very small, essentially infinitesimal. By so restricting it, a whole lot of cumbersome terms are dropped. (Compare with the parallel situation of (2-13) simplified to (2-14).) Fortunately, the small independent variable case will be all we need for our work with the SM.

The Gell-Mann $SU(3)$ matrix is

Gell-Mann SU(3) matrix useful in QFT, $|\alpha_i| << 1$

$$N(\alpha_i)=i\begin{bmatrix} -i+\alpha_3+\frac{\alpha_8}{\sqrt{3}} & \alpha_1-i\alpha_2 & \alpha_4-i\alpha_5 \\ \alpha_1+i\alpha_2 & -i-\alpha_3+\frac{\alpha_8}{\sqrt{3}} & \alpha_6-i\alpha_7 \\ \alpha_4+i\alpha_5 & \alpha_6+i\alpha_7 & -i-\frac{2\alpha_8}{\sqrt{3}} \end{bmatrix} \qquad |\alpha_i|\ll 1 \tag{2-28}$$

Note the similarity between (2-28) for $SU(3)$ and (2-25) for $SU(2)$. However, (2-25) is good globally, whereas (2-28) is only good locally, essentially because $SU(2)$ is a simpler theory than $SU(3)$.

Note that $N(0) = I$. Note also (where "*HOT*" means higher order terms) that

$$Det\,N \text{ for } |\alpha_i|\ll 1$$
$$\approx 1-i\left(-\alpha_3+\frac{\alpha_8}{\sqrt{3}}\right)-i\left(-\frac{2\alpha_8}{\sqrt{3}}\right)-i\left(\alpha_3+\frac{\alpha_8}{\sqrt{3}}\right)-i\left(-\frac{2\alpha_8}{\sqrt{3}}\right)-i\left(\alpha_3+\frac{\alpha_8}{\sqrt{3}}\right)-i\left(-\alpha_3+\frac{\alpha_8}{\sqrt{3}}\right)+HOT\approx 1, \tag{2-29}$$

and

$$\underset{|\alpha_i|\ll 1}{N^{\dagger}N\approx}\begin{bmatrix} i+\alpha_3+\frac{\alpha_8}{\sqrt{3}} & \alpha_1-i\alpha_2 & \alpha_4-i\alpha_5 \\ \alpha_1+i\alpha_2 & i-\alpha_3+\frac{\alpha_8}{\sqrt{3}} & \alpha_6-i\alpha_7 \\ \alpha_4+i\alpha_5 & \alpha_6+i\alpha_7 & i-\frac{2\alpha_8}{\sqrt{3}} \end{bmatrix}\begin{bmatrix} -i+\alpha_3+\frac{\alpha_8}{\sqrt{3}} & \alpha_1-i\alpha_2 & \alpha_4-i\alpha_5 \\ \alpha_1+i\alpha_2 & -i-\alpha_3+\frac{\alpha_8}{\sqrt{3}} & \alpha_6-i\alpha_7 \\ \alpha_4+i\alpha_5 & \alpha_6+i\alpha_7 & -i-\frac{2\alpha_8}{\sqrt{3}} \end{bmatrix}$$

$$\approx\begin{bmatrix} 1+i\alpha_3+i\frac{\alpha_8}{\sqrt{3}}-i\alpha_3-i\frac{\alpha_8}{\sqrt{3}}+HOT & i\alpha_1+\alpha_2-i\alpha_1-\alpha_2+HOT & i\alpha_4+\alpha_5-i\alpha_4-\alpha_5+HOT \\ -i\alpha_1+\alpha_2+i\alpha_1-\alpha_2+HOT & 1-i\alpha_3+i\frac{\alpha_8}{\sqrt{3}}+i\alpha_3-i\frac{\alpha_8}{\sqrt{3}}+HOT & i\alpha_6+\alpha_7-i\alpha_6-\alpha_7+HOT \\ -i\alpha_4+\alpha_5+i\alpha_4-\alpha_5+HOT & -i\alpha_6+\alpha_7+i\alpha_6-\alpha_7+HOT & 1-i\frac{2\alpha_8}{\sqrt{3}}+i\frac{2\alpha_8}{\sqrt{3}}+HOT \end{bmatrix} \tag{2-30}$$

$$\approx\begin{bmatrix} 1 & 0 & 0 \\ 0 & 1 & 0 \\ 0 & 0 & 1 \end{bmatrix}.$$

So, for small values of the 8 parameters, N is special unitary and equals the identity when all parameters are zero.

Note also that taking all imaginary terms in N as zero, we should get an $SO(3)$ group. Doing so with (2-28), we get A of (2-14) (with a relabeling of variables such that $\theta_3 = -\alpha_2$, $\theta_2 = \alpha_5$, and $\theta_1 = -\alpha_7$.)

2.2.8 Direct, Outer, and Tensor Products

The Concept

In group theory applications, one commonly runs into a type of multiplication known as the direct product, which is also common in linear algebra.

Consider an example where elements of two groups are represented by matrices, and although they could have the same dimension, in our case the first matrix A has dimension 3, and the second matrix B, dimension 2. Note the symbol X in (2-31) represents direct product.

$$\mathbf{A}\times\mathbf{B} \xrightarrow[\text{represented by matrices}]{\text{direct product of groups}} A\times B = \begin{bmatrix} A_{11} & A_{12} & A_{13} \\ A_{21} & A_{22} & A_{23} \\ A_{31} & A_{32} & A_{33} \end{bmatrix} \times \begin{bmatrix} B_{11} & B_{12} \\ B_{21} & B_{22} \end{bmatrix}$$

$$= \begin{bmatrix} A_{11}\begin{bmatrix} B_{11} & B_{12} \\ B_{21} & B_{22} \end{bmatrix} & A_{12}\begin{bmatrix} B_{11} & B_{12} \\ B_{21} & B_{22} \end{bmatrix} & A_{13}\begin{bmatrix} B_{11} & B_{12} \\ B_{21} & B_{22} \end{bmatrix} \\ A_{21}\begin{bmatrix} B_{11} & B_{12} \\ B_{21} & B_{22} \end{bmatrix} & A_{22}\begin{bmatrix} B_{11} & B_{12} \\ B_{21} & B_{22} \end{bmatrix} & A_{23}\begin{bmatrix} B_{11} & B_{12} \\ B_{21} & B_{22} \end{bmatrix} \\ A_{31}\begin{bmatrix} B_{11} & B_{12} \\ B_{21} & B_{22} \end{bmatrix} & A_{32}\begin{bmatrix} B_{11} & B_{12} \\ B_{21} & B_{22} \end{bmatrix} & A_{33}\begin{bmatrix} B_{11} & B_{12} \\ B_{21} & B_{22} \end{bmatrix} \end{bmatrix} \qquad (2\text{-}31)$$

Essentially, this is a component-wise operation wherein components live side-by-side and do not mix. The direct product is also called the Cartesian product. In index notation,

Direct product definition

$$A\times B \xrightarrow[\text{notation}]{\text{in index}} = A_{ij}B_{k'l'}\,. \qquad (2\text{-}32)$$

Direct product indices = indices of constituent matrices

The result (2-32) has four different indices, 2 indices for A_{ij} plus 2 indices for $B_{k'l'}$, with different dimensions for each index. The i index dimension is 3; the j dimension is 3; the k' dimension is 2; and the l' dimension is 2. The total number of components is 3 X 3 X 2 X 2 = 36. Said another way, it is the total number of components of A times the total number of components of B, i.e., 9 X 4 = 36.

Number of components = product of number of components of constituent matrices

Note that, in this example, we use primes for the indices of the $B_{k'l'}$ in (2-32) to make it clear that generally the $B_{k'l'}$ components live in a completely different world from the components of A_{ij}. That is, the factors with unprimed subscripts in each element in (2-32) are usually (but not always) of different character from the factors with primed subscripts. Keep this in mind in future work, whether or not we employ primed subscripts to differentiate component factors.

We have already used the direct product concept in Vol. 1 (see (4-123), pg. 115, reproduced as (2-33) below), and there it was called the *outer* product (of spinor fields).

$$\underbrace{\psi\bar{\psi}}_{\substack{\text{not writing}\\ \text{out spinor}\\ \text{indices}}} = \underbrace{\psi_\alpha\bar{\psi}_\beta}_{\substack{\text{with spinor}\\ \text{indices}\\ \text{written}}} = \psi_\alpha\psi^\dagger_\delta\gamma^0_{\delta\beta} = X_{\alpha\beta} = \text{outer product, a matrix quantity in spinor space} \qquad (2\text{-}33)$$

Similar example from prior work

Outer product of two vectors = a matrix

Here, the outer product is symbolized by having the adjoint (complex conjugate transpose) factor on the RHS. (The inner product is symbolized by having it on the LHS and yields a scalar, rather than a matrix quantity.) Spinors are essentially vectors (4D in QFT) in spinor space.

Note that a direct product can be carried out between many different types of mathematical objects. (2-31) [i.e., (2-32)] is a direct product of matrix *group* representations. (2-33) is a direct product of *spinors* (vectors in spinor space). For historical reasons, the latter is more commonly called the outer product.

Direct and outer prods similar in concept, but former term used for matrices, the latter typically for vectors

The general direct product principle, regardless of the mathematical entities involved, is this. We get a separate component in the result for each possible multiplication of one component in the first entity times one component in the second.

In tensor analysis, tensors are commonly represented by matrices, and one often runs into what could be called a direct product of tensors, which looks just like (2-31) [i.e., (2-32)]. However, that term rarely seems to be used, and the almost universally used one is tensor product. For tensors, ⊗ is typically employed instead of X.

Tensor product like direct product but doesn't go by that name, and usually employs different symbol

Note that a vector is a 1st rank tensor. (A scalar is a zeroth rank tensor. What we usually think of as a tensor is a 2nd rank tensor.) In that context, the operation of what we might expect to be referred to as the direct product of two vectors is instead commonly labeled the outer product or the tensor product. For two vectors **w** and **y**, it is denoted by **w** ⊗ **y**.

Note further that tensor products do not generally commute. **w** ⊗ **y** does not equal **y** ⊗ **w**, except for special cases.

The various terminologies are probably a little confusing. (They were for me while learning it.) But generally, the term direct product is conventionally applied to groups and matrices, and employs the symbol X. The term tensor product is typically used for any tensors, including vectors, and commonly employs the symbol ⊗. As noted, another, very common term for the tensor product of vectors, in particular, is outer product.

Summary of the nomenclature: direct product, tensor product, outer product

A Hypothetical Example

Now consider **A** and **B** being operators that operate on vectors **v** in a vector space. Imagine the quantities represented by vectors have two characteristics. They are colored, and they are charged. The colors are red, green, blue and the charges are + and –. We represent a given vector **v** as a tensor product (which is an outer product) of a 3D vector **w** representing color (*r, g, b* symbolizing the particular color component in the vector) and a 2D vector **y** representing charge (*p* symbolizing the positively charged component; *n*, the negatively charged).

A hypothetical example for illustrative purposes

$$\mathbf{v}=\mathbf{w}\otimes\mathbf{y}\xrightarrow[\text{by column and row matrices}]{\text{tensor product represented}} v=\begin{bmatrix} r\\ g\\ b\end{bmatrix}\begin{bmatrix} p & n\end{bmatrix}=\begin{bmatrix} rp & rn\\ gp & gn\\ bp & bn\end{bmatrix} \quad\text{or}\quad v_{jl}=w_j y_l . \qquad (2\text{-}34)$$

In one sense, the vector **v** is a matrix (2 columns, 3 rows), but in the abstract vector space sense it is a vector in a vector space, where that space is comprised of 3X2 matrices.

The **A** operator here is related to color and thus acts only on the **w** part of **v**. It is blind to the **y** part. The **B** operator is related to charge and acts only on the **y** part. So, given (2-31) and (2-32),

$$(\mathbf{A}\times\mathbf{B})\mathbf{v}=(\mathbf{A}\times\mathbf{B})(\mathbf{w}\otimes\mathbf{y})=\mathbf{Aw}\otimes\mathbf{By}=\mathbf{v}''\xrightarrow[\text{notation}]{\text{in index}} A_{ij}B_{k'l'}v_{jl'}=A_{ij}B_{k'l'}w_j y_{l'}=v''_{ik'} \quad (2\text{-}35)$$

A particular operator commonly acts on only one of the constituent vectors of a tensor product element

---

Do **Problem 10** to show (2-35) in terms of matrices.

You are probably already considering the color part of the (state) vector **v** above in terms of the strong force, and that, in fact, is why I choose color as a characteristic for this example. Similarly, the same state represented symbolically by **v** may have a particular weak interaction charge.

When we get to weak interactions, we will see that operators like **B** act on a two-component state (such as a quark or lepton). We can imagine a particular **B** operator, a weak charge operator, acting on the state that would yield an eigenvalue equal to its weak charge.

With strong interactions, we will see that operators like **A** act on a 3-component state (such as a quark). For example, a quark state having a given color *r* (red) would be in a color eigenstate (with zero values for *g* and *b* in w_j), and we could imagine a particular **A** operator, a color operator, acting on it that would yield an eigenvalue corresponding to red. This is all very rough around the edges, and not completely accurate, since we are trying to convey the general idea as simply as possible. Many details and modifications will be needed (in pages ahead) to make it correct and more complete.

Direct Products and the Standard Model

In fact, the famous *SU*(3)X*SU*(2)X*U*(1) relation of the SM symbolizes the action of three groups (strong/color interaction operators in *SU*(3), weak interaction operators in *SU*(2), and QED operators in *U*(1)) whose operators act on fields (vectors in the mathematical sense here). Each field has a separate part (indices) in it for each of the three interaction types. And each such part of the field is acted upon only by operators associated with that particular interaction. More on this, with examples, later in this chapter.

SM: direct product of 3 group reps SU(3)XSU(2)XU(1)

To Summarize

The direct product of group matrix representations acts on the tensor product of vector spaces. Each original group acts independently on each individual vector space.

An aside

An aside on spin and group theory

The following will not be relevant for our work, but I mention it in case you run across it in other places (such as the references in the footnotes on pg. 8). Do not spend too much time scratching your head over this for now, but save it to come back to if and when you run into it elsewhere.

In some applications, the two parts of the state vector, such as **w** and **y** in (2-34), respond to the same operator(s). For example, a spin 1 particle interacting with a spin ½ particle in NRQM can both be operated on by a spin operator. When two such particles interact to form a bound system, that total system has six possible spin eigenstates (four states with $J_{tot} = 3/2, J_z = 3/2, 1/2, -1/2, -3/2$, and two with $J_{tot} = 1/2, J_z = 1/2, -1/2$). The state vector of the system is the outer product of the spin 1 state multiplied by the spin ½ state. Both parts of the system state vector relate to spin and both parts are acted on by a spin operator.

The two parts of the system state vector have, respectively, 3 spin components (spin 1 particle has eigenstates $J_z = 1, 0, -1$) and 2 spin components (spin ½ particle has eigenstates $J_z = 1/2, -1/2$). The spin operator acts in the 3D space of the spin 1 particle and also in the 2D space of the spin ½ particle.

In that case, instead of a 3X2 state vector matrix with six components representing the system, one can formulate the math using a six-component column vector for the system. And then the spin operator for the system becomes a 6X6 matrix, instead of a 3X3 matrix group direct product multiplied by a 2X2 matrix group.

This is commonly done and can be confusing when one considers direct products defined as in (2-31) (equivalently, (2-32)), as we do here.

End of aside

Things to Note

In NRQM, we have already used the concept of separate operators acting independently on state vectors, where we have free particle states like

NRQM example of different operators operating on different parts of state vector (i.e., on an outer (tensor) product)

$$\text{NRQM spin up state} \quad \psi_{state} = Ae^{-i(Et-\mathbf{k}\cdot\mathbf{x})}\begin{bmatrix}1\\0\end{bmatrix}. \qquad (2\text{-}36)$$

The Hamiltonian operator $H = i\frac{\partial}{\partial t}$ acts on the $e^{-i(Et-\mathbf{k}\cdot\mathbf{x})}$ part of the wave function and does nothing to the 2-component spinor part. The spin operator S_z operates on the spinor part, but not the $e^{-i(Et-\mathbf{k}\cdot\mathbf{x})}$ part.

We commonly write the outer (tensor) products of two vectors as a column vector on the left times a row vector on the right, as in (2-34). An inner product, conversely, is denoted by a row vector on the left with a column vector on the right, as most readers have been doing for a long time.

Often write inner product as row on left, column on right; outer product as column on left, row on right

$$\mathbf{w}_1 \cdot \mathbf{w}_2 \xrightarrow[\text{row on left and column on right}]{\text{inner product represented by}} = \begin{bmatrix} r_1 & g_1 & b_1 \end{bmatrix}\begin{bmatrix} r_2 \\ g_2 \\ b_2 \end{bmatrix} = r_1r_2 + g_1g_2 + b_1b_2 \qquad (2\text{-}37)$$

However, as noted in the solutions book, in the answer to Problem 10, the row vs column vector methodology can become cumbersome, and even a bit confusing, at times. Of all the choices, the most foolproof notation is the index notation, as on the RHS of (2-34) and (2-35).

But index notation is most foolproof

Additionally, be aware that the tensor product is defined for tensors of all ranks, not just rank 1 (vectors). For example, for two 2nd rank tensors expressed as matrices T_{ij} and S_{kl}, the tensor product is $T_{ij}S_{kl} = Z_{ijkl}$, a rank 4 tensor.

Finally, and likely to cause even more confusion, the literature is not 100% consistent with use of the term "direct product". Further, take care that the symbol ⊗ is often used for it, and sometimes X is used for tensor product.

2.2.9 Summary of Types of Operations Involving Groups

Wholeness Chart 2-2 lists the types of operations associated with groups that we have covered.

Wholeness Chart 2-2. Types of Operations Involving Groups and Vector Spaces

Operation	What	Type	Relevance	In Matrix Representation
A ∘ B	Group operation (binary)	Between 2 elements **A** and **B** of the group	Defining characteristic of the group	Matrix multiplication, AB
Av	Group action on a vector space	A group element **A** operates on a vector **v**	Some, but not all groups, may do this.	Matrix multiplication with column vector, Av
AXB	Direct product of group elements	Combining groups (**A** & **B** here symbolize entire groups **A** & **B**.)	Larger group formed from 2 smaller ones	Larger group $C_{ijk'l'} = A_{ij}B_{k'l'}$ has indices of A and indices of B^*
w ⊗ y	Outer (tensor) product of vectors	Combining vectors	Composite formed from 2 vectors	Composite v_{ij} has index of w_i and index of y_j

* The direct product in group matrix representations, in some applications (see "An Aside" section on pg. 20), can instead be re-expressed as a matrix with the same number of indices as each of A and B (in the example, two indices), but of dimension equal to the product of the dimensions of A and B.

2.2.10 Overview of Types of Groups

Infinite vs finite groups

The types of groups we have encountered are summarized in Wholeness Chart 2-3, along with one (first row) we have yet to mention, infinite groups, which simply have an infinite number of group elements. One example is all real numbers with the group operation being addition. Another is continuous 2D rotations (see (2-6)), which is infinite because there are an infinite number of angles θ through which we can rotate (even when θ is constrained to $0 \le \theta < 2\pi$, since θ is continuous). As a result, Lie groups are infinite groups, as they have one or more continuously varying parameters. On the other hand, a finite group has a finite number of elements. One example is shown in (2-7), which has only four group members.

Note the various types of groups are not mutually exclusive. For example, we could have an $SO(n)$, direct product, Abelian, Lie group. Or many other different combinations of group types.

Wholeness Chart 2-3. Overview of Types of Groups

Type of Group	Characteristic	Symbols	Matrix Representation
Infinite (vs. finite)	Group has an infinite number of elements.		Example: 2D rotation matrices as function of θ
Abelian (vs Non-abelian)	All elements commute	**AB = BA**	Some groups of matrices Abelian, but generally no.
Lie (vs Non-Lie)	Elements continuous smooth functions of continuous, smooth variable(s) θ_i	**A=A(θ_i)**	Example: rotation matrices as function of rotation angle(s)
Orthogonal, *O*(*n*)	Under **A**, magnitude of vector unchanged. All group elements real.	|**Av**| = |**v**|	$A^{-1} = A^{T}$ $\lvert\text{Det } A\rvert = 1$
Special Orthogonal, *SO*(*n*)	As in block above	As in block above	As in block above, but Det $A = 1$
Unitary, *U*(*n*)	Under **A**, magnitude of vector unchanged. At least some group elements complex.	|**Av**| = |**v**|	$A^{-1} = A^{\dagger}$ $\lvert\text{Det } A\rvert = 1$
Special Unitary, *SU*(*n*)	As in block above	As in block above	As in block above, but Det $A = 1$
Direct product	Composite group is formed by direct product of two groups	**C = AXB**	Example: $C_{ijk'l'} = A_{ij}B_{k'l'}$

2.2.11 Same Physical Phenomenon Characterized by Different Groups

It is interesting that certain natural phenomena can be characterized by different groups. For example, consider 2D rotation. Mathematically, we can characterize rotation by the *SO*(2) group, represented by (2-6). (Some other parametrizations are shown in (2-9) and (2-10).) This group rotates a vector, such as the position vector (x,y), through an angle θ.

Different mathematical groups can characterize the same physical phenomenon

But we can also characterize 2D rotation via

$$U(\theta) = e^{i\theta} \quad \left(\text{a representation of } U(1)\right), \qquad \text{Repeat of (2-17)}$$

One example

where the unitary group (2-17) rotates a complex number $x + iy$ though an angle θ. (See Problem 8.)

Note, the *SO*(2) group and the *U*(1) group above are different groups (here characterizing the same real world phenomenon), and *not* different representations of the same group[1].

SO(2) *and U*(1) *are different groups, not different representations of the same group*

That is why we prefer to say a particular group is a *characterization* of a given natural phenomenon and not a "representation" of the phenomenon. (The word "characterization" is employed by me in this text to help avoid confusion with the term "representation", but it is not generally used by others.)

In a similar way, which we will look at very briefly later on, both *SO*(3) and *SU*(2) can characterize 3D rotation. In fact, *SU*(2) is a preferred way of handling spin (which is a 3D angular momentum vector) for spin ½ particles in NRQM, as seen from different orientations (z axis up, x axis up, y axis up, or other orientations). Many QM textbooks show this.[2]

2.2.12 Subgroups

A subgroup is a subset of elements of a group, wherein the subset elements under the group binary operation satisfy the properties of a group. For one, all binary operations between elements of the subset result in another element in the subset. That is, the subgroup has closure, within itself.

A subgroup is a group unto itself inside a larger group

As examples, the group *SO*(n) is a subgroup of *O*(n), and *SU*(n) is a subgroup of *U*(n). So is rotation in 2D a subgroup of rotation in 3D, where rotation in 2D can be represented by an *SO*(3) 3X3 matrix with unity as one diagonal component and zeros for the other components of the row and column that diagonal component is in. (See (2-12).) And (2-7) is a subgroup of (2-6). As a relativistic example, the set of 3D rotations *SO*(3) is a subgroup of the Lorentz group *SO*(3,1). And the Lorentz group is itself a subgroup of the Poincaré group, which consists of the Lorentz group plus translations (in 4D).

The most important example for us is the group covering the standard model, i.e., *SU*(3)X*SU*(2)X *U*(1), which is a direct product of its subgroups *SU*(3), *SU*(2), and *U*(1),

Other than working individually with the subgroups of the SM, we will not be doing much with subgroups, but mention them in passing, as you will no doubt see the term in the literature.

Note that a matrix subgroup is not the same thing as a submatrix. The latter is obtained by removing one or more rows and/or columns from a larger (parent) matrix, so it would have fewer rows and/or columns than its parent. A matrix subgroup, on the other hand, is comprised of elements of the larger matrix group, so each matrix of the subgroup must have the same number of rows and columns as the matrices of the larger group.

2.2.13 Most Texts Treat Group Theory More Formally Than This One

We have purposefully not used formal mathematical language in our development of group theory, in keeping with the pedagogic principles delineated in the preface of Vol. 1. In short, I think that, for most of us, it is easier to learn a theory introduced via concrete examples than via more abstract presentations, as in some other texts.

But, for those who may consult other books, Table 2-1 shows some symbols you will run into in more formal treatments, and what they mean in terms of what we have done here.

For example, where we said that the closure property of a group means the result of the group operation between any two elements in the group is also an element of the group, the formal notation for this would be

---

[1] In mathematics lingo, *SO*(2) and *U*(1) are said to be "isomorphic". A group isomorphism is a function between two groups that sets up a one-to-one correspondence between the elements of the groups in a way that respects the given group operations.

[2] For one, see Merzbacher, E. *Quantum Mechanics*, 2nd ed, (Wiley, 1970), pg. 271 and Chap. 16.

$$\forall \mathbf{A},\mathbf{B} \in G, \quad \mathbf{A} \circ \mathbf{B} \in G\,. \tag{2-38}$$

Note, however, that different authors can use different conventions. For example, the symbol $\subseteq$ is sometimes used to designate "is a sub*group* of". So, terminology usage requires some vigilance.

Table 2-1.
Some Symbols Used in Formal Group Theory

Brief look at more formal treatment of group theory

Symbol	Use	Meaning
$\in$	$\mathbf{A} \in G$	**A** is a member of group (or set) G
$\notin$	$\mathbf{A} \notin G$	**A** is not a member of group (or set) G
$\subseteq$	$\mathbf{A} \subseteq S$	**A** is a subset of set S
$\leq$	$\mathbf{A} \leq G$	**A** is a subgroup of group G
$\forall$	$\forall\, \mathbf{A}$	For any and all elements **A**
$\mathbb{R}$		Set of real numbers (for some authors, **R**)
$\mathbb{C}$		Set of complex numbers (for some, **C**)
$\mathbb{R}^3$		3D space of real numbers (for some, $\mathbf{R}^3$)

2.3 Lie Algebras

Lie algebra intro

Different from a group, a Lie algebra, like any algebra, has *two binary operations* (in a matrix representation, matrix addition plus a matrix multiplication type operation) between elements of the algebra and also a *scalar operation* (which will be multiplication of matrix elements by scalars in our applications). But, as we will see, a *Lie* algebra, in addition, can be used to generate an associated Lie group, in ways that prove to be advantageous, particularly in QFT. An *SO*(2) group can be generated from its particular associated Lie algebra; an *SU*(2) group, from its particular associated Lie algebra; an *SU*(3) group, from its Lie algebra; etc.

The set element multiplication type operation used for matrix Lie algebras is not simple matrix multiplication (which won't actually work) such as *AB*, but a matrix commutation operation. That is,

2nd binary operation for the algebra is commutation

$$\text{common } 2^{\text{nd}} \text{ operation in physics} \rightarrow \left[A,B\right] = AB - BA = C \qquad C \text{ an element of the algebra,} \tag{2-39}$$

where (2-39) is commonly called the Lie bracket. (In more abstract Lie algebras, the Lie bracket does not have to be a commutator. In the special case we are dealing with, it is.)

As with Lie groups, a precise mathematical definition of Lie algebras, particularly abstract (vs matrix) ones, is beyond the level of this book. For our purposes, where all our groups and algebras have elements composed of matrices, the following heuristic simplification will do.

Suitable definition of Lie algebra for our purposes

A Lie algebra is an algebra with binary operations between elements, consisting of addition and commutation, that can generate an associated Lie group (in a manner yet to be shown).

In practice, this will mean, because of their connection to Lie groups (with the continuous, smooth nature of their elements dependence on parameters) that Lie algebra elements vary smoothly and continuously as functions of continuous, smooth parameters (scalars).

Our goal now is to deduce the relationship between Lie algebras and Lie groups, and then show how it applies to certain areas of physics.

2.3.1 Relating a Lie Group to a Lie Algebra: Simple Example of SO(2)

Consider the *SO*(2) 2D rotation group representation of (2-6), reproduced below for convenience.

One parameter, real Lie group (2D rotation)

$$\hat{M}(\theta) = \begin{bmatrix} \cos\theta & -\sin\theta \\ \sin\theta & \cos\theta \end{bmatrix} \qquad \text{Repeat of (2-6)}$$

We can express $\hat{M}(\theta)$ as a Taylor expansion around $\theta = 0$.

$$\hat{M}(\theta) = \hat{M}(0) + \theta \hat{M}'(0) + \frac{\theta^2}{2!}\hat{M}''(0) + \frac{\theta^3}{3!}\hat{M}'''(0) + \ldots\,, \tag{2-40}$$

which for small θ ($|\theta| << 1$) becomes (where the factor of i is inserted at the end because it will make things easier in the future)

$$\hat{M}(\theta) \approx \hat{M}(0) + \theta\hat{M}'(0) = \begin{bmatrix} \cos 0 & -\sin 0 \\ \sin 0 & \cos 0 \end{bmatrix} + \theta \begin{bmatrix} -\sin 0 & -\cos 0 \\ \cos 0 & -\sin 0 \end{bmatrix},$$
$$= I + \theta \begin{bmatrix} 0 & -1 \\ 1 & 0 \end{bmatrix} = I + i\theta\hat{X} \qquad |\theta| \ll 1. \tag{2-41}$$

Taylor expansion

where[1]

$$\hat{X} = -i\hat{M}'(0) = -i\begin{bmatrix} 0 & -1 \\ 1 & 0 \end{bmatrix} = \begin{bmatrix} 0 & i \\ -i & 0 \end{bmatrix}. \tag{2-42}$$

The single matrix $\hat{X}$ and the continuous scalar field θ represent a Lie algebra, where the elements are all of form $\theta\hat{X}$, the first operation is matrix addition, and the second operation is matrix commutation (Commutation, as mentioned earlier, is necessary, and we will look at why a little later.)

Lie algebra for SO(2) has one matrix $\hat{X}$, and elements $\theta\hat{X}$

Do **Problem 11** to show that (2-42) is an algebra and gain a valuable learning experience. If you have some trouble, it will help to continue reading to the end of Sect. 2.3.3, and then come back to this.

A Lie algebra such as this one (i.e., $\theta\hat{X}$) is often called the tangent space of the group (here, $\hat{M}(\theta)$) because the derivative of a function, when plotted, is tangent to that function at any given point. $\hat{X}$ in the present case is the first derivative of $\hat{M}$ at $\theta = 0$. Actually, it is *i* times $\hat{X}$. We have inserted the imaginary factor because doing so works best in physics, especially in quantum theory, where operators are Hermitian. Here, $\hat{X}$ is Hermitian, i.e., $\hat{X}^\dagger = \hat{X}$, and we will see the consequences of that later in this chapter. Mathematicians typically don't use the *i* factor, but the underlying idea is the same.

Generating *SO*(2) Lie Group from Its Lie Algebra

Given (2-41), and knowing $\hat{X}$, we can actually reproduce the group $\hat{M}$ from the Lie algebra $\theta\hat{X}$. We show this using (2-40) with (2-6) and its derivatives, along with (2-42). That is,

From $\hat{X}$, can generate SO(2) group via expansion

$$\hat{M}(\theta) = \underbrace{\begin{bmatrix} \cos 0 & -\sin 0 \\ \sin 0 & \cos 0 \end{bmatrix}}_{I} + \theta\underbrace{\begin{bmatrix} -\sin 0 & -\cos 0 \\ \cos 0 & -\sin 0 \end{bmatrix}}_{i\hat{X}} + \frac{\theta^2}{2!}\underbrace{\begin{bmatrix} -\cos 0 & \sin 0 \\ -\sin 0 & -\cos 0 \end{bmatrix}}_{-I} + \frac{\theta^3}{3!}\underbrace{\begin{bmatrix} \sin 0 & \cos 0 \\ -\cos 0 & \sin 0 \end{bmatrix}}_{-i\hat{X}} + .. \tag{2-43}$$

Realizing that every even derivative factor in (2-40) is the identity matrix (times either 1 or – 1), and every odd derivative factor is $i\hat{X}$ (times 1 or – 1), we can obtain the group element for any θ directly from the Lie algebra matrix $\hat{X}$, as illustrated by re-expressing (2-43) as

$$\hat{M}(\theta) = \begin{bmatrix} 1 - \frac{\theta^2}{2!} + \frac{\theta^4}{4!} .. & -\theta + \frac{\theta^3}{3!} - \frac{\theta^5}{5!} + .. \\ \theta - \frac{\theta^3}{3!} + \frac{\theta^5}{5!} .. & 1 - \frac{\theta^2}{2!} + \frac{\theta^4}{4!} + .. \end{bmatrix} = \begin{bmatrix} \cos\theta & -\sin\theta \\ \sin\theta & \cos\theta \end{bmatrix}. \tag{2-44}$$

Exponentiation of Lie Algebra to Generate *SO*(2) Lie Group

More directly, as shown below, we can simply exponentiate $\theta\hat{X}$ to get $\hat{M}$. From (2-40) and (2-43), where we note $\hat{X}\hat{X} = I$,

$$\begin{bmatrix} \cos\theta & -\sin\theta \\ \sin\theta & \cos\theta \end{bmatrix} = \hat{M}(\theta) = \hat{M}(0) + \theta\hat{M}'(0) + \frac{\theta^2}{2!}\hat{M}''(0) + \frac{\theta^3}{3!}M'''(0) + ...$$
$$= I + \theta(i\hat{X}) + \frac{\theta^2}{2!}(-I) + \frac{\theta^3}{3!}(-i\hat{X}\,I) + ... \tag{2-45}$$
$$= I + (i\theta\hat{X}) + \frac{(i\theta\hat{X})^2}{2!} + \frac{(i\theta\hat{X})^3}{3!} + ... = e^{i\theta\hat{X}}.$$

or via $SO(2) = e^{i\theta\hat{X}}$

---

[1] Some authors, usually non-physicists, define $\hat{X}$ as simply $\hat{M}'(0)$ without a factor of *i*.

In essence, $\hat{X}$ can generate $\hat{M}$ (via exponentiation as in the last part of (2-45)). The matrix $\hat{X}$ is called the generator of the group $\hat{M}$ (or of the Lie algebra). It is a basis vector in the Lie algebra vector space. Actually, it is *the* basis vector in this case, as there is only one basis matrix, and that is $\hat{X}$.

$\hat{X}$ called the generator of the Lie group/algebra

Note that the exponentiation to find $\hat{M}$ can be expressed via addition with $\hat{X}$ and I, as in the LHS of the last row of (2-45).

As an aside, inserting the i in the last step of (2-41) to define $\hat{X}$ as in (2-42) led to (2-45).

Key point

Knowing the generator of the Lie algebra, we can construct (generate) the associated Lie group simply by exponentiation of the generator (times the scalar field). Knowing the associated Lie algebra is (almost) the same as knowing the group.

Knowing generator essentially = knowing the group

2.3.2 Starting with a Different Parametrization of SO(2)

Let's repeat the process of the prior section, but start with a different parametrization of the same $SO(2)$ group, i.e., (2-9), which we repeat below for convenience.

$$\hat{M}(x)=\begin{bmatrix}\sqrt{1-x^2} & -x \\ x & \sqrt{1-x^2}\end{bmatrix} \qquad \text{Repeat of (2-9)}$$

Different form of SO(2)

We calculate the generator in similar fashion as before.

$$\hat{X}=-i\hat{M}'(x)_{x=0}=-i\begin{bmatrix}\dfrac{-2x}{2\sqrt{1-x^2}} & -1 \\ 1 & \dfrac{-2x}{2\sqrt{1-x^2}}\end{bmatrix}_{x=0}=-i\begin{bmatrix}0 & -1\\ 1 & 0\end{bmatrix}=\begin{bmatrix}0 & i \\ -i & 0\end{bmatrix} \tag{2-46}$$

It has same generator as other form

(2-46) is the same as (2-42), but generally different parametrizations of the same group can have different generators, even though the underlying abstract associated Lie algebra is the same. We will see an example of this shortly in a problem related to $SO(3)$.

But generally, different forms of group have different generators

We could, if we wished, obtain the original group by expanding (2-9), similar to what we did in (2-45), and the terms in the expansion would involve only the matrices I and $\hat{X}$.

2.3.3 Lie Algebra for a Three Parameter Lie Group: SO(3) Example

The 2D rotation example above was a one parameter (i.e., θ) Lie group. Consider the 3D rotation group matrix representation A of (2-11) to (2-13) with the three parameters θ_1, θ_2, and θ_3. The multivariable Taylor expansion, parallel to (2-45), is

3D rotation = SO(3) group

3 parameters

$$\begin{aligned}&A(\theta_1,\theta_2,\theta_3)=A_1(\theta_1)A_2(\theta_2)A_3(\theta_3)=e^{i\theta_1\hat{X}_1}e^{i\theta_2\hat{X}_2}e^{i\theta_3\hat{X}_3}=\\ &\left(A_1\big|_{\theta_1=0}+\theta_1 A_1'\big|_{\theta_1=0}+\tfrac{1}{2!}\theta_1^2 A_1''\big|_{\theta_1=0}+\dots\right)\left(A_2\big|_{\theta_2=0}+\theta_2 A_2'\big|_{\theta_2=0}+\tfrac{1}{2!}\theta_2^2 A_2''\big|_{\theta_2=0}+\dots\right)\times\\ &\qquad\qquad\left(A_3\big|_{\theta_3=0}+\theta_3 A_3'\big|_{\theta_3=0}+\tfrac{1}{2!}\theta_3^2 A_3''\big|_{\theta_3=0}+\dots\right)=\\ &\left(I+\left(i\theta_1\hat{X}_1\right)+\frac{\left(i\theta_1\hat{X}_1\right)^2}{2!}+\dots\right)\left(I+\left(i\theta_2\hat{X}_2\right)+\frac{\left(i\theta_2\hat{X}_2\right)^2}{2!}+\dots\right)\left(I+\left(i\theta_3\hat{X}_3\right)+\frac{\left(i\theta_3\hat{X}_3\right)^2}{2!}+\dots\right).\end{aligned} \tag{2-47}$$

Taylor expansion in 3 parameters

Parallel to what we did for the one parameter case, we can find three generators

$$\hat{X}_i=-i\left.\frac{\partial A_i}{\partial\theta_i}\right|_{\theta_i=0} \qquad i=1,2,3 \text{ (no sum)}, \tag{2-48}$$

which, more explicitly expressed (taking first derivatives of components in (2-12)), are

$$\hat{X}_1=i\begin{bmatrix}0&0&0\\0&0&1\\0&-1&0\end{bmatrix} \qquad \hat{X}_2=i\begin{bmatrix}0&0&-1\\0&0&0\\1&0&0\end{bmatrix} \qquad \hat{X}_3=i\begin{bmatrix}0&1&0\\-1&0&0\\0&0&0\end{bmatrix}. \tag{2-49}$$

3 generators $\hat{X}_i$

Note, for future reference, the commutation relations between the $\hat{X}_i$. For example,

$$\left[\hat{X}_1,\hat{X}_2\right]=i^2\left(\begin{bmatrix}0&0&0\\0&0&1\\0&-1&0\end{bmatrix}\begin{bmatrix}0&0&-1\\0&0&0\\1&0&0\end{bmatrix}-\begin{bmatrix}0&0&-1\\0&0&0\\1&0&0\end{bmatrix}\begin{bmatrix}0&0&0\\0&0&1\\0&-1&0\end{bmatrix}\right)$$
$$=-\begin{bmatrix}0&0&0\\1&0&0\\0&0&0\end{bmatrix}+\begin{bmatrix}0&1&0\\0&0&0\\0&0&0\end{bmatrix}=\begin{bmatrix}0&1&0\\-1&0&0\\0&0&0\end{bmatrix}=-i\hat{X}_3\,. \qquad (2\text{-}50)$$

In general, where ε_{ijk} is the Levi-Civita symbol, and as you can prove to yourself by calculating the other two commutations involved (or just take my word for it),

Commutation relations for generators

$$\left[\hat{X}_i,\hat{X}_j\right]=-i\varepsilon_{ijk}\hat{X}_k \qquad i,j,k \text{ each take on value } 1,2, \text{ or } 3\,. \qquad (2\text{-}51)$$

Scalar field multiplication of the θ_i by their respective $\hat{X}_i$, under matrix addition and matrix commutation, comprise a Lie algebra. We now show that this is indeed an algebra.

Demonstrating (2-49) with θ_i yield elements of an algebra

$\hat{X}_i$ and θ_i yield an algebra

An algebra has two binary operations (matrix addition and commutation of two matrices here) and one scalar multiplication operation. We need to check that the elements in a set created using (2-49) as a basis and employing these operations satisfy the requirements for an algebra, as given in Wholeness Chart 2-1, pg. 9.

Showing it

We consider the set for which every possible (matrix) element in the set can be constructed by addition of the $\hat{X}_i$, each multiplied by a scalar. That is, for any element of the set $\hat{X}_a$

$$\hat{X}_a=\theta_{a1}\hat{X}_1+\theta_{a2}\hat{X}_2+\theta_{a3}\hat{X}_3=\theta_{ai}\hat{X}_i. \qquad \hat{X}_i \text{ denote the three elements of (2-49)}$$

For example, the special case $\hat{X}_1$ has $\theta_{11}=1$, and $\theta_{12}=\theta_{13}=0$. More general elements can have any real values for the θ_{ai}.

First, we look at the 1st binary operation of matrix addition with scalar field multiplication.

1st binary operation satisfies group properties

Closure: It may be obvious that by our definition of set elements above, every addition of any two elements yields an element of the set. To show it explicitly, for any two elements, we can write

$$\hat{X}_a=\theta_{a1}\hat{X}_1+\theta_{a2}\hat{X}_2+\theta_{a3}\hat{X}_3 \qquad \hat{X}_b=\theta_{b1}\hat{X}_1+\theta_{b2}\hat{X}_2+\theta_{b3}\hat{X}_3,$$

$$\hat{X}_c=\hat{X}_a+\hat{X}_b=(\theta_{a1}+\theta_{b1})\hat{X}_1+(\theta_{a2}+\theta_{b2})\hat{X}_2+(\theta_{a3}+\theta_{b3})\hat{X}_3==\theta_{c1}\hat{X}_1+\theta_{c2}\hat{X}_2+\theta_{c3}\hat{X}_3$$

So, the sum of any two elements of the set is in the set, and we have closure.

Associative: $\hat{X}_a+(\hat{X}_b+\hat{X}_c)=(\hat{X}_a+\hat{X}_b)+\hat{X}_c$. Matrix addition is associative.

Identity: For $\theta_{di}=0$ with i values $=1,2,3$, $\hat{X}_b+\hat{X}_d=\hat{X}_b$. So, $\hat{X}_d=[0]_{3X3}$ is the identity element for matrix addition.

Inverse: Each element $\hat{X}_a$ of the set has an inverse $(-\hat{X}_a)$, since $\hat{X}_a+(-\hat{X}_a)=[0]_{3X3}$ (the identity element).

Commutation: $\hat{X}_a+\hat{X}_b=\hat{X}_b+\hat{X}_a$. Matrix addition is commutative.

1st binary operation commutative, so we have a vector space (the vectors are matrices)

Thus, under addition and scalar multiplication, the set of all elements $\hat{X}_a$ comprises a vector space and satisfies the requirements for one of the operations of an algebra.

Second, we look at the 2nd binary operation of matrix commutation with scalar field multiplication.

2nd binary operation satisfies closure requirement of an algebra

Closure: $[\hat{X}_a,\hat{X}_b]=[\theta_{ai}\hat{X}_i,\theta_{bj}\hat{X}_j],=\theta_{ai}\theta_{bj}[\hat{X}_i,\hat{X}_j]$, which from (2-51) yields the following.

$$[\hat{X}_a,\hat{X}_b]=-i\theta_{ai}\theta_{bj}\varepsilon_{ijk}\hat{X}_k=-i\theta_{ck}\hat{X}_k \qquad \text{where } \theta_{ck}=\theta_{ai}\theta_{bj}\varepsilon_{ijk}.$$

At this point, recalling that all of the scalars, such as θ_{ai} and θ_{bj}, are *real*, and ε_{ijk} is real too, we *cannot* write $[\hat{X}_a,\hat{X}_b]=\hat{X}_c$, because of the imaginary factor i. That is, the RHS of the previous relation $(-i\theta_{ck}\hat{X}_k)$ is real and *not* an element of the set (which has all imaginary elements), so there is no closure. However, we can easily fix the situation by defining the binary commutation operation to include a factor of i, as follows.

Definition of 2nd binary operation: $i[\hat{X}_a, \hat{X}_b]$ (2-52)

Given (2-51), we find (2-52) yields $i[\hat{X}_a, \hat{X}_b] = \theta_{ai}\theta_{bj}\,\varepsilon_{ijk}\hat{X}_k = \theta_{ck}\hat{X}_k$, which is in the set. Therefore, under the 2nd operation of (2-52), there is closure.

Third, we look at both binary operations together.

Distributive: *From* Wholeness Chart 2-1, **A**∘(**B** ✢ **C**) = (**A**∘**B**) ✢ (**A**∘**C**) for us, is distributive, *if* we have $i[A, B+C] = i[A,B] + i[A,C]$ or simply $[A, B+C] = [A,B] + [A,C]$.

2nd binary operation distributive over 1st: satisfies requirement of an algebra

Now $[A, B+C] = [\hat{X}_a, \hat{X}_b + \hat{X}_c] = \theta_{ai}\hat{X}_i\,(\theta_{bj}\hat{X}_j + \theta_{cj}\hat{X}_j) - (\theta_{bj}\hat{X}_j + \theta_{cj}\hat{X}_j)\,\theta_{ai}\hat{X}_i$

or $= \theta_{ai}\hat{X}_i\theta_{bj}\hat{X}_j - \theta_{bj}\hat{X}_j\theta_{ai}\hat{X}_i + \theta_{ai}\hat{X}_i\theta_{cj}\hat{X}_j - \theta_{cj}\hat{X}_j\theta_{ai}\hat{X}_i$

$= [\hat{X}_a,\hat{X}_b] + [\hat{X}_a,\hat{X}_c] = [A,B] + [A,C]$.

So, the commutation operation is distributive over the addition operation.

Conclusion: The set of the $\hat{X}_a$'s under matrix addition, the matrix commutation operation (2-52), and scalar field multiplication is an algebra. It is a Lie algebra because we can use it to generate a Lie group (via (2-47)). Note that every element in the set is a smooth, continuous function of the smooth, continuous (real) variables θ_{ai}.

So, we have an algebra, a Lie algebra

Further, regarding the 2nd binary operation, one sees from the analysis below that this particular algebra is non-associative, non-unital, and non-Abelian.

Associative: General relation **A**∘(**B**∘**C**) = (**A**∘**B**)∘**C**. For us, it is associative, *if* we have $i\,[A, i[B,C]] = i\,[i[A,B],C]$ or simply $[A,[B,C]] = [[A,B],C]$.

Now: $[A, [B,C]] = [\hat{X}_a, [\hat{X}_b,\hat{X}_c]] = \hat{X}_a(\hat{X}_b\hat{X}_c - \hat{X}_c\hat{X}_b) - (\hat{X}_b\hat{X}_c - \hat{X}_c\hat{X}_b)\hat{X}_a$

$= \hat{X}_a\hat{X}_b\hat{X}_c - \hat{X}_a\hat{X}_c\hat{X}_b - \hat{X}_b\hat{X}_c\hat{X}_a + \hat{X}_c\hat{X}_b\hat{X}_a$

$= \theta_{ai}\theta_{bj}\theta_{ck}(\hat{X}_i\hat{X}_j\hat{X}_k - \hat{X}_i\hat{X}_k\hat{X}_j - \hat{X}_j\hat{X}_k\hat{X}_i + \hat{X}_k\hat{X}_j\hat{X}_i)$

And: $[[A,B], C] = [[\hat{X}_a,\hat{X}_b],\hat{X}_c] = (\hat{X}_a\hat{X}_b - \hat{X}_b\hat{X}_a)\,\hat{X}_c - \hat{X}_c\,(\hat{X}_a\hat{X}_b - \hat{X}_b\hat{X}_a)$

$= \hat{X}_a\hat{X}_b\hat{X}_c - \hat{X}_b\hat{X}_a\hat{X}_c - \hat{X}_c\hat{X}_a\hat{X}_b + \hat{X}_c\hat{X}_b\hat{X}_a$

$= \theta_{ai}\theta_{bj}\theta_{ck}(\hat{X}_i\hat{X}_j\hat{X}_k - \hat{X}_j\hat{X}_i\hat{X}_k - \hat{X}_k\hat{X}_i\hat{X}_j + \hat{X}_k\hat{X}_j\hat{X}_i)$

These relations are not equal. The middle terms differ because the $\hat{X}_i$ do not commute. So, the second binary operation is non-associative and we say that this Lie algebra is non-associative.

This algebra is non-associative

Identity: An element I would be the identity element under commutation relation (2-52), if and only if, $i\,[I, \hat{X}_a] = \hat{X}_a$ where $\hat{X}_a$ is any element in the set. As shown by doing Problem 12, there is no such I. Since no identity element exists, this algebra is non-unital.

and non-unital

Inverse: If there is no identity element, there is no meaning for an inverse.

Commutative: General relation **A**∘**B** = **B**∘**A** needed for all elements, for the binary operation ∘. For us, ∘ is commutation, so we need commutation of the commutation operation. That is, we need, in general, $[A,B] = [B,A]$. Thus, as one example, $i[\hat{X}_1, \hat{X}_2] = i[\hat{X}_2, \hat{X}_1]$. But from (2-51), or simply from general knowledge of commutation, this is not true (we are off by a minus sign), so there are elements in the set that do not commute under the 2nd binary operation (2-52) (which is itself a commutation relation). This 2nd binary operation is non-Abelian, and thus, so is the algebra.

and non-Abelian

End of demo

Do **Problem 12** to show there is no identity element for the 2nd operation (2-52) in the *SO*(3) Lie algebra.

Do **Problem 13** to see why we took matrix commutation as our second binary operation for the Lie group, rather than the simpler alternative of matrix multiplication.

The commutation relations embody the structure of the Lie algebra and Lie group, and tell us almost everything one needs to know about both the algebra and the group. Because of this "structuring", the $-\varepsilon_{ijk}$ of (2-51) are often called the structure constants. We will see that other groups have their own particular (different) structure constants, but in every case, they tell us the properties of the algebra and associated group.

Constants in the generator commutation relations are called "structure constants". These structure the group (contain the key info about the group)

Another Choice of Parametrization

Consider an alternative parametrization for $SO(3)$. where instead of θ_i, we use $\theta'_i = -\theta_i$. This could be considered rotation (of a vector for example) in the cw direction around each axis.

Do **Problem 14** to find the generator commutation relations in $SO(3)$. where $\theta_i \rightarrow \theta'_i = -\theta_i$ in (2-12).

With this alternative set of parameters, we find different generators $\hat{X}'_i$, where

$$\left[\hat{X}'_i, \hat{X}'_j\right] = i\varepsilon_{ijk}\hat{X}'_k \qquad \text{for parametrization } \theta'_i = -\theta_i \tag{2-53}$$

We have a different set of structure constants, which differ from those in (2-51) by a minus sign. So, the structure "constants" are not really constant in the sense that they can change for different choices of parametrization. The fact that they were the same for different choices of parametrization in $SO(2)$ [see Section 2.3.2, pg. 25], was a coincidence. More generally, they are not the same.

But structure "constants" change with different parametrizations

In what follows, we will stick with the original parametrization of θ_i, as in (2-12).

Quick intermediate summary for $SO(3)$

For $SO(3)$, and our original choice of parametrization, the

$$\text{generators obey } \left[\hat{X}_i, \hat{X}_j\right] = -i\varepsilon_{ijk}\hat{X}_k \quad i,j,k \text{ each take on a value } 1, 2, \text{ or } 3, \quad \text{repeat of (2-51)}$$

and, for any parametrization, the Lie algebra operations are addition and

$$\text{the 2}^{\text{nd}} \text{ binary operation is } \quad i[\hat{X}_a, \hat{X}_b]\,. \qquad \text{repeat of (2-52)}$$

Summary of $SO(3)$:
1) 3 generators
2) 3 commutation rels
3) binary operations: addition & commutation

2.3.4 Generating SO(3) from Its Lie Algebra

As with the $SO(2)$ Lie group, one can generate the $SO(3)$ group from its generators, via the expansion in the last line of (2-47). We won't go through all the algebra involved, as the steps for each factor parallel those for $SO(2)$, and the actual doing of it is fairly straightforward (though tedious).

2.3.5 Exponentiation Relating SO(3) Lie Group to Its Lie Algebra

General Case is Tricky

For a one parameter Lie group such as (2-6) in θ, the relationship between it and the associated Lie algebra $\theta\hat{X}$ (see (2-42)) was simple exponentiation (2-45). One can generate the group via $M(\theta) = e^{i\theta\hat{X}}$. For a Lie group of more than one parameter, however, things get a little trickier, because one must use the Baker-Campbell-Hausdorff relationship for exponentiation of operators,

$$e^X e^Y = e^{X+Y+\frac{1}{2}[X,Y]+\frac{1}{12}\left(\left[X,[X,Y]\right]+\left[Y,[Y,X]\right]\right)+\ldots}, \tag{2-54}$$

where we imply the infinite series of nested commutators after the second commutator relation[1]. If X and Y commute (as numbers do), we get the familiar addition of exponents relation. When they don't, such as with many operators, things get more complicated.

In our example of $SO(3)$ (2-47), one might naively expect to obtain the Lie group from the Lie algebra using the exponentiation relation on the RHS of (2-55), but due to (2-54) one cannot.

Exponential addition law for operators makes exponentiation of generators to get $SO(3)$ Lie group not simple

$$A(\theta_1,\theta_2,\theta_3) = A_1(\theta_1)A_2(\theta_2)A_3(\theta_3) = e^{i\theta_1\hat{X}_1}e^{i\theta_2\hat{X}_2}e^{i\theta_3\hat{X}_3} \neq e^{i(\theta_1\hat{X}_1+\theta_2\hat{X}_2+\theta_3\hat{X}_3)} = e^{i\theta_i\hat{X}_i} \tag{2-55}$$

So, if you have a particular Lie algebra element $\theta_i\hat{X}_i$ (some sum of the generators), you do not use the RHS of (2-55) to generate the Lie group (2-47). You have to use the relationship in the middle of

---

[1] To be precise, (2-54) only holds if X and Y are "sufficiently small", where defining that term mathematically would take us too far afield. Simply note that all operators we will work with will be sufficiently small.

(2-55). Conversely, if you have a Lie group element in terms of three θ_i, such as A on the LHS of (2-55), you cannot assume the associated Lie algebra element to be exponentiated is $\theta_i\hat{X}_i$.

To get the Lie algebra element, call it $\hat{\theta}_i\hat{X}_i$, associated with a given Lie group element in terms of θ_i, we need to use (2-54). That is, we need to find the $\hat{\theta}_i$ values in

$$A(\theta_1,\theta_2,\theta_3) = A_1(\theta_1)A_2(\theta_2)A_3(\theta_3) = e^{i\hat{\theta}_i\hat{X}_i} \qquad \hat{\theta}_i = \hat{\theta}_i(\theta_j). \tag{2-56}$$

As a simple example, consider the case where $\theta_3 = 0$, so the total group action amounts to a rotation through θ_2 followed by a rotation through θ_1. (Operations act from right to left in (2-56).) Using (2-54), we find

$$\begin{aligned} A(\theta_1,\theta_2,\theta_3=0) &= A_1(\theta_1)A_2(\theta_2) = e^{i\theta_1\hat{X}_1}e^{i\theta_2\hat{X}_2} \\ &= e^{i\theta_1\hat{X}_1+i\theta_2\hat{X}_2+\frac{1}{2}\left[i\theta_1\hat{X}_1,i\theta_2\hat{X}_2\right]+\frac{1}{12}\left(\left[i\theta_1\hat{X}_1,\left[i\theta_1\hat{X}_1,i\theta_2\hat{X}_2\right]\right]+\left[i\theta_2\hat{X}_2,\left[i\theta_2\hat{X}_2,i\theta_1\hat{X}_1\right]\right]\right)+\dots} \\ &= e^{i\hat{\theta}_i\hat{X}_i} . \end{aligned} \tag{2-57}$$

So, using the defining commutation relation of the Lie algebra (2-51), we find

$$\begin{aligned} i\hat{\theta}_i\hat{X}_i &= i\theta_1\hat{X}_1 + i\theta_2\hat{X}_2 + \tfrac{1}{2}\left[i\theta_1\hat{X}_1,i\theta_2\hat{X}_2\right] + \tfrac{1}{12}\left[i\theta_1\hat{X}_1,\left[i\theta_1\hat{X}_1,i\theta_2\hat{X}_2\right]\right] + \tfrac{1}{12}\left[i\theta_2\hat{X}_2,\left[i\theta_2\hat{X}_2,i\theta_1\hat{X}_1\right]\right].. \\ &= i\theta_1\hat{X}_1 + i\theta_2\hat{X}_2 + \tfrac{1}{2}\theta_1\theta_2\left(i\hat{X}_3\right) + i\tfrac{1}{12}\theta_1^2\theta_2\left[\hat{X}_1,i\hat{X}_3\right] + i\tfrac{1}{12}\theta_1\theta_2^2\left[\hat{X}_2,-i\hat{X}_3\right] + \dots \\ &= i\theta_1\hat{X}_1 + i\theta_2\hat{X}_2 + i\tfrac{1}{2}\theta_1\theta_2\hat{X}_3 - i\tfrac{1}{12}\theta_1^2\theta_2\hat{X}_2 - i\tfrac{1}{12}\theta_1\theta_2^2\hat{X}_1 + .. \end{aligned} \tag{2-58}$$

But we can still generate the group from $\hat{X}_i$ using the generator commutation relations

At second order, $\hat{\theta}_i\hat{X}_i \approx \theta_1\hat{X}_1 + \theta_2\hat{X}_2 + \frac{1}{2}\theta_1\theta_2\hat{X}_3$, so $\hat{\theta}_1 \approx \theta_1$, $\hat{\theta}_2 \approx \theta_2$, $\hat{\theta}_3 \approx \frac{1}{2}\theta_1\theta_2$. In principle, we can find the $\hat{\theta}_i$ at any order by using all terms in (2-58) up to that order. And for cases where $\theta_3 \neq 0$, one just repeats the process one more time using the results of (2-58) with (2-54) and the third operator in the exponent $\theta_3\hat{X}_3$.

Do **Problem 15** to obtain the third order $\hat{\theta}_i$ values for our example.

A key thing to notice is that any two group elements $A(\theta_{A1}, \theta_{A2}, \theta_{A3})$ and $B(\theta_{B1}, \theta_{B2}, \theta_{B3})$ of form like (2-56), when multiplied together via the group operation of matrix multiplication, are also in the group, i.e., $AB = C$, where C is in the group. That is, due to the commutation relations (2-51) used in (2-54) we will always get a result equal to the exponentiation of $\underline{\theta}_i\hat{X}_i$, i.e., $C = e^{i\underline{\theta}_i\hat{X}_i}$, where the $\underline{\theta}_i$ can be determined. That is, every group operation on group elements yields a group element, and that group element has an associated Lie algebra element $\underline{\theta}_i\hat{X}_i$. All of this is only because each of the commutation relations (2-51) used in (2-54) [and thus, (2-58)] yields one of the Lie algebra basis matrices $\hat{X}_i$.

Group property of AB=C (with A, B ,C in group) still holds

<u>Infinitesimal Scalars θ_i Case is Simpler</u>

For small θ_i in (2-58), at lowest order $\hat{\theta}_i\hat{X}_i \approx \theta_1\hat{X}_1 + \theta_2\hat{X}_2$, so $\hat{\theta}_1 \approx \theta_1$, $\hat{\theta}_2 \approx \theta_2$, $\hat{\theta}_3 \approx \theta_3 = 0$. It is common to simply consider the group and the algebra to be local (small values of θ_i), so orders higher than the lowest are negligible, and one can simply identify $\hat{\theta}_i \approx \theta_i$. Then, we find (2-57) becomes

$$A(\theta_1,\theta_2,\theta_3=0) = A_1(\theta_1)A_2(\theta_2) = e^{i\theta_1\hat{X}_1}e^{i\theta_2\hat{X}_2} \approx e^{i\theta_1\hat{X}_1+i\theta_2\hat{X}_2} \qquad |\theta_1|,|\theta_2| \ll 1 , \tag{2-59}$$

and for the more general case,

$$A(\theta_1,\theta_2,\theta_3) = A_1(\theta_1)A_2(\theta_2)A_3(\theta_3) = e^{i\theta_1\hat{X}_1}e^{i\theta_2\hat{X}_2}e^{i\theta_3\hat{X}_3} \approx e^{i\theta_i\hat{X}_i} \approx I + i\theta_i\hat{X}_i \qquad |\theta_i| \ll 1 \tag{2-60}$$

Exponentiation to get SO(3) group is simpler in infinitesimal case

In principle, we can generate the global (finite) Lie group by taking $\theta_i \rightarrow d\theta_i$ in (2-60) and carrying out step-wise integration. And of course, we can always generate the finite group with the first part of (2-55), $A(\theta_1,\theta_2,\theta_3) = e^{i\theta_1\hat{X}_1}e^{i\theta_2\hat{X}_2}e^{i\theta_3\hat{X}_3}$.

2.3.6 Summary of SO(n) Lie Groups and Algebras

The first three rows of Wholeness Chart 2-5, pg. 37, summarize special orthogonal Lie groups, their associated Lie algebras, and the use of exponentiation of the latter to generate the former. Note, we have introduced the symbol $\hat{N}$ (with a caret) as a surrogate for what we have been calling the 3D rotation matrix A. The symbol A is common in the literature for that matrix, whereas in the chart, we use carets for real matrices and no carets for complex ones (which are yet to be treated herein).

Summary of special orthogonal Lie groups in top part of Wholeness Chart 2-5

The number of generators for an $SO(n)$ Lie algebra is deduced from what we found for $SO(2)$ and $SO(3)$. The result can be proven, but is not central to our work, so we will not go through that proof here. Also, we use the symbol $\hat{\theta}_i(\theta_j)$ for both $SO(3)$ and $SO(n)$, in order to emphasize the parallel between the two cases, though in general, the functional form of $\hat{\theta}_i$ will be different in each case.

You may wish to follow along with the rest of the chart as we develop $SU(2)$ and $SU(3)$ theory in the next sections.

2.3.7 The SU(2) Associated Lie Algebra

We find the Lie algebra generators for the $SU(2)$ Lie group (2-20) [with notation of (2-25)], repeated below for convenience, in similar fashion to what we did for $SO(2)$ and $SO(3)$.

$$M=\begin{bmatrix} a & b \\ -b^* & a^* \end{bmatrix}=\begin{bmatrix} \alpha_0+i\alpha_3 & \alpha_2+i\alpha_1 \\ -\alpha_2+i\alpha_1 & \alpha_0-i\alpha_3 \end{bmatrix}=i\begin{bmatrix} -i\alpha_0+\alpha_3 & \alpha_1-i\alpha_2 \\ \alpha_1+i\alpha_2 & -i\alpha_0-\alpha_3 \end{bmatrix} \qquad \text{Repeat of (2-25)}$$

That is, from the multivariable Taylor expansion

Expanding SU(2) group of 3 independent parameters to get generators

$$M(\alpha_i)=\underbrace{M(0,0,0)}_{I}+\alpha_1\left.\frac{\partial M}{\partial\alpha_1}\right|_{\alpha_i=0}+\alpha_2\left.\frac{\partial M}{\partial\alpha_2}\right|_{\alpha_i=0}+\alpha_3\left.\frac{\partial M}{\partial\alpha_3}\right|_{\alpha_i=0}$$
$$+\left.\frac{(\alpha_1)^2}{2!}\frac{\partial^2 M}{\partial\alpha_1^2}\right|_{\alpha_i=0}+\left.\frac{\alpha_1\alpha_2}{2!}\frac{\partial^2 M}{\partial\alpha_1\partial\alpha_2}\right|_{\alpha_i=0}+\dots, \qquad (2\text{-}61)$$

the generators are

$$X_1=-i\left.\frac{\partial M}{\partial\alpha_1}\right|_{\alpha_i=0} \qquad X_2=-i\left.\frac{\partial M}{\partial\alpha_2}\right|_{\alpha_i=0} \qquad X_3=-i\left.\frac{\partial M}{\partial\alpha_3}\right|_{\alpha_i=0}. \qquad (2\text{-}62)$$

Evaluating (2-62) for (2-25), we find, with (2-23),

The 3 generators X_i

$$X_1=-i\left.\frac{\partial M}{\partial\alpha_1}\right|_{\alpha_i=0}=-i\left(\frac{\partial}{\partial\alpha_1}\begin{bmatrix} \alpha_0+i\alpha_3 & \alpha_2+i\alpha_1 \\ -\alpha_2+i\alpha_1 & \alpha_0-i\alpha_3 \end{bmatrix}\right)_{\alpha_i=0}=-i\begin{bmatrix} 0 & i \\ i & 0 \end{bmatrix}=\begin{bmatrix} 0 & 1 \\ 1 & 0 \end{bmatrix}$$
$$X_2=-i\left.\frac{\partial M}{\partial\alpha_2}\right|_{\alpha_i=0}=\begin{bmatrix} 0 & -i \\ i & 0 \end{bmatrix} \qquad X_3=-i\left.\frac{\partial M}{\partial\alpha_3}\right|_{\alpha_i=0}=\begin{bmatrix} 1 & 0 \\ 0 & -1 \end{bmatrix}, \qquad (2\text{-}63)$$

The 3 generators X_i are the Paul matrices

which are the Pauli matrices, and which have the commutation relations

with the Pauli matrices commutation relations

$$\left[X_i,X_j\right]=i2\varepsilon_{ijk}X_k \xrightarrow[\text{symbols}]{\text{more common}} \left[\sigma_i,\sigma_j\right]=i2\varepsilon_{ijk}\sigma_k \quad \text{or} \quad \left[\frac{\sigma_i}{2},\frac{\sigma_j}{2}\right]=i\varepsilon_{ijk}\frac{\sigma_k}{2}. \qquad (2\text{-}64)$$

We will not take the time to show that the X_i along with the three scalar field multipliers comprise an algebra under the binary operations of addition and commutation. We have done that twice before for other algebras and should be able to simply accept it here. I assure you it is indeed an algebra.

Note that had we defined M with $\alpha_i \rightarrow ½\alpha_i$ (i = 1,2,3), we would have found $X_i \rightarrow ½X_i = ½\sigma_i$. Then, the commutation relations would have been as in the RHS of (2-64), and we would have the structure constants ε_{ijk}. So, ε_{ijk} are the structure constants if we take our Lie algebra X_i as $½\sigma_i$ (which is common in QFT); $2\varepsilon_{ijk}$ are the structure constants if we take our Lie algebra X_i as σ_i.

SU(2) generators can be taken as $½\sigma_i$, with structure constants ε_{ijk}

The Lie algebra X_i for the $SU(2)$ group has the same number of generators, and for $X_i = ½\sigma_i$, has the same commutation relation we found for one parametrization of the $SO(3)$ group [(2-53)]. The two different groups $SU(2)$ and $SO(3)$ have similar structure and are similar in many ways. For one,

SU(2) happens to be similar to what we found for SO(3)

which we won't get into in depth here[1], as it doesn't play much role in the standard model of QFT, the 3-dimensional (pseudo) vector of angular momentum can be treated under either the 3D rotation group $SO(3)$ or (non-relativistically) under the 2D $SU(2)$ group. As you may have run into in other studies, spin angular momentum in NRQM is often analyzed using a 2D complex column vector with the top component representing spin up (along z axis) and the lower component representing spin down. The Pauli matrices, via their operations on the column vector, play a key role in all of that.

Take caution that $SO(3)$ and $SU(2)$ are *not* different representations of the same group, even though they may, under certain parametrizations, share the same structure constants (same commutation relations). They are different groups, but they can characterize the same physical phenomenon. This is similar to the relationship between 2D rotation group $SO(2)$ and the $U(1)$ group we looked at in Problem 8.

SU(2) and SO(3) are different groups, but can both characterize 3D rotation. They have similar Lie algebra structures

2.3.8 Generating the SU(2) Group from the SU(2) Lie Algebra

Do Problem 16 to prove to yourself that the X_i above generate the $SU(2)$.

From the results of Problem 16, we see that (2-61) can be expressed as

Expressing SU(2) elements in terms of generators and independent parameters

$$M(\alpha_i) = I + i\alpha_1 X_1 + i\alpha_2 X_2 + i\alpha_3 X_3 - \frac{\alpha_1^2}{2!}I - \frac{\alpha_2^2}{2!}I - \frac{\alpha_3^2}{2!}I + \frac{\alpha_1\alpha_2}{2!}[0] + \ldots \qquad \text{(2-65)}$$

2.3.9 Exponentiation of the SU(2) Lie Algebra

Finite Parameter Case

One can obtain the Lie group from the Lie algebra via the expansion (2-61) along with (2-62), expressed in (2-65). One can also obtain it in a second (related) way, which involves exponentiation, in a manner similar to what we saw earlier with $SO(2)$ and $SO(3)$. However, we would find doing so to be a mathematical morass, so we will simply draw parallels to what we saw with earlier groups.

Exponentiating Lie algebra to get SU(2) Lie group is a nightmare

We illustrate how it is done in principle

Consider a general Lie algebra element

$$X = \alpha_1 X_1 + \alpha_2 X_2 + \alpha_3 X_3 , \qquad \text{(2-66)}$$

where one could exponentiate it as

$$e^{iX} = e^{i(\alpha_1 X_1 + \alpha_2 X_2 + \alpha_3 X_3)} . \qquad \text{(2-67)}$$

and where we note, in passing, that (see (2-54) and (2-55))

$$e^{i(\alpha_1 X_1 + \alpha_2 X_2 + \alpha_3 X_3)} \neq e^{i\alpha_1 X_1} e^{i\alpha_2 X_2} e^{i\alpha_3 X_3} . \qquad \text{(2-68)}$$

We would like to explore whether (2-67) equals (2-25) [equivalently, (2-65)],

$$\begin{aligned} e^{i(\alpha_1 X_1 + \alpha_2 X_2 + \alpha_3 X_3)} &\stackrel{?}{=} M(\alpha_i) \\ &= I + i\alpha_1 X_1 + i\alpha_2 X_2 + i\alpha_3 X_3 - \frac{\alpha_1^2}{2!}I - \frac{\alpha_2^2}{2!}I - \frac{\alpha_3^2}{2!}I + \frac{\alpha_1\alpha_2}{2!}[0] + \ldots \end{aligned} \qquad \text{(2-69)}$$

By expanding the top row LHS of (2-69) around $\alpha_i = 0$, we could see whether or not it matches the expansion of M in (2-69), 2nd row. We will not go through all that tedium, but draw instead on our knowledge of the other multiple parameter case $SO(3)$, where we found the equal sign with the question mark in (2-69) is actually a $\neq$ sign. If we wished, however, we could, with a copious amount of labor, find a matrix function to exponentiate that would give us M. That is, similar to (2-56),

$$M(\alpha_i) = e^{i\beta_i X_i} \neq e^{i(\alpha_1 X_1 + \alpha_2 X_2 + \alpha_3 X_3)} \qquad \beta_i = \beta_i(\alpha_i) . \qquad \text{(2-70)}$$

Do Problem 17 to help in what comes next.

---

[1] See footnote references on pg. 8 or almost any text on group theory.

Infinitesimal Parameter Case

However, for small values of $|\alpha_i|$, (<< 1), as can be found by doing Problem 17,

$$e^{i(\alpha_1 X_1+\alpha_2 X_2+\alpha_3 X_3)} \approx I+i\alpha_1 X_1+i\alpha_2 X_2+i\alpha_3 X_3 \approx M(\alpha_i) \quad |\alpha_i| \ll 1. \tag{2-71}$$

Exponentiation in infinitesimal case is simpler

As with prior cases, one could generate the global (finite) $SU(2)$ group by step-wise integration over infinitesimal α_i.

2.3.10 Another Parametrization of SU(2)

(2-72) below is a different parametrization of $SU(2)$ with three different parameters.

Do **Problem 18** to prove it.

Another parametrization of SU(2)

$$M(\phi_1,\phi_2,\phi_3)=M(\phi_i)=\begin{bmatrix} \cos\phi_1 e^{i\phi_2} & \sin\phi_1 e^{i\phi_3} \\ -\sin\phi_1 e^{-i\phi_3} & \cos\phi_1 e^{-i\phi_2}\end{bmatrix} \qquad a=\cos\phi_1 e^{i\phi_2} \qquad b=\sin\phi_1 e^{i\phi_3} \tag{2-72}$$

Note that (2-6) is a special case (subgroup, actually) of (2-72) where $\phi_2 = \phi_3 = 0$ (and here, $\phi_1 = -\theta$).

We find the generators for (2-72) in the same way as we did for (2-25).

The Lie algebra generators of (2-73) below turn out to be somewhat different from the earlier example (2-63) in that they are switched around, and X_3 is not found via the simple derivative with all $\phi_i = 0$, as we had earlier. There are subtleties of group theory involved here, and we don't want to go off on too much of a tangent from our fundamental goal, so we will leave it at that.

As we noted earlier, different parametrizations generally lead to different generators. In any vector space (such as our Lie algebra) one can have different basis vectors (matrices are the vectors here). So, the generators for any matrix Lie group depend on what choice we make for the independent parameters.

The generators for this form

$$X_1=-i\left.\frac{\partial M}{\partial\phi_1}\right|_{\phi_i=0}=-i\begin{bmatrix} -\sin\phi_1 e^{i\phi_2} & \cos\phi_1 e^{i\phi_3} \\ -\cos\phi_1 e^{-i\phi_3} & -\sin\phi_1 e^{-i\phi_2}\end{bmatrix}_{\phi_i=0}=-i\begin{bmatrix}0 & 1\\ -1 & 0\end{bmatrix}=\begin{bmatrix}0 & -i\\ i & 0\end{bmatrix}$$

$$X_2=-i\left.\frac{\partial M}{\partial\phi_2}\right|_{\phi_i=0}=-i\begin{bmatrix} i\cos\phi_1 e^{i\phi_2} & 0 \\ 0 & -i\cos\phi_1 e^{-i\phi_2}\end{bmatrix}_{\phi_i=0}=-i\begin{bmatrix}i & 0\\ 0 & -i\end{bmatrix}=\begin{bmatrix}1 & 0\\ 0 & -1\end{bmatrix}$$

$$-i\left.\frac{\partial M}{\partial\phi_3}\right|_{\phi_i=0}=-i\begin{bmatrix} 0 & i\sin\phi_1 e^{i\phi_3} \\ i\sin\phi_1 e^{-i\phi_3} & 0\end{bmatrix}_{\phi_i=0}=\begin{bmatrix}0 & 0\\ 0 & 0\end{bmatrix} \quad \text{Not relevant}$$

$$-i\left.\frac{\partial^2 M}{\partial\phi_1\partial\phi_3}\right|_{\phi_i=0}=-i\begin{bmatrix} 0 & i\cos\phi_1 e^{i\phi_3} \\ i\cos\phi_1 e^{-i\phi_3} & 0\end{bmatrix}_{\phi_i=0}=\begin{bmatrix}0 & 1\\ 1 & 0\end{bmatrix} \quad \text{needed in expansion of } M$$

$$X_3=-i\left.\frac{\partial M(\phi_1,0,\pi/2)}{\partial\phi_1}\right|_{\phi_1=0}=-i\begin{bmatrix} -\sin\phi_1 & \cos\phi_1 e^{i\pi/2} \\ -\cos\phi_1 e^{-i\pi/2} & -\sin\phi_1\end{bmatrix}_{\phi_1=0}=\begin{bmatrix}0 & 1\\ 1 & 0\end{bmatrix} \quad \text{3rd generator}$$

(2-73)

For small values of the parameters in (2-72),

$$M(\phi_1,\phi_2,\phi_3)\approx\begin{bmatrix}1+i\phi_2 & \phi_1\\ -\phi_1 & 1-i\phi_2\end{bmatrix} \qquad |\phi_i| \ll 1 \tag{2-74}$$

and we can express the group matrix as

$$M(\phi_i)\approx I+if_j(\phi_i)X_j \approx e^{if_j(\phi_i)X_j} \approx e^{i(\phi_1 X_1+\phi_2 X_2)} \qquad |\phi_i| \ll 1 \quad f_1=\phi_1, f_2=\phi_2, f_3=0 \tag{2-75}$$

However, we cannot express $M(\phi_i)$ globally as either a simple finite sum or as an exponentiation of $i(\phi_1 X_1 + \phi_2 X_2)$ as in (2-75). We could express it globally as an infinite sum, similar in concept to (2-61). We could also express it globally (after a whole lot of algebra) as an exponentiation of some function $\tilde{\beta}_j(\phi_i)$ of the ϕ_i as in

$$M(\phi_i) = e^{i\tilde{\beta}_j(\phi_i)X_j} \qquad \text{any size } \phi_i, \ \tilde{\beta}_j(\phi_i) \text{ deduced to make it work}. \qquad (2\text{-}76)$$

Bottom line:

(Finite parameters α_i) For an $SU(2)$ parametrization of particular form (2-25), we can obtain a global expression of the group 1) as a finite sum of terms linear in the generators and the identity, and 2) as a (complicated) exponentiation of the generators (see (2-70)).

(Other finite parameters e.g., ϕ_i) For other parametrizations, 1) above needs an infinite sum of terms.

(Any infinitesimal parameters) In any parametrization, we can find a local expression of the group as 1) a finite sum of terms linear in the generators and the identity or 2) as an exponential [see (2-75) and (2-71)]. This is the usual approach to the Lie algebra as the tangent space (local around the identity) to the Lie group. Locally, finding the group from the Lie algebra via exponentiation is relatively easy. Globally, it is generally horrendous.

Note that, in general, the generators for different parametrizations can be different, and thus so are the structure constants. However, we can find linear combinations of the generators from one parametrization that equal the generators from another parametrization. In other words, the vector space of the abstract Lie algebra for a given Lie group is the same for any parametrization, even though we generally get different generators (basis vectors [= matrices]) for different parametrizations.

(2-72) has value in analyzing spin. (2-25) has value in QFT. Different parametrizations work better for different applications.

This form good for spin; prior form better for QFT

Digression for Brief Look at Spin and $SU(2)$ in NRQM

However, we will now digress briefly to show how (2-72) can be used for spin analysis. Recall the wave function in NRQM had a two-component column vector representing spin.

Brief look at how this form of SU(2) can handle spin

$$|\psi\rangle_{\substack{spin\\up}} = Ae^{-ikx}\begin{bmatrix}1\\0\end{bmatrix} \text{ spin in } +z \text{ direction} \qquad |\psi\rangle_{\substack{spin\\down}} = Ae^{-ikx}\begin{bmatrix}0\\1\end{bmatrix} \text{ spin in } -z \text{ direction} \qquad (2\text{-}77)$$

Consider the case where we rotate the spin down particle to a spin up orientation (or conversely, rotating our observing coordinate system in the opposite direction). In physical space we have rotated the z axis 180° and could use the $SO(3)$ rotation group (2-47) to rotate the 3D (pseudo) vector for spin angular momentum through 180°. However, for the manner in which we represent spin in (2-77), that would not work, as spin there is represented by a two-component vector, not a three-component one. But, consider the $SU(2)$ parametrization (2-72) where, in this case, $\phi_2 = \phi_3 = 0$, and the ϕ_1 is a rotation about the x axis (which effectively rotates the z axis around the x axis). We actually need to take $\phi_1 = \phi/2$, where ϕ is the actual physical angle of rotation, in order to make it work, as we are about to see.

Then note what (2-72) does to the spin down wave function on the RHS of (2-77).

3D rotation in SU(2) via $\phi_1 = \phi_{3D}/2$.

$$M|\psi\rangle_{\substack{spin\\down}} = \begin{bmatrix}\cos\phi_1 & \sin\phi_1\\ -\sin\phi_1 & \cos\phi_1\end{bmatrix}Ae^{-ikx}\begin{bmatrix}0\\1\end{bmatrix} = Ae^{-ikx}\underbrace{\begin{bmatrix}\cos\frac{\phi}{2} & \sin\frac{\phi}{2}\\ -\sin\frac{\phi}{2} & \cos\frac{\phi}{2}\end{bmatrix}}_{\text{for } \phi=180^\circ}\begin{bmatrix}0\\1\end{bmatrix}$$

$$= Ae^{-ikx}\begin{bmatrix}0 & 1\\ -1 & 0\end{bmatrix}\begin{bmatrix}0\\1\end{bmatrix} = Ae^{-ikx}\begin{bmatrix}1\\0\end{bmatrix} = |\psi\rangle_{\substack{spin\\up}}. \qquad (2\text{-}78)$$

So, we see that the spinor (two-component column vector) lives in a 2D vector space, on which the elements of the $SU(2)$ group operate. And rotations in 3D space can be characterized, in a 2D complex space, by the $SU(2)$ group. Because ϕ_1 in the 2D complex space of $SU(2)$ ranges over 360°, while ϕ in the physical space of $SO(3)$ ranges over 720°, we say $SU(2)$ is the double cover of $SO(3)$.

As noted earlier this has wide ranging application in analyzing spin, but we will leave further treatment of this topic to other sources, such as those cited previously.

End of digression

We do note that the matrix operation of (2-78) is sometimes referred to as a raising operation as it raises the lower component into the upper component slot. Conversely, when an operation transfers an upper component to a lower component slot, it is called a lowering operation. We will run into these concepts again in QFT.

Raising operation: column vector component up one level. Lowering operation: down one

2.3.11 Summary of Parametrizations and Characterizations

Wholeness Chart 2-4 summarizes what we have found for different relationships between Lie groups and their respective Lie algebras.

Wholeness Chart 2-4. Lie Group Parametrizations and Characterizations

Lie Group Relationships	Examples	Matrices	Lie Algebra
Same group, different parametrizations	$SU(2)$ of (2-25) and (2-72)	Different forms for matrices, but same dimension	Same abstract Lie algebra. May (or may not) have same basis matrices (generators) with same structure constants
Different groups characterizing same physical world phenomenon	$SO(3)$ and $SU(2)$, both for 3D rotation. $SO(2)$ and $U(1)$, both for 2D rotation	Different forms for matrices, different dimensions	Different abstract Lie algebra. May (or may not) have same structure constants

2.3.12 Shortcut Way to Generate the First Parametrization

Note that because of its particular form, our first parametrization (2-25) of the $SU(2)$ representation can be found rather easily from the Lie algebra simply by adding the generators and the identity matrix, multiplied by their associated parameters. That is,

1st SU(2) form is easy to generate from the generators

$$\begin{aligned} M &= \begin{bmatrix} \alpha_0 + i\alpha_3 & \alpha_2 + i\alpha_1 \\ -\alpha_2 + i\alpha_1 & \alpha_0 - i\alpha_3 \end{bmatrix} = \alpha_0 \begin{bmatrix} 1 & 0 \\ 0 & 1 \end{bmatrix} + i\alpha_1 \begin{bmatrix} 0 & 1 \\ 1 & 0 \end{bmatrix} + i\alpha_2 \begin{bmatrix} 0 & -i \\ i & 0 \end{bmatrix} + i\alpha_3 \begin{bmatrix} 1 & 0 \\ 0 & -1 \end{bmatrix} \\ &= \alpha_0 I + i\alpha_1 X_1 + i\alpha_2 X_2 + i\alpha_3 X_3 = \sqrt{1-\alpha_1^2-\alpha_2^2-\alpha_3^2}\, I + i\alpha_1 X_1 + i\alpha_2 X_2 + i\alpha_3 X_3 \qquad (2\text{-}79) \\ &= I + i\alpha_1 X_1 + i\alpha_2 X_2 + i\alpha_3 X_3 - \frac{\alpha_1^2}{2!} I - \frac{\alpha_2^2}{2!} I - \frac{\alpha_3^2}{2!} I + \frac{\alpha_1 \alpha_2}{2!}[0] + \dots , \end{aligned}$$

where the last line, in which we expand the dependent variable α_0 in terms of the independent variables, is simply our original expansion (2-61), which in terms of the generators is (2-65).

So, in this particular parametrization, going back and forth between the Lie group and the Lie algebra (plus the identity matrix) is relatively easy.

However, it is not so easy and simple with the second parametrization (2-72). In the expansion of $M(\phi_i)$ (which we didn't do), one gets terms of form $\phi_i X_i$ in the infinite summation, but the original matrix had functions of $\sin\phi_1$, $\cos\phi_1$, $e^{\pm i\phi_2}, e^{\pm i\phi_3}$ multiplied by one another. That gets complicated in a hurry.

2nd SU(2) form is hard to generate from the generators

As noted, in NRQM, we deal with the 2nd parametrization (2-72). In QFT, we deal with the 1st. So, in this sense, QFT is easier. (But, probably only in that sense.)

2.3.13 The SU(3) Lie Algebra

We repeat (2-28), the most suitable form of $SU(3)$ for our purposes, below

$$N = i\begin{bmatrix} -i+\alpha_3+\frac{\alpha_8}{\sqrt{3}} & \alpha_1 - i\alpha_2 & \alpha_4 - i\alpha_5 \\ \alpha_1 + i\alpha_2 & -i-\alpha_3+\frac{\alpha_8}{\sqrt{3}} & \alpha_6 - i\alpha_7 \\ \alpha_4 + i\alpha_5 & \alpha_6 + i\alpha_7 & -i-\frac{2\alpha_8}{\sqrt{3}} \end{bmatrix}$$

Repeat of (2-28)

$$= \begin{bmatrix} 1 & 0 & 0 \\ 0 & 1 & 0 \\ 0 & 0 & 1 \end{bmatrix} + i\begin{bmatrix} \alpha_3+\frac{\alpha_8}{\sqrt{3}} & \alpha_1 - i\alpha_2 & \alpha_4 - i\alpha_5 \\ \alpha_1 + i\alpha_2 & -\alpha_3+\frac{\alpha_8}{\sqrt{3}} & \alpha_6 - i\alpha_7 \\ \alpha_4 + i\alpha_5 & \alpha_6 + i\alpha_7 & -\frac{2\alpha_8}{\sqrt{3}} \end{bmatrix} \qquad |\alpha_i| \ll 1.$$

The Gell-Mann matrices λ_i, which are convenient to work with in $SU(3)$ theory, are

$$\lambda_1 = \begin{bmatrix} 0 & 1 & 0 \\ 1 & 0 & 0 \\ 0 & 0 & 0 \end{bmatrix} \quad \lambda_2 = \begin{bmatrix} 0 & -i & 0 \\ i & 0 & 0 \\ 0 & 0 & 0 \end{bmatrix} \quad \lambda_3 = \begin{bmatrix} 1 & 0 & 0 \\ 0 & -1 & 0 \\ 0 & 0 & 0 \end{bmatrix} \quad \lambda_4 = \begin{bmatrix} 0 & 0 & 1 \\ 0 & 0 & 0 \\ 1 & 0 & 0 \end{bmatrix}$$
$$\lambda_5 = \begin{bmatrix} 0 & 0 & -i \\ 0 & 0 & 0 \\ i & 0 & 0 \end{bmatrix} \quad \lambda_6 = \begin{bmatrix} 0 & 0 & 0 \\ 0 & 0 & 1 \\ 0 & 1 & 0 \end{bmatrix} \quad \lambda_7 = \begin{bmatrix} 0 & 0 & 0 \\ 0 & 0 & -i \\ 0 & i & 0 \end{bmatrix} \quad \lambda_8 = \frac{1}{\sqrt{3}}\begin{bmatrix} 1 & 0 & 0 \\ 0 & 1 & 0 \\ 0 & 0 & -2 \end{bmatrix}. \tag{2-80}$$

Using these λ_i matrices, we can construct (or generate) N

Then, with (2-80) and (2-28), we find

$$\begin{aligned} N &= I + i(\alpha_1\lambda_1 + \alpha_2\lambda_2 + \alpha_3\lambda_3 + \alpha_4\lambda_4 + \alpha_5\lambda_5 + \alpha_6\lambda_6 + \alpha_7\lambda_7 + \alpha_8\lambda_8) \\ &= I + i\alpha_i\lambda_i \qquad |\alpha_i| \ll 1, \end{aligned} \tag{2-81}$$

which parallels the second line of (2-79). By doing Problem 19, you can show that the λ_i are the Lie algebra generators of N.

Do **Problem 19** to show that λ_i are the Lie algebra generators of N. This problem is important for a sound understanding of $SU(3)$.

Then do **Problem 20** to help in what comes next.

It turns out, if one cranks all the algebra using (2-80), that the following commutation relations exist between the Lie algebra generators (basis vector matrices), similar to (2-64) in $SU(2)$ theory,

$SU(3)$ generators' commutation relations

$$[\lambda_i, \lambda_j] = i2f_{ijk}\lambda_k \qquad \text{or} \qquad \left[\frac{\lambda_i}{2}, \frac{\lambda_j}{2}\right] = i\, f_{ijk}\frac{\lambda_k}{2}, \tag{2-82}$$

and where repeated indices, as usual, indicate summation. Similar to what we found with $SU(2)$ in (2-64), we can take our $SU(3)$ generators as ½λ_i just as readily as we can take them to be λ_i. For the former, the structure constants are f_{ijk}; for the latter, $2f_{ijk}$. The f_{ijk} are not, however, as simple as the structure constants for $SU(2)$, which took on the values ±1 of the Levi-Civita symbol ε_{ijk}.

$SU(3)$ generators can be taken as ½λ_i, with structure constants f_{ijk}

So that you are not confused, note that some authors define another tensor $F_i = \lambda_i/2$, use that to construct N, and refer when needed to the explicit form of the Gell-Mann matrices λ_i. The choice is, of course, conventional, but I think it easier and clearer to stick to one set of matrices (the λ_i), and that is what we do herein.

The f_{ijk} do turn out to be totally anti-symmetric, like the ε_{ijk}.

f_{ijk} are fully anti-symmetric

$$f_{ijk} = -f_{jik} = f_{jki} = -f_{kji} = f_{kij} = -f_{ikj}, \tag{2-83}$$

and they take on the specific values shown in Table 2-2 (some of which you can check via your solution to Problem 20).

Table 2-2
Values of *SU*(3) Structure Constants

ijk	123	147	156	246	257	345	367	458	678	Others
f_{ijk}	1	$\frac{1}{2}$	$-\frac{1}{2}$	$\frac{1}{2}$	$\frac{1}{2}$	$\frac{1}{2}$	$-\frac{1}{2}$	$\frac{\sqrt{3}}{2}$	$\frac{\sqrt{3}}{2}$	0

Note the form of (2-28) parallels that of (2-25). The form of (2-28) meshes with the conventional, widely used definition of (2-80).

As with $SU(2)$, we will not go through all the steps to show the λ_i with the α_i comprise a Lie algebra, as they parallel what we did for $SO(2)$ and $SO(3)$. One can see that the λ_i matrices and the scalars α_i form a vector space, and that with commutation as the second binary operation, defined via (2-82), there is closure.

The λ_i with α_i satisfy requirements of a Lie algebra

2.3.14 Generating the SU(3) Group from the SU(3) Lie Algebra

We include this section to keep the parallel with the $SU(2)$ group (Section 2.3.8 on pg. 31), although it should be fairly obvious that for $|\alpha_i| \ll 1$, we can generate the $SU(3)$ group from its Lie algebra via (2-81).

2.3.15 Exponentiation of the SU(3) Lie Algebra

General Case

One can generate the $SU(3)$ Lie group from its Lie algebra via a second, related way, which involves exponentiation, in a manner similar to what we saw earlier with $SO(2)$, $SO(3)$, and $SU(2)$. As we found before, doing so would be an algebraic nightmare, so we will once again simply outline the steps that would be taken.

Due to Baker-Campbell-Hausdorf, exponentiating Lie algebra to get SU(3) Lie group would be a mathematical morass

Consider a general $SU(3)$ Lie algebra element

$$\Lambda = i\alpha_i\lambda_i \qquad \text{any size } |\alpha_i|, \tag{2-84}$$

where, as we found with other groups earlier,

$$e^{i\alpha_i\lambda_i} \neq N(\alpha_i) \qquad \text{any size } |\alpha_i|. \tag{2-85}$$

A Taylor expansion of the LHS of (2-85) would not give us the RHS.

However, again as discussed before for other groups, we could, with a whole lot of effort, deduce other parameters, call them β_i, for which

But in principle, we could do it with $e^{i\beta_i\lambda_i}$ and eight β_i as functions of eight α_i.

$$e^{i\beta_i\lambda_i} = N(\alpha_i) \qquad \beta_i = \beta_i(\alpha_i). \tag{2-86}$$

The functions $\beta_i(\alpha_i)$ here are, in general, different functions from $\beta_i(\alpha_i)$ of (2-70), as they must be since there are different numbers of independent variables in the two cases. We use the same symbol to emphasize the parallels between $SU(2)$ and $SU(3)$. We won't be taking the time and effort to find the β_i of (2-86) here.

Infinitesimal Case

But, as with other groups, the non-equal sign in (2-85) becomes approximately equal for small parameters, i.e., $\alpha_i \to 0$, as the higher-order terms are dwarfed by the lower-order ones. And essentially, the equal sign replaces the non-equal sign for infinitesimal values of α_i. So,

Infinitesimal case is easier, as then $\beta_i = \alpha_i$.

$$\beta_i \to \alpha_i \qquad \text{as } \alpha_i \to 0, \tag{2-87}$$

and

$$e^{i\beta_i\lambda_i} \approx e^{i\alpha_i\lambda_i} \approx N(\alpha_i) \approx I + i\alpha_i\lambda_i \qquad |\alpha_i| \ll 1. \tag{2-88}$$

Generalizing Exponentiation

Wholeness Chart 2-5 below summarizes, and extrapolates to groups of degree n, what we have learned about exponentiation of the Lie algebra generators to generate the group.

Note that we have deduced the number of generators for a group of degree n, for each group type, from what we've learned about 2nd and 3rd degree groups. The number of generators for an nXn $SO(n)$ matrix equals the number of components above and to the right of the diagonal, i.e., one half of the total number of off-diagonal components. The number of generators for an nXn $SU(n)$ matrix equals the total number of components minus one.

Summary of n degree Lie groups shown in Wholeness Chart 2-5.

In the chart, and generally from here on throughout this text, we employ carets for real matrices and no carets for complex ones. Take caution, however, that this is not a commonly used convention. In particular, A is typically used to denote the 3D physical rotation matrix, but in the chart, we use $\hat{N}$.

And as noted earlier for special orthogonal groups, in order to draw clear parallels, we use the same symbol (i.e., $\hat{\theta}_i$ for real groups, β_i for complex ones) as functions of the independent variables in the 3 and n degree cases, but the functional dependence is generally different. That is $\beta_i(\alpha_i)$ is a different function of α_i for 3D matrices than it is for nD matrices (for $n \neq 3$).

Wholeness Chart 2-5. Exponentiation of Lie Algebra Generators

Lie Algebra	**Generators** Preferred Parametrization	**Lie Group Representation** Finite Parameters	 Infinitesimal Parameters	
$SO(2)$	$\hat{X}$ of (2-42)	$\hat{M} = e^{i\theta\hat{X}}$	$\hat{M} = e^{i\theta\hat{X}}$ $= I + i\theta\hat{X}$	
$SO(3)$	3 $\hat{X}_i$ matrices of (2-49) $i = 1, 2, 3$	$\hat{N}(\theta_i) = A(\theta_1,\theta_2,\theta_3) = e^{i\hat{\theta}_i\hat{X}_i}$ $= e^{i\theta_1\hat{X}_1}e^{i\theta_2\hat{X}_2}e^{i\theta_3\hat{X}_3}$ $\hat{\theta}_i = \hat{\theta}_i(\theta_j)$	$\hat{N} = e^{i\hat{\theta}_i\hat{X}_i} = e^{i\theta_i\hat{X}_i}$ $= I + i\theta_i\hat{X}_i$ $= I + i\hat{\theta}_i\hat{X}_i$	$\hat{\theta}_i = \theta_i$
$SO(n)$	$\frac{n^2-n}{2}$ matrices $\hat{Y}_i$ $i = 1, 2, \ldots, \frac{n^2-n}{2}$	$\hat{P}(\theta_1,\theta_2,\ldots,\theta_n) = e^{i\hat{\theta}_i\hat{Y}_i}$ $= e^{i\theta_1\hat{Y}_1}e^{i\theta_2\hat{Y}_2}\ldots e^{i\theta_n\hat{Y}_n}$ $\hat{\theta}_i = \hat{\theta}_i(\theta_j)$	$\hat{P} = e^{i\hat{\theta}_i\hat{Y}_i} = e^{i\theta_i\hat{Y}_i}$ $= I + i\theta_i\hat{Y}_i$ $= I + i\hat{\theta}_i\hat{Y}_i$	$\hat{\theta}_i = \theta_i$
$SU(2)$	3 Pauli matrices* $X_i = \sigma_i$ of (2-63) $i = 1, 2, 3$	$M(\alpha_i) = e^{i\beta_i X_i}$ $\beta_i = \beta_i(\alpha_i)$	$M = e^{i\beta_i X_i} = e^{i\alpha_i X_i}$ $= I + i\alpha_i X_i$ $= I + i\beta_i X_i$	$\beta_i = \alpha_i$
$SU(3)$	8 Gell-Mann matrices* $X_i = \lambda_i$ of (2-80) $i = 1, 2, \ldots 8$	$N(\alpha_i) = e^{i\beta_i X_i}$ $\beta_i = \beta_i(\alpha_j)$	$N = e^{i\beta_i X_i} = e^{i\alpha_i X_i}$ $= I + i\alpha_i X_i$ $= I + i\beta_i X_i$	$\beta_i = \alpha_i$
$SU(n)$	$n^2 - 1$ matrices Y_i $i = 1, 2, \ldots, n^2 - 1$	$P(\alpha_i) = e^{i\beta_i Y_i}$ $\beta_i = \beta_i(\alpha_j)$	$P = e^{i\beta_i Y_i} = e^{i\alpha_i Y_i}$ $= I + i\alpha_i Y_i$ $= I + i\beta_i Y_i$	$\beta_i = \alpha_i$

* We can, instead, as is common in QFT, take $X_i = ½\sigma_i$ in $SU(2)$ and $X_i = ½\lambda_i$ in $SU(3)$. This would simply mean our arbitrary parameters β_i and α_i above would be multiplied by 2.

2.3.16 Other Parametrizations of SU(3)

As with $SU(2)$, there are other parametrizations of $SU(3)$ [than (2-28)], but we won't delve into those at this point.

Other forms of SU(3) exist, but we won't look at them here

2.4 "Rotations" in Complex Space and Associated Symmetries

2.4.1 Conceptualizing SU(n) Operations on Vectors

As we have noted in Vol. 1 (pg. 27, Box 2-3, and top half of pg. 199), a unitary operator operating on a generally complex state vector keeps the "length" (absolute magnitude) of that vector unchanged. This is parallel to real orthogonal matrices operating on real column vectors.

For real matrices and vectors, the operation of the matrix on the vector corresponds to a rotation of the vector (for the active operation interpretation), or alternatively, to a rotation of the coordinate axes from which the (stationary) vector is viewed (passive interpretation). (See Vol. 1, Section 6.1.2, pg. 164.)

SU(n) group operations complex, but analogous to rotations in real spaces

For complex vector spaces and complex matrix operators acting on them, one can conceptualize, in an abstract sense, a unitary operation as a "rotation" in complex space of the complex vector (for the active interpretation), or alternatively, as a "rotation" of the abstract vector space coordinate "axes" from which the (stationary) vector is viewed (passive interpretation). Having this perspective in the back of your mind can often help in following the mathematical procedures involved in carrying out such operations.

2.4.2 *Symmetries Under SU(n)* Operations

We should know that a real scalar invariant, such as the length of a vector, remains unchanged under a special orthogonal transformation, i.e., a rotation. That is, for nD space, where $\hat{P}$ (see (2-6) for the 2D case) is the rotation operation and $[v]$ is the column vector form of the abstract vector **v**,

SO(n) group operations = rotations with vector length invariant

$$|\mathbf{v}|^2 = \mathbf{v}\bullet\mathbf{v} = [v]^T[v] \quad [v'] = [\hat{P}][v] \quad \rightarrow \quad |\mathbf{v}'|^2 = \mathbf{v}'\bullet\mathbf{v}' = [v']^T[v'] = [v]^T[\hat{P}]^T[\hat{P}][v] \quad (2\text{-}89)$$

For an orthogonal matrix, $[\hat{P}]^T = [\hat{P}]^{-1}$, so

$$|\mathbf{v}'|^2 = [v]^T[\hat{P}]^T[\hat{P}][v] = [v]^T[v] = |\mathbf{v}|^2, \quad (2\text{-}90)$$

and the length of the vector after rotation is unchanged, perhaps not such a surprising result to most readers. The key points are that 1) the length of the vector is invariant under the transformation (we say the scalar length is invariant or symmetric) and 2) we can carry this concept over to complex spaces.

So, now, consider an nD complex column vector represented by $[w]$ and an $SU(n)$ matrix operator P. (See (2-28) and Problem 7 for $n = 2$). We have

SU(n) group operations keep complex vector magnitude invariant, i.e., like "rotations"

$$|\mathbf{w}|^2 = \mathbf{w}\bullet\mathbf{w} = [w]^\dagger[w] \quad [w'] = [P][w] \quad \rightarrow \quad |\mathbf{w}'|^2 = \mathbf{w}'\bullet\mathbf{w}' = [w']^\dagger[w'] = [w]^\dagger[P]^\dagger[P][w]. \quad (2\text{-}91)$$

For a unitary matrix, $[P]^\dagger = [P]^{-1}$, so

$$|\mathbf{w}'|^2 = [w]^\dagger[P]^\dagger[P][w] = [w]^\dagger[w] = |\mathbf{w}|^2, \quad (2\text{-}92)$$

and the "length" (absolute magnitude) of the vector after the unitary operation is unchanged. As this is the hallmark of pure rotation of a vector, i.e., with no stretching/compression, in real vector spaces, it is natural to think of a special unitary operation as a kind of "rotation" in complex vector space.

Bottom line: The absolute magnitude of a real (complex) vector is symmetric (invariant) under an orthogonal (unitary) transformation.

Reminder note: Recall from Chap. 6, Sect. 6.1.2, of Vol. 1. (pgs.164-166) that for a scalar to be symmetric it must have both 1) the same functional form in transformed (primed for us) variables as it had in the original (non-prime for us) variables, and 2) the same value after the transformation.

A symmetric scalar (like vector length) has same value and functional form after transformation

As examples, in (2-89) and (2-90), and also in (2-91) and (2-92), we had

$$|\mathbf{v}|^2 = \underbrace{v_1v_1 + v_2v_2 + \dots + v_nv_n}_{\text{original functional form}} = \underbrace{v_1'v_1' + v_2'v_2' + \dots + v_n'v_n'}_{\text{transformed functional form same as original}} = |\mathbf{v}'|^2 \quad \text{equal sign means same numeric value}, \quad (2\text{-}93)$$

$$|\mathbf{w}|^2 = \underbrace{w_1^*w_1 + w_2^*w_2 + \dots + w_n^*w_n}_{\text{original functional form}} = \underbrace{w_1'^*w_1' + w_2'^*w_2' + \dots + w_n'^*w_n'}_{\text{transformed functional form same as original}} = |\mathbf{w}'|^2 \quad \text{equal sign means same numeric value}. \quad (2\text{-}94)$$

2.4.3 Applications in Quantum Theory

In General

In theories of quantum mechanics, a state vector can be expressed as a superposition of orthogonal basis state vectors, e.g.,

$$\underbrace{|\psi\rangle}_{\substack{\text{general}\\ \text{state}}} = A_1 \underbrace{|\psi_1\rangle}_{\substack{\text{basis}\\ \text{state}}} + A_2 \underbrace{|\psi_2\rangle}_{\substack{\text{basis}\\ \text{state}}} + A_3|\psi_3\rangle + \dots A_n|\psi_n\rangle = A_i|\psi_i\rangle . \tag{2-95}$$

For the usual normalization, the probability of measuring (observing) the system to be in any given state i is $|A_i|^2$ and the "length" (absolute magnitude) of the state vector is unity, i.e.,

QM complex. States are SU(n) vectors with total (normalized) probability ("length") =1

$$\text{total probability} = \text{"length" of state vector} = |A_1|^2 + |A_2|^2 + |A_3|^2 + \dots |A_n|^2 = \sum_i |A_i|^2 = 1 . \tag{2-96}$$

The complex vector space state (2-95) can be represented as a column vector i.e.,

$$|\psi\rangle = \psi_{state} \xrightarrow[\text{basis}]{\text{in chosen}} = \begin{bmatrix} \langle\psi_1|\psi\rangle \\ \langle\psi_2|\psi\rangle \\ \vdots \\ \langle\psi_n|\psi\rangle \end{bmatrix} = \begin{bmatrix} A_1 \\ A_2 \\ \vdots \\ A_n \end{bmatrix}, \tag{2-97}$$

where in NRQM and RQM the vector space is Hilbert space.

A unitary operator, represented by the matrix P, operating on the state vector yields

$$P|\psi\rangle = |\psi'\rangle = \begin{bmatrix} P_{11} & P_{12} & \cdots & P_{1n} \\ P_{21} & P_{22} & \cdots & P_{2n} \\ \vdots & \vdots & \ddots & \vdots \\ P_{n1} & P_{n2} & \cdots & P_{nn} \end{bmatrix} \begin{bmatrix} A_1 \\ A_2 \\ \vdots \\ A_n \end{bmatrix} = P_{ij}A_j = A'_i = \begin{bmatrix} A'_1 \\ A'_2 \\ \vdots \\ A'_n \end{bmatrix}, \tag{2-98}$$

where, because of unitarity [parallel to (2-92)],

Under unitary transformation, total probability (state vector "length") remains = 1

$$\langle\psi'|\psi'\rangle = \sum_i |A'_i|^2 = 1 = \sum_i |A_i|^2 = \langle\psi|\psi\rangle . \tag{2-99}$$

The absolute value of the state vector, its "length" (= total probability for the system to be found in *some* state), remains equal to one. This result generalizes to states in all branches of quantum mechanics, NRQM, RQM, and QFT.

Quantum Field Theory

Consider a unitary transformation, such as the S operator of QFT (Vol. 1, Sects. 7.4 to 7.5.2, pgs. 194-199), which describes a transition in Fock space (complex vector space with basis vectors being multiparticle eigenstates) from one state vector (components A_i) to another state vector (components A_i'). This parallels (2-95) to (2-99) for Fock space rather than Hilbert space. (See Vol. 1, Wholeness Chart 3-2, pg. 68, for a comparison of these two spaces.) In Vol. 1, Fig. 7-2, pg. 199, the S_{fi} there correspond to the A_i' here. The total probability before and after the transformation remains unchanged, i.e., the total probability for the system to be in some state remains equal to one.

In QFT, under unitary S operator (interactions), total probability (state "length") remains = 1

In QFT, the Lagrangian (density) is an invariant scalar under external (global) transformations (Vol. 1, pg. 173), i.e., Poincaré (Lorentz plus 4D translation) transformations. It is called a world scalar, or a Lorentz scalar (Vol. 1, Sect. 2.5.1, point 11, pgs. 24-25). This particular invariance constitutes an external symmetry.

But the Lagrangian also has internal (local) symmetries, which leave it invariant under certain other transformations in other abstract spaces. These are also called gauge symmetries. (See Vol. 1, pgs. 178, 290-298.) In QED, we found the Lagrangian was symmetric (invariant) under the gauge transformations shown in the cited pages.

In QFT, scalar $\mathcal{L}$ invariant (symmetric) under external (Poincaré) and certain internal (gauge) transformations

As a simple example, consider the fermion mass term in $\mathcal{L}$, $m\bar{\psi}\psi$, under the gauge transformation

Example of an invariant term in $\mathcal{L}$ under a certain gauge transformation

$$\psi \to \psi' = e^{-i\alpha}\psi , \tag{2-100}$$

$$m\bar{\psi}\psi \to m\bar{\psi}'\psi' = m\left(\bar{\psi}e^{i\alpha}\right)\left(e^{-i\alpha}\psi\right) = m\bar{\psi}\psi . \tag{2-101}$$

That term in $\mathcal{L}$ is symmetric with respect to the transformation (2-100), and as we found in Chaps. 6 and 11 of Vol. 1, when we include the concomitant transformation on the photon field, so are all the other terms, collectively, in $\mathcal{L}$. ψ here is known as a gauge field.

With reference to (2-17) and Prob. 8, the transformation (2-100) is a $U(1)$ transformation. The QED Lagrangian is symmetric under (particular) $U(1)$ transformations. We have yet to discuss weak and strong interactions, which we will find are symmetric under particular $SU(2)$ and $SU(3)$ transformations, respectively, which is why the SM is commonly referred to by the symbology

$\mathcal{L}$ invariance under U(1), SU(2), and SU(3) transformations underlies SM

$$SU(3)\times SU(2)\times U(1) \qquad [\text{the standard model}]. \tag{2-102}$$

There is some oversimplification here, as we will find the weak and QED interactions are intertwined in non-trivial ways, but that is the general idea.

Now consider certain terms in the free Lagrangian, where we note the subscripts e and ν_e refer to electron and electron neutrino fields, respectively, and the superscript L refers to something called left-handed chirality, the exact definition of which we leave to later chapters on weak interaction theory. For now, just consider it a label for a particular vector space we are interested in, which we will find later on in this book to be related to the weak interaction. For $\gamma^{\mu}\partial_{\mu} = \not{\partial}$,

$$\mathcal{L}_{\substack{two\\terms}} = \bar{\psi}^L_{\nu_e}\gamma^{\mu}\partial_{\mu}\psi^L_{\nu_e} + \bar{\psi}^L_e\gamma^{\mu}\partial_{\mu}\psi^L_e = \begin{bmatrix}\bar{\psi}^L_{\nu_e} & \bar{\psi}^L_e\end{bmatrix}\begin{bmatrix}\not{\partial}\psi^L_{\nu_e}\\ \not{\partial}\psi^L_e\end{bmatrix} = \begin{bmatrix}\bar{\psi}^L_{\nu_e} & \bar{\psi}^L_e\end{bmatrix}\not{\partial}\begin{bmatrix}\psi^L_{\nu_e}\\ \psi^L_e\end{bmatrix}, \tag{2-103}$$

We have cast the usual scalar terms just after the first equal sign into a two-component, complex, column vector (2D vector in complex space). In this case, it is composed of fields, not states.

Now, let's see how (2-103) transforms under a typical $SU(2)$ transformation, which we label M and recall that $M^{\dagger} = M^{-1}$. M here is a global transformation, i.e., not a function of spacetime x^{μ}.

Example of symmetry of two terms in $\mathcal{L}$, under a global SU(2) transformation

$$\begin{aligned}\mathcal{L}'_{\substack{two\\terms}} &= \underbrace{\begin{bmatrix}\bar{\psi}'^L_{\nu_e} & \bar{\psi}'^L_e\end{bmatrix}\not{\partial}\begin{bmatrix}\psi'^L_{\nu_e}\\ \psi'^L_e\end{bmatrix}}_{\text{transformed functional form}} = \underbrace{\begin{bmatrix}\bar{\psi}^L_{\nu_e} & \bar{\psi}^L_e\end{bmatrix}M^{\dagger}}_{\text{transformed row vec}}\not{\partial}\;\underbrace{M\begin{bmatrix}\psi^L_{\nu_e}\\ \psi^L_e\end{bmatrix}}_{\text{transformed col vec}} \\ &= \begin{bmatrix}\bar{\psi}^L_{\nu_e} & \bar{\psi}^L_e\end{bmatrix}\not{\partial}\underbrace{M^{\dagger}M}_{M^{-1}}\begin{bmatrix}\psi^L_{\nu_e}\\ \psi^L_e\end{bmatrix} = \underbrace{\begin{bmatrix}\bar{\psi}^L_{\nu_e} & \bar{\psi}^L_e\end{bmatrix}\not{\partial}\begin{bmatrix}\psi^L_{\nu_e}\\ \psi^L_e\end{bmatrix}}_{\text{orignal functional form}} = \underbrace{\mathcal{L}_{\substack{two\\terms}}}_{\text{original value}}\end{aligned} \tag{2-104}$$

The transformed terms in (2-104) equal the original terms in (2-103), in both functional form and value. These terms are symmetric with respect to any global $SU(2)$ transformation.

We note that we have used a global transformation in order to make a point in the simplest possible way. The actual $SU(2)$ transformations in the SM are local (gauge) transformations, for which it is considerably more complicated to demonstrate invariance (but which we will do later in this book.)

Nature has decided which particular terms form the 2-component vectors

One may wonder why we chose the particular two $\mathcal{L}$ terms of (2-103) to form our two component complex vector, instead of others, like perhaps an electron field and a muon field, or a right chirality (RC in this book, though many texts use RH) neutrino field and a left chirality (LC) electron field. The answer is simply that the form of (2-103) plays a fundamental role in weak interaction theory. Nature chose that particular two component vector as the one for which $SU(2)$ weak symmetry holds. But, we are getting ahead of ourselves. Much detail on this is yet to come in later chapters.

These are LC electron field and its associated neutrino field

Note that quarks share the same $SU(2)$ symmetry with leptons, provided we form our two component quark weak field vector from the up and down quark fields. That is,

And LC up and down quarks

$$\mathcal{L}_{\substack{two\ other\\terms}} = \bar{\psi}^L_u\not{\partial}\psi^L_u + \bar{\psi}^L_d\not{\partial}\psi^L_d = \begin{bmatrix}\bar{\psi}^L_u & \bar{\psi}^L_d\end{bmatrix}\not{\partial}\begin{bmatrix}\psi^L_u\\ \psi^L_d\end{bmatrix}, \tag{2-105}$$

where (2-105) is invariant under the transformation M, in the same way we showed in (2-104).

Similarly, 3-component color vectors exist for SU(3)

Similar logic applies for $SU(3)$ transformations related to the strong (color) interaction. If quarks have three different eigenstates of color (red, green, and blue), we can represent terms for them in the Lagrangian as in (2-106). Note that up quarks can have any one of the three colors as eigenstates, and likewise, down quarks can have any one of the three. The same holds true for RC vs LC fields. So, in (2-106), we don't use the up/down subscripts for quark fields, nor the L superscript, as the results apply to both kinds of chirality and both components of the quark weak (2D) field vector in (2-105).

Consider then, $$\mathcal{L}_{\substack{three\ quark\\terms}} = \bar{\psi}_r \not{\partial} \psi_r + \bar{\psi}_g \not{\partial} \psi_g + \bar{\psi}_b \not{\partial} \psi_b = \begin{bmatrix} \bar{\psi}_r & \bar{\psi}_g & \bar{\psi}_b \end{bmatrix} \not{\partial} \begin{bmatrix} \psi_r \\ \psi_g \\ \psi_b \end{bmatrix}. \qquad (2\text{-}106)$$

The three terms in (2-106) are symmetric under a global $SU(3)$ transformation.

Do **Problem 21** to prove it. Then do **Problem 22.**

Note that fermions differ in type [flavor] (electron, neutrino, up quark, down quark, etc.), chirality (LC vs RC), and color (R, G, or B). Each of these can be designated via a sub or superscript index. For example, an up, LC, green quark field could be written as (note use of capital letter Ψ)

Some symbols for LC vs RC, color

$$\begin{gathered}\text{up, green, LC quark} \rightarrow \Psi^L_{ug} \qquad \text{generally} \rightarrow \Psi^h_{fa} \text{ where here } h = L; f = u; a = g \\ h = L,R \quad f = u,d \text{ (for quarks) or } \nu_e, e \text{ (for leptons)} \quad a = r,g,b \text{ (quarks); } 0 \text{ (leptons)}\end{gathered} \qquad (2\text{-}107)$$

Leptons have no a component, as they are colorless (i.e., do not feel the strong force). And if, instead (which will often prove helpful), we want to use numbers for the two and three-component vectors to be acted upon by $SU(2)$ and $SU(3)$ matrices, we can take

$$f = 1,2 \text{ (for quarks); } 1,2 \text{ (for leptons)} \qquad a = 1,2,3 \text{ (quarks); nothing (leptons)} \qquad (2\text{-}108)$$

Then, (2-107) is essentially an outer (tensor) product (Sect. 2.2.8, pg. 18) of a 2D vector with a 3D vector. An $SU(2)$ transformation on it would only affect the f indices and not the a indices. An $SU(3)$ transformation would only affect the a indices and not the f indices. That is, for quark fields, where M is an $SU(2)$ operation and N is an $SU(3)$ operation, we can write

Fields are outer products of 2D, 3D, spin, and spacetime elements

$$M_{mf} N_{na} \Psi^h_{fa} = \Psi'^h_{mn} \qquad f = 1,2 \quad a = 1,2,3 \quad m = 1,2 \quad n = 1,2,3. \qquad (2\text{-}109)$$

In the weak and strong interactions chapters, we will see how these $SU(2)$ and $SU(3)$ symmetries, via Noether's theorem (parallel to what we saw in QED for $U(1)$ symmetries) lead to weak and strong charge conservation (at least prior to symmetry breaking for weak charge). In addition, associated local transformation symmetries will lead to the correct interaction terms in the Lagrangian.

Full Expressions of Typical Fields in QFT

Finally, we note in (2-110) what typical fields might look like, if we expressed them in terms of column vectors, where the subscripts W and S refer to the weak and strong interactions, respectively. The subscripts u and e on the creation/destruction operators refer to up quark field and electron field, respectively. The down quark would have unity in the lower $SU(2)$ column vector component and zero in the upper. The neutrino field has unity in the top slot; the electron field, unity in the bottom one. The M matrix of (2-109) would operate on the two-component vector and nothing else; the N matrix on the three-component vector and nothing else. All of the operators we worked with in QED, such as electric charge, momentum, etc. would be formed from the ψ_u (or ψ_e) part, as in QED.

Examples of typical field outer products

$$\text{LC green up quark} \quad \Psi^L_{ug} = \Psi^L_{12} = \underbrace{\sum_{r,\mathbf{p}} \sqrt{\frac{m}{VE_\mathbf{p}}} \left(c_{ur}(\mathbf{p}) u_r(\mathbf{p}) e^{-ipx} + d^\dagger_{ur}(\mathbf{p}) v_r(\mathbf{p}) e^{ipx} \right)}_{\text{general solution to Dirac equation for up quark, } \psi_u} \begin{bmatrix} 1 \\ 0 \end{bmatrix}_W \begin{bmatrix} 0 \\ 1 \\ 0 \end{bmatrix}_S = \psi_u \begin{bmatrix} 1 \\ 0 \end{bmatrix}_W \begin{bmatrix} 0 \\ 1 \\ 0 \end{bmatrix}_S$$

$$\text{LC electron} \quad \Psi^L_e = \Psi^L_2 = \underbrace{\sum_{r,\mathbf{p}} \sqrt{\frac{m}{VE_\mathbf{p}}} \left(c_{er}(\mathbf{p}) u_r(\mathbf{p}) e^{-ipx} + d^\dagger_{er}(\mathbf{p}) v_r(\mathbf{p}) e^{ipx} \right)}_{\text{general solution for electron field, } \psi_e} \begin{bmatrix} 0 \\ 1 \end{bmatrix}_W = \psi_e \begin{bmatrix} 0 \\ 1 \end{bmatrix}_W \qquad (2\text{-}110)$$

u_r and v_r are also column vectors (in 4D spinor space) not explicitly shown in (2-110). As shown later, $d^\dagger_{ur}$ and $d^\dagger_{er}$ here create antiparticles with "anti-color" (for quarks) *and* opposite (RC) chirality.

Note further that it will be easier in the future if we use slightly different notation (parallel to that in (2-103) to (2-105)), whereby the last terms in both rows of (2-110) are written as

Alternative symbolism

$$\psi_e \begin{bmatrix} 0 \\ 1 \end{bmatrix}_W = \begin{bmatrix} 0 \\ \psi_e \end{bmatrix}_W = \begin{bmatrix} 0 \\ \psi^L_e \end{bmatrix} \qquad \psi_u \begin{bmatrix} 1 \\ 0 \end{bmatrix}_W \begin{bmatrix} 0 \\ 1 \\ 0 \end{bmatrix}_S = \begin{bmatrix} \psi_u \\ 0 \end{bmatrix}_W \begin{bmatrix} 0 \\ 1 \\ 0 \end{bmatrix}_S = \begin{bmatrix} \psi^L_u \\ 0 \end{bmatrix} \begin{bmatrix} 0 \\ 1 \\ 0 \end{bmatrix}_S. \qquad (2\text{-}111)$$

Summary and Conclusions

Wholeness Chart 2-6 summarizes transformations and symmetries for various spaces.

Symmetries of the Lagrangian exist under $SU(n)$ transformations and these, as we will find, have profound implications for QFT.

Wholeness Chart 2.6. Symmetries in Various Spaces

Space	Operator	Example	Transformed	Sym?
Euclidean, 2D	$SO(2)$ rotation	vector length, $\|\mathbf{v}\|$	$\|\mathbf{v}'\| = \|\mathbf{v}\|$	Y
	"	function (circle) $x^2 + y^2 = r^2$	function $x'^2 + y'^2 = r^2$	Y
	"	function (ellipse) $x^2 + 3y^2 = r^2$	function $x'^2 + 3y'^2 \neq r^2$	N
Physical, 3D	$SO(3)$ rotation	vector length $\|\mathbf{v}\|$	$\|\mathbf{v}'\| = \|\mathbf{v}\|$	Y
	"	laws of nature (functional form)	laws in primed coordinates = laws in unprimed coordinates	Y
	"	vector component v^i	$v'^i \neq v^i$	N
	Non-orthogonal transform	vector length $\|\mathbf{v}\|$	$\|\mathbf{v}'\| \neq \|\mathbf{v}\|$	N
Minkowski, 4D	$SO(3,1)$ Lorentz transform	4D position vec. length = Δs	$\Delta s' = \Delta s$	Y
	"	mass m	$m^2 = p^\mu p_\mu = p'^\mu p'_\mu$	Y
	"	laws of nature (functional form)	Laws in primed coordinates = laws in unprimed coordinates	Y
	"	4-momentum component p^μ	$p'^\mu \neq p^\mu$	N
	Non-Lorentz (non 4D orthogonal) transform	4D position vec. length = Δs	$\Delta s' \neq \Delta s$	N
Hilbert, *n*D n "axes" = single particle eigenstates	$SU(n)$ ("rotation")	state vector length = total probability = 1	$\Sigma \|A'_i\|^2 = \Sigma \|A_i\|^2 = 1$ i.e., $\langle\psi'\|\psi'\rangle = \langle\psi\|\psi\rangle = 1$	Y
	"	Schrödinger eq (SE)	SE primed = SE unprimed	Y
	Non-unitary transform	state vector length	$\langle\psi'\|\psi'\rangle = \Sigma \|A'_i\|^2 \neq \Sigma \|A_i\|^2 = \langle\psi\|\psi\rangle$	N
Fock, *n*D n "axes" = multiparticle eigenstates	Complex "rotation", such as S operator	state vector length = total probability = 1	$\langle\psi'\|\psi'\rangle = \Sigma \|A'_i\|^2 = \Sigma \|A_i\|^2 = \langle\psi\|\psi\rangle$	Y
	"	K-G, Dirac, etc. field equations functs. of ϕ, ψ, etc	same field equations in terms of primed fields, ϕ', ψ', etc.	Y
	Non-unitary transform	state vector length	$\langle\psi'\|\psi'\rangle = \Sigma \|A'_i\|^2 \neq \Sigma \|A_i\|^2 = \langle\psi\|\psi\rangle$	N
fields operating on those states	$U(1)$ ("rotation") such as $\psi \rightarrow \psi' = \psi e^{-i\alpha}$	free fermion Lagrangian (density) $\mathcal{L}_0^{1/2}$	$\mathcal{L}'_0{}^{1/2}$ is same function of ψ' as $\mathcal{L}_0^{1/2}$ is of ψ	Y
Weak, 2D 2 "axes" = LC e^- & ν	$SU(2)$ ("rotation")	$\mathcal{L}_0$ functs. of ψ_e, ψ_ν, etc.	$\mathcal{L}_0'$ same functs. of ψ_e', ψ_ν', etc.	Y
	"	field eqs. in ψ_e, ψ_ν, etc	same field eqs. in ψ_e', ψ_ν', etc.	Y
Color, 3D 3 "axes" = RGB	$SU(3)$ ("rotation")	$\mathcal{L}_0$ functions of quark, lepton, boson fields	$\mathcal{L}_0'$ same functions of primed fields as $\mathcal{L}_0$ is of unprimed fields	Y
	"	field eqs. in unprimed fields	same field eqs. in primed fields.	Y

2.5 Singlets and Multiplets

2.5.1 Physical Space and Orthogonal Transformation Groups

In the ordinary 3D space of our experience, we know of two different types of entities, vectors and scalars. In an n dimensional real space, vectors have n components, but regardless of what n is, scalars always have one component. An orthogonal transformation on the vector will change the components of the vector. But an orthogonal transformation leaves the value of a scalar unchanged. It is as if the orthogonal operation acting on the scalar is simply the 1D identity "matrix", i.e., the number one.

For an $SO(n)$ group operating on n dimensional vectors, the matrix representation of the group is an nXn matrix. But for the group operating on scalars, the same group is represented by unity, a 1X1 matrix. The nXn matrix is one representation of the group (that acts on vectors). The 1X1 matrix is another representation of the same group (that acts on scalars).

It may seem confusing that we deal with 1X1 matrices for an $SO(n)$ group, instead of nXn matrices. To help, consider the $SO(2)$ group of (2-6), where **v** symbolizes a vector and $\hat{\mathbf{s}}$ symbolizes a scalar,

Rotations SO(2) change vector component values, but not scalar value

$$SO(2):\;\; \hat{\mathbf{M}}(\theta)\mathbf{v} \xrightarrow[\text{rep}]{\text{matrix}} \begin{bmatrix} \cos\theta & -\sin\theta \\ \sin\theta & \cos\theta \end{bmatrix}\begin{bmatrix} v_1 \\ v_2 \end{bmatrix} = \begin{bmatrix} v_1' \\ v_2' \end{bmatrix} \qquad \hat{\mathbf{M}}(\theta)\hat{\mathbf{s}} \xrightarrow[\text{rep}]{\text{matrix}} [1]\hat{s} = \hat{s}' = \hat{s}\,. \tag{2-112}$$

We say the 2X2 matrix in (2-112) is a <u>2D representation</u> of the group, and the unit "matrix" is the <u>1D representation</u> of the same group. The former acts on vectors, the latter on scalars. In group theory lingo (as opposed to typical physics lingo), the vector is called a <u>doublet</u>, and the scalar a <u>singlet</u>.

2D vector = doublet, scalar = singlet, SO(2) operation has 2D & 1D reps

The matrix generators for each rep (we will start using "<u>rep</u>" sometimes as shorthand for "representation") are different, but they represent the same group. For the 2D rep, the generator is the $\hat{X}$ matrix of (2-42), which commutes with itself. Since the 1D rep is essentially the 1D identity matrix (the number one), and the group can be expressed as $e^{i\theta\hat{X}}$ (see Wholeness Chart 2-5, pg. 37), for any θ, the generator $\hat{X}$ in 1D is zero, since $1 = e^{i\theta 0}$. This commutes with itself, and thus, the commutation relations are the same, a criterion for different reps of the same group (with the same parametrization).

Same commutator relation for 2D & 1D reps

To express it slightly differently, consider the vector space of real numbers, which are 1X1 column vectors. Transform these vectors under an $SO(2)$ transformation, which operates on 1X1 vectors and so must be represented by 1X1 matrices. Such a transformation matrix must have determinant of one, so the only entry in the 1X1 matrix has to be one.

In 3D space, the vector is called a <u>triplet</u>, and the scalar a singlet. The three generators $\hat{X}_i$ of (2-49) for the 3D rep obey the commutation relations (2-51). For the 1D rep, each of the $\hat{X}_i$ is zero. (See Wholeness Chart 2-5 where the $\hat{\theta}_i$ are arbitrary and the group operator must be unity.) Thus, in the 1D rep, the generators also obey (2-51) [rather trivially]. The commutation relation holds in both reps of the $SO(3)$ group.

For SO(3), 3D vector = triplet, scalar = singlet, operation has 3D and 1D reps

Same commutators in both reps

The important point is that singlets are unchanged by $SO(n)$ group operations. In a nutshell, because $Det\,\hat{M} = 1$, and for 1D, $\hat{M}$ is 1X1, $\hat{M}$ must = 1. In practice, we simply remember that when an SO group operates on a scalar (a singlet), the scalar (singlet) remains unchanged.

Singlet unchanged under SO(n) action

2.5.2 Complex Space and Unitary Transformation Groups

As we've seen before, $SU(n)$ operations are the complex cousins of $SO(n)$ operations. Just as we had nD <u>multiplets</u> (vectors) and 1D singlets (scalars) upon which special orthogonal matrices operated in real n dimensional space, so we have nD multiplets and 1D singlets upon which special unitary groups operate in complex n dimensional space.

For SU(n), vector = multiplet, scalar = singlet, operation has nD and 1D reps

As an example, consider the $SU(2)$ group with 2D rep of (2-25), where **w** symbolizes a complex doublet and **s** symbolizes a complex singlet.

$$SU(2):\mathbf{Mw} \xrightarrow[\text{rep}]{\text{matrix}} \begin{bmatrix} \alpha_0 + i\alpha_3 & \alpha_2 + i\alpha_1 \\ -\alpha_2 + i\alpha_1 & \alpha_0 - i\alpha_3 \end{bmatrix}\begin{bmatrix} w_1 \\ w_2 \end{bmatrix} = \begin{bmatrix} w_1' \\ w_2' \end{bmatrix} \qquad \mathbf{Ms} \xrightarrow[\text{rep}]{\text{matrix}} [1]s = s' = s \tag{2-113}$$

For singlet, 1D "matrix" rep = 1

For the 1D case, **M** is the 1D identity matrix, and since $M = e^{i\beta_i X_i}$ (Wholeness Chart 2-5, pg. 37) for all β_i, all three $X_i = 0$. The commutation relations hold trivially. And the singlet is invariant.

Same commutators

Similar logic holds for *SU*(3) with the 3D complex vectors as triplets and complex scalars are singlets. In the 1D rep, the group takes the form of a 1D identity matrix, the generators are all zero, and a singlet is unchanged under the group action. These conclusions hold for any *SU*(*n*).

For 1D rep in SU(n), generators all = 0; singlet unchanged under action of SU(n)

As a foreshadowing of things to come, we will see later on that right-chiral fermions are weak [*SU*(2)] singlets and left-chiral ones form weak doublets. Similarly, each lepton is a color [*SU*(3)] singlet and quarks form color triplets. But don't think too much more about this right now.

Some fields in QFT are singlets, some are multiplet components

2.5.3 Other Multiplets

There are yet other dimensional representations for any given group. For example, 3D and 4D representations of *SO*(2) are

$$\hat{\underline{M}} = \begin{bmatrix} \cos\theta & -\sin\theta & \\ \sin\theta & \cos\theta & \\ & & 1 \end{bmatrix} \qquad \hat{\underset{\sim}{M}} = \begin{bmatrix} \cos\theta & -\sin\theta & & \\ \sin\theta & \cos\theta & & \\ & & \cos\theta & -\sin\theta \\ & & \sin\theta & \cos\theta \end{bmatrix}. \tag{2-114}$$

Reps can have dimensions other than n and 1

These share similarities with (2-6) and, respectively, would act on an *SO*(2) triplet and an *SO*(2) quadruplet.

The generators for (2-114) are, respectively,

$$\hat{\underline{X}} = \begin{bmatrix} 0 & i & \\ -i & 0 & \\ & & 0 \end{bmatrix} \qquad \hat{\underset{\sim}{X}} = \begin{bmatrix} 0 & i & & \\ -i & 0 & & \\ & & 0 & i \\ & & -i & 0 \end{bmatrix}, \tag{2-115}$$

which share similarities with (2-42). The commutation relation for the generator of each of the matrices in (2-114) is the same as for the generators of (2-6). (In this trivial case of only one generator, the generator commutes with itself.)

Similarly, there can be different dimension representations for any given group. For the same parametrization, the commutation relations for every such representation will be the same.

Reps of different dimensions have same commutators for generators

In the simplest view, for a given choice of parameters, the representation comprising an *n*X*n* matrix is called the fundamental (or standard) representation. There are more sophisticated ways, steeped in mathematical jargon, to define the fundamental rep, but this definition, though perhaps not fully precise, should suffice for our purposes. The rep associated with the singlet is called the trivial representation.

Fundamental rep, simple definition = nXn matrix rep

We will not do much, if anything, in this book with representations of dimensions other than *n* and 1 (acting on *n*-plet and singlet) for any group of degree *n*.

2.5.4 General Points for Multiplets and Associated Reps

In both real and complex spaces, we know that components of a multiplet change when acted upon by the (*SO*(*n*) or *SU*(*n*)) group, but the singlet is invariant. We can also recognize that a matrix representation of a group does not have to be of dimension *n*. For the same group, matrix reps of different dimensions having the same parametrizations all have the same structure constants (same commutation relations).

Singlets invariant under SU(n) transformations

Commonly used names for fundamental reps of certain groups are shown in Table 2-3. Note we show the formal mathematical symbols (see Table 2-1, pg. 23) for the vector spaces in parentheses.

Table 2-3. Common Names of Some Fundamental Representations

Group	Vector Space	Name
SU(2)	2D complex ($\mathbb{C}^2$)	spinor
SO(3)	3D real ($\mathbb{R}^3$)	vector
SO(3,1)	4D relativistic, real ($\mathbb{R}^4$)	four-vector

Common names for fundamental reps

2.5.5 Singlets and Multiplets in QFT

As you may have guessed, what we called 2-component and 3-component vectors in (2-103) to (2-111) are more properly called doublets and triplets. In the referenced equations they were $SU(2)$ LC (weak interaction) field doublets and $SU(3)$ color (strong interaction) field triplets.

In QFT, as we know, fields create and destroy states. So, we will find that the individual fields, which are the components in doublets and triplets, create and destroy particles in (generally) multiparticle states. In the SM, we will deal with both fields and states, just as we have in all the QFT we have studied to date.

In QFT, components of multiplet fields create and destroy multiparticle states

2.6 Matrix Operators on States vs Fields

In Section 2.4.3, we began to apply what we had learned of group theory to quantum mechanics. We reviewed how the action of a unitary operator (represented by a matrix) on a state left the magnitude of that state unchanged, even though the component parts of the state changed.

We then discussed the action of unitary operators on *fields* (as opposed to states) in QFT. We showed examples of this for $U(1)$ in (2-101), $SU(2)$ in (2-104), and $SU(3)$ in Problem 21. The question can then arise as to whether, in QFT, we take our vectors (multiplets upon which group operators operate), to be fields or states. The answer is a little subtle.

In QFT, do group operators act on field multiplets or state multiplets?

2.6.1 The Spin Operator on States and Fields

We start by referencing Vol. 1, Sect. 4.1.10, pg. 93. There we show the RQM spin operator, where for simplicity we only discuss the z direction component of the total spin operator, acting on a particular RQM (single particle) spinor state at rest with spin in the z direction. See Vol. 1, (4-40).

To help answer: spin operator example from Vol. 1

$$\text{For RQM} \qquad \Sigma_3\left|\psi_{\substack{spin\,up\\ \mathbf{p}=0}}\right\rangle = \frac{1}{2}\begin{bmatrix}1&&&\\&-1&&\\&&1&\\&&&-1\end{bmatrix}\begin{pmatrix}1\\0\\0\\0\end{pmatrix}e^{-ipx} = \frac{1}{2}\begin{pmatrix}1\\0\\0\\0\end{pmatrix}e^{-ipx} = \frac{1}{2}\left|\psi_{\substack{spin\,up\\ \mathbf{p}=0}}\right\rangle \tag{2-116}$$

However, in QFT, as we discussed in detail in Vol. 1, Sect. 4.9, pgs. 113-115, we need to take into account that a state may be multiparticle. We found that by defining our spin operator (which acts on states) to be the LHS of (2-117) below, we got the RHS. See Vol. 1, (4-110) and (4-119).

$${}_{\text{QFT}}\Sigma_3 = \int_V \psi^\dagger \Sigma_3 \psi\, d^3x \;\rightarrow\; {}_{\text{QFT}}\Sigma_3 = \sum_{r,\mathbf{p}} \frac{m}{E_\mathbf{p}}\left(u_r^\dagger(\mathbf{p})\Sigma_3 u_r(\mathbf{p})N_r(\mathbf{p}) + v_r^\dagger(\mathbf{p})\Sigma_3 v_r(\mathbf{p})\bar{N}_r(\mathbf{p})\right) \tag{2-117}$$

QFT spin operator on states includes number operators and spinors

So, for a state with an at rest spin up electron and an at rest spin up positron, we get, where all number operators yield zero except for the electron $N_r(\mathbf{p}=0)$ and positron $\bar{N}_r(\mathbf{p}=0)$, they each yield 1.

$$\begin{aligned}
{}_{\text{QFT}}\Sigma_3\left|\psi_{\substack{spin\,up\\ \mathbf{p}=0}},\bar{\psi}_{\substack{spin\,up\\ \mathbf{p}=0}}\right\rangle &= \sum_{r,\mathbf{p}}\frac{m}{E_\mathbf{p}}\left(u_r^\dagger(\mathbf{p})\Sigma_3 u_r(\mathbf{p})N_r(\mathbf{p}) + v_r^\dagger(\mathbf{p})\Sigma_3 v_r(\mathbf{p})\bar{N}_r(\mathbf{p})\right)\left|\psi_{\substack{spin\,up\\ \mathbf{p}=0}},\bar{\psi}_{\substack{spin\,up\\ \mathbf{p}=0}}\right\rangle \\
&= \left(\underbrace{\begin{pmatrix}1&0&0&0\end{pmatrix}\frac{1}{2}\begin{bmatrix}1&&&\\&-1&&\\&&1&\\&&&-1\end{bmatrix}\begin{pmatrix}1\\0\\0\\0\end{pmatrix}}_{\text{spin up for particle}} + \underbrace{\begin{pmatrix}0&0&1&0\end{pmatrix}\frac{1}{2}\begin{bmatrix}1&&&\\&-1&&\\&&1&\\&&&-1\end{bmatrix}\begin{pmatrix}0\\0\\1\\0\end{pmatrix}}_{\text{spin up for anti-particle}}\right)\left|\psi_{\substack{spin\,up\\ \mathbf{p}=0}},\bar{\psi}_{\substack{spin\,up\\ \mathbf{p}=0}}\right\rangle \\
&= \left(\tfrac{1}{2}+\tfrac{1}{2}\right)\left|\psi_{\substack{spin\,up\\ \mathbf{p}=0}},\bar{\psi}_{\substack{spin\,up\\ \mathbf{p}=0}}\right\rangle = \left|\psi_{\substack{spin\,up\\ \mathbf{p}=0}},\bar{\psi}_{\substack{spin\,up\\ \mathbf{p}=0}}\right\rangle.
\end{aligned} \tag{2-118}$$

The eigenvalue, representing total spin, is one, the sum of the spins for both particles, so it all works. In QFT, the (multiparticle) state is just represented by a symbol, the ket symbol like that in (2-118), and we generally don't think of it as having structure, such as spinor column vectors. The definition (2-117) provides the spinor column vectors in the operator itself. In any operation on a ket of given spin, these column vectors in the QFT spin operator give us the needed structure that leads to the correct result. We don't worry about spinor structure in the kets. It is included in the spin operator.

If we had used a single particle in (2-118), we would have found the spin eigenvalue of ½. Note, as in (2-119) below, that if we had used the original operator of (2-116) and operated with it on the QFT field (instead of the state) we would also get an eigenvalue of ½.

$$\text{On a quantum field } \Sigma_3 \psi_{\substack{spinup\\ \mathbf{p}=0}} = \frac{1}{2}\begin{bmatrix} 1 & & & \\ & -1 & & \\ & & 1 & \\ & & & -1 \end{bmatrix}\left(c_r(\mathbf{p})\begin{pmatrix}1\\0\\0\\0\end{pmatrix} e^{-ipx} + d_r^\dagger(\mathbf{p})\begin{pmatrix}0\\0\\1\\0\end{pmatrix} e^{ipx}\right)$$

$$= \frac{1}{2}\left(c_r(\mathbf{p})\begin{pmatrix}1\\0\\0\\0\end{pmatrix} e^{-ipx} + d_r^\dagger(\mathbf{p})\begin{pmatrix}0\\0\\1\\0\end{pmatrix} e^{ipx}\right) = \frac{1}{2}\psi_{\substack{spinup\\ \mathbf{p}=0}} . \qquad (2\text{-}119)$$

Σ_3 operating on spin up field yields same eigenvalue as $\int \psi^\dagger \Sigma_3 \psi\, d^3x$ on one particle spin up state

Conclusion: Σ_3 acting on a quantum field of given spin yields the same eigenvalue as ${}_{QFT}\Sigma_3 = \int \psi^\dagger \Sigma_3 \psi\, d^3x$ operating on a single particle state of the same spin.

2.6.2 Other Operators on States vs Fields

Similar for other QFT operators

Somewhat similar logic works for other operators in QFT. Commonly, our operators, like M in (2-104), operate on fields directly, as we show in (2-104). If we want the corresponding operation to apply to (multiparticle) states, we parallel what we did for spin in going from (2-116) to (2-117). That is,

$$\text{Operation on quantum fields } M\Psi^L = M\begin{bmatrix}\psi^L_{\nu_e}\\ \psi^L_e\end{bmatrix} = M\psi^L_{\nu_e}\begin{bmatrix}1\\0\end{bmatrix} + M\psi^L_e\begin{bmatrix}0\\1\end{bmatrix}$$

$$\text{Corresponding operation on state } \left(\int_V \Psi^{L\dagger} M \Psi^L\, d^3x\right)\left| n_{\nu_e}\, \psi^L_{\nu_e}, n_e\, \psi^L_e\right\rangle \qquad (2\text{-}120)$$

Generally, SU(n) group operators in QFT act on field multiplets, but we can find associated operators for states

where the ket here could have any number n_{ν_e} of neutrinos and any number n_e of electrons, where the neutrinos and electrons could have any spin and momentum values, and where M only operates on the two-component column vector shown and not the spacetime or 4D spinor column vector factors in the fields.

Let's consider the particular form for M of

$$M = \begin{bmatrix}1 & 0\\ 0 & -1\end{bmatrix}. \qquad (2\text{-}121)$$

Hopefully, in parallel with (2-117), and recalling (2-110), 2nd row, we can intuitively deduce that

$$\int_V \Psi^{L\dagger} M \Psi^L\, d^3x = \sum_{r,\mathbf{p}}\begin{pmatrix} (1\ \ 0)M\begin{pmatrix}1\\0\end{pmatrix} N^L_{\nu_e,r}(\mathbf{p}) + (0\ \ 1)M\begin{pmatrix}0\\1\end{pmatrix} N^L_{e,r}(\mathbf{p}) \\ -(1\ \ 0)M\begin{pmatrix}1\\0\end{pmatrix} \overline{N}^R_{\nu_e,r}(\mathbf{p}) - (0\ \ 1)M\begin{pmatrix}0\\1\end{pmatrix} \overline{N}^R_{e,r}(\mathbf{p})\end{pmatrix} \qquad (2\text{-}122)$$

Do **Problem 23** to prove (2-122). Or take less time by just looking at the solution booklet answer.

Now let's consider an example where we have a single electron ket ($n_{\nu_e} = 0$, $n_e = 1$ in (2-120)).

$$\left(\int_V \Psi^{L\dagger} M \Psi^L\, d^3x\right)\left|\psi^L_e\right\rangle = \sum_{r,\mathbf{p}} (0\ \ 1)\begin{bmatrix}1 & 0\\ 0 & -1\end{bmatrix}\begin{pmatrix}0\\1\end{pmatrix} N_{e,r}(\mathbf{p})\left|\psi^L_e\right\rangle = -\left|\psi^L_e\right\rangle. \qquad (2\text{-}123)$$

For M acting directly on the electron field, we get

$$M\Psi^L_e = M\begin{bmatrix}0\\ \psi^L_e\end{bmatrix} = M\psi^L_e\begin{bmatrix}0\\1\end{bmatrix} = \psi^L_e\begin{bmatrix}1 & 0\\ 0 & -1\end{bmatrix}\begin{bmatrix}0\\1\end{bmatrix} = -\psi^L_e\begin{bmatrix}0\\1\end{bmatrix} = -\begin{bmatrix}0\\ \psi^L_e\end{bmatrix} = -\Psi^L_e . \qquad (2\text{-}124)$$

The eigenvalues in (2-123) and (2-124) are the same, i.e., – 1. We generalize this result below.

Conclusion #1: Eigenvalues from any unitary operator M operating on field multiplets Ψ are the same as those from the associated operator $\int \Psi^{\dagger} M \Psi \, d^3x$ operating on single particle states.

Eigenvalues for an operator M acting on field multiplets same as $\int \Psi^{\dagger} M \Psi \, d^3x$ on one particle states.

Conclusion #2: In group theory applied to QFT, elements of the group (such as M in our example above) act on multiplets composed of fields. The group theory vectors comprise fields, not states.

SU(n) group elements in QFT act on field multiplets, not states

We will see this material again, in a different, more rigorous way later in the book, when we get to four-currents and conservation laws in weak and strong interaction theories.

2.6.3 Summary

This material is summarized and expanded on in Wholeness Chart 2-13, pg. 58b.

2.7 Cartan Subalgebras and Eigenvectors

2.7.1 SU(2) Cartan Subalgebra

Note that one of the $SU(2)$ generators (2-63) is diagonal (and like M of (2-121) in the example above).

$$X_3 = \begin{bmatrix} 1 & 0 \\ 0 & -1 \end{bmatrix}. \tag{2-125}$$

Observe the operation of this generator matrix on particular $SU(2)$ doublets.

$$X_3 \begin{bmatrix} C_1 \\ 0 \end{bmatrix} = \begin{bmatrix} 1 & 0 \\ 0 & -1 \end{bmatrix} \begin{bmatrix} C_1 \\ 0 \end{bmatrix} = \begin{bmatrix} C_1 \\ 0 \end{bmatrix} \qquad X_3 \begin{bmatrix} 0 \\ C_2 \end{bmatrix} = \begin{bmatrix} 1 & 0 \\ 0 & -1 \end{bmatrix} \begin{bmatrix} 0 \\ C_2 \end{bmatrix} = -\begin{bmatrix} 0 \\ C_2 \end{bmatrix}. \tag{2-126}$$

Diagonal generators operating on single component multiplets yield eigenvalues

---

Do **Problem 24** to show the effect of X_3 on a more general $SU(2)$ doublet.

The first doublet in (2-126) has an eigenvalue, under the transformation X_3, of +1; the second, of −1. We can use these eigenvalues to label each type of doublet. If d symbolizes a given doublet, one with $C_1 = 1$ and $C_2 = 0$, it would be d_{+1}; for $C_1 = 0$ and $C_2 = 1$, d_{-1}. A general doublet, with any values for the constants C_i, would be $d_i = C_1 d_{+1} + C_2 d_{-1}$. As another foreshadowing of things to come, consider the particular QFT weak interaction field doublets of (2-127), where a constant factor of $g/2$ (where the constant g is discussed more below) is used by convention. Note this can be considered as simply taking $X_3 = ½\sigma_i$ instead of σ_i, with a constant factor of g.

$$\begin{aligned} \frac{g}{2} X_3 \Psi_e^L &= \frac{g}{2} \begin{bmatrix} 1 & 0 \\ 0 & -1 \end{bmatrix} \begin{bmatrix} 0 \\ \psi_e^L \end{bmatrix} = -\frac{g}{2} \begin{bmatrix} 0 \\ \psi_e^L \end{bmatrix} = -\frac{g}{2} \Psi_e^L \\ \frac{g}{2} X_3 \Psi_{\nu_e}^L &= \frac{g}{2} \begin{bmatrix} 1 & 0 \\ 0 & -1 \end{bmatrix} \begin{bmatrix} \psi_{\nu_e}^L \\ 0 \end{bmatrix} = \frac{g}{2} \begin{bmatrix} \psi_{\nu_e}^L \\ 0 \end{bmatrix} = \frac{g}{2} \Psi_{\nu_e}^L \end{aligned} \tag{2-127}$$

Can be used to assign component eigenvalues (quantum numbers) to weak doublet states

If the doublet is a LC electron, we get an eigenvalue for the $\frac{g}{2} X_3$ operation of $-\frac{g}{2}$. If it is a LC neutrino, we get $+\frac{g}{2}$. From what we learned in Section 2.6 (eigenvalues for a group operator on a field are the same as those for the associated operator on a single particle state), we can use these eigenvalues to label the associated particles.

Note that for singlets, in the 1D rep, the generator X_3 is zero (as are the other generators). So,

$$\frac{g}{2} X_3 \psi_e^R = \frac{g}{2}[0] \psi_e^R = 0 \qquad \frac{g}{2} X_3 \psi_{\nu_e}^R = \frac{g}{2}[0] \psi_{\nu_e}^R = 0. \tag{2-128}$$

For singlets, eigenvalues = 0

It will turn out, when we get to weak interactions, that the operator $\frac{g}{2} X_3$ corresponds to the weak charge operator. Weak doublets and singlets are eigenstates of that operator, with eigenvalues that correspond to the weak charge of the respective particles.

Recall that RC fermions do not feel the weak force, and from (2-128), we see that RC fermions have zero weak charge. From (2-127), LC electrons have weak charge $-\frac{g}{2}$; LC neutrinos, $+\frac{g}{2}$.

Weak operator eigenvalues = weak charges

Due to the negative signs in (2-122) on the antiparticle number operators, an antiparticle ket would have the opposite sign for weak charge (as antiparticles have for any type charge in the SM). The constant g reflects the inherent strength of the weak interaction, as greater value for it means a higher weak charge. It is called the weak coupling constant, though it is actually on the order of the e/m coupling constant e. The weak interaction is weak for another reason we will delve into later.

In practice, when discussing weak charge, the coupling constant g is commonly ignored. That is, one generally says the weak charge of the LC electron is – ½; of the LC neutrino, + ½. We just need to keep in mind that it is actually g times +½ or – ½.

Weak charge includes weak coupling g, but g commonly omitted in weak charge designations

Note that diagonal matrices commute. Note also that the diagonal matrix X_3, by itself (in either the 2D or 1D rep), forms a Lie algebra. It conforms to all the properties we list for an algebra in Wholeness Chart 2-1, pg. 9, with the second operation as commutation, closure, and the structure constant of the parent algebra (or really any structure constant since all elements in the algebra commute). Hence, X_3 comprises a subalgebra (within the entire algebra for $SU(2)$), and is called a Cartan subalgebra, after its discoverer Élie Cartan. A Cartan sub-algebra, in the general case, comprises a sub-algebra of a parent Lie algebra, wherein the sub-algebra comprises all commuting elements of the parent algebra (for matrices, all diagonal matrices).

Subset of diagonal matrices called Cartan subalgebra

The Cartan subalgebra provides us with a means for labeling the vectors (upon which the parent algebra acts) with vector eigenvalues under operation of the Cartan generators. And the eigenvectors, in turn, provide a basis with which we can construct any general (non-eigenstate) vector.

Wholeness Chart 2-7a lists the weak charge eigenvalues for the first generation (first family) of fermions. As one might expect, the other two known lepton generations (muon/muon neutrino and tau/tau neutrino) are directly parallel. That is, they form the same multiplets, with the same weak charges, just as they had parallel e/m charges in electrodynamics.

Quarks form weak doublets and singlets as well, and they parallel those for leptons. As might be expected, the LC up quark typically occupies the upper slot in the doublet; and the LC down quark, the lower slot. Just as with leptons, there are second and third generations of quarks, which, like their lepton cousins, play a far smaller role in creation than their first generation counterparts. The second of these is comprised of the charmed and strange quarks; the third, of the top and bottom quarks. As a mnemonic, remember that the more positive "quality" (up, charmed, top) for quarks gets the upper slot (positive weak charge) in the doublet, while the more negative one (down, strange, bottom), the lower (negative weak charge) slot.

You may wish to go back and compare what we said about charges in Section 2.2.8 (pg. 18) subheading A Hypothetical Example with this present section.

Aside on Group Theory Lingo

In formal group theory, the X_3 eigenvalues of the LC electron neutrino (up quark) and the LC electron (down quark) doublets are called weights. Weight I is + ½, and weight II is – ½. We will not be using this terminology in this book.

Wholeness Chart 2-7a. *SU*(2) Cartan Subalgebra Generator Eigenvalues for 1st Generation Fermions

Leptons

Particle	Multiplet	Weak Charge
RC electron neutrino	singlet $\psi^R_{\nu_e}$	0
RC electron	singlet ψ^R_e	0
LC electron neutrino	doublet $\begin{bmatrix}\psi^L_{\nu_e}\\0\end{bmatrix}=\Psi^L_{\nu_e}$	$+\frac{1}{2}$
LC electron	doublet $\begin{bmatrix}0\\\psi^L_e\end{bmatrix}=\Psi^L_e$	$-\frac{1}{2}$

Quarks

Particle	Multiplet	Weak Charge
RC up quark	singlet ψ^R_u	0
RC down quark	singlet ψ^R_d	0
LC up quark	doublet $\begin{bmatrix}\psi^L_u\\0\end{bmatrix}=\Psi^L_u$	$+\frac{1}{2}$
LC down quark	doublet $\begin{bmatrix}0\\\psi^L_d\end{bmatrix}=\Psi^L_d$	$-\frac{1}{2}$

2.7.2 SU(3) Cartan Subalgebra

The $SU(3)$ generators (2-80) have two diagonal matrices,

SU(3) Cartan subalgebra

$$\lambda_3 = \begin{bmatrix} 1 & 0 & 0 \\ 0 & -1 & 0 \\ 0 & 0 & 0 \end{bmatrix} \qquad \lambda_8 = \frac{1}{\sqrt{3}} \begin{bmatrix} 1 & 0 & 0 \\ 0 & 1 & 0 \\ 0 & 0 & -2 \end{bmatrix} . \tag{2-129}$$

Consider their operations on a typical strong (color) triplet state, such as that created and destroyed by the green quark field of (2-111), where g_S is the strong coupling constant. As with $SU(2)$, it is conventional to divide g_S by 2 (or equivalently, simply take $X_i = ½\lambda_i$ instead of λ_i, with constant g_S) for finding eigenvalues,

Finding the two eigenvalues of an SU(3) color triplet state

$$\frac{g_S}{2}\lambda_3 \begin{bmatrix} \psi_u^L \\ 0 \end{bmatrix} \begin{bmatrix} 0 \\ 1 \\ 0 \end{bmatrix}_S = \frac{g_S}{2} \begin{bmatrix} 1 & 0 & 0 \\ 0 & -1 & 0 \\ 0 & 0 & 0 \end{bmatrix} \begin{bmatrix} \psi_u^L \\ 0 \end{bmatrix} \begin{bmatrix} 0 \\ 1 \\ 0 \end{bmatrix}_S = \begin{bmatrix} \psi_u^L \\ 0 \end{bmatrix} \frac{g_S}{2} \begin{bmatrix} 1 & 0 & 0 \\ 0 & -1 & 0 \\ 0 & 0 & 0 \end{bmatrix} \begin{bmatrix} 0 \\ 1 \\ 0 \end{bmatrix}_S = -\frac{g_S}{2} \begin{bmatrix} \psi_u^L \\ 0 \end{bmatrix} \begin{bmatrix} 0 \\ 1 \\ 0 \end{bmatrix}_S$$

$$\frac{g_S}{2}\lambda_8 \begin{bmatrix} \psi_u^L \\ 0 \end{bmatrix} \begin{bmatrix} 0 \\ 1 \\ 0 \end{bmatrix}_S = \frac{g_S}{2\sqrt{3}} \begin{bmatrix} 1 & 0 & 0 \\ 0 & 1 & 0 \\ 0 & 0 & -2 \end{bmatrix} \begin{bmatrix} \psi_u^L \\ 0 \end{bmatrix} \begin{bmatrix} 0 \\ 1 \\ 0 \end{bmatrix}_S = \frac{g_S}{2\sqrt{3}} \begin{bmatrix} \psi_u^L \\ 0 \end{bmatrix} \begin{bmatrix} 0 \\ 1 \\ 0 \end{bmatrix}_S . \tag{2-130}$$

The up green quark has two eigenvalues from the two diagonal matrices. We will label them

SU (3) eigenvalues of green quark

$$\varepsilon_3 = -\frac{g_S}{2} \qquad \varepsilon_8 = +\frac{g_S}{2\sqrt{3}} . \tag{2-131}$$

Any quantum field with those two eigenvalues will be a green quark field. The shortcut way, rather than using two different eigenvalues to label a strong interaction triplet eigenvector, is to conglomerate them into one, i.e., color (R, G, or B).

Do **Problem 25** to gain practice with finding different quantum numbers (eigenvalues) for quark and lepton states other than the green quark.

With the results of Problem 25, we can build Wholeness Chart 2-7b. Antiparticle states have eigenvalues of opposite signs from particle states. As with weak charges, the strong interaction eigenvalues are usually expressed without the coupling constant factor of g_S. And as before, the value of that constant reflects the inherent strength of the strong interaction. And it is, as one might expect, significantly greater than the weak or electromagnetic coupling constants.

SU (3) eigenvalues commonly expressed without g_S factor

In parallel with $SU(2)$, $SU(3)$ singlets have zero color charge. These are the leptons, which do not feel the strong force. With regard to color, there is no difference between RC and LC quarks or leptons.

SU (3) eigenvalues for singlets = 0

Wholeness Chart 2-7b. *SU*(3) Cartan Subalgebra Generator Eigenvalues for Fermions

Color	Multiplet	ε_3	ε_8
R quark	triplet $\begin{bmatrix} 1 \\ 0 \\ 0 \end{bmatrix}$	$\frac{1}{2}$	$\frac{1}{2\sqrt{3}}$
G quark	triplet $\begin{bmatrix} 0 \\ 1 \\ 0 \end{bmatrix}$	$-\frac{1}{2}$	$\frac{1}{2\sqrt{3}}$
B quark	triplet $\begin{bmatrix} 0 \\ 0 \\ 1 \end{bmatrix}$	0	$-\frac{1}{\sqrt{3}}$
Colorless (all leptons)	singlet	0	0

What About a Single Color (Strong Charge) Operator?

One could ask why we have two separate operators ½λ_3 and ½λ_8, with two separate eigenvalues, for the single variable of color. Why not a single 3X3 matrix operator, as we implied (for pedagogic reasons) on pg. 19? For example, the operator

$$\frac{g_S}{2}\left(\lambda_3+\sqrt{3}\lambda_8\right)=\frac{g_S}{2}\left(\begin{bmatrix}1&0&0\\0&-1&0\\0&0&0\end{bmatrix}+\begin{bmatrix}1&0&0\\0&1&0\\0&0&-2\end{bmatrix}\right)=g_s\begin{bmatrix}1&0&0\\0&0&0\\0&0&-1\end{bmatrix} \tag{2-132}$$

would have a red eigenvalue (ignoring the g_S factor) of +1; a green eigenvalue of 0, and a blue eigenvalue of –1.

But that implies the green charge is zero, so there would be no attraction, or repulsion, of it by a red or blue quark, and we know that is not the case. There is, in fact, no choice of a single diagonal matrix for which we would have equal magnitude eigenvalues of *r,g,b* quarks, but different signs for the three (as there are only two signs.). There are ample reasons, besides not being the accepted convention, why using a single diagonal matrix as a color operator would not be advantageous.

Another Aside on Group Theory Lingo

As in *SU*(2), mathematicians call the sets of eigenvalues in Wholeness Chart 2-7b "weights". For *SU*(3), there are three weights. Weight I = $\left(\frac{1}{2},\frac{1}{2\sqrt{3}}\right)$, weight II = $\left(-\frac{1}{2},\frac{1}{2\sqrt{3}}\right)$, weight III = $\left(0,\frac{-1}{\sqrt{3}}\right)$. But again, we will not be using this terminology in this book. We mention it because you will run into it in other texts.

2.7.3 Cartan Subalgebras and Observables

Almost all observables are built out of Cartan subalgebra elements of certain Lie algebras. Recall that the eigenvalues for particular operators are observables. We see with our examples from *SU*(2) and *SU*(3) above, and *U*(1) from Vol. 1, that we identify (and thus distinguish between) particles by their e/m, weak and strong operator eigenvalues. These eigenvalues are the charges associated with the respective interactions (electric, weak, and strong charges). For the weak and strong interactions, these are the eigenvalues of the Cartan subalgebra operator(s). In the strong interaction case, we streamline by labeling certain eigenvalue pairs as particular colors.

Cartan subalgebra elements are operators corresponding to observables

2.7.4 SU(n) Cartan Subalgebras

We will not delve into special unitary groups which act on vectors (multiplets) in spaces of dimension greater than 3. However, such spaces play a key role in many advanced theories, so we sum up the general principles we have uncovered as applied to special unitary operations in complex spaces of any dimension.

For an *SU*(*n*) Lie group in matrix representation, we will have $n^2 - 1$ generators in the associated Lie algebra. One can choose a basis where $n - 1$ of these are simultaneously diagonal. These diagonal matrices commute and form a subalgebra called the Cartan subalgebra. Each vector (multiplet) that is an eigenvector of all of these Cartan (diagonal) generators has $n - 1$ eigenvalues associated with those generators. These eigenvalues can be used to label the *n* independent vectors (multiplets).

SU(*n*) *has* $n - 1$ *generators in Cartan subalgebra*

2.8 Group Theory Odds and Ends

2.8.1 Graphic Analogy to Lie Groups and Algebras

We noted earlier that the vector space of a Lie algebra is commonly known as the tangent space to its associated Lie group. This is because, essentially, the basis vectors (the generators, which are matrices for us) of the vector space, are first derivatives (with respect to particular parameters), and a first derivative is tangent to its underlying function.

Fig. 2-1 illustrates this in a graphic, and heuristic, way, where the Lie group is represented as the surface of a sphere, and the Lie algebra, as a tangent plane to that sphere at the identity of the group. One can imagine more extended analogies, in higher dimensional spaces, wherein there are more than two generators X_i.

In the figure, in order for $e^{i\alpha_i X_i}$ to represent an element of the group, we need to restrict the parameters so $|\alpha_i| << 1$, i.e., we have to be very close to the identity. Each set of α_i corresponds to a different point (group element) on the sphere surface. As the α_i change, one moves along that surface. Group elements not infinitesimally close to the identity can, in principle (as summarized in Wholeness Chart 2-5, pg. 37), be represented by $e^{i\beta_j(\alpha_i)X_j}$, where the β_j are dependent on the α_i.

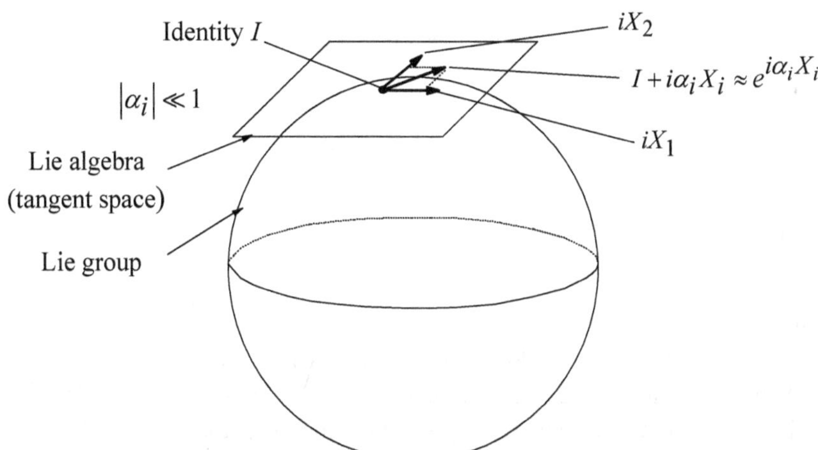

Figure 2-1. Schematic Analogy for Lie Groups and Algebras

2.8.2 Hermitian Generators and Unitarity

Note from the summary in Wholeness Chart 2-8 that all the generators we have looked at are Hermitian. This is a general rule for $SO(n)$ and $SU(n)$ groups.

Wholeness Chart 2-8. Lie Groups and Their Lie Algebra Generators

Lie Group	Matrix Rep	Generators for Fundamental Representation	Exponentiated Form
$SO(2)$	$\hat{M}$ (2-6)	$\hat{X} = \begin{bmatrix} 0 & i \\ -i & 0 \end{bmatrix}$	$e^{i\theta\hat{X}}$
$SO(3)$	$\hat{N}(=A)$ (2-11)	$\hat{X}_1 = \begin{bmatrix} 0 & 0 & 0 \\ 0 & 0 & i \\ 0 & -i & 0 \end{bmatrix}$ $\hat{X}_2 = \begin{bmatrix} 0 & 0 & -i \\ 0 & 0 & 0 \\ i & 0 & 0 \end{bmatrix}$ $\hat{X}_3 = \begin{bmatrix} 0 & i & 0 \\ -i & 0 & 0 \\ 0 & 0 & 0 \end{bmatrix}$	$e^{i\hat{\theta}_i\hat{X}_i}$
$SO(n)$	$\hat{P}$	$\hat{Y}_i$ (imaginary, Hermitian, and traceless)	$e^{i\hat{\theta}_i\hat{Y}_i}$
$SU(2)$	M (2-25)	$X_1 = \begin{bmatrix} 0 & 1 \\ 1 & 0 \end{bmatrix}$ $X_2 = \begin{bmatrix} 0 & -i \\ i & 0 \end{bmatrix}$ $X_3 = \begin{bmatrix} 1 & 0 \\ 0 & -1 \end{bmatrix}$	$e^{i\beta_i X_i}$
$SU(3)$	N (2-26)	$\lambda_1 = \begin{bmatrix} 0 & 1 & 0 \\ 1 & 0 & 0 \\ 0 & 0 & 0 \end{bmatrix}$ $\lambda_2 = \begin{bmatrix} 0 & -i & 0 \\ i & 0 & 0 \\ 0 & 0 & 0 \end{bmatrix}$ $\lambda_3 = \begin{bmatrix} 1 & 0 & 0 \\ 0 & -1 & 0 \\ 0 & 0 & 0 \end{bmatrix}$ $\lambda_4 = \begin{bmatrix} 0 & 0 & 1 \\ 0 & 0 & 0 \\ 1 & 0 & 0 \end{bmatrix}$ $\lambda_5 = \begin{bmatrix} 0 & 0 & -i \\ 0 & 0 & 0 \\ i & 0 & 0 \end{bmatrix}$ $\lambda_6 = \begin{bmatrix} 0 & 0 & 0 \\ 0 & 0 & 1 \\ 0 & 1 & 0 \end{bmatrix}$ $\lambda_7 = \begin{bmatrix} 0 & 0 & 0 \\ 0 & 0 & -i \\ 0 & i & 0 \end{bmatrix}$ $\lambda_8 = \frac{1}{\sqrt{3}}\begin{bmatrix} 1 & 0 & 0 \\ 0 & 1 & 0 \\ 0 & 0 & -2 \end{bmatrix}$	$e^{i\beta_i\lambda_i}$
$SU(n)$	P	Y_i (complex, Hermitian, and traceless)	$e^{i\beta_i Y_i}$

Do **Problem 26** to prove that all generators of $SO(n)$ and $SU(n)$ groups are Hermitian.

If you did the suggested problem, you saw that in order to generate a unitary group matrix (or orthogonal group matrix in the case of real matrices) via exponentiation of a Lie algebra element, the generators must be Hermitian, i.e., $Y_i^\dagger = Y_i$.

Note also that the four cases we studied all have traceless generators. This too is a general rule for generators of all special orthogonal and special unitary groups.

SU(n) groups have Hermitian, traceless generators

Do **Problem 27** to prove it.

2.8.3 Terminology: Vector Spaces, Dimensions, Dual Spaces, and More

Wholeness Chart 2-9, pg. 53, summarizes all of this section and more.

Use of Terms "Vector Space" and "Dimension"

Don't confuse the use of the term "vector space" when applied to matrix operation of a group on a vector versus the use of the same term when applied to Lie algebras. In the former case, the vector space is the home of the vector (or multiplet in group theory language) upon which the matrix acts. In the latter case, the vector space is the space of matrices (which are vectors in this sense) whose basis matrices (basis vectors) are the generators.

Vector space on which group matrices operate different from Lie algebra vector space

The dimension of a given representation equals the dimension of the vector space in which it acts. The $SU(2)$ representation with generators X_i shown in Wholeness Chart 2-8 operates on 2D vectors (doublets), i.e., it comprises 2x2 matrices and has dimension 2. However if an $SU(2)$ representation operates on a singlet, a 1D entity, such as in (2-128), it has dimension 1.

Dimension of a rep = dimension of vectors operated on

The dimension of a Lie algebra, on the other hand, equals the number of generators for the associated Lie group, since each generator is a basis vector (a matrix in this case) for the vector space of the Lie algebra. An $SU(n)$ Lie algebra has $n^2 - 1$ generators, so its dimension is $n^2 - 1$.

Dimension of a Lie algebra = number of generators

Further, and making it even more confusing, the dimension of a Lie group is commonly taken to be the dimension of its Lie algebra. So, an $SU(3)$ Lie group would have dimension 8, but its matrix representation in the fundamental rep (acting on three component vectors) would have dimension 3.

Dimension of a Lie group same as dimension of its algebra

So, be careful not to confuse the two uses of the term "vector space" and the two uses of the term "dimension" in Lie group theory.

Dual Vectors

Dual vector is a mathematical term for what we, in certain applications, have called the complex conjugate transpose of a vector. More generally, it is the entity with which a vector forms an inner product to generate a real scalar, whose value equals the square of the absolute value ("length" or magnitude) of the vector.

Dual vector inner product with associated vector = square of vector magnitude

Examples include the dual $\mathbf{r}^\dagger$ of the position vector $\mathbf{r}$ in any dimension, the bra $\langle\psi|$ as the dual of the ket in QM, the covariant 4D spacetime position vector as the dual of the contravariant 4D position vector, and the complex conjugate transpose as the dual of the weak doublet.

$$r_i = \begin{bmatrix} x \\ y \end{bmatrix} \rightarrow \underbrace{r_i^\dagger = r_i^{\mathrm{T}} = \begin{bmatrix} x & y \end{bmatrix}}_{\text{dual vector}} \quad r_i^\dagger r_i = |\mathbf{r}|^2 \qquad |\phi\rangle \rightarrow \underbrace{\langle\phi|}_{\text{dual vector}} \quad \langle\phi|\phi\rangle = \begin{matrix}\text{NRQM probability,} \\ \text{QFT norm (real scalar)}\end{matrix}$$

$$x^\mu = \begin{bmatrix} x^0 \\ x^1 \\ x^2 \\ x^3 \end{bmatrix} \rightarrow \underbrace{x_\mu = \begin{bmatrix} x_0 & x_1 & x_2 & x_3 \end{bmatrix}}_{\text{dual vector}} \qquad x_\mu x^\mu = \text{spacetime interval squared (real scalar)} \tag{2-133}$$

$$\begin{bmatrix} \psi_{\nu_e}^L \\ \psi_e^L \end{bmatrix} \rightarrow \underbrace{\begin{bmatrix} \bar{\psi}_{\nu_e}^L & \bar{\psi}_e^L \end{bmatrix}}_{\text{dual vector}} \qquad \begin{bmatrix} \bar{\psi}_{\nu_e}^L & \bar{\psi}_e^L \end{bmatrix} \begin{bmatrix} \psi_{\nu_e}^L \\ \psi_e^L \end{bmatrix} = \bar{\psi}_{\nu_e}^L \psi_{\nu_e}^L + \bar{\psi}_e^L \psi_e^L \quad \text{(real scalar)}$$

To be precise, the components of the dual vector for the weak doublet are adjoints, i.e., they are complex conjugate transposes in spinor space post multiplied by the γ^0 matrix. (See Vol. 1, Sect. 4.1.6, pg. 91.)

See Jeevanjee (footnote on pg. 8) for more on dual vectors.

Other Terminology

The order of a group is the number of elements in the group. The order of a matrix (even a matrix representing a group), on the other hand, is *m*X*n*, where *m* is the number of rows and *n* is the number of columns. For a square *n*X*n* matrix, the order is simply stated as *n*, and is often used interchangeably with the term dimension (of the matrix).

Order of group = num of elements

Order of square matrix same as its dimension = num of rows (or columns)

Wholeness Chart 2-9. Various Terms in Group, Matrix, and Tensor Theory

Term	Used with	Meaning	Examples
Degree	groups	n in $SO(n)$, $SU(n)$, $U(n)$, etc.	$SO(3)$ has degree 3 $SU(2)$ has degree 2
Vector space	matrix (group representation)	space of vectors upon which matrix representation acts	$SO(2)$: $M\begin{bmatrix} x \\ y \end{bmatrix}$ 2D space of x, y axes
	Lie algebra	space of matrix generators (= basis vectors in the space)	$SU(2)$: space spanned by X_1, X_2, X_3 generators
Dimension	Lie group	same as Lie algebra below	see Lie algebra below
	matrix (can be a group rep)	number of components in a vector upon which square matrix acts	$M\begin{bmatrix} x \\ y \end{bmatrix}$ M has dimension 2
	Lie algebra	number of generators (= number of independent parameters)	$SU(2)$ has dimension 3 $SU(n)$ has dimension $n^2 - 1$
Dual vector space	vector space	separate space of vecs: each inner product with original vec = vec magnitude squared	$\mathbf{r}^\dagger$, $\langle\psi\|$, x_μ
Order	group	number of elements in underlying set	any Lie group = ∞
	matrix	*m*X*n* where *m* = rows, *n* = columns (for square matrix, same as dimension)	2X3 matrix has order 2X3 3X3 matrix has order 3
	tensor	same as rank of tensor below	see rank of tensor below
Sub	group	subset of elements in a parent group that by itself satisfies properties of a group (including closure within the subgroup)	2D rotations is a subgroup of 3D rotations
	matrix	submatrix obtained by removing rows and/or columns from parent matrix	$\begin{bmatrix} a & b \\ d & e \end{bmatrix}$ submatrix of $\begin{bmatrix} a & b & c \\ d & e & f \\ g & h & i \end{bmatrix}$
	algebra	subset of elements in a parent algebra that by itself satisfies properties of an algebra (including closure within the subalgebra)	Cartan subalgebra (diagonal matrices) for the Lie algebra of any $SU(n)$ group
Rank	matrix	number of independent vectors (columns)	identity matrix in 2D: rank = 2
	Lie algebra	number of generators in Cartan sub-algebra	$SU(3)$ has rank 2; $U(1)$, rank 1 $SU(n)$ has rank $n - 1$
	tensor	number of indices	rank of tensor T_{ijk} is 3

2.8.4 Spinors in QFT and the Lorentz Group

In Vol. 1, pg. 171, we noted that when our reference frame undergoes a Lorentz transformation, the 4D spinors (spinor space vectors) transform too. The manner in which they transform was signified there by the symbol *D*, which is a particular 4D matrix group (spinor space operator). *D* is the spinor space *representation* of the Lorentz group. It is a representation of that group in spinor space, as opposed to the usual way we see the Lorentz group represented, as the Lorentz transformation in 4D spacetime. With considerable time and effort, which we will not take here (but which is shown in the citations of the footnote on the aforenoted page in Vol. 1), one can deduce the precise form of the spinor space representation from the spacetime representation.

Lorentz transformation has a rep in spinor space

2.8.5 Casimir Operators

An operator for a Lie algebra that commutes with all generators is called a Casimir operator. The identity operator multiplied by any constant is a Casimir operator, for example. However, it must be (aside from the arbitrary constant) constructed from the generators. An example from *SU*(2) is

Casimir operator commutes with all generators

$$X_1X_1 + X_2X_2 + X_3X_3 = I + I + I = 3I \qquad (2\text{-}134)$$

We will not be doing anything further with Casimir operators herein, but mention them because they are usually part of other developments of group theory you will run into.

2.8.6 Jacobi Identity

We note in passing that many texts define a Lie algebra using what is called the Jacobi identity,

$$\left[[X,Y],Z\right] + \left[[Y,Z],X\right] + \left[[Z,X],Y\right] = 0 \ , \qquad (2\text{-}135)$$

to define the second binary operation in the algebra, rather than the more straightforward way (at least for our special case) of simply defining that operation as a commutator. Each bracket in (2-135) is considered a Lie bracket, and it constitutes the second operation. Formally, the Lie brackets do not have to be commutators, they just have to satisfy (2-135), which commutators do.

Jacobi identity is formal, most general way of expressing 2nd Lie algebra operation

---

Do **Problem 28** if you wish to show that for the brackets in the Jacobi identity signifying commutation, then the identity is satisfied. (This will not be relevant for our work, so it is not critical to do this problem.)

We will not do anything more with the Jacobi identity. We mention it only because you will no doubt run into it in other texts. The bottom line is that the Jacobi identity is simply the formal way of defining the second operation for a Lie algebra. For our purposes, this is commutation.

2.8.7 Reducible and Irreducible Group Representations

Most presentations of group theory discuss what are known as reducible (or irreducible) representations of groups. We are going to postpone treatment of that topic until Part 4 of this book, however, as it will not be really relevant before we get to strong interactions.

2.8.8 Abelian (Commuting) vs Non-Abelian Operators

Recognize that if a matrix is diagonal, then exponentiation of it results in a diagonal matrix as well. So, an element of the Cartan subalgebra exponentiated will yield a diagonal element of the associated Lie group. Diagonal matrices commute.

If all elements of a particular Lie group commute, then all of its Lie algebra generators will commute as well. As we have mentioned, a commuting group is called an Abelian group; a non-commuting group, non-Abelian. The *SU*(2) and *SU*(3) groups are non-Abelian and their Lie algebras are non-Abelian, as well. The *U*(1) group of (2-17) is Abelian. Cartan subalgebras are Abelian.

SU(2), SU(3) Lie algebras non-Abelian; U(1) and Cartan subalgebras Abelian

2.8.9 Raising and Lowering Operators vs Eigenvalue Operators

As we have mentioned before (pg. 34), some operators change components of a multiplet and some can leave a multiplet unchanged, but typically multiplied by a constant.

In the latter case the constants are eigenvalues and the multiplet is an eigenvector of the operator. Examples of such operators include (see Wholeness Chart 2-8) X_3 of *SU*(2), as well as λ_3 and λ_8 of

2 kinds of matrices: diag → eigenvalues; non-diag → raise or lower multiplet components

SU(3), acting on multiplets with only one non-zero component. See (2-127) and (2-130). Such operators are generally diagonal and members of the Cartan subalgebra.

<u>*SU*(2)</u>

Other, non-diagonal operators like the *SU*(2) generators X_1 and X_2 of (2-63) raise and lower components. For examples,

$$X_1 v = \begin{bmatrix} 0 & 1 \\ 1 & 0 \end{bmatrix}\begin{bmatrix} 0 \\ C_2 \end{bmatrix} = \begin{bmatrix} C_2 \\ 0 \end{bmatrix} \quad X_1 \underset{\sim}{v} = \begin{bmatrix} 0 & 1 \\ 1 & 0 \end{bmatrix}\begin{bmatrix} C_1 \\ 0 \end{bmatrix} = \begin{bmatrix} 0 \\ C_1 \end{bmatrix} \quad X_2 v = \begin{bmatrix} 0 & -i \\ i & 0 \end{bmatrix}\begin{bmatrix} 0 \\ C_2 \end{bmatrix} = -i\begin{bmatrix} C_2 \\ 0 \end{bmatrix}. \qquad (2\text{-}136)$$

However, since, for example, X_1 can either raise or lower components of a doublet, it is generally preferred to call entities such as

SU(2) *raising and lowering operators*

$$\frac{X_1 + iX_2}{2} = \frac{1}{2}\begin{bmatrix} 0 & 1 \\ 1 & 0 \end{bmatrix} + i\frac{1}{2}\begin{bmatrix} 0 & -i \\ i & 0 \end{bmatrix} = \begin{bmatrix} 0 & 1 \\ 0 & 0 \end{bmatrix} \qquad \frac{X_1 - iX_2}{2} = \frac{1}{2}\begin{bmatrix} 0 & 1 \\ 1 & 0 \end{bmatrix} - i\frac{1}{2}\begin{bmatrix} 0 & -i \\ i & 0 \end{bmatrix} = \begin{bmatrix} 0 & 0 \\ 1 & 0 \end{bmatrix} \qquad (2\text{-}137)$$

the <u>raising operator</u> (LHS of (2-137)) and <u>lowering operator</u> (RHS of (2-137)).

<u>*SU*(3)</u>

All of the *SU*(3) generators other than λ_3 and λ_8 will similarly raise and lower components of triplets in 3D complex space.

Fig. 2-2 is a plot of the *SU*(3) eigenvalues listed in Wholeness Chart 2-7b, pg. 49. Bars over color symbols indicate anti-particles (with opposite color charges, and thus opposite eigenvalues). Note that the tips of the vectors signifying the quarks lie on a circle and all leptons sit at the origin, as they are *SU*(3) singlets, and thus have zero for each *SU*(3) eigenvalue.

Plot of fermion SU(3) *eigenvalues*

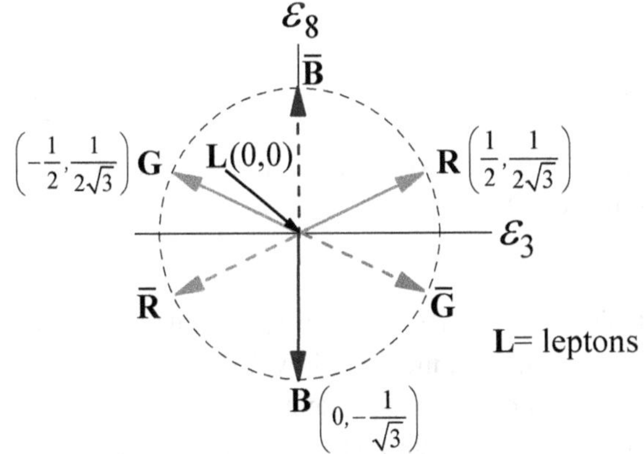

Figure 2-2. Quark and Lepton *SU*(3) Eigenvalue Plot

Note that the raising operator (see (2-80))

$$\frac{\lambda_1}{2} + i\frac{\lambda_2}{2} = \frac{1}{2}\begin{bmatrix} 0 & 1 & 0 \\ 1 & 0 & 0 \\ 0 & 0 & 0 \end{bmatrix} + i\frac{1}{2}\begin{bmatrix} 0 & -i & 0 \\ i & 0 & 0 \\ 0 & 0 & 0 \end{bmatrix} = \begin{bmatrix} 0 & 1 & 0 \\ 0 & 0 & 0 \\ 0 & 0 & 0 \end{bmatrix} \qquad (2\text{-}138)$$

operating on a green quark triplet

G → *R raising operator*

$$\left(\frac{\lambda_1}{2} + i\frac{\lambda_2}{2}\right)\begin{bmatrix} 0 \\ 1 \\ 0 \end{bmatrix} = \begin{bmatrix} 0 & 1 & 0 \\ 0 & 0 & 0 \\ 0 & 0 & 0 \end{bmatrix}\begin{bmatrix} 0 \\ 1 \\ 0 \end{bmatrix} = \begin{bmatrix} 1 \\ 0 \\ 0 \end{bmatrix} \qquad (2\text{-}139)$$

raises the triplet component from the second slot to the first, i.e., it turns it into a red quark. A comparable lowering operator does the reverse, turning a red quark into a green one.

R → *G lowering operator*

$$\left(\frac{\lambda_1}{2} - i\frac{\lambda_2}{2}\right)\begin{bmatrix} 1 \\ 0 \\ 0 \end{bmatrix} = \begin{bmatrix} 0 & 0 & 0 \\ 1 & 0 & 0 \\ 0 & 0 & 0 \end{bmatrix}\begin{bmatrix} 1 \\ 0 \\ 0 \end{bmatrix} = \begin{bmatrix} 0 \\ 1 \\ 0 \end{bmatrix} \qquad (2\text{-}140)$$

In terms of Fig. 2-2, the raising operator of (2-138) rotates the green quark vector of (2-139) clockwise 120°. The lowering operator of (2-140) rotates the red quark in the opposite direction by the same amount.

SU(3) raising and lowering operators rotate color states in Fig. 2-2

Do **Problem 29** to find other *SU*(3) raising and lowering operators.

All of the non-diagonal *SU*(3) generators play a role in raising and lowering components in quark triplets. The diagonal *SU*(3) generators determine the eigenvalues for quark triplet eigenstates.

Bottom line: There are two types of operators. One raises and/or lowers multiplet components. The other merely multiplies eigen-multiplets by an eigenvalue. Similar reasoning applies to operators in *SU*(n) for any n, as well as any *SO*(n). We could, of course, have operators which are linear combinations of those two types.

2.8.10 Colorless Composite States

Consider a quark combination, such as one would find in any baryon, of RGB. The eigenvalues of the three parts of the composite add to zero. Graphically, vector addition in Fig. 2-2 of the three color eigenstates sums to the origin. The composite is colorless. It has zero for both eigenvalues.

Hadrons are colorless (vectors in Fig. 2-2 add to zero)

Similarly, a meson is comprised of a quark and an antiquark of the same color (actually anti-color). The eigenvalues sum to zero, and the two vectors in Fig. 2-2 vector sum to the origin.

One can figuratively think of the strong force as a tendency for points on the outer circle of Fig. 2-2 to attract points on the opposite side of the circle. We will, of course, speak more about this when we get to a more formal treatment of strong interactions.

and have total eigenvalues for $\lambda_3/2$ and $\lambda_8/2 = 0$

2.8.11 Group Action of S Operator versus SU(n) Operators

Note that the S operator (see (2-4)) acts on *states*, whereas the *SU*(n) operators act on *field* multiplets. The S operator is really a group of operators and the vector space on which it operates is composed of quantum states. The *U*(1), *SU*(2), and *SU*(3) groups we have been looking at, on the other hand, act on vectors (multiplets) composed of quantum fields.

S operator acts on states; SU(n) operators act on fields

2.8.12 Unitary vs Special Unitary

The most general complex square matrix P of dimension n has n^2 different complex number components. Each of these has two real numbers, one of which is multiplied by i, so, there are $2n^2$ real numbers, call them real variables, in all. P equals a sum of $2n^2$ independent matrices (for which half of them could be real and half imaginary), each such matrix multiplied by an independent real variable.

Progression from general matrix to unitary to special summarized in Wholeness Chart 2-10

$$P=\begin{bmatrix} p_{Re11}+ip_{Im11} & p_{Re12}+ip_{Im12} & \cdots & p_{Re1n}+ip_{Im1n} \\ p_{Re21}+ip_{Im21} & p_{Re22}+ip_{Im22} & \cdots & \vdots \\ \vdots & \vdots & \ddots & \vdots \\ p_{Ren1}+ip_{Imn1} & \cdots & \cdots & p_{Renn}+ip_{Imnn} \end{bmatrix}$$
$$= p_{Re11}\begin{bmatrix} 1 & 0 & \cdots & 0 \\ 0 & 0 & \cdots & \vdots \\ \cdots & \cdots & \ddots & \vdots \\ 0 & \cdots & \cdots & 0 \end{bmatrix} + p_{Im11}\begin{bmatrix} i & 0 & \cdots & 0 \\ 0 & 0 & \cdots & \vdots \\ \cdots & \cdots & \ddots & \vdots \\ 0 & \cdots & \cdots & 0 \end{bmatrix} + p_{Re12}\begin{bmatrix} 0 & 1 & \cdots & 0 \\ 0 & 0 & \cdots & \vdots \\ \cdots & \cdots & \ddots & \vdots \\ 0 & \cdots & \cdots & 0 \end{bmatrix} + \ldots\ldots + p_{Imnn}\begin{bmatrix} 0 & 0 & \cdots & 0 \\ 0 & 0 & \cdots & \vdots \\ \cdots & \cdots & \ddots & \vdots \\ 0 & \cdots & \cdots & i \end{bmatrix}. \tag{2-141}$$

General complex matrix as sum of $2n^2$ basis (matrix) vectors with $2n^2$ indep real variables

Consider what happens when we impose the restriction that P is unitary.

$$\begin{gathered} P \text{ a general } n\mathrm{X}n \text{ matrix with } 2n^2 \text{ independent real variables} \\ P^{-1}P = P^{\dagger}P = I \rightarrow n^2 \text{ constraint equations} \rightarrow 2n^2 - n^2 = n^2 \text{ independent real variables} \end{gathered} \tag{2-142}$$

Unitary → n^2 constraint eqs on real variables → n^2 independent real variables

The imposition of unitarity yields n^2 separate scalar equations, one for each component of I, in terms of the $2n^2$ real variables. These constrain the variables so only n^2 of them are independent. P is then equal to a sum of the same $2n^2$ independent matrices, for which half of them could be real and half imaginary, but half of the variables in front of the matrices are dependent, and half independent.

We could, instead, combine the real and imaginary matrices to give us half as many complex matrices. We also have the freedom to choose whatever variables we wish as independent, as long as the remaining variables satisfy the unitary constraint equations. In effect, by doing these things, we are just changing our basis vectors (matrices here) in the n dimensional complex space P lives in.

→ n^2 complex basis (matrix) vectors

The matrices can then be combined in such a way that we can construct P from a sum of these n^2 generally complex matrices, each multiplied by an independent real variable.

Now consider that P is also special.

$$\text{Det}\,P = 1 \quad \rightarrow \quad \text{one scalar constraint equation} \rightarrow \text{one variable becomes dependent}\,, \qquad (2\text{-}143)$$

Making group special imposes one more constraint

so, we have $n^2 - 1$ independent variables with the same number of independent generators (basis vectors in the matrix space of the algebra).

Wholeness Chart 2-10 summarizes these results. Compare them with Wholeness Charts 2-5 and 2-8 on pgs. 37 and 51, respectively.

Wholeness Chart 2-10. Imposition of Unitarity and Special-ness on a General Matrix

Matrix P of Order n	**Scalar Constraint Equations**	**Independent Real Variables**	**Number of Matrices as Basis for P**	**P as Sum of Matrices (infinitesimal)**
Most general P	0	$2n^2$	$2n^2$ complex matrices Z_i	$P = p_i Z_i$ $i = 1,.., 2n^2$
Unitary $P^{-1} = P^{\dagger}$	n^2	n^2	n^2 generally complex	$P = I + i\alpha_i Y_i$ $i = 1,.., n^2$
Also, special Det $P = 1$	1	$n^2 - 1$	$n^2 - 1$ plus identity matrix	$P = I + i\alpha_i Y_i$ $i = 1,.., n^2 - 1$

2.8.13 A Final Note on Groups vs Tensors

I hope this section does not confuse you, as up to here, we have carefully discriminated between groups and tensors.

Tensors represent a lot of things in our physical world. Examples (of rank 2) include the stress and strain tensors (continuum mechanics), the vector rotation and moment of inertia tensors (classical mechanics), the electromagnetic tensor $F^{\mu}{}_{\nu}$ (see Vol. 1, pg. 138), and the stress-energy tensor of general relativity $T^{\mu}{}_{\nu}$. All of these examples are grounded in the physical world of space and time, and when expressed as matrices, all have real components. They act on real vectors and yield other (related) real vectors.

Some examples of tensors

However, in a more general sense, tensor components could be complex, and are in fact found in applications in QM, which, as we know, is a theory replete with complex numbers and complex state vectors.

A common definition of tensors delineates how they transform under a change in coordinate system. Hopefully, you have seen that before, as we cannot digress to consider it in depth here. (See Jeevanjee (footnote on pg. 8) for more details, if needed.)

Tensor action same in physical world, different components in different coordinate systems

(2-6), for fixed θ, is an expression of a tensor in an orthonormal coordinate system. This tensor rotates a 2D physical space vector through an angle θ. If we change our coordinate system (transform it so the x axis is no longer horizontal), we will transform the components of (2-6) and the components of the vector it operates on. But, in physical space, the same physical vector **v** will rotate from its original physical position by the same angle θ. Only the coordinate values are different. The actual physical operation is the same. For (2-6) to be a tensor, that same physical operation has to be performed regardless of what coordinate system one has transformed (2-6) to (what component values one has for the transformed matrix).

(2-6), which rotates a vector, is a tensor (for a fixed value of θ). So is each of (2-7). So is each of (2-13) for any given fixed set of values for θ_i. A set of such tensors, such as the elements of (2-7), or (2-6) with $0 \leq \theta < 2\pi$, or (2-13) with $0 \leq \theta_i < 2\pi$, can form a group.

Here is the point. The group $SO(2)$ comprises a set of rotation tensors in 2D. Each element in the group is a tensor. The same goes for $SO(3)$. And, if we extrapolate to complex spaces, the same thing can be considered true of any $SU(n)$. The group elements can be thought of as complex tensors[1].

Group members of SO(n) and SU(n) are tensors

Recall that groups can have many different types of members, from ordinary numbers to vectors to matrices, etc. In this chapter, we have examined groups made up of tensor elements.

BUT, the language can then get cumbersome. Earlier (Wholeness Chart 2-2, pg. 21), we distinguished between the direct product of group elements and the tensor product of vectors. But, in our cases, the direct product of group elements is, technically speaking, a tensor product of tensors. Confusing? Yes, certainly.

To circumvent this confusion as best we can, in this text, we will henceforth avoid calling our group elements tensors. We will treat group elements as represented by matrices, and call the vectors they operate on multiplets, vectors, or occasionally, rank 1 tensors. We only mention the connection of rank 2 tensors to group elements in this section. I do this because I, the author, was once confused by the similarity between groups and tensors, and I believe others must have similar confusion, as well.

But we will refer to them as "matrices", and avoid the term "tensors"

2.8.14 More Advanced Aspects of Group Theory

We note that we have presented group theory in the simplest possible way, and have done more "showing" than "proving". In particular, strict mathematical derivations of abstract Lie algebras are intimately involved with the concept of differentiation on manifolds (which we won't define here) and manifold tangent spaces. Matrix Lie algebras, as opposed to abstract ones, can be derived and defined using parametrizations and exponentiation. For matrix groups, the two approaches (abstract and matrix) are equivalent, though the latter is generally considered easier to grok (English vernacular for "understand"). Though we have focused in this chapter on matrix Lie groups and the matrix approach, we have related that to differentiation of group matrices and thus indirectly to differentiable manifolds. (See Fig. 2-1, pg. 51, and recall the Lie algebra is the tangent space of the Lie group.)

Additionally, you may have heard of other types of groups associated, for example, with string (or *M*) theory, such as E_8 and $O(32)$ groups. We will not be delving into any of these in this book. In fact, $U(1)$ and special unitary groups $SU(n)$ are all we will be concerned with from here on out, and only those in the latter group with $n = 2$ or 3.

There is much more to group theory, but we've developed what we need for the SM

Finally, as you may be aware, there are many more aspects and levels to group theory than we have covered here. These include additional terminology, advanced theory, and numerous applications[2]. We have developed what we need for the task at hand, i.e., understanding the SM.

---

[1] For veterans of tensor analysis only: To be precise, a group element in this context is a mixed tensor, as it transforms a vector to another vector in the same vector space (not to a vector in the dual vector space.) For real vectors and orthonormal bases, there is no need to bring up this distinction, as there is no difference between the original vector space and its dual space.

[2] See the references on pg. 8. Also, for a thorough, pedagogic treatment of even more advanced group theory, such as that needed for study of grand unified theories, see Marina von Steinkirch's *Introduction to Group Theory for Physicists* (2011). She wrote it as a grad student, for grad students, as an elaboration on H. Georgi's far more terse *Lie Algebras in Particle Physics* (1999). Free legal download available at www.astro.sunysb.edu/steinkirch/books/group.pdf.

2.8.15 Summary of Operators, Fields, States, and Their Interrelationships

This sub-section was added in the second revision (July 2025) of this volume.

Wholeness Chart 2-13 below summarizes the relationships between operators, fields, and states in QFT, some of which was addressed in Sect. 2.6, pg. 45. It may help to view that chart as you read the following paragraphs.

There can be some confusion as to whether we take our fields (such as ψ) or our states (such as $|e^-\rangle$) as vectors in a space. It can actually be either, so one needs to be careful about context. In one context, the fields are considered Hilbert space vectors, so operators such as spin Σ_3, the Pauli matrices σ_i, and the Gell-Mann matrices λ_i operate on vectors in that space, for example, $\Sigma_3\,\psi$.

But, the fields, such as ψ, in turn, operate on separate entities, the quantum states (to create and destroy particles in those states). In this second context, the quantum states are taken as vectors in Fock space (a generalization of Hilbert space), and the field operators act directly on those states, such as $\psi|e^-\rangle$.

Beyond this, we have yet other non-field type operators, such as the Hamiltonian H, electric charge Q, 3-momentum $\mathbf{P}$, etc., which operate on states, as in $H|e^-\rangle$.

Consider typical case examples for these three kinds of operations. If ψ is in a spin eigenstate, say spin ½ up, then $\Sigma_3\,\psi = \frac{1}{2}\psi$. If we consider only the destruction part of ψ, i.e., ψ^+, then $\psi^+\,|e^-\rangle = |0\rangle$. And if E is the definite energy of an electron, then $H|e^-\rangle = E|e^-\rangle$.

One needs to learn to be somewhat agile in use of the terms "operator" and "vector" when working in QFT, and glean their meaning from context. In particular, in the literature, flipping back and forth between vector and operator contexts for fields, such as ψ, is common, so beware.

Finally, as noted in Sect. 2.6, pg. 45, we can have operators, such as spin Σ_3, which operate on fields, that have parallel (in this case, $_{QFT}\Sigma_3$) operators, which operate on states. See (2-119) and (2-118); the conclusion below (2-119); (2-120); and the discussion following (2-120). This point is summarized in the two examples of Wholeness Chart 2-13. In later chapters, we will see other similar cases (in electroweak and strong theory) for particular values of the M matrix in (2-120).

This space on the rest of this page is intentional, so all of Wholeness Chart 2-13 can be included on a single page, the next one.

Wholeness Chart 2-13. Quantum Field Theory Vector Spaces

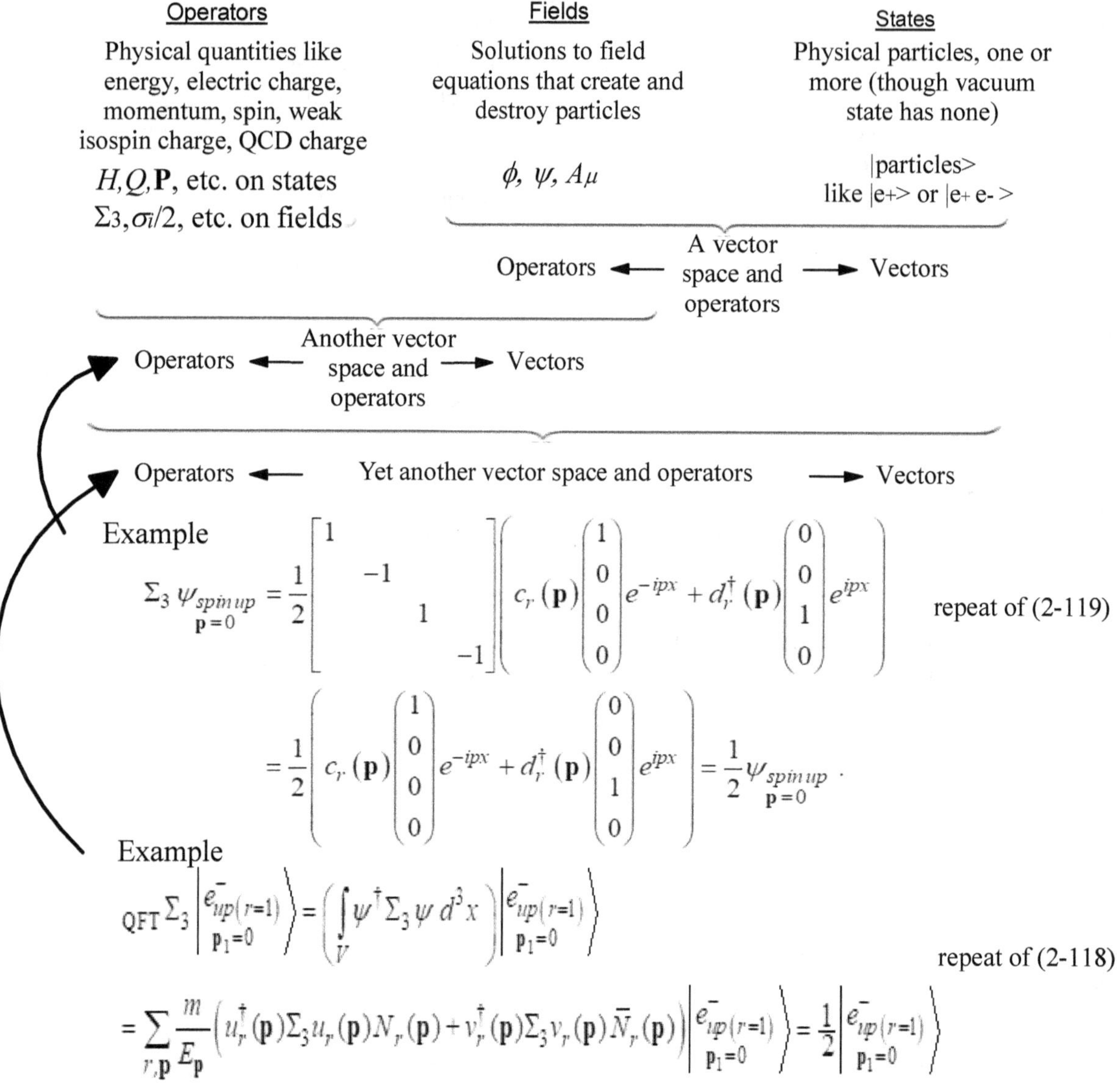

$$\Sigma_3\,\psi_{\substack{spin\,up\\ \mathbf{p}=0}} = \frac{1}{2}\begin{bmatrix}1&&&\\&-1&&\\&&1&\\&&&-1\end{bmatrix}\left(c_r(\mathbf{p})\begin{pmatrix}1\\0\\0\\0\end{pmatrix}e^{-ipx} + d_r^\dagger(\mathbf{p})\begin{pmatrix}0\\0\\1\\0\end{pmatrix}e^{ipx}\right)$$

$$= \frac{1}{2}\left(c_r(\mathbf{p})\begin{pmatrix}1\\0\\0\\0\end{pmatrix}e^{-ipx} + d_r^\dagger(\mathbf{p})\begin{pmatrix}0\\0\\1\\0\end{pmatrix}e^{ipx}\right) = \frac{1}{2}\psi_{\substack{spin\,up\\ \mathbf{p}=0}}\,.$$

$${}_{\text{QFT}}\Sigma_3\left|\begin{matrix}e^-_{up(r=1)}\\ \mathbf{p}_1=0\end{matrix}\right\rangle = \left(\int_V \psi^\dagger\Sigma_3\psi\, d^3x\right)\left|\begin{matrix}e^-_{up(r=1)}\\ \mathbf{p}_1=0\end{matrix}\right\rangle$$

$$= \sum_{r,\mathbf{p}}\frac{m}{E_{\mathbf{p}}}\left(u_r^\dagger(\mathbf{p})\Sigma_3 u_r(\mathbf{p})N_r(\mathbf{p}) + v_r^\dagger(\mathbf{p})\Sigma_3 v_r(\mathbf{p})\bar{N}_r(\mathbf{p})\right)\left|\begin{matrix}e^-_{up(r=1)}\\ \mathbf{p}_1=0\end{matrix}\right\rangle = \frac{1}{2}\left|\begin{matrix}e^-_{up(r=1)}\\ \mathbf{p}_1=0\end{matrix}\right\rangle$$

2.9 Chapter Summary

There are several ways to summarize this chapter. The first is the graphic overview in Section 2.9.1. Second is the verbal summary of Section 2.9.2. Third is in chart form and found in Section 2.9.3. To get a more detailed summary simply review the other wholeness charts in the chapter. Being structured, graphic, and concise yet extensive, they paint a good overview of group theory as it applies to QFT.

2.9.1 Pictorial Overview

Wholeness Chart 2-11. is a schematic "big picture" view of this chapter and needs little in the way of explanation.

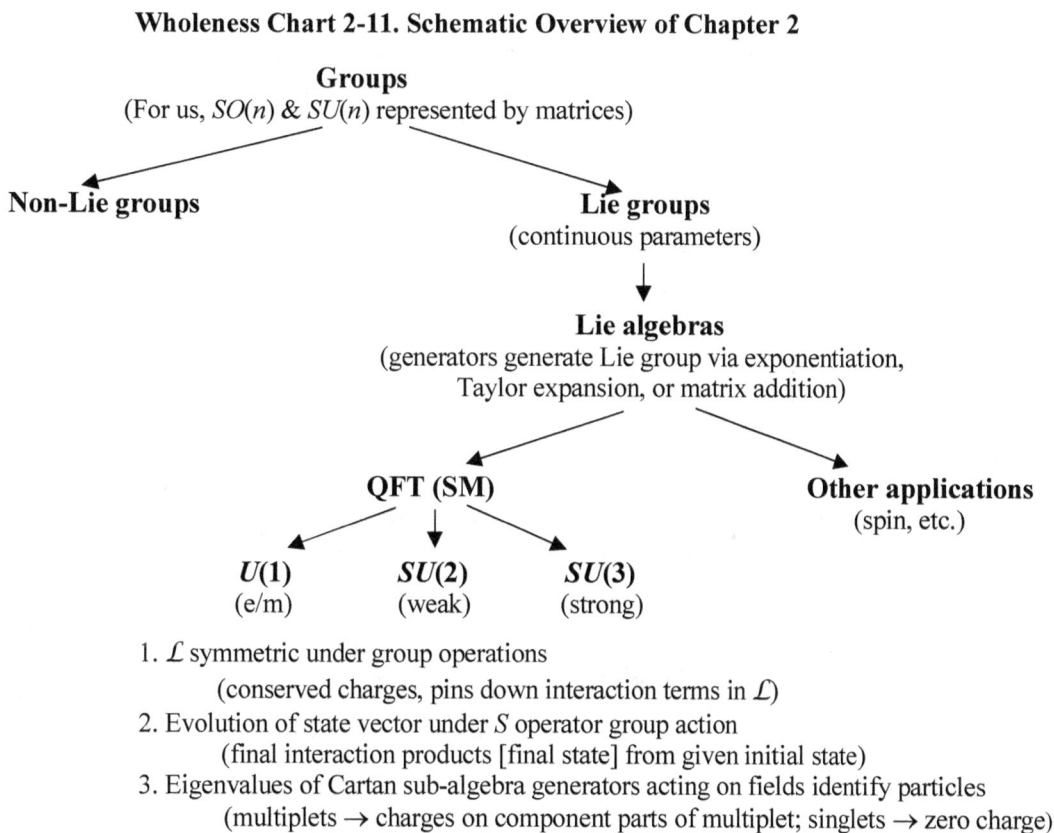

Wholeness Chart 2-11. Schematic Overview of Chapter 2

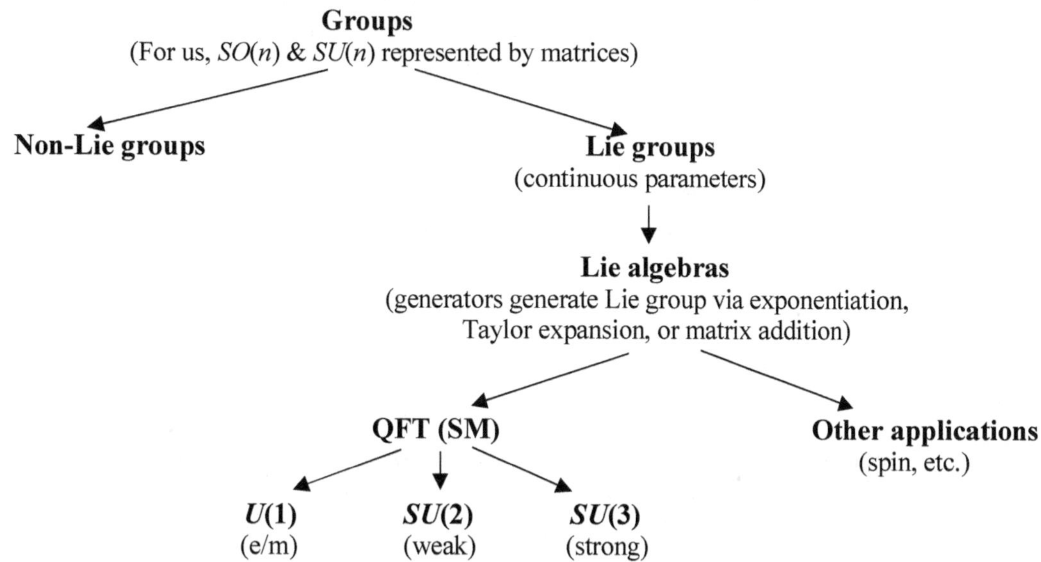

1. $\mathcal{L}$ symmetric under group operations
   (conserved charges, pins down interaction terms in $\mathcal{L}$)
2. Evolution of state vector under S operator group action
   (final interaction products [final state] from given initial state)
3. Eigenvalues of Cartan sub-algebra generators acting on fields identify particles
   (multiplets $\rightarrow$ charges on component parts of multiplet; singlets $\rightarrow$ zero charge)

2.9.2 Verbal Summary

A group can be represented as a set of matrices and a binary operation between the matrices that satisfy the properties of a group, such as closure, being invertible, etc. The group may act on vectors in a vector space, and in our applications will virtually always do so.

There are many classifications of groups (orthogonal, unitary, special, Abelian, Lie, direct product, finite, and more) and a particular group could belong to one or several of these classifications. For unitary (orthogonal) groups, the complex conjugate transpose (transpose) of any element equals the inverse of that element. For special groups, the determinant of every element equals unity. Special orthogonal groups [$SO(n)$] rotate vectors (active transformation) or coordinate systems (passive transformation) in real spaces. Special unitary groups [$SU(n)$] can be conceptualized as comparable "rotations" in complex spaces.

An algebra can be represented as a set of matrices with scalar multiplication and two binary operations that satisfy certain properties such as closure.

Lie groups have all elements that vary as continuous, smooth functions of one or more continuous, smooth variables (parameters). Lie algebras have elements that satisfy the properties of an algebra, generate an associated Lie group, and, as a result, are continuous, smooth functions of one or more continuous, smooth variables (parameters).

The basis elements of a Lie algebra can generate the Lie group

1) globally, and locally, as factors in the terms of a Taylor expansion about the identity,
2) globally and locally, via exponentiation (which is simplified locally), or
3) locally (generally, but in some particular cases globally, as well), via matrix addition of the identity plus the Lie algebra basis elements times the independent parameters.

Such Lie algebra elements are called the generators of the group. The generators can generate the group via exponentiation, though in the global case, finding the form of the associated real variables used in the exponentiation can be very complicated for all but the simplest matrix groups. For infinitesimal values of the parameters, however, the exponentiation simplifies, as we only deal with first order terms in the expansion of the exponential.

Lie algebras we will deal with comprise matrices and the operations of matrix addition and commutation. Structure constants relate such commutation to the particular matrices (within the Lie algebra) the commutation results in.

A symmetry under any group operation (matrix multiplication for us) means some value, as well as the functional form of that value, remains unchanged under that operation.

For a given group $SU(n)$, there will generally be multiplets (vectors) of n dimensions and singlets (scalars) of dimension one, upon which the group elements operate. Diagonal Lie algebra elements of the associated Lie group commute, and when operating on multiplets with one non-zero component, yield associated eigenvalues. The set of such diagonal elements is called the Cartan subalgebra. Operation of such elements on singlets yields eigenvalues of zero. The dimension of a representation of the group equals the number of components in the multiplet (or singlet) upon which it acts. The same group, when acting on different dimension vectors, is represented by different matrices, i.e., it has different representations.

In the standard model (SM), the weak force is embodied by the $SU(2)$ group; the strong force, the $SU(3)$ group; and the electromagnetic force, the $U(1)$ group. Cartan subalgebra elements in these spaces acting on quantum field multiplets yield eigenvalues that are the same as those of associated operators acting on single particle states. For $SU(2)$ and $SU(3)$, these eigenvalues, alone or in sets, correspond to weak and strong (color) charges. Certain symmetries in each of the $U(1)$, $SU(2)$, and $SU(3)$ groups dictate the nature of the interactions for each of the various forces. Much of the rest of this book is devoted to exploring these symmetries and their associated interactions.

2.9.3 Summary in Chart Form

Wholeness Chart 2-12. Table Overview of Chapter 2

Entity	**Definition/Application**	**Comment**
Group	Elements satisfy criteria for a group	
Lie Group	Elements = continuous smooth function(s) of continuous smooth parameter(s) α_i	
$SU(n)$ Lie Group	Det $P = 1$ $\quad P^{\dagger}P = I$	$P(n)$ = matrix rep of $SU(n)$
Generators = Bases of Lie Algebra	$Y_i = -i\dfrac{\partial P}{\partial \alpha_i} \qquad i = 1, \ldots, n^2 - 1$	
Commutation relations	$\left[Y_i, Y_j\right] = i c_{ijk} Y_k$	Usually different structure constants c_{ijk} for different parameters α_i.
Lie Algebra, 2nd Operation	$i\,[Y_i, Y_j]$	Elements satisfy algebra criteria. 1st binary operation is addition. Both operations have closure.
Group from Algebra For $\lvert\alpha_i\rvert << 1 \rightarrow$	$P = e^{i\beta_i(\alpha_j)Y_i}$ $P \approx e^{i\alpha_i Y_i} \approx I + i\alpha_i Y_i \quad \delta P \approx i\alpha_i Y_i$	Can also construct P from Taylor expansion with Y_i used therein.
Cartan subalgebra	Diagonal generators of Y_i (symbol Y_i^{Cart} here) Acting on vector space → eigenvectors & eigenvalues	All elements commute.
QFT	Y_i^{Cart} eigenvalues of field multiplets (vectors in a vector space) correspond to weak & strong (color) charges.	Later chapters to show these do, in fact, correspond to such charges.
	P & Y_i act on vector space of quantum fields. S operator acts on Fock (vector) space of (multiparticle) states. $\int \Psi^{\dagger} P \Psi d^3x$ & $\int \Psi^{\dagger} Y_i \Psi d^3x$ act on Fock (vector) space of states	Y_i acting on field multiplet yields same eigenvalue(s) as $\int \Psi^{\dagger} Y_i \Psi d^3x$ acting on associated single particle state.
	Symmetries of $\mathcal{L}$ for quantum field multiplet transformations under $SU(n)$ operators yield 1) charge conservation (via Noether's theorem) and 2) the form of the interaction $\mathcal{L}$.	To be shown in later chapters.

2.10 Appendix: Proof of Particular Determinant Relationship

In Problem 27 one is asked to start with relation (2-144) from matrix theory, where A and B are matrices.

$$\text{If } B = e^{A}\text{, then } Det\ B = e^{Trace\ A} \tag{2-144}$$

Here we prove that relation.

We can diagonalize A in the first relation of (2-144) via a similarity transformation, and B will be diagonalized at the same time.

$$B_{diag} = e^{A_{diag}} \tag{2-145}$$

For any component of the matrix in (2-145) with row and column number k,

$$B^{kk}_{diag} = e^{A^{kk}_{diag}} \tag{2-146}$$

Then, taking determinants, we have

$$Det\ B_{diag} = B^{11}_{diag} B^{22}_{diag} B^{33}_{diag} .... = e^{A^{11}_{diag}} e^{A^{22}_{diag}} e^{A^{33}_{diag}} ... = e^{\left(A^{11}_{diag} + A^{22}_{diag} + A^{33}_{diag} ...\right)} = e^{Trace\ A_{diag}} . \tag{2-147}$$

The Baker-Campbell-Hausdorf relation is not relevant for the next to last equal sign because all exponents are mere numbers. Since both the trace and the determinant are invariant under a similarity transformation, we have, therefore,

$$Det\ B = e^{Trace\ A} . \tag{2-148}$$

2.11 Problems

1. Give your own examples of a group, field, vector space, and algebra.
2. Is a commutative group plus a second binary operation a field?
   Is a commutative group plus a scalar operation with a scalar field a vector space?
   Is a vector space plus a second binary operation an algebra?
   Does a field plus a scalar operation with a scalar field comprise an algebra? Is this algebra unital?
3. Show that 3D spatial vectors, under addition, form a vector space. Then show that QM states do as well. What do we call the space of QM states?
4. Show that the matrices $\hat{N}$ below are an orthogonal group $O(2)$ that is not special orthogonal, i.e., not $SO(2)$.

$$\hat{N}(\theta) = \begin{bmatrix} -cos\,\theta & sin\theta \\ sin\theta & cos\,\theta \end{bmatrix}$$

   Show the operation of $\hat{N}$ on the vector [1,0] graphically. Graph the operation on the same vector for $\theta = 0$. Do these graphs help in understanding why special orthogonal transformations are more appropriate to represent the kinds of phenomena we see in nature? Explain your answer.
5. Write down a 2D matrix that cannot be expressed using (2-6) and a 3D matrix than cannot be expressed using (2-12) and (2-13).
6. Show that U of (2-17) forms a group under matrix multiplication.
7. Show that any unitary operation U (not just (2-17)) operating on a vector (which could be a quantum mechanical state) leaves the magnitude of the vector unchanged.

8. Does $e^{i\theta}$ [which is a representation of the unitary group $U(1)$] acting on a complex number characterize the same thing as the $SO(2)$ group representation (2-6) acting on a 2D real vector? Explain your answer mathematically. (Hint: Express the components of a 2D real vector as the real and imaginary parts of a complex scalar. Then, compare the effect of $e^{i\theta}$ on that complex scalar to the effect of (2-6) on the 2D real vector.) Note that $U(1)$ and $SO(2)$ are *different groups*. We do *not* say that $e^{i\theta}$ here is a representation of $SO(2)$. $U(1)$ and $SO(2)$ can describe the same physical world phenomenon, but they are *not* different representations (of a particular group), as the term "representation" is employed in group theory.

9. Show that M of (2-20) obeys the group closure property under the group operation of matrix multiplication.

10. Show (2-35) in terms of matrices.

11. Show that $\hat{X} = \begin{bmatrix} 0 & i \\ -i & 0 \end{bmatrix}$ along with a scalar field multiplier θ comprise an algebra. Use Wholeness Chart 2-1 as an aid. Note that every $\theta\hat{X}$ is in the set of elements comprising the algebra, and the operations are matrix addition and matrix commutation. This is considered a trivial Lie algebra. Why do you think it is considered such?

12. Show there is no identity element for the 2nd operation (2-52) in the $SO(3)$ Lie algebra. (Hint: The identity element has to work for every element in the set, so you only have to show there is no identity for a single element of your choice.)

13. Why did we take a matrix commutation relation as our second binary operation for the algebra for our $SO(3)$ Lie group, rather than the simpler alternative of matrix multiplication? (Hint: Examine closure.)

14. Find the generators in $SO(3)$ for the parametrization $\theta'_i = -\theta_i$ in (2-12). Then, find the commutation relations for those generators.

15. Obtain the $\hat{\theta}_i$ values for (2-58) up to third order.

16. Show that $X_1 = \begin{bmatrix} 0 & 1 \\ 1 & 0 \end{bmatrix}, X_2 = \begin{bmatrix} 0 & -i \\ i & 0 \end{bmatrix}$, and $X_3 = \begin{bmatrix} 1 & 0 \\ 0 & -1 \end{bmatrix}$ generate the three-parameter $SU(2)$ group. (Hint: Use (2-61), along with (2-20), (2-23), and the derivative of (2-23) to get M, and prove that all elements shown in the text in that expansion can be obtained with the generators and the identity matrix. Then, presume that all other elements not shown can be deduced in a similar way, with similar results.) Then sum up the second order terms in the expansion to see if it gives you, to second order, the group matrix (2-20).

17. For $SU(2)$ with $|\alpha_1|, |\alpha_2|, |\alpha_3| << 1$, show $e^{i(\alpha_1 X_1 + \alpha_2 X_2 + \alpha_3 X_3)} \approx I + i\alpha_1 X_1 + i\alpha_2 X_2 + i\alpha_3 X_3 \approx M$.

18. Show that $M(\theta_1, \theta_2, \theta_3) = \begin{bmatrix} \cos\theta_1 e^{i\theta_2} & \sin\theta_1 e^{i\theta_3} \\ -\sin\theta_1 e^{-i\theta_3} & \cos\theta_1 e^{-i\theta_2} \end{bmatrix}$ is an $SU(2)$ Lie group. That is, verify that the group is both unitary and special.

19. Although this is already explicit in (2-81), confirm that the λ_i, in (2-80), are the generators of the matrix N of (2-28). (Hint: Expand N in terms of α_i. Comparing with (2-61) to (2-63) and (2-79) may help.)

20. Show that $[\lambda_1, \lambda_2] = i\,2\lambda_3$ and that $[\lambda_6, \lambda_7] = i\sqrt{3}\lambda_8 - i\lambda_3$.

21. Show that the three terms $\mathcal{L}_{\substack{three\ quark\\ terms}} = \bar{\psi}_r \not{\partial} \psi_r + \bar{\psi}_g \not{\partial} \psi_g + \bar{\psi}_b \not{\partial} \psi_b = \begin{bmatrix} \bar{\psi}_r & \bar{\psi}_g & \bar{\psi}_b \end{bmatrix} \not{\partial} \begin{bmatrix} \psi_r \\ \psi_g \\ \psi_b \end{bmatrix}$ are symmetric under an $SU(3)$ transformation that is independent of space and time coordinates x^μ.

22. Show that for a non-unitary group acting on any arbitrary vector, the vector magnitude is generally not invariant.

23. Prove (2-122). This is time consuming, so you might just want to look directly at the solution in the solutions booklet.

24. Show the effect of the diagonal $SU(2)$ generator on a doublet where neither component is zero. What do you conclude from the result?

25. Find the strong interaction quantum numbers for the down red quark, the up blue quark, the LC electron, and the RC electron neutrino states.

26. For any unitary matrix P, $P^\dagger P = I$. If $P = e^{i\beta_i Y_i}$, show that all Y_i are Hermitian. For any orthogonal matrix $\hat{P}$, $\hat{P}^T \hat{P} = I$. If $\hat{P} = e^{i\hat{\theta}_i \hat{Y}_i}$, show that all $\hat{Y}_i$ are purely imaginary and Hermitian.

27. Given what was stated in Problem 26, and using the relation from matrix theory below, show that all special orthogonal and special unitary generators must be traceless. The relation below is derived in the appendix of this chapter.

From matrix theory: For $B = e^A$, Det $B = e^{Trace\ A}$.

28. Show that if the brackets in the Jacobi identity $\left[[X,Y],Z\right] + \left[[Y,Z],X\right] + \left[[Z,X],Y\right] = 0$ are commutators, then the identity is satisfied automatically.

29. Find the $SU(3)$ raising operator that turns a blue quark into a green one. Then find the lowering operator that does the reverse, i.e., changes a green quark into a blue one. In terms of Fig. 2-2, what does the action of each of these operators do? What does the action of the $\frac{1}{2}(\lambda_1 + i\lambda_2)$ operator do to a blue quark?

Chapter 3

Relevant Mathematical Topics

"Young man, in mathematics, you don't understand things. You just get used to them."
John von Neumann

3.0 Preliminaries

I chose the quote above because I disagree so strongly with it. And I suspect, so do many of you reading this book. If concepts are presented in a clear, pedagogic way, they can be understood by reasonably capable students. And in fact, they must be understood, if one wishes to make any kind of mark in any field, like physics, whose foundation is mathematical.

Our goal in this chapter, and more so in this entire book, is to understand. As I've said before, I will do my best to help you do so, and if possible, you will help me and others by suggesting ways to make it all even clearer and easier for others, who come later. (See "Provide Feedback" link on book website. URL opposite pg. 1.)

3.0.1 Background

Certain topics in mathematics play key roles in the path integral (PI) approach to QFT. However, these topics can also be employed in the canonical quantization approach, though they are not typically used there. This can lead, and has led, to student confusion about these topics as somehow being an inextricable part of the path integral method, and that method alone. They are not.

This chapter: new techniques for canonical approach

In this chapter, we will introduce these concepts in the context of canonical quantization, primarily for QED. In Part 2 of this book, we will use them in developing QED via path integrals. Later on, we will see them again, as they are a foundational part of the path integral development of strong and weak interactions, though for the latter theory they are not essential and typically not covered in introductory courses.

But later will apply these techniques to PI approach

Note that those anxious to get to electroweak interactions without delay can skip this chapter and the next one now, then return to them before beginning strong interactions in Chap. 12. Chaps. 5 through 11, on electroweak theory, only make cursory reference to Chaps. 3 and 4.

Strictly speaking, this chapter, much like Chap. 2, is not purely mathematical. It also applies certain mathematical structures to QFT. Because those structures underlie all three of the SM interactions as well as both the canonical and PI approaches, I felt it best to include them in Part 1 of this book, which presents overarching principles that permeate all of the material to be covered in Parts 2, 3, and 4.

3.0.2 Chapter Overview

In this chapter, we will introduce and develop the mathematics of

- Green functions,
- Grassmann variables, and
- the generating functional.

The three main topics in this chapter

You have probably already learned about Green functions in prior courses, but the meaning of the term in QFT is somewhat different from that of the usual mathematics course. We will explore those differences, and in so doing, refine our understanding of these widely used functions. With the QFT

From QFT Green function can deduce transition amplitude

form of the Green function in hand, we can deduce the transition amplitude for a given interaction. That is where its value lies.

Grassmann variables are a particular type of variable that are used in the process of finding relevant Green functions. In that case, they are actually Grassmann *fields*, and are treated as *classical*, rather than quantum fields. Despite being classical in nature, and not operators, they have an unusual property such fields typically do not have. They anti-commute. If you are suspecting that these fields may be related to fermions, you would be correct. More on this later.

Grassmann variables are classical and anti-commute

Grassmann fields are used with something called the generating functional to determine a particular Green function. The generating functional *generates* the Green function by taking derivatives of it with respect to fields, some of which are Grassmann fields. That Green function is then used to find the transition amplitude.

From generating functional can deduce Green functions

We show this sequence of steps graphically in Wholeness Chart 3-1 below.

Wholeness Chart 3-1. Finding Transition Amplitude from the Generating Functional

find generating functional (via canonical or PI approach)

↓

take derivatives with respect to fields (including Grassmann fields)

↓

Green function

↓

transition amplitude

From generating functional to transition amplitudes

Our goal in this chapter is to find the generating functional (via canonical quantization) for any particular interaction. From it, as we will see, we can obtain the transition amplitude.

In Part 2 of this book, we will find the form of the generating functional for the path integral approach. Even though the forms of the Green function are different in the two approaches, we will show they are effectively equivalent and thus, yield the same transition amplitude.

Generating functional and QFT Green functions have different forms in canonical and PI approaches

There is a summary of the material in this chapter in Wholeness Chart 3-7 beginning on page 102. I recommend following that, step-by-step, as you proceed through the chapter.

This chapter: canonical approach

3.1 Green Functions

3.1.1 Usual Green Function Definition/Derivation from Mathematics

The usual Green function technique described in most mathematics texts entails solving an equation of form (3-1), where $u(x)$ is the solution, L is an operator, and $f(x)$ is a known function.

$$Lu(x) = f(x) \tag{3-1}$$

The usual math Green function helps solve differential equations

The Green function for this equation is $G(x,s)$, which solves

$$LG(x,s) = \delta(x-s). \tag{3-2}$$

The solution $u(x)$ to (3-1) is then (see proof below)

$$u(x) = \int G(x,s) f(s) ds. \tag{3-3}$$

Proof

Putting (3-3) into (3-1), then using (3-2), we have

$$Lu(x) = L\int G(x,s)f(s)ds = \int LG(x,s)f(s)ds = \int \delta(x-s)f(s)ds = f(x). \tag{3-4}$$

End of proof

3.1.2 Usual Math Definition of Green Function for QFT Field Equations

The Klein-Gordon (Scalar) Equation Green Function

Consider the familiar scalar field equation for Klein-Gordon fields,

$$\left(\partial_\alpha \partial^\alpha + m^2\right)\phi = 0. \tag{3-5}$$

The Green function for (3-5) is found from (3-6) (where a minus sign on the RHS doesn't change the essence of the issue and is conventional for scalars in QFT, we use y in place of the s of (3-2), and x and y here represent 4D spacetime positions).

Can use usual math Green function with scalar field equation

$$\left(\partial_\alpha \partial^\alpha + m^2\right) G(x,y) = -\delta(x-y) \tag{3-6}$$

To solve (3-6), convert it to momentum space via a Fourier transform[1] over x.

$$\left(-k^2 + m^2\right)\hat{G}(k,y) = -e^{iky}. \tag{3-7}$$

The solution to (3-7) is

$$\hat{G}(k,y) = \frac{e^{iky}}{k^2 - m^2}. \tag{3-8}$$

But for the factor of $-e^{iky}$, $\hat{G}$ is the inverse of the K-G equation operator in momentum space.

Converting (3-8) back to position space via a reverse Fourier transform, we get

When we do, we get Green function = scalar propagator

$$G(x,y) = \frac{1}{(2\pi)^4}\int d^4k \frac{e^{-ik(x-y)}}{k^2 - m^2 + i\varepsilon} = \Delta_F(x-y) = G(x-y) \text{ with } \Delta_F(k) = \frac{1}{k^2 - m^2 + i\varepsilon}, \tag{3-9}$$

which, when multiplied by i, is just the scalar Feynman propagator, and for which the (x,y) argument can be expressed $(x-y)$. Note that we add the infinitesimal quantity $i\varepsilon$ to the denominator of the integrand in (3-9) for reasons delineated in Vol. 1, pg. 76. Essentially, the integrand would otherwise blow up at $|k| = m$. After doing the work one needs to do with (3-9) (effectively integrating it in momentum space, as we did in Vol. 1 to obtain transition amplitudes), the ε can be taken to zero (and we actually get a suitable result).

<u>The Photon Green Function</u>

In analogy to the Klein-Gordon (scalar) case, instead of (3-5), we have the massless spin 1 field equation, the Maxwell equation for photons,

$$\left(\partial_\alpha \partial^\alpha\right) A^\mu = 0. \tag{3-10}$$

Do **Problem 1** to show the Maxwell equation Green function is as shown in (3-11).

$$G^{\mu\nu}(x-y) = \frac{-g^{\mu\nu}}{(2\pi)^4}\int d^4k \frac{e^{-ik(x-y)}}{k^2 + i\varepsilon} \quad \text{with } D_F^{\mu\nu}(k) = \frac{-g^{\mu\nu}}{k^2 + i\varepsilon} \tag{3-11}$$

Similarly, for massless spin 1 field equation, we get Green function = photon propagator

<u>The Spin ½ Green Function</u>

Do **Problem 2** to show the Dirac equation Green function is as shown in (3-12).

Similarly, the Green function for the Dirac equation $\left(i\gamma^\alpha \partial_\alpha - m\right)\psi = 0$ is

And for spin ½ field equation, Green function = spinor propagator

$$G(x-y) = \frac{1}{(2\pi)^4}\int_{-\infty}^{+\infty} \frac{e^{-ip(x-y)}}{\not{p} - m + i\varepsilon} d^4p = S_F(x-y)$$
$$\text{with } S_F(p) = \frac{1}{\not{p} - m + i\varepsilon} = \frac{\not{p} + m}{p^2 - m^2 + i\varepsilon}. \tag{3-12}$$

<u>General Conclusion for Green Functions of Usual Mathematics and QFT Field Equations</u>

The Green function (defined as negative of the usual math definition and multiplied by i) for each particle type field equation equals the Feynman propagator. So, we can see why propagators are sometimes called Green functions.

Little wonder that propagators are often called Green functions

---

[1] Our convention is the one typically used in QFT. $\hat{f}(k) = \int d^4x\, f(x) e^{\pm ikx}$ where $kx = k^\mu x_\mu$, and $f(x) = (2\pi)^{-4}\int d^4k\, \hat{f}(k) e^{\mp ikx}$, where it is irrelevant which exponential signs one uses, as long as the two transforms have opposite signs therein. Note also that from Fourier transform tables (or taking the time to prove it yourself) $\widehat{\partial_\mu f(x)} = \mp ik_\mu \hat{f}(k)$.

3.1.3 Simplified Overview of Green Function Methodology of QFT

A Somewhat Different (Extended) Approach to Green Functions

When students first see the more extended use of Green functions in QFT, which goes well beyond merely finding propagators, there is typically confusion, as the methodology therein seems quite different from the usual math Green functions, as described in the foregoing. So, we will consider two different approaches to Green functions, the usual math one (see above) and, what we will term the "QFT Green function methodology" (see below). The QFT Green function methodology results are more extensive, in that they can be applied to entire interactions, not simply a sole propagator (though as we will see, they also include, as a subset, the sole propagator, the usual math Green function).

QFT Green function methodology different from, and goes beyond, usual math methodology

It applies to entire interactions, not just propagators

Brief Overview of QFT Green Function Methodology: In Words

The QFT Green function methodology approach provides a final result, the Green function, that can be readily transformed into a Feynman amplitude for a given interaction. The transformation involves straightforward substitution, as explained in words and pictures in the following.

As it turns out, the QFT methodology yields a Green function consisting of propagators (e.g., $S_F(p)$), and only propagators. That is, unlike Feynman amplitudes, there are no external line expressions such as $u_1(\mathbf{p})$. One obtains the appropriate Feynman amplitude simply by substituting an external line expression for each incoming/outgoing particle (that appears as a propagator in a Green function) in the particular interaction of interest. (e.g., Substitute $u_1(\mathbf{p})$ for the Green function $S_F\,(p)$.)

QFT Green functions composed only of propagators

There are other subtleties, such as getting the signs right on the 4-momenta, but that is the technique in its essence. Thereby, one turns a Green function into a Feynman amplitude.

Substituting external particles for certain propagators yields transition amplitude

Brief Overview of QFT Methodology Green Functions: In Pictures

The Green function is a mathematical entity, just as the transition amplitude is a mathematical entity. And just as transition amplitudes can be represented pictorially by Feynman diagrams, so can Green functions be represented by diagrams. Fig. 3-1 displays the relationship between a typical diagram associated with a Green function and a corresponding Feynman diagram.

Showing this with diagrams

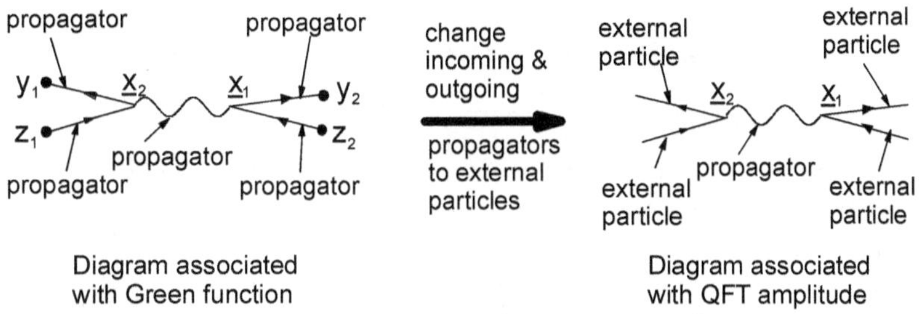

Figure 3-1. Changing a Green Function Diagram to External Particles Feynman Diagram
(for one type of Bhabha Scattering)

Note from the LH diagram in Fig. 3-1 that, as mentioned above, a Green function (and pictorially, a Green function diagram) is made up wholly of propagators. In the LH diagram, the incoming electron and positron of Bhabha scattering (only lowest order and only one of two ways it can occur are shown) are represented by propagators. The outgoing electron and positron in the RH diagram are represented in the LH diagram by propagators, as well. Small dots on the end of the lines in diagrams indicate the lines represent propagators, rather than what we formerly thought of as external particles.

In Green function lingo, the propagators that become external particles are called legs. A Green function with n legs is called an n-point Green function. For example, the Green function corresponding to the diagram on the LHS of Fig. 3-1 is a four-point Green function.

Legs → external particles. n legs = n-point Green function

To convert the diagram (and mathematically, the Green function) to an external particle diagram (and mathematically, to a transition amplitude), we change each leg to an external particle. Note carefully, though, that by convention, time in all leg 4-momenta in a Green function is defined to be directed inward in the Green function diagram, even though that makes no sense physically. The top LH diagram in Fig. 3-2 is the form of Green function diagram one will find in other texts, where you have to keep in mind that time is flowing inward on all legs. In the top RH diagram, we represent the same thing, but with all leg time directions toward the right. This will help us in visualizing the Feynman diagrams we wish to convert the Green function diagram into, but be aware that drawing it in this way is not done elsewhere.

Time direction of all legs defined as inward

Now here is a key feature of Green functions. Depending on what interaction we wish to analyze, we will choose different legs in the Green function to represent different external particles in a particular Feynman amplitude. This is illustrated in the second row of Fig. 3-2 where the particular interaction of interest dictates which legs convert to incoming (time inward) particles and which convert to outgoing (time outward) particles.

Converting different legs in Green function to different initial/ final particles yields different interactions

We see that a single QFT Green function can allow us to analyze a number of different interactions. The first Feynman diagram in the second row represents Bhabha scattering (1st way, as in Vol. 1, pg. 221, Fig. 8-2). The second diagram in that row represents Møller scattering (1st way, Vol. 1, pg. 229, Fig. 8-4). The third diagram, positron Møller scattering (1st way). Take care that subscript numbers on Green function legs have nothing to do with subscript numbers in Feynman diagrams. Final particles in the latter diagrams have primes; initial particles, no prime.

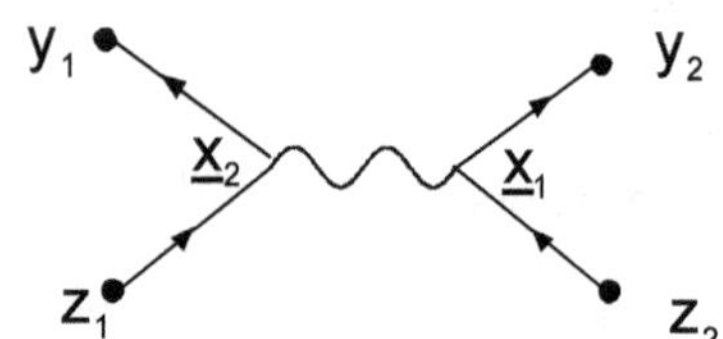

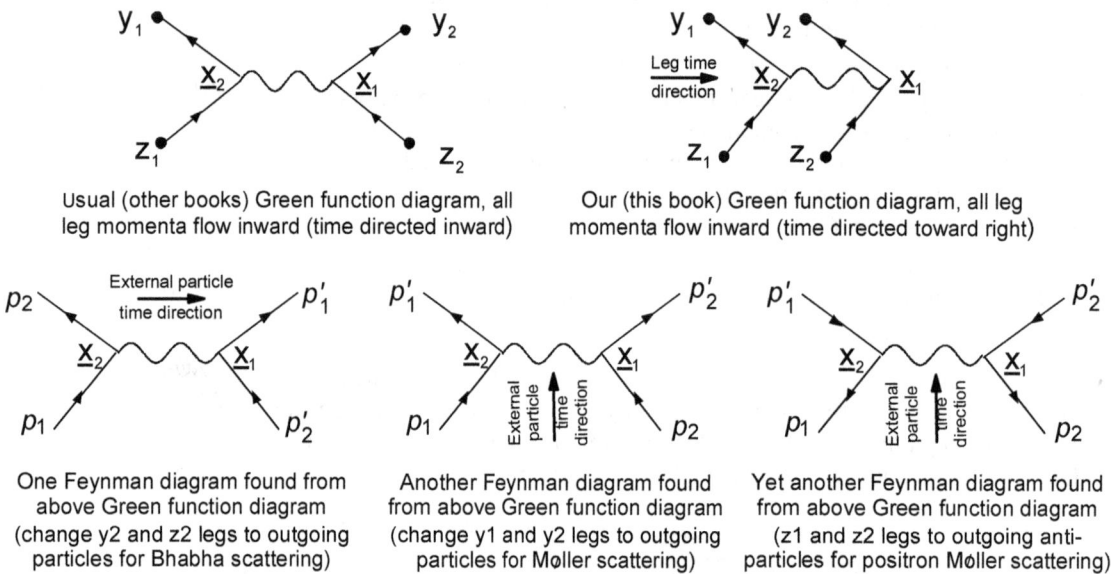

Usual (other books) Green function diagram, all leg momenta flow inward (time directed inward)

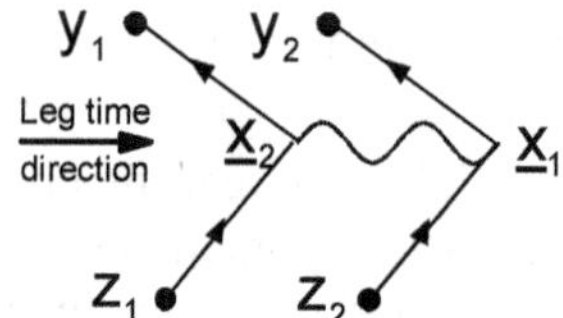

Our (this book) Green function diagram, all leg momenta flow inward (time directed toward right)

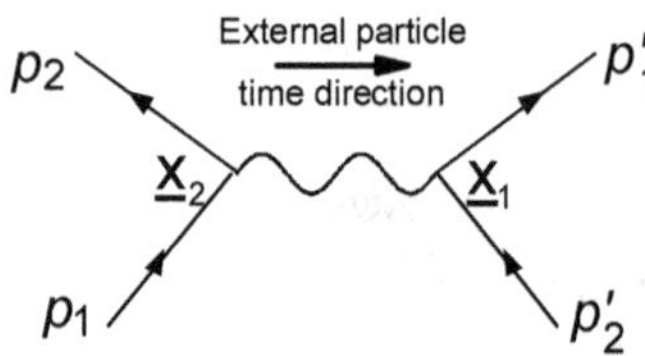

One Feynman diagram found from above Green function diagram (change y2 and z2 legs to outgoing particles for Bhabha scattering)

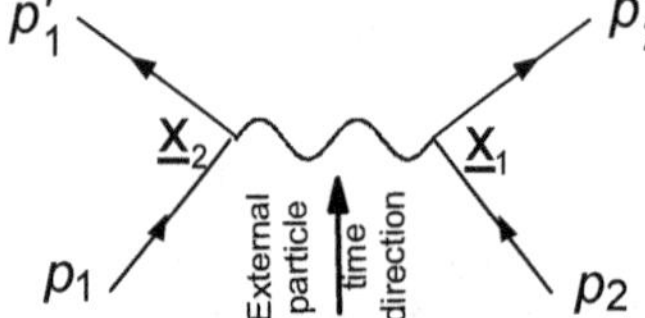

Another Feynman diagram found from above Green function diagram (change y1 and y2 legs to outgoing particles for Møller scattering)

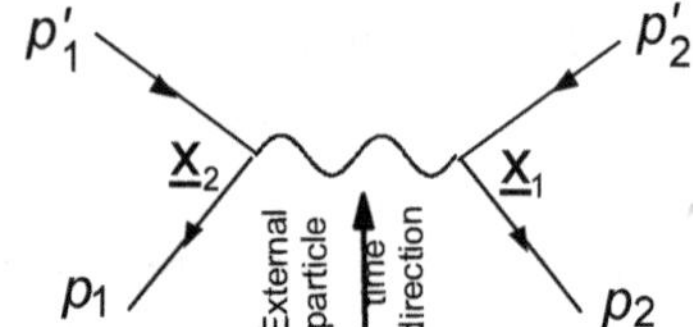

Yet another Feynman diagram found from above Green function diagram (z1 and z2 legs to outgoing anti-particles for positron Møller scattering)

Figure 3-2. Schematic of Converting Green Functions to Feynman Amplitudes

One can think of the conversion as simply swinging whichever legs we want in the top RH diagram of Fig. 3-2 to be final (outgoing) particles, such that they are outgoing with respect to time, while keeping the internal arrows on each such line fixed on the line. Effectively, this is the same as changing, for each final particle, the sign of the four-momenta from that of the corresponding propagator leg in the Green function. This is because 3-momentum entails a time derivative, so changing the sign of time, changes the sign of the 3-momentum; and because in our fields, we have a factor of e^{-iEt}, so changing the sign of t is the same as changing the sign on E instead. This is somewhat heuristic for now, but, as we will see, the math behind it all supports this perspective.

Momenta on outgoing particles of opposite sign from that of corresponding legs

By way of further explanation, in getting the last diagram in Fig. 3-2 we initially have the primed (outgoing) momenta for the lower legs, and the non-primed (incoming) momenta for the upper legs, with time passing from up to down. Then, we flipped the diagram 180°, so time flows upward instead.

One of the powers of the Green function approach is that a single function contains the information we need to evaluate a number of different interactions. Fig. 3-3 (with diagrams like the usual ones of other texts) shows us what, given the foregoing paragraphs, should not be too big of a surprise. That is, a Green function is comprised of many (an infinite number including higher order) terms (in pictorial form, diagrams), just as we once learned a transition amplitude is composed of many (an infinite number of) higher order terms (in pictorial form, diagrams).

One Green function has info for more than one interaction

Plus has all higher order terms in it

Note, in particular, that the Green function diagram of Fig. 3-2 represents only part of the total Green function, i.e., in Fig. 3-3, only the first diagram. By converting legs to external particles, the Green function with four fermion legs can yield the Feynman amplitudes for every possible interaction having four external fermionic particles. And it does this for all orders at once.

Green function with certain legs yields all interactions, to all orders, with those particular legs as external particles

Green functions with various different legs can represent all possible types of interactions. For example, the Green function with two fermionic and two photonic legs can yield all possible interactions with two external fermions and two external photons. Green functions hold a whole lot of valuable information in a compact form.

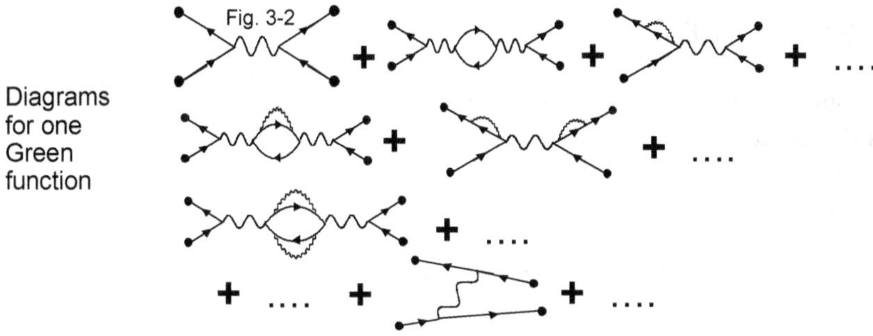
Fig. 3-2
Diagrams for one Green function
All lines are propagators. Same legs (incoming and outgoing propagators) in all diagrams. All orders in one Green function.

Figure 3-3. Schematic of the Green Function with Four Fermion Legs

All of the Feynman-like diagrams of Fig. 3-3 (shown in the usual form of other books) can typically be combined into one diagram, using a shaded circle to represent all possible internal propagator configurations of a connected diagram for the same given legs (external particles in an external particle transition amplitude). Three of these are shown in Fig. 3-4 (again using the usual form, not our special form, as in Fig. 3-2), where the topmost example represents Fig. 3-3. The second example in Fig. 3-4, with two photon legs, is just a dressed photon propagator, i.e., the bare propagator together with all its self-energy modifications. The third diagram with three legs, two fermion and one photon, represents the dressed vertex.

Shaded circle will represent all possible ways and orders of interaction of particular legs

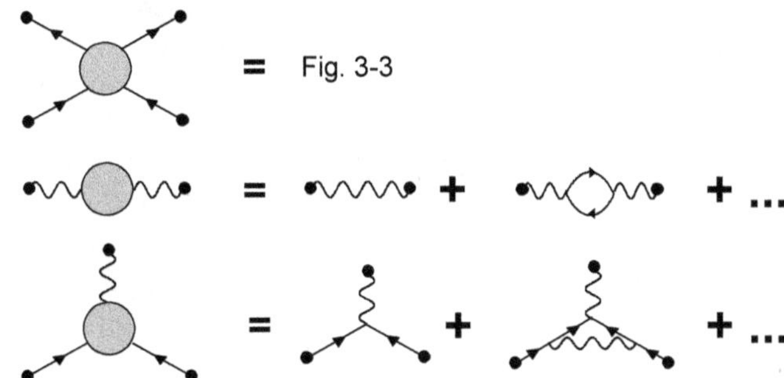
= Fig. 3-3

Figure 3-4. Symbolism for Combining Sub-diagrams into One Diagram

The shaded circle symbols are also used in external particle Feynman diagrams (with external particles rather than propagator legs) to represent the sum of all connected sub-diagrams with the same external particles.

So, What Good Is It?

One may ask what value the Green function methodology has, as we already know how to construct amplitudes for any given interaction (at least for QED, at this point). In certain contexts, it has several key advantages.

Green functions may relate less directly to observable quantities than transition amplitudes do, but they are easier to calculate, and the latter are readily obtained from them. Additionally, as we will see, transition amplitudes for several different interactions can be derived from a single Green function, allowing us to more readily discern relationships between these interactions.

And as we will also eventually see, QCD shares certain similarities with QED, but the more complicated form of $SU(3)$ compared to $U(1)$ makes strong force field theory far more challenging. It turns out that the development of QCD using Green functions (in tandem with the PI approach) is considerably easier than it otherwise would be.

Additionally, advanced topics in QFT that require going beyond low-order perturbation solution methods (such as in QCD) are, almost without exception, formulated in terms of Green functions.

Green functions offer advantages in QCD and other non-perturbative theories

Beyond that, they teach us, once again, that the physical, phenomenal world can be represented in different, yet consonant, mathematical ways.

3.1.4 The Math Behind the Green Function Methodology of QFT

Mathematical Form of the QFT Green Function

In this sub-section, we first present, in (3-13), the defining expression for Green functions in the QFT methodology, written $G^{\mu\nu...}(x_1, x_2,...y_1,...z_1,...)$. We then explore that relation and present examples, with different legs (incoming/outgoing particles in the associated Feynman diagrams), in order to show how Green functions, as discussed in prior sections, arise from (3-13). Note that for most people, it takes some time getting used to Green functions, so if you scratch your head for a while, be consoled that you are not alone. In time, with some effort, things should become clearer.

Math definition of QFT Green function for canonical approach

$$\text{Green function } G^{\mu\nu\cdots}(x_1,x_2,...y_1,... z_1,...)=\frac{\langle 0|T\left\{SA^\mu(x_1)A^\nu(x_2)...\psi(y_1)...\bar{\psi}(z_1)...\right\}|0\rangle}{\langle 0|S|0\rangle} \tag{3-13}$$

S is the familiar *S* operator (*not* the action, which often uses the same symbol) from the canonical quantization approach to QFT; A^μ, ψ, and $\bar{\psi}$ are the usual interaction picture QED quantum (operator) fields from that same approach; and *T* indicates time ordering. Kets and bras are interaction picture states. (3-13) is the position space form of the Green function, which has a concomitant momentum-space form obtained via the usual Fourier transform methodology (to be shown).

One field (not counting fields in S) for each leg

Take care to note that there is one field in (3-13) for each leg (incoming/outgoing particle in an interaction), and nothing (not counting *S*) for other particles (internal lines in a Feynman diagram). We label the arguments for the photon field x_i; for ψ, y_i; and for $\bar{\psi}$, z_i.

For now, we are not going to worry about the denominator in (3-13), but focus on the numerator. Shortly, we will show how the denominator arises mathematically and ends up being little more than a phase factor. The critical part of the Green function is the numerator.

The *S* operator (see Vol. 1, Wholeness Chart 8-4, pg. 248, last row and pg. 249, 3rd row) is (where we underline dummy integration variables to distinguish them from the $x_1, x_2, \ldots$ in the Green function (3-13))

Recalling S operator general form

$$\begin{aligned} S &= Te^{-i\int\mathcal{H}_I(\underline{x})d^4\underline{x}} = Te^{i\int\mathcal{L}_I(\underline{x})d^4\underline{x}} = \sum_{n=0}^{\infty}\frac{i^n}{n!}\int T\left\{\mathcal{L}_I(\underline{x_1})...\mathcal{L}_I(\underline{x_n})\right\}d^4\underline{x_1}...d^4\underline{x_n} \\ &= I + i\int\mathcal{L}_I(\underline{x_1})d^4\underline{x_1} - \tfrac{1}{2!}\int\int T\left\{\mathcal{L}_I(\underline{x_1})\mathcal{L}_I(\underline{x_2})\right\}d^4\underline{x_1}d^4\underline{x_2} + \ldots \end{aligned} \tag{3-14}$$

For QED (see Vol. 1, pg. 215, (8-2)), (3-14) becomes

Expanding QED S operator

$$S = I + ie\int\left(\bar{\psi}\gamma^\alpha A_\alpha\psi\right)_{\underline{x_1}}d^4\underline{x_1} - \tfrac{e^2}{2!}\int\int T\left\{\left(\bar{\psi}\gamma^\alpha A_\alpha\psi\right)_{\underline{x_1}}\left(\bar{\psi}\gamma^\beta A_\beta\psi\right)_{\underline{x_2}}\right\}d^4\underline{x_1}d^4\underline{x_2} + \ldots \tag{3-15}$$

We can re-express (3-13) by inserting (3-15) into it, to get

Expanding QED S operator inside Green function

$$\begin{aligned} G^{\mu\nu\cdots}(x_1,x_2,...y_1,...z_1,...) &= \frac{\langle 0|T\left\{A^\mu(x_1)A^\nu(x_2)...\psi(y_1)...\bar{\psi}(z_1)...\right\}|0\rangle}{\langle 0|S|0\rangle} + \\ &\frac{ie\langle 0|T\left\{\int\left(\bar{\psi}\gamma^\alpha A_\alpha\psi\right)_{\underline{x_1}}d^4\underline{x_1}A^\mu(x_1)A^\nu(x_2)...\psi(y_1)...\bar{\psi}(z_1)...\right\}|0\rangle}{\langle 0|S|0\rangle} + \\ &\frac{-\tfrac{e^2}{2!}\langle 0|T\left\{\int\int\left\{\left(\bar{\psi}\gamma^\alpha A_\alpha\psi\right)_{\underline{x_1}}\left(\bar{\psi}\gamma^\beta A_\beta\psi\right)_{\underline{x_2}}\right\}d^4\underline{x_1}d^4\underline{x_2}A^\mu(x_1)A^\nu(x_2)...\psi(y_1)...\bar{\psi}(z_1)...\right\}|0\rangle}{\langle 0|S|0\rangle} \\ &+\ldots \end{aligned} \tag{3-16}$$

How Only Propagators Remain in Any Green Function

Recall from Wick's theorem (3-17) (Vol. 1, pg. 204, (7-78)) how we can convert the time ordering of (3-16) into normal ordering, and along the way we obtain terms with propagators in them. (*A,B,C* etc. are quantum fields such as A^μ, ψ, and $\bar{\psi}$.)

$$T\left\{(AB...)_{x_1}\,......(AB...)_{x_n}\right\}=N\left\{(AB...)_{x_1}\,......(AB...)_{x_n}\right\} \tag{3-17}$$

$$+N\left\{(AB..)_{x_1}(AB..)_{x_2}...\right\}+N\left\{(AB..)_{x_1}(AB..)_{x_2}...\right\}+...+N\left\{...(A...Z)_{x_{n-1}}(A...Z)_{x_n}\right\}$$

$$+N\left\{(ABC...)_{x_1}(ABC...)_{x_2}...\right\}+N\left\{(ABC...)_{x_1}(ABC...)_{x_2}...\right\}+....$$

+ (all normal ordered terms with three non-equal times contractions)

+ etc.

Recalling Wick's theorem

Using (3-17) in (3-16) we will end up with, thanks to normal ordering, a lot of terms having destruction operators on the RHS. Any of those operating on the vacuum ket $|0\rangle$ in the numerator of (3-16) will result in zero. We will also get terms with all creation operators and each of those will produce a ket state that does not match (i.e. is orthogonal to) the vacuum state bra.

So, the only terms surviving are those with only propagators (i.e., contractions) in them and no operators, since propagators are numeric. (That is, in the sense that they are not creation or destruction operators. We ignore the sense in which they have spinor or spacetime indices, which makes them a little more than just numbers.) These terms each have a number sandwiched between a vacuum bra and a vacuum ket, so we can take the number outside the bracket, with $\langle 0|0\rangle = 1$.

Using Wick theorem, only terms surviving in Green function are propagators

Conclusion: We can forget about all terms arising in (3-16) except for those having only propagators and no operators in them. They are the only terms surviving in (3-16).

Thus, we see how the claim made earlier that Green functions comprise only propagators is true for the relation defined by (3-13). It remains to be shown that these equal our familiar transition amplitudes when we substitute external particle relations for the incoming/outgoing propagators (i.e., the legs).

The Denominator of the Green Function

The Green function in position space (3-13) has a bra and a ket representing the same initial vacuum state, i.e., at $t = -\infty$. This can be re-written more explicitly with notation specifying that the vacuum under consideration is the initial vacuum as

Look at denominator in Green function

$$G^{\mu\nu...}\left(x_1,x_2,...y_1,...z_1,...\right)=\frac{\left\langle 0,t=-\infty\middle|T\left\{SA^{\mu}\left(x_1\right)A^{\nu}\left(x_2\right)...\psi\left(y_1\right)...\bar{\psi}\left(z_1\right)...\right\}\middle|0,t=-\infty\right\rangle}{\left\langle 0,t=-\infty\middle|S\middle|0,t=-\infty\right\rangle} \tag{3-18}$$

Re-write Green function more explicitly

Examining the denominator of (3-18), we note that the operator S takes a given state (ket) at $t = -\infty$ to its final state (ket) at $t = +\infty$. Any terms in S that create particle states different from the vacuum would end up dropping out of the denominator, since the resulting ket would be orthogonal to the bra vacuum state. Similarly, any terms in S that destroy the vacuum (leaving zero) would drop out. The *total* S operator itself can never destroy the vacuum (and leave zero in the denominator), as one can see from its form in the above referenced wholeness chart in Vol. 1. S comprises e raised to a complex power, and when expanded, has the identity ($S^{(0)} = I$) as its lowest order term (which cannot annihilate the vacuum).

It is reasonable to presume that we cannot measure any difference between the vacuum at $t = -\infty$ and the vacuum at $t = +\infty$. Hence, the two can only differ by a phase factor, which is not detectable physically. (Phase factors disappear in probability and expectation value calculations). Thus,

$$\left|0,t=+\infty\right\rangle = e^{i\phi}\left|0,t=-\infty\right\rangle, \tag{3-19}$$

Denominator can only have inner product of final vacuum ket with initial vacuum bra

and the denominator of (3-18) becomes

$$\left\langle 0,t=-\infty\middle|S\middle|0,t=-\infty\right\rangle = \left\langle 0,t=-\infty\middle|\middle|0,t=+\infty\right\rangle = \left\langle 0,t=-\infty\middle|e^{i\phi}\middle|0,t=-\infty\right\rangle = e^{i\phi}. \tag{3-20}$$

Two vacuums can only differ by a phase, so, denominator just a phase factor

As we claimed earlier, the denominator is, at most, a phase factor, and phase factors are irrelevant when we calculate probability. The phase ϕ could, in fact, be zero, in which case the denominator (3-20) would be one, but we work here with the most general case.

Green Functions in 4-Momentum Space

Just as we can express our transition amplitudes in position space (see, for example, Vol. 1, pg. 221, (8-26)) or in momentum space (Vol. 1, pg. 223, (8-34)), so we can express a Green function in either space. In (3-13) (S in symbol form) and (3-16) (S expressed explicitly), it was displayed in position space. It is defined below in momentum space as (3-21), where q_i are 4-momenta, all of which (by mathematical convention, even though unrealistic physically) flow inwards, and where here, and often in the future, we employ a single integral sign as space-saving, less complicated shorthand for multiple ones. (See footnote on pg. 66 for the Fourier transform convention typically employed in QFT, though that was only for a single variable x, i.e., single k in momentum space.) We will shortly do an example using (3-21) to get some "hands-on" experience with it and gain better understanding for each of the factors in it, including the delta function (which might seem a little strange at this point).

$$\boxed{\begin{aligned}&\hat{G}^{\mu\nu\ldots}\left(q_1,q_2,\ldots q_n\right)(2\pi)^4\,\delta^{(4)}\left(q_1+q_2+\ldots+q_n\right)\\&\quad\equiv\int G^{\mu\nu\ldots}\left(x_1,x_2,\ldots x_n\right)\left(e^{-iq_1x_1}e^{-iq_2x_2}\ldots e^{-iq_nx_n}\right)dx_1dx_2\ldots dx_n\end{aligned}} \qquad (3\text{-}21)$$

Green function in momentum space

3.1.5 Interaction Example Using a Green Function: Bhabha Scattering

In Bhabha scattering we have an incoming electron, an incoming positron, an outgoing electron, and an outgoing positron. So, in (3-13), we take our legs to be the four fields corresponding to those particles. See Fig. 3-1, pg. 67, and Fig. 3-2, pg. 68. Note that we have no photon legs for this case, so we have no $A^\mu(x_i)$ factors and no superscripts on G. Our four-point Green function is

Green function (in position space) for 4 fermion legs (will lead to Bhabha scattering amplitude)

$$G\left(y_1,y_2,z_1,z_2\right)=\frac{\langle 0|T\left\{S\,\psi(y_1)\psi(y_2)\bar{\psi}(z_1)\,\bar{\psi}(z_2)\right\}|0\rangle}{\langle 0|S|0\rangle}\quad(\text{for Bhabha scattering}). \qquad (3\text{-}22)$$

Note, we want to create the electron particle (which will be the beginning of a propagator in the Green function) at z_1, so we use the quantum field $\bar{\psi}(z_1)$ for that. To create a positron at y_1, we use $\psi(y_1)$; to destroy an electron at y_2, $\psi(y_2)$; and to destroy a positron at z_2, $\bar{\psi}(z_2)$.(Recall that in a Green function, all particles are virtual.)

As shown earlier, the denominator in (3-22) only represents a phase factor, and in finding probabilities, we work with the square of the absolute value of transition amplitudes, where phase factors drop out. So, we will concentrate on the numerator of (3-22). Using (3-15) in (3-22), we have

the numerator of $G\left(y_1,y_2,z_1,z_2\right)=$

$$\langle 0|T\left\{\left(\begin{aligned}&I+ie\int\left(\bar{\psi}\gamma^\alpha A_\alpha\psi\right)_{\underline{x_1}}d^4\underline{x_1}\\&-\tfrac{e^2}{2!}\int\!\!\int T\left\{\left(\bar{\psi}\gamma^\alpha A_\alpha\psi\right)_{\underline{x_1}}\left(\bar{\psi}\gamma^\beta A_\beta\psi\right)_{\underline{x_2}}\right\}d^4\underline{x_1}d^4\underline{x_2}+\ldots\end{aligned}\right)\psi(y_1)\psi(y_2)\bar{\psi}(z_1)\,\bar{\psi}(z_2)\right\}|0\rangle \qquad (3\text{-}23)$$

$$=\langle 0|T\left\{\begin{aligned}&\psi(y_1)\psi(y_2)\bar{\psi}(z_1)\,\bar{\psi}(z_2)+ie\int\left(\bar{\psi}\gamma^\alpha A_\alpha\psi\right)_{\underline{x_1}}d^4\underline{x_1}\,\psi(y_1)\psi(y_2)\bar{\psi}(z_1)\,\bar{\psi}(z_2)\\&-\tfrac{e^2}{2!}\int\!\!\int T\left\{\left(\bar{\psi}\gamma^\alpha A_\alpha\psi\right)_{\underline{x_1}}\left(\bar{\psi}\gamma^\beta A_\beta\psi\right)_{\underline{x_2}}\right\}d^4\underline{x_1}d^4\underline{x_2}\,\psi(y_1)\psi(y_2)\bar{\psi}(z_1)\,\bar{\psi}(z_2)+\ldots\end{aligned}\right\}|0\rangle.$$

Expand numerator of Green function

From Wick's theorem (3-17) and our knowledge that, due to normal ordering, only terms with propagators and no operators survive, we get (3-23) equal to

$$\langle 0|\left\{\begin{aligned}&\psi(y_1)\psi(y_2)\bar{\psi}(z_1)\bar{\psi}(z_2)\;+\;\psi(y_1)\psi(y_2)\bar{\psi}(z_1)\bar{\psi}(z_2)\\&-\tfrac{e^2}{2!}\int\!\!\int\;(\bar{\psi}\gamma^\alpha A_\alpha\psi)_{\underline{x_1}}(\bar{\psi}\gamma^\beta A_\beta\psi)_{\underline{x_2}}\psi(y_1)\psi(y_2)\bar{\psi}(z_1)\bar{\psi}(z_2)d^4\underline{x_1}d^4\underline{x_2}\\&-\tfrac{e^2}{2!}\int\!\!\int\;(\bar{\psi}\gamma^\alpha A_\alpha\psi)_{\underline{x_1}}(\bar{\psi}\gamma^\beta A_\beta\psi)_{\underline{x_2}}\psi(y_1)\psi(y_2)\bar{\psi}(z_1)\bar{\psi}(z_2)d^4\underline{x_1}d^4\underline{x_2}\;+\;\ldots\ldots\end{aligned}\right\}|0\rangle \qquad (3\text{-}24)$$

Apply Wick theorem

Note that the second term in the penultimate row of (3-23) dropped out in going to (3-24) because there is no way we can make a photon propagator out of a single A_α field. Note also that all quantities inside the bra and ket of (3-24) are numbers, not operators, so we can move those numbers outside the bracket with $\langle 0|0\rangle = 1$, and just deal with the numbers inside the large parentheses.

Term in e^1 drops out

The first term in (3-24) is simply two unconnected propagators that don't interact. That term represents a positron at y_1 traveling to z_2 and an electron at z_1 traveling to y_2. For the S operator approach, such a term corresponded to no interaction between the electron and positron, which is one of the ways the particles could behave. The incoming electron and positron remain unchanged and are in the same state outgoing as they were incoming. Such a happening has a probability, just as the interaction of the two has a probability[2].

First term in e^0 = two unconnected propagators

Here that term does not represent external particles, however, but propagators. But recall that we obtain the Feynman amplitudes by substituting external particle relations for the legs in a Green function. That is what we would do here, so we would end up with the behavior described in the foregoing paragraph, with the Green function diagrams for this term of Fig. 3-5 (in the usual form of other texts). These diagrams are not connected, and we will discuss the ramifications of that shortly.

No interactions therein, so not relevant for finding amplitude

Figure 3-5. Disconnected Green Function Diagrams of First Term in Top Row in (3-24)

Same logic for other term in e^0

All of the reasoning applied to the first term in (3-24) also applies to the second term.

Looking now at the third term in (3-24), we see it represents the LHS of Fig. 3-1 (which equals the top row of Fig. 3-2). That is, the contractions of (3-24), are the virtual particles (propagators) of the LHS of Fig. 3-1 (top row in Fig. 3-2). The factor in front of that third term in (3-24) is the same factor we have when we evaluate Feynman amplitudes. We can surmise, when we replace the incoming and outgoing propagators with external particle relations (i.e., take the appropriate S_F to the appropriate $u_r(\mathbf{p}_1)$, $v_r(\mathbf{p}_2)$, $\bar{u}_r(\mathbf{p}'_1)$, or $\bar{v}_r(\mathbf{p}'_2)$), then that term becomes a Feynman amplitude with external particles, and at that point, represents the RHS of Fig. 3-1 (lower diagrams of Fig. 3-2).

Term in e^2 similar to second order Feynman amplitude except propagators instead of external particles

---

To start seeing that explicitly, do **Problem 3** to show (3-25).

With the third term in (3-24) having contractions cast as propagators, (see Vol. 1, pg. 202, (7-75) and Appendix A of this chapter),

$$G^{\,numer}_{\,3rd\,term}\left(y_1,y_2,z_1,z_2\right)=-\frac{e^2}{2}\int\left(-iS_F\left(y_2-\underline{x}_1\right)\right)\gamma^{\alpha}iS_F\left(\underline{x}_1-z_2\right) \times iD_{F\alpha\beta}\left(\underline{x}_1-\underline{x}_2\right)\left(-iS_F\left(y_1-\underline{x}_2\right)\right)\gamma^{\beta}iS_F\left(\underline{x}_2-z_1\right)d^4\underline{x}_1 d^4\underline{x}_2 . \quad (3\text{-}25)$$

Convert contractions to propagators in position space

To aid in what follows, we can represent (3-25) as the Green function diagram(s) of Fig. 3-6, which is really just a repeat of Fig. 3-2 for the upper row and lower LH diagram there. This will give us a sense of all legs (which will be external particles in the related Feynman diagram) having time directed inwards, which, as we noted earlier, is the rule in Green function analyses. Note that the time arrows in Fig. 3-6 only apply to the legs/external particles. Internal events $\underline{x}_1$ and $\underline{x}_2$ can occur in any time order, and in fact do, when we integrate the transition amplitude over all time and space. Note also that, as with Fig. 3-2, the first and second diagrams are really the same thing, but it can help in what we are about to do to think of it as represented as the middle diagram in Fig. 3-6, as that may be more helpful in visualizing all leg momenta flowing inward.

Math form of leg 4-momenta implies all external particles would have momenta inward

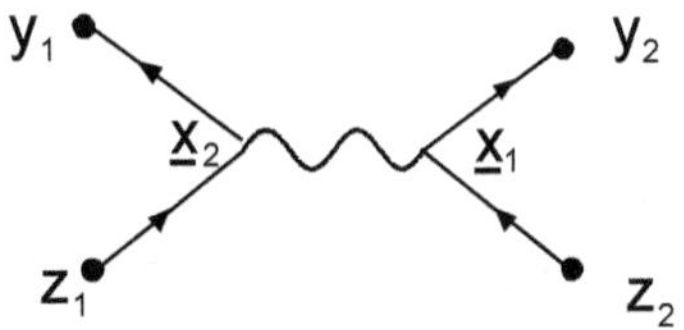

Usual Green function diagram, all leg momenta flow inward (time flow inward)

Our Green function diagram, all leg momenta flow inward (time toward right)

Associated Feynman diagram, physically directed external momenta (time)

Figure 3-6. First Way for Bhabha Scattering: Green Function and Feynman Diagrams

Picturing Green function with all leg time inward vs Feynman amplitude external momenta with final particles time outward

Using (3-25) in (3-21), we find the momentum space equivalent of this term in the Green function.

---

[2] However, when we integrate over all space and time in the standard approach to QFT Feynman amplitude calculations, the particles then have to interact, as they must eventually over infinite time. Hence this term doesn't come into play in that approach. See Vol. 1, pg. 441, which explains the rationale for integration over all time and space.

$$\hat{G}^{numer}_{3rd\,term}\left(p_{y1},p_{y2},p_{z2},p_{z1}\right)(2\pi)^4\,\delta^{(4)}\left(p_{y1}+p_{y2}+p_{z2}+p_{z1}\right)$$
$$=\int G_{3rd\,term}\left(y_1,y_2,z_1,z_2\right)\left(e^{-ip_{y1}y_1}e^{-ip_{y2}y_2}e^{-ip_{z1}z_1}e^{-ip_{z2}z_2}\right)dy_1dy_2dz_1dz_2$$
$$=-\frac{e^2}{2}\int\underbrace{i\left(\frac{1}{(2\pi)^4}\int S_F(p_a)e^{-ip_a(y_2-\underline{x}_1)}\,d^4p_a\right)}_{S_F(y_2-\underline{x}_1)}\gamma^\alpha\underbrace{i\left(\frac{1}{(2\pi)^4}\int S_F(p_b)e^{-ip_b(\underline{x}_1-z_2)}\,d^4p_b\right)}_{S_F(\underline{x}_1-z_2)}$$
$$\times\underbrace{i\frac{1}{(2\pi)^4}\int D_{F\alpha\beta}(k)e^{-ik(\underline{x}_1-\underline{x}_2)}\,d^4k}_{D_{F\alpha\beta}(\underline{x}_1-\underline{x}_2)}\underbrace{i\left(\frac{1}{(2\pi)^4}\int S_F(p_c)e^{-ip_c(y_1-\underline{x}_2)}d^4p_c\right)}_{S_F(y_1-\underline{x}_2)}$$
$$\times\gamma^\beta\underbrace{i\left(\frac{1}{(2\pi)^4}\int S_F(p_d)e^{-ip_d(\underline{x}_2-z_1)}\,d^4p_d\right)}_{S_F(\underline{x}_2-z_1)}\left(e^{-ip_{y1}y_1}e^{-ip_{y2}y_2}e^{-ip_{z1}z_1}e^{-ip_{z2}z_2}\right)d^4\underline{x}_1d^4\underline{x}_2dy_1dy_2dz_1dz_2\,. \tag{3-26}$$

Converting position space Green function to momentum space

Cranking the math

There are a lot of integrals in (3-26). Six with respect to space and five for propagators with respect to 4-momentum. To start, let's isolate the factors with dy_1 in them and carry out that integration.

$$\int e^{-ip_cy_1}e^{-ip_{y1}y_1}dy_1=(2\pi)^4\,\delta^{(4)}\left(p_{y1}+p_c\right) \tag{3-27}$$

Using (3-27) in the integration over p_c factor of (3-26) results in

$$\frac{1}{(2\pi)^4}\int S_F(p_c)e^{ip_c\underline{x}_2}(2\pi)^4\,\delta^{(4)}\left(p_{y1}+p_c\right)d^4p_c=S_F\left(-p_{y1}\right)e^{-ip_{y1}\underline{x}_2}\,. \tag{3-28}$$

Next, isolate the factors in dy_2 and carry out that integration to get

$$\int e^{-ip_ay_2}e^{-ip_{y2}y_2}dy_2=(2\pi)^4\,\delta^{(4)}\left(p_{y2}+p_a\right), \tag{3-29}$$

and the integration over p_a factor of (3-26) results in

$$\frac{1}{(2\pi)^4}\int S_F(p_a)e^{ip_a\underline{x}_1}(2\pi)^4\,\delta^{(4)}\left(p_{y2}+p_a\right)d^4p_a=S_F\left(-p_{y2}\right)e^{-ip_{y2}\underline{x}_1}\,. \tag{3-30}$$

Similarly, isolating the factors in dz_1 and dz_2 and integrating leads to

$$\int e^{ip_dz_1}e^{-ip_{z1}z_1}dz_1=(2\pi)^4\,\delta^{(4)}\left(p_{z1}-p_d\right)$$
$$\frac{1}{(2\pi)^4}\int S_F(p_d)e^{-ip_d\underline{x}_2}(2\pi)^4\,\delta^{(4)}\left(p_{z1}-p_d\right)d^4p_d=S_F\left(p_{z1}\right)e^{-ip_{z1}\underline{x}_2} \tag{3-31}$$

$$\int e^{ip_bz_2}e^{-ip_{z2}z_2}dz_2=(2\pi)^4\,\delta^{(4)}\left(p_{z2}-p_b\right)$$
$$\frac{1}{(2\pi)^4}\int S_F(p_b)e^{-ip_b\underline{x}_1}(2\pi)^4\,\delta^{(4)}\left(p_{z2}-p_b\right)d^4p_b=S_F\left(p_{z2}\right)e^{-ip_{z2}\underline{x}_1}\,. \tag{3-32}$$

Putting (3-28) to (3-32) into (3-26) yields

$$\hat{G}^{numer}_{3rd\,term}\left(p_{y1},p_{y2},p_{z2},p_{z1}\right)(2\pi)^4\,\delta^{(4)}\left(p_{y1}+p_{y2}+p_{z2}+p_{z1}\right)$$
$$=-\frac{e^2}{2}\int iS_F\left(-p_{y2}\right)e^{-ip_{y2}\underline{x}_1}\gamma^\alpha iS_F\left(p_{z2}\right)e^{-ip_{z2}\underline{x}_1}i\left(\frac{1}{(2\pi)^4}\int D_{F\alpha\beta}(k)e^{-ik(\underline{x}_1-\underline{x}_2)}\,d^4k\right) \tag{3-33}$$
$$\times\, iS_F\left(-p_{y1}\right)e^{-ip_{y1}\underline{x}_2}\gamma^\beta iS_F\left(p_{z1}\right)e^{-ip_{z1}\underline{x}_2}d^4\underline{x}_1d^4\underline{x}_2\,.$$

Now, isolate the factors with $\underline{x}_2$ in them and carry out the integration over $\underline{x}_2$.

$$\int e^{ik\underline{x}_2} e^{-ip_{y1}\underline{x}_2} e^{-ip_{z1}\underline{x}_2} d^4\underline{x}_2 = (2\pi)^4 \delta^{(4)}\left(k - (p_{y1} + p_{z1})\right) \tag{3-34}$$

So, k must equal the sum of the incoming momenta from the particles on the LHS in the diagram, which makes sense and is what we are used to from finding transition amplitudes in QED.

Using (3-34) in the integration over k in (3-33) yields

$$\frac{1}{(2\pi)^4}\int D_{F\alpha\beta}(k) e^{-ik\underline{x}_1} (2\pi)^4 \delta^{(4)}\left(k - (p_{y1} + p_{z1})\right) d^4k = D_{F\alpha\beta}(p_{y1} + p_{z1}) e^{-i(p_{y1}+p_{z1})\underline{x}_1} . \tag{3-35}$$

Now carry out the integration in (3-33) for all the factors in $\underline{x}_1$.

$$\int e^{-ip_{y2}\underline{x}_1} e^{-ip_{z2}\underline{x}_1} e^{-i(p_{y1}+p_{z1})\underline{x}_1} d^4\underline{x}_1 = (2\pi)^4 \delta^{(4)}\left(p_{y1} + p_{z1} + p_{z2} + p_{y2}\right) \tag{3-36}$$

Resulting delta function means sum of all Green function leg (inward) momenta = 0

So, $p_{z2} + p_{y2}$, the sum of momenta of the RH propagators in the LH diagram of Fig. 3-6, must be of opposite sign from k, the sum of the momenta $p_{y1} + p_{z1}$ of the LH propagators in the same diagram. This is because the time flow of legs in a Green function is assumed to be inwards in the diagram. Hence, if k is the incoming momentum from the LH side in the LH diagram, it must equal the negative of the incoming momentum on the RH side of the same diagram (which is physically the outgoing momentum).

Thus, Feynman amplitude outgoing momenta = negative of Green function momenta for those legs

Plugging (3-35) and (3-36) into (3-33), we see why the delta function and factor of $(2\pi)^4$ was included in the definition (3-21). The delta functions on both sides of the equation, along with the $(2\pi)^4$ factor, cancel, and we get

$$\begin{aligned}\hat{G}^{numer}_{3rd\,term}(p_{y1}, p_{y2}, p_{z2}, p_{z1}) = -\frac{e^2}{2}\left(-iS_F(-p_{y2})\right)\gamma^\alpha iS_F(p_{z2}) \\ \times iD_{F\alpha\beta}(p_{y1}+p_{z1})\left(-iS_F(-p_{y1})\right)\gamma^\beta iS_F(p_{z1}).\end{aligned} \tag{3-37}$$

Final result: momentum space form of this term in Green function

The minus signs for the momenta of the y_1 and y_2 propagators are consistent with what we learned in Vol. 1 (Wholeness Chart 8-1, pg. 234). The mathematical expression of a positron propagator employs the negative of the physical four momentum. (In this context, "physical" refers to the Green function diagram momentum, not the real-world momentum.) In the LH and middle diagrams of Fig. 3-6, those momenta, p_{y1} and p_{y2}, are flowing inward, and that is how we should think of them. The minus signs in front of them in (3-37) are just due to a mathematical quirk that arises for antiparticles.

The fourth term in (3-24), via the logic of Vol. 1, pg. 258, Box 9-1, turns out to be equivalent to the third term. In brief, the coordinates in the integrations are dummy variables that can be interchanged, and along with the anti-commutation relations for fermions, the two terms turn out to be the same thing. Adding them together gives us a single term like the third one, but without the factor of ½.

Next term in e^2 is mathematically equivalent to prior one

$$\begin{aligned}\hat{G}^{numer}_{3rd+4th\,term}(p_{y1}, p_{y2}, p_{z2}, p_{z1}) = -e^2\left(-iS_F(-p_{y2})\right)\gamma^\alpha iS_F(p_{z2}) \\ \times iD_{F\alpha\beta}(p_{y1}+p_{z1})\left(-iS_F(-p_{y1})\right)\gamma^\beta iS_F(p_{z1})\end{aligned} \tag{3-38}$$

Combining the two terms

We could have arrived at (3-38) more quickly using our experience (Vol. 1, Chap. 8) in converting transition amplitudes from expressions in position space to expressions in momentum space. That is, instead of all the steps in going from (3-24) to (3-38), we could have simply extrapolated our knowledge of Feynman rules (Vol. 1, Box 8-3, pg. 236) to express the third and fourth terms in (3-24) directly in momentum space as in (3-38).

However, since our Green function propagator legs in (3-24) have 4-momentum flowing inward, i.e., from y_2 to $\underline{x}_1$ and from z_2 to $\underline{x}_1$, but the Feynman amplitude has them in the opposite direction, we need a sign reversal to account for that. So, we need to make the changes of (3-39) below to get the transition amplitude. In summary, for our example, we can quickly go from the Green function in spacetime to the Feynman amplitude in momentum space as follows.

First Way Bhabha Scattering Occurs

$$S_F(\underline{x}_2 - z_1) \xrightarrow[\text{space}]{\text{momentum}} S_F(p_{z1}) \to u_{r_1}(\mathbf{p}_1) \quad -S_F(y_1 - \underline{x}_2) \xrightarrow[\text{space}]{\text{momentum}} -S_F(-p_{y1}) \to \bar{v}_{r_2}(\mathbf{p}_2)$$
$$S_F(\underline{x}_1 - z_2) \to S_F(p_{z2}) \to v_{r_2'}(-\mathbf{p}_{z2}) = v_{r_2'}(\mathbf{p}_2') \quad -S_F(y_2 - \underline{x}_1) \to -S_F(-p_{y2}) \to \bar{u}_{r_1'}(-\mathbf{p}_{y2}) = \bar{u}_{r_1'}(\mathbf{p}_1')$$
$$p_{y1},\, p_{z1},\, p_{y2},\, p_{z2} = \text{Green function 4-momenta, all incoming} \tag{3-39}$$
$$p_1',\, p_2'\ (p_1,\, p_2) = \text{outgoing (incoming) Feynman amplitude particle 4-momenta}$$
$$p_1 = p_{z1},\quad p_2 = p_{y1} \quad p_1' = -p_{y2},\quad p_2' = -p_{z2}$$

Conversions needed to turn Green function term into a Feynman sub-amplitude

In doing so, be careful with the numeric subscripts in position space vs momentum space. In position space, numerical subscripts i = 1,2,... in x_i, y_i, z_i refer to particular field legs. In momentum space the numerical subscripts 1,2,... refer to particular external particle momenta. These are, in general, not related. In (3-39), we have employed the convention used in Vol. 1, and commonly in QFT. That is, incoming particles in Feynman amplitudes are numbered without primes and outgoing particles are numbered with primes.

Making the changes of (3-39), we find that we turn (3-38) into the Feynman amplitude,

$$\underset{\substack{+4th \\ term}}{\hat{G}^{numer}_{3rd}}(p_{y1}, p_{y2}, p_{z2}, p_{z1}) \to \underset{\substack{+4th \\ term}}{\mathcal{M}^{numer}_{3rd}} = \underbrace{-e^2 \bar{u}_{r_1'}(\mathbf{p}_1')\gamma^\alpha v_{r_2'}(\mathbf{p}_2') i D_{F\alpha\beta}(p_1 + p_2) \bar{v}_{r_2}(\mathbf{p}_2)\gamma^\beta u_{r_1}(\mathbf{p}_1)}_{\text{Feynman sub-amplitude } \mathcal{M}^{(2)}_{B1}} \tag{3-40}$$
$$\underbrace{S^{(2)}_{B1}}_{\substack{\text{transition} \\ \text{amplitude}}} = \sqrt{\frac{m}{VE_{\mathbf{p}_1}}}\sqrt{\frac{m}{VE_{\mathbf{p}2}}}\sqrt{\frac{m}{VE_{\mathbf{p}_1'}}}\sqrt{\frac{m}{VE_{\mathbf{p}_2'}}}(2\pi)^4 \delta^{(4)}(p_1 + p_2 - (p_1' + p_2'))\mathcal{M}^{(2)}_{B1}\,.$$

The result: Feynman (sub) amplitude for (one way of) Bhabha scattering at second order

Compare (3-40) to Vol. 1, (8-34), pg. 223 and we see the momentum space form of this term in the Green function for the particle allocations of (3-39) corresponds to the first way Bhabha scattering can occur. As mentioned earlier, the denominator of the Green function is merely a phase factor, which will drop out when we calculate probabilities, so can be ignored.

We ignore denominator as it is only a phase factor that drops out in probability calculations

Second Way for Bhabha Scattering

Now, consider the second way it can occur as in the last diagram of Fig. 3-3. This is represented by two more second order terms in (3-24), which differ only in employment of dummy integration variables $\underline{x}_1$ and $\underline{x}_2$, so can be combined into one, yielding

$$\underset{\substack{more \\ terms}}{G^{numer}_{two}}(y_1, y_2, z_1, z_2) = -e^2 \int (\bar{\psi}\gamma^\alpha A_\alpha \psi)_{\underline{x}_1} (\bar{\psi}\gamma^\beta A_\beta \psi)_{\underline{x}_2} \psi(y_1)\psi(y_2)\bar{\psi}(z_1)\bar{\psi}(z_2) d^4\underline{x}_1 d^4\underline{x}_2 \tag{3-41}$$

Other terms in Green function will yield amplitude for second way of Bhabha scattering

From this, similar to what we did before, we get

$$G^{numer}_{two\,more\,terms}(y_1, y_2, z_1, z_2) = -e^2 \int (-iS_F(y_1 - \underline{x}_1))\gamma^\alpha iS_F(\underline{x}_1 - z_2)$$
$$\times iD_{F\alpha\beta}(\underline{x}_1 - \underline{x}_2)(-iS_F(y_2 - \underline{x}_2))\gamma^\beta iS_F(\underline{x}_2 - z_1) d^4\underline{x}_1 d^4\underline{x}_2\,. \tag{3-42}$$

From those, one gets position space propagator form for those terms

Diagrammatically, this can be represented as either the LH or middle part of Fig. 3-7. Once again, the middle part should be easier as, in it, momentum flow for the legs is in the direction we are used to seeing in Feynman diagrams, which serve as our guides in constructing the math behind the diagrams. The associated Feynman diagram is on the RHS.

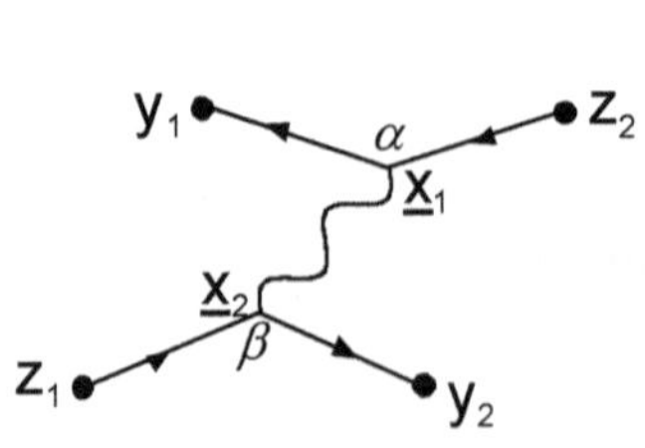

Usual Green function diagram, all leg momenta flow inward (time flow inward)

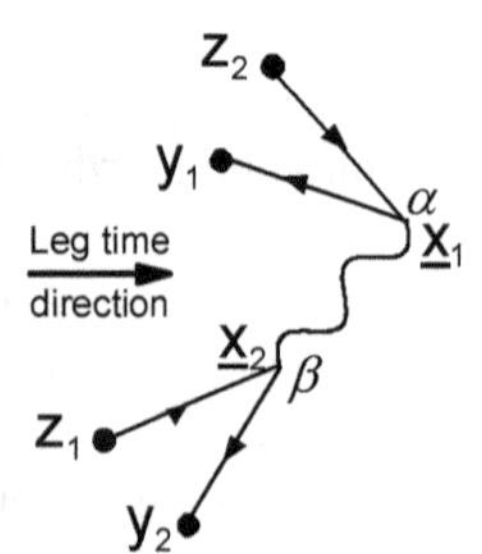

Our Green function diagram, all leg momenta flow inward (time toward right)

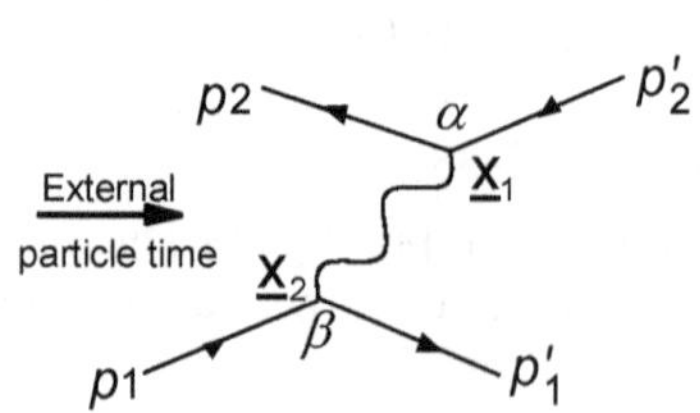

Associated Feynman diagram, physically directed momenta (time)

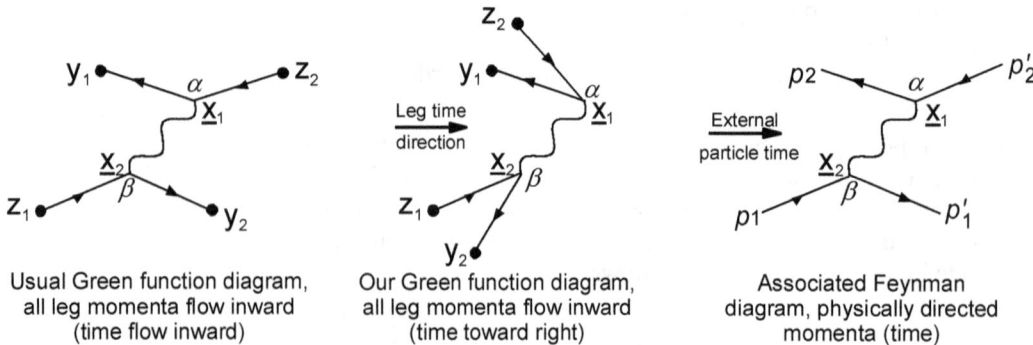

Comparing Green function diagrams for these terms to Feynman diagrams with time directions explicit

Figure 3-7. Second Way for Bhabha Scattering: Green Function and Feynman Diagrams

We will not go through all the steps to convert (3-42) to momentum space form, as it parallels what we did for the first way Bhabha scattering can occur. From that experience, looking at (3-42) plus the middle diagram in Fig. 3-7, and recognizing that all momenta of legs in Green functions flow inwards, we can extrapolate to get (3-43).

$$\begin{aligned}\hat{G}^{numer}_{two\,more\,terms}\left(p_{y1},p_{y2},p_{z2},p_{z1}\right)=&-e^2\left(-iS_F\left(-p_{y1}\right)\right)\gamma^\alpha iS_F\left(p_{z2}\right)\\ &\times iD_{F\alpha\beta}\left(p_{y2}+p_{z1}\right)\left(-iS_F\left(-p_{y2}\right)\right)\gamma^\beta iS_F\left(p_{z1}\right)\end{aligned} \tag{3-43}$$

From our experience, we can short cut the math to get momentum space form

This differs from (3-38) only by an exchange of p_{y1} with p_{y2}. Again, for 4-momenta of legs flowing inwards, the propagators with negative signs in the argument represent anti-particles. And since physical momentum (as in the RHS of Fig. 3-7), for the y_2 particle is the negative of p_{y2}, the argument of the photon propagator makes sense physically.

To obtain the Feynman amplitude, we make the changes of (3-44).

<u>Second Way Bhabha Scattering Occurs</u>

Conversions needed to turn this Green function term into a Feynman sub-amplitude

$$\begin{gathered}S_F\left(\underline{x_2}-z_1\right)\to S_F\left(p_{z1}\right)\to u_{r_1}\left(\mathbf{p}_1\right)\qquad -S_F\left(y_2-\underline{x_2}\right)\to -S_F\left(-p_{y2}\right)\to\bar{u}_{r_1'}\left(-\mathbf{p}_{y2}\right)=\bar{u}_{r_1'}(\mathbf{p}_1')\\ S_F\left(\underline{x_1}-z_2\right)\to S_F\left(p_{z2}\right)\to v_{r2'}\left(-\mathbf{p}_{z2}\right)=v_{r2'}\left(\mathbf{p}_2'\right)\quad -S_F\left(y_1-\underline{x_1}\right)\to -S_F\left(-p_{y1}\right)\to\bar{v}_{r2}\left(\mathbf{p}_{y1}\right)=\bar{v}_{r2}(\mathbf{p}_2)\\ p_{y1},\,p_{z1},\,p_{y2},\,p_{z2}=\text{Green function 4-momenta, incoming}\\ p_1',p_2'\left(p_1,p_2\right)=\text{ outgoing (incoming) Feynman amplitude particle 4-momenta}\\ p_1=p_{z1},\quad p_2=p_{y1}\quad p_1'=-p_{y2},\quad p_2'=-p_{z2}\end{gathered} \tag{3-44}$$

We thus find (3-45) below, where the sign change in the Feynman sub-amplitude arises because we would have to adjust it to match the same normal order as the Feynman amplitude for the first way of Bhabha scattering (3-40), and each exchange of adjacent fermion operators (which have become spinors in (3-45)) would bring in a minus sign. (See Feynman rule #9 in Vol. 1, pg. 514.) That is, we keep the relation order we have in (3-45), as it signifies the order the hidden spinor indices function, but recognize that we would have to make spinor switches to match the order in (3-40). There would be three such switches, giving us a factor of -1, which changes the minus sign in (3-43) to a plus sign.

$$\begin{aligned}\underset{\substack{more\\terms}}{\hat{G}^{numer}_{two}}\left(p_{y1},p_{y2},p_{z2},p_{z1}\right)\to \underset{\substack{more\\terms}}{\mathcal{M}_{two}} &= \underbrace{e^2\bar{v}_{r_2}(\mathbf{p}_2)\gamma^\alpha v_{r_2'}(\mathbf{p}_2')iD_{F\,\alpha\beta}\overbrace{\left(p_2'-p_2\right)}^{=p_1-p_1'}\bar{u}_{r_1'}(\mathbf{p}_1')\gamma^\beta u_{r_1}(\mathbf{p}_1)}_{\text{Feynman sub-amplitude }\mathcal{M}^{(2)}_{B2}}\\ \underbrace{S^{(2)}_{B2}}_{\substack{\text{transition}\\ \text{sub-amplitude}}} &=\sqrt{\frac{m}{VE_{\mathbf{p}_1}}}\sqrt{\frac{m}{VE_{\mathbf{p}_2}}}\sqrt{\frac{m}{VE_{\mathbf{p}_1'}}}\sqrt{\frac{m}{VE_{\mathbf{p}_2'}}}\left(2\pi\right)^4\delta^{(4)}\left(p_1+p_2-\left(p_1'+p_2'\right)\right)\mathcal{M}^{(2)}_{B2}\end{aligned} \tag{3-45}$$

The result: Feynman (sub) amplitude for (second way of) Bhabha scattering at second order

(3-45) is what we found in Vol. 1, pg. 224, (8-47). Adding the transition amplitudes of (3-40) and (3-45) gives us the total Bhabha scattering transition amplitude (to second order.)

Be forewarned that almost all of the literature on Green functions in QFT show Green function diagrams like the LHS of Fig. 3-1 (pg. 67), the LHS of Fig. 3-6, and the LHS of Fig. 3-7, and not like I have shown them here in the middle of Fig. 3-6 and the middle of Fig. 3-7. In other texts, one is supposed to keep in mind that momentum always flows inward. But that can be confusing when working with the signs in the transition from Green functions to Feynman amplitudes. So, I have drawn such figures as we have herein with time moving from left to right to help conceptually. Do not expect to find Green function diagrams looking like that elsewhere.

Our Green function also has higher order terms, which would yield higher order amplitude corrections

Note that (3-24) also contains higher order terms (from other terms in S) which would be associated with other, higher order diagrams, some of which are shown in Fig. 3-3.

Beyond those terms, however, (3-24) contains yet more terms, one of which we show in (3-46).

$$\langle 0|\left\{-\frac{e^2}{2!}\int(\bar{\psi}\gamma^\alpha A_\alpha\psi)_{\underline{x}_1}(\bar{\psi}\gamma^\beta A_\beta\psi)_{\underline{x}_2}\psi(y_1)\psi(y_2)\bar{\psi}(z_1)\bar{\psi}(z_2)\,d^4\underline{x_1}\,d^4\underline{x_2}\right\}|0\rangle \tag{3-46}$$

Plus, other second order terms

The corresponding Green function diagram (in the usual form, not our special form) for (3-46) is

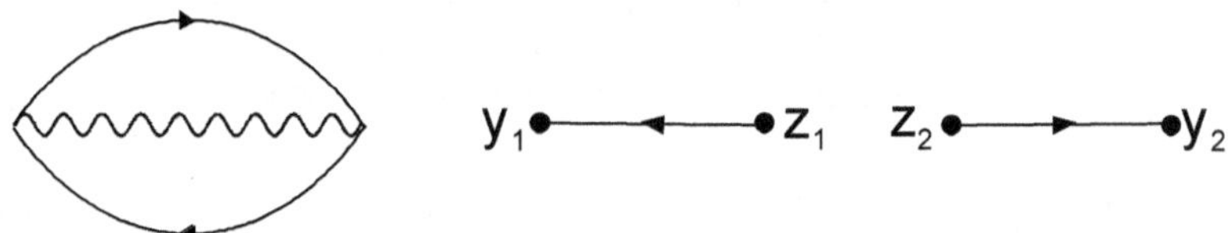

Diagrams for one of these (disconnected) terms

Fig. 3-8. Disconnected Green Function Diagrams of (3-46), Part of (3-24)

And, of course, there are more disconnected diagrams in (3-24).

Show some more such diagrams by doing **Problem 4**.

Note a general principle from (3-46) and Fig. 3-8. If a Green function term has separate factors, each with a different set of spacetime coordinates, and the factors are not connected via a contraction, then each of those factors represents a Green function diagram disconnected from the others.

3.1.6 Visualizing Why the QFT Green Function Works

To understand the QFT Green function a little more intuitively, consider the bra corresponding to the ket of (3-19)

$$\langle 0,t=+\infty| = \langle 0,t=-\infty|e^{-i\phi} \quad \text{or} \quad \langle 0,t=-\infty| = e^{i\phi}\langle 0,t=+\infty| \tag{3-47}$$

and substitute that into the numerator of (3-18) along with (3-20) into the denominator. We get

Re-writing Green function with final state bras

$$G^{\mu\ldots}(x_1,\ldots y_1,\ldots z_1,\ldots) = \frac{e^{i\phi}\langle 0,t=+\infty|T\left\{SA^{\mu}(x_1)\ldots\psi(y_1)\ldots\bar{\psi}(z_1)\ldots\right\}|0,t=-\infty\rangle}{e^{i\phi}}, \tag{3-48}$$

where the phase factors cancel.

(3-48) acts more like the amplitudes we are familiar with. That is, it takes an initial state at an earlier time ($t = -\infty$, here) to a later time ($t = +\infty$, here). In prior work, to find an amplitude, we generally started with a state of certain particles, which was acted on by the operators in S to produce a final state of particles. The time ordered S operator worked fine for that.

Here in (3-48), on the other hand, we start with a vacuum state at $t = -\infty$ and end with a vacuum state at $t = +\infty$. So, to get anything meaningful, we have to create some particles. The fields $A^{\mu}(x_1)\ldots\psi(y_1)\ldots\bar{\psi}(z_1)\ldots$ can be used for that. Which ones we use depends on the problem at hand. If we want to examine a problem with an initial electron and positron, we would use $\bar{\psi}(z_1)$ to create an electron at z_1 and $\psi(y_1)$ to create a positron at y_1.

Then, the S operator would take those fields forward in time and cause them to interact (e.g., to annihilate creating a virtual photon which then transforms into an outgoing electron and positron, as in Bhabha scattering). But then we have to get back to the vacuum again (at $t = +\infty$). So, we use additional fields from $A^{\mu}(x_1)\ldots\psi(y_1)\ldots\bar{\psi}(z_1)\ldots$ to destroy the resulting electron and positron and leave the vacuum. In our example, we would have a field therein of form $\bar{\psi}(z_1')$ to destroy the positron at $z_1{}'$ and $\psi(y_1{}')$ to destroy the electron at $y_1{}'$.

In this form, Green function much like transition amplitude

But note, the S operator is a unitary operator that can be expressed as a sequence of unitary operations, i.e., using U to represent these operations (see Vol. 1, Chap. 18, where that symbol is also employed)

$$S = U(+\infty,-\infty) = U(+\infty,y_1')U(y_1',z_1')U(z_1',y_1)U(y_1,z_1)U(z_1,-\infty). \tag{3-49}$$

Since S and the other operators in (3-48) are time ordered, each of the U operators in S of (3-49) operates in the appropriate time sequence. In our prior example, $U(z_1, -\infty)$ takes the initial vacuum to the time at z_1, where $\bar{\psi}(z_1)$ creates an electron. $U(y_1, z_1)$ then takes that state to y_1, where $\psi(y_1)$ creates a positron (y_1 and z_1 could be the same 4D point), etc. until we arrive at the final vacuum state.

(3-13) [the same as (3-18)] is just a different mathematical way to express the amplitude associated with this unfolding of events. But in it, the final bra is different in that it represents the identical vacuum state of the ket, both being at $t = -\infty$. As noted earlier, the denominator of (3-13) [the same as (3-18)] effectively cancels out the difference in the final kets of the two methodologies.

3.1.7 Other Interaction Examples Using Green Functions

Møller Scattering

Note that, as depicted graphically in Fig. 3-2, pg. 68, lower middle diagram, the same Green function (3-24) could also represent Møller (electron-electron) scattering, as the four legs in (3-24) could create electron propagators beginning at z_1 and z_2, then end with two electron propagators at y_1 and y_2. This parallels what we learned about transition amplitudes for these two processes in QED. (See Vol. 1, Wholeness Chart 8-4, pg. 250 for $S_{(B)}^{(2)}$ and pg. 225.)

Same Green function yields amplitudes for different interactions (here, any with 4 external fermions)

That same Green function could, in addition, represent positron-positron scattering, as in the last diagram of Fig. 3-2.

The Photon Propagator

Consider the photon propagator, with Feynman diagrams shown in Fig. 3-9 to the first two orders

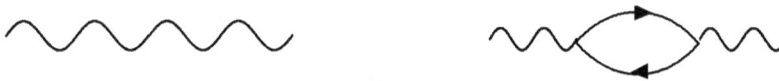

Figure 3-9. The Photon Propagator to First Two Orders

Now consider the Green function (3-13) with only two photon legs (a two-point Green function), where again we focus on the numerator.

Green function with only two photon legs

$$G^{\mu\nu}(x_1,x_2)=\frac{\langle 0|T\left\{SA^{\mu}(x_1)A^{\nu}(x_2)\right\}|0\rangle}{\langle 0|S|0\rangle}$$

$$=\frac{\langle 0|T\left\{A^{\mu}(x_1)A^{\nu}(x_2)\right\}|0\rangle}{\langle 0|S|0\rangle}+\frac{ie\langle 0|T\left\{\int\left(\bar{\psi}\gamma^{\alpha}A_{\alpha}\psi\right)_{\underline{x}_1}d^4\underline{x}_1A^{\mu}(x_1)A^{\nu}(x_2)\right\}|0\rangle}{\langle 0|S|0\rangle} \tag{3-50}$$

$$\frac{-\frac{e^2}{2!}\langle 0|T\left\{\int\left\{\left(\bar{\psi}\gamma^{\alpha}A_{\alpha}\psi\right)_{\underline{x}_1}\left(\bar{\psi}\gamma^{\beta}A_{\beta}\psi\right)_{\underline{x}_2}\right\}d^4\underline{x}_1d^4\underline{x}_2A^{\mu}(x_1)A^{\nu}(x_2)\right\}|0\rangle}{\langle 0|S|0\rangle}+\ldots$$

In converting the time ordering to normal ordering via Wick's theorem, the second term in the middle line of (3-50) will drop out, as it has an odd number of photon fields, and only with an even number can a non-zero term result. The third term, similar to what we have seen before, is identical in all but dummy integration variables from a fourth term (not shown). They sum to give a term like the 3rd one without the factor of ½. Thus, we end up with

$$\text{numerator of } G^{\mu\nu}(x_1,x_2)=$$

$$\langle 0|\left\{\begin{matrix}\underbrace{A^{\mu}(x_1)A^{\nu}(x_2)}-e^2\int\left(\bar{\psi}\gamma^{\alpha}A_{\alpha}\psi\right)_{\underline{x}_1}\left(\bar{\psi}\gamma^{\beta}A_{\beta}\psi\right)_{\underline{x}_2}A^{\mu}(x_1)A^{\nu}(x_2)d^4\underline{x}_1d^4\underline{x}_2\\+\ \ldots\end{matrix}\right\}|0\rangle \tag{3-51}$$

Again, this equals simply the quantities inside the large parentheses, since they are numbers and $\langle 0|0\rangle = 1$.

Note we are not converting the leg propagators to external particles, since we are dealing here simply with a propagator (photon propagator) and not an external particle interaction.

We should be able to see that the terms shown in (3-51) are simply the lowest order propagator (first term) and the second order correction (second term), shown diagrammatically in Fig. 3-9. These are just the relation derived in the standard QFT canonical approach for the photon propagator to second order.

Thus, we see that the Green function for two photon legs is simply the photon propagator. We showed this to second order and can surmise that if we included yet higher order terms in (3-50), we would get the higher order terms for the photon propagator as found in standard QFT without Green functions calculations.

We get the photon propagator

3.1.8 Green Functions That Vanish

Note that unless the Green function (3-13) has the same number of ψ legs as $\bar{\psi}$ legs, it will equal zero, because otherwise the ket would either be destroyed or not match the bra. This is true since S always has the same number of each of those types of fermion field operator. This means Green functions corresponding to the two LHS graphs of Fig. 3-10 equal zero. And it means electron number must be conserved, which it is in the real world.

Green functions with no fermion legs and an odd number of photon legs, such as those represented by the two RHS graphs of Fig. 3-10 (in the usual form, not our special form), will also vanish. This is a result of Furry's theorem. See Vol. 1, pg. 341. For each such configuration, there will be another configuration with opposite sign on the transition amplitude and thus, the Green function, so the two configurations will cancel one another.

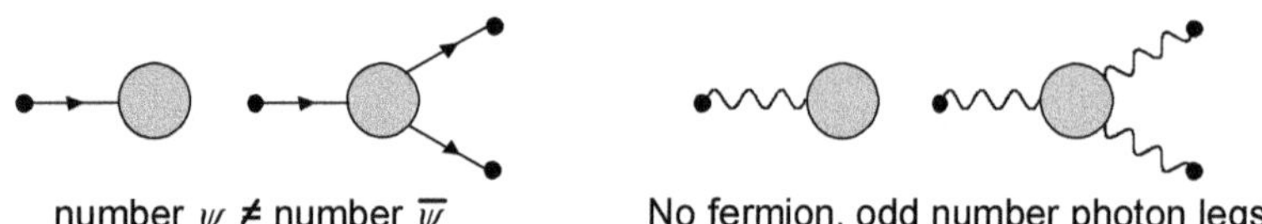

Figure 3-10. Examples of Vanishing Green Functions

Bottom line: The Green function vanishes when

1) the number of ψ legs ≠ number of $\bar{\psi}$ legs, or
2) there are no fermion legs and an odd number of photon legs

Cases where Green function is zero

3.1.9 Connected Green Functions

For any QED case with more than three legs (such as (3-24) with four), we will have connected and disconnected graphs. We can represent this diagrammatically for (3-24) as shown (in the usual form) in Fig. 3-11.

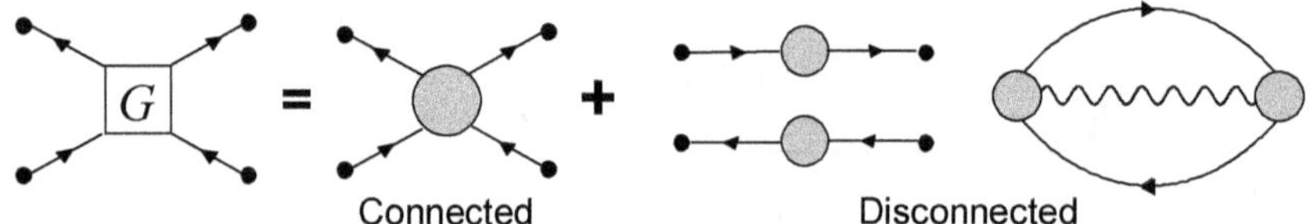

Green functions have connected and disconnected parts

Figure 3-11. Decomposition of the Green Function (3-24) into Connected and Disconnected Parts

Disconnected diagrams represent two or more independent processes, each with its own transition amplitude, as if it acted alone, except the two amplitudes are multiplied together. So, there is nothing really new therein beyond those two individual transition amplitudes. Feynman rules do not, in fact, even apply to disconnected diagrams. As a result, focus is usually directed to the connected Green function G_C,

$$G_C^{\mu\nu\ldots}(x_{1,}\ldots) = G^{\mu\nu\ldots}(x_{1,}\ldots) - \sum \text{disconnected diagrams} \qquad (3\text{-}52)$$

Meaningful analyses use connected Green function

Disconnected Vacuum Processes in the Green Function Denominator and Numerator

As noted on pg. 71, the denominator of the Green function (3-18) just changes the vacuum from possibly one phase to another. If we expanded it, we would see it represents vacuum bubble processes. (See Vol. 1, pgs. 234-235.) This leads to a quantity in the denominator.

We will not expound on it in depth here, as it is not critical, involves a lengthy derivation, and would take us a bit afield, but it can be shown[3] that the disconnected vacuum bubble processes in the numerator lead to a multiplicative factor there. One can see this, for example, in (3-46) where the vacuum bubble contribution to the math is a multiplicative factor. It turns out that the quantity in the denominator and the multiplicative factor in the numerator are equal, which may feel right intuitively, since they are both vacuum bubble contributions. The bottom line: they cancel each other out,

Vacuum bubble contributions in numerator and denominator cancel

[3] If you really need to know how, see F. Mandl and G. Shaw, *Quantum Field Theory*, 2nd ed., Wiley (2010), pgs. 248-252.

effectively meaning we can ignore vacuum bubble processes in the Green function. This provides justification for keeping the denominator in the definition (3-13).

Disconnected, Connected, and Vacuum Bubble Diagrams

Although in the preceding we showed vacuum bubbles as part of disconnected diagrams, note that we can have them in either connected or disconnected diagrams. Likewise, we can have no vacuum bubbles in either connected or disconnected diagrams. There is no inherent connection between vacuum bubbles and disconnected diagrams. They are independent concepts. Wholeness Chart 3-2 illustrates this independence.

Wholeness Chart 3-2. Comparing Connected, Disconnected, and Vacuum Bubble Green Function Diagrams

	Connected		Disconnected	
	No Vacuum Bubble	Vacuum Bubbles		No Vacuum Bubble
$\hat{G}_{numerator}$	$\langle 0\vert T\{S\psi\psi\bar{\psi}\,\bar{\psi}A^{\mu}A^{\nu}\}\vert 0\rangle$	$\langle 0\vert T\{S\}\vert 0\rangle$	$\langle 0\vert T\{S\psi\psi\bar{\psi}\,\bar{\psi}A^{\mu}A^{\nu}\}\vert 0\rangle$	$\langle 0\vert T\{S\psi\psi\bar{\psi}\,\bar{\psi}A^{\mu}A^{\nu}\}\vert 0\rangle$
Example diagrams				
	Each block above represents one term in the associated expansion of $\hat{G}_{numerator}$ in the block above that. Diagrams shown are usual form (like other texts), not like our special form (this text) diagrams.			

The same Green function numerator $\hat{G}_{numerator}$ (2nd, 4th, and 5th columns), when expanded, has many terms, where each term may have connected and/or disconnected parts (factors), For the terms with disconnected parts (factors), some have vacuum bubbles and some do not. The denominator of the Green function, which happens to equal the particular numerator example chosen of the 3rd column, cancels all disconnected vacuum bubble diagrams, but only those (with vacuum bubbles). To focus on just the connected diagrams, we still need to carry out (3-52), to eliminate the disconnected parts that do not have vacuum bubbles.

Vacuum/non-vacuum bubbles independent of connected/ disconnected

There are no purely vacuum bubble parts for the Green function shown in the 2nd, 4th, and 5th columns, since that has (six) external (non-vacuum) particles (legs) built into it.

Bottom line: To deduce transition amplitudes for given external particles, we need only the connected, non-vacuum bubble parts of the Green function for those external particles.

External particle interactions → connected, non-vacuum bubble

An aside: Note from the chart above that the denominator of the Green function $\langle 0|S|0\rangle = e^{i\phi}$ of (3-13) and (3-48) (S is time ordered inherently) represents the effect of the vacuum, i.e., the vacuum bubbles.

3.1.10 Another Example

Consider a different (four-point) Green function, where we continue our convention in this chapter of using spacetime argument symbolism of x_i for A^{μ}, y_i for ψ, and z_i for $\bar{\psi}$,

A different Green function example

$$G^{\mu\nu}\left(x_1,x_2,y_1,z_1\right)=\frac{\langle 0|T\left\{A^{\mu}(x_1)A^{\nu}(x_2)\psi(y_1)\bar{\psi}(z_1)S\right\}|0\rangle}{\langle 0|S|0\rangle}. \tag{3-53}$$

In (3-53), we have purposely placed S at the end of the string of operators to make a point. In moving S to the left side of the relation, we would exchange the operators in S sequentially with the operator legs. Every boson-boson exchange, and every boson-fermion exchange, would leave the relation unchanged because those fields commute. Each exchange for fermions with fermions would produce a minus sign factor, because fermion fields anti-commute. But there are two fermion legs, so such exchanges, in total, would produce no change in the relation. (See Vol. 1, pgs. 72 and 118.) In any non-zero Green function, the total number of fermion legs must be even, as we need the same number of ψ legs as $\bar{\psi}$ legs. So, we can always place S at the beginning or end of the expression.

S operator OK at beginning or end of {} brackets

We note this because you will often see a Green function in the literature presented in a form like (3-53), with S on the right.

That said, let us turn attention to (3-53).

Do **Problem 5** to show (3-54).

The lowest order contribution to the numerator of (3-53) is

$$G^{(0)\mu\nu}_{\text{numerator}}\left(x_1,x_2,y_1,z_1\right)=\underbrace{A^{\mu}\left(x_1\right)A^{\nu}\left(x_2\right)}\underbrace{\psi\left(y_1\right)\bar{\psi}\left(z_1\right)}=iD_F^{\mu\nu}\left(x_1-x_2\right)iS_F\left(y_1-z_1\right), \tag{3-54}$$

Lowest order term is disconnected, of independent photon & fermion propagators

which can be seen graphically as two disconnected propagators, one fermionic and one photonic.

In general, like all Green functions, (3-53) can be decomposed into connected and disconnected parts, as shown graphically in Fig. 3-12 (again, in the usual form, not our special form).

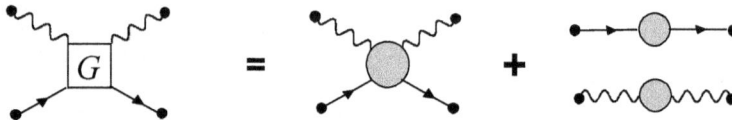

Figure 3-12. Decomposition of Green Function (3-53) Into Connected and Disconnected Parts

---

Do **Problem 6** to show (3-55).

The second order part of the connected Green function portion of (3-53) in momentum space is found by calculating the connected Green function in position space, $G_c^{\mu\nu\cdots}\left(x_{1,}\ldots\right)$, then Fourier transforming that to momentum space. Or, more rapidly, using Feynman rules instead of carrying out all the integration operations in the transform. The result, which we express in lowered indices because it will eventually prove easier to compare with what we have done before, is

$$\begin{aligned}\hat{G}^{(2)}_{c\;\mu\nu}\left(k_{x1},k_{x2},p_{y1},p_{z1}\right)= & \\ & -e^2\left(-iS_F\left(-p_{y1}\right)\right)iD_{F\mu\alpha}\left(k_{x2}\right)\gamma^{\alpha}iS_F\left(p_{z1}+k_{x1}\right)\gamma^{\beta}iD_{F\beta\nu}\left(k_{x1}\right)iS_F\left(p_{z1}\right) \\ & -e^2\left(-iS_F\left(-p_{y1}\right)\right)iD_{F\nu\alpha}\left(k_{x1}\right)\gamma^{\alpha}iS_F\left(p_{z1}+k_{x2}\right)\gamma^{\beta}iD_{F\beta\mu}\left(k_{x2}\right)iS_F\left(p_{z1}\right),\end{aligned} \tag{3-55}$$

2nd order terms in this connected Green function in momentum space

which has Green function diagrams on the LHS (and middle) of Fig. 3-13.

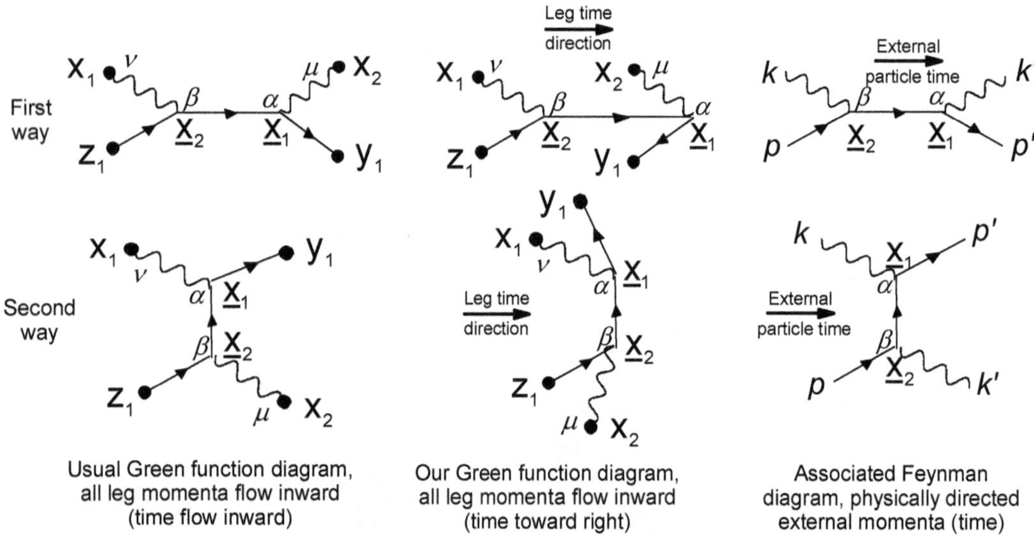

Figure 3-13. Compton Scattering Feynman Diagrams from 2nd Order Green Function (3-55)

The minus signs on p_{y1} in (3-55) come in, as before, because the mathematics just works out such that the argument of antiparticle propagators equals the negative of the physical 4-momentum (and here the Green function diagram is considered to represent the "physical" situation).

Green function diagrams and associated Feynman diagrams for Compton scattering

Note that (3-55) can be expressed in short hand notation, given that the order of photon propagators is unimportant. $D_{\mu\alpha F}(k) = D_{\alpha\mu F}(k)$. That is, looking at the second term in (3-55) and switching dummy variables α and β, we can express that as

$$\begin{aligned}&-e^2\left(-iS_F\left(-p_{y1}\right)\right)i\underbrace{D_{F\nu\beta}}_{D_{F\beta\nu}}\left(k_{x1}\right)\gamma^\beta iS_F\left(p_{z1}+k_{x2}\right)\gamma^\alpha i\underbrace{D_{F\alpha\mu}}_{D_{F\mu\alpha}}\left(k_{x2}\right)iS_F\left(p_{z1}\right)\\&=-e^2\left(-iS_F\left(-p_{y1}\right)\right)iD_{F\mu\alpha}\left(k_{x2}\right)\gamma^\beta iS_F\left(p_{z1}+k_{x2}\right)\gamma^\alpha iD_{F\beta\nu}\left(k_{x1}\right)iS_F\left(p_{z1}\right).\end{aligned} \tag{3-56}$$

Then (3-55) becomes

$$\begin{aligned}&\hat{G}^{(2)}_{c\,\mu\nu}\left(k_{x1},k_{x2},p_{y1},p_{z1}\right)\\&\quad=-e^2\left(-iS_F\left(-p_{y1}\right)\right)iD_{F\mu\alpha}\left(k_{x2}\right)\gamma^\alpha iS_F\left(p_{z1}+k_{x1}\right)\gamma^\beta iD_{F\beta\nu}\left(k_{x1}\right)iS_F\left(p_{z1}\right)\\&\qquad\quad-e^2\left(-iS_F\left(-p_{y1}\right)\right)iD_{F\mu\alpha}\left(k_{x2}\right)\gamma^\beta iS_F\left(p_{z1}+k_{x2}\right)\gamma^\alpha iD_{F\beta\nu}\left(k_{x1}\right)iS_F\left(p_{z1}\right)\\&\quad=\left(-iS_F\left(-p_{y1}\right)\right)iD_{F\mu\alpha}\left(k_{x2}\right)\Gamma^{\alpha\beta}\left(k_{x1},k_{x2},p_{y1},p_{z1}\right)iD_{F\beta\nu}\left(k_{x1}\right)iS_F\left(p_{z1}\right).\end{aligned} \tag{3-57}$$

Streamlined Green function terms using vertex function

where

$$\Gamma^{\alpha\beta}\left(k_{x1},k_{x2},p_{y1},p_{z1}\right)=-e^2\left(\gamma^\alpha iS_F\left(p_{z1}+k_{x1}\right)\gamma^\beta+\gamma^\beta iS_F\left(p_{z1}+k_{x2}\right)\gamma^\alpha\right). \tag{3-58}$$

(3-58) is called a vertex function, which readers of Vol. 1 have seen before on pgs. 459-460 therein. These are helpful in streamlining notation when a given interaction can be represented by more than one Feynman diagram (where, of course, the incoming and outgoing particles/propagators are the same in each diagram). Green functions (and associated vertex functions), on the other hand, are more general, in that any legs can be converted to initial or final particles, depending on the interaction we wish to analyze.

The Feynman Amplitude for Compton Scattering

Consider the Compton scattering case, which is represented graphically on the RHS of Fig. 3-13.

$$\gamma(k,r)+e^-(p,s)\rightarrow\gamma(k',r')+e^-(p',s'), \tag{3-59}$$

We obtain the Feynman amplitude from the Green function (3-55) with the following changes.

$$p_{z1}=p \qquad k_{x1}=k \qquad k_{x2}=-k' \qquad p_{y1}=-p', \tag{3-60}$$

and for second order

$$iS_F\left(p_{z1}\right)\rightarrow u_s(\mathbf{p}) \;\; iD_{\beta\nu F}\left(k_{x1}\right)\rightarrow\varepsilon_{\beta,r}(\mathbf{k}) \;\; iD_{\mu\alpha F}\left(k_{x2}\right)\rightarrow\varepsilon_{\alpha,r'}(\mathbf{k}') \;\; -iS_F\left(-p_{y1}\right)\rightarrow\bar{u}_{s'}(\mathbf{p}'). \tag{3-61}$$

Changes to get Compton amplitude at second order

Thus, (3-57) becomes

$$\begin{aligned}\mathcal{M}^{(2)}_{Compton}\left(k,k',p',p\right)&=\underbrace{\bar{u}_{s'}(\mathbf{p}')\varepsilon_{\alpha,r'}(\mathbf{k}')\Gamma^{\alpha\beta}\left(k,-k',-p',p\right)\varepsilon_{\beta,r}(\mathbf{k})u_s(\mathbf{p})}_{\mathcal{M}^{(2)}_{C1}+\mathcal{M}^{(2)}_{C2}}\\\mathcal{M}^{(2)}_{C1}&=-e^2\bar{u}_{s'}(\mathbf{p}')\varepsilon_{\alpha,r'}(\mathbf{k}')\gamma^\alpha iS_F\left(q=p+k\right)\gamma^\beta\varepsilon_{\beta,r}(\mathbf{k})u_s(\mathbf{p})\\\mathcal{M}^{(2)}_{C2}&=-e^2\bar{u}_{s'}(\mathbf{p}')\varepsilon_{\alpha,r'}(\mathbf{k}')\gamma^\beta iS_F\left(q=p-k'\right)\gamma^\alpha\varepsilon_{\beta,r}(\mathbf{k})u_s(\mathbf{p}),\end{aligned} \tag{3-62}$$

Compton scattering Feynman amplitude at second order

the Compton scattering Feynman amplitude. (Compare with Vol. 1, pg. 228, (8-64) and (8-65) where dummy indices α,β here are μ,ν there, and positioning of photon polarization vectors is irrelevant.)

Higher Order Relations

For higher orders in perturbation theory, relations (3-61) have to be modified using the relations in Vol. 1, pg. 370, Wholeness Chart 14-4, Column IX, as follows.

$$\begin{aligned}&iS_F\left(p_{z1}\right)\rightarrow\left(Z_f^{nth}\right)^{1/2}u_s(\mathbf{p}) \qquad iD_{\beta\nu F}\left(k_{x1}\right)\rightarrow\left(Z_\gamma^{nth}\right)^{1/2}\varepsilon_{\beta,r}(\mathbf{k})\\&iD_{\mu\alpha F}\left(k_{x2}\right)\rightarrow\left(Z_\gamma^{nth}\right)^{1/2}\varepsilon_{\alpha,r'}(\mathbf{k}') \qquad -iS_F\left(-p_{y1}\right)\rightarrow\left(Z_f^{nth}\right)^{1/2}\bar{u}_{s'}(\mathbf{p}').\end{aligned} \tag{3-63}$$

Changes to get Compton amplitude at higher order

For second order, (3-63) reduces to (3-61).

3.1.11 Crossing

Note that, as discussed in Section 3.1.3, the single Green function (3-53) [equivalently for second order, (3-55)] can represent several different interactions, such as

$$\begin{aligned}&\gamma+e^-\to\gamma+e^-\ \ (\text{Compton scattering}) \qquad \gamma+e^+\to\gamma+e^+\ \ (\text{Compton positron scattering})\\&\gamma+\gamma\to e^++e^-\ \ (\text{fermion pair production}) \quad e^++e^-\to\gamma+\gamma\ \ (\text{fermion pair annihilation}).\end{aligned} \qquad (3\text{-}64)$$

One Green function can yield multiple Feynman amplitudes

We simply need to take different initial and final particles when we transition from the Green function to the Feynman amplitude.

Interactions linked in this way are called crossed reactions; and the relations between them, crossing relations. Crossing is typically emphasized in the Green function methodology, but it is actually inherent in the S operator methodology of Vol. 1, as well. See Wholeness Chart 8-4, therein, pg. 250, where (3-65) is one term in the S operator.

Different interactions from same Green function called crossed reactions

$$S_B^{(2)}=-e^2\int d^4x_1d^4x_2N\left\{\left(\bar{\psi}A\psi\right)_{x_1}\left(\bar{\psi}A\psi\right)_{x_2}\right\} \qquad (3\text{-}65)$$

(3-65) can give rise to a transition amplitude, i.e., an S matrix element, (and concomitantly, a Feynman amplitude) for any one of (3-64).

3.1.12 Comparing the Standard Math Green Function with QFT Green Function Methodologies

Green Function Free Photon Propagator

From the result of (3-51), we concluded that

$$G_c^{\mu\nu}(x_1,x_2)=\underbrace{D_F^{\mu\nu}(x_1-x_2)}_{\text{no loops}}+\underbrace{(\text{higher order})}_{\text{fermion-antifermion loops}}\ , \qquad (3\text{-}66)$$

Two photon leg connected Green function = photon propagator

where

$$D_F^{\mu\nu}(k)=\frac{-g^{\mu\nu}}{k^2+i\varepsilon}. \qquad (3\text{-}67)$$

Green Function by Two Different Approaches = Propagator

So now we see the connection between the usual math Green function approach and the QFT Green function methodology. Using the former with the QFT field equations, we get the no loops term in (3-66), i.e., (3-11) containing the familiar photon Feynman propagator, and for that reason alone, we can see why it is common to call the propagator a "Green function". It is the usual math Green function for Maxwell's equation.

But there is another reason. The QFT Green function methodology also gives rise to the propagator as we have known it (apart from a factor of i), $iD_F^{\mu\nu}$, in (3-66). The full propagator (including higher order terms) of (3-66) is a Green function, as found via the QFT Green function methodology.

Both usual math and QFT Green functions yield propagators

Greater Extent of QFT Green Function Methodology

As shown above, we can use the QFT Green function methodology to obtain the math form of the propagators. However, the QFT Green function methodology has a wider range of application. At its core is (3-13), and that relation can be used to find a Green function for any interaction, in addition to just those of the free propagators (such as we did for Bhabha scattering in the first example above.)

But QFT Green functions have wider application

3.1.13 Green Functions Odds and Ends

Order and Points in Green Functions

Don't confuse the nth order of a term in a Green function with the n-points of the entire Green function. "n" is commonly used to refer to both.

The number of points n of a particular Green function equals the number of external legs. For example, in (3-53), (3-54), and (3-55), the points $n = 4$. For any such Green function, the order of one of its terms is the number of vertices in that term (which equals the power to which e is raised). In (3-54), the order n is zero. In (3-55), the order n is two. In (3-53), all orders are included, so it is

n-point Green function vs n^{th} order of term in Green function

inappropriate to talk of its order. The order of a Green function term equals the order of the S operator used to obtain that term.

Other Notation for Green Function

There is an alternative notation for Green functions using $\langle\rangle$ type brackets that can be convenient, and is often used. It is

Alternative symbolism for Green function uses <> brackets

$$\left\langle A^{\mu}(x_1)A^{\nu}(x_2)\ldots\psi(y_1)\ldots\bar{\psi}(z_1)\ldots\right\rangle = G^{\mu\nu\ldots}(x_1,x_2,\ldots y_1,\ldots z_1,\ldots) \quad \text{(symbolism, canonical or IP)}$$

$$= \frac{\langle 0|T\left\{SA^{\mu}(x_1)A^{\nu}(x_2)\ldots\psi(y_1)\ldots\bar{\psi}(z_1)\ldots\right\}|0\rangle}{\langle 0|S|0\rangle} \quad (\text{canonical expression}). \qquad (3\text{-}68)$$

Note that in Chap. 4 we will study the path integral way to represent Green functions, which differs from the canonical expression we have been dealing with. The $\langle\rangle$ notation of (3-68) is used for both.

Outer Product of Fermion Leg Fields

Observe that the adjoint Fermi fields $\bar{\psi}$ in our definition of Green functions (3-13), repeated in (3-68), are placed to the right of the ψ fields indicating the products involved are outer products (in spinor space), not inner products.

Legs in Green function form outer products

Time and Momentum in Green Functions

In Figs. 3-2 (pg. 68), 3-6 (pg. 73), 3-7 (pg. 76), and 3-13 (pg. 82) we showed *time* in the Green function diagram legs directed inward. Other texts depict leg *momentum* flowing inward. If the legs are truly external particles, then the two perspectives are equivalent, as momentum for real particles is directed from an earlier event to a later event.

We show time direction in Green function diagrams; others show momentum

However, to be precise, Green function legs are propagators (prior to conversion to Feynman diagrams/amplitudes), not real particles, and as we found in Vol. 1 (see, for example, pgs. 243-246), propagators can have 3-momentum directed in any direction (not just along the apparent direction of travel). And they can carry either positive or negative energy. In this context, it is not really consistent to talk of propagators (legs) having momentum directed inward, though it is commonly done. And I, with all this in mind (and not wanting to confuse readers), have sort of swept this distinction between time and momentum direction under the rug earlier in the chapter.

But propagator momentum can be in any direction

What is really meant by all this is that leg 4-momenta in Green function math expressions switch signs for *final* (not *initial*) legs when we convert those legs to external (real) particles, in order to find the correct math form of the related Feynman amplitude.

In this book, we do this by effectively switching the direction of time (for the final legs) when we convert a Green function to a Feynman amplitude. Mathematically, we discussed this on pg. 68. From a more conceptual standpoint, if we reverse the time direction, 3-momentum will flow in the opposite direction. And what seems like repulsion (positive energy carried by virtual particles) becomes attraction (negative energy carried by them), as described in the above referenced pages from Vol. 1.

Earlier in the chapter, I used the term "momentum flow inward" which I intend to imply that 3-momentum, whatever direction it is pointing for a virtual particle (propagator), that vector direction, while remaining constant, flows over time from the position at the earlier time to that at the later.

Given the above reasoning, it appears that thinking of time directed inward for Green function legs may be more correct than thinking of momentum directed inward. Additionally, it seems simpler to me conceptually (as in the above cited figures). So, I prefer thinking in terms of changing the direction of time in the legs, rather than the direction of momentum.

Thinking in terms of time flow may be more consistent and easier

But again, if we think in terms of the legs as representing external (real) particles, where 3-momentum must point from the creation event to the destruction event and energy is always positive, then the two perspectives are equivalent.

3.1.14 Overview of Green Functions in Canonical Form

Bottom Line: The Shortcut Method

Steps from a Green function to associated Feynman amplitudes

Step 1. If we know the legs in our Green function, we can quickly sketch a figure like the LH side of Fig. 3-6 (pg. 73), the middle diagram of Fig. 3-7 (pg. 76), or the LH side of Fig. 3-13 (pg. 82), with all leg momenta (time flow) inward.

Step 2. Given what we've been through in this chapter, we should then be able to make the jump directly to constructing the momentum space Green function, composed solely of propagators, as in (3-38), (3-43), or (3-55), without explicitly going through the lengthy math behind it all.

Step 3. From there, to get the Feynman amplitude for a given interaction, we simply exchange the leg propagators for external particle spinors (fermions) and/or external polarization vectors (photons), as in (3-39), (3-44), or (3-61). In doing so, we need to keep in mind that the final (outward) particles have opposite sign for momentum from those in the corresponding Green function legs. Comparing the Green function diagram to the associated Feynman diagram can help keep all of this straight.

Schematic Overview

Wholeness Chart 3-3 provides an overview of Green functions in the canonical approach.

Wholeness Chart 3-3. Summary of QED Green Functions for Canonical Approach

	Canonical Approach	**Path Integral**	**Comments**
Green functions	$G^{\mu\nu\dots}(x_1,x_2,\dots y_1,\dots z_1,\dots)=\langle A^\mu(x_1)A^\nu(x_2)\dots\psi(y_1)\dots\bar{\psi}(z_1)\dots\rangle$		Common Green function notation.
	$=\dfrac{\langle 0\vert T\{SA^\mu(x_1)\dots\psi(y_1)\dots\bar{\psi}(z_1)\dots\}\vert 0\rangle}{\langle 0\vert S\vert 0\rangle}$ Ignore denominator for connected Green functions. (Cancels vacuum bubble factor in numerator.)	To be covered later in book.	Green function will be all propagators. Time inward for all legs $A^\mu(x_1)\dots\psi(y_1)\dots\bar{\psi}(z_1)\dots$ All orders included. n-point Green function = n number of legs.
	Example Green function diagram (2nd order) 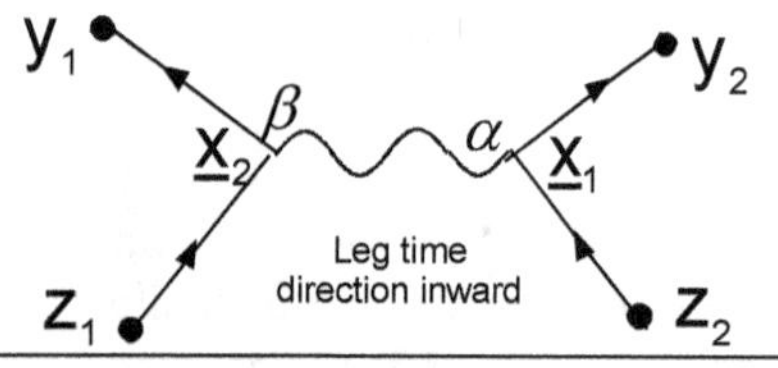		Different terms in Green function represented by different Green function diagrams, each of a particular order with a particular way to connect legs to vertices.
	Convert above to momentum space.		Can do via 1) lengthy math, or 2) if you are experienced, directly
Feynman amplitudes	For each external particle, change the associated propagator to external line (e.g., $S_F(p)\rightarrow u(\mathbf{p})$) to get amplitude. Need final particles $p_{Feyn}=-p_{Green}$.		Need to change appropriate sign of final momenta in arguments of internal propagators, as well.
	Above example transformed to Feynman diagram 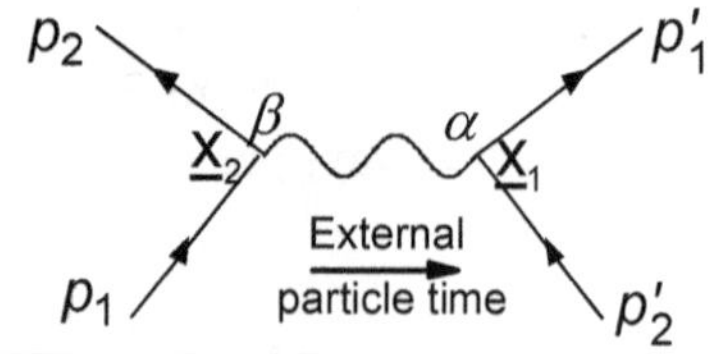		Need to include different ways for same initial and final particles to interact.
Significance	Knowing $G^{\mu\dots}$ for a case with particular legs, one knows the Feynman amplitudes for all interactions, to all orders, where those legs are external particles.		

3.2 Grassmann Fields

Following Wholeness Chart 3-4, pg. 91, may help as you progress through this section.

3.2.1 Defining Characteristic

Grassmann numbers (or Grassmann variables), though just numeric (classical, not quantum, operators), are defined to be anti-commuting. This seems strange at first, and may take some getting used to, but they do serve vital functions in many areas of math and physics. Their anti-commuting nature gives rise to different properties than we are used to for many operations, such as the lack of interchangeability of the order of derivatives.

Grassmann numbers and fields are classical, but anti-commute

Grassmann fields are simply numeric (classical) *fields* with the same basic property. They anti-commute. They are closely associated with fermions and are used as a calculational device for obtaining Green functions from the generating functional (which we have yet to study). Grassmann fields will be needed to obtain Green functions that have fermionic fields in them.

3.2.2 Grassmann Algebras

Recall that an algebra has two operations between its elements, and for Grassmann numbers (variables, fields), these are addition and multiplication, where the generators of the algebra anti-commute. Thus, sets of numbers (variables, fields) employing these operations, and satisfying the criteria of an algebra, are called Grassmann algebras, after the 19th century mathematician H. Grassmann, who invented them (but never saw them much used or appreciated in his lifetime).

Definition of Grassmann Algebra

A Grassmann algebra is an abstract algebra, any of whose elements f can be constructed from a field ("field" in the group theory sense of Chapter 2, not in the field theory sense of Section 3.2.1) of ordinary numbers $p_0, p_i, p_{ij}, p_{ijk}, \dots p_{ijk\dots n}$, and its n generators $\theta_1, \theta_2, \dots \theta_i, \dots \theta_j, \dots, \theta_n$ having form

Grassmann algebra definition

$$f\left(p_0, p_i, p_{ij}, \dots p_{ij\dots n}\right) = p_0 + p_i\theta_i + p_{ij}\theta_i\theta_j + \dots + p_{ij..n}\theta_i\theta_j\dots\theta_n, \tag{3-69}$$

(where, as always, repeated indices mean summation) and wherein, the multiplication of generators has the additional property

$$\theta_i\theta_j = -\theta_j\theta_i, \text{ i.e., the generators all anti-commute } \left[\theta_i, \theta_j\right]_+ = 0. \tag{3-70}$$

End of definition

A Grassmann algebra has significant differences from the $SU(n)$ Lie algebras we studied in Chapter 2, not the least of which is in having elements formed from linear combinations of products of generators. However, it is still an algebra, as it satisfies all of the criteria for one, as one can verify.

We generalize the anti-commutation relations (3-70) for a string of generators, where $k-1$ is the number of times we commute the kth generator in the string to bring it to the left hand side, as

Generalizing anti-commutation relation for Grassmann generators

$$\theta_1\theta_2\dots\theta_{k-1}\theta_k\theta_{k+1}\dots\theta_w = (-1)^{k-1}\,\theta_k\left(\theta_1\theta_2\dots\theta_{k-1}\theta_{k+1}\dots\theta_w\right). \tag{3-71}$$

Do **Problem 7** to show (3-72) and (3-73).

Restrictions on Generator Products and $i, j, \dots n$

If you did Problem 7, you found the important result

Square of Grassmann generator = 0

$$\theta_i^2 = 0, \tag{3-72}$$

and that (3-69) can be expressed with fewer terms, yet still in most general form, with the constraint

and this lets us streamline form for Grassmann algebra element

$$i < j < \dots < n. \tag{3-73}$$

An Example Where $n = 2$

The most general case for (3-69) with $n = 2$ has form, where we introduce shorthand notation for the argument of an element f,

Grassmann algebra example with two generators

$$f\left(p_0, p_1, p_2, p_{12}\right) = f(p) = p_0 + p_1\theta_1 + p_2\theta_2 + p_{12}\theta_1\theta_2 \quad \left(= p_0 + p_1\theta_1 + p_2\theta_2 - p_{12}\theta_2\theta_1\right). \tag{3-74}$$

The binary operation of multiplication means

$$\begin{aligned} f(p)f(q) &= f(p_0,p_1,p_2,p_{12})f(q_0,q_1,q_2,q_{12}) \\ &= (p_0 + p_1\theta_1 + p_2\theta_2 + p_{12}\theta_1\theta_2)(q_0 + q_1\theta_1 + q_2\theta_2 + q_{12}\theta_1\theta_2) \end{aligned} \tag{3-75}$$

is an element of the algebra. It is also associative and distributive, so this is

$$\begin{aligned} f(p)f(q) &= p_0q_0 + (p_0q_1 + p_1q_0)\theta_1 + \underbrace{p_1q_1\theta_1^2}_{0} + (p_0q_2 + p_2q_0)\theta_2 + \underbrace{p_2q_2\theta_2^2}_{0} \\ &\quad + (p_0q_{12} + p_1q_2 - p_2q_1 + p_{12}q_0)\theta_1\theta_2 + \underbrace{p_1q_{12}\theta_1^2\theta_2}_{0} + \underbrace{p_2q_{12}\theta_2\underbrace{\theta_1\theta_2}_{-\theta_2\theta_1}}_{0} \\ &\quad + \underbrace{p_{12}q_1\theta_1\theta_2\theta_1}_{0} + \underbrace{p_{12}q_2\theta_1\theta_2\theta_2}_{0} + \underbrace{p_{12}q_{12}\theta_1\theta_2\theta_1\theta_2}_{0} \\ &= p_0q_0 + (p_0q_1 + p_1q_0)\theta_1 + (p_0q_2 + p_2q_0)\theta_2 + (p_0q_{12} + p_1q_2 - p_2q_1 + p_{12}q_0)\theta_1\theta_2 . \end{aligned} \tag{3-76}$$

Note that anti-commutation of elements $f(p)$ and $f(q)$ would require, among other things, that $p_0q_0 + q_0p_0 = 0$, which is generally not the case. Similarly, commutation would require that $p_1q_2(\theta_1\theta_2 - \theta_2\theta_1) = 2p_1q_2\theta_1\theta_2 = 0$, which is also generally not the case.

So, different elements of the algebra, $f(p)$ and $f(q)$, do not, in general, commute or anti-commute with each other, even though their generators anti-commute. This conclusion generalizes to algebras with any number n of generators, not just 2.

Different elements in the algebra do not have to anti-commute even though the generators do

3.2.3 "Differentiation" with Respect to θ_i

Keep in mind that Grassmann generators θ_i are abstract entities, defined solely by their operational properties, so it doesn't really make sense to consider differentiation with respect to them. However, one can define what is (loosely) called differentiation with respect to θ_i as an operator symbolized by $\partial/\partial\theta_i$ and having the properties

$$\frac{\partial\theta_i}{\partial\theta_j} = \delta_{ij} \quad i,j = 1,\ldots,n \quad \text{and} \quad \frac{\partial a}{\partial\theta_i} = 0 \quad a = \text{ an ordinary number, such as } p_i, p_{ij}, \text{etc.} . \tag{3-77}$$

Properties of "differentiation" with respect to a generator

The LHS of (3-77), at least symbolically, looks similar to ordinary differentiation.

Differentiation for products of two Grassmann generators is defined as

$$\frac{\partial(\theta_1\theta_2)}{\partial\theta_1} = \underbrace{\frac{\partial\theta_1}{\partial\theta_1}}_{1}\theta_2 + \theta_1\underbrace{\frac{\partial\theta_2}{\partial\theta_1}}_{0} = \theta_2 \qquad \frac{\partial(\theta_2\theta_1)}{\partial\theta_1} = \frac{\partial(-\theta_1\theta_2)}{\partial\theta_1} = -\frac{\partial(\theta_1\theta_2)}{\partial\theta_1} = -\theta_2 . \tag{3-78}$$

We can generalize this result to any number n of generators using (3-71).

$$\frac{\partial(\theta_1\theta_2\ldots\theta_{k-1}\theta_k\theta_{k+1}\ldots\theta_n)}{\partial\theta_k} = (-1)^{k-1}\left(\frac{\partial\theta_k}{\partial\theta_k}\right)\theta_1\theta_2\ldots\theta_{k-1}\theta_{k+1}\ldots\theta_n = (-1)^{k-1}\theta_1\theta_2\ldots\theta_{k-1}\theta_{k+1}\ldots\theta_n . \tag{3-79}$$

Differentiation of a product of generators

To be complete, we mention that (3-78) and (3-79) define what is known as a "left derivative", because it moves the generator whose derivative we take to the left hand side. We will not have need for the "right derivative".

It will aid us shortly if we further generalize (3-79) to (3-80) (where only one term will survive).

$$\begin{aligned} \frac{\partial(\theta_1\ldots\theta_{j-1}\theta_j\theta_{j+1}\ldots\theta_n)}{\partial\theta_k} &= (-1)^0\underbrace{\left(\frac{\partial\theta_1}{\partial\theta_k}\right)}_{\delta_{1k}}(\theta_2\ldots\theta_{j-1}\theta_j\theta_{j+1}\ldots\theta_n) + (-1)^1\underbrace{\left(\frac{\partial\theta_2}{\partial\theta_k}\right)}_{\delta_{2k}}(\theta_1\theta_3\ldots\theta_{j-1}\theta_j\theta_{j+1}\ldots\theta_n) \\ &\quad + \ldots + (-1)^{j-1}\underbrace{\left(\frac{\partial\theta_j}{\partial\theta_k}\right)}_{\delta_{jk}}(\theta_1\ldots\theta_{j-1}\theta_{j+1}\ldots\theta_n) + \ldots + (-1)^{n-1}\underbrace{\left(\frac{\partial\theta_n}{\partial\theta_k}\right)}_{\delta_{nk}}(\theta_1\theta_3\ldots\theta_{j-1}\theta_j\theta_{j+1}\ldots\theta_{n-1}) . \end{aligned} \tag{3-80}$$

Most general expression of the differentiation

Do **Problem 8** to understand how (3-81) arises.

If you did Problem 8, you showed that (3-81) below is valid for a Grassmann algebra with two generators. Hopefully, we can then accept the generalized version for any number of generators,

Anti-commutation relations of Grassmann generators and differentiation with respect to them

$$\left[\frac{\partial}{\partial\theta_i},\frac{\partial}{\partial\theta_j}\right]_+ = 0 \qquad \left[\theta_i,\frac{\partial}{\partial\theta_j}\right]_+ = \delta_{ij}\,. \tag{3-81}$$

3.2.4 Grassmann Algebras with a Continuous Set of Generators

Grassmann Fields

Grassmann fields: continuous rather than discrete generators

The above results for a finite set of n generators θ_i can be extended to an infinite, continuous set of generators $\theta(x)$ with anti-commutation relations

Grassmann fields anti-commutation relations

$$\boxed{\left[\theta(x),\,\theta(y)\right]_+ = 0}\,. \tag{3-82}$$

$\theta(x)$ is called a Grassmann field.

Grassmann Functionals

Similar to (3-69),we can define

$$F[\theta] = f_0 + \int f_1(x)\theta(x)d^4x + \int f_2(x,y)\theta(x)\theta(y)d^4xd^4y + ... \tag{3-83}$$

Grassmann functionals: similar to, but not exactly like, usual functionals

$F[\theta]$ is called a Grassmann functional, though one should take particular care regarding it, as it is not really an ordinary functional, which generally (See Vol. 1, pg. 489, footnote) comprises a mapping of a function (such as $\theta(x)$ might be here) to a scalar number (such as F might be here) usually via a definite integral. For an "ordinary" functional, $\theta(x)$ would be an ordinary function of x, but here it is a Grassmann field, which does not have ordinary properties. (3-83) is thus purely a formal expression, though we refer to it as a "functional".

Functional Differentiation with Respect to Grassmann Fields

We *define* functional differentiation $\delta/\delta\theta(y)$ via extension of (3-77), to have the properties

Definition: functional differentiation for Grassmann fields

$$\boxed{\frac{\delta\theta(x)}{\delta\theta(y)} = \delta^{(4)}(x-y) \qquad \frac{\delta f}{\delta\theta(y)} = 0 \qquad f = \text{an ordinary function of } x \text{ and/or } y,} \tag{3-84}$$

and from (3-80),

Differentiation of a product of generator fields: most general expression

$$\begin{aligned}\frac{\delta}{\delta\theta(x)}\left(\theta(x_1)\theta(x_2)...\theta(x_j)...\theta(x_n)\right) &= \delta^{(4)}(x-x_1)(-1)^0\left(\theta(x_2)...\theta(x_j)...\theta(x_n)\right)\\ &+\delta^{(4)}(x-x_2)(-1)^1\left(\theta(x_1)\theta(x_3)...\theta(x_j)...\theta(x_n)\right)+...\\ &+\delta^{(4)}(x-x_j)(-1)^{j-1}\left(\theta(x_1)...\theta(x_{j-1})\theta(x_{j+1})...\theta(x_n)\right)+...\\ &+\delta^{(4)}(x-x_n)(-1)^{n-1}\left(\theta(x_1)...\theta(x_j)...\theta(x_{n-1})\right).\end{aligned} \tag{3-85}$$

The LHS of (3-84) may seem a little strange, in that it implies the derivative of a Grassmann field $\theta(x)$ with respect to itself, i.e., $\theta(y)$ with $y = x$, is infinity, rather than one, as we might expect. However, it all works because it is always used inside integrals, where the delta function forces $x = y$ in the integrand upon integration, which has the same effect as the LHS being equal to unity when $x = y$, and zero otherwise.

Further, from extension of (3-81), we have

Anti-commutation relations for Grassmann fields and their derivatives

$$\left[\frac{\delta}{\delta\theta(x)},\frac{\delta}{\delta\theta(y)}\right]_+ = 0 \qquad \left[\theta(x),\frac{\delta}{\delta\theta(y)}\right]_+ = \delta^{(4)}(x-y)\,. \tag{3-86}$$

In Practice, Commonly Paired Independent Grassmann Fields

In QFT, we will typically work with fairly simple (relatively speaking) functionals, and these will usually have two associated, but *independent*, Grassmann fields (think "related to ψ and $\bar{\psi}$"), which occur in pairs. We will denote these by $\theta(x)$ and $\tilde{\theta}(x)$, and their independence leads to the relations

Paired Grassmann fields (that are independent)

$$\frac{\delta\theta(x)}{\delta\tilde{\theta}(y)}=0 \qquad \frac{\delta\tilde{\theta}(x)}{\delta\theta(y)}=0\,. \tag{3-87}$$

Differentiation with paired Grassmann fields

Anti-commutators for Independent Fields

Anti-commutators between two independent fields, and their functional derivatives, are defined to be zero. Useful examples are

$$\left[\theta(x),\tilde{\theta}(y)\right]_+=0 \quad \left[\frac{\delta}{\delta\theta(x)},\frac{\delta}{\delta\tilde{\theta}(y)}\right]_+=0 \quad \left[\theta(x),\frac{\delta}{\delta\tilde{\theta}(y)}\right]_+=0 \quad \left[\tilde{\theta}(x),\frac{\delta}{\delta\theta(y)}\right]_+=0\,. \tag{3-88}$$

Anti-commutation relations for paired Grassmann fields

Example Functionals

A common functional form, where a function of x and y, $K(x,y)$, is known as the kernel, and where we introduce the short-hand notation $[AKB]$, which in the present case becomes $\left[\theta K\tilde{\theta}\right]$ is

$$\boxed{[AKB]=\int A(x)K(x,y)B(y)d^4x\,d^4y} \xrightarrow[\text{case}]{\text{special}} \left[\theta K\tilde{\theta}\right]=\int\theta(x)K(x,y)\tilde{\theta}(y)d^4x\,d^4y\,. \tag{3-89}$$

Example of a common functional (and its notation) employing Grassmann fields

From (3-84), (3-87), and (3-88) in (3-89), where the minus sign in the last row is from (3-85),

$$\begin{aligned}\frac{\delta}{\delta\tilde{\theta}(z)}\left[\theta K\tilde{\theta}\right]&=\int\underbrace{\left(\frac{\delta}{\delta\tilde{\theta}(z)}\theta(x)\right)}_{0}K(x,y)\tilde{\theta}(y)d^4x\,d^4y+\int\theta(x)\underbrace{\left(\frac{\delta}{\delta\tilde{\theta}(z)}K(x,y)\right)}_{0}\tilde{\theta}(y)d^4x\,d^4y\\&\quad+\int\underbrace{\left(-\frac{\delta}{\delta\tilde{\theta}(z)}\tilde{\theta}(y)\right)}_{-\delta^{(4)}(z-y)}\theta(x)K(x,y)d^4x\,d^4y=-\int\theta(x)K(x,z)d^4x\,.\end{aligned} \tag{3-90}$$

Differentiation of that example functional

An example that will prove helpful, but is a bit more complicated,

using $e^X=1+X+\frac{X^2}{2}+\frac{X^3}{3!}+\dots$, is

$$\begin{aligned}e^{i\left[\theta K\tilde{\theta}\right]}&=1+i\int\theta(x_1)K(x_1,y_1)\tilde{\theta}(y_1)d^4x_1d^4y_1\\&\quad-\frac{1}{2}\int\left(\theta(x_1)K(x_1,y_1)\tilde{\theta}(y_1)\right)\left(\theta(x_2)K(x_2,y_2)\tilde{\theta}(y_2)\right)d^4x_1d^4y_1d^4x_2d^4y_2-\dots\\&=\sum_{n=0}^{\infty}\frac{i^n}{n!}\int\left(\theta(x_1)K(x_1,y_1)\tilde{\theta}(y_1)\right)\dots\left(\theta(x_n)K(x_n,y_n)\tilde{\theta}(y_n)\right)d^4x_1d^4y_1\dots d^4x_nd^4y_n\,.\end{aligned} \tag{3-91}$$

Another example more like what we will actually work with in QFT

The order of the bracketed quantities $\left(\theta(x_i)K(x_i,y_i)\tilde{\theta}(y_i)\right)$ (no sum) in each integral of (3-91) is unimportant, since each has two Grassmann fields that anti-commute. We get a factor of $(-1)^4$ each time we exchange such bracketed quantities, so those bracketed quantities commute. By the same logic, they also commute with the functional differential operators $\delta/\delta\theta(z)$ and $\delta/\delta\tilde{\theta}(y)$.

Do **Problem 9** to deduce (3-92).

Note that, much like more pedestrian exponential forms,

$$\frac{\delta}{\delta\theta(z)}e^{i\left[\theta K\tilde{\theta}\right]}=i\frac{\delta\left[\theta K\tilde{\theta}\right]}{\delta\theta(z)}e^{i\left[\theta K\tilde{\theta}\right]} \qquad \frac{\delta}{\delta\tilde{\theta}(z)}e^{i\left[\theta K\tilde{\theta}\right]}=i\frac{\delta\left[\theta K\tilde{\theta}\right]}{\delta\tilde{\theta}(z)}e^{i\left[\theta K\tilde{\theta}\right]}\,. \tag{3-92}$$

Differentiation of that example

3.2.5 Summary of Grassmann Numbers and Fields

Wholeness Chart 3-4 summarizes the key relations for Grassmann versus ordinary variables.

Wholeness Chart 3-4. Summary of Properties: Grassmann Numbers and Fields

	Numbers		**Fields (Non-Operator)**	
	Ordinary	**Grassmann**	**Ordinary**	**Grassmann**
Commutation	$[x,y]=0$		$[\phi(x),\phi(y)]=0$	
Anti-commutation		$[\theta_1,\theta_2]_+=0$		$[\theta(x),\theta(y)]_+=0$
Square	$x^2=xx$	$\theta_1^2=0$	$\phi^2(x)=\phi(x)\phi(x)$	$\theta^2(x)=0$
Derivatives	Variable Derivatives		Functional Derivatives	
Delta Relations	$\frac{\partial x_j}{\partial x_k}=\delta_{jk}$	$\frac{\partial \theta_j}{\partial \theta_k}=\delta_{jk}$	$\frac{\delta\phi(x)}{\delta\phi(y)}=\delta^{(4)}(x-y)$	$\frac{\delta\theta(x)}{\delta\theta(y)}=\delta^{(4)}(x-y)$
Products	$\frac{\partial(x_1x_2)}{\partial x_1}=x_2$ $\frac{\partial(x_1x_2)}{\partial x_2}=x_1$ $\frac{\partial(x_1..x_j..x_n)}{\partial x_k}=$ sum of terms like $\delta_{jk}x_1..x_{j-1}x_{j+1}..x_n$ all but one = 0	$\frac{\partial(\theta_1\theta_2)}{\partial\theta_1}=\theta_2$ $\frac{\partial(\theta_1\theta_2)}{\partial\theta_2}=-\theta_1$ $\frac{\partial(\theta_1..\theta_j.\theta_n)}{\partial\theta_k}=$ sum of terms like $\delta_{jk}\times(-1)^{j-1}\theta_1..\theta_{j-1}\theta_{j+1}..\theta_n$ all but one = 0	$\frac{\delta(\phi(x_1)..\phi(x_j)..\phi(x_n))}{\delta\phi(x)}=$ sum of terms like $\delta^{(4)}(x-x_j)\times\phi(x_1)..\phi(x_{j-1})\phi(x_{j+1})..\phi(x_n)$	we won't be using $\frac{\delta\{\theta(x)\theta(y)\}}{\delta\theta(x)}$ or $\frac{\delta\{\theta(x)\theta(y)\}}{\delta\theta(y)}$, so ignore $\frac{\delta(\theta(x_1)..\theta(x_j)..\theta(x_n))}{\delta\theta(x)}=$ sum of terms like $\delta^{(4)}(x-x_j)\times(-1)^{j-1}\theta(x_1)..\theta(x_{j-1})\theta(x_{j+1})..\theta(x_n)$
Commutation	$\left[\frac{\partial}{\partial x_i},\frac{\partial}{\partial x_j}\right]=0$ $\left[x_i,\frac{\partial}{\partial x_j}\right]=\delta_{ij}$		$\left[\frac{\delta}{\delta\phi(x)},\frac{\delta}{\delta\phi(y)}\right]=0$ $\left[\phi(x),\frac{\delta}{\delta\phi(y)}\right]=\delta^{(4)}(x-y)$	
Anti-commutation		$\left[\frac{\partial}{\partial\theta_i},\frac{\partial}{\partial\theta_j}\right]_+=0$ $\left[\theta_i,\frac{\partial}{\partial\theta_j}\right]_+=\delta_{ij}$		$\left[\frac{\delta}{\delta\theta(x)},\frac{\delta}{\delta\theta(y)}\right]_+=0$ $\left[\theta(x),\frac{\delta}{\delta\theta(y)}\right]_+=\delta^{(4)}(x-y)$
Paired Grassmann Fields				$\frac{\delta\theta(x)}{\delta\tilde\theta(y)}=0$ $\frac{\delta\tilde\theta(x)}{\delta\theta(y)}=0$ All anti-commutators between paired fields and derivatives = 0. (See (3-88))
Example: Derivative of Functional				$[\theta K\tilde\theta]=\int\theta(x)K(x,y)\tilde\theta(y)d^4xd^4y$ $\frac{\delta}{\delta\theta(z)}e^{i[\theta K\tilde\theta]}=i\frac{\delta[\theta K\tilde\theta]}{\delta\theta(z)}e^{i[\theta K\tilde\theta]}$ $=i\left(\int K(z,y)\tilde\theta(y)d^4y\right)e^{i[\theta K\tilde\theta]}$

3.3 The Generating Functional

A useful way of finding Green functions employs use of something called the generating functional defined below in (3-93). We will show how this is done after explaining what we mean by the symbols in (3-93).

3.3.1 Definition of Generating Functional (Canonical Approach)

The generating functional in the canonical approach is defined by

$$\boxed{Z[J_k,\sigma,\bar{\sigma}]=\frac{\langle 0|S'|0\rangle}{\langle 0|S|0\rangle}}, \tag{3-93}$$

Definition of generating functional and symbols therein

where $\mathcal{L}_I$ below is the interaction Lagrangian, $\mathcal{L}_S$ in (3-93) is a new entity we introduce below, called the source Lagrangian, and S' is

$$\boxed{\begin{aligned} S' &= T\left\{e^{i\int \mathcal{L}_I'(\underline{x})d^4\underline{x}}\right\}=\sum_{n=0}^{\infty}\frac{i^n}{n!}\int_{-\infty}^{\infty}\cdots\int_{-\infty}^{\infty}T\{\mathcal{L}_I'(\underline{x_1})\mathcal{L}_I'(\underline{x_2})\ldots\mathcal{L}_I'(\underline{x_n})\}d^4\underline{x_1}d^4\underline{x_2}\ldots d^4\underline{x_n} \\ \mathcal{L}_I' &= \mathcal{L}_I+\mathcal{L}_S=\mathcal{L}_I+\underbrace{J_\kappa(\underline{x})A^\kappa(\underline{x})+\bar{\sigma}(\underline{x})\psi(\underline{x})+\bar{\psi}(\underline{x})\sigma(\underline{x})}_{\mathcal{L}_S}\,. \end{aligned}} \tag{3-94}$$

Note the newly introduced fields $J_\kappa(\underline{x})\ldots\bar{\sigma}(\underline{x})\ldots\sigma(\underline{x})$ in $\mathcal{L}_S$ above are fictitious classical fields (just functions, not operators) called source fields, whereas the fields $A^\mu(\underline{x})\ldots\psi(\underline{x})\ldots\bar{\psi}(\underline{x})$ are quantum (operator) fields. Note further that $J_\kappa(x)$ is NOT the fermion 4-current of QFT, i.e., it is not $\bar{\psi}\gamma_\kappa\psi$. It is a new (fictitious) entity that will be used as a helpful tool. Still further, note that the term "source" in source field here does *not* imply such fields are a source for electric, magnetic, or other fields, as the term is commonly used in electromagnetism. I believe the word is used here since, as we will see, it is a *de facto* source for Green function generation.

Source fields are fictitious and classical, and will help in finding Green functions

The $\bar{\sigma}(\underline{x})$ and $\sigma(\underline{x})$ fields are Grassmann fields, and so now, we are beginning to see why we studied such fields in Section 3.2. Further, these two fields are paired Grassmann fields (symbolized in general before by θ and $\tilde{\theta}$), as in (3-87), (3-88), and the penultimate row of Wholeness Chart 3-4.

Fermion source fields are Grassmann fields

In what follows, we don't solve for, or use, algebraic forms of the source fields $J_\kappa(\underline{x})\ldots.\bar{\sigma}(\underline{x})\ldots.\sigma(\underline{x})$. We just employ those source fields as symbols that aid us in developing useful relations. They are part of the "trick" we use to deduce Green functions, as we will shortly see.

Finally, note that the spinor/Grassmann field products in (3-94) are inner products, e.g., showing the spinor indices, $\bar{\sigma}(\underline{x})\psi(\underline{x})=\bar{\sigma}_\alpha(\underline{x})\psi_\alpha(\underline{x})$, where, as is customary, they are usually not shown. In essence, $\bar{\sigma}$ and σ are spinors in their own right, though different from spinor fields like $\bar{\psi}$ and ψ.

3.3.2 Definition and Example of Functional Differentiation of a Functional

Definition: Functional Differentiation with Respect to J_μ

Similar to what we have already seen for Grassmann fields, the (non-Grassmann, classical) field J_K is defined to have the following functional differentiation property.

$$\boxed{\frac{\delta J_\kappa(x'')}{\delta J_\mu(x)}=\delta^\mu_\kappa\delta^{(4)}(x''-x) \qquad \frac{\delta f}{\delta J_\mu(x)}=0 \qquad f=\text{any function other than } J_\mu} \tag{3-95}$$

Functional differentiation for photon source field similar to Grassmann field

This is similar to (3-84), except that was for Grassmann fields, and (3-95) is for ordinary (non-Grassmann, commuting) fields.

Example: Functional Differentiation of a Functional with Respect to J_μ

As an example of applying (3-95), consider the functional

$$F[J_\kappa]=\int f(\underline{x_1})J_\kappa(\underline{x_1})d\underline{x_1}, \tag{3-96}$$

Functional differentiation of a functional

which is a functional because it has a different numeric value for each different form of the function $J_\kappa(\underline{x}_1)$ (with the assumption that $f(\underline{x}_1)$ is a fixed, known function). Now take a functional derivative of (3-96) with respect to J_μ, using (3-95), and we get

$$\frac{\delta}{\delta J_\mu(x_1)}\overbrace{\left(\int f(\underline{x}_1)J_\kappa(\underline{x}_1)d\underline{x}_1\right)}^{\text{functional } F[J_\kappa]}=\left(\int f(\underline{x}_1)\frac{\delta J_\kappa(\underline{x}_1)}{\delta J_\mu(x_1)}d\underline{x}_1\right)=\left(\int f(\underline{x}_1)\delta_\kappa^\mu\delta(x_1-\underline{x}_1)d\underline{x}_1\right)=\delta_\kappa^\mu f(x_1). \quad (3\text{-}97)$$

3.3.3 Functional Derivatives of the Generating Functional

Functional derivative of the generating functional with respect to photon source field

From (3-93), with (3-94) and (3-95), along with the usual rule for taking the derivatives of an exponential function (2nd to 3rd lines below),

$$\frac{1}{i}\frac{\delta Z[J_k,\sigma,\bar\sigma]}{\delta J_\mu(x_1)}=\frac{\langle 0|\frac{1}{i}\frac{\delta S'}{\delta J_\mu(x_1)}|0\rangle}{\langle 0|S|0\rangle}=\frac{\langle 0|\frac{1}{i}\frac{\delta}{\delta J_\mu(x_1)}T\left\{e^{i\int\mathcal{L}_I'(\underline{x})d^4\underline{x}}\right\}|0\rangle}{\langle 0|S|0\rangle}$$

$$=\frac{\langle 0|\frac{1}{i}\frac{\delta}{\delta J_\mu(x_1)}T\left\{e^{i\int\{\mathcal{L}_I(\underline{x})+\mathcal{L}_S(\underline{x})\}d^4\underline{x}}\right\}|0\rangle}{\langle 0|S|0\rangle}=\frac{\langle 0|\frac{1}{i}T\left\{\frac{\delta}{\delta J_\mu(x_1)}\left\{e^{i\int\left\{\begin{matrix}\mathcal{L}_I(\underline{x})+J_\kappa(\underline{x})A^\kappa(\underline{x})\\+\bar\sigma(\underline{x})\psi(\underline{x})+\bar\psi(\underline{x})\sigma(\underline{x})\end{matrix}\right\}d^4\underline{x}}\right\}\right\}|0\rangle}{\langle 0|S|0\rangle}$$

$$=\frac{\langle 0|\frac{1}{i}T\left\{\left(\frac{\delta}{\delta J_\mu(x_1)}i\int\left\{\begin{matrix}\mathcal{L}_I(\underline{x})+J_\kappa(\underline{x})A^\kappa(\underline{x})\\+\bar\sigma(\underline{x})\psi(\underline{x})+\bar\psi(\underline{x})\sigma(\underline{x})\end{matrix}\right\}d^4\underline{x}\right)\left\{e^{i\int\left\{\begin{matrix}\mathcal{L}_I(\underline{x})+J_\kappa(\underline{x})A^\kappa(\underline{x})\\+\bar\sigma(\underline{x})\psi(\underline{x})+\bar\psi(\underline{x})\sigma(\underline{x})\end{matrix}\right\}d^4\underline{x}}\right\}\right\}|0\rangle}{\langle 0|S|0\rangle} \quad (3\text{-}98)$$

$$=\frac{\langle 0|T\left\{\left(\int\frac{\delta J_\kappa(\underline{x})}{\delta J_\mu(x_1)}A^\kappa(\underline{x})d^4\underline{x}\right)\left\{e^{i\int\{\mathcal{L}_I(\underline{x})+J_\kappa(\underline{x})A^\kappa(\underline{x})+\bar\sigma(\underline{x})\psi(\underline{x})+\bar\psi(\underline{x})\sigma(\underline{x})\}d^4\underline{x}}\right\}\right\}|0\rangle}{\langle 0|S|0\rangle}$$

$$=\frac{\langle 0|T\left\{\left(\int\delta_\kappa^\mu\delta^{(4)}(x_1-\underline{x})A^\kappa(\underline{x})d^4\underline{x}\right)\left\{e^{i\int\{\mathcal{L}_I(\underline{x})+J_\kappa(\underline{x})A^\kappa(\underline{x})+\bar\sigma(\underline{x})\psi(\underline{x})+\bar\psi(\underline{x})\sigma(\underline{x})\}d^4\underline{x}}\right\}\right\}|0\rangle}{\langle 0|S|0\rangle}$$

$$=\frac{\langle 0|T\left\{A^\mu(x_1)\left\{e^{i\int\{\mathcal{L}_I(\underline{x})+J_\kappa(\underline{x})A^\kappa(\underline{x})+\bar\sigma(\underline{x})\psi(\underline{x})+\bar\psi(\underline{x})\sigma(\underline{x})\}d^4\underline{x}}\right\}\right\}|0\rangle}{\langle 0|S|0\rangle}=\frac{\langle 0|T\left\{A^\mu(x_1)S'\right\}|0\rangle}{\langle 0|S|0\rangle}.$$

Now suppose in the last line of (3-98), we take the source fields equal to zero, i.e., $J_\kappa=\bar\sigma=\sigma=0$. Then we get

Setting source fields to zero in the result

$$\left.\frac{\langle 0|T\left\{A^\mu(x_1)e^{i\int\{\mathcal{L}_I(\underline{x})+J_\kappa(\underline{x})A^\kappa(\underline{x})+\bar\sigma(\underline{x})\psi(\underline{x})+\bar\psi(\underline{x})\sigma(\underline{x})\}d^4\underline{x}}\right\}|0\rangle}{\langle 0|S|0\rangle}\right|_{\substack{J_\kappa=\bar\sigma_\alpha\\=\sigma_\alpha=0}}$$
$$=\frac{\langle 0|T\left\{A^\mu(x_1)e^{i\int\mathcal{L}_I(\underline{x})d^4\underline{x}}\right\}|0\rangle}{\langle 0|S|0\rangle}=\frac{\langle 0|T\left\{A^\mu(x_1)S\right\}|0\rangle}{\langle 0|S|0\rangle}, \quad (3\text{-}99)$$

which, from (3-13), is just a one-point Green function with a single photon leg. By taking the functional derivative of the generating functional with respect to J_μ, and setting the classical fields in the result equal to zero, we get a Green function for a single external photon.

This, as we know from Section 3.1.8 (pg. 80), will vanish and so, it may not have much value, other than illustration of the functional derivative methodology. However, note what happens when we start with the last line of (3-98) and take another derivative like we did in the first line.

$$\frac{1}{i}\frac{\delta}{\delta J_\nu(x_2)}\frac{\langle 0|T\{A^\mu(x_1)S'\}|0\rangle}{\langle 0|S|0\rangle}=\frac{\langle 0|\frac{1}{i}T\left\{A^\mu(x_1)\left(\frac{\delta}{\delta J_\nu(x_2)}S'\right)\right\}|0\rangle}{\langle 0|S|0\rangle}$$

$$=\frac{\langle 0|\frac{1}{i}T\left\{A^\mu(x_1)\left(\frac{\delta}{\delta J_\nu(x_2)}\left\{e^{i\int\left\{\mathcal{L}_I(\underline{x})+J_\kappa(\underline{x})A^\kappa(\underline{x})+\bar{\sigma}(\underline{x})\psi(\underline{x})+\bar{\psi}(\underline{x})\sigma(\underline{x})\right\}d^4\underline{x}}\right\}\right)\right\}|0\rangle}{\langle 0|S|0\rangle}$$

$$=\frac{\langle 0|\frac{1}{i}T\left\{A^\mu(x_1)\left(i\int\left\{\frac{\delta J_\kappa(\underline{x})}{\delta J_\nu(x_2)}\right\}A^\kappa(\underline{x})d^4\underline{x}\right)\left\{e^{i\int\left\{\mathcal{L}_I(\underline{x})+J_\kappa(\underline{x})A^\kappa(\underline{x})+\bar{\sigma}(\underline{x})\psi(\underline{x})+\bar{\psi}(\underline{x})\sigma(\underline{x})\right\}d^4\underline{x}}\right\}\right\}|0\rangle}{\langle 0|S|0\rangle} \qquad \text{(3-100)}$$

$$=\frac{\langle 0|T\left\{A^\mu(x_1)\left(\int\delta^\nu_\kappa\delta^{(4)}(x_2-\underline{x})A^\kappa(\underline{x})d^4\underline{x}\right)\left\{e^{i\int\left\{\mathcal{L}_I(\underline{x})+J_\kappa(\underline{x})A^\kappa(\underline{x})+\bar{\sigma}(\underline{x})\psi(\underline{x})+\bar{\psi}(\underline{x})\sigma(\underline{x})\right\}d^4\underline{x}}\right\}\right\}|0\rangle}{\langle 0|S|0\rangle}$$

$$=\frac{\langle 0|T\left\{A^\mu(x_1)A^\nu(x_2)\left\{e^{i\int\left\{\mathcal{L}_I(\underline{x})+J_\kappa(\underline{x})A^\kappa(\underline{x})+\bar{\sigma}(\underline{x})\psi(\underline{x})+\bar{\psi}(\underline{x})\sigma(\underline{x})\right\}d^4\underline{x}}\right\}\right\}|0\rangle}{\langle 0|S|0\rangle}=\frac{\langle 0|T\{A^\mu(x_1)A^\nu(x_2)S'\}|0\rangle}{\langle 0|S|0\rangle}.$$

Taking another functional derivative with respect to photon source field

Now take the source fields in the last line of (3-100) equal to zero, and we get

$$\frac{\langle 0|T\left\{A^\mu(x_1)A^\nu(x_2)e^{i\int\mathcal{L}_I(\underline{x})d^4\underline{x}}\right\}|0\rangle}{\langle 0|S|0\rangle}=\frac{\langle 0|T\{A^\mu(x_1)A^\nu(x_2)S\}|0\rangle}{\langle 0|S|0\rangle}, \qquad \text{(3-101)}$$

the two-point Green function with two photon legs (3-50), which we showed did not vanish, and in fact, equals the photon propagator.

Taking source fields to zero yields a two-point Green function – generated from Z, the generating functional

---

Do **Problem 10** to deduce (3-102) and (3-103).

From Prob. 10, where the subscript "0" means all source fields equal zero, we have

$$\left(\frac{1}{i}\frac{\delta}{\delta\sigma(z_1)}\frac{\langle 0|T\{A^\mu(x_1)A^\nu(x_2)S'\}|0\rangle}{\langle 0|S|0\rangle}\right)_0=\frac{\langle 0|T\{A^\mu(x_1)A^\nu(x_2)\bar{\psi}(z_1)S\}|0\rangle}{\langle 0|S|0\rangle}, \qquad \text{(3-102)}$$

which vanishes via the logic of Section 3.1.8 (pg. 80). But taking another derivative yields

$$\left(\frac{1}{i}\frac{\delta}{\delta\bar{\sigma}(y_1)}\frac{\langle 0|T\{A^\mu(x_1)A^\nu(x_2)\bar{\psi}(z_1)S'\}|0\rangle}{\langle 0|S|0\rangle}\right)_0$$

$$=\frac{\langle 0|T\{A^\mu(x_1)A^\nu(x_2)\bar{\psi}(z_1)\psi(y_1)S\}|0\rangle}{\langle 0|S|0\rangle}=(-1)\frac{\langle 0|T\{A^\mu(x_1)A^\nu(x_2)\psi(y_1)\bar{\psi}(z_1)S\}|0\rangle}{\langle 0|S|0\rangle}, \qquad \text{(3-103)}$$

Two more derivatives of Z with respect to Grassmann fields yields a four-point Green function

where (3-103) is the (non-zero and quite useful) four-point Green function (3-53). Note the minus sign arising when we took the derivative with respect to $\bar{\sigma}$, since we had to switch positions of the ψ operator field and the $\bar{\psi}$ operator field (which anti-commute) to obtain the standard form (3-103) of the Green function. Each time we take a derivative with respect to $\bar{\sigma}$, we pick up another such minus sign.

We conclude that, in general, we can obtain any Green function we like by taking functional derivatives of the generating functional with respect to the appropriate (fictitious) source fields associated with the type of external legs in that particular Green function. For two external photons, we take two derivatives, one with respect to J_μ and one with respect to J_ν. For two external photons

Can get any Green function we like via derivatives of Z, one for each leg

and two external fermions, we take those same two derivatives plus one with respect to σ, and one with respect to $\bar{\sigma}$. Details and examples follow in the next sections.

3.3.4 Relation Between Green Function and Generating Functional

The relationship (3-104) below generalizes the special case of Section 3.3.3 and shows specifically how to obtain any Green function one wishes from the generating functional. Note that any Green function where the number of ψ factors (from $\bar{\sigma}$ derivatives) does not equal the number of $\bar{\psi}$ factors (from σ derivatives) vanishes.

Formula for deriving any Green function from Z

$$G^{\mu\ldots}\left(x_1,\ldots y_1,\ldots z_1,\ldots\right)=\left(-1\right)^{\bar{n}}\left(\frac{1}{i}\right)^{n}\left.\frac{\delta^n Z\left[J_k,\sigma,\bar{\sigma}\right]}{\delta J_\mu\left(x_1\right)\ldots\,\delta\bar{\sigma}\left(y_1\right)\ldots\,\delta\sigma\left(z_1\right)}\right|_0 \tag{3-104}$$

n = number of total derivatives, no sum on n; $\bar{n}$ = derivatives with respect to $\bar{\sigma}$ fields (also = derivatives with respect to σ fields)

Take care that we have to take the fermionic functional derivatives in the order shown in the above definition (all σ derivatives before all $\bar{\sigma}$ derivatives) or the minus signs won't always work out correctly. This is because the derivatives of paired fields anti-commute (see (3-88)), so switching their order switches signs.

3.3.5 Another Example: Four External Fermions

We can apply (3-104) using four derivatives with respect to fermion source fields to get the four fermion leg Green function of (3-22). Given what we have learned about how functional derivatives act on the generating functional, we can pretty much just write down the results (where the change in the last line is explained in the comment after (3-53)).

Finding four-point Green function that can give us Bhabha scattering amplitude

$$\begin{aligned}G\left(y_1,y_2,z_1,z_2\right)&=\left(-1\right)^2\left(\frac{1}{i}\right)^4\left.\frac{\delta^4 Z\left[J_k,\sigma,\bar{\sigma}\right]}{\delta\bar{\sigma}\left(y_1\right)\delta\bar{\sigma}\left(y_2\right)\delta\sigma\left(z_1\right)\delta\sigma\left(z_2\right)}\right|_0\\
&=\left.\frac{\left\langle 0\right|T\left\{\dfrac{\delta^4 S'}{\delta\bar{\sigma}\left(y_1\right)\delta\bar{\sigma}\left(y_2\right)\delta\sigma\left(z_1\right)\delta\sigma\left(z_2\right)}\right\}\left|0\right\rangle}{\left\langle 0\right|S\left|0\right\rangle}\right|_0\\
&=\left.\frac{\left\langle 0\right|T\left\{\dfrac{\delta^4}{\delta\bar{\sigma}\left(y_1\right)\delta\bar{\sigma}\left(y_2\right)\delta\sigma\left(z_1\right)\delta\sigma\left(z_2\right)}e^{i\int\left\{\begin{smallmatrix}\mathcal{L}_I\left(\underline{x}\right)+J_\kappa\left(\underline{x}\right)A^\kappa\left(\underline{x}\right)\\+\bar{\sigma}_\alpha\left(\underline{x}\right)\psi_\alpha\left(\underline{x}\right)+\bar{\psi}_\alpha\left(\underline{x}\right)\sigma_\alpha\left(\underline{x}\right)\end{smallmatrix}\right\}d^4\underline{x}}\right\}\left|0\right\rangle}{\left\langle 0\right|S\left|0\right\rangle}\right|_0\\
&=\frac{\left\langle 0\right|T\left\{\psi\left(y_1\right)\psi\left(y_2\right)\bar{\psi}\left(z_1\right)\bar{\psi}\left(z_2\right)S\right\}\left|0\right\rangle}{\left\langle 0\right|S\left|0\right\rangle}=\frac{\left\langle 0\right|T\left\{S\psi\left(y_1\right)\psi\left(y_2\right)\bar{\psi}\left(z_1\right)\bar{\psi}\left(z_2\right)\right\}\left|0\right\rangle}{\left\langle 0\right|S\left|0\right\rangle}.\end{aligned} \tag{3-105}$$

Do **Problem 11** to gain more practice in rapidly going from the generating functional to a desired Green function.

3.3.6 A Different Form for the Generating Functional

We will later find Z for PI approach and show it equals Z for canonical approach

In Part 2 of this book we will study the form of the generating functional Z used in the path integral approach, as opposed to the canonical approach form (3-93) (with (3-94)) of this chapter. We will want to show the two are equivalent. To do this, we will need to recast the canonical generating functional into a form more suitable for comparison with the PI generating functional.

To do this we need another form for Z

First example: Free photon case

The first step in finding the new form for Z is to find the special case generating functional of the free photon, which consists of a single photon propagating, but not interacting with anything. So there are no internal fermion loops (no interactions with fermions at all) at higher order such as we saw in Fig. 3-9, pg. 79, and (3-51). A Green function for this case with two photon legs would lead to a free photon propagator (without internal loops), where critically, the Green function would have no interaction terms. In other words, we take the interaction Lagrangian $\mathcal{L}_I = 0$ in both the Green function and the generating functional that generates it.

Wholeness Chart 3-5 provides an overview of what we seek to accomplish, where we use the symbol $\underline{Z_0}$ as the free field generating functional with the subscript "0" signifying free field, no interactions. We will be going through the steps in the LHS of the chart in the following pages. The RHS will be covered in Part 2 of this book. You may wish to refer back to this chart as we proceed though those steps here and later on. Don't worry if you don't fully understand the chart now. You will (hopefully, if you work hard) as we proceed.

Free generating functional signified by Z_0

Wholeness Chart 3-5. A Different Form for *Z* for Comparing Later On with Path Integral *Z*

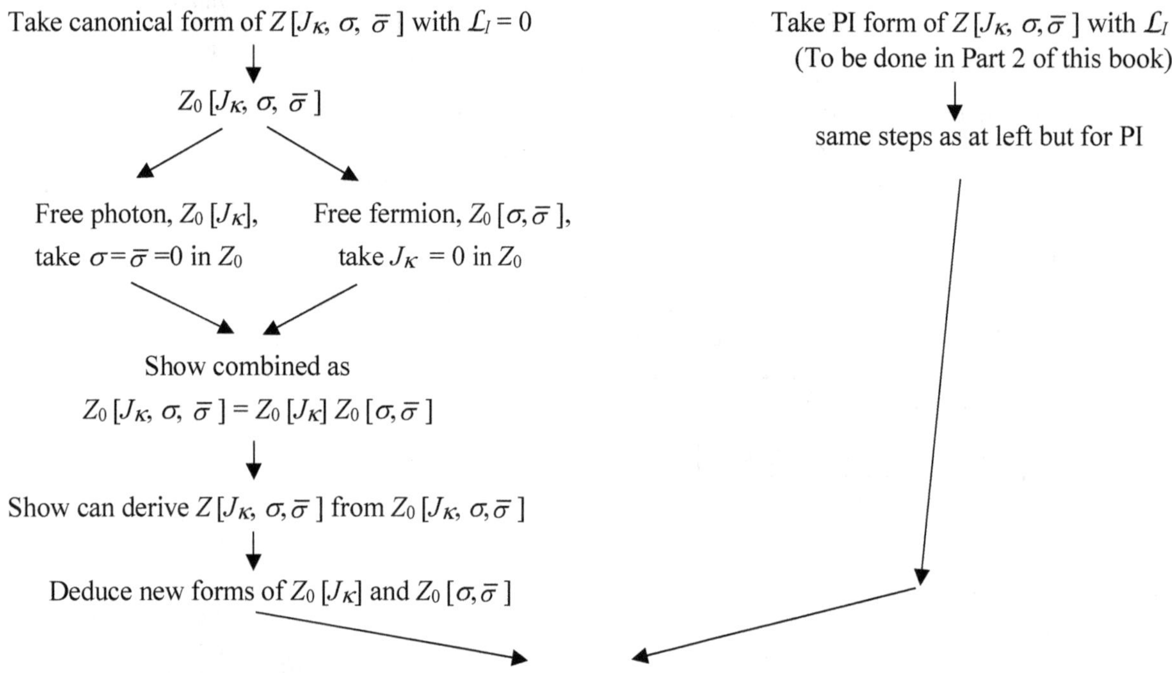
Take canonical form of $Z[J_K, \sigma, \bar{\sigma}]$ with $\mathcal{L}_I = 0$
$Z_0[J_K, \sigma, \bar{\sigma}]$
Free photon, $Z_0[J_K]$, take $\sigma = \bar{\sigma} = 0$ in Z_0
Free fermion, $Z_0[\sigma, \bar{\sigma}]$, take $J_K = 0$ in Z_0
Show combined as $Z_0[J_K, \sigma, \bar{\sigma}] = Z_0[J_K]\, Z_0[\sigma, \bar{\sigma}]$
Show can derive $Z[J_K, \sigma, \bar{\sigma}]$ from $Z_0[J_K, \sigma, \bar{\sigma}]$
Deduce new forms of $Z_0[J_K]$ and $Z_0[\sigma, \bar{\sigma}]$
Take PI form of $Z[J_K, \sigma, \bar{\sigma}]$ with $\mathcal{L}_I = 0$
(To be done in Part 2 of this book)
same steps as at left but for PI
Compare to see new forms of $Z_0[J_K]$ and $Z_0[\sigma, \bar{\sigma}]$ from PI same as those from canonical,
so, Z_0, and thus Z, the same in both, and the two approaches are equivalent.

Free photon propagator Z_0 from general Z definition

For the free field case, from (3-14), $S \rightarrow S_0 = 1$, and the general definition for generating functional (3-93) becomes

$$Z_0\left[J_k, \sigma, \bar{\sigma}\right] = \frac{\langle 0|S_0'|0\rangle}{\langle 0|S_0|0\rangle} = \frac{\langle 0|S_0'|0\rangle}{\langle 0|1|0\rangle} = \langle 0|S_0'|0\rangle \tag{3-106}$$

Free source S' operator with $\mathcal{L}_I = 0$ signified by S'_0

In (3-94), the interaction Lagrangian $\mathcal{L}_I = 0$, so $\mathcal{L}_I' = \mathcal{L}_S$. Thus,

$$S_0' = T\left\{e^{i\int \mathcal{L}_S(\underline{x})d^4\underline{x}}\right\} = \sum_{n=0}^{\infty} \frac{i^n}{n!}\int T\left\{\mathcal{L}_S(\underline{x_1})\mathcal{L}_S(\underline{x_2})\ldots\mathcal{L}_S(\underline{x_n})\right\} d^4\underline{x_1} d^4\underline{x_2}\ldots d^4\underline{x_n}\,. \tag{3-107}$$

Substituting the value for $\mathcal{L}_S$ in (3-94) into (3-107), we have

$$S_0' = \sum_{n=0}^{\infty}\frac{i^n}{n!}\int T\left\{\begin{matrix}\left(J_\mu(\underline{x_1})A^\mu(\underline{x_1}) + \bar{\sigma}(\underline{x_1})\psi(\underline{x_1}) + \bar{\psi}(\underline{x_1})\sigma(\underline{x_1})\right)\\ \times\left(J_\nu(\underline{x_2})A^\nu(\underline{x_2}) + \bar{\sigma}(\underline{x_2})\psi(\underline{x_2}) + \bar{\psi}(\underline{x_2})\sigma(\underline{x_2})\right)\times\ldots\\ \times\left(J_\rho(\underline{x_n})A^\rho(\underline{x_n}) + \bar{\sigma}(\underline{x_n})\psi(\underline{x_n}) + \bar{\psi}(\underline{x_n})\sigma(\underline{x_n})\right)\end{matrix}\right\} d^4\underline{x_1}\ldots d^4\underline{x_n} \tag{3-108}$$

S'_0 expressed as an expansion

Because we seek a relationship for the free photon field, we now ignore the fermion fields, i.e., take $\sigma = \bar{\sigma} = 0$. Later, we will generalize our result to include both fermions and photons. We also simplify by only looking at the first few terms in (3-108). Doing that, we get

$$S'_0\Big|_{\sigma=\bar\sigma=0} = \sum_{n=0}^{\infty}\frac{i^n}{n!}\int T\left\{J_\mu(\underline{x_1})A^\mu(\underline{x_1})J_\nu(\underline{x_2})A^\nu(\underline{x_2})J_\rho(\underline{x_3})A^\rho(\underline{x_3})\ldots\right\}d^4\underline{x_1}d^4\underline{x_2}\ldots d^4\underline{x_n}$$
$$=1+i\int T\left\{J_\mu(\underline{x_1})A^\mu(\underline{x_1})\right\}d\underline{x_1}-\frac{1}{2}\int T\left\{J_\mu(\underline{x_1})A^\mu(\underline{x_1})J_\nu(\underline{x_2})A^\nu(\underline{x_2})\right\}d^4\underline{x_1}d^4\underline{x_2}+\ldots \tag{3-109}$$

S'_0 with Grassmann source fields $\sigma = 0$ is for free photon field

Restricting our effective S'_0 to this for use in (3-106), we have (recall we need an even number of photon fields in the second line of (3-109) to give rise to a contraction in (3-106))

$$\text{effective terms in } S'_0\Big|_{\sigma=\bar\sigma=0} = 1-\frac{1}{2}\int J_\mu(\underline{x_1})\underbrace{A^\mu(\underline{x_1})A^\nu(\underline{x_2})}J_\nu(\underline{x_2})d^4\underline{x_1}d^4\underline{x_2}+\ldots$$
$$=1-\frac{1}{2}\int J_\mu(\underline{x_1})iD_F^{\mu\nu}(\underline{x_1}-\underline{x_2})J_\nu(\underline{x_2})d^4\underline{x_1}d^4\underline{x_2}+\ldots \tag{3-110}$$

Thus, from (3-106) (where Z_0 is now only a function of J_K without fermion sources),

Series expansion of generating functional $Z_0[J_K]$ for free photon field

$$Z_0\left[J_k,\sigma,\bar\sigma\right]_{\sigma=\bar\sigma=0} \equiv Z_0\left[J_k\right]=\langle 0|S'_0|0\rangle=\langle 0|0\rangle S'_{0\,\sigma=\bar\sigma=0\ \text{effective}}=S'_{0\,\sigma=\bar\sigma=0\ \text{effective}}$$
$$=1-\frac{i}{2}\int J_\mu(\underline{x_1})D_F^{\mu\nu}(\underline{x_1}-\underline{x_2})J_\nu(\underline{x_2})d^4\underline{x_1}d^4\underline{x_2}+\begin{pmatrix}\text{higher order terms with}\\ \text{more than two factors of } J\end{pmatrix}. \tag{3-111}$$

Compare the last part of this to

$$e^X=1+X+\frac{X^2}{2}+\frac{X^3}{3!}+\ldots, \tag{3-112}$$

and we can surmise (with *H.O.T.* meaning "higher order terms")_

$Z_0[J_K]$ converted to exponential form

$$e^{-\frac{i}{2}\int J_\mu(\underline{x_1})D_F^{\mu\nu}(\underline{x_1}-\underline{x_2})J_\nu(\underline{x_2})d^4\underline{x_1}d^4\underline{x_2}}$$
$$=1-\frac{i}{2}\int J_\mu(\underline{x_1})D_F^{\mu\nu}(\underline{x_1}-\underline{x_2})J_\nu(\underline{x_2})d^4\underline{x_1}d^4\underline{x_2}+(H.O.T.). \tag{3-113}$$

With the definition (3-114) below, which is the same as the LHS of (3-89) applied here to boson source fields and the photon propagator,

$\left[J_\sigma D_F^{\sigma\rho}J_\rho\right]$ symbol defined

$$\text{symbol}\left[J_\sigma D_F^{\sigma\rho}J_\rho\right]\equiv\int J_\sigma(\underline{x_1})D_F^{\sigma\rho}(\underline{x_1}-\underline{x_2})J_\rho(\underline{x_2})d^4\underline{x_1}d^4\underline{x_2}, \tag{3-114}$$

we see that, at least to lowest order, for only photons and no fermions, (3-111) is

$Z_0[J_K]$ expressed using new symbol

$$Z_0\left[J_k,\sigma,\bar\sigma\right]_{\sigma=\bar\sigma=0}\equiv Z_0\left[J_k\right]=e^{-\frac{i}{2}\int J_\sigma(\underline{x_1})D_F^{\sigma\rho}(\underline{x_1}-\underline{x_2})J_\rho(\underline{x_2})d^4\underline{x_1}d^4\underline{x_2}}=e^{-\frac{i}{2}\left[J_\sigma D_F^{\sigma\rho}J_\rho\right]}. \tag{3-115}$$

Hopefully, we can accept that if we carried out the above steps and those of the appendix for the higher order terms, the relation (3-115) still holds. It does.

Showing Z_0 Yields Green Function for Free Photon Propagator

From the relation (3-104), where the Green function we seek has no fermions ($\bar n = 0$) and two photon legs ($n = 2$),

Using $Z_0[J_K]$ to get related Green function

$$G_0^{\mu\nu}(x_1,x_2)=\left(\frac{1}{i}\right)^2\frac{\delta^2 Z_0\left[J_k\right]}{\delta J_\mu(x_1)\delta J_\nu(x_2)}\Bigg|_0. \tag{3-116}$$

Inserting (3-115) into (3-116), we find, to lowest order,

$$\left(\frac{1}{i}\right)^2 \frac{\delta^2 Z_0[J_k]}{\delta J_\mu(x_1)\delta J_\nu(x_2)}\bigg|_0 = -\frac{\delta^2 e^{-\frac{i}{2}\left[J_\sigma D_F^{\sigma\rho} J_\rho\right]}}{\delta J_\mu(x_1)\delta J_\nu(x_2)}\bigg|_0$$

$$= \underbrace{-\frac{\delta^2(1)}{\delta J_\mu(x_1)\delta J_\nu(x_2)}\bigg|_0}_{=0} + \underbrace{\left(\frac{\delta^2}{\delta J_\mu(x_1)\delta J_\nu(x_2)}\frac{i}{2}\overbrace{\int J_\sigma(\underline{x_1})D_F^{\sigma\rho}(\underline{x_1}-\underline{x_2})J_\rho(\underline{x_2})d^4\underline{x_1}d^4\underline{x_2}}^{\left[J_\sigma D_F^{\sigma\rho} J_\rho\right]}\right)_0}_{\text{Term X}} + \ldots \tag{3-117}$$

We see term X in (3-117) is

$$\text{Term X} = \left(\frac{\delta}{\delta J_\nu(x_2)}\left(\frac{\delta}{\delta J_\mu(x_1)}\frac{i}{2}\int J_\sigma(\underline{x_1})D_F^{\sigma\rho}(\underline{x_1}-\underline{x_2})J_\rho(\underline{x_2})d^4\underline{x_1}d^4\underline{x_2}\right)\right)_0$$

$$= \frac{i}{2}\left(\frac{\delta}{\delta J_\nu(x_2)}\left(\begin{array}{l}\int \frac{\delta J_\sigma(\underline{x_1})}{\delta J_\mu(x_1)}D_F^{\sigma\rho}(\underline{x_1}-\underline{x_2})J_\rho(\underline{x_2})d^4\underline{x_1}d^4\underline{x_2} \\ \qquad + \int J_\sigma(\underline{x_1})D_F^{\sigma\rho}(\underline{x_1}-\underline{x_2})\frac{\delta J_\rho(\underline{x_2})}{\delta J_\mu(x_1)}d^4\underline{x_1}d^4\underline{x_2}\end{array}\right)\right)_0$$

$$= \frac{i}{2}\left(\frac{\delta}{\delta J_\nu(x_2)}\left(\begin{array}{l}\int \delta_\sigma^\mu\delta(x_1-\underline{x_1})D_F^{\sigma\rho}(\underline{x_1}-\underline{x_2})J_\rho(\underline{x_2})d^4\underline{x_1}d^4\underline{x_2} \\ \qquad + \int J_\sigma(\underline{x_1})D_F^{\sigma\rho}(\underline{x_1}-\underline{x_2})\delta_\rho^\mu\delta(x_1-\underline{x_2})d^4\underline{x_1}d^4\underline{x_2}\end{array}\right)\right)_0 \tag{3-118}$$

$$= \frac{i}{2}\left(\frac{\delta}{\delta J_\nu(x_2)}\left(\begin{array}{l}\int D_F^{\mu\rho}(x_1-\underline{x_2})J_\rho(\underline{x_2})d^4\underline{x_2} \\ \qquad + \int J_\sigma(\underline{x_1})D_F^{\sigma\mu}(\underline{x_1}-x_1)d^4\underline{x_1}\end{array}\right)\right)_{J=0} \quad \left(\begin{array}{l}\text{get next row via sym in} \\ D_F^{\sigma\mu} \text{ indices \& switch} \\ \text{dummy index } \rho\to\sigma,\end{array}\right)$$

$$= \frac{i}{2}\left(\frac{\delta}{\delta J_\nu(x_2)}\left(\begin{array}{l}\int D_F^{\mu\sigma}(x_1-\underline{x_2})J_\sigma(\underline{x_2})d^4\underline{x_2} \\ \qquad + \int D_F^{\mu\sigma}(x_1-\underline{x_1})J_\sigma(\underline{x_1})d^4\underline{x_1}\end{array}\right)\right)_{J=0} \quad \left(\begin{array}{l}\text{next row via switching} \\ \text{dummy integration} \\ \text{variable } \underline{x_1}\to\underline{x_2}\end{array}\right)$$

$$= i\left(\frac{\delta}{\delta J_\nu(x_2)}\left(\int D_F^{\mu\sigma}(x_1-\underline{x_2})J_\sigma(\underline{x_2})d^4\underline{x_2}\right)\right)_{J=0}.$$

So,

$$\text{Term X} = i\left(\int D_F^{\mu\sigma}(x_1-\underline{x_2})\frac{\delta J_\sigma(\underline{x_2})}{\delta J_\nu(x_2)}d^4\underline{x_2}\right)_{J=0}$$
$$= i\left(\int D_F^{\mu\sigma}(x_1-\underline{x_2})\delta_\sigma^\nu\delta(x_2-\underline{x_2})d^4\underline{x_2}\right)_{J=0} = iD_F^{\mu\nu}(x_1-x_2). \tag{3-119}$$

Thus, from (3-116) and (3-117),

$$G_0^{\mu\nu}(x_1,x_2) = G_0^{\mu\nu}(x_1-x_2) = \left(\frac{1}{i}\right)^2 \frac{\delta^2 Z_0[J_k]}{\delta J_\mu(x_1)\delta J_\nu(x_2)}\bigg|_0 = iD_F^{\mu\nu}(x_1-x_2). \tag{3-120}$$

$Z_0[J_K]$ yields free photon propagator Green function

Result (3-120) shows that the usual Green function of mathematics for Maxwell's equation (3-11) equals the free photon Green function from the QFT Green function methodology using the generating functional.

Note there are no higher order terms in (3-120), since from (3-111) and (3-113), terms with three or more factors of J in them will still have at least one factor of J in them after the differentiation. And when we set all such J equal to zero in (3-120), these higher order terms go to zero.

Extrapolating Free Field Photon Derivation to Free Fermions

The above procedure for photons can be carried out in parallel for fermions, using Grassmann source fields, to obtain

Free fermion propagator $Z_0[\sigma,\bar{\sigma}]$ found in similar fashion

$$Z_0[J_k,\sigma,\bar{\sigma}]_{J_\kappa=0} \equiv Z_0[\sigma,\bar{\sigma}] = e^{-i[\bar{\sigma}S_F\sigma]}. \tag{3-121}$$

Do **Problem 12** to show (3-121) and that the Green function obtained from it is the spinor propagator.

Note the exponent in (3-121) does not have the factor of ½ found in the exponent of (3-115).

Generalizing to Include Free Photons and Fermions

Since free fields do not interact with one another, and thus are independent mathematically, we can use

Z_0 for free photons and free fermions combines Z_0 of each separately

$$\boxed{Z_0[J_k,\sigma,\bar{\sigma}] = Z_0[J_k]Z_0[\sigma,\bar{\sigma}] = e^{-\frac{i}{2}[J_\alpha D_F^{\alpha\rho} J_\rho]}e^{-i[\bar{\sigma}S_F\sigma]}} \tag{3-122}$$

in (3-104) to obtain either the free photon or the free fermion Green function. For example, using (3-122) in (3-104) and only two derivatives with respect to photonic source fields, as in (3-123) below, yields (3-120). Similarly, using (3-122) in (3-104) with only two Grassmann field derivatives gives us a Green function equivalent to the free fermion propagator.

$$\begin{aligned} G_0^{\mu\nu}(x_1,x_2) = G_0^{\mu\nu}(x_1-x_2) = (-1)^0\left(\frac{1}{i}\right)^2 \left.\frac{\delta^2 Z_0[J_k,\sigma,\bar{\sigma}]}{\delta J_\mu(x_1)\delta J_\nu(x_2)}\right|_0 = iD_F^{\mu\nu}(x_1-x_2) \\ G_0(y_1,z_1) = G_0(y_1-z_1) = (-1)^1\left(\frac{1}{i}\right)^2 \left.\frac{\delta^2 Z_0[J_\kappa,\sigma,\bar{\sigma}]}{\delta\bar{\sigma}(y_1)\delta\sigma(z_1)}\right|_0 = iS_F(y_1-z_1). \end{aligned} \tag{3-123}$$

Extrapolating from Free to Interacting Fields

We can use Z_0 to find Z for interactions

We now wish to extend Z_0 of (3-122), which is for free fields, to a more general relation Z for interacting fields (of different form from (3-93)).

To this end, first consider the operator, which is known as the interaction operator,

To do so, we'll use the interaction operator $I_\delta(x)$

$$I_\delta(x) = \left(-\frac{1}{i}\frac{\delta}{\delta\sigma(x)}\right)\gamma_\mu\left(\frac{1}{i}\frac{\delta}{\delta\bar{\sigma}(x)}\right)\left(\frac{1}{i}\frac{\delta}{\delta J_\mu(x)}\right). \tag{3-124}$$

It turns out, as we will prove, that exponentiating the integral of (3-124) (along with a factor of *ie*) and having it operate on (3-122) gives us Z, the generating functional including interactions. Specifically, where, as discussed earlier, we can ignore $\langle 0|S|0\rangle$ in the denominator for connected interactions,

Relation to generate Z from $I_\delta(x)$ and Z_0 (proof to follow)

$$\boxed{\begin{aligned} Z[J_k,\sigma,\bar{\sigma}] &= \frac{e^{ie\int I_\delta(x)d^4x}Z_0[J_k,\sigma,\bar{\sigma}]}{\langle 0|S|0\rangle} \qquad \left(\begin{array}{c}\text{can ignore } \langle 0|S|0\rangle \text{ for} \\ \text{connected interactions}\end{array}\right) \\ &= \frac{e^{ie\int\left(-\frac{1}{i}\frac{\delta}{\delta\sigma(x)}\right)\gamma_\mu\left(\frac{1}{i}\frac{\delta}{\delta\bar{\sigma}(x)}\right)\left(\frac{1}{i}\frac{\delta}{\delta J_\mu(x)}\right)d^4x}\, e^{-\frac{i}{2}[J_\alpha D_F^{\alpha\rho}J_\rho]}e^{-i[\bar{\sigma}S_F\sigma]}}{\langle 0|S|0\rangle}. \end{aligned}} \tag{3-125}$$

Proof of (3-125)

Recall that

$$S' = T\left\{e^{i\int(\mathcal{L}_I(\underline{x})+\mathcal{L}_S(\underline{x}))d^4\underline{x}}\right\} \qquad \text{(repeat of (3-94))}$$

$$S'_0 = T\left\{e^{i\int\mathcal{L}_S(\underline{x})d^4\underline{x}}\right\} = T\left\{e^{i\int\{J_\kappa(\underline{x})A^\kappa(\underline{x})+\bar{\sigma}(\underline{x})\psi(\underline{x})+\bar{\psi}(\underline{x})\sigma(\underline{x})\}d^4\underline{x}}\right\} \qquad \text{(repeat of (3-107)).}$$

With the above, we have the following.

Step 1: Calculate $I_\delta(\underline{x})$ acting on S'_0.

$$
\begin{aligned}
&I_\delta(x)S'_0 \\
&=\left(-\frac{1}{i}\frac{\delta}{\delta\sigma(x)}\right)\gamma_\mu\left(\frac{1}{i}\frac{\delta}{\delta\bar{\sigma}(x)}\right)\left(\frac{1}{i}\frac{\delta}{\delta J_\mu(x)}\right)T\left\{e^{i\int\left(J_\kappa(\underline{x})A^\kappa(\underline{x})+\bar{\sigma}(\underline{x})\psi(\underline{x})+\bar{\psi}(\underline{x})\sigma(\underline{x})\right)d^4\underline{x}}\right\} \\
&=\left(-\frac{1}{i}\frac{\delta}{\delta\sigma(x)}\right)\gamma_\mu\left(\frac{1}{i}\frac{\delta}{\delta\bar{\sigma}(x)}\right)T\left\{A^\mu(x)e^{i\int J_\kappa(\underline{x})A^\kappa(\underline{x})+\bar{\sigma}(\underline{x})\psi(\underline{x})+\bar{\psi}(\underline{x})\sigma(\underline{x})d^4\underline{x}}\right\} \\
&=\left(-\frac{1}{i}\frac{\delta}{\delta\sigma(x)}\right)\gamma_\mu T\left\{\psi(x)A^\mu(x)e^{i\int J_\kappa(\underline{x})A^\kappa(\underline{x})+\bar{\sigma}(\underline{x})\psi(\underline{x})+\bar{\psi}(\underline{x})\sigma(\underline{x})d^4\underline{x}}\right\} \qquad (3\text{-}126) \quad \textit{Step 1} \\
&=T\{\underbrace{\bar{\psi}(x)\gamma_\mu\psi(x)A^\mu(x)}_{\mathcal{L}_I(x)/e}\ \underbrace{e^{i\int J_\kappa(\underline{x})A^\kappa(\underline{x})+\bar{\sigma}(\underline{x})\psi(\underline{x})+\bar{\psi}(\underline{x})\sigma(\underline{x})d^4\underline{x}}}_{S'_0}\ \}
\end{aligned}
$$

$$
\text{or} \quad eI_\delta(x)S'_0=T\{\mathcal{L}_I(x)S'_0\}.
$$

Step 2: Use Step 1 to calculate $e^{ie\int I_\delta(x)d^4x}S'_0$.

$$
\begin{aligned}
&e^{ie\int I_\delta(x)d^4x}S'_0=\left(I+ie\int I_\delta(x_1)d^4x_1-\frac{e^2}{2}\int I_\delta(x_1)d^4x_1\int I_\delta(x_2)d^4x_2+\ldots\right)S'_0 \qquad \textit{Step 2} \\
&=T\left\{\left(I+i\int\mathcal{L}_I(x_1)d^4x_1-\tfrac{1}{2}\int\mathcal{L}_I(x_1)d^4x_1\int\mathcal{L}_I(x_2)d^4x_2+\ldots\right)S'_0\right\} \qquad (3\text{-}127) \\
&=T\left\{e^{i\int\mathcal{L}_I(x)d^4x}\underbrace{e^{i\int\{J_\kappa(\underline{x})A^\kappa(\underline{x})+\bar{\sigma}(\underline{x})\psi(\underline{x})+\bar{\psi}(\underline{x})\sigma(\underline{x})\}d^4\underline{x}}}_{S'_0}\right\}=T\left\{e^{i\int\mathcal{L}_I(x)d^4x}e^{i\int\mathcal{L}_S d^4\underline{x}}\right\}=S'.
\end{aligned}
$$

Step 3: Take VEV of Step 2 to find (3-125)

Take the vacuum expectation value of the first and last parts of (3-127) to get the generating functionals (3-93) and (3-106) *Step 3*

$$
\begin{aligned}
&\langle 0|S'|0\rangle=e^{ie\int I_\delta(x)d^4x}\langle 0|S'_0|0\rangle \\
\text{or}\quad &\langle 0|S|0\rangle Z[J_k,\sigma,\bar{\sigma}]=e^{ie\int I_\delta(x)d^4x}Z_0[J_k,\sigma,\bar{\sigma}].
\end{aligned} \qquad (3\text{-}128)
$$

So, along with (3-122), we get

$$
\begin{aligned}
Z[J_k,\sigma,\bar{\sigma}]&=\frac{e^{ie\int\left(-\frac{1}{i}\frac{\delta}{\delta\sigma(x)}\right)\gamma_\mu\left(\frac{1}{i}\frac{\delta}{\delta\bar{\sigma}(x)}\right)\left(\frac{1}{i}\frac{\delta}{\delta J_\mu(x)}\right)d^4x}\,e^{-\frac{i}{2}\left[J_\alpha D_F^{\alpha\rho}J_\rho\right]_e-i\left[\bar{\sigma}S_F\sigma\right]_e}}{\langle 0|S|0\rangle} \qquad \text{(last line of (3-125)).} \\
&=\underbrace{e^{ie\int\left(-\frac{1}{i}\frac{\delta}{\delta\sigma(x)}\right)\gamma_\mu\left(\frac{1}{i}\frac{\delta}{\delta\bar{\sigma}(x)}\right)\left(\frac{1}{i}\frac{\delta}{\delta J_\mu(x)}\right)d^4x}\,e^{-\frac{i}{2}\left[J_\alpha D_F^{\alpha\rho}J_\rho\right]_e-i\left[\bar{\sigma}S_F\sigma\right]_e}}_{\text{for connected interactions}},
\end{aligned}
$$

Where we note that for connected Green functions, we can ignore the factor $\langle 0|S|0\rangle$, as was discussed earlier in the chapter.

End of Proof

3.4 Odds and Ends

3.4.1 Alternative Nomenclature

Note that Green functions are often called correlation functions. Why this name? We explain in Appendix B.

Also, the numerator of (3-93) is sometimes called the generating functional, and the entire relation, including the denominator, as the *normalized* generating functional. For us, however, (3-93) is simply the generating functional (and it is already normalized in the sense some others use.)

3.4.2 Math Terms Summary and Clarification

A functional derivative is a derivative with respect to a function. (See (3-84), (3-90), (3-95), and (3-98).)

A functional F is a function of a function in the sense that $F[f(x)]$ has a different numeric value for each different function $f(x)$. In our cases, a functional is almost always a definite integral of form $F[f]=\int_{f_a}^{f_b} g(x)f(x)dx$, where $g(x)$ is a fixed function (which could equal 1), but f can take on different dependences on x. For each such different dependence, F has a (generally) different numeric value.

A functional integral (also called a path integral) is an integral of a functional (such as $F[f(x)]$) over all possible functions $f(x)$, for example $\int F\mathcal{D}f(x)$, as explained in Chapter 18 of Vol. 1. We will return to this in Part 2 of this book. For an ordinary integral, the domain over which integration is carried out is a region of space. For a functional integral, the domain is a space of functions.

3.4.3 Green Functions in Other Pictures

We have been working, since Chap. 7 of Vol. 1 (see pgs. 187-194), in the interaction picture. But the literature sometimes expresses Green functions in the Heisenberg picture. We review the differences in Appendix C. Though that is not a prerequisite for the rest of this book, it may help avoid confusion when consulting other sources.

3.4.4 LSZ Reduction Formula

While it will not be a topic of much focus in this book, and is not needed for the remainder of it, you may elsewhere run into the LSZ reduction formula (after its discoverers, H. Lehmann, K. Symanzik, and W. Zimmermann). We briefly review it in Appendix D, for those who are interested.
Bottom line: The LSZ reduction formula is a mathematical formula for converting (reducing) Green functions into transition amplitudes, and that formula can be used instead of our simple substitution method, as illustrated, for example, in (3-39).

3.5 Chapter Summary

Wholeness Chart 3-7 is a pretty self-contained summary of this chapter. Wholeness Charts 3-1 (pg. 65), 3-2, (pg. 81), 3-3 (pg.86), 3-4 (pg. 91) and 3-5 (pg. 96) summarize various underlying components of Wholeness Chart 3-7.

A key result of this chapter, which is a sort of grand summary of Wholeness Chart 3-7, is shown in Wholeness Chart 3-6. From the free field generating functions for the photon and Dirac spinor we can deduce the entire theory of QED. That is, from them, using (3-122), we can obtain the full free field generating functional $Z_0[J_\kappa,\sigma,\bar{\sigma}]$; from that, using (3-125), we obtain the interacting generating functional Z; from Z, using (3-104), we obtain the Green function of our choice $G^{\mu\dots}$; and from $G^{\mu\dots}$, by substituting external particles for leg propagators, we obtain the transition amplitude for the interaction of our choice S_{fi}.

Wholeness Chart 3-6. Deducing the Whole Theory from the Free Generating Functionals

$$Z_0[J_\kappa] \text{ and } Z_0[\sigma,\bar{\sigma}] \rightarrow Z_0 \rightarrow Z \rightarrow G^{\mu\dots} \rightarrow S_{fi}$$

Wholeness Chart 3-7. Summary of QED Green Functions and Generating Functional

	Canonical Approach	Path Integral Approach	Comments
S operator	$S=\sum_{n=0}^{\infty}\frac{i^n}{n!}\int T\left\{\mathcal{L}_I(\underline{x}_1)...\mathcal{L}_I(\underline{x}_n)\right\}d^4\underline{x}_1...d^4\underline{x}_n$ $=T\left\{e^{i\int\mathcal{L}_I(\underline{x})d^4\underline{x}}\right\}$	This column to be filled in later in book	See (3-14)
Transition amplitude	$S_{fi}=\langle f\|S\|i\rangle=\delta_{if}$ $+\prod_{\mathbf{p}''}\sqrt{\frac{m}{VE_{\mathbf{p}''}}}\prod_{\mathbf{k}''}\sqrt{\frac{1}{2V\omega_{\mathbf{k}''}}}(2\pi)^4\delta(p_f-p_i)\mathcal{M}_{fi}$		Vol. 1, (19-5), pg. 514
Green functions	Green function will be all propagators. For each external particle, change associated propagator to external line (e.g., $S_F(p)\to u(\mathbf{p})$) to get amplitude.		All Green function legs time inward. Need final $p\to -p$ to get amplitude
	$G^{\mu...}(x_1,...y_1,...z_1,...)=$ $\frac{\langle 0\|T\left\{SA^{\mu}(x_1)...\psi(y_1)...\bar{\psi}(z_1)...\right\}\|0\rangle}{\langle 0\|S\|0\rangle}$ Ignore denominator above for connected Feynman diagrams		(3-13) Analogous results hold for other field theories (e.g., weak, strong interactions).
Generating functional Z	Z used to find Green functions		
	$Z\left[J_k,\sigma,\bar{\sigma}\right]=\frac{\langle 0\|S'\|0\rangle}{\langle 0\|S\|0\rangle}$ where ↓ $S'=T\left\{e^{i\int\mathcal{L}_I'(\underline{x})d^4\underline{x}}\right\}$ $\mathcal{L}_I'=\mathcal{L}_I+\mathcal{L}_S$ where $\mathcal{L}_S=J_{\kappa}A^{\kappa}+\bar{\sigma}\psi+\bar{\psi}\sigma$		(3-93) $\mathcal{L}_S$ called "source Lagrangian". $J_{\kappa},\bar{\sigma},\sigma$ sources are fictitious and classical fields, not operators. Analogous results for other field theories.
Green function from generating functional	$G^{\mu..}=(-1)^{\bar{n}}\left(\frac{1}{i}\right)^{n}\frac{\delta^n Z[J_k,\sigma,\bar{\sigma}]}{\delta J_{\mu}(x_1)..\delta\bar{\sigma}(y_1)..\delta\sigma(z_1)}\Bigg\|_0$		(3-104) $\|_0$ means $J=\bar{\sigma}=\sigma=0$ after taking derivatives
	Defs: n = tot num source fields = total num derivatives = <u>points</u> of Green function $\bar{n}$ = tot num of σ source fields = tot num derivatives wrt σ <u>order</u> = number of vertices in graph (Green funct has all orders in it due to $\mathcal{L}_I$)		
	Plug and chug Z (for canonical approach) in above to get $G^{\mu...}$ of canonical		Yet to show two approaches equiv
Significance	Knowing $G^{m...}$ for a case with particular fields $A^{\mu},\psi,\bar{\psi}$, one knows the amplitude for that case. Knowing Z, one knows all possible $G^{\mu...}$. Z contains the whole theory.		

We will later prove the equivalence of the canonical and PI approaches by using the blocks below.			
Free field case	Take $\mathcal{L}_I = 0$ in canonical Z above $\rightarrow Z_0$ $Z_0\left[J_k,\sigma,\bar{\sigma}\right] = Z_0\left[J_k\right]Z_0\left[\sigma,\bar{\sigma}\right]$		(3-122) Show by substitution & chugging math.
Interacting fields case	$Z\left[J_k,\sigma,\bar{\sigma}\right] = \dfrac{e^{ie\int I_\delta(x)d^4x} Z_0\left[J_k,\sigma,\bar{\sigma}\right]}{\langle 0 \vert S \vert 0 \rangle}$ $I_\delta(x) = \left(-\dfrac{1}{i}\dfrac{\delta}{\delta\sigma(x)}\right)\gamma_\mu\left(\dfrac{1}{i}\dfrac{\delta}{\delta\bar{\sigma}(x)}\right)\left(\dfrac{1}{i}\dfrac{\delta}{\delta J_\mu(x)}\right)$		(3-125) To show, find $I_\delta(x)S'$, then take VEV.
Equivalent free fields expression	$Z_0\left[J_k\right] = e^{-\frac{i}{2}\left[J_\mu D_F^{\mu\nu} J_\nu\right]}$ $Z_0\left[\sigma,\bar{\sigma}\right] = e^{-i\left[\bar{\sigma} S_F \sigma\right]}$ $\left[AKB\right] = \int A(\underline{x})K(\underline{x},\underline{y})B(\underline{y})d^4\underline{x}d^4\underline{y}$		(3-115) & (3-121) (3-89)
Alternative symbolism for Green function			
	$\left\langle A^\mu(x_1)..\psi(y_1)..\bar{\psi}(z_1)..\right\rangle = G^{\mu.}(x_1,.y_1,.\,z_1,.)$	← Same for PI	(3-68)
	Free fields case: $\left\langle A^\mu(x_1)..\psi(y_1)..\bar{\psi}(z_1)..\right\rangle_0$	← Same for PI	

End of Part 1, Mathematical Preliminaries

This concludes the laying of the mathematical groundwork for our studies of the standard model, both via the path integral approach and the canonical approach.

3.6 Appendices

3.6.1 Appendix A. Rules for Interchanging Fermion Fields and Propagators

If you did Problems 3 and/or 6, you may have felt a little confused regarding sign changes when we switch adjacent fermion fields. That is, when do we switch signs, and when do we not?

Below is a summary. Note that when we have only contractions, there is no normal ordering, as contractions are only numbers.

Wholeness Chart 3-8. Fermion Sign Switching in Transition Amplitudes and Green Functions

	Only Fields	**Field and One End of Contraction**	**Only Contractions**
The Rule	Switch signs	Switch signs	No sign change
Example	$N\{..\psi_{x_1}\psi_{x_2}\} = -N\{..\psi_{x_2}\psi_{x_1}\}$	$N\{..\bar{\psi}_{x_1}\psi_{x_1}\bar{\psi}_{x_2}\psi_{x_2}\} = -N\{..\bar{\psi}_{x_1}\psi_{x_1}\psi_{x_2}\bar{\psi}_{x_2}\}$	$..\bar{\psi}_{x_1}\psi_{x_1}\bar{\psi}_{x_2}\psi_{x_2} = ..\bar{\psi}_{x_1}\psi_{x_1}\psi_{x_2}\bar{\psi}_{x_2}$

3.6.2 Appendix B. Correlation Functions

Consider the function f of **x** and **y**, where ϕ and ϕ' are two different functions,

$$f(\mathbf{x},\mathbf{y}) = \phi(\mathbf{x})\phi'(\mathbf{y}) \tag{3-129}$$

If, when **x** and **y** are close to each other, the values of ϕ and ϕ' are almost (or at) their respective (positive) maxima, the value of f will be high. If ϕ is high, but instead, $\phi' = 0$, then $f = 0$. In the first case, the correlation between ϕ and ϕ' is high. In the second case low. If ϕ were high, but ϕ' were a large negative value, then $f < 0$, and correlation between ϕ and ϕ' is even lower. If both ϕ and ϕ' are near (or at) their respective largest negative minima, $f > 0$ and large, so, ϕ and ϕ' are highly correlated. In a nutshell, the more alike ϕ and ϕ' are when **x** and **y** are close, the more correlated the two functions are, and the higher the value for f.

The same logic can be applied for any **x** and **y**, not just when they are close. Suppose for example, when **x** = 10 and **y** = 25, the two ϕ and ϕ' are at their respective positive value maxima. Then f would be high, and the two functions are highly correlated between those particular spatial locations. If at **x** = 10 and **y** = 38, ϕ' is at a minimum, then, they would not be well correlated between those two locations, and f would be much lower.

So, in general, a function like f, which is a product of two functions of separate independent variables, has a high value when the two function values are close to each other and a low value when they are not. So, f is a correlation function.

The concept can be extended to 4D and more than two functions (like $\phi_1, \phi_2, \ldots, \phi_n$) of more than two independent variables (like $x_1, x_2, \ldots, x_m$). Hence, propagators, which are two-point Green functions (two functions of two independent variables), such as the first term in (3-51), $iD_F^{\mu\nu}(x_1 - x_2)$, represent correlations in spacetime between the fields $A^\mu(x_1)$ and $A^\nu(x_2)$, and thus, are correlation functions. Higher point Green functions would be correlation functions between several fields for events located at several different places in spacetime, not just two.

3.6.3 Appendix C. Green Functions and Transition Amplitudes in Other Pictures

Recall that in the IP, states evolve via the interaction part H_I^I of the interaction picture full Hamiltonian $H^I = H_0^I + H_I^I$, where superscripts denote the picture and subscripts denote the free and interaction parts. Operators (such as our quantum fields ϕ, ψ, etc.) evolve via the free Hamiltonian H_0^I. We built our entire understanding of QFT in Vol. 1 on the foundation of the IP. (See Vol. 1, Wholeness Charts 7-1, pg. 188, and 8-4, pgs. 248-249.)

We could, in principle, have built, instead, on the Schrödinger picture (SP), in which states evolve via the full Hamiltonian $H^S = H_0^S + H_I^S$, and operators (for the usual case without explicit Hamiltonian time dependence) are frozen (time independent). Or we could, again in principle, have built on the Heisenberg picture (HP), in which states are frozen (time independent) and operators evolve via the full Hamiltonian $H^H = H_0^H + H_I^H$.

For the SP and HP, we say "in principle" because those routes would have been a whole lot harder than the IP, and that is the reason we (and others) prefer the IP methodology. However, in the literature, one often encounters Green functions expressed in the Heisenberg picture (HP), so we now briefly summarize both Green functions and transition amplitudes in all three pictures. As this topic is a bit tangential to our direction in this text, we do not treat it in depth, so the presentation is streamlined and may not be as pedagogic as it otherwise might be. My apologies for that.

First, note from Wholeness Chart 3-9, the major reason why finding transition amplitudes in the SP would have been so difficult. The initial (multiparticle, generally) state evolves via the full Hamiltonian H^S, and because interactions are included, deducing what that state becomes at any given time is not easy. In that non-free case, the states are not simple complex sinusoids, but something far more complicated. Additionally, they interact with one another, which makes them even more complicated mathematically, and thus, generally intractable to express in closed form.

In the HP, the states are fixed, and all the dynamics occurs in the fields. The fields do not obey the free field equations, but the full field equations (including interactions), so they are not simple complex sinusoids (as they are in the IP where they obey the free field equations). That means our creation and destruction operators like a, $a^\dagger$, c_r, etc. in the fields A^μ, ψ, etc. are functions of time (i.e., they change as the system evolves). Thus, if we create an initial state at an initial time, we create certain kinds of particles, but destroying that state later on would happen with destruction operators that have evolved since that creation. The transition amplitude S_{fi} would depend on the way in which those operators have evolved.

Srednicki[4], among others, develops the HP in detail for determining transition amplitudes and Green functions. Here, we simply layout an overview outline of how that is done.

We start by symbolically representing the destruction and creation operators time dependence for the simplest case, a real scalar field, as

$$a_{\mathbf{k}}(t) \text{ and } a_{\mathbf{k}}^{\dagger}(t), \tag{3-130}$$

In the IP, these operators are independent of time. With some effort, one can work out the time dependence, but we won't do that here.

We can then consider creating an initial state at $t = -\infty$ out of the vacuum via

$$|i\rangle_H = a_{\mathbf{k}}^{\dagger}(t=-\infty)|0\rangle_H, \tag{3-131}$$

and returning to the vacuum at $t = +\infty$ via

$$C|0\rangle_H = a_{\mathbf{k}}(t=+\infty)|i\rangle_H = a_{\mathbf{k}}(t=+\infty)a_{\mathbf{k}}^{\dagger}(t=-\infty)|0\rangle_H. \tag{3-132}$$

The time dependence in $a_{\mathbf{k}}$ arises from interaction term(s) in the Hamiltonian, and so, its action on a state carries the changes that can arise via interactions between $t = -\infty$ and $t = +\infty$. Since things change due to the interaction, the final vacuum state is not exactly the same as the initial one. We represent this change by C.

Writing an expectation value for such a process, with an arbitrary number of initial and final particles, we have

$$_H\langle 0|\underbrace{a_{\mathbf{k}'_2}(t=+\infty)a_{\mathbf{k}'_1}(t=+\infty)\dots a_{\mathbf{k}_2}^{\dagger}(t=-\infty)a_{\mathbf{k}_1}^{\dagger}(t=-\infty)\dots|0\rangle_H}_{C|0\rangle_H} = C. \tag{3-133}$$

This can be re-written as

$$\underbrace{_H\langle 0|a_{\mathbf{k}'_2}(t=+\infty)a_{\mathbf{k}'_1}(t=+\infty)\dots}_{\langle f|}\underbrace{a_{\mathbf{k}_2}^{\dagger}(t=-\infty)a_{\mathbf{k}_1}^{\dagger}(t=-\infty)\dots|0\rangle_H}_{|i\rangle} = C = S_{fi} = {}_H\langle f|i\rangle_H, \tag{3-134}$$

where we re-characterize the expectation value in terms of an initial non-vacuum state and what is typically a different final non-vacuum state, and commonly, $C = S_{fi} \neq 0$. Take care that often it is not stated that the expression is in the HP, and that can be very confusing, as the similar expression in the IP is always equal to zero (for different initial and final states).

Wholeness Chart 3-9. Transition Amplitudes and Green Functions in Three Pictures

	IP	**SP**	**HP**
S_{fi}	$S_{fi} = {}_I\langle f\|F\rangle_I = {}_I\langle f\|S\|i\rangle_I =$ ${}_I\langle f\|\underbrace{e^{-iT\int H_I^I dt}\|i\rangle_I}_{\|F\rangle_I=\sum_{f'} S_{f'i}\|f'\rangle_I}$	$S_{fi} = {}_S\langle f\|F\rangle_S =$ ${}_S\langle f\|\underbrace{e^{-iT\int H^S dt}\|i\rangle_S}_{\|F\rangle_S=\sum_{f'} S_{f'i}\|f'\rangle_S}$	$S_{fi} = {}_H\langle f\|i\rangle_H =$ $\underbrace{\langle 0\|A^{\mu}(+\infty)\psi(+\infty)\dots}_{\langle f\|}\underbrace{A^{\nu}(-\infty)\bar{\psi}(-\infty)\|0\rangle}_{\|i\rangle}$
$G^{\mu\dots}$	$\dfrac{{}_I\langle 0\|T\left\{e^{-i\int H_I^I dt}A^{\mu}(x_1)\dots\psi(y_1)\dots\right\}\|0\rangle_I}{\underbrace{{}_I\langle 0\|e^{-iT\int H_I^I dt}\|0\rangle_I}_{S}}$	$\dfrac{{}_S\langle 0\|T\left\{e^{-i\int H^S dt}A^{\mu}(x_1)\dots\psi(y_1)\dots\right\}\|0\rangle_S}{{}_S\langle 0\|e^{-iT\int H^S dt}\|0\rangle_S}$	$\dfrac{{}_H\langle 0\|T\left\{A^{\mu}(x_1)\dots\psi(y_1)\dots\right\}\|0\rangle_H}{{}_H\langle 0\|0\rangle_H}$

Note some key aspects of (3-134). First, to make it all equal the transition amplitude, some terms and multiplying factors have been tucked into the creation and destruction operators, which we will not take the time and space to deduce.

---

[4] Srednicki, M., *Quantum Field Theory* (Cambridge 2007), Chap. 5.

Second, from all that we have done previously, the RHS of (3-134) looks strange, as we would have thought, given an interaction occurring, that it should equal zero, since we are considering the case where the initial and final states are orthogonal (different eigenstates in Fock space where the initial state evolved into the final state)[5]. However, that is how it is typically represented in the literature. Again, note that it is often not stated that the expression is in the HP, and that can be confusing.

Third, (3-134) can be extended to spinor and vector fields, for any interactions between those fields, as in QED or other theories.

Finally, (3-134) is expressed in momentum space, and the 4D position space equivalent is

$$\underbrace{{}_H\langle 0|\phi_{\mathbf{k}'_2}(\mathbf{x},t=+\infty)\phi_{\mathbf{k}'_1}(\mathbf{x},t=+\infty)\ldots}_{\langle f|}\underbrace{\phi_{\mathbf{k}_2}(\mathbf{x},t=-\infty)\phi_{\mathbf{k}_1}(\mathbf{x},t=-\infty)\ldots|0\rangle_H}_{|i\rangle}=S_{fi}={}_H\langle f|i\rangle_H \quad (3\text{-}135)$$

For the QED fields, this looks like the first block under the HP heading in Wholeness Chart 3-9, where there, and in the next row therein, we have omitted some of the QED fields, due to lack of space.

In that last row of the chart, hopefully one can see the parallels between the three pictures. Whatever operator in the top row came before the initial state $|i\rangle$ in the expectation value relation for S_{fi} is stuck into the Green function numerator before the leg fields. For the IP, this was S. For the HP, it is 1.

Although this appendix is neither detailed nor complete, hopefully, it can serve as an aid to avoiding confusion when one encounters expressions for the Green function and other things in other sources, which use a picture such as the HP, rather than the IP.

3.6.4 Appendix D. The LSZ Reduction Formula

A topic you will run into in the literature, but which will not be focused on in this book, is something known as the LSZ reduction formula. As it is a bit of a sidelight to the main text, we will not provide a proof, but merely present it, with rationale for why it works. A proof can be found in Schwartz[6], as well as other texts.

Bottom line: The LSZ reduction formula converts (reduces) Green functions into transition amplitudes. It is a formal mathematical means which can be used instead of our seemingly simpler and more straightforward substitutions of external particle relations for leg propagators as, for example, in (3-39).

I apologize for the briefness of this section, realize that it does not do justice to the subject, and recognize that it does not meet the pedagogic standards of the rest of this text. But we have other fish to fry, for which I'd prefer to devote more ink.

Let's start with an example, the 2nd order part of the Green function (in momentum space) of (3-57) with two fermion and two photon legs,

$$\hat{G}^{(2)}_{c\ \mu\nu}=\left(-iS_F\left(-p_{y1}\right)\right)iD_{F\mu\alpha}\left(k_{x2}\right)\Gamma^{\alpha\beta}\left(k_{x1},k_{x2},p_{y1},p_{z1}\right)iD_{F\beta\nu}\left(k_{x1}\right)iS_F\left(p_{z1}\right), \quad \text{repeat of (3-57)}$$

where $\Gamma^{\alpha\beta}$ is defined in (3-58) and essentially sums the two different ways the internal fermion propagator shows up in Compton scattering. The propagators shown explicitly in (3-57) are for the legs, which will be external particles in the transition amplitude.

Now re-write (3-57), using the momentum space forms for the propagators, where we follow a common practice of not writing out the infinitesimal $i\varepsilon$ term in the denominators, as

$$\hat{G}^{(2)}_{c\ \mu\nu}=\frac{-i}{-\not{p}_{y1}-m}\frac{ig_{\mu\alpha}}{k^2_{x2}}\Gamma^{\alpha\beta}\frac{ig_{\beta\nu}}{k^2_{x1}}\frac{i}{\not{p}_{z1}-m}. \quad (3\text{-}136)$$

Then, note we can get the Feynman amplitude part of the transition amplitude for Compton scattering from (3-136) by multiplying it by appropriate factors, i.e.,

---

[5] We are considering the case where it is possible for the initial state to transition into the final state and $C = S_{fi}$ is not zero. For a case where it is not possible for the initial state to become the final state, we would have $C = S_{fi} = 0$.

[6] Schwartz, M. *Quantum Field Theory and the Standard Model*, Cambridge (2014), pg. 69-72.

$$\mathcal{M}^{(2)}_{Compton}\left(k,k',p',p\right)$$

$$= \bar{u}_{s'}(\mathbf{p}')\frac{-\not{p}_{y1}-m}{-i}\varepsilon^{\mu}_{r'}(\mathbf{k}')\frac{k^2_{x2}}{i}\underbrace{\left(\frac{-i}{-\not{p}_{y1}-m}\frac{ig_{\mu\alpha}}{k^2_{x2}}\Gamma^{\alpha\beta}\frac{ig_{\beta\nu}}{k^2_{x1}}\frac{i}{\not{p}_{z1}-m}\right)}_{\hat{G}^{(2)}_{C\,\mu\nu}}\frac{k^2_{x1}}{i}\varepsilon^{\nu}_{r}(\mathbf{k})\frac{\not{p}_{z1}-m}{i}u_s(\mathbf{p}) \quad (3\text{-}137)$$

$$= \bar{u}_{s'}(\mathbf{p}')\varepsilon_{\alpha,r'}(\mathbf{k}')\Gamma^{\alpha\beta}\left(k,-k',-p',p\right)\varepsilon_{\beta,r}(\mathbf{k})u_s(\mathbf{p}) = \mathcal{M}^{(2)}_{C1}+\mathcal{M}^{(2)}_{C2},$$

where we have made the substitutions of (3-60), repeated below, which switches signs on the particular incoming 4-momenta in the Green function that are outgoing in the Feynman amplitude.

$$p_{z1}=p \qquad k_{x1}=k \qquad k_{x2}=-k' \qquad p_{y1}=-p' \,. \qquad \text{repeat of (3-60)}$$

(3-137) is an example of the LSZ formula used for the specific case of Compton scattering. Basically, it entails multiplication of the Green function by the inverse of the leg propagators along with the appropriate spinor and polarization factors. On face value, it may seem trivial, and hardly an improvement on our straightforward substitution method, but has uses in advanced QFT.

In the general case, with n (n') incoming (outgoing) electrons, m (m') incoming (outgoing) positrons, and l total incoming plus outgoing photons, we can express the LSZ reduction formula as

$$\mathcal{M}=\left(\prod_{i'}^{n'}\bar{u}_{i',s_{i'}}(\mathbf{p}_{i'})\frac{\not{p}_{i'}-m}{i}\right)\left(\prod_{i'''}^{m'}\frac{\not{p}_{i'''}-m}{i}v_{i''',s_{i'''}}(\mathbf{p}_{i'''})\right)\left(\prod_{j}^{l}\varepsilon^{\mu_j}_{r_j}(\mathbf{k}'_j)\frac{k_j^2}{i}\right)$$
$$\times\hat{G}_{C\,\mu_1\mu_2\ldots\mu_l}\left(\prod_{i}^{n}\frac{\not{p}_{i}-m}{i}u_{i,s_i}(\mathbf{p}_i)\right)\left(\prod_{i''}^{m}\frac{\not{p}_{i''}-m}{i}\bar{v}_{i'',s_{i''}}(\mathbf{p}_{i''})\right). \qquad (3\text{-}138)$$

There is subtle sign issue with (3-138), reflected in our example of (3-57), where we had a minus sign come in with the factor of $-iS_F(-p_{y1})$. This arose because of the relation shown in Vol. 1, pg. 202, (7-75), where

$$\underbrace{\psi_\alpha(x_1)\bar{\psi}_\beta(x_2)}=-\underbrace{\bar{\psi}_\beta(x_2)\psi_\alpha(x_1)}=iS_{F\alpha\beta}(x_1-x_2) \qquad \text{repeat of Vol. 1, (7-75).} \quad (3\text{-}139)$$

To fully account for this, we would need a factor of $(-1)^Z$ in (3-138), where Z is the number of $-iS_F$ propagators we have in the Green function due to the ordering of fermionic fields in converting a contraction to a Dirac propagator. But since, in calculating probabilities from our amplitude, the sign of the amplitude drops out, we can safely ignore this. And ignore it we did in (3-138).

There is a comparable, more complicated, expression for the LSZ formula in position space for use with the Green function in position space (where, for example, $k^2 \to -\partial^\mu\partial_\mu$), but we do not get into that here. If you run into it in the literature, the section on page 124 in Chap. 4 herein titled "Third relation: Operator inverses" will help in understanding it.

Finally, be forewarned that the LSZ formula is often expressed with the HP form of the Green function (see Wholeness Chart 3-9), and that can be confusing, if you are not prepared for it.

3.7 Problems

1. Find the Green function for the Maxwell equation $-\left(\partial_\alpha\partial^\alpha\right)A^\mu=0$. Note that the Maxwell equation actually has a minus sign in front, which is often ignored (but which we need to keep for this problem). The minus sign arises from the Lagrangian, which has form $\mathcal{L}^1_0=-\frac{1}{4}F_{\mu\nu}F^{\mu\nu}=-\frac{1}{2}\left(\partial_\nu A_\mu\partial^\nu A^\mu-\partial_\mu A_\nu\partial^\nu A^\mu\right)$. Plugging $\mathcal{L}^1_0$ into the Euler-Lagrange equation gives us the Maxwell equation above (with the minus sign). (Hint: Consider the case where instead of zero on the RHS, we had a function $f^\mu(x)$, and the solution is found via $A^\mu(x)=\int G^{\mu\nu}(x,y)f_\nu(y)dy=\int G^{\mu\nu}(x,y)g_{\nu\rho}f^\rho(y)dy$.)

2. Find the Green function for the Dirac equation $-(i\gamma^{\alpha}\partial_{\alpha} - m)\psi = 0$. Note that, like Maxwell's equation (see Prob. 1), the Dirac equation, when derived from the QED Lagrangian and the Euler-Lagrange equation actually has a minus sign in front (which doesn't matter normally, as it equals zero, but does matter here). See Vol. 1, Chap. 4, Prob. 22. Hint: Other than the factor e^{iky}, the Green function in momentum space is the inverse of the Dirac equation operator in momentum space. This problem is actually very simple, once you see it.

3. Derive (3-25), and in the process, confirm that the spinor indices therein, when written out explicitly and not hidden, are correct. Hints: Start from (3-24). Using (7-75) of Vol. 1, pg. 202 and Appendix A of this chapter will help.

4. Show a set of disconnected diagrams arising from a term in the 4-point Green function with all fermion legs other than the ones shown explicitly in this chapter.

5. Show that the lowest order part of (3-53) is (3-54), and sketch the Green function diagrams for that lowest order contribution.

6. Show that the second order part of (3-53) is (3-55).

7. Use (3-70) to show (3-72) (i.e., $\theta_i^2 = 0$) and that the subscripts in (3-69) obey (3-73), i.e., $i < j < \ldots < n$.

8. For a Grassmann algebra with $n = 2$, show

$$\left[\frac{\partial}{\partial\theta_1}, \frac{\partial}{\partial\theta_2}\right]_+ = 0 \qquad \left[\theta_1, \frac{\partial}{\partial\theta_1}\right]_+ = 1 \qquad \left[\theta_1, \frac{\partial}{\partial\theta_2}\right]_+ = 0$$

and then use the results to justify the general relation (3-81).

9. Derive the LH relation in (3-92) from (3-91). Then use it to glean the RH relation.

10. From the last line of (3-100), derive (3-102) and then, (3-103).

11. Starting with the generating functional for QED, find the Green function for i) four fermion legs and one photon leg, and ii) four photon legs. Show a Feynman diagram for an associated transition amplitude one could find for each of these cases. Hint: For ii), see Vol. 1, Sect. 13.3, pgs. 328-329.

12. Show that $Z_0[\sigma,\bar{\sigma}] = e^{-i[\bar{\sigma}S_F\sigma]}$ and that the Green function generated by it is the spinor propagator.

Part Two

QED Revisited: Path Integrals

"Whatever you think, it's more than that."
The Incredible String Band

Chapter 4 The Many Paths Approach to QED

Chapter 4

The Many Paths Approach to QED

Visiting Einstein one day, I could not resist telling him about Feynman's new way to express quantum theory. "Feynman has found a beautiful picture to understand the probability amplitude for a dynamical system to go from one specified configuration at one time to another specified configuration at a later time. He treats on a footing of absolute equality every conceivable history that leads from the initial state to the final one, no matter how crazy the motion in between. The contributions of these histories differ not at all in amplitude, only in phase... This prescription reproduces all of standard quantum theory. How could one ever want a simpler way to see what quantum theory is all about! Doesn't this marvelous discovery make you willing to accept the quantum theory, Professor Einstein?" He replied in a serious voice, "I still cannot believe that God plays dice. But maybe", he smiled, "I have earned the right to make my mistakes".

John Wheeler

4.0 Preliminaries

4.0.1 Prerequisites

Prerequisites for this chapter

A prerequisite for this chapter is Chap. 18 of Vol. 1, or its equivalent. That chapter develops NRQM from the path integral (functional integral, many paths, sum over histories) point of view and then, at the end, extrapolates that quantum particle perspective to a single quantum field.

A second prerequisite is familiarity with the canonical quantization approach to QFT, as expressed in Chaps. 1 to 17 of Vol.1, and in Chap. 3 of this book, or their equivalents.

4.0.2 Chapter Overview

In this chapter, we will extend the quantum field theory of a single quantum field deduced from the path integral (PI) approach (PIA) in Chap. 18, Vol. 1, to multiple quantum fields that interact in QED, and deduce practical ways to evaluate the relevant functional integrals.

Overview of key areas covered in this chapter

More specifically, we will do the following.

- Discuss the basic premise of the PI approach to QFT and list the associated postulates.
- Evaluate a key functional integral that will prove invaluable.
- Deduce the generating functional for the PI approach, and see how it equals that of the canonical approach.
- Note that Green functions, and thus transition amplitudes, since they are derivable from the generating functional, must be the same in both approaches, and therefore, cross sections, decays, and bound state solutions are the same, as well.

We'll show generating functional of PI approach same as canonical

A major part of the chapter will be devoted to the third point above. The generating functional lets us determine virtually anything we wish to know about particle interactions, and thus, in a nutshell, contains essentially the entirety of QFT within it. So, by showing the PI approach yields the same generating functional as the canonical approach, we validate the PI approach. Both approaches will then make the same experimental predictions, and these, as we know, have been validated via real world testing many times over.

4.1 Overview of the Path Integral Approach

4.1.1 Quick Review of PI for NRQM

Feynman's postulates for NRQM PI approach

Feynman's original postulates for functional quantization in NRQM, (Vol. 1, pg. 497) are

1. A particle is assumed classical in the sense that it can be considered a point-like object, with both its position and its 3-momentum well defined along each individual path, so those values determine the Lagrangian at any point and time along any given path. However, the particle is assumed quantum mechanical in that, like a wave function, it has a phase (at the point).
2. The phase value at any final event is $iS/\hbar$, where the action S is calculated along a particular path beginning with a particular initial event. (The quantity $e^{iS/\hbar}$ is referred to herein as a "phasor".)
3. The probability density for the final event is given by the square of the magnitude of a typically complex amplitude.
4. That amplitude is found by adding together the phasor values at that final event from all paths between the initial and final events, including classically impossible paths. The amplitude of the resultant summation must then be normalized relative to all other possible final events, and it is this normalized form of the amplitude that is referred to in 3.

Phasors are not weighted when summing over all paths – all have same amplitude

Note there is no weighting of the various path phasors. That is, the different individual paths in the summation do not have different amplitudes (see Vol. 1, (18-24) and Fig. 18-3, pg. 495). The correlation with the classical result comes from destructive interference among the paths far from the classical path, and constructive interference among the paths close to the classical path.

The amplitude for the limiting case of an infinite number of discrete paths goes over to a functional integral (see Vol. 1, Wholeness Chart 18-2, pg. 498) of form (where T is the time interval from x_i to x_f, L is the full Lagrangian at each point along a particular path of the quantum particle, $\mathcal{D}$ implies a functional integral over all paths, and C is a constant of proportionality that must be fixed by other means in order to normalize the result)

Amplitude proportional to limit of sum of phasors over paths, as number of paths → ∞

$$U\left(x_i,x_f;T\right)=C\lim_{N\to\infty}\sum_{j=1}^{N}e^{iS_j/\hbar}=C\int_{x_i}^{x_f}e^{i\int_0^T\frac{L}{\hbar}dt}\mathcal{D}x(t)\quad\text{(amplitude for NRQM particle)},\quad(4\text{-}1)$$

$$\text{where}\quad |U|^2=\begin{array}{c}\text{probability density of measuring particle at}\\ \text{final event (for correct normalization factor } C).\end{array}\quad(4\text{-}2)$$

That limit is a path integral

In earlier (non PI) studies of NRQM, we found (4-2) by squaring the absolute value of the normalized particle wave function at the final event.

As noted in Vol. 1, for RQM, the same logic is used to obtain (4-1), except that a relativistic Lagrangian is used.

Actually evaluating path integrals is challenging

Actually finding (4-2) is another matter, however, beyond mere theory. How does one carry out a functional integral, an integration over all possible paths of an integrand that itself contains an integration? As noted in Chap. 18 of Vol. 1, for meaningful situations, numerical solutions are generally the only practical answer, and they are not trivial.

4.1.2 Extending NRQM/RQM PI Approach to Fields

The PI Approach to QFT in Words

PI in QFT superimposes paths of fields instead of particles – each field configuration over all space not just one point

In the PI approach to fields, we superimpose all imaginable paths of *fields*, instead of *particles.* ("Path" of a field means the evolution in time of a 3D field configuration, not a 1D line, as with the path of a zero dimensional particle.) That is, we superimpose (add) all possible different ways individual fields could evolve *at every point* in space through time between the initial and final field configurations, *not just one point* as in postulate #1 above for particles.

Imagine a volume in which we normally could think of a single free field, spread throughout the volume, moving and changing over time. Then consider *all* imaginable ways (*not* just the physically realizable way that must obey the stationary action principle) that field could evolve from the initial to final configuration. Then, for each of those ways, calculate the phasor value at each moment in time for each point in the field. At each such point at each moment in time, add all the phasor values from all the infinite number of ways together. The phasor values far from the physically realizable one destructively interfere; while the phasor values near the physically realizable one, constructively interfere. The result, amazingly, turns out to be phasor values for what the single, physically realizable, field would have evolved as. For the final time, this "sum of all phasors" result determines the transition amplitude.

The same logic applies to a finite set of interacting fields (such as an electron field and a positron field). Each such field (electron or positron) could be treated as a superposition over the entire volume, at each point in space and time, of an infinite number of all imaginable field evolutions from the initial field configuration to the final one. Doing so, we get the same result as from physically realizable fields in the canonical approach. (Such as the probability for Bhabha scattering.) This is the basic principle of the PI approach to QFT.

So, the QFT functional quantization postulates, corresponding to the above for NRQM/RQM, are

Postulates for QFT PI approach

1. A field is assumed classical in the sense that it can be considered well defined, with both the field itself and its conjugate momentum having specific values at each point in space and time along the path of the field ("path" implies evolution of a field configuration over a region of space here). Those values determine the Lagrangian density $\mathcal{L}$ at each point and time along any given path. However, the field is assumed quantum mechanical in that it has a phase, as well as an amplitude, at every point and every time along the path.

Similar to NRQM but for fields, not particles

2. The action S of a particular field at a given time is calculated along a particular path (over a volume) starting with a particular initial field configuration. (The quantity $e^{iS/\hbar}$, where $S = \int \mathcal{L} dVdt$, is referred to herein as a "phasor".)
3. The probability of finding the final field configuration is given by the square of the magnitude of a typically complex amplitude.
4. That amplitude is found by adding together the phasor values at the final time from all imaginable paths between the initial and final field configurations, including classically impossible paths. The amplitude of the resultant summation must then be normalized, and it is this normalized form of the amplitude that is referred to in 3.

This process is referred to as "path integration" (for fields this time, not particles as in Vol. 1).

Key Points to Be Aware of

Note that PI fields are not operators, but simply functions of space and time. They do not obey commutation relations, so one does not think of them as quantum fields that create and destroy states, but as somewhat similar in many ways to classical fields or good, old-fashioned wave functions. They evolve much like the states we grew to know and love in the Schrödinger picture. If we were to deal with a single evolving such field (not a superposition of a large number of them), or a finite set of such single fields interacting, (such as the electron and positron case above), then we would virtually be doing RQM in the Schrödinger picture, where the states are "fields" in the classical sense, i.e., they are spread out in time and space.

PI fields are Heisenberg picture fields that are not operators, but just functions

That is, states in the Schrödinger picture of RQM evolve in much the same way as fields in the Heisenberg picture of QFT. Both obey the same wave equations we are familiar with (Klein-Gordon, Dirac, and Maxwell equations with interaction terms included) and are controlled by the full Hamiltonian, not just the free part of the Hamiltonian, as fields are in the interaction picture. (We showed Heisenberg picture fields obey these equations in the Appendix A of Vol. 1, Chap. 3).

Heisenberg fields evolve much like states in Schrödinger picture both via full H

So, we could imagine a canonical theory in which we develop Heisenberg fields, where the states are frozen (unchanging), and the fields evolve governed by the full Hamiltonian. But this is very hard, indeed, intractable. The interaction picture is much easier. So, using that picture is how the canonical approach to QFT was successfully developed.

Enter Richard Feynman

But then Richard Feynman comes along with his PI approach. It turns out to be easier to solve than the canonical version in the Heisenberg picture, but, as in the Heisenberg picture, it uses the full Hamiltonian for field evolution. The major difference for Feynman, however, is that he does not choose to follow the physically realizable evolution of a single field (or a finite set of such fields interacting), but rather the evolution of a whole bunch (infinite number) of fields (over all imaginable paths) superimposed. Remarkably, this not only actually works, but is far easier (though not "easy").

Bottom line: In the QFT PI approach, we use Heisenberg fields, i.e., the full Hamiltonian. States are frozen, i.e., only fields evolve. We find probabilities of interactions from field behavior alone.

All the "action" (colloquially, as in the sense of "activity", and also in the math sense of the time integral of the Lagrangian) is in the fields, and unlike the canonical approach, we don't have to think much about states as something separate in the theory. In a sense, fields represent the states, and we could almost think of the fields as representing SM particles, though technically, they are Heisenberg SM fields.

As an aside, note that this throws a little bit of a monkey wrench into the Vol. 1, Chap. 1, Wholeness Chart 1-2, pg. 7, second row, where we stated that solutions to the field equations in QFT are operators. That is only actually true for the canonical approach, which is what that chapter, and almost all of that book, was focused on.

A Naïve Method for the PI Approach in QFT

Given the above considerations, one could make an educated guess on the best way to proceed in order to obtain valid transition amplitudes in the PI approach to QFT. This would be reasonably straightforward, but it leads to very real practical problems. In particular, doing closed form calculations is simply not feasible.

Before exploring a less obvious, but more successful, means to the PI transition amplitude, we will review (below) how we might extrapolate NRQM/RQM PI methods to QFT via the naïve and seemingly straightforward (but ultimately inadequate) route.

Wholeness Chart 18-4 of Vol. 1, pgs. 506-507, denotes correlations in going over from a particle theory (like NRQM/RQM) to a field theory (like QFT). These are, where ϕ is a scalar field,

Extrapolating NRQM/RQM PI approach to QFT

	Particle Theory		Field Theory
Dependent variable	$x(t)$	$\rightarrow$	$\phi(x^1,x^2,x^3,t)=\phi(x^\mu)$
Independent variable(s)	t	$\rightarrow$	$x^1,x^2,x^3,t \;=\; x^\mu$.

(4-3)

Using these to correlate with (4-1) (see Vol. 1, Wholeness Chart 18-5, pg. 508), one could postulate (18-64) of Vol. 1, i.e., that the transition amplitude for a single quantum field is

PI transition amplitude for a single field in QFT (assuming our postulates hold, which we later validate)

$$U(i,f;T)=C\int_{\phi_i}^{\phi_f} e^{i\int_0^T \frac{\mathcal{L}}{\hbar}d^4x}\mathcal{D}\phi(x^\mu) \qquad \text{(QFT transition amplitude for single field)}. \tag{4-4}$$

$$\text{For free scalar field, } \mathcal{L}=\partial_\alpha\phi^\dagger\partial^\alpha\phi-\mu^2\phi^\dagger\phi \qquad \phi=\phi(x^\mu).$$

Whereas (4-2) is a probability density (a value at a point), since (the top row of) (4-4) is a value for a field over all of the space through which the field travels rather than a point in space, the square of its absolute value can be expected to represent the total probability (of finding the field in that particular configuration).

The functional integral of (4-4) is over all possible field configurations $\phi(x^\mu)$ in the volume at hand over the time T from the initial field configuration to the final one. We can conceptualize this, but actually carrying out the math is challenging.

We can then intuit how (4-4) could be extended to more than one field. That is, where $\mathcal{L}$ is now the Lagrangian density involving all fields and their interactions, and we switch to natural units,

PI transition amplitude for multiple fields in QFT

$$U(i,f;T)=C\int_i^f e^{i\int_0^T \mathcal{L}d^4x}\mathcal{D}\phi_1\left(x_1^\mu\right)\mathcal{D}\phi_2\left(x_2^\mu\right)\ldots\mathcal{D}\phi_n\left(x_n^\mu\right) \qquad \text{(multiple fields, natural units)}. \tag{4-5}$$

For QED, this might look like (where the symbol A means the photon field A^ν, streamlined spacetime position symbols without the μ superscript, as in Chap. 3, and the meanings shown for $\mathcal{D}A$, $\mathcal{D}\bar{\psi}$, and $\mathcal{D}\psi$)

PI transition amplitude for QED (in QFT)

$$U(i,f;T)=C\int_i^f e^{i\int_0^T \mathcal{L}d^4x}\mathcal{D}A\mathcal{D}\bar{\psi}\mathcal{D}\psi \qquad \text{(QED)}$$
$$\mathcal{D}A=\mathcal{D}A_1(x_1)\mathcal{D}A_2(x_2)\ldots \quad \mathcal{D}\bar{\psi}=\mathcal{D}\bar{\psi}_1(z_1)\mathcal{D}\bar{\psi}_2(z_2)\ldots \quad \mathcal{D}\psi=\mathcal{D}\psi_1(y_1)\mathcal{D}\psi_2(y_2)\ldots, \tag{4-6}$$

where the explicit form of the QED Lagrangian density (as shown in Vol. 1, top of pg. 298) is

$$\mathcal{L} = \mathcal{L}_0 + \mathcal{L}_I = \underbrace{-\tfrac{1}{2}\left(\partial_\beta A_\nu \partial^\beta A^\nu - \partial_\nu A_\beta \partial^\beta A^\nu\right)}_{-\frac{1}{4}F_{\nu\beta}F^{\nu\beta}} + \bar{\psi}\left(i\gamma^\nu \partial_\nu - m\right)\psi + \underbrace{e\bar{\psi}\gamma^\nu \psi A_\nu}_{\mathcal{L}_I \text{ (interaction included)}} . \tag{4-7}$$

QED Lagrangian

Note a few things with (4-6). First, the main integral between i and f means the path integral is from initial field configurations $A^\mu_{1i},\dots,\bar{\psi}_{1i},\dots,\psi_{1i}\dots$ to the final field configurations $A^\mu_{1f},\dots,\bar{\psi}_{1f},\dots,\psi_{1f}\dots$. This integral sign is commonly written in the literature without limits, i.e., as simply $\int$. We will do the same, but keep in mind what it really means.

Path integral is from initial (fixed) to final (fixed) field configurations

Second, the integral in the exponential between initial time 0 and final time T implies the fields can only cover all imaginable paths that move forward in time from time 0 to time T. However, one can surmise that the PI superposition concept can be extended to all times, i.e., from time $-\infty$ to time $+\infty$. The initial and final times (and field configurations) would still be fixed, but as the fields travel the myriad number of imaginable paths to get from the initial configuration to the final one, they would traverse all events over all time and space.

But PI paths between initial and final field configurations vary over all time

In other words, some of these paths could take the initial field configurations backward in time, and then later forward in time, in order to reach the final configuration. In this scenario, the fields moving backward in time add destructively and make no contribution to the final result.

Since the integral in question (in the exponential in (4-6)) is already over all space ($V \to \infty$), by increasing our time interval from $-\infty$ to $+\infty$, we have manipulated the form of that integral to be similar to ones we have worked with before in the canonical approach. Thus, the streamlined notation transition amplitude, as generally expressed in the literature, and as we will henceforth express it, is

PI transition amplitude over all T and V is S_{fi}

$$S_{fi} = \lim_{\substack{0\to -\infty \\ T\to +\infty}} U(i,f;T) = C\int_i^f e^{i\int_{-\infty}^{+\infty}\mathcal{L}d^4x}\mathcal{D}A\mathcal{D}\bar{\psi}\mathcal{D}\psi = \underbrace{C\int e^{i\int \mathcal{L}d^4x}\mathcal{D}A\mathcal{D}\bar{\psi}\mathcal{D}\psi}_{\text{streamlined notation}} . \tag{4-8}$$

<u>Comparison with the Canonical Approach</u>

Note how this compares with the canonical transition amplitude (Vol. 1, (19-2) & (19-3), pg. 513),

$$S_{fi} = \langle f|S|i\rangle = \langle f|Te^{i\int \mathcal{L}_I(x)d^4x}|i\rangle . \tag{4-9}$$

Compare to canonical transition amplitude S_{fi} (which is over all T and V and which we will see is same as PI)

For different initial and final field configurations in (4-8) (hidden in the not-shown limits of the streamlined integral sign symbol in the functional integral over the fields), we would get different values for S_{fi}. These field configurations would correlate with our initial and final external particles from the canonical approach.

Wholeness Chart 4-1 lists some of the major differences between the canonical approach (<u>CA</u>) and the PI approach (<u>PIA</u>).

Wholeness Chart 4-1. Some Differences between the CA and PIA

List of differences between canonical and PI approaches

Canonical Approach	Path Integral Approach
All physically realizable paths between initial and final particle states	All imaginable paths between initial and final field configurations
Paths obey stationary action principle	Paths don't obey stationary action principle
Integrate each internal event (vertex) over all spacetime	Integrate all imaginable paths throughout all spacetime
Interaction picture with $\mathcal{L}_I$ paramount	Heisenberg picture with full $\mathcal{L}$ paramount
Field commutation relations	All fields commute; no such relations
Fields are creation/destruction operators	Fields are just (classical) functions, not operators
Wick's theorem applies	Wick's theorem not applicable (Only for operators)
Time ordering in transition amplitude	No time ordering in transition amplitude
Time only forward	Time forward and backward

Fig. 4-1, though heuristic, can help us visualize what is happening in the PIA by comparing it to the CA. In the CA, we had a transition amplitude where the fields created and destroyed states, and all fields and particles followed physically realizable paths. The example shown is the first way for Bhabha scattering.

In the PIA, on the other hand, we have (non-operator) fields (two initially in the figure) that are in an initial configuration and evolve via all possible paths to the final configuration (of two final fields). In Fig. 4-1, we show three of an infinite number of all imaginable paths. Those field evolutions far from physically realizable destructively interfere. Those close to the physically realizable evolution constructively interfere. Note that in the course of evolving from the initial to the final configuration, some of the field paths travel, for a while, backward in time. And the extension in time covered by all the different fields is from $-\infty$ to $+\infty$, even though the initial and final configurations are at finite times. That is, in position space, the whole thing starts at a given (finite) initial time, and ends at a given (finite) final time, but in between, the fields evolve over all time.

Illustration of major difference between CA and PIA

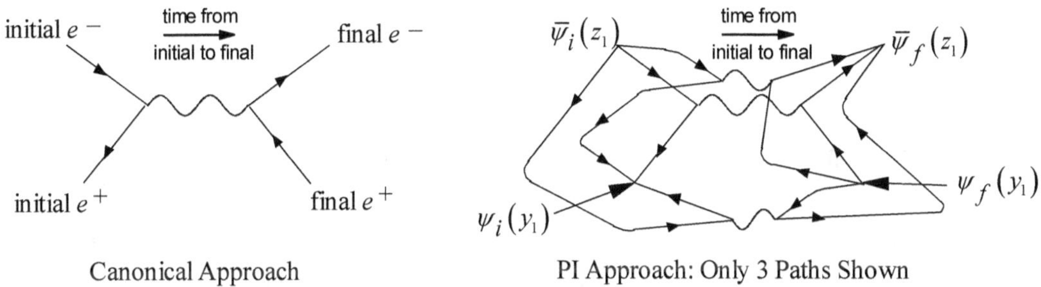

Figure 4-1. Heuristic Graphical Comparison of CA and PIA Transition Amplitudes

As noted, Fig. 4-1 is heuristic. It depicts fields, which are spread out in space, as located at a particular point in space. This is not so different from the Feynman diagrams we have used so much, where, as discussed in Vol. 1, pg. 238, a line would normally represent a point moving through time, but in actuality symbolizes a wave spread out through all of space.

Initial fields travel paths over all space and time to become final fields

The point is that our initial and final fields, such as $\psi_i(y_1)$ for example, are really functions of space and time, y_1 in this example. They are spread throughout space and not located at a single point as Fig. 4-1 seems to depict them. And their paths, as they evolve, are not really along lines in spacetime, but changing at all points in space as time changes (backward and forward, as it happens).

All intermediate paths interfere to leave physically realizable path

Nevertheless, Fig. 4-1 can provide us with some insight on the import of the math in (4-8). Just bear in mind how symbolic it all is.

The Problem with the Naïve Method

Problem: evaluating path integrals directly is intractable

If we thought evaluation of (4-4), with a single field, was challenging, consider (4-8), which for three field types, is well beyond that. In fact, evaluating the integral with the interaction part of $\mathcal{L}$ included (which we need for interactions), is essentially impossible, at least in closed form. So, though the theory may seem appealing, its implementation is anything but.

The Solution

Solution: Generating functional $\rightarrow$ Green function $\rightarrow S_{fi}$, and generating functional can be evaluated

Recall from Chap. 3 that given the generating functional, we can find Green functions, and thus transition amplitudes, for every possible particle interaction. So, if instead of the transition amplitude considered above, we were able to evaluate the generating functional in the PI approach, we would then have the entire theory in our hands. It turns out that evaluating the generating functional is far easier (but again, not really "easy") than evaluating the transition amplitude directly.

And that is the way the PI approach to QFT is developed. By focusing on the generating functional, rather than the transition amplitude. It is far more tractable that way. And that is the main reason we took so much time in Chap. 3 developing the theory therein. While it is little more than an interesting sidelight for the canonical approach, it turns out to be essential for the PI approach.

So, we will focus on PI generating functional

The rest of this chapter centers on finding the PI generating functional and showing it is equivalent to the canonical generating functional. This tells us that PI approach is indeed valid, and that the two approaches are but two different itineraries to get to Rome.

4.2 A Key Relation for the PI Approach

4.2.1 Our Direction in This Section

Functional integrals are generally intractable. Elegant in principle. Not so in practice.

But there is one specific type of functional integral we can evaluate (with a little development). And fortunately, that one type can be utilized in a manner that lets us determine the PI generating functional, and from it, all possible Green functions. And then, from them, all possible transition amplitudes. In this section, we deduce that most valuable of integral types.

There is one type of PI we can evaluate and with it, we can evaluate the PI generating functional

4.2.2 The Key Functional Integral

We start with the well-known (at least in math circles) integral (4-10) (see Vol. 1, (18-46), pg. 503, where "b" there is taken as "b'" here, and where certain convergence issues are addressed). We employ the usual notation where an integral sign without limits implies integration from $-\infty$ to $+\infty$).

Determining that key PI relation

$$\int e^{-ax^2+b'x}dx = \sqrt{\frac{\pi}{a}}e^{b'^2/4a} \qquad Re(a)\geq 0\,, \tag{4-10}$$

Integral over x from integral table

We then make the following changes to the constants,

$$a \rightarrow \frac{-ik}{2} \qquad\qquad b' \rightarrow ib\,. \tag{4-11}$$

With those changes, we end up with the integral

$$\int e^{\frac{i}{2}kx^2+ibx}dx = \sqrt{\frac{\pi}{-ik/2}}e^{-b^2/(-4ik/2)} = \sqrt{\frac{i2\pi}{k}}e^{-ib^2/2k}\,. \tag{4-12}$$

Modified via substitution of variables

We are interested in using (4-12) to evaluate the integral (4-13) below, where, as always, repeated indices mean summation,

$$\int e^{\frac{i}{2}x_iK_{ij}x_j+ib_ix_i}dx_1dx_2...dx_n\,, \tag{4-13}$$

Extending to multiple independent x_i

which is simply a multidimensional integral over n independent (orthogonal) dimensions (axes) x_i. K_{ij} couples the ith dimension with the jth. Stronger coupling (higher K_{ij} value) would have stronger impact on the result of the double integration over x_i and x_j. K_{ij} is, effectively, a weighting.

K_{ij} couples ith dimension to jth

Do **Problem 1** to show (4-14).

By doing Problem 1, or by looking it up in integral tables, we have the relation

$$\int e^{\frac{i}{2}x_iK_{ij}x_j+ib_ix_i}dx_1dx_2...dx_n = \sqrt{\frac{(i2\pi)^n}{Det\,K_n}}e^{-\frac{i}{2}b_iK_{ij}^{-1}b_j} \quad \begin{matrix} Det\,K_n = \text{determinant of } K_{ij}\;\; i,j=1,...,n \\ K_{ij}\text{ symmetric} \qquad \text{no sum on } n\end{matrix}\,, \tag{4-14}$$

Integral for n independent x_i

where we see the coupling between dimensions, represented by K_{ij}, is, as we noted above, a key element in the final number one gets for the integral.

Now employ a different real variable ϕ_i in (4-14).

Change x_i value to field value ϕ_i in tiny volume

$$\text{For } x_i \rightarrow \phi_i\,, \tag{4-15}$$

$$\int_{-\infty}^{+\infty}\int_{-\infty}^{+\infty}\cdots\int_{-\infty}^{+\infty} e^{\frac{i}{2}\phi_iK_{ij}\phi_j+ib_i\phi_i}d\phi_1d\phi_2...d\phi_n = \sqrt{\frac{(i2\pi)^n}{Det\,K_n}}e^{-\frac{i}{2}b_iK_{ij}^{-1}b_j} \qquad (\text{no sum on } n), \tag{4-16}$$

Integral using ϕ_i over n tiny volumes

where integration limits are shown explicitly to help pedagogically as we think about this later. We will consider ϕ_1 to represent the part of the single field ϕ (which is spread out over all space) inside at the center of a 1st (typically small) volume, which we label as "1"; the part ϕ_2 at the center of a different 2nd volume we label as "2"; and ϕ_i for that part of the field at the center of the ith volume. In the course of the integration, each ϕ_i varies inside its own volume over all different possible values, i.e., from $-\infty$ to $+\infty$. This variation of the field ϕ_i inside each volume i is considered independent of the variation of the field in all other volumes, i.e., the ϕ_i are all independent, just as each x_i in (4-14) was independent of all the others. Note, the ϕ_i are not separate fields, but the values of the single field

ϕ at different points (the centers of each volume) in 3D physical space. They are independent, so, in a mathematical sense, could be considered different "axes" in an abstract space, but physically, they are values at points in the 3D world we live in.

Similar to before, K_{ij} couples ϕ_i in the ith volume to ϕ_j in the jth volume. Whatever that dependence happens to be will play a key role in determining the value of the integral. And all these volumes together equal the entire volume considered for the problem at hand. This entire volume can be taken as infinity (the whole of 3D space), as we are used to doing in the canonical approach.

Integrate over all ϕ_i values inside each ith tiny volume

That means the field values in the integration take on any and all values, and thus we are effectively (approximately in this discrete case) adding all possible configurations for the values of ϕ_i throughout space. If we extend our 3D volumes to 4D spacetime, then we talk about summing (approximately) all paths (configurations) for the field ϕ throughout space as it evolves through time. Hence, if we take the limit where our (now 4D) volumes become infinitesimal, and n goes to infinity,

That means all n integrations cover all values for ϕ throughout the whole volume

$$\lim_{n\to\infty} d\phi_1 d\phi_2 ... d\phi_n = \mathcal{D}\phi\,, \tag{4-17}$$

So, for 4D & $n \to \infty$, we have a path integral

and we are talking about a path integral over all spacetime. In that case, (4-16) becomes (where we now use the single integral sign but think of it like it is in (4-16))

$$\int e^{\frac{i}{2}\int \phi(x)K(x,y)\phi(y)dxdy + i\int b(x)\phi(x)dx} \underbrace{\underbrace{d\phi_1 d\phi_2 ... d\phi_n}_{n\to\infty}}_{\mathcal{D}\phi} = C\underbrace{e^{-\frac{i}{2}\int b(x)K^{-1}(x,y)b(y)dxdy}}_{=1 \text{ for } b=0}\,, \tag{4-18}$$

And that gives us a result for this particular path integral

$$\text{where} \qquad C = \lim_{n\to\infty}\sqrt{\frac{(i2\pi)^n}{Det\, K_n}} \qquad (\text{no sum on } n). \tag{4-19}$$

Though it does give us a weird constant in front

The square root factor in (4-19) has infinity in the numerator and possibly in the denominator as well, since in the limit, the matrix K_{ij} gets an infinite number of components. We will later find that this infinite factor drops out of calculations we will wish to make, so for now, we'll just carry it along as if it weren't an issue.

Fortunately, this later drops out of calculations

Note that (4-18) is a path integral, but one from $t_i = -\infty$ to $t_f = +\infty$, i.e., without initial and final field configurations specified, so it is a little different from the ones we've seen before.

As the next step of our development, we note that differential and other operators in an integral can be represented by matrices in the sum of small elements whose limit is taken to get the integral. For example, to keep things simple, think temporarily of the exponent in (4-18) as a (1D) line integral rather than that of a 4D volume. Consider the matrix of (4-20), where Δx is the length of each tiny line segment into which the exponent in (4-18) would be broken up.

Showing that we can take K(x,y) as an operator

$$K_{ij}\phi_j = \frac{1}{\Delta x}\begin{bmatrix} 1 & -1 & & \\ & 1 & -1 & \\ & & \ddots & \\ & & & 1 \end{bmatrix}\begin{bmatrix} \phi_1 \\ \phi_2 \\ \vdots \\ \phi_n \end{bmatrix} = \begin{bmatrix} \frac{\phi_1 - \phi_2}{\Delta x} \\ \frac{\phi_2 - \phi_3}{\Delta x} \\ \vdots \end{bmatrix} \xrightarrow[n\to\infty]{\text{in limit}} \begin{bmatrix} \frac{\partial \phi_1}{\partial x} \\ \frac{\partial \phi_2}{\partial x} \\ \vdots \end{bmatrix} \quad \text{so} \quad K_{ij} \to \frac{\partial}{\partial x} = K \tag{4-20}$$

The particular matrix shown in (4-20) becomes, in the limit, a first derivative with respect to x. Had we worked in reverse, starting with the first derivative of the continuous case, we could have found the matrix that, in the discrete case, approximates that derivative. And we could do that for other operations, as well. Whatever operator we are interested in for a continuous formulation, there will be corresponding matrix representing it in the discrete formulation.

So, in the limit of going from the discrete case of (4-16) to the integral of (4-18), we could have the particular matrix in the discrete case that corresponds to a particular operator in a continuum. We can conclude that (4-18) is valid for K as either a matrix (of infinite dimension) or an operator[1].

[1] The matrix K must be symmetric for (4-16) to hold, but (4-20) is not. It turns out that a matrix representing the second derivative actually does meet this requirement, and everything we will do using the relationships derived here will involve second derivatives. We are using (4-20) only to illustrate the point that matrices can approximate operators.

As a final generalization of (4-18), we can extend it to quantities with spacetime (or other) indices, by changing the symbol for the field from ϕ to A^ν, since the former is typically reserved for scalars.

Generalizing from a real scalar like ϕ to a real entity with indices like A_ν

$$\int e^{\overbrace{\frac{i}{2}\int A_\mu(\underline{x})K^\mu{}_\nu(\underline{x},\underline{y})A^\nu(\underline{y})d\underline{x}d\underline{y}}^{\left[A_\mu K^\mu{}_\nu A^\nu\right]}+i\int b_\nu(\underline{x})A^\nu(\underline{x})d\underline{x}}\mathcal{D}A(x)=C\,e^{-\frac{i}{2}\overbrace{\int b_\mu(\underline{x})K^{-1\mu}{}_\nu(\underline{x},\underline{y})b^\nu(\underline{y})d\underline{x}d\underline{y}}^{\left[b_\mu K^{-1\mu}{}_\nu b^\nu\right]}}. \tag{4-21}$$

Our final result for real spin 1 photon fields

For spinors, we have (4-22), which is derived in the appendix of this chapter. In it, as usual, the spinor indices are hidden, and we note that, due to the nature of spinors and Grassmann variables, the potentially divergent constant C' is different from C in (4-21). We will see that this constant drops out of any relations we will have interest in (at least for QED).

$$\int e^{i\overbrace{\int\bar{\psi}(\underline{x})K(\underline{x},\underline{y})\psi(\underline{y})d\underline{x}d\underline{y}}^{\left[\bar{\psi}K\psi\right]}+i\int\bar{\sigma}(\underline{x})\psi(\underline{x})d\underline{x}+i\int\bar{\psi}(\underline{x})\sigma(\underline{x})d\underline{x}}\mathcal{D}\bar{\psi}\mathcal{D}\psi=C'e^{-i\overbrace{\int\bar{\sigma}(\underline{x})K^{-1}(\underline{x},\underline{y})\sigma(\underline{y})d\underline{x}d\underline{y}}^{\left[\bar{\sigma}K^{-1}\sigma\right]}}, \tag{4-22}$$

Corresponding result for spinors and Grassmann variables

We have put the above relations in boxes because they, as we will see, are the key to unlocking the power of the PI approach, but you don't need to memorize them.

4.3 Green Functions in the PI Approach

4.3.1 Intuiting the Form of the PI Green Function

The Math

As we develop the theory from this point on, you should follow along with Wholeness Chart 4-4, pg. 126, at the end of the chapter, which shows in summary form what we are about to do and compares it to what we did in Chap. 3. (We fill in the blanks of Wholeness Chart 3-7 at the end of Chap. 3.)

Follow along with Wholeness Chart 4-4

In this section, we will simply present the PI approach (PIA) Green function, which we will eventually prove is equivalent to the canonical approach (CA) Green function. However, before doing so, we present some justification for why it would have the given form.

PIA Green function presented below

In the first column of Wholeness Chart 4-4, we see that in the CA, when we pass from the transition amplitude to the Green function, we have a denominator that is simply the transition amplitude from the infinite past vacuum to the infinite future vacuum, and a numerator that is the same except for the insertion of the leg field operators. But the real action occurs between what eventually will be considered the initial and final coordinates of the legs (in (3-13), the arguments $x_1, x_2, \dots y_1, \dots z_1, \dots$ of the leg fields). The rest is integrated (inside S).

So, we might expect the PIA Green function to have a parallel relationship with the PIA transition amplitude (4-8). That is, where legs in the PIA are parallel to legs in the CA, we have the leg-free transition amplitude in the denominator and a numerator with leg fields inserted. And we do.

Extrapolating parallels from CA to PIA

The PIA Green function, with $\mathcal{L} = \mathcal{L}_0 + \mathcal{L}_I$, is

$$G^{\mu\nu..}(x_1,x_2,\dots,y_1,\dots,z_1,\dots)=\frac{\int\left\{e^{i\int\mathcal{L}d^4x}A^\mu(x_1)A^\nu(x_2)\dots\psi(y_1)\dots\bar{\psi}(z_1)\dots\right\}\mathcal{D}A\mathcal{D}\bar{\psi}\mathcal{D}\psi}{\int e^{i\int\mathcal{L}d^4x}\mathcal{D}A\mathcal{D}\bar{\psi}\mathcal{D}\psi}, \tag{4-23}$$

PIA Green function (we prove valid later)

As in CA, PIA legs fixed initially and finally while other integration goes on

where the fields that are functions of the 4D coordinates $x_1, x_2, \dots y_1, \dots z_1, \dots$ are, as in the CA, called legs and are fixed at the initial and final times, with integration over everything else. In this process, the leg fields, which are functions of those coordinates, are effectively the initial and final field configurations which stay fixed, and between which, in the PIA, all imaginable field evolutions (not just physically realizable ones) are considered superimposed.

Legs are essentially initial and final field configurations

To visualize this, think of several fields spread out over a 3D volume (which we commonly take as infinite), each at an initial time $t_{x_1\,initial}, t_{x_2\,initial}, \dots, t_{y_1\,initial} \dots$. A moment later, the fields have evolved and have different configurations in the same volume. A moment after that, the configurations are yet different again, until one arrives at the final configurations of the fields at final times. Such an evolution of the fields is considered one path. The path integral of (4-23) effectively adds all such paths, i.e., all imaginable paths.

Green functions integrate all paths for given legs

The above two paragraphs were simplified, in order to make a point, as we should recall that in Green functions, all legs are considered to have time flowing forward, and thus all legs, to be strictly precise, are initial field configurations. When we use the Green function (4-23) to obtain transition amplitudes, we need to reverse the time flow on what we want to be the final legs, so they are final fields, not initial ones.

Legs are not points, but fields over all space with particular fixed values at initial and final times

Note that each 3D coordinate x_i (or y_i or z_i) does not actually represent a single point, but all of space, as, for example, $A^{\mu}(x_1)$ represents a field which is a function of x_1. (Recall that x_1 is really x_1^{ν}, where we hide the spacetime index, and the subscript is associated with a particular photon leg field.) Parallel logic holds for the other leg fields. Essentially, G^{μ} $(x_1, \ldots y_1, \ldots z_1, \ldots)$ has spacetime dependence on $x_1, \ldots y_1, \ldots z_1, \ldots$ It is different at different events $x_1, \ldots y_1, \ldots z_1, \ldots$. When we convert it into the momentum space Green function $\hat{G}(q_1, q_2, \ldots, q_n)$ (see (3-21)), we integrate over all those events $x_1, \ldots y_1, \ldots z_1, \ldots$ And it is from that momentum space version of the Green function that we can deduce the momentum space version of the transition amplitude.

When we convert to momentum space, legs become a function of momentum

In getting (4-23), we have used our intuition, with the CA as a guide. It turns out, as has been proven decades ago now, that (4-23) is indeed the Green function for the PIA. However, we will prove it is so, rather than just intuit it, as we progress through this chapter.

Heuristic Graphics Again

Look back again at the RHS of Fig. 4-1, pg. 115. The only fixed things there were the initial and final field configurations. Everything else varied over all space and time, as expressed mathematically in the transition amplitude of (4-8). All imaginable relations between the fields, all the ways they could interact with one another, play out in the integration, except, in our example, those values of $\psi_i(y_1), \bar{\psi}_i(z_1), \psi_f(y_1)$, and $\bar{\psi}_f(z_1)$, which stay fixed.

The same thing is true for the Green function (4-23). Applied to our example, the fields $\psi_i(y_1), \bar{\psi}_i(z_1), \psi_f(y_1)$, and $\bar{\psi}_f(z_1)$ stay fixed, while all else varies. The RHS of Fig. 4-1 could represent our PIA Green function, as easily as it can represent our PIA transition amplitude.

Fig. 4-1 RHS shows initial and final fields fixed, same as legs are in PI Green function

Converting from PIA Green Functions to PIA Transition Amplitudes

So, just as we had rules for converting CA Green functions to CA transition amplitudes (like changing leg propagators to initial and final external particles), so we can have rules for changing PIA Green functions to PIA transition amplitudes.

The basic rule in the PIA is this. Change legs in the PIA Green function to chosen initial and final conditions in the PIA transition amplitude. As in the CA, which legs you choose for initial fields, and which for final fields, determine which of the several interactions the Green function can lead to we will actually analyze.

This is summarized in Wholeness Chart 4-2.

Wholeness Chart 4-2. Going from Green Functions to Transition Amplitudes for CA and PIA

CA and PIA rules for taking Green function → S_{fi}

Approach	In Green Function	e.g.	Change to	In Transition Amplitude	e.g.
CA	Legs	$S_F(\mathbf{p})$	→	Initial & final particles	$u_r(\mathbf{p})$
PIA	Legs	$\psi(y_1)$	→	Initial & final fields	$\psi_i(y_1)$

4.3.2 The Plan for Proving It

Note from all we did in Chap. 3, summarized in Wholeness Chart 4-4, pg. 126, that if we have the generating functional Z, we can get the Green function. And we can find the generating functional $Z[J_k, \sigma, \bar{\sigma}]$ for the interaction case from the generating functional for the non-interacting case, $Z_0[J_k, \sigma, \bar{\sigma}] = Z_0[J_k] Z_0[\sigma, \bar{\sigma}]$. In essence, we are working backwards in Wholeness Chart 4-4.

Still need to prove PIA Green function equivalent to CA one

That is, by starting with $Z_0[J_k]$ and $Z_0[\sigma,\bar{\sigma}]$, we can find Z, and from Z, we can find G^{μ}. Then from G^{μ}, we can find S_{fi}. We showed this for the CA.

The key: show $Z_0[J_\kappa]$ and $Z_0[\sigma,\bar{\sigma}]$ in PIA equivalent to CA, as we can derive all the rest from them

Now, if we can prove the PIA has the same $Z_0[J_k]$ and $Z_0[\sigma,\bar{\sigma}]$ as the CA, then the two approaches will have the same Z. And thus, using the same relation (3-104), the same Green function G^{μ} will result. And thus, yield the same S_{fi}. If, in doing this, we find the form of G^{μ} to be that of (4-23), then we will have proven the equivalence of (4-23), the Green function of the PIA, to (3-13), the Green function of the CA.

It will be easiest to do this if we start by tentatively assuming (4-23) is true, use it to determine Z, and then, from that, deduce $Z_0[J_k,\sigma,\bar{\sigma}]$, which equals $Z_0[J_k]Z_0[\sigma,\bar{\sigma}]$. If $Z_0[J_k]$ and $Z_0[\sigma,\bar{\sigma}]$ are the same as we found for the CA, then we have proven the two approaches are equivalent and (4-23) is the same as (3-13). This is how we will proceed in the following sections, and it is displayed schematically in Wholeness Chart 4-3.

Wholeness Chart 4-3. Showing Our Guess for Path Integral Green Function is Correct

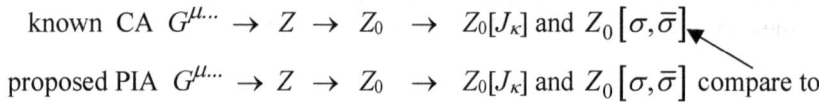

The plan for showing PIA equivalent to CA

known CA $G^{\mu\ldots} \rightarrow Z \rightarrow Z_0 \rightarrow Z_0[J_\kappa]$ and $Z_0[\sigma,\bar{\sigma}]$

proposed PIA $G^{\mu\ldots} \rightarrow Z \rightarrow Z_0 \rightarrow Z_0[J_\kappa]$ and $Z_0[\sigma,\bar{\sigma}]$ compare to

Note we actually derive quantities by starting on the RHS and moving to the LHS, in opposite direction to the arrows. But, given any quantity to the left of an arrow, we can deduce (by trial and error) what quantity to the right of the arrow would yield that to the left.

The reason we are doing this is because, as it will turn out, we can use the one type of functional integral ((4-21) for photons and (4-22) for spinors) that we know how to work out to help us find $Z_0[J_\kappa]$ and $Z_0[\sigma,\bar{\sigma}]$. We cannot do this with any of the other functional integrals in the other quantities shown in the second row of Wholeness Chart 4-3. But in being able to evaluate those two PIA quantities, we are then able to obtain valid expressions for all the other PIA quantities.

4.4 Generating Functional in the PI Approach

4.4.1 The Interacting Generating Functional Z

We can make a guess (or at least I, the author can, knowing what lies ahead) for the PIA generating functional, including interactions, as

$$Z[J_k,\sigma,\bar{\sigma}]=\frac{\int e^{i\int(\mathcal{L}+\mathcal{L}_S)d^4\underline{x}}\mathcal{D}A\mathcal{D}\bar{\psi}\mathcal{D}\psi}{\int e^{i\int\mathcal{L}d^4\underline{x}}\mathcal{D}A\mathcal{D}\bar{\psi}\mathcal{D}\psi}=\frac{\int e^{i\int\left(\mathcal{L}_0+\mathcal{L}_I+J_\kappa A^\kappa+\bar{\sigma}\psi+\bar{\psi}\sigma\right)d^4\underline{x}}\mathcal{D}A\mathcal{D}\bar{\psi}\mathcal{D}\psi}{\int e^{i\int\left(\mathcal{L}_0+\mathcal{L}_I\right)d^4\underline{x}}\mathcal{D}A\mathcal{D}\bar{\psi}\mathcal{D}\psi}. \qquad (4\text{-}24)$$

The PIA form of the generating functional (yet to be validated)

where the source Lagrangian $\mathcal{L}_S$ and its source fields are as defined in Chap. 3, pg. 92.

Then, we can see if (4-24) inserted into the relation (3-104), which we repeat below (but give it a new equation number (4-25) for this chapter), gives us (4-23).

$$G^{\mu..}=(-1)^{\bar{n}}\left(\frac{1}{i}\right)^n\frac{\delta^n Z[J_k,\sigma,\bar{\sigma}]}{\delta J_\mu(x_1)\ldots\delta\bar{\sigma}(y_1)\ldots\delta\sigma(z_1)\ldots}\bigg|_0 \quad \text{no sum on } n \qquad \text{(same as (3-104))(4-25)}$$

Use same relationship between generating functional and Green function as in CA

So, let's do it. Let's plug (4-24) into (4-25) and see if we get (4-23).

$$G^{\mu..}(x_1,\ldots,y_1,\ldots,z_1,\ldots)$$

$$=(-1)^{\bar{n}}\left(\frac{1}{i}\right)^n \frac{\delta^n}{\delta J_\mu(x_1)\ldots\delta\bar{\sigma}(y_1)\ldots\delta\sigma(z_1)\ldots}\left.\frac{\int e^{i\int(\mathcal{L}+J_\kappa A^\kappa+\bar{\sigma}\psi+\bar{\psi}\sigma)d^4\underline{x}}\mathcal{D}A\mathcal{D}\bar{\psi}\mathcal{D}\psi}{\int e^{i\int\mathcal{L}d^4\underline{x}}\mathcal{D}A\mathcal{D}\bar{\psi}\mathcal{D}\psi}\right|_0 \quad (4\text{-}26)$$

Start to proving the same relationship works in the PIA

$\bar{n}$ = derivatives wrt $\bar{\sigma}$ fields, n = derivatives wrt all fields, no sum on n.

From what we learned in Chap. 3 about functional derivatives, we hopefully can see that (4-26) yields (4-23) (repeated below). Note that each derivative (n of them) yields a factor of i in the numerator, and each derivative with respect to $\bar{\sigma}$, ($\bar{n}$ of them) yields a factor of -1.

$$G^{\mu..}(x_1,\ldots,y_1,\ldots,z_1,\ldots)=\frac{\int\left\{e^{i\int\mathcal{L}d^4\underline{x}}A^\mu(x_1)\ldots\psi(y_1)\ldots\bar{\psi}(z_1)\right\}\mathcal{D}A\mathcal{D}\bar{\psi}\mathcal{D}\psi}{\int e^{i\int\mathcal{L}d^4\underline{x}}\mathcal{D}A\mathcal{D}\bar{\psi}\mathcal{D}\psi}, \quad \text{repeat of } (4\text{-}23)$$

We get Green function from Z in same way in both approaches

If (4-23) is not obvious to you from (4-26), do **Problem 2** to show it explicitly.

So, if (4-24) is the correct form (i.e., equivalent to CA case) for the PIA generating functional, then (4-23) is the correct form for the PIA Green function. The Green function in both cases is derived from Z via the same relation (4-25).

4.4.2 The Free Generating Functional Z_0

From here, it should help to follow Wholeness Chart 3-5. pg. 96, which lays out the last few steps in each row of Wholeness Chart 4-3 in greater detail.

As with the CA, we define the PIA free generating functional Z_0 as Z with $\mathcal{L}_I = 0$. From (4-24),

$$Z_0\left[J_k,\sigma,\bar{\sigma}\right]=\frac{\int e^{i\int(\mathcal{L}_0+\mathcal{L}_S)d^4\underline{x}}\mathcal{D}A\mathcal{D}\bar{\psi}\mathcal{D}\psi}{\int e^{i\int\mathcal{L}_0 d^4\underline{x}}\mathcal{D}A\mathcal{D}\bar{\psi}\mathcal{D}\psi}=\frac{\int e^{i\int(\mathcal{L}_0+J_\kappa A^\kappa+\bar{\sigma}\psi+\bar{\psi}\sigma)d^4\underline{x}}\mathcal{D}A\mathcal{D}\bar{\psi}\mathcal{D}\psi}{\int e^{i\int\mathcal{L}_0 d^4\underline{x}}\mathcal{D}A\mathcal{D}\bar{\psi}\mathcal{D}\psi}. \quad (4\text{-}27)$$

PIA Z_0 found using $\mathcal{L}_I = 0$ in Z, as in CA

Note that (4-27) can be broken into two factors.

$$Z_0\left[J_k,\sigma,\bar{\sigma}\right]=\frac{\int e^{i\int\left(\mathcal{L}_0^1+\mathcal{L}_0^{1/2}+J_\kappa A^\kappa+\bar{\sigma}\psi+\bar{\psi}\sigma\right)d^4\underline{x}}\mathcal{D}A\mathcal{D}\bar{\psi}\mathcal{D}\psi}{\int e^{i\int\left(\mathcal{L}_0^1+\mathcal{L}_0^{1/2}\right)d^4\underline{x}}\mathcal{D}A\mathcal{D}\bar{\psi}\mathcal{D}\psi} \quad \left(\begin{matrix}\text{superscripts for}\\ \text{spin 1 and spin }\frac{1}{2}\end{matrix}\right)$$

$$=\underbrace{\frac{\left(\int e^{i\int\left(\mathcal{L}_0^1+J_\kappa A^\kappa\right)d^4\underline{x}}\mathcal{D}A\right)}{\left(\int e^{i\int\mathcal{L}_0^1 d^4\underline{x}}\mathcal{D}A\right)}}_{Z_0\left[J_k\right]}\underbrace{\frac{\left(\int e^{i\int\left(\mathcal{L}_0^{1/2}+\bar{\sigma}\psi+\bar{\psi}\sigma\right)d^4\underline{x}}\mathcal{D}\bar{\psi}\mathcal{D}\psi\right)}{\left(\int e^{i\int\mathcal{L}_0^{1/2}d^4\underline{x}}\mathcal{D}\bar{\psi}\mathcal{D}\psi\right)}}_{Z_0\left[\sigma,\bar{\sigma}\right]} \quad (4\text{-}28)$$

$$=Z_0\left[J_k\right]Z_0\left[\sigma,\bar{\sigma}\right].$$

PIA $Z_0\left[J_k,\sigma,\bar{\sigma}\right]=Z_0\left[J_k\right]Z_0\left[\sigma,\bar{\sigma}\right]$ as in CA

(4-28) is just as we found in the CA (3-122) for the free generating functional there.

4.4.3 Finding Z from the Free Generating Functional Z_0

We next seek to prove the same relationship exists between Z_0 and Z in the PIA as we found in the CA with (3-125). That is, we want to prove (4-29), where N_I is a factor to be determined that plays the same role as $\langle 0|S|0\rangle$ plays in (3-125), or equivalently, the second row of (3-128).

$$\boxed{\begin{aligned}Z\left[J_k,\sigma,\bar{\sigma}\right]&=\frac{1}{N_I}e^{ie\int I_\delta(x)d^4x}Z_0\left[J_k,\sigma,\bar{\sigma}\right]\\ I_\delta(x)&=\left(-\frac{1}{i}\frac{\delta}{\delta\sigma(x)}\right)\gamma_\mu\left(\frac{1}{i}\frac{\delta}{\delta\bar{\sigma}(x)}\right)\left(\frac{1}{i}\frac{\delta}{\delta J_\mu(x)}\right).\end{aligned}} \quad (4\text{-}29)$$

PIA Z found from PIA Z_0 same way as in CA

Proof follows

Proof of (4-29)

Step 1. Calculate $I_\delta(x)$ acting on Z_0.

The proof

$$I_\delta(x)Z_0[J_k,\sigma,\bar{\sigma}]$$

Step 1: Find $I_\delta(x)Z_0$ by plugging & chugging

$$=\left(-\frac{1}{i}\frac{\delta}{\delta\sigma(x)}\right)\gamma_\mu\left(\frac{1}{i}\frac{\delta}{\delta\bar{\sigma}(x)}\right)\left(\frac{1}{i}\frac{\delta}{\delta J_\mu(x)}\right)\frac{\int e^{i\int(\mathcal{L}_0+J_\kappa A^\kappa+\bar{\sigma}\psi+\bar{\psi}\sigma)d^4\underline{x}}\mathcal{D}A\mathcal{D}\bar{\psi}\mathcal{D}\psi}{\int e^{i\int\mathcal{L}_0 d^4\underline{x}}\mathcal{D}A\mathcal{D}\bar{\psi}\mathcal{D}\psi}$$

$$=\left(-\frac{1}{i}\frac{\delta}{\delta\sigma(x)}\right)\gamma_\mu\left(\frac{1}{i}\frac{\delta}{\delta\bar{\sigma}(x)}\right)\frac{\int A^\mu(x)e^{i\int(\mathcal{L}_0+J_\kappa A^\kappa+\bar{\sigma}\psi+\bar{\psi}\sigma)d^4\underline{x}}\mathcal{D}A\mathcal{D}\bar{\psi}\mathcal{D}\psi}{\int e^{i\int\mathcal{L}_0 d^4\underline{x}}\mathcal{D}A\mathcal{D}\bar{\psi}\mathcal{D}\psi} \tag{4-30}$$

$$=\frac{\int\left(\bar{\psi}\gamma_\mu\psi A^\mu\right)_x e^{i\int(\mathcal{L}_0+J_\kappa A^\kappa+\bar{\sigma}\psi+\bar{\psi}\sigma)d^4\underline{x}}\mathcal{D}A\mathcal{D}\bar{\psi}\mathcal{D}\psi}{\int e^{i\int\mathcal{L}_0 d^4\underline{x}}\mathcal{D}A\mathcal{D}\bar{\psi}\mathcal{D}\psi}$$

$$=\frac{1}{e}\frac{\int\mathcal{L}_I(x)e^{i\int(\mathcal{L}_0+J_\kappa A^\kappa+\bar{\sigma}\psi+\bar{\psi}\sigma)d^4\underline{x}}\mathcal{D}A\mathcal{D}\bar{\psi}\mathcal{D}\psi}{\int e^{i\int\mathcal{L}_0 d^4\underline{x}}\mathcal{D}A\mathcal{D}\bar{\psi}\mathcal{D}\psi}\qquad\left(\text{In }\frac{1}{e},\ e\text{ is the QED coupling constant}\right).$$

Step 2. Use Step 1 to calculate $e^{ie\int I_\delta(x)d^4x}Z_0[J_k,\sigma,\bar{\sigma}]$.

Step 2: Use step 1 to find $e^{ie\int I_\delta(x)d^4x}Z_0=Z$ by plug & chug

$$e^{ie\int I_\delta(x)d^4x}Z_0[J_k,\sigma,\bar{\sigma}]=\frac{e^{ie\int I_\delta(x)d^4x}\int e^{i\int(\mathcal{L}_0+J_\kappa A^\kappa+\bar{\sigma}\psi+\bar{\psi}\sigma)d^4\underline{x}}\mathcal{D}A\mathcal{D}\bar{\psi}\mathcal{D}\psi}{\int e^{i\int\mathcal{L}_0 d^4\underline{x}}\mathcal{D}A\mathcal{D}\bar{\psi}\mathcal{D}\psi}$$

$$=\frac{\int\left(I+ie\int I_\delta(x_1)d^4x_1-\frac{e^2}{2}\int I_\delta(x_1)d^4x_1\int I_\delta(x_2)d^4x_2+\dots\right)e^{i\int(\mathcal{L}_0+J_\kappa A^\kappa+\bar{\sigma}\psi+\bar{\psi}\sigma)d^4\underline{x}}\mathcal{D}A\mathcal{D}\bar{\psi}\mathcal{D}\psi}{\int e^{i\int\mathcal{L}_0 d^4\underline{x}}\mathcal{D}A\mathcal{D}\bar{\psi}\mathcal{D}\psi}$$

$$=\frac{\int\left(I+i\int\mathcal{L}_I(x_1)d^4x_1-\frac{1}{2}\int\mathcal{L}_I(x_1)d^4x_1\int\mathcal{L}_I(x_2)d^4x_2+\dots\right)e^{i\int(\mathcal{L}_0+J_\kappa A^\kappa+\bar{\sigma}\psi+\bar{\psi}\sigma)d^4\underline{x}}\mathcal{D}A\mathcal{D}\bar{\psi}\mathcal{D}\psi}{\int e^{i\int\mathcal{L}_0 d^4\underline{x}}\mathcal{D}A\mathcal{D}\bar{\psi}\mathcal{D}\psi}$$

$$=\frac{\int e^{i\int\mathcal{L}_I(x)d^4x}e^{i\int(\mathcal{L}_0+J_\kappa A^\kappa+\bar{\sigma}\psi+\bar{\psi}\sigma)d^4\underline{x}}\mathcal{D}A\mathcal{D}\bar{\psi}\mathcal{D}\psi}{\int e^{i\int\mathcal{L}_0 d^4\underline{x}}\mathcal{D}A\mathcal{D}\bar{\psi}\mathcal{D}\psi}=\frac{\int e^{i\int(\mathcal{L}_0+\mathcal{L}_I+J_\kappa A^\kappa+\bar{\sigma}\psi+\bar{\psi}\sigma)d^4\underline{x}}\mathcal{D}A\mathcal{D}\bar{\psi}\mathcal{D}\psi}{\int e^{i\int\mathcal{L}_0 d^4\underline{x}}\mathcal{D}A\mathcal{D}\bar{\psi}\mathcal{D}\psi} \tag{4-31}$$

$$=Z[J_k,\sigma,\bar{\sigma}]\frac{\int e^{i\int(\mathcal{L}_0+\mathcal{L}_I)d^4\underline{x}}\mathcal{D}A\mathcal{D}\bar{\psi}\mathcal{D}\psi}{\int e^{i\int\mathcal{L}_0 d^4\underline{x}}\mathcal{D}A\mathcal{D}\bar{\psi}\mathcal{D}\psi}=Z[J_k,\sigma,\bar{\sigma}]N_I=\underbrace{e^{ie\int I_\delta(x)d^4x}Z_0[J_k,\sigma,\bar{\sigma}]}_{\text{From 1st row of Step 2}},$$

Proven

which, with N_I determined, is (4-29) and parallels (3-125), and the 2nd row of (3-128). QED.

End Proof

4.4.4 What We Have Shown and What We Need to Show

What We Have Shown

We have shown PIA theory deduced in same way from Z_0 as in CA

Note that we have shown the second row of Wholeness Chart 4-3, pg. 120. That is, just as we showed in Chap. 3 for the CA, we have here shown that in the PIA, for given Z_0, we can calculate Z; from Z, we can calculate the Green function, and from that, we can get the transition amplitude.

But note carefully, that in deducing all this, we assumed our PIA Green function (4-23) is the same as our CA Green function (3-13). We have not yet proven that, i.e., we have not proven that the second row in Wholeness Chart 4-3 is equivalent to the first.

But we have not shown they are equivalent yet

What We Need to Show

If we can show that the $Z_0[J_k]$ and $Z_0[\sigma,\bar{\sigma}]$ of the PIA (end of second row in Wholeness Chart 4-3) are identical to the $Z_0[J_k]$ and $Z_0[\sigma,\bar{\sigma}]$ of the CA (end of first row), then, since the entire theory of either approach can be produced in the same way from those entities, we will have proven that the two approaches are identical. That is what we do in Sect. 4.4.5.

If show the respective $Z_0[J_k]$ and $Z_0[\sigma,\bar{\sigma}]$ are equivalent, that shows both entire approaches are

One Result from When We Show Z_0 is the Same in the CA and PIA (Section added in book revision)

If $Z_0[J_k,\sigma,\bar{\sigma}]$ is the same in the CA and PIA (which we will prove), we have an important result. Consider, from (3-94) and (4-24), that

$$\underbrace{Z[0,0,0]=1}_{\text{CA}} \qquad\qquad \underbrace{Z[0,0,0]=1}_{\text{PIA}}. \tag{4-31)+1}$$

(3-128) [equivalent to (3-125)] and the last row of (4-31) [equivalent to the 1st row of (4-29)] are

$$\underbrace{Z[J_k,\sigma,\bar{\sigma}]\langle 0|S|0\rangle = e^{ie\int I_\delta(x)d^4x}Z_0[J_k,\sigma,\bar{\sigma}]}_{\text{CA}} \quad \underbrace{Z[J_k,\sigma,\bar{\sigma}]N_I = e^{ie\int I_\delta(x)d^4x}Z_0[J_k,\sigma,\bar{\sigma}]}_{\text{PIA}}. \tag{4-31)+2}$$

Neither S nor N_I is a function of the source fields J_k,σ, or $\bar{\sigma}$, so they must be the same for any value of those fields, including $J_k=\sigma=\bar{\sigma}=0$, when $Z=1$. Thus, from (4-31)+2 we have

$$\underbrace{\overbrace{Z[0,0,0]}^{1}\langle 0|S|0\rangle = e^{ie\int I_\delta(x)d^4x}Z_0[J_k,\sigma,\bar{\sigma}]_{J_k=\sigma=\bar{\sigma}=0}}_{\text{CA}} \quad \underbrace{\overbrace{Z[0,0,0]}^{1}N_I = e^{ie\int I_\delta(x)d^4x}Z_0[J_k,\sigma,\bar{\sigma}]_{J_k=\sigma=\bar{\sigma}=0}}_{\text{PIA}}. \tag{4-31)+3}$$

Conclusion: If Z_0 is the same in the CA and PIA, then $\langle 0|S|0\rangle = N_I$, and that means, from (4-31)+2, that Z is the same in both, as well.

4.4.5 Proving the Free Generating Functionals in Both Approaches are Equivalent

We seek to show that the following, which we found true in the CA (i.e., (3-115) and (3-121)), is true in the PIA.

$$Z_0[J_k]=e^{-\frac{i}{2}\left[J_\mu D_F^{\mu\nu}J_\nu\right]}=e^{-\frac{i}{2}\int J_\mu(\underline{x})D_F^{\mu\nu}(\underline{x},\underline{y})J_\nu(\underline{y})d^4\underline{x}\,d^4\underline{y}}$$
$$Z_0[\sigma,\bar{\sigma}]=e^{-i[\bar{\sigma}S_F\sigma]}=e^{-i\int\bar{\sigma}(\underline{x})S_F(\underline{x},\underline{y})\sigma(\underline{y})d^4\underline{x}\,d^4\underline{y}}, \tag{4-32}$$

where the integrals are of form (4-21) and (4-22).

We need to show $Z_0[J_k]$ & $Z_0[\sigma,\bar{\sigma}]$ equal relations at left, which are same as CA

Tools for Proving (4-32)

We will use (4-21) and (4-22) as well as several other relations in the course of proving (4-32). We first derive those extra relations below.

To help us show it, we'll need several relations that are proven below

First relation: An integration by parts expression

$$\text{We know } \underbrace{\partial_\mu}_{d}\Big(\underbrace{A_\nu}_{u}\,\underbrace{\partial^\mu A^\nu}_{v}\Big)=\underbrace{\partial_\mu A_\nu}_{du}\,\underbrace{\partial^\mu A^\nu}_{v}+\underbrace{A_\nu}_{u}\,\underbrace{\partial_\mu\partial^\mu A^\nu}_{dv}. \tag{4-33}$$

(4-33) and the standard result from the integral of a divergence yields

$$\int\partial_\mu\left(A_\nu\,\partial^\mu A^\nu\right)d^4x=\left(A_\nu\,\partial^\mu A^\nu\right)\Big|_{-\infty}^{+\infty}=\int\partial_\mu A_\nu\,\partial^\mu A^\nu\,d^4x+\int A_\nu\,\partial_\mu\partial^\mu A^\nu\,d^4x. \tag{4-34}$$

With the reasonable assumption that the field and/or its derivative vanish at infinity, we have

$$\int\partial_\mu A_\nu\,\partial^\mu A^\nu\,d^4x=-\int A_\nu\,\partial_\mu\partial^\mu A^\nu\,d^4x. \tag{4-35}$$

First relation we'll need

End of one integration by parts expression.

Second relation: Additional integration by parts expression.

$$\int \partial_\mu \left(A_\nu \partial^\nu A^\mu \right) d^4x = \int \left(\partial_\mu A_\nu \right)\left(\partial^\nu A^\mu \right) d^4x + \int A_\nu \partial_\mu \partial^\nu A^\mu d^4x$$
$$= A_\nu \partial^\nu A^\mu \Big|_{-\infty}^{+\infty} = 0 \ \left(\text{assuming } A_\nu \text{ or } \partial^\nu A^\mu = 0 \text{ at } \infty \right). \tag{4-36}$$

So, $$\int \left(\partial_\mu A_\nu \right)\left(\partial^\nu A^\mu \right) d^4x = -\int A_\nu \partial^\nu \partial_\mu A^\mu d^4x \ \left(= 0 \text{ in Lorenz gauge where } \partial_\mu A^\mu = 0 \right). \tag{4-37}$$

Second relation we'll need

End of additional integration by parts expression.

Third and fourth relations: Operator inverses.

Consider the Klein-Gordon equation operator, the negative of which we will symbolize by *K*.

$$\left(\partial_\alpha \partial^\alpha + m^2 \right) \phi = 0 \qquad -\left(\partial_\alpha \partial^\alpha + m^2 \right) = K \qquad -K\phi = 0 \tag{4-38}$$

If *K* were a discrete matrix, its inverse K^{-1} would be found via $K_{ij} K_{jk}^{-1} = I_{ik}$, but in the continuum limit, this becomes

$$KK^{-1} = \delta(x-y) = -\left(\partial_\alpha \partial^\alpha + m^2 \right) K^{-1}. \tag{4-39}$$

But this is the relation we found in (3-6) where $K^{-1} = G(x,y)$, which, as we showed in (3-9), yields

$$K^{-1}(x,y) = \Delta_F(x-y) = \frac{1}{(2\pi)^4} \int d^4k \frac{e^{-ik(x-y)}}{k^2 - m^2 + i\varepsilon}, \tag{4-40}$$

and this means that (*i* times) the inverse of the operator *K* of (4-38) (the negative of the Klein-Gordon operator) is simply the scalar propagator.

This can readily be extended to spin 1 fields, i.e., for a massive field,

$$\left(\partial_\alpha \partial^\alpha + m^2 \right) A^\mu = 0 \qquad -\left(\partial_\alpha \partial^\alpha + m^2 \right) = K \qquad -KA^\mu = 0, \tag{4-41}$$

Third relation is inverse of Maxwell equation operator

where $K^{-1\mu\nu} = G^{\mu\nu}$ of (3-11),

$$K^{-1\mu\nu}(x-y) = \frac{-g^{\mu\nu}}{(2\pi)^4} \int d^4k \frac{e^{-ik(x-y)}}{k^2 - m^2 + i\varepsilon} \xrightarrow{\text{for } m=0} = D_F^{\mu\nu}(x-y). \tag{4-42}$$

which is this, at left

So, (*i* times) the inverse of the negative of the Maxwell equation operator is the photon propagator.

Similarly, as noted in Chap. 3, for the Dirac equation operator,

$$\left(i\gamma^\alpha \partial_\alpha - m \right) \psi = 0 \qquad \left(i\gamma^\alpha \partial_\alpha - m \right) = K_S \qquad K_S \psi = 0, \tag{4-43}$$

Similar relation is inverse of Dirac equation operator

where $K_S^{-1} = G_S$,

$$K_S^{-1}(x-y) = \frac{1}{(2\pi)^4} \int \frac{e^{-ip(x-y)} \left(\not{p} + m \right)}{p^2 - m^2 + i\varepsilon} d^4p = S_F(x-y), \tag{4-44}$$

which is this, at left

and (*i* times) the inverse of the Dirac equation operator K_S is the spinor propagator.
End of operator inverses derivation.

<u>Proving (4-32)</u>

Proving PIA $Z_0[J_k]$ same as in CA

Consider the PIA form of (4-28),

$$Z_0[J_k] = \frac{\int e^{i\int \left(\mathcal{L}_0^1 + J_\kappa A^\kappa \right) d^4\underline{x}} \mathcal{D}A}{\int e^{i\int \mathcal{L}_0^1 d^4\underline{x}} \mathcal{D}A} = \frac{\int e^{i\int \left(\frac{1}{2}\left(-\partial_\mu A_\nu \partial^\mu A^\nu + \partial_\mu A_\nu \partial^\nu A^\mu \right) + J_\kappa A^\kappa \right) d^4\underline{x}} \mathcal{D}A}{\int e^{i\int \frac{1}{2}\left(-\partial_\mu A_\nu \partial^\mu A^\nu + \partial_\mu A_\nu \partial^\nu A^\mu \right) d^4\underline{x}} \mathcal{D}A}. \tag{4-45}$$

Use (4-35) with real A^ν and (4-37) with the Lorenz gauge in (4-45) to obtain

Using relations we derived earlier

$$Z_0[J_k] = \frac{\int e^{i\int \left(\frac{1}{2} A_\nu \partial_\mu \partial^\mu A^\nu - \frac{1}{2} A_\nu \partial^\nu \partial_\mu A^\mu + J_\kappa A^\kappa \right) d^4\underline{x}} \mathcal{D}A}{\int e^{i\int \left(\frac{1}{2} A_\nu \partial_\mu \partial^\mu A^\nu - \frac{1}{2} A_\nu \partial^\nu \partial_\mu A^\mu \right) d^4\underline{x}} \mathcal{D}A} = \frac{\int e^{i\int \left(\frac{1}{2} A_\nu \partial_\mu \partial^\mu A^\nu + J_\kappa A^\kappa \right) d^4\underline{x}} \mathcal{D}A}{\int e^{i\int \frac{1}{2} A_\nu \partial_\mu \partial^\mu A^\nu d^4\underline{x}} \mathcal{D}A}. \tag{4-46}$$

Now use (4-21), essentially the only form of the path integral for vector fields we know how to evaluate, with real $A^\nu, K = \partial_\mu \partial^\mu$, and real $b_\nu = J_\nu$. Then,

$$Z_0[J_k] = \frac{\lim\limits_{n\to\infty}\sqrt{\frac{(i2\pi)^n}{Det\, K_n}}\, e^{-\frac{i}{2}\int J_\mu(\underline{x}) K^{-1\mu\nu}(\underline{x},\underline{y}) J_\nu(\underline{y}) d^4\underline{x} d^4\underline{y}}}{\lim\limits_{n\to\infty}\sqrt{\frac{(i2\pi)^n}{Det\, K_n}}} \quad (\text{no sum on } n), \qquad (4\text{-}47)$$

Weird constant drops out, as promised

where the infinite constants cancel (as we promised). Along with (4-42), (4-47) becomes

$$Z_0[J_k] = e^{-\frac{i}{2}\int J^\mu(\underline{x}) D_{F\mu\nu}(\underline{x},\underline{y}) J^\nu(\underline{y}) d^4\underline{x} d^4\underline{y}} = e^{-\frac{i}{2}\left[J_\mu D_F^{\mu\nu} J_\nu\right]}, \qquad (4\text{-}48)$$

Proven: PIA $Z_0[J_K]$ same as CA $Z_0[J_K]$

which is the same $Z_0[J_k]$ as we found in the CA, (3-115).

In similar fashion, considering the PIA form of $Z_0[\sigma,\bar{\sigma}]$ in (4-28), we would find

Proving: PIA $Z_0[\sigma,\bar{\sigma}]$ same as in CA

$$Z_0[\sigma,\bar{\sigma}] = \frac{\left(\int e^{i\int\left(\mathcal{L}_0^{1/2} + \bar{\sigma}\psi + \bar{\psi}\sigma\right)d^4\underline{x}}\mathcal{D}\bar{\psi}\mathcal{D}\psi\right)}{\left(\int e^{i\int \mathcal{L}_0^{1/2} d^4\underline{x}}\mathcal{D}\bar{\psi}\mathcal{D}\psi\right)} = e^{-i\int \bar{\sigma}(\underline{x}) S_F(\underline{x},\underline{y})\sigma(\underline{y}) d^4\underline{x} d^4\underline{y}} = e^{-i\left[\bar{\sigma} S_F \sigma\right]}. \qquad (4\text{-}49)$$

Proven (if you do Prob 3): PIA $Z_0[\sigma,\bar{\sigma}]$ same as in CA

To prove (4-49), do **Problem 3**.

(4-49) is the same as $Z_0[\sigma,\bar{\sigma}]$ in the CA, (3-121), so from (4-28), so is

$$\text{PIA} \quad Z_0[J_k,\sigma,\bar{\sigma}] = Z_0[J_k] Z_0[\sigma,\bar{\sigma}] = e^{-\frac{i}{2}\left[J_\mu D_F^{\mu\nu} J_\nu\right]} e^{-i\left[\bar{\sigma} S_F \sigma\right]} = \quad \text{CA} \quad Z_0[J_k,\sigma,\bar{\sigma}]. \quad (4\text{-}50)$$

So, PIA $Z_0[J_k,\sigma,\bar{\sigma}]$ same as in CA

End of proof

4.5 Chapter Summary: What Have We Proven and Why?

4.5.1 The Proposed PIA and the Known CA are Equivalent

As we have shown in this chapter and Chap. 3 (as summarized in Wholeness Chart 4-3, pg. 120 and in more detail in Wholeness Chart 4-4, pg. 126), if we know Z_0, we know the entire theory. Z_0 gives us Z; Z gives us $G^{\mu\nu\ldots}$; and $G^{\mu\nu\ldots}$ gives us S_{fi}, the transition amplitude. And the steps to get any particular S_{fi} we seek are the same in both the CA and the PIA. (Note that Wholeness Chart 4-4 adds the path integral approach column to Wholeness Chart 3-7, at the end of Chap. 3.)

Since Z_0 is same in both PIA and CA, and rest of the theory is derived in same way from Z_0, then the two approaches are equivalent

As shown above, Z_0 is the same thing in both the CA and the PIA. Then, the Z, $G^{\mu\nu\ldots}$, and S_{fi} we find, via the same procedures shown in this chapter and Chap. 3, are the same.

All the relations we have deduced for Z_0, Z, $G^{\mu\nu\ldots}$, and S_{fi} in the PIA are equivalent to those we found in the CA. The two approaches describe one and the same thing.

4.5.2 The PIA Actually Works

Our proposed PIA has been validated because it equals the known-to-be valid CA. So, we have proven Feynman's original proposal that integrating over an infinite number of all imaginable field paths will give us the same result as the physically realizable path alone. Field paths away from the physically realizable path destructively interfere. Those close constructively interfere. It all works.

And thus, we have validated Feynman's original hypothesis for the many paths approach to fields

4.5.3 How This Will Help in the Future

Since we have the same free generating functional in both the CA and PIA, and so, we end up with the same transition amplitudes, the PIA seems, on the face of it, to offer us little in the calculation of those amplitudes. However, it proves very useful in theoretical developments of certain more complicated theories such as QCD, as we will see later in this book. Analogous results hold for those other theories that go beyond QED.

The PIA will help us with QCD and other theories more complicated than QED

Wholeness Chart 4-4. Comparison of Two Approaches to QED Green Functions and Generating Functional

	Canonical Approach (CA)	Path Integral Approach (PIA)	Comments
S operator	$S=\sum_{n=0}^{\infty}\frac{i^n}{n!}\int T\left\{\mathcal{L}_I(\underline{x}_1)...\mathcal{L}_I(\underline{x}_n)\right\}d^4\underline{x}_1...d^4\underline{x}_n$ $=T\left\{e^{i\int\mathcal{L}_I(\underline{x})d^4\underline{x}}\right\}$	N/A	See (3-14) Fields $A^\mu,\psi,\bar{\psi}$ are operators in CA, functions in PIA
Transition amplitude	$S_{fi}=\langle f\|S\|i\rangle=\langle f\|Te^{i\int\mathcal{L}_I(\underline{x})d^4\underline{x}}\|i\rangle$	$S_{fi}=C\underbrace{\int_{A_i,\bar{\psi}_i,\psi_i}^{A_f,\bar{\psi}_f,\psi_f}}_{\text{common symbol }\int}e^{i\int\mathcal{L}d^4x}\mathcal{D}A\mathcal{D}\bar{\psi}\mathcal{D}\psi$	Vol. 1, (19-5), pg. 514 & (4-8)
Green functions	Green function will be all propagators. For each external particle, change associated propagator to external line (e.g., $S_F(p)\rightarrow u(\mathbf{p})$) to get amplitude.	← Ditto. Get same final result in PIA via different route.	All Green function legs time inward. Need final $p\rightarrow -p$ to get amplitude
	$G^{\mu...}(x_1,...y_1,...z_1,...)=$ $\dfrac{\langle 0\|T\left\{SA^\mu(x_1)...\psi(y_1)...\bar{\psi}(z_1)...\right\}\|0\rangle}{\langle 0\|S\|0\rangle}$ Ignore denominator above for connected Feynman diagrams	$G^{\mu..}(x_1,.y_1,.z_1,.)=$ $\dfrac{\int\left\{e^{i\int\mathcal{L}d^4x}A^\mu(x_1)..\psi(y_1)..\bar{\psi}(z_1)\right\}\mathcal{D}A\mathcal{D}\bar{\psi}\mathcal{D}\psi}{\int e^{i\int\mathcal{L}d^4x}\mathcal{D}A\mathcal{D}\bar{\psi}\mathcal{D}\psi}$	(3-13) & (4-23) Analogous results hold for other field theories (e.g., weak, strong interactions).
Generating functional	Z used to find Green functions	← Ditto	
	$Z[J_k,\sigma,\bar{\sigma}]=\dfrac{\langle 0\|S'\|0\rangle}{\langle 0\|S\|0\rangle}$ where ↓ $S'=T\left\{e^{i\int\mathcal{L}_I'(\underline{x})d^4\underline{x}}\right\}$ $\mathcal{L}_I'=\mathcal{L}_I+\mathcal{L}_S$ where $\mathcal{L}_S=J_\kappa A^\kappa+\bar{\sigma}\psi+\bar{\psi}\sigma$	$Z[J_k,\sigma,\bar{\sigma}]=\dfrac{\int e^{i\int(\mathcal{L}+\mathcal{L}_S)d^4x}\mathcal{D}A\mathcal{D}\bar{\psi}\mathcal{D}\psi}{\int e^{i\int(\mathcal{L})d^4x}\mathcal{D}A\mathcal{D}\bar{\psi}\mathcal{D}\psi}$	(3-93) & (4-24) $\mathcal{L}_S$ called "source Lagrangian". $J_\kappa,\bar{\sigma},\sigma$ sources are fictitious and classical fields, not operators. Analogous results for other field theories.
Green function from generating functional	$G^{\mu..}=(-1)^{\bar{n}}\left(\frac{1}{i}\right)^n\left.\dfrac{\delta^n Z[J_k,\sigma,\bar{\sigma}]}{\delta J_\mu(x_1)..\delta\bar{\sigma}(y_1)..\delta\sigma(z_1)}\right\|_0$	← Ditto, but for PIA Z	(3-104) & (4-26) $\|_0$ means $J=\bar{\sigma}=\sigma=0$ after taking derivatives
	Defs: n = tot num source fields = total num derivatives = <u>points</u> of Green function $\bar{n}$ = tot num of σ source fields = tot num derivatives with respect to σ <u>order</u> = number of vertices in graph (Green function has all orders in it due to $\mathcal{L}_I$)		
	Plug and chug Z (for CA) in above to prove one gets $G^{\mu...}$ of CA	← Ditto, but for PIA Z and $G^{\mu...}$	
Significance	Knowing $G^{\mu..}$ for a case with particular fields $A^\mu,\psi,\bar{\psi}$, one knows the amplitude for that case. Knowing Z, one knows all possible $G^{\mu...}$. Z contains the whole theory.		

Proving the equivalence of the canonical and PI approaches by using the blocks below.			
Free field case	Take $\mathcal{L}_I = 0$ in canonical Z above → Z_0 $Z_0[J_k,\sigma,\bar{\sigma}] = Z_0[J_k]Z_0[\sigma,\bar{\sigma}]$	← Ditto, but PIA form of Z_0	(3-122) & (4-28) Show by substitution & chugging math.
Interacting fields case	$Z[J_k,\sigma,\bar{\sigma}] = \dfrac{e^{ie\int I_\delta(x)d^4x} Z_0[J_k,\sigma,\bar{\sigma}]}{\langle 0\|S\|0\rangle}$ $I_\delta(x) = \left(-\dfrac{1}{i}\dfrac{\delta}{\delta\sigma(x)}\right)\gamma_\mu\left(\dfrac{1}{i}\dfrac{\delta}{\delta\bar{\sigma}(x)}\right)\left(\dfrac{1}{i}\dfrac{\delta}{\delta J_\mu(x)}\right)$	$Z[J_k,\sigma,\bar{\sigma}] = \dfrac{e^{ie\int I_\delta(x)d^4x} Z_0[J_k,\sigma,\bar{\sigma}]}{N_I}$ but for PIA form of Z_0 ← Ditto.	(3-125) & (4-29) Show by substitution & chugging math.
	For same Z_0 in CA and PIA, $\langle 0\|S\|0\rangle = N_I$		(4-31)+3
Equivalent free fields expression	$Z_0[J_k] = e^{-\frac{i}{2}\left[J_\mu D_F^{\mu\nu} J_\nu\right]}$ $Z_0[\sigma,\bar{\sigma}] = e^{-i[\bar{\sigma}S_F\sigma]}$ $[AKB] = \int A(\underline{x})K(\underline{x},\underline{y})B(\underline{y})d^4\underline{x}d^4\underline{y}$	← Ditto	(3-115) & (3-121), (4-48) & (4-49) (3-89)
Significance	1) Free field generating functional is easy (relatively) to calculate & one can get interacting case from it. 2) Because free field generating functional is same for CA and PIA, their interacting generating functionals are as well, so the two approaches are equivalent. (Green functions and thus, transition amplitudes also equivalent.)		

Alternative symbolism for Green function			
	$\langle A^\mu(x_1)..\psi(y_1)..\bar{\psi}(z_1)..\rangle = G^{\mu..}(x_1,.y_1,.z_1,.)$	← Ditto	(3-68)
	Free fields case: $\langle A^\mu(x_1)..\psi(y_1)..\bar{\psi}(z_1)..\rangle_0$	← Ditto	

End of Part 2, Path Integrals in QED

This concludes our introduction to the path integral approach and its application in QED.

4.6 Appendix: Evaluating the Fermionic Path Integral

4.6.1 Quick Review of Grassmann Algebra

In Chap. 3, we looked at Grassmann algebras and differentiation with respect to generators of the algebra, which we have denoted by θ_i. See pgs. 87-91. Here we will look at integration over those generators.

Of the properties shown in Chap. 3, we recall

$$\left[\theta_i,\theta_j\right]_+ = 0 \qquad \theta_i^2 = 0 \tag{4-51}$$

and for $n = 2$ generators, the most general element f of the algebra, where p_i are ordinary numbers, is

$$f\left(p_0,p_1,p_2\right) = p_0 + p_1\theta_1 + p_2\theta_2 + p_{12}\theta_1\theta_2 = p_0 + p_1\theta_1 + p_2\theta_2 - p_{12}\theta_2\theta_1 . \tag{4-52}$$

For any n number of generators, it is

$$f\left(p_0,p_i,p_{ij},\dots p_{ij\dots n}\right) = p_0 + p_i\theta_i + p_{ij}\theta_i\theta_j + \dots + p_{ij..n}\theta_i\theta_j\dots\theta_n \qquad i<j<n . \tag{4-53}$$

In physics, our Grassmann generators will be $\theta_1 = \psi_1(x_1)$, $\theta_2 = \psi_2(x_2)$, etc., but for now we will take a more general approach and use the symbols θ_i.

4.6.2 Integration with Respect to Grassmann Variables

As we noted in Chap. 3 for "differentiation" with respect to Grassmann variables, such variables are abstract entities, so we will need to define what we mean operationally by the term "integration" with respect to them, just as we did with "differentiation".

To start, we will want our integration to be linear, so that (where X and Y are functions of the θ_i)

$$\int (aX + bY)\,d\theta_1 \ldots d\theta_n = a\int X\,d\theta_1 d\ldots d\theta_n + b\int Y\,d\theta_1 d\ldots d\theta_n \,. \tag{4-54}$$

For a single generator in (4-54), we have, given (4-53), the most general integral

$$\int (p_0 + p_1\theta)\,d\theta = p_0\int d\theta + p_1\int \theta\,d\theta \,. \tag{4-55}$$

We will want integration to yield a normal complex number, not a Grassmann number, so we will need to define

$$p_0\int d\theta = 0\,, \tag{4-56}$$

and, by convention, we define a normalization

$$\int \theta\,d\theta = 1\,. \tag{4-57}$$

This makes (4-55) look a lot like differentiation.

$$\int (p_0 + p_1\theta)\,d\theta = p_1 \qquad\qquad \frac{d}{d\theta}(p_0 + p_1\theta) = p_1\,, \tag{4-58}$$

so, integration and differentiation do the same thing for Grassmann numbers. In general,

$$\int X d\theta_1 d\theta_2 \ldots d\theta_n = \frac{\partial}{\partial\theta_1}\frac{\partial}{\partial\theta_2}\cdots\frac{\partial}{\partial\theta_n}X \tag{4-59}$$

which leads to

$$\int (\theta_n \ldots \theta_1)\,d\theta_1 d\theta_2 \ldots d\theta_n = \int \theta_n \ldots \left(\int \theta_2 \left(\int \theta_1\,d\theta_1\right) d\theta_2\right) \ldots d\theta_n = 1\,. \tag{4-60}$$

Nested integrals like (4-60) are evaluated from the inside out. And since θ_i anti-commute with one another, we take the $d\theta_i$ to do so, as well. For example,

$$\int \theta_2\theta_1\,d\theta_1 d\theta_2 = -\int \theta_1\theta_2\,d\theta_1 d\theta_2 = -\int \theta_2\theta_1\,d\theta_2 d\theta_1\,, \tag{4-61}$$

which is consistent with the way derivatives with respect to Grassmann variables behave.

Bottom line: For Grassmann variables, integration behaves like differentiation. In general, we can evaluate an integral via (4-59).

4.6.3 Path Integrals and Grassmann Variables

To prove (4-62), the first relation we will use in this section, do **Problem 4**.

Consider

$$e^{-\theta_1 A_{12}\theta_2} = 1 - A_{12}\theta_1\theta_2\,, \tag{4-62}$$

which, if you did Problem 4, you know is an exact, not approximate, relation. It is good for any θ_i, not just small ones. Then, an integral of (4-62) is

$$\int e^{-\theta_1 A_{12}\theta_2}\,d\theta_1 d\theta_2 = \int \left(1 - A_{12}\theta_1\theta_2\right) d\theta_1 d\theta_2 = \int A_{12}\theta_2\theta_1\,d\theta_1 d\theta_2 = A_{12}\,. \tag{4-63}$$

Relabeling, to make the next step easier, where the overbar simply denotes a different independent Grassmann variable,

$$\int e^{-\bar{\theta}_1 A_{11}\theta_1}\,d\bar{\theta}_1 d\theta_1 = \int \left(1 - A_{11}\bar{\theta}_1\theta_1\right) d\bar{\theta}_1 d\theta_1 = \int A_{11}\theta_1\bar{\theta}_1\,d\bar{\theta}_1\theta_1 = A_{11}\,. \tag{4-64}$$

Now, consider a number n of θ_i Grassmann variables plus another n independent Grassmann variables $\bar{\theta}_i$ in the integral

$$\int e^{-\bar{\theta}_i A_{ij}\theta_j} d\bar{\theta}_1 d\theta_1 \dots d\bar{\theta}_n d\theta_n = \int e^{-\bar{\theta}_1 A_{11}\theta_1} e^{-\bar{\theta}_1 A_{12}\theta_2} \dots e^{-\bar{\theta}_n A_{nn}\theta_n} d\bar{\theta}_1 d\theta_1 \dots d\bar{\theta}_n d\theta_n$$
$$= \int \left(1-\bar{\theta}_1 A_{11}\theta_1\right)\left(1-\bar{\theta}_1 A_{12}\theta_2\right)\dots\left(1-\bar{\theta}_n A_{nn}\theta_n\right) d\bar{\theta}_1 d\theta_1 \dots d\bar{\theta}_n d\theta_n . \quad (4\text{-}65)$$

This is simply n^2 repetitions of (4-63) and (4-64), and in expanding the factors of the integrand, no terms with any duplicate Grassmann variables will survive. And due to (4-56), only terms with exactly $n\,\theta_i$ and $n\,\bar{\theta}_i$ will survive.

Consider (4-65) for $n = 2$, where the second line below results because only terms with all the Grassmann variables, and no duplicates, will survive the integration,

$$\int e^{-\bar{\theta}_i A_{ij}\theta_j} d\bar{\theta}_1 d\theta_1 d\bar{\theta}_2 d\theta_2 = \int \left(1-\bar{\theta}_1 A_{11}\theta_1\right)\left(1-\bar{\theta}_1 A_{12}\theta_2\right)\left(1-\bar{\theta}_2 A_{21}\theta_1\right)\left(1-\bar{\theta}_2 A_{22}\theta_2\right) d\bar{\theta}_1 d\theta_1 d\bar{\theta}_2 d\theta_2$$
$$= \int \left(\bar{\theta}_1 A_{11}\theta_1 \bar{\theta}_2 A_{22}\theta_2 + \bar{\theta}_1 A_{12}\theta_2 \bar{\theta}_2 A_{21}\theta_1\right) d\bar{\theta}_1 d\theta_1 d\bar{\theta}_2 d\theta_2 . \quad (4\text{-}66)$$

Evaluating the last integral form in (4-66) (noting the sign changes that arise from anti-commutation) results in

$$\int e^{-\bar{\theta}_i A_{ij}\theta_j} d\bar{\theta}_1 d\theta_1 d\bar{\theta}_2 d\theta_2 = \int \bar{\theta}_1 A_{11}\theta_1 \bar{\theta}_2 A_{22}\theta_2 \, d\bar{\theta}_1 d\theta_1 d\bar{\theta}_2 d\theta_2 + \int \bar{\theta}_2 A_{21}\theta_1 \, \bar{\theta}_1 A_{12}\theta_2 d\bar{\theta}_1 d\theta_1 d\bar{\theta}_2 d\theta_2$$
$$= -\int A_{11}\theta_1 \bar{\theta}_2 A_{22}\theta_2 \underbrace{\bar{\theta}_1 d\bar{\theta}_1}_{\int \bar{\theta}_1 d\bar{\theta}_1 = 1} d\theta_1 d\bar{\theta}_2 d\theta_2 - \int \bar{\theta}_2 A_{21}\theta_1 A_{12}\theta_2 \underbrace{\bar{\theta}_1 d\bar{\theta}_1}_{\int \bar{\theta}_1 d\bar{\theta}_1 = 1} d\theta_1 d\bar{\theta}_2 d\theta_2$$
$$= -\int A_{11}\bar{\theta}_2 A_{22}\theta_2 \underbrace{\theta_1 d\theta_1}_{\int \theta_1 d\theta_1 = 1} d\bar{\theta}_2 d\theta_2 + \int \bar{\theta}_2 A_{21} A_{12}\theta_2 \underbrace{\theta_1 d\theta_1}_{\int \theta_1 d\theta_1 = 1} d\bar{\theta}_2 d\theta_2 \quad (4\text{-}67)$$
$$= \int A_{11}A_{22}\theta_2 \bar{\theta}_2 \, d\bar{\theta}_2 d\theta_2 - \int A_{21}A_{12}\theta_2 \bar{\theta}_2 d\bar{\theta}_2 d\theta_2 = \int A_{11}A_{22}\theta_2 \, d\theta_2 - \int A_{21}A_{12}\theta_2 d\theta_2$$
$$= A_{11}A_{22} - A_{21}A_{12} = Det\, A .$$

The same result can be obtained using (4-59).

To see a similar result for $n = 3$, do **Problem 5**.

The results of (4-67) and Problem 5 can be generalized to any n as

$$\int e^{-\bar{\theta}_i A_{ij}\theta_j} d\bar{\theta}_1 d\theta_1 \dots d\bar{\theta}_n d\theta_n = Det\, A_n , \quad (4\text{-}68)$$

where the subscript n is a reminder that matrix A has dimension n; that is A_{ij}, with $i, j = 1, 2, \dots, n$. Note this differs from our results for a real, ordinary variable ϕ (or A_ν) of (4-18) (or (4-21)), where there we had the square root of a determinant in the denominator.

Now, consider independent (fixed) Grassmann numbers η_i and $\bar{\eta}_i$ (which are *not* variables over which we will integrate) and the integral

$$\int e^{-\bar{\theta}_i A_{ij}\theta_j + \bar{\theta}_i \eta_i + \bar{\eta}_i \theta_i} d\bar{\theta}_1 d\theta_1 \dots d\bar{\theta}_n d\theta_n . \quad (4\text{-}69)$$

We want to re-express (4-69), but in doing so, we will need the following relation.

$$\bar{\theta}_i \eta_i = \bar{\theta}_j \eta_j = \bar{\theta}_j \delta_{jl} \eta_l = \bar{\theta}_j A_{jk} A^{-1}_{kl} \eta_l \quad (4\text{-}70)\text{a}$$

$$\eta_i \bar{\theta}_i = \eta_j \bar{\theta}_j = \eta_j \delta_{jl} \bar{\theta}_l = \eta_j A^{-1}_{jk} A_{kl} \bar{\theta}_l \quad (4\text{-}70)\text{b}$$

With (4-70), (4-69) becomes

$$\int e^{-\bar{\theta}_i A_{ij}\theta_j+\bar{\theta}_i\eta_i+\bar{\eta}_i\theta_i}d\bar{\theta}_1 d\theta_1 \ldots d\bar{\theta}_n d\theta_n=\int e^{-\bar{\theta}_i A_{ij}\theta_j+\bar{\theta}_j A_{jk}A_{kl}^{-1}\eta_l+\bar{\eta}_i A_{ij}^{-1}A_{jk}\theta_k}d\bar{\theta}_1 d\theta_1 \ldots d\bar{\theta}_n d\theta_n$$
$$=e^{\bar{\eta}_i A_{ij}^{-1}\eta_j}\int e^{-\bar{\theta}_j A_{jk}\theta_k+\bar{\theta}_j A_{jk}A_{kl}^{-1}\eta_l+\bar{\eta}_i A_{ij}^{-1}A_{jk}\theta_k-\bar{\eta}_i A_{ij}^{-1}\eta_j}d\bar{\theta}_1 d\theta_1 \ldots d\bar{\theta}_n d\theta_n \qquad (4\text{-}71)$$
$$=e^{\bar{\eta}_i A_{ij}^{-1}\eta_j}\int e^{-\left(\bar{\theta}_j-\bar{\eta}_i A_{ij}^{-1}\right)A_{jk}\left(\theta_k-A_{kl}^{-1}\eta_l\right)}d\bar{\theta}_1 d\theta_1 \ldots d\bar{\theta}_n d\theta_n .$$

Now do the following substitution of variables in the last row of (4-71).

$$\bar{\theta}'_j=\bar{\theta}_j-\bar{\eta}_i A_{ij}^{-1} \qquad d\bar{\theta}'_j=d\left(\bar{\theta}_j-\bar{\eta}_i A_{ij}^{-1}\right)=d\bar{\theta}_j$$
$$\theta'_k=\theta_k-A_{kl}^{-1}\eta_l \qquad d\theta'_k=d\left(\theta_k-A_{kl}^{-1}\eta_l\right)=d\theta_k \qquad (4\text{-}72)$$

With (4-72) and (4-68), (4-71) becomes

$$\int e^{-\bar{\theta}_i A_{ij}\theta_j+\bar{\theta}_i\eta_i+\bar{\eta}_i\theta_i}d\bar{\theta}_1 d\theta_1 \ldots d\bar{\theta}_n d\theta_n=e^{\bar{\eta}_i A_{ij}^{-1}\eta_j}\int e^{-\bar{\theta}'_j A_{jk}\theta'_k}d\bar{\theta}'_1 d\theta'_1 \ldots d\bar{\theta}'_n d\theta'_n$$
$$=\left(Det\, A_n\right)e^{\bar{\eta}_i A_{ij}^{-1}\eta_j} . \qquad (4\text{-}73)$$

Now, consider, as we did with the ϕ_i in (4-16), that each subscript designates a small volume in which the θ_i and $\bar{\theta}_i$ are measured and each integration is over $d\bar{\theta}_{\underline{i}}d\theta_{\underline{i}}$ inside that volume. Then the entire integration, as in (4-17) and (4-18), includes all possible values for θ_i and $\bar{\theta}_i$ over the entire large volume. That is, it includes all paths for the θ_i and $\bar{\theta}_i$. If we take the limit where the small volumes go to infinitesimal volumes as $n \to \infty$, then we can represent (4-73) as

$$\lim_{n\to\infty}\int e^{-\bar{\theta}_i A_{ij}\theta_j+\bar{\theta}_i\eta_i+\bar{\eta}_i\theta_i}d\bar{\theta}_1 d\theta_1 \ldots d\bar{\theta}_n d\theta_n$$
$$=\int e^{\int\left(-\bar{\theta}(\underline{x})A(\underline{x},\underline{y})\theta(\underline{y})+\bar{\theta}(\underline{x})\eta(\underline{x})+\bar{\eta}(\underline{x})\theta(\underline{x})\right)d\underline{x}d\underline{y}}D\bar{\theta}D\theta=\left(Det\, A\right)e^{\left[\bar{\eta}A^{-1}\eta\right]} \qquad (4\text{-}74)$$
$$Det\, A=\lim_{n\to\infty}Det\, A_n .$$

As we noted with the path integral over ϕ_i, A can be an operator (which is approximated in the discrete case by a matrix).

Now, if we replace θ with ψ, η with $i\sigma$, and A with $-iK$, we get (4-22) with $C' = Det\,(-iK)$.

4.7 Problems

1. Starting with (4-12), deduce (4-14). Assume K_{ij} is a symmetric matrix that can be diagonalized. Hint: Consider the transformation of the exponent such that the matrix/operator K_{ij} has diagonal form, i.e., with eigenvalues on the diagonal. x_i and b_i are also transformed. Use primes to indicate the transformed quantities and call the eigenvalues λ_i. Note that we get one factor under a square root sign for each value of the index i, and it has λ_i in the denominator. The determinant of the matrix is simply the product of all the eigenvalues, and this is invariant, i.e., the same in the primed and unprimed coordinates. Continue with similar logic for the exponential part with b_i.
2. Using (4-26), prove (4-23).
3. Prove (4-49).
4. Prove via Taylor expansion that for θ_i as Grassmann variables, $e^{-\theta_1 A_{12}\theta_2}=1-A_{12}\theta_1\theta_2$ is an exact relationship and not just an approximation.
5. Evaluate (4-65) for $n = 3$ and show that it equals the determinant of A.

Part Three
Electroweak Interactions

"Particle physics may appear complex, but ... [its] simplicity is of the kind that lies on the far side of difficulty, confusion, and complexity. The journey is difficult, but the view once achieved is magnificent."

Joseph Conlon in *Why String Theory?*

Chapter 5 Electroweak Toolkit

Chapter 6 The High Energy Symmetric Universe

Chapter 7 The Low Energy Universe: Broken Symmetry

Chapter 8 Electroweak Scattering and Decay

Chapter 9 Electroweak Symmetries: Continuous & Discrete

Chapter 10 Neutrinos

Chapter 11 Additional Topics in Electroweak Theory

Chapter 5

Electroweak Toolkit

"Live as if you were to die tomorrow. Learn as if you were to live forever."
Mahatma Gandhi

5.0 Preliminaries

We now return to the canonical approach to QFT until we note otherwise.

5.0.1 Chapter Overview

In this chapter, as background for electroweak interaction theory, we will investigate four subjects,

- alternative spinor representations,
- chirality,
- spinor spacetime transformations (specifically boosts), and
- massive spin 1 (vector) fields.

Four topics in this chapter

In Vol. 1, pg. 86, while solving the Dirac equation, we noted in passing that there were different sets of matrices and spinors (different representations) that would satisfy the basic conditions ((4-6) in Vol. 1) that Dirac determined his approach must meet. Then, and since, we have dealt exclusively with one of these, the standard (or Dirac-Pauli) representation (or rep). The matrices and solutions to the Dirac equation in this rep are shown in Vol. 1, (4-10) (equivalently, (4-11)) on pg. 87 and (4-20) on pg. 89, and repeated in this chapter as (5-2) and (5-3).

1) Spinor reps:
- standard
- Weyl
- Majorana

In this chapter we will explore two other spinor reps, known as the Weyl representation and the Majorana representation, then discuss the advantages they offer.

The second topic we will cover in this chapter is something called chirality, which we mentioned, though not actually defined, in Chap. 2 (pgs. 40-44), but will finally define a few pages hence. Chirality is a characteristic of spinor fields intimately connected with the weak force, and is often confused with helicity, though they are quite different. Confusion arises primarily from two causes: i) helicity and chirality each have two fundamental forms, and historically, though unfortunately, these two forms have been labeled right-hand (R) and left-hand (L) for both helicity and chirality; and ii) for spinor fermions traveling at speed c (hypothetical massless fermions), helicity and chirality become effectively the same thing. That is, at speeds below c, an L helicity fermion field is different from an L chirality fermion field, but for speed equal to c, an L helicity fermion is identical to an L chirality fermion. R helicity and R chirality behave in a similar way.

2) Chirality, related to the weak interaction; different from, but often confused with, helicity

We will see how all this works out in due course, though it would have been a lot easier for a lot of people, if chirality had originally been labeled "inside" and "outside", or "light" and "dark", or "happy" and "sad", instead of "left" and "right". The main point, however, is that the weak interaction "senses" chirality, not helicity or any of a number of other things.

3) Spinor spacetime transformations

The third area we will explore is the transformation of spinor fields under rotations and boosts.

The fourth topic of this chapter comprises massive vector fields, which are similar in many ways to the massless photon field we worked so much with in QED, but have significant differences. Such massive fields are governed by the Proca equation, which is a generalization of the Maxwell equation in its 4-vector potential (i.e., $A^{\mu}(x)$) form to include fields with mass. As you have probably already heard, the force-mediating bosons of present day (low energy) weak interactions, the W^+, W^-, and Z particles, are massive and have spin 1. These particles are called intermediate vector bosons (IVB),

4) Massive vector fields: For us, W^+, W^-, and Z

where the modifier "intermediate" is an anachronism left over from the 1960s, when it was being conjectured that they had masses in between those of particular (composite) vector bosons known at the time (which, unlike IVBs, are now known to be made of two quarks.)

The electromagnetic force, mediated by massless photons, extends, in principle, to infinity (via the inverse square law in the static case). Massive vector boson forces, on the other hand, drop off with distance much more rapidly than the massless photon vector boson. Their effect is dampened by their mass, in ways we will soon see (and much like we found for the Yukawa potential for massive scalars in Prob. 2, Chap. 16, of Vol. 1).

5.1 Spinor Representations

We start our exploration of spinor representations by recalling the Dirac equation,

$$\left(i\gamma^{\mu}_{\alpha\beta}\partial_{\mu} - mI_{\alpha\beta}\right)\psi_{\beta} = 0 \quad \text{or} \quad \left(i\gamma^{\mu}\partial_{\mu} - m\right)\psi = 0 \quad \text{or} \quad \left(i\not{\partial} - m\right)\psi = 0\,, \tag{5-1}$$

Dirac equation

and then, in the next sub-section, the rep for, and solution to, that equation we are familiar with, i.e., the standard rep spinor ψ.

5.1.1 The Standard (Dirac-Pauli) Rep

In Vol. 1, we elected the standard spinor rep by choosing the particular set of matrices in (5-1) of

$${}_S\gamma^0 = \begin{bmatrix} 1 & & & \\ & 1 & & \\ & & -1 & \\ & & & -1 \end{bmatrix} \quad {}_S\gamma^1 = \begin{bmatrix} & & & 1 \\ & & 1 & \\ & -1 & & \\ -1 & & & \end{bmatrix} \quad {}_S\gamma^2 = \begin{bmatrix} & & & -i \\ & & i & \\ & i & & \\ -i & & & \end{bmatrix} \quad {}_S\gamma^3 = \begin{bmatrix} & & 1 & \\ & & & -1 \\ -1 & & & \\ & 1 & & \end{bmatrix} \quad \begin{matrix}\text{Vol. 1,} \\ \text{(4-10)},\end{matrix} \tag{5-2)a}$$

Dirac matrices in standard rep

where we add a subscript "S" to distinguish the standard rep from others we are about to introduce. It is common to cast these matrices in streamlined form, where σ_i are the Pauli matrices, as

$${}_S\gamma^0 = \begin{bmatrix} I & 0 \\ 0 & -I \end{bmatrix} \quad {}_S\gamma^1 = \begin{bmatrix} 0 & \sigma_1 \\ -\sigma_1 & 0 \end{bmatrix} \quad {}_S\gamma^2 = \begin{bmatrix} 0 & \sigma_2 \\ -\sigma_2 & 0 \end{bmatrix} \quad {}_S\gamma^3 = \begin{bmatrix} 0 & \sigma_3 \\ -\sigma_3 & 0 \end{bmatrix} \quad \begin{matrix}\text{Vol. 1,} \\ \text{(4-11)}\end{matrix}\,. \tag{5-2)b}$$

An even more streamline, and popular, notation is

$${}_S\gamma^0 = \begin{bmatrix} I & 0 \\ 0 & -I \end{bmatrix} \quad {}_S\gamma^i = \begin{bmatrix} 0 & \sigma_i \\ -\sigma_i & 0 \end{bmatrix}. \tag{5-2)c}$$

Standard rep Dirac matrices streamlined form

We then found the solutions (bases for the general solution ψ) to (5-1), given (5-2),

$${}_S\psi^{(1)} = \underbrace{\sqrt{\frac{E+m}{2m}}\begin{pmatrix} 1 \\ 0 \\ \dfrac{p^3}{E+m} \\ \dfrac{p^1+ip^2}{E+m} \end{pmatrix}}_{\substack{\text{spinor } {}_Su_1 = \text{ part} \\ \text{of solution in 4D} \\ \text{spinor space}}} \underbrace{e^{-ipx}}_{\substack{\text{4D} \\ \text{space-} \\ \text{time} \\ \text{part}}} = {}_Su_1e^{-ipx} \qquad {}_S\psi^{(2)} = \underbrace{\sqrt{\frac{E+m}{2m}}\begin{pmatrix} 0 \\ 1 \\ \dfrac{p^1-ip^2}{E+m} \\ \dfrac{-p^3}{E+m} \end{pmatrix}}_{\text{spinor } {}_Su_2} e^{-ipx} = {}_Su_2e^{-ipx}$$

$${}_S\psi^{(3)} = \underbrace{\sqrt{\frac{E+m}{2m}}\begin{pmatrix} \dfrac{p^3}{E+m} \\ \dfrac{p^1+ip^2}{E+m} \\ 1 \\ 0 \end{pmatrix}}_{\text{spinor } {}_Sv_2} e^{ipx} = {}_Sv_2e^{ipx} \qquad {}_S\psi^{(4)} = \underbrace{\sqrt{\frac{E+m}{2m}}\begin{pmatrix} \dfrac{p^1-ip^2}{E+m} \\ \dfrac{-p^3}{E+m} \\ 0 \\ 1 \end{pmatrix}}_{\text{spinor } {}_Sv_1} e^{ipx} = {}_Sv_1e^{ipx}\,, \quad \begin{matrix}\text{Vol. 1,} \\ \text{(4-20)}\end{matrix} \tag{5-3}$$

Independent (basis) solutions to Dirac equation in standard rep

where you might want to take note of the conventional way of subscript numbering v_1 and v_2, in a seemingly opposite way of numbering u_1 and u_2. Doing it this way actually makes things easier later.

Keep in mind the general form ψ of the solution to the Dirac equation, in this rep,

$$
\begin{aligned}
{}_S\psi &= \sum_{\mathbf{p}} \sqrt{\frac{m}{VE_{\mathbf{p}}}} \left(c_1(\mathbf{p})\, {}_S\psi^{(1)} + c_2(\mathbf{p})\, {}_S\psi^{(2)} + d_2^{\dagger}(\mathbf{p})\, {}_S\psi^{(3)} + d_1^{\dagger}(\mathbf{p})\, {}_S\psi^{(4)} \right) \\
&= \sum_{r,\mathbf{p}} \sqrt{\frac{m}{VE_{\mathbf{p}}}} \left(c_r(\mathbf{p})\, {}_S u_r e^{-ipx} + d_r^{\dagger}(\mathbf{p})\, {}_S v_r e^{ipx} \right) \\
&= \sum_{\mathbf{p}} \sqrt{\frac{m}{VE_{\mathbf{p}}}} \left(\begin{aligned} & c_1(\mathbf{p}) \sqrt{\frac{E_{\mathbf{p}}+m}{2m}} \begin{pmatrix} 1 \\ 0 \\ \frac{p^3}{E_{\mathbf{p}}+m} \\ \frac{p^1+ip^2}{E_{\mathbf{p}}+m} \end{pmatrix} e^{-ipx} + c_2(\mathbf{p}) \sqrt{\frac{E_{\mathbf{p}}+m}{2m}} \begin{pmatrix} 0 \\ 1 \\ \frac{p^1-ip^2}{E_{\mathbf{p}}+m} \\ \frac{-p^3}{E_{\mathbf{p}}+m} \end{pmatrix} e^{-ipx} \\ & + d_2^{\dagger}(\mathbf{p}) \sqrt{\frac{E_{\mathbf{p}}+m}{2m}} \begin{pmatrix} \frac{p^3}{E_{\mathbf{p}}+m} \\ \frac{p^1+ip^2}{E_{\mathbf{p}}+m} \\ 1 \\ 0 \end{pmatrix} e^{ipx} + d_1^{\dagger}(\mathbf{p}) \sqrt{\frac{E_{\mathbf{p}}+m}{2m}} \begin{pmatrix} \frac{p^1-ip^2}{E_{\mathbf{p}}+m} \\ \frac{-p^3}{E_{\mathbf{p}}+m} \\ 0 \\ 1 \end{pmatrix} e^{ipx} \end{aligned} \right) \\
&= \sum_{\mathbf{p}} \sqrt{\frac{m}{VE_{\mathbf{p}}}} \sqrt{\frac{E_{\mathbf{p}}+m}{2m}} \begin{pmatrix} c_1(\mathbf{p}) e^{-ipx} + \left(d_2^{\dagger}(\mathbf{p}) \frac{p^3}{E_{\mathbf{p}}+m} + d_1^{\dagger}(\mathbf{p}) \frac{p^1-ip^2}{E_{\mathbf{p}}+m} \right) e^{ipx} \\ c_2(\mathbf{p}) e^{-ipx} + \left(d_2^{\dagger}(\mathbf{p}) \frac{p^1+ip^2}{E_{\mathbf{p}}+m} - d_1^{\dagger}(\mathbf{p}) \frac{p^3}{E_{\mathbf{p}}+m} \right) e^{ipx} \\ \left(c_1(\mathbf{p}) \frac{p^3}{E_{\mathbf{p}}+m} + c_2(\mathbf{p}) \frac{p^1-ip^2}{E_{\mathbf{p}}+m} \right) e^{-ipx} + d_2^{\dagger}(\mathbf{p}) e^{ipx} \\ \left(c_1(\mathbf{p}) \frac{p^1+ip^2}{E_{\mathbf{p}}+m} - c_2(\mathbf{p}) \frac{p^3}{E_{\mathbf{p}}+m} \right) e^{-ipx} + d_1^{\dagger}(\mathbf{p}) e^{ipx} \end{pmatrix} = \begin{bmatrix} {}_S\psi_1 \\ {}_S\psi_2 \\ {}_S\psi_3 \\ {}_S\psi_4 \end{bmatrix} = {}_S\psi . \qquad (5\text{-}4)
\end{aligned}
$$

General solution to Dirac equation in standard rep

Note that the numeric subscripts for the four components in spinor space ${}_S\psi_i$ (i = 1,2,3,4) are a different thing from the numeric subscripts on destruction and creation operators $c_1(\mathbf{p}), c_2(\mathbf{p}), d_2^{\dagger}(\mathbf{p}), d_1^{\dagger}(\mathbf{p})$, the numeric superscripts on 3-momentum components p^i, and the superscripts in parentheses denoting the four basis vectors in spinor space ${}_S\psi^{(1)}$, ${}_S\psi^{(2)}$, ${}_S\psi^{(3)}$, ${}_S\psi^{(4)}$.

<u>Special Case ($v << c$) where Standard Rep Advantageous</u>

For non-relativistic speeds, where $E_{\mathbf{p}} \approx m >> |\mathbf{p}|$, (5-3) reduce to

$$
E_{\mathbf{p}} \approx m >> |\mathbf{p}| \qquad
\begin{aligned}
{}_S\psi^{(1)} &\approx \begin{pmatrix} 1 \\ 0 \\ 0 \\ 0 \end{pmatrix} e^{-ipx} = {}_S u_1 e^{-ipx} & \quad {}_S\psi^{(2)} &\approx \begin{pmatrix} 0 \\ 1 \\ 0 \\ 0 \end{pmatrix} e^{-ipx} = {}_S u_2 e^{-ipx} \\
{}_S\psi^{(3)} &\approx \begin{pmatrix} 0 \\ 0 \\ 1 \\ 0 \end{pmatrix} e^{ipx} = {}_S v_2 e^{ipx} & \quad {}_S\psi^{(4)} &\approx \begin{pmatrix} 0 \\ 0 \\ 0 \\ 1 \end{pmatrix} e^{ipx} = {}_S v_1 e^{ipx} ,
\end{aligned} \qquad (5\text{-}5)
$$

Standard rep basis solutions for $v << c$

which correspond, respectively, to spin up particle/field, spin down particle/field, spin up antiparticle/field, and spin down antiparticle/field. We can see this mathematically by operating on each of the solutions in (5-5) with the spin operator (Vol. 1, (4-39), pg. 93) in the standard rep,

$$ {}_S\Sigma_3 = \frac{1}{2}\begin{bmatrix} 1 & & & \\ & -1 & & \\ & & 1 & \\ & & & -1 \end{bmatrix} \quad \left(\text{natural units}\right). \tag{5-6} $$

Spin operator in standard rep

(5-6) acting on each of the solutions (5-5) yields, in succession, spin up (eigenvalue +½), spin down (eigenvalue –½), spin up (eigenvalue +½), spin down (eigenvalue –½). This is a nice rep to work in when energy levels are non-relativistic. Things get simplified.

Recall from Chap. 2 that (5-6) acting on an eigen-field ψ of the spin operator yields the same eigenvalue as $\int {}_S\psi^\dagger {}_S\Sigma_3\, {}_S\psi\, d^3x$ acting on the single fermion eigenstate $|\psi\rangle$. This was also noted in Vol. 1, Sect. 4.9.1, pgs. 113-115.

We won't have too much further to say about the standard rep, as we have been working with it since Chap. 4 of Vol. 1 and should be reasonably familiar with it.

5.1.2 Transforming to a Different Rep

Note that, with a unitary transformation in 4D spinor space, we can transform (5-1) into any number of different forms (different matrices and different solution forms).

$$ U_{\rho\alpha}\big(\ i\gamma^\mu_{\alpha\beta}\underbrace{U^\dagger_{\beta\delta}U_{\delta\eta}}_{I_{\beta\eta}}\partial_\mu - mI_{\alpha\beta}\underbrace{U^\dagger_{\beta\delta}U_{\delta\eta}}_{I_{\beta\eta}}\ \big)\psi_\eta = 0 $$

$$ \rightarrow\ \big(\ \underbrace{iU_{\rho\alpha}\gamma^\mu_{\alpha\beta}U^\dagger_{\beta\delta}}_{\gamma'^\mu_{\rho\delta}}\partial_\mu - \underbrace{mU_{\rho\alpha}I_{\alpha\beta}U^\dagger_{\beta\delta}}_{I_{\rho\delta}}\ \big)\underbrace{U_{\delta\eta}\psi_\eta}_{\psi'_\delta} = 0 \tag{5-7} $$

$$ \rightarrow\ \left(i\gamma'^\mu_{\rho\delta}\partial_\mu - mI_{\rho\delta}\right)\psi'_\delta = 0 \qquad \rightarrow\ \left(i\gamma'^\mu\partial_\mu - m\right)\psi' = 0. $$

Transforming the Dirac equation to different γ^μ and ψ

The primed ψ' is a solution to the Dirac equation formulated with primed matrices γ'^μ. Actually, for given $U_{\delta\eta}$, there are four independent (primed) solutions to (5-7), but they will look different from (5-3), just as the primed matrices will look different from (5-2). The primed quantities comprise a different spinor space representation, i.e., a different rep. In short, any 4X4 unitary matrix can transform the standard rep to a different rep, via (with spinor indices suppressed)

$$ U\psi = \psi' \qquad\qquad U\gamma^\mu U^\dagger = \gamma'^\mu, \tag{5-8} $$

where ψ (and ψ') can be any one of four basis solutions to the Dirac equation in the corresponding representation, or a general solution made up of a superposition of such basis solutions.

The bottom line: There are innumerable ways to express the Dirac equation, with innumerable sets of different matrices and concomitant solutions, provided we transform quantities with a unitary matrix as in (5-8). Different choices for U will lead to different gamma matrices and therefore, different solutions ψ.

Infinite number of U, i.e., ways to transform Dirac equation

The value of this: As we have probably already learned in our physics careers, we can often turn a difficult-to-solve problem into an easier one, if we transform the governing equations into a different form. Two particular choices for U (and thus choices for the forms of γ^μ and ψ) that make theory and calculation easier, and which are most often used in QFT, give us what are called the Weyl and Majorana reps. We develop and explore these in the next two sections.

We'll pick particular choices for U that help us with certain problems in QFT

Relation to Reps in Group Theory

I mention in passing that the gamma matrices, under matrix multiplication together with their anti-commutation relations, form, in a complicated way we will not delve into, an algebra (see Chap. 2) called the Dirac algebra. This is one kind of a more general class of algebras known formally as Clifford algebras, a nomenclature you will probably run into in the literature. Like any group or algebra, our particular type of Clifford algebra (the Dirac algebra) can be represented in different ways (for us, by different matrices). So, when we talk of different spinor reps, we are using the term "rep" in the same sense we did for group theory.

Different gamma matrix sets in different reps of Dirac algebra

5.1.3 The Weyl Rep

Named after its originator, Herman Weyl (pronounced "vile"), the Weyl representation is, as we will see, convenient for handling massless spin ½ fermions, and thus, historically, it was applied to neutrinos, in the days when they were considered massless. It still has uses today as an approximation for neutrinos and other highly relativistic particles (which approach the speed of light, where (rest) mass is negligible), in supersymmetric theories, and later for us, as an aid in understanding chirality.

Weyl rep: useful when $m = 0$

Transformation from the Standard to the Weyl Rep

The Weyl rep (subscript "W") is found using the following unitary matrix.

Transformation from standard to Weyl rep

$$ {}_{S\text{-}W}U = \frac{1}{\sqrt{2}}\begin{bmatrix} 1 & & -1 & \\ & 1 & & -1 \\ 1 & & 1 & \\ & 1 & & 1 \end{bmatrix} = \frac{1}{\sqrt{2}}\begin{bmatrix} I & -I \\ I & I \end{bmatrix}. \tag{5-9} $$

Do **Problems 1** and **2** to find (5-10) and (5-11).

Gamma Matrices in the Weyl Rep

The gamma matrices in the Weyl rep, from (5-8) and (5-9), are

$$ {}_W\gamma^0 = \begin{bmatrix} & & 1 & \\ & & & 1 \\ 1 & & & \\ & 1 & & \end{bmatrix} \quad {}_W\gamma^1 = \begin{bmatrix} & & & 1 \\ & & 1 & \\ & -1 & & \\ -1 & & & \end{bmatrix} \quad {}_W\gamma^2 = \begin{bmatrix} & & & -i \\ & & i & \\ & i & & \\ -i & & & \end{bmatrix} \quad {}_W\gamma^3 = \begin{bmatrix} & & 1 & \\ & & & -1 \\ -1 & & & \\ & 1 & & \end{bmatrix}, \tag{5-10a} $$

Weyl rep gamma matrices

or streamlined, as

$$ {}_W\gamma^0 = \begin{bmatrix} 0 & I \\ I & 0 \end{bmatrix} \quad {}_W\gamma^i = \begin{bmatrix} 0 & \sigma_i \\ -\sigma_i & 0 \end{bmatrix}. \tag{5-10b} $$

As an aside, we can note that all but the γ^0 matrix turn out to be the same as in the standard rep.

Basis Solutions in the Weyl Rep

The eigensolutions to the Dirac equation in the Weyl rep, from (5-8), are

$$ {}_W\psi^{(1)} = \sqrt{\frac{E+m}{4m}}\begin{pmatrix} 1-\frac{p^3}{E+m} \\ -\frac{p^1+ip^2}{E+m} \\ 1+\frac{p^3}{E+m} \\ \frac{p^1+ip^2}{E+m} \end{pmatrix} e^{-ipx} = {}_W u_1 e^{-ipx} \qquad {}_W\psi^{(2)} = \sqrt{\frac{E+m}{4m}}\begin{pmatrix} -\frac{p^1-ip^2}{E+m} \\ 1+\frac{p^3}{E+m} \\ \frac{p^1-ip^2}{E+m} \\ 1-\frac{p^3}{E+m} \end{pmatrix} e^{-ipx} = {}_W u_2 e^{-ipx} $$

Basis spinors in Weyl rep

$$ {}_W\psi^{(3)} = \sqrt{\frac{E+m}{4m}}\begin{pmatrix} \frac{p^3}{E+m}-1 \\ \frac{p^1+ip^2}{E+m} \\ \frac{p^3}{E+m}+1 \\ \frac{p^1+ip^2}{E+m} \end{pmatrix} e^{ipx} = {}_W v_2 e^{ipx} \qquad {}_W\psi^{(4)} = \sqrt{\frac{E+m}{4m}}\begin{pmatrix} \frac{p^1-ip^2}{E+m} \\ \frac{-p^3}{E+m}-1 \\ \frac{p^1-ip^2}{E+m} \\ \frac{-p^3}{E+m}+1 \end{pmatrix} e^{ipx} = {}_W v_1 e^{ipx}. \tag{5-11} $$

Similar to (5-4), the general solution in the Weyl rep has form

$$ {}_W\psi = \begin{bmatrix} {}_W\psi_1 \\ {}_W\psi_2 \\ {}_W\psi_3 \\ {}_W\psi_4 \end{bmatrix} = \sum_{\mathbf{p}} \sqrt{\frac{m}{VE_{\mathbf{p}}}} \left(c_1(\mathbf{p})\, {}_W\psi^{(1)} + c_2(\mathbf{p})\, {}_W\psi^{(2)} + d_2^\dagger(\mathbf{p})\, {}_W\psi^{(3)} + d_1^\dagger(\mathbf{p})\, {}_W\psi^{(4)} \right). \quad (5\text{-}12) $$

General solution in Weyl rep

Creation and Destruction Operators in Weyl Rep

Since the transformation (5-9) does not act on $c_1(\mathbf{p})$ in ${}_s\psi$, it plays the same role as in the standard rep, i.e., it destroys a particle having the 3-momentum **p** and spin value $r = 1$. But in the Weyl rep we represent that with a different mathematical form of the spinor field, ${}_W\psi^{(1)}$. Likewise, $d_1^\dagger(\mathbf{p})$ creates the corresponding antiparticle. Other terms in (5-12) follow suit. The same result would be obtained by applying our basic QFT postulates to ${}_W\psi$ and following the same steps as we did in Vol 1.

*Same **p**, r states created & destroyed, but different spinor form for them*

Special Case (Relativistic Speeds $v \to c$) where Weyl Rep Advantageous

Now, consider a highly relativistic case, where $E >> m$ and with the momentum in the x^3 direction, so $p^1 = p^2 = 0$ and $E \approx p^3$ (natural units) in (5-11) and (5-12).

$$ {}_W\psi^{(1)} \approx \sqrt{\frac{E}{4m}} \begin{pmatrix} 1-\frac{E}{E} \\ 0 \\ 1+\frac{E}{E} \\ 0 \end{pmatrix} e^{-ipx} = \sqrt{\frac{E}{m}} \begin{pmatrix} 0 \\ 0 \\ 1 \\ 0 \end{pmatrix} e^{-ipx} \qquad {}_W\psi^{(2)} \approx \sqrt{\frac{E}{4m}} \begin{pmatrix} 0 \\ 1+\frac{E}{E} \\ 0 \\ 1-\frac{E}{E} \end{pmatrix} e^{-ipx} = \sqrt{\frac{E}{m}} \begin{pmatrix} 0 \\ 1 \\ 0 \\ 0 \end{pmatrix} e^{-ipx} $$

$$ {}_W\psi^{(3)} \approx \sqrt{\frac{E}{4m}} \begin{pmatrix} \frac{E}{E}-1 \\ 0 \\ \frac{E}{E}+1 \\ 0 \end{pmatrix} e^{ipx} = \sqrt{\frac{E}{m}} \begin{pmatrix} 0 \\ 0 \\ 1 \\ 0 \end{pmatrix} e^{ipx} \qquad {}_W\psi^{(4)} \approx \sqrt{\frac{E}{4m}} \begin{pmatrix} 0 \\ -\frac{E}{E}-1 \\ 0 \\ -\frac{E}{E}+1 \end{pmatrix} e^{ipx} = -\sqrt{\frac{E}{m}} \begin{pmatrix} 0 \\ 1 \\ 0 \\ 0 \end{pmatrix} e^{ipx}. \quad (5\text{-}13) $$

Special case: v almost = c, so E >> m

With (5-13), (5-12) becomes

$$ {}_W\psi = \begin{bmatrix} {}_W\psi_1 \\ {}_W\psi_2 \\ {}_W\psi_3 \\ {}_W\psi_4 \end{bmatrix} \approx \sum_{\mathbf{p}} \sqrt{\frac{1}{V}} \left(c_1(\mathbf{p}) \begin{pmatrix} 0 \\ 0 \\ 1 \\ 0 \end{pmatrix} e^{-ipx} + c_2(\mathbf{p}) \begin{pmatrix} 0 \\ 1 \\ 0 \\ 0 \end{pmatrix} e^{-ipx} + d_2^\dagger(\mathbf{p}) \begin{pmatrix} 0 \\ 0 \\ 1 \\ 0 \end{pmatrix} e^{ipx} - d_1^\dagger(\mathbf{p}) \begin{pmatrix} 0 \\ 1 \\ 0 \\ 0 \end{pmatrix} e^{ipx} \right). \quad (5\text{-}14) $$

This is an obvious simplification, and below, we discuss how these solutions can help in analysis.

Value of the Weyl Rep

The usefulness of the Weyl rep lies in how it alters the form of the Dirac equation and how that helps us in the analysis of highly relativistic spin ½ fermions. So, let's express that equation in the Weyl rep.

$$ \left(i_W\gamma^\mu\partial_\mu - m\right){}_W\psi = 0 \quad \rightarrow \quad \left(i_W\gamma^0\partial_0 + i_W\gamma^1\partial_1 + i_W\gamma^2\partial_2 + i_W\gamma^3\partial_3 - mI\right){}_W\psi = 0 $$

$$ \rightarrow \left(i\left(\begin{bmatrix} & & 1 & \\ & & & 1 \\ 1 & & & \\ & 1 & & \end{bmatrix}\partial_0 + \begin{bmatrix} & & & 1 \\ & & 1 & \\ & -1 & & \\ -1 & & & \end{bmatrix}\partial_1 + \begin{bmatrix} & & & -i \\ & & i & \\ & i & & \\ -i & & & \end{bmatrix}\partial_2 + \begin{bmatrix} & & 1 & \\ & & & -1 \\ -1 & & & \\ & 1 & & \end{bmatrix}\partial_3 \right) - m\begin{bmatrix} 1 & & & \\ & 1 & & \\ & & 1 & \\ & & & 1 \end{bmatrix} \right) \begin{bmatrix} {}_W\psi_1 \\ {}_W\psi_2 \\ {}_W\psi_3 \\ {}_W\psi_4 \end{bmatrix} = \begin{bmatrix} 0 \\ 0 \\ 0 \\ 0 \end{bmatrix} \quad (5\text{-}15) $$

Writing out Dirac equation in Weyl rep with matrices expressed

$$ \rightarrow \quad i \begin{bmatrix} \partial_0\,{}_W\psi_3 + \partial_1\,{}_W\psi_4 - i\partial_2\,{}_W\psi_4 + \partial_3\,{}_W\psi_3 + im\,{}_W\psi_1 \\ \partial_0\,{}_W\psi_4 + \partial_1\,{}_W\psi_3 + i\partial_2\,{}_W\psi_3 - \partial_3\,{}_W\psi_4 + im\,{}_W\psi_2 \\ \partial_0\,{}_W\psi_1 - \partial_1\,{}_W\psi_2 + i\partial_2\,{}_W\psi_2 - \partial_3\,{}_W\psi_1 + im\,{}_W\psi_3 \\ \partial_0\,{}_W\psi_2 - \partial_1\,{}_W\psi_1 - i\partial_2\,{}_W\psi_1 + \partial_3\,{}_W\psi_2 + im\,{}_W\psi_4 \end{bmatrix} = \begin{bmatrix} 0 \\ 0 \\ 0 \\ 0 \end{bmatrix}, $$

which is four coupled partial differential equations in four scalar unknowns. Not such a simple thing to solve or work with.

But now consider that we have either i) a massless fermion ($m = 0$), or ii) a highly relativistic state where speed approaches c and energy and momenta values are far greater than m_w, so we can ignore mass. Then, (5-15) becomes (where we move the bottom two rows of (5-15) to the top)

$$\begin{bmatrix} \partial_0 {}_w\psi_1 - \partial_1 {}_w\psi_2 + i\partial_2 {}_w\psi_2 - \partial_3 {}_w\psi_1 \\ \partial_0 {}_w\psi_2 - \partial_1 {}_w\psi_1 - i\partial_2 {}_w\psi_1 + \partial_3 {}_w\psi_2 \end{bmatrix} = \begin{bmatrix} 0 \\ 0 \end{bmatrix}$$
$$\begin{bmatrix} \partial_0 {}_w\psi_3 + \partial_1 {}_w\psi_4 - i\partial_2 {}_w\psi_4 + \partial_3 {}_w\psi_3 \\ \partial_0 {}_w\psi_4 + \partial_1 {}_w\psi_3 + i\partial_2 {}_w\psi_3 - \partial_3 {}_w\psi_4 \end{bmatrix} = \begin{bmatrix} 0 \\ 0 \end{bmatrix}, \qquad (5\text{-}16)$$

For $m = 0$, Dirac equation in Weyl rep uncouples into two sets of equations, each set involving two scalar components

which comprises two separate sets of differential equations which are not coupled to each other, with each set having two equations in two scalar unknowns. As shorthand, in terms of the Pauli matrices, we can write these as

$$\left(\partial_0 \begin{bmatrix} 1 & \\ & 1 \end{bmatrix} - \partial_1 \begin{bmatrix} & 1 \\ 1 & \end{bmatrix} - \partial_2 \begin{bmatrix} & -i \\ i & \end{bmatrix} - \partial_3 \begin{bmatrix} 1 & \\ & -1 \end{bmatrix}\right) \begin{bmatrix} {}_w\psi_1 \\ {}_w\psi_2 \end{bmatrix} = \begin{bmatrix} 0 \\ 0 \end{bmatrix}$$
$$\left(\partial_0 \begin{bmatrix} 1 & \\ & 1 \end{bmatrix} + \partial_1 \begin{bmatrix} & 1 \\ 1 & \end{bmatrix} + \partial_2 \begin{bmatrix} & -i \\ i & \end{bmatrix} + \partial_3 \begin{bmatrix} 1 & \\ & -1 \end{bmatrix}\right) \begin{bmatrix} {}_w\psi_3 \\ {}_w\psi_4 \end{bmatrix} = \begin{bmatrix} 0 \\ 0 \end{bmatrix} \qquad (5\text{-}17)$$
$$\rightarrow \quad (I\partial_0 - \underbrace{\sigma_i\partial_i}_{\boldsymbol{\sigma}\cdot\boldsymbol{\nabla}})\underbrace{{}_w\psi^L}_{\begin{bmatrix} {}_w\psi_1 \\ {}_w\psi_2 \end{bmatrix}} = 0 \qquad (I\partial_0 + \underbrace{\sigma_i\partial_i}_{\boldsymbol{\sigma}\cdot\boldsymbol{\nabla}})\underbrace{{}_w\psi^R}_{\begin{bmatrix} {}_w\psi_3 \\ {}_w\psi_4 \end{bmatrix}} = 0 .$$

called Weyl equations, solutions called L and R Weyl spinors, ψ^L and ψ^R

These are known as the Weyl equations and ${}_w\psi^L$ and ${}_w\psi^R$ as Weyl spinors, where we will shortly show why the L and R superscript notation is used. ${}_w\psi^L$ and ${}_w\psi^R$ have only two components each and are subsets of the usual 4-component spinor we are familiar with. Note that if we have an L (or R) Weyl spinor, we can simply ignore two components and only have to solve a 2X2 matrix equation (one of those in the last line of (5-17)). For this reason, the terms L Weyl spinor and R Weyl spinor can refer either to the two component spinors of (5-17) or the 4-component spinors of (5-18) below, depending on context. That is, we can represent the 4-component spinor field in the Weyl rep as

2 equations in 2 unknowns easier to solve than 4 equations in 4 unknowns

$$\underbrace{{}_W\psi = \begin{bmatrix} {}_W\psi^L \\ {}_W\psi^R \end{bmatrix} = \begin{bmatrix} {}_W\psi_1 \\ {}_W\psi_2 \\ {}_W\psi_3 \\ {}_W\psi_4 \end{bmatrix}}_{\text{Weyl rep spinor field}} \qquad \underbrace{\begin{bmatrix} {}_W\psi^L \\ 0 \end{bmatrix} = \begin{bmatrix} {}_W\psi_1 \\ {}_W\psi_2 \\ 0 \\ 0 \end{bmatrix}}_{L \text{ Weyl spinor}} = \underbrace{{}_W\psi^L}_{\substack{\text{often} \\ \text{used} \\ \text{symbol}}} \qquad \underbrace{\begin{bmatrix} 0 \\ {}_W\psi^R \end{bmatrix} = \begin{bmatrix} 0 \\ 0 \\ {}_W\psi_3 \\ {}_W\psi_4 \end{bmatrix}}_{R \text{ Weyl spinor}} = \underbrace{{}_W\psi^R}_{\substack{\text{often} \\ \text{used} \\ \text{symbol}}} . \qquad (5\text{-}18)$$

The Dirac Equation in the Weyl Rep for $m \neq 0$

From (5-15) with mass not vanishing, we have, instead of (5-17), the more general form of the Dirac equation in the Weyl rep,

$$i\left(\partial_0 \begin{bmatrix} 1 & \\ & 1 \end{bmatrix} - \partial_1 \begin{bmatrix} & 1 \\ 1 & \end{bmatrix} - \partial_2 \begin{bmatrix} & -i \\ i & \end{bmatrix} - \partial_3 \begin{bmatrix} 1 & \\ & -1 \end{bmatrix}\right) \begin{bmatrix} {}_w\psi_1 \\ {}_w\psi_2 \end{bmatrix} - m \begin{bmatrix} 1 & \\ & 1 \end{bmatrix} \begin{bmatrix} {}_w\psi_3 \\ {}_w\psi_4 \end{bmatrix} = \begin{bmatrix} 0 \\ 0 \end{bmatrix}$$
$$i\left(\partial_0 \begin{bmatrix} 1 & \\ & 1 \end{bmatrix} + \partial_1 \begin{bmatrix} & 1 \\ 1 & \end{bmatrix} + \partial_2 \begin{bmatrix} & -i \\ i & \end{bmatrix} + \partial_3 \begin{bmatrix} 1 & \\ & -1 \end{bmatrix}\right) \begin{bmatrix} {}_w\psi_3 \\ {}_w\psi_4 \end{bmatrix} - m \begin{bmatrix} 1 & \\ & 1 \end{bmatrix} \begin{bmatrix} {}_w\psi_1 \\ {}_w\psi_2 \end{bmatrix} = \begin{bmatrix} 0 \\ 0 \end{bmatrix}$$
$$\rightarrow \quad i(I\partial_0 - \sigma_i\partial_i)\underbrace{{}_w\psi^L}_{\begin{bmatrix} {}_w\psi_1 \\ {}_w\psi_2 \end{bmatrix}} - mI\underbrace{{}_w\psi^R}_{\begin{bmatrix} {}_w\psi_3 \\ {}_w\psi_4 \end{bmatrix}} = 0 \qquad i(I\partial_0 + \sigma_i\partial_i)\underbrace{{}_w\psi^R}_{\begin{bmatrix} {}_w\psi_3 \\ {}_w\psi_4 \end{bmatrix}} - mI\underbrace{{}_w\psi^L}_{\begin{bmatrix} {}_w\psi_1 \\ {}_w\psi_2 \end{bmatrix}} = 0 \qquad (5\text{-}19)$$
$$\rightarrow \quad i(\partial_0 - \boldsymbol{\sigma}\cdot\boldsymbol{\nabla})\,{}_w\psi^L - m\,{}_w\psi^R = 0 \quad i(\partial_0 + \boldsymbol{\sigma}\cdot\boldsymbol{\nabla})\,{}_w\psi^R - m\,{}_w\psi^L = 0 .$$

For $m \neq 0$, not uncoupled, but still designate top 2 components ψ^L and bottom two ψ^R

The coupled equations (5-19) become uncoupled when $m = 0$ (and approximately so, in the highly relativistic regime, when we can ignore mass).

Do **Problem 3** to find the spin operator in the Weyl rep, (5-20).

The Spin Operator in the Weyl Rep

To see, at least in part, why the L and R notation is used, first note the spin operator in the Weyl rep, which, if you did Problem 3, you found turns out to be the same as the spin operator of the standard rep, i.e.,

Transform spin operator in standard rep to get spin operator in Weyl rep

$$ {}_W\Sigma_3 = \frac{1}{2}\begin{bmatrix} 1 & & & \\ & -1 & & \\ & & 1 & \\ & & & -1 \end{bmatrix} \qquad \left(\text{natural units}\right). \tag{5-20}$$

Why the L and R Notation

Now, consider the highly relativistic case of (5-14), where $E >> m$, and consider we only have a field with ${}_W\psi^L$ in it.

$$\text{Consider } {}_W\psi = \begin{bmatrix} {}_W\psi^L \\ 0 \end{bmatrix} = \begin{bmatrix} {}_W\psi_1 \\ {}_W\psi_2 \\ 0 \\ 0 \end{bmatrix}$$

$$\approx \sum_{\mathbf{p}} \sqrt{\frac{1}{V}} \left(0 + c_2(\mathbf{p}) \begin{pmatrix} 0 \\ 1 \\ 0 \\ 0 \end{pmatrix} e^{-ipx} + 0 + d_1^\dagger(\mathbf{p}) \begin{pmatrix} 0 \\ -1 \\ 0 \\ 0 \end{pmatrix} e^{ipx} \right) \tag{5-21}$$

$$= \sum_{\mathbf{p}} \sqrt{\frac{1}{V}} \left(c_2(\mathbf{p}) \begin{pmatrix} 0 \\ 1 \\ 0 \\ 0 \end{pmatrix} e^{-ipx} + d_1^\dagger(\mathbf{p}) \begin{pmatrix} 0 \\ -1 \\ 0 \\ 0 \end{pmatrix} e^{ipx} \right).$$

Now, operate on (5-21) with (5-20),

$${}_W\Sigma_3 \begin{bmatrix} {}_W\psi^L \\ 0 \end{bmatrix} = \frac{1}{2}\begin{bmatrix} 1 & & & \\ & -1 & & \\ & & 1 & \\ & & & -1 \end{bmatrix} \begin{bmatrix} {}_W\psi_1 \\ {}_W\psi_2 \\ 0 \\ 0 \end{bmatrix} \qquad \left({}_W\psi_1 \approx 0 \text{ for special case of } v \to c\right)$$

$$\approx \frac{1}{2}\begin{bmatrix} 1 & & & \\ & -1 & & \\ & & 1 & \\ & & & -1 \end{bmatrix} \left(\sum_{\mathbf{p}} \sqrt{\frac{1}{V}} \left(c_2(\mathbf{p}) \begin{pmatrix} 0 \\ 1 \\ 0 \\ 0 \end{pmatrix} e^{-ipx} + d_1^\dagger(\mathbf{p}) \begin{pmatrix} 0 \\ -1 \\ 0 \\ 0 \end{pmatrix} e^{ipx} \right) \right) \tag{5-22}$$

$$= -\frac{1}{2} \sum_{\mathbf{p}} \sqrt{\frac{1}{V}} \left(c_2(\mathbf{p}) \begin{pmatrix} 0 \\ 1 \\ 0 \\ 0 \end{pmatrix} e^{-ipx} + d_1^\dagger(\mathbf{p}) \begin{pmatrix} 0 \\ -1 \\ 0 \\ 0 \end{pmatrix} e^{ipx} \right) = -\frac{1}{2}\begin{bmatrix} {}_W\psi_1 = 0 \\ {}_W\psi_2 \\ 0 \\ 0 \end{bmatrix} = -\frac{1}{2}\begin{bmatrix} {}_W\psi^L \\ 0 \end{bmatrix},$$

For highly relativistic case, ψ^L has L helicity

and we find the L field has spin eigenvalue of –½, i.e., left-hand spin in the x^3 direction. Since the 3-momentum in this special case is in the x^3 direction, we have a left-hand helicity field. That is part of the reason the L is used for this two-component field part of the whole Weyl rep four-component field.

Recall: Spin Aligns with **p** for Relativistic Speeds

In Vol, 1, pgs. 95-96, we discussed how spin aligns with the velocity (or 3-momentum) direction when $v = c$, and gets very close to alignment for v close to c. This is what we have seen above. At such speeds, spin is aligned with (or closely aligned with) **p**. In the case above, the alignment was along the x^3 axis (in the negative direction there) and so was 3-momentum (but in positive direction). This is true for a particle/field seen in a frame where the particle/field has high velocity. In its own rest frame, the spin would generally (in most cases) point in some other direction away from x^3.

Do **Problem 4** to show that the R Weyl field has positive (right-hand) spin in the x^3 direction in the special highly relativistic case we are dealing with here and thus, has right-hand helicity.

For highly relativistic case, ψ^R has R helicity

Things to Note About L and R Weyl Spinors

Other axis orientations at relativistic speeds

If we have the special case of 1) highly relativistic speed and 2) **p** aligned with x^3 axis, we have seen that the L and R Weyl fields, ${}_W\psi^L$ and ${}_W\psi^R$, have left-hand and right-hand helicity, respectively. If we re-orient our coordinate axes, so the x^3 axis is not aligned with **p**, then the physical L (R) field itself still has the same left-hand (right-hand) helicity. (Its spin is still aligned along the direction of **p**). Our expressions for the spinor field and the spin operator will change and be far more complicated, but if we cranked the math, we would eventually find that the spin and the 3-momentum yield the same helicities in the second coordinate system as in the original.

*In coordinate axes with **p** not aligned along x^3, for $v = c$, ψ^L still L helicity; ψ^R still R helicity, but math more complicated*

At relativistic speeds, the spin aligns with the 3-momentum, so in that case, if we align our x^3 coordinate axis in the direction of **p**, everything simplifies, as the spin must also align with that axis (with right-hand or left-hand helicity). The p^1 and p^2 are zero, and we get the simple relation (5-14).

Conclusion: A L (R) Weyl field ${}_W\psi^L$ (${}_W\psi^R$) at highly relativistic speeds is in an L (R) helicity eigenstate, regardless of whether the 3-momentum is aligned with the x^3 axis or not. Mathematical relationships are simplest when they are so aligned, but the L (R) helicity quality (at high speeds) is a characteristic of the field, unrelated to the particular coordinate system we describe it in.

Non-relativistic speeds

We know the following on physical grounds from the logic of Vol. 1, pgs. 95-96.

Conclusion: For $v << c$, spin does not generally (except in a special case) align with **p**, so ${}_W\psi^L$ and ${}_W\psi^R$ are generally not L and R helicity eigenstates.

For non-relativistic case, ψ^L usually not L helicity; ψ^R usually not R helicity

---

Do **Problem 5** to show the above conclusion mathematically.

Creation and Destruction Operators

From (5-21) and what we noted earlier about the creation and destruction properties of $c_2(\mathbf{p})$ and $d_1^\dagger(\mathbf{p})$, we can conclude that for $v = c$, ${}_W\psi^L$ will either destroy a Weyl particle or create a Weyl antiparticle, both of which will have L helicity . Similarly, for $v = c$, ${}_W\psi^R$ will either destroy a Weyl particle or create a Weyl antiparticle, both of which will have R helicity. For lower speeds, particles and antiparticles are created and destroyed by ${}_W\psi$, but it is not so easy to characterize them.

ψ^L destroys (creates) particles (antiparticles); for v=c, both with L helicity

Kinematics vs Dynamics (Interactions)

Charged fermion spin effectively acts like a classical current loop, and gives rise to a magnetic moment. See Vol. 1, pgs. 411-412. So, a fermion and an anti-fermion with the same spin have opposite magnetic moments, since the charges are opposite, and thus, the current in the effective loop flows in the opposite direction, i.e., it has opposite sign. Recall that magnetic moment comes into play in dynamics, i.e., in interactions. So, a fermion and anti-fermion are kinematically identical (considering spin), but dynamically opposite (considering magnetic moment).

Hence, at highly relativistic speeds, the anti-particle creation operator $d_1^\dagger(\mathbf{p})$ of ${}_W\psi^L$, creates an antiparticle with spin in the same direction as a particle destroyed by $c_2(\mathbf{p})$, but with magnetic moment in the opposite direction. Helicity is the same, but the magnetic moment of one is the negative of that of the other. We will soon see this concept again in our discussion of chirality.

For v=c, ψ^L destroys particles and creates antiparticles with opposite mag moments

Weyl Fermions vs Weyl Spinors

Back in the days when neutrinos were thought to be massless, the Weyl rep was the rep of choice to evaluate them in, since the Dirac equation decouples in that rep. So, neutrinos created and destroyed by the ${}_W\psi^L$ and ${}_W\psi^R$ were named Weyl fermions (massless spin ½ particles). No known particle is a Weyl fermion, except in an approximate sense, as neutrinos are typically highly relativistic.

Hypothetical massless fermions called Weyl fermions, always in helicity eigenstates

However, Weyl spinors (though incomplete in a sense) can be very helpful in other ways because they can be used as building blocks of any fermion field, since adding them together yields the full fermion field, i.e.,

$$ {}_W\psi = \begin{bmatrix} {}_W\psi^L \\ {}_W\psi^R \end{bmatrix} = \begin{bmatrix} {}_W\psi^L \\ 0 \end{bmatrix} + \begin{bmatrix} 0 \\ {}_W\psi^R \end{bmatrix} = \underbrace{{}_W\psi^L + {}_W\psi^R}_{\text{alternative notation}} \quad . \tag{5-23} $$

Note that since, for massive fields, neither ${}_W\psi^L$ nor ${}_W\psi^R$ alone is a solution of the Dirac equation, they do not individually represent actual physical fields, and they do not individually destroy/create real particles, but only 'parts' (certain spinor space components) of real particles. We will refer to the fields ${}_W\psi^L$ and ${}_W\psi^R$ as Weyl spinors, or Weyl fields, which destroy/create actual particles, i.e., Weyl fermions, only in the massless case.

Massive ${}_W\psi^L$, ${}_W\psi^R$ (Weyl spinors/fields) are components of physical field ${}_W\psi$

Do **Problem 6** for more practice with the Weyl rep.

Summary of the Weyl Rep

Wholeness Chart 5-1 is a concise summary of the key aspects of the Weyl rep.

Wholeness Chart 5-1. Overview of Weyl Representation

	Highly Relativistic, $v = c$	**Non-highly relativistic, $v \ll c$**
Field form	${}_W\psi = \begin{bmatrix} {}_W\psi^L \\ {}_W\psi^R \end{bmatrix} = \begin{bmatrix} {}_W\psi_1 \\ {}_W\psi_2 \\ {}_W\psi_3 \\ {}_W\psi_4 \end{bmatrix}$ (5-18)	Same as at left
Dirac equation	${}_W\psi^L$ and ${}_W\psi^R$ equations uncouple 2 sets, 2 eq. in 2 unknowns in each set See (5-17)	${}_W\psi^L$ and ${}_W\psi^R$ equations coupled 4 eq., 4 unknowns See (5-15)
Helicity	${}_W\psi^L$ (${}_W\psi^R$) always in L (R) helicity eigenstate	${}_W\psi^L$ (${}_W\psi^R$) generally not in L (R) helicity eigenstate (is in special case: **p** aligned with spin)
Simplest coordinate system to use	x^3 aligned with **p**	Same as at left
Creation/destruction & spin (kinematics)	${}_W\psi^L$ destroys L helicity particle & creates L helicity antiparticle ${}_W\psi^R$ destroys R helicity particle & creates R helicity antiparticle	$\begin{bmatrix} {}_W\psi^L \\ {}_W\psi^R \end{bmatrix}$ destroys/creates antiparticles/particles, but generally not in so simple form
Creation/destruction & magnetic moment (dynamics)	${}_W\psi^L$ destroys particle of X mag moment creates antiparticle of $-X$ mag moment ${}_W\psi^R$ destroys particle of $-X$ mag moment creates antiparticle of X mag moment	Generally, not in so simple form

5.1.4 The Majorana Rep

Named after its originator, Ettore Majorana, a truly great theoretical physicist who disappeared mysteriously at the age of 32, the Majorana rep has advantages when dealing with (hypothetical at this point) fermions that are their own antiparticles. Such particles are known as Majorana particles, or here, more specifically, Majorana fermions. Certain advanced theories posit that neutrinos are actually Majorana fermions, and experiments to prove them so, or not so, have yet (as of this writing) to bear fruit. For the record, the kinds of fermions we have been dealing with up to this point are considered to have distinctly different antineutrinos, and are called Dirac fermions. We will discuss this more later, when we get to the study of neutrinos, which turn out to be the only fermions that could be of the Majorana type.

Majorana fermions are their own antiparticles

easier to work with them in Majorana rep

We note now, to avoid possible future confusion, that Majorana fermions can be expressed in any of the standard, Weyl, or Majorana reps. It is just easier to work with them in the Majorana rep.

The Transformation from the Standard to the Majorana Rep

The unitary transformation from the standard to the Majorana rep is

Transformation from standard to Majorana rep

$$_{S-M}U = \frac{1}{\sqrt{2}}\begin{bmatrix} 1 & & & -i \\ & 1 & i & \\ & -i & -1 & \\ i & & & -1 \end{bmatrix} = \frac{1}{\sqrt{2}}\begin{bmatrix} I & \sigma_2 \\ \sigma_2 & -I \end{bmatrix}. \tag{5-24}$$

Do **Problem 7** to find (5-25).

The Majorana Rep Gamma Matrices

$$_M\gamma^0 = \begin{bmatrix} & & & -i \\ & & i & \\ & -i & & \\ i & & & \end{bmatrix} \quad _M\gamma^1 = \begin{bmatrix} i & & & \\ & -i & & \\ & & i & \\ & & & -i \end{bmatrix} \quad _M\gamma^2 = \begin{bmatrix} & & & i \\ & & -i & \\ & -i & & \\ i & & & \end{bmatrix} \quad _M\gamma^3 = \begin{bmatrix} & -i & & \\ -i & & & \\ & & & -i \\ & & -i & \end{bmatrix} \tag{5-25}$$

In streamlined form,

Gamma matrices in Majorana rep are all purely imaginary

$$_M\gamma^0 = \begin{bmatrix} 0 & \sigma_2 \\ \sigma_2 & 0 \end{bmatrix} \quad _M\gamma^1 = \begin{bmatrix} i\sigma_3 & \\ & i\sigma_3 \end{bmatrix} \quad _M\gamma^2 = \begin{bmatrix} 0 & -\sigma_2 \\ \sigma_2 & 0 \end{bmatrix} \quad _M\gamma^3 = \begin{bmatrix} -i\sigma_1 & \\ & -i\sigma_1 \end{bmatrix}. \tag{5-26}$$

Dirac Equation and Its Solutions in Majorana Rep

Note that (5-25) (equivalently, (5-26)) comprise all purely imaginary matrices. This means that in the Dirac equation (5-7), expressed in the Majorana rep as

Dirac eq in Majorana rep is real, not complex

$$\left(i_M\gamma^\mu\partial_\mu - m\right)_M\psi = 0\,, \tag{5-27}$$

all terms are real. And that means we can have purely real solutions to that equation. Of course, we could have purely imaginary solutions as well, or complex solutions where the real and imaginary parts are each an independent solution. But the reason for using the Majorana rep is that it facilitates handling of real field solutions, which we can recall (Vol. 1, pgs. 65 and 148) create and destroy particles that are their own antiparticles.

So, it can have purely real solutions

The real photon field

Just as real Maxwell eq had real photon field solutions

$$A^\mu = \sum_{r,\mathbf{k}} \frac{1}{\sqrt{2V\omega_\mathbf{k}}}\left(\varepsilon_r^\mu(\mathbf{k})a_r(\mathbf{k})e^{-ikx} + \varepsilon_r^\mu(\mathbf{k})a_r^\dagger(\mathbf{k})e^{ikx}\right) \tag{5-28}$$

is an example, though that is a bosonic field, not a fermionic field. Note that the second term is the complex conjugate of the first, so their sum is real. And the photons that (5-28) creates and destroys are antiparticles of themselves. This would not be the case if we had $b_r^\dagger$ instead of $a_r^\dagger$.

Different Possible Gamma Matrices for Real Dirac Equation

There are different possible gamma matrices, other than (5-26), which result in a real Dirac equation, and thus different real solutions, as well. Any orthogonal matrix transforming (5-26) and its associated solutions (i.e., where U of (5-8) is real) would work. But we will focus on solution forms

We work with best of many possible real gamma matrix reps

that will serve our ultimate purpose best, and for that, (5-26) works well. However, be aware that other authors may use slightly different matrices from (5-26) for the Majorana rep[1].

Do **Problem 8** to find (5-29).

Direct Transformation Basis Solutions in the Majorana Rep

Note we use a prime to denote the basis in the Majorana rep that we find from directly applying (5-24) to ${}_S\psi$, because eventually we will not find that basis the most useful one. More on that below.

Carrying out the aforenoted transformation, we find

$$
{}_M\psi'^{(1)} = \underbrace{\sqrt{\frac{E+m}{4m}}\begin{pmatrix} 1 - i\frac{p^1+ip^2}{E+m} \\ i\frac{p^3}{E+m} \\ -\frac{p^3}{E+m} \\ i - \frac{p^1+ip^2}{E+m} \end{pmatrix}}_{{}_M u_1(\mathbf{p})} e^{-ipx} \qquad {}_M\psi'^{(2)} = \underbrace{\sqrt{\frac{E+m}{4m}}\begin{pmatrix} i\frac{p^3}{E+m} \\ 1 + i\frac{p^1-ip^2}{E+m} \\ -i - \frac{p^1-ip^2}{E+m} \\ \frac{p^3}{E+m} \end{pmatrix}}_{{}_M u_2(\mathbf{p})} e^{-ipx}
$$

$$
{}_M\psi'^{(3)} = \underbrace{\sqrt{\frac{E+m}{4m}}\begin{pmatrix} \frac{p^3}{E+m} \\ \frac{p^1+ip^2}{E+m} + i \\ -i\frac{p^1+ip^2}{E+m} - 1 \\ i\frac{p^3}{E+m} \end{pmatrix}}_{{}_M v_2(\mathbf{p})} e^{ipx} \qquad {}_M\psi'^{(4)} = \underbrace{\sqrt{\frac{E+m}{4m}}\begin{pmatrix} \frac{p^1-ip^2}{E+m} - i \\ -\frac{p^3}{E+m} \\ i\frac{p^3}{E+m} \\ i\frac{p^1-ip^2}{E+m} - 1 \end{pmatrix}}_{{}_M v_1(\mathbf{p})} e^{ipx}. \tag{5-29}
$$

Basis spinors in Majorana rep found via direct transformation

As we can see in (5-29), the bases we have found for the Majorana rep are complex, not real. And as we mentioned, the value of the Majorana rep is that we can have real solutions, which would represent Majorana fields (whose antiparticles are also its particles). So, there must be a way to construct a real solution from some linear combination of (5-29).

These bases complex, but we seek real solution

Finding a Real Majorana Solution from a Complex Basis

Finding a real solution

In seeking a real solution composed of a linear combination of the complex basis solutions in (5-29), we observe that ${}_M\psi'^{(1)}$ and ${}_M\psi'^{(4)}$ are closely related in their components, in that most of the terms (inside the brackets) differ by a simple factor of i. Similarly, there is a close relationship between ${}_M\psi'^{(2)}$ and ${}_M\psi'^{(3)}$. In fact, we can first verify the following.

$$
\left({}_M\psi'^{(1)}\right)^* = i\,{}_M\psi'^{(4)}. \tag{5-30}
$$

A relationship that will help us find that real solution

Thus, we ask if the question mark on the first equal sign below has a yes answer.

---

[1] See, for example, P. B. Pal, Dirac, Majorana, and Weyl fermions, *Am. J. Phys.* **79** (5), May 2011, pgs. 485-498, arXiv:1006.1718.

$$\left({}_M\psi'^{(1)}\right)^* \stackrel{?}{=} i\left({}_M\psi'^{(4)}\right) \rightarrow \sqrt{\frac{E+m}{4m}}\begin{pmatrix} 1-i\dfrac{p^1+ip^2}{E+m} \\ i\dfrac{p^3}{E+m} \\ -\dfrac{p^3}{E+m} \\ i-\dfrac{p^1+ip^2}{E+m}\end{pmatrix}^* e^{ipx} \stackrel{?}{=} \sqrt{\frac{E+m}{4m}}\, i \begin{pmatrix} \dfrac{p^1-ip^2}{E+m}-i \\ -\dfrac{p^3}{E+m} \\ i\dfrac{p^3}{E+m} \\ i\dfrac{p^1-ip^2}{E+m}-1\end{pmatrix} e^{ipx}$$

$$\begin{pmatrix} 1+\dfrac{ip^1+p^2}{E+m} \\ \dfrac{-ip^3}{E+m} \\ \dfrac{-p^3}{E+m} \\ -i+\dfrac{-p^1+ip^2}{E+m}\end{pmatrix} \stackrel{?}{=} \begin{pmatrix} \dfrac{ip^1+p^2}{E+m}+1 \\ \dfrac{-ip^3}{E+m} \\ \dfrac{-p^3}{E+m} \\ \dfrac{-p^1+ip^2}{E+m}-i\end{pmatrix}, \qquad ?\text{ answer is yes.} \tag{5-31}$$

So, using (5-31), we can obtain a real field from

$${}_M\psi_{r=1} = \overbrace{\sum_{\mathbf{p}}\sqrt{\frac{m}{VE_{\mathbf{p}}}}\left(c_1(\mathbf{p})\,{}_M\psi'^{(1)} + c_1^\dagger(\mathbf{p})\,i\,{}_M\psi'^{(4)}\right)}^{\text{linear combination of primed basis spinors}} = \overbrace{\sum_{\mathbf{p}}\sqrt{\frac{m}{VE_{\mathbf{p}}}}\left(c_1(\mathbf{p})\,{}_M\psi'^{(1)} + \left(c_1(\mathbf{p})\,{}_M\psi'^{(1)}\right)^*\right)}^{\text{real}}, \tag{5-32}$$

Using that relationship to find a real solution as linear combination of basis spinors

where we have employed the accepted convention of using the dagger on $c_1(\mathbf{p})$ to represent complex conjugation, as it is not a column vector in spinor space, so the transpose part is meaningless. As shown in the first part of (5-33), we can then define a new basis of spinors (unprimed) as constants times the old (primed) basis spinors (constant 1 in the first case, i in the second). Thus, we have the Majorana field constructed from the primed basis spinors and also expressed in the unprimed basis as

$${}_M\psi_{r=1} = \overbrace{\sum_{\mathbf{p}}\sqrt{\frac{m}{VE_{\mathbf{p}}}}\left(c_1(\mathbf{p})\underbrace{{}_M\psi'^{(1)}}_{\substack{\text{define}\\ \text{as } {}_M\psi^{(1)}}} + c_1^\dagger(\mathbf{p})\underbrace{i\left({}_M\psi'^{(4)}\right)}_{\substack{\text{define}\\ \text{as } {}_M\psi^{(4)}}}\right)}^{\text{linear combination of primed basis spinors}} = \overbrace{\sum_{\mathbf{p}}\sqrt{\frac{m}{VE_{\mathbf{p}}}}\left(c_1(\mathbf{p})\,{}_M\psi^{(1)} + c_1^\dagger(\mathbf{p})\,{}_M\psi^{(4)}\right)}^{\text{linear combination of unprimed basis spinors}}$$

$$= \sum_{\mathbf{p}}\sqrt{\frac{m}{VE_{\mathbf{p}}}}\left(c_1(\mathbf{p})\underbrace{\sqrt{\frac{E_{\mathbf{p}}+m}{4m}}\begin{pmatrix} 1+\dfrac{-ip^1+p^2}{E_{\mathbf{p}}+m} \\ \dfrac{ip^3}{E_{\mathbf{p}}+m} \\ \dfrac{-p^3}{E_{\mathbf{p}}+m} \\ i+\dfrac{-p^1-ip^2}{E_{\mathbf{p}}+m}\end{pmatrix}}_{{}_Mu_1(\mathbf{p})} e^{-ipx} + c_1^\dagger(\mathbf{p})\underbrace{\sqrt{\frac{E_{\mathbf{p}}+m}{4m}}\begin{pmatrix} \dfrac{ip^1+p^2}{E_{\mathbf{p}}+m}+1 \\ \dfrac{-ip^3}{E_{\mathbf{p}}+m} \\ \dfrac{-p^3}{E_{\mathbf{p}}+m} \\ \dfrac{-p^1+ip^2}{E_{\mathbf{p}}+m}-i\end{pmatrix}}_{i\,{}_Mv_1(\mathbf{p}) = {}_Mu_1^*(\mathbf{p})} e^{ipx}\right). \tag{5-33}$$

Spin r = 1: real solution in Majorana rep

In RQM, $c_1(\mathbf{p})$ and $c_1^\dagger(\mathbf{p})$ are numeric coefficients; in QFT, they are destruction and creation operators. Here, $c_1^\dagger(\mathbf{p})$ will create a particle of 3-momentum $\mathbf{p}$ and spin state $r = 1$, and $c_1(\mathbf{p})$ will destroy the same particle. Since ${}_M\psi_{r=1}$ is real, the particle is neutrally charged (Vol. 1, pg. 65) and is its own antiparticle.

Do **Problem 9** to find (5-34)

Similarly, from (5-29), we can find ${}_M\psi_{r=2}$, if we realize that

$$\left({}_M\psi'^{(2)}\right)^* = -i\,{}_M\psi'^{(3)}. \tag{5-34}$$

Parallel to (5-32) and (5-33), using (5-34), we can find a real field that can be made of a linear combination of bases spinors (5-29)

$${}_M\psi_{r=2} = \sum_{\mathbf{p}}\sqrt{\frac{m}{VE_{\mathbf{p}}}}\Big(c_2(\mathbf{p})\underbrace{{}_M\psi'^{(2)}}_{\substack{\text{define as}\\ {}_M\psi^{(2)}}} + c_2^\dagger(\mathbf{p})\underbrace{\left(-i_M\psi'^{(3)}\right)}_{\substack{\text{define as}\\ {}_M\psi^{(3)}}}\Big) = \sum_{\mathbf{p}}\sqrt{\frac{m}{VE_{\mathbf{p}}}}\underbrace{\left(c_2(\mathbf{p})\,{}_M\psi'^{(2)} + c_2^\dagger(\mathbf{p})\left({}_M\psi'^{(2)}\right)^*\right)}_{\text{real}}$$

Spin r = 2: real solution in Majorana rep

$$= \sum_{\mathbf{p}}\sqrt{\frac{m}{VE_{\mathbf{p}}}}\left(c_2(\mathbf{p})\,{}_M\psi^{(2)} + c_2^\dagger(\mathbf{p})\,{}_M\psi^{(3)}\right)$$

$$= \sum_{\mathbf{p}}\sqrt{\frac{m}{VE_{\mathbf{p}}}}\left(\underbrace{c_2(\mathbf{p})\sqrt{\frac{E+m}{4m}}\begin{pmatrix}\frac{ip^3}{E+m}\\ 1+\frac{ip^1+p^2}{E+m}\\ -i+\frac{-p^1+ip^2}{E+m}\\ \frac{p^3}{E+m}\end{pmatrix}}_{{}_M u_2(\mathbf{p})}e^{-ipx} + c_2^\dagger(\mathbf{p})\underbrace{\sqrt{\frac{E+m}{4m}}\begin{pmatrix}\frac{-ip^3}{E+m}\\ 1+\frac{-ip^1+p^2}{E+m}\\ i+\frac{-p^1-ip^2}{E+m}\\ \frac{p^3}{E+m}\end{pmatrix}}_{-i_M v_2(\mathbf{p})={}_M u_2^*(\mathbf{p})}e^{ipx}\right). \tag{5-35}$$

This is the spin $r = 2$ equivalent of (5-33). It creates and destroys neutral particles that are their own antiparticles but with different spin state.

A Complete Real Field Solution

Thus, a general Majorana (real) fermion field in the Majorana rep comprises the sum of (5-33) and (5-35), where we introduce notation that distinguishes between the character of the field and the rep that field is being expressed in,

Complete real solution (field) in Majorana rep

$$\begin{matrix}\text{Majorana field}\rightarrow MF\\ \text{Majorana rep}\rightarrow M\end{matrix}\psi = \sum_{\mathbf{p},r}\sqrt{\frac{m}{VE_{\mathbf{p}}}}\left(c_r(\mathbf{p})\,{}_M u_r(\mathbf{p})e^{-ipx} + c_r^\dagger(\mathbf{p})\,{}_M u_r^*(\mathbf{p})e^{ipx}\right). \tag{5-36}$$

creates and destroys Majorana particles

(5-36) will create and destroy particle states that are their own antiparticles.

Majorana Field in a Different Rep

Regardless of which rep (Majorana, standard, Weyl) we express a Majorana field in, it will still create and destroy states that physically are their own antiparticles. It's a bit similar to a gauge, in that we can use different underlying math, but the things we predict we will measure in the physical world must be the same for all such different mathematical schemes. The $c_r(\mathbf{p})$ and $c_r^\dagger(\mathbf{p})$ operators destroy and create particles of physical momentum $\mathbf{p}$ and spin state r, but the mathematical form of the spinors in the standard rep is different from the form they have in the Majorana rep (or the Weyl rep).

Deducing Majorana field in the standard rep

But then, we might ask, what does a Majorana field look like in another rep? We will answer that question for the standard rep and then generalize that answer to other reps.

If the transformation from the standard rep to the Majorana rep is (5-24), then the transformation in the opposite direction, from the Majorana to standard rep, is the inverse of that,

$$_{M-S}U = _{S-M}U^{-1} = _{S-M}U^{\dagger} = \frac{1}{\sqrt{2}}\begin{bmatrix} 1 & & & -i \\ & 1 & i & \\ & -i & -1 & \\ i & & & -1 \end{bmatrix}^{\dagger} = \frac{1}{\sqrt{2}}\begin{bmatrix} 1 & & & -i \\ & 1 & i & \\ & -i & -1 & \\ i & & & -1 \end{bmatrix}, \tag{5-37}$$

Transformation from Majorana rep to standard rep

which just happens to be the same as the original $_{S\text{-}M}U$.

Transforming (5-36) to the standard rep (where we carry it out explicitly in Appendix A), we have

$$^{MF}_{S}\psi = _{M-S}U\,^{MF}_{M}\psi = \sum_{\mathbf{p},r}\sqrt{\frac{m}{VE_{\mathbf{p}}}}\left(c_r(\mathbf{p})\underbrace{_{M-S}U_M u_r(\mathbf{p})}_{=_S u_r(\mathbf{p})}e^{-ipx} + c_r^{\dagger}(\mathbf{p})\underbrace{_{M-S}U\,_M u_r^{*}(\mathbf{p})}_{\neq _S u_r^{*}(\mathbf{p})}e^{ipx}\right)$$
$$\neq \sum_{\mathbf{p},r}\sqrt{\frac{m}{VE_{\mathbf{p}}}}\left(c_r(\mathbf{p})\,_S u_r(\mathbf{p})e^{-ipx} + c_r^{\dagger}(\mathbf{p})\,_S u_r^{*}(\mathbf{p})e^{ipx}\right). \tag{5-38}$$

Majorana field in standard rep does not have simple form we might naively expect

Note that a Majorana field in the standard rep does not have the form, as one might naively expect, of the second row in (5-38). The second term does not transform in the manner needed to yield $_S u_r^{*}(\mathbf{p})$ in the second term in the standard rep. To get that term, we would have needed the transformation

$$_S u_r^{*}(\mathbf{p}) = _{M-S}U^{*}\,_M u_r^{*}(\mathbf{p}) \quad \text{since} \quad _S u_r(\mathbf{p}) = _{M-S}U\,_M u_r(\mathbf{p}), \tag{5-39}$$

but $_{M\text{-}S}U^{*}$ is not the transformation between those reps, $_{M\text{-}S}U$ is.

Said another way, $_S u_r^{*}(\mathbf{p})$ does not solve the Dirac equation in the standard rep (it is not one of the four basis spinors (5-3)).

The bottom line: A Majorana field takes the simplest form in the Majorana rep.

Majorana field simplest form in Majorana rep

Do Fermions That Are Their Own Antiparticles Exist?

We reiterate that there are no known fermions that are their own antiparticles (Majorana fermions). It is possible that neutrinos may be Majorana in nature, and experimentalists have been trying for some time to determine if they are or not. We will have more to say about this when we study neutrinos later in the book.

Only neutrinos could be Majorana fermions, but we don't know

5.1.5 Spinor Fields and Inner Products in Different Reps

Transforming Spinor Space Scalars to a Different Rep

Note that any quantity that is a scalar with respect to spinor space will be the same in any rep. This is much like world scalars in relativity, which have the same value and same form regardless of what reference frame they are observed in. A different rep is, in an abstract way, like observing the same physical quantity but from a different observer's perspective.

Spinor space scalars are rep invariant

As an example of a spinor scalar under transformation, consider

$$\bar{\psi}\psi = \psi^{\dagger}\gamma^{0}\psi = \psi^{\dagger}U^{\dagger}U\gamma^{0}U^{\dagger}U\psi = \psi^{\dagger}U^{\dagger}\left(U\gamma^{0}U^{\dagger}\right)U\psi = \psi'^{\dagger}\gamma'^{0}\psi' = \bar{\psi}'\psi', \tag{5-40}$$

Such as $\bar{\psi}\psi$

which has the same form and (because of the equal signs) the same value in both the non-primed and the primed rep. It is invariant (symmetric) under a spinor rep transformation.

Do **Problems 10 and 11** to find (5-41) and (5-42).

Similarly, where H is the Hamiltonian and Q is the charge operator (which operate on states),

$$\bar{\psi}\gamma^{\mu}\psi = \bar{\psi}'\gamma'^{\mu}\psi' \qquad \bar{\psi}\,\partial\!\!\!/\,\psi = \bar{\psi}'\,\partial\!\!\!/'\,\psi' \qquad e\bar{\psi}A\!\!\!/\,\psi = e\bar{\psi}'A\!\!\!/'\psi'$$
$$H = H' \qquad Q = Q'. \tag{5-41}$$

and other examples, including H and Q,

And the spinor basis inner product relations from Vol. 1, pg. 90, (4-25), are invariant, as well.

$$u_r^{\dagger}(\mathbf{p})u_s(\mathbf{p}) = u_r'^{\dagger}(\mathbf{p})u_s'(\mathbf{p}) = v_r^{\dagger}(\mathbf{p})v_s(\mathbf{p}) = v_r'^{\dagger}(\mathbf{p})v_s'(\mathbf{p}) = \frac{E}{m}\delta_{rs} \qquad (5\text{-}42)$$
$$u_r^{\dagger}(\mathbf{p})v_s(-\mathbf{p}) = u_r'^{\dagger}(\mathbf{p})v_s'(-\mathbf{p}) = 0.$$

as well as spinor inner products,

And thus, the transition amplitude S_{fi}, which is a scalar, is the same for any rep, as well.

and the transition amplitude

Transforming Spinor Space Matrix Operators

Changing to a different rep does, however, change spinors (column matrices) and their complex conjugate transposes (row matrices), as well as spinor matrix operators, to (generally) different forms. For example, consider the spin operator (5-6) of standard rep form transformed to the Majorana rep. Using (5-6) and (5-24), along with (2-64) and the anti-commutation relations for Pauli matrices $\left[\sigma_i,\sigma_j\right]_+ = 2\delta_{ij}I$, we have

Spinors and spinor operators do (generally) change in different reps

$$_M\Sigma_3 = {}_{S-M}U\;{}_S\Sigma_3\;{}_{S-M}U^{\dagger} = \frac{1}{\sqrt{2}}\begin{bmatrix} 1 & & & -i \\ & 1 & i & \\ & -i & -1 & \\ i & & & -1 \end{bmatrix} \frac{1}{2}\begin{bmatrix} 1 & & & \\ & -1 & & \\ & & 1 & \\ & & & -1 \end{bmatrix} \frac{1}{\sqrt{2}}\begin{bmatrix} 1 & & & -i \\ & 1 & i & \\ & -i & -1 & \\ i & & & -1 \end{bmatrix}$$
$$= \frac{1}{4}\begin{bmatrix} I & \sigma_2 \\ \sigma_2 & -I \end{bmatrix}\begin{bmatrix} \sigma_3 & \\ & \sigma_3 \end{bmatrix}\begin{bmatrix} I & \sigma_2 \\ \sigma_2 & -I \end{bmatrix} = \frac{1}{4}\begin{bmatrix} I & \sigma_2 \\ \sigma_2 & -I \end{bmatrix}\begin{bmatrix} \sigma_3 & \sigma_3\sigma_2 \\ \sigma_3\sigma_2 & -\sigma_3 \end{bmatrix} = \frac{1}{4}\begin{bmatrix} \sigma_3+\sigma_2\sigma_3\sigma_2 & \sigma_3\sigma_2-\sigma_2\sigma_3 \\ \sigma_2\sigma_3-\sigma_3\sigma_2 & \sigma_2\sigma_3\sigma_2+\sigma_3 \end{bmatrix} \qquad (5\text{-}43)$$
$$= \frac{1}{4}\begin{bmatrix} \sigma_3-\sigma_3\sigma_2\sigma_2 & -i2\sigma_1 \\ i2\sigma_1 & \sigma_3-\sigma_3\sigma_2\sigma_2 \end{bmatrix} = \frac{1}{4}\begin{bmatrix} 0 & -i2\sigma_1 \\ i2\sigma_1 & 0 \end{bmatrix} = \frac{1}{2}\begin{bmatrix} & & & -i \\ & & -i & \\ & i & & \\ i & & & \end{bmatrix}.$$

For example, the spin operator is different in standard and Majorana representations

Weyl Fields and Majorana Fields are the Same Physical Fields in Any Rep

Keep in mind that the L (R) Weyl field when $v = c$, will destroy particles and create antiparticles with left-hand (right-hand) helicity, regardless of which rep we wish to express it in (and regardless of what coordinate system we use). An L (R) Weyl field is always an L (R) Weyl field.

Similarly, a Majorana field will create and destroy particles that are their own antiparticles, regardless of which rep we express it in (and regardless of what coordinate system we use). A Majorana field is always a Majorana field.

Though having different forms in different reps, Majorana and Weyl fields are the same fields (have same physical properties in any rep)

5.1.6 Summary of Three Different Spinor Reps

Wholeness Chart 5-2 is a quick synopsis of the advantage of each of the three reps we have studied in this Section 5.1. We will have a more detailed summary after we study chirality in Section 5.2.

Wholeness Chart 5-2. Advantages of Different Reps

Different reps make different things simple

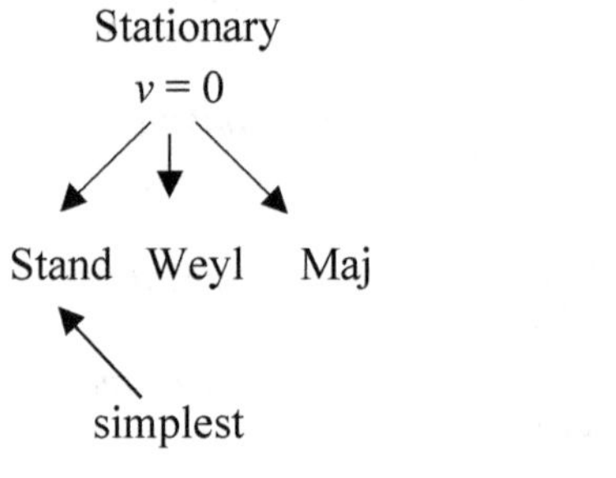

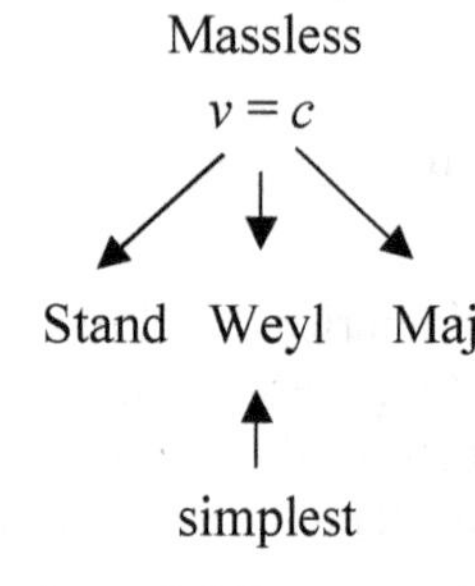

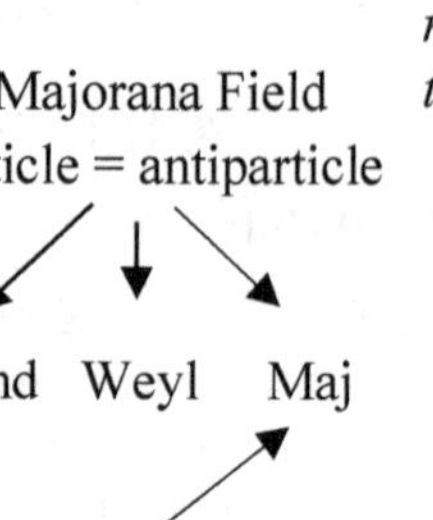

Standard (Weyl) useful for approximate case $v \approx 0$ ($v \to c$)

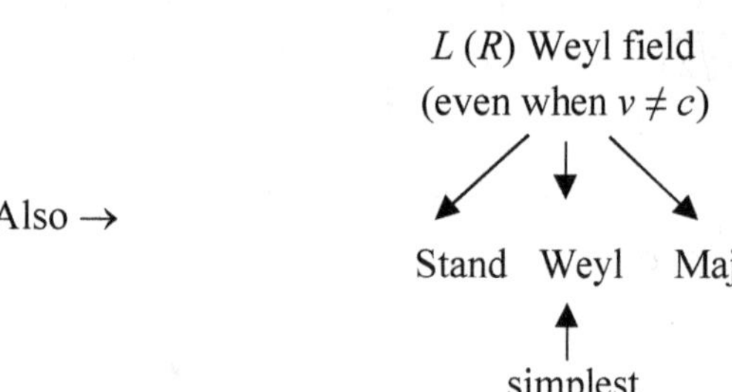

5.2 Chirality

We've mentioned chirality more than once in the preceding pages of this book, but now, finally, we have the tools in place to define it precisely.

Time for a precise definition of chirality

5.2.1 Chirality and the Weak Force

At *high energies*, at which all particles are effectively massless, it turns out that Nature has decided that the weak force is felt by L Weyl fermions, but not by R Weyl fermions. L Weyl fermions interact via the weak force. R Weyl fermions do not. No one knows why. It just happens to be that way.

L Weyl field senses weak force; R does not (at least at high energy)

At *low energies*, when particle mass comes into play, this becomes a bit more complicated, as we will discuss in later chapters.

<u>A New Gamma Matrix, γ^5</u>

The first step in understanding chirality comprises defining a new (to us, in this book) gamma matrix, called γ^5, whose relevance we will learn about, and whose use we will become accustomed to. It is defined in terms of the gamma matrices we already know. (We will shortly show (5-44) is the same in any rep.)

Defining γ^5, a new gamma matrix

$$\gamma^5 \equiv i\gamma^0\gamma^1\gamma^2\gamma^3 \tag{5-44}$$

We can gain insight into this in the easiest way, if we work in the Weyl rep, where (5-44) becomes (see (5-10))

$$_W\gamma^5 = i\,_W\gamma^0\,_W\gamma^1\,_W\gamma^2\,_W\gamma^3 = i\,_W\gamma^0\,_W\gamma^1 \begin{bmatrix} &&&-i\\ &&i&\\ &i&&\\ -i&&& \end{bmatrix}\begin{bmatrix} &&1&\\ &&&-1\\ -1&&&\\ &1&& \end{bmatrix} = i\,_W\gamma^0\,_W\gamma^1\begin{bmatrix} &-i&&\\ -i&&&\\ &&&-i\\ &&-i& \end{bmatrix}$$

$$= i\,_W\gamma^0 \begin{bmatrix} &&&1\\ &&1&\\ &-1&&\\ -1&&& \end{bmatrix}\begin{bmatrix} &-i&&\\ -i&&&\\ &&&-i\\ &&-i& \end{bmatrix} = i\,_W\gamma^0\begin{bmatrix} &&-i&\\ &&&-i\\ i&&&\\ &i&& \end{bmatrix} = i\begin{bmatrix} &&1&\\ &&&1\\ 1&&&\\ &1&& \end{bmatrix}\begin{bmatrix} &&-i&\\ &&&-i\\ i&&&\\ &i&& \end{bmatrix} \tag{5-45}$$

$$= i\begin{bmatrix} i&&&\\ &i&&\\ &&-i&\\ &&&-i \end{bmatrix} = \begin{bmatrix} -1&&&\\ &-1&&\\ &&1&\\ &&&1 \end{bmatrix} = \begin{bmatrix} -I&\\ &I \end{bmatrix}.$$

Form of γ^5 in the Weyl rep

Now, consider (5-45) acting on the L Weyl field expressed in the Weyl rep (5-18), where, by convention, the symbol ${}_W\psi^L$ can be used to represent either a four-component spinor, or the top two components of that spinor, depending on context.

$$ {}_W\gamma^5\,{}_W\psi^L = \begin{bmatrix} -1 & & & \\ & -1 & & \\ & & 1 & \\ & & & 1 \end{bmatrix} \begin{bmatrix} {}_W\psi_1 \\ {}_W\psi_2 \\ 0 \\ 0 \end{bmatrix} = - \begin{bmatrix} {}_W\psi_1 \\ {}_W\psi_2 \\ 0 \\ 0 \end{bmatrix} = -{}_W\psi^L \quad \text{or} \quad \begin{bmatrix} -I & \\ & I \end{bmatrix} \begin{bmatrix} {}_W\psi^L \\ 0 \end{bmatrix} = - \begin{bmatrix} {}_W\psi^L \\ 0 \end{bmatrix}. \tag{5-46}$$

L Weyl field has γ^5 eigenvalue of -1

The eigenvalue for ${}_W\gamma^5$ acting on ${}_W\psi^L$ is -1. For the R Weyl field, it is $+1$.

$$ {}_W\gamma^5\,{}_W\psi^R = \begin{bmatrix} -1 & & & \\ & -1 & & \\ & & 1 & \\ & & & 1 \end{bmatrix} \begin{bmatrix} 0 \\ 0 \\ {}_W\psi_3 \\ {}_W\psi_4 \end{bmatrix} = \begin{bmatrix} 0 \\ 0 \\ {}_W\psi_3 \\ {}_W\psi_4 \end{bmatrix} = {}_W\psi^R \quad \text{or} \quad \begin{bmatrix} -I & \\ & I \end{bmatrix} \begin{bmatrix} 0 \\ {}_W\psi^R \end{bmatrix} = \begin{bmatrix} 0 \\ {}_W\psi^R \end{bmatrix}. \tag{5-47}$$

R Weyl field has γ^5 eigenvalue of $+1$

Note that the Weyl spinor eigenvalues for the operator γ^5 are the same regardless of whether the Weyl field is massless or not, and regardless of how the spin is aligned relative to the 3-momentum. They are also the same regardless of what coordinate system, or inertial reference frame, we wish to employ.

These eigenvalues the same whether $v = c$, or not, unlike helicity

γ^5 is the Chirality Operator

Now, guess what? γ^5 is the chirality operator. The L (R) Weyl field is always in an eigenstate of γ^5 with eigenvalue -1 ($+1$). In fact, we could use the eigenvalues to designate the Weyl fields, calling the L field -1; and the R field, $+1$. What is most common, however, is that the L (R) Weyl field is referred to as the L (R) chiral field.

γ^5 is the chirality operator; Weyl fields are chiral fields

γ^5 can also be considered a reflection (or inversion) operator as it flips (inverts) the signs of the first two components of ${}_W\psi$, which is essentially a reflection of the first two spinor space "axes".

The Weak Interaction Senses L Chirality

At *high energies*, only the left chiral field feels the weak force. The right one doesn't. We will show how this works mathematically, shortly, and what happens at low energy, in due course. First, we need to understand a little more about γ^5, and discuss how to project the L (R) chiral/Weyl field portion out of a general spinor field. Note that, unless otherwise stated, all that we do in the remainder of this chapter with Weyl fields is for massless, or highly relativistic, fields.

Only L Weyl/chiral field feels weak force (at high energy)

5.2.2 Key Operator Properties in Chiral Field Theory

Properties of γ^5

Transforming γ^5 to another rep, we have

$$\begin{aligned} \gamma'^5 &= i\gamma'^0\gamma'^1\gamma'^2\gamma'^3 = i\left(U\,{}_W\gamma^0U^\dagger\right)\left(U\,{}_W\gamma^1U^\dagger\right)\left(U\,{}_W\gamma^2U^\dagger\right)\left(U\,{}_W\gamma^3U^\dagger\right) \\ &= i\,U\,{}_W\gamma^0\,{}_W\gamma^1\,{}_W\gamma^2\,{}_W\gamma^3U^\dagger = U\,{}_W\gamma^5U^\dagger , \end{aligned} \tag{5-48}$$

γ^5 transforms in usual way of other gamma matrices

which is what we would intuitively suspect, anyway.

From (5-45), it should be obvious that, at least in the Weyl rep,

$$\left(\gamma^5\right)^\dagger = \gamma^5 \qquad \left(\gamma^5\right)^2 = I \qquad \left(\gamma^5\right)^\dagger\gamma^5 = \left(\gamma^5\right)^2 = I . \tag{5-49}$$

Some key properties of γ^5 (true in any rep)

The first relation in (5-49) is invariant (the same in any rep) via a matrix theory rule that Hermitian matrices remain Hermitian under unitary transformations. But, we can prove it using (5-48).

$$\left(\gamma'^5\right)^\dagger = \left(U\,{}_W\gamma^5U^\dagger\right)^\dagger = U\left({}_W\gamma^5\right)^\dagger U^\dagger = U\,{}_W\gamma^5U^\dagger = \gamma'^5 . \tag{5-50}$$

Show the other relations in (5-49) are also rep invariant by doing **Problem 12**.

In general, relationships between matrices expressed in the form of equations are invariant under unitary transformations, and we will simply incorporate this rule in what follows. The gamma matrix expressions we derive henceforth, unless stated to the contrary, will be valid in any spinor rep.

Relations between matrices valid in any rep

Another useful relation is

$$\gamma^5\gamma^\mu = -\gamma^\mu\gamma^5 \quad \text{or} \quad \left[\gamma^5,\gamma^\mu\right]_+ = 0, \tag{5-51}$$

One such relationship (valid in any rep)

which can be proven by simple substitution in the Weyl rep.

Do **Problem 13** to prove (5-51) is true for a couple of the four gamma matrices.

Other relations for γ^5 that will eventually be useful, including their forms in the standard and Majorana reps, can be found in Appendix B.

The Weyl Field Projection Operators

Consider the following operator, which we can construct out of the identity (represented in short hand as 1) and the chirality operator γ^5, expressed here in the Weyl rep (because it is easiest),

$$P_L = \tfrac{1}{2}\left(1-\gamma^5\right) = \tfrac{1}{2}\left(\begin{bmatrix}1&&&\\&1&&\\&&1&\\&&&1\end{bmatrix} - \begin{bmatrix}-1&&&\\&-1&&\\&&1&\\&&&1\end{bmatrix}\right) = \tfrac{1}{2}\begin{bmatrix}2&&&\\&2&&\\&&0&\\&&&0\end{bmatrix} = \begin{bmatrix}1&&&\\&1&&\\&&0&\\&&&0\end{bmatrix}. \tag{5-52}$$

Then, note the effect of (5-52) on a general spinor state (in the Weyl rep, as is (5-52)).

$$P_L\,{}_W\psi = \tfrac{1}{2}\left(1-\gamma^5\right){}_W\psi = \begin{bmatrix}1&&&\\&1&&\\&&0&\\&&&0\end{bmatrix}\begin{bmatrix}{}_W\psi_1\\{}_W\psi_2\\{}_W\psi_3\\{}_W\psi_4\end{bmatrix} = \begin{bmatrix}{}_W\psi_1\\{}_W\psi_2\\0\\0\end{bmatrix} = {}_W\psi^L. \tag{5-53}$$

L chiral projection operator P_L projects L chiral field from ψ

The P_L operator projects out the L chiral/Weyl field. Similarly,

$$P_R\,{}_W\psi = \underbrace{\tfrac{1}{2}\left(1+\gamma^5\right)}_{P_R}{}_W\psi = \begin{bmatrix}0&&&\\&0&&\\&&1&\\&&&1\end{bmatrix}\begin{bmatrix}{}_W\psi_1\\{}_W\psi_2\\{}_W\psi_3\\{}_W\psi_4\end{bmatrix} = \begin{bmatrix}0\\0\\{}_W\psi_3\\{}_W\psi_4\end{bmatrix} = {}_W\psi^R, \tag{5-54}$$

R chiral projection operator P_R projects R chiral field from ψ

and P_R projects out the R chiral field.

Note that P_L (P_R) projects out the L (R) chiral field in any rep. Recall the physical field is unchanged from rep to rep, though the expressions of the spinor space matrices and column/row spinors can change. P_L (P_R) and ψ will have different components in different reps, but in every rep P_L (P_R) acting on ψ will yield ψ^L (ψ^R), expressed in that rep. It is just easier to see how it does that in the Weyl rep, and that is why we used that rep above.

P_L (P_R) projects L (R) chiral field in any rep

Some Properties of P_L (P_R)

Note that using the middle relation of (5-49), in any rep, we have

$$\left(P_L\right)^2 = \tfrac{1}{2}\left(1-\gamma^5\right)\tfrac{1}{2}\left(1-\gamma^5\right) = \tfrac{1}{4}\left(1-2\gamma^5+\left(\gamma^5\right)^2\right) = \tfrac{1}{4}\left(1-2\gamma^5+1\right) = \tfrac{1}{2}\left(1-\gamma^5\right) = P_L, \tag{5-55}$$

which makes sense, since $\left(P_L\right)^2\psi = P_L\left(P_L\psi\right) = P_L\psi^L = \psi^L = P_L\psi$. Generally, again in any rep,

$$\left(P_L\right)^2 = P_L \qquad \left(P_R\right)^2 = P_R \qquad \left(P_L\right)^\dagger = P_L \qquad \left(P_R\right)^\dagger = P_R \qquad P_L P_R = 0. \tag{5-56}$$

Key properties of projection operators

Do **Problem 14** to show the last four relations in (5-56).

Further, where we use (5-51), (5-56), and from (5-10), $\left(\gamma^0\right)^\dagger = \gamma^0$,

$$\bar{\psi}P_R \equiv \psi^\dagger\gamma^0 P_R = \psi^\dagger\gamma^0\tfrac{1}{2}\left(1+\gamma^5\right) = \psi^\dagger\tfrac{1}{2}\left(1-\gamma^5\right)\gamma^0$$
$$= \psi^\dagger P_L\gamma^0 = \psi^\dagger P_L^\dagger\gamma^0 = \left(P_L\psi\right)^\dagger\gamma^0 = \psi^{L\dagger}\gamma^0 = \bar{\psi}^L. \tag{5-57}$$

R projection operator effect on adjoint spinor field

P_R acting on $\bar{\psi}$ from the right yields $\bar{\psi}^L$, which creates left chiral particles in a ket.

Do **Problem 15** to show $\bar{\psi}P_L = \bar{\psi}^R$.

L projection operator effect on adjoint spinor field

Creation and Destruction Properties of *L* (*R*) Fields

ψ^L destroys particles and creates their opposites, anti-particles. But, as far as the weak force is concerned, the *R* chiral (RC) particle is the opposite of the *L* chiral (LC) particle. So, ψ^L destroys LC particles and creates their opposite, RC antiparticles.

Recall from Wholeness Chart 5-1, pg. 141, that even though, for the massless case, ψ^L creates particles and destroys antiparticles with the same kinematics (helicities), the particles and antiparticles have opposite dynamics (magnetic moments). So, with regard to the dynamics of the weak force, it should not be too surprising that ψ^L destroys particles and creates antiparticles with opposite weak interaction characteristics.

Summary of Creation and Destruction Properties of Weyl/Chiral Fields

Wholeness Chart 5-3. Summary of Chiral Field Effects on States

Creation and destruction operator effects of chiral fields

Chiral field	Effect
$\psi^L = P_L\psi = \frac{1}{2}\left(1-\gamma^5\right)\psi$	Destroys LC particles, creates RC antiparticles
$\psi^R = P_R\psi = \frac{1}{2}\left(1+\gamma^5\right)\psi$	" RC " , " LC "
$\bar{\psi}^L = \bar{\psi}P_R = \bar{\psi}\frac{1}{2}\left(1+\gamma^5\right)$	Creates LC particles, destroys RC antiparticles
$\bar{\psi}^R = \bar{\psi}P_L = \bar{\psi}\frac{1}{2}\left(1-\gamma^5\right)$	" RC " , " LC "

Key Mnemonic

As an aide to remembering the essence of Wholeness Chart 5-3:

Aide to keeping it all straight in your mind

- *L* (or *R*) superscript → refers to particle chirality (never antiparticle)
- No bar on ψ → destroys particles
- Bar on $\bar{\psi}$ → creates particles

Think of *L* and *R* superscripts as standing for particles, where no bar means destroying them, and bar, creating them (as it has always been for us before). Then, for other effects of the particular field (last part of second column in Wholeness Chart 5-3), 1) take particle → antiparticle, 2) exchange creation ↔ destruction, and 3) exchange LC ↔ RC.

The shaded rows in Wholeness Chart 5-3, with fields having L superscript, are the ones we will use most, as they are the ones involved in weak interactions. I recommend copying the chart and taping it to the wall where you study QFT, as you may find yourself referring to it often.

Symbols for Chirality and Helicity

In other texts, helicity and chirality are commonly both symbolized by *L* and *R*. Herein, in the non-sub/superscript text, we use either LC and RC, or alternatively *L* and *R*, for chirality; and LH (RH) for helicity. Weyl fields are always LC or RC, but only LH or RH, when $v = c$ (massless or highly relativistic cases). When we use *L* or *R* as superscripts on ψ, it will always refer to chirality. If we mean helicity, we will superscript with *LH* or *RH*.

In this book: (as superscripts) L and R = chirality; LH and RH = helicity

5.2.3 Typical Weak Interaction Lagrangian Term

Recall (Vol. 1, pg. 186, (7-20)) that the interaction term in the QED Lagrangian, corresponding to a vertex in a Feynman diagram, is

$$_{QED}\mathcal{L}_I = e\bar{\psi}\gamma^\mu\psi A_\mu = e\, j^\mu A_\mu \quad \text{where lepton e/m 4-current} = j^\mu = \bar{\psi}\gamma^\mu\psi \tag{5-58}$$

We will jump ahead of ourselves a little now by expressing a typical interaction term in the weak interaction Lagrangian at high energy, where $W_{3\mu}$ is one of the three intermediate vector bosons that act as virtual bosons mediating the weak force, comparable to the photon for QED. The term looks a lot like (5-58) and, where the fermion fields are electrons in this case, is

Example Lagrangian term for weak interaction

$$\text{one term in high energy Lagrangian} \quad _{Weak}\mathcal{L}_I \propto \bar{\psi}^L\gamma^\mu\psi^L W_{3\mu}\,. \tag{5-59}$$

Note the left chiral fields are the ones interacting. The right chiral part of the field ψ is not involved. The $W_{3\mu}$ field only interacts with the L chiral field, and not the R chiral field.

With the expressions of Wholeness Chart 5-3, (5-51), and (5-56), we can re-write (5-59) in terms of our projection operators as

That term expressed with 1) projection operators and 2) a new kind of 4-current

$$\begin{aligned}\bar{\psi}^L\gamma^\mu\psi^L W_{3\mu} &= \bar{\psi}P_R\gamma^\mu P_L\psi W_{3\mu} = \bar{\psi}\tfrac{1}{2}\left(1+\gamma^5\right)\gamma^\mu\tfrac{1}{2}\left(1-\gamma^5\right)\psi W_{3\mu} = \bar{\psi}\gamma^\mu\tfrac{1}{2}\left(1-\gamma^5\right)\tfrac{1}{2}\left(1-\gamma^5\right)\psi W_{3\mu}\\ &= \bar{\psi}\gamma^\mu\left(P_L\right)^2\psi W_{3\mu} = \bar{\psi}\gamma^\mu P_L\psi W_{3\mu} = \underbrace{\bar{\psi}\gamma^\mu\tfrac{1}{2}\left(1-\gamma^5\right)\psi}_{j'^\mu_W} W_{3\mu} = j'^\mu_W W_{3\mu}.\end{aligned} \tag{5-60}$$

j'^μ_W is comparable to the e/m lepton 4-current of (5-58), and in fact, has (5-58) as one of its terms (along with a factor of ½).

$$j'^\mu_W = \bar{\psi}^L\gamma^\mu\psi^L = \bar{\psi}\gamma^\mu\tfrac{1}{2}\left(1-\gamma^5\right)\psi = \underbrace{\tfrac{1}{2}\bar{\psi}\gamma^\mu\psi}_{j^\mu} - \underbrace{\tfrac{1}{2}\bar{\psi}\gamma^\mu\gamma^5\psi}_{j^{5\mu}} \tag{5-61}$$

j^μ, the spinor (or lepton) 4-current we have seen before (Vol. 1, pg. 112, (4-97), is often called the <u>vector current</u>, and $j^{5\mu}$, the <u>axial (or axial vector) current</u>, because of their transformation properties (which we discuss in a later chapter.)

The actual weak interaction Lagrangian has more terms with 4-currents in it than we show here, and gets more complicated, so we postpone an in-depth look at it to a later chapter. But, hopefully, (5-60) gives you some idea of where we are headed with all this, and how it parallels what we already know about QED.

5.2.4 Chirality Summary

Wholeness Chart 5-4 summarizes the introduction to chirality of this section.

Overview of our intro to chirality

Wholeness Chart 5-4. Development of Chiral Operators and Fields

Spinor fields physically the same in all reps & spinor matrix equations have same form in all reps

↓

So, pick the easiest rep to work in = Weyl rep

↓

Find γ^5 (chirality operator) $= i\gamma^0\gamma^1\gamma^2\gamma^3$ in Weyl rep

↓

Show chiral fields (Weyl fields) ψ^L and ψ^R are eigenspinors of γ^5 in Weyl rep
so true in any rep

↓

Construct projection operators $P_L = \tfrac{1}{2}\left(1-\gamma^5\right)$ & $P_R = \tfrac{1}{2}\left(1+\gamma^5\right)$, then show $P_L\psi = \psi^L$ & $P_R\psi = \psi^R$ in Weyl rep
so true in any rep

↓

Note typical term in weak interaction Lagrangian in Weyl rep $\bar{\psi}^L\gamma^\mu\psi^L W_{3\mu} = \bar{\psi}\gamma^\mu\tfrac{1}{2}\left(1-\gamma^5\right)\psi W_{3\mu}$
so true in any rep

5.3 Spinors and Boosts

5.3.1 Idiosyncrasies of Spinor State Boosts in RQM

Boosting a spinor (going from $\mathbf{v} = 0$ to $\mathbf{v} \neq 0$) is not represented mathematically as simply as one might naively think. To see this most easily, consider an RQM general spin state, where we have a particle, but no antiparticle, of 3-momentum $\mathbf{p}$ in the standard rep (see (5-4), and upper case C_i represent numeric coefficients (in RQM), not operators (as with lower case c_i in QFT),

Boosting a spinor not quite as one might expect

$$\psi_{state} = \sqrt{\frac{m}{VE_{\mathbf{p}}}}\left(C_1(\mathbf{p})u_1(\mathbf{p})e^{-ipx} + C_2(\mathbf{p})u_2(\mathbf{p})e^{-ipx}\right)$$

$$= \sqrt{\frac{m}{VE_{\mathbf{p}}}}\sqrt{\frac{E_{\mathbf{p}}+m}{2m}}\left(C_1(\mathbf{p})\begin{pmatrix}1\\0\\\frac{p^3}{E_{\mathbf{p}}+m}\\\frac{p^1+ip^2}{E_{\mathbf{p}}+m}\end{pmatrix} + C_2(\mathbf{p})\begin{pmatrix}0\\1\\\frac{p^1-ip^2}{E_{\mathbf{p}}+m}\\\frac{-p^3}{E_{\mathbf{p}}+m}\end{pmatrix}\right)e^{-ipx}. \qquad (5\text{-}62)$$

RQM fermion general state

When $\mathbf{v} = 0$, this is

For $\mathbf{p} = 0$

$$\psi_{state} = \underbrace{\sqrt{\frac{m}{VE_{\mathbf{p}}}}\sqrt{\frac{E_{\mathbf{p}}+m}{2m}}}_{\sqrt{1/V}}\left(C_1(0)\begin{pmatrix}1\\0\\0\\0\end{pmatrix} + C_2(0)\begin{pmatrix}0\\1\\0\\0\end{pmatrix}\right)e^{-ipx} \qquad (E_{\mathbf{p}} = m, \text{ for } \mathbf{p} = 0). \qquad (5\text{-}63)$$

Under a spatial rotation and/or boost, as we saw in Vol. 1, pg. 171, (6-22), we need to operate on ψ_{state} with the spinor matrix D (defined below and formally known as the representation of the Lorentz group in spinor space), to get the new (primed) state. That is,

$$\psi'_{state} = D\psi_{state}, \quad \text{where } D = e^{-i\left(L^k\Theta^k + M^k Q^k\right)}, \text{ and} \qquad (5\text{-}64)$$

$$L^k = -\tfrac{i}{2}\varepsilon_{ij}{}^{k}\gamma^i\gamma^j, \quad \Theta^k = \underbrace{\left(\theta^1,\theta^2,\theta^3\right)}_{\substack{\text{rotation}\\\text{angles}}}, \quad M^k = -\tfrac{i}{2}\gamma^0\gamma^k, \quad Q^k = \underbrace{\left(v^1,v^2,v^3\right)}_{\substack{\text{boost velocity}\\\text{components}}} \quad v \ll c. \qquad (5\text{-}65)$$

RQM fermion spacetime transformation

We don't need to worry now about details of (5-64) and (5-65).[1] What we do need to understand is that, for a boost (and no rotation to keep it simple) we don't merely change the p^i values in the u_i of (5-62) from zero to the boosted values to get ψ'. We need to use (5-64). Otherwise, we would have

Boost is not simply new p^i value in u_i

$$\text{wrong} \rightarrow \quad \psi'_{state} = \sqrt{\frac{m}{VE_{\mathbf{p}}}}\left(C_1(0)u_1(\mathbf{p})e^{-ipx} + C_2(0)u_2(\mathbf{p})e^{-ipx}\right), \text{ instead of} \qquad (5\text{-}66)$$

$$\text{right} \rightarrow \quad \psi'_{state} = \sqrt{\frac{m}{VE_{\mathbf{p}}}}\left(C'_1(\mathbf{p})u_1(\mathbf{p})e^{-ipx} + C'_2(\mathbf{p})u_2(\mathbf{p})e^{-ipx}\right) \qquad (5\text{-}67)$$

Boost uses D; and get new C_i from that

That is, $C_i(0) \neq C_i'(\mathbf{p})$. As a general state is boosted, the amplitudes C_i of the basis spinors u_i change. And they change in a different manner than the basis spinors do. That is, the state ψ_{state} changes by (5-64); the basis spinors change according to their defined values of (5-62). If we want to express ψ' in terms of the basis spinors, as in (5-67), then we need to find the new amplitudes $C_i'(\mathbf{p})$ for the boosted general state.

Conclusion. We cannot take an unboosted general spin state and find the boosted general spin state just by inputting the new $\mathbf{p}$ instead of $\mathbf{p} = 0$ into the relations for $u_1(\mathbf{p})$ and $u_2(\mathbf{p})$ (along with the new $E_{\mathbf{p}}$). The spinor basis states do not, on their own, take into account the full effect of the boost. The coefficients change, as well.

u_i and C_i both change under boost, but in different ways

This is shown heuristically in Vol. 1, pg. 99, Fig. 4-2.

---

[1] Small v will work for us, as we focus on infinitesimal transformations, and here $v \ll 1$. For more on (5-65), and for large v, see www.quantumfieldtheory.info/Spinor_Lorentz_transf.pdf., as well as the footnote on pg. 171 of Vol. 1.

5.3.2 Finding the Boosted General Spin State (finding $C_1'(\mathbf{p})$ and $C_2'(\mathbf{p})$)

The steps for determining $C_1'(\mathbf{p})$ and $C_2'(\mathbf{p})$ for a boosted particle whose unboosted general spin state is known are as follows.

Steps to find boosted C_i'

1. Start with the unboosted ($\mathbf{p} = 0$) general spin state in spinor space.

$$\psi_{state} = \sqrt{\frac{m}{VE_{\mathbf{p}}}}\left(C_1(\mathbf{p}=0)u_1(\mathbf{p}=0)e^{-ipx} + C_2(\mathbf{p}=0)u_2(\mathbf{p}=0)e^{-ipx}\right) \tag{5-68}$$

2. Apply D, the Lorentz group representation of the boost (5-64), to (5-68) to get the boosted state ψ'_{state}.
3. Determine $u_1(\mathbf{p})$ and $u_2(\mathbf{p})$ from equation (5-62).
4. Use the results from steps 2 and 3, along with normalization conditions (see Vol. 1, pg. 90, (4-25)) to solve for boosted $C_1'(\mathbf{p})$ and $C_2'(\mathbf{p})$. That is, from

$$\underbrace{\psi'_{state}}_{\substack{\text{known}\\ \text{from \#2}}} = \sqrt{\frac{m}{VE_{\mathbf{p}}}}\left(C_1'(\mathbf{p})\underbrace{u_1(\mathbf{p})}_{\substack{\text{known}\\ \text{from \#3}}}e^{-ipx} + C_2'(\mathbf{p})\underbrace{u_2(\mathbf{p})}_{\substack{\text{known}\\ \text{from \#3}}}e^{-ipx}\right), \tag{5-69}$$

find the new coefficients via

$$\int\sqrt{\frac{m}{VE_{\mathbf{p}}}}e^{ipx}u_1^{\dagger}(\mathbf{p})\psi'_{state}\,d^3x = C_1'(\mathbf{p}) \qquad \int\sqrt{\frac{m}{VE_{\mathbf{p}}}}e^{ipx}u_2^{\dagger}(\mathbf{p})\psi'_{state}\,d^3x = C_2'(\mathbf{p}). \tag{5-70}$$

Assuming we started with the usual normalization

$$\left|C_1(0)\right|^2 + \left|C_2(0)\right|^2 = 1, \tag{5-71}$$

the above steps will lead to

$$\left|C_1'(\mathbf{p})\right|^2 + \left|C_2'(\mathbf{p})\right|^2 = 1. \tag{5-72}$$

5.3.3 Boosted Fields in QFT

In QFT, the coefficients $C_1(\mathbf{p})$ and $C_2(\mathbf{p})$ of RQM's ψ_{state} go over into the destruction operators $c_1(\mathbf{p})$ and $c_2(\mathbf{p})$ of QFT's field ψ, and we also include the antiparticle creation operators, as in the first row of (5-4). The adjoint field $\bar{\psi}$ has the reverse creation and destruction operator properties. $c_1(\mathbf{p})$ still destroys a particle of spin state $r = 1$ and 3-momentum $\mathbf{p}$. Similar for $c_2(\mathbf{p})$, but for $r = 2$.

Boosted fields in QFT handled similarly

In our work to date, we have generally focused on fields creating/destroying states in either the $r = 1$ or the $r = 2$ spin basis state, corresponding to $u_1(\mathbf{p})$ or $u_2(\mathbf{p})$, and not a superposition of the two (a field that creates/destroys a general state). If we were to consider boosts and superpositions of such fields/states, we would have to take into account the analysis of Sections 5.3.1 and 5.3.2.

We will not delve into this further, at this point, as we have other fish to fry, and need to move on.

5.4 Massive Spin 1 (Vector) Fields

As you probably already know, the three intermediate vector bosons that carry the weak force (that act as virtual particles in weak interactions) in our contemporary universe have mass. In that, they differ from the photon of the QED interaction, which is massless. Before we delve into weak interaction theory, it will help if we understand some characteristics of such massive spin 1 particles/fields. To do so, we will start by comparing them to photons.

A look at massive spin 1 bosons

5.4.1 Qualitative Look at Massless Spin 1 (Photon) Degrees of Freedom

Being massless, and thus traveling at the speed of light, photons must have their spin aligned either in, or opposite to, their direction of travel. We showed this in Vol. 1, Box 4-2, pg. 95 and Fig. 4-1, pg. 96.

Photons have *spin in $\pm\mathbf{k}$ direction*

Additionally, their field components must be transverse to that direction. Thus, a classical photon can have no component of its field A_μ in the direction it is traveling, i.e., it can only have transverse (spatial) components.

They also have only transverse field components

In Vol. 1, pgs. 151-154, we discussed the Gupta-Bleuler approach in QFT whereby only transverse components of massless fields are measurable. If longitudinal and time-like components did exist, they would cancel one another, not be detectable, and so are effectively non-existent in our world. We can thus consider, for all intents and purposes, that massless fields have only transverse field components.

To summarize

Real (not virtual) photons are constrained by having

1) spin pointing in, or opposite to, their direction of travel, and

2) only field components transverse to their direction of travel.

5.4.2 Qualitative Look at Massive Spin 1 Degrees of Freedom

However, massive particles with spin, and in particular for us with spin 1, do not have to have their spin aligned with their direction of travel, since they travel at less than the speed of light. Spin can be off at virtually any angle.

Massive bosons not so constrained (spin in any direction), i.e., they have more degrees of freedom

Similarly, their field components are not constrained to be solely transverse, i.e., they can have a longitudinal component. So, we would expect massive spin 1 particles to be less constrained, and thus to have more degrees of freedom, than photons do. And we would be correct.

To summarize

Massive particles can have

1) spin pointing in any direction, and

2) field components in any direction

However, we need to look more closely at the underlying math to fully understand those degrees of freedom and the constraints (or lack thereof) on them.

5.4.3 Mathematical Look at Massive Spin 1 Degrees of Freedom

Massive Spin 1 Field Equation

Before proceeding, it may help to go back and review Vol. 1, pg. 141, Wholeness Chart 5-1, right hand column. There you can find Maxwell's equation in terms of the 4-potential A^μ,

$$\partial^\alpha \partial_\alpha A^\mu(x) - \partial^\mu \underbrace{\left(\partial_\nu A^\nu(x)\right)}_{=0 \text{ in Lorenz gauge}} = 0 . \qquad \text{(Maxwell equation)} \qquad (5\text{-}73)$$

Photon field equation before adopting a gauge

where the Lorenz condition (gauge) turns the second term to zero (and gives us the more familiar form of the Maxwell equation). The corresponding equation for a massive spin 1 field, known as the Proca equation, parallels (5-73), but has an additional mass term that parallels that of the Klein-Gordon equation. We next express that equation, (5-74) below, for Z^μ, the 4-potential for the Z intermediate vector boson, one of the three weak massive spin 1 fields operating in our present day universe. The other two such bosons are governed by the same equation, and for them, the same conclusions will apply as we shall draw for the Z.

$$\partial^\alpha \partial_\alpha Z^\mu(x) - \partial^\mu \left(\partial_\nu Z^\nu(x)\right) + m_Z^2 Z^\mu(x) = 0 . \qquad \text{(Proca equation)} \qquad (5\text{-}74)$$

Massive spin 1 field equation adds mass term

Constraints and Degrees of Freedom

When we say "degrees of freedom" (d.o.f.) we are referring to the 4 spacetime components of A^μ or Z^μ. It is easiest to think about these in the "photon and axes aligned coordinate system" of Vol. 1, pg. 142, Fig. 5-1(d), where **k** of the photon is aligned with the x^3 axis. For the photon, there are only two measurable components, and for the chosen coordinate system, these are represented by $A^1(x^\alpha)$ and $A^2(x^\alpha)$, which are perpendicular to each other and to **k** (the x^3 axis here). We say the photon is constrained to only two degrees of freedom.

Degree of freedom (d.o.f.) = free spacetime component of field

It will help if we review why this is so for 1) the classical photon, and 2) the QFT photon. Chap. 5 of Vol. 1 can help here, if the review below is too terse. Following those reviews, we will look at the analogous situation for massive spin 1 fields.

Synopsis of the Classical Photon

For the classical photon, the Lorenz condition (which simplifies (5-73)),

$$\partial_\nu A^\nu(x) = 0\,, \tag{5-75}$$

1st constraint on classical photon (Lorenz gauge) → 1 fewer d.o.f.

comprises one equation in four scalar unknowns (the components of A^ν). It is one constraint and therefore limits the free unknowns (components, degrees of freedom) to three. Any one of them can be considered to depend on the other three.

Now, empirically, we know that two of the components are perpendicular to **k** (the x^3 axis in the easiest to use coordinate system). And classically, we don't detect any A^0 component, i.e., none in the time direction. So, we take that component equal to zero, which is a second constraint. Doing that, reduces (5-75) to what is known as the Coulomb gauge

$$\partial_i A^i(x) = 0 \qquad i = 1,2,3 \quad \text{since} \quad A^0(x) = 0\,. \tag{5-76}$$

2nd constraint on classical photon (Coulomb gauge) → yet 1 fewer d.o.f.

This can be re-written, after taking derivatives, where the k_i arise from derivatives of the exponential in ikx of $A^i(x)$, as

$$k_i A^i(x) = \mathbf{k}\bullet\mathbf{A} = 0\,, \tag{5-77}$$

meaning the 3D photon field **A** is perpendicular to **k**. The photon field is transverse to its direction of travel. Equation (5-76), one equation in three unknowns, constrains what would otherwise be three degrees of freedom (free components) to two.

These constraints leave 2 d.o.f. & make photon field transverse, as we observe in the world

The Coulomb gauge (i.e., $A^0 = 0$) is equivalent to assuming that there is no component in the **k** direction (for our coordinate system, $A^3 = 0$) and two independent components perpendicular to **k**.

2nd constraint equivalent to assuming no longitudinal component

Synopsis of the QFT Photon

As shown in Vol. 1, Chap. 5, Sect. 5.8, pgs. 150-154, the Gupta-Bleuler weak Lorenz condition leads to the conclusion that the expectation value for measuring the energy of the scalar (timelike) photon component A^0 cancels that of the longitudinal (in **k** direction) photon component (A^3 in our chosen coordinate system). We can never measure either, so physically, the photon field we measure is transverse.

In QFT, Gupta-Bleuler condition instead of Lorenz → timelike component cancels longitudinal→measure only transverse comps

Technically, though, that is only one constraint, as the scalar and longitudinal components could well exist, but just not be measured. To close the loop, we could assume, as we did in the classical case, that the scalar photon component is zero, i.e., $A^0 = 0$. This means, via Gupta-Bleuler, that the longitudinal component must also be zero, leaving us with two free components (A^1 and A^2 in our chosen coordinate system), which are transverse to **k**.

Could add constraint $A^3 = 0$ to ensure only (two) transverse components

Note that the assumption $A^0 = 0$, due to Gupta-Bleuler, is equivalent to the assumption that there is no component in the **k** direction (for our coordinate system, $A^3 = 0$).

Bottom Line for Photons

Classically, or quantum mechanically, the photon field has four degrees of freedom (spacetime components), but we can apply two gauge conditions to reduce these to two. That makes things easier and helps match theory with experiment. The two constraint equations are 1) the Lorenz condition and 2) either i) the timelike component of A^μ is zero, or ii) its effective equivalent, the longitudinal component of A^μ is zero.

Classical or QFT, two gauge constraints → photon has only two d.o.f., which are transverse components

Caveat

Keep in mind that these results apply only to real (on shell) photons. Photon propagators typically involve independent (non-zero) values for all four photon components. This is one more way that virtual particles are not bound by the usual rules for real particles.

But photon propagator is virtual particle & has all 4 components

Massive Spin 1 Fields Degrees of Freedom

Note what happens if we take the divergence of (5-74)

$$\partial_\mu\left(\partial^\alpha\partial_\alpha Z^\mu(x) - \partial^\mu\left(\partial_\nu Z^\nu(x)\right) + m_Z^2 Z^\mu(x)\right) = 0\,. \tag{5-78}$$

Do **Problem 16** to show (5-79).

We get

$$\partial_\mu Z^\mu(x) = 0 . \qquad (5\text{-}79)$$

Massive spin 1 field already constrained without applying a gauge

This looks like the Lorenz condition (5-75) used in (5-73), but note carefully that here it is not a gauge we choose to impose on the field (to make things easier, as it were). If you did Problem 16, you saw that this relation arises due to the mass term in (5-78). We *have* to use (5-79) for massive fields. It is not an option (not a gauge), as we had for massless spin 1 fields. (5-79) is a constraint equation and it means we have no more than three independent components (degrees of freedom), no matter what.

which means only 3 d.o.f. from the "get-go"

Now, we know that, physically, a massive particle can have its spatial field **Z** aligned in any direction, not just transverse to the direction of **k**. So, we can't take any of the three spatial components, i.e., Z^i, identically equal to zero, as we did for massless fields. We have to surmise that the constraint of equation (5-79) can be interpreted to mean we have three free spatial components (which then fix the timelike component).

But massive field can have any spin direction and a longitudinal field component

Resultant Form of Proca Equation

Given (5-79), (5-74) becomes

$$\boxed{\partial^\alpha \partial_\alpha Z^\mu(x) + m_Z^2 Z^\mu(x) = 0} \qquad \text{(Proca equation, without zero term)}, \quad (5\text{-}80)$$

More meaningful and useful form of Proca equation

and this is the form of the Proca equation used throughout the literature. Note though, that this form, which, but for the mass term, is the same as the form we commonly use for Maxwell's equation, is not the result of imposing a gauge condition. The common form of Maxwell's equation is due to such an imposed gauge (Lorenz gauge). The Proca equation is not. It just is what it is, with no sleight of hand.

Bottom Line for Massive Spin 1 Fields

There is no gauge freedom for real massive spin 1 particles. They must have the field equation (5-80), and they must have three (not two as with massless spin 1 particles) degrees of freedom (free components), which experiment dictates we take as the three spatial components. Such particles are not restricted to having components transverse to their direction of travel.

Massive spin 1 field uses no gauges and has 3 d.o.f., 3 spatial field components

5.5 Chapter Summary

5.5.1 Spinor Representations

With a unitary spinor space transformation U, we can transform the field solutions of the Dirac equation ($\psi' = U\psi$) and the gamma matrices ($\gamma'^\mu = U\gamma^\mu U^\dagger$) to an unlimited number of different forms (representations or reps), though the relations between those matrices and solutions remain the same, i.e., the Dirac equation still looks like $\left(i\gamma'^\mu\partial_\mu - m\right)\psi' = 0$.

Other than the standard rep, which we have used since early on in Vol. 1, the Weyl and Majorana reps can prove quite useful in QFT. In particular, Weyl, or chiral, fields, which interact via the weak force, take on their simplest form in the Weyl rep. And Majorana fields, for which antiparticles equal particles, take their simplest form in the Majorana rep. So, we can deduce key relations for Weyl and Majorana fields in those respective reps (and thus in the simplest ways), and those relations will then be valid in any rep.

The Weyl rep is overviewed in Wholeness Chart 5-1, pg. 141. All three major reps are overviewed in Wholeness Chart 5-2, pg. 148.

5.5.2 Chirality

Chirality, often confused with helicity, is that characteristic of spinor fields that plays a role in the weak interaction. While L and R chiral (Weyl) spinor fields are always eigenstates of chirality, they only become eigenstates of helicity when their velocity magnitude equals the speed of light.

Wholeness Chart 5-4, pg. 152, summarizes the development of chiral fields and the associated operators. The particular effect of each chiral field and its adjoint (creation/destruction properties) is listed in Wholeness Chart 5-3, pg. 151.

5.5.3 Boosting Spinors

One cannot obtain a boosted spinor field by simply using the boosted values for **p** in u_i and v_i. One has to use the spinor transformation, the representation of the Lorentz group in spinor space, which is symbolized by D and defined in (5-64) and (5-65).

Steps for obtaining coefficients for a boosted RQM spinor state are listed in Sect. 5.3.2, pg. 154.

5.5.4 Massive Spin 1 Fields

Massive spin 1 fields, such as the mediators of the weak force, are constrained by their field equation to have three degrees of freedom (d.o.f.), i.e., three independent spatial field components. There is no freedom to impose gauges on such fields, and they do not have to be purely transverse to their 3-momentum direction. The spin and spatial 3-vector of such a field can be in any direction.

5.6 Appendices

5.6.1 Appendix A: Expressing Majorana Field in the Standard Rep

Expressing (5-38) term-by-term, we have

$$ {}^{MF}_{\;S}\psi = {}_{M\text{-}S}U\,{}^{MF}_{\;M}\psi = \sum_{\mathbf{p}} \sqrt{\frac{m}{VE_{\mathbf{p}}}} \begin{pmatrix} c_1(\mathbf{p})\,{}_{M\text{-}S}U_M u_1(\mathbf{p})e^{-ipx} + c_1^\dagger(\mathbf{p})\,{}_{M\text{-}S}U_M u_1^*(\mathbf{p})e^{ipx} \\ + c_2(\mathbf{p})\,{}_{M\text{-}S}U_M u_2(\mathbf{p})e^{-ipx} + c_2^\dagger(\mathbf{p})\,{}_{M\text{-}S}U_M u_2^*(\mathbf{p})e^{ipx} \end{pmatrix}. \tag{5-81}$$

Looking at the first term in (5-81) and using (5-37), (5-29), and (5-3), we get

$$ c_1(\mathbf{p})\,{}_{M\text{-}S}U_M u_1(\mathbf{p})e^{-ipx} = c_1(\mathbf{p})\frac{1}{\sqrt{2}} \begin{bmatrix} 1 & & & -i \\ & 1 & i & \\ & -i & -1 & \\ i & & & -1 \end{bmatrix} \sqrt{\frac{E_{\mathbf{p}}+m}{4m}} \begin{pmatrix} 1+\dfrac{-ip^1+p^2}{E_{\mathbf{p}}+m} \\ \dfrac{ip^3}{E_{\mathbf{p}}+m} \\ \dfrac{-p^3}{E_{\mathbf{p}}+m} \\ i+\dfrac{-p^1-ip^2}{E_{\mathbf{p}}+m} \end{pmatrix} e^{-ipx} $$

$$ = c_1(\mathbf{p})\frac{1}{\sqrt{2}}\sqrt{\frac{E_{\mathbf{p}}+m}{4m}} \begin{pmatrix} 1+\dfrac{-ip^1+p^2}{E_{\mathbf{p}}+m}+1+\dfrac{ip^1-p^2}{E_{\mathbf{p}}+m} \\ \dfrac{ip^3}{E+m}+\dfrac{-ip^3}{E+m} \\ \dfrac{p^3}{E_{\mathbf{p}}+m}+\dfrac{p^3}{E_{\mathbf{p}}+m} \\ i+\dfrac{p^1+ip^2}{E_{\mathbf{p}}+m}-i+\dfrac{p^1+ip^2}{E_{\mathbf{p}}+m} \end{pmatrix} e^{-ipx} = c_1(\mathbf{p})\frac{2}{\sqrt{2}}\underbrace{\sqrt{\frac{E_{\mathbf{p}}+m}{4m}} \begin{pmatrix} 1 \\ 0 \\ \dfrac{p^3}{E_{\mathbf{p}}+m} \\ \dfrac{p^1+ip^2}{E_{\mathbf{p}}+m} \end{pmatrix}}_{{}_S u_1(\mathbf{p})} e^{-ipx} \tag{5-82}$$

$$ = c_1(\mathbf{p})\,{}_S u_1(\mathbf{p})e^{-ipx}. $$

For the second term in (5-81), we get

$$c_1^\dagger(\mathbf{p})\,{}_{M\text{-}S}U_{\;M}u_1^*(\mathbf{p})e^{ipx}=c_1^\dagger(\mathbf{p})\frac{1}{\sqrt{2}}\begin{bmatrix}1&&&-i\\&1&i&\\&-i&-1&\\i&&&-1\end{bmatrix}\sqrt{\frac{E_{\mathbf{p}}+m}{4m}}\begin{pmatrix}1+\dfrac{-ip^1+p^2}{E_{\mathbf{p}}+m}\\\dfrac{ip^3}{E_{\mathbf{p}}+m}\\\dfrac{-p^3}{E_{\mathbf{p}}+m}\\i-\dfrac{p^1+ip^2}{E_{\mathbf{p}}+m}\end{pmatrix}^*e^{ipx}$$

$$=c_1^\dagger(\mathbf{p})\frac{1}{\sqrt{2}}\sqrt{\frac{E_{\mathbf{p}}+m}{4m}}\begin{bmatrix}1&&&-i\\&1&i&\\&-i&-1&\\i&&&-1\end{bmatrix}\begin{pmatrix}\dfrac{ip^1+p^2}{E_{\mathbf{p}}+m}+1\\\dfrac{-ip^3}{E_{\mathbf{p}}+m}\\\dfrac{-p^3}{E_{\mathbf{p}}+m}\\\dfrac{-p^1+ip^2}{E_{\mathbf{p}}+m}-i\end{pmatrix}e^{ipx}=c_1^\dagger(\mathbf{p})\frac{1}{\sqrt{2}}\sqrt{\frac{E_{\mathbf{p}}+m}{4m}}\begin{pmatrix}\dfrac{ip^1+p^2}{E_{\mathbf{p}}+m}+1+\dfrac{ip^1+p^2}{E_{\mathbf{p}}+m}-1\\\dfrac{-ip^3}{E_{\mathbf{p}}+m}+\dfrac{-ip^3}{E_{\mathbf{p}}+m}\\\dfrac{-p^3}{E_{\mathbf{p}}+m}+\dfrac{p^3}{E_{\mathbf{p}}+m}\\\dfrac{-p^1+ip^2}{E_{\mathbf{p}}+m}+i+\dfrac{p^1-ip^2}{E_{\mathbf{p}}+m}+i\end{pmatrix}e^{ipx}$$

$$=c_1^\dagger(\mathbf{p})\frac{2}{\sqrt{2}}\sqrt{\frac{E_{\mathbf{p}}+m}{4m}}\begin{pmatrix}\dfrac{ip^1+p^2}{E_{\mathbf{p}}+m}\\\dfrac{-ip^3}{E_{\mathbf{p}}+m}\\0\\i\end{pmatrix}e^{ipx}=c_1^\dagger(\mathbf{p})i\underbrace{\sqrt{\frac{E_{\mathbf{p}}+m}{2m}}\begin{pmatrix}\dfrac{p^1-ip^2}{E_{\mathbf{p}}+m}\\\dfrac{-p^3}{E_{\mathbf{p}}+m}\\0\\1\end{pmatrix}}_{{}_Sv_1(\mathbf{p})}e^{ipx}=c_1^\dagger(\mathbf{p})i\,{}_Sv_1(\mathbf{p})e^{ipx}. \quad (5\text{-}83)$$

Putting (5-82) and (5-83) into (5-81), along with similar expressions for the $r=2$ terms, we have

$${}^{MF}_{S}\psi={}_{M\text{-}S}U\,{}^{MF}_{M}\psi=\sum_{\mathbf{p}}\sqrt{\frac{m}{VE_{\mathbf{p}}}}\begin{pmatrix}c_1(\mathbf{p})\,{}_Su_1(\mathbf{p})e^{-ipx}+\underbrace{ic_1^\dagger(\mathbf{p})}_{\substack{\text{special case}\\\text{of }{}_Sd_1^\dagger(\mathbf{p})}}\,{}_Sv_1(\mathbf{p})e^{ipx}\\+c_2(\mathbf{p})\,{}_Su_2(\mathbf{p})e^{-ipx}\underbrace{-ic_2^\dagger(\mathbf{p})}_{\substack{\text{special case}\\\text{of }{}_Sd_2^\dagger(\mathbf{p})}}\,{}_Sv_2(\mathbf{p})e^{ipx}\end{pmatrix}, \quad (5\text{-}84)$$

the standard rep form of the Majorana field, which carries the restriction that the anti-particle operators (usually denoted with $d_r(\mathbf{p})$ symbol) are defined as shown (complex conjugates of the $c_r(\mathbf{p})$). That restriction ensures that ${}^{MF}_{S}\psi$, like ${}^{MF}_{M}\psi$, will create and destroy particles that are the same as their antiparticles, even though (5-84) is not real, but complex. Real fields create and destroy particles that are their own antiparticles (are Majorana particles), but being real is only sufficient for creation and destruction of Majorana particles, not necessary.

Deriving our charge operator, just as we did in Vol. 1, pg. 111, we find, with the anti-particle definitions of (5-84),

$$Q = -e\sum_{r,\mathbf{p}}\left(N_r(\mathbf{p}) - \bar{N}_r(\mathbf{p})\right) = -e\sum_{r,\mathbf{p}}\left(c_r^\dagger(\mathbf{p})c_r(\mathbf{p}) - d_r^\dagger(\mathbf{p})d_r(\mathbf{p})\right)$$
$$\underbrace{= -e\sum_{r,\mathbf{p}}\left(c_r^\dagger(\mathbf{p})c_r(\mathbf{p}) - c_r^\dagger(\mathbf{p})c_r(\mathbf{p})\right) = 0}_{\text{For Majorana field}}, \tag{5-85}$$

the electric charge on a Majorana particle is zero, the only value it can have if it is to be identical to its own antiparticle.

However, the form of (5-36), where the Majorana field is real when expressed in the Majorana rep, is simpler, and easier to deal with, than that same field (5-84) expressed in the standard rep. Nevertheless, when we deal with a Majorana field in any rep, the operator coefficients will create and destroy particles that are their own antiparticles.

5.6.2 Appendix B. Useful γ^5 Relations

For future reference, we collect in one place, pertinent γ^5 relations.

$$\left(\gamma^5\right)^\dagger = \gamma^5 \qquad \left(\gamma^5\right)^2 = I \qquad \left(\gamma^5\right)^\dagger\gamma^5 = \left(\gamma^5\right)^2 = I \qquad \gamma^5\gamma^\mu = -\gamma^\mu\gamma^5 \quad \text{repeat of (5-49) and (5-51)}$$
$$\bar{\psi}^L\gamma^\mu\psi^L = \bar{\psi}\gamma^\mu\tfrac{1}{2}\left(1-\gamma^5\right)\psi \qquad \bar{\psi}^R\gamma^\mu\psi^R = \bar{\psi}\gamma^\mu\tfrac{1}{2}\left(1+\gamma^5\right)\psi \qquad \text{from (5-60)} \tag{5-86}$$

We cite, without proof, trace relations (5-87), though their derivations can be found on the website for this book (URL opposite pg 1). As with integral tables and similar relations for gamma matrices cited in Vol. 1, Chap. 4, Appendix A, we can use them with confidence that those who have gone before us have worked them out correctly. $\varepsilon^{\alpha\beta\gamma\delta}$ is the Levi-Civita tensor of rank 4.

$$\operatorname{Tr}\gamma^5 = \operatorname{Tr}\left(\gamma^5\gamma^\alpha\right) = \operatorname{Tr}\left(\gamma^5\gamma^\alpha\gamma^\beta\right) = \operatorname{Tr}\left(\gamma^5\gamma^\alpha\gamma^\beta\gamma^\gamma\right) = 0 \qquad \operatorname{Tr}\left(\gamma^5\gamma^\alpha\gamma^\beta\gamma^\gamma\gamma^\delta\right) = -i4\varepsilon^{\alpha\beta\gamma\delta}. \tag{5-87}$$

For all three reps, including the standard and Majorana ones,

$${}_S\gamma^5 = \begin{bmatrix} & I \\ I & \end{bmatrix} = \begin{bmatrix} & & 1 & \\ & & & 1 \\ 1 & & & \\ & 1 & & \end{bmatrix} \quad {}_W\gamma^5 = \begin{bmatrix} -I & \\ & I \end{bmatrix} = \begin{bmatrix} -1 & & & \\ & -1 & & \\ & & 1 & \\ & & & 1 \end{bmatrix} \quad {}_M\gamma^5 = \begin{bmatrix} \sigma_2 & \\ & -\sigma_2 \end{bmatrix} = \begin{bmatrix} & -i & & \\ i & & & \\ & & & i \\ & & -i & \end{bmatrix}. \tag{5-88}$$

Do **Problem 17** to show the first and third expressions in (5-88).

5.7 Problems

1. Show that (5-9) is unitary. Then, use (5-8) with (5-2) and (5-9) to deduce (5-10). Find ${}_W\gamma^0$ two ways, using gamma matrices and the unitary matrix in the 4X4 form, then in the 2X2 compact form with submatrices being themselves 2X2. Then, find the other three Weyl rep gamma matrices using only the 2X2 compact form.
2. Using (5-3), (5-8), and (5-9) find the four basis solutions (5-11) to the Dirac equation in the Weyl rep.
3. Find the spin operator for the Weyl rep.
4. Show that the *R* Weyl field has positive (right hand) spin in the x^3 direction in the relativistic case where the x^3 axis is aligned with the 3-momentum vector and thus, has right-hand helicity.
5. Show that for $v \ll c$, ${}_W\psi^L$ and ${}_W\psi^R$ are generally not in *L* and *R* helicity eigenstates. That is, for low velocities, fields that are eigenstates of *L* and *R* helicity do *not* correspond to pure ${}_W\psi^L$ and ${}_W\psi^R$ fields, respectively, as in the highly relativistic case. Do the analysis in the most convenient coordinate system, where x^3 is aligned with 3-momentum $\mathbf{p}$ (so $p^1 = p^2 = 0$). Will the conclusion be valid in other coordinate systems?
6. For 3-momentum in the negative x^3 direction at relativistic speed, find the basis spinors for ${}_W\psi$, the spin direction and helicity for ${}_W\psi^L$.

7. Find the Majorana rep gamma matrices. Use the 2X2 matrix formulation with the 2X2 submatrices of (5-24). Recall that different Pauli matrices anti-commute, that commutation $[\sigma_i, \sigma_j] = i2\varepsilon_{ijk}\sigma_k$, and that the same Pauli matrix multiplied by itself yields the identity matrix. See (5-2)b.

8. Find the four basis solutions to the Dirac equation in the Majorana rep one gets using the transformation between reps.

9. Show that $\left({}_M\psi'^{(2)}\right)^* = -i\,{}_M\psi'^{(3)}$.

10. Show that $\bar{\psi}'\gamma'^1\psi' = \bar{\psi}\gamma^1\psi$, $\bar{\psi}'\gamma'^\mu\psi' = \bar{\psi}\gamma^\mu\psi$, $\bar{\psi}'\not{\partial}'\psi' = \bar{\psi}\not{\partial}\psi$, $e\bar{\psi}'\partial_\mu A^\mu\psi' = e\bar{\psi}\partial_\mu A^\mu\psi$, where primes indicate a different spinor rep than non-primes. Show also that for the Hamiltonian and charge operators, $H' = H$ and $Q' = Q$.

11. Show that $u_r'^\dagger(\mathbf{p})u_s'(\mathbf{p}) = u_r^\dagger(\mathbf{p})u_s(\mathbf{p}) = v_r'^\dagger(\mathbf{p})v_s'(\mathbf{p}) = v_r^\dagger(\mathbf{p})v_s(\mathbf{p}) = \frac{E}{m}\delta_{rs}$, where primes indicate a different spinor rep than non-primes, and also that $u_r'^\dagger(\mathbf{p})v_s'(-\mathbf{p}) = u_r^\dagger(\mathbf{p})v_s(-\mathbf{p}) = 0$.

12. Prove that the relations $\left(\gamma^5\right)^2 = I$ and $\left(\gamma^5\right)^\dagger\gamma^5 = \left(\gamma^5\right)^2 = I$ are true in any spinor rep.

13. Show that $\gamma^5\gamma^0 = -\gamma^0\gamma^5$ and $\gamma^5\gamma^1 = -\gamma^1\gamma^5$.

14. Prove that $\left(P_R\right)^2 = P_R$, $\left(P_L\right)^\dagger = P_L$, $\left(P_R\right)^\dagger = P_R$, and $P_L P_R = 0$. Justify the first and fourth of these, as we did $(P_L)^2$ after (5-55).

15. Show $\bar{\psi}P_L = \bar{\psi}^R$.

16. Take the divergence of the Proca equation for Z^μ (5-74) to show that $\partial_\mu Z^\mu = 0$.

17. Show that in the standard and Majorana reps, ${}_S\gamma^5 = \begin{bmatrix} & I \\ I & \end{bmatrix} = \begin{bmatrix} & & 1 & \\ & & & 1 \\ 1 & & & \\ & 1 & & \end{bmatrix}$ and ${}_M\gamma^5 = \begin{bmatrix} \sigma_2 & \\ & -\sigma_2 \end{bmatrix} = \begin{bmatrix} & -i & & \\ i & & & \\ & & & i \\ & & -i & \end{bmatrix}$.

Chapter 6

The High Energy Symmetric Universe

"I wonder if wondering is the meaning of life?
Art Paul, world renowned graphics designer

6.0 Preliminary Remarks

You have almost certainly heard the story. In the early universe, things (the Lagrangian, specifically) were symmetric, and all particles were massless. When the electroweak ("e/w" for short) part of that symmetry broke, the Higgs field bestowed masses onto most particles (but not the photon or the gluons). The period prior to the breaking, from 10^{-36} to 10^{-12} seconds after the Big Bang, is known as the electroweak epoch, which given the usual use of the term "epoch" to delineate exceptionally long periods of time, is a tad humorous.

e/w $\mathcal{L}$ symmetry broke at end of e/w epoch and gave particles mass

In this and the next chapter, we will start to get a handle on the mathematics behind this grand story of electroweak symmetry and its breaking.

Before that symmetry breaking took place, however, the mathematics describing the universe was simpler (though still far from simple) than afterwards. So, in this chapter, we will investigate the interactions of the fields that eventually (after "all" that time of the electroweak epoch) gave rise to what we know today as the weak and electromagnetic fields. With that under our belts, we will then explore, in the next chapter, how the symmetry broke, what that meant, and the implications of it all.

But before that, easier math to describe

This chapter covers that math

6.1 Background for Electroweak Symmetry and Its Breaking

There are a number of concepts related to electroweak symmetry and its breaking that one should have command of prior to delving deeply into the subject. We review and/or introduce each such concept in a separate subsection below.

6.1.1 Complex vs Real Bosons

Recall the usual Lagrangian for a free *complex* scalar boson field, which we studied in Vol. 1 (see pg. 49, (3-32)),

$$\mathcal{L} = \partial^{\mu}\phi^{\dagger}\partial_{\mu}\phi - \mu^{2}\phi^{\dagger}\phi \qquad \text{(Lagrangian for free complex scalar field)}. \tag{6-1}$$

For a *real* scalar boson field (Vol. 1, Chap. 3, pg. 83, Prob. 15), on the other hand, where $\phi^{\dagger} = \phi$, the Lagrangian looks a bit different, i.e., it has an additional factor of ½ in front, as in

Form of e/w $\mathcal{L}$ slightly different for real and complex scalars

$$\mathcal{L} = \tfrac{1}{2}\partial^{\mu}\phi\partial_{\mu}\phi - \tfrac{1}{2}\mu^{2}\phi\phi \qquad \text{(Lagrangian for free real scalar field)}. \tag{6-2}$$

The additional factor of ½ comes in because we must substitute (6-2) into the Euler-Lagrange equation to get the real boson field equation of motion wherein derivatives there are taken with respect to ϕ and $\phi_{,\mu}$ and said entities ϕ and $\phi_{,\mu}$ are squared in (6-2). In (6-1), to get the field equation of motion for ϕ, we take derivatives with respect to $\phi^{\dagger}$ and $\phi^{\dagger}{}_{,\mu}$, and said entities are not squared in (6-1). Handling it in this way leaves the same field equation for ϕ, whether it is real or complex. Otherwise the two would be off by a factor of 2.

6.1.2 Potential Energy Terms

It should help considerably if you now read Sect. 6.12, pg. 206, which was added as a pedagogic aid in the second revision of this text in July 2025.

In QFT, particularly in introductory courses, we don't usually consider potential energy in the Lagrangian. To derive classical potentials like the Coulomb potential, we consider QFT interactions via Møller scattering (without potential energy terms in the Lagrangian) to obtain the relevant Feynman amplitudes (Vol. 1, pgs. 404-410), and then, from them, the effective potential. The characteristics of the virtual particles mediating the interaction give rise to the corresponding classical potential measured in experiments.

QFT typically does without potential energy term $\mathcal{V}$,

deduces classical potential from QFT interactions

However, in classical field theory, and thus in QFT, one can work with a potential energy, to be precise, a potential energy *density* in the Lagrangian (density). (It is often simply called the potential.) This will be done when we examine the Higgs field, coming up shortly. For now, simply note that we could have a scalar field Lagrangian of form (where $\mu = 0$ in this example)

But we can and, in e/w theory, do have a potential $\mathcal{V}$

$$\mathcal{L} = \partial^\mu \phi^\dagger \partial_\mu \phi - \mathcal{V}\left(\phi, \phi^\dagger\right) \qquad \left(\text{massless complex scalar field}\right). \tag{6-3}$$

Note that potential energy density $\mathcal{V}$ is generally a function of the field (and its complex conjugate) and not derivatives of the field. This mimics what we know from classical theory (Vol. 1, Wholeness Chart 18-4, pgs. 506-507), where the parallel between particles and fields, i.e., $x^i(t)$ (particle) $\rightarrow \phi(t,x^i)$ (field), is commonly employed to extrapolate field theory from particle theory. For example,

$$\begin{gathered} V(x) \quad \rightarrow \quad \mathcal{V}\left(\phi^\dagger, \phi\right) \\ \left[\text{e.g., spring-mass system} = \tfrac{1}{2}kx^2\right] \rightarrow \left[\text{e.g.,} = \tfrac{1}{2}\mu\phi^2\left(\text{real}\right) \text{ or } \mu\phi^\dagger\phi\left(\text{complex}\right)\right], \end{gathered} \tag{6-4}$$

$\mathcal{V}$ typically a function of fields (not derivatives)

where we note the parallels between the mass term of QFT and the potential energy of classical theory. In QFT, in a sense, mass is somewhat akin to classical potential energy in that, relativistically, it can be converted into kinetic energy, just as the potential energy in a spring can be converted into kinetic energy.

And of course, just like we can have different systems in classical theory with different forms for potential energy (such as gravitational potential energy around a planet, where $V = -GmM/r$), we can have different forms for $\mathcal{V}$ as a function of ϕ in field theory.

Different systems have different $\mathcal{V}$

If we wish, at least mathematically, we could consider (6-1) [or (6-2)] as if it were a massless field with potential $\mu^2\phi^\dagger\phi$ [or $\frac{1}{2}\mu^2\phi\phi$ for a real field]. That is,

$$\mathcal{L} = \partial^\mu \phi^\dagger \partial_\mu \phi - \mathcal{V} \qquad \text{where } \mathcal{V} = \mu^2\phi^\dagger\phi = \mu^2\left|\phi\right|^2 . \tag{6-5}$$

Usual scalar mass term in $\mathcal{L}$ like a potential $\mathcal{V}$

Note that in (6-5), $\mathcal{V}$ must be positive (for real μ). Consider, on the other hand, a Lagrangian where μ is imaginary. That would make $\mathcal{V}$ negative. This case, with an imaginary μ, is a key part of electroweak symmetry breaking, as we shall see.

6.1.3 Different Ways to Represent ϕ

Note that we typically express a complex scalar field as

$$\phi(x) = \sum_{\mathbf{k}} \frac{1}{\sqrt{2V\omega_{\mathbf{k}}}}(a(\mathbf{k})e^{-ikx} + b^\dagger(\mathbf{k})e^{ikx}) \qquad \phi^\dagger(x) = \sum_{\mathbf{k}'} \frac{1}{\sqrt{2V\omega_{\mathbf{k}'}}}(b(\mathbf{k}')e^{-ik'x} + a^\dagger(\mathbf{k}')e^{ik'x}) . \tag{6-6}$$

We want to express complex scalar field differently from usual

However, we will find it advantageous later on to express ϕ as the sum of its real part and its imaginary part. That is, where ϕ_1 and ϕ_2 are real,

$$\phi = \phi_1 + i\phi_2 = \sum_{\mathbf{k}}\left(\phi_{1\mathbf{k}} + i\phi_{2\mathbf{k}}\right). \tag{6-7}$$

As sum of real and imaginary parts

From the left expression of (6-6), where for simplicity we only examine a single $\mathbf{k}$ value term,

$$Re\{\phi_{\mathbf{k}}\} = \phi_{1\mathbf{k}} = \frac{1}{\sqrt{2V\omega_{\mathbf{k}}}}\left(\frac{a(\mathbf{k})e^{-ikx} + a^{\dagger}(\mathbf{k})e^{ikx}}{2} + \frac{b^{\dagger}(\mathbf{k})e^{ikx} + b(\mathbf{k})e^{-ikx}}{2}\right)$$
$$Im\{\phi_{\mathbf{k}}\} = i\phi_{2\mathbf{k}} = \frac{1}{\sqrt{2V\omega_{\mathbf{k}}}}\left(\frac{a(\mathbf{k})e^{-ikx} - a^{\dagger}(\mathbf{k})e^{ikx}}{2} + \frac{b^{\dagger}(\mathbf{k})e^{ikx} - b(\mathbf{k})e^{-ikx}}{2}\right) \tag{6-8}$$

Thus, we can re-express $\phi_{1\mathbf{k}}$ and $\phi_{2\mathbf{k}}$ as

$$\phi_{1\mathbf{k}} = \frac{1}{\sqrt{2V\omega_{\mathbf{k}}}}\left(\frac{a(\mathbf{k}) + b(\mathbf{k})}{2}e^{-ikx} + \frac{a^{\dagger}(\mathbf{k}) + b^{\dagger}(\mathbf{k})}{2}e^{ikx}\right) = \frac{1}{\sqrt{2V\omega_{\mathbf{k}}}}\left(\alpha(\mathbf{k})e^{-ikx} + \alpha^{\dagger}(\mathbf{k})e^{ikx}\right)$$
$$\phi_{2\mathbf{k}} = \frac{1}{\sqrt{2V\omega_{\mathbf{k}}}}\left(\frac{a(\mathbf{k}) - b(\mathbf{k})}{2i}e^{-ikx} + \frac{-a^{\dagger}(\mathbf{k}) + b^{\dagger}(\mathbf{k})}{2i}e^{ikx}\right) = \frac{1}{\sqrt{2V\omega_{\mathbf{k}}}}\left(\beta(\mathbf{k})e^{-ikx} + \beta^{\dagger}(\mathbf{k})e^{ikx}\right), \tag{6-9}$$

Result: real and imaginary parts of complex scalar field

where $\alpha(\mathbf{k})$ and $\beta(\mathbf{k})$ are the operators for the real and imaginary parts of the complex scalar field, respectively, whose definition in terms of $a(\mathbf{k})$, $b(\mathbf{k})$, and their complex conjugates can be gleaned from (6-9). ϕ_1 and ϕ_2 each thus acts much like a photon field, which is real and has similar form to the RHS of each row of (6-9).

Coefficient operators $\alpha(\mathbf{k})$ and $\beta(\mathbf{k})$ create and destroy particles $|\phi_1\rangle$ and $|\phi_2\rangle$

In electroweak symmetry breaking, one often refers to the fields ϕ_1 and $i\phi_2$ as real and imaginary components of ϕ, but their specific forms, as in (6-9), are not shown in any other text I am aware of. The key thing to note is that given the usual postulates of QFT, one can deduce that the $\alpha(\mathbf{k})$ and $\beta(\mathbf{k})$ operators are destruction operators and their complex conjugates are creation operators. These destroy and create particle states $|\phi_1\rangle$ and $|\phi_2\rangle$ in a manner which we are used to. Note that ϕ_2, with operators $\beta(\mathbf{k})$ and $\beta^{\dagger}(\mathbf{k})$ would destroy and create states $|\phi_2\rangle$.

6.1.4 Plotting $\mathcal{V}$

In electroweak symmetry breaking one commonly sees plots of $\mathcal{V}$ vs ϕ in terms of the real and imaginary field components of ϕ in (6-7). One example is (6-5),

$$\mathcal{V} = \mu^2\phi^{\dagger}\phi = \mu^2|\phi|^2 = \mu^2(\phi_1 - i\phi_2)(\phi_1 + i\phi_2) = \mu^2(\phi_1^2 + \phi_2^2), \tag{6-10}$$

which can be plotted as a paraboloid over the ϕ_1-ϕ_2 plane.

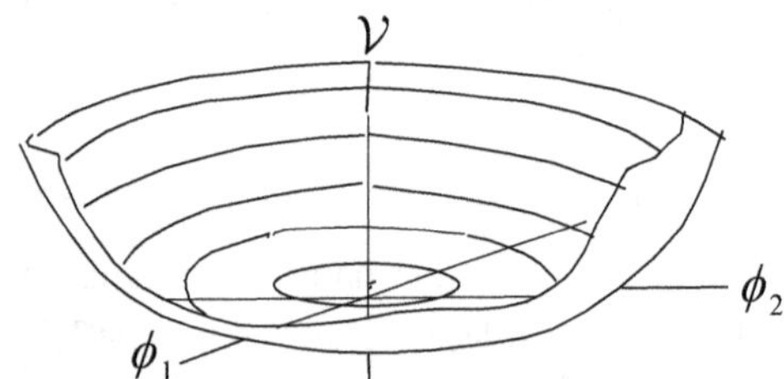

Figure 6-1. Plot of $\mathcal{V}$ vs ϕ for Real μ

Plot of the usual scalar mass term as a potential

As a student I was perplexed by the use of such graphs in QFT where ϕ is not a variable, but an operator, as in (6-7) with operators $\alpha(\mathbf{k})$ and $\beta(\mathbf{k})$ of (6-9). That is, ϕ in QFT does not vary. It does not have greater or lesser values extending outward from ϕ=0, as appears to be shown in Fig. 6-1.

I know of no other text that addresses this question, but here, offer the following resolution.

From (6-10), using (6-9), we have

$$\mathcal{V} = \mu^2(\phi_1^2 + \phi_2^2) = \frac{\mu^2}{2V}\sum_{\mathbf{k}}\sum_{\mathbf{k}'}\frac{1}{\sqrt{\omega_{\mathbf{k}}\omega_{\mathbf{k}'}}}\begin{pmatrix}\alpha(\mathbf{k})\alpha(\mathbf{k}')e^{-ikx}e^{-ik'x} + \alpha(\mathbf{k})\alpha^{\dagger}(\mathbf{k}')e^{-ikx}e^{ik'x} \\ + \alpha^{\dagger}(\mathbf{k})\alpha(\mathbf{k}')e^{ikx}e^{-ik'x} + \alpha^{\dagger}(\mathbf{k})\alpha^{\dagger}(\mathbf{k}')e^{ikx}e^{ik'x}\end{pmatrix}$$
$$+ \frac{\mu^2}{2V}\sum_{\mathbf{k}}\sum_{\mathbf{k}'}\frac{1}{\sqrt{\omega_{\mathbf{k}}\omega_{\mathbf{k}'}}}\begin{pmatrix}\beta(\mathbf{k})\beta(\mathbf{k}')e^{-ikx}e^{-ik'x} + \beta(\mathbf{k})\beta^{\dagger}(\mathbf{k}')e^{-ikx}e^{ik'x} \\ + \beta^{\dagger}(\mathbf{k})\beta(\mathbf{k}')e^{ikx}e^{-ik'x} + \beta^{\dagger}(\mathbf{k})\beta^{\dagger}(\mathbf{k}')e^{ikx}e^{ik'x}\end{pmatrix}, \tag{6-11}$$

$$\text{and} \quad V_{\phi} = \int \mathcal{V} d^3x. \tag{6-12}$$

With a substantial amount of algebra[1] (which would take us far afield of our present objective, so I hope you can take my word for it), we would find the effective total potential V_ϕ of

$$V_\phi = \mu^2 \sum_{\mathbf{k}} \frac{1}{\omega_{\mathbf{k}}}\left(\alpha^\dagger(\mathbf{k})\alpha(\mathbf{k}) + \tfrac{1}{2} + \beta^\dagger(\mathbf{k})\beta(\mathbf{k}) + \tfrac{1}{2}\right) = \mu^2 \sum_{\mathbf{k}} \frac{1}{\omega_{\mathbf{k}}}\left(N_\alpha(\mathbf{k}) + N_\beta(\mathbf{k}) + 1\right), \quad (6\text{-}13)$$

where our old friends the number operators show up again. As is the usual (but yet to be fully justified) practice in QFT, we drop the term "1". This is the same as assuming (6-11) is normal ordered[2], whereby the "1" term never shows up in (6-13).

Dividing (6-13) by the volume V, we get an effective (representing measurable) $\mathcal{V}$ potential energy density operator for the field,

$$\mathcal{V} = \frac{V_\phi}{V} = \frac{\mu^2}{V} \sum_{\mathbf{k}} \frac{1}{\omega_{\mathbf{k}}}\left(N_\alpha(\mathbf{k}) + N_\beta(\mathbf{k})\right), \quad (6\text{-}14)$$

which operates on kets. So, we have expectation values (expected measured values)

Expectation value of $\mathcal{V}$ as a function of particle number densities

$$\overline{\mathcal{V}} = \langle n_{\mathbf{k}}\phi_{\mathbf{k}}, n_{\mathbf{k}'}\phi_{\mathbf{k}'},..|\mathcal{V}|n_{\mathbf{k}}\phi_{\mathbf{k}}, n_{\mathbf{k}'}\phi_{\mathbf{k}'},..\rangle = \frac{\mu^2}{V}\sum_{\mathbf{k}} \frac{1}{\omega_{\mathbf{k}}}\left(n_\alpha(\mathbf{k}) + n_\beta(\mathbf{k})\right) = \mu^2 \sum_{\mathbf{k}} \frac{1}{\omega_{\mathbf{k}}} \underbrace{\left(\mathcal{n}_\alpha(\mathbf{k}) + \mathcal{n}_\beta(\mathbf{k})\right)}_{\frac{n_\alpha(\mathbf{k})}{V} + \frac{n_\beta(\mathbf{k})}{V}}, \quad (6\text{-}15)$$

where script $\mathcal{n}$ represents particle number density and non-script n, particle number.

<u>The main point</u>

The main point of all this is that when we plot $\mathcal{V}$ vs ϕ as in Fig. 6-1, what we are actually representing is, in fact, $\overline{\mathcal{V}}$, the expectation value for potential energy density, vs the number of particles expected to be measured per unit volume, i.e., the particle number density.

That is, in Fig. 6-1, we are actually plotting

$$\overline{\mathcal{V}} = \sum_{\mathbf{k}} \overline{\mathcal{V}}_{\mathbf{k}} = \sum_{\mathbf{k}} \frac{\mu^2}{\omega_{\mathbf{k}}}\left(\mathcal{n}_\alpha(\mathbf{k}) + \mathcal{n}_\beta(\mathbf{k})\right), \quad (6\text{-}16)$$

where we can visualize each $\overline{\mathcal{V}}_{\mathbf{k}}$ plotted versus two number density axes labeled $\mathcal{n}_\alpha(\mathbf{k})$ and $\mathcal{n}_\beta(\mathbf{k})$.

However, note that $\mathcal{V}$ in (6-5) is expressed as a function of the bilinear operator term $\phi^\dagger\phi$, and $\overline{\mathcal{V}}_{\mathbf{k}}$ of (6-16), which is derived from $\mathcal{V}$, is a function of $\mathcal{n}_\alpha(\mathbf{k})$ and $\mathcal{n}_\beta(\mathbf{k})$. If we were to plot $\overline{\mathcal{V}}_{\mathbf{k}}$ vs $\mathcal{n}_\alpha(\mathbf{k})$ and $\mathcal{n}_\beta(\mathbf{k})$, we would effectively be plotting $\overline{\mathcal{V}}_{\mathbf{k}}$ vs $|\phi_{\mathbf{k}}|^2 = \phi_{1\mathbf{k}}^2 + \phi_{2\mathbf{k}}^2$, i.e., over the $\phi_{1\mathbf{k}}^2$ and $\phi_{2\mathbf{k}}^2$ axes, not the $\phi_{1\mathbf{k}}$ and $\phi_{2\mathbf{k}}$ axes. $\mathcal{n}_\alpha(\mathbf{k})$ and $\mathcal{n}_\beta(\mathbf{k})$ correspond to $\phi_{1\mathbf{k}}^2$ and $\phi_{2\mathbf{k}}^2$, not $\phi_{1\mathbf{k}}$ and $\phi_{2\mathbf{k}}$.

Our plot is expectation value of $\mathcal{V}$ vs square roots of particle number densities

So, if we plot $\overline{\mathcal{V}}_{\mathbf{k}}$ against $\phi_{1\mathbf{k}}$ and $\phi_{2\mathbf{k}}$, to appear like Fig. 6-1, then we essentially are plotting $\overline{\mathcal{V}}_{\mathbf{k}}$ vs the square root of the particle number densities $\mathcal{n}_\alpha(\mathbf{k})$ and $\mathcal{n}_\beta(\mathbf{k})$ (for all particles having the same $\mathbf{k}$). That is, the horizontal axes in Fig. 6-1, with everything converted to expectation values, assuming the vertical axis labels $\overline{\mathcal{V}}_{\mathbf{k}}$, would be labeled $\sqrt{\mathcal{n}_\alpha(\mathbf{k})}$ and $\sqrt{\mathcal{n}_\beta(\mathbf{k})}$.

Of course, matters are complicated further in that we have to add all $\overline{\mathcal{V}}_{\mathbf{k}}$ to get $\overline{\mathcal{V}}$, but the same general principle holds for each $\overline{\mathcal{V}}_{\mathbf{k}}$. In essence, we would need a multi-dimensional space with a pair of orthogonal axes for each $\mathbf{k}$, and a single axis for $\overline{\mathcal{V}}$, which would be dependent on the number density values of all of the other independent axes. The greater the values on all those other axes, the greater the value of $\overline{\mathcal{V}}$. Fig. 6-1 is a symbolic oversimplification of that, but it will serve us well in what is to come.

---

[1] See Vol. 1, pgs. 53-54, for how this is done for the full Hamiltonian $H = \int \mathcal{H}\, d^3x$. Equation (3-53) therein effectively parallels (6-13) and (6-14) herein, and leads to the expectation value of the potential in (6-15). The extra bilinear operator terms in Vol. 1 (3-53) drop out when we take expectation values, so (6-13) herein is really an effective potential density operator giving the same result in (6-15) as the complete expression of Vol. 1 (3-53) would have.

[2] See Vol. 1, pgs. 60, 203, 209.

Bottom line of the main point

In plots you will see of $\mathcal{V}$ vs ϕ, of which QFT is replete, the distance from the centerline axis is related to the particle number density. Further out, more dense. Closer in, less dense. At the center, zero particle number density.

6.1.5 Defining the Vacuum

We generally think of the vacuum as the state with zero real particles, i.e., zero particle density, which would make the centerline of Fig. 6-1 the vacuum. This is also the lowest energy state (lowest $\mathcal{V}$).

The vacuum, generally
- *no particles*
- *lowest energy*

However, we will soon see a case where the zero-particle state is not the lowest energy state. Don't worry about this for now, as it will become clearer a little later.

This will be modified shortly

6.1.6 Transformations

Complex characterization of ϕ

Complex ϕ, as used conventionally in electroweak theory, comprises (6-7) along with a factor of $1/\sqrt{2}$ (which helps in the long run), i.e.,

Characterizing a scalar field ϕ as a complex math entity

$$\text{For } U(1) \text{ as employed in QFT} \quad \phi = \frac{1}{\sqrt{2}}(\phi_1 + i\phi_2) \quad \phi_1, \phi_2 \text{ real}. \tag{6-17}$$

This is a complex vector space. It can be transformed under the action of a $U(1)$ group (transformation). Recall from Chap. 2 (pgs. 15 and 22), that such a group can be represented as

$$U(1) = \left\{e^{i\alpha}\right\} \tag{6-18}$$

where here we take α = constant. Transformation of (6-17) under this group takes the form

Associated U(1) transformation

$$\begin{aligned} &\phi' = U\phi = e^{i\alpha}\phi = (1 + i\alpha + \ldots)\phi \xrightarrow[\text{i.e., small } \alpha]{\text{infinitesimal}} \phi' \approx (1+i\alpha)\phi = \phi + \delta\phi \\ &\phi^{\dagger\prime} = \phi^\dagger U^\dagger = \phi^\dagger e^{-i\alpha} = \phi^\dagger(1 - i\alpha + \ldots) \xrightarrow[\text{i.e., small } \alpha]{\text{infinitesimal}} \phi^{\dagger\prime} \approx \phi^\dagger(1 - i\alpha) = \phi^\dagger + \delta\phi^\dagger . \end{aligned} \tag{6-19}$$

For infinitesimal transformations,

Change in ϕ for infinitesimal U(1) transformation

$$\delta\phi = i\alpha\phi \qquad \delta\phi^\dagger = -i\alpha\phi^\dagger . \tag{6-20}$$

Do **Problem 1** to show the scalar Lagrangian (6-1) is symmetric under the $U(1)$ transformation (6-19).

Recall from Chap. 2, if we were to have a set of $SU(2)$ transformations (a group), the generators would be 2X2 matrices. For an $SU(3)$ transformation, 3X3 matrices. Here, for $U(1)$, the generator is simply a 1X1 matrix, to be precise, the number one. The quantity $\underline{\alpha}$ is a parameter by which the $U(1)$ generator is multiplied, just as we had parameters by which we multiplied the $SU(2)$ and $SU(3)$ generators in group theory. Note that in some literature, α is called the generator, but this is not strictly correct.

U(1) generator is the number 1

α is the parameter

Observe that this transformation (6-18) is effectively a rotation about an axis perpendicular to ϕ, at the point where $\phi = 0$ in Fig. 6-1.

Essentially a rotation in 2D complex plane

Real characterization of ϕ

Alternatively, we can characterize the same field and the same (rotation) transformation solely in terms of real entities, i.e., in a representation of $SO(2)$, which is a real group, where,

Alternative characterization: ϕ as a real column vector

$$\text{for } SO(2), \; \phi = \frac{1}{\sqrt{2}}\begin{bmatrix}\phi_1 \\ \phi_2\end{bmatrix}. \tag{6-21}$$

Note that you may find some literature referring to (6-21) as a different *representation of U(1)*, but in this text, we consider it a different group, i.e., $SO(2)$, rather than $U(1)$. As discussed in Chap. 2, every element of one group corresponds to exactly one element of the other, as do group operations between elements. (In mathematics lingo, the two different groups are "isomorphic"). Further, for an active transformation (the vector rotates, not the coordinate system),

$$\phi' = \underbrace{\begin{bmatrix} \cos\alpha & -\sin\alpha \\ \sin\alpha & \cos\alpha \end{bmatrix}}_{SO(2)} \begin{bmatrix} \frac{\phi_1}{\sqrt{2}} \\ \frac{\phi_2}{\sqrt{2}} \end{bmatrix} \xrightarrow[\text{i.e., small } \alpha]{\text{infinitesimal}} \phi' \approx \begin{bmatrix} 1 & -\alpha \\ \alpha & 1 \end{bmatrix} \begin{bmatrix} \frac{\phi_1}{\sqrt{2}} \\ \frac{\phi_2}{\sqrt{2}} \end{bmatrix} = \begin{bmatrix} \frac{\phi_1}{\sqrt{2}} - \alpha\frac{\phi_2}{\sqrt{2}} \\ \frac{\phi_2}{\sqrt{2}} + \alpha\frac{\phi_1}{\sqrt{2}} \end{bmatrix} = \phi + \delta\phi \,, \quad (6\text{-}22)$$

Associated SO(2) transformation

where the matrix on the left in (6-22) is the simple rotation matrix in 2D. Thus, both forms (6-19) and (6-22) of the transformation are effectively rotations about the vertical axis in Fig. 6-1. Note,

$$\delta\phi = \frac{1}{\sqrt{2}} \begin{bmatrix} -\alpha\phi_2 \\ \alpha\phi_1 \end{bmatrix}. \quad (6\text{-}23)$$

Change in ϕ for infinitesimal SO(2) transformation

The Scalar Lagrangian in Terms of Real and Complex Reps

We can express the scalar Lagrangian (6-1) we know so well in either the complex rep of $U(1)$ or the real rep of $SO(2)$. In the complex rep of $U(1)$, it is

$$\begin{aligned} \mathcal{L} &= \partial^\mu \phi^\dagger \partial_\mu \phi - \mu^2 \phi^\dagger \phi = \frac{1}{\sqrt{2}}\left(\partial^\mu \phi_1 - i\partial^\mu \phi_2\right)\frac{1}{\sqrt{2}}\left(\partial_\mu \phi_1 + i\partial_\mu \phi_2\right) - \mu^2 \frac{1}{\sqrt{2}}\left(\phi_1 - i\phi_2\right)\frac{1}{\sqrt{2}}\left(\phi_1 + i\phi_2\right) \\ &= \frac{1}{2}\left(\partial^\mu \phi_1 \partial_\mu \phi_1 + \partial^\mu \phi_2 \partial_\mu \phi_2\right) - \frac{\mu^2}{2}\left(\phi_1^2 + \phi_2^2\right). \end{aligned} \quad (6\text{-}24)$$

$\mathcal{L}$ in complex U(1) rep

In the real rep of $SO(2)$, it is.

$$\begin{aligned} \mathcal{L} &= \partial^\mu \phi^{\mathrm{T}} \partial_\mu \phi - \mu^2 \phi^{\mathrm{T}} \phi = \frac{1}{\sqrt{2}}\partial^\mu \begin{bmatrix} \phi_1 & \phi_2 \end{bmatrix} \frac{1}{\sqrt{2}}\partial_\mu \begin{bmatrix} \phi_1 \\ \phi_2 \end{bmatrix} - \mu^2 \frac{1}{\sqrt{2}}\begin{bmatrix} \phi_1 & \phi_2 \end{bmatrix}\frac{1}{\sqrt{2}}\begin{bmatrix} \phi_1 \\ \phi_2 \end{bmatrix} \\ &= \frac{1}{2}\begin{bmatrix} \partial^\mu \phi_1 & \partial^\mu \phi_2 \end{bmatrix}\begin{bmatrix} \partial_\mu \phi_1 \\ \partial_\mu \phi_2 \end{bmatrix} - \frac{\mu^2}{2}\begin{bmatrix} \phi_1 & \phi_2 \end{bmatrix}\begin{bmatrix} \phi_1 \\ \phi_2 \end{bmatrix} = \frac{1}{2}\left(\partial^\mu \phi_1 \partial_\mu \phi_1 + \partial^\mu \phi_2 \partial_\mu \phi_2\right) - \frac{\mu^2}{2}\left(\phi_1^2 + \phi_2^2\right). \end{aligned} \quad (6\text{-}25)$$

$\mathcal{L}$ in real SO(2) rep same as in U(1) rep

As seen in the last parts of (6-24) and (6-25), the Lagrangian in both reps is the same.

Do **Problem 2** to show the scalar Lagrangian of (6-1) is symmetric under the $SO(2)$ real rep transformation of (6-22).

Infinitesimal Transformation of the Lagrangian

We will generally find it easier to do infinitesimal transformations. If they are symmetric, we can conclude that integrating them will yield finite transformations that are also symmetric.

Easier to determine symmetry of $\mathcal{L}$ under infinitesimal transformations

So, if we prove the Lagrangian is symmetric for infinitesimal changes in the parameters of a given group (α in our $U(1)$ case), we will know it is symmetric under finite changes in the group parameters, i.e. under finite transformations.

That is, for small variations in our fields under the group action, we get a small variation in $\mathcal{L}$. If that variation is zero, $\mathcal{L}$ is symmetric under the group action. Specifically, where ϕ is symbolic of any field in $\mathcal{L}$, and δ represents infinitesimal change,

$$\text{if } \delta\phi \neq 0, \text{ but } \mathcal{L}' = \mathcal{L} + \delta\mathcal{L} = \mathcal{L} \quad (\text{i.e., } \delta\mathcal{L} = 0) \quad \text{then } \mathcal{L} \text{ is symmetric}. \quad (6\text{-}26)$$

If $\delta\mathcal{L} = 0$, $\mathcal{L}$ is symmetric (invariant) under the transformation

In the future, we will be satisfied that $\mathcal{L}$ is symmetric, if we can show $\delta\mathcal{L} = 0$.

At Different Locations

Note that for the field ϕ, from either (6-20) or (6-23),

$$\left.\begin{aligned} &\text{at } \phi = 0, \ \delta\phi = 0 \rightarrow U(1)\phi = 0 \\ &\text{at } \phi \neq 0, \ \delta\phi \neq 0 \rightarrow U(1)\phi \neq 0 \end{aligned}\right\} \begin{aligned} &\text{And also, } \delta\mathcal{L} = 0 \text{ at all } \phi_1 \text{ and } \phi_2 \text{ under} \\ &\text{the } U(1) \text{ transformation about the vacuum} \end{aligned} \quad (6\text{-}27)$$

Under the transformation, the fields ϕ_1 and ϕ_2 change everywhere except when they represent the vacuum, the lowest energy state that also has zero particle density. Under the same transformation, the Lagrangian is invariant everywhere in ϕ_1-ϕ_2 space. The potential $\mathcal{V}$ is part of the Lagrangian, and it remains invariant under the transformation, as well. This can be visualized as rotating the horizontal axes in Fig. 6-1 about the vertical axis through the angle α. The plot looks unchanged.

$\mathcal{V}$ in Fig. 6-1 is symmetric

6.1.7 Conclusions for Scalars We Are Familiar With

For the scalar Lagrangian we have been working with since early on in Vol. 1,

Key takeaways for scalar $\mathcal{L}$ and its $\mathcal{V}$

- the $\mu^2\phi^\dagger\phi$ term can be considered either i) a mass term, or ii) a potential energy density term $\mathcal{V}$ (the potential, for short)
- the potential is invariant (symmetric) under the action of the $U(1)$ group (transformation)
- the whole scalar Lagrangian (6-24) is invariant under the action of the global $U(1)$ group
- a plot of $\mathcal{V}$ vs ϕ_1 and ϕ_2 is symbolic for plotting the expectation $\overline{\mathcal{V}}$ vs (the square root of) the particle number density, and
- the vacuum, as we know it in our studies to date, has $\mathcal{V} = 0$ and particle density = 0 (not counting the ½ quanta of the zero-point energy).

6.2 The Scalar Field Potential at High Energy

All of Sect. 6.1 dealt with a scalar field as we would deal with it in our present-day universe. We did it all that way for pedagogic reasons – to gain some comfort with certain concepts in the simplest possible way.

Prior was for scalar $\mathcal{V}$ in contemporary universe

However, note carefully, that the picture changes radically if we consider things prior to the end of the electroweak epoch. In particular, the potential $\mathcal{V}$ has a different form than (6-5) and a different shape from that of Fig. 6-1. This has radical ramifications for our theory, as we are about to discover.

Now we'll look at scalar $\mathcal{V}$ in e/w epoch

6.2.1 Background

The route to a successful high energy electroweak theory, which made the correct predictions as the universe's energy density plummeted, was long and arduous, with many players making critical contributions. Along the way, some key figures, such as Jeffrey Goldstone, Peter Higgs, and others produced models that brought us closer. Two of these, the Goldstone model and the Higgs model, were major stepping stones. The final result, whose key architects were Sheldon Glashow, Abdus Salam, and Steven Weinberg, resulted in the 1979 Nobel Prize, and is known, appropriately, as the Glashow/Salam/Weinberg or GSW model.

History lesson: 3 e/w models

Goldstone, Higgs, & GSW

GSW nailed it

In the next chapter we will study each of these three models in depth, as they all have something to teach us. And we will see what they predict as energy levels fall in the early universe, how the symmetry of the Lagrangian is broken in the process, and what that would leave us with today.

But in this chapter, after a brief overview of the Goldstone and Higgs models at high energy, we will go directly to the GSW model, which is now known to be the correct one, and examine its characteristics and behavior in the electroweak epoch. We want to get to know its symmetry well, before we dive into the complex process of how that symmetry broke as that epoch drew to a close.

6.2.2 The Scalar (Higgs) Field in Different Models

As we have alluded to before, and as has you have no doubt heard, the Higgs field is the key player in electroweak symmetry breaking, and it turns out to be a scalar field. It is employed in each of the three different models (the Higgs model isn't the only one with the Higgs field), though it can take on different mathematical forms.

Higgs scalar field part of all 3 models

In both the Goldstone and Higgs models, the Higgs field is a complex scalar field as in (6-17). Thus, in group theory context, it is an element in a 1-D vector space of complex scalars upon which the $U(1)$ group may operate. The difference between the two models is related to their behavior under a $U(1)$ transformation, and thus (see Vol. 1, pgs. 290-297), to their interaction properties.

Recall the common way to represent the $U(1)$ group, and the meaning of "global" and "local" therein.

$$U(1) = e^{i\alpha} \qquad \alpha = \text{constant} \rightarrow \text{global transformation} \qquad \alpha = \alpha\left(x^\mu\right) \rightarrow \text{local transformation}\,. \quad (6\text{-}28)$$

Local vs global U(1) transformation

Then recall that in QED (see referenced pages above), the free field Lagrangian is symmetric under a global transformation, but not a local one. And we can only keep the total (free plus interaction) Lagrangian symmetric under a local transformation set if it has the correct interaction terms. As we

QED symmetric under local U(1) $\rightarrow$ correct interaction terms in $\mathcal{L}$

have no doublets or triplets (See Chap. 2 herein) in QED, the entire theory is symmetric under the $U(1)$ group action (global, but not local, in the free theory, local and global in the interaction theory).

Goldstone e/w model: only global symmetry, no interactions, and unobserved scalar

The Goldstone model for the electroweak interactions has global $U(1)$ symmetry of the free Lagrangian but is limited, in that it does not include interactions. Further, it predicts a massless scalar field, which is not observed. We will see these things explicitly in the next chapter.

Higgs e/w model: local symmetry & interactions, but massive photon

The Higgs model overcame these problems, i.e., it included interactions without a massless scalar and had local $U(1)$ symmetry. But, it did not yield the correct results at contemporary energy levels. In particular, it predicted a massive photon field, and that was a death knell.

GSW e/w model: local symmetry, different Higgs form, correct predictions

Both the Goldstone and Higgs models posited the Higgs scalar field to be a $U(1)$ singlet. A major part of the breakthrough by the GSW theory was to consider it to be an $SU(2)$ doublet. This led to a symmetry of $SU(2)$ (for the doublet) combined with $U(1)$ (for other fields in the SM), i.e., a combination $SU(2)\text{X}U(1)$ symmetry. Importantly, this symmetry was valid for a local transformation, not just a global one, and as a result, produced the correct interactions and a massless photon, at low energies.

GSW has Higgs as SU(2) doublet, not U(1) singlet

So, now the key question is "what does the GSW Higgs doublet look like?". If you are guessing this means we have, effectively, two Higgs fields, one for each component of the doublet, you would be correct. We will delve into the mathematical form of that doublet in detail in the following section.

We will understand all this better, as we progress, but for now, simply try to get a feeling for it, based on what we already know about local symmetries (from Vol 1) and groups (transformations) from Chap.2. The essence of it all is summarized in Wholeness Chart 6-1.

Wholeness Chart 6-1. Differences Between Goldstone, Higgs, and Glashow/Salam/Weinberg Models

Summary of the 3 models

	Goldstone Model	Higgs Model	GSW Model
Higgs fields	Complex scalar singlet $\phi=\frac{1}{\sqrt{2}}(\phi_1+i\phi_2)$	Same as at left	Complex scalar doublet Φ See (6-29)
Interactions?	No	Yes	Yes
Symmetry	Global $U(1)$ sym of $\mathcal{L}$	Local $U(1)$ sym of $\mathcal{L}$	Local $SU(2)\text{X}U(1)$ sym of $\mathcal{L}$
Shortcomings	No interactions, massless scalar	Massive photon	None for SM

6.2.3 The Scalar Higgs Field in the GSW Model

GSW postulates

The GSW model is based on

1) two complex scalar Higgs fields ϕ_a and ϕ_b (instead of one ϕ), which, as it turns out, can be expressed advantageously as an $SU(2)$ doublet, the Higgs doublet[1] (more on this below)

1. Higgs SU(2) doublet

$$\Phi(x)=\frac{1}{\sqrt{2}}\begin{bmatrix}\phi_a\\ \phi_b\end{bmatrix}=\frac{1}{\sqrt{2}}\begin{bmatrix}\phi_1+i\phi_2\\ \phi_3+i\phi_4\end{bmatrix}\qquad \phi_1,\phi_2,\phi_3,\phi_4 \text{ real, and} \qquad (6\text{-}29)$$

2. $\mathcal{L}$ has local SU(2)XU(1) symmetry

2) a Lagrangian density with local $SU(2)\text{X}U(1)$ symmetry (more on this below).

6.2.4 Higgs Potential in the GSW Model

The Higgs potential in electroweak theory is different from (6-10). It adds an extra term that is quartic in the scalar (Higgs) field, in addition to the one that is quadratic.

---

[1] Note that we define (6-29) with the normalization factor $1/\sqrt{2}$, whereas some authors, like Mandl and Shaw, leave this out at this point (and include it later in defining the ϕ_i). I believe it is easier to follow the derivation of the theory using the form employed here, as it parallels that typically used in the Goldstone and Higgs derivations.

First, the Goldstone & Higgs Model Potential

In the Goldstone and Higgs models, the potential takes the form, where we take note that, though it seems weird at this point, μ is imaginary,

$$\mathcal{V} = \mu^2\phi^\dagger\phi + \lambda\left(\phi^\dagger\phi\right)^2 \qquad\qquad \mu^2 < 0,\ \lambda \text{ real and} > 0$$
$$= \mu^2 \frac{1}{\sqrt{2}}(\phi_1 - i\phi_2)\frac{1}{\sqrt{2}}(\phi_1 + i\phi_2) + \lambda\left(\frac{1}{\sqrt{2}}(\phi_1 - i\phi_2)\frac{1}{\sqrt{2}}(\phi_1 + i\phi_2)\right)^2 \qquad (6\text{-}30)$$
$$= \frac{\mu^2}{2}\left(\phi_1^2 + \phi_2^2\right) + \frac{\lambda}{4}\left(\phi_1^2 + \phi_2^2\right)^2 = -\frac{|\mu^2|}{2}\left(\phi_1^2 + \phi_2^2\right) + \frac{\lambda}{4}\left(\phi_1^2 + \phi_2^2\right)^2.$$

Goldstone and Higgs $\mathcal{V}$ for singlet, scalar Higgs field ϕ

Plotting (6-30), we get Fig. 6-2, which is often called the "Mexican hat" diagram. It arises from the quartic ϕ^4 term in (6-30). You may have seen "ϕ^4 theory" introduced at early stages of other texts. This is where it arises in QFT, and it ends up being a key feature of electroweak theory.

Mexican hat $\mathcal{V}$ plot of Goldstone & Higgs

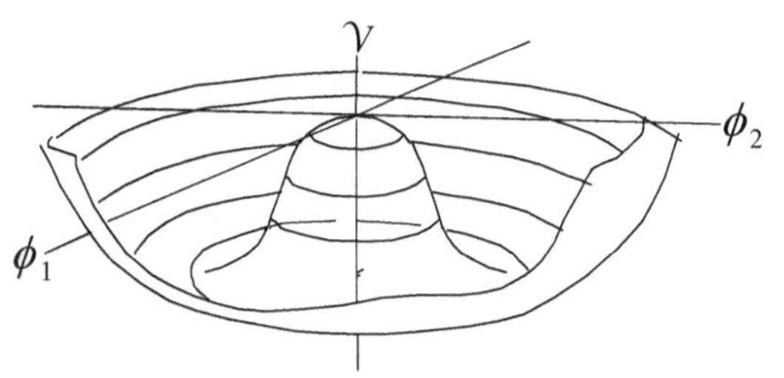

Here, zero particle density state is not the lowest energy state

Figure 6-2. The Higgs (Mexican Hat) Potential for the Goldstone and Higgs Models

Note that instead of a minimum of $\mathcal{V}$, the $\phi_1 = \phi_2 = 0$ location is now a (local) maximum (due to the imaginary μ value). As we will see in the next chapter, the universe starts there, but quickly rolls off that maximum into the trough, the lowest level potential available. But for now, we are focused on the high energy universe, which keeps $\mathcal{V}$ high (at its local maximum energy), in the center of Fig. 6-2).

High energy state in center is e/w epoch

low energy in trough is current universe

This chapter → high energy center

Now, the GSW Potential

As mentioned above, the GSW Higgs potential is similar to that of the Goldstone and Higgs models, except that the one complex scalar field ϕ used there becomes two complex scalar fields, designated herein by $\phi_a = \phi_1 + i\phi_2$ and $\phi_b = \phi_3 + i\phi_4$. So instead of (6-30), we get

$$\mathcal{V} = \frac{\mu^2}{2}\left(\phi_a^\dagger\phi_a + \phi_b^\dagger\phi_b\right) + \frac{\lambda}{4}\left(\phi_a^\dagger\phi_a + \phi_b^\dagger\phi_b\right)^2 = \frac{\mu^2}{2}\left(\phi_1^2 + \phi_2^2 + \phi_3^2 + \phi_4^2\right) + \frac{\lambda}{4}\left(\phi_1^2 + \phi_2^2 + \phi_3^2 + \phi_4^2\right)^2 \qquad (6\text{-}31)$$
$$= -\frac{|\mu^2|}{2}\left(\phi_1^2 + \phi_2^2 + \phi_3^2 + \phi_4^2\right) + \frac{\lambda}{4}\left(\phi_1^2 + \phi_2^2 + \phi_3^2 + \phi_4^2\right)^2 \qquad \mu^2 < 0,\ \lambda > 0.$$

The GSW $\mathcal{V}$ → four real ϕ_i fields instead of two

Note we can re-write (6-31) in a more streamlined way, using the doublet form of (6-29), as

$$\mathcal{V} = \frac{\mu^2}{2}\begin{bmatrix}\phi_1 - i\phi_2 & \phi_3 - i\phi_4\end{bmatrix}\begin{bmatrix}\phi_1 + i\phi_2 \\ \phi_3 + i\phi_4\end{bmatrix} + \frac{\lambda}{4}\left(\begin{bmatrix}\phi_1 - i\phi_2 & \phi_3 - i\phi_4\end{bmatrix}\begin{bmatrix}\phi_1 + i\phi_2 \\ \phi_3 + i\phi_4\end{bmatrix}\right)^2 \qquad (6\text{-}32)$$
$$= \mu^2\Phi^\dagger\Phi + \lambda\left(\Phi^\dagger\Phi\right)^2 = -|\mu^2|\Phi^\dagger\Phi + \lambda\left(\Phi^\dagger\Phi\right)^2.$$

Streamlined doublet notation for GSW $\mathcal{V}$

So, we can imagine a 5D plot of (6-31), similar to the 3D plot of Fig. 6-2 with $\mathcal{V}$ on the vertical axis plotted against four mutually perpendicular axes for the four ϕ_i. Except for the number of dimensions, the behavior will be similar.

Visualize as 5D version of Fig. 6-2

The universe will tend to "fall off" the peak and into the valley, the region of lowest potential, and from our perspective there, as depicted in Fig. 6-2, things are no longer symmetric. From the center of the plot, on top of the Mexican hat shape, as we rotate, nothing seems changed. But from the trough, as we rotate our viewpoint, we see low lying terrain in one direction turn to a ramp upward in another, and then back again to low lying in another. Graphically, the loss of symmetry is apparent. It also turns up in the mathematics, as we will see later on.

When universe falls off central maximum, $\mathcal{V}$, and hence $\mathcal{L}$, no longer symmetric

For now, however, we are focused on the high energy regime where $\mathcal{V}$ is a (local) maximum.

6.3 The High Energy Lagrangian

6.3.1 Ignoring Quarks for Now

To keep things as simple as possible at the beginning, we will ignore quark interactions (strong, electroweak, and with the Higgs field). Once one has an understanding of the lepton behavior, incorporation of quarks into electroweak theory becomes much easier, even straightforward.

We'll ignore quarks to start

6.3.2 Our Direction in this Chapter

The General Approach

Our Lagrangian at high energy needs to have massless particles as that is needed for local symmetry to hold and thus, for a viable interaction theory[1]. We will see this more clearly later, so just take my word for it, for now. When the universe's energy falls and symmetry breaks, leptons, vector bosons, and the Higgs field acquire mass. The key, we will see, is that these fields are not quite the same fields after symmetry breaking as they were before.

Only massless fields in $\mathcal{L}$ at high energy

We'll see how mass arises in next chapter

We will first investigate the free Lagrangian (subscript "0") in the high energy theory. Then, we will extrapolate from methods used in QED to go from the free to the interaction Lagrangian, still at high energy. Finally, in the next chapter, we will explore how the GSW theory, i.e., its Lagrangian, changes as we go from high to low energies. See Wholeness Chart 6-2.

Start with free Lagrangian $\mathcal{L}_0$

Wholeness Chart 6-2. Lagrangians We Will Examine (in This Order)

$$\underbrace{\mathcal{L}_0}_{\substack{\text{at high energy}\\ \text{(electroweak epoch,}\\ \text{this chapter)}}} \quad\rightarrow\quad \underbrace{\mathcal{L} = \mathcal{L}_0 + \mathcal{L}_I}_{\substack{\text{at high energy}\\ \text{(electroweak epoch,}\\ \text{this chapter)}}} \quad\rightarrow\quad \underbrace{\mathcal{L} = \mathcal{L}_0 + \mathcal{L}_I}_{\substack{\text{at low energy}\\ \text{(present epoch,}\\ \text{next chapter)}}}$$

Use that to find full $\mathcal{L}$ at high energy & (next chapter) at low energy

Group Theory to be Applied

In the remainder of this chapter, we assume the reader is familiar with group theory as presented in Chap. 2, specifically with Sect. 2.2.8 (pgs. 18-21) dealing with direct, tensor and outer products, and Sects. 2.5 to 2.7.1 (beginning on pg. 43) on singlets, multiplets, Cartan subalgebras, and their eigenvalues.

6.3.3 Vector Bosons at High Energy

We need to understand one more thing before looking closely into the high energy electroweak Lagrangian. That, like the trick of using a Higgs doublet instead of a singlet, was one more key that helped lead to a successful theory, and it was deduced by Steven Weinberg, one of the theory's principal discoverers. Here is that trick.

At contemporary energy levels, we know of the photon field, which mediates the electromagnetic interaction, and the intermediate vector boson fields, the W^+, its antiparticle the W^-, and the Z, which mediate the weak interaction. (Sometimes the W^+, which is positively charged, is simply labeled W, and the W^-, as $W^\dagger$, since the antiparticle is the complex conjugate of the particle.)

However, at high energy, two of these fields, the photon and the Z blend mix together into two other different, independent fields, symbolized by B_μ and $W_{3\mu}$. The B_μ is known as the hypercharge field, and the $W_{3\mu}$ is one of the three high energy weak intermediate vector bosons. The relationships between these high energy and low energy fields are given by

A_μ (photon) and Z_μ fields of our epoch were mixed as B_μ and $W_{3\mu}$ fields in e/w epoch

$$\overbrace{\begin{bmatrix} B_\mu \\ W_{3\mu} \end{bmatrix}}^{\text{high energy}} = \begin{bmatrix} \cos\theta_W & -\sin\theta_W \\ \sin\theta_W & \cos\theta_W \end{bmatrix} \overbrace{\begin{bmatrix} A_\mu \\ Z_\mu \end{bmatrix}}^{\text{low energy}} \qquad \text{or} \qquad \begin{bmatrix} A_\mu \\ Z_\mu \end{bmatrix} = \begin{bmatrix} \cos\theta_W & \sin\theta_W \\ -\sin\theta_W & \cos\theta_W \end{bmatrix} \begin{bmatrix} B_\mu \\ W_{3\mu} \end{bmatrix}$$

$$B_\mu = A_\mu \cos\theta_W - Z_\mu \sin\theta_W \qquad\qquad A_\mu = B_\mu \cos\theta_W + W_{3\mu} \sin\theta_W \qquad (6\text{-}33)$$

$$W_{3\mu} = A_\mu \sin\theta_W + Z_\mu \cos\theta_W \qquad\qquad Z_\mu = -B_\mu \sin\theta_W + W_{3\mu} \cos\theta_W ,$$

B_μ is hypercharge field, $W_{3\mu}$ one of high energy weak fields

[1] For example, in QED, a massive photon would make $\mathcal{L}$ non-symmetric. See Vol 1, pg. 295-296.

where Weinberg calls θ_W the weak mixing angle, but others call it the Weinberg mixing angle. Its experimentally determined value is $\theta_W \approx 28.13^o$.

Why does it work this way? No one knows, but it does.

The point is, when we construct the high energy Lagrangian, we will be using the B_μ and $W_{3\mu}$ fields, not the A_μ and Z_μ fields. When, in the next chapter, we deduce the low energy Lagrangian from the high energy one, we will substitute the LHS relations of (6-33) to obtain the fields whose particles we measure today, the photon and Z fields. But at high energy, it is the B_μ and $W_{3\mu}$ fields that one would detect most readily with measuring instruments (if one were alive and functioning in the electroweak epoch.)

In high energy theory, we deal with B_μ and $W_{3\mu}$, not A_μ and Z_μ

Similarly, the other two present-day fields, W^+_μ and W^-_μ, mix in the high energy universe to yield different fields, $W_{1\mu}$ and $W_{2\mu}$, functioning at that energy level. The relations between these fields are

$$\underbrace{W_{1\mu}}_{\substack{\text{high}\\\text{energy}}} = \frac{W^+_\mu + W^-_\mu}{\sqrt{2}} \quad \underbrace{W_{2\mu}}_{\substack{\text{high}\\\text{energy}}} = i\frac{W^+_\mu - W^-_\mu}{\sqrt{2}} \qquad \underbrace{W^+_\mu}_{\substack{\text{low}\\\text{energy}}} = \frac{W_{1\mu} - iW_{2\mu}}{\sqrt{2}} \quad \underbrace{W^-_\mu}_{\substack{\text{low}\\\text{energy}}} = \frac{W_{1\mu} + iW_{2\mu}}{\sqrt{2}}. \tag{6-34}$$

Additionally, high energy $W_{1\mu}$ & $W_{2\mu}$ mixture of low energy W^+_μ & W^-_μ

From the LHS of (6-34), since $W^-_\mu = \left(W^+_\mu\right)^\dagger$, we can see that $W_{1\mu}$ and $W_{2\mu}$ are real fields. From (6-33), since A_μ is real, and we know (now, from experiment) that Z_μ is real, then B_μ and $W_{3\mu}$ must be real, as well. This is kind of a nice result, as it means that, at high energy, we can deal with force mediating boson fields that are real, and hence simpler than complex fields. Further, as noted earlier, all fields at high energy are massless.

B_μ and three $W_{i\mu}$ fields are all real and are the fields we deal with at high energy

6.3.4 Lepton Interaction Terms in the High Energy Lagrangian

At the beginning of the universe, we have two electroweak fields, the weak high energy field and the weak hypercharge field mediated, respectively, by vector bosons W_i^μ and B^μ.

The Hypercharge Lagrangian Terms

It turns out that the math for the one B^μ gauge field is similar to that of the familiar photon field A^μ. That is, it can be described by a theory based on $U(1)$ type transformations (symmetry under $e^{i\alpha}$ gauge transformations on fields ψ^L and ψ^R). In fact, it is analogous in all regards. In QED theory, we had an interaction term

$$_{QED}\mathcal{L}_I = e\bar{\psi}\gamma^\mu\psi A_\mu, \tag{6-35}$$

Review of QED $\mathcal{L}_I$

where $-e$ is the charge of the electron, and e is the electromagnetic coupling constant. In hypercharge theory, where g' is the weak hypercharge coupling constant (or just hypercharge coupling constant), we have a typical term

$$\underset{\text{one term}}{_{hypercharge}\mathcal{L}_I} = g'Y\bar{\psi}^L_e\gamma^\mu\psi^L_e B_\mu, \quad Y = -\tfrac{1}{2} \text{ for } \psi^L_e \left(\text{other values for other fields}\right). \tag{6-36}$$

Similar form for B_μ part of e/w $\mathcal{L}_I$

The LC electron has a Y value of – ½ , where $g'Y$ is its hypercharge, although commonly, one refers to hypercharge as simply Y, and drops the g'. Terms for the LC neutrino, the RC electron, and the RC neutrino each have their own value for Y, and these are respectively, – ½, –1, and 0. These multiplied by g' have the values they do for the same reason we had the factor of $+e$ in (6-35). Those values lead to the correct experimental results. No one knows (as of this writing) of any deeper reason for them.

Y = hypercharge

Different Y for different field types

Terms parallel to (6-36) exist for the 2nd lepton generation (muons and muon neutrinos) and the 3rd lepton generation (taus and tau neutrinos), and as you might guess, their Y values equal those of their parallel particles in the 1st generation. For example, the LC muon has the same hypercharge as the LC electron, and so on, with the other particles.

2nd & 3rd generations have same Y values

The High Energy Weak Interaction Lagrangian Terms

The Lagrangian interaction terms for the three W_i^μ with fermions are a little more complicated. It turns out that certain combinations of spinor fields naturally group together. In particular, as noted in Chap. 2, it has been found that grouping the electron and electron neutrino fields together as components in an $SU(2)$ doublet leads to meaningful symmetries (and thus relevant interactions that

$W_{i\mu}$ interactions a bit more complicated

match experiment) as in (6-37) below. (Note, as with the Higgs doublet symbolized by the capital letter Φ, we use capital Ψ for lepton doublets.)

LC SU(2) doublet for electron & its associated neutrino

$$\Psi_e^L = \begin{pmatrix} \psi_{\nu_e}^L \\ \psi_e^L \end{pmatrix} \qquad \bar{\Psi}_e^L = \left(\bar{\psi}_{\nu_e}^L \quad \bar{\psi}_e^L \right) \qquad \text{electron doublet} \qquad (6\text{-}37)$$

Thus, we will find a term in the free part of the Lagrangian like the LHS in (6-38) below that can be more compactly expressed, using (6-37), as the RHS of (6-38).

Streamlined form for certain terms in $\mathcal{L}$ using the doublet

$$\bar{\psi}_{\nu_e}^L \gamma^\mu \partial_\mu \psi_{\nu_e}^L + \bar{\psi}_e^L \gamma^\mu \partial_\mu \psi_e^L = \bar{\Psi}_e^L \gamma^\mu \partial_\mu \Psi_e^L = \bar{\Psi}_e^L \not{\partial} \Psi_e^L . \qquad (6\text{-}38)$$

Similarly, it turns out that the interaction terms in the Lagrangian with the W_3^μ field (see (5-59)) are similar to the interaction terms of (6-35) and (6-36), except that here we have the <u>weak coupling constant g</u> (divided by 2, which is conventional and helps things work out a little better later on), as in the top LHS of (6-39) below. The compact form, using the $SU(2)$ doublet, is shown in the bottom row.

Further use of doublet for interaction terms with a factor of g/2

g is weak coupling constant

$$\underset{\text{two terms}}{\text{weak at high energy}}\,\mathcal{L}_I = -\frac{g}{2}\bar{\psi}_{\nu_e}^L \not{W}_3 \psi_{\nu_e}^L + \frac{g}{2}\bar{\psi}_e^L \not{W}_3 \psi_e^L = -\frac{g}{2}\left(\bar{\psi}_{\nu_e}^L \quad \bar{\psi}_e^L \right)\begin{bmatrix} 1 & 0 \\ 0 & -1 \end{bmatrix} \not{W}_3 \begin{pmatrix} \psi_{\nu_e}^L \\ \psi_e^L \end{pmatrix}$$
$$= -\frac{g}{2}\bar{\Psi}_e^L \begin{bmatrix} 1 & 0 \\ 0 & -1 \end{bmatrix} \not{W}_3 \Psi_e^L = -\frac{g}{2}\bar{\Psi}_e^L \sigma_3 \not{W}_3 \Psi_e^L . \qquad (6\text{-}39)$$

Note the σ_3 matrix is the third Pauli matrix. It turns out that (6-39) can be generalized to incorporate all three W_i^μ fields into the Lagrangian using all the Pauli matrices, as follows, where repeated i indices means summation, as usual. Again, we are simply postulating the interaction form for the fields of (6-40) below, though we know that form has been validated empirically. (Note that some authors use the symbol τ_i instead of σ_i for the Pauli matrices.)

Compact form for all weak interaction e & ν_e terms in $\mathcal{L}$

$$\text{Weak interaction terms in Lagrangian have form } -\frac{g}{2}\bar{\Psi}_e^L \sigma_i \not{W}_i \Psi_e^L \quad \sigma_i = \text{Pauli matrix } i = 1,2,3 \ . (6\text{-}40)$$

Pauli matrices

$$\sigma_1 = \begin{bmatrix} 0 & 1 \\ 1 & 0 \end{bmatrix} \quad \sigma_2 = \begin{bmatrix} 0 & -i \\ i & 0 \end{bmatrix} \quad \sigma_3 = \begin{bmatrix} 1 & 0 \\ 0 & -1 \end{bmatrix} \qquad (6\text{-}41)$$

We review the following identities for Pauli matrices (that can be proven by simple substitution of (6-41) into (6-42)), which will be useful in the future. ε_{ijk} is the Levi-Civita permutation tensor. The last row in (6-42) is merely a generalization of the 2nd and 3rd terms in the first row.

Relations between Pauli matrices

$$\sigma_i^\dagger = \sigma_i \qquad \sigma_1\sigma_1 = \sigma_2\sigma_2 = \sigma_3\sigma_3 = I \qquad \sigma_i\sigma_j = i\varepsilon_{ijk}\sigma_k \quad (\text{for } i \neq j)$$
$$\sigma_i\sigma_j = \delta_{ij}I + i\varepsilon_{ijk}\sigma_k \qquad \left[\sigma_i, \sigma_j\right] = 2i\varepsilon_{ijk}\sigma_k \qquad \left[\sigma_i, \sigma_j\right]_+ = 2\delta_{ij}I \quad (\text{for any } i, j, k) \qquad (6\text{-}42)$$

(6-39), with σ_3, reflects interactions between electron fields and other electron fields, and between electron type neutrinos and other electron type neutrinos, mediated by the W_3 boson. Interactions between electron and electron type neutrino fields are provided via the other two Pauli matrices in (6-40) and mediated by the W_1 and W_2 bosons. For example, where we keep in mind that the σ_i are completely different animals from the γ^μ (part of $\not{W}_1$), which operate in different abstract spaces, so they commute,

Example of some $\mathcal{L}$ terms for e and ν_e interactions

$$-\frac{g}{2}\bar{\Psi}_e^L \sigma_1 \not{W}_1 \Psi_e^L = -\frac{g}{2}\left(\bar{\psi}_{\nu_e}^L \quad \bar{\psi}_e^L \right)\begin{bmatrix} 0 & 1 \\ 1 & 0 \end{bmatrix} \not{W}_1 \begin{pmatrix} \psi_{\nu_e}^L \\ \psi_e^L \end{pmatrix} = -\frac{g}{2}\bar{\psi}_{\nu_e}^L \not{W}_1 \psi_e^L - \frac{g}{2}\bar{\psi}_e^L \not{W}_1 \psi_{\nu_e}^L . \qquad (6\text{-}43)$$

We can further generalize (6-40) (again, this is a postulate that ends up working) to include all three lepton families as the

Compact form for 3 generations, all weak interaction terms

$$\text{form of all lepton weak interaction terms in } \mathcal{L}\text{:} \quad -\frac{g}{2}\bar{\Psi}_l^L \sigma_i \not{W}_i \Psi_l^L \qquad (\text{sum over } l \text{ and } i) . \quad (6\text{-}44)$$

Note we have transitioned into the use of a two dimensional space, with the column matrix doublet of (6-37) acting as a 2D vector and the 2X2 Pauli matrices acting as operators in that 2D space. We have moved into the arena of group theory. The group is $SU(2)$.

We are now using 2D space of SU(2)

Weak Isospin Nomenclature

Recall that spin in NRQM was represented in the wave function as a two-component spinor (much simpler than the four-component spinor of relativistic theories), which reflected spin up or spin down. That two component spinor, in light of group theory, is an $SU(2)$ doublet. We may not have called it that back when we studied this material, but that is what it is. The Pauli matrices, the generators of the $SU(2)$ group, played a role. The eigenvectors of σ_3 (the Cartan sub-algebra) represented spin up and spin down (in z direction). The eigenvectors of σ_1 represented spin in the $+x$ and $-x$ directions, with parallel relations for σ_2 in the $+y$ and $-y$ directions.

Here in weak theory, we find an $SU(2)$ doublet once again in (6-37). Because the treatment of this doublet will parallel what is already known about spin in NRQM, such doublets are called weak isospin doublets (or sometimes, either just isospin doublets or weak doublets). The entire weak interaction theory at high energy is called weak isospin theory.

Weak theory SU(2) doublets like NRQM spin SU(2) doublets

Called weak isospin

6.3.5 SU(2)XU(1)

The weak interaction (isospin) terms in the Lagrangian comprise fields arranged as doublets and singlets that behave as such entities do in standard $SU(2)$ group theory. A transformation in the associated 2D space acts like a 2X2 matrix multiplication on doublets, whereby the components of the doublet generally change. A singlet is unchanged (invariant) under an $SU(2)$ transformation. The Pauli matrices are the generators of the $SU(2)$ algebra for the $SU(2)$ group in this space for the 2D rep.

So, we have some fields (components of $SU(2)$ doublets, as we will see) that are players in the $SU(2)$ domain of weak interactions. And we have some fields ($SU(2)$ singlets as it will turn out) that are unaffected by the weak interaction and are not players in that interaction. Likewise, some fields feel the hypercharge force of the $U(1)$ domain, and others do not.

Some fields part of SU(2) doublets; some, SU(2) singlets

Hence, each field can be considered as an outer (or tensor) product of different parts. The left chirality electron field ψ_e^L has a part of it that responds to the $SU(2)$ weak force, and a part that responds to the $U(1)$ hypercharge force. An $SU(2)$ operator affects the $SU(2)$ part; a $U(1)$ operator affects the $U(1)$ part. We can think of the two parts as being multiplied together in the field via an outer product. This is where the "X", which we learned in Chap. 2 stands for direct product, comes from in $SU(2)XU(1)$. $SU(2)$ operators are direct product multiplied by $U(1)$ operators, and they operate on vectors (doublets and singlets) that are outer products of $SU(2)$ and $U(1)$ vectors.

Each field an outer product of SU(2) part and U(1) part

Now, some fields act like they don't have one of these parts. For example, the RC electron field ψ_e^R doesn't respond to the $SU(2)$ weak force. But, in group theory terms, that field is actually an $SU(2)$ singlet, and as we saw in Chap. 2 (Wholeness Chart 2-7a, pg. 48) such a field has weak charge of zero. So, it does not interact via the weak force. We can, therefore, in practice, consider it as a non-player in $SU(2)$ type interactions. Note that it does not have any $SU(2)$ index label. Like scalars in 2D or 3D physical space, it does not have components, and has no component label.

Some fields, like RC electron, don't respond to SU(2) force & have no effective SU(2) part

The RC electron field does, however, have a weak hypercharge of -1, as noted in the paragraph after (6-36). It is a player in $U(1)$ type interactions.

When we construct a Lagrangian (at high energy) below, it will reflect these characteristics.

6.3.6 The Free Lagrangian at High Energy (Excluding Quarks)

The free Lagrangian at high energy ((6-45) below), ignoring quarks for now, has a form one might intuitively expect. The leptons look familiar from QED, the vector gauge boson terms are similar to what we saw for photons in Vol.1, eqs. (11-6) on pg. 288, and (11-41) on pg. 297 (where $F^{\mu\nu} = \partial^\nu A^\mu - \partial^\mu A^\nu$), and the scalar Higgs has the usual form for scalars, except for the doublet format and the quartic λ term, which we examined earlier. We use the streamlined, and meaningful in terms of group theory, doublet notation for the leptons (6-37) and Higgs (6-29).

Note that some authors use $F_i^{\mu\nu}$, where we use $W_i^{\mu\nu}$. We use different notation to avoid confusion with photons in QED.

High energy, free electroweak Lagrangian $\mathcal{L}_0 = \mathcal{L}_0^L + \mathcal{L}_0^B + \mathcal{L}_0^H$

$$\mathcal{L}_0^L = i\left(\bar{\Psi}_l^L \partial\!\!\!/ \Psi_l^L + \bar{\psi}_l^R \partial\!\!\!/ \psi_l^R + \bar{\psi}_{\nu_l}^R \partial\!\!\!/ \psi_{\nu_l}^R\right) \qquad \text{(leptons, massless)}$$

$$\mathcal{L}_0^B = -\tfrac{1}{4} W_i^{\mu\nu} W_{i\,\mu\nu} - \tfrac{1}{4} B^{\mu\nu} B_{\mu\nu} \qquad \text{(gauge bosons)} \qquad (6\text{-}45)$$

$$\text{where } W_i^{\mu\nu} \equiv \partial^\nu W_i^\mu - \partial^\mu W_i^\nu \qquad B^{\mu\nu} \equiv \partial^\nu B^\mu - \partial^\mu B^\nu$$

$$\mathcal{L}_0^H = \left(\partial^\mu \Phi\right)^\dagger \left(\partial_\mu \Phi\right) - \mu^2 \Phi^\dagger \Phi - \lambda \left(\Phi^\dagger \Phi\right)^2 \qquad \text{(Higgs boson).}$$

Parts of e/w $\mathcal{L}_0$ (free Lagrangian)

6.3.7 The Interaction Lagrangian at High Energy

We first review, in brief, how interaction theory can be deduced from free field theory in QED.

Minimal Substitution in QED

As shown in Vol. 1. pg. 297, in QED, one can obtain the interaction Lagrangian from the free Lagrangian via minimal substitution, which is summarized below in (6-46).

Review of minimal substitution in QED using QED covariant derivative

Minimal substitution, $U(1)$ theory for QED: $\partial_\mu \rightarrow D_\mu = \partial_\mu - ieA_\mu$

$$\underbrace{\bar{\psi} i\gamma^\mu \partial_\mu \psi - m\bar{\psi}\psi}_{\text{free fermion } \mathcal{L}_0^{1/2}} \xrightarrow[\text{substitution}]{\text{minimal}} \bar{\psi} i\gamma^\mu D_\mu \psi - m\bar{\psi}\psi = \underbrace{\bar{\psi} i\gamma^\mu \partial_\mu \psi - m\bar{\psi}\psi}_{\text{free fermion } \mathcal{L}_0^{1/2}} + \underbrace{e\bar{\psi}\gamma^\mu \psi A_\mu}_{\text{interaction } \mathcal{L}_I} \qquad (6\text{-}46)$$

The extra term in the QED covariant derivative D_μ in (6-46) (i times the charge on the electron [the coupling constant times −1] multiplied by the boson field [photon] that mediates the e/m interaction) gave us an $\mathcal{L}$ with i) local symmetry and ii) the correct electromagnetic interaction term.

Minimal Substitution in Electroweak Theory

Apply same idea of minimal substitution to e/w theory

In seeking overall guiding principles from which to deduce any theory, including QFT, one looks to extrapolate principles that worked well in a less complex, well-developed area of that theory to a more complex, undeveloped one. And so the founders of QFT reasoned with regard to minimal substitution._ If it worked with QED, the same idea may work with weak and strong interactions. And indeed, it did. The $U(1)$ mathematics of QED were extended to $SU(2)XU(1)$ for electroweak theory; and to $SU(3)$ for QCD (which we will study later in the book).

So, in e/w theory, for the $U(1)$ part, we do much the same thing as we did in QED, except that we will use B^μ instead of A^μ and $g'Y$ (the hypercharge) instead of $-e$ (the electromagnetic charge). For the $SU(2)$ part, we use W_i^μ instead of A^μ, g/2 instead of $-e$, and factors of the $SU(2)$ generators σ_i in order to give interaction terms like (6-44). The $U(1)$ generator is simply the identity matrix. $SU(2)$ generators are the Pauli matrices σ_i. Thus, e/w minimal substitution employs the electroweak covariant derivative of (6-47).

e/w covariant derivative for e/w minimal substitution

$$\partial^\mu \rightarrow D^\mu \text{ where } \boxed{\begin{array}{c} D^\mu = \partial^\mu + i\frac{g}{2}\sigma_j W_j^\mu + ig'YB^\mu \text{ for fermions, } B \text{ boson, Higgs} \\ g \rightarrow -g \text{ for } W \text{ bosons.} \end{array}} \qquad (6\text{-}47)$$

The different signs on the term with g parallels the different values for Y in the term with g'. Both reflect different ways and strengths by which particular particles interact, as they give rise to various different forms for terms in the Lagrangian. These values are dictated to us by Nature. Taking them as they are in (6-47) gives rise to the correct theory, i.e., an $\mathcal{L}$ with i) local symmetry and ii) the correct e/w interaction terms.

Key objective: local symmetry of $\mathcal{L}$.

Means to that end: covariant derivative

The guiding principle is really local symmetry, and the covariant derivative, when properly defined, gives us that. As you might imagine, getting the proper form for (6-47), with the correct values for Y and more, was a key part in the process of deducing the correct theory.

In summary, to get electroweak interaction theory at high energy, we do two things.

Postulates to deduce e/w interaction theory $\mathcal{L}$ from free theory $\mathcal{L}_0$

1) Use minimal substitution of the e/w covariant derivative $\partial^\mu \rightarrow D^\mu$ (6-47) into (6-45), and
2) postulate particular interaction (coupling) terms between the Higgs and the lepton fields.

Doing this, we find the full Lagrangian at high energy of (6-48), where ${}_AY_{ll}$ and ${}_BY_{l\nu_l}$ are coupling constants between the Higgs and the leptons. (6-48) has a border around it because I consider it the easiest way to remember the high energy $\mathcal{L}$. Further below we elaborate on each term in (6-48).

High energy, total (free + interacting) electroweak Lagrangian $\mathcal{L} = \mathcal{L}^L + \mathcal{L}^B + \mathcal{L}^H + \mathcal{L}^{LH}$

$$\mathcal{L}^L = i\left(\bar{\Psi}_l^L \not{D} \Psi_l^L + \bar{\psi}_l^R \not{D} \psi_l^R + \bar{\psi}_{\nu_l}^R \not{D} \psi_{\nu_l}^R\right) \qquad \text{(leptons, massless)}$$

$$\mathcal{L}^B = -\tfrac{1}{4} G_i^{\mu\nu} G_{i\,\mu\nu} - \tfrac{1}{4} B^{\mu\nu} B_{\mu\nu} \qquad \text{(gauge bosons)}$$

$$\mathcal{L}^H = \left(D^\mu \Phi\right)^\dagger \left(D_\mu \Phi\right) - \mu^2 \Phi^\dagger \Phi - \lambda \left(\Phi^\dagger \Phi\right)^2 \qquad \text{(Higgs boson)} \tag{6-48}$$

$$\mathcal{L}^{LH} = -\,{}_AY_{ll'} \bar{\Psi}_l^L \Phi \psi_{l'}^R - {}_BY_{l\nu_{l'}} \bar{\Psi}_l^L \tilde{\Phi} \psi_{\nu_{l'}}^R + h.c. \qquad \text{(lepton-Higgs coupled)}$$

where D_μ, $\tilde{\Phi}$ defined in (6-47), (6-62), $G_i^{\mu\nu} = D^\nu W_i^\mu - D^\mu W_i^\nu = \partial^\nu W_i^\mu - \partial^\mu W_i^\nu + g\varepsilon_{ijk} W_j^\mu W_k^\nu$.

Total e/w $\mathcal{L}$ at high energy, compact form

Below shows expanded forms for these terms

For $\mathcal{L}^L$:

There are three lepton families (l = 1,2,3 for e, μ, τ), for which the $SU(2)$ singlets are

$$\psi_l^R \text{ and } \psi_{\nu_l}^R \tag{6-49}$$

and the $SU(2)$ doublets are

$$\Psi_l^L = \begin{pmatrix} \psi_{\nu_l}^L \\ \psi_l^L \end{pmatrix} \qquad \bar{\Psi}_l^L = \begin{pmatrix} \bar{\psi}_{\nu_l}^L & \bar{\psi}_l^L \end{pmatrix}. \tag{6-50}$$

Since the high energy RC leptons (6-49) do not feel the weak force, in our minimal substitution for them, we will need to leave out terms in W_j^μ, so we will not get terms in the Lagrangian coupling ψ_l^R and $\psi_{\nu_l}^R$ with W_j^μ.

Said another, more formal (and complicated) way, while the $SU(2)$ generators for $SU(2)$ doublets are the Pauli matrices, the $SU(2)$ generators for $SU(2)$ singlets are zero. (See Chap. 2, pg. 43.) Hence, for the second term in (6-47) acting on singlets, we would need the singlet generators, not the doublet generators σ_i. Since those are zero, that term drops out of our gauge covariant derivative (6-47) for singlets (6-49). Said in simpler language, the $SU(2)$ operators do not act on RC fields.

Thus, in (6-48), we have, with

$$\text{weak hypercharge } Y = -1 \text{ for } \psi_l^R \qquad Y = 0 \text{ for } \psi_{\nu_l}^R \qquad Y = -\tfrac{1}{2} \text{ for } \Psi_l^L\,, \tag{6-51}$$

$$\begin{aligned}
i\bar{\psi}_l^R \not{D} \psi_l^R &= i\bar{\psi}_l^R \gamma_\mu D^\mu \psi_l^R = i\bar{\psi}_l^R \gamma_\mu \left(\partial^\mu + ig'YB^\mu\right) \psi_l^R \\
&= i\bar{\psi}_l^R \gamma_\mu \partial^\mu \psi_l^R - g'Y\bar{\psi}_l^R \gamma_\mu B^\mu \psi_l^R = i\bar{\psi}_l^R \not{\partial} \psi_l^R + g'\bar{\psi}_l^R \not{B} \psi_l^R \\[1em]
i\bar{\psi}_{\nu_l}^R \not{D} \psi_{\nu_l}^R &= i\bar{\psi}_{\nu_l}^R \gamma_\mu D^\mu \psi_{\nu_l}^R = i\bar{\psi}_{\nu_l}^R \gamma_\mu \left(\partial^\mu + ig'YB^\mu\right) \psi_{\nu_l}^R \\
&= i\bar{\psi}_{\nu_l}^R \gamma_\mu \partial^\mu \psi_{\nu_l}^R - g'Y\bar{\psi}_{\nu_l}^R \gamma_\mu B^\mu \psi_{\nu_l}^R = i\bar{\psi}_{\nu_l}^R \not{\partial} \psi_{\nu_l}^R \\[1em]
i\bar{\Psi}_l^L \not{D} \Psi_l^L &= i\bar{\Psi}_l^L \gamma_\mu D^\mu \Psi_l^L = i\bar{\Psi}_l^L \gamma_\mu \left(\partial^\mu + i\tfrac{g}{2}\sigma_j W_j^\mu + ig'YB^\mu\right) \Psi_l^L \\
&= i\bar{\Psi}_l^L \gamma_\mu \partial^\mu \Psi_l^L - \tfrac{g}{2} \bar{\Psi}_l^L \gamma_\mu \sigma_j W_j^\mu \Psi_l^L - g'Y\bar{\Psi}_l^L \gamma_\mu B^\mu \Psi_l^L \\
&= i\bar{\Psi}_l^L \not{\partial} \Psi_l^L - \tfrac{g}{2} \bar{\Psi}_l^L \not{W}_j \sigma_j \Psi_l^L + \tfrac{g'}{2} \bar{\Psi}_l^L \not{B} \Psi_l^L\,.
\end{aligned} \tag{6-52}$$

Terms for leptons interacting with W and B bosons

For $\mathcal{L}^B$

It is assumed that the B^μ and the W_i^μ particles do not interact directly with one another, so there are no products between them. Hence, $G_i^{\mu\nu}$ has no B^μ field in it; and $B^{\mu\nu}$ has no W_i^μ fields in it.

For the W_i^μ fields

Applying minimal substitution of (6-47) to the free Lagrangian terms of W_i^μ in (6-45), where we can drop the non-interacting part of the covariant derivative (i.e., the part in B^μ), we have

$$\begin{aligned} G_i^{\mu\nu} = D^\nu W_i^\mu - D^\mu W_i^\nu &= \partial^\nu W_i^\mu - i\frac{g}{2}\sigma_j W_j^\nu W_i^\mu - \partial^\mu W_i^\nu + i\frac{g}{2}\sigma_j W_j^\mu W_i^\nu \\ &= \partial^\nu W_i^\mu - \partial^\mu W_i^\nu - i\frac{g}{2}\sigma_j\left(W_j^\nu W_i^\mu - W_j^\mu W_i^\nu\right). \end{aligned} \qquad (6\text{-}53)$$

$G_i^{\mu\nu}$ factor of W bosons terms

It will simplify things if we contract (6-53) with the Pauli matrices σ_i.

$$\sigma_i G_i^{\mu\nu} = \sigma_i\left(\partial^\nu W_i^\mu - \partial^\mu W_i^\nu\right) \underbrace{- i\frac{g}{2}\sigma_i\sigma_j W_j^\nu W_i^\mu + i\frac{g}{2}\sigma_i\sigma_j W_j^\mu W_i^\nu}_{\boxed{A}} \qquad (6\text{-}54)$$

and evaluate $\boxed{A}$ in (6-54) using $\sigma_i\sigma_j = \delta_{ij}I + i\varepsilon_{ijk}\sigma_k$ from the second row of (6-42).

$$\begin{aligned} \boxed{A} &= -i\frac{g}{2}\left(\delta_{ij}I + i\varepsilon_{ijk}\,\sigma_k\right)W_j^\nu W_i^\mu + i\frac{g}{2}\left(\delta_{ij}I + i\varepsilon_{ijk}\,\sigma_k\right)W_j^\mu W_i^\nu \\ &= \underbrace{i\frac{g}{2}\left(-W_i^\nu W_i^\mu + W_i^\mu W_i^\nu\right)}_{=0,\text{ field commutes with itself}} \underbrace{+\frac{g}{2}\varepsilon_{kij}\sigma_k W_j^\nu W_i^\mu - \frac{g}{2}\varepsilon_{kij}\sigma_k W_j^\mu W_i^\nu}_{\text{exchange dummies } i,j,k} \\ &= \frac{g}{2}\left(\varepsilon_{ijk}\sigma_i W_k^\nu W_j^\mu - \varepsilon_{ijk}\sigma_i W_k^\mu W_j^\nu\right) = \frac{g}{2}\sigma_i\varepsilon_{ijk}\left(W_k^\nu W_j^\mu - W_k^\mu W_j^\nu\right). \end{aligned} \qquad (6\text{-}55)$$

With $\boxed{A}$ of (6-55) in (6-54), σ_i can be factored out, as the relationship must hold for all possible values of the fields. This leaves

$$\begin{aligned} G_i^{\mu\nu} &= \partial^\nu W_i^\mu - \partial^\mu W_i^\nu + \frac{g}{2}\Big(-\varepsilon_{ijk}W_k^\mu W_j^\nu \underbrace{+\ \varepsilon_{ijk}W_k^\nu W_j^\mu}_{\text{switch dummies } j,k}\Big) \\ &= W_i^{\mu\nu} + \frac{g}{2}\left(-\varepsilon_{ijk}W_k^\mu W_j^\nu - \varepsilon_{ijk}W_j^\nu W_k^\mu\right) = W_i^{\mu\nu} - \frac{g}{2}\overbrace{\varepsilon_{ijk}\left(W_k^\mu W_j^\nu + W_j^\nu W_k^\mu\right)}^{\boxed{B}}. \end{aligned} \qquad (6\text{-}56)$$

Due to our second postulate for QFT (Vol. 1, pg. 4, (1-8) and pg. 31, bold framed box in Wholeness Chart 2-5) different boson fields (such as W_1^μ and W_2^μ) commute with each other and with derivatives of the different fields (such as W_1^μ and $\partial_\nu W_2^\mu$). Thus, $\boxed{B}$ and $G_i^{\mu\nu}$ of (6-56) become

$$\boxed{B} = \varepsilon_{ijk}\left(W_k^\mu W_j^\nu + W_j^\nu W_k^\mu\right) = 2\varepsilon_{ijk}W_k^\mu W_j^\nu = -2\varepsilon_{ijk}W_j^\mu W_k^\nu\,, \qquad (6\text{-}57)$$

$$G_i^{\mu\nu} = W_i^{\mu\nu} + g\varepsilon_{ijk}W_j^\mu W_k^\nu\,. \qquad (6\text{-}58)$$

Equivalent, more useful form for $G_i^{\mu\nu}$ factor

I note in passing that virtually every other QFT text simply presents (6-58) without relating it to the covariant derivative. That (6-58) can, in fact, be derived from the covariant derivative means the theory is inherently more elegant, as all non-Higgs coupling interaction terms can be deduced from one overarching principle, and not simply inserted *ad hoc* into the theory.

Note further that sign in the definition of $W_i^{\mu\nu}$ is conventional, varies with author, and was chosen here to match $F^{\mu\nu}$ of Vol. 1. Had we defined it instead as $W_i^{\mu\nu} = -\partial^\nu W_i^\mu + \partial^\mu W_i^\nu$, then we would not have needed the minus sign on g in (6-47), and the theory would have appeared even more elegant.

Do **Problems 3** and **4** to show (6-59) and (6-60).

For the B^μ field,

$$\text{Free } B^{\mu\nu} = \partial^\nu B^\mu - \partial^\mu B^\nu \quad \rightarrow \quad \text{Interaction } B^{\mu\nu} = \partial^\nu B^\mu - \partial^\mu B^\nu\,. \qquad (6\text{-}59)$$

B boson terms (free, don't self-interact)

For the term $-\frac{1}{4}G_i^{\mu\nu}G_{i\mu\nu}$ in the interaction Lagrangian (6-48), from Prob. 4,

$$-\frac{1}{4}G_i^{\mu\nu}G_{i\,\mu\nu} = -\frac{1}{4}W_i^{\mu\nu}W_{i\,\mu\nu} + g\varepsilon_{ijk}W_{i\mu}W_{j\nu}\partial^\mu W_k^\nu - \frac{1}{4}g^2\varepsilon_{ijk}\varepsilon_{ilm}W_j^\mu W_k^\nu W_{l\mu}W_{m\nu}\,. \qquad (6\text{-}60)$$

W bosons terms (free plus self-interacting)

Recall that each term in the Lagrangian is represented in Feynman diagrams as a vertex. (6-60) implies we will have vertices with three and four W_i^μ particles. (6-60) also implies, since the Euler-Lagrange equation takes single derivatives of each term in the Lagrangian, that the field equation is non-linear (more than one factor of the W_i^μ fields in at least some terms).

Each term in $\mathcal{L}$ is a vertex → W bosons have 3 & 4 particle vertices, & e/w theory is non-linear

For $\mathcal{L}^H$

The Higgs field is, as mentioned earlier, represented in the GSW theory as an $SU(2)$ doublet. And it is coupled, in the interaction theory, to the weak and hypercharge fields. Thus, we adopt minimal substitution for the doublet, where the hypercharge for the Higgs is ½. As with fermions, no one knows why the Higgs has this particular hypercharge. It does, however, like the fermion hypercharges discussed earlier, support an $\mathcal{L}$ that is symmetric under $SU(2)XU(1)$ (discussed immediately following), and as we have noted before, seems to be an underlying principle to which Nature adheres.

Hence, in (6-48), we have

$$D^\mu\Phi = \left(\partial^\mu + \frac{i}{2} g\sigma_j W_j^\mu + ig'YB^\mu\right)\Phi \qquad Y = \frac{1}{2} \text{ for } \Phi . \tag{6-61}$$

Higgs doublet minimal substitution

For $\mathcal{L}^{LH}$

The symbol $\tilde{\Phi}$ is used as shorthand in (6-48). It is defined (compare to (6-29)) as

$$\tilde{\Phi} = \frac{1}{\sqrt{2}}\begin{pmatrix} \phi_b^* \\ -\phi_a^* \end{pmatrix} = \frac{1}{\sqrt{2}}\begin{pmatrix} \phi_3 - i\phi_4 \\ -\phi_1 + i\phi_2 \end{pmatrix}. \tag{6-62}$$

Definition of $\tilde{\Phi}$ for Lepton-Higgs coupling

The "*h.c.*" in (6-48) stands for hermitian conjugate (complex conjugate transpose) of the other terms, and is a common shorthand used for terms in the Lagrangian.

There is nothing really to derive here from the free Lagrangian, as we are simply postulating that the term $\mathcal{L}^{LH}$ be inserted as part of the interaction Lagrangian. The matrices ${}_AY_{ll'}$ and ${}_BY_{l\nu_{l'}}$ couple the lepton and Higgs fields, are called Yukawa matrices, and are unrelated to hypercharge Y (which by coincidence uses the same letter symbol Y, though without subscripts). The components of those matrices are called Yukawa couplings. Note, in passing, that the Yukawa matrices couple lepton fields of different generations. There is much going on in $\mathcal{L}^{LH}$, and we will have a lot more to say about it in the next chapter, so, for now, don't strain too much on trying to interpret it.

Lepton-Higgs coupling called Yukawa coupling

6.3.8 Why This Form for $\mathcal{L}$?

The Lagrangian takes the form (6-48) because that form is symmetric. Further, Gerardus 't Hooft has shown that the e/w Lagrangian must be symmetric to be renormalizable. We need a symmetric $\mathcal{L}$.

e/w $\mathcal{L}$ is symmetric (as we will show later)

Review of Local Symmetry in QED

Recall from QED, that if the Lagrangian is symmetric under some set of local transformations of its fields, then the interactions that show up in that particular Lagrangian mirror those in the real world. As an overview of that symmetry (see Vol. 1, pgs. 293-298), note the full QED Lagrangian is

$$\mathcal{L} = -\tfrac{1}{4}F_{\nu\beta}F^{\nu\beta} + \bar{\psi}\left(i\gamma^\nu\partial_\nu - m\right)\psi + e\bar{\psi}\gamma^\nu\psi A_\nu . \tag{6-63}$$

The local set of transformations, where the transformation on ψ is $U(1)$, is

$$\psi \to \psi' = e^{-i\alpha(x^\mu)}\psi \qquad A_\nu \to A'_\nu = A_\nu - \tfrac{1}{e}\partial_\nu\alpha(x^\mu). \tag{6-64}$$

Review of local symmetry of QED $\mathcal{L}$

Substituting (6-64) into (6-63), we get the primed Lagrangian (with transformed [primed] fields) having the same form as the unprimed Lagrangian. (See the reference for details to get the last = sign.)

$$\mathcal{L} \xrightarrow[A_\nu \to A'_\nu = A_\nu - \frac{1}{e}\partial_\nu\alpha(x^\mu)]{\psi\to\psi' = e^{-i\alpha(x^\mu)}\psi} \mathcal{L}' = -\tfrac{1}{4}F'_{\nu\beta}F'^{\nu\beta} + \bar{\psi}'\left(i\gamma^\nu\partial_\nu - m\right)\psi' + e\bar{\psi}'\gamma^\nu\psi' A'_\nu = \mathcal{L}. \tag{6-65}$$

Insisting on local U(1) symmetry led to correct QED interaction terms

Thus, the Lagrangian is symmetric under the transformation set. And the particular interaction term (last term in (6-63)) had to be just what it is, in order to have symmetry *and* in order to represent the real interaction Nature prescribes. In QFT, we generalize that rule to all SM interactions. Local symmetry dictates real world interactions (plus ensures renormalizability).

Insisting on local e/w SU(2)XU(1) symmetry leads to correct e/w interaction terms

Local Symmetry for Electroweak Interactions

The transformation set under which (6-48) is symmetric (invariant) is shown below, where the symbol x is shorthand for x^μ. Note that there are far more fields involved than in QED, and so, there are considerably more fields to be transformed than in QED. Additionally, the transformations on the fermions are not simply of type $U(1)$, but include both $U(1)$ and $SU(2)$, i.e., we are talking a $SU(2)XU(1)$ symmetry transformation. When a set of transformations, such as these, must be carried out together, they are collectively called coupled transformations.

The e/w coupled transformation set that works is shown below

Demonstrating that the invariance of $\mathcal{L}$ holds for electroweak theory with the particular coupled transformation set we are about to show is lengthy. We discuss this in depth after presenting those transformations explicitly in the following subsections.

We will demonstrate symmetry after presenting that transformation set

6.3.9 Local Finite Transformations

$U(1)$ transformations in electroweak theory parallel those of QED in (6-64), but instead of A^μ, we will have the fields B^μ and W_i^μ, and instead of a parameter $\alpha(x)$ in the local transformation $e^{-i\alpha(x)}$, we will use other parameters labeled $f(x)$ and $\omega_i(x)$. The group theory generator for $U(1)$ is the identity, and the (single) parameter for $U(1)$ we use in QED is $\alpha(x)$. α effects a rotation in the complex plane through an angle α.

$SU(2)$ transformations can be visualized heuristically as "rotations" in a 2D complex space having complex values on the axes. The generators for this space in QFT are the Pauli matrices σ_i, as we learned in Chap. 2. So, the "rotations" will be effected by the operator $e^{i\omega_i(x)\sigma_i/2}$, where each ω_i reflects the amount by which a doublet is "rotated" about the ith "axis". These are explicitly expressed in Wholeness Chart 6-3.

Finite local e/w SU(2)XU(1) symmetry transformation set

Wholeness Chart 6-3. Finite *SU*(2) and *U*(1) Local Transformations in Electroweak Theory

<u>SU(2)</u>

$$\left.\begin{aligned}
&\Psi_l^L(x) \to \Psi_l'^L = e^{i\omega_i(x)\sigma_i/2}\Psi_l^L = \left(1+\tfrac{i}{2}\omega_i(x)\sigma_i+....\right)\Psi_l^L \\
&\bar{\Psi}_l^L(x) \to \bar{\Psi}_l'^L = \bar{\Psi}_l^L e^{-i\omega_i(x)\sigma_i/2} = \bar{\Psi}_l^L\left(1-\tfrac{i}{2}\omega_i(x)\sigma_i+....\right) \\
&\psi_{l\,\&\nu_l}^R(x) \to \psi_{l\,\&\nu_l}'^R = \psi_{l\,\&\nu_l}^R \\
&\bar{\psi}_{l\,\&\nu_l}^R(x) \to \bar{\psi}_{l\,\&\nu_l}'^R = \bar{\psi}_{l\,\&\nu_l}^R
\end{aligned}\right\} \tag{6-66}$$

$$W_i^\mu(x)\sigma_i \to W_i'^\mu\sigma_i = e^{i\omega_j(x)\sigma_j/2}\left(W_i^\mu\sigma_i\right)e^{-i\omega_k(x)\sigma_k/2} \tag{6-67}$$

$$B^\mu \to B'^\mu = B^\mu \tag{6-68}$$

$$\left.\begin{aligned}
&\Phi(x) \to \Phi' = e^{i\omega_i(x)\sigma_i/2}\Phi = \left(1+\tfrac{i}{2}\omega_i(x)\sigma_i+...\right)\Phi \\
&\tilde{\Phi}(x) \to \tilde{\Phi}' = e^{i\omega_i(x)\sigma_i/2}\tilde{\Phi} = \left(1+\tfrac{i}{2}\omega_i(x)\sigma_i+...\right)\tilde{\Phi}
\end{aligned}\right\} \tag{6-69}$$

<u>U(1)</u>

$$\left.\begin{aligned}
&\Psi_l^L(x) \to \Psi_l'^L = e^{iYf(x)}\Psi_l^L = \left(1+iYf(x)+....\right)\Psi_l^L \\
&\bar{\Psi}_l^L(x) \to \bar{\Psi}_l'^L = \bar{\Psi}_l^L e^{-iYf(x)} = \bar{\Psi}_l^L\left(1-iYf(x)+....\right) \\
&\psi_l^R(x) \to \psi_l'^R = e^{iYf(x)}\psi_l^R = \left(1+iYf(x)+....\right)\psi_l^R \\
&\bar{\psi}_l^R(x) \to \bar{\psi}_l'^R = \bar{\psi}_l^R e^{-iYf(x)} = \bar{\psi}_l^R\left(1-iYf(x)+....\right) \\
&\psi_{\nu_l}^R(x) \to \psi_{\nu_l}'^R = \psi_{\nu_l}^R \qquad \bar{\psi}_{\nu_l}^R(x) \to \bar{\psi}_{\nu_l}'^R = \bar{\psi}_{\nu_l}^R
\end{aligned}\right\} \tag{6-70}$$

$$W_i^\mu(x) \to W_i'^\mu = W_i^\mu - \frac{1}{g}\partial^\mu\omega_i(x) \tag{6-71}$$

$$B^\mu \to B'^\mu = B^\mu - \frac{1}{g'}\partial^\mu f(x) \tag{6-72}$$

$$\left.\begin{aligned}
&\Phi(x) \to \Phi' = e^{iYf(x)}\Phi = \left(1+iYf(x)+...\right)\Phi \\
&\tilde{\Phi}(x) \to \tilde{\Phi}' = e^{iYf(x)}\tilde{\Phi} = \left(1+iYf(x)+...\right)\tilde{\Phi}
\end{aligned}\right\} \tag{6-73}$$

6.3.10 Local Infinitesimal Transformations

Far easier to evaluate infinitesimal transformations & same conclusions

It is far easier to examine the invariance of $\mathcal{L}$ for infinitesimal transformations, i.e., for the above where ω_i and f are very small. One can then, in principle, integrate to obtain the finite transformation case. But, if $\mathcal{L}$ is symmetric under the infinitesimal transformation, then it should also be so under the finite transformation.

If small change in $\mathcal{L}$ is zero (i.e. $\delta\mathcal{L} = 0$) under transformation set, then $\mathcal{L}$ is symmetric

That is, if we have small changes in the Lagrangian under the transformation, then

$$\mathcal{L} \rightarrow \mathcal{L}' = \mathcal{L} + \delta\mathcal{L} \qquad (\delta\mathcal{L} = \text{ change in } \mathcal{L} \text{ due to small transformations in the fields}). \qquad (6\text{-}74)$$

If $\delta\mathcal{L} = 0$, then $\mathcal{L}'=\mathcal{L}$, and the Lagrangian is symmetric.

So, we take the easy (most efficient) way out, and examine the Lagrangian symmetry under the infinitesimal transformation set that is the small parameter limit of the finite set (6-66) to (6-73). The relevant relations are then as in (6-75) to (6-82) below.

After looking over (6-75) below and understanding where if comes from, do **Problem 5** to show (6-76).

Wholeness Chart 6-4. Infinitesimal $SU(2)$ and $U(1)$ Local Transformations in Electroweak Theory

Infinitesimal local e/w SU(2) and U(1) separate symmetry transformations

<u>SU(2)</u>

$$\left.\begin{aligned}
&\Psi_l^L(x) \rightarrow \Psi_l'^L \approx \left(1+\tfrac{i}{2}\omega_i(x)\sigma_i\right)\Psi_l^L = \Psi_l^L + \delta\Psi_l^L && \delta\Psi_l^L = \tfrac{i}{2}\omega_i(x)\sigma_i\Psi_l^L \\
&\bar{\Psi}_l^L(x) \rightarrow \bar{\Psi}_l'^L \approx \bar{\Psi}_l^L\left(1-\tfrac{i}{2}\omega_i(x)\sigma_i\right) = \bar{\Psi}_l^L + \delta\bar{\Psi}_l^L && \delta\bar{\Psi}_l^L = -\tfrac{i}{2}\omega_i(x)\bar{\Psi}_l^L\sigma_i \\
&\psi_{l\,\&\nu_l}^R(x) \rightarrow \psi_{l\,\&\nu_l}'^R = \psi_{l\,\&\nu_l}^R && \delta\psi_{l\,\&\nu_l}^R = 0 \\
&\bar{\psi}_{l\,\&\nu_l}^R(x) \rightarrow \bar{\psi}_{l\,\&\nu_l}'^R = \bar{\psi}_{l\,\&\nu_l}^R && \delta\bar{\psi}_{l\,\&\nu_l}^R = 0
\end{aligned}\right\} \qquad (6\text{-}75)$$

$$W_i^\mu(x) \rightarrow W_i'^\mu \approx W_i^\mu - \varepsilon_{ijk}\omega_j(x)W_k^\mu = W_i^\mu + \delta W_i^\mu \qquad \delta W_i^\mu = -\varepsilon_{ijk}\omega_j(x)W_k^\mu \qquad (6\text{-}76)$$

$$B^\mu \rightarrow B'^\mu = B^\mu \qquad \delta B^\mu = 0 \qquad (6\text{-}77)$$

$$\left.\begin{aligned}
&\Phi(x) \rightarrow \Phi' \approx \left(1+\tfrac{i}{2}\omega_i(x)\sigma_i\right)\Phi = \Phi + \delta\Phi && \delta\Phi = \tfrac{i}{2}\omega_i(x)\sigma_i\Phi \\
&\tilde{\Phi}(x) \rightarrow \tilde{\Phi}' \approx \left(1+\tfrac{i}{2}\omega_i(x)\sigma_i\right)\tilde{\Phi} = \tilde{\Phi} + \delta\tilde{\Phi} && \delta\tilde{\Phi} = \tfrac{i}{2}\omega_i(x)\sigma_i\tilde{\Phi}
\end{aligned}\right\} \qquad (6\text{-}78)$$

<u>U(1)</u>

$$\left.\begin{aligned}
&\Psi_l^L(x) \rightarrow \Psi_l'^L \approx \left(1+iYf(x)\right)\Psi_l^L = \Psi_l^L + \delta\Psi_l^L && \delta\Psi_l^L = iYf(x)\Psi_l^L \\
&\bar{\Psi}_l^L(x) \rightarrow \bar{\Psi}_l'^L \approx \bar{\Psi}_l^L\left(1-iYf(x)\right) = \bar{\Psi}_l^L + \delta\bar{\Psi}_l^L && \delta\bar{\Psi}_l^L = -iYf(x)\bar{\Psi}_l^L \\
&\psi_l^R(x) \rightarrow \psi_l'^R \approx \left(1+iYf(x)\right)\psi_l^R = \psi_l^R + \delta\psi_l^R && \delta\psi_l^R = iYf(x)\psi_l^R \\
&\bar{\psi}_l^R(x) \rightarrow \bar{\psi}_l'^R \approx \bar{\psi}_l^R\left(1-iYf(x)\right) = \bar{\psi}_l^R + \delta\bar{\psi}_l^R && \delta\bar{\psi}_l^R = -iYf(x)\bar{\psi}_l^R \\
&\psi_{\nu_l}^R(x) \rightarrow \psi_{\nu_l}'^R = \psi_{\nu_l}^R \qquad \bar{\psi}_{\nu_l}^R(x) \rightarrow \bar{\psi}_{\nu_l}'^R = \bar{\psi}_{\nu_l}^R && \delta\psi_{\nu_l}^R = \delta\bar{\psi}_{\nu_l}^R = 0
\end{aligned}\right\} \qquad (6\text{-}79)$$

$$W_i^\mu(x) \rightarrow W_i'^\mu \approx W_i^\mu - \frac{1}{g}\partial^\mu\omega_i(x) = W_i^\mu + \delta W_i^\mu \qquad \delta W_i^\mu = -\frac{1}{g}\partial^\mu\omega_i(x) \qquad (6\text{-}80)$$

$$B^\mu \rightarrow B'^\mu = B^\mu - \frac{1}{g'}\partial^\mu f(x) = B^\mu + \delta B^\mu \qquad \delta B^\mu = -\frac{1}{g'}\partial^\mu f(x) \qquad (6\text{-}81)$$

$$\left.\begin{aligned}
&\Phi(x) \rightarrow \Phi' \approx \left(1+iYf(x)\right)\Phi = \Phi + \delta\Phi && \delta\Phi = iYf(x)\Phi \\
&\tilde{\Phi}(x) \rightarrow \tilde{\Phi}' \approx \left(1+iYf(x)\right)\tilde{\Phi} = \tilde{\Phi} + \delta\tilde{\Phi} && \delta\tilde{\Phi} = iYf(x)\tilde{\Phi}
\end{aligned}\right\} \qquad (6\text{-}82)$$

6.3.11 Local SU(2)XU(1) Infinitesimal Transformations

Under $SU(2)XU(1)$ transformations combining the local, infinitesimal $SU(2)$ and $U(1)$ transformations of (6-75) to (6-82), the differences in the fields (the terms with δ in front of them) add. Thus,

Wholeness Chart 6-5. Infinitesimal $SU(2)XU(1)$ Local Transformations in Electroweak Theory

Infinitesimal local e/w SU(2)XU(1) symmetry transformations together as a set

$\underline{SU(2)XU(1)}$

$$\left.\begin{aligned}
&\Psi_l^L \to \Psi_l'^L \approx \left(1+\tfrac{i}{2}\omega_i(x)\sigma_i + iYf(x)\right)\Psi_l^L = \Psi_l^L + \delta\Psi_l^L && \delta\Psi_l^L = \tfrac{i}{2}\omega_i(x)\sigma_i\Psi_l^L + iYf(x)\Psi_l^L \\
& && Y=-\tfrac{1}{2} \updownarrow \\
&\bar{\Psi}_l^L \to \bar{\Psi}_l'^L \approx \bar{\Psi}_l^L\left(1-\tfrac{i}{2}\omega_i(x)\sigma_i - iYf(x)\right) = \bar{\Psi}_l^L + \delta\bar{\Psi}_l^L && \delta\bar{\Psi}_l^L = -\tfrac{i}{2}\omega_i(x)\bar{\Psi}_l^L\sigma_i - iYf(x)\bar{\Psi}_l^L \\
&\psi_l^R \to \psi_l'^R = \psi_l^R + iYf(x)\psi_l^R && \delta\psi_l^R = iYf(x)\psi_l^R \quad \leftarrow Y=-1 \\
&\bar{\psi}_l^R \to \bar{\psi}_l'^R = \bar{\psi}_l^R - iYf(x)\bar{\psi}_l^R && \delta\bar{\psi}_l^R = -iYf(x)\bar{\psi}_l^R \quad \leftarrow Y=-1 \\
&\psi_{\nu_l}^R(x) \to \psi_{\nu_l}'^R = \psi_{\nu_l}^R \qquad \bar{\psi}_{\nu_l}^R(x) \to \bar{\psi}_{\nu_l}'^R = \bar{\psi}_{\nu_l}^R && \delta\psi_{\nu_l}^R = \delta\bar{\psi}_{\nu_l}^R = 0
\end{aligned}\right\} \quad (6\text{-}83)$$

$$W_i^\mu \to W_i'^\mu \approx W_i^\mu - \frac{1}{g}\partial^\mu\omega_i(x) - \varepsilon_{ijk}\omega_j(x)W_k^\mu = W_i^\mu + \delta W_i^\mu \qquad \delta W_i^\mu = -\frac{1}{g}\partial^\mu\omega_i(x) - \varepsilon_{ijk}\omega_j(x)W_k^\mu \quad (6\text{-}84)$$

$$B^\mu \to B'^\mu = B^\mu - \frac{1}{g'}\partial^\mu f(x) \qquad \delta B^\mu = -\frac{1}{g'}\partial^\mu f(x) \quad (6\text{-}85)$$

$$\left.\begin{aligned}
&\Phi \to \Phi' \approx \left(1+\tfrac{i}{2}\omega_i(x)\sigma_i + iYf(x)\right)\Phi = \Phi + \delta\Phi && \delta\Phi = \tfrac{i}{2}\omega_i(x)\sigma_i\Phi + iYf(x)\Phi \quad \leftarrow Y=\tfrac{1}{2} \\
&\tilde{\Phi} \to \tilde{\Phi}' \approx \left(1+\tfrac{i}{2}\omega_i(x)\sigma_i + iYf(x)\right)\tilde{\Phi} = \tilde{\Phi} + \delta\tilde{\Phi} && \delta\tilde{\Phi} = \tfrac{i}{2}\omega_i(x)\sigma_i\tilde{\Phi} + iYf(x)\tilde{\Phi} \quad \leftarrow Y=-\tfrac{1}{2}
\end{aligned}\right\} \quad (6\text{-}86)$$

Note that some authors define what we call ω_i as $g\omega_i$ and f as what we call $g'f$. The choice is conventional, as it doesn't really matter, provided one stays consistent throughout, since the ω_i and f are arbitrary variables, anyway.

6.3.12 Showing $\mathcal{L}$ is Symmetric

To show $\mathcal{L}$ is symmetric, we need to show $\mathcal{L}'$ (same form as $\mathcal{L}$, but with primed fields) is equal to $\mathcal{L}$. In the infinitesimal case, where A, B, C, ... are fields, this can be expressed as

$$\mathcal{L}' = \mathcal{L}(A',B',C',\ldots) = \mathcal{L}(A,B,C,\ldots) + \delta\mathcal{L} = \mathcal{L}(A,B,C,\ldots) \quad (\text{for symmetry } \delta\mathcal{L}=0), \quad (6\text{-}87)$$

$$\delta\mathcal{L} = \frac{\partial\mathcal{L}(A,B,C,\ldots)}{\partial A}\delta A + \frac{\partial\mathcal{L}(A,B,C,\ldots)}{\partial B}\delta B + \frac{\partial\mathcal{L}(A,B,C,\ldots)}{\partial C}\delta C + \ldots . \quad (6\text{-}88)$$

The δA, δB, etc. for different fields in e/w theory are shown in (6-83) to (6-86). For fermions, the ordering of the spinor fields (which reflects the order of the hidden spinor indices) needs to be kept after the derivatives in (6-88) are taken. That is, showing the δA, δB, etc. in (6-88) to the right of the derivative is only symbolic. For spinor fields, without the indices expressed, we need to put the δA, δB, etc. in the place where we took the derivative of A, B, etc.

(6-88) can be simplified by writing it as

Summary of how to determine $\delta\mathcal{L}$

$$\delta\mathcal{L} = \mathcal{L}(\delta A,B,C,\ldots) + \mathcal{L}(A,\delta B,C,\ldots) + \mathcal{L}(A,B,\delta C,\ldots) + \ldots \quad (=0 \text{ for symmetry}), \quad (6\text{-}89)$$

as long as we have a separate term for each factor of a given field, such as A, that appears in any given term of $\mathcal{L}$. For example, if we have a term in AA, then $\delta A^2 = \delta(AA) = (\delta A)A + A(\delta A)$.

Do **Problem 6** to show that, using (6-89), the QED Lagrangian of (6-63) under the infinitesimal form of the $U(1)$ transformation set (6-64) has $\delta\mathcal{L} = 0$, and thus, is symmetric.

So, to prove the e/w Lagrangian is symmetric, the fastest, simplest (though still not fast nor simple) way is to find $\delta\mathcal{L}$ under the infinitesimal (local) $SU(2)XU(1)$ transformations of (6-83) to (6-86), and show it is equal to zero. That is, just plug (6-83) to (6-86) into (6-89) for $\mathcal{L}$ of (6-48). If the result is zero, we have shown $\mathcal{L}$ to be symmetric.

This is what we do in Appendix A, pg. 196, that no other text I am aware of carries out in detail. Typically, when developing e/w theory, it is simply just stated that (6-48) is symmetric under a set of infinitesimal (local) $SU(2)XU(1)$ transformations, with perhaps a few terms, such as the Higgs field terms, actually demonstrated to be symmetric.

In Appendix A, we show $\delta\mathcal{L} = 0$

It takes some time and effort to go through every step in Appendix A, and if you wish to simply accept that it is true, I, the author, can empathize with your position. However, if you really want to understand how e/w theory upholds the symmetry that is talked about so much, you will need to digest every step of that appendix. The choice is yours.

As a bit of encouragement, the symmetry under $SU(3)$ color transformations for QCD, when we get to it, is much simpler, I reckon, by an order of magnitude or so.

6.4 Summary of Postulates for High Energy Electroweak Theory

Postulates for high energy e/w interaction theory

The basic postulates from which we derive electroweak theory are these.

1. Free fields in the high energy e/w $\mathcal{L}$ are represented by massless fields having terms similar in form to those in QED.
2. The scalar potential in $\mathcal{L}$ has form $\mathcal{V} = \mu^2\Phi^\dagger\Phi + \lambda\left(\Phi^\dagger\Phi\right)^2 \quad \mu^2 < 0,\ \lambda > 0$.
3. Interaction theory is obtained via minimal substitution in the free field $\mathcal{L}$ plus 4 and 5 below.
4. Postulated extra terms are added to $\mathcal{L}$ to couple the Higgs field with lepton fields.
5. Fields associated with observed gauge bosons of present-day universe are linear combinations of the fields used in minimal substitution.

With these postulates, one finds a symmetric theory whose interactions match experiment.

6.5 A Return to Charges: Weak Hypercharge, Weak Isospin, Electric

Looking more closely at e/w charges

6.5.1 The Simpler View

Recall in QED, the Lagrangian interaction term (we only had one then … how simple!) had form

$$\text{QED} \qquad \mathcal{L}_I = e\bar{\psi}\gamma^\mu\psi A_\mu \qquad \rightarrow \qquad \mathcal{H}_I = -e\bar{\psi}\gamma^\mu\psi A_\mu\ , \tag{6-90}$$

Review of how charge showed up in QED $\mathcal{L}$

where the charge on the electron is $-e$. In other words, we could read the charge on the fermion off as the negative of the factor in front of the Lagrangian term describing the interaction vertex between two electrons and a photon.

In high energy electroweak theory, we can do a similar thing. Consider an e/w interaction term in (6-48) comparable to (6-90) such as the middle term in the last row of (6-52) for $l = e$ and $j = 3$,

Using parallel with QED to deduce e/w charges

$$-\frac{g}{2}\bar{\Psi}^L_e\gamma^\mu\sigma_3\Psi^L_e W_{3\mu} = -\frac{g}{2}\begin{pmatrix}\bar{\psi}^L_{\nu_e} & \bar{\psi}^L_e\end{pmatrix}\gamma^\mu\begin{bmatrix}1 & 0\\ 0 & -1\end{bmatrix}\begin{pmatrix}\psi^L_{\nu_e}\\ \psi^L_e\end{pmatrix}W_{3\mu}$$
$$= -\frac{g}{2}\bar{\psi}^L_{\nu_e}\gamma^\mu\psi^L_{\nu_e}W_{3\mu} + \frac{g}{2}\bar{\psi}^L_e\gamma^\mu\psi^L_e W_{3\mu}. \tag{6-91}$$

Comparing (6-91) to (6-90) we could surmise the <u>weak isospin charge</u> (sometimes just called the "<u>weak charge</u>") on the LC electron is $-g/2$, and on the LC electron neutrino $+g/2$. And we would be correct. In similar fashion, we infer the weak isospin charges and weak hypercharges on all particles, and thus construct Wholeness Chart 6-6 below.

e/w isospin charges on LC e and ν_e deduced

As one more example, consider the RC fields ψ^R_e and $\psi^R_{\nu_e}$. They have no terms of form like (6-91) , so their weak isospin charges are zero. So, where we generalize from the electron e and its neutrino to all 3 families with $l = e, \mu, \tau$,

e/w isospin charges on RC e and ν_e deduced

for physical weak isospin charge gI_3^W,

$$I_3^W = +\tfrac{1}{2} \text{ for } \psi_{\nu_l}^L \qquad I_3^W = -\tfrac{1}{2} \text{ for } \psi_l^L \qquad I_3^W = 0 \text{ for } \psi_l^R \qquad I_3^W = 0 \text{ for } \psi_{\nu_l}^R . \tag{6-92}$$

Summary of e/w isospin charges I_3^W for 3 generations

Note that it is common, as a sort of shorthand notation, to call I_3^W the weak isospin charge, even though the actual charge one would measure with physical instruments would be gI_3^W. Note some authors use T_3 or another symbol for this. The symbol I_3^W is chosen as the "*I*" represents isospin; the "*W*", weak; and the "3", the third Paul matrix σ_3, which is used to determine the value in front of the respective terms in (6-91).

Similar to (6-91), we get other terms in the Lagrangian in terms of g', like

$$\frac{g'}{2}\bar{\psi}_e^L \gamma^\mu \psi_e^L B_\mu , \tag{6-93}$$

which, similar to what we did above, we can compare to (6-90) and deduce a charge associated with LC electrons mediated by the B_μ gauge boson of $-g'/2$. We call this charge the weak hypercharge (sometimes just "hypercharge"), and speak of that charge on the LC electron as equal to $-½$, even though the actual physically measurable weak hypercharge is $-g'/2$. In this context, we use the label Y to represent weak hypercharge, and for the LC electron $Y = -½$.

We can deduce hypercharge Y as we did isospin charge, from terms in $\mathcal{L}$

Similarly, N_Q is commonly deemed the electric charge, even though the actual physically measurable electric charge is $Q = e\,N_Q$. (Some authors use Q for what we consider N_Q to be.)

With foresight, and for our choice of factors in minimal substitution using (6-47), it turns out we find a consistent theory, where local symmetries hold, and we get the correct interactions, if

$$N_Q = I_3^W + Y , \tag{6-94}$$

Relation between electric, isospin, & hypercharges

which is known as the Gell-Mann-Nishijima relation. This just falls out of the theory, as it is composed.

Be aware that some authors use a different convention for the factors in minimal substitution and, for consistency with that convention, need to define $N_Q = I_3^W + ½Y$.

Note that the RC neutrinos have no charge of any kind and so do not interact at high energy via the weak isospin or weak hypercharge interactions. Being leptons, they also do not interact via the QCD color force and have zero color charge.

RC neutrinos have no charge of any kind → don't interact in SM

6.5.2 Plotting Charges

From Wholeness Chart 6-6 one can note that in QED only the antiparticles had opposite electric charge from particles, but for the weak $SU(2)$ interaction, we have particles (neutrinos, specifically) that have opposite weak isospin charge of other particles (electrons). Note also that the antiparticle of a LC electron is a RC positron, and vice versa. The antiparticle of a LC neutrino is a RC antineutrino, and vice versa. All charges have reversed sign in antiparticles compared to particles.

You can check the relation (6-94) to see that it is reflected in Wholeness Chart 6-6.

Wholeness Chart 6-6. Charges: Electric, Weak Isospin, and Hypercharge

Plots of 3 kinds of charges for different particles

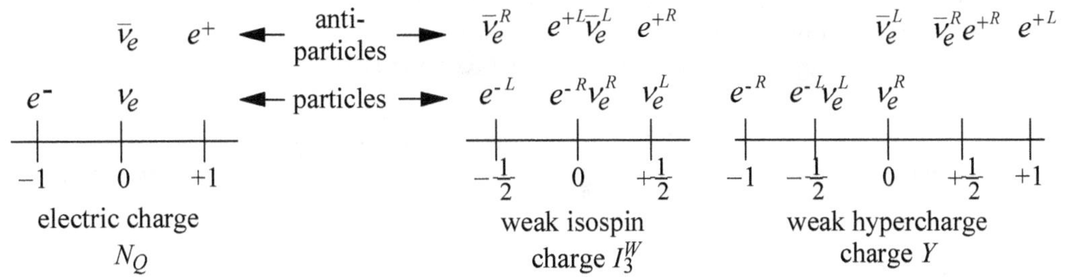

These charges are listed in a table in Wholeness Chart 6-10, pg. 196.

6.5.3 The More Sophisticated, but More Complicated, View

Actually, there is a more elegant (but perhaps less easy to grok) method for determining charge on all particles similar to what we did in QED. (See Vol. 1, pgs. 112-113, 173-176.) That is, using Noether's theorem, we find the conserved, weak 4-current operators (often just called "4-currents") for the relevant fields. Then we integrate the $\mu = 0$ component over all space to get the charge operator. We then operate on a given ket (representing a given particle) with the charge operator and get the eigenvalue, which is the charge of that particle. This more sophisticated method with conserved 4-currents also proves conservation of weak isospin charge and hypercharge at high energy.

We will later deduce 4-currents to find these same charges

We will save further work on e/w 4-currents for a later chapter.

6.5.4 Back to Charges and the Cartan Subalgebra

In Chap. 2 (pg. 47) we stated without proof that the Cartan subalgebra matrix element for $SU(2)$, i.e., σ_3, acting on doublets and singlets had eigenvalues equal to the weak isospin charge in units of $g/2$. (We called it just "weak charge" in Chap. 2 for pedagogic reasons and because that briefer nomenclature is sometimes used.) But this is just what we found in (6-91) above. That is

$$\frac{g}{2}\begin{bmatrix}1 & 0\\ 0 & -1\end{bmatrix}\begin{pmatrix}\psi^L_{\nu_e}\\ \psi^L_e\end{pmatrix} \rightarrow \frac{g}{2}\begin{bmatrix}1 & 0\\ 0 & -1\end{bmatrix}\begin{pmatrix}\psi^L_{\nu_e}\\ 0\end{pmatrix} = \underbrace{\frac{g}{2}}_{\substack{\text{eigenvalue =}\\ \text{weak charge}}}\begin{pmatrix}\psi^L_{\nu_e}\\ 0\end{pmatrix} \qquad \frac{g}{2}\begin{bmatrix}1 & 0\\ 0 & -1\end{bmatrix}\begin{pmatrix}0\\ \psi^L_e\end{pmatrix} = \underbrace{-\frac{g}{2}}_{\substack{\text{eigenvalue =}\\ \text{weak charge}}}\begin{pmatrix}0\\ \psi^L_e\end{pmatrix}. \tag{6-95}$$

Cartan subalgebra method of Chap. 2 yields same e/w charges as above

The eigenvalue for the LC neutrino acted on by the matrix $\frac{g}{2}\sigma_3$ is $\frac{g}{2}$; for the LC electron, $-\frac{g}{2}$. As we stated in Chap. 2, these are the weak (isospin) charges, the same as what we deduced from (6-91). This occurs because, in minimal substitution, we get matrix terms in the first row of (6-91), with the σ_3 matrix sandwiched between LC row and column doublets. When we matrix multiply those matrix terms, we get non-matrix (scalar product) terms, as in the second row of (6-91), which upon comparison with similar terms in QED (as in (6-90)) allow us to deduce the weak charges.

For the $SU(2)$ singlet representation, as we stated in Chap. 2, the generators of the algebra are all zero, so the Cartan subalgebra matrix, for the singlet rep, is zero. Zero acting on any singlet has an eigenvalue of zero, i.e., no weak charge, as we saw in (6-92).

Bottom line: The generators of the Cartan subalgebra for $SU(2)$ acting on weak doublets and singlets yield eigenvalues equal to the weak isospin charge (in units of $g/2$) of the respective fields.

6.6 Typical High Energy Interactions

Let's take a look at how various terms in the interaction Lagrangian (6-48) play a role in determining transition amplitudes and associated Feynman diagrams.

Looking at high energy e/w interactions

Recall that when one converts the Lagrangian to the Hamiltonian via the Legendre transformation (Vol. 1, (2-14), pg. 18), one can then use the interaction Hamiltonian in the interaction picture (IP), along with the Dyson-Wicks expansion, to obtain transition amplitudes. (See Vol. 1, Wholeness Chart 8-4, pgs. 248-251.) In doing so, terms in the integrand of the integral for the transition amplitude arise having factors with fields at particular spacetime points connected by contractions (propagators) to factors with fields at other spacetime points. Each such grouping of fields at the same spacetime point represents, in a Feynman diagram, a vertex connecting incoming and outgoing particles at that point. (See Vol. 1, pg. 250.) Thus, each term in the Lagrangian represents a vertex in a Feynman diagram.

Overview of using $\mathcal{L}$ to find transition amplitudes

So, let's short circuit all those many pages of algebra and calculus for e/w theory and cut straight to the chase. We'll look at terms in the e/w Lagrangian and consider them as Feynman diagram vertices connected by virtual particle propagators.

Each term in $\mathcal{L}$ represents a vertex in Feynman diagram

6.6.1 First Example

As an example, the last two terms in (6-52) are interaction terms, representing the interactions between the LC fermion doublet and, respectively, the W fields and the B fields. These two terms are part of the very first term in the total e/w Lagrangian $\mathcal{L}$ of (6-48). Specifically, they are part of the interaction Lagrangian $\mathcal{L}_I$.

1st example LC doublet terms in $\mathcal{L}$

two (interaction) terms in high energy e/w $\mathcal{L} = -\frac{g}{2}\bar{\Psi}_l^L \not{W}_j \sigma_j \Psi_l^L + \frac{g'}{2}\bar{\Psi}_l^L \not{B} \Psi_l^L$. (6-96)

Let's focus on the term with the W fields for $l = e$ (the electron family)

$$-\frac{g}{2}\bar{\Psi}_e^L \not{W}_j \sigma_j \Psi_e^L = -\frac{g}{2}\begin{pmatrix}\bar{\psi}_{\nu_e}^L & \bar{\psi}_e^L\end{pmatrix}\gamma^\mu W_{j\mu}\sigma_j \begin{pmatrix}\psi_{\nu_e}^L \\ \psi_e^L\end{pmatrix}$$

$$= -\frac{g}{2}\begin{pmatrix}\bar{\psi}_{\nu_e}^L & \bar{\psi}_e^L\end{pmatrix}\gamma^\mu \underbrace{\begin{bmatrix}0 & 1 \\ 1 & 0\end{bmatrix}}_{\sigma_1}\begin{pmatrix}\psi_{\nu_e}^L \\ \psi_e^L\end{pmatrix} W_{1\mu} - \frac{g}{2}\begin{pmatrix}\bar{\psi}_{\nu_e}^L & \bar{\psi}_e^L\end{pmatrix}\gamma^\mu \underbrace{\begin{bmatrix}0 & -i \\ i & 0\end{bmatrix}}_{\sigma_2}\begin{pmatrix}\psi_{\nu_e}^L \\ \psi_e^L\end{pmatrix} W_{2\mu}$$
$$-\frac{g}{2}\begin{pmatrix}\bar{\psi}_{\nu_e}^L & \bar{\psi}_e^L\end{pmatrix}\gamma^\mu \underbrace{\begin{bmatrix}1 & 0 \\ 0 & -1\end{bmatrix}}_{\sigma_3}\begin{pmatrix}\psi_{\nu_e}^L \\ \psi_e^L\end{pmatrix} W_{3\mu} \quad (6\text{-}97)$$

We'll focus on W boson mediated terms

$$= -\frac{g}{2}\bar{\psi}_{\nu_e}^L \gamma^\mu \psi_e^L W_{1\mu} - \frac{g}{2}\bar{\psi}_e^L \gamma^\mu \psi_{\nu_e}^L W_{1\mu} + i\frac{g}{2}\bar{\psi}_{\nu_e}^L \gamma^\mu \psi_e^L W_{2\mu} - i\frac{g}{2}\bar{\psi}_e^L \gamma^\mu \psi_{\nu_e}^L W_{2\mu}$$
$$-\frac{g}{2}\bar{\psi}_{\nu_e}^L \gamma^\mu \psi_{\nu_e}^L W_{3\mu} + \frac{g}{2}\bar{\psi}_e^L \gamma^\mu \psi_e^L W_{3\mu}. \quad (6\text{-}98)$$

And further focus on just two of those terms, each of which represents a vertex

Consider the last two terms in (6-98) as they would manifest, after the Dyson-Wicks expansion, as one of the terms in the S operator (that has a contraction on the $W_{3\mu}$ field) i.e.,

$$\text{one of the terms in } S = S_{W_3}^{(2)} = -\frac{g^2}{4}\int d^4x_1 d^4x_2 N\left\{\left(\bar{\psi}_{\nu_e}^L \not{W}_3 \psi_{\nu_e}^L\right)_{x_1}\left(\bar{\psi}_e^L \not{W}_3 \psi_e^L\right)_{x_2}\right\}. \quad (6\text{-}99)$$

They will give us a (2nd order) term in the S operator

The contraction between the W particles is found in the same way we found propagators in Vol. 1 for scalars, spinors, and photons. We will eventually just extrapolate those results to get a mathematical expression for the weak boson propagators, but for now, we are just trying to get a feeling for what the terms in the Lagrangian represent, and will simply use symbols for the exact expression.

(6-99) can give us 2nd order transition amplitudes (S matrix elements) for more than one interaction type, depending on the initial and final particles involved. One of the possible interactions associated with (6-99) would be for an initial LC electron, an initial RC positron, a final LC neutrino, and a final RC antineutrino. For these, we would have a transition amplitude (an S_{fi} matrix element) of

$$\text{one of the elements in } S_{fi} = \left\langle \nu_e^L \bar{\nu}_e^R \middle| S_{W_3}^{(2)} \middle| e^{-L} e^{+R}\right\rangle \quad \text{for } i = e^{-L} \text{ and } e^{+R} \quad f = \nu_e^L \text{ and } \bar{\nu}_e^R, \quad (6\text{-}100)$$

And that gives rise to different S_{fi}, one of which we examine

which can be represented in a 2nd order Feynman diagram by Fig. 6-3.

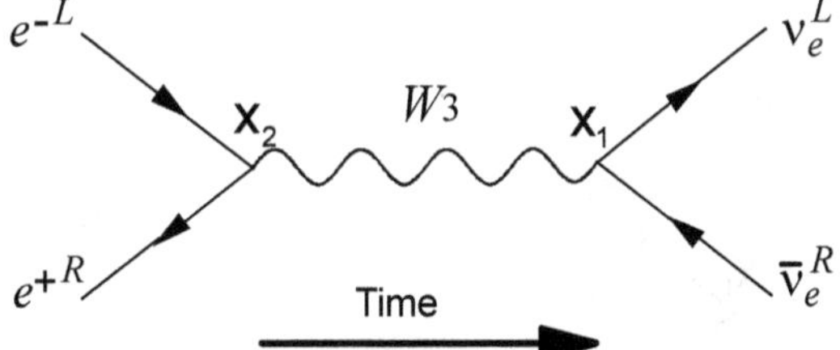

That one has this Feynman diagram

Figure 6-3. One Feynman Diagram from (6-99)

We could, as we did in Chap. 8 of Vol. 1, find the transition amplitude for this interaction by carrying out the lengthy calculations involved in substituting (6-99) into (6-100) and cranking all the pages of concomitant algebra. However, as Feynman did for QED, so he (and others) did for e/w interactions. That is, he found a set of rules, Feynman rules as they are called, for e/w interactions, of which Fig. 6-3 represents one such interaction. We will wait to detail these rules, however, to a later chapter, as they are more relevant to interactions we see today, at low energy levels, than to those at the high energy levels we are working with in this chapter.

We could find S_{fi} via long calculation or via Feynman rules for e/w theory

Feynman rules to be shown in later chapter

All charge types conserved in this example

As an aside, note how electric charge is conserved in this example interaction. And so are both weak isospin charge and hypercharge, as you can check using Wholeness Chart 6-6, pg. 183, and Fig. 6-3, or just by realizing that particles and their antiparticles (which have opposite chirality) have opposite signs on all types of charges. So, we'll have zero total charge initially and finally.

6.6.2 Second Example

2nd example of W boson mediated interaction

Now, consider the first and third terms in (6-98),

$$-\frac{g}{2}\bar{\psi}^L_{\nu_e}\gamma^\mu\psi^L_e W_{1\mu} + i\frac{g}{2}\bar{\psi}^L_{\nu_e}\gamma^\mu\psi^L_e W_{2\mu} = -\frac{g}{2}\bar{\psi}^L_{\nu_e}\gamma^\mu\psi^L_e\left(W_{1\mu} - iW_{2\mu}\right). \tag{6-101}$$

From (6-34), this is equal to

$W_{1\mu}$ and $W_{2\mu}$ terms combine into one term in W_μ^+

$$= -\frac{g}{\sqrt{2}}\bar{\psi}^L_{\nu_e}\gamma^\mu\psi^L_e\frac{\left(W_{1\mu} - iW_{2\mu}\right)}{\sqrt{2}} = -\frac{g}{\sqrt{2}}\bar{\psi}^L_{\nu_e}\gamma^\mu\psi^L_e W^+_\mu. \tag{6-102}$$

As an aside, I implied originally that W_μ^+ was for low energy, but I did that for pedagogic reasons, so you wouldn't get confused. W_μ^+, as a combination of the $W_{1\mu}$ and $W_{2\mu}$, can be used as a surrogate for them at any energy level, though we will soon see it is most advantageous at low energy levels.

Similarly, the second and fourth terms in (6-98), along with (6-34) give us

Two other $W_{1\mu}$ and $W_{2\mu}$ terms combine into one term in W_μ^-

$$-\frac{g}{2}\bar{\psi}^L_e\gamma^\mu\psi^L_{\nu_e}W_{1\mu} - i\frac{g}{2}\bar{\psi}^L_e\gamma^\mu\psi^L_{\nu_e}W_{2\mu} = -\frac{g}{\sqrt{2}}\bar{\psi}^L_e\gamma^\mu\psi^L_{\nu_e}\frac{\left(W_{1\mu} + iW_{2\mu}\right)}{\sqrt{2}} = -\frac{g}{\sqrt{2}}\bar{\psi}^L_e\gamma^\mu\psi^L_{\nu_e}W^-_\mu. \tag{6-103}$$

Can express whole $\mathcal{L}$ in terms of W_μ^+ & W_μ instead of $W_{1\mu}$ & $W_{2\mu}$

It turns out that all terms in the Lagrangian in $W_{1\mu}$ and $W_{2\mu}$ can be converted into terms in W_μ^+ and W_μ^-, and that the latter are more convenient and useful. Note that W_μ^+ and W_μ^- are complex fields, so unlike the $W_{i\mu}$, which are real, they can carry electric charge. The W_μ^+ has positive electric charge, the W_μ^-, negative, as will become more apparent as we progress below.

(6-102) and (6-103) are two terms in the Lagrangian, which, in the Dyson-Wicks expansion will give us several terms in the S operator, one of which is

One of the resulting terms in S operator

$$\text{one of the terms in } S = S^{(2)}_{\substack{W^+W^-\\ e\&\nu}} = -\frac{g^2}{2}\int d^4x_1 d^4x_2 N\left\{\left(\bar{\psi}^L_e \underline{W^- \psi^L_{\nu_e}\right)_{x_1}\left(\bar{\psi}^L_{\nu_e}W^+}\psi^L_e\right)_{x_2}\right\}. \tag{6-104}$$

(6-104), like (6-99), can give us 2nd order transition amplitudes (S matrix elements), for more than one interaction type, depending on the initial and final particles involved. One of these, for example, is

And that gives rise to different S_{fi}, one of which is this

$$\text{one of the elements in } S_{fi} = \left\langle e^{-L}\bar{\nu}^R_e\nu^L_e\right|S^{(2)}_{\substack{W^+W^-\\ e\&\nu}}\left|e^{-L}\right\rangle \quad \text{for } i = e^{-L} \quad f = \nu^L_e,\, \bar{\nu}^R_e, \text{ and } e^{-L}, \tag{6-105}$$

which can be represented in a Feynman diagram by Fig. 6-4.

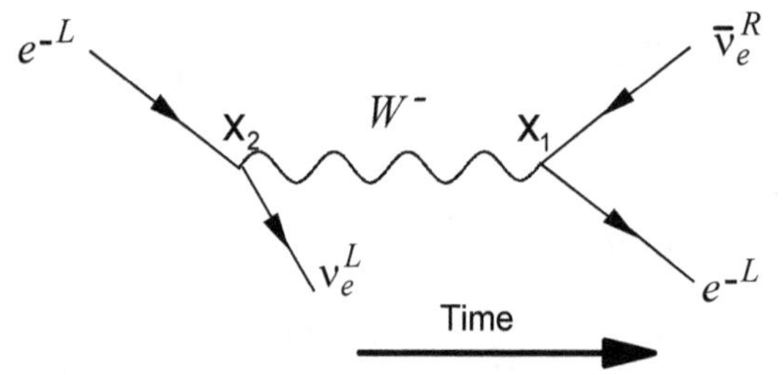

That one has this Feynman diagram

Figure 6-4. One Feynman Diagram from (6-105)

And again, all charge types conserved

Here we have an initial LC electron emitting a neutrino and an antineutrino.

Note again, the conservation of all types of charge. Electric charge should be obvious. Weak isospin charge I_3^W is –½ for the initial LC electron. The final particles have isospin charges of +½ (LC neutrino), –½ (LC electron), and –½ (RC antineutrino), for a net of –½, initially and finally. Weak hypercharge Y is –½ initially and finally, –½ –½ + ½ = –½.

We can surmise, assuming charge conservation at each vertex, that the W^- boson has $N_Q = -1$, $I_3^W = -1$ and $Y = 0$. In Fig. 6-3, the W_3 would have zero for all types of charges.

6.6.3 Third Example

Consider now terms like those in (6-96) for another generation of leptons, say, the muon generation. And restrict ourselves again to only the $SU(2)$ terms. Then for $l = \mu$, we have

$$-\frac{g}{2}\bar{\Psi}_l^L W_j \sigma_j \Psi_l^L \xrightarrow[\text{only}]{\text{muon family}} -\frac{g}{2}\bar{\Psi}_\mu^L W_j \sigma_j \Psi_\mu^L . \tag{6-106}$$

3rd example, this time with muons involved

These terms parallel those in (6-97) and (6-98), except that we have μ subscripts instead of e. And these terms (6-106) are in the Lagrangian in addition to those of (6-98).

Via the same steps we took from (6-98) to (6-104), we would then get

$$\text{one of the terms in } S = S^{(2)}_{\substack{W^+W^- \\ \mu\&e}} = -\frac{g^2}{2}\int d^4x_1 d^4x_2 N\left\{\left(\bar{\psi}_e^L W^- \psi_{\nu_e}^L\right)_{x_1}\left(\bar{\psi}_{\nu_\mu}^L W^+ \psi_\mu^L\right)_{x_2}\right\}, \tag{6-107}$$

An S operator term with electron and muon generations involved

and so, where we have an initial muon, we get

$$\text{one of the elements in } S_{fi} = \left\langle e^{-L}\bar{\nu}_e^R \nu_\mu^L \middle| S^{(2)}_{\substack{W^+W^- \\ \mu\&e}} \middle| \mu^{-L}\right\rangle \quad \text{for } i = \mu^{-L} \quad f = \nu_\mu^L, \bar{\nu}_e^R, \text{ and } e^{-L}. \tag{6-108}$$

which gives rise to different S_{fi}, one of which is this

The Feynman diagram for this looks like

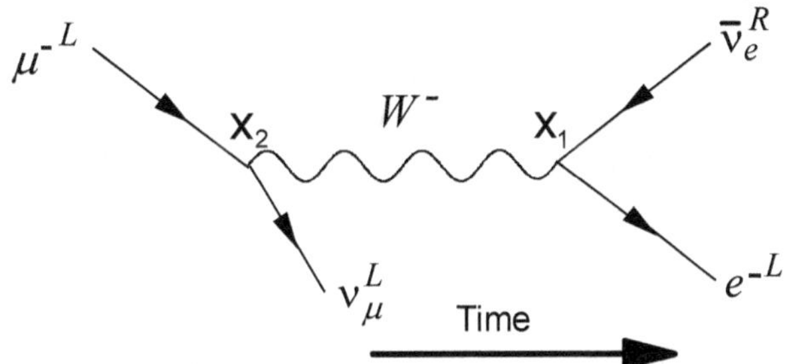

That one has this Feynman diagram

Figure 6-5. One Feynman Diagram from (6-108)

We have the decay, via weak interaction, of a muon into an electron and two neutrinos (one of them actually an anti-neutrino), with charge conservation at every step. Charges here parallel those of Fig. 6-4, since the LC muon has all the same charges as the LC electron.

And again, all charge types conserved

6.6.4 Other Examples

As you may imagine, since there are so many terms in the e/w interaction Lagrangian (as opposed to only one in QED per generation), we have a plethora of different types of vertices, and a plethora of different types of interactions, only a bare three of which have we looked at explicitly above.

Many other examples of many different e/w interactions

In all of these different interactions, both types of charge at high energy, weak isospin and hypercharge, are conserved, and that implies, via (6-94), that electric charge will be as well. We will, however, have a wrinkle to add in to this conclusion, at low energy, when we study symmetry breaking.

All charges will be conserved, at least at high energy

Do **Problem 9** to work out another e/w interaction on your own.

6.6.5 Higgs Coupling Terms

We note that the terms $\mathcal{L}^{LH}$ in (6-48) couple the Higgs (doublet) field Φ to the leptons. At high energy that means we would have a particular vertex in Feynman diagrams for each individual term in $\mathcal{L}^{LH}$. In other words, the Higgs interacts with all of the leptons in a way similar to what we have shown in the foregoing for interactions involving W bosons and various leptons.

We will not explore that in depth herein, though, from what we have seen, one should have some feeling for how that would progress. The real action involving the Higgs takes place after symmetry breaking, at low energy, and that is what we will examine closely in the next chapter.

We will not look at Higgs interactions until the next chapter

but they will behave in similar fashion, with each $\mathcal{L}$ term $\rightarrow$ a vertex

6.6.6 Feynman Rules for Electroweak Theory

As we have already noted, but I wish to emphasize here, we do not have to go through the lengthy and tedious process of evaluating transition amplitudes from first principles. We can, instead, rely on certain established rules (yet to be stated herein), Feynman rules like we had in QED, to cut to the chase and quickly formulate whatever transition amplitude we are interested in. We simply look at the terms in the Lagrangian to see what vertices each represents in a Feynman diagram. We then consider which of these vertices is involved in a particular interaction, deduce the propagators that would connect those vertices, and go to the e/w Feynman rules to construct our transition amplitude S_{fi}.

As noted, S_{fi} found via long calculation or via Feynman rules

This is just the shortcut we learned in QED, but in e/w theory there are far more different types of vertices, and more propagator types. E/w theory is more complicated, but in essence, follows the same logic and procedures as QED.

We will present Feynman rules for the low energy theory in a later chapter. For high energy work, such as one might do when studying the first instants after the Big Bang, one can extrapolate those rules to the high energy case. We will not, however, do that here, as it is peripheral to our main objectives.

Feynman rules shortcut in a later chapter

6.7 Quarks and Electroweak Theory

We will now take a brief look at quarks in the e/w interaction. We will ignore, for the present time, the quark strong interaction.

And now, a brief look at quarks in e/w theory

6.7.1 The Quark Lagrangian

Virtually everything we have said for leptons applies to quarks. They have terms in $\mathcal{L}$ parallel to the terms for leptons.

Can apply what we learned for leptons virtually exactly

The Free High Energy Lagrangian Including Quarks

In the free high energy Lagrangian (6-45), we simply insert the free quark Lagrangian $\mathcal{L}_0^Q$ having the same form as the free lepton Lagrangian $\mathcal{L}_0^L$ as in (6-109) below. The q = 1,2,3 subscript represents LC doublets for the three generations (defined below); the q_u = 1,2,3, the RC singlets for the three generations of up, charm, and top quark; and the q_d = 1,2,3, for the three generations of down, strange, and bottom quarks. Subscripts *u, d, c, s, t, b*, respectively, stand for up, down, charm, strange, top, and bottom quarks.

High energy, free electroweak Lagrangian $\quad \mathcal{L}_0 = \mathcal{L}_0^L + \mathcal{L}_0^Q + \mathcal{L}_0^B + \mathcal{L}_0^H$

$$\mathcal{L}_0^Q = i\left(\bar{\Psi}_q^L \partial\!\!\!/ \Psi_q^L + \bar{\psi}_{q_u}^R \partial\!\!\!/ \psi_{q_u}^R + \bar{\psi}_{q_d}^R \partial\!\!\!/ \psi_{q_d}^R\right) \qquad \text{(quarks, massless)} \tag{6-109}$$

Add quark terms to high energy free $\mathcal{L}$

where

$$\Psi_{q=1}^L = \begin{pmatrix} \psi_u^L \\ \psi_d^L \end{pmatrix} \quad \Psi_{q=2}^L = \begin{pmatrix} \psi_c^L \\ \psi_s^L \end{pmatrix} \quad \Psi_{q=3}^L = \begin{pmatrix} \psi_t^L \\ \psi_b^L \end{pmatrix} \quad \text{LC quark doublets}$$

$$\psi_{q_u=1}^R = \psi_u^R \quad \psi_{q_u=2}^R = \psi_c^R \quad \psi_{q_u=3}^R = \psi_t^R \quad \text{RC quark singlets} \tag{6-110}$$

$$\psi_{q_d=1}^R = \psi_d^R \quad \psi_{q_d=2}^R = \psi_s^R \quad \psi_{q_d=3}^R = \psi_b^R \quad \text{RC quark singlets.}$$

The LC up type and down type quarks, in all three generations, form doublets mirroring the LC lepton doublets, in all three generations. The RC up and down type quarks form singlets mirroring the RC neutrino and e, μ, τ lepton singlets, in all three generations.

Quark doublets and singlets mirror lepton doublets and singlets

The Total (Free plus Interacting) High Energy Lagrangian Including Quarks

Lepton $SU(2)XU(1)$ Interactions

To get the interaction terms between the quarks and the W_i^μ, as well as the B^μ, we do what we did for leptons in getting (6-48) from (6-45). We replace the ordinary derivative in (6-109) with the covariant derivative. The only difference is in the Y values the quarks take on.

$\partial^\mu \rightarrow D^\mu$ for quarks, where

$$\boxed{\begin{array}{l} D^\mu = \partial^\mu + i\frac{g}{2}\sigma_j W_j^\mu + ig'YB^\mu \quad \text{for quarks} \\ Y = \frac{1}{6} \text{ LC quarks}, \ \frac{2}{3}\text{RC } u,c,t, \ -\frac{1}{3}\text{RC } d,s,b \end{array}} \qquad (6\text{-}111)$$

Covariant derivative for quarks mirrors that for leptons, but different Y

Quark-Higgs Interactions

We also need terms in the Lagrangian coupling quarks to the Higgs field like the form shown in (6-48) for $\mathcal{L}^{LH}$. These are labeled $\mathcal{L}^{QH}$ and are shown in (6-112).

All Quark Interaction Terms in the Lagrangian

So the terms we need to add to (6-48) for quarks are

$$\boxed{\begin{array}{ll} \mathcal{L}^Q = i\left(\bar{\Psi}_q^L \not{D} \Psi_q^L + \bar{\psi}_{q_u}^R \not{D} \psi_{q_u}^R + \bar{\psi}_{q_d}^R \not{D} \psi_{q_d}^R\right) & \text{(quarks, massless)} \\ \mathcal{L}^{QH} = -{}_C Y_{qq_d} \bar{\Psi}_q^L \Phi \psi_{q_d}^R - {}_D Y_{qq_u} \bar{\Psi}_q^L \tilde{\Phi} \psi_{q_u}^R + h.c. & \text{(quark-Higgs coupled)} \end{array}} \qquad (6\text{-}112)$$

Terms for quarks to add to high energy total e/w $\mathcal{L}$

where the quarks have their own Yukawa matrices ${}_C Y_{qq_d}$ and ${}_D Y_{qq_u}$ (more on this later), and the total e/w Lagrangian is (6-113). Note the ${}_C Y$ and ${}_D Y$ matrices have nothing to do with hypercharge Y.

$$\boxed{\begin{array}{c} \textbf{High energy, total (free + interacting) e / w Lagrangian, quarks and leptons} \\ \mathcal{L} = \mathcal{L}^L + \mathcal{L}^Q + \mathcal{L}^B + \mathcal{L}^H + \mathcal{L}^{LH} + \mathcal{L}^{QH} \end{array}} \qquad (6\text{-}113)$$

Total high energy e/w $\mathcal{L}$ including leptons and quarks

Quark Isospin Charges and Hypercharges

The quark hypercharges are simply their Y values in (6-111).

Quark hypercharges, like lepton ones, = Y

Their weak isospin charges are found in the same way as we did for leptons. Either by looking at the terms in the Lagrangian and comparing to similar terms in QED, or by operating on the doublets and singlets with the Cartan subalgebra operator (multiplied by $g/2$).

Quark isospin charges deduced like lepton ones, from terms in $\mathcal{L}$

In the first case, since all terms in quarks mirror comparable terms with leptons, those mirrored terms will have the same coefficients in front, and thus, the same charges. That is, the last term in (6-91) will be mirrored by a quark term as in

$$\frac{g}{2}\bar{\psi}_e^L \gamma^\mu \psi_e^L W_{3\mu} \xrightarrow{\text{mirrored by}} \frac{g}{2}\bar{\psi}_d^L \gamma^\mu \psi_d^L W_{3\mu}, \qquad (6\text{-}114)$$

Quark isospin charges mirror leptons

so, the LC down quark will have the same $I_3^W = -\frac{1}{2}$ as the LC electron. And so on, for other quarks.

In the second case, we have the mirror quark doublets, as in

$$\Psi_{q=1}^L = \begin{pmatrix} \psi_u^L \\ \psi_d^L \end{pmatrix} \Leftrightarrow \Psi_e^L = \begin{pmatrix} \psi_{\nu_e}^L \\ \psi_e^L \end{pmatrix} \quad \Psi_{q=2}^L = \begin{pmatrix} \psi_c^L \\ \psi_s^L \end{pmatrix} \Leftrightarrow \Psi_\mu^L = \begin{pmatrix} \psi_{\nu_\mu}^L \\ \psi_\mu^L \end{pmatrix} \quad \Psi_{q=3}^L = \begin{pmatrix} \psi_t^L \\ \psi_b^L \end{pmatrix} \Leftrightarrow \Psi_\tau^L = \begin{pmatrix} \psi_{\nu_\tau}^L \\ \psi_\tau^L \end{pmatrix} \qquad (6\text{-}115)$$

Quark doublets mirror lepton ones

so, operation by the Cartan subalgebra matrix (σ_3 in the doublet rep) will yield the same eigenvalues for LC quarks as (6-95) did for LC leptons. RC quarks, like all RC leptons, are singlets, so have zero as their Cartan generator in the singlet rep, and thus, zero weak isospin eigenvalues, i.e., $I_3^W = 0$.

And same charges found using Cartan subalgebra operation

Do **Problem 10** to confirm quark electric charges, in units of e, are +2/3 for up type quarks and – 1/3 for down type.

6.7.2 Symmetry of Quark Terms in Electroweak Lagrangian

Since (6-112) has the same form as the $\mathcal{L}^L$ and $\mathcal{L}^{LH}$ terms in (6-48), and the latter are symmetric under (6-83) to (6-86), then (6-112) must be symmetric, as well. So, (6-113) is symmetric.

Total e/w $\mathcal{L}$, including quarks, has local SU(2)XU(1) symmetry

6.7.3 Typical Quark Interactions

As noted earlier, due to having precisely the same form, terms in $\mathcal{L}^Q$ of the interaction Lagrangian for quarks and terms in $\mathcal{L}^L$ for leptons will look identical, except for a switch in the subscripts denoting quarks or leptons. Just keep in mind that the LC up type quarks in the quark doublet correspond to LC neutrinos in the lepton double, and the LC down type quarks, to the LC electron, muon, and tau.

We only have to switch lepton subscripts for quark ones in S and S_{fi} lepton derivations

Thus, if we were to follow the steps of (6-106) to (6-108) where, instead of lepton subscripts, we had quark subscripts, we would find, parallel to (6-107),

which, as one example, gives us a term in S like this one

$$\text{one of the terms in } S = S^{(2)}_{\substack{W^+W^- \\ u\&d}} = -\frac{g^2}{2}\int d^4x_1 d^4x_2 N\left\{\left(\bar{\psi}_e^L \not{W}^- \psi_{\nu_e}^L\right)_{x_1}\left(\bar{\psi}_u^L \not{W}^+ \psi_d^L\right)_{x_2}\right\}, \tag{6-116}$$

and parallel to (6-108),

And from that, an S_{fi} like this one

$$\text{one of the elements in } S_{fi} = \left\langle e^{-L}\bar{\nu}_e^R u^L \middle| S^{(2)}_{u\&d} \middle| d^L \right\rangle \quad \text{for } i = d^L \quad f = u^L, \bar{\nu}_e^R, \text{ and } e^{-L}. \tag{6-117}$$

This represents an initial LC down quark transitioning (decaying) into a LC up quark, a LC electron, and a RC electron antineutrino can be represent diagrammatically as Fig. 6-6. During the electroweak epoch, nucleons had yet to form since quarks were so close together, they were essentially free and not bound. They, along with the strong force mediator gluons (which we will study later in the book), formed what is known as a quark-gluon plasma. Within this plasma, a lot of different interactions took place, one of which is that of (6-117) and Fig. 6-6.

With a Feynman diagram like this one

showing a down quark decaying to an up quark along with an electron and an antineutrino

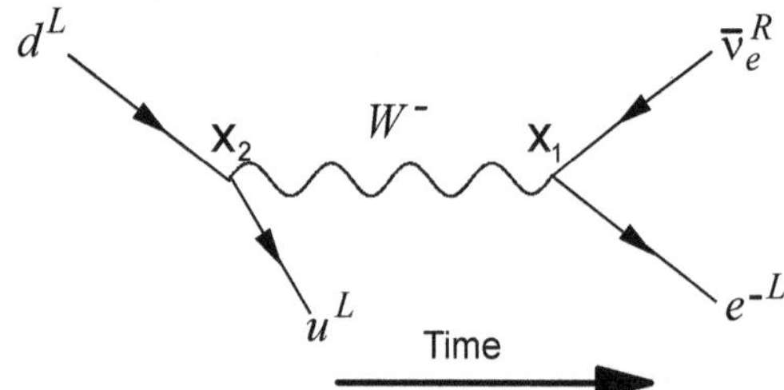

Figure 6-6. Feynman Diagram for Down Quark Decay Via (6-117)

And again, all charges are conserved

Note again that electric charge is conserved at every step. The down quark has –1/3 for electric charge, the W^-, –1, and the up quark +2/3, so the first vertex conserves charge. Similarly so at the second vertex. Weak isospin charge is –½ for the initial down quark. The final particles have isospin charges of +½ (LC up quark), –½ (LC electron), and –½ (RC antineutrino), for a net final isospin charge of –½, i.e., conservation of that charge. If electric charge and weak isospin charge are conserved, then by (6-94) so is hypercharge, though you can check it step-by-step, if you like.

As a foreshadowing of things to come, in low energy theory, the quarks will be bound into nucleons. The neutron, in particular, will be comprised of one up and two down quarks. When one of the down quarks decays, as in Fig. 6-6, to an up quark, we are left with two ups and a down, i.e., a proton. Thus, Fig. 6-6, at low energy, is effectively our Feynman diagram for decay of a neutron into a proton. But for now, we return our focus to the high energy case.

6.8 Other Things to Note

6.8.1 Mass Terms in the Lagrangian

Mass terms destroy the symmetry of $\mathcal{L}$, but we need to show it

All the terms in the e/w $\mathcal{L}$ of (6-48) (with massless fields) are symmetric under the infinitesimal (local) $SU(2)XU(1)$ transformations of (6-83) to (6-86). What would happen if we had an extra massive term?

Let's examine that for the field with the very simplest transformation, the B^μ field, whose mass terms would parallel that of a scalar field ϕ (or A^μ if it had mass, as in Vol. 1, pg. 296, (11-39)) in QED. From (6-85)

Proof B_μ mass term destroys $\mathcal{L}$ symmetry

$$\begin{aligned} m_B^2 B^\mu B_\mu \rightarrow \; & \delta\left(m_B^2 B^\mu B_\mu\right) = m_B^2\left(\delta B^\mu\right)B_\mu + m_B^2 B^\mu\left(\delta B_\mu\right) \\ & = m_B^2\left(-\frac{1}{g'}\partial^\mu f(x)\right)B_\mu + m_B^2 B^\mu\left(-\frac{1}{g'}\partial_\mu f(x)\right) = -2m_B^2 B_\mu \frac{1}{g'}\partial^\mu f(x) \neq 0. \end{aligned} \tag{6-118}$$

This term is not invariant for any non-zero $f(x)$, so if we had a massive B^μ field, our theory would not be symmetric, and thus, according to the principle that Nature seems to adhere to (symmetric theories represent the interactions found in our real physical world), we would have an incorrect theory.

Similar results can be shown for other fields, but as they are more complicated, we will save time by just extrapolating (6-118) to, and drawing conclusions for, those other fields. Interested parties can, however, do Problem 11 and/or see in Appendix B how fermion mass terms are not symmetric under an $SU(2)$ transformation (though they would be under a $U(1)$ transformation, as they were in QED).

Proof fermion mass terms destroy $\mathcal{L}$ symmetry

Do **Problem 11** to show that if the $W_1{}^\mu$ field were massive, $\mathcal{L}$ would not be symmetric.

Proof W boson mass term destroys $\mathcal{L}$ symmetry

Bottom line

The e/w $\mathcal{L}$ is not symmetric under $SU(2)$X$U(1)$ if it contains a mass term for any B^μ, $W_i{}^\mu$, or fermion field.

6.8.2 Being Clear on Commutation and Non-Commutation

Take care with the context in which the term "non-commutation" is used. With regard to any bosonic field in QFT, we have non-commutation between the field (such as ϕ), and its conjugate momentum (such as π_S), as in,

$$\left[\phi_r(\mathbf{x},t),\pi_s(\mathbf{y},t)\right]=i\delta_{rs}\delta(\mathbf{x}-\mathbf{y}) \tag{6-119}$$

and this is true regardless of whether the field is of a $U(1)$, $SU(2)$, $SU(3)$ or other special unitary group theory. That is, (6-119) is a basic postulate of all of QFT for any boson field. (See Vol. 1, pg. 30.) For fermions, we have anti-commutation in place of (6-119).

Be clear on non-commutation for field and its conjugate momentum vs SU(n) non-commutation

However, with regard to the transformations of fields in Hilbert space, there are some transformation types (like those in a $U(1)$ theory such as $e^{i\alpha}$ and $e^{i\beta}$) that commute. There are others (like those in an $SU(2)$ theory such as σ_1 and σ_2) that do not. $U(1)$ theories have non-commutation of fields and their conjugate momenta, but commutation of their transformations in $U(1)$ space. $SU(n)$ theories are non-commutative for both.

6.8.3 Yang-Mills Fields

You will no doubt see the W_i^μ fields referred to as Yang-Mills fields. They are the massless fields associated with an $SU(2)$ theory, whose generators (the Pauli matrices in this case) are non-Abelian (do not commute).

In early 1954, Chen Ning Yang and Robert Mills extended the concept of gauge theory for Abelian groups, e.g. quantum electrodynamics, to non-Abelian groups to try and provide an explanation for strong interactions. The quanta (particle states) of a Yang–Mills field must be massless in order to maintain gauge invariance of the Lagrangian.

In general, any massless set of fields (in $SU(2)$, the set comprises the three fields W_i^μ) in a non-Abelian $SU(n)$ theory that leaves the Lagrangian invariant under an $SU(n)$ transformation is comprised of Yang-Mills fields. The associated theory is called a Yang-Mills theory.

Yang-Mills fields
- *non-Abelian ($SU(n)$)*
- *massless*
- *invariant $\mathcal{L}$*

6.8.4 Non-Abelian Fields and Non-linearity

Non-commuting $SU(n)$ Operators in the Lagrangian

We saw that the non-Abelian group $SU(2)$ gives rise to extra terms in the interaction Lagrangian (see (6-48) with (6-60)), with more than two of the same interaction type field (the $W_i{}^\mu$ fields) as factors in those terms. That is, the terms are greater than bilinear in the fields. Abelian groups, like $U(1)$, do not have such terms in $\mathcal{L}$. See QED theory (e/m $U(1)$) and (6-59) of hypercharge $U(1)$ theory.

We can see, as an example, from our evaluation of $G_i^{\mu\nu}$ beginning with (6-53), if we did not have the σ_j factor in the covariant derivative, we would have extra terms at the end of that line with a factor of $W_i^\nu W_i^\mu - W_i^\mu W_i^\nu = W_i^\mu W_i^\nu - W_i^\mu W_i^\nu = 0$. The non-Abelian generators σ_i gave us the extra non-zero terms of (6-53).

The same conclusion is reached for $SU(3)$, and in fact for any $SU(n)$ theory. Such non-Abelian theories contain generators of a given interaction type that do not commute. And these give rise, after minimal substitution, to self-interaction terms (that are greater than bilinear in $\mathcal{L}$).

Non-commuting generators give rise to $\mathcal{L}$ terms with more than two factors of same type boson fields

Non-commuting $SU(n)$ Operators in $\mathcal{L}$ and Non-Linear QFT Theories

So, they give rise to non-linear field equations

When we substitute the Lagrangian $\mathcal{L}$ into the Euler-Lagrange equation, we take derivatives of $\mathcal{L}$ with respect to the field and with respect to the derivatives of the field. With only bilinear terms in $\mathcal{L}$, this results in a field equation that is linear in the field. For trilinear, quadrilinear, etc. terms, it results in a field equation that is non-linear in the field i.e., it has at least one term with more than one factor of the field in it. In a non-linear equation, the dependent variable (field for us) affects itself, i.e., it is self-interacting.

Non-linearity means self-interacting

Quantum chromodynamics (QCD) turns out to be an $SU(3)$ theory, with eight (non-commuting) matrices in a 3D space as generators, parallel to the three Pauli matrices in electroweak $SU(2)$. Thus, QCD has eight different gauge fields comparable to the three W_i^μ fields of $SU(2)$ electroweak theory. And thus, QCD has similar extra terms arising in the QCD $\mathcal{L}$ that are not bilinear in the QCD gauge fields. So, QCD is a non-Abelian, and non-linear, field theory, as well.

The bottom line: Non-Abelian field theories are non-linear, or self-interacting, theories. Abelian field theories are linear, and not self-interacting. Yang-Mills theories are non-linear theories. See Wholeness Chart 6-7 for a summary overview.

Wholeness Chart 6-7. Non-Abelian Theories Have Non-Linear Field Equations

Summary: Abelian vs non-Abelian theories

$U(1)$ Abelian	**$SU(n)$ Non-Abelian**
1 field, e.g., B^μ in high energy e/w theory	>1 field, e.g., 3 W_i^μ in high energy e/w theory
↓	↓
Identity matrix is only generator, commutes with itself	Non-commuting generator matrices
↓	↓
Minimal substitution	Minimal substitution
↓	↓
Only bilinear boson field terms in $\mathcal{L}$	Trilinear and/or higher boson field terms in $\mathcal{L}$
↓	↓
Linear field equation e.g., terms only in B^μ	Non-linear field equation e.g., term(s) in $W_i^\mu W_j^\mu$

Caveat

The above is the usual way the nonlinear issue is presented. However, note that in QED interaction theory we have a Lagrangian term $e\bar{\psi}A_\mu\gamma^\mu\psi$, which is trilinear in fields and gives rise to coupled field equations that are nonlinear, as shown in Vol. 1, (7-18) and (7-19), pg. 186, repeated below.

$$\partial^\alpha\partial_\alpha A^\mu = -e\bar{\psi}\gamma^\mu\psi \qquad \text{and} \qquad \left(i\gamma^\mu\partial_\mu - m\right)\psi = -e\gamma^\mu A_\mu\psi\,. \tag{6-120}$$

So, in effect the $U(1)$ QED interaction theory (that lacks non-Abelian generators) actually is nonlinear. But, the free QED theory (photons alone and fermions alone, non-interacting) is linear. (Take the RHS of each above equation as zero.) In $SU(n)$ theories on the other hand, the bosons alone field equations are nonlinear. Their Lagrangian terms are greater than bilinear in the boson fields, and so the bosons only field equations will have terms with two or more factors of those boson fields in them.

$U(1)$ linearity refers to photons not interacting with fermions, and not self-interacting

Hence, when you hear talk about Abelian theories as linear and non-Abelian as nonlinear, it is with reference to the bosons alone field equations.

Note the subtlety in nomenclature. We use the term "free" in QFT to refer to the field equations and Lagrangian *before* minimal substitution, where the field equations are indeed always linear. The term "bosons only" or "bosons alone" refers to boson field equations and boson parts of the Lagrangian without coupling to fermions, but *after* minimal substitution, where the associated field equations for $SU(n)$ theories, with $n > 1$, are nonlinear.

$SU(n)$ nonlinearity refers to bosons not interacting with fermions, yet self-interacting

6.8.5 SU(2) and U(1) Symmetries Hold Independently and are Unique

SU(2) and U(1) symmetries hold separately

Note that because each of the $\omega_i(x)$ and $f(x)$ in transformations (6-83) to (6-86) can be varied independently of each other, each of the $SU(2)$ and $U(1)$ transformation sets acting alone leaves the Lagrangian invariant. For example, consider $f = 0$. The Lagrangian is still symmetric under the $SU(2)$ transformation provided by non-zero values for the ω_i. That is, it is invariant under just the $SU(2)$ transformation part of (6-83) to (6-86), ignoring the $U(1)$ part. The same logic works in reverse. $\mathcal{L}$ is symmetric for those transformations of the fields where $f \neq 0$ and $\omega_i = 0$. So, $\mathcal{L}$ is invariant under just the $U(1)$ transformation without the $SU(2)$ transformation.

The $SU(2)$ and $U(1)$ transformations hold independently.

Further, they appear unique. There are no other known transformations that uphold the symmetry.

and they are unique

6.8.6 Insight into the Term "Gauge Symmetry"

Note what we get when we transform ("rotate" in complex 2D space) a doublet in $SU(2)$.

$$\begin{pmatrix} \psi'^{L}_{\nu_e} \\ \psi'^{L}_{e} \end{pmatrix} = e^{i\omega_i(x)\sigma_i/2} \begin{pmatrix} \psi^{L}_{\nu_e} \\ \psi^{L}_{e} \end{pmatrix} = \begin{bmatrix} A & B \\ C & D \end{bmatrix} \begin{pmatrix} \psi^{L}_{\nu_e} \\ \psi^{L}_{e} \end{pmatrix} = \begin{pmatrix} A\psi^{L}_{\nu_e} + B\psi^{L}_{e} \\ C\psi^{L}_{\nu_e} + D\psi^{L}_{e} \end{pmatrix}. \tag{6-121}$$

We could build a viable theory by transforming ("rotating") to another basis for our fields

but we choose easiest to work in

Our new (primed) fields $\psi'^{L}_{\nu_e}$ and ψ'^{L}_{e} are linear combinations of our original (unprimed) fields $\psi^{L}_{\nu_e}$ and ψ^{L}_{e}. If we did this throughout our Lagrangian, as part of a set of transformations on the fields in the Lagrangian, we would get a theory in terms of these new fields, not the original ones. And this theory would work because it would still give us correct measurable quantities in experiments (like cross sections, etc.). The theory with the new fields would be a whole lot more complicated, but ultimately would make the same predictions.

This is what we mean by different gauges – same measured quantities, different fields

Recall the fields themselves are not measurable. They are behind the scenes, as it were. So, the things we can measure remain unchanged, but things we cannot (like the fields themselves) change. This is a characteristic of a gauge theory. The gauge (i.e., field) changes, but the experiments do not. This is what is meant when people talk of QFT as being a gauge symmetric theory.

As an aside, note that the primed fields do not create and destroy particles in eigenstates of electric charge since each is a linear combination of fields with zero (neutrino) and -1 (electron) charges on their respective particles. You can begin to imagine how complicated using the primed fields would make our theory. So, we choose to build our theory on fields that create and destroy particles in charge eigenstates. If you think QFT is hard as it is, imagine what it would be like learning it if we used the primed fields in (6-121) instead of the unprimed ones.

6.8.7 A Note on Boson Transformations

Transformations of W_i^{μ} and B^{μ} not truly U(1)

I do not recall seeing this point expressed elsewhere, but note that (6-71) and (6-72) are not actually $U(1)$ transformations for W_i^{μ} and B^{μ}, respectively. They are listed under that heading herein, and the entire symmetry transformation set (6-66) to (6-73) (and its infinitesimal expressions) is commonly referred to as $SU(2)$X$U(1)$. But, since the transformations (6-71) and (6-72) do not satisfy $U^{\dagger}U = I$, this is, strictly speaking, incorrect.

However, entire transformation set called SU(2)XU(1)

However, (6-71) and (6-72) are Lie group transformations and are dependent on the same Lie group parameters f and ω_i as the $U(1)$ and $SU(2)$ transformations on fermions. This has given rise to the common practice of simply deeming the entire transformation set (6-66) to (6-73) as $SU(2)$X$U(1)$.

6.8.8 Higgs Interpreted as 5th Force

Higgs interaction can be interpreted as a 5th force

Note that the $\mathcal{L}^{LH}$ of (6-48) comprises terms coupling two lepton fields with the Higgs field. And thus, each of those terms represents a vertex in a Feynman diagram, similar in ways to the vertex of QED theory. Here, the Higgs boson parallels the role played there by the photon (boson). And in parallel fashion, one could consider the Higgs to be the mediator of a particular force, just as the photon is the mediator of the e/m force. In essence, although you won't hear it mentioned much, one could interpret this as a separate interaction type beyond the usual three of the SM plus gravity, with the coupling constants being the components of the flavor matrices ${}_AY_{ll'}$ and ${}_BY_{l\nu_{l'}}$. That is, the Higgs interaction could be considered a fifth force. We will have a bit more to say about this in Chap. 11.

6.8.9 Right Chiral Neutrinos May Not Exist

Since RC neutrinos don't interact via SM forces, we don't know if they even exist

RC neutrinos have no electric, weak, or hypercharge. Since they are leptons, we know they do not interact via the strong force, so also have zero strong force charge. Hence, they simply do not interact with other particles via any of the three standard model forces. Neutrino masses are so small, and gravitational coupling is so weak, that it is virtually impossible to detect whether they interact gravitationally. Since we have no way to detect RC neutrinos directly, we really don't know if they even exist. It is certainly possible there are no such animals.

However, since 1998, experiments have detected neutrino oscillations, about which you have probably heard, and these lend strong indirect support to the existence of RC neutrinos. We will study how and why this is so in a later chapter.

6.9 Chapter Summary

6.9.1 History and Background of the Electroweak Theory

Historically, two early models of the electroweak interactions, the Goldstone and Higgs models, paved the way for the successful Glashow-Salaam-Weinberg (GSW) model. In the GSW model, the Higgs scalar field is represented by an $SU(2)$ doublet Φ, and the Higgs potential $\mathcal{V}$ looks, graphically, like a Mexican hat (though with $\mathcal{V}$ plotted versus four (real) variables ϕ_i.) The plot one typically sees of $\mathcal{V}$ vs fields ϕ_i represents, in actuality, a plot of the expectation value of $\mathcal{V}$ vs particle number densities (more correctly, the square root of these densities).

Additionally, relations discovered by Steven Weinberg between the low energy photon & Z^μ weak mediator field, and the high energy weak mediator W_3 & hypercharge mediator B^μ fields play a critical role in relating theoretical results to experiments in the present-day universe. See (6-33).

Low energy e/w theory (during and after symmetry breaking) will be covered in the next chapter.

6.9.2 Finding Solutions in Principle and in Practice

Note that one can use $\mathcal{L}$ to find the interacting field equations, which are a set of coupled differential equations in several unknowns (one for each component of each field) and which are further complicated by being non-linear. In principle, one simply solves this set of equations to find the mathematical description of each field, interacting with all the others, as time evolves. In practice, this is essentially impossible.

The answer in QED and e/w theory is perturbation theory, using the interaction picture and the interaction Lagrangian, as summarized in Part 4 of Wholeness Chart 6-8, and as we have been doing in Vol. 1 and the present volume.

6.9.3 Development of the Theory

Wholeness Chart 6-8 summarizes the key points of this chapter.

Wholeness Chart 6-8. Overview of the Development of High Energy GSW Theory

Step	**Relevant Points & Relations**	**Ref.**
Approach	1) Start with free, massless Lagrangian $\mathcal{L}_0$ with Higgs potential $\mathcal{V}$ of Mexican hat form, Higgs field as a doublet, weak isospin fields W_i^μ, weak hypercharge fields B^μ, lepton fields, and quark fields, 2) apply minimal substitution in $\mathcal{L}_0$ with $SU(2)XU(1)$ generators plus add terms coupling Higgs to fermions to get total e/w $\mathcal{L}$, 3) show result is symmetric under local $SU(2)XU(1)$ transformations, 4) using $\mathcal{L}$, find S_{fi}	
1) $\mathcal{L}_0$	$\mathcal{L}_0 = \mathcal{L}_0^L + \mathcal{L}_0^Q + \mathcal{L}_0^B + \mathcal{L}_0^H$ $\mathcal{L}_0^L = i\left(\bar{\Psi}_l^L \partial\!\!\!/ \Psi_l^L + \bar{\psi}_l^R \partial\!\!\!/ \psi_l^R + \bar{\psi}_{\nu_l}^R \partial\!\!\!/ \psi_{\nu_l}^R\right)$ (leptons, massless) $\mathcal{L}_0^Q = i\left(\bar{\Psi}_q^L \partial\!\!\!/ \Psi_q^L + \bar{\psi}_{q_u}^R \partial\!\!\!/ \psi_{q_u}^R + \bar{\psi}_{q_d}^R \partial\!\!\!/ \psi_{q_d}^R\right)$ (quarks, massless) $\mathcal{L}_0^B = -\frac{1}{4} W_i^{\mu\nu} W_{i\,\mu\nu} - \frac{1}{4} B^{\mu\nu} B_{\mu\nu}$ (gauge bosons, massless) $\mathcal{L}_0^H = \left(\partial^\mu \Phi\right)^\dagger \left(\partial_\mu \Phi\right) \underbrace{- \mu^2 \Phi^\dagger \Phi - \lambda\left(\Phi^\dagger \Phi\right)^2}_{-\mathcal{V}}$ $\mu^2 < 0$ $\lambda > 0$ (Higgs boson) $W_i^{\mu\nu} = \partial^\nu W_i^\mu - \partial^\mu W_i^\nu$ $B^{\mu\nu} = \partial^\nu B^\nu - \partial^\mu B^\nu$.	(6-45) & (6-109)
2a) $\partial^\mu \rightarrow D^\mu$	$D^\mu = \partial^\mu + i\frac{g}{2}\sigma_j W_j^\mu + ig'YB^\mu$ for leptons, quarks, B boson, Higgs Y differs for different leptons $g \rightarrow -g$ for W bosons	(6-47) & (6-111)
2b) Higgs-fermion coupling	$\mathcal{L}^{LH} + \mathcal{L}^{QH}$ added in below	(6-48) & (6-112)
2) $\mathcal{L}$ (e/w)	$\mathcal{L}^L = i\left(\bar{\Psi}_l^L D\!\!\!\!/ \Psi_l^L + \bar{\psi}_l^R D\!\!\!\!/ \psi_l^R + \bar{\psi}_{\nu_l}^R D\!\!\!\!/ \psi_{\nu_l}^R\right)$ (leptons, massless) $\mathcal{L}^Q = i\left(\bar{\Psi}_q^L D\!\!\!\!/ \Psi_q^L + \bar{\psi}_{q_u}^R D\!\!\!\!/ \psi_{q_u}^R + \bar{\psi}_{q_d}^R D\!\!\!\!/ \psi_{q_d}^R\right)$ (quarks, massless) $\mathcal{L}^B = -\frac{1}{4} G_i^{\mu\nu} G_{i\,\mu\nu} - \frac{1}{4} B^{\mu\nu} B_{\mu\nu}$ (gauge bosons, massless) $\mathcal{L}^H = \left(D^\mu \Phi\right)^\dagger \left(D_\mu \Phi\right) - \mu^2 \Phi^\dagger \Phi - \lambda\left(\Phi^\dagger \Phi\right)^2$ (Higgs boson) $\mathcal{L}^{LH} = -{}_A Y_{ll'}\left(\bar{\Psi}_l^L \psi_{l'}^R \Phi + \Phi^\dagger \bar{\psi}_{l'}^R \Psi_l^L\right) - {}_B Y_{l\nu_{l'}}\left(\bar{\Psi}_l^L \psi_{\nu_{l'}}^R \tilde{\Phi} + \tilde{\Phi}^\dagger \bar{\psi}_{\nu_{l'}}^R \Psi_l^L\right)$ (lepton-Higgs coupled) $\mathcal{L}^{QH} = -{}_C Y_{qq_d}\left(\bar{\Psi}_q^L \psi_{q_d}^R \Phi + \Phi^\dagger \bar{\psi}_{q_d}^R \Psi_q^L\right) - {}_D Y_{qq_u}\left(\bar{\Psi}_q^L \psi_{q_u}^R \tilde{\Phi} + \tilde{\Phi}^\dagger \bar{\psi}_{q_u}^R \Psi_q^L\right)$ (quark-Higgs coupled) $-\frac{1}{4} G_i^{\mu\nu} G_{i\,\mu\nu} = -\frac{1}{4} W_i^{\mu\nu} W_{i\,\mu\nu} + g\varepsilon_{ijk} W_{i\mu} W_{j\nu} \partial^\mu W_k^\nu - \frac{1}{4} g^2 \varepsilon_{ijk}\varepsilon_{ilm} W_j^\mu W_k^\nu W_{l\mu} W_{m\nu}$	(6-48), (6-112) & (6-60)
3) $\mathcal{L}$ symmetry	Local, infinitesimal $SU(2)XU(1)$ transformations shown pg. 181 Appendix A shows $\delta\mathcal{L} = 0$ under local $SU(2)XU(1)$ group transformation for massless fields Guiding principle of Nature: Demanding $\mathcal{L}$ locally symmetric means interaction terms correct	(6-83) to (6-86)
4) Finding S_{fi}	Use Interaction Picture Two Ways i) $\mathcal{L}_I \rightarrow \mathcal{H}_I \rightarrow S$ operator $\rightarrow$ Dyson-Wicks expansion $\rightarrow \langle f \| S \| i \rangle = S_{fi} \rightarrow$ evaluate integrals ii) Use Feynman diagrams and rules, where each term in $\mathcal{L}_I$ is a vertex ii) is simpler, and we'll do in a later chapter	
Notes	Weak isospin charge, hypercharge, and electric charge conserved. Non-abelian theories ($SU(n)$ with $n > 1$) have terms in $\mathcal{L}$ with more than two of the same boson mediator factors and thus, have non-linear free field equations.	

6.9.4 Masses and Symmetry

It is not necessary to remember every detail of Wholeness Chart 6-9. Just the simple idea that the e/w Lagrangian would not be symmetric if it contained any mass terms for any field.

Wholeness Chart 6-9. Mass Terms in High Energy Electroweak $\mathcal{L}$ and Symmetry

	High Energy Symmetry	
Mass Term Field Type	***U*(1)**	***SU*(2)**
Bosons		
Real Spin 1 W_i^μ	No	No
Real Spin 1 B^μ	No	No
Fermions		
Leptons (3 generations)	Yes	No
Quarks (3 generations)	Yes	No

6.9.5 Summary of Fermion Charges

Wholeness Chart 6-10 collates all the information we have developed regarding the different types of charges.

Wholeness Chart 6-10. Various Charges for Elementary Particles ($N_Q = I_3^W + Y$)

Fermion Type	Left-chiral Fermions	Electric Charge N_Q	Weak Isospin I_3^W	Weak Hyper-charge Y	Right-chiral Fermions	Electric Charge N_Q	Weak Isospin I_3^W	Weak Hyper-charge Y
Leptons	$\nu_e^L, \nu_\mu^L, \nu_\tau^L$	0	+ 1/2	– 1/2	$\nu_e^R, \nu_\mu^R, \nu_\tau^R$	0	0	0
	$e^{-L}, \mu^{-L}, \tau^{-L}$	– 1	– 1/2	– 1/2	$e^{-R}, \mu^{-R}, \tau^{-R}$	– 1	0	– 1
Quarks	u^L, c^L, t^L	+ 2/3	+ 1/2	+ 1/6	u^R, c^R, t^R	+ 2/3	0	+ 2/3
	d^L, s^L, b^L	– 1/3	– 1/2	+ 1/6	d^R, s^R, b^R	– 1/3	0	– 1/3

6.10 Appendices

Appendix A. Showing the Symmetry of the High Energy Lagrangian

6.10.1 Expanded Form of High Energy Lagrangian

To show the invariance of the high energy Lagrangian under the transformation set of (6-83) to (6-86), pg. 181, we need to work with the expanded version of the Lagrangian of (6-48) including the terms shown in (6-52) to (6-62). We repeat these below, for easy reference.

For $\mathcal{L}^L$

$$\mathcal{L}^L = i\bar{\Psi}_l^L \not{D}\Psi_l^L + i\bar{\psi}_l^R \not{D}\psi_l^R + i\bar{\psi}_{\nu_l}^R \not{D}\psi_{\nu_l}^R$$

$$\begin{aligned} i\bar{\Psi}_l^L \not{D}\Psi_l^L &= i\bar{\Psi}_l^L \left(\not{\partial} + \tfrac{i}{2}g\sigma_j \not{W}_j - \tfrac{i}{2}g'\not{B}\right)\Psi_l^L = i\bar{\Psi}_l^L \not{\partial}\Psi_l^L - \frac{g}{2}\bar{\Psi}_l^L \sigma_j \not{W}_j \Psi_l^L + \frac{g'}{2}\bar{\Psi}_l^L \not{B}\Psi_l^L \\ i\bar{\psi}_l^R \not{D}\psi_l^R &= i\bar{\psi}_l^R\left(\not{\partial} - ig'\not{B}\right)\psi_l^R = i\bar{\psi}_l^R \not{\partial}\psi_l^R + g'\bar{\psi}_l^R \not{B}\psi_l^R \\ i\bar{\psi}_{\nu_l}^R \not{D}\psi_{\nu_l}^R &= i\bar{\psi}_{\nu_l}^R \not{\partial}\psi_{\nu_l}^R \end{aligned} \qquad (6\text{-}122)$$

$$\mathcal{L}^L = i\bar{\Psi}_l^L \not{\partial}\Psi_l^L + i\bar{\psi}_l^R \not{\partial}\psi_l^R + i\bar{\psi}_{\nu_l}^R \not{\partial}\psi_{\nu_l}^R - \frac{g}{2}\bar{\Psi}_l^L \sigma_j \not{W}_j \Psi_l^L + \frac{g'}{2}\bar{\Psi}_l^L \not{B}\Psi_l^L + g'\bar{\psi}_l^R \not{B}\psi_l^R$$

For $\mathcal{L}^B$

$$\mathcal{L}^B = -\tfrac{1}{4}B^{\mu\nu}B_{\mu\nu} - \tfrac{1}{4}G_i^{\mu\nu}G_{i\,\mu\nu}$$

$$B^{\mu\nu} = \partial^\nu B^\mu - \partial^\mu B^\nu \qquad G_i^{\mu\nu} = \partial^\nu W_i^\mu - \partial^\mu W_i^\nu \underbrace{+ g e_{ijk} W_j^\mu W_k^\nu}_{\substack{= g e_{ijk} W_k^\nu W_j^\mu \\ = -g e_{ijk} W_j^\nu W_k^\mu}} \tag{6-123}$$

For $\mathcal{L}^H$

$$\mathcal{L}^H = \left(D^\mu \Phi\right)^\dagger \left(D_\mu \Phi\right) - \mu^2 \Phi^\dagger \Phi - \lambda \left(\Phi^\dagger \Phi\right)^2$$
$$D_\mu \Phi = \left(\partial_\mu + \tfrac{i}{2} g \sigma_k W_{k\,\mu} + i\tfrac{g'}{2} B_\mu\right)\Phi \tag{6-124}$$

For $\mathcal{L}^{LH}$

$$\mathcal{L}^{LH} = -{}_A Y_{ll'} \bar{\Psi}_l^L \psi_{l'}^R \Phi - {}_A Y_{ll'}^\dagger \Phi^\dagger \bar{\psi}_{l'}^R \Psi_l^L - {}_B Y_{l\nu_{l'}} \bar{\Psi}_l^L \psi_{\nu_{l'}}^R \tilde{\Phi} - {}_B Y_{l\nu_l'}^\dagger \tilde{\Phi}^\dagger \bar{\psi}_{\nu_{l'}}^R \Psi_l^L$$
$$\tilde{\Phi} = \frac{1}{\sqrt{2}}\begin{pmatrix} \phi_b^* \\ -\phi_a^* \end{pmatrix} = \frac{1}{\sqrt{2}}\begin{pmatrix} \phi_3 - i\phi_4 \\ -\phi_1 + i\phi_2 \end{pmatrix} \tag{6-125}$$

Using the above (6-122) to (6-125), as in (6-48),

$$\mathcal{L} = \mathcal{L}^L + \mathcal{L}^B + \mathcal{L}^H + \mathcal{L}^{LH}\,. \tag{6-126}$$

6.10.2 The Approach

As discussed in Sect. 6.3.12, pg. 181, under any infinitesimal transformation of the underlying fields, we have

$$\mathcal{L}' = \mathcal{L} + \delta\mathcal{L}\,, \tag{6-127}$$

where $\delta\mathcal{L}$ is the variation in the Lagrangian due to the variation in the underlying fields. If $\delta\mathcal{L} = 0$, then $\mathcal{L}' = \mathcal{L}$, and the transformation is symmetric. So, all we need to do is show $\delta\mathcal{L} = 0$ for (6-126). We will separately examine each of the terms in (6-126) to demonstrate that this does, indeed, happen.

6.10.3 Showing the Symmetry Term by Term in the High Energy Lagrangian

For $\mathcal{L}^L$

From the last row of (6-122),

$$\begin{aligned}\mathcal{L}'^{\,L} &= \mathcal{L}^L + \delta\mathcal{L}^L \\ &= i\left(\bar{\Psi}_l^L + \delta\bar{\Psi}_l^L\right)\not{\partial}\left(\Psi_l^L + \delta\Psi_l^L\right) + i\left(\bar{\psi}_l^R + \delta\bar{\psi}_l^R\right)\not{\partial}\left(\psi_l^R + \delta\psi_l^R\right) + i\left(\bar{\psi}_{\nu_l}^R + \delta\bar{\psi}_{\nu_l}^R\right)\not{\partial}\left(\psi_{\nu_l}^R + \delta\psi_{\nu_l}^R\right) \\ &\quad - \tfrac{g}{2}\left(\bar{\Psi}_l^L + \delta\bar{\Psi}_l^L\right)\sigma_j\left(\not{W}_j + \delta\not{W}_j\right)\left(\Psi_l^L + \delta\Psi_l^L\right) + \tfrac{g'}{2}\left(\bar{\Psi}_l^L + \delta\bar{\Psi}_l^L\right)\left(\not{B} + \delta\not{B}\right)\left(\Psi_l^L + \delta\Psi_l^L\right) \\ &\quad + g'\left(\bar{\psi}_l^R + \delta\bar{\psi}_l^R\right)\left(\not{B} + \delta\not{B}\right)\left(\psi_l^R + \delta\psi_l^R\right),\end{aligned} \tag{6-128}$$

where for part of our chosen transformation set, $\delta\psi_{\nu_l}^R = \delta\bar{\psi}_{\nu_l}^R = 0$. Because the variation δ is infinitesimal, we can ignore all orders in δ higher than the first. Thus,

$$\begin{aligned}\delta\mathcal{L}^L &\approx i\bar{\Psi}_l^L\not{\partial}\left(\delta\Psi_l^L\right) + i\left(\delta\bar{\Psi}_l^L\right)\not{\partial}\Psi_l^L + i\bar{\psi}_l^R\not{\partial}\left(\delta\psi_l^R\right) + i\left(\delta\bar{\psi}_l^R\right)\not{\partial}\psi_l^R \\ &\quad - \tfrac{g}{2}\bar{\Psi}_l^L\sigma_j\not{W}_j\left(\delta\Psi_l^L\right) - \tfrac{g}{2}\bar{\Psi}_l^L\sigma_j\left(\delta\not{W}_j\right)\Psi_l^L - \tfrac{g}{2}\left(\delta\bar{\Psi}_l^L\right)\sigma_j\not{W}_j\Psi_l^L \\ &\quad + \tfrac{g'}{2}\bar{\Psi}_l^L\not{B}\left(\delta\Psi_l^L\right) + \tfrac{g'}{2}\bar{\Psi}_l^L\left(\delta\not{B}\right)\Psi_l^L + \tfrac{g'}{2}\left(\delta\bar{\Psi}_l^L\right)\not{B}\Psi_l^L \\ &\quad + g'\bar{\psi}_l^R\not{B}\left(\delta\psi_l^R\right) + g'\bar{\psi}_l^R\left(\delta\not{B}\right)\psi_l^R + g'\left(\delta\bar{\psi}_l^R\right)\not{B}\psi_l^R\,.\end{aligned} \tag{6-129}$$

Let's start with one of the easier parts of (6-129), the first and last terms of the next to last row of (6-129). It turns out, they cancel. That is, using (6-83),

$$\begin{aligned}&\tfrac{g'}{2}\bar{\Psi}_l^L \not{B}\left(\delta\Psi_l^L\right)+\tfrac{g'}{2}\left(\delta\bar{\Psi}_l^L\right)\not{B}\Psi_l^L\\&=\tfrac{g'}{2}\bar{\Psi}_l^L \not{B}\left(\tfrac{i}{2}\omega_i\sigma_i\Psi_l^L-\tfrac{i}{2}f\Psi_l^L\right)+\tfrac{g'}{2}\left(-\tfrac{i}{2}\omega_i\bar{\Psi}_l^L\sigma_i+\tfrac{i}{2}f\bar{\Psi}_l^L\right)\not{B}\Psi_l^L\\&=\tfrac{g'}{2}\bar{\Psi}_l^L\left(\tfrac{i}{2}\omega_i\sigma_i-\tfrac{i}{2}f\right)\not{B}\Psi_l^L+\tfrac{g'}{2}\bar{\Psi}_l^L\left(-\tfrac{i}{2}\omega_i\sigma_i+\tfrac{i}{2}f\right)\not{B}\Psi_l^L=0\,.\end{aligned}\tag{6-130}$$

Do **Problem 7** to show the first and last terms of the last row in (6-129) cancel. (Hint: This is a piece of cake.)

The first and last terms of the second row in (6-129) become, where we use (6-42) in the last line of (6-131),

$$\begin{aligned}&-\tfrac{g}{2}\bar{\Psi}_l^L\sigma_j\not{W}_j\left(\delta\Psi_l^L\right)-\tfrac{g}{2}\left(\delta\bar{\Psi}_l^L\right)\sigma_j\not{W}_j\Psi_l^L\\&=-\tfrac{g}{2}\bar{\Psi}_l^L\sigma_j\not{W}_j\left(\tfrac{i}{2}\omega_i\sigma_i\Psi_l^L-\tfrac{i}{2}f\Psi_l^L\right)-\tfrac{g}{2}\left(-\tfrac{i}{2}\omega_i\bar{\Psi}_l^L\sigma_i+\tfrac{i}{2}f\bar{\Psi}_l^L\right)\sigma_j\not{W}_j\Psi_l^L\\&=-i\tfrac{g}{4}\omega_i\bar{\Psi}_l^L\sigma_j\sigma_i\not{W}_j\Psi_l^L\underbrace{+i\tfrac{g}{4}f\bar{\Psi}_l^L\sigma_j\not{W}_j\Psi_l^L}_{\text{cancels}}+i\tfrac{g}{4}\omega_i\bar{\Psi}_l^L\sigma_i\sigma_j\not{W}_j\Psi_l^L\underbrace{-i\tfrac{g}{4}f\bar{\Psi}_l^L\sigma_j\not{W}_j\Psi_l^L}_{\text{cancels}}\\&=i\tfrac{g}{4}\omega_i\bar{\Psi}_l^L\left(\sigma_i\sigma_j-\sigma_j\sigma_i\right)\not{W}_j\Psi_l^L=i\tfrac{g}{4}\bar{\Psi}_l^L\left(i2\varepsilon_{ijk}\sigma_k\right)\omega_i\not{W}_j\Psi_l^L=-\tfrac{g}{2}\bar{\Psi}_l^L\varepsilon_{ijk}\omega_i\not{W}_j\sigma_k\Psi_l^L\,.\end{aligned}\tag{6-131}$$

The middle term of the second row in (6-129), using (6-84), is

$$\begin{aligned}&-\tfrac{g}{2}\bar{\Psi}_l^L\sigma_j\left(\delta\not{W}_j\right)\Psi_l^L\\&=-\tfrac{g}{2}\bar{\Psi}_l^L\sigma_j\left(-\tfrac{1}{g}\not{\partial}\omega_j-\varepsilon_{jki}\omega_k\not{W}_i\right)\Psi_l^L=\tfrac{1}{2}\bar{\Psi}_l^L\sigma_j\left(\not{\partial}\omega_j\right)\Psi_l^L+\tfrac{g}{2}\bar{\Psi}_l^L\varepsilon_{kij}\omega_k\not{W}_i\sigma_j\Psi_l^L\\&=\tfrac{1}{2}\bar{\Psi}_l^L\sigma_j\left(\not{\partial}\omega_j\right)\Psi_l^L+\tfrac{g}{2}\bar{\Psi}_l^L\varepsilon_{ijk}\omega_i\not{W}_j\sigma_k\Psi_l^L\,,\end{aligned}\tag{6-132}$$

the last term of which cancels with the final result in (6-131), leaving us with the first term in the last row of (6-132) equal to the entire second row of (6-129).

Assimilating all of the results from (6-130) to (6-132) into (6-129), we have

$$\begin{aligned}\delta\mathcal{L}^L=&\overbrace{i\bar{\Psi}_l^L\not{\partial}\left(\delta\Psi_l^L\right)}^{\boxed{1}}+\overbrace{i\left(\delta\bar{\Psi}_l^L\right)\not{\partial}\Psi_l^L}^{\boxed{2}}+\overbrace{i\bar{\psi}_l^R\not{\partial}\left(\delta\psi_l^R\right)}^{\boxed{3}}+\overbrace{i\left(\delta\bar{\psi}_l^R\right)\not{\partial}\psi_l^R}^{\boxed{4}}\\&+\overbrace{\tfrac{1}{2}\bar{\Psi}_l^L\sigma_j\left(\not{\partial}\omega_j\right)\Psi_l^L}^{\boxed{5}}+\overbrace{\tfrac{g'}{2}\bar{\Psi}_l^L\left(\delta\not{B}\right)\Psi_l^L}^{\boxed{6}}+\overbrace{g'\bar{\psi}_l^R\left(\delta\not{B}\right)\psi_l^R}^{\boxed{7}}\,.\end{aligned}\tag{6-133}$$

Let's look at terms $\boxed{1}$ and $\boxed{2}$.

$$\begin{aligned}&\overbrace{i\bar{\Psi}_l^L\not{\partial}\left(\delta\Psi_l^L\right)}^{\boxed{1}}+\overbrace{i\left(\delta\bar{\Psi}_l^L\right)\not{\partial}\Psi_l^L}^{\boxed{2}}\\&=i\bar{\Psi}_l^L\not{\partial}\left(\tfrac{i}{2}\omega_i\sigma_i\Psi_l^L\right)+i\bar{\Psi}_l^L\not{\partial}\left(-\tfrac{i}{2}f\Psi_l^L\right)+i\bar{\Psi}_l^L\left(-\tfrac{i}{2}\omega_i\sigma_i\right)\not{\partial}\Psi_l^L+i\bar{\Psi}_l^L\left(\tfrac{i}{2}f\right)\not{\partial}\Psi_l^L\\&=-\tfrac{1}{2}\bar{\Psi}_l^L\left(\not{\partial}\omega_i\right)\sigma_i\Psi_l^L\underbrace{-\tfrac{1}{2}\bar{\Psi}_l^L\omega_i\sigma_i\not{\partial}\Psi_l^L}_{\text{cancels with 5th term}}+\tfrac{1}{2}\bar{\Psi}_l^L\left(\not{\partial}f\right)\Psi_l^L\\&\qquad\underbrace{+\tfrac{1}{2}\bar{\Psi}_l^L f\not{\partial}\Psi_l^L}_{\text{cancels with last term}}+\underbrace{\tfrac{1}{2}\bar{\Psi}_l^L\omega_i\sigma_i\not{\partial}\Psi_l^L}_{\text{cancels with 2nd term}}\underbrace{-\tfrac{1}{2}\bar{\Psi}_l^L f\not{\partial}\Psi_l^L}_{\text{cancels with 4th term}}\\&=-\tfrac{1}{2}\bar{\Psi}_l^L\left(\not{\partial}\omega_i\right)\sigma_i\Psi_l^L+\tfrac{1}{2}\bar{\Psi}_l^L\left(\not{\partial}f\right)\Psi_l^L\,.\end{aligned}\tag{6-134}$$

Now the $\boxed{5}$ and $\boxed{6}$ terms.

$$\overbrace{\tfrac{1}{2}\bar{\Psi}_l^L\sigma_j\left(\not{\partial}\omega_j\right)\Psi_l^L}^{\boxed{5}}+\overbrace{\tfrac{g'}{2}\bar{\Psi}_l^L\left(\delta\not{B}\right)\Psi_l^L}^{\boxed{6}}=\tfrac{1}{2}\bar{\Psi}_l^L\sigma_j\left(\not{\partial}\omega_j\right)\Psi_l^L+\tfrac{g'}{2}\bar{\Psi}_l^L\left(-\tfrac{1}{g'}\not{\partial}f\right)\Psi_l^L\,.\tag{6-135}$$

(6-135) cancels (6-134), and leaves us with

$$\delta\mathcal{L}^{L} = \overbrace{+i\bar{\psi}_l^R \not\partial\left(\delta\psi_l^R\right)}^{\boxed{3}} \overbrace{+i\left(\delta\bar{\psi}_l^R\right)\not\partial\psi_l^R}^{\boxed{4}} \overbrace{+g'\bar{\psi}_l^R\left(\delta\not B\right)\psi_l^R}^{\boxed{7}}, \qquad (6\text{-}136)$$

which becomes

$$\begin{aligned}\delta\mathcal{L}^{L} &= i\bar{\psi}_l^R \not\partial\left(-if\psi_l^R\right) + i\left(if\bar{\psi}_l^R\right)\not\partial\psi_l^R + g'\bar{\psi}_l^R\left(-\tfrac{1}{g'}\not\partial f\right)\psi_l^R \\ &= \underbrace{\bar{\psi}_l^R\left(\not\partial f\right)\psi_l^R}_{\text{cancels with last term}} \quad \underbrace{+\bar{\psi}_l^R f\not\partial\psi_l^R - \bar{\psi}_l^R f\not\partial\psi_l^R}_{=0} \quad \underbrace{-\bar{\psi}_l^R\left(\not\partial f\right)\psi_l^R}_{\text{cancels with 1st term}} = 0\,.\end{aligned} \qquad (6\text{-}137)$$

Result of this section: We have just shown that the $\mathcal{L}^L$ part of the high energy e/w $\mathcal{L}$ is invariant under the transformation set of (6-83) to (6-86).

For $\mathcal{L}^B$

Re-stating (6-123) for convenience,

$$\mathcal{L}^B = \overbrace{-\tfrac{1}{4}B^{\mu\nu}B_{\mu\nu}}^{\boxed{9}} \overbrace{-\tfrac{1}{4}G_i^{\mu\nu}G_{i\,\mu\nu}}^{\boxed{10}} \qquad \text{repeat of} \quad (6\text{-}123)$$

$$B^{\mu\nu} = \partial^\nu B^\mu - \partial^\mu B^\nu \qquad G_i^{\mu\nu} = \partial^\nu W_i^\mu - \partial^\mu W_i^\nu - g\varepsilon_{ijk}W_j^\nu W_k^\mu$$

For the $\boxed{9}$ term in (6-123), we find from (6-85) that B^μ only transforms under the $U(1)$ part of the transformation, so the transformation parallels the symmetric transformation of the QED Lagrangian term for the free part of the photon field A^μ, as shown in Vol. 1, (11-36), pg. 294. So, from that, we recognize the $\boxed{9}$ term contributes zero to $\delta\mathcal{L}^B$, and we only need to examine term $\boxed{10}$.

$$\delta\mathcal{L}^B = \text{ term } \boxed{10} \text{ variation } = -\tfrac{1}{4}\left(\left(\delta G_i^{\mu\nu}\right)G_{i\,\mu\nu} + G_i^{\mu\nu}\left(\delta G_{i\,\mu\nu}\right)\right) \quad \text{(sum on all indices)} \qquad (6\text{-}138)$$

We need to find $\delta G_i^{\mu\nu}$ to evaluate (6-138). We know $G_i^{\mu\nu}$ from (6-123) and from that

$$\delta G_i^{\mu\nu} = \partial^\nu\left(\delta W_i^\mu\right) - \partial^\mu\left(\delta W_i^\nu\right) - g\varepsilon_{ijk}\left(\delta W_j^\nu\right)W_k^\mu - g\varepsilon_{ijk}W_j^\nu\left(\delta W_k^\mu\right) \qquad (6\text{-}139)$$

With (6-84), (6-139) becomes

$$\begin{aligned}\delta G_i^{\mu\nu} &= \partial^\nu\left(-\tfrac{1}{g}\partial^\mu\omega_i - \varepsilon_{ijk}\omega_j W_k^\mu\right) - \partial^\mu\left(-\tfrac{1}{g}\partial^\nu\omega_i - \varepsilon_{ijk}\omega_j W_k^\nu\right) \\ &\quad - g\varepsilon_{ijk}\left(-\tfrac{1}{g}\partial^\nu\omega_j - \varepsilon_{jmn}\omega_m W_n^\nu\right)W_k^\mu - g\varepsilon_{ijk}W_j^\nu\left(-\tfrac{1}{g}\partial^\mu\omega_k - \varepsilon_{kmn}\omega_m W_n^\mu\right) \\ &= \underbrace{-\tfrac{1}{g}\partial^\nu\partial^\mu\omega_i}_{\text{cancels}} - \varepsilon_{ijk}\partial^\nu\left(\omega_j W_k^\mu\right) \underbrace{+\tfrac{1}{g}\partial^\mu\partial^\nu\omega_i}_{\text{cancels}} + \varepsilon_{ijk}\partial^\mu\left(\omega_j W_k^\nu\right) \\ &\quad + \varepsilon_{ijk}\left(\partial^\nu\omega_j\right)W_k^\mu + g\varepsilon_{ijk}\varepsilon_{jmn}\omega_m W_n^\nu W_k^\mu + \varepsilon_{ijk}\left(\partial^\mu\omega_k\right)W_j^\nu + g\varepsilon_{ijk}\varepsilon_{kmn}\omega_m W_j^\nu W_n^\mu\end{aligned} \qquad (6\text{-}140)$$

$$\begin{aligned} &= \underbrace{-\varepsilon_{ijk}\left(\partial^\nu\omega_j\right)W_k^\mu}_{\substack{\text{cancels}\\ \text{with}}} - \varepsilon_{ijk}\omega_j\left(\partial^\nu W_k^\mu\right) \underbrace{+\varepsilon_{ijk}\left(\partial^\mu\omega_j\right)W_k^\nu}_{\substack{\text{cancels}\\ \text{with}}} + \varepsilon_{ijk}\omega_j\left(\partial^\mu W_k^\nu\right) \\ &\quad \overbrace{+\varepsilon_{ijk}\left(\partial^\nu\omega_j\right)W_k^\mu} + g\varepsilon_{ijk}\varepsilon_{jmn}\omega_m W_n^\nu W_k^\mu \underbrace{\overbrace{+\varepsilon_{ijk}\left(\partial^\mu\omega_k\right)W_j^\nu}}_{-\varepsilon_{ijk}\left(\partial^\mu\omega_j\right)W_k^\nu} + g\varepsilon_{ijk}\varepsilon_{kmn}\omega_m W_j^\nu W_n^\mu \\ &= \underbrace{-\varepsilon_{ijk}\omega_j\left(\partial^\nu W_k^\mu - \partial^\mu W_k^\nu\right)}_{\boxed{10a}} \underbrace{+g\varepsilon_{ijk}\varepsilon_{jmn}\omega_m W_n^\nu W_k^\mu}_{\boxed{10b}} \underbrace{+g\varepsilon_{ijk}\varepsilon_{kmn}\omega_m W_j^\nu W_n^\mu}_{\boxed{10c}}.\end{aligned} \qquad (6\text{-}141)$$

So, slightly re-arranged,

$$\delta G_i^{\mu\nu} = \underbrace{-\varepsilon_{ijk}\omega_j\left(\partial^\nu W_k^\mu - \partial^\mu W_k^\nu\right)}_{\boxed{10a}} \underbrace{+g\varepsilon_{ijk}\left(\varepsilon_{jmn}\omega_m W_n^\nu\right)W_k^\mu}_{\boxed{10b}} \underbrace{+g\varepsilon_{ijk}W_j^\nu\left(\varepsilon_{kmn}\omega_m W_n^\mu\right)}_{\boxed{10c}}. \tag{6-142}$$

We will re-express the terms labeled $\boxed{10b}$ and $\boxed{10c}$ in (6-142), but to do this, we first need to deduce a particular mathematical relationship we will employ to do so.

Digression to deduce a needed math relationship

A well-known relation between cross products of three 3D vectors **a**, **b**, and **c** is

$$\mathbf{a}\times(\mathbf{b}\times\mathbf{c})+\mathbf{b}\times(\mathbf{c}\times\mathbf{a})+\mathbf{c}\times(\mathbf{a}\times\mathbf{b})=0, \tag{6-143}$$

where (6-143) can be re-arranged as

$$(\mathbf{a}\times\mathbf{b})\times\mathbf{c}+\mathbf{b}\times(\mathbf{a}\times\mathbf{c})=\mathbf{a}\times(\mathbf{b}\times\mathbf{c}). \tag{6-144}$$

Now re-write (6-144) in terms of vector indices as

$$\varepsilon_{ijk}\left(\varepsilon_{jmn}a_m b_n\right)c_k+\varepsilon_{ijk}b_j\left(\varepsilon_{kmn}a_m c_n\right)=\varepsilon_{ijk}a_j\left(\varepsilon_{kmn}b_m c_n\right). \tag{6-145}$$

Then, make the substitutions

$$a\to\omega \qquad b\to W^\nu \qquad c\to W^\mu \tag{6-146}$$

in (6-145), to get the relationship we seek to use in (6-142)

$$\underbrace{g\varepsilon_{ijk}\left(\varepsilon_{jmn}\omega_m W_n^\nu\right)W_k^\mu}_{\boxed{10b}} \underbrace{+g\varepsilon_{ijk}W_j^\nu\left(\varepsilon_{kmn}\omega_m W_n^\mu\right)}_{\boxed{10c}} = g\varepsilon_{ijk}\omega_j\left(\varepsilon_{kmn}W_m^\nu W_n^\mu\right). \tag{6-147}$$

End of digression

Thus, from (6-147), (6-142) becomes

$$\begin{aligned}\delta G_i^{\mu\nu} &= \underbrace{-\varepsilon_{ijk}\omega_j\left(\partial^\nu W_k^\mu-\partial^\mu W_k^\nu\right)}_{\boxed{10a}} + g\varepsilon_{ijk}\omega_j\left(\varepsilon_{kmn}W_m^\nu W_n^\mu\right)\\ &= -\varepsilon_{ijk}\omega_j\left(\partial^\nu W_k^\mu-\partial^\mu W_k^\nu - g\left(\varepsilon_{kmn}W_m^\nu W_n^\mu\right)\right),\end{aligned} \tag{6-148}$$

or from (6-123),

$$\delta G_i^{\mu\nu} = -\varepsilon_{ijk}\omega_j G_k^{\mu\nu}. \tag{6-149}$$

Hence, (6-138) becomes

$$\begin{aligned}\delta\mathcal{L}^B &= -\tfrac{1}{4}\Big(\left(-\varepsilon_{ijk}\omega_j G_k^{\mu\nu}\right)G_{i\,\mu\nu} + \underbrace{G_i^{\mu\nu}\left(-\varepsilon_{ijk}\omega_j G_{k\,\mu\nu}\right)}_{\text{dummies } i\to k,\ k\to i}\Big)\\ &= \tfrac{1}{4}\left(\varepsilon_{ijk}\omega_j G_k^{\mu\nu}G_{i\,\mu\nu}+\varepsilon_{kji}\omega_j G_{k\,\mu\nu}G_i^{\mu\nu}\right) = \tfrac{1}{4}\left(\varepsilon_{ijk}\omega_j G_k^{\mu\nu}G_{i\,\mu\nu}-\varepsilon_{ijk}\omega_j G_k^{\mu\nu}G_{i\,\mu\nu}\right)=0.\end{aligned} \tag{6-150}$$

$\boxed{\text{Result of this section}}$: $\mathcal{L}^B$ is invariant under the transformation set of (6-83) to (6-86).

For $\mathcal{L}^H$

Re-stating (6-124) here for convenience,

$$\begin{aligned}\mathcal{L}^H &= \left(D^\mu\Phi\right)^\dagger\left(D_\mu\Phi\right)-\mu^2\Phi^\dagger\Phi-\lambda\left(\Phi^\dagger\Phi\right)^2\\ D_\mu\Phi &= \left(\partial_\mu+\tfrac{i}{2}g\sigma_k W_{k\,\mu}+i\tfrac{g'}{2}B_\mu\right)\Phi\end{aligned} \qquad \text{repeat of (6-124)}$$

We first show the invariance of $\Phi^\dagger\Phi$.

$$\left(\Phi^\dagger\Phi\right)'\approx\left(\Phi^\dagger+\delta\Phi^\dagger\right)\left(\Phi+\delta\Phi\right)\approx\Phi^\dagger\Phi+\left(\delta\Phi^\dagger\right)\Phi+\Phi^\dagger\left(\delta\Phi\right), \tag{6-151}$$

So,

$$\delta\left(\Phi^{\dagger}\Phi\right)\approx\left(\delta\Phi^{\dagger}\right)\Phi+\Phi^{\dagger}\left(\delta\Phi\right). \tag{6-152}$$

From (6-86),

$$\begin{aligned}\delta\left(\Phi^{\dagger}\Phi\right)&\approx\left(-\tfrac{i}{2}\omega_i\Phi^{\dagger}\sigma_i^{\dagger}-\tfrac{i}{2}f\Phi^{\dagger}\right)\Phi+\Phi^{\dagger}\left(\tfrac{i}{2}\omega_i\sigma_i\Phi+\tfrac{i}{2}f\Phi\right)\qquad\left(\text{where }\sigma_i^{\dagger}=\sigma_i\right)\\&=-\tfrac{i}{2}\omega_i\Phi^{\dagger}\sigma_i\,\Phi-\tfrac{i}{2}f\Phi^{\dagger}\Phi+\tfrac{i}{2}\omega_i\Phi^{\dagger}\sigma_i\Phi+\tfrac{i}{2}f\Phi^{\dagger}\Phi=0,\end{aligned} \tag{6-153}$$

and $\Phi^{\dagger}\Phi$ is unchanged under the transformation, so we can ignore the last two terms in (6-124).

As an aside, note that instead of looking at infinitesimal transformations, it would actually have been simpler to express the transformation in its finite form, represented by U. Doing that, we have $\Phi'^{\dagger}\Phi' = (\Phi^{\dagger}U^{\dagger})(U\Phi)=\Phi^{\dagger}\Phi$.

Thus,

$$\delta\mathcal{L}^{H}=\delta\left(\left(D^{\mu}\Phi\right)^{\dagger}\left(D_{\mu}\Phi\right)\right)=\left(\delta\left(D^{\mu}\Phi\right)^{\dagger}\right)\left(D_{\mu}\Phi\right)+\left(D^{\mu}\Phi\right)^{\dagger}\left(\delta\left(D_{\mu}\Phi\right)\right). \tag{6-154}$$

We need to show (6-154), the variation of the kinetic Higgs term in the Lagrangian under the given transformation set, is zero. I have been unable to find this derivation anywhere in other texts, and it is almost invariably simply ignored. The only text where I have seen it even mentioned simply shrugs it off by saying the kinetic Higgs term "obviously shares this property [of invariance with other terms in the Lagrangian]". As you will see in the following, this is anything but obvious. In fact, it is the most problematic and involved proof of invariance of any term in the electroweak Lagrangian. We evaluate it via two different methods, then compare and contrast those two methods.

Method 1

Given that $D_{\mu}\Phi$ is a doublet, we could define its variation to be of form (6-86), which is for the doublet Φ. That is, we are presuming $D_{\mu}\Phi$ varies just the same way Φ does, i.e.,

$$\begin{aligned}\delta\left(D_{\mu}\Phi\right)&=\left(\tfrac{i}{2}\omega_i\sigma_i+\tfrac{i}{2}f\right)D_{\mu}\Phi=\tfrac{i}{2}\omega_i\sigma_i\left(D_{\mu}\Phi\right)+\tfrac{i}{2}f\left(D_{\mu}\Phi\right)\\\delta\left(D_{\mu}\Phi\right)^{\dagger}&=\left(D_{\mu}\Phi\right)^{\dagger}\left(-\tfrac{i}{2}\omega_i\sigma_i^{\dagger}-\tfrac{i}{2}f\right)=-\tfrac{i}{2}\omega_i\left(D_{\mu}\Phi\right)^{\dagger}\sigma_i-\tfrac{i}{2}f\left(D_{\mu}\Phi\right)^{\dagger}.\end{aligned} \tag{6-155}$$

Substituting (6-155) into (6-154), we have

$$\begin{aligned}\delta\mathcal{L}^{H}&=-\tfrac{i}{2}\omega_i\left(D^{\mu}\Phi\right)^{\dagger}\sigma_i\left(D_{\mu}\Phi\right)-\tfrac{i}{2}f\left(D^{\mu}\Phi\right)^{\dagger}\left(D_{\mu}\Phi\right)\\&\quad+\tfrac{i}{2}\omega_i\left(D^{\mu}\Phi\right)^{\dagger}\sigma_i\left(D_{\mu}\Phi\right)+\tfrac{i}{2}f\left(D^{\mu}\Phi\right)^{\dagger}\left(D_{\mu}\Phi\right)=0.\end{aligned} \tag{6-156}$$

Method 2

However, one could argue (probably persuasively) that Method 1 differs from what should be considered the standard way of determining a variation. That is, one should express $D_{\mu}\Phi$ in terms of all the various fields therein, then apply the appropriate infinitesimal transformations to each field, as we noted in (6-89) of Sect. 6.3.12, and as we did for $\delta\mathcal{L}^{L}$ above. If that equals the first row of (6-155), then the second row of (6-155) is valid, and thus, so is (6-156). If it does not, then we need to re-think the assumption of Method 1.

So, expressing the variation of the second row of (6-124), we have

$$\begin{aligned}\underbrace{\delta\left(D_{\mu}\Phi\right)}_{\text{Method 2}}&=\underbrace{\partial_{\mu}\left(\delta\Phi\right)}_{\boxed{12}}\underbrace{+\tfrac{i}{2}g\sigma_k\left(\delta W_{k\mu}\right)\Phi}_{\boxed{13}}\underbrace{+\tfrac{i}{2}g\sigma_kW_{k\mu}\left(\delta\Phi\right)}_{\boxed{14}}\underbrace{+\tfrac{i}{2}g'\left(\delta B_{\mu}\right)\Phi}_{\boxed{15}}\underbrace{+\tfrac{i}{2}g'B_{\mu}\left(\delta\Phi\right)}_{\boxed{17}}\\&=\underbrace{\partial_{\mu}\left(\tfrac{i}{2}\omega_i\sigma_i\Phi+\tfrac{i}{2}f\Phi\right)}_{\boxed{12}}\underbrace{+\tfrac{i}{2}g\sigma_k\left(-\tfrac{1}{g}\partial_{\mu}\omega_k-\varepsilon_{kmn}\omega_mW_{n\mu}\right)\Phi}_{\boxed{13}}\\&\quad\underbrace{+\tfrac{i}{2}g\sigma_kW_{k\mu}\left(\tfrac{i}{2}\omega_i\sigma_i\Phi+\tfrac{i}{2}f\Phi\right)}_{\boxed{14}}\underbrace{+\tfrac{i}{2}g'\left(-\tfrac{1}{g'}\partial_{\mu}f\right)\Phi}_{\boxed{15}}\underbrace{+\tfrac{i}{2}g'B_{\mu}\left(\tfrac{i}{2}\omega_i\sigma_i\Phi+\tfrac{i}{2}f\Phi\right)}_{\boxed{17}}\end{aligned} \tag{6-157}$$

$$
\begin{aligned}
&=\underbrace{\partial_\mu\left(\tfrac{i}{2}\omega_i\sigma_i\Phi\right)}_{\boxed{12a}}+\underbrace{\partial_\mu\left(\tfrac{i}{2}f\Phi\right)}_{\boxed{12b}}+\underbrace{\tfrac{i}{2}g\sigma_k\left(-\tfrac{1}{g}\partial_\mu\omega_k\right)\Phi}_{\boxed{13a}}+\underbrace{\tfrac{i}{2}g\sigma_k\left(-\varepsilon_{kmn}\omega_m W_{n\,\mu}\right)\Phi}_{\boxed{13b}} \\
&\quad+\underbrace{\tfrac{i}{2}g\sigma_k W_{k\,\mu}\left(\tfrac{i}{2}\omega_i\sigma_i\Phi\right)}_{\boxed{14a}}+\underbrace{\tfrac{i}{2}g\sigma_k W_{k\,\mu}\left(\tfrac{i}{2}f\Phi\right)}_{\boxed{14b}}+\underbrace{\tfrac{i}{2}g'\left(-\tfrac{1}{g'}\partial_\mu f\right)\Phi}_{\boxed{15}}+\underbrace{\tfrac{i}{2}g'B_\mu\left(\tfrac{i}{2}\omega_i\sigma_i\Phi\right)}_{\boxed{17a}}+\underbrace{\tfrac{i}{2}g'B_\mu\left(\tfrac{i}{2}f\Phi\right)}_{\boxed{17b}}
\end{aligned}
\qquad (6\text{-}158)
$$

$$
\begin{aligned}
&=\underbrace{\tfrac{i}{2}\sigma_k\left(\partial_\mu\omega_k\right)\Phi}_{\substack{\boxed{12a-1}\\ \text{cancels }\boxed{13a}}}+\underbrace{\tfrac{i}{2}\omega_i\sigma_i\left(\partial_\mu\Phi\right)}_{\boxed{12a-2}}+\underbrace{\tfrac{i}{2}\left(\partial_\mu f\right)\Phi}_{\substack{\boxed{12b-1}\\ \text{cancels }\boxed{15}}}+\underbrace{\tfrac{i}{2}f\left(\partial_\mu\Phi\right)}_{\boxed{12b-2}}\underbrace{-\tfrac{i}{2}\sigma_k\left(\partial_\mu\omega_k\right)\Phi}_{\substack{\boxed{13a}\\ \text{cancels }\boxed{12a-1}}}\underbrace{-\tfrac{i}{2}g\varepsilon_{kmn}\omega_m W_{n\,\mu}\sigma_k\Phi}_{\boxed{13b}} \\
&\quad\underbrace{-\tfrac{g}{4}\omega_i W_{k\,\mu}\sigma_k\sigma_i\Phi}_{\boxed{14a}}\underbrace{-\tfrac{g}{4}f\sigma_k W_{k\,\mu}\Phi}_{\boxed{14b}}\underbrace{-\tfrac{i}{2}\left(\partial_\mu f\right)\Phi}_{\substack{\boxed{15}\\ \text{cancels }\boxed{12b-1}}}\underbrace{-\tfrac{g'}{4}\omega_i B_\mu\sigma_i\Phi}_{\boxed{17a}}\underbrace{-\tfrac{g'}{4}fB_\mu\Phi}_{\boxed{17b}}.
\end{aligned}
\qquad (6\text{-}159)
$$

Or finally, via Method 2,

$$
\begin{aligned}
\underbrace{\delta\left(D_\mu\Phi\right)}_{\text{Method 2}}&=\underbrace{\tfrac{i}{2}\omega_i\sigma_i\left(\partial_\mu\Phi\right)}_{\boxed{12a-2}}\underbrace{+\tfrac{i}{2}f\left(\partial_\mu\Phi\right)}_{\boxed{12b-2}}\underbrace{-\tfrac{i}{2}g\varepsilon_{kmn}\omega_m W_{n\,\mu}\sigma_k\Phi}_{\boxed{13b}}\underbrace{-\tfrac{g}{4}\omega_i W_{k\,\mu}\sigma_k\sigma_i\Phi}_{\boxed{14a}}\underbrace{-\tfrac{g}{4}fW_{k\,\mu}\sigma_k\Phi}_{\boxed{14b}} \\
&\quad\underbrace{-\tfrac{g'}{4}\omega_i B_\mu\sigma_i\Phi}_{\boxed{17a}}\underbrace{-\tfrac{g'}{4}fB_\mu\Phi}_{\boxed{17b}}.
\end{aligned}
\qquad (6\text{-}160)
$$

Comparing Method 1 to Method 2

For Method 1 comparison, we express the first row of (6-155) in terms of the fields, operators, and variables in $D_\mu\Phi$ to obtain

$$
\begin{aligned}
\underbrace{\delta\left(D_\mu\Phi\right)}_{\text{Method 1}}&=\left(\tfrac{i}{2}\omega_i\sigma_i+\tfrac{i}{2}f\right)D_\mu\Phi=\tfrac{i}{2}\omega_i\sigma_i\left(D_\mu\Phi\right)+\tfrac{i}{2}f\left(D_\mu\Phi\right) \\
&=\tfrac{i}{2}\omega_i\sigma_i\left(\partial_\mu+\tfrac{i}{2}g\sigma_k W_{k\,\mu}+i\tfrac{g'}{2}B_\mu\right)\Phi+\tfrac{i}{2}f\left(\partial_\mu+\tfrac{i}{2}g\sigma_k W_{k\,\mu}+i\tfrac{g'}{2}B_\mu\right)\Phi \\
&=\tfrac{i}{2}\omega_i\sigma_i\partial_\mu\Phi-\tfrac{g}{4}\omega_i W_{k\,\mu}\sigma_i\sigma_k\Phi-\tfrac{g'}{4}\omega_i\sigma_i B_\mu\Phi+\tfrac{i}{2}f\partial_\mu\Phi-\tfrac{g}{4}f\sigma_k W_{k\,\mu}\Phi-\tfrac{g'}{4}fB_\mu\Phi.
\end{aligned}
\qquad (6\text{-}161)
$$

Re-arranging order and numbering terms, we have

$$
\begin{aligned}
\underbrace{\delta\left(D_\mu\Phi\right)}_{\text{Method 1}}&=\underbrace{\tfrac{i}{2}\omega_i\sigma_i\partial_\mu\Phi}_{\boxed{112a-2}}\underbrace{+\tfrac{i}{2}f\partial_\mu\Phi}_{\boxed{112b-2}}-\tfrac{g}{4}\omega_i W_{k\,\mu}\sigma_i\sigma_k\Phi\underbrace{-\tfrac{g}{4}fW_{k\,\mu}\sigma_k\Phi}_{\boxed{114b}}\underbrace{-\tfrac{g'}{4}\omega_i\sigma_i B_\mu\Phi}_{\boxed{117a}}\underbrace{-\tfrac{g'}{4}fB_\mu\Phi}_{\boxed{117b}} \\
&=\underbrace{\tfrac{i}{2}\omega_i\sigma_i\partial_\mu\Phi}_{\boxed{112a-2}}\underbrace{+\tfrac{i}{2}f\partial_\mu\Phi}_{\boxed{112b-2}}-\tfrac{g}{4}\omega_i W_{k\,\mu}\left(\sigma_k\sigma_i+i2\varepsilon_{ikj}\sigma_j\right)\Phi\underbrace{-\tfrac{g}{4}fW_{k\,\mu}\sigma_k\Phi}_{\boxed{114b}}\underbrace{-\tfrac{g'}{4}\omega_i B_\mu\sigma_i\Phi}_{\boxed{117a}}\underbrace{-\tfrac{g'}{4}fB_\mu\Phi}_{\boxed{117b}} \\
&=\underbrace{\tfrac{i}{2}\omega_i\sigma_i\partial_\mu\Phi}_{\boxed{112a-2}}\underbrace{+\tfrac{i}{2}f\partial_\mu\Phi}_{\boxed{112b-2}}\underbrace{-\tfrac{g}{4}\omega_i W_{k\,\mu}\sigma_k\sigma_i\Phi}_{\boxed{114a}}\underbrace{-i\tfrac{g}{2}\varepsilon_{ikj}\omega_i W_{k\,\mu}\sigma_j\Phi}_{\boxed{113b}}\underbrace{-\tfrac{g}{4}fW_{k\,\mu}\sigma_k\Phi}_{\boxed{114b}}\underbrace{-\tfrac{g'}{4}\omega_i B_\mu\sigma_i\Phi}_{\boxed{117a}}\underbrace{-\tfrac{g'}{4}fB_\mu\Phi}_{\boxed{117b}} \\
&=\underbrace{\tfrac{i}{2}\omega_i\sigma_i\partial_\mu\Phi}_{\boxed{112a-2}}\underbrace{+\tfrac{i}{2}f\partial_\mu\Phi}_{\boxed{112b-2}}\underbrace{-i\tfrac{g}{2}\varepsilon_{jik}\omega_i W_{k\,\mu}\sigma_j\Phi}_{\substack{\boxed{113b}\\ \underbrace{}_{-i\frac{g}{2}\varepsilon_{kmn}\omega_m W_{n\mu}\sigma_k\Phi}}}\underbrace{-\tfrac{g}{4}\omega_i W_{k\,\mu}\sigma_k\sigma_i\Phi}_{\boxed{114a}}\underbrace{-\tfrac{g}{4}fW_{k\,\mu}\sigma_k\Phi}_{\boxed{114b}}\underbrace{-\tfrac{g'}{4}\omega_i B_\mu\sigma_i\Phi}_{\boxed{117a}}\underbrace{-\tfrac{g'}{4}fB_\mu\Phi}_{\boxed{117b}}
\end{aligned}
\qquad (6\text{-}162)
$$

Comparing (6-160) to (6-162), we see that Method 1 and Method 2 yield the same results, and that justifies our assumption in Method 1 that $D_\mu\Phi$ transforms in the same way as Φ. Hence, the two methods are equivalent, and in either, we find $\delta\mathcal{L}^H = 0$.

Result of this section : $\mathcal{L}^H$ is invariant under the transformation set of (6-83) to (6-86), and additionally, $D_\mu\Phi$ transforms in the same way as Φ.

For $\mathcal{L}^{LH}$

We repeat (6-125) below for convenience, and we write out the *h.c.* parts explicitly.

$$\mathcal{L}^{LH} = -{}_AY_{ll'}\bar{\Psi}_l^L\Phi\psi_{l'}^R \underbrace{-{}_AY_{ll'}^\dagger\bar{\psi}_{l'}^R\Phi^\dagger\Psi_l^L}_{h.c.\text{ of term at left}} - {}_BY_{l\nu_{l'}}\bar{\Psi}_l^L\tilde{\Phi}\psi_{\nu_{l'}}^R \underbrace{-{}_BY_{l\nu_{l'}}^\dagger\bar{\psi}_{\nu_{l'}}^R\tilde{\Phi}^\dagger\Psi_l^L}_{h.c.\text{ of term at left}}$$

$$\tilde{\Phi} = \frac{1}{\sqrt{2}}\begin{pmatrix}\phi_b^* \\ -\phi_a^*\end{pmatrix} = \frac{1}{\sqrt{2}}\begin{bmatrix}\phi_3 - i\phi_4 \\ -\phi_1 + i\phi_2\end{bmatrix} \qquad \text{repeat of (6-125)}$$

$$\delta\mathcal{L}^{LH} = -\delta\left({}_AY_{ll'}\bar{\Psi}_l^L\Phi\psi_{l'}^R + {}_AY_{ll'}^\dagger\bar{\psi}_{l'}^R\Phi^\dagger\Psi_l^L\right) - \delta\left({}_BY_{l\nu_{l'}}\bar{\Psi}_l^L\tilde{\Phi}\psi_{\nu_{l'}}^R + {}_BY_{l\nu_{l'}}^\dagger\bar{\psi}_{\nu_{l'}}^R\tilde{\Phi}^\dagger\Psi_l^L\right) \qquad (6\text{-}163)$$

Because the Yukawa matrices are Hermitian, we can re-write (6-163) as

$$\delta\mathcal{L}^{LH} = -{}_AY_{ll'}\delta\left(\bar{\Psi}_l^L\Phi\psi_{l'}^R + \bar{\psi}_{l'}^R\Phi^\dagger\Psi_l^L\right) \underbrace{-{}_BY_{l\nu_{l'}}\delta\left(\bar{\Psi}_l^L\tilde{\Phi}\psi_{\nu_{l'}}^R + \bar{\psi}_{\nu_{l'}}^R\tilde{\Phi}^\dagger\Psi_l^L\right)}_{\text{ignore for now}} \qquad (6\text{-}164)$$

The part of (6-163) we are not ignoring for now that is being varied (inside the parentheses) is

$$\delta\left(\bar{\Psi}_l^L\Phi\psi_{l'}^R + \bar{\psi}_{l'}^R\Phi^\dagger\Psi_l^L\right) = \begin{pmatrix}\left(\delta\bar{\Psi}_l^L\right)\Phi\psi_{l'}^R + \bar{\Psi}_l^L(\delta\Phi)\psi_{l'}^R + \bar{\Psi}_l^L\Phi\left(\delta\psi_{l'}^R\right) \\ +\left(\delta\bar{\psi}_{l'}^R\right)\Phi^\dagger\Psi_l^L + \bar{\psi}_{l'}^R\left(\delta\Phi^\dagger\right)\Psi_l^L + \bar{\psi}_{l'}^R\Phi^\dagger\left(\delta\Psi_l^L\right)\end{pmatrix}$$

$$= \begin{pmatrix}\left(-\frac{i}{2}\omega_i\bar{\Psi}_l^L\sigma_i + \frac{i}{2}f\bar{\Psi}_l^L\right)\Phi\psi_{l'}^R + \bar{\Psi}_l^L\left(\frac{i}{2}\omega_i\sigma_i\Phi + \frac{i}{2}f\Phi\right)\psi_{l'}^R + \bar{\Psi}_l^L\Phi\left(-if\psi_{l'}^R\right) \\ +\left(if\bar{\psi}_{l'}^R\right)\Phi^\dagger\Psi_l^L + \bar{\psi}_{l'}^R\left(-\frac{i}{2}\omega_i\Phi^\dagger\sigma_i - \frac{i}{2}f\Phi^\dagger\right)\Psi_l^L + \bar{\psi}_{l'}^R\Phi^\dagger\left(\frac{i}{2}\omega_i\sigma_i\Psi_l^L - \frac{i}{2}f\Psi_l^L\right)\end{pmatrix}. \qquad (6\text{-}165)$$

$$\begin{aligned} &= -\tfrac{i}{2}\omega_i\bar{\Psi}_l^L\sigma_i\Phi\psi_{l'}^R + \tfrac{i}{2}f\bar{\Psi}_l^L\Phi\psi_{l'}^R + \tfrac{i}{2}\omega_i\bar{\Psi}_l^L\sigma_i\Phi\psi_{l'}^R + \tfrac{i}{2}f\bar{\Psi}_l^L\Phi\psi_{l'}^R - if\bar{\Psi}_l^L\Phi\psi_{l'}^R \\ &\quad + if\bar{\psi}_{l'}^R\Phi^\dagger\Psi_l^L - \tfrac{i}{2}\omega_i\bar{\psi}_{l'}^R\Phi^\dagger\sigma_i\Psi_l^L - \tfrac{i}{2}f\bar{\psi}_{l'}^R\Phi^\dagger\Psi_l^L + \tfrac{i}{2}\omega_i\bar{\psi}_{l'}^R\Phi^\dagger\sigma_i\Psi_l^L - \tfrac{i}{2}f\bar{\psi}_{l'}^R\Phi^\dagger\Psi_l^L\,. \end{aligned} \qquad (6\text{-}166)$$

In the first row of (6-166), the 1st and 3rd terms cancel. So do the 2nd, 4th, and 5th terms. In the second row, the 1st, 3rd, and 5th terms cancel, as do the 2nd and 4th.. So, in (6-163),

$$\delta\left(\bar{\Psi}_l^L\Phi\psi_{l'}^R + \bar{\psi}_{l'}^R\Phi^\dagger\Psi_l^L\right) = 0 \qquad (6\text{-}167)$$

Note in passing that a singlet is not a two component object in *SU*(2) space like a doublet, so it commutes (like any scalar in Euclidean space) with the 2X2 Pauli matrices.

Do **Problem 8** to show the part of (6-164) we initially ignored above is also zero.

Result of this section : $\mathcal{L}^{LH}$ is invariant under the transformation set of (6-83) to (6-86).

6.10.4 Final Conclusion

The entire high energy electroweak Lagrangian is symmetric under the transformation set (6-83) to (6-86), i.e.,

$$\text{high energy electroweak } \delta\mathcal{L} = 0\,. \qquad (6\text{-}168)$$

Appendix B. How Fermion Mass Terms Break Symmetry in the Lagrangian

Consider what happens to the Lagrangian (6-48), which is symmetric without any mass terms, if we add in mass terms, of the usual form, for fermions.

For the LC fermions, we need to cast the terms in doublet form, since our transformation (6-83) includes $SU(2)$ terms in the Pauli matrices. Additionally, we need to note a key relation, which we will repeat in the main part (not appendix) of the next chapter, as it pops up often in QFT,

$$\begin{aligned}\bar{\psi}_e^L\psi_e^R+\bar{\psi}_e^R\psi_e^L&=\left(\bar{\psi}_e\tfrac{1}{2}\left(1+\gamma^5\right)\right)\left(\tfrac{1}{2}\left(1+\gamma^5\right)\psi_e\right)+\left(\bar{\psi}_e\tfrac{1}{2}\left(1-\gamma^5\right)\right)\left(\tfrac{1}{2}\left(1-\gamma^5\right)\psi_e\right)\\&=\tfrac{1}{4}\bar{\psi}_e\left(1+2\gamma^5+\underbrace{\left(\gamma^5\right)^2}_{=1}\right)\psi_e+\tfrac{1}{4}\bar{\psi}_e\left(1-2\gamma^5+\left(\gamma^5\right)^2\right)\psi_e\\&=\tfrac{1}{4}\bar{\psi}_e\left(2+2\gamma^5\right)\psi_e+\tfrac{1}{4}\bar{\psi}_e\left(2-2\gamma^5\right)\psi_e=\bar{\psi}_e\psi_e.\end{aligned}\tag{6-169}$$

So, for mass terms of any particular fermion field f, we have

$$m_f\bar{\psi}_f\psi_f=m_f\bar{\psi}_f^L\psi_f^R+m_f\bar{\psi}_f^R\psi_f^L.\tag{6-170}$$

Thus,

$$\begin{aligned}m_{\nu_e}\bar{\psi}_{\nu_e}\psi_{\nu_e}+m_e\bar{\psi}_e\psi_e&=\bar{\psi}_{\nu_e}^L m_{\nu_e}\psi_{\nu_e}^R+\bar{\psi}_{\nu_e}^R m_{\nu_e}\psi_{\nu_e}^L+\bar{\psi}_e^L m_e\psi_e^R+\bar{\psi}_e^R m_e\psi_e^L\\&=\begin{pmatrix}\bar{\psi}_{\nu_e}^L & \bar{\psi}_e^L\end{pmatrix}\begin{pmatrix}m_{\nu_e} & \\ & m_e\end{pmatrix}\begin{pmatrix}\psi_{\nu_e}^R\\ \psi_e^R\end{pmatrix}+\begin{pmatrix}\bar{\psi}_{\nu_e}^R & \bar{\psi}_e^R\end{pmatrix}\begin{pmatrix}m_{\nu_e} & \\ & m_e\end{pmatrix}\begin{pmatrix}\psi_{\nu_e}^L\\ \psi_e^L\end{pmatrix}\\&=\bar{\Psi}_e^L\begin{pmatrix}m_{\nu_e} & \\ & m_e\end{pmatrix}\begin{pmatrix}\psi_{\nu_e}^R\\ \psi_e^R\end{pmatrix}+\begin{pmatrix}\bar{\psi}_{\nu_e}^R & \bar{\psi}_e^R\end{pmatrix}\begin{pmatrix}m_{\nu_e} & \\ & m_e\end{pmatrix}\Psi_e^L.\end{aligned}\tag{6-171}$$

Since the $U(1)$ and $SU(2)$ parts of the transformation work independently, we only need to show that either the $U(1)$ part or the $SU(2)$ part yields $\delta\mathcal{L}\neq 0$, i.e., a non-symmetric Lagrangian. So, with the $SU(2)$ part of (6-83) into our conjectured mass terms (6-171), where we keep in mind that the RC field are singlets under $SU(2)$, so they are invariant under that transformation, we have

$$\begin{aligned}&\delta\left(\bar{\Psi}_e^L\begin{pmatrix}m_{\nu_e} & \\ & m_e\end{pmatrix}\begin{pmatrix}\psi_{\nu_e}^R\\ \psi_e^R\end{pmatrix}+\begin{pmatrix}\bar{\psi}_{\nu_e}^R & \bar{\psi}_e^R\end{pmatrix}\begin{pmatrix}m_{\nu_e} & \\ & m_e\end{pmatrix}\Psi_e^L\right)\\&=\left(\delta\bar{\Psi}_e^L\right)\begin{pmatrix}m_{\nu_e} & \\ & m_e\end{pmatrix}\begin{pmatrix}\psi_{\nu_e}^R\\ \psi_e^R\end{pmatrix}+\begin{pmatrix}\bar{\psi}_{\nu_e}^R & \bar{\psi}_e^R\end{pmatrix}\begin{pmatrix}m_{\nu_e} & \\ & m_e\end{pmatrix}\left(\delta\Psi_e^L\right)\\&=\left(-\tfrac{i}{2}\omega_i(x)\bar{\Psi}_e^L\sigma_i\right)\begin{pmatrix}m_{\nu_e} & \\ & m_e\end{pmatrix}\begin{pmatrix}\psi_{\nu_e}^R\\ \psi_e^R\end{pmatrix}+\begin{pmatrix}\bar{\psi}_{\nu_e}^R & \bar{\psi}_e^R\end{pmatrix}\begin{pmatrix}m_{\nu_e} & \\ & m_e\end{pmatrix}\left(\tfrac{i}{2}\omega_i(x)\sigma_i\Psi_e^L\right).\end{aligned}\tag{6-172}$$

Each value of i is independent, since $\omega_i(x)$ varies independently for each i. If for any i, we have $\delta\mathcal{L}\neq 0$, then $\mathcal{L}$ cannot be symmetric, as none of the other terms in i can come to the rescue and cancel the non-zero term(s). So, let's look at $i = 1$. Then, that part of (6-172) is

$$\begin{aligned}&=-\tfrac{i}{2}\omega_1(x)\bar{\Psi}_e^L\begin{pmatrix} & 1\\ 1 & \end{pmatrix}\begin{pmatrix}m_{\nu_e} & \\ & m_e\end{pmatrix}\begin{pmatrix}\psi_{\nu_e}^R\\ \psi_e^R\end{pmatrix}+\tfrac{i}{2}\omega_1(x)\begin{pmatrix}\bar{\psi}_{\nu_e}^R & \bar{\psi}_e^R\end{pmatrix}\begin{pmatrix}m_{\nu_e} & \\ & m_e\end{pmatrix}\begin{pmatrix} & 1\\ 1 & \end{pmatrix}\Psi_e^L\\&=\tfrac{i}{2}\omega_1(x)\left(\underbrace{-\begin{pmatrix}\bar{\psi}_{\nu_e}^L & \bar{\psi}_e^L\end{pmatrix}\begin{pmatrix} & 1\\ 1 & \end{pmatrix}\begin{pmatrix}m_{\nu_e} & \\ & m_e\end{pmatrix}\begin{pmatrix}\psi_{\nu_e}^R\\ \psi_e^R\end{pmatrix}}_{\boxed{A}}\underbrace{+\begin{pmatrix}\bar{\psi}_{\nu_e}^R & \bar{\psi}_e^R\end{pmatrix}\begin{pmatrix}m_{\nu_e} & \\ & m_e\end{pmatrix}\begin{pmatrix} & 1\\ 1 & \end{pmatrix}\begin{pmatrix}\psi_{\nu_e}^L\\ \psi_e^L\end{pmatrix}}_{\boxed{B}}\right).\end{aligned}\tag{6-173}$$

In order for $\delta\mathcal{L}=0$, we need $\boxed{A}+\boxed{B}=0$.

$$\boxed{A}+\boxed{B}=-\left(\bar{\psi}_{\nu_e}^{L}\;\;\bar{\psi}_{e}^{L}\right)\begin{pmatrix} & 1\\ 1 & \end{pmatrix}\begin{pmatrix} m_{\nu_e}\psi_{\nu_e}^{R}\\ m_e\psi_e^{R}\end{pmatrix}+\left(\bar{\psi}_{\nu_e}^{R}\;\;\bar{\psi}_{e}^{R}\right)\begin{pmatrix} m_{\nu_e} & \\ & m_e\end{pmatrix}\begin{pmatrix}\psi_e^{L}\\ \psi_{\nu_e}^{L}\end{pmatrix}$$

$$=-\left(\bar{\psi}_{\nu_e}^{L}\;\;\bar{\psi}_{e}^{L}\right)\begin{pmatrix} m_e\psi_e^{R}\\ m_{\nu_e}\psi_{\nu_e}^{R}\end{pmatrix}+\left(\bar{\psi}_{\nu_e}^{R}\;\;\bar{\psi}_{e}^{R}\right)\begin{pmatrix} m_{\nu_e}\psi_e^{L}\\ m_e\psi_{\nu_e}^{L}\end{pmatrix}=\underbrace{-\bar{\psi}_{\nu_e}^{L}m_e\psi_e^{R}}_{\boxed{1}}\underbrace{-\bar{\psi}_{e}^{L}m_{\nu_e}\psi_{\nu_e}^{R}}_{\boxed{2}}\underbrace{+\bar{\psi}_{\nu_e}^{R}m_{\nu_e}\psi_e^{L}}_{\boxed{3}}\underbrace{+\bar{\psi}_{e}^{R}m_e\psi_{\nu_e}^{L}}_{\boxed{4}}. \qquad (6\text{-}174)$$

Terms with the same mass must cancel in order to have a zero result. Looking at two of these,

$$\boxed{1}+\boxed{4}=-\bar{\psi}_{\nu_e}^{L}m_e\psi_e^{R}+\bar{\psi}_{e}^{R}m_e\psi_{\nu_e}^{L}\neq 0, \qquad (6\text{-}175)$$

and that is all we need to show to prove that for the entire $\mathcal{L}$, $\delta\mathcal{L}\neq 0$, and $\mathcal{L}$ is not symmetric. (A similar result holds for the neutrino mass terms, and in fact for all fermion mass terms in all three generations.)

As an aside, fermion mass terms (but not boson mass terms) are symmetric under $U(1)$.

$$\delta\left(m\bar{\psi}_l\psi_l\right)=m\left(\delta\bar{\psi}_l\right)\psi_l+m\bar{\psi}_l\left(\delta\psi_l\right)=m\left(-iYf(x)\bar{\psi}_l\right)\psi_l+m\bar{\psi}_l\left(iYf(x)\psi_l\right)=0. \qquad (6\text{-}176)$$

In QED, we had a fermion mass term and it did not destroy the $U(1)$ symmetry of e/m.

Bottom line: Fermion mass terms in $\mathcal{L}$ are not symmetric under $SU(2)$ transformations (though they are under $U(1)$). Boson mass terms (as we showed in the main part of this chapter and for QED in Vol. 1) are not symmetric under $U(1)$, and that is a major reason the photon, as well as the B^{μ} field, must be massless.

6.11 Problems

1. Show $\mathcal{L}=\partial^{\mu}\phi^{\dagger}\partial_{\mu}\phi-\mu^2\phi^{\dagger}\phi$ is symmetric under the global $U(1)$ transformation in the (complex) rep of $\phi'=U(1)\phi=e^{i\alpha}\phi=(1+i\alpha+\ldots)\phi$. Do this for both the finite transformation (α large) and the infinitesimal ($\alpha \ll 1$) transformation.
2. Show that $\mathcal{L}$ of Prob. 1 is symmetric under the global $SO(2)$ transformation in the (real) rep of $SO(2)=\begin{bmatrix}\cos\alpha & -\sin\alpha\\ \sin\alpha & \cos\alpha\end{bmatrix}=\begin{bmatrix}1-\ldots & -\alpha-\ldots\\ \alpha+\ldots & 1-\ldots\end{bmatrix}$. Do this for both the finite transformation (α large) and the infinitesimal ($\alpha \ll 1$) transformation.
3. Show that the interaction $B^{\mu\nu}$ is the same as the free $B^{\mu\nu}$.
4. Show that $-\frac{1}{4}G_i^{\mu\nu}G_{i\,\mu\nu}=-\frac{1}{4}W_i^{\mu\nu}W_{i\,\mu\nu}+g\varepsilon_{ijk}W_{i\mu}W_{j\nu}\partial^{\mu}W_k^{\nu}-\frac{1}{4}g^2\varepsilon_{ijk}\varepsilon_{ilm}W_j^{\mu}W_k^{\nu}W_{l\mu}W_{m\nu}$. Hint: Use (6-57), i.e., $\varepsilon_{ijk}\left(W_k^{\mu}W_j^{\nu}+W_j^{\nu}W_k^{\mu}\right)=-2\varepsilon_{ijk}W_j^{\mu}W_k^{\nu}$, where the same relation is valid after lowering indices. Also, be aware, as noted after (6-56), that different boson fields (such as $W_1{}^{\mu}$ and $W_2{}^{\mu}$) commute with each other and with derivatives of the different fields (such as $W_1{}^{\mu}$ and $\partial_{\nu}W_2{}^{\mu}$).
5. Derive (6-76) from (6-67) using the first relation in the second row of (6-42).
6. Show that, using $\delta\mathcal{L}=\mathcal{L}(\delta A,B,C,\ldots)+\mathcal{L}(A,\delta B,C,\ldots)+\mathcal{L}(A,B,\delta C,\ldots)+\ldots$, in the QED Lagrangian of $\mathcal{L}=-\frac{1}{4}F_{\nu\beta}F^{\nu\beta}+\bar{\psi}\left(i\gamma^{\nu}\partial_{\nu}-m\right)\psi+e\bar{\psi}\gamma^{\nu}\psi A_{\nu}$ under the infinitesimal form of the $U(1)$ transformation set $\psi\rightarrow\psi'=e^{-i\alpha(x^{\mu})}\psi$ and $A_{\nu}\rightarrow A'_{\nu}=A_{\nu}-\frac{1}{e}\partial_{\nu}\alpha(x^{\mu})$ has $\delta\mathcal{L}=0$, and thus $\mathcal{L}$ is symmetric.
7. Show that the first and last terms in the last row of (6-129) cancel.
8. Show that ${}_{B}Y_{l\nu_{l'}}\delta\left(-\bar{\Psi}_l^{L}\tilde{\Phi}\psi_{\nu_{l'}}^{R}-\bar{\psi}_{\nu_{l'}}^{R}\tilde{\Phi}^{\dagger}\Psi_l^{L}\right)=0$, where δ represents the infinitesimal electroweak transformation set we use in this chapter.

9. From the $SU(2)$ terms in the interaction e/w Lagrangian, deduce the form of the corresponding 2nd order (tree level) S operator term and the associated S matrix term for the decay of the tau particle into an electron and neutrinos. Draw the relevant Feynman diagram.
10. Given the Y values for quarks in (6-111), the weak isospin charges for quark doublets parallel to those for leptons of (6-95), the knowledge that quark singlets have weak isospin charge of zero, and (6-94), show that all up type quarks have electric charge +2/3, and all down type quarks have electric charge –1/3.
11. Show that if the $\mathcal{L}$ had a $W_1{}^{\mu}$ field with mass, having a mass term of form $m_W^2 W_1^{\mu} W_{1\mu}$, then $\mathcal{L}$ would not be symmetric. Note: You only have to look at the $U(1)$ transformation.

6.12 Potentials in QFT (Added July 2025)

6.12.1 Review of Potentials in QED: Volume 1

For the vast majority of the theory and problems in QFT, there is no external potential field (no external forces acting on particles). In those cases, what we consider classically to be a potential field is, on the QFT level, a collection of virtual particles being exchanged between the real particles in macroscopic objects. See, for example, the end of Chap. 8 in Vol. 1, Sect. 8.10, pgs. 243-246.

However, there are cases where we consider what, in QFT jargon is called an "external field", and that is essentially a potential field, as understood classically, but represented in QFT as a fixed source (not a particle, *per se*) that emits and absorbs virtual photons. See, in above reference, from the bottom of pg. 415 to top of pg. 416 (sub-section titled "Changes to Feynman Rules for External Fields") and the Appendix of Chap. 16 (titled *Deriving Feynman Rules for Static, External (Potential) Fields*), pg. 430 to 431a.

Wholeness Chart 6-11. Potential Fields in QED

Model Employed	How Often Used	Potential Field in QFT?	Result	See *SFQFT* Vol. 1
Particles interacting with one another	~98% of the time	No	Macroscopic potential field	Sect. 8.10, pgs. 243-246 graphically; most of book generally
Particle interacting with external field	~2% of the time	Yes	Macroscopic potential field	bottom pg. 415 to top pg. 416 and Chap. 16 Appendix, pgs. 430-431a

6.12.2 The Higgs Potential: Volume 2

Sect. 6.12.1 focused on a potential as affecting a particle, where the potential is external to the particle. This could occur, in QFT math, two ways. One, as an interaction between particles, where one particle feels the effect of the other via a virtual particle exchange. (First row below column titles in above chart.) Each particle moves under the interaction with the other. The second way is interaction with an external field. This is also mediated by a virtual particle emitted from the external field. The external field is considered stationary.

There is a third way we talk about potential, as we do in this (and subsequent) chapter(s).

Classically, this can be thought of much like a trampoline. The gravitational potential pulls downward, more in the middle than the sides. This is an external potential. But as the trampoline stretches, potential energy builds up within the trampoline itself, like in a spring. The trampoline, like a stretched spring, in its tendency to move back to equilibrium with gravity, can cause something on it (like a gymnast) to move, and in so doing, potential energy changes to kinetic energy. So, a classical field can have potential energy stored inside it (like stretching in this example). It can also react to an external potential (like the gravity potential field in this examples).

In QFT, as we see in this chapter, the Higgs field can be considered to have its own internal potential energy, which will naturally tend toward the lowest potential energy state, like the trampoline tending to return to the unstretched state, or a ball rolling into the center of a bowl. This is actually modeled as potential energy *density*, since we are working with a field, and thus, we deal with energy densities (potential, kinetic, etc.) at each point in space for the field. This potential energy density for the Higgs field is what we mean by the Higgs potential, and what we study in this chapter.

Chapter 7

The Low Energy Universe: Broken Symmetry

"Nothing in physics seems so hopeful as the idea that it is possible for a theory to have a high degree of symmetry that was hidden from us in everyday life. The physicist's task is to find this deeper symmetry."
Steven Weinberg

7.0 Preliminaries

Lagrangian e/w terms change at low energy and fields gain mass

In this chapter we investigate what happens to the high energy Lagrangian of the previous chapter as energy density decreases from its levels at the end of the electroweak epoch down to where it is today. We will see, mathematically, how the elegant symmetry of the Lagrangian we showed at high energy is broken, and in the process, how various fields acquire mass, or more specifically, how mass terms spring up in the Lagrangian. And then we will investigate weak and electromagnetic interactions at contemporary energy levels, including an overview of the methodology involved in calculating their probabilities. The actual calculations are, as might be expected, quite extensive, so we will save them for a later chapter.

7.0.1 Chapter Overview

Topics in this chapter

In this chapter, we will

- look at three different e/w models: Goldstone, Higgs, GSW (Glashow-Salaam-Weinberg),
- discern why GSW is the correct model, and
- examine, in detail, the GSW model at contemporary energy levels.

As part of the above, we will

- explore the effect of the Higgs potential on the universe as temperature drops,
- follow the concomitant impact on the Lagrangian including
  - its loss of symmetry, and
  - the changes in form of certain of its terms,
- see how fields acquire mass, and
- investigate subtleties involving particle mass states and e/w interactions.

Wholeness Chart 7-12 by Latham Boyle on pg. 252b, is an excellent one page, bottom line summary of this material. You may wish to follow along with that as you progress through the chapter.

7.0.2 Background for Electroweak Symmetry Breaking

How to determine mass from terms in the Lagrangian

As energy levels fall, the Lagrangian changes form. In particular, mass terms arise where there were no such terms to begin with. So, we need to be able to look at the resulting Lagrangian and discern which terms are mass terms and how much mass is represented in each such term. To this end, we review below the general forms of the Lagrangian mass terms for bosons and fermions.

Scalar Mass Terms

For complex scalar fields

Note that in the usual Lagrangian for a free *complex* scalar boson field,

$$\mathcal{L} = \partial^{\mu}\phi^{\dagger}\partial_{\mu}\phi - \mu^{2}\phi^{\dagger}\phi \qquad \text{(Lagrangian for free complex scalar field)}, \qquad (7\text{-}1)$$

the mass μ squared is the coefficient of the $-\phi^{\dagger}\phi$ term, and we can make that a general rule for similar such terms for any complex boson field. The coefficient of such a term (which is called a "mass term") for any given boson field is the mass squared for that field.

For a *real* scalar boson field, on the other hand, the mass term looks a little different, i.e., it has an additional factor of ½ in front (and $\phi^{\dagger}=\phi$), as in

For real scalar fields

$$\mathcal{L}=\tfrac{1}{2}\partial^{\mu}\phi\partial_{\mu}\phi-\tfrac{1}{2}\mu^{2}\phi\phi \qquad \text{(Lagrangian for free real scalar field)} \; . \tag{7-2}$$

As an example of the mass of a boson mass term we are already familiar with, the Lagrangian in QED has no term of form $A^{\mu}A_{\mu}$ (comparable to the term with $\phi\phi$ in (7-2)). So, the coefficient for that term is zero, which means the photon is massless.

Bottom line

Consequently, as a general principle, if we have a Lagrangian for any scalar field, we can simply read off the mass of the associated bosonic particle from the mass term, as in (7-1) and (7-2).

Fermion Mass Term

For Dirac particles, the free Lagrangian has the form

For fermion fields

$$\mathcal{L}=i\bar{\psi}\partial\!\!\!/\psi-m\bar{\psi}\psi \; . \tag{7-3}$$

So, given the Lagrangian for any spinor field, we can simply read off the associated particle mass as the coefficient of the $-\bar{\psi}\psi$ term.

Vector Mass Term

Do **Problem 1** to show that the mass term for spin 1 fields is as shown in (7-4).

For vector fields

$$\mathcal{L}_{0}^{1}=-\tfrac{1}{4}\left(\partial^{\nu}Z^{\mu}-\partial^{\mu}Z^{\nu}\right)\left(\partial_{\nu}Z_{\mu}-\partial_{\mu}Z_{\nu}\right)+\tfrac{1}{2}m_{Z}^{2}Z^{\mu}Z_{\mu} \qquad Z^{\mu}=\text{real massive vector field} \; . \tag{7-4}$$

Note that the sign of the mass terms for vectors is positive; for scalars, negative. For complex vector fields, as with complex scalar fields, there is no factor of ½.

7.1 Goldstone Model

7.1.1 Assumptions for the Goldstone Model

An early attempt to describe weak interactions: the Goldstone model

Wholeness Chart 7-1 on pg. 215 summarizes the Goldstone model, which you may not understand quite yet, but could wish to follow along with as we develop the model. That model is based on

1) a complex scalar Higgs field ϕ,
2) global $U(1)$ symmetry with no interactions, and
3) a particular form for potential energy density $\mathcal{V}$ as in

The Goldstone potential in $\mathcal{L}$ has ϕ to 4th power term

$$\mathcal{L}=\partial^{\mu}\phi^{\dagger}\partial_{\mu}\phi\underbrace{-\mu^{2}\phi^{\dagger}\phi-\lambda\left(\phi^{\dagger}\phi\right)^{2}}_{-\mathcal{V}} \qquad \text{Complex } \phi=\tfrac{1}{\sqrt{2}}\left(\phi_{1}+i\phi_{2}\right) \quad \phi_{1},\phi_{2} \text{ real}$$

$$=\tfrac{1}{2}\left(\partial^{\mu}\phi_{1}\right)\left(\partial_{\mu}\phi_{1}\right)+\tfrac{1}{2}\left(\partial^{\mu}\phi_{2}\right)\left(\partial_{\mu}\phi_{2}\right)\underbrace{-\tfrac{\mu^{2}}{2}\left(\phi_{1}^{2}+\phi_{2}^{2}\right)-\tfrac{\lambda}{4}\left(\phi_{1}^{2}+\phi_{2}^{2}\right)^{2}}_{-\mathcal{V}} \; . \tag{7-5}$$

As noted in Chap. 6, in this model, μ^2 is negative (μ is imaginary) and the potential density is

$$\mathcal{V}=\mu^{2}\phi^{\dagger}\phi+\lambda\left(\phi^{\dagger}\phi\right)^{2}=-\frac{\left|\mu^{2}\right|}{2}\left(\phi_{1}^{2}+\phi_{2}^{2}\right)+\tfrac{\lambda}{4}\left(\phi_{1}^{2}+\phi_{2}^{2}\right)^{2} \quad \mu^{2}<0 \quad \lambda>0 \; . \tag{7-6}$$

7.1.2 The True and False Vacuums

A closer look at that potential

We plotted the potential of (7-6) in Chap. 6, repeated below as Fig. 7-1, and noted that it is commonly referred to as the "Mexican hat" potential. Note that for small ϕ_1 and ϕ_2, the λ term is negligible and the shape approaches an upside-down paraboloid (dependence on the squares of ϕ_1 and ϕ_2). For large ϕ_1 and ϕ_2, the λ term dominates and $\mathcal{V}$ approximates a 4th power dependence on ϕ_1 and ϕ_2. The $\phi_1=\phi_2=0$ location (where $\mathcal{V}=0$), which we referred to in Chap. 6 as the high energy location, is generally referred to as the false vacuum, for reasons we are about to delineate.

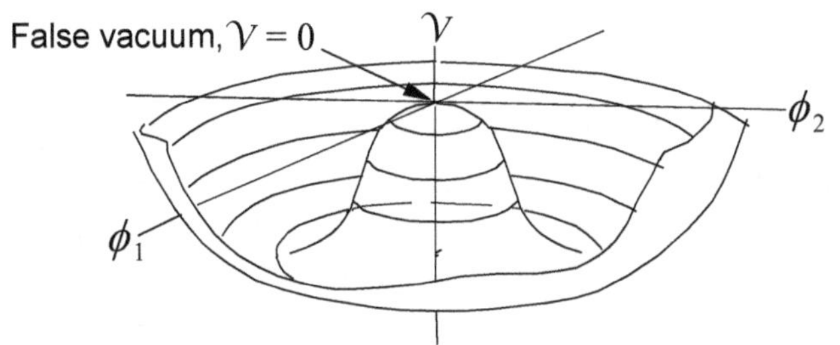

Plot of the Goldstone potential

Figure 7-1. $\mathcal{V}$ vs Higgs Field ϕ in Goldstone Model (Mexican Hat Potential)

In classical field theory, the false vacuum point is unstable for a classical potential shaped like that of Fig. 7-1. But recall from Chap. 6 that $\mathcal{V}$ in Fig. 7-1 really represents the expectation value for the potential $\overline{\mathcal{V}}$, which is what we would measure classically, on average. And $\overline{\mathcal{V}}$ must behave like classical potentials do. The universe with a field ϕ having a potential like that shown in Fig. 7-1 will be unstable and tend to move to lower $\mathcal{V}$. In the graph, it will tend to slide down the surface sloping away from the vertical centerline and settle somewhere in the circular trough surrounding the centerline. Which particular location it ends up in circumferentially is random. See the LHS of Fig. 7-2.

The universe slides from unstable false vacuum to stable true vacuum

We can now see why we called the top of the Mexican hat the false vacuum. It does not represent the lowest energy state. What we call the true vacuum (for which there are many possible locations in the trough of the hat) is the lowest energy state. Our definition of vacuum in this case corresponds to the lowest energy state, *not* to the absence of real particles, since the true vacuum has a non-zero density of ϕ particles.

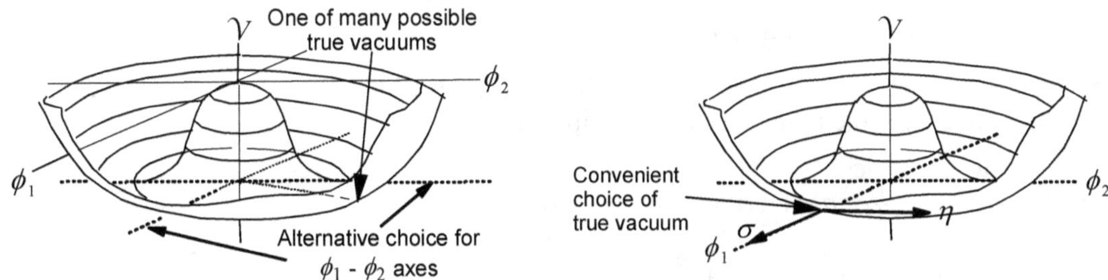

Figure 7-2. Possible True Vacuums and Preferred True Vacuum

We align our axes to simplify analysis

Due to symmetry, there is no loss in generality in choosing as our true vacuum the location shown in the RHS of Fig. 7-2. Also, where we place our coordinate axes is always arbitrary. So, as usual, we place them where it will be (as we will see) easiest to do our math. We will shortly discuss the new σ - η axes coordinate system, where the σ axis is taken to be aligned with the ϕ_1 axis.

7.1.3 Characteristics of the Minimum Potential (True Vacuum)

Location of the Minimum

Our chosen minimum of $\mathcal{V}$ for $\phi_2 = 0$ is located at a real, positive value for ϕ_1 denoted by the symbol v. This allows us to find an expression relating μ and v, from (7-6), as follows.

$$\left.\frac{\partial \mathcal{V}}{\partial \phi_1}\right|_{\substack{\phi_1=v\\ \phi_2=0}} = 0 = \left(\mu^2\phi_1 + \tfrac{\lambda}{4}2\left(\phi_1^2+\phi_2^2\right)2\phi_1\right)_{\substack{\phi_1=v\\ \phi_2=0}} \quad\rightarrow\quad \mu^2 = -\lambda\phi_1^2\Big|_{\phi_1=v} = -\lambda v^2 < 0\,, \qquad \text{(7-7)}$$

Thus, the effective "mass" μ of ϕ (for λ positive) is imaginary, which contrasts with what we know about typical particles having real mass. And the location of the minimum is

Determining location of true vacuum, at minimum $\mathcal{V}$

$$\phi_1\Big|_{min} = v = \sqrt{\frac{-\mu^2}{\lambda}} > 0\,. \qquad \text{(7-8)}$$

Defining New Fields at the Minimum

With an eye to future analyses, we define new fields σ and η as

$$\sigma = \phi_1 - v \qquad \eta = \phi_2 \qquad \rightarrow \qquad \phi_1 = \sigma + v \qquad \phi_2 = \eta\,, \tag{7-9}$$

Defining new fields at the minimum (at the true vacuum)

shown graphically in Fig. 7-2. We then substitute the RHS of (7-9) into the potential $\mathcal{V}$ in (7-6) to get

$$\begin{aligned}\mathcal{V} &= \tfrac{\mu^2}{2}\left(\phi_1^2+\phi_2^2\right)+\tfrac{\lambda}{4}\left(\phi_1^2+\phi_2^2\right)^2 = \tfrac{\mu^2}{2}\left(\sigma+v\right)^2+\tfrac{\mu^2}{2}\eta^2+\tfrac{\lambda}{4}\left(\left(\sigma+v\right)^2+\eta^2\right)^2 \\ &= \tfrac{\mu^2}{2}\left(\sigma^2+\eta^2\right)+\mu^2\sigma v+\tfrac{\mu^2}{2}v^2+\tfrac{\lambda}{4}\left(\sigma^2+2\sigma v+v^2+\eta^2\right)^2 .\end{aligned} \tag{7-10}$$

Using the RHS of (7-7) for μ^2, this becomes

$$\mathcal{V} = \underbrace{-\tfrac{\lambda v^2}{2}\left(\sigma^2+\eta^2\right)-\lambda v^3\sigma-\tfrac{\lambda v^4}{2}}_{\boxed{X}}+\tfrac{\lambda}{4}\left(\sigma^2+2\sigma v+v^2+\eta^2\right)^2 \tag{7-11}$$

Expanding $\mathcal{V}$ in terms of the new fields

Working out the square at the end of (7-11), we find

$$\begin{aligned}\mathcal{V} = \underbrace{\lambda\left(-\overset{\boxed{1}}{\frac{v^2\sigma^2}{2}}-\overset{\boxed{2}}{\frac{v^2\eta^2}{2}}\overset{\boxed{3}}{-v^3\sigma}-\frac{v^4}{2}\right)}_{\boxed{X}} &+\lambda\Big(\tfrac{1}{4}\sigma^4+\tfrac{1}{4}\sigma^2 2\sigma v+\overset{\boxed{1a}}{\tfrac{1}{4}\sigma^2v^2}+\tfrac{1}{4}\sigma^2\eta^2 \\ &+\tfrac{1}{4}\sigma^2 2\sigma v+\tfrac{1}{4}4\sigma^2v^2+\overset{\boxed{3a}}{\tfrac{1}{4}2\sigma v v^2}+\tfrac{1}{4}2\sigma v\eta^2 \\ &+\overset{\boxed{1b}}{\tfrac{1}{4}\sigma^2v^2}+\overset{\boxed{3b}}{\tfrac{1}{4}2\sigma v v^2}+\tfrac{1}{4}v^4+\overset{\boxed{2a}}{\tfrac{1}{4}\eta^2v^2} \\ &+\tfrac{1}{4}\sigma^2\eta^2+\tfrac{1}{4}2\sigma v\eta^2+\overset{\boxed{2b}}{\tfrac{1}{4}v^2\eta^2}+\tfrac{1}{4}\eta^4\Big).\end{aligned} \tag{7-12}$$

The terms with the same numbers (e.g., 1,2,3) in boxes above them cancel. So, we are left with (7-13) (where the terms with letters over them have the same form and can be combined)

$$\begin{aligned}\mathcal{V} &= \lambda\begin{pmatrix}\overset{\boxed{A}}{-\tfrac{1}{2}v^4}+\tfrac{1}{4}\sigma^4+\overset{\boxed{B}}{\tfrac{1}{2}\sigma^3v}+\overset{\boxed{C}}{\tfrac{1}{4}\sigma^2\eta^2}+\overset{\boxed{B}}{\tfrac{1}{2}\sigma^3v}+\sigma^2v^2+\overset{\boxed{D}}{\tfrac{1}{2}\sigma v\eta^2} \\ \overset{\boxed{A}}{+\tfrac{1}{4}v^4}+\overset{\boxed{C}}{\tfrac{1}{4}\sigma^2\eta^2}+\overset{\boxed{D}}{\tfrac{1}{2}\sigma v\eta^2}+\tfrac{1}{4}\eta^4\end{pmatrix} \\ &= \lambda\left(\overset{\boxed{A}}{-\frac{v^4}{4}}+\frac{\sigma^4}{4}+\overset{\boxed{B}}{\sigma^3v}+\overset{\boxed{C}}{\frac{\sigma^2\eta^2}{2}}+\overset{\boxed{D}}{\sigma v\eta^2}+\sigma^2v^2+\frac{\eta^4}{4}\right).\end{aligned} \tag{7-13}$$

Rearranging (7-13), we have

$$\begin{aligned}\mathcal{V} &= \lambda\left(\overset{\boxed{A}}{-\frac{v^4}{4}}+\sigma^2v^2+\overset{\boxed{B}}{\sigma^3v}+\overset{\boxed{D}}{\sigma v\eta^2}+\tfrac{1}{4}\left(\sigma^4+\overset{\boxed{C}}{2\sigma^2\eta^2}+\eta^4\right)\right) \\ &= -\tfrac{1}{4}\lambda v^4+\tfrac{1}{2}\left(2\lambda v^2\right)\sigma^2+\left(\lambda v\right)\sigma\left(\sigma^2+\eta^2\right)+\tfrac{\lambda}{4}\left(\sigma^2+\eta^2\right)^2 .\end{aligned} \tag{7-14}$$

The result: $\mathcal{V}$ in terms of the new fields

7.1.4 Interpretation of Terms in $\mathcal{V}$ in Light of the New Fields

The first term in (7-14) and the Vacuum Expectation Value of $\mathcal{V}$

The first term in the last row of (7-14) is a constant, that represents a contribution to the energy density of the vacuum at the true vacuum ($\sigma = \eta = 0$), i.e., after symmetry breaking, by the scalar field. Thus, the vacuum energy density remnant left in the vacuum after symmetry breaking that one would expect to measure (the vacuum expectation value or VEV) is the negative value

$$\text{true vaccum VEV of } \mathcal{V} = \left\langle 0_{true}\right|-\tfrac{1}{4}\lambda v^4\left|0_{true}\right\rangle = -\tfrac{1}{4}\lambda v^4 . \tag{7-15}$$

Expectation value of energy density $\mathcal{V}$ at minimum in terms of the new fields

This is the vertical distance in Fig. 7-2 from the false vacuum to the true vacuum and is called the Higgs condensate. Commonly, condensation is a phase transition of a gas to a liquid involving a fluid that transitions to a lower energy state, and the analogy gives rise to the name.

Aside on the VEV of ϕ_1

We can see, from (7-9) and (7-16) below that v in the $-\frac{1}{4}\lambda v^4$ term is the VEV of the field ϕ_1 at the true vacuum.

$$\text{VEV of } \phi_1 = \langle 0_{true}|\phi_1|0_{true}\rangle = \langle 0_{true}|\sigma + v|0_{true}\rangle = \langle 0_{true}|v|0_{true}\rangle = v \qquad (7\text{-}16)$$

The field gets a VEV at true vacuum as well

Typically, in QFT, fields have zero VEVs, i.e., one cannot measure the field directly. This is because a typical quantum field contains only construction and destruction operators acting alone and no number operators (no bilinear combinations of operators), so

$$\text{VEV of typical field } \phi_{typical} = \langle 0|\phi_{typical}|0\rangle = 0\,. \qquad (7\text{-}17)$$

though, as we recall, fields normally have no VEV

But in our case, there is a constant term v in our field (along with a typical quantum field σ), so the Higgs field ϕ_1 acquires a VEV as a result of the symmetry breaking.

End of Aside and Back to the VEV of $\mathcal{V}$

We now can see that (7-15) is proportional to the fourth power of the VEV of the field ϕ_1 (i.e. of v) of (7-16), and linearly dependent on the positive constant λ. One can think of (7-15) as a constant value for the energy density of the Higgs field pervading all space, from the point of view of the true vacuum. However, as we mentioned earlier, much like what is done in other parts of QFT, we ignore constant vacuum energy, no matter how large, and no matter whether it is negative or positive (negative in this case).

The large VEV for $\mathcal{V}$ is negative and generally just ignored

The other terms in (7-14)

The other terms in the bottom row of (7-14) have been arranged so constants are on the left of each term and fields on the right. All of these terms will end up in the Lagrangian density (7-5), with opposite signs. There, the 2nd term in the bottom row of (7-14) acts like a mass term in the free Lagrangian for the σ field, where $2\lambda v^2$ is mass squared. The 3rd and 4th terms are not bilinear (as free terms in the Lagrangian are), but tri and quadrilinear and thus, represent interactions between fields (including the interaction of the σ with itself via σ^3 and σ^4 terms).

For bilinear terms of a certain field (such as σ here) in $\mathcal{L}$, the field equation is linear in that field (via the Euler-Lagrange equation). For cubic or quartic terms in $\mathcal{L}$, the field equation is nonlinear.

Note the mass of our new, *real* field σ, the true vacuum Higgs field, is $\sqrt{2\lambda v^2}$. The mass of the η field is zero, as there is no term having form η^2. We can think of the mass term as a term that is non-zero when small oscillations are made about the zero point ($\sigma = 0$ here) in the direction of the field in the Mexican hat diagram (in the σ axis direction here). Note that the σ field has real, not imaginary mass. Graphically, a real boson mass has an upward curvature (2nd derivative positive) and an imaginary mass (like ϕ if we interpret the ϕ^2 term as a mass term) has a downward curvature (2nd derivative negative).

Reading off the true vacuum Higgs mass from a term in $\mathcal{V}$

The original field ϕ is also commonly called the Higgs field, but, more precisely, it is the false vacuum Higgs field. (We may be getting a little ahead of ourselves, as we need a more elaborate symmetry breaking model than this in order to match reality, but the Higgs particle found at CERN in 2012 was the true vacuum Higgs, not the false vacuum Higgs.)

Two versions of the Higgs field; ϕ at false vacuum, σ at true vacuum

For the η field, small oscillations in the η axis direction produce no change in $\mathcal{V}$, as the curve is essentially flat in that direction. So, the η mass is zero. The η field is known as a Goldstone (or a Nambu-Goldstone) boson field (after its discoverers Yoichiro Nambu and Jeffrey Goldstone). In general, any *massless* scalar boson that results from a broken global symmetry is a Nambu-Goldstone boson.

The η field is massless, called a Nambu-Goldstone boson

7.1.5 Further Comment on Vacuum Energy

Note we have two sources for large vacuum energy density that we ignore, the zero-point-energy (ZPE) for the Higgs (and other particles) that arises in similar manner to what we found in general, in Vol.1, for bosons, and, as well, the negative Higgs condensate term $-\frac{1}{4}\lambda v^4$ in (7-15).

Two kinds of large vacuum energy: ZPE and Higgs condensate

In practice, the true vacuum is often simply assumed to have zero potential energy density, and in that case, the false vacuum would have positive potential energy density. Graphically, it depends on where we wish to place our horizontal axes in Fig. 7-2. If we place the plane of those axes at the false vacuum (solid lines), the true vacuum has negative potential energy density and the false vacuum, zero. If we place that plane at the true vacuum (dotted lines), then the true vacuum has zero potential energy density and the false vacuum has a positive one.

By switching axes, we can make true vacuum energy zero and false vacuum energy positive

There is no universally accepted choice of location, height wise, for the horizontal axes, though, as noted, in practice we typically assume present day vacuum energy is zero, and in doing so, we are, in effect, adopting the dotted line axes placement in Fig. 7-2.

There is no universally adopted choice for the location of the axes

7.1.6 Comparing False and True Vacuums

<u>Degeneracy</u>

Note from Fig. 7-2 that there are many possible true vacuums, but only one false vacuum. One says the true vacuum state is degenerate, but the false vacuum state is not.

True vacuum is degenerate; false vacuum is not

<u>Rotational Symmetry Transformation about False Vacuum (Vertical Axis in Fig. 7-1)</u>

Consider the rotational symmetry of the field ϕ about the false vacuum. Mathematically, we can characterize that as a transformation in a complex space or in a real space, as follows.

$\mathcal{V}$ (and $\mathcal{L}$) symmetric about false vacuum

1) Complex characterization

The global $U(1)$ transformation for the Goldstone model can be expressed mathematically in a complex representation of ϕ as shown in the first line of (7-5), i.e., where

Complex characterization of ϕ

$$\text{Complex } \phi = \frac{1}{\sqrt{2}}\left(\phi_1 + i\phi_2\right) \quad \phi_1, \phi_2 \text{ real}, \tag{7-18}$$

as (with α = constant for global symmetry)

$$\begin{aligned} \phi' &= U(1)\phi = e^{i\alpha}\phi = (1 + i\alpha + ...)\phi \quad \xrightarrow[\text{i.e., small } \alpha]{\text{infintesimal}} \quad \phi' \approx (1 + i\alpha)\phi = \phi + \delta\phi \\ \phi'^{\dagger} &= \phi^{\dagger}U^{\dagger}(1) = \phi^{\dagger}e^{-i\alpha} = \phi^{\dagger}(1 - i\alpha + ...) \quad \xrightarrow[\text{i.e., small } \alpha]{\text{infintesimal}} \quad \phi'^{\dagger} \approx \phi^{\dagger}(1 - i\alpha) = \phi^{\dagger} + \delta\phi^{\dagger}. \end{aligned} \tag{7-19}$$

So, for infinitesimal transformations

$$\delta\phi = i\alpha\phi \qquad \delta\phi^{\dagger} = -i\alpha\phi^{\dagger}. \tag{7-20}$$

Do **Problem 2** to show (7-5) is symmetric under the transformation (7-19).

For $\phi = 0$, there are no ϕ particles (zero expectation value for particle number density along the ϕ_1 and ϕ_2 axes as described in Chap. 6, pg. 165 under heading "The Main Point"), so even though ϕ is a field (and not a state), $\phi = 0$ represents the zero ϕ particle state (i.e., the false vacuum state) of $|0_{false}\rangle$. Thus, we must have $\alpha\phi = 0 = \delta\phi$. The transformation $U(1)$ leaves $\phi = 0$ at $\phi = 0$, i.e., unchanged.

$U(1)$ acting on ϕ at false vacuum leaves ϕ unchanged

2) Real characterization

Alternatively, we can characterize the same field under the same physical transformation solely in terms of real entities, i.e., as an $SO(2)$ transformation, as follows. (The astute reader may notice a sign change below in α from what we did earlier, but this is ultimately irrelevant, as we can define the direction of α as we wish, plus it has opposite signs for active vs passive transformations. We here choose the sign commonly employed in the literature.)

Real characterization of ϕ

$$\text{Real } \phi = \frac{1}{\sqrt{2}}\begin{bmatrix} \phi_1 \\ \phi_2 \end{bmatrix}, \tag{7-21}$$

$$\phi' = \begin{bmatrix} \cos\alpha & \sin\alpha \\ -\sin\alpha & \cos\alpha \end{bmatrix} \begin{bmatrix} \frac{\phi_1}{\sqrt{2}} \\ \frac{\phi_2}{\sqrt{2}} \end{bmatrix} \quad \xrightarrow[\text{i.e., small } \alpha]{\text{infintesimal}} \quad \phi' \approx \begin{bmatrix} 1 & \alpha \\ -\alpha & 1 \end{bmatrix} \begin{bmatrix} \frac{\phi_1}{\sqrt{2}} \\ \frac{\phi_2}{\sqrt{2}} \end{bmatrix} = \begin{bmatrix} \frac{\phi_1}{\sqrt{2}} + \alpha\frac{\phi_2}{\sqrt{2}} \\ \frac{\phi_2}{\sqrt{2}} - \alpha\frac{\phi_1}{\sqrt{2}} \end{bmatrix} = \phi + \delta\phi. \tag{7-22}$$

The matrix on the LHS of (7-22) is the simple rotation matrix in 2D. Thus, both forms of the transformation are effectively rotations about the vertical axis in Fig. 7-1. Note,

$$\delta\phi = \begin{bmatrix} \delta\phi_1 \\ \delta\phi_2 \end{bmatrix} = \frac{1}{\sqrt{2}}\begin{bmatrix} \alpha\phi_2 \\ -\alpha\phi_1 \end{bmatrix}. \tag{7-23}$$

SO(2) acting on ϕ at false vacuum leaves ϕ unchanged

Do **Problem 3** to show (7-5) is symmetric under the transformation (7-22).

---

3) Effect of rotation about false vacuum at different ϕ

Note that for the field ϕ, from (7-20) (or equivalently, in real space, from (7-23)),

$$\left.\begin{array}{l} \text{at } \phi=0,\ \delta\phi=0 \rightarrow U(1)\phi=0 \\ \text{at } \phi\neq 0,\ \delta\phi\neq 0 \rightarrow U(1)\phi\neq 0 \end{array}\right\} \begin{array}{l}\text{And also, } \delta\mathcal{L}=0 \text{ at all } \phi_1 \text{ and } \phi_2 \text{ under} \\ \text{the } U(1) \text{transformation about false vacuum}\end{array} \quad (7\text{-}24)$$

U(1) acting on complex ϕ (or SO(2) acting on real ϕ) leaves $\mathcal{L}$ unchanged for all ϕ; but ϕ changes when $\phi \neq 0$

Rotation about False Vacuum with Respect to True Vacuum Real Fields σ and η

Only real characterization shown

The above results can be expressed in terms of the true vacuum fields σ and η, instead of ϕ (or equivalently, ϕ_1 and ϕ_2). We define the true vacuum fields as the *SO*(2) doublet Σ, which here is *not* the spin operator, as in (7-25).

Same group action, but now about false vacuum axis in terms of true vacuum real fields σ and η

$$\text{At } \phi=\frac{1}{\sqrt{2}}\begin{bmatrix}\phi_1\\ \phi_2\end{bmatrix}=\begin{bmatrix}0\\0\end{bmatrix} \quad \left(\text{in terms of } \sigma \text{ and } \eta \rightarrow \Sigma=\frac{1}{\sqrt{2}}\begin{bmatrix}\sigma\\ \eta\end{bmatrix}=\frac{1}{\sqrt{2}}\begin{bmatrix}\phi_1-v\\ \phi_2\end{bmatrix}=\frac{1}{\sqrt{2}}\begin{bmatrix}-v\\0\end{bmatrix}\right) \quad (7\text{-}25)$$

$$\text{At } \phi=\frac{1}{\sqrt{2}}\begin{bmatrix}v\\0\end{bmatrix} \quad \left(\text{in terms of } \sigma \text{ and } \eta \rightarrow \Sigma=\frac{1}{\sqrt{2}}\begin{bmatrix}\sigma\\ \eta\end{bmatrix}=\frac{1}{\sqrt{2}}\begin{bmatrix}0\\0\end{bmatrix}\right) \quad (7\text{-}26)$$

From (7-25), (7-26), and (7-23), we have

$$\left.\begin{array}{l} \text{at } \Sigma=\frac{1}{\sqrt{2}}\begin{bmatrix}-v\\0\end{bmatrix},\quad \phi=\frac{1}{\sqrt{2}}\begin{bmatrix}0\\0\end{bmatrix},\quad \delta\phi=0\rightarrow\delta\phi_1=\delta\phi_2=0\rightarrow\delta\Sigma=0 \\ \text{at } \Sigma=\frac{1}{\sqrt{2}}\begin{bmatrix}0\\0\end{bmatrix},\quad \phi=\frac{1}{\sqrt{2}}\begin{bmatrix}v\\0\end{bmatrix},\quad \delta\phi\neq 0\rightarrow\delta\phi_1,\delta\phi_2\neq 0\rightarrow\delta\Sigma\neq 0 \end{array}\right\} \begin{array}{c}\text{And also, } \delta\mathcal{L}=0 \text{ at all} \\ \sigma \text{ and } \eta \text{ under the} \\ U(1)\text{ transformation} \\ \text{about false vacuum}\end{array} .(7\text{-}27)$$

$\mathcal{L}$ still invariant (symmetric) about false vacuum when expressed in terms of σ and η

Rotational Transformation about True Vacuum

Now imagine we consider a rotation about a vertical axis passing through the true vacuum of the RHS of Fig. 7-2, instead of the false vacuum. We will not do the math, but it should be fairly apparent (hopefully) that $\mathcal{V}$, and hence $\mathcal{L}$, will not be symmetric under this rotation, whether we express $\mathcal{L}$ in terms of σ and η, or in terms of ϕ_1 and ϕ_2.

Bottom line: $\mathcal{L}$ is symmetric about the false vacuum, but not about the true vacuum. This is true regardless of which fields we choose to express $\mathcal{L}$ in terms of.

$\mathcal{L}$ symmetric about false vacuum, not about true vacuum, expressed in terms of any fields

7.1.7 Conclusions

ϕ is an operator and $\mathcal{V}$ is expressed in terms of operators, but, as noted, when we refer to them in the context of the Goldstone model, (and other models to come), we can think of them as representing expectation values (square root of the number of ϕ type particles, and the potential, respectively) and thus treat them as if they were actual numbers, not operators. The same holds for σ and η.

False vacuum

Note that, at $\phi = 0$, i) the second derivative of $\mathcal{V}$ (see (7-6)) with respect to ϕ_1 or ϕ_2 is negative (for $\mu^2 < 0$), ii) $\delta\phi = 0$ means there is no degeneracy of ϕ, iii) symmetry is unbroken ($\mathcal{V}$ is symmetric in ϕ_1 and ϕ_2 and in terms of σ and η, i.e., rotate the horizontal axes in Fig. 7-1 about the vertical axis and all looks just the same), and iv) the VEV of the field $\phi\,(=(\phi_1+i\phi_2)/\sqrt{2}\,)$ is zero.

Bottom line for the false vacuum

Note also that at $\phi = 0$, there are no particles (zero expectation value for particle number density along the ϕ_1 and ϕ_2 axes) so even though ϕ is a field (and not a state) $\phi = 0$ represents the zero ϕ particle state (the false vacuum state of $|0_{false}\rangle$. Thus, we must have $\alpha\phi = 0$. This is often, somewhat confusingly in my experience, referred to via the jargon "the generator α destroys the vacuum" (turns it to zero). This should be, I submit, "α times the generator (the identity) destroys the vacuum".

True vacuum

At $\phi = v/\sqrt{2}$, i) the second derivative of $\mathcal{V}$ (see (7-14)) with respect to σ is positive for $m_\sigma^2 > 0$ and with respect to η is zero, ii) $\delta\phi \neq 0$ means there is degeneracy of ϕ, iii) symmetry is broken ($\mathcal{V}$ is unsymmetric in terms of ϕ_1 and ϕ_2 and in terms of σ and η), and iv) the VEV of the field ϕ is not zero.

Bottom line for the true vacuum

Since $\delta\phi \neq 0$, then $i\alpha \neq 0$, and α times the $U(1)$ generator 1 does not destroy the true vacuum.

7.1.8 The Goldstone Theorem

The famous theorem associated with the Goldstone model is shown below and repeated in the last row of Wholeness Chart 7-1 (with explanations for some of the jargon referenced by letter superscripts and in parenthetical remarks in the chart). Note the symmetry in the Goldstone theorem is explicitly for *global*, not local, symmetry transformations, and those global transformations can be more general than what we have looked at here.

Goldstone theorem for global symmetry

Here we have examined global $U(1)$ symmetry, but the Goldstone theorem also holds for any other global symmetry. These could be $SU(2)$, $SU(3)$, etc. symmetries, which would have matrices (2X2, 3X3, etc.) instead of (the 1X1 identity) 1 as the generators of the transformation. In this Section 7.1, we have not examined more general cases than $U(1)$ symmetry, but from working through the implications of that symmetry, we should be able to accept the Goldstone theorem implications for these other types of global symmetry, as well.

Goldstone theorem good for U(1), SU(2), SU(3), etc.

Goldstone Theorem

The Goldstone theorem applies to the general case of an infinitesimal transformation $\varepsilon_i T_i$ (in the foregoing special $U(1)$ case, $\varepsilon_i = i\alpha$, and the matrix generator T_i = the number 1). It is as follows.

For any global symmetry of $\mathcal{L}^a$ which is not realized[b] in the spectrum of physical states[c],

Parts of Goldstone theorem with jargon explanations

i) there must exist a massless scalar particle (Nambu-Goldstone boson),

ii) there must be a degeneracy of physical states[d], and

iii) the corresponding generator of the transformation[e] does not annihilate the (true) vacuum or more generally $T_i\Phi \neq 0$[f] for the true[g] vacuum.

a) i.e., $\delta\mathcal{L} = 0$ for $\Phi \rightarrow \Phi' = \Phi + \delta\Phi = \Phi + \varepsilon_i T_i \Phi$, where Φ is a field singlet or multiplet

b) i.e., is broken ($\delta\mathcal{L} \neq 0$), such as at the true vacuum

c) i.e., range of particles we observe in our universe (at the true vacuum where we are now)

d) i.e., of the (true) vacuum

e) i.e. identity 1 for the $U(1)$ transformation shown earlier, T_i, more generally

f) i.e., $\delta\Phi \neq 0$

g) i.e., realized vacuum (where we are at in the present day).

7.1.9 Summary and Issues for the Goldstone Model

Problems with the Goldstone Model

1. It predicts a massless scalar (the η field) particle (a Nambu-Goldstone boson), which is not observed in nature.
2. It has no interactions in it, so it can't describe the real world.

Goldstone model lacks interactions and predicts field not observed in nature

Usefulness of Goldstone model

The Goldstone model has the advantages of 1) being a good lead-in to more realistic symmetry breaking models because it is easier to understand, and 2) using the same form for the potential $\mathcal{V}$ as is used in more elaborate models (including the GSW one that is part of the standard model of elementary particle physics), so the same mathematical procedure used in it can be transferred directly to, and employed in, those other models.

But has historical importance and leads to more elaborate, correct theory

Bottom line simplified language summary of the Goldstone model results

Any *global* symmetry (i.e., no gauge interactions) in a scalar field potential that is unstable will spontaneously break and lead to a massless scalar field (Goldstone boson) that is not observed in the real world. The global symmetry will also lead to a massive scalar field.

It says any unstable global $\mathcal{L}$ symmetry in a scalar field breaks to a massless (unobserved) scalar

Overview of the Goldstone Model

Wholeness Chart 7-1 summarizes the key elements of the Goldstone model on a single page.

Wholeness Chart 7-1. Goldstone Model

	At False Vacuum		**At True Vacuum**	
Pictorially	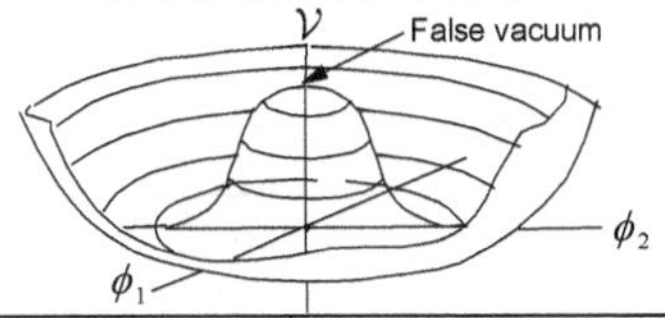 $\mathcal{V}$ False vacuum ϕ_1 ϕ_2		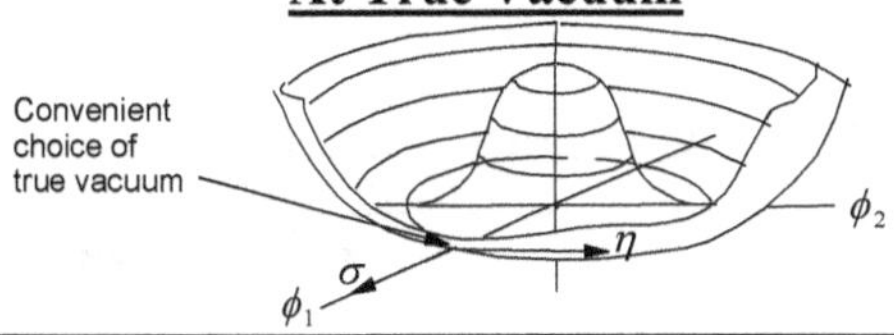 Convenient choice of true vacuum σ η ϕ_1 ϕ_2	
	Complex Rep	**Real Rep**	**Complex Rep**	**Real Rep**
Scalar Field (2 DOFs)	$\phi(x)=\frac{1}{\sqrt{2}}(\phi_1+i\phi_2)$	$\phi(x)=\frac{1}{\sqrt{2}}\begin{bmatrix}\phi_1\\ \phi_2\end{bmatrix}$	$\Sigma=\phi-\frac{v}{\sqrt{2}}=\frac{1}{\sqrt{2}}(\phi_1-v+i\phi_2)$ $=\frac{1}{\sqrt{2}}(\sigma+i\eta)$	$\Sigma=\frac{1}{\sqrt{2}}\begin{bmatrix}\phi_1-v\\ \phi_2\end{bmatrix}=\frac{1}{\sqrt{2}}\begin{bmatrix}\sigma\\ \eta\end{bmatrix}$
$\mathcal{L}$	$\partial^\mu\phi^\dagger\partial_\mu\phi-\mu^2\lvert\phi\rvert^2-\lambda\lvert\phi\rvert^4$ $\mu^2<0,\ \lambda>0$	$\frac{1}{2}(\partial^\mu\phi_1)^2+\frac{1}{2}(\partial^\mu\phi_2)^2$ $-\frac{\mu^2}{2}(\phi_1^2+\phi_2^2)-\frac{\lambda}{4}(\phi_1^2+\phi_2^2)^2$	Real rep more illuminating	$\left.\frac{1}{2}(\partial^\mu\sigma)^2-\frac{1}{2}(2\lambda v^2)\sigma^2+\frac{1}{2}(\partial^\mu\eta)^2\right\}\mathcal{L}_0$ $\left.-(\lambda v)\sigma(\sigma^2+\eta^2)-\frac{\lambda}{4}(\sigma^2+\eta^2)^2\right\}\mathcal{L}_I$
$\mathcal{V}$	$\mu^2\lvert\phi\rvert^2+\lambda\lvert\phi\rvert^4$	$\frac{\mu^2}{2}(\phi_1^2+\phi_2^2)+\frac{\lambda}{4}(\phi_1^2+\phi_2^2)^2$		$\frac{1}{2}(2\lambda v^2)\sigma^2+(\lambda v)\sigma(\sigma^2+\eta^2)+\frac{\lambda}{4}(\sigma^2+\eta^2)^2$
$\mathcal{H}$	$\partial^0\phi^\dagger\partial_0\phi+\nabla\phi^\dagger\bullet\nabla\phi+\mathcal{V}$ $=\mathcal{V}$ for min $\mathcal{H}$	(form not so important) $=\mathcal{V}$ for min $\mathcal{H}$		$=\mathcal{V}$ for min $\mathcal{H}$
Note	$\mathcal{V}$ ($\mathcal{H}$ & $\mathcal{L}$) sym functions of ϕ_1,ϕ_2 about false vac		$\mathcal{V}$ ($\mathcal{H}$&$\mathcal{L}$) not sym functs of σ,η about true vac	
Classical Interp	Unstable at $\phi=0$	As at left	Stable in σ at $\sigma=0$. (Not unstable in η)	
QFT Interpret	Unstable at $\phi=0$ & VEV $\langle_{false}0\lvert\phi\rvert 0_{false}\rangle=0$	As at left	Not unstable at $\sigma=0$. Small changes $\rightarrow$ σ particles created. $\langle_{true}0\lvert\phi\rvert 0_{true}\rangle=\langle_{true}0\lvert\frac{1}{\sqrt{2}}(\sigma+v)\rvert 0_{true}\rangle=\frac{v}{\sqrt{2}}\neq 0$	
Particle Masses	For small ϕ, coeff of ϕ^2 $m_\phi^2=\mu^2<0\rightarrow m_\phi$ imag	For small ϕ_1, ϕ_2, as at left	small σ, η, as at right	For small σ, η, coeffs of $\sigma^2/2$, $\eta^2/2$ $m_\sigma^2=2\lambda v^2>0\quad m_\eta^2=0$
Slope of $\mathcal{V}$	Curved downward $\rightarrow$ imaginary effective mass		Curved upward $\rightarrow$ real mass. Flat $\rightarrow$ zero mass	
Degeneracy	For $\mathcal{V}$ (& $\mathcal{L}$&$\mathcal{H}$) stationary, a unique $\phi=0$. No degeneracy (of false vacuum).		For $\mathcal{V}$ (&$\mathcal{H}$) min ($\mathcal{L}$ max), many possible Σ, i.e., degeneracy of (true) vacuum. (But once universe settles in a true vacuum state, it stays there.)	
***U*(1) Global Symmetry Transform** (α =constant & $\delta\mathcal{L}=\delta\mathcal{V}=0$)	$\phi'=U(1)\phi=e^{i\alpha}\phi$ (finite) Complex plane rotation $\left.\begin{matrix}\phi=0,\ U(1)\phi=0\\ \phi\neq 0,\ U(1)\phi\neq 0\end{matrix}\right\}\delta\mathcal{L}=0$ $\delta\phi=i\alpha\phi$ (infinitesimal)	$\phi'=\begin{bmatrix}\cos\alpha & \sin\alpha\\ -\sin\alpha & \cos\alpha\end{bmatrix}\begin{bmatrix}\frac{\phi_1}{\sqrt{2}}\\ \frac{\phi_2}{\sqrt{2}}\end{bmatrix}$ Real plane rotation As at left for $\delta\mathcal{L}$ $\delta\phi=\frac{1}{\sqrt{2}}\begin{bmatrix}\alpha\phi_2\\ -\alpha\phi_1\end{bmatrix}$ (infinitesimal)	Same transformation about false vacuum, i.e., as at $\phi=\frac{1}{\sqrt{2}}\begin{bmatrix}\phi_1\\ \phi_2\end{bmatrix}=\begin{bmatrix}0\\0\end{bmatrix}\left(\text{or }\Sigma=\frac{1}{\sqrt{2}}\begin{bmatrix}\sigma\\ \eta\end{bmatrix}=\frac{1}{\sqrt{2}}\begin{bmatrix}-v\\0\end{bmatrix}\right)$. But at $\phi=\frac{1}{\sqrt{2}}\begin{bmatrix}v\\0\end{bmatrix}\left(\text{or }\Sigma=\frac{1}{\sqrt{2}}\begin{bmatrix}\sigma\\ \eta\end{bmatrix}=\frac{1}{\sqrt{2}}\begin{bmatrix}0\\0\end{bmatrix}\right)$, then $\delta\phi=i\alpha\phi\left(=\frac{1}{\sqrt{2}}\begin{bmatrix}0\\ -\alpha v\end{bmatrix}\right)\neq 0$ (but still $\delta\mathcal{L}=0$)	
Conclusions for $\alpha\phi$=0 & $\delta\mathcal{L}$=0	At $\phi=0$, $\delta\phi=0$ means 1. 2nd deriv of $\mathcal{V}$ wrspt ϕ_1 or ϕ_2 is neg $\rightarrow$ imag "mass" 2. No degeneracy of ϕ 3. Sym unbroken about false vac ($\mathcal{V}$ sym in ϕ_1 & ϕ_2) 4. VEV of field ϕ is zero. 5. Generator times α annihil vac ($\alpha\phi=0$ at $\phi=0$)		At $\phi=v/\sqrt{2}$, $\delta\phi\neq 0$ means 1. 2nd deriv of $\mathcal{V}$ wrspt η is 0 $\rightarrow$ zero mass for η 2. Degeneracy of ϕ 3. Sym broken about true vac ($\mathcal{V}$ not sym in σ & η) 4. VEV of field ϕ is not zero. (Field gets a VEV.) 5. Generator times α does not annihil vac ($\alpha\phi\neq 0$)	
Goldstone Theorem	For general case of infinitesimal transformation $\varepsilon_i T_i$ (in above, $\varepsilon_i=i\alpha$; matrix T_i = number 1): For any global symmetry of $\mathcal{L}$ (i.e., $\delta\mathcal{L}=0$ for $\Phi\rightarrow\Phi+\varepsilon_i T_i\Phi$) which is not realized (i.e. is broken) in the spectrum of physical states (i.e., range of particles we measure in our universe), i) there must exist a massless scalar particle (Nambu-Goldstone boson), ii) there must be degeneracy of physical states (i.e., of the true vacuum), iii) the corresponding generator of the transformation (i.e. 1 for infinitesimal transformation above) does not annihilate the vacuum or more generally $T_i\Phi\neq 0$ (i.e., $\delta\Phi\neq 0$) for the true (realized) vacuum.			

7.2 Higgs Model

The difference between the Goldstone and Higgs models is simply that the latter includes interactions, as shown in Wholeness Chart 7-2 below. Note the potential $\mathcal{V}$ is the same.

Higgs model like Goldstone except local symmetry and interactions

Wholeness Chart 7-2. Difference Between Goldstone and Higgs Models

	Goldstone Model	Higgs Model
Higgs fields	complex scalar ϕ	same as at left
Interactions?	No (global $U(1)$ sym of $\mathcal{L}$)	Yes (local $U(1)$ sym of $\mathcal{L}$)

$$\mathcal{L} = \left|D_\mu\phi\right|^2 - \tfrac{1}{4}F_{\mu\nu}F^{\mu\nu} - \mu^2\left|\phi\right|^2 - \lambda\left|\phi\right|^4 \qquad \phi = \tfrac{1}{\sqrt{2}}\left(\phi_1 + i\phi_2\right) \quad \phi_1, \phi_2 \text{ real}$$
$$= \left(D^\mu\phi\right)^\dagger\left(D_\mu\phi\right) - \tfrac{1}{4}F_{\mu\nu}F^{\mu\nu} \underbrace{-\tfrac{\mu^2}{2}\left(\phi_1^2 + \phi_2^2\right) - \tfrac{\lambda}{4}\left(\phi_1^2 + \phi_2^2\right)^2}_{-\mathcal{V}} \tag{7-28}$$

Same $\mathcal{V}$ as Goldstone

where
$$D^\mu = \partial^\mu + iqA^\mu \qquad F^{\mu\nu} = \partial^\nu A^\mu - \partial^\mu A^\nu, \tag{7-29}$$

A^μ here is a suitable U(1) gauge field that could be like a photon field

The Higgs model is summarized in Wholeness Chart 7-3, pg. 219, which you may want to follow along with as we progress in this section.

As with the Goldstone model, we choose the minimum of $\mathcal{V}$ for $\phi_2 = 0$ to be located at a positive value for ϕ_1 denoted by the symbol v. This allows us to find an expression relating μ and v, as follows.

$$\left.\frac{\partial\mathcal{V}}{\partial\phi_1}\right|_{\substack{\phi_1=v\\ \phi_2=0}} = 0 = \left.\left(\mu^2\phi_1 + \tfrac{\lambda}{4}2\left(\phi_1^2+\phi_2^2\right)2\phi_1\right)\right|_{\substack{\phi_1=v\\ \phi_2=0}} \rightarrow \quad \mu^2 = -\lambda\phi_1^2\Big|_{\phi_1=v} = -\lambda v^2 < 0\,, \tag{7-30}$$

And thus, as before, for $\lambda > 0$, the "mass" μ of ϕ is imaginary.

7.2.1 $\mathcal{V}$ of $\mathcal{L}$ Expressed via New Fields σ and η

So, just as we did in the Goldstone model in (7-9), we define new fields

Express with same fields σ and η, as in Goldstone

$$\sigma = \phi_1 - v \qquad \eta = \phi_2 \qquad \rightarrow \qquad \phi_1 = \sigma + v \qquad \phi_2 = \eta \tag{7-31}$$

and with (7-31) (the same as (7-9)) into the potential $\mathcal{V}$ of (7-28) (the same as in (7-5)), we get the same $\mathcal{V}$ we had in the Goldstone case (the same as (7-14)),

$$\mathcal{V} = -\tfrac{1}{4}\lambda v^4 + \tfrac{1}{2}\left(2\lambda v^2\right)\sigma^2 + \left(\lambda v\right)\sigma\left(\sigma^2+\eta^2\right) + \tfrac{\lambda}{4}\left(\sigma^2+\eta^2\right)^2. \tag{7-32}$$

Same $\mathcal{V}$ in terms of fields σ and η

7.2.2 Kinetic Terms in $\mathcal{L}$ Expressed via New Fields σ and η

We use the gauge covariant derivative definition of (7-29) to determine the interaction terms for the Higgs. Specifically, we find the covariant derivative terms in $\mathcal{L}$ of (7-28) in terms of ϕ_1 and ϕ_2.

Use covariant derivative to get interactions

$$\left|D_\mu\phi\right|^2 = \left(D^\mu\phi\right)^\dagger\left(D_\mu\phi\right) = \left(\partial^\mu\phi + iqA^\mu\phi\right)^\dagger\left(\partial_\mu\phi + iqA_\mu\phi\right) = \left(\partial^\mu\phi^\dagger - iqA^\mu\phi^\dagger\right)\left(\partial_\mu\phi + iqA_\mu\phi\right)$$
$$= \left(\partial^\mu\phi^\dagger\right)\partial_\mu\phi + iqA_\mu\left(\partial^\mu\phi^\dagger\right)\phi - iqA^\mu\phi^\dagger\partial_\mu\phi + q^2A^\mu A_\mu\phi^\dagger\phi \tag{7-33}$$
$$= \frac{1}{2}\left[\begin{array}{l}\left(\partial^\mu\left(\phi_1 - i\phi_2\right)\right)\partial_\mu\left(\phi_1 + i\phi_2\right) + iqA_\mu\left(\partial^\mu\left(\phi_1 - i\phi_2\right)\right)\left(\phi_1 + i\phi_2\right)\\ \qquad - iqA^\mu\left(\phi_1 - i\phi_2\right)\partial_\mu\left(\phi_1 + i\phi_2\right) + q^2A^\mu A_\mu\left(\phi_1 - i\phi_2\right)\left(\phi_1 + i\phi_2\right)\end{array}\right]$$

$$= \frac{1}{2}\left[\begin{array}{l}\left|\partial^\mu\phi_1\right|^2 + \left|\partial^\mu\phi_2\right|^2 \overset{\boxed{1}}{+iqA_\mu\left(\partial^\mu\phi_1\right)\phi_1} \overset{\boxed{A}}{-qA_\mu\left(\partial^\mu\phi_1\right)\phi_2} \overset{\boxed{B}}{+qA_\mu\left(\partial^\mu\phi_2\right)\phi_1} \overset{\boxed{2}}{+iqA_\mu\left(\partial^\mu\phi_2\right)\phi_2}\\ \underset{\boxed{1}}{-iqA_\mu\phi_1\partial^\mu\phi_1} \underset{\boxed{B}}{+qA_\mu\phi_1\partial^\mu\phi_2} \underset{\boxed{A}}{-qA_\mu\phi_2\partial^\mu\phi_1} \underset{\boxed{2}}{-iqA_\mu\phi_2\partial^\mu\phi_2} + q^2A^\mu A_\mu\phi_1^2 + q^2A^\mu A_\mu\phi_2^2\end{array}\right]. \tag{7-34}$$

The terms with numbers over them cancel. The terms with letters over them are equal if ϕ_1 and ϕ_2 commute, and they do because they are different fields. This leaves us with

Interaction plus kinetic terms expressed via ϕ field

$$\left|D_\mu \phi\right|^2 = \tfrac{1}{2}\left[\left|\partial^\mu \phi_1\right|^2 + \left|\partial^\mu \phi_2\right|^2 - 2qA_\mu\left(\partial^\mu \phi_1\right)\phi_2 + 2qA_\mu\left(\partial^\mu \phi_2\right)\phi_1 + q^2 A^\mu A_\mu \phi_1^2 + q^2 A^\mu A_\mu \phi_2^2\right]. \quad (7\text{-}35)$$

With the RHS of (7-31), and realizing v is a constant, (7-35) becomes

$$\begin{aligned}\left|D^\mu \phi\right|^2 &= \tfrac{1}{2}\left(\partial^\mu \sigma\right)\left(\partial_\mu \sigma\right) + \tfrac{1}{2}\left(\partial^\mu \eta\right)\left(\partial_\mu \eta\right) - qA_\mu\left(\partial^\mu \sigma\right)\eta + qA_\mu\left(\partial^\mu \eta\right)(\sigma + v) \\ &\quad + \tfrac{q^2}{2}A^\mu A_\mu(\sigma+v)^2 + \tfrac{q^2}{2}A^\mu A_\mu \eta^2 \\ &= \tfrac{1}{2}\left(\partial^\mu \sigma\right)\left(\partial_\mu \sigma\right) + \tfrac{1}{2}\left(\partial^\mu \eta\right)\left(\partial_\mu \eta\right) - qA_\mu\left(\partial^\mu \sigma\right)\eta + qA_\mu\left(\partial^\mu \eta\right)\sigma + qvA_\mu\left(\partial^\mu \eta\right) \\ &\quad + \tfrac{q^2}{2}A^\mu A_\mu \sigma^2 + q^2 v A^\mu A_\mu \sigma + \tfrac{q^2}{2}v^2 A^\mu A_\mu + \tfrac{q^2}{2}A^\mu A_\mu \eta^2 .\end{aligned} \quad (7\text{-}36)$$

Re-arranging so the bilinear (quadratic in fields), and linear in q, terms come first, we have

Interaction plus kinetic terms expressed via σ and η fields

$$\begin{aligned}\left|D^\mu \phi\right|^2 &= \tfrac{1}{2}\left(\partial^\mu \sigma\right)\left(\partial_\mu \sigma\right) + \tfrac{1}{2}\left(\partial^\mu \eta\right)\left(\partial_\mu \eta\right) + qvA_\mu\left(\partial^\mu \eta\right) \\ &\quad - qA_\mu\left(\partial^\mu \sigma\right)\eta + qA_\mu\left(\partial^\mu \eta\right)\sigma + \tfrac{q^2}{2}A^\mu A_\mu \sigma^2 + q^2 v A^\mu A_\mu \sigma + \tfrac{q^2}{2}v^2 A^\mu A_\mu + \tfrac{q^2}{2}A^\mu A_\mu \eta^2 .\end{aligned} \quad (7\text{-}37)$$

7.2.3 $\mathcal{L}$ expressed via New Fields σ and η

With (7-32) and (7-37), the last row of (7-28) becomes

$$\begin{aligned}\mathcal{L} &= \tfrac{1}{2}\left(\partial^\mu \sigma\right)\left(\partial_\mu \sigma\right) + \tfrac{1}{2}\left(\partial^\mu \eta\right)\left(\partial_\mu \eta\right) + qvA_\mu\left(\partial^\mu \eta\right) - \tfrac{1}{4}F_{\mu\nu}F^{\mu\nu} \\ &\quad - qA_\mu\left(\partial^\mu \sigma\right)\eta + qA_\mu\left(\partial^\mu \eta\right)\sigma + \tfrac{q^2}{2}A^\mu A_\mu \sigma^2 + q^2 v A^\mu A_\mu \sigma + \tfrac{q^2}{2}v^2 A^\mu A_\mu + \tfrac{q^2}{2}A^\mu A_\mu \eta^2 \\ &\quad \underbrace{+\tfrac{1}{4}\lambda v^4 - \tfrac{1}{2}\left(2\lambda v^2\right)\sigma^2 - (\lambda v)\sigma\left(\sigma^2+\eta^2\right) - \tfrac{\lambda}{4}\left(\sigma^2+\eta^2\right)^2}_{-\mathcal{V}} .\end{aligned} \quad (7\text{-}38)$$

Dropping the constant term $\frac{1}{4}\lambda v^4$ once again, and re-arranging, (7-38) becomes

Entire Higgs model $\mathcal{L}$ expressed via σ and η fields

$$\begin{aligned}\mathcal{L} &= \underbrace{\tfrac{1}{2}\left(\partial^\mu \sigma\right)\left(\partial_\mu \sigma\right) - \tfrac{1}{2}\left(2\lambda v^2\right)\sigma^2 - \tfrac{1}{4}F_{\mu\nu}F^{\mu\nu} + \tfrac{q^2 v^2}{2}A^\mu A_\mu + \tfrac{1}{2}\left(\partial^\mu \eta\right)\left(\partial_\mu \eta\right)}_{=\mathcal{L}_0 \text{ (bilinear terms, with no mixed products, represent free fields)}} \underbrace{+\, qvA_\mu\left(\partial^\mu \eta\right)}_{\substack{\text{interaction} \\ \text{of } A_\mu \text{ and } \eta}} \\ &\quad \underbrace{\begin{aligned}&-qA_\mu\left(\partial^\mu \sigma\right)\eta + qA_\mu\left(\partial^\mu \eta\right)\sigma - (\lambda v)\sigma\left(\sigma^2+\eta^2\right) - \tfrac{\lambda}{4}\left(\sigma^2+\eta^2\right)^2 \\ &\qquad + \tfrac{q^2}{2}A^\mu A_\mu \sigma^2 + q^2 v A^\mu A_\mu \sigma + \tfrac{q^2}{2}A^\mu A_\mu \eta^2 .\end{aligned}}_{\text{cubic and quartic terms in the fields, represent interactions}}\end{aligned} \quad (7\text{-}39)$$

But we can't just read mass off of the terms, because different fields are coupled

Note that the last term in the top row of (7-39) couples the A^μ and η fields, so they are no longer independent fields (at lowest order). Hence, we cannot simply surmise that the 4th term describes a field A^μ of squared mass q^2v^2. A^μ has mass, we just can't assume it can be read off from the coefficient of $A^\mu A_\mu$, as our rule for determining such mass was deduced from Lagrangians for vector fields with no bilinear terms coupling such a vector field to another field.

7.2.4 Degrees of Freedom and the Unitary Gauge

Degrees of Freedom

Number of DOF = number of independent components of a field

Massless vectors, like photon, have 2 DOF

Note that the photon field, which is a massless vector field, for a given **k**, has two degrees of freedom (DOF). Any general photon can be considered composed of two independent fields (DOF), each with a polarization aligned at a right angle to the other and also at a right angle to the direction of **k.** (See Vol. 1, pgs. 141-144, 150-155.) In 4 D, one might naively expect a 4D vector to have four degrees of freedom, since there would be four independent components, but as shown in the reference just cited, the longitudinal and time-like components for photons precisely cancel one another in any calculation of expectation values. As a result, neither is ever measured. The bottom line: massless vector fields effectively have two independent components, two DOF.

A real scalar field, such as ϕ_1 or ϕ_2, has only one DOF. A scalar is a single number classically. (A complex scalar field would have two DOF, the real part being one and the (independent) imaginary part being the other.)

Real scalar: 1 DOF
Complex " : 2 DOF

A massive vector field, it turns out, has three degrees of freedom, as we showed in Chap. 5, Sect. 5.4, pgs. 154-157. As a result of having mass, it cannot travel at the speed of light, and thus all polarization components do not have to be perpendicular to **k**, as was the case for photons. This frees up one degree of freedom compared to massless vector fields/particles. Its independent components can align with 3 independent directions in 3D.

Massive vector field: 3 DOF

Symmetry Breaking Appears to Change the Number of Degrees of Freedom

Note that before symmetry breaking, we have four DOF: one for ϕ_1; one for ϕ_2; and two for the massless A^μ.

However, after symmetry breaking, we have five: one for σ; one for η; and three for the massive A^μ. The symmetry breaking seems to have added a degree of freedom. But that cannot be a physical reality. We cannot simply redefine our fields and by so doing, change the nature of reality, the total number of degrees of freedom expressed by the fields.

Higgs model symmetry breaking seems to add one DOF

The answer lies at the heart of gauge theory. There are certain things we can measure, resulting from behaviors of underlying fields, but the fields themselves are not measurable. They are gauge fields. So, as long as we develop a theory where the measurables are unchanged, we can manipulate the underlying (unmeasurable) gauge field(s) in any convenient way. This is what we did in classical e/m and QED. We chose the Lorenz gauge because it made our math easiest.

But physical reality won't permit adding a DOF mathematically

In our case, we do a similar thing, keeping in mind that nature must be described with only four degrees of freedom in its fields (using fields defined relative to the false vacuum), so any theory with five has an extra, superfluous, one.

7.2.5 The Unitary Gauge

Since, using the fields defined relative to the true vacuum, we have an extra degree of freedom, we can constrain our DOF in some way, with some constraint equation between them, that will simplify our work.

So, we can constrain one DOF, i.e., we can choose a gauge condition

The simplest such gauge (constraint) choice turns out to be the constraint equation $\eta = 0$. This is known as the unitary gauge. With it, (7-39) takes the much simpler form, with A^μ no longer coupled in bilinear fashion with a non-zero field η,

The simplest gauge is the unitary gauge, where we set $\eta = 0$

$$\mathcal{L} = \underbrace{\tfrac{1}{2}(\partial^\mu\sigma)(\partial_\mu\sigma) - \tfrac{1}{2}(2\lambda v^2)\sigma^2 - \tfrac{1}{4}F_{\mu\nu}F^{\mu\nu} + \frac{(qv)^2}{2}A^\mu A_\mu}_{\mathcal{L}_0}$$
$$\underbrace{-(\lambda v)\sigma^3 - \tfrac{\lambda}{4}\sigma^4 + \tfrac{q^2}{2}A^\mu A_\mu\sigma^2 + q^2 v A^\mu A_\mu\sigma}_{\mathcal{L}_I} . \qquad (7\text{-}40)$$

This choice eliminates bilinear coupling terms of different fields

The free field terms are in the top row of (7-40) and the interaction terms on the bottom row. A^μ is now independent *at lowest order* from other fields, so we can read its mass off directly (4th term). From the 2nd term, $m_\sigma = \sqrt{2\lambda v^2}$. The A^μ field now has a mass whereas it did not have one before symmetry breaking. As an aside, recall (Vol. 1, pg. 296 and also Chap. 6 herein) that if the photon had mass in QED, we would not have symmetry of $\mathcal{L}$. Thus, the A^μ mass term breaks symmetry in the Higgs model.

So, now we can read off the mass of the Higgs field σ

7.2.6 Summary and Issues for the Higgs Model

Problems with the Higgs Model

The Higgs model yields a massive photon and so cannot be correct. Further, it doesn't include any weak force terms, which, of course, are part of our universe.

Problems: a massive photon and no weak interactions

Advantages of the Higgs Model

The Higgs model advances what we learned in the Goldstone model to fields with interactions, and thus helps us when we proceed to the Glashow/Salam/Weinberg model, which does a pretty good job of describing our world.

But it gets us one step closer to a better model

Wholeness Chart 7-3. Higgs Model

	At False Vacuum		**At True Vacuum**	
Pictorially	Same as Goldstone figure		Same as Goldstone figure	
	Characterization		**Characterization**	
	Complex	**Real**	**Complex**	**Real**
Higgs Field (same as Goldstone)	$\phi(x)=\frac{1}{\sqrt{2}}(\phi_1+i\phi_2)$	$\phi(x)=\frac{1}{\sqrt{2}}\begin{bmatrix}\phi_1\\ \phi_2\end{bmatrix}$	$\Sigma=\phi-\frac{v}{\sqrt{2}}=\frac{1}{\sqrt{2}}(\phi_1-v+i\phi_2)$ $=\frac{1}{\sqrt{2}}(\sigma+i\eta)$	$\Sigma=\frac{1}{\sqrt{2}}\begin{bmatrix}\phi_1-v\\ \phi_2\end{bmatrix}=\frac{1}{\sqrt{2}}\begin{bmatrix}\sigma\\ \eta\end{bmatrix}$
$\mathcal{L}$	$(D^\mu\phi)^\dagger D_\mu\phi-\frac{1}{4}F_{\mu\nu}F^{\mu\nu}$ $-\mu^2\lvert\phi\rvert^2-\lambda\lvert\phi\rvert^4$ $D^\mu=\partial^\mu+iqA^\mu$	$\frac{1}{2}\lvert D^\mu\phi_1\rvert^2+\frac{1}{2}\lvert D^\mu\phi_2\rvert^2-\frac{1}{4}F_{\mu\nu}F^{\mu\nu}$ $-\frac{\mu^2}{2}(\phi_1^2+\phi_2^2)-\frac{\lambda}{4}(\phi_1^2+\phi_2^2)^2$	Real more illuminating	For transformation to σ,η coordinates $\left.\begin{matrix}\frac{1}{2}\lvert\partial^\mu\sigma\rvert^2-\frac{1}{2}(2\lambda v^2)\sigma^2+\frac{1}{2}\lvert\partial^\mu\eta\rvert^2\\ -\frac{1}{4}F_{\mu\nu}F^{\mu\nu}+\frac{1}{2}(qv)^2A_\mu A^\mu\end{matrix}\right\}\mathcal{L}_0$ $+qvA_\mu(\partial^\mu\eta)$+cubic & quartic terms $\}\mathcal{L}_I$
$\mathcal{V}$	Same as Goldstone model	Same as Goldstone model		Same as Goldstone model
U(1) Local Symmetry Transform	Infinitesimal form: $\delta\phi=i\alpha(x)\phi$ $\delta A_\mu=-\frac{1}{q}\partial_\mu\alpha(x)$ $\mathcal{L}_{false}$ sym fn of ϕ & A_μ about false vacuum	2D rotation in ϕ_1, ϕ_2 space, different at each x^μ $\mathcal{L}_{false}$ sym fn of ϕ_1, ϕ_2, A_μ about false vacuum		$\mathcal{L}_{true}$ not symmetric function of σ, η, A_μ about true vacuum
Difficulties with Form for $\mathcal{L}$				Product term in A_μ & $\partial^\mu\eta$ means A_μ & η not indep normal modes & cannot equate factors of terms in $\mathcal{L}_0$ with mass. $\mathcal{L}_{false}$ has 4 DOF. $\mathcal{L}_{true}$ has 5 (1 σ, 1 η, 3 from massive A_μ). Change of variable cannot alter DOF. A non-physical field has arisen.
Fixing Up with Unitary Gauge				Choose unitary gauge $\eta=0$ to restore 4 DOF and make σ & A_μ independent field modes (at 2nd order).
$\mathcal{L}$ Under Unitary Gauge				$\left.\begin{matrix}\frac{1}{2}\lvert\partial^\mu\sigma\rvert^2-\frac{1}{2}(2\lambda v^2)\sigma^2\\ -\frac{1}{4}F_{\mu\nu}F^{\mu\nu}+\frac{1}{2}(qv)^2A_\mu A^\mu\end{matrix}\right\}\mathcal{L}_0$ $\left.\begin{matrix}-(\lambda v)\sigma^3-\frac{\lambda}{4}\sigma^4\\ +\frac{q^2}{2}A^\mu A_\mu\sigma^2+q^2vA^\mu A_\mu\sigma\end{matrix}\right\}\mathcal{L}_I$
Masses				For small σ & A_μ, $m_\sigma=\sqrt{2\lambda v^2}\quad m_\gamma=\lvert qv\rvert$
Result				At true vacuum: 1 real, massive scalar field σ (Higgs) 1 real, massive vector field A_μ (photon)
Summary	Global symmetry (Goldstone model) $\rightarrow$ massless Nambu-Goldstone scalar boson η (not observed) Local symmetry (Higgs model) $\rightarrow$ massive A_μ (not what we observe) and no η.			
Conclusion	Neither the Goldstone nor Higgs model matches reality.			

7.3 Glashow/Salam/Weinberg Model

7.3.1 Background

Ignoring Quarks for Now

As we did in Chap. 6, we will initially ignore quarks and add them in later, once the reader has gained an understanding of the behavior of leptons under symmetry breaking, which quite parallels the quark behavior.

Treat just leptons now, quarks later

Summary Overview of What We are about to Do

Symmetry breaking in the Glashow/Weinberg/Salam (GSW) model is summarized in Wholeness Chart 7-8, pg. 240. It should help to follow that as we progress through the development of the model.

Key Fermion Mass Term Relation

Do **Problem 4** to show (7-41) and (7-42).

We will need the first of the following relations to deduce the true vacuum Lagrangian in the GSW model. You can do Prob. 4 to prove it or find the derivation done for you in Chap. 6, Appendix B.

$$\bar{\psi}_l \psi_l = \bar{\psi}_l^L \psi_l^R + \bar{\psi}_l^R \psi_l^L \qquad (\text{no sum on } l) \qquad (7\text{-}41)$$

$$\bar{\psi}_l \gamma^\mu \psi_l = \bar{\psi}_l^R \gamma^\mu \psi_l^R + \bar{\psi}_l^L \gamma^\mu \psi_l^L \qquad (\text{no sum on } l). \qquad (7\text{-}42)$$

Two relations we will use repeatedly

Recalling Mixing of False Vacuum and True Vacuum Bosons

In Chap. 6, we noted the mixing between the false vacuum boson fields and the true vacuum boson fields, where θ_W is the Weinberg mixing angle,

$$\underbrace{\begin{bmatrix} B_\mu \\ W_{3\mu} \end{bmatrix}}_{\text{false vacuum}} = \begin{bmatrix} \cos\theta_W & -\sin\theta_W \\ \sin\theta_W & \cos\theta_W \end{bmatrix} \underbrace{\begin{bmatrix} A_\mu \\ Z_\mu \end{bmatrix}}_{\text{true vacuum}} \qquad \text{or} \qquad \begin{bmatrix} A_\mu \\ Z_\mu \end{bmatrix} = \begin{bmatrix} \cos\theta_W & \sin\theta_W \\ -\sin\theta_W & \cos\theta_W \end{bmatrix} \begin{bmatrix} B_\mu \\ W_{3\mu} \end{bmatrix}$$

$$B_\mu = A_\mu \cos\theta_W - Z_\mu \sin\theta_W \qquad A_\mu = B_\mu \cos\theta_W + W_{3\mu} \sin\theta_W \qquad (7\text{-}43)$$

$$W_{3\mu} = A_\mu \sin\theta_W + Z_\mu \cos\theta_W \qquad Z_\mu = -B_\mu \sin\theta_W + W_{3\mu} \cos\theta_W,$$

Recalling mixing via the Weinberg angle and

and

$$\underbrace{W_{1\mu}}_{\substack{\text{false}\\\text{vacuum}}} = \frac{W_\mu^+ + W_\mu^-}{\sqrt{2}} \qquad \underbrace{W_{2\mu}}_{\substack{\text{false}\\\text{vacuum}}} = i\frac{W_\mu^+ - W_\mu^-}{\sqrt{2}} \qquad \underbrace{W_\mu^+}_{\substack{\text{true}\\\text{vacuum}}} = \frac{W_{1\mu} - iW_{2\mu}}{\sqrt{2}} \qquad \underbrace{W_\mu^-}_{\substack{\text{true}\\\text{vacuum}}} = \frac{W_{1\mu} + iW_{2\mu}}{\sqrt{2}} \qquad (7\text{-}44)$$

W_μ^+ sometimes symbolized by W_μ, with W_μ^- symbolized by $W_\mu^\dagger$.

the superpositions of real fields W_1 and W_2 to yield complex fields W^+ and W^-

GSW Model vs Goldstone and Higgs Models

As noted in Chap. 6, the GSW model is based on

Differences in GSW from Higgs model

1) two complex scalar Higgs fields ϕ_a and ϕ_b (instead of one ϕ), which can be expressed as a doublet

$$\Phi(x) = \frac{1}{\sqrt{2}}\begin{bmatrix} \phi_a \\ \phi_b \end{bmatrix} = \frac{1}{\sqrt{2}}\begin{bmatrix} \phi_1 + i\phi_2 \\ \phi_3 + i\phi_4 \end{bmatrix} \qquad \phi_1, \phi_2, \phi_3, \phi_4 \text{ real, and} \qquad (7\text{-}45)$$

Higgs field here is an SU(2) doublet

2) a Lagrangian density with local $SU(2)XU(1)$ symmetry

and symmetry is local SU(2)XU(1), not just U(1)

Note that we define (7-45) with the normalization factor $\frac{1}{\sqrt{2}}$, whereas some authors, like Mandl and Shaw, leave this out at this point (and include it in later in defining the ϕ_i). I believe it is easier to follow the derivation of the theory using the form employed here, as it parallels that used in the Goldstone and Higgs derivations.

The difference between the Goldstone, Higgs, and GSW models is shown in Wholeness Chart 7-4 below.

Wholeness Chart 7-4 Difference Between Goldstone, Higgs, and Glashow/Salam/Weinberg Models

Comparison of the three models

	Goldstone Model	Higgs Model	GSW Model
Higgs fields	Complex scalar singlet ϕ $\phi = \frac{1}{\sqrt{2}}(\phi_1 + i\phi_2)$	Same as at left	Complex scalar doublet Φ See (7-45)
Interactions?	No (global $U(1)$ sym of $\mathcal{L}$)	Yes (local $U(1)$ sym of $\mathcal{L}$)	Yes (local $SU(2)\text{X}U(1)$ sym of $\mathcal{L}$)

Recalling the False Vacuum Electroweak Lagrangian

In Chap. 6, we found a false vacuum (high energy) symmetric Lagrangian of form (7-46) below, a repeat of (6-48).

False Vacuum, Total Electroweak Lepton Lagrangian $\mathcal{L} = \mathcal{L}^L + \mathcal{L}^B + \mathcal{L}^H + \mathcal{L}^{LH}$

Recalling the false vacuum $\mathcal{L}$

$$\begin{aligned}
&\mathcal{L}^L = i\left(\bar{\Psi}_l^L \not{D} \Psi_l^L + \bar{\psi}_l^R \not{D} \psi_l^R + \bar{\psi}_{\nu_l}^R \not{D} \psi_{\nu_l}^R\right) && \text{(leptons, massless)} \\
&\mathcal{L}^B = -\tfrac{1}{4} G_i^{\mu\nu} G_{i\,\mu\nu} - \tfrac{1}{4} B^{\mu\nu} B_{\mu\nu} && \text{(gauge bosons)} \qquad (7\text{-}46) \\
&\mathcal{L}^H = \left(D^\mu \Phi\right)^\dagger \left(D_\mu \Phi\right) \underbrace{- \mu^2 \Phi^\dagger \Phi - \lambda\left(\Phi^\dagger \Phi\right)^2}_{-\mathcal{V}} && \text{(Higgs boson)} \\
&\mathcal{L}^{LH} = -\, {}_A Y_{ll'} \bar{\Psi}_l^L \Phi \psi_{l'}^R - {}_B Y_{l\nu_{l'}} \bar{\Psi}_l^L \tilde{\Phi} \psi_{\nu_{l'}}^R + h.c. && \text{(lepton-Higgs coupled)},
\end{aligned}$$

where

$$\Psi_l^L = \begin{pmatrix} \psi_{\nu_l}^L \\ \psi_l^L \end{pmatrix} \qquad (7\text{-}47)$$

$$D^\mu = \partial^\mu + \tfrac{i}{2} g \sigma_j W_j^\mu + i g' Y B^\mu \qquad \not{D} = \gamma^\mu D_\mu, \qquad (7\text{-}48)$$

$$G_i^{\mu\nu} = D^\nu W_i^\mu - D^\mu W_i^\nu = \partial^\nu W_i^\mu - \partial^\mu W_i^\nu + g \varepsilon_{ijk} W_j^\mu W_k^\nu, \qquad (7\text{-}49)$$

$$B^{\mu\nu} = D^\nu B^\mu - D^\mu B^\nu = \partial^\nu B^\mu - \partial^\mu B^\nu \qquad (7\text{-}50)$$

$$\Phi(x) = \frac{1}{\sqrt{2}}\begin{bmatrix} \phi_a \\ \phi_b \end{bmatrix} = \frac{1}{\sqrt{2}}\begin{bmatrix} \phi_1 + i\phi_2 \\ \phi_3 + i\phi_4 \end{bmatrix} \qquad \tilde{\Phi} = \frac{1}{\sqrt{2}}\begin{pmatrix} \phi_b^* \\ -\phi_a^* \end{pmatrix} = \frac{1}{\sqrt{2}}\begin{pmatrix} \phi_3 - i\phi_4 \\ -\phi_1 + i\phi_2 \end{pmatrix} \qquad \phi_1, \phi_2, \phi_3, \phi_4 \text{ real.} \qquad (7\text{-}51)$$

Higgs Potential in GSW Model

As noted in Chap. 6 ((6-31) and (6-32)), the Higgs potential in the GSW model parallels that of the Goldstone and Higgs models except the one complex scalar field ϕ used there becomes two complex scalar fields $\phi_a = \phi_1 + i\phi_2$ and $\phi_b = \phi_3 + i\phi_4$ here. So instead of (7-6), we get

$\mathcal{V}$ in GSW model with Φ field doublet via ϕ_i fields

$$\begin{aligned}
\mathcal{V} &= \mu^2 \Phi^\dagger \Phi + \lambda\left(\Phi^\dagger \Phi\right)^2 = \frac{\mu^2}{2}\begin{bmatrix} \phi_1 - i\phi_2 & \phi_3 - i\phi_4 \end{bmatrix}\begin{bmatrix} \phi_1 + i\phi_2 \\ \phi_3 + i\phi_4 \end{bmatrix} + \frac{\lambda}{4}\left(\begin{bmatrix} \phi_1 - i\phi_2 & \phi_3 - i\phi_4 \end{bmatrix}\begin{bmatrix} \phi_1 + i\phi_2 \\ \phi_3 + i\phi_4 \end{bmatrix}\right)^2 \qquad (7\text{-}52) \\
&= \frac{\mu^2}{2}\left(\phi_1^2 + \phi_2^2 + \phi_3^2 + \phi_4^2\right) + \frac{\lambda}{4}\left(\phi_1^2 + \phi_2^2 + \phi_3^2 + \phi_4^2\right)^2 \qquad \mu^2 < 0,\ \lambda > 0.
\end{aligned}$$

7.3.2 Higgs Fields in Terms of Other Real Fields

Just as we did earlier in the Goldstone and Higgs models, we examine first the potential $\mathcal{V}$ from the Φ field in (7-46) and (7-52), which is symmetric in Φ (and in $\phi_1, \phi_2, \phi_3, \phi_4$). We then, again similar to earlier procedures, designate new fields that are functions of the old ones. That is, where η_1, η_2, σ, and η_3, are those new (real) fields, normalized by convention with a $\sqrt{2}$ factor, and v is a real constant,

Φ field doublet via new fields σ, η_1, η_2, η_3 and constant v

$$\phi_1 = \eta_1;\ \phi_2 = \eta_2;\ \phi_3 = (\sigma + v);\ \phi_4 = \eta_3 \quad \rightarrow \quad \Phi(x) = \frac{1}{\sqrt{2}}\begin{bmatrix} \phi_1 + i\phi_2 \\ \phi_3 + i\phi_4 \end{bmatrix} = \frac{1}{\sqrt{2}}\begin{bmatrix} \eta_1 + i\eta_2 \\ (\sigma + v) + i\eta_3 \end{bmatrix}. \tag{7-53}$$

It turns out that when we substitute (7-53) into (7-52), and move to the minimum of $\mathcal{V}$ at the true vacuum, we end up with 1) $\mathcal{L}$ being not symmetric in η_1, η_2, σ, and η_3, and 2) additional degrees of freedom (independent components of fields) in $\mathcal{L}$ from what we had in $\mathcal{L}$ for $\phi_1, \phi_2, \phi_3, \phi_4$. This parallels the Higgs model case, where we had one extra degree of freedom at the true vacuum with the new fields versus the false vacuum with the original fields. Recall that the additional degree of freedom in the Higgs model arose from the transformation of two massless bosons at false vacuum to one massless boson and one massive boson at the true vacuum. The change in massive bosons by one meant one additional degree of freedom. Here we start with four real massless boson fields instead of two, and like in Higgs model, we will find that we end up with only one massless boson at the true vacuum. Because here in the GSW model we have three more massive particles than originally, we pick up three additional degrees of freedom.

We will not show this gaining of mass for three of the bosons explicitly for the GSW model, as it is cumbersome to do so. We can gain some confidence, however, by recognizing that our presumption of it will (as we will see) lead to a valid theory, supported by experiment.

This results in 3 additional DOF

Recall that we cannot measure fields directly, and our measurables cannot be affected by how we designate our fields. That is, our fields are gauge fields. See Vol. 1, pgs. 177-178. So, our having three (unphysical) extra degrees of freedom means that we are free to constrain those degrees of freedom in any way we like. That is, we can prescribe a gauge condition for each, as is convenient, without affecting any measurables.

We eliminate these with 3 constraint equations on the fields

So, of course, we want to pick the gauge conditions that give us the simplest way to analyze the case at hand. That turns out to be what is called the unitary gauge (the same term as used in the Higgs model), i.e.,

$$\eta_1 = 0 \qquad \eta_2 = 0 \qquad \eta_3 = 0 \qquad \text{unitary gauge}. \tag{7-54}$$

These are $\eta_i = 0$, called the unitary gauge,

and thus,

$$\Phi(x) = \frac{1}{\sqrt{2}}\begin{bmatrix} 0 \\ \phi_3 \end{bmatrix} = \frac{1}{\sqrt{2}}\begin{bmatrix} 0 \\ \sigma + v \end{bmatrix} \quad \text{in unitary gauge}. \tag{7-55}$$

which simplifies the Higgs doublet

7.3.3 Higgs Field at True Vacuum

To find the true vacuum (where $\mathcal{V}$ is a minimum), we use (7-55) in (7-52) and take the derivative of (7-52) with respect to ϕ_3.

$$\frac{\partial \mathcal{V}}{\partial \phi_3} = 0 = \frac{\partial}{\partial \phi_3}\left(\mu^2 \Phi^\dagger \Phi + \lambda (\Phi^\dagger \Phi)^2\right) = \frac{\partial}{\partial \phi_3}\left(\frac{\mu^2}{2}\phi_3^2 + \frac{\lambda}{4}\phi_3^4\right) = \mu^2 \phi_3 + \lambda \phi_3^3 \quad \text{(at true vacuum)}$$
$$\rightarrow -\mu^2 = \lambda \phi_{3,true\,vac}^2 \quad \rightarrow \quad \phi_{3,true\,vac} = \sqrt{\frac{-\mu^2}{\lambda}}. \tag{7-56}$$

Finding the minimum of $\mathcal{V}$ in the GSW model

The last expression in (7-56) is the location on the positive ϕ_3 axis where the Higgs field potential is a minimum.

With an eye to what comes later, we want to define σ as zero at the Higgs potential minimum. Thus in (7-55), we take

$$v = \sqrt{\frac{-\mu^2}{\lambda}}. \tag{7-57}$$

Taking our constant v equal to this minimum

Note from (7-57), that if v is real (which is assumed, since we took ϕ_3 as positive at that location), then μ is imaginary. Note also, that by taking $\sigma = 0$ at the true vacuum (along with the unitary gauge where $\eta_i = 0$), we are defining the σ coordinate axis origin as being at the true vacuum, as in the righthand side of Fig. 7-2, pg. 209.

σ is then zero at the minimum, i.e., at the true vacuum

Key Point

So, σ looks to us, living at the true vacuum, like the Higgs field of relevance to us. It has zero particle number density at the true vacuum, which we usually think of as a characteristic of a vacuum. In the present epoch, we think of σ as the Higgs field, not Φ (or, in the unitary gauge, ϕ_3).

And so, σ is effectively our Higgs field in our present-day universe

7.3.4 Finally, the GSW Lagrangian at True Vacuum

Finding the e/w (Lepton) Lagrangian for the True Vacuum Fields

Substituting (7-55), (7-57), (7-43), and (7-44) into (7-46), we get the present day e/w Lagrangian (7-58) below. All but the $\mathcal{L}^{LH}$ part of (7-46) is done in the appendix, beginning on pg. 251. The $\mathcal{L}^{LH}$ part is not as straightforward as the rest, with a particularly interesting and meaningful wrinkle, so, rather than do that in the appendix, we evaluate it in Sect. 7.3.5 below.

Substitute Higgs and gauge boson relations into $\mathcal{L}$ at false vacuum

Just as I did in Chap. 6 for the appendix there explicitly showing the false vacuum e/w Lagrangian symmetry, I heartily recommend that you study the appendix in this chapter step-by-step. It is long, but when you finish, you will see how each term in the true vacuum e/w Lagrangian arises. And that is something you will thereafter, and forever, feel confident about. No other text I am aware of shows this derivation in its completeness. Others just state the final result, and perhaps, show how two or three of the terms arise. Personally, I found that very unfulfilling. Not until I worked through all of this myself, did I truly feel I understood e/w symmetry breaking. And so, I advise you to gain that same understanding, at far less price in time and effort than I, myself, had to expend to reach that end.

All but one part of this is done in the appendix

The final result, culled from (7-167), (7-185), and (7-212) in the appendix, is

True Vacuum, Total Electroweak Lepton Lagrangian $\mathcal{L} = \mathcal{L}_0 + \mathcal{L}_I$

$$
\begin{aligned}
\mathcal{L}_0 = {} & \bar{\psi}_l \big(i\partial\!\!\!/ - \underbrace{\tfrac{v g_l}{\sqrt{2}}}_{m_l}\big)\psi_l + \bar{\psi}_{\nu_l} i\partial\!\!\!/ \psi_{\nu_l} + \underbrace{\mathcal{L}^{m_\nu}}_{\text{neutrino mass terms}} && \text{(leptons)} \\
& -\tfrac{1}{4}F_{\mu\nu}F^{\mu\nu} && \text{(real photon field)} \\
& -\tfrac{1}{2}W^{\dagger}_{\mu\nu}W^{\mu\nu} + \underbrace{\left(\tfrac{1}{2}vg\right)^2}_{m_W^2} W^{\dagger}_{\mu}W^{\mu} && \text{(complex } W \text{ fields)} \\
& -\tfrac{1}{4}Z_{\mu\nu}Z^{\mu\nu} + \tfrac{1}{2}\underbrace{\left(\frac{1}{2}\frac{vg}{\cos\theta_W}\right)^2}_{m_Z^2} Z_\mu Z^\mu && \text{(real } Z \text{ field)} \\
& +\tfrac{1}{2}(\partial_\mu\sigma)(\partial^\mu\sigma) - \tfrac{1}{2}\underbrace{(-2\mu^2)}_{m_H^2}\sigma^2 && \text{(real true vac Higgs } \sigma) \\
\mathcal{L}_I = {} & \mathcal{L}^{LB} + \mathcal{L}^{LH} + \mathcal{L}^{BB} + \mathcal{L}^{HH} + \mathcal{L}^{HB} && \text{(interaction terms)},
\end{aligned} \tag{7-58}
$$

The result of all that substitution and algebra is $\mathcal{L}$ in terms of true vacuum fields, of which σ is one

where

$$
Z_{\mu\nu} = \partial_\nu Z_\mu - \partial_\mu Z_\nu \qquad W_{\mu\nu} = \partial_\nu W^+_\mu - \partial_\mu W^+_\nu \qquad W^{\dagger}_{\mu\nu} = \partial_\nu W^-_\mu - \partial_\mu W^-_\nu , \tag{7-59}
$$

and the interaction terms are found in the afore cited (7-167), (7-185), and (7-212), and also, in (7-97) below, which is yet to be studied. The lepton mass terms are also yet to be studied.

Note that the sign of the mass terms for vectors is positive; for scalars, negative. For complex vector fields, as with complex scalar fields, there is no factor of ½.

The complete Lagrangian for true vacuum fields, including interaction terms, is expressed in the chapter summary on pg. 251, in relations (7-143) to (7-151).

Some Things to Note

Note that for a *complex* vector field ($W_\mu = W_\mu^+$ here), mass is, as we discussed after (7-4), the square root of the factor in front of the bilinear expression in the Lagrangian of that single type of complex boson field. For a *real* vector field (Z^μ and σ here), mass is the square root of that same factor aside from a factor of ½, as in (7-4). The photon field has no such term, meaning its mass is zero, one indication that the methodology used by GSW does indeed give us a valid theory.

The mass of the contemporary Higgs can be expressed either in terms of μ or v and λ, via (7-57).

Reading off the mass of the Higgs σ from its mass term in $\mathcal{L}$

$$m_H = \sqrt{2}\left|\mu\right| = \sqrt{2}v\sqrt{\lambda} \ . \qquad (7\text{-}60)$$

Prior to the discovery of the Higgs particle in 2012, v, g, and θ_W had been determined from scattering experiments, so it was possible to predict the correct (measured) masses of the W's and the Z particles. But λ was still unknown, so it was not possible to get a handle on m_H. Note that m_H is the tree level mass for the Higgs, but that is modified by higher order corrections (higher order Feynman diagrams/amplitudes). See Section 7.4.15. The masses for the W and Z particles also change due to higher order corrections, but theorists took those into account in predicting the experimentally measured values.

Massless fields at false vacuum have become massive fields at true vacuum

Consider that we started in (7-46) with all massless fields (because massive fields mean the Lagrangian cannot be symmetric), and after symmetry breaking (of the Φ field, as easily seen in the Mexican hat figure, but also of $\mathcal{L}$ itself, mathematically), we now have fields with mass terms. The symmetry breaking bore the fruit of massive particles, corresponding to those in our present-day universe.

Neutrinos being QED immune leads to this identity

Note that, as found in (7-158), in order for neutrinos to be electrically neutral, the appropriate terms linking neutrinos to photons have to drop out, and the only way that can happen is if

$$g \sin\theta_W = g' \cos\theta_W \ . \qquad (7\text{-}61)$$

GSW photon coupling to electrons term

Additionally, as found in (7-161), the only term coupling leptons to photons has form

$$g \sin\theta_W \bar{\psi}\gamma^\mu \psi A_\mu \qquad \left(\text{compare to } e\bar{\psi}\gamma^\mu\psi A_\mu \text{ in QED}\right), \qquad (7\text{-}62)$$

which means, in order to obtain QED at the true vacuum, we must have (from (7-62) and (7-61))

The above force this relation onto the GSW model

$$\boxed{e = g \sin\theta_W = g' \cos\theta_W} \ . \qquad (7\text{-}63)$$

<u>The Most Pertinent Terms for Us</u>

Lepton-boson terms in $\mathcal{L}$ are central part of e/w theory

The terms we will be most focused on, when we calculate things like decay and scattering, are those in $\mathcal{L}^{LB}$, i.e., the couplings between the leptons familiar to us and the bosons mediating interactions between them, such as the Ws, the Z, and the photon. (The lepton-photon coupling we studied pretty thoroughly in QED.) As found in (7-167) of the appendix, and using (5-86),

Lepton-boson terms $\mathcal{L}^{LB}$

$$\begin{aligned}\mathcal{L}^{LB} = \; & \underbrace{e\bar{\psi}_l A\!\!\!/ \psi_l}_{\text{QED interaction}} - \frac{g}{2\sqrt{2}}\bar{\psi}_{\nu_l} W\!\!\!\!/^{\,+}\left(1-\gamma^5\right)\psi_l - \frac{g}{2\sqrt{2}}\bar{\psi}_l W\!\!\!\!/^{\,-}\left(1-\gamma^5\right)\psi_{\nu_l} \\ & - \frac{g}{4\cos\theta_W}\bar{\psi}_{\nu_l} Z\!\!\!/\left(1-\gamma^5\right)\psi_{\nu_l} + \frac{g}{4\cos\theta_W}\bar{\psi}_l Z\!\!\!/\left(1-4\sin^2\theta_W - \gamma^5\right)\psi_l \\ = \; & e\bar{\psi}_l A\!\!\!/ \psi_l - \frac{g}{\sqrt{2}}\bar{\psi}^L_{\nu_l} W\!\!\!\!/^{\,+}\psi^L_l - \frac{g}{\sqrt{2}}\bar{\psi}^L_l W\!\!\!\!/^{\,-}\psi^L_{\nu_l} - \frac{g}{2\cos\theta_W}\bar{\psi}^L_{\nu_l} Z\!\!\!/ \psi^L_{\nu_l} \\ & + \frac{g}{2\cos\theta_W}\bar{\psi}^L_l Z\!\!\!/ \psi^L_l - \underbrace{g\sin\theta_W}_{e} \tan\theta_W \bar{\psi}_l Z\!\!\!/ \psi_l \, .\end{aligned} \qquad (7\text{-}64)$$

Note, in passing, that the last term in (7-64) can be expressed, via (7-42), as

Alternative form for charged lepton and Z field terms

$$-e\tan\theta_W \bar{\psi}_l Z\!\!\!/ \psi_l = -e\tan\theta_W \bar{\psi}^L_l Z\!\!\!/ \psi^L_l - e\tan\theta_W \bar{\psi}^R_l Z\!\!\!/ \psi^R_l \, . \qquad (7\text{-}65)$$

There are other terms linking quarks into all of this, and we will have additional parts of the Lagrangian for them. More on this later.

A number of terms for Higgs interaction with fermions and bosons

For Higgs interaction calculations, such as determining how to detect it at CERN, one needs to consider terms in $\mathcal{L}^{LH}$, $\mathcal{L}^{HB}$, and $\mathcal{L}^{HH}$ (see (7-66) below and (7-212) of the appendix), along with the interaction terms for quarks, represented by the symbol $\mathcal{L}^{QH}$ (and presented later). This is not trivial work.

7.3.5 Lepton Mass Eigenstates

Higgs-Lepton Terms

Consider the lepton-Higgs coupling of (7-46),

Exploring the Higgs-Lepton terms in the low energy $\mathcal{L}$

$$\mathcal{L}^{LH} = -{}_AY_{ll'}\bar{\Psi}_l^L\Phi\psi_{l'}^R - {}_BY_{l\nu_{l'}}\bar{\Psi}_l^L\tilde{\Phi}\psi_{\nu_{l'}}^R + h.c., \tag{7-66}$$

where the l and l' indices represent lepton flavors, where l or $l' = 1$ represents e; $= 2$, μ; and $= 3$, τ, and matrices ${}_AY$ and ${}_BY$ are Yukawa matrices.

Charged Lepton Mass Terms

Now, for the first term after the equal sign in (7-66), first, carry out the operations in $SU(2)$ (weak isospin) space, in the unitary gauge (7-55),

$$-{}_A Y_{ll'}\bar{\Psi}_l^L\Phi\psi_{l'}^R = -{}_A Y_{ll'}\left(\bar{\psi}_{\nu_l}^L\ \bar{\psi}_l^L\right)\frac{1}{\sqrt{2}}\begin{pmatrix}0\\ \sigma+v\end{pmatrix}\psi_{l'}^R = -\frac{1}{\sqrt{2}}(\sigma+v)\,{}_AY_{ll'}\bar{\psi}_l^L\psi_{l'}^R. \tag{7-67}$$

Expanding one of those charged lepton terms in the unitary gauge

Then, express the operations in flavor space (which is an $SU(3)$ space for flavor different from the color $SU(3)$ space of Chap. 2) of the result, i.e.,

$$(7\text{-}67) = -\frac{1}{\sqrt{2}}(\sigma+v)\left(\bar{\psi}_e^L\ \bar{\psi}_\mu^L\ \bar{\psi}_\tau^L\right)\underbrace{\begin{bmatrix}{}_AY_{ee} & {}_AY_{e\mu} & {}_AY_{e\tau}\\ {}_AY_{\mu e} & {}_AY_{\mu\mu} & {}_AY_{\mu\tau}\\ {}_AY_{\tau e} & {}_AY_{\tau\mu} & {}_AY_{\tau\tau}\end{bmatrix}}_{{}_AY}\begin{pmatrix}\psi_e^R\\ \psi_\mu^R\\ \psi_\tau^R\end{pmatrix}, \tag{7-68}$$

(7-68) represents nine terms in the Lagrangian, that is, where here and throughout, summation is not carried out over repeated indices e, μ, and τ (that is, we treat them like numbers),

$$-{}_A Y_{ll'}\bar{\Psi}_l^L\Phi\psi_{l'}^R = -\frac{1}{\sqrt{2}}(\sigma+v)\begin{pmatrix}\bar{\psi}_{eA}^LY_{ee}\psi_e^R + \bar{\psi}_{eA}^LY_{e\mu}\psi_\mu^R + \bar{\psi}_{eA}^LY_{e\tau}\psi_\tau^R\\ +\bar{\psi}_{\mu A}^LY_{\mu e}\psi_e^R + \bar{\psi}_{\mu A}^LY_{\mu\mu}\psi_\mu^R + \bar{\psi}_{\mu A}^LY_{\mu\tau}\psi_\tau^R\\ +\bar{\psi}_{\tau A}^LY_{\tau e}\psi_e^R + \bar{\psi}_{\tau A}^LY_{\tau\mu}\psi_\mu^R + \bar{\psi}_{\tau A}^LY_{\tau\tau}\psi_\tau^R\end{pmatrix}. \tag{7-69}$$

This will give us terms with the field σ in them plus terms with the constant v in them, i.e.,

This gives us low energy Higgs-lepton terms, and lepton mass-like terms

$$\text{vertex interaction terms} = -\frac{\sigma}{\sqrt{2}}\bar{\psi}_{eA}^LY_{ee}\psi_e^R - \frac{\sigma}{\sqrt{2}}\bar{\psi}_{eA}^LY_{e\mu}\psi_\mu^R - \frac{\sigma}{\sqrt{2}}\bar{\psi}_{eA}^LY_{e\tau}\psi_\tau^R - \ldots = -\frac{\sigma}{\sqrt{2}}\bar{\psi}_{lA}^LY_{ll'}\psi_{l'}^R \tag{7-70}$$

$$\text{mass-like terms} = -\frac{v}{\sqrt{2}}\bar{\psi}_{eA}^LY_{ee}\psi_e^R - \frac{v}{\sqrt{2}}\bar{\psi}_{eA}^LY_{e\mu}\psi_\mu^R - \frac{v}{\sqrt{2}}\bar{\psi}_{eA}^LY_{e\tau}\psi_\tau^R - \ldots = -\frac{v}{\sqrt{2}}\bar{\psi}_{lA}^LY_{ll'}\psi_{l'}^R. \tag{7-71}$$

The mass-like terms of (7-71) seem strange, as some of them have two different fields, for example, the second term therein with an electron and a muon field. But mass terms in the Lagrangian we have seen before have always had two fields of the same flavor field in them, like the first term in (7-71). To get only one type of field in each mass-like term, we can transform the 3-vectors and the matrix ${}_AY_{ll'}$ of (7-71), such that the matrix is diagonal.

But, some mass-like terms mix flavors, unlike usual mass terms

---

Do **Problem 6** to show (7-72) below.

Aside

From the theory of linear algebra in complex spaces, or by doing Prob. 6, we learn that we can transform a (nonsingular) complex matrix Y, not necessarily Hermitian or unitary, into a real diagonal matrix with all positive components via two (particular, not just any two) unitary transformations, represented by the U and K matrices in

Transforming a general complex matrix into a real diagonal matrix

$$\text{real diagonal } M = UYK^\dagger \xrightarrow[\text{out explicitly}]{\text{indices written}} M_{ll'} = U_{ln\,A}Y_{ns}K_{sl'}^\dagger \quad \text{all } M_{ll'} > 0 \ \left(l = l' \text{ as diagonal}\right). \tag{7-72}$$

For the special case of Y Hermitian, we should recall that we would have $K = U$.

End of Aside

In (7-71), we consider transformations like (7-72), where a wavy underline indicates entities expressed in the new basis, and we keep in mind that gamma matrices act in a different space from flavor space matrices U and K, so they commute with U and K (and their hermitian conjugates),

$$-\frac{v}{\sqrt{2}}\bar{\psi}^L_{l\,A}Y_{ll'}\psi^R_{l'} = -\frac{v}{\sqrt{2}}\psi^{L\dagger}_m \underbrace{\gamma^0 U^{-1}_{ml}}_{U^{-1}_{ml}\gamma^0} U_{ln\,A}Y_{ns}K^{-1}_{sl'}K_{l't}\psi^R_t = -\frac{v}{\sqrt{2}}\underbrace{\psi^{L\dagger}_m U^{\dagger}_{ml}\gamma^0}_{\underset{\sim}{\psi}^{L\dagger}_l\gamma^0=\underset{\sim}{\bar{\psi}}^L_l}\underbrace{U_{ln\,A}Y_{ns}K^{\dagger}_{sl'}}_{{}_A\underset{\sim}{Y}_{ll'}=M_{ll'}}\underbrace{K_{l't}\psi^R_t}_{\underset{\sim}{\psi}^R_{l'}} \qquad (7\text{-}73)$$
$$= -\frac{v}{\sqrt{2}}\underset{\sim}{\bar{\psi}}^L_l M_{ll'}\underset{\sim}{\psi}^R_{l'} = -\frac{v}{\sqrt{2}}\underset{\sim}{\bar{\psi}}^L_1 M_{11}\underset{\sim}{\psi}^R_1 - \frac{v}{\sqrt{2}}\underset{\sim}{\bar{\psi}}^L_2 M_{22}\underset{\sim}{\psi}^R_2 - \frac{v}{\sqrt{2}}\underset{\sim}{\bar{\psi}}^L_3 M_{33}\underset{\sim}{\psi}^R_3 .$$

The mass-like terms with Y matrix transformed to a diagonal matrix → only terms with same flavor factors

Thus, we have transformed the expression (7-71), repeated at the beginning of (7-73), which mixes different flavors in terms therein, into the last part of (7-73), in which each term has only factors of the same type.

For future use, we summarize the transformations used as

Summarizing action of K and U matrices

$$\underset{\sim}{\psi}^R_{l'} = K_{l't}\psi^R_t \qquad \underset{\sim}{\psi}^L_l = U_{lm}\psi^L_m \qquad M_{ll'} = {}_A\underset{\sim}{Y}_{ll'} = U_{ln\,A}Y_{ns}K^{\dagger}_{sl'}$$
$$\psi^R_t = K^{\dagger}_{tl'}\underset{\sim}{\psi}^R_{l'} \qquad \psi^L_m = U^{\dagger}_{ml}\underset{\sim}{\psi}^L_l \qquad {}_AY_{ns} = U^{\dagger}_{nl\,A}\underset{\sim}{Y}_{ll'}K_{l's} = U^{\dagger}_{nl}M_{ll'}K_{l's} . \qquad (7\text{-}74)$$

In the new basis, the fields are linear combinations of original basis fields

Note that, for example, $\quad \underset{\sim}{\psi}^R_1 = K_{1e}\psi^R_e + K_{1\mu}\psi^R_\mu + K_{1\tau}\psi^R_\tau , \qquad (7\text{-}75)$

i.e., our new field $\underset{\sim}{\psi}^R_1$ is a linear combination of the old ones. Similarly, $\underset{\sim}{\psi}^R_2$, $\underset{\sim}{\psi}^R_3$, $\underset{\sim}{\bar{\psi}}^L_1$, $\underset{\sim}{\bar{\psi}}^L_2$ and $\underset{\sim}{\bar{\psi}}^L_3$ are linear combinations of the original fields, as well.

The same transformations U and K used in (7-73) can be used in (7-70), with similar results, i.e., a matrix diagonalization leaving only terms with the same (new basis) type leptons.

Now, consider the term in (7-66) that is the hermitian conjugate (symbolized by *h.c.*) of (7-73), where we can diagonalize that via the same transformations U and K.

$$\left(-\frac{v}{\sqrt{2}}\bar{\psi}^L_{l\,A}Y_{ll'}\psi^R_{l'}\right)^{\dagger} = -\frac{v}{\sqrt{2}}\bar{\psi}^R_{l'\,A}Y^{\dagger}_{ll'}\psi^L_{l} = -\frac{v}{\sqrt{2}}\underbrace{\bar{\psi}^R_s K^{\dagger}_{sl}}_{\underset{\sim}{\bar{\psi}}^R_l}\underbrace{K_{lt\,A}Y^{\dagger}_{tn}U^{\dagger}_{nl'}}_{{}_A\underset{\sim}{Y}^{\dagger}_{ll'}=M^{\dagger}_{ll'}=M_{ll'}}\underbrace{U_{l'm}\psi^L_m}_{\underset{\sim}{\psi}^L_{l'}} \qquad (7\text{-}76)$$
$$= -\frac{v}{\sqrt{2}}\underset{\sim}{\bar{\psi}}^R_l M_{ll'}\underset{\sim}{\psi}^L_{l'} = -\frac{v}{\sqrt{2}}\underset{\sim}{\bar{\psi}}^R_1 M_{11}\underset{\sim}{\psi}^L_1 - \frac{v}{\sqrt{2}}\underset{\sim}{\bar{\psi}}^R_2 M_{22}\underset{\sim}{\psi}^L_2 - \frac{v}{\sqrt{2}}\underset{\sim}{\bar{\psi}}^R_3 M_{33}\underset{\sim}{\psi}^L_3 .$$

Repeating above steps for hermitian conjugate term in $\mathcal{L}$

Consider the first terms in the last part of (7-73) and the last part of (7-76), along with (7-41),

$$-\frac{v}{\sqrt{2}}\underset{\sim}{\bar{\psi}}^L_1 M_{11}\underset{\sim}{\psi}^R_1 - \frac{v}{\sqrt{2}}\underset{\sim}{\bar{\psi}}^R_1 M_{11}\underset{\sim}{\psi}^L_1 = -\frac{v}{\sqrt{2}}M_{11}\left(\underset{\sim}{\bar{\psi}}^L_1\underset{\sim}{\psi}^R_1 + \underset{\sim}{\bar{\psi}}^R_1\underset{\sim}{\psi}^L_1\right) = -\underbrace{\frac{v}{\sqrt{2}}M_{11}}_{\substack{\text{mass of new} \\ \text{basis field } \psi_1}}\underset{\sim}{\bar{\psi}}_1\underset{\sim}{\psi}_1 . \qquad (7\text{-}77)$$

Adding original term to its h.c. yields the traditional mass type term

Conclusion: If we use the new basis (with wavy underlines), we get nice, neat mass terms in our Lagrangian, with no mixing of different type fields in those mass terms. And we can read the mass directly off of the terms as

$$\text{mass of } i^{th} \text{ charged lepton field in new basis} = \frac{v}{\sqrt{2}}M_{ii} \quad (\text{no sum, } i=1,2,3). \qquad (7\text{-}78)$$

Mass of each charged lepton in new basis

How About Other Charged Lepton Terms in the Lagrangian?

Note the kinetic term in (7-58), and transform it using the same transformations as above. Using (7-42), one gets (where the equality of the first and last parts of the top line is something you should make a point of committing to memory)

$$i\bar{\psi}_l\not{\partial}\psi_l = i\bar{\psi}_l\not{\partial}\tfrac{1}{2}\left(1-\gamma^5\right)\psi_l + i\bar{\psi}_l\not{\partial}\tfrac{1}{2}\left(1+\gamma^5\right)\psi_l = i\bar{\psi}^L_l\not{\partial}\psi^L_l + i\bar{\psi}^R_l\not{\partial}\psi^R_l$$
$$= i\underset{\sim}{\bar{\psi}}^L_m U^{\dagger}_{ml}U_{ln}\not{\partial}\underset{\sim}{\psi}^L_n + i\underset{\sim}{\bar{\psi}}^R_m K^{\dagger}_{ml}K_{ln}\not{\partial}\underset{\sim}{\psi}^R_n = i\underset{\sim}{\bar{\psi}}^L_l\not{\partial}\underset{\sim}{\psi}^L_l + i\underset{\sim}{\bar{\psi}}^R_l\not{\partial}\underset{\sim}{\psi}^R_l = i\underset{\sim}{\bar{\psi}}_l\not{\partial}\underset{\sim}{\psi}_l . \qquad (7\text{-}79)$$

Charged lepton kinetic terms unchanged in form in new basis

The kinetic term looks the same in either basis.

So, let's look at interaction terms with leptons in them, i.e., those in (7-64), and see what our transformations U and K give us. First, the QED interaction term,

$$e\bar{\psi}_l\not{A}\psi_l = e\,A_\mu\bar{\psi}_l\gamma^\mu\psi_l = eA_\mu\bar{\psi}^L_l\gamma^\mu\psi^L_l + eA_\mu\bar{\psi}^R_l\gamma^\mu\psi^R_l$$
$$= eA_\mu\underset{\sim}{\bar{\psi}}^L_m U^{\dagger}_{ml}U_{ln}\gamma^\mu\underset{\sim}{\psi}^L_n + eA_\mu\underset{\sim}{\bar{\psi}}^R_m K^{\dagger}_{ml}K_{ln}\gamma^\mu\underset{\sim}{\psi}^R_n \qquad (7\text{-}80)$$
$$= eA_\mu\underset{\sim}{\bar{\psi}}^L_l\gamma^\mu\underset{\sim}{\psi}^L_l + eA_\mu\underset{\sim}{\bar{\psi}}^R_l\gamma^\mu\underset{\sim}{\psi}^R_l = eA_\mu\underset{\sim}{\bar{\psi}}_l\gamma^\mu\underset{\sim}{\psi}_l .$$

QED term unchanged in new basis, as well

The QED interaction term (7-80) has the same form in either basis, just as the kinetic term does.

Likewise for lepton-Z field interaction terms

Do **Problem 7** to show that charged lepton interaction terms with the Z vector boson have the same form in both bases, as well.

We are going to wait just a bit to look at terms with leptons interacting with the W bosons.

We'll investigate lepton-W terms a bit later. All other terms unchanged

Neutrino Mass Terms

All of the above logic we used for the e, μ, and τ field terms in $\mathcal{L}^{LH}$ of (7-66) can be directly applied to neutrinos. Instead of the M matrix being the transformation of the ${}_AY$ matrix, as in (7-72), we have

Repeating above diagonalization to new basis for neutrinos

$$\text{real diagonal } {}_\nu M = T_B Y L^\dagger \xrightarrow[\text{out explicitly}]{\text{indices written}} {}_\nu M_{l\nu_{l'}} = T_{ln}\,{}_BY_{ns}\,L^\dagger_{s\nu_{l'}}, \qquad (7\text{-}81)$$

where the matrices T and L are different from U and K, because ${}_BY$ is different from ${}_AY$ (and so, ${}_\nu M$ will be different from M, where the subscript here indicates it refers to a neutrino term).

Repeating the steps we went through for the first term in (7-66), for the second term therein, we get, first,

$$-{}_BY_{l\nu_{l'}}\bar{\Psi}^L_l\tilde{\Phi}\psi^R_{\nu_{l'}} = -{}_BY_{l\nu_{l'}}\begin{pmatrix}\bar{\psi}^L_{\nu_l} & \bar{\psi}^L_l\end{pmatrix}\frac{1}{\sqrt{2}}\begin{pmatrix}\sigma+v\\0\end{pmatrix}\psi^R_{\nu_{l'}} = -\frac{1}{\sqrt{2}}(\sigma+v)\,{}_BY_{\nu_l\nu_{l'}}\bar{\psi}^L_{\nu_l}\psi^R_{\nu_{l'}}, \qquad (7\text{-}82)$$

and then, parallel to (7-73), (7-76), and (7-77),

$$-{}_BY_{l\nu_{l'}}\bar{\Psi}^L_l\tilde{\Phi}\psi^R_{\nu_{l'}} + h.c. = \underbrace{-\frac{v}{\sqrt{2}}\bar{\underset{\sim}{\psi}}_{\nu_l}\,{}_\nu M_{\nu_l\nu_{l'}}\underset{\sim}{\psi}_{\nu_{l'}}}_{\text{diagonal matrix mass term}} \quad \underbrace{-\frac{\sigma}{\sqrt{2}}\bar{\underset{\sim}{\psi}}_{\nu_l}\,{}_\nu M_{\nu_l\nu_{l'}}\underset{\sim}{\psi}_{\nu_{l'}}}_{\text{interaction term}}$$

$$\overbrace{-\underbrace{\frac{v}{\sqrt{2}}\,{}_\nu M_{\nu_1\nu_1}}_{\substack{m_{\nu_1},\text{ mass of 1st}\\ \text{new basis field}}}\bar{\underset{\sim}{\psi}}_{\nu_1}\underset{\sim}{\psi}_{\nu_1} - \underbrace{\frac{v}{\sqrt{2}}\,{}_\nu M_{\nu_2\nu_2}}_{\substack{m_{\nu_2},\text{ mass of 2nd}\\ \text{new basis field}}}\bar{\underset{\sim}{\psi}}_{\nu_2}\underset{\sim}{\psi}_{\nu_2} - \underbrace{\frac{v}{\sqrt{2}}\,{}_\nu M_{\nu_3\nu_3}}_{\substack{m_{\nu_3},\text{ mass of 3rd}\\ \text{new basis field}}}\bar{\underset{\sim}{\psi}}_{\nu_3}\underset{\sim}{\psi}_{\nu_3}} \,. \qquad (7\text{-}83)$$

So, we have the

$$\text{mass of } i^{th} \text{ neutrino field in new basis} = \frac{v}{\sqrt{2}}\,{}_\nu M_{ii} \quad (\text{no sum, } i=1,2,3), \qquad (7\text{-}84)$$

Mass of each neutrino in new basis

where

$$\begin{matrix} \underset{\sim}{\psi}^R_{\nu_l} = L_{\nu_l\nu_t}\psi^R_{\nu_t} & \underset{\sim}{\psi}^L_{\nu_l} = T_{\nu_l\nu_t}\psi^L_{\nu_t} & {}_\nu M_{\nu_l\nu_{l'}} = {}_B\underset{\sim}{Y}_{\nu_l\nu_{l'}} = T_{\nu_l n}\,{}_BY_{ns}\,L^\dagger_{s\nu_{l'}} \\ \psi^R_{\nu_t} = L^\dagger_{\nu_t\nu_l}\underset{\sim}{\psi}^R_{\nu_l} & \psi^L_{\nu_t} = T^\dagger_{\nu_t\nu_l}\underset{\sim}{\psi}^L_{\nu_l} & Y_{ns} = T^\dagger_{n\nu_l}\,{}_B\underset{\sim}{Y}_{\nu_l\nu_{l'}}L_{\nu_{l'}s} = T^\dagger_{n\nu_l}\,{}_\nu M_{\nu_l\nu_{l'}}L_{\nu_{l'}s} \end{matrix} \,. \qquad (7\text{-}85)$$

Summarizing action of L and T matrices

Other Terms with Neutrinos

Kinetic and interaction terms will follow suit with the charged leptons, as well. They will take the same form for both the original (non-underlined) and the wavy underline basis, except for the interactions with W bosons, which we have yet to discuss.

All other neutrino terms without W interactions unchanged in new basis

Distinguishing Between Bases

The wavy underlined fields basis is called the mass basis, and the fields therein, the mass eigenstates. The original basis fields are called flavor eigenstates, or charge eigenstates, and they form the flavor basis. Note the word "state" in this context is not used in the sense of states vs fields, but in the sense of an eigenstate in an eigenstate problem.

Wavy underline = mass basis with mass eigenstates

Original basis, no underline = flavor basis with flavor eigenstates

We will have much to say about these bases, but first, we need to examine the ramifications of using the mass basis for terms containing leptons and W bosons.

The W Boson-Lepton Interaction Terms

From (7-64), we have

Examining terms with W fields in $\mathcal{L}^{LB}$

$$W \text{ boson terms in } \mathcal{L}^{LB} = -\frac{g}{\sqrt{2}}\bar{\psi}^L_{\nu_l}\mathcal{W}^+\psi^L_l - \frac{g}{\sqrt{2}}\bar{\psi}^L_l\mathcal{W}^-\psi^L_{\nu_l} \qquad (7\text{-}86)$$

Let's convert that to the mass eigenstate basis using (7-74) and (7-85), and see what we have.

Transforming fields to mass basis

$$-\frac{g}{\sqrt{2}}\bar{\psi}^L_{\nu_l}W^+\psi^L_l-\frac{g}{\sqrt{2}}\bar{\psi}^L_l W^-\psi^L_{\nu_l}=-\frac{g}{\sqrt{2}}\bar{\psi}^L_{\nu_t}W^+\underbrace{T_{\nu_t l}U^\dagger_{lm}}_{P^\dagger_{\nu_t m}}\psi^L_m-\frac{g}{\sqrt{2}}\bar{\psi}^L_{l'}W^-\underbrace{U_{l'l}T^\dagger_{l\nu_t}}_{P_{l'\nu_t}}\psi^L_{\nu_t}$$

$$\text{with more common subscript symbols} = -\frac{g}{\sqrt{2}}\bar{\psi}^L_{\nu_l}W^+\underbrace{T_{\nu_l s}U^\dagger_{sl'}}_{P^\dagger_{\nu_l l'}}\psi^L_{l'}-\frac{g}{\sqrt{2}}\bar{\psi}^L_{l'}W^-\underbrace{U_{l's}T^\dagger_{s\nu_l}}_{P_{l'\nu_l}}\psi^L_{\nu_l}\,. \tag{7-87}$$

Note that, since T and U are unitary, the matrix P is unitary, because (writing as matrix multiplication with flavor indices suppressed)

Combine U and $T^\dagger$ matrices into unitary matrix $P=UT^\dagger$

$$P=UT^\dagger \qquad\qquad P^\dagger P=TU^\dagger UT^\dagger=TT^\dagger=I\,. \tag{7-88}$$

P is called, after its discoverers, the Pontecorvo-Maki-Nakagawa-Sakata mixing matrix, or more simply, the PMNS matrix. Note that a number of authors use the symbol U for this, instead of P.

Called PMNS matrix

Writing Out the W Boson-Lepton Interaction Terms with the PMNS Matrix

So, in the mass basis, (7-86) can be written as

$\mathcal{L}$ terms with PMNS matrix written out

$$W\text{ boson terms in }\mathcal{L}^{LB}=-\frac{g}{\sqrt{2}}\begin{pmatrix}\bar{\psi}^L_{\nu_1} & \bar{\psi}^L_{\nu_2} & \bar{\psi}^L_{\nu_3}\end{pmatrix}W^+\underbrace{\begin{bmatrix}P^*_{11} & P^*_{21} & P^*_{31}\\ P^*_{12} & P^*_{22} & P^*_{32}\\ P^*_{13} & P^*_{23} & P^*_{33}\end{bmatrix}}_{P^\dagger}\begin{pmatrix}\psi^L_1\\ \psi^L_2\\ \psi^L_3\end{pmatrix}$$

$$-\frac{g}{\sqrt{2}}\begin{pmatrix}\bar{\psi}^L_1 & \bar{\psi}^L_2 & \bar{\psi}^L_3\end{pmatrix}W^-\underbrace{\begin{bmatrix}P_{11} & P_{12} & P_{13}\\ P_{21} & P_{22} & P_{23}\\ P_{31} & P_{32} & P_{33}\end{bmatrix}}_{\text{PMNS matrix }P}\begin{pmatrix}\psi^L_{\nu_1}\\ \psi^L_{\nu_2}\\ \psi^L_{\nu_3}\end{pmatrix}. \tag{7-89}$$

Choice of Basis

Our theory, as we have noted before (Vol. 1, pgs. 177-178 and elsewhere), is a gauge theory. That is, we can redefine our fields in any way, as long as that redefinition doesn't change predictions for things that we could measure in the real world. We never measure the field, itself (Vol. 1, pgs. 189-190). It has zero expectation value. So, in our present case, we can redefine our fields in any way that does not change any calculation result, such as transition amplitudes. And so, we will want to choose a basis that makes our problem easiest to solve.

Because QFT is a gauge theory, we can pick any basis we like

This works because all types of charges are the same in different generations, so a linear combination of charged fields in a given basis (such as electron field plus a muon field) gives us a field that has the same charges as each of the original fields. So, if we like, we can take the new field as part of a new basis, and even, if we choose, call it a flavor field (such as "electron") even though it is different from what we called that flavor field (electron in our example) in the original basis. The new "electron" flavor field in the new basis will have all the same charges as the original basis electron field had, so it is not unreasonable to just call it an electron. Similarly, with neutrinos. Confusing? Yes, but that is what is done in QFT.

A new basis field is a combination of 3 flavor fields, all having the same charge

So, it could be called a flavor field, itself

For example, we can construct a new (primed) set of basis fields via some unitary matrix D,

$$\psi'^L_l=D_{lm}\psi^L_m\xrightarrow[\text{new basis fields}]{\text{for the first of the}}\psi'^L_a=D_{ae}\psi^L_e+D_{a\mu}\psi^L_\mu+D_{a\tau}\psi^L_\tau\xrightarrow[\text{our new }e]{\text{call }a}\psi'^L_a=\psi'^L_e\,. \tag{7-90}$$

And we could even call it by the same name as one of the original flavors

Key point: We can consider a field in the new basis a flavor field, if we like, and give it any name we like. For example, if it is made up of what, in the original flavor basis, we called fields labeled e, μ, τ, we can even call the new field e (electron).

With all this in mind, one can choose a basis that is some transformation set of either the mass basis, the original flavor basis, or some combination of the two, provided such transformation set does not, in the final analysis, change any physical predictions.

But, here's the rub. In the literature, two such different bases (each a combination of the original flavor and mass bases) are used, one for handling leptons, the other for handling quarks. And, as can be a source of great consternation, the basis chosen is virtually never specified. So, you can see a treatment of this topic in one manner for leptons, and think you are starting to understand it, only to look at a different reference for quarks (that uses a different basis, but doesn't say so), where the treatment seems irreconcilably different from the prior treatment.

Caution: basis used for leptons different from that for quarks

It took me a long time to unravel all of this, by myself. I pass on the following, which I hope will save you, the reader, the frustrations I felt in trying to make sense of this issue. It is not simple, and should take you some time to fathom, but in high likelihood, it will make your path toward understanding shorter and less painful than it otherwise would be.

<u>A Hybrid Basis: Basis I</u>

For leptons, can be convenient to use a hybrid basis, Basis I

As we showed above, except for the mass terms and the lepton terms with W^+ or W^- factors, we could use either the mass basis or the original flavor basis and get the same terms in the Lagrangian. This concept is used in one of the conventional choices of basis, which we will call <u>Basis I</u>, and which we show in the following.

In hybrid Basis I, new charged leptons identical to mass eigenstates

We label new basis charged leptons e, μ, τ

For the charged leptons (electron, muon, tau), we define a new (primed) basis where the new states are the same as the mass eigenstates, and we assign our original flavor names to the new basis states.

$$\psi'^L_{l'} = \underset{\sim}{\psi}^L_l \;\left(\psi'^L_e = \underset{\sim}{\psi}^L_1 ; \psi'^L_\mu = \underset{\sim}{\psi}^L_2 ; \psi'^L_\tau = \underset{\sim}{\psi}^L_3\right) \quad \psi'^R_{l'} = \underset{\sim}{\psi}^R_l \;\left(\psi'^R_e = \underset{\sim}{\psi}^R_1 ; \psi'^R_\mu = \underset{\sim}{\psi}^R_2 ; \psi'^R_\tau = \underset{\sim}{\psi}^R_3\right), \tag{7-91}$$

In this basis, we will get mass terms looking like (7-77), with the labeling of (7-91),

$$\text{charged lepton mass terms in Basis I} = -\frac{v}{\sqrt{2}} M_{ee}\,\bar{\psi}'_e \psi'_e \; -\frac{v}{\sqrt{2}} M_{\mu\mu}\,\bar{\psi}'_\mu \psi'_\mu \; -\frac{v}{\sqrt{2}} M_{\tau\tau}\,\bar{\psi}'_\tau \psi'_\tau \,. \tag{7-92}$$

There is no mixing of flavors in the mass terms for charged leptons and they have definite mass.

In this basis, e, μ, τ fields have definite mass, i.e., mass matrix is diagonal

The above will work fine, since P is unitary, and by the same logic shown earlier, all other charged lepton terms in the Lagrangian without W^+ or W^- factors (all terms not mixing charged leptons with neutrinos) will look the same in the original flavor basis, the mass basis, and this new basis, Basis I.

To take care of the terms with W^+ and W^- factors, we define the LC neutrino new flavor eigenstates in terms of the PMNS matrix and the mass eigenstates. That is, we take

But, LC neutrinos related to mass eigenstates via PMNS matrix

We label new basis neutrinos ν_e, ν_μ, ν_τ

$$\begin{pmatrix} \psi'^L_{\nu_e} \\ \psi'^L_{\nu_\mu} \\ \psi'^L_{\nu_\tau} \end{pmatrix} = \underbrace{\begin{bmatrix} P_{e1} & P_{e2} & P_{e3} \\ P_{\mu 1} & P_{\mu 2} & P_{\mu 3} \\ P_{\tau 1} & P_{\tau 2} & P_{\tau 3} \end{bmatrix}}_{\text{PMNS matrix } P} \begin{pmatrix} \underset{\sim}{\psi}^L_{\nu_1} \\ \underset{\sim}{\psi}^L_{\nu_2} \\ \underset{\sim}{\psi}^L_{\nu_3} \end{pmatrix} \qquad \begin{array}{l} \text{or } \psi'^L_{\nu_{l'}} = P_{\nu_{l'}\nu_l} \underset{\sim}{\psi}^L_{\nu_l} \\ \text{with } \underset{\sim}{\psi}^L_{\nu_l} = \left(P_{\nu_{l'}\nu_l}\right)^\dagger \psi'^L_{\nu_{l'}} = P^*_{\nu_l \nu_{l'}} \psi'^L_{\nu_{l'}} \,. \end{array} \tag{7-93}$$

RC neutrinos cannot be detected because they don't interact via SM forces, so we are at liberty to define them in Basis I any way we wish. To keep things simple, we define them as equal to their mass eigenstates, i.e., similar to (7-91),

RC neutrinos in this basis identical to mass eigenstates

$$\psi'^R_{\nu_{l'}} = \underset{\sim}{\psi}^R_{\nu_l} \qquad \left(\psi'^R_{\nu_e} = \underset{\sim}{\psi}^R_{\nu_1} ; \psi'^R_{\nu_\mu} = \underset{\sim}{\psi}^R_{\nu_2} ; \psi'^R_{\nu_\tau} = \underset{\sim}{\psi}^R_{\nu_3}\right). \tag{7-94}$$

The mass terms for neutrinos in Basis I, therefore, look like (7-95) below, where we start with (7-83) and use (7-41), then (7-93) and (7-94) to get the penultimate line.

Neutrino mass terms in this basis not diagonal

$$\begin{aligned} \underbrace{\mathcal{L}^{m_\nu}}_{\substack{\text{neutrino} \\ \text{mass terms}}} &= -\frac{v}{\sqrt{2}} \bar{\underset{\sim}{\psi}}_{\nu_1}\, {}_\nu M_{\nu_1\nu_1} \underset{\sim}{\psi}_{\nu_1} - \frac{v}{\sqrt{2}} \bar{\underset{\sim}{\psi}}_{\nu_2}\, {}_\nu M_{\nu_2\nu_2} \underset{\sim}{\psi}_{\nu_2} - \frac{v}{\sqrt{2}} \bar{\underset{\sim}{\psi}}_{\nu_3}\, {}_\nu M_{\nu_3\nu_3} \underset{\sim}{\psi}_{\nu_3} \\ &= -\frac{v}{\sqrt{2}} \bar{\underset{\sim}{\psi}}_{\nu_l}\, {}_\nu M_{\nu_l\nu_{l'}} \underset{\sim}{\psi}_{\nu_{l'}} = -\frac{v}{\sqrt{2}} \bar{\underset{\sim}{\psi}}^R_{\nu_l}\, {}_\nu M_{\nu_l\nu_{l'}} \underset{\sim}{\psi}^L_{\nu_{l'}} - \frac{v}{\sqrt{2}} \bar{\underset{\sim}{\psi}}^L_{\nu_l}\, {}_\nu M_{\nu_l\nu_{l'}} \underset{\sim}{\psi}^R_{\nu_{l'}} \qquad l,l'=1,2,3 \\ &= -\frac{v}{\sqrt{2}} \bar{\psi}'^R_{\nu_l} \underbrace{{}_\nu M_{\nu_l\nu_{l'}} P^\dagger_{\nu_{l'}\nu_m}}_{\text{non-diagonal}} \psi'^L_{\nu_m} - \frac{v}{\sqrt{2}} \bar{\psi}'^L_{\nu_l} \underbrace{P_{\nu_l m}\, {}_\nu M_{m\nu_{l'}}}_{\text{non-diagonal}} \psi'^R_{\nu_{l'}} \qquad l,l',m=e,\mu,\tau \\ &= -\frac{v}{\sqrt{2}} \bar{\psi}'^R_\nu\, {}_\nu M\, P^\dagger \psi'^L_\nu - \frac{v}{\sqrt{2}} \bar{\psi}'^L_\nu P\, {}_\nu M\, \psi'^R_\nu \qquad \text{shorthand notation}. \end{aligned} \tag{7-95}$$

LC neutrinos in Basis I are not in mass eigenstates. They have indefinite mass.

The terms with a factor of W^+ or W^- are shown in (7-89), which given (7-93), can be expressed as

All terms in $\mathcal{L}$ unchanged except neutrino mass terms and W-lepton interaction terms

$$\begin{aligned}\text{W boson terms in } \mathcal{L}^{LB} \text{ Basis I} &= -\frac{g}{\sqrt{2}}\begin{pmatrix}\bar{\psi}^L_{\underset{\sim}{\nu_1}} & \bar{\psi}^L_{\underset{\sim}{\nu_2}} & \bar{\psi}^L_{\underset{\sim}{\nu_3}}\end{pmatrix}\underbrace{\begin{bmatrix}P^*_{e1} & P^*_{\mu 1} & P^*_{\tau 1}\\ P^*_{e2} & P^*_{\mu 2} & P^*_{\tau 2}\\ P^*_{e3} & P^*_{\mu 3} & P^*_{\tau 3}\end{bmatrix}}_{P^\dagger} W^+ \begin{pmatrix}\psi'^L_e\\ \psi'^L_\mu\\ \psi'^L_\tau\end{pmatrix}\\ &\quad -\frac{g}{\sqrt{2}}\begin{pmatrix}\bar{\psi}'^L_e & \bar{\psi}'^L_\mu & \bar{\psi}'^L_\tau\end{pmatrix} W^- \underbrace{\begin{bmatrix}P_{e1} & P_{e2} & P_{e3}\\ P_{\mu 1} & P_{\mu 2} & P_{\mu 3}\\ P_{\tau 1} & P_{\tau 2} & P_{\tau 3}\end{bmatrix}}_{\text{PMNS matrix } P}\begin{pmatrix}\psi^L_{\underset{\sim}{\nu_1}}\\ \psi^L_{\underset{\sim}{\nu_2}}\\ \psi^L_{\underset{\sim}{\nu_3}}\end{pmatrix} \qquad (7\text{-}96)\\ &= -\frac{g}{\sqrt{2}}\begin{pmatrix}\bar{\psi}'^L_{\nu_e} & \bar{\psi}'^L_{\nu_\mu} & \bar{\psi}'^L_{\nu_\tau}\end{pmatrix} W^+ \begin{pmatrix}\psi'^L_e\\ \psi'^L_\mu\\ \psi'^L_\tau\end{pmatrix} - \frac{g}{\sqrt{2}}\begin{pmatrix}\bar{\psi}'^L_e & \bar{\psi}'^L_\mu & \bar{\psi}'^L_\tau\end{pmatrix} W^- \begin{pmatrix}\psi'^L_{\nu_e}\\ \psi'^L_{\nu_\mu}\\ \psi'^L_{\nu_\tau}\end{pmatrix}\end{aligned}$$

or

$$\begin{aligned}\text{W boson terms in } \mathcal{L}^{LB} \text{ Basis } I &= -\frac{g}{\sqrt{2}}\bar{\psi}'^L_{\nu_e} W^+ \psi'^L_e - \frac{g}{\sqrt{2}}\bar{\psi}'^L_{\nu_\mu} W^+ \psi'^L_\mu - \frac{g}{\sqrt{2}}\bar{\psi}'^L_{\nu_\tau} W^+ \psi'^L_\tau\\ &\quad -\frac{g}{\sqrt{2}}\bar{\psi}'^L_e W^- \psi'^L_{\nu_e} - \frac{g}{\sqrt{2}}\bar{\psi}'^L_\mu W^- \psi'^L_{\nu_\mu} - \frac{g}{\sqrt{2}}\bar{\psi}'^L_\tau W^- \psi'^L_{\nu_\tau}.\end{aligned} \qquad (7\text{-}97)$$

W-lepton interaction terms in Basis I

In this basis we have, from (7-92), that the new basis charged lepton flavor states have definite masses, but from (7-95), the LC neutrino flavor states have indefinite mass (as they are superpositions of mass eigenstates). All terms will look like those in the original flavor basis, except the charged lepton mass terms will have a diagonal flavor matrix (different flavors are not mixed in the charged lepton mass terms), and neutrino mass terms will have a different non-diagonal matrix ($P_\nu M$ of (7-95) instead of $_BY$ of (7-82) with neutrino flavors mixing).

All terms in $\mathcal{L}$ look like original flavor basis, except mass terms

To help us visualize what is going on, we can write out the full expressions for the LC neutrino fields in (7-97) as

$$\begin{gathered}\psi'^L_{\nu_e} = P_{e1}\psi^L_{\underset{\sim}{\nu_1}} + P_{e2}\psi^L_{\underset{\sim}{\nu_2}} + P_{e3}\psi^L_{\underset{\sim}{\nu_3}} \qquad \psi'^L_{\nu_\mu} = P_{\mu 1}\psi^L_{\underset{\sim}{\nu_1}} + P_{\mu 2}\psi^L_{\underset{\sim}{\nu_2}} + P_{\mu 3}\psi^L_{\underset{\sim}{\nu_3}}\\ \psi'^L_{\nu_\tau} = P_{\tau 1}\psi^L_{\underset{\sim}{\nu_1}} + P_{\tau 2}\psi^L_{\underset{\sim}{\nu_2}} + P_{\tau 3}\psi^L_{\underset{\sim}{\nu_3}},\end{gathered} \qquad (7\text{-}98)$$

But neutrino flavor state is a superposition of mass states

and thus, they do not have definite masses. If you measure the new LC electron flavor neutrino mass, you could get any one of three values. The mass $m_{\nu 1}$ of $\psi^L_{\underset{\sim}{\nu_1}}$; the mass $m_{\nu 2}$ of $\psi^L_{\underset{\sim}{\nu_2}}$; or the mass $m_{\nu 3}$ of $\psi^L_{\underset{\sim}{\nu_3}}$. Over many measurements, you would get an average, or expectation, value for the LC electron neutrino mass. That value would depend on the various values of P_{e1}, P_{e2}, and P_{e3}. We will have a lot more to say about neutrino masses in a later chapter devoted solely to neutrinos.

Neutrino flavor state has no definite mass

Note notation change: We note that, once a basis is chosen, the primes are dropped. That is, in the literature, one sees fields, regardless of the basis chosen, without primes (or wavy underlines). Often (but not always), the flavor letter labels e, μ, and τ are used for the new basis states and the numbers 1,2,3 for the mass eigenstates.

Primes and wavy underlines not used in the literature

In Fig. 7-3, on the LHS, we show an interaction that, in this choice of basis, cannot occur, i.e., there is no term in the second row of (7-97) for an incoming muon neutrino, and an outgoing electron at a vertex. (Note that the W^- field destroys a W^- particle and creates a W^+ particle.) Electrons in Basis I can only be produced from electron neutrinos, but not muon or tau neutrinos. In general, there is no generation mixing at a vertex.

Basis I

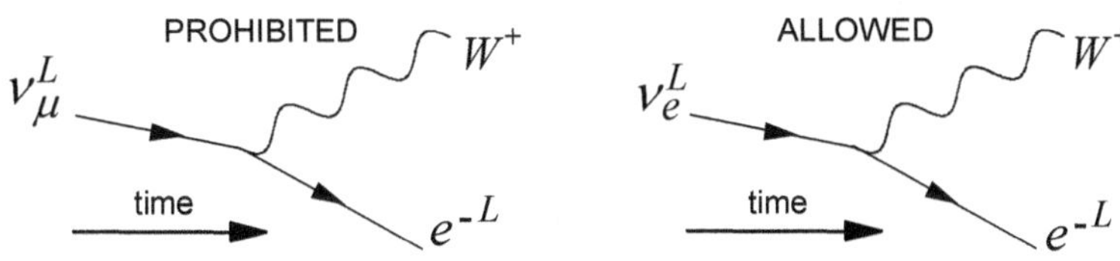

ν_μ^L and ν_e^L have indefinite mass

Figure 7-3. In Basis I: No Generation Changing at a Vertex, Neutrinos Not in Mass Eigenstates

Allowed and unallowed vertices in Basis I

We note in passing that Basis I is the basis of choice for evaluating neutrino oscillations, which we mentioned in Chap. 6, and which we will treat in depth in a later chapter.

Basis I is typically used to evaluate neutrino oscillations

A Different Basis: Basis II

Instead of Basis I, we could choose the following (double primed) basis, which we label Basis II. In Basis II, we take all the new flavor basis states to be identical to mass eigenstates. So, here, we have that

(7-91) and (7-94) are also in Basis II but, instead of (7-93),

$$\begin{pmatrix} \psi''^L_{\nu_e} \\ \psi''^L_{\nu_\mu} \\ \psi''^L_{\nu_\tau} \end{pmatrix} = \begin{pmatrix} \underset{\sim}{\psi}^L_{\nu_1} \\ \underset{\sim}{\psi}^L_{\nu_2} \\ \underset{\sim}{\psi}^L_{\nu_3} \end{pmatrix} . \tag{7-99}$$

A different basis, Basis II, where all fields (including neutrinos) taken as mass eigenstates

All terms will look identical in this Basis II from what we had in the original flavor basis, except for the mass terms, which will be diagonal, and for the terms with factors of W^+ and W^- in them. In other words, Basis II is simply the under wavy line basis, where we take each mass eigenstate as a new flavor state. Mass eigenstate 1 for both charged leptons and neutrinos is considered the electron generation; 2, the muon generation; and 3, the tau generation. They, instead of the last row of (7-96), are then as follows.

All terms in $\mathcal{L}$ look like original flavor basis, except mass terms, which now don't mix flavors, plus W-lepton terms

W boson terms in $\mathcal{L}^{LB}$ Basis II

$$= -\frac{g}{\sqrt{2}} \begin{pmatrix} \bar{\psi}''^L_{\nu_e} & \bar{\psi}''^L_{\nu_\mu} & \bar{\psi}''^L_{\nu_\tau} \end{pmatrix} \underbrace{\begin{bmatrix} P^*_{ee} & P^*_{\mu e} & P^*_{\tau e} \\ P^*_{e\mu} & P^*_{\mu\mu} & P^*_{\tau\mu} \\ P^*_{e\tau} & P^*_{\mu\tau} & P^*_{\tau\tau} \end{bmatrix}}_{P^\dagger} W\!\!\!/^{+} \begin{pmatrix} \psi''^L_{e} \\ \psi''^L_{\mu} \\ \psi''^L_{\tau} \end{pmatrix}$$
$$-\frac{g}{\sqrt{2}} \begin{pmatrix} \bar{\psi}''^L_{e} & \bar{\psi}''^L_{\mu} & \bar{\psi}''^L_{\tau} \end{pmatrix} W\!\!\!/^{-} \underbrace{\begin{bmatrix} P_{ee} & P_{e\mu} & P_{e\tau} \\ P_{\mu e} & P_{\mu\mu} & P_{\mu\tau} \\ P_{\tau e} & P_{\tau\mu} & P_{\tau\tau} \end{bmatrix}}_{\text{PMNS matrix } P} \begin{pmatrix} \psi''^L_{\nu_e} \\ \psi''^L_{\nu_\mu} \\ \psi''^L_{\nu_\tau} \end{pmatrix} . \tag{7-100}$$

W-lepton terms

All fields will be in mass eigenstates, but the charged lepton-neutrino vertex will now allow generation changing. For example, from the second row of (7-100), we have, comparable to (7-97), from (7-89) with (7-99),

All fields, including neutrinos, now in mass eigenstates

$$\text{3 of the terms in (7-100)} = -\frac{g}{\sqrt{2}}\bar{\psi}''^L_e P_{ee} W\!\!\!/^{-} \psi''^L_{\nu_e} - \frac{g}{\sqrt{2}}\bar{\psi}''^L_e P_{e\mu} W\!\!\!/^{-} \psi''^L_{\nu_\mu} - \frac{g}{\sqrt{2}}\bar{\psi}''^L_e P_{e\tau} W\!\!\!/^{-} \psi''^L_{\nu_\tau} . \tag{7-101}$$

These will allow an outgoing electron, and an outgoing W^+, from any one of an incoming electron neutrino, muon neutrino, or tau neutrino. See Fig. 7-4. The probabilities for each incoming flavor of neutrino are determined by the P_{ij} values in (7-101).

But now, unlike Basis I, flavors mix at vertices having a W field

Basis II

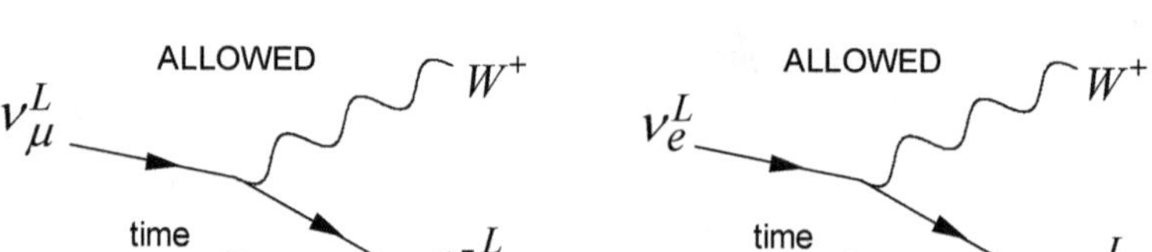

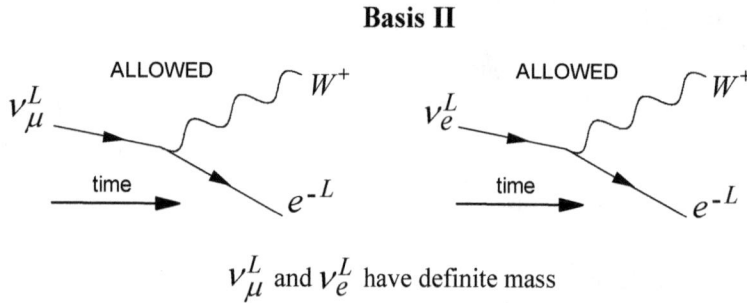

ν_μ^L and ν_e^L have definite mass

Figure 7-4. In Basis II: Generation Changing at a Vertex, Neutrinos in Mass Eigenstates

Figs. 7-3 and 7-4 may seem a little confusing and difficult to reconcile, but keep in mind that the neutrinos having the same symbol in each are actually different entities. In Fig. 7-3, the electron neutrino is actually a linear combination of 3 neutrino mass eigenstates, so it includes what we label in Fig. 7-4 of Basis II as $\nu_\mu{}^L$, as well as what we label in Basis II as $\nu_e{}^L$ and $\nu_\tau{}^L$, which are all mass eigenstates. That is, the LHS of Fig. 7-4 comes in as part of the allowed interaction in Fig. 7-3.

As noted, Basis I is the preferred basis for analyzing neutrino oscillations. We would get the same final results in either basis, but Basis I is easier in this case, and it is the one that is being employed, between the lines, in popular accounts one reads of such oscillations. Details on this in a later chapter.

Basis I easiest to analyze neutrino oscillations

As we are about to see, the Basis II approach is the one used for quarks. The use of that approach for quarks and Basis I for leptons can be a source of great confusion, when one tries to fathom what one would expect to be rather straightforward correlations between the two types of fermions.

Basis II easiest to analyze quark interactions

Summary of the Two Bases

Note again, that only in this book will you find Basis I and Basis II labeled in that way, so don't expect to see it elsewhere in the literature.

Labels "Basis I" and "Basis II" not used outside of this text

Wholeness Chart 7-5 summarizes the differences and similarities between the two different bases. The shaded box highlights the only case in which the mass eigenstate for a field is not used, i.e., for LC neutrino fields.

Wholeness Chart 7-5. Comparison of Two Possible Lepton Basis Choices

(Basis I Virtually Always Used for Leptons)

	Basis I		**Basis II**		**Both I and II**
	Basis?	Generation Change with $W^{\pm}$?	Basis?	Generation Change with $W^{\pm}$?	Other Generation Change?
LC charged lepton fields (e^L, μ^L, τ^L)	mass (definite mass)	No	mass (definite mass)	Yes	No
RC charged lepton fields (e^R, μ^R, τ^R)	mass (definite mass)	No such vertex	mass (definite mass)	No such vertex	No
LC neutrinos	$\psi'^L_{\nu_{l'}} = P_{\nu_{l'}\nu_l}\psi^L_{\nu_l}$ (no definite mass)	No	mass (definite mass)	Yes	No
RC neutrinos	mass (definite mass)	No such vertex	mass (definite mass)	No such vertex	No

From the chart we see what we had already stated. Basis II is simply the mass basis for all fields. Basis I differs from it in that it uses different eigenstates for the LC neutrino fields.

Bottom line In both Bases I and II, except for the following, $\mathcal{L}$ looks like it did in the original (flavor) eigen basis, and we can think of the fields expressed therein as flavor fields e, μ, τ, ν_e, ν_μ, ν_τ (even though they are different from what we originally called flavor fields, but since they have the same charges that is OK). The exceptions are

Leptons in $\mathcal{L}$ labeled e, μ, τ, ν_e, ν_μ, ν_τ

in Basis I, the neutrino mass terms mix generations (similar to our original flavor basis) but the charged lepton mass terms do not (unlike original basis), whereas

in Basis II, no mass terms mix generations (unlike original basis), but the W-lepton field terms do mix generations (unlike original basis).

Summary of Basis I and Basis II

In essence, Basis I behaves like our original flavor basis (which mixed generations in the neutrino mass terms but didn't in the W-lepton terms) except that the e, μ, τ, mass terms do not mix generations. Basis II acts differently than our original basis by not mixing generations in any mass terms, but mixing them in the W-lepton terms.

The Diagonal Mass Matrices

The symbols typically used for the diagonal mass matrices (7-78) and (7-84) are

$$M=\begin{bmatrix} g_e & & \\ & g_\mu & \\ & & g_\tau \end{bmatrix} \qquad m_e=\frac{v}{\sqrt{2}}g_e \quad m_\mu=\frac{v}{\sqrt{2}}g_\mu \quad m_\tau=\frac{v}{\sqrt{2}}g_\tau \tag{7-102}$$

Common symbols for charged lepton diagonal mass matrices

$$\text{in general}\quad {}_\nu M=\begin{bmatrix} g_{\nu_1} & & \\ & g_{\nu_2} & \\ & & g_{\nu_3} \end{bmatrix} \qquad m_1=\frac{v}{\sqrt{2}}g_{\nu_1} \quad m_2=\frac{v}{\sqrt{2}}g_{\nu_2} \quad m_3=\frac{v}{\sqrt{2}}g_{\nu_3}$$

$$\text{in Basis II}\quad {}_\nu M=\begin{bmatrix} g_{\nu_e} & & \\ & g_{\nu_\mu} & \\ & & g_{\nu_\tau} \end{bmatrix} \qquad \underbrace{m_{\nu_e}=\frac{v}{\sqrt{2}}g_{\nu_e}}_{=\frac{v}{\sqrt{2}}g_{\nu_1}} \quad \underbrace{m_{\nu_\mu}=\frac{v}{\sqrt{2}}g_{\nu_\mu}}_{=\frac{v}{\sqrt{2}}g_{\nu_2}} \quad \underbrace{m_{\nu_\tau}=\frac{v}{\sqrt{2}}g_{\nu_\tau}}_{=\frac{v}{\sqrt{2}}g_{\nu_3}}. \tag{7-103}$$

Common symbols for neutrino diagonal mass matrices

These are the symbols we use in (7-58). The g symbols are called Yukawa couplings. They, along with the field VEV at true vacuum v, determine the masses of the associated particles. One of the great mysteries of the SM is why the Yukawa couplings have the values they do. Models to explain them, beyond the SM, have been proposed, but none have, as of this writing, provided an unequivocal answer to the mystery.

g_l, g_{ν_l} called Yukawa couplings

PMNS Matrix Component Values

The most general 3X3 complex matrix contains 18 possible real, independent numbers, as each component has form $a + ib = Ae^{i\theta}$. (Either a and b, or magnitude A and phase angle θ, are real and independent.) If it is unitary, as the PMNS matrix is, those 18 real numbers are constrained by the 9 scalar equations in $U^\dagger U=I$, so a 3X3 unitary matrix has 9 independent real numbers, some of which can be thought of as constants like A, and some as phase angles, like θ.

Determining independent real numbers in PMNS matrix

It turns out that five of these real numbers can be taken as phase angles[1] that can be shifted from the PMNS matrix into the fields. That is, we will see the same predictions from the theory (which match what is measured in the physical world) if we take each of those phase angles out of the PMNS matrix and stick them into the associated fields. For example, in (7-93), where we ignore the wavy underlining of the fields (which, as mentioned, is generally dropped in the literature with numbers 1,2,3 symbolizing mass eigenstates), one of the terms can be converted, as follows.

$$P_{e1}\psi^L_{\nu_1}=A_{e1}e^{i\theta_1}\psi^L_{\nu_1}=A_{e1}e^{i\theta_1}\underbrace{e^{-i\theta_1}\underline{\psi}^L_{\nu_1}}_{\psi^L_{\nu_1}}=A_{e1}\underline{\psi}^L_{\nu_1}\ \xrightarrow[\text{and } A_{e1}=\text{new } P_{e1}]{\text{taking } \underline{\psi}^L_{\nu_1}=\text{new } \psi^L_{\nu_1}}\ =\text{new } P_{e1}\psi^L_{\nu_1}. \tag{7-104}$$

[1] Valle, J.W.F., Neutrino Physics Overview", *Journal of Physics: Conference Series*, **53**(1), 473-505 (2006). arXiv:hep-ph/0608101.

In a similar manner, four other phase angles can be removed from the PMNS matrix without changing theory predictions. And thus, there remain only four independent real numbers whose values can affect predictions. Three of these can be taken as mixing angles in flavor space, similar to the Weinberg angle θ_W of (7-43). Each such angle can be visualized as a rotation in 3D flavor space about one of the axes in that space. The remaining real number is a phase angle, designated by δ. When all is said and done, we have

4 independent components, one of which is a phase angle

PMNS matrix as a series of 2D rotations plus a phase

$$P=\begin{bmatrix}1 & & \\ & \cos\theta_{23} & \sin\theta_{23} \\ & -\sin\theta_{23} & \cos\theta_{23}\end{bmatrix}\begin{bmatrix}\cos\theta_{13} & & \sin\theta_{13}e^{-i\delta} \\ & 1 & \\ -\sin\theta_{13}e^{i\delta} & & \cos\theta_{13}\end{bmatrix}\begin{bmatrix}\cos\theta_{12} & \sin\theta_{12} & \\ -\sin\theta_{12} & \cos\theta_{12} & \\ & & 1\end{bmatrix}$$

$$=\begin{bmatrix}\cos\theta_{12}\cos\theta_{13} & \sin\theta_{12}\cos\theta_{13} & \sin\theta_{13}e^{-i\delta} \\ -\sin\theta_{12}\cos\theta_{23}-\cos\theta_{12}\sin\theta_{23}\sin\theta_{13}e^{i\delta} & \cos\theta_{12}\cos\theta_{23}-\sin\theta_{12}\sin\theta_{23}\sin\theta_{13}e^{i\delta} & \sin\theta_{23}\cos\theta_{13} \\ \sin\theta_{12}\sin\theta_{23}-\cos\theta_{12}\cos\theta_{23}\sin\theta_{13}e^{i\delta} & -\cos\theta_{12}\sin\theta_{23}-\sin\theta_{12}\cos\theta_{23}\sin\theta_{13}e^{i\delta} & \cos\theta_{23}\cos\theta_{13}\end{bmatrix}. \quad (7\text{-}105)$$

The three angles θ_{ij} and the phase δ have been found experimentally, and as of 2020[1], are

Experimental values for the 4 independent numbers

$$\theta_{12}=33.44^{\circ}\begin{pmatrix}+0.78^{\circ}\\-0.75^{\circ}\end{pmatrix}\quad \theta_{23}=49.0^{\circ}\begin{pmatrix}+1.1^{\circ}\\-1.4^{\circ}\end{pmatrix}\quad \theta_{13}=8.57^{\circ}\begin{pmatrix}+0.13^{\circ}\\-0.12^{\circ}\end{pmatrix}\quad \delta=195^{\circ}\begin{pmatrix}+51^{\circ}\\-25^{\circ}\end{pmatrix}. \quad (7\text{-}106)$$

The 3σ ranges for the magnitudes of the components of P are

Absolute values of components in PMNS matrix

$$\text{PMNS matrix } P \text{ component magnitudes } =\begin{bmatrix}0.801\leftrightarrow 0.845 & 0.513\leftrightarrow 0.579 & 0.143\leftrightarrow 0.156 \\ 0.233\leftrightarrow 0.507 & 0.461\leftrightarrow 0.694 & 0.631\leftrightarrow 0.778 \\ 0.261\leftrightarrow 0.526 & 0.471\leftrightarrow 0.701 & 0.611\leftrightarrow 0.761\end{bmatrix}. \quad (7\text{-}107)$$

If the off-diagonal components in (7-107) were close to zero, in Basis I, from (7-98), we would expect neutrino mass eigenstates to be almost identical to the flavor eigenstates, or in Basis II, from (7-100), very little generation changing at a vertex. Instead, those components are comparable in magnitude to the diagonal components, and so, we expect generous amounts of each mass eigenstate in each new flavor in Basis I, and considerable generation mixing in Basis II.

As off diagonals are not close to zero, neutrino mass eigenstates are not close to flavor eigenstates (Basis I)

7.3.6 Quarks in the GSW Theory

As you might expect, just as we found at high energy (the false vacuum) in the previous chapter, all that we have found for leptons in e/w theory is directly paralleled by quarks. That is, we just treat the quark doublets and singlets like their associated lepton doublets and singlets, as in the following correspondences.

Quarks parallel leptons

$$\Psi^L_{q=1}=\begin{pmatrix}\psi^L_u\\ \psi^L_d\end{pmatrix}\Leftrightarrow\Psi^L_e=\begin{pmatrix}\psi^L_{\nu_e}\\ \psi^L_e\end{pmatrix}\quad \Psi^L_{q=2}=\begin{pmatrix}\psi^L_c\\ \psi^L_s\end{pmatrix}\Leftrightarrow\Psi^L_\mu=\begin{pmatrix}\psi^L_{\nu_\mu}\\ \psi^L_\mu\end{pmatrix}\quad \Psi^L_{q=3}=\begin{pmatrix}\psi^L_t\\ \psi^L_b\end{pmatrix}\Leftrightarrow\Psi^L_\tau=\begin{pmatrix}\psi^L_{\nu_\tau}\\ \psi^L_\tau\end{pmatrix} \quad (7\text{-}108)$$

Doublet and singlet parallels, quarks and leptons

$$\begin{matrix}\psi^R_{q_u=1}=\psi^R_u\Leftrightarrow\psi^R_{\nu_e} & \psi^R_{q_u=2}=\psi^R_c\Leftrightarrow\psi^R_{\nu_\mu} & \psi^R_{q_u=3}=\psi^R_t\Leftrightarrow\psi^R_{\nu_\tau}\\ \psi^R_{q_d=1}=\psi^R_d\Leftrightarrow\psi^R_e & \psi^R_{q_d=2}=\psi^R_s\Leftrightarrow\psi^R_\mu & \psi^R_{q_d=3}=\psi^R_b\Leftrightarrow\psi^R_\tau .\end{matrix} \quad (7\text{-}109)$$

Then, all we have to do is take our lepton results at true vacuum and make the substitutions in front of the arrows of (7-108) and (7-109). We summarize the result in the next few subsections.

Take care with the notation, where we use q as a subscript for the LC doublet (both LC up type and down type quarks), but q_u for the RC up type quarks and q_d for the RC down type quarks.

Quark Lagrangian Terms at True Vacuum

So, we get quark terms at low energy, which we must add to (7-58) to get the total Lagrangian (7-110) below, where the q subscript on individual fields denotes all six quarks. (For doublets, q represents the three generations of quark doublets.)

[1] For the most recent experimental values for these and much more see Particle Data Group https://pdg.lbl.gov/.

True Vacuum, Total Electroweak Lagrangian $\mathcal{L} = \mathcal{L}_0 + \mathcal{L}_I$

$$\text{quark part of } \mathcal{L}_0 = \underbrace{\bar{\psi}_{q_d}\big(i\not{\partial} - \underbrace{\tfrac{vg_{q_d}}{\sqrt{2}}}_{m_{q_d}}\big)\psi_{q_d} + \bar{\psi}_{q_u}\big(i\not{\partial} - \underbrace{\tfrac{vg_{q_u}}{\sqrt{2}}}_{m_{q_u}}\big)\psi_{q_u}}_{\text{alternative notation} \rightarrow \bar{\psi}_q\left(i\not{\partial} - \underbrace{\tfrac{vg_q}{\sqrt{2}}}_{m_q}\right)\psi_q} \tag{7-110}$$

$$\mathcal{L}_I = \underbrace{\mathcal{L}^{LB} + \mathcal{L}^{LH}}_{\text{lepton interactions}} + \mathcal{L}^{BB} + \mathcal{L}^{HH} + \mathcal{L}^{HB} \underbrace{+\mathcal{L}^{QB} + \mathcal{L}^{QH}}_{\text{quark interactions}} ,$$

Total e/w $\mathcal{L}$ at low energy, including quarks and leptons

where we discuss the quark mass and interaction terms in the following subsections.

Quark-Higgs Terms

In the flavor basis used for minimal substitution, we get terms representing quark interactions with the Higgs parallel to (7-66),

Quark-Higgs terms in $\mathcal{L}$ behave like lepton-Higgs terms

$$\text{false vacuum } \mathcal{L}^{QH} = -{}_CY_{qq_d}\bar{\Psi}_q^L\Phi\psi_{q_d}^R - {}_DY_{qq_u}\bar{\Psi}_q^L\tilde{\Phi}\psi_{q_u}^R + h.c. \tag{7-111}$$

When we insert the Higgs doublet expressed in the true vacuum Higgs field σ and the Higgs VEV v in the unitary gauge we get, comparable to (7-70) and (7-71),

$$\text{weak vertex terms } \frac{\sigma}{\sqrt{2}}\bar{\psi}_d^L{}_CY_{dd}\psi_d^R + \frac{\sigma}{\sqrt{2}}\bar{\psi}_d^L{}_CY_{ds}\psi_s^R + \frac{\sigma}{\sqrt{2}}\bar{\psi}_d^L{}_CY_{db}\psi_b^R + \ldots = \frac{\sigma}{\sqrt{2}}\bar{\psi}_{q_d}^L\,{}_CY_{q_dq_d'}\,\psi_{q_d'}^R \tag{7-112}$$

$$\text{mass-like terms } \frac{v}{\sqrt{2}}\bar{\psi}_d^L{}_CY_{dd}\psi_d^R + \frac{v}{\sqrt{2}}\bar{\psi}_d^L{}_CY_{ds}\psi_s^R + \frac{v}{\sqrt{2}}\bar{\psi}_d^L{}_CY_{db}\psi_b^R + \ldots = \frac{v}{\sqrt{2}}\bar{\psi}_{q_d}^L\,{}_CY_{q_dq_d'}\,\psi_{q_d'}^R . \tag{7-113}$$

At low energy, quarks interact with Higgs σ plus have mass-like terms

When we diagonalize the two matrices, we'll get terms like

$$\text{in mass basis } \mathcal{L}^{QH} = \frac{\sigma}{\sqrt{2}}\bar{\psi}_{\underset{\sim}{d}}^L{}_C\underset{\sim}{Y}_{dd}\psi_{\underset{\sim}{d}}^R + \frac{\sigma}{\sqrt{2}}\bar{\psi}_{\underset{\sim}{s}}^L{}_C\underset{\sim}{Y}_{ss}\psi_{\underset{\sim}{s}}^R + \frac{\sigma}{\sqrt{2}}\bar{\psi}_{\underset{\sim}{b}}^L{}_C\underset{\sim}{Y}_{bb}\psi_{\underset{\sim}{b}}^R \tag{7-114}$$

$$+\frac{v}{\sqrt{2}}\bar{\psi}_{\underset{\sim}{d}}^L{}_C\underset{\sim}{Y}_{dd}\psi_{\underset{\sim}{d}}^R + \frac{v}{\sqrt{2}}\bar{\psi}_{\underset{\sim}{s}}^L{}_C\underset{\sim}{Y}_{ss}\psi_{\underset{\sim}{s}}^R + \frac{v}{\sqrt{2}}\bar{\psi}_{\underset{\sim}{b}}^L{}_C\underset{\sim}{Y}_{bb}\psi_{\underset{\sim}{b}}^R , \tag{7-115}$$

We diagonalize the quark mass matrix, as we did with leptons

where, as with leptons, the wavy underlines imply the mass eigenstate basis. (7-114) represents interactions between down type quarks and the low energy Higgs. (7-115) are the mass terms where

$$\begin{aligned} {}_C\underset{\sim}{Y} = M_{down\,type} &= \begin{bmatrix} g_d & & \\ & g_s & \\ & & g_b \end{bmatrix} & m_d &= \frac{v}{\sqrt{2}}g_d & m_s &= \frac{v}{\sqrt{2}}g_s & m_b &= \frac{v}{\sqrt{2}}g_b \\ {}_D\underset{\sim}{Y} = M_{up\,type} &= \begin{bmatrix} g_u & & \\ & g_c & \\ & & g_t \end{bmatrix} & m_u &= \frac{v}{\sqrt{2}}g_u & m_c &= \frac{v}{\sqrt{2}}g_c & m_t &= \frac{v}{\sqrt{2}}g_t . \end{aligned} \tag{7-116}$$

Symbols for components of diagonal mass matrix

These are the symbols we use in (7-110), where the g values are the Yukawa couplings for quarks.

Quark Mass Eigenstates

When we diagonalize the quark Y (Yukawa) matrices, we end up with a situation much like we had with the PMNS matrix for leptons. That is, all terms except those with a W field will look the same in the (mass) basis, i.e., the basis where the quark Yukawa matrices are diagonal. The terms with W fields in them parallel (7-89) and are

$$\begin{aligned}
&\text{true vacuum } W \text{ terms in } \mathcal{L}^{QB} = -\frac{g}{\sqrt{2}}\left(\bar{\psi}_u^L \; \bar{\psi}_c^L \; \bar{\psi}_t^L\right)W^+\begin{pmatrix}\psi_d^L\\ \psi_s^L\\ \psi_b^L\end{pmatrix} - \frac{g}{\sqrt{2}}\left(\bar{\psi}_d^L \; \bar{\psi}_s^L \; \bar{\psi}_b^L\right)W^-\begin{pmatrix}\psi_u^L\\ \psi_c^L\\ \psi_t^L\end{pmatrix}\\
&= -\frac{g}{\sqrt{2}}\left(\bar{\underset{\sim}{\psi}}_u^L \; \bar{\underset{\sim}{\psi}}_c^L \; \bar{\underset{\sim}{\psi}}_t^L\right)W^+\underbrace{\begin{bmatrix}V_{ud} & V_{us} & V_{ub}\\ V_{cd} & V_{cs} & V_{cb}\\ V_{td} & V_{ts} & V_{tb}\end{bmatrix}}_{\text{CKM matrix } V}\begin{pmatrix}\underset{\sim}{\psi}_d^L\\ \underset{\sim}{\psi}_s^L\\ \underset{\sim}{\psi}_b^L\end{pmatrix} \qquad (7\text{-}117)\\
&\quad -\frac{g}{\sqrt{2}}\left(\bar{\underset{\sim}{\psi}}_d^L \; \bar{\underset{\sim}{\psi}}_s^L \; \bar{\underset{\sim}{\psi}}_b^L\right)W^-\underbrace{\begin{bmatrix}V_{ud}^* & V_{cd}^* & V_{td}^*\\ V_{us}^* & V_{cs}^* & V_{ts}^*\\ V_{ub}^* & V_{cb}^* & V_{tb}^*\end{bmatrix}}_{V^\dagger}\begin{pmatrix}\underset{\sim}{\psi}_u^L\\ \underset{\sim}{\psi}_c^L\\ \underset{\sim}{\psi}_t^L\end{pmatrix}.
\end{aligned}$$

We end up with CKM matrix for quarks parallel to PMNS matrix for leptons

where the quark equivalent V of the PMNS matrix is called the Cabibbo-Kobayashi-Maskawa (CKM) matrix. (It may be confusing, but this is the common convention, since from direct comparison with the PMNS matrix, one might expect the CKM matrix to be $V^\dagger$, not V.)

Note notation change: We now, as we did with leptons, drop the wavy underline and keep in mind that every field shown in the resulting Lagrangian is a mass eigenstate field.

Change to common notation with no wavy underlines

So, all fields in (7-110) are expressed in the mass basis, parallel to what we termed Basis II for leptons. We will call this Basis IIq, where the "q" stands for quarks. This is the basis that is universally used for quarks. If we do refer to the alternative basis, parallel to Basis I for leptons, where not all quark fields are expressed as mass eigenstate fields, we will call that basis Basis Iq, though we will generally ignore Basis Iq, as it is not used in the literature.

Quark convention parallel to lepton Basis II , called Basis IIq

Quark-Gauge Boson Terms

The terms representing the low energy quark interactions with the W, Z, and photon fields parallel (7-64) for leptons (except that here we use all mass basis fields, i.e., we use Basis IIq) and turn out to be (after some algebra, similar to what we did for leptons). That is,

In $\mathcal{L}$ with Basis IIq, all quark fields are mass eigenstates, denoted by u,d,c,s,t,b

$$\begin{aligned}
\mathcal{L}^{QB} &= \underbrace{-\tfrac{2}{3}e\bar{\psi}_{q_u}A\psi_{q_u} + \tfrac{1}{3}e\bar{\psi}_{q_d}A\psi_{q_d}}_{\text{QED quark interactions}}\\
&\quad -\frac{g}{2\sqrt{2}}\bar{\psi}_{q_u}W^+\left(1-\gamma^5\right)V_{q_u q_d}\psi_{q_d} - \frac{g}{2\sqrt{2}}\bar{\psi}_{q_d}W^-\left(1-\gamma^5\right)V^\dagger_{q_d q_u}\psi_{q_u}\\
&\quad -\frac{g}{4\cos\theta_W}\bar{\psi}_{q_u}Z\left(1-\frac{8\sin^2\theta_W}{3}-\gamma^5\right)\psi_{q_u} + \frac{g}{4\cos\theta_W}\bar{\psi}_{q_d}Z\left(1-\frac{4\sin^2\theta_W}{3}-\gamma^5\right)\psi_{q_d} \qquad (7\text{-}118)\\
&= -\tfrac{2}{3}e\bar{\psi}_{q_u}A\psi_{q_u} + \tfrac{1}{3}e\bar{\psi}_{q_d}A\psi_{q_d} - \frac{g}{\sqrt{2}}\bar{\psi}^L_{q_u}W^+V_{q_u q_d}\psi^L_{q_d} - \frac{g}{\sqrt{2}}\bar{\psi}^L_{q_d}W^-V^\dagger_{q_d q_u}\psi^L_{q_u}\\
&\quad -\frac{g}{2\cos\theta_W}\bar{\psi}^L_{q_u}Z\psi^L_{q_u} + \frac{g}{2\cos\theta_W}\bar{\psi}^L_{q_d}Z\psi^L_{q_d}\\
&\quad +\tfrac{2}{3}\underbrace{g\sin\theta_W}_{e}\tan\theta_W\bar{\psi}_{q_u}Z\psi_{q_u} - \tfrac{1}{3}\underbrace{g\sin\theta_W}_{e}\tan\theta_W\bar{\psi}_{q_d}Z\psi_{q_d}.
\end{aligned}$$

Quark-gauge boson vertex terms in $\mathcal{L}$

Note the quark QED terms in (7-118) have factors in front representing their respective charges (+2/3 for up type quarks and –1/3 for down types), just as we had the factor e for a charged lepton field and zero (no such term) for neutrinos.

Quark-photon terms have factor representing quark electric change

As you might be guessing, we don't need to memorize all the terms in the Lagrangian. We can look them up, when we need them.

Choice of Basis

The basis we work in for quarks, as noted above, is the one where all fields are mass eigenstates. We summarize the two possible quark bases in Wholeness Chart 7-6 below, which parallels

Summary of Basis Iq vs Basis IIq, even though we never use Basis Iq

Wholeness Chart 7-5 for leptons. Note that even though we chose the original mass eigenstates, we are now free to call them by flavor state names, even though they are not the same as the original flavor state fields. Yes, this can be confusing. But this is how it is done.

Again, be clear that Basis Iq is never used (that I have seen) for quarks.

And perhaps now you can see the bewilderment that can arise when Basis I (of Wholeness Chart 7-5) is used commonly for neutrino oscillation calculations, but Basis IIq (of Wholeness Chart 7-6) is used for quark interaction calculations.

In the chart, shaded boxes represent the only cases where there is generation changing at a vertex.

Wholeness Chart 7-6. Comparison of Two Possible Quark Basis Choices

(Basis IIq Always Used for Quarks)

	Basis Iq		Basis IIq		Both Iq and IIq
	Basis?	Generation Change with $W^{\pm}$?	Basis?	Generation Change with $W^{\pm}$?	Other Generation Change?
LC down type quark fields (d^L, s^L, b^L)	mass (definite mass)	No	mass (definite mass)	Yes	No
RC down type quark fields (d^R, s^R, b^R)	mass (definite mass)	No such vertex	mass (definite mass)	No such vertex	No
LC up type quark fields (u^L, c^L, t^L)	$\psi'^L_{q_{d'}} = V_{q_{d'}q_d}\psi^L_{\sim q_d}$ (no definite mass)	No	mass (definite mass)	Yes	No
RC up type quark fields (u^R, c^R, t^R)	mass (definite mass)	No such vertex	mass (definite mass)	No such vertex	No

Sample Quark Interaction

Similar to leptons in Basis II, quarks in Basis IIq can change generations at a vertex that has a W field, one example of which is portrayed in Fig. 7-5.

Basis IIq

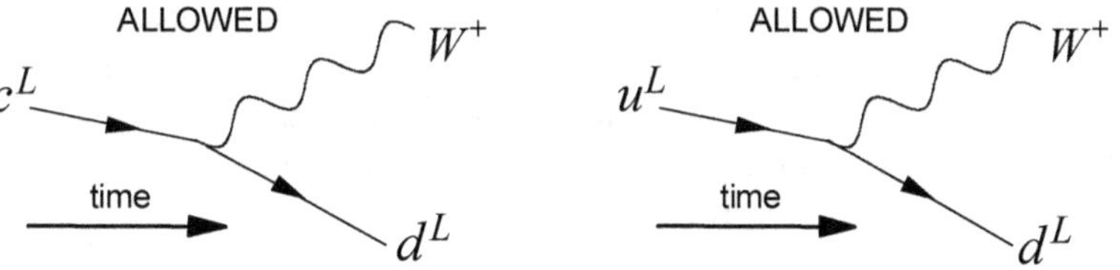

Generation mixing at W-quark vertex a basic feature of Basis IIq

all quarks have definite mass

Figure 7-5. Quark Generation Changing at a Vertex

Do **Problem 8** to determine the term in the Lagrangian that corresponds to each side of Fig. 7-5.

CKM Matrix Component Values

Form of CKM matrix like PMNS matrix except different constants

In the same way that the PMNS matrix has four independent real parameters, one of which is a phase angle, so does the CKM matrix. So,

$$V = \begin{bmatrix} \cos\phi_{12}\cos\phi_{13} & \sin\phi_{12}\cos\phi_{13} & \sin\phi_{13}e^{-i\alpha} \\ -\sin\phi_{12}\cos\phi_{23}-\cos\phi_{12}\sin\phi_{23}\sin\phi_{13}e^{i\alpha} & \cos\phi_{12}\cos\phi_{23}-\sin\phi_{12}\sin\phi_{23}\sin\phi_{13}e^{i\alpha} & \sin\phi_{23}\cos\phi_{13} \\ \sin\phi_{12}\sin\phi_{23}-\cos\phi_{12}\cos\phi_{23}\sin\phi_{13}e^{i\alpha} & -\cos\phi_{12}\sin\phi_{23}-\sin\phi_{12}\cos\phi_{23}\sin\phi_{13}e^{i\alpha} & \cos\phi_{23}\cos\phi_{13} \end{bmatrix}. \quad (7\text{-}119)$$

The best known experimentally determined values for the CKM angles, as of 2020, are

Experimental values for the CKM constants

$$\phi_{12} = 13.04 \pm 0.05^\circ \qquad \phi_{13} = 0.201 \pm 0.011^\circ \qquad \phi_{23} = 2.38 \pm 0.06^\circ \qquad \alpha = 68.75 \pm 4.58^\circ. \quad (7\text{-}120)$$

The magnitudes of the components, to the accuracies of (7-120), are

Magnitudes of CKM matrix components

$$\text{CKM matrix } V \text{ component magnitudes} = \begin{bmatrix} 0.974 & 0.227 & 0.004 \\ 0.226 & 0.973 & 0.041 \\ 0.009 & 0.040 & 0.999 \end{bmatrix}. \quad (7\text{-}121)$$

Note that the off-diagonal components are much smaller than the PMNS matrix (7-107) ones. This means there is not much quark generation mixing at vertices, as in the LHS of Fig. 7-5.

Not much generation mixing at W vertices

Historical Notes

The CKM matrix has been known since 1973, when the picture of three generations of quarks was coming into focus. It was preceded, however, by a two-generation version introduced by Nicola Cabibbo in 1963. In that version the mass and flavor eigenstates were related by

Cabibbo matrix for only 2 quark generations

$$\begin{pmatrix} d \\ s \end{pmatrix} = V_C \begin{pmatrix} \underset{\sim}{d} \\ \underset{\sim}{s} \end{pmatrix} = \begin{bmatrix} \cos\theta_c & \sin\theta_c \\ -\sin\theta_c & \cos\theta_c \end{bmatrix} \begin{pmatrix} \underset{\sim}{d} \\ \underset{\sim}{s} \end{pmatrix}, \quad (7\text{-}122)$$

where θ_C is known as the Cabibbo angle and V_C as the Cabibbo matrix.

In 2008, Kobayashi and Maskawa shared one half of the Nobel Prize in Physics for their discovery of the matrix that for many years was called the KM matrix, rather than the CKM matrix. Many physicists felt it was unfair that Cabibbo was not also honored, as his prior work was the foundation upon which the CKM matrix was built.

Given that all off diagonal components in (7-121) are almost zero, except for the ones relating down to strange quarks, the Cabibbo matrix can be a good approximation to the full three generation CKM matrix.

Cabibbo matrix good approximation of CKM

The PMNS matrix was introduced in 1962 by Ziro Maki, Masami Nakagawa and Shoichi Sakata, to explain the neutrino oscillations conjectured at that time by Bruno Pontecorvo. These oscillations were not detected until 1998, and as mentioned, we will look more closely at them in a later chapter.

PMNS matrix suggested in 1960s, found viable in 1998

7.3.7 Masses of Particles

Fig. 7-6 is a log plot depicting the various masses of SM particles as known in 2025. Quark masses shown are for bare quarks, which actually aren't found in nature. When they are bound into a hadron (3 quarks) or a meson (quark-antiquark pair), their effective mass changes, due to the binding energy holding the quarks together, which via the mass-energy equivalence adds to the mass of the composite (hadron or meson) particle[1]. Neutrino masses are not known precisely (as of 2025).

Wholeness Chart 7-7 lists the masses numerically. We have included the gluon, which is the massless gauge particle mediator of the strong force, even though we have not studied it yet. The neutrinos shown in the chart are flavor eigenstates (of indefinite mass) with the expectation value (the average mass over many measurements) shown. Neutrinos in the figure, on the other hand, are labeled as mass eigenstates. Neutrino mass values are not known to very high accuracy as of 2025, although it is known that the heaviest neutrino mass must be less than .45 eV.

---

[1] It turns out, in strong interactions, that the color force potential (energy), unlike the Coulomb potential (but much like the potential of a spring holding two masses together), is positive. So, when the strong force binds quarks, the presence of that potential energy adds to the total mass-energy of the composite particle, and makes it more than the sum of the mass-energies of the individual quarks. With the Coulomb potential, as in the hydrogen atom, the negative potential energy makes the total mass-energy of the composite less than the sum of the mass-energies of the parts.

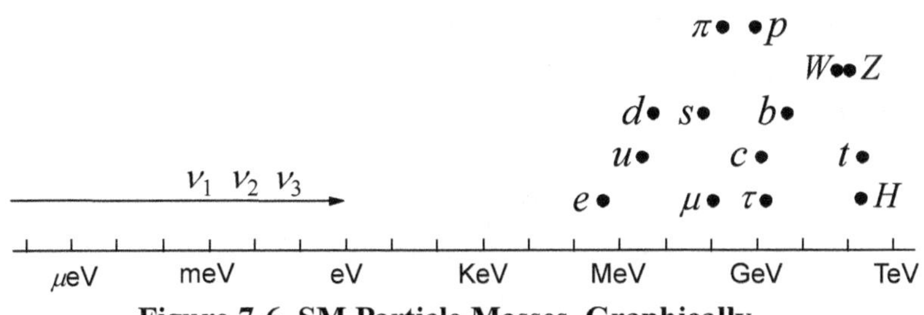

Log plot of particle masses

Figure 7-6. SM Particle Masses, Graphically

The extraordinary range of the masses is noteworthy. Neutrino masses are over 500,000 times smaller than the electron mass, and the electron mass is more than 300,000 times smaller than that of the top quark. As noted earlier, with reference to the Yukawa couplings that dictate the mass values, no one has, as of this writing, discovered any underlying theoretical reason for them being what they are, though we will look at a conjectured mechanism for the small neutrino masses later in the book.

Enormous range of particle masses

No accepted theory for why these values

Wholeness Chart. 7-7. SM Particle Masses Numerically

<u>Leptons</u>		<u>Quarks</u>	Bare	Effective	<u>Bosons</u>		<u>Composite</u>	
e^-	.511 MeV	u	2.2 MeV	.312 GeV in proton	photon	0	proton (≈ neutron)	.935 GeV
μ	.106 GeV	d	4.6 MeV	.312 GeV in proton	$W^{\pm}$	80.4 GeV	charged pi meson	.140 GeV
τ	1.777 GeV	c	1.28 GeV		Z	91.2 GeV	neutral pi meson	.135 GeV
ν_e	Expectation value < 1 eV	s	.096 GeV		gluon	0		
ν_μ		t	173.1 GeV		Higgs	125.1 GeV		
ν_τ		b	4.18 GeV					

7.3.8 Measurement of Flavor or Mass? (Added in June 2022 revision)

Note that a typical neutrino detection device is composed of normal matter, i.e., electrons and nuclei. From Fig. 7-3, pg. 231, we see that only electron neutrinos could interact with such a device, not muon or tau neutrinos. The latter two types would be undetectable.

I recall, as a student studying the material of this chapter, wondering that if mass eigenstates and flavor eigenstates are different, which of the two do we actually measure in experiment? From the prior paragraph, we can understand that neutrino measurement devices generally detect a given flavor, and commonly in the early days of neutrino research, this was the electron neutrino flavor.

Neutrino experiments detect flavor eigenstates, not mass eigenstates (The two are different for neutrinos)

So, each detection could have one of three different masses. Any number we would assign as mass of an electron neutrino is then a statistical average, over many mass measurements, of those three mass values. Such measurements are extremely difficult, however, so neutrino masses are not known to high precision. (More on this in Chap. 10.)

The situation with quarks is different, since we use Basis IIq, visualized in Fig. 7-5, pg. 237, as opposed to Basis I, used for neutrinos and visualized in Fig. 7-3. While we can't measure quarks directly, we can think in terms of a measurement with a normal matter detector of first-generation particles, but, unlike the neutrino case, these detectors can interact with any of the three generations. Charge measurements can't discern between flavors, but mass measurements do. Different hadrons with the same charge, like the proton (*uud*) and the Σ^+ (*uus*) are distinguished between from their mass differences. Unlike with neutrinos, quark flavor eigenstates are the same as quark mass eigenstates. So, quark generation discernment in experiment is done via their respective masses.

Quarks discerned in experiments via mass eigenstates (which for quarks are the same as flavor eigenstates)

7.3.9 Summary of the GSW Model

Wholeness Chart 7-8 summarizes the symmetry breaking of the GSW model from the false vacuum to the true vacuum, and shows how key terms in the Lagrangian change under the transition.

Summary of GSW model

Wholeness Chart 7-8. Glashow/Salam/Weinberg Model

	At False Vacuum Complex Characterization	Real	**At True Vacuum** Complex	Real Characterization
Higgs Field (isospin doublet)	$\Phi(x)=\frac{1}{\sqrt{2}}\begin{bmatrix}\phi_a\\ \phi_b\end{bmatrix}=\frac{1}{\sqrt{2}}\begin{bmatrix}\phi_1+i\phi_2\\ \phi_3+i\phi_4\end{bmatrix}$	$\Phi=\frac{1}{\sqrt{2}}\begin{bmatrix}\phi_1\\ \phi_2\\ \phi_3\\ \phi_4\end{bmatrix}$	$\Sigma=\frac{1}{\sqrt{2}}\begin{bmatrix}\eta_1+i\eta_2\\ \sigma+i\eta_3\end{bmatrix}$	$\Sigma(x)=\frac{1}{\sqrt{2}}\begin{bmatrix}\eta_1\\ \eta_2\\ \sigma\\ \eta_3\end{bmatrix}=\frac{1}{\sqrt{2}}\begin{bmatrix}\phi_1\\ \phi_2\\ \phi_3-v\\ \phi_4\end{bmatrix}$ $\phi_3=v$ is min of $\mathcal{V}$. $v=\sqrt{\frac{-\mu^2}{\lambda}}$
$\mathcal{L}$ (only leptons shown, quarks are parallel)	$\mathcal{L}_{false}=\mathcal{L}^L+\mathcal{L}^B+\mathcal{L}^H+\mathcal{L}^{LH}$ (massless) $\mathcal{L}^L=i\left(\bar{\Psi}_l^L\not{D}\Psi_l^L+\bar{\psi}_l^R\not{D}\psi_l^R+\bar{\psi}_{\nu_l}^R\not{D}\psi_{\nu_l}^R\right)$ $\mathcal{L}^B=-\frac{1}{4}G_i^{\mu\nu}G_{i\mu\nu}-\frac{1}{4}B^{\mu\nu}B_{\mu\nu}$ $\mathcal{L}^H=\left(D^\mu\Phi\right)^\dagger\left(D_\mu\Phi\right)-\mu^2\Phi^\dagger\Phi-\lambda\left(\Phi^\dagger\Phi\right)^2$ $\mathcal{L}^{LH}=-{}_AY_{ll'}\bar{\Psi}_l^L\Phi\psi_{l'}^R-{}_BY_{l\nu_{l'}}\bar{\Psi}_l^L\tilde{\Phi}\psi_{\nu_{l'}}^R$ $+ h.c.$ $D^\mu=\partial^\mu+\frac{i}{2}g\sigma_jW_j^\mu+ig'YB^\mu$	Use algebra on $\mathcal{L}$ relations at left to obtain theory in terms of real ϕ_i	Real rep more illumina-ting	For $\mathcal{L}_{true}$: 1. Subst in $\mathcal{L}_{false}$ (translates axis to $\mathcal{V}_{min}$) $\phi_{1,2,4}=\eta_{1,2,4}$ $\phi_3=(\sigma+v)$ 2. Substitute $B_\mu=\cos\theta_W A_\mu-\sin\theta_W Z_\mu$ $W_{3\mu}=\sin\theta_W A_\mu+\cos\theta_W Z_\mu$ $W_{1\mu}=\frac{W_\mu+W_\mu^\dagger}{\sqrt{2}}$ $W_{2\mu}=i\frac{W_\mu-W_\mu^\dagger}{\sqrt{2}}$ 3. Result: Complicated $\mathcal{L}_{true}$ with a) non independent (non normal) modes b) 3 more DOF than $\mathcal{L}_{false}$ 4. Resolve using unitary gauge below.
Local Sym Transform *SU*(2)X*U*(1)	$\delta\Psi_l^L=\frac{i}{2}\omega_i\sigma_i\Psi_l^L+iYf\Psi_l^L$ $Y=-\frac{1}{2}$ $\delta\psi_l^R=iYf\psi_l^R$ $Y=-1$ $\delta W_i^\mu=-\frac{1}{g}\partial^\mu\omega_i-\varepsilon_{ijk}\omega_jW_k^\mu$ $\delta B^\mu=-\frac{1}{g'}\partial^\mu f$ $\delta\Phi=\frac{i}{2}\omega_i\sigma_i\Phi+iYf\Phi$ $Y=\frac{1}{2}$			$\mathcal{L}_{true}$ not symmetric in η_1, η_2, σ, η_3 at true vacuum (broken symmetry)
Unitary Gauge				$\eta_1=\eta_2=\eta_3=0$ (3 DOF less, norm modes)
$\mathcal{L}$ **in Unitary Gauge**				Via steps 1, 2, 4 above $\mathcal{L}_{true}=\mathcal{L}_0+\mathcal{L}_I$ becomes $\mathcal{L}_0=\bar{\psi}_l\left(i\not{\partial}-\frac{vg_l}{\sqrt{2}}\right)\psi_l+\bar{\psi}_{\nu_l}i\not{\partial}\psi_{\nu_l}+\mathcal{L}^{m_\nu}$ $-\frac{1}{4}F_{\mu\nu}F^{\mu\nu}-\frac{1}{2}W_{\mu\nu}^\dagger W^{\mu\nu}+\left(\frac{1}{2}vg\right)^2W_\mu^\dagger W^\mu$ $-\frac{1}{4}Z_{\mu\nu}Z^{\mu\nu}+\frac{1}{2}\left(\frac{1}{2}\frac{vg}{\cos\theta_W}\right)^2Z_\mu Z^\mu$ $+\frac{1}{2}\left(\partial_\mu\sigma\right)\left(\partial^\mu\sigma\right)-\frac{1}{2}\left(-2\mu^2\right)\sigma^2$ $\mathcal{L}_I=\mathcal{L}^{LB}+\mathcal{L}^{LH}+\mathcal{L}^{BB}+\mathcal{L}^{HH}+\mathcal{L}^{HB}$
Determining Masses				Inspection of above: $m_l=\frac{vg_l}{\sqrt{2}}$ $m_\gamma=0$ $m_W=\frac{1}{2}vg$ $m_Z=\frac{1}{2}\frac{vg}{\cos\theta_W}$ $m_H=\sqrt{-2\mu^2}=\sqrt{2\lambda v^2}$
Note				1. Zero photon mass ($m_\gamma=0$ as no $A^\mu A_\mu$ term) 2. Know g, θ_W, v in 1970s via exper→ predict $m_{Z,W}$ 3. λ only in $\mathcal{L}^{HH}$. Not predict m_H. Need to measure.
Determining Couplings				Interactn terms: e^- with $A^\mu\rightarrow e=g\sin\theta_W$; ν with $A^\mu\rightarrow 0$; e^-& ν with $W\rightarrow g/2\sqrt{2}$, with $Z\rightarrow$ $g/4\cos\theta_W$

7.4 Other Things to Note

7.4.1 Possible Changes and Differences in Mexican Hat Shape

It is possible that the Mexican hat shape shown repeatedly herein for the Higgs field potential energy density is not static but changes over time, i.e., in (7-52), $\mu = \mu(t)$, or $\mu = \mu(t)$ and $\lambda = \lambda(t)$. One example, illustrated in 2D only for simplicity, could be like that shown in Fig. 7-7.

Higgs potential could change in time, not known

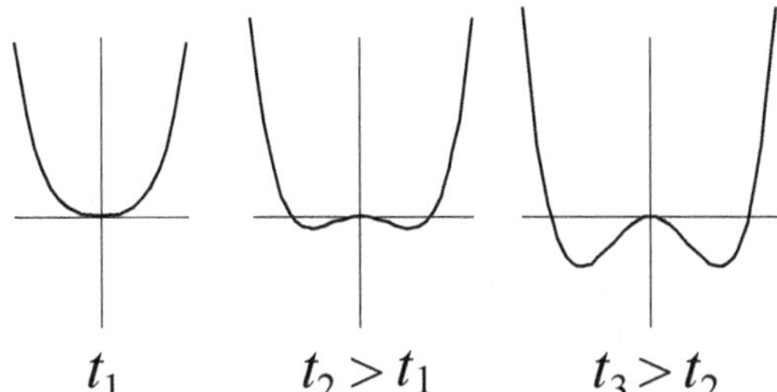

Figure 7-7. Possible Change in Shape of Higgs Potential with Time (Only 2D shown)

In the case of Fig. 7-7, at t_1 the field would be in a stable configuration. However, at later times, the false vacuum would be unstable.

It is also possible, that the false vacuum might be slightly, but not strongly stable, as in Fig. 7-8. In such case, we would expect the universe, at some point, tunneled out of the small indention in the middle of the curve and fell into the stable true vacuum state.

Higgs potential could be slightly stable at false vacuum, not known

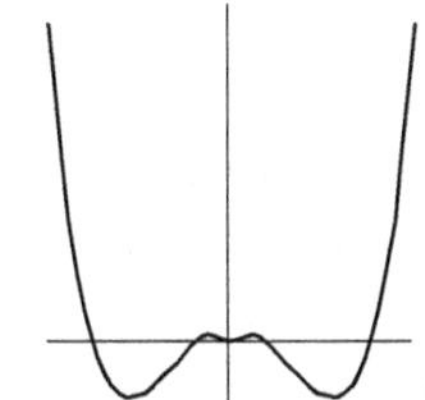

Figure 7-8. Possible Slightly Stable False Vacuum

As symbolized in Figs. 7-1 and 7-2 on pg. 209, the universe would be in equilibrium at the false vacuum, but not be stable. A slight fluctuation in the ϕ particle density would displace the universe off of the false vacuum (ϕ particle density not precisely zero) and begin the universe's descent into the true vacuum, which is in equilibrium, but not unstable. For the case of Fig. 7-8, larger fluctuations in the underlying particle density would be necessary to initiate symmetry breaking. Fluctuations in the horizontal direction (i.e., in the density of ϕ particles) would need to exceed the local maxima to begin the "roll down" into the valley.

It is also possible that the true vacuum is only slightly stable, and perhaps its outer side does not extend upward indefinitely, but merely humps before turning downward. In such case, our universe could one day tunnel out of the valley we now find ourselves in, and blow all we know to smithereens.

7.4.2 Particle Creation

As the universe slides down the slope of the Mexican hat potential from the false vacuum to the true vacuum, it moves radially outward from the central axis of Fig 7-2, pg. 209. As noted in Chap. 6, this means the ϕ particle density one would expect to measure increases. Hence, the reduction in potential energy is accompanied by an increase in particles, i.e., an increase in mass-energy.

Sliding down Mexican hat creates Higgs particles

Typically, these $|\phi\rangle$ particles decay (via the coupling with other particles) into other standard model particles. Thus, the process of spontaneous symmetry breaking populates the universe with more and more particles (until it stabilizes in the true vacuum.)

These decay into other particles, populating the universe with particles

7.4.3 GSW Model and Unification of E/M with Weak force

Before I first saw the GSW model some decades ago, I had heard a lot of talk about unifying the two forces, electromagnetic and weak, into a single electroweak force. This seemed so elegant and profound to me.

However, when I actually studied and learned it, I realized, with some disappointment, that there was more hype than substance in all the talk. At the true vacuum, there are two different types of interactions, electromagnetic and weak, mediated by two different types of gauge boson fields, the e/m photon A^μ and the weak intermediate vector bosons Z^μ, W^μ, and $W^{\mu\dagger}$. In a true unification, we would expect these two to merge at the false vacuum (high energy) into a single field. However, at the GSW false vacuum we still have two kinds of interactions, mediated by the two types of gauge boson fields, the hypercharge field B^μ and the isospin fields W_i^μ. Yes, the former (true vacuum) fields are linear combinations of the latter (false vacuum) fields. But in both cases, we have two different kinds of fields. In other words, there is no true unification (into a single field type).

Not true unification at false vacuum, still two kinds of fields

In some sense, one might argue there is some unification, as we now know the two fields are related, i.e., intertwined, so to speak. The A^μ field, for instance, is a combination of what, at high energy, is the weak and hypercharge fields. Ditto for the Z^μ. So, each type of low energy field is a result of combining both types of high energy fields. But, it is still two types of fields transforming to two types of fields.

But unification in some sense, as low energy weak and e/m fields combos of same fields W_i and B

Grand unified theories (GUTs), on the other hand, actually unify the three force fields (weak, strong, e/m) into a single type of force at high energy. The difference is that the GSW theory has been verified by experiment, and appears unique, whereas the extant GUT theories, as of this writing, fall short in these regards. Similarly, superstring (or M) theories appear to unite all four forces (above plus gravity) into a single type force, but with, as yet, no experimental verification.

In contrast, hypothetical GUT theories have true unification

Nevertheless, the GSW theory is a giant step for science, as it correctly describes the current properties, and evolution, of the electromagnetic and weak interactions.

7.4.4 Electroweak Symmetry Breaking is Apparent, Not Real.

The Lagrangian actually never loses its symmetry. It only appears to. It is always symmetric with respect to the high energy, false vacuum. If we express $\mathcal{L}$ in terms of the ϕ_i, instead of the η_i and σ, we get the original high energy $\mathcal{L}$. It is never really lost, and it is symmetric about the false vacuum.

$\mathcal{L}$ never really loses its symmetry, it is just hidden at low energy

At the true vacuum, we merely express $\mathcal{L}$ in terms of the fields η_i and σ, which have zero particle number expectation values (ignoring ZPE) at the true (present day, low energy) vacuum. There, $\mathcal{L}$ is not symmetric (in terms of the ϕ_i and also in terms of the η_i and σ). But it is only because our point of view is different that we seem to have lost symmetry.

By analogy, consider yourself at a point in the middle of a huge bowl. You rotate yourself about the vertical axis through that point and the bowl still looks the same to you. It is symmetric. Now move up the side of the bowl to another point and again rotate yourself about the vertical axis through this point. The bowl obviously looks to you like it is oriented differently. It does not maintain its appearance during the rotation. It appears unsymmetric. But, it is still a symmetric bowl. It just depends on the location you rotate yourself about.

Similar logic holds for $\mathcal{L}$ about two different points in our Hilbert space of fields. The symmetry is not really broken (although it is commonly stated that way), it is hidden.

Electroweak symmetry breaking is one form of what is known as spontaneous symmetry breaking, which occurs without outside influences (spontaneously) and also is hidden, i.e., only apparent. With explicit symmetry breaking, on the other hand, symmetry is truly broken.

7.4.5 Summary of Charge Types and Amounts on SM Particles

Wholeness Chart 7-9 lists the electric charge, weak isospin charge, and weak hypercharge for the electroweak bosons and SM fermions. The fermion part of this chart is a repeat of Wholeness Chart 6-10 of Chap. 6.

Table of e/w theory charges for all SM particles

We can now see the reason for the letter designations of the weak fields. W_i is obviously for "weak". W^+ is the positively charged (both electric and weak isospin) weak force carrier; W^-, its negatively charged antiparticle. Z stands for "zero" charge (all of electric, weak isospin, and hypercharge).

Do **Problem 9** to deduce the weak isospin charge and hypercharge for the false vacuum Higgs ϕ_3.

Wholeness Chart 7-9. Various Charges for Elementary Particles ($N_Q = I_3^W + Y$)

Fermion Type	LC Fermions	Electric Charge N_Q	Weak Isospin I_3^W	Hyper-charge Y	RC Fermions	Electric Charge N_Q	Weak Isospin I_3^W	Hyper-charge Y
Leptons	$\nu_e^L, \nu_\mu^L, \nu_\tau^L$	0	+ 1/2	– 1/2	$\nu_e^R, \nu_\mu^R, \nu_\tau^R$	0	0	0
	$e^{-L}, \mu^{-L}, \tau^{-L}$	– 1	– 1/2	– 1/2	$e^{-R}, \mu^{-R}, \tau^{-R}$	– 1	0	– 1
Quarks	u^L, c^L, t^L	+ 2/3	+ 1/2	+ 1/6	u^R, c^R, t^R	+ 2/3	0	+ 2/3
	d^L, s^L, b^L	– 1/3	– 1/2	+ 1/6	d^R, s^R, b^R	– 1/3	0	– 1/3

Original $\mathcal{L}$ Gauge Fields	Boson	Electric Charge N_Q	Weak Isospin I_3^W	Hyper-charge Y
Weak	W_1	Not in electric or weak isospin charge eigenstates		0
	W_2			0
	W_3	0	0	0
Hypercharge	B	0	0	0
Higgs	ϕ_3	0	– 1/2	+ 1/2

Contemporary $\mathcal{L}$ Gauge Fields	Boson	Electric Charge N_Q	Weak Isospin I_3^W	Hyper-charge Y
Weak	W^+	+ 1	+ 1	0
	W^-	– 1	– 1	0
	Z	0	0	0
E/m	γ	0	0	0
Higgs	σ	0	– 1/2	+ 1/2

7.4.6 Phenomenology and Terminology: "The Vacuum Eats Charge"

Although weak (isospin and hypercharge) charges are conserved in interactions like those we looked at in Chap. 6, an interesting effect arises due to the mass term for chiral fields, in which weak charges are not actually conserved.

The low energy Lagrangian (7-58) (and also in Wholeness Chart 7-8) has the electron field mass term (see (7-41)), due to symmetry breaking of the Higgs field

$$\text{One term in } \bar{\psi}_l \frac{v g_l}{\sqrt{2}} \psi_l = \bar{\psi}_e m_e \psi_e = m_e \bar{\psi}_e \psi_e = m_e \bar{\psi}_e^L \psi_e^R + m_e \bar{\psi}_e^R \psi_e^L \qquad (\text{no sum on } e). \quad (7\text{-}123)$$

This term came from $\mathcal{L}^{LH}$ at high energy (see Wholeness Chart 7-8).

Consider the second term after the last equal sign in (7-123). It can be represented by a Feynman diagram like Fig. 7-9, where an LC electron is destroyed and an RC electron is created. Since an LC electron has weak isospin charge of –½ and an RC electron has zero, a half unit of negative isospin charge has been lost. Similarly, a half unit of negative hypercharge has been gained. This is sometimes referred to as the "vacuum eating charge". It is due to the mass term and only occurs there.

Phrase "vacuum eats charge" refers to weak isospin charge when LC particle changes to RC, or vice versa

Note that from the 1st term after the last equal sign in (7-123), we get the opposite effect. An RC electron changes to an LC electron and an additional –½ unit of weak isospin charge is added, a half unit of negative hypercharge lost.

That charge seems to not be conserved in individual interactions

The net effect of many such interactions sums to zero overall charge created or destroyed, even though the individual interactions do not conserve charge.

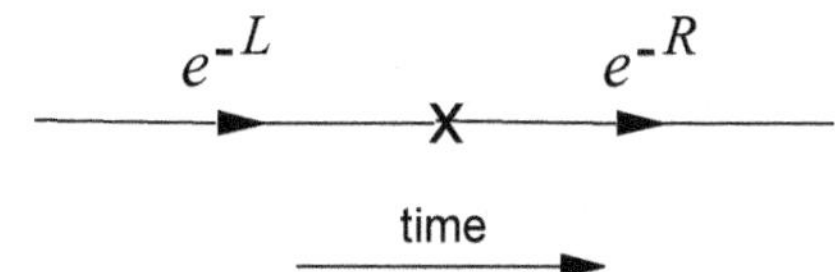

Figure 7-9. Weak Isospin Charge is "Eaten" (Not Conserved)

In reality, if we were to consider things from the point of view of the false vacuum, instead of the X in Fig. 7-9 (which represents the mass that came from the constant v in $\phi_b = \phi_3 + i\phi_4 - v$ when symmetry broke), we would have the Higgs particle from the field ϕ_b being created. That is, instead of the X, i.e., instead of the m_e in (7-123), we would have the ϕ_b field (times the Yukawa coupling g_e), as in

$$-g_e \bar{\psi}_e^R \phi_b \psi_e^L . \tag{7-124}$$

(7-124) represents a Feynman diagram vertex, one of whose manifestations is shown in Fig. 7-10.

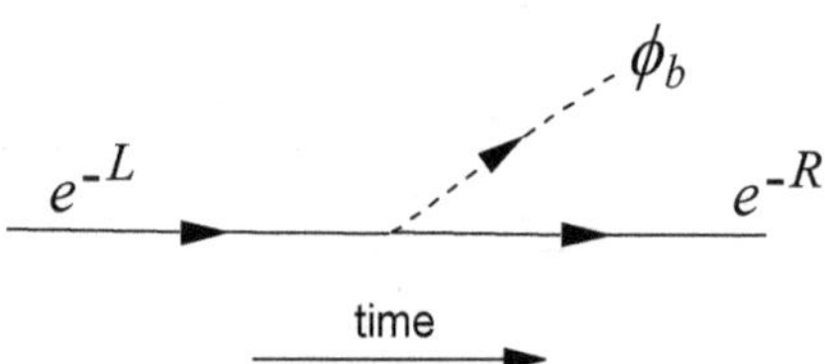

But from perspective of false vacuum, that charge is conserved

Figure 7-10. Interaction of Electron and High Energy Higgs

The diagram shows an LC electron in (of isospin –½ and hypercharge –½) with two particles out, a high energy Higgs particle (of isospin –½ and hypercharge +½) and an RC electron (of isospin 0 and hypercharge –1). Each charge type before equals the sum of charges of the same type after. The high energy Higgs ϕ_b carries off weak isospin charge and hypercharge.

In other words, it is only an illusion from the point of view of the true vacuum that charge is not conserved. From the point of view of the false vacuum, it is conserved.

There are, of course, similar terms to (7-123), and similar behavior, for muons and taus.

Note that at the false vacuum, $\mathcal{L}$ is symmetric and charge (both isospin and hypercharge) is conserved. At the true vacuum, $\mathcal{L}$ is not symmetric (due to the mass terms), and charge is not conserved (due to the mass terms). This is all connected by Noether's theorem (Vol. 1, pgs. 172-177, 290-297), which we will apply to electroweak theory in a later chapter.

7.4.7 Gauge Boson Fields Don't Actually Change with Energy Level

We commonly talk of the B^μ and W_i^μ as the high energy (or false vacuum) fields, and the A^μ, $W^{\pm\mu}$, and Z^μ fields as the low energy (or true vacuum) fields. However, this is not really true, as via (7-43) and (7-44), each field of one set is really a linear combination of the fields in the other set, and this is true at any energy level. Our theory is a gauge theory, and the fields are gauge fields, which are unmeasurable directly and can be different, yet predict the same measured results. So, we could actually work with either set at high energy, or either set at low energy.

Since QFT is a gauge theory, we could use B & W_i fields at low energy

However, analysis is much simpler if we use the B^μ and W_i^μ set at high energy (to deduce the form of $\mathcal{L}$, and show its invariance, for example), and the A^μ, $W^{\pm\mu}$, and Z^μ set at low energy (as they reflect the properties of particles we envision intuitively as interacting at low energies.) As we've noted before, don't confuse the form of the theory with the phenomenology it predicts. Different forms of the former can yield the same for the latter.

But, theory is easier to relate to our present world using photon, $W^\pm$, and Z fields, instead

7.4.8 Weak Interactions at Low Energy Can be Right Chiral

As we have seen, the photon and Z particles seen at contemporary energy levels are really linear combinations of the B and W_3 particles at high energy. In particular, from (7-43)

$$B_\mu = A_\mu \cos\theta_W - Z_\mu \sin\theta_W . \tag{7-125}$$

We know the B_μ serves as a gauge boson mediating part of the weak force between right chiral fermions and the term reflecting this in the Lagrangian is, as we found before with (7-42),

$$i\bar{\psi}_l \not{D}\psi_l \xrightarrow[\text{the terms}]{\text{one of}} i\bar{\psi}_l \left(-ig'B_\mu\gamma^\mu\right)\psi_l = g'\bar{\psi}_l^R \not{B}\psi_l^R + g'\bar{\psi}_l^L \not{B}\psi_l^L . \tag{7-126}$$

But the B field is partly composed of the photon and Z fields, as in (7-125), so the next to last term in (7-126) becomes

$$g'\bar{\psi}_l^R \not{B} \psi_l^R = g'\bar{\psi}_l^R \left(\not{A}\cos\theta_W - \not{Z}\sin\theta_W\right)\psi_l^R = g'\cos\theta_W \bar{\psi}_l^R \not{A}\psi_l^R - g'\sin\theta_W \bar{\psi}_l^R \not{Z}\psi_l^R . \qquad (7\text{-}127)$$

From (7-63),

$$g'\cos\theta_W = e . \qquad (7\text{-}128)$$

Solving (7-128) for g' and plugging into the last term in (7-127) for l=e, we obtain,

$$-e\tan\theta_W \bar{\psi}_e^R \gamma^\mu Z_\mu \psi_e^R \qquad (\text{term in low energy Lagrangian, } l = e), \qquad (7\text{-}129)$$

which, in fact, is what we already showed in the last term of (7-65). (I repeated the development above for pedagogic reasons.)

The point is that the term (7-129) couples the weak force carrying Z to the right chiral electron. In a Feynman diagram, this looks like Fig. 7-11.

At low energy, RC fields actually do take part in weak interactions, contrary to what one might think

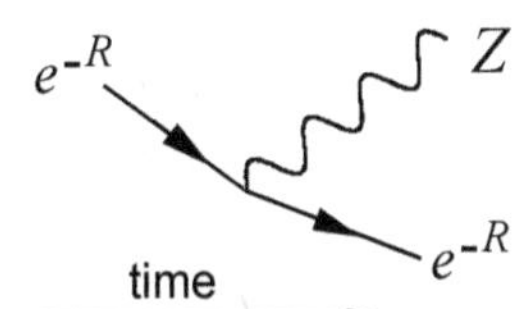

Figure 7-11. Right Chiral Electron Feels the Weak Z Force

Bottom line: Even though one hears everywhere that the weak force only couples to left chiral fermions, and not to right chiral ones, that is only true at high energy for the weak force mediated by the W_i gauge bosons. It is not true at low energy for the weak force mediated by the Z boson.

Only at high energy, is it strictly correct to limit weak interactions to LC fields

This was the reason why we earlier cautioned that certain statements made about only LC particles responding to the weak force were only true for the high energy fields W_i^μ.

7.4.9 No Flavor Changing Neutral Currents

Similar to what we saw in Sect. 5.2.3, pg. 151, in (7-129), it is common to call the factor (7-130) in the last term in the second row of (7-64), for l=e,

$$\begin{aligned} j^\mu_{eZ} &= \frac{g}{4\cos\theta_W}\bar{\psi}_e\gamma^\mu\left(1 - 4\sin^2\theta_W - \gamma^5\right)\psi_e \\ &= \frac{g}{2\cos\theta_W}\bar{\psi}_e\gamma^\mu\tfrac{1}{2}\left(1-\gamma^5\right)\psi_e - \underbrace{g\sin\theta_W}_{e}\tan\theta_W\,\bar{\psi}_e\gamma^\mu\psi_e , \end{aligned} \qquad (7\text{-}130)$$

a four-current, or more loosely, just a current. We can then re-write

Example of a neutral 4-current

$$\text{last term, second row of (7-64), (for } l\text{=}e) = j^\mu_{eZ} Z_\mu . \qquad (7\text{-}131)$$

(7-130) is called a neutral current because, for example, the destruction of an incoming electron by ψ_e is matched by the creation of an electron of same e/m and weak charge by $\bar{\psi}_e$. The charge effect on anything else its fields might interact with is zero, i.e., neutral. And in fact, that is what happens in Fig. 7-11, which represents the effect of the last term in (7-130) on the RC part of an electron. A neutral Z particle is created. In general, a neutral current neither leads to production nor destruction of a charged particle at a vertex.

On the other hand, for the term in (7-64)

$$-\frac{g}{\sqrt{2}}\bar{\psi}^L_{\nu_e}\gamma^\mu\psi^L_e W^+_\mu \qquad (7\text{-}132)$$

we would have a charged current

$$j^\mu_{eW} = -\frac{g}{\sqrt{2}}\bar{\psi}^L_{\nu_e}\gamma^\mu\psi^L_e = -\frac{g}{\sqrt{2}}\bar{\psi}_{\nu_e}\gamma^\mu\frac{1}{2}\left(1-\gamma^5\right)\psi_e , \qquad (7\text{-}133)$$

Example of a charged 4-current

which gives rise to a Feynman diagram vertex like Fig. 7-12, where we can see that a charged W particle is created via the action of (7-132). In general, a charged current leads to creation or destruction of a charged particle at a vertex. (In (7-132), W^+ destroys a W^+ particle or creates a W^- one.)

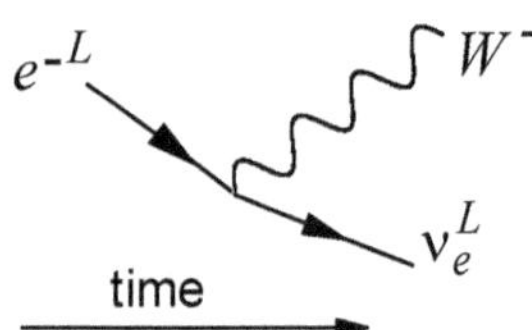

Figure 7-12. Diagram of Charged Current Effect at a Vertex

Note the flavor in Fig. 7-11 of the fermion does not change, whereas in Fig. 7-12, it does. This turns out to be a well-known general rule, stated below.

There are no flavor changing neutral 4-currents

Bottom line: There are no flavor changing neutral currents.

7.4.10 Note on Derivation of Weak Four-Currents

Four-currents for weak interactions can be derived from the symmetry of $\mathcal{L}$, and Noether's theorem. (See Vol. 1, Chapters 6 and 11, where it was done for QED.) We delve further into this in a later chapter.

4-currents can be derived via Noether's theorem

7.4.11 Running Coupling Constant

The interaction terms in $\mathcal{L}$ of the Glashow/Salam/Weinberg model (7-58) give rise to radiative type corrections to the effective Higgs field coupling, via adding up relevant Feynman diagram sub-amplitudes, which vary as the energy level changes. This is similar to QED (see Vol. 1, pgs. 304-317) where the QED coupling constant varies with energy level, due to the contributions of Feynman diagrams beyond tree level. Thus, we find our $SU(2)$ and $U(1)$ coupling constants g and g' become functions of energy level (represented by p)

Weak isospin and hypercharge couplings vary with energy, just as e(p) does in QED

$$g = g(p) \qquad g' = g'(p) \,. \tag{7-134}$$

Details are beyond the scope of this book

We will not delve more deeply into this issue here. It is quite complex.

7.4.12 Renormalization

Renormalization of electroweak theory was long an intractable problem, but was brilliantly solved in the 1970s by Gerardus 't Hooft and Martinus Veltman, and in 1999, they received a Nobel Prize for their work. We do not consider what they did in depth here, but simply note, very briefly, the approach they took.

't Hooft and Veltman realized that renormalization is easier to prove by choosing a different gauge (other than the unitary gauge), in which η_1, η_2, and $\eta_3 \neq 0$. This results in unphysical (ghost) fields we don't observe. Ghost fields are linear combinations of the 12 degrees of freedom of the other fields of the unitary gauge [1 massive Higgs σ(1 DOF), 3 massive W and Z (9 DOF), and 1 massless photon (2 DOF)]. Conceptually, and practically for calculating cross sections, that approach is more complicated. However, it provides advantages that facilitate proof of the renormalizabilty of the theory.

e/w renormalization beyond our present scope, but achieved using a different gauge

For more on this subject, see, for example, Peskin & Schroeder[1] and Itzykson and Zuber[2].

7.4.13 Any Candidate Higgs-like Field Must Be a Real Scalar with GSW Potential

For physics to be the same everywhere we would need the Higgs spin, if such existed, to be the same everywhere. But the only spin that would be the same everywhere is spin zero. Otherwise, we would have Higgs spin pointing in different directions, randomly, at different locations in space, and that would mean interactions that were different at different locations. At high energy this wouldn't make much difference, as Higgs interactions would simply be like other interactions of other particles. But once the e/w symmetry breaks, at contemporary energy levels, when the Higgs bestows masses on particles, the effect could lead to varying masses for the same particle at different points in space.

[1] M.E. Peskin and D. V. Schroeder, *An Introduction to Quantum Field Theory*, (Perseus 1995). pgs. 352-363 and 383-388.

[2] C. Itzykson and J.B. Zuber, *Quantum Field Theory*, (McGraw-Hill 1985), pgs. 594 and 617.

And the masses themselves would not be scalars, but spinors or vectors. Hence, the Higgs needs to be a scalar.

Further, a complex Higgs, such as (7-53), in a non-unitary gauge, unlike (7-55), would entail a complex constant instead of the real constant v. And this, via the mass terms in (7-58), would lead to complex, not real, masses, which, of course, have not been observed.

Only a real, scalar Higgs field with $\mathcal{V}$ as in GSW model works

Additionally, any term with more than four factors (as in (7-52)) of a scalar field, each factor having dimension 1 (as we showed at the end of Chap. 1), would need a coupling constant (comparable to μ^2 and λ) with negative mass dimension in order to give the term the dimension 4, as all terms in $\mathcal{L}$ must have. This leads, for reasons beyond our present scope, to non-renormalizable theories, and additionally, to undesirable effects in the ultraviolet range. For more on this topic, see https://edu.itp.phys.ethz.ch/hs12/qft1/Chapter07.pdf. Also see Chap. 16 herein, pgs. 475 and 491.

Bottom line: Only a real, scalar Higgs with potential of form (7-52) yields a valid e/w theory.

7.4.14 Higgs Spin Experimentally

Experiment → Higgs must be spin 0 or 2

CERN experiments have shown the true vacuum Higgs decays, via one of its decay channels (one of the ways it decays), into two photons. Conservation of linear momentum in the Higgs center of mass (COM) frame means the outgoing photons must have equal and opposite 3-momentum. Since the spin of massless particles must be aligned with their 3-momentum, they share the same spin axis. So, the photon spins are aligned in either the same direction or opposite directions. They either add or subtract, and since photon spin equals one, their total spin must be either zero or 2. Conservation of angular momentum then means the original Higgs must have the same spin as the total spin of the two final photons together, i.e., either total spin 2 or total spin 0. As of 2025, experimental evidence has strongly favored spin 0.

7.4.15 Mass of the Higgs and Gauge Hierarchy

Radiative corrections to the Higgs mass

Radiative corrections also affect the Higgs couplings and thus, particle masses, particularly the Higgs mass itself. For example, in (7-70), we would get a term (where we ultimately drop the wavy underlining symbolism)

$$\frac{\sigma}{\sqrt{2}}\bar{\psi}^{L}_{1A}Y_{11}\psi_{1}^{R}=\frac{\sigma}{\sqrt{2}}\bar{\psi}^{L}_{eA}Y_{ee}\psi_{e}^{R}=\frac{\sigma}{\sqrt{2}}\bar{\underset{\sim}{\psi}}{}^{L}_{e}M_{ee}\underset{\sim}{\psi}{}_{e}^{R}=\frac{\sigma}{\sqrt{2}}\bar{\underset{\sim}{\psi}}{}^{L}_{e}g_{e}\underset{\sim}{\psi}{}_{e}^{R}=\frac{g_{e}}{\sqrt{2}}\sigma\bar{\underset{\sim}{\psi}}{}^{L}_{e}\underset{\sim}{\psi}{}_{e}^{R}, \qquad (7\text{-}135)$$

which leads to a vertex with a Higgs and two charged electron family members (like electrons or positrons) such as either of the vertices shown in the RHS of Fig. 7-13, where H represents the Higgs field σ and dashed lines are typically used for the Higgs. Two such vertices can be linked via a lepton loop, such as that shown on the RHS of the figure.

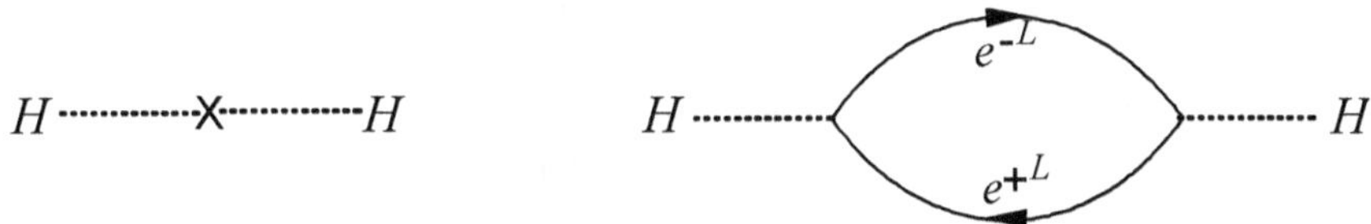

Figure 7-13. A Higgs Mass Term Diagram and a Radiative Correction Diagram

Radiative corrections imply Higgs mass should be enormous, but it isn't

The key thing is that the loop diagram contributes to the amplitude for the Higgs mass term. It adds another piece besides the term like that of (7-2) with $m_H = \mu$ in it, represented by the LHS of Fig. 7-13. Recall, we determine the mass of a real scalar ϕ by the factor by which ϕ^2 is multiplied. So, this radiative correction loop contributes to the effective mass we would measure in an experiment.

This discrepancy is known as the "gauge hierarchy problem"

Other fermions, such as the muon, tau, and quarks contribute radiative corrections, as well, and in addition, there are higher order (more than one loop) terms. Still further, there are also terms where gauge bosons interact with the Higgs, and even the Higgs field interacting with itself. When all of this is calculated straightforwardly, one gets an enormous number for the Higgs mass, on the order of 10^{16} GeV or more. But the Higgs mass is a little more than 10^2 GeV. This is known as the gauge hierarchy problem. Theory and experiment don't match, and in fact, are not remotely close. The "hierarchy" part refers to the mass/energy levels, whose numerical values are hierarchical in magnitude.

Note that, for the gauge hierarchy problem, the higher loop evaluation is done via the cutoff method of regularization, which we found did not give the correct result in QED. Using dimensional regularization for the Higgs calculation, there are still serious mass issues, but they are different[1].

The two leading candidate theories to solve this problem are 1) supersymmetry (SUSY) and 2) extra dimensions. SUSY, however, has not found any experimental support at the time of this writing (2025), at CERN or elsewhere. Without a natural solution like these, one ends up with a need for some very unnatural, very fine-tuned cancellation of sub-amplitudes (plus and minus amplitude values on the order of 10^{16} GeV cancelling to leave a term on the order of 10^2 GeV.)

7.4.16 LC and RC Fields with Mass Don't Actually Exist

I have so far presented fields, such as the electron or neutrino, at low energy (and thus having mass) as LC or RC fields. I did this for pedagogic reasons, as I think, in the early stages of learning about all this stuff, it is easier to grok that way.

Looking at LC and RC fields at low energy with more scrutiny

But, in truth, such fields do not actually exist physically and are simply a fiction we use as a shorthand, as explained below.

Free Lagrangian and Chiral Fields

Consider the usual Dirac field term in the free Lagrangian at low energy, expressed as chiral fields,

$$i\bar{\psi}\gamma^{\mu}\partial_{\mu}\psi = i\bar{\psi}\gamma^{\mu}\partial_{\mu}\frac{1}{2}\left(1-\gamma^{5}\right)\psi + i\bar{\psi}\gamma^{\mu}\partial_{\mu}\frac{1}{2}\left(1+\gamma^{5}\right)\psi = i\bar{\psi}^{L}\gamma^{\mu}\partial_{\mu}\psi^{L} + i\bar{\psi}^{R}\gamma^{\mu}\partial_{\mu}\psi^{R}\,. \tag{7-136}$$

where the fields (particles created and destroyed by them) have mass, so there is another term, which can be expressed in terms of chiral fields, using

$$-m\bar{\psi}\psi = -m\bar{\psi}^{L}\psi^{R} - m\bar{\psi}^{R}\psi^{L}\,. \tag{7-137}$$

Substituting the LHS of (7-136) and (7-137) into the Euler-Lagrange equation for $\bar{\psi}$ yields the Dirac equation.

$$\frac{\partial}{\partial\bar{\psi}}\left(i\bar{\psi}\gamma^{\mu}\partial_{\mu}\psi - m\bar{\psi}\psi\right) = 0 \qquad \rightarrow \; i\gamma^{\mu}\partial_{\mu}\psi - m\psi = 0\,. \tag{7-138}$$

But note what happens when we substitute the RHS of those equations in the Euler-Lagrange equation for $\bar{\psi}^{L}$. We get

$$\begin{aligned}&\frac{\partial}{\partial\bar{\psi}^{L}}\left(i\bar{\psi}^{L}\gamma^{\mu}\partial_{\mu}\psi^{L} + i\bar{\psi}^{R}\gamma^{\mu}\partial_{\mu}\psi^{R} - m\bar{\psi}^{L}\psi^{R} - m\bar{\psi}^{R}\psi^{L}\right) = 0\\ &\rightarrow\; i\gamma^{\mu}\partial_{\mu}\psi^{L} - m\psi^{R} = 0,\end{aligned} \tag{7-139}$$

which is half of the Dirac equation as expressed in (5-19) of Chap. 5, i.e., two of the four equations coupled in 4 scalar unknowns. Repeating the process but taking derivatives with respect to $\bar{\psi}^{R}$, yields

$$i\gamma^{\mu}\partial_{\mu}\psi^{R} - m\psi^{L} = 0\,, \tag{7-140}$$

the other half of the Dirac equation (the other two coupled equations) as shown in (5-19).

Conclusion

LC and RC fields don't actually solve Dirac eq at low energy, only at high

Neither ψ^L nor ψ^R alone solves the Dirac equation, so left (or right) chiral particles cannot exist physically (unless they are massless).

Bottom Line

Written as in the RHS of (7-136) and (7-137), it appears that we are writing the Lagrangian in terms of a field, the LC field (with mass), that does not solve the Dirac equation.

So not strictly correct physically to consider LC and RC fields as actual fields at low energy

So, in interactions below the e/w symmetry breaking scale, there really are no physical fields with left (or right) chirality, although things are often treated that way. The notation (and thinking/visualizing) of having actual physical LC fields interacting weakly and RC fields not interacting weakly can be helpful, but is misleading. In reality, the full fermion field reacts in a way where the L Weyl field portion links up via the weak force, but the R portion does not.

---

[1] For more on this, see www.quantumfieldtheory.info/Potential_solution_to_vacuum_energy_and_gauge_hierarchy.pdf, R. Klauber, "Mechanism for Resolving Gauge Hierarchy and Large Vacuum Energy" (2018) http://arxiv.org/abs/1802.03277, and Grange, Mathiot, Mutet, and Werner, *Phys Rev D* 88, 12, 125015 (2013), https://arxiv.org/abs/1312.5278 .

<u>More Correct Interpretation</u>

So, it is more correct to write (a typical interaction term)

$$-\frac{g}{\sqrt{2}}\bar{\psi}_{\nu_e}\gamma^{\mu}\frac{1}{2}\left(1-\gamma^5\right)\psi_e W_{\mu}^{+}=-\frac{g}{\sqrt{2}}\bar{\psi}_{\nu_e}\gamma^{\mu}P_L\psi_e W_{\mu}^{+}=-\frac{g}{\sqrt{2}}\bar{\psi}_{\nu_e}\gamma^{\mu}\left(P_L\right)^2\psi_e W_{\mu}^{+} \qquad (7\text{-}141)$$

than

$$-\frac{g}{\sqrt{2}}\bar{\psi}^L_{\nu_e}\gamma^{\mu}\psi^L_e W_{\mu}^{+}, \qquad (7\text{-}142)$$

although (7-142) is commonly used. We need to keep in mind that expressions like (7-142) really mean the LC part of the field ψ is sensitive to the weak interaction, not that there exists a LC massive field. The P_L operator in (7-141) filters out the LC parts of the spinor fields, such that only those parts play a role in the weak interaction.

In reality, it is the LC part of a general field ψ that interacts weakly

<u>Feynman Diagrams</u>

The Feynman diagram corresponding to (7-142) is shown in Fig. 7-14 (where the W^+ field destroys a W^+ particle or creates a W^- particle). The left side of the figure is correct at the false vacuum, since LC particles are massless and hence have a physical reality. The right side of the figure is the correct Feynman diagram at the true vacuum, where fermions have mass, and hence, LC particles do not exist physically. However, the left side diagram is commonly used as shorthand at the true vacuum.

To be precise, we should use general fields with LC projection operator

though it is common practice to simply use the LC projected part of the field as a field itself

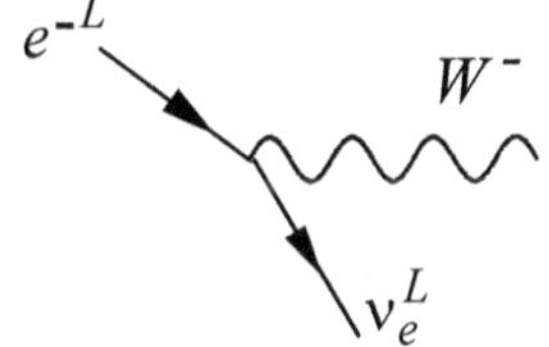

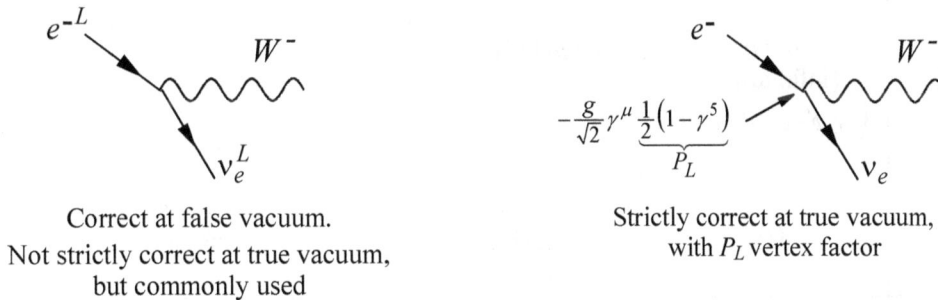

Correct at false vacuum.
Not strictly correct at true vacuum, but commonly used

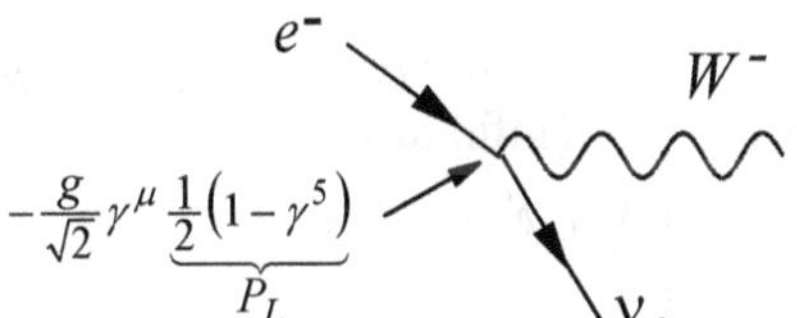

Strictly correct at true vacuum, with P_L vertex factor

Figure 7-14. Examples of Correct and Incorrect, but Commonly Used, Feynman Diagrams

7.4.17 Why is the Higgs Particle Called the "God Particle"?

You may have heard the Higgs particle referred to as the "God particle". It is called this, at least in part, for the following reason.

Prior to symmetry breaking, no particles had mass. So, all traveled at the speed of light. But particles traveling at the speed of light are not going to coalesce into nuclei and atoms, as they are simply moving too fast. So, nothing we know of as our world could ever form. Once the particles gain mass, via the coupling to the Higgs field, they no longer travel at the speed of light, so they effectively slow down enough to join together to form the building blocks of our universe. Without the Higgs, we would have no universe as we know it. It is the foundation of all, the source of all. Hence, the allusion to God, the source of everything according to the world's major religions.

Higgs gave particles mass, so slowed them down from light speed

Only then could creation arise, so Higgs like God

Additionally, when a (massless) particle travels at the speed of light, time does not progress on that particle. In effect, time is frozen and nothing happens. From this perspective, the Higgs gives rise to time itself. Without any mass, no particles would experience the passage of time. The "God particle", by bestowing mass, bestows time on the participants in the evolution of the universe, something for which many would credit God[1].

Also, no time passes on a massless particle, so the Higgs, in a sense, gave birth to time

7.5 Chapter Summary

7.5.1 Overview of Three Symmetry Breaking Models

Wholeness Chart 7-10 is an overview of the three symmetry breaking models, Goldstone, Higgs, and GSW.

---

[1] See *The God Particle*, Leon M. Lederman and Dick Teresi, Dell Publishing (1993).

Wholeness Chart 7-10. Electroweak Symmetry Breaking: Overview of Three Models

	Single Real Scalar Field	**Complex Scalar Field Singlet (Isoscalar = isospin singlet)**		**Complex Scalar Field Doublet (Isospinor = isospin doublet)**	
		Goldstone Model	**Higgs Model**	**(Not treated)**	**Glashow/Salam/Weinberg**
Higgs Field	$\phi(x) = \text{real}$	$\phi(x) = \frac{1}{\sqrt{2}}(\phi_1 + i\phi_2)$	as at left		$\Phi(x) = \frac{1}{\sqrt{2}}\begin{bmatrix}\phi_1 + i\phi_2 \\ \phi_3 + i\phi_4\end{bmatrix}$
Symmetry Transform Type	Global reflection $\mathcal{L}(\phi) = \mathcal{L}(-\phi)$	Global $U(1)$ (i.e., no interactions)	Local $U(1)$ (i.e., interactions)	Global $SU(2)XU(1)$	Local $SU(2)XU(1)$ (i.e., interactions)
$\mathcal{V}(\phi)$ [or $\mathcal{H}(\phi)$] (same symmetry as $\mathcal{L}$)	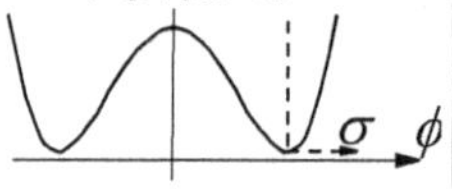 $\mathcal{V}$ or $\mathcal{H}$ or $-\mathcal{L}$ σ ϕ	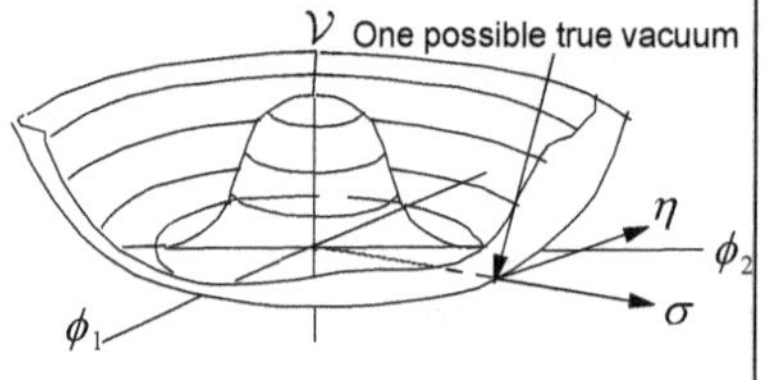 $\mathcal{V}$ One possible true vacuum η ϕ_2 σ ϕ_1	As at left		Similar in 5D (4 ϕ_i,1 $\mathcal{V}$) to Goldstone. Have true vacuum at min $\mathcal{V}$.
Symmetry Transform	$\phi(x) \to -\phi(x)$ (discrete)	$\delta\phi = i\alpha\phi$ (infinitesimal) (or $\phi' = e^{i\alpha}\phi$ finite)	$\delta\phi = i\alpha(x)\phi$ $\delta A^\mu = -\frac{1}{e}\partial^\mu\alpha(x)$ $[\& \not\partial \to \not D \text{ in } \mathcal{L}_0]$		$\delta\Phi = \frac{i}{2}\omega_i\sigma_i\Phi + iYf\Phi$ $\delta B^\mu = -\frac{1}{g'}\partial^\mu f$ $\delta W_i^\mu = -\frac{1}{g'}\partial^\mu\omega_i - \varepsilon_{ijk}\omega_j W_k^\mu$ $\delta\Psi_l^L = \frac{i}{2}\omega_i\sigma_i\Psi_l^L + iYf\Psi_l^L$
Degrees of Freedom, False Vacuum	1 (ϕ)	2 (ϕ_1, ϕ_2)	4 (ϕ_1, ϕ_2, 2 massless A^μ polariz states)		12 (ϕ_1, ϕ_2, ϕ_3, ϕ_4, 2 for massless B^μ, 2 each for 3 massless W_i^μ)
New Bosons = Lin Combins of Old		Easiest to align σ axis with ϕ_1 axis	As at left		$B^\mu, W_i^\mu \to W^{\pm\mu}, Z^\mu, A^\mu$ $\phi_{1,2,4} = \eta_{1,2,3}$ $\phi_3 = \sigma + v$ Still 12 DOFs, still massless
Degrees of Freedom, True Vacuum	1 (σ)	2 (σ, η) $m_\eta = 0$ $m_\sigma^2 > 0$	5 (σ,η, 3 from massive A^μ) 1 non physical field $m_\eta = 0$ $m_\sigma^2 > 0$		15 (1 each from η_1,η_2,η_3,σ, 9 from massive $W^{\pm\mu}$,Z^μ, 2 from massless A^μ) 3 non physical fields
Unitary Gauge			Eliminate one DOF by choosing $\eta = 0$		Get what is observed physically by choosing $\eta_1 = \eta_2 = \eta_3 = 0$. Eliminate 3 DOFs.
Result			1 massive Higgs σ 1 massive A^μ 4 DOFs		1 massive Higgs σ 3 massive $W^{\pm\mu}$,Z^μ 1 massless A^μ 12 DOFs

7.5.2 The Entire Electroweak Lagrangian at True Vacuum

Following, in one place, are all the terms in the low energy e/w Lagrangian. Each interaction term represents a possible vertex in an interaction. This one page is something you might want to make a paper copy of and pin to the wall where you study.

Total Electroweak True Vacuum Lagrangian

All fields shown are flavor fields (e.g., $\nu_e, e, \mu, \nu_\mu, \tau, \nu_\tau, u, d, c, s, t, b$)

$$\mathcal{L} = \mathcal{L}_0 + \mathcal{L}_I = \mathcal{L}_0 + \mathcal{L}^{LB} + \mathcal{L}^{QB} + \mathcal{L}^{LH} + \mathcal{L}^{QH} + \mathcal{L}^{HH} + \mathcal{L}^{HB} + \mathcal{L}^{BB} \tag{7-143}$$

$$\begin{aligned}
\mathcal{L}_0 = {} & \bar{\psi}_q\big(i\not{\partial} - \underbrace{\tfrac{vg_q}{\sqrt{2}}}_{m_q}\big)\psi_q + \bar{\psi}_l\big(i\not{\partial} - \underbrace{\tfrac{vg_l}{\sqrt{2}}}_{m_l}\big)\psi_l + \bar{\psi}_{\nu_l} i\not{\partial}\psi_{\nu_l} \underbrace{- \tfrac{v}{\sqrt{2}}\bar{\psi}^R_{\nu_l}\, {}_\nu M_{\nu_l i} P^\dagger_{i\nu_{l'}}\, \psi^L_{\nu_{l'}} - \tfrac{v}{\sqrt{2}}\bar{\psi}^L_{\nu_l} P_{\nu_l i}\, {}_\nu M_{i\nu_{l'}} \psi^R_{\nu_{l'}}}_{\text{non-diagonal neutrino mass matrix } P_\nu M} \\
& -\tfrac{1}{4}F_{\mu\nu}F^{\mu\nu} - \tfrac{1}{2}W^\dagger_{\mu\nu}W^{\mu\nu} + \underbrace{\left(\tfrac{1}{2}vg\right)^2}_{m_W^2} W^\dagger_\mu W^\mu - \tfrac{1}{4}Z_{\mu\nu}Z^{\mu\nu} + \tfrac{1}{2}\underbrace{\left(\tfrac{1}{2}\tfrac{vg}{\cos\theta_W}\right)^2}_{m_Z^2} Z_\mu Z^\mu \\
& + \tfrac{1}{2}\left(\partial_\mu\sigma\right)\left(\partial^\mu\sigma\right) - \tfrac{1}{2}\underbrace{\left(-2\mu^2\right)}_{m_H^2}\sigma^2
\end{aligned} \tag{7-144}$$

$$\begin{aligned}
\mathcal{L}^{LB} = {} & e\bar{\psi}_l \not{A}\psi_l - \tfrac{g}{\sqrt{2}}\bar{\psi}^L_{\nu_l}\not{W}^+\psi^L_l - \tfrac{g}{\sqrt{2}}\bar{\psi}^L_l\not{W}^-\psi^L_{\nu_l} \\
& - \frac{g}{2\cos\theta_W}\bar{\psi}^L_{\nu_l}\not{Z}\psi^L_{\nu_l} + \frac{g}{2\cos\theta_W}\bar{\psi}^L_l\not{Z}\psi^L_l - \underbrace{g\sin\theta_W}_{e}\tan\theta_W\bar{\psi}_l\not{Z}\psi_l
\end{aligned} \tag{7-145}$$

$$\begin{aligned}
\mathcal{L}^{QB} = {} & -\tfrac{2}{3}e\bar{\psi}_{q_u}\not{A}\psi_{q_u} + \tfrac{1}{3}e\bar{\psi}_{q_d}\not{A}\psi_{q_d} - \tfrac{g}{\sqrt{2}}\bar{\psi}^L_{q_u}\not{W}^+V_{q_u q_d}\psi^L_{q_d} - \tfrac{g}{\sqrt{2}}\bar{\psi}^L_{q_d}\not{W}^-V^\dagger_{q_d q_u}\psi^L_{q_u} \\
& - \frac{g}{2\cos\theta_W}\bar{\psi}^L_{q_u}\not{Z}\psi^L_{q_u} + \frac{g}{2\cos\theta_W}\bar{\psi}^L_{q_d}\not{Z}\psi^L_{q_d} \\
& + \tfrac{2}{3}\underbrace{g\sin\theta_W}_{e}\tan\theta_W\bar{\psi}_{q_u}\not{Z}\psi_{q_u} - \tfrac{1}{3}\underbrace{g\sin\theta_W}_{e}\tan\theta_W\bar{\psi}_{q_d}\not{Z}\psi_{q_d}\,.
\end{aligned} \tag{7-146}$$

$$\mathcal{L}^{LH} = -\frac{g_{\nu_l}}{\sqrt{2}}\sigma\bar{\psi}^R_{\nu_l}\, {}_\nu M_{\nu_l i}P^\dagger_{i\nu_{l'}}\, \psi^L_{\nu_{l'}} - \frac{g_{\nu_l}}{\sqrt{2}}\sigma\bar{\psi}^L_{\nu_l} P_{\nu_l i\nu} M_{i\nu_{l'}}\psi^R_{\nu_{l'}} - \frac{g_l}{\sqrt{2}}\sigma\bar{\psi}_l\,\psi_l \tag{7-147}$$

$$\mathcal{L}^{QH} = -\frac{g_{q_u}}{\sqrt{2}}\sigma\bar{\psi}_{q_u}\,\psi_{q_u} - \frac{g_{q_d}}{\sqrt{2}}\sigma\bar{\psi}_{q_d}\,\psi_{q'_d} \tag{7-148}$$

$$\mathcal{L}^{HH} = -\frac{\lambda}{4}\sigma^4 - \lambda v\sigma^3 \tag{7-149}$$

$$\mathcal{L}^{HB} = \frac{g^2}{4}W^{-\mu}W^+_\mu\sigma^2 + \frac{vg^2}{2}W^{-\mu}W^+_\mu\sigma + \frac{g^2}{8\cos^2\theta_W}Z^\mu Z_\mu\sigma^2 + \frac{vg^2}{4\cos^2\theta_W}Z^\mu Z_\mu\sigma \tag{7-150}$$

$$\begin{aligned}
\mathcal{L}^{BB} = {} & \\
& ig\cos\theta_W\left\{\left(W^-_\alpha W^+_\beta - W^-_\beta W^+_\alpha\right)\partial^\alpha Z^\beta + \left(\partial_\alpha W^+_\beta - \partial_\beta W^+_\alpha\right)W^{-\beta}Z^\alpha - \left(\partial_\alpha W^-_\beta - \partial_\beta W^-_\alpha\right)W^{+\beta}Z^\alpha\right\} \\
& + ie\left\{\left(W^-_\alpha W^+_\beta - W^-_\beta W^+_\alpha\right)\partial^\alpha A^\beta + \left(\partial_\alpha W^+_\beta - \partial_\beta W^+_\alpha\right)W^{-\beta}A^\alpha - \left(\partial_\alpha W^-_\beta - \partial_\beta W^-_\alpha\right)W^{+\beta}A^\alpha\right\} \\
& + g^2\cos^2\theta_W\left(W^+_\alpha W^-_\beta Z^\alpha Z^\beta - W^+_\beta W^{-\beta}Z_\alpha Z^\alpha\right) + e^2\left(W^+_\alpha W^-_\beta A^\alpha A^\beta - W^+_\beta W^{-\beta}A_\alpha A^\alpha\right) \\
& + eg\cos\theta_W\left\{W^+_\alpha W^-_\beta\left(Z^\alpha A^\beta + A^\alpha Z^\beta\right) - 2W^+_\beta W^{-\beta}A_\alpha Z^\alpha\right\} \\
& + \tfrac{1}{2}g^2 W^-_\alpha W^+_\beta\left(W^{-\alpha}W^{+\beta} - W^{+\alpha}W^{-\beta}\right).
\end{aligned} \tag{7-151}$$

We will be primarily concerned with (7-145) and (7-146), the interactions of leptons and quarks with the gauge bosons $W^\pm$, Z, and γ (photon), though the photon interactions were studied thoroughly in Vol. 1 (QED).

Note that the PMNS matrix P is incorporated in the neutrino mass of (7-144) and Higgs-neutrino terms of (7-147). The CKM matrix V is incorporated in the quark-W interaction terms of (7-146).

7.5.3 Other Summary Prior Wholeness Charts

Certain wholeness charts and figures in this chapter are good summaries of key aspects of the GSW theory. We won't repeat them here, but merely cite which they are, and what page they can be found on.

Particle mass summary: Wholeness Chart 7-7 and Fig. 7-6, pg. 239.

GSW model: Wholeness Chart 7-8, pg. 240.

Particle charges: Wholeness Chart 7-9, pg. 243.

7.5.4 Leptons, Quarks, and Mass Matrices

Nutshell Summary of Mass Eigenstates and Diagonalization

All of our work related to mass matrices can be summarized in a nutshell, as follows.

- For quarks, we diagonalize the mass matrix, and
- for leptons, we diagonalize the interaction (flavor) matrix (with the *W* bosons).

Thus,

- for quarks, the interaction (flavor) matrix (with the W bosons) is non-diagonal; and
- for leptons, the mass matrix is non-diagonal.

Detailed Summary of Mass vs Flavor Eigenstates

See the following page for Wholeness Chart 7-11, a detailed, yet single page, overview of mass vs flavor eigenstates.

7.5.5 Grand Summary Chart

Wholeness Chart 7-12, two pages hence, is a terrific overview of the standard model. This chart was created by Professor Latham Boyle, and a free full color version can be downloaded legally at https://upload.wikimedia.org/wikipedia/commons/2/2f/Standard_Model_Of_Particle_Physics--Most_Complete_Diagram.png. I suggest you do so, print it out, and stick it to the wall near the desk where you study.

7.5.6 Calculations

The goal of all that we are doing in this book is, of course, to calculate probabilities for particular interactions and from them, deduce scattering cross-sections and particle decay rates. See a summary of this for QED (without the decay part) on pg. 513 of Vol. 1.

That is, just as we did in Vol. 1, Chaps. 7 and 8 for QED, we could start here with our e/w $\mathcal{L}_I$, deduce $\mathcal{H}_I$, use that to find the S operator, employ the Dyson-Wicks expansion, evaluate very unwieldy integrals, and determine the transition amplitude S_{fi}. The square of the absolute value of that gives the probability of the initial state i transitioning (scattering) into the final state f.

But, just as in QED, we can short cut that very long, tedious process by employing Feynman diagrams and Feynman rules to determine the e/w S matrix, whose elements $S_{fi.}$ are transition amplitudes for various electroweak interactions.

We are not going to explore the long tedious route here. We will simply, in the next chapter, vault over the tedium and instead, go directly to the Feynman rules for electroweak theory.

Wholeness Chart 7-11. Mass vs Flavor Eigenstate Bases in QFT

$\mathcal{L}$ Term	**Original Flavor Basis**	**Mass Eigenstates Basis**	**Basis I (New Flavors)**	**Basis II (Yet Different Flavors)**
Kinetic	$i\bar{\psi}\,\partial\!\!\!/\,\psi$	$i\bar{\underline{\psi}}\,\partial\!\!\!/\,\underline{\psi}$ (same)	$i\bar{\psi}'\,\partial\!\!\!/\,\psi'$ (same)	$i\bar{\psi}''\,\partial\!\!\!/\,\psi''$ (same)
Non-W interactions	e.g., $e\bar{\psi}A\!\!\!/\,\psi$	$e\bar{\underline{\psi}}A\!\!\!/\,\underline{\psi}$ (same)	$e\bar{\psi}'A\!\!\!/\,\psi'$ (same)	$e\bar{\psi}''A\!\!\!/\,\psi''$ (same)
Mass terms	Need to add $h.c.$ terms to below $-\frac{v}{\sqrt{2}}\bar{\psi}^L_{\nu_l}\,{}_B Y_{\nu_l\nu_{l'}}\psi^R_{\nu_{l'}} = -\frac{v}{\sqrt{2}}\times$ $\begin{bmatrix}\bar{\psi}^L_{\nu_e} & \bar{\psi}^L_{\nu_\mu} & \bar{\psi}^L_{\nu_\tau}\end{bmatrix}\begin{bmatrix}Y_{\nu_e\nu_e} & Y_{\nu_e\nu_\mu} & Y_{\nu_e\nu_\tau}\\ Y_{\nu_\mu\nu_e} & Y_{\nu_\mu\nu_\mu} & Y_{\nu_\mu\nu_\tau}\\ Y_{\nu_\tau\nu_e} & Y_{\nu_\tau\nu_\mu} & Y_{\nu_\tau\nu_\tau}\end{bmatrix}_B\begin{bmatrix}\psi^L_{\nu_e}\\ \psi^L_{\nu_\mu}\\ \psi^L_{\nu_\tau}\end{bmatrix}$ Non-diag neutrino mass matrix $\frac{v}{\sqrt{2}}\,{}_B Y$ $-\frac{v}{\sqrt{2}}\bar{\psi}^L_l\,{}_A Y_{ll'}\psi^R_{l'} + h.c. = -\frac{v}{\sqrt{2}}\times$ $\begin{bmatrix}\bar{\psi}^L_e & \bar{\psi}^L_\mu & \bar{\psi}^L_\tau\end{bmatrix}\begin{bmatrix}Y_{ee} & Y_{e\mu} & Y_{e\tau}\\ Y_{\mu e} & Y_{\mu\mu} & Y_{\mu\tau}\\ Y_{\tau e} & Y_{\tau\mu} & Y_{\tau\tau}\end{bmatrix}_A\begin{bmatrix}\psi^L_e\\ \psi^L_\mu\\ \psi^L_\tau\end{bmatrix}$ Non-diag charged lepton mass $\frac{v}{\sqrt{2}}\,{}_A Y$	$m_{\nu_i} = \frac{vg_{\nu_i}}{\sqrt{2}}$ $\quad -m_{\nu_i}\bar{\underline{\psi}}_{\nu_i}\underline{\psi}_{\nu_i} =$ $-\begin{bmatrix}\bar{\underline{\psi}}_{\nu_1} & \bar{\underline{\psi}}_{\nu_2} & \bar{\underline{\psi}}_{\nu_3}\end{bmatrix}\begin{bmatrix}m_{\nu_1} & & \\ & m_{\nu_2} & \\ & & m_{\nu_3}\end{bmatrix}\begin{bmatrix}\underline{\psi}_{\nu_1}\\ \underline{\psi}_{\nu_2}\\ \underline{\psi}_{\nu_3}\end{bmatrix}$ Diagonal neutrino mass matrix $m_l = \frac{vg_l}{\sqrt{2}}$ $\quad m_l\bar{\underline{\psi}}_l\underline{\psi}_l =$ $-\begin{bmatrix}\bar{\underline{\psi}}_1 & \bar{\underline{\psi}}_2 & \bar{\underline{\psi}}_3\end{bmatrix}\begin{bmatrix}m_1 & & \\ & m_2 & \\ & & m_3\end{bmatrix}\begin{bmatrix}\underline{\psi}_1\\ \underline{\psi}_2\\ \underline{\psi}_3\end{bmatrix}$ Diagonal charged lepton mass matrix	Need to add $h.c.$ terms to below $-\frac{v}{\sqrt{2}}\bar{\psi}'^R_{\nu_l}\,{}_\nu M_{\nu_l i}P^\dagger_{i\nu_{l'}}\psi'^L_{\nu_{l'}} = -\frac{v}{\sqrt{2}}\times$ $\begin{bmatrix}\bar{\psi}'^L_{\nu_e} & \bar{\psi}'^L_{\nu_\mu} & \bar{\psi}'^L_{\nu_\tau}\end{bmatrix}\begin{bmatrix}MP^\dagger_{\nu_e\nu_e} & MP^\dagger_{\nu_e\nu_\mu} & MP^\dagger_{\nu_e\nu_\tau}\\ MP^\dagger_{\nu_\mu\nu_e} & MP^\dagger_{\nu_\mu\nu_\mu} & MP^\dagger_{\nu_\mu\nu_\tau}\\ MP^\dagger_{\nu_\tau\nu_e} & MP^\dagger_{\nu_\tau\nu_\mu} & MP^\dagger_{\nu_\tau\nu_\tau}\end{bmatrix}\begin{bmatrix}\psi'^L_{\nu_e}\\ \psi'^L_{\nu_\mu}\\ \psi'^L_{\nu_\tau}\end{bmatrix}$ Non-diag neutrino mass matrix $-\frac{v}{\sqrt{2}}MP^\dagger$ $m_l = \frac{vg_l}{\sqrt{2}}$ $\quad m_l\bar{\psi}'_l\psi'_l =$ $-\begin{bmatrix}\bar{\psi}'_e & \bar{\psi}'_\mu & \bar{\psi}'_\tau\end{bmatrix}\begin{bmatrix}m_e & & \\ & m_\mu & \\ & & m_\tau\end{bmatrix}\begin{bmatrix}\psi'_e\\ \psi'_\mu\\ \psi'_\tau\end{bmatrix}$ Diagonal charged lepton mass matrix	$m_q = \frac{vg_q}{\sqrt{2}}$ $\quad -m_q\bar{\psi}''_q\psi''_q =$ $-\begin{bmatrix}\bar{\psi}''_u & \bar{\psi}''_c & \bar{\psi}''_t\end{bmatrix}\begin{bmatrix}m_u & & \\ & m_c & \\ & & m_t\end{bmatrix}\begin{bmatrix}\psi''_u\\ \psi''_c\\ \psi''_t\end{bmatrix}$ Diagonal up type quark (or neutrino) mass matrix $m_q = \frac{vg_q}{\sqrt{2}}$ $\quad -m_q\bar{\psi}''_q\psi''_q =$ $-\begin{bmatrix}\bar{\psi}''_d & \bar{\psi}''_s & \bar{\psi}''_b\end{bmatrix}\begin{bmatrix}m_d & & \\ & m_s & \\ & & m_b\end{bmatrix}\begin{bmatrix}\psi''_d\\ \psi''_s\\ \psi''_b\end{bmatrix}$ Diagonal down type quark (or charged lepton) mass matrix
Mass?	Mass indefinite for all. Flavor definite for all.	Mass definite for all. Flavor indefinite for all.	Mass definite for e, μ, τ, indefinite for ν. Flavor definite for all (but defined differently than original flavor).	Mass definite for all. Flavor definite for all (but defined differently than original flavor)..
W-boson/ lepton interactions	No mixing of generations. e.g., no e^- to ν_μ at W vertex. $-\frac{g}{\sqrt{2}}\bar{\psi}^L_l W\!\!\!\!/^-\psi^L_{\nu_l} =$ $-\begin{bmatrix}\bar{\psi}^L_e & \bar{\psi}^L_\mu & \bar{\psi}^L_\tau\end{bmatrix}\begin{bmatrix}1 & & \\ & 1 & \\ & & 1\end{bmatrix}W\!\!\!\!/^-\begin{bmatrix}\psi^L_{\nu_e}\\ \psi^L_{\nu_\mu}\\ \psi^L_{\nu_\tau}\end{bmatrix}$	Mixing of generations. e.g., 1 to 2 can occur at W vertex. $-\frac{g}{\sqrt{2}}\bar{\underline{\psi}}^L_{l'}R_{l'\nu_i}W\!\!\!\!/^-\underline{\psi}^L_{\nu_i} = -\frac{g}{\sqrt{2}}\times$ $\begin{pmatrix}\bar{\underline{\psi}}^L_1 & \bar{\underline{\psi}}^L_2 & \bar{\underline{\psi}}^L_3\end{pmatrix}\begin{bmatrix}R_{11} & R_{12} & R_{13}\\ R_{21} & R_{22} & R_{23}\\ R_{31} & R_{32} & R_{33}\end{bmatrix}W\!\!\!\!/^-\begin{pmatrix}\underline{\psi}^L_{\nu_1}\\ \underline{\psi}^L_{\nu_2}\\ \underline{\psi}^L_{\nu_3}\end{pmatrix}$	No mixing of generations. e.g., no e^- to ν_μ at W vertex. $-\frac{g}{\sqrt{2}}\bar{\psi}'^L_l W\!\!\!\!/^-\psi'^L_{\nu_l} =$ $-\begin{bmatrix}\bar{\psi}'^L_e & \bar{\psi}'^L_\mu & \bar{\psi}'^L_\tau\end{bmatrix}\begin{bmatrix}1 & & \\ & 1 & \\ & & 1\end{bmatrix}W\!\!\!\!/^-\begin{bmatrix}\psi'^L_{\nu_e}\\ \psi'^L_{\nu_\mu}\\ \psi'^L_{\nu_\tau}\end{bmatrix}$	Mixing of generations. e.g., d to c quark can occur at W vertex. $-\frac{g}{\sqrt{2}}\bar{\psi}''^L_{q_d}V^\dagger_{q_d q_u}W\!\!\!\!/^-\psi''^L_{q_u} = -\frac{g}{\sqrt{2}}\times$ $\begin{bmatrix}\bar{\psi}''^L_d & \bar{\psi}''^L_s & \bar{\psi}''^L_b\end{bmatrix}\begin{bmatrix}V^\dagger_{du} & V^\dagger_{dc} & V^\dagger_{dt}\\ V^\dagger_{su} & V^\dagger_{sc} & V^\dagger_{st}\\ V^\dagger_{bu} & V^\dagger_{bc} & V^\dagger_{bt}\end{bmatrix}W\!\!\!\!/^-\begin{bmatrix}\psi''^L_u\\ \psi''^L_c\\ \psi''^L_t\end{bmatrix}$
Used for	Introducing/developing SM of QFT	Intermediate step to bases at right	Leptons	Quarks

Wholeness Chart 7-12. The Standard Model in a Nutshell

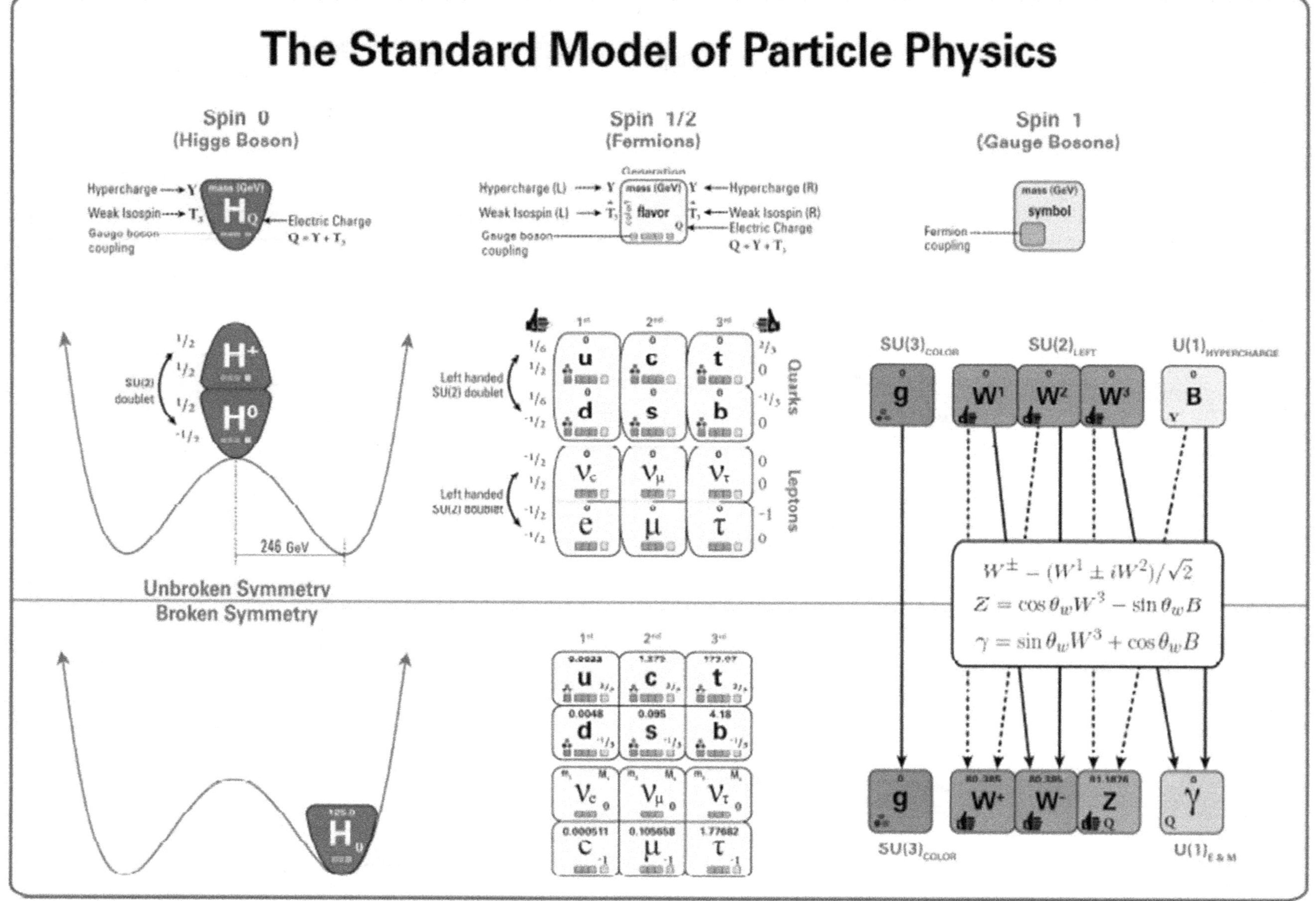

Credti: Latham Boyle, CC BY-SA 4.0 https://creativecommons.org/licenses/by-sa/4.0 , via Wikimedia Commons
https://upload.wikimedia.org/wikipedia/commons/2/2f/Standard_Model_Of_Particle_Physics--Most_Complete_Diagram.png

7.6 Appendix. Transforming the Lagrangian to the True Vacuum

We look at each term in the false vacuum $\mathcal{L}$ of (7-46) expressed in the fields of the true vacuum to get the true vacuum $\mathcal{L}$. As we do, we will substitute the Higgs field Φ, with unitary gauge (7-55),

$$\Phi(x)=\frac{1}{\sqrt{2}}\begin{bmatrix}0\\ \phi_3\end{bmatrix}=\frac{1}{\sqrt{2}}\begin{bmatrix}0\\ \sigma+v\end{bmatrix} \quad \text{in unitary gauge} \qquad \text{repeat of (7-55)}$$

into (7-46) to get $\mathcal{L}$ in terms of σ and v. That form of $\mathcal{L}$ will be the effective Lagrangian for us in our present epoch, where we consider the Higgs field to be σ, instead of Φ.

We will also substitute the Weinberg mixing relations (7-43), and the W mixing relations (7-44),

$$B_\mu = A_\mu \cos\theta_W - Z_\mu \sin\theta_W \qquad W_{3\mu} = A_\mu \sin\theta_W + Z_\mu \cos\theta_W , \qquad \text{repeat of (7-43)}$$

$$W_{1\mu}=\frac{W_\mu^+ + W_\mu^-}{\sqrt{2}} \quad W_{2\mu}=i\frac{W_\mu^+ - W_\mu^-}{\sqrt{2}} \quad \text{or} \quad W_\mu^+=\frac{W_{1\mu}-iW_{2\mu}}{\sqrt{2}} \quad W_\mu^-=\frac{W_{1\mu}+iW_{2\mu}}{\sqrt{2}} , \qquad \text{repeat of (7-44)}$$

W_μ^+ sometimes symbolized by W_μ, with W_μ^- symbolized by $W_\mu^\dagger$.

so $\mathcal{L}$ is expressed entirely in terms of the relevant spin 1 fields (Ws, Z, photon) for the current epoch.

7.6.1 $\mathcal{L}^L$ in Terms of True Vacuum Fields

In Chap. 6 ((6-52) and pgs. 182-187), we looked at the lepton part of the Lagrangian (in this chapter, the first row of (7-46)),

$$\mathcal{L}^L = i\left(\bar{\Psi}_l^L \not{D}\Psi_l^L + \bar{\psi}_l^R \not{D}\psi_l^R + \bar{\psi}_{\nu_l}^R \not{D}\psi_{\nu_l}^R\right), \tag{7-152}$$

and after substituting for the covariant derivative, found interaction terms in $\mathcal{L}$. Each individual term is represented, in a Feynman diagram, as a vertex. We repeat that result first, then insert (7-43) and (7-44) for our final, present day, $\mathcal{L}^L$.

From (7-152), along with (7-48),

$$\mathcal{L}^L = \underbrace{i\bar{\psi}_l^R \not{\partial}\psi_l^R + g'\bar{\psi}_l^R \not{B}\psi_l^R}_{i\bar{\psi}_l^R \not{D}\psi_l^R} + \underbrace{i\bar{\psi}_{\nu_l}^R \not{\partial}\psi_{\nu_l}^R}_{i\bar{\psi}_{\nu_l}^R \not{D}\psi_{\nu_l}^R} + \underbrace{i\bar{\Psi}_l^L \not{\partial}\Psi_l^L - \frac{g}{2}\bar{\Psi}_l^L \not{W}_j \sigma_j \Psi_l^L + \frac{g'}{2}\bar{\Psi}_l^L \not{B}\Psi_l^L}_{i\bar{\Psi}_l^L \not{D}\Psi_l^L} \tag{7-153}$$

Now, with (7-43) and (7-44),

$$\begin{aligned}\mathcal{L}^L = {} & i\bar{\psi}_l^R \not{\partial}\psi_l^R + i\bar{\psi}_{\nu_l}^R \not{\partial}\psi_{\nu_l}^R + g'\bar{\psi}_l^R\left(\not{A}\cos\theta_W - \not{Z}\sin\theta_W\right)\psi_l^R + i\bar{\Psi}_l^L \not{\partial}\Psi_l^L \\ & \underbrace{-\frac{g}{2}\bar{\Psi}_l^L \not{W}_1\begin{bmatrix}0&1\\1&0\end{bmatrix}\Psi_l^L - \frac{g}{2}\bar{\Psi}_l^L \not{W}_2\begin{bmatrix}0&-i\\i&0\end{bmatrix}\Psi_l^L}_{-\frac{g}{2}\bar{\Psi}_l^L\begin{bmatrix}0 & \not{W}_1 - i\not{W}_2\\ \not{W}_1 + i\not{W}_2 & 0\end{bmatrix}\Psi_l^L} - \frac{g}{2}\bar{\Psi}_l^L\left(\not{A}\sin\theta_W + \not{Z}\cos\theta_W\right)\begin{bmatrix}1 & \\ & -1\end{bmatrix}\Psi_l^L \\ & + \frac{g'}{2}\bar{\Psi}_l^L\left(\not{A}\cos\theta_W - \not{Z}\sin\theta_W\right)\Psi_l^L\end{aligned} \tag{7-154}$$

$$\begin{aligned}\mathcal{L}^L = {} & i\bar{\psi}_l^R \not{\partial}\psi_l^R + i\bar{\psi}_{\nu_l}^R \not{\partial}\psi_{\nu_l}^R + g'\bar{\psi}_l^R\left(\not{A}\cos\theta_W - \not{Z}\sin\theta_W\right)\psi_l^R + i\begin{bmatrix}\bar{\psi}_{\nu_l}^L & \bar{\psi}_l^L\end{bmatrix}\not{\partial}\begin{bmatrix}\psi_{\nu_l}^L\\ \psi_l^L\end{bmatrix} \\ & -\frac{g}{\sqrt{2}}\begin{bmatrix}\bar{\psi}_{\nu_l}^L & \bar{\psi}_l^L\end{bmatrix}\begin{bmatrix}0 & \not{W}^+\\ \not{W}^- & 0\end{bmatrix}\begin{bmatrix}\psi_{\nu_l}^L\\ \psi_l^L\end{bmatrix} - \frac{g}{2}\begin{bmatrix}\bar{\psi}_{\nu_l}^L & \bar{\psi}_l^L\end{bmatrix}\left(\not{A}\sin\theta_W + \not{Z}\cos\theta_W\right)\begin{bmatrix}1 & \\ & -1\end{bmatrix}\begin{bmatrix}\psi_{\nu_l}^L\\ \psi_l^L\end{bmatrix} \\ & + \frac{g'}{2}\begin{bmatrix}\bar{\psi}_{\nu_l}^L & \bar{\psi}_l^L\end{bmatrix}\left(\not{A}\cos\theta_W - \not{Z}\sin\theta_W\right)\begin{bmatrix}\psi_{\nu_l}^L\\ \psi_l^L\end{bmatrix}.\end{aligned} \tag{7-155}$$

It is straightforward to multiply all the factors in the terms of (7-155) to get

$$\begin{aligned}\mathcal{L}^L = & \, i\bar{\psi}_l^R \not\partial \psi_l^R + i\bar{\psi}_{\nu_l}^R \not\partial \psi_{\nu_l}^R + g' \cos\theta_W \bar{\psi}_l^R \not{A} \psi_l^R - g' \sin\theta_W \bar{\psi}_l^R \not{Z} \psi_l^R + i\bar{\psi}_{\nu_l}^L \not\partial \psi_{\nu_l}^L + i\bar{\psi}_l^L \not\partial \psi_l^L \\ & - \tfrac{g}{\sqrt{2}} \bar{\psi}_{\nu_l}^L \not{W}^+ \psi_l^L - \tfrac{g}{\sqrt{2}} \bar{\psi}_l^L \not{W}^- \psi_{\nu_l}^L - \tfrac{g}{2} \bar{\psi}_{\nu_l}^L \left(\not{A} \sin\theta_W + \not{Z} \cos\theta_W\right) \psi_{\nu_l}^L \\ & \qquad\qquad + \tfrac{g}{2} \bar{\psi}_l^L \left(\not{A} \sin\theta_W + \not{Z} \cos\theta_W\right) \psi_l^L \\ & + \frac{g'}{2} \bar{\psi}_{\nu_l}^L \left(\not{A} \cos\theta_W - \not{Z} \sin\theta_W\right) \psi_{\nu_l}^L + \frac{g'}{2} \bar{\psi}_l^L \left(\not{A} \cos\theta_W - \not{Z} \sin\theta_W\right) \psi_l^L .\end{aligned} \tag{7-156}$$

Or finally, to separate out every term that will act as a vertex,

$$\begin{aligned}\mathcal{L}^L = & \left. i\bar{\psi}_l^R \not\partial \psi_l^R + i\bar{\psi}_{\nu_l}^R \not\partial \psi_{\nu_l}^R + i\bar{\psi}_l^L \not\partial \psi_l^L + i\bar{\psi}_{\nu_l}^L \not\partial \psi_{\nu_l}^L \right\} \quad \text{free} \\ & \left. \overbrace{+g' \cos\theta_W \bar{\psi}_l^R \not{A} \psi_l^R}^{\boxed{A}} \; \overbrace{-g' \sin\theta_W \bar{\psi}_l^R \not{Z} \psi_l^R}^{\boxed{B}} \right\} \quad \text{RC interacting} \\ & \left. \begin{aligned} & - \tfrac{g}{\sqrt{2}} \bar{\psi}_{\nu_l}^L \not{W}^+ \psi_l^L - \tfrac{g}{\sqrt{2}} \bar{\psi}_l^L \not{W}^- \psi_{\nu_l}^L \overbrace{- \frac{g \sin\theta_W}{2} \bar{\psi}_{\nu_l}^L \not{A} \psi_{\nu_l}^L}^{\boxed{C}} \overbrace{- \frac{g \cos\theta_W}{2} \bar{\psi}_{\nu_l}^L \not{Z} \psi_{\nu_l}^L}^{\boxed{D}} \\ & \qquad \overbrace{+ \frac{g \sin\theta_W}{2} \bar{\psi}_l^L \not{A} \psi_l^L}^{\boxed{E}} \overbrace{+ \frac{g \cos\theta_W}{2} \bar{\psi}_l^L \not{Z} \psi_l^L}^{\boxed{F}} \\ & \overbrace{+ \frac{g' \cos\theta_W}{2} \bar{\psi}_l^L \not{A} \psi_l^L}^{\boxed{G}} \overbrace{- \frac{g' \sin\theta_W}{2} \bar{\psi}_l^L \not{Z} \psi_l^L}^{\boxed{H}} \\ & \qquad \overbrace{+ \frac{g' \cos\theta_W}{2} \bar{\psi}_{\nu_l}^L \not{A} \psi_{\nu_l}^L}^{\boxed{J}} \overbrace{- \frac{g' \sin\theta_W}{2} \bar{\psi}_{\nu_l}^L \not{Z} \psi_{\nu_l}^L}^{\boxed{K}} \end{aligned} \right\} \quad \text{LC interacting} .\end{aligned} \tag{7-157}$$

Note that the terms labeled $\boxed{C}$ and $\boxed{J}$ above reflect interaction between neutrinos and photons, i.e., neutrinos feeling the electromagnetic interaction. But, we know they don't, so those terms have to cancel if we are to have a valid theory. They cancel IF, and only if,

$$g \sin\theta_W = g' \cos\theta_W . \tag{7-158}$$

Thus, we take (7-158) as a valid relationship between g, g', and θ_W, i.e, as a valid constraint on the theory. And so,

$$\boxed{C} + \boxed{J} = 0 . \tag{7-159}$$

The forms of $\boxed{A}$, $\boxed{E}$, and $\boxed{G}$ will add to give us

$$\begin{aligned} & \overbrace{+g' \cos\theta_W \bar{\psi}_l^R \not{A} \psi_l^R}^{\boxed{A}} \; \overbrace{+ \frac{g \sin\theta_W}{2} \bar{\psi}_l^L \not{A} \psi_l^L}^{\boxed{E}} \; \overbrace{+ \frac{g' \cos\theta_W}{2} \bar{\psi}_l^L \not{A} \psi_l^L}^{\boxed{G}} \\ & = g \sin\theta_W \bar{\psi}_l^R \not{A} \psi_l^R + g \sin\theta_W \bar{\psi}_l^L \not{A} \psi_l^L = g \sin\theta_W \left(\bar{\psi}_l^R \not{A} \psi_l^R + \bar{\psi}_l^L \not{A} \psi_l^L \right). \end{aligned} \tag{7-160}$$

From (7-42), (7-160) becomes

$$\boxed{A} + \boxed{E} + \boxed{G} = g \sin\theta_W \bar{\psi}_l \not{A} \psi_l . \tag{7-161}$$

Since the interaction term in the QED Lagrangian (between electron fields and photon fields) is

$$\mathcal{L}_{QED} = e \bar{\psi}_l \not{A} \psi_l , \tag{7-162}$$

(7-161) can only be a valid low energy term in the Lagrangian if (where the last part comes from (7-158))

$$\boxed{e = g \sin\theta_W = g' \cos\theta_W .} \tag{7-163}$$

(7-163) is a fundamental relationship that is used often in e/w theory. It can help to commit it to memory.

Thus, (7-161), in its most streamlined and relevant form, is

$$\boxed{A}+\boxed{E}+\boxed{G}=e\bar{\psi}_l A\!\!\!/\psi_l . \tag{7-164}$$

Now look at $\boxed{D}$ and $\boxed{K}$.

$$\begin{aligned}
\boxed{D}+\boxed{K} &= -\frac{g\cos\theta_W}{2}\bar{\psi}^L_{\nu_l}Z\!\!\!/\psi^L_{\nu_l}-\frac{g'\sin\theta_W}{2}\bar{\psi}^L_{\nu_l}Z\!\!\!/\psi^L_{\nu_l}=-\left(\frac{g\cos\theta_W}{2}+\frac{g'\sin\theta_W}{2}\right)\bar{\psi}^L_{\nu_l}Z\!\!\!/\psi^L_{\nu_l}\\
&=-\left(\frac{g\cos\theta_W}{2}+\frac{e}{\cos\theta_W}\frac{\sin\theta_W}{2}\right)\bar{\psi}^L_{\nu_l}Z\!\!\!/\psi^L_{\nu_l}=-\left(\frac{g\cos\theta_W}{2}+\frac{g\sin\theta_W}{\cos\theta_W}\frac{\sin\theta_W}{2}\right)\bar{\psi}^L_{\nu_l}Z\!\!\!/\psi^L_{\nu_l}\\
&=-\frac{g}{2\cos\theta_W}\left(\cos^2\theta_W+\sin^2\theta_W\right)\bar{\psi}^L_{\nu_l}Z\!\!\!/\psi^L_{\nu_l}=-\frac{g}{2\cos\theta_W}\bar{\psi}^L_{\nu_l}Z\!\!\!/\psi^L_{\nu_l}\\
&=-\frac{g}{2\cos\theta_W}\bar{\psi}_{\nu_l}Z\!\!\!/\tfrac{1}{2}\left(1-\gamma^5\right)\psi_{\nu_l}=-\frac{g}{4\cos\theta_W}\bar{\psi}_{\nu_l}Z\!\!\!/\left(1-\gamma^5\right)\psi_{\nu_l}.
\end{aligned} \tag{7-165}$$

Finally, $\boxed{B}$, $\boxed{F}$, and $\boxed{H}$ give us (where three lines from the end, we use (5-86) of Chap. 5)

$$\begin{aligned}
\boxed{B}+\boxed{F}+\boxed{H} &= -g'\sin\theta_W\bar{\psi}^R_l Z\!\!\!/\psi^R_l+\frac{g\cos\theta_W}{2}\bar{\psi}^L_l Z\!\!\!/\psi^L_l-\frac{g'\sin\theta_W}{2}\bar{\psi}^L_l Z\!\!\!/\psi^L_l\\
&=-\frac{g\sin\theta_W}{\cos\theta_W}\sin\theta_W\bar{\psi}^R_l Z\!\!\!/\psi^R_l+\left(\frac{g\cos\theta_W}{2}-\frac{g\sin\theta_W}{\cos\theta_W}\frac{\sin\theta_W}{2}\right)\bar{\psi}^L_l Z\!\!\!/\psi^L_l\\
&=-\frac{g}{\cos\theta_W}\sin^2\theta_W\bar{\psi}^R_l Z\!\!\!/\psi^R_l+\frac{1}{2}\frac{g}{\cos\theta_W}\left(\cos^2\theta_W-\sin^2\theta_W\right)\bar{\psi}^L_l Z\!\!\!/\psi^L_l\\
&=-\frac{g}{\cos\theta_W}\sin^2\theta_W\bar{\psi}^R_l Z\!\!\!/\psi^R_l+\frac{1}{2}\frac{g}{\cos\theta_W}\left(1-2\sin^2\theta_W\right)\bar{\psi}^L_l Z\!\!\!/\psi^L_l\\
&=-\frac{g}{\cos\theta_W}\sin^2\theta_W\left(\bar{\psi}^R_l Z\!\!\!/\psi^R_l+\bar{\psi}^L_l Z\!\!\!/\psi^L_l\right)+\frac{1}{2}\frac{g}{\cos\theta_W}\bar{\psi}^L_l Z\!\!\!/\psi^L_l\\
&=-\frac{g}{\cos\theta_W}\sin^2\theta_W\bar{\psi}_l Z\!\!\!/\psi_l+\frac{1}{2}\frac{g}{\cos\theta_W}\bar{\psi}^L_l Z\!\!\!/\psi^L_l\\
&=-\frac{g}{\cos\theta_W}\sin^2\theta_W\bar{\psi}_l Z\!\!\!/\psi_l+\frac{1}{2}\frac{g}{\cos\theta_W}\bar{\psi}_l Z\!\!\!/\tfrac{1}{2}\left(1-\gamma^5\right)\psi_l\\
&=\frac{g}{4\cos\theta_W}\bar{\psi}_l Z\!\!\!/\left(1-4\sin^2\theta_W-\gamma^5\right)\psi_l.
\end{aligned} \tag{7-166}$$

Result of this section

Thus, (7-157), with (7-159), (7-164), (7-165), and (7-166), becomes

$$\underbrace{\mathcal{L}^L}_{\substack{\text{false}\\ \text{vacuum}\\ \text{fields}}}\xrightarrow[\text{vacuum fields}]{\text{substitute true}}\underbrace{\overbrace{i\bar{\psi}^R_l\partial\!\!\!/\psi^R_l+i\bar{\psi}^L_l\partial\!\!\!/\psi^L_l}^{i\bar{\psi}_l\partial\!\!\!/\psi_l}+\overbrace{i\bar{\psi}^R_{\nu_l}\partial\!\!\!/\psi^R_{\nu_l}+i\bar{\psi}^L_{\nu_l}\partial\!\!\!/\psi^L_{\nu_l}}^{i\bar{\psi}_{\nu_l}\partial\!\!\!/\psi_{\nu_l}}}_{\text{free field, kinetic terms (in }\mathcal{L}^L_0)}\quad\underbrace{+e\bar{\psi}_l A\!\!\!/\psi_l}_{\substack{\text{QED interaction}\\ \text{term (in }\mathcal{L}^{LB})}}$$

$$\underbrace{-\frac{g}{2\sqrt{2}}\bar{\psi}_{\nu_l}W\!\!\!/^{+}\left(1-\gamma^5\right)\psi_l-\frac{g}{2\sqrt{2}}\bar{\psi}_l W\!\!\!/^{-}\left(1-\gamma^5\right)\psi_{\nu_l}}_{\text{interaction terms (in }\mathcal{L}^{LB})}\quad\underbrace{-\frac{g}{4\cos\theta_W}\bar{\psi}_{\nu_l}Z\!\!\!/\left(1-\gamma^5\right)\psi_{\nu_l}}_{\text{interaction term (in }\mathcal{L}^{LB})} \tag{7-167}$$

$$\underbrace{+\frac{g}{4\cos\theta_W}\bar{\psi}_l Z\!\!\!/\left(1-4\sin^2\theta_W-\gamma^5\right)\psi_l .}_{\text{iinteraction terms (in }\mathcal{L}^{LB})}$$

7.6.2 $\mathcal{L}^B$ in Terms of True Vacuum Fields

From the second row of (7-46),

$$\mathcal{L}^B = -\tfrac{1}{4}G_i^{\mu\nu}G_{i\,\mu\nu} - \tfrac{1}{4}B^{\mu\nu}B_{\mu\nu}\,. \tag{7-168}$$

Weak Isospin Boson Terms $-\tfrac{1}{4}G_i^{\mu\nu}G_{i\,\mu\nu}$

If you did Prob. 4 in Chap. 6 (or if you looked at the answer in the solutions booklet) you found, for the first term after the equal sign in (7-168),

$$-\tfrac{1}{4}G_i^{\mu\nu}G_{i\,\mu\nu} = \overbrace{\underbrace{-\tfrac{1}{4}\left(\partial^\nu W_i^\mu - \partial^\mu W_i^\nu\right)\left(\partial_\nu W_{i\,\mu} - \partial_\mu W_{i\nu}\right)}_{\text{free field, kinetic terms }\left(\text{in } \mathcal{L}_0^B\right)}}^{-\frac{1}{4}W_i^{\mu\nu}W_{i\,\mu\nu}} \underbrace{+ g\varepsilon_{ijk}W_{i\mu}W_{j\nu}\partial^\mu W_k^\nu - \tfrac{1}{4}g^2\varepsilon_{ijk}\varepsilon_{ilm}W_j^\mu W_k^\nu W_{l\mu}W_{m\nu}}_{\text{interaction terms }\left(\text{in } \mathcal{L}^{BB}\right)}\,. \tag{7-169}$$

Free Field, Kinetic Terms for W^+ and W^- Fields from (7-169)

$$\begin{aligned}
-\tfrac{1}{4}W_i^{\mu\nu}W_{i\,\mu\nu} &= -\tfrac{1}{4}\left(\partial^\nu W_i^\mu - \partial^\mu W_i^\nu\right)\left(\partial_\nu W_{i\,\mu} - \partial_\mu W_{i\nu}\right) \\
&= -\tfrac{1}{4}\left(\partial^\nu W_i^\mu\right)\partial_\nu W_{i\,\mu} + \tfrac{1}{4}\left(\partial^\nu W_i^\mu\right)\partial_\mu W_{i\nu} + \underbrace{\tfrac{1}{4}\left(\partial^\mu W_i^\nu\right)\partial_\nu W_{i\,\mu}}_{\frac{1}{4}\left(\partial^\nu W_i^\mu\right)\partial_\mu W_{i\nu}} \underbrace{- \tfrac{1}{4}\left(\partial^\mu W_i^\nu\right)\partial_\mu W_{i\nu}}_{-\frac{1}{4}\left(\partial^\nu W_i^\mu\right)\partial_\nu W_{i\,\mu}} \\
&= -\tfrac{1}{2}\left(\partial^\nu W_i^\mu\right)\partial_\nu W_{i\,\mu} + \tfrac{1}{2}\left(\partial^\nu W_i^\mu\right)\partial_\mu W_{i\nu} \\
&= -\tfrac{1}{2}\left(\partial^\nu W_1^\mu\right)\partial_\nu W_{1\mu} + \tfrac{1}{2}\left(\partial^\nu W_1^\mu\right)\partial_\mu W_{1\nu} - \tfrac{1}{2}\left(\partial^\nu W_2^\mu\right)\partial_\nu W_{2\,\mu} + \tfrac{1}{2}\left(\partial^\nu W_2^\mu\right)\partial_\mu W_{2\nu} \\
&\quad - \tfrac{1}{2}\left(\partial^\nu W_3^\mu\right)\partial_\nu W_{3\mu} + \tfrac{1}{2}\left(\partial^\nu W_3^\mu\right)\partial_\mu W_{3\nu}\,.
\end{aligned} \tag{7-170}$$

Then using the W field expressions of (7-43) and (7-44) in (7-170), we find it is

$$\begin{aligned}
&= \underbrace{-\tfrac{1}{2}\left\{\partial^\nu\left(\frac{W^{+\mu}+W^{-\mu}}{\sqrt{2}}\right)\right\}\left\{\partial_\nu\left(\frac{W_\mu^+ + W_\mu^-}{\sqrt{2}}\right)\right\}}_{\boxed{A}} \underbrace{+\tfrac{1}{2}\left\{\partial^\nu\left(\frac{W^{+\mu}+W^{-\mu}}{\sqrt{2}}\right)\right\}\left\{\partial_\mu\left(\frac{W_\nu^+ + W_\nu^-}{\sqrt{2}}\right)\right\}}_{\boxed{B}} \\
&\quad \underbrace{+\tfrac{1}{2}\left\{\partial^\nu\left(\frac{W^{+\mu}-W^{-\mu}}{\sqrt{2}}\right)\right\}\left\{\partial_\nu\left(\frac{W_\mu^+ - W_\mu^-}{\sqrt{2}}\right)\right\}}_{\boxed{C}} \underbrace{-\tfrac{1}{2}\left\{\partial^\nu\left(\frac{W^{+\mu}-W^{-\mu}}{\sqrt{2}}\right)\right\}\left\{\partial_\mu\left(\frac{W_\nu^+ - W_\nu^-}{\sqrt{2}}\right)\right\}}_{\boxed{D}} \\
&\quad \underbrace{-\tfrac{1}{2}\left\{\left(\partial^\nu\left(A^\mu \sin\theta_W + Z^\mu\cos\theta_W\right)\right)\left(\partial_\nu\left(A_\mu\sin\theta_W + Z_\mu\cos\theta_W\right)\right)\right\}}_{\boxed{E}} \\
&\quad \underbrace{+\tfrac{1}{2}\left\{\left(\partial^\nu\left(A^\mu \sin\theta_W + Z^\mu\cos\theta_W\right)\right)\left(\partial_\mu\left(A_\nu\sin\theta_W + Z_\nu\cos\theta_W\right)\right)\right\}}_{\boxed{F}}.
\end{aligned} \tag{7-171}$$

Note that

$$\boxed{A}+\boxed{B}+\boxed{C}+\boxed{D} = -\tfrac{1}{4}W_1^{\mu\nu}W_{1\mu\nu} - \tfrac{1}{4}W_2^{\mu\nu}W_{2\,\mu\nu}, \tag{7-172}$$

and

$$\boxed{E}+\boxed{F} = -\tfrac{1}{4}W_3^{\mu\nu}W_{3\mu\nu}\,. \tag{7-173}$$

Expanding $\boxed{A}$ and $\boxed{C}$, we find

$$\boxed{A}+\boxed{C}$$

$$
\begin{aligned}
&=-\tfrac{1}{2}\left\{\partial^{\nu}\left(\frac{W^{+\mu}+W^{-\mu}}{\sqrt{2}}\right)\right\}\left\{\partial_{\nu}\left(\frac{W_{\mu}^{+}+W_{\mu}^{-}}{\sqrt{2}}\right)\right\}+\tfrac{1}{2}\left\{\partial^{\nu}\left(\frac{W^{+\mu}-W^{-\mu}}{\sqrt{2}}\right)\right\}\left\{\partial_{\nu}\left(\frac{W_{\mu}^{+}-W_{\mu}^{-}}{\sqrt{2}}\right)\right\}\\
&=-\tfrac{1}{4}\left\{\partial^{\nu}\left(W^{+\mu}+W^{-\mu}\right)\right\}\left\{\partial_{\nu}\left(W_{\mu}^{+}+W_{\mu}^{-}\right)\right\}+\tfrac{1}{4}\left\{\partial^{\nu}\left(W^{+\mu}-W^{-\mu}\right)\right\}\left\{\partial_{\nu}\left(W_{\mu}^{+}-W_{\mu}^{-}\right)\right\}\\
&=\underbrace{-\tfrac{1}{4}\left(\partial^{\nu}W^{+\mu}\right)\partial_{\nu}W_{\mu}^{+}}_{\text{cancels term below}}\underbrace{-\tfrac{1}{4}\left(\partial^{\nu}W^{+\mu}\right)\partial_{\nu}W_{\mu}^{-}-\tfrac{1}{4}\left(\partial^{\nu}W^{-\mu}\right)\partial_{\nu}W_{\mu}^{+}}_{\text{indentical to terms below}}\underbrace{-\tfrac{1}{4}\left(\partial^{\nu}W^{-\mu}\right)\partial_{\nu}W_{\mu}^{-}}_{\text{cancels term below}}\\
&\quad+\tfrac{1}{4}\left(\partial^{\nu}W^{+\mu}\right)\partial_{\nu}W_{\mu}^{+}-\tfrac{1}{4}\left(\partial^{\nu}W^{+\mu}\right)\partial_{\nu}W_{\mu}^{-}-\tfrac{1}{4}\left(\partial^{\nu}W^{-\mu}\right)\partial_{\nu}W_{\mu}^{+}+\tfrac{1}{4}\left(\partial^{\nu}W^{-\mu}\right)\partial_{\nu}W_{\mu}^{-}\\
&=-\tfrac{1}{2}\left(\partial^{\nu}W^{+\mu}\right)\partial_{\nu}W_{\mu}^{-}-\tfrac{1}{2}\left(\partial^{\nu}W^{-\mu}\right)\partial_{\nu}W_{\mu}^{+}.
\end{aligned}
\tag{7-174}
$$

In similar fashion,

$$\boxed{B}+\boxed{D}$$

$$
\begin{aligned}
&=\tfrac{1}{2}\left\{\partial^{\nu}\left(\frac{W^{+\mu}+W^{-\mu}}{\sqrt{2}}\right)\right\}\left\{\partial_{\mu}\left(\frac{W_{\nu}^{+}+W_{\nu}^{-}}{\sqrt{2}}\right)\right\}-\tfrac{1}{2}\left\{\partial^{\nu}\left(\frac{W^{+\mu}-W^{-\mu}}{\sqrt{2}}\right)\right\}\left\{\partial_{\mu}\left(\frac{W_{\nu}^{+}-W_{\nu}^{-}}{\sqrt{2}}\right)\right\}\\
&=\tfrac{1}{4}\left(\partial^{\nu}W^{+\mu}+\partial^{\nu}W^{-\mu}\right)\left(\partial_{\mu}W_{\nu}^{+}+\partial_{\mu}W_{\nu}^{-}\right)-\tfrac{1}{4}\left(\partial^{\nu}W^{+\mu}-\partial^{\nu}W^{-\mu}\right)\left(\partial_{\mu}W_{\nu}^{+}-\partial_{\mu}W_{\nu}^{-}\right)\\
&=\tfrac{1}{4}\left(\partial^{\nu}W^{+\mu}\right)\partial_{\mu}W_{\nu}^{+}+\tfrac{1}{4}\left(\partial^{\nu}W^{+\mu}\right)\partial_{\mu}W_{\nu}^{-}+\tfrac{1}{4}\left(\partial^{\nu}W^{-\mu}\right)\partial_{\mu}W_{\nu}^{+}+\tfrac{1}{4}\left(\partial^{\nu}W^{-\mu}\right)\partial_{\mu}W_{\nu}^{-}\\
&\quad-\tfrac{1}{4}\left(\partial^{\nu}W^{+\mu}\right)\partial_{\mu}W_{\nu}^{+}+\tfrac{1}{4}\left(\partial^{\nu}W^{+\mu}\right)\partial_{\mu}W_{\nu}^{-}+\tfrac{1}{4}\left(\partial^{\nu}W^{-\mu}\right)\partial_{\mu}W_{\nu}^{+}-\tfrac{1}{4}\left(\partial^{\nu}W^{-\mu}\right)\partial_{\mu}W_{\nu}^{-}\\
&=\tfrac{1}{2}\left(\partial^{\nu}W^{+\mu}\right)\partial_{\mu}W_{\nu}^{-}+\tfrac{1}{2}\left(\partial^{\nu}W^{-\mu}\right)\partial_{\mu}W_{\nu}^{+}.
\end{aligned}
\tag{7-175}
$$

From (7-174) and (7-175)

$$
\begin{aligned}
&\underbrace{\boxed{A}+\boxed{C}+\boxed{B}+\boxed{D}}_{-\frac{1}{4}W_{1}^{\mu\nu}W_{1\mu\nu}-\frac{1}{4}W_{2}^{\mu\nu}W_{2\mu\nu}}\\
&\quad=-\tfrac{1}{2}\left(\partial^{\nu}W^{+\mu}\right)\partial_{\nu}W_{\mu}^{-}-\tfrac{1}{2}\left(\partial^{\nu}W^{-\mu}\right)\partial_{\nu}W_{\mu}^{+}+\tfrac{1}{2}\left(\partial^{\nu}W^{+\mu}\right)\partial_{\mu}W_{\nu}^{-}+\tfrac{1}{2}\left(\partial^{\nu}W^{-\mu}\right)\partial_{\mu}W_{\nu}^{+},
\end{aligned}
\tag{7-176}
$$

With an eye to what we are about to do, we move the second term to the end, and then switch some dummy variables, to obtain

$$
\begin{aligned}
&=-\tfrac{1}{2}\left(\partial^{\nu}W^{+\mu}\right)\partial_{\nu}W_{\mu}^{-}+\underbrace{\tfrac{1}{2}\left(\partial^{\nu}W^{+\mu}\right)\partial_{\mu}W_{\nu}^{-}}+\underbrace{\tfrac{1}{2}\left(\partial^{\nu}W^{-\mu}\right)\partial_{\mu}W_{\nu}^{+}}\underbrace{-\tfrac{1}{2}\left(\partial^{\nu}W^{-\mu}\right)\partial_{\nu}W_{\mu}^{+}}\\
&=-\tfrac{1}{2}\left(\partial^{\nu}W^{+\mu}\right)\partial_{\nu}W_{\mu}^{-}+\tfrac{1}{2}\left(\partial^{\mu}W^{+\nu}\right)\partial_{\nu}W_{\mu}^{-}+\tfrac{1}{2}\left(\partial_{\mu}W_{\nu}^{-}\right)\partial^{\nu}W^{+\mu}-\tfrac{1}{2}\left(\partial_{\mu}W_{\nu}^{-}\right)\partial^{\mu}W^{+\nu}.
\end{aligned}
\tag{7-177}
$$

The two different W fields commute, as all different boson fields do, to leave us with

$$=-\tfrac{1}{2}\left(\left(\partial_{\nu}W_{\mu}^{-}\right)\partial^{\nu}W^{+\mu}-\left(\partial_{\nu}W_{\mu}^{-}\right)\partial^{\mu}W^{+\nu}-\left(\partial_{\mu}W_{\nu}^{-}\right)\partial^{\nu}W^{+\mu}+\left(\partial_{\mu}W_{\nu}^{-}\right)\partial^{\mu}W^{+\nu}\right). \tag{7-178}$$

Now consider the quantity defined as

$$W^{\mu\nu}=\partial^{\nu}W^{+\mu}-\partial^{\mu}W^{+\nu}, \tag{7-179}$$

so,

$$
\begin{aligned}
-\tfrac{1}{2}W_{\mu\nu}^{\dagger}W^{\mu\nu}&=-\tfrac{1}{2}\left(\partial_{\nu}W_{\mu}^{-}-\partial_{\mu}W_{\nu}^{-}\right)\left(\partial^{\nu}W^{+\mu}-\partial^{\mu}W^{+\nu}\right)\\
&=-\tfrac{1}{2}\left(\left(\partial_{\nu}W_{\mu}^{-}\right)\partial^{\nu}W^{+\mu}-\left(\partial_{\nu}W_{\mu}^{-}\right)\partial^{\mu}W^{+\nu}-\left(\partial_{\mu}W_{\nu}^{-}\right)\partial^{\nu}W^{+\mu}+\left(\partial_{\mu}W_{\nu}^{-}\right)\partial^{\mu}W^{+\nu}\right).
\end{aligned}
\tag{7-180}
$$

Comparing (7-180) to (7-178), we see that in (7-171)

$$\boxed{A}+\boxed{C}+\boxed{B}+\boxed{D}=-\tfrac{1}{4}W_1^{\mu\nu}W_{1\mu\nu}-\tfrac{1}{4}W_2^{\mu\nu}W_{2\mu\nu}=\underbrace{\overbrace{-\tfrac{1}{2}\left(\partial_\nu W_\mu^- -\partial_\mu W_\nu^-\right)\left(\partial^\nu W^{+\mu}-\partial^\mu W^{+\nu}\right)}^{-\frac{1}{2}W^\dagger_{\mu\nu}W^{\mu\nu}}}_{\text{free, kinetic term in } \mathcal{L}_0^B}. \quad (7\text{-}181)$$

Free Field, Kinetic Terms for Z and Photon Fields from (7-168) and (7-169)

The last term in (7-168), expanded below using (7-43) as (7-182), and the terms $\boxed{E}$ and $\boxed{F}$ in (7-171) all have true vacuum fields Z^μ and A^μ in them and so, need to be considered together.

$$\begin{aligned}-\tfrac{1}{4}B^{\mu\nu}B_{\mu\nu}&=-\tfrac{1}{4}\left(\partial^\nu B^\mu-\partial^\mu B^\nu\right)\left(\partial_\nu B_\mu-\partial_\mu B_\nu\right)\\&=-\tfrac{1}{4}\left(\partial^\nu B^\mu\right)\partial_\nu B_\mu+\tfrac{1}{4}\left(\partial^\nu B^\mu\right)\partial_\mu B_\nu+\tfrac{1}{4}\left(\partial^\mu B^\nu\right)\partial_\nu B_\mu-\tfrac{1}{4}\left(\partial^\mu B^\nu\right)\partial_\mu B_\nu\\&=\underbrace{-\tfrac{1}{4}\left(\partial^\nu\left(A^\mu\cos\theta_W-Z^\mu\sin\theta_W\right)\right)\partial_\nu\left(A_\mu\cos\theta_W-Z_\mu\sin\theta_W\right)}_{\boxed{G}}\\&\quad\underbrace{+\tfrac{1}{4}\left(\partial^\nu\left(A^\mu\cos\theta_W-Z^\mu\sin\theta_W\right)\right)\partial_\mu\left(A_\nu\cos\theta_W-Z_\nu\sin\theta_W\right)}_{\boxed{H}}\\&\quad\underbrace{+\tfrac{1}{4}\left(\partial^\mu\left(A^\nu\cos\theta_W-Z^\nu\sin\theta_W\right)\right)\partial_\nu\left(A_\mu\cos\theta_W-Z_\mu\sin\theta_W\right)}_{\boxed{J}}\\&\quad\underbrace{-\tfrac{1}{4}\left(\partial^\mu\left(A^\nu\cos\theta_W-Z^\nu\sin\theta_W\right)\right)\partial_\mu\left(A_\nu\cos\theta_W-Z_\nu\sin\theta_W\right)}_{\boxed{K}}.\end{aligned}\quad (7\text{-}182)$$

Do **Problem 5** to show (7-183).

If you did Problem 5, you found, where the quantities in boxes refer to (7-171) and (7-182),

$$\begin{aligned}\boxed{E}+\boxed{F}+\boxed{G}+\boxed{H}+\boxed{J}+\boxed{K}&=-\tfrac{1}{4}W_3^{\mu\nu}W_{3\mu\nu}-\tfrac{1}{4}B^{\mu\nu}B_{\mu\nu}\\&=\underbrace{-\tfrac{1}{4}\left(\partial_\nu A_\mu-\partial_\mu A_\nu\right)\left(\partial^\nu A^\mu-\partial^\mu A^\nu\right)}_{-\frac{1}{4}F_{\mu\nu}F^{\mu\nu}}\underbrace{-\tfrac{1}{4}\left(\partial_\nu Z_\mu-\partial_\mu Z_\nu\right)\left(\partial^\nu Z^\mu-\partial^\mu Z^\nu\right)}_{-\frac{1}{4}Z_{\mu\nu}Z^{\mu\nu}}.\end{aligned}\quad (7\text{-}183)$$

Interaction Terms in (7-169)

Plugging (7-43) and (7-44) into the interaction terms part of (7-169) represents an algebraic marathon of considerable proportion. It also represents little in the way of learning, and so, constitutes a massive investment in time and effort, for little real reward.

Additionally, the resulting (large number of) terms in the Lagrangian are not used in any but the most esoteric computations in e/w theory. We certainly will not be using them herein. They are higher order corrections in a theory where higher order corrections are far smaller, and far less impactful, than in QED.

The above two paragraphs are an attempt at justifying my reluctance to spend several pages working through algebra, even though I try very hard to explicitly write out for you all derivations, step-by-step, of QFT relationships expressed in this volume and Vol. 1. Instead, in this case, I will just write out the result, which I have lifted from another book[1], and never actually worked out myself. Perhaps, like me, you can just accept it.

That result is

---

[1] F. Mandl and G. Shaw, *Quantum Field Theory*, 2nd ed., Wiley (2010), pg. 421.

$$
\begin{aligned}
\mathcal{L}^{BB} = & ig\cos\theta_W\left\{\left(W^-_\alpha W^+_\beta - W^-_\beta W^+_\alpha\right)\partial^\alpha Z^\beta + \left(\partial_\alpha W^+_\beta - \partial_\beta W^+_\alpha\right)W^{-\beta}Z^\alpha - \left(\partial_\alpha W^-_\beta - \partial_\beta W^-_\alpha\right)W^{+\beta}Z^\alpha\right\} \\
& + ie\left\{\left(W^-_\alpha W^+_\beta - W^-_\beta W^+_\alpha\right)\partial^\alpha A^\beta + \left(\partial_\alpha W^+_\beta - \partial_\beta W^+_\alpha\right)W^{-\beta}A^\alpha - \left(\partial_\alpha W^-_\beta - \partial_\beta W^-_\alpha\right)W^{+\beta}A^\alpha\right\} \\
& + g^2\cos^2\theta_W\left(W^+_\alpha W^-_\beta Z^\alpha Z^\beta - W^+_\beta W^{-\beta}Z_\alpha Z^\alpha\right) \\
& + e^2\left(W^+_\alpha W^-_\beta A^\alpha A^\beta - W^+_\beta W^{-\beta}A_\alpha A^\alpha\right) \\
& + eg\cos\theta_W\left\{W^+_\alpha W^-_\beta\left(Z^\alpha A^\beta + A^\alpha Z^\beta\right) - 2W^+_\beta W^{-\beta}A_\alpha Z^\alpha\right\} \\
& + \tfrac{1}{2}g^2 W^-_\alpha W^+_\beta\left(W^{-\alpha}W^{+\beta} - W^{+\alpha}W^{-\beta}\right).
\end{aligned}
\qquad (7\text{-}184)
$$

Find yourself longing for the simple days of QED, with only one interaction term?

Result of this section

Thus, (7-168) with (7-169), (7-181), (7-183), and (7-184) becomes

$$
\underbrace{\mathcal{L}^B}_{\substack{\text{false}\\\text{vacuum}\\\text{fields}}} \xrightarrow[\text{vacuum fields}]{\text{substitute true}} \left.\begin{aligned}
&\overbrace{-\tfrac{1}{4}\left(\partial_\nu A_\mu - \partial_\mu A_\nu\right)\left(\partial^\nu A^\mu - \partial^\mu A^\nu\right)}^{-\frac{1}{4}F_{\mu\nu}F^{\mu\nu}}\overbrace{-\tfrac{1}{4}\left(\partial_\nu Z_\mu - \partial_\mu Z_\nu\right)\left(\partial^\nu Z^\mu - \partial^\mu Z^\nu\right)}^{-\frac{1}{4}Z_{\mu\nu}Z^{\mu\nu}} \\
&\overbrace{-\tfrac{1}{2}\left(\partial_\nu W^-_\mu - \partial_\mu W^-_\nu\right)\left(\partial^\nu W^{+\mu} - \partial^\mu W^{+\nu}\right)}^{-\frac{1}{2}W^\dagger_{\mu\nu}W^{\mu\nu}}
\end{aligned}\right\}\begin{matrix}\text{free field,}\\\text{kinetic}\\\text{terms}\\(\text{in }\mathcal{L}^B_0)\end{matrix} \quad (7\text{-}185)
$$

$$
\underbrace{+\mathcal{L}^{BB}}_{\substack{\text{21 nonlinear terms,}\\\text{each a vertex}}} .
$$

7.6.3 $\mathcal{L}^H$ in Terms of True Vacuum Fields

Relation We'll Use

From (7-43) and (7-44)

$$
\begin{aligned}
W^\mu_j W_{j\mu} = & W^\mu_1 W_{1\mu} + W^\mu_2 W_{2\mu} + W^\mu_3 W_{3\mu} = \left(\frac{W^{+\mu}+W^{-\mu}}{\sqrt{2}}\right)\left(\frac{W^+_\mu + W^-_\mu}{\sqrt{2}}\right) \\
& + \left(i\frac{W^{+\mu}-W^{-\mu}}{\sqrt{2}}\right)\left(i\frac{W^+_\mu - W^-_\mu}{\sqrt{2}}\right) + \left(A^\mu\sin\theta_W + Z^\mu\cos\theta_W\right)\left(A_\mu\sin\theta_W + Z_\mu\cos\theta_W\right) \\
= & \tfrac{1}{2}\Big(W^{+\mu}W^+_\mu + W^{+\mu}W^-_\mu + W^{-\mu}W^+_\mu + W^{-\mu}W^-_\mu \underbrace{-W^{+\mu}W^+_\mu}_{\text{cancels 1st term}} + W^{+\mu}W^-_\mu + W^{-\mu}W^+_\mu \underbrace{-W^{-\mu}W^-_\mu}_{\text{cancels 4th term}}\Big) \qquad (7\text{-}186) \\
& + A^\mu A_\mu\sin^2\theta_W + 2A^\mu Z_\mu\sin\theta_W\cos\theta_W + Z^\mu Z_\mu\cos^2\theta_W \\
= & W^{+\mu}W^-_\mu + W^{-\mu}W^+_\mu + A^\mu A_\mu\sin^2\theta_W + 2A^\mu Z_\mu\sin\theta_W\cos\theta_W + Z^\mu Z_\mu\cos^2\theta_W .
\end{aligned}
$$

Or, finally,

$$
W^\mu_j W_{j\mu} = \underbrace{2W^{-\mu}W^+_\mu}_{W^\mu_1 W_{1\mu} + W^\mu_2 W_{2\mu}} \underbrace{+ A^\mu A_\mu\sin^2\theta_W + 2A^\mu Z_\mu\sin\theta_W\cos\theta_W + Z^\mu Z_\mu\cos^2\theta_W}_{W^\mu_3 W_{3\mu}} . \qquad (7\text{-}187)
$$

We will shortly use the above expression.

Higgs Part of Lagrangian

From the third row of (7-46) (in terms of the false vacuum Higgs field Φ),

$$
\mathcal{L}^H = \left(D^\mu\Phi\right)^\dagger\left(D_\mu\Phi\right) - \mathcal{V} = \left(D^\mu\Phi\right)^\dagger\left(D_\mu\Phi\right) - \mu^2\Phi^\dagger\Phi - \lambda\left(\Phi^\dagger\Phi\right)^2 . \qquad (7\text{-}188)
$$

To see this in terms of the Higgs field σ at true vacuum, we need to substitute (7-55) into (7-188).

The Covariant Derivative Term

Let's focus first on the term with covariant derivatives, and referring to (6-61), we have

$$\left(D^\mu\Phi\right)^\dagger\left(D_\mu\Phi\right)=\left(\partial^\mu\Phi^\dagger-\tfrac{i}{2}g\Phi^\dagger\sigma_jW_j^\mu-\tfrac{i}{2}g'B^\mu\Phi^\dagger\right)\left(\partial_\mu\Phi+\tfrac{i}{2}g\sigma_kW_{k\mu}\Phi+\tfrac{i}{2}g'B_\mu\Phi\right) \quad (7\text{-}189)$$

$$\begin{aligned}&=\partial^\mu\Phi^\dagger\partial_\mu\Phi+\overbrace{\left(\partial^\mu\Phi^\dagger\right)\tfrac{i}{2}g\sigma_kW_{k\mu}\Phi}^{\boxed{1a}}+\overbrace{\left(\partial^\mu\Phi^\dagger\right)\tfrac{i}{2}g'B_\mu\Phi}^{\boxed{2a}}\\
&\quad\overbrace{-\tfrac{i}{2}g\Phi^\dagger\sigma_jW_j^\mu\partial_\mu\Phi}^{\boxed{1b}}-\tfrac{i}{2}g\Phi^\dagger\sigma_jW_j^\mu\tfrac{i}{2}g\sigma_kW_{k\mu}\Phi\overbrace{-\tfrac{i}{2}g\Phi^\dagger\sigma_jW_j^\mu\tfrac{i}{2}g'B_\mu\Phi}^{\boxed{3}}\\
&\quad\overbrace{-\tfrac{i}{2}g'B^\mu\Phi^\dagger\partial_\mu\Phi}^{\boxed{2b}}\overbrace{-\tfrac{i}{2}g'B^\mu\Phi^\dagger\tfrac{i}{2}g\sigma_kW_{k\mu}\Phi}^{=\boxed{3}}-\tfrac{i}{2}g'B^\mu\Phi^\dagger\tfrac{i}{2}g'B_\mu\Phi.\end{aligned} \quad (7\text{-}190)$$

For Φ in the unitary gauge (7-55), Φ is real. So, the terms labeled $\boxed{1a}$, $\boxed{1b}$, $\boxed{2a}$, and $\boxed{2b}$ cancel. The terms labeled $\boxed{3}$ add. Thus,

$$\begin{aligned}&\left(D^\mu\Phi\right)^\dagger\left(D_\mu\Phi\right)\\
&\quad=\partial^\mu\Phi^T\partial_\mu\Phi+\underbrace{\tfrac{1}{4}g^2\Phi^T\sigma_j\sigma_kW_j^\mu W_{k\mu}\Phi}_{\frac{1}{4}g^2\Phi^T\left(\delta_{jk}I+i\varepsilon_{jki}\sigma_i\right)W_j^\mu W_{k\mu}\Phi}+\tfrac{1}{2}gg'\Phi^T\sigma_j\Phi W_j^\mu B_\mu+\tfrac{1}{4}(g')^2\Phi^T\Phi B^\mu B_\mu.\end{aligned} \quad (7\text{-}191)$$

In the underbracket, the term with the factor of $\varepsilon_{jki}\sigma_iW_j^\mu W_{k\mu}$ vanishes, since $W_j^\mu W_{k\mu}-W_k^\mu W_{j\mu}$ equals zero.

Now substitute (7-55) into (7-191) to get

$$\begin{aligned}\left(D^\mu\Phi\right)^\dagger\left(D_\mu\Phi\right)&=\partial^\mu\Phi^T\partial_\mu\Phi+\tfrac{1}{4}g^2\Phi^T\Phi W_j^\mu W_{j\mu}+\tfrac{1}{2}gg'\Phi^T\sigma_j\Phi W_j^\mu B_\mu+\tfrac{1}{4}(g')^2\Phi^T\Phi B^\mu B_\mu\\
&=\left(\partial^\mu\tfrac{1}{\sqrt{2}}\begin{bmatrix}0&\sigma+v\end{bmatrix}\right)\left(\partial_\mu\tfrac{1}{\sqrt{2}}\begin{bmatrix}0\\\sigma+v\end{bmatrix}\right)+\tfrac{1}{4}g^2\tfrac{1}{\sqrt{2}}\begin{bmatrix}0&\sigma+v\end{bmatrix}\tfrac{1}{\sqrt{2}}\begin{bmatrix}0\\\sigma+v\end{bmatrix}W_j^\mu W_{j\mu}\\
&\quad+\tfrac{1}{2}gg'\tfrac{1}{\sqrt{2}}\begin{bmatrix}0&\sigma+v\end{bmatrix}\sigma_j\tfrac{1}{\sqrt{2}}\begin{bmatrix}0\\\sigma+v\end{bmatrix}W_j^\mu B_\mu+\tfrac{1}{4}(g')^2\tfrac{1}{\sqrt{2}}\begin{bmatrix}0&\sigma+v\end{bmatrix}\tfrac{1}{\sqrt{2}}\begin{bmatrix}0\\\sigma+v\end{bmatrix}B^\mu B_\mu\end{aligned} \quad (7\text{-}192)$$

$$\begin{aligned}&=\tfrac{1}{2}\partial^\mu\sigma\partial_\mu\sigma+\tfrac{1}{8}g^2(\sigma+v)^2W_j^\mu W_{j\mu}\\
&\qquad+\underbrace{\tfrac{1}{4}gg'\underbrace{\begin{bmatrix}0&\sigma+v\end{bmatrix}\sigma_j\begin{bmatrix}0\\\sigma+v\end{bmatrix}}_{=0\text{ for }j=1\text{ or }2}W_j^\mu}_{=-(\sigma+v)^2W_3^\mu}B_\mu+\tfrac{1}{8}(g')^2(\sigma+v)^2B^\mu B_\mu.\end{aligned} \quad (7\text{-}193)$$

Use (7-187) in the second term in (7-193), and (7-43) elsewhere, to find (7-193) is

$$\begin{aligned}&=\tfrac{1}{2}\partial^\mu\sigma\partial_\mu\sigma+\tfrac{1}{8}g^2(\sigma+v)^2\begin{pmatrix}2W^{-\mu}W_\mu^++A^\mu A_\mu\sin^2\theta_W\\+2A^\mu Z_\mu\sin\theta_W\cos\theta_W+Z^\mu Z_\mu\cos^2\theta_W\end{pmatrix}\\
&\qquad-\tfrac{1}{4}gg'(\sigma+v)^2\left(A^\mu\sin\theta_W+Z^\mu\cos\theta_W\right)\left(A_\mu\cos\theta_W-Z_\mu\sin\theta_W\right)\\
&\qquad+\tfrac{1}{8}(g')^2(\sigma+v)^2\left(A^\mu\cos\theta_W-Z^\mu\sin\theta_W\right)\left(A_\mu\cos\theta_W-Z_\mu\sin\theta_W\right).\end{aligned} \quad (7\text{-}194)$$

$$
\begin{aligned}
&= \tfrac{1}{2}\partial^\mu\sigma\,\partial_\mu\sigma + \tfrac{1}{8}g^2(\sigma+v)^2\begin{pmatrix} 2W^{-\mu}W_\mu^+ + A^\mu A_\mu \sin^2\theta_W \\ \quad + 2A^\mu Z_\mu \sin\theta_W \cos\theta_W + Z^\mu Z_\mu \cos^2\theta_W \end{pmatrix} \\
&\quad - \tfrac{1}{4}gg'(\sigma+v)^2\left(\sin\theta_W\cos\theta_W\left(A^\mu A_\mu - Z^\mu Z_\mu\right) - A^\mu Z_\mu \sin^2\theta_W + A^\mu Z_\mu \cos^2\theta_W\right) \qquad (7\text{-}195) \\
&\quad + \tfrac{1}{8}(g')^2(\sigma+v)^2\begin{pmatrix} A^\mu A_\mu \cos^2\theta_W \\ \quad - 2A^\mu Z_\mu \sin\theta_W\cos\theta_W + Z^\mu Z_\mu \sin^2\theta_W \end{pmatrix}
\end{aligned}
$$

With (7-163), (7-195) becomes

$$
\begin{aligned}
&= \tfrac{1}{2}\partial^\mu\sigma\,\partial_\mu\sigma + \tfrac{1}{8}\left(\frac{e^2}{\sin^2\theta_W}\right)(\sigma+v)^2\begin{pmatrix} 2W^{-\mu}W_\mu^+ + A^\mu A_\mu \sin^2\theta_W \\ \quad + 2A^\mu Z_\mu \sin\theta_W\cos\theta_W + Z^\mu Z_\mu\cos^2\theta_W \end{pmatrix} \\
&\quad - \tfrac{1}{4}\left(\frac{e^2}{\sin\theta_W\cos\theta_W}\right)(\sigma+v)^2\begin{pmatrix} \sin\theta_W\cos\theta_W\left(A^\mu A_\mu - Z^\mu Z_\mu\right) \\ \quad - A^\mu Z_\mu\sin^2\theta_W + A^\mu Z_\mu\cos^2\theta_W \end{pmatrix} \qquad (7\text{-}196) \\
&\quad + \tfrac{1}{8}\left(\frac{e^2}{\cos^2\theta_W}\right)(\sigma+v)^2\begin{pmatrix} A^\mu A_\mu\cos^2\theta_W \\ \quad - 2A^\mu Z_\mu\sin\theta_W\cos\theta_W + Z^\mu Z_\mu\sin^2\theta_W \end{pmatrix}
\end{aligned}
$$

$$
\begin{aligned}
&= \tfrac{1}{2}\partial^\mu\sigma\,\partial_\mu\sigma + \tfrac{1}{8}e^2(\sigma+v)^2\left(2\frac{W^{-\mu}W_\mu^+}{\sin^2\theta_W} + A^\mu A_\mu + 2A^\mu Z_\mu\frac{\cos\theta_W}{\sin\theta_W} + Z^\mu Z_\mu\frac{\cos^2\theta_W}{\sin^2\theta_W}\right) \\
&\quad + \tfrac{1}{4}e^2(\sigma+v)^2\left(-A^\mu A_\mu + Z^\mu Z_\mu + A^\mu Z_\mu\frac{\sin\theta_W}{\cos\theta_W} - A^\mu Z_\mu\frac{\cos\theta_W}{\sin\theta_W}\right) \qquad (7\text{-}197) \\
&\quad + \tfrac{1}{8}e^2(\sigma+v)^2\left(A^\mu A_\mu - 2A^\mu Z_\mu\frac{\sin\theta_W}{\cos\theta_W} + Z^\mu Z_\mu\frac{\sin^2\theta_W}{\cos^2\theta_W}\right).
\end{aligned}
$$

Key Point

The important thing to notice here is that the terms in $A^\mu A_\mu$ cancel. If they didn't, because v is a constant, we would get terms in $\mathcal{L}$ of form

$$
C\tfrac{1}{2}e^2v^2A^\mu A_\mu \qquad C = \text{a constant} \quad m_\gamma = \sqrt{C}\,ev, \qquad (7\text{-}198)
$$

so, the photon mass m_γ would not be zero. We would have a massive photon, and the theory would be headed for the trash barrel. The GSW model gives us a massless photon, which we need to have, and that is one of its great successes.

Note that (7-158) led to cancellation of the photon mass term, another corroboration of that result.

Result of Covariant Derivative Term

The terms in $A^\mu Z_\mu$ cancel in (7-197), leaving us with

$$
\begin{aligned}
\left(D^\mu\Phi\right)^\dagger\left(D_\mu\Phi\right) &= \tfrac{1}{2}\partial^\mu\sigma\,\partial_\mu\sigma + \tfrac{1}{8}e^2(\sigma+v)^2\left(2\frac{W^{-\mu}W_\mu^+}{\sin^2\theta_W}\right) + \tfrac{1}{8}e^2(\sigma+v)^2 Z^\mu Z_\mu\underbrace{\left(ctn^2\theta_W\right)}_{\frac{1}{\sin^2\theta_W}-1} \\
&\quad + \underbrace{\tfrac{1}{4}e^2(\sigma+v)^2 Z^\mu Z_\mu}_{\text{cancels "}-1\text{" parts of prior and next terms}} + \tfrac{1}{8}e^2(\sigma+v)^2 Z^\mu Z_\mu\underbrace{\left(\tan^2\theta_W\right)}_{\frac{1}{\cos^2\theta_W}-1} \qquad (7\text{-}199)
\end{aligned}
$$

$$=\tfrac{1}{2}\partial^{\mu}\sigma\partial_{\mu}\sigma+\tfrac{1}{4}g^{2}\sin^{2}\theta_{W}\left(\sigma+v\right)^{2}\frac{W^{-\mu}W_{\mu}^{+}}{\sin^{2}\theta_{W}}+\tfrac{1}{8}e^{2}\left(\sigma+v\right)^{2}Z^{\mu}Z_{\mu}\underbrace{\left(\frac{1}{\sin^{2}\theta_{W}}+\frac{1}{\cos^{2}\theta_{W}}\right)}_{\frac{\cos^{2}\theta_{W}+\sin^{2}\theta_{W}}{\sin^{2}\theta_{W}\cos^{2}\theta_{W}}} \tag{7-200}$$

$$=\tfrac{1}{2}\partial^{\mu}\sigma\partial_{\mu}\sigma+\tfrac{1}{4}g^{2}\left(\sigma+v\right)^{2}W^{-\mu}W_{\mu}^{+}+\tfrac{1}{8}\left(g^{2}\sin^{2}\theta_{W}\right)\left(\sigma+v\right)^{2}Z^{\mu}Z_{\mu}\frac{1}{\sin^{2}\theta_{W}\cos^{2}\theta_{W}}$$

$$=\tfrac{1}{2}\partial^{\mu}\sigma\partial_{\mu}\sigma+\tfrac{1}{4}g^{2}\left(\sigma+v\right)^{2}W^{-\mu}W_{\mu}^{+}+\tfrac{1}{8}\frac{g^{2}}{\cos^{2}\theta_{W}}\left(\sigma+v\right)^{2}Z^{\mu}Z_{\mu}. \tag{7-201}$$

Expanding (7-201), we obtain

$$\begin{aligned}\left(D^{\mu}\Phi\right)^{\dagger}\left(D_{\mu}\Phi\right)= &\underbrace{\tfrac{1}{2}\partial^{\mu}\sigma\partial_{\mu}\sigma}_{\substack{\text{free, kinetic}\\ \text{term (in } \mathcal{L}_0^{H})}}\ \underbrace{+\tfrac{1}{4}g^{2}W^{-\mu}W_{\mu}^{+}\sigma^{2}+\tfrac{1}{2}vg^{2}W^{-\mu}W_{\mu}^{+}\sigma}_{\text{interaction terms (in } \mathcal{L}^{HB})}\ \underbrace{+\tfrac{1}{4}\overbrace{v^{2}g^{2}}^{m_W^2}W^{-\mu}W_{\mu}^{+}}_{\substack{\text{free, mass}\\ \text{term (in } \mathcal{L}_0^{B})}}\\ &\underbrace{+\tfrac{1}{8}\frac{g^{2}}{\cos^{2}\theta_{W}}Z^{\mu}Z_{\mu}\sigma^{2}+\tfrac{1}{4}\frac{vg^{2}}{\cos^{2}\theta_{W}}Z^{\mu}Z_{\mu}\sigma}_{\text{interaction terms (in } \mathcal{L}^{HB})}\ \underbrace{+\tfrac{1}{2}\overbrace{\left(\tfrac{1}{4}\frac{v^{2}g^{2}}{\cos^{2}\theta_{W}}\right)}^{m_Z^2}Z^{\mu}Z_{\mu}}_{\text{free, mass term (in } \mathcal{L}_0^{B})},\end{aligned} \tag{7-202}$$

where the interaction terms (W with Higgs σ and Z with Higgs σ) and the mass terms are designated. From (7-4), we see that the W mass and the Z mass are given, respectively, by

$$m_{W}=\tfrac{1}{2}vg \qquad m_{Z}=\frac{1}{2}\frac{vg}{\cos\theta_{W}}=\frac{m_{W}}{\cos\theta_{W}} \qquad \left(v=\sqrt{\frac{-\mu^{2}}{\lambda}}>0\right). \tag{7-203}$$

The interaction terms are non-linear terms, i.e., they lead to non-linear terms in the field equations. The mass terms will end up in the free part of the Lagrangian under $\mathcal{L}^{B}$.

<u>The $\mathcal{V}$ Terms</u>

With (7-52),

$$\mathcal{V}=\mu^{2}\Phi^{\dagger}\Phi+\lambda\left(\Phi^{\dagger}\Phi\right)^{2}=\frac{\mu^{2}}{2}\begin{bmatrix}\phi_1-i\phi_2 & \phi_3-i\phi_4\end{bmatrix}\begin{bmatrix}\phi_1+i\phi_2\\ \phi_3+i\phi_4\end{bmatrix}+\frac{\lambda}{4}\left(\begin{bmatrix}\phi_1-i\phi_2 & \phi_3-i\phi_4\end{bmatrix}\begin{bmatrix}\phi_1+i\phi_2\\ \phi_3+i\phi_4\end{bmatrix}\right)^{2}, \tag{7-204}$$

which, with the unitary gauge of (7-55), becomes

$$\mathcal{V}=\frac{\mu^{2}}{2}\begin{bmatrix}0 & \sigma+v\end{bmatrix}\begin{bmatrix}0\\ \sigma+v\end{bmatrix}+\frac{\lambda}{4}\left(\begin{bmatrix}0 & \sigma+v\end{bmatrix}\begin{bmatrix}0\\ \sigma+v\end{bmatrix}\right)^{2}=\frac{\mu^{2}}{2}\left(\sigma+v\right)^{2}+\frac{\lambda}{4}\left(\sigma+v\right)^{4}. \tag{7-205}$$

From (7-57),

$$\lambda=\frac{-\mu^{2}}{v^{2}}. \tag{7-206}$$

Expanding (7-205) with (7-206), we get

$$\begin{aligned}\mathcal{V}&=\frac{\mu^{2}}{2}\left(\sigma+v\right)^{2}-\frac{\mu^{2}}{4v^{2}}\left(\sigma+v\right)^{4}=\frac{\mu^{2}}{2}\left(\sigma^{2}+2\sigma v+v^{2}\right)-\frac{\mu^{2}}{4v^{2}}\left(\sigma^{2}+2\sigma v+v^{2}\right)\left(\sigma^{2}+2\sigma v+v^{2}\right)\\ &=\frac{\mu^{2}}{2}\left(\sigma^{2}+2\sigma v+v^{2}\right)-\frac{\mu^{2}}{4v^{2}}\left(\sigma^{4}+2\sigma^{3}v+\sigma^{2}v^{2}+2\sigma^{3}v+4\sigma^{2}v^{2}+2\sigma v^{3}+\sigma^{2}v^{2}+2\sigma v^{3}+v^{4}\right)\\ &=\frac{\mu^{2}}{2}\left(\sigma^{2}+2\sigma v+v^{2}\right)-\frac{\mu^{2}}{4v^{2}}\left(\sigma^{4}+4\sigma^{3}v+6\sigma^{2}v^{2}+4\sigma v^{3}+v^{4}\right).\end{aligned} \tag{7-207}$$

Or

$$\mathcal{V}=\overset{\boxed{A1}}{\frac{\mu^2\sigma^2}{2}}+\overset{\boxed{B1}}{\mu^2\sigma v}+\overset{\boxed{C1}}{\frac{\mu^2 v^2}{2}}-\frac{\mu^2\sigma^4}{4v^2}-\frac{\mu^2\sigma^3}{v}-\overset{\boxed{A2}}{\frac{3\mu^2\sigma^2}{2}}-\overset{\boxed{B2}}{\mu^2\sigma v}-\overset{\boxed{C2}}{\frac{\mu^2 v^2}{4}}. \tag{7-208}$$

Combining terms with the same letters over them in (7-208),we get (where we use (7-206) to get the RHS)

$$\mathcal{V}=\overset{\boxed{A}}{-\mu^2\sigma^2}+\overset{\boxed{C}}{\frac{\mu^2 v^2}{4}}-\frac{\mu^2\sigma^4}{4v^2}-\frac{\mu^2\sigma^3}{v}=\overset{\boxed{A}}{-\tfrac{1}{2}\left(2\mu^2\sigma^2\right)}\overset{\boxed{C}}{-\frac{\lambda v^4}{4}}+\tfrac{1}{4}\lambda\sigma^4+\lambda v\sigma^3. \tag{7-209}$$

Note that the $\boxed{C}$ term is a constant, the energy density left in the vacuum from the Higgs symmetry breaking. This term is often considered to represent a <u>condensate</u> of the Higgs field (the <u>Higgs condensate</u>), since a phase transition as temperature lowers, such as liquification of water vapor, generally entails a condensation of the substance making the transition.

$$\text{Higgs condensate energy density}=-\frac{\lambda v^4}{4}. \tag{7-210}$$

With reference to Fig. 7-2, pg. 209, if we take our coordinate axes as the solid lines (which we are doing in this derivation), then the potential energy density at the true vacuum is the negative number (7-210). If, on the other hand, we take our coordinate axes as the dotted lines in the figure, then the potential at the true vacuum is zero, but the potential at the false vacuum is the negative of (7-210), i.e., it is positive.

However, as noted before, this constant term (7-210) is generally just ignored, for no real valid theoretical reason. It is another example, other than ZPE, of the huge mismatch between theory and observation, with regard to the vacuum. In ignoring (7-210), we are implicitly taking our coordinate axes as the dotted lines in Fig. 7-2.

Returning to (7-209) and using (7-188) with the Higgs condensate taken as zero, we have

$$\underbrace{\mathcal{L}^H}_{\substack{\text{false}\\\text{vacuum}\\\text{fields}}}\xrightarrow[\text{vacuum fields}]{\text{substitute true}}\underbrace{\left(D^\mu\Phi\right)^\dagger\left(D_\mu\Phi\right)}_{\text{found earlier}}\underbrace{-\tfrac{1}{2}\underbrace{\overset{\boxed{A}}{\left(-2\mu^2\right)}}_{m_H^2}\sigma^2}_{\substack{\text{free, mass}\\\text{term (in }\mathcal{L}_0^H)}}\underbrace{-\tfrac{1}{4}\lambda\sigma^4-\lambda v\sigma^3}_{\substack{\text{self-interaction}\\\text{terms (in }\mathcal{L}^{HH})}}, \tag{7-211}$$

where, as noted in (7-2), the factor labeled m_H. represents the Higgs mass, which since μ is imaginary, is real and positive. The other two terms represent higher order (nonlinear, self-interacting) Higgs field terms and comprise $\mathcal{L}^{HH}$.

Using (7-202) with (7-211) gives us our final result for $\mathcal{L}^H$ converted to the true vacuum.

Result of this Section

$$\underbrace{\mathcal{L}^H}_{\substack{\text{false}\\\text{vacuum}\\\text{fields}}}\xrightarrow[\substack{\text{vacuum}\\\text{fields}}]{\substack{\text{substitute}\\\text{true}}}\underbrace{\tfrac{1}{2}\partial^\mu\sigma\partial_\mu\sigma-\tfrac{1}{2}\overbrace{\left(-2\mu^2\right)}^{m_H^2}\sigma^2}_{\substack{\text{free field, kinetic \&}\\\text{mass terms (in }\mathcal{L}_0^H)}}\underbrace{+\overbrace{\frac{v^2g^2}{4}}^{m_W^2}W^{-\mu}W_\mu^+ +\tfrac{1}{2}\overbrace{\left(\frac{v^2g^2}{4\cos^2\theta_W}\right)}^{m_Z^2}Z^\mu Z_\mu}_{\text{free field mass terms (in }\mathcal{L}_0^B)}$$

$$\underbrace{+\frac{g^2}{4}W^{-\mu}W_\mu^+\sigma^2+\frac{vg^2}{2}W^{-\mu}W_\mu^+\sigma+\frac{g^2}{8\cos^2\theta_W}Z^\mu Z_\mu\sigma^2+\frac{vg^2}{4\cos^2\theta_W}Z^\mu Z_\mu\sigma}_{\text{nonlinear interaction terms (in }\mathcal{L}^{HB})} \tag{7-212}$$

$$\underbrace{-\tfrac{\lambda}{4}\sigma^4-\lambda v\sigma^3}_{\substack{\text{self interaction, nonlinear}\\\text{terms (in }\mathcal{L}^{HH})}}.$$

7.7 Problems

1. Show that for Z^μ a real, massive vector field, whose kinetic term in the Lagrangian has form $-\frac{1}{4}\left(\partial^\nu Z^\mu - \partial^\mu Z^\nu\right)\left(\partial_\nu Z_\mu - \partial_\mu Z_\nu\right)$, the mass term in the Lagrangian has form $\frac{1}{2}m_Z^2 Z^\mu Z_\mu$.
2. Show (7-5) is symmetric under the transformation (7-19).
3. Show (7-5) is symmetric under the transformation (7-22).
4. Deduce $\bar{\psi}_l \psi_l = \bar{\psi}_l^L \psi_l^R + \bar{\psi}_l^R \psi_l^L$ and $\bar{\psi}_l \gamma^\mu \psi_l = \bar{\psi}_l^R \gamma^\mu \psi_l^R + \bar{\psi}_l^L \gamma^\mu \psi_l^L$, where there is no sum on l.
5. Prove, where the block letters refer to (7-171) and (7-182), that

$$\boxed{E} + \boxed{F} + \boxed{G} + \boxed{H} + \boxed{J} + \boxed{K}$$
$$= \boxed{E} + \boxed{F} - \frac{1}{4}B^{\mu\nu}B_{\mu\nu} = \underbrace{-\frac{1}{4}\left(\partial_\nu A_\mu - \partial_\mu A_\nu\right)\left(\partial^\nu A^\mu - \partial^\mu A^\nu\right)}_{-\frac{1}{4}F_{\mu\nu}F^{\mu\nu}} \underbrace{-\frac{1}{4}\left(\partial_\nu Z_\mu - \partial_\mu Z_\nu\right)\left(\partial^\nu Z^\mu - \partial^\mu Z^\nu\right)}_{-\frac{1}{4}Z_{\mu\nu}Z^{\mu\nu}}.$$

6. Prove that for a (non-singular) complex matrix Y (not necessarily unitary or hermitian), one can find unitary matrices U and K that diagonalize Y via $M = UYK^\dagger$, where M is real and diagonal with all positive components.

   Hint: Start by recognizing that $YY^\dagger$ is an hermitian matrix, since $(YY^\dagger)^\dagger = YY^\dagger$. It is also positive definite as for any non-zero vector x, $x^\dagger(YY^\dagger)x = [x^\dagger Y][Y^\dagger x]$ is the product of a complex vector times its complex conjugate, and thus, is positive. Then, consider it can be diagonalized with a unitary matrix U, where we call the resulting diagonal matrix M^2, whose components are the eigenvalues of $YY^\dagger$. That is, $U(YY^\dagger)U^\dagger = M^2$. Hermitian, positive definite matrices, like $YY^\dagger$, must have real, positive eigenvalues, so M^2 has all real positive values on its diagonal.

   Note $M^2 = MM^\dagger$, where M can be diagonal and have all real, positive components, though that is not necessary. Nevertheless, for the proof, choose M to be diagonal with each component the positive square root of the corresponding diagonal component of M^2. Then define $K = M^{-1}UY$. Then show first that K is unitary, then, that $UYK^\dagger = M$, where M (as we defined it originally) is diagonal with all real, positive components. It will help if you note that from $M^2 = U(YY^\dagger)U^\dagger$, $M^{-2} = (U(YY^\dagger)U^\dagger)^{-1} = (U^\dagger)^{-1}(YY^\dagger)^{-1}U^{-1} = U((Y^\dagger)^{-1}Y^{-1})U^\dagger$.

7. Show that charged lepton interaction terms with the Z vector boson have the same form in both the mass and flavor bases.
8. Find the term in the Lagrangian that corresponds to each side of Fig. **7-5**, pg. 237.
9. Deduce the weak isospin charge and the weak hypercharge for the Higgs particle ϕ_3. There are two ways to find the isospin charge. First, from using the lepton-Higgs coupled part of the high energy Lagrangian with the unitary gauge (7-54) and examining charge conservation at a vertex. Second, by using the $SU(2)$ weak isospin charge operator on Φ in the unitary gauge. Only the first way works for the hypercharge. Note that because the Higgs is real, it has zero electric charge, which is born out by the relation in the heading of Wholeness Chart 7-9, pg. 243.

Chapter 8

Experiment: Scattering and Decay

"Soon I knew the craft of experimental physics was beyond me - it was the sublime quality of patience - patience in accumulating data, patience with recalcitrant equipment - which I sadly lacked."
Abdus Salam

8.0 Preliminaries

Yet, Nobel laureate Salam certainly made his mark as a world-class theorist. His comment reminds us all of the vital contributions made by the great experimentalists of our age and of ages past. Without them, there would be no theory.

For me, personally, the quote resonates deeply, as I too was ill qualified for experimental work. Thankfully, for myself and others, I realized that truth early on.

8.0.1 Background

Steps to Our Goal

Recall, from pg. 511 of Vol. 1, the process to our overall goals.

Big picture of the QFT process and its goals

2nd quantization postulates $\rightarrow$ QFT theory $\rightarrow$ transition amplitude S_{fi} calculation
$\rightarrow$ probability ($= |S_{fi}|^2$) $\rightarrow$ scattering, decay, other experimental results $\rightarrow$ confirmation of theory

Obtaining Transition Amplitudes

There are two ways to go from QFT theory to a particular transition amplitude S_{fi}.

Two ways to S_{fi}

- In the interaction picture: Find $\mathcal{L}_I$ and from it, $\mathcal{H}_I$. Then use $\mathcal{H}_I$ in the Dyson expansion along with Wick's theorem to get unwieldly relations for S_{fi} with numerous complicated integrals. Evaluate those integrals and convert to momentum space.
- Use Feynman diagrams and Feynman rules.

We'll take the easier, Feynman's rules

We are going to skip the first of these and go directly to Feynman's far easier and quicker method.

From Transition Amplitudes to Cross Sections and Decay Rates

To find decay rates and scattering cross sections from S_{fi}, we will borrow heavily from the methodologies and relations developed in Vol. 1, Chaps. 16 and 17. We will not rederive the general, more universal, results that were obtained there.

We'll apply evaluation methods from QED to find e/w decays and cross sections from S_{fi}

8.0.2 Chapter Overview

So, in this chapter, we will apply the experimentally related calculation techniques we developed in Vol. 1 for QED to what we have learned so far in this volume for electroweak interactions, and in so doing, postdict certain well-known empirical results. In essence, we will

The 3 major steps in this chapter

- introduce Feynman's rules for electroweak theory,
- employ them to find transition amplitudes S_{fi} of interest, and
- with those S_{fi} relations, determine decay rates and cross sections.

8.1 Feynman Rules for Electroweak Interactions

8.1.1 The Rules

Feynman rules for e/w theory in chapter summary

The Feynman rules we will deal with will be for the low energy (true vacuum) fields of Chap. 7, not the high energy (false vacuum) fields of Chap. 6. These represent the essence of this chapter, and so, should be in the chapter summary. As they are lengthy, we do not repeat them here. You may wish to put a bookmark on the page where they start (pg. 290), for reference as you continue through the chapter. We will be referring to them often.

Tree level calculations give good results in many cases

We will focus on that

Note that these rules can be used for tree level calculations using contemporary epoch, measured energy level values for the running coupling constants g and g'. In this book, we are not doing in depth study of the very complex subject of e/w renormalization nor of the concomitant dependence of weak coupling constants on energy level. As we will see, we can still do very relevant determinations of decay times and scattering cross sections for weak interactions using lowest order computations.

8.1.2 The Massive Vector Boson Propagator

Massive vector propagator derived in Appendix A

Feynman rule #12, which arises from the propagator for a massive vector field, has an element, the second term in the numerator, that no doubt looks a little strange and unexpected. We deduce the propagator for vector fields with mass in Appendix A. It is straightforward to extrapolate that result to rule #12, as it parallels the reduction of the massless vector propagator to rule #2.

8.1.3 Examples of Feynman Diagrams and Vertex Factors

As examples

Two examples of vertex factors

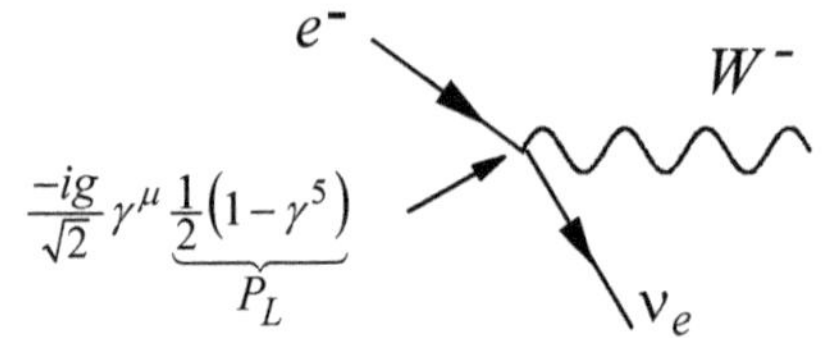

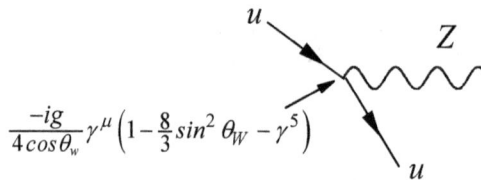

Vertex factor for Feynman rule #15-2 Vertex factor for Feynman rule #15-11

Figure 8-1. Examples of Feynman Diagrams Showing Feynman Rules Vertex Factors

Note the vertex factor in the LHS of Fig. 8-1 has a factor of i in it that was not in the similar diagram on the RHS of Fig. 7-13. pg. 249, in Chap. 7. This is because Fig. 8-1 reflects the Feynman rules vertex factor, whereas Fig. 7-13 was reflecting the factor in the associated interaction term in the Lagrangian. In general, such terms differ by a factor of i.

8.2 Particle Decay

Decay easier than scattering, so we'll do it first

It is easier to analyze particle decay than scattering, so we will focus on that first. Recall that in QED we had scattering, but no particle decay. We need weak interactions for that.

8.2.1 General Expression for Fermion Decay

Step 1: derive the most general decay rate relation

Before calculating the decay rate for any specific particle, we will derive a general expression, which will be the starting point for any particular particle rate of decay we wish to find explicitly.

Consider a single initial particle P decaying into N final particles, primed and labeled 1 to N in

$$P \rightarrow P_1' + P_2' + \ldots + P_N' . \tag{8-1}$$

Now consider the transition amplitude (8-121), Part A of Feynman's rules, for the case of (8-1), where i means initial, f means final, and l means lepton

General decay amplitude

$$S_{fi} = \delta_{fi} + \left((2\pi)^4 \delta^{(4)}\left(\sum p_f - \sum p_i\right) \frac{1}{(2VE)^{1/2}} \prod_f \frac{1}{\left(2VE_f\right)^{1/2}} \prod_l (2m_l)^{1/2} \right) \mathcal{M} . \tag{8-2}$$

The square of the absolute value of (8-2) is the probability of decay, so if we divide that by time T, we get the probability of decay per unit time, w. Note that $\delta_{fi} = 0$, since the initial and final states are not equal, so $f \neq i$.

$$w = \frac{|S_{fi}|^2}{T} = \frac{\delta_{fi}}{T} + \frac{1}{VT}\left((2\pi)^4 \delta^{(4)}\left(\sum p_f - \sum p_i\right)\right)^2 \frac{1}{2E}\left(\prod_f \frac{1}{2VE_f}\right)\left(\prod_l 2m_l\right)|\mathcal{M}|^2 \quad (8\text{-}3)$$

Decay rate for specific final 4-momenta p_f

As we did in Vol. 1, we take the time interval T and the volume V as infinite. As noted there, this makes the problem easier to solve and yields the same cross sections as taking T and V finite. In this case, it means the decay rates will be the same, either way. As we saw in Vol. 1, (17-44), pg. 454, for $\sum p_f = \sum p_i$ in our integrals, we can take

$$(2\pi)^4 \delta^{(4)}\left(\sum p_f - \sum p_i\right) = VT \qquad \text{for} \quad V,T \to \infty . \quad (8\text{-}4)$$

Hence, one of the delta functions in (8-3) cancels with VT in the denominator and the term with δ_{fi} in it drops out, leaving

$$w = (2\pi)^4 \delta^{(4)}\left(\sum p_f - \sum p_i\right)\frac{1}{2E}\left(\prod_f \frac{1}{2VE_f}\right)\left(\prod_l 2m_l\right)|\mathcal{M}|^2 . \quad (8\text{-}5)$$

Simplified general decay rate for specific p_f

(8-5) provides the decay rate with one set of values for all of the final particle momenta $\mathbf{p}'_f$. It is effectively a probability (of decay per unit time) for a particle to decay where the final products have certain precise values for momenta. So, w is essentially a probability per momenta values of the final particles, i.e., w is a probability density over momentum space. To get the total probability of decay (per unit time) we will need to integrate over all possible final momenta. Hence, w is vanishingly small for any one set of $\mathbf{p}'_f$, as the possible values are continuous, and hence there are an infinite number of such sets of values.

We need to obtain a decay rate over a range of values, $\mathbf{p}'_f$ to $\mathbf{p}'_f + d\,\mathbf{p}'_f$, which we can integrate to get the total decay rate. The decay rate over $\mathbf{p}'_f$ to $\mathbf{p}'_f + d\,\mathbf{p}'_f$ is called the differential decay rate $d\Gamma_d$, where the subscript d indicates a particular decay process. To get it, we need to find the number of multiparticle final states between $\mathbf{p}'_f$ and $\mathbf{p}'_f + d\,\mathbf{p}'_f$.

Need decay rate for differential range of p_f

We found this in Vol. 1, relation (17-49), pg. 455. (See also Fig. 17-12, pg. 444.)

$$\text{number of final states between } \mathbf{p}'_f \text{ and } \mathbf{p}'_f + d\mathbf{p}'_f \quad \text{equals} \quad dN_f = \prod_f \frac{Vd\mathbf{p}'_f}{(2\pi)^3} . \quad (8\text{-}6)$$

For this, need number of states in differential range of p_f

So, the rate of decay over this infinitesimal range of final 3-momenta is the differential decay rate

$$\begin{aligned} d\Gamma_d &= w dN_f = w\prod_f \frac{Vd\mathbf{p}'_f}{(2\pi)^3} \\ &= (2\pi)^4 \delta^{(4)}\left(\sum p_f - \sum p_i\right)\frac{1}{2E}\left(\prod_f \frac{1}{2VE_f}\right)\left(\prod_l 2m_l\right)|\mathcal{M}_d|^2 \prod_f \frac{Vd\mathbf{p}'_f}{(2\pi)^3}, \end{aligned} \quad (8\text{-}7)$$

With that, get differential decay rate over this range, $d\Gamma_d$

Or, more succinctly, our final, most general result,

$$\text{for decay process } d, \quad d\Gamma_d = (2\pi)^4 \delta^{(4)}\left(\sum p_f - \sum p_i\right)\frac{1}{2E}\left(\prod_l 2m_l\right)\left(\prod_f \frac{d\mathbf{p}'_f}{(2\pi)^3 2E_f}\right)|\mathcal{M}_d|^2 . \quad (8\text{-}8)$$

Simplest form of $d\Gamma_d$, our most general decay rate relation

To go further, we would need a specific particle and deduce its amplitude $\mathcal{M}$ for use in (8-8). We would then integrate (8-8) over all possible final state momenta and spins to get the total decay rate Γ_d for a particular decay process. In that integration the delta function is eliminated and conservation of 4-momenta is imposed on the decay process.

Integrate $d\Gamma_d$ to get total decay rate Γ_d for particular decay process

8.2.2 Particle Lifetimes

Decay Branches

Note that a given particle, be it elementary (like a muon) or composite (like a neutron) could decay in different ways to different final particles. These are called different decay modes or decay branches or decay channels. For example, three possible decay modes (branches or channels) for the muon are

A particle can decay in different ways called modes or branches or channels

$$\mu^- \rightarrow e^- + \nu_\mu + \bar{\nu}_e \qquad \mu^- \rightarrow e^- + \nu_\mu + \bar{\nu}_e + \gamma \qquad \mu^- \rightarrow e^- + \nu_\mu + \bar{\nu}_e + e^+ + e^- . \tag{8-9}$$

As it turns out, the first of these occurs 98.6% of the time.

Decay Rates

For each branch, we would calculate a different branch decay rate Γ_m, where m would equal 1,2, or 3 in the case of (8-9), each number representing a different mode (branch) of decay for the same particle.

Each mode has its own total decay rate Γ_m

Consider a hypothetical particle having two decay branches, one averaging 10 decays per second, the other at 1 decay per second. We would then expect a total of 11 total decays/second. The lesson: total decay rate Γ for a given particle is simply the sum of the individual branch decay rates for that particle.

Total decay rate for a particle = sum of decay rates for all branches

$$\text{Total decay rate (including all modes) for given particle} = \Gamma = \sum_m \Gamma_m . \tag{8-10}$$

Particle Lifetime from Decay Rate

The inverse of the total decay rate (decays/second) is simply the expected lifetime (seconds/decay), i.e., the mean time interval to decay, τ.

Expected particle lifetime = inverse of its total decay rate

$$\text{particle lifetime} \quad \tau = \frac{1}{\Gamma} = \frac{1}{\sum_m \Gamma_m} . \tag{8-11}$$

Branching Ratios

The branching ratio B_m is the percentage of the time we would see that particular branch (the mth branch). In the case of our two-decay mode example above, the second branch would occur 1/11 of the time, and we would have $B_2 \approx 0.0909$. In general, the branching ratio is

Branching ratio B_m = % of time seeing particular branch (mode) of decay

$$B_m = \frac{\Gamma_m}{\Gamma} = \frac{\Gamma_m}{\sum_{m'} \Gamma_{m'}} . \tag{8-12}$$

For muon decay (8-9), the branching ratio of the first mode would be 0.986.

Note that some authors use the symbol Γ to represent an individual branch (where we use Γ_m), and $\Sigma\Gamma$ to represent total decay mode (where we use Γ).

Particle Half-lives from Lifetime

Do **Problem 1** to find the half-life for a particle is that of (8-13).

$$\text{particle half-life} = t_{1/2} = .693\,\tau \qquad \tau = \text{lifetime} = \text{expected (mean) life.} \tag{8-13}$$

Particle lifetime vs half life

8.2.3 Muon Decay

For Fixed Spins

We can apply (8-8) to the muon, where we will approximate the total decay rate by ignoring the second and third modes of (8-9), as they contribute very little. We will first need to evaluate the Feynman amplitude for (8-8).

Step 2: applying general decay rate relation to muon decay case

The Feynman diagram for the primary mode, where we note that a propagator combines both the particle and antiparticle character into one expression (as on RHS below), is

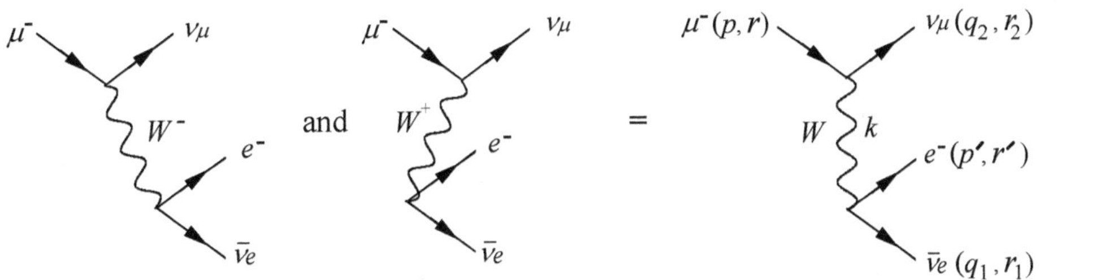

Most common muon decay mode

Figure 8-2. Primary Decay Mode of Muon

Using Feynman rules 4, 12, and 15-2, we obtain the amplitude, which includes a W propagator

Feynman amplitude for this decay mode

$$\mathcal{M}_{m=1} = -\frac{g^2}{8}\left(\bar{u}_{r'}(\mathbf{p}')\gamma^\mu\left(1-\gamma^5\right)v_{r_1}(\mathbf{q}_1)\right)\frac{i\left(-g_{\mu\nu}+k_\mu k_\nu / m_W^2\right)}{k^2-m_W^2+i\varepsilon}\left(\bar{u}_{r_2}(\mathbf{q}_2)\gamma^\nu\left(1-\gamma^5\right)u_r(\mathbf{p})\right). \quad (8\text{-}14)$$

where

As usual, 4-momenta conserved at vertices

$$k = p - q_2 = p' + q_1. \quad (8\text{-}15)$$

Note that because the muon mass m_μ is far larger than the muon neutrino mass, at contemporary energy levels, $p >> q_2$, and so from (8-15), $k \approx p$. Hence, $k^2 \approx p^2 = m_\mu^2$. Since m_W is almost a thousand times larger than m_μ, the denominator of (8-14) can be approximated by m_W^2 alone with an error of only about 1 part in 10^6. Similar logic applies to the $k_\mu k_\nu / m_W^2$ term. This means we can approximate and simplify (8-14) as

Feynman amplitude approximation, resulting from large W mass

$$\mathcal{M}_{m=1} \approx -\frac{iG}{\sqrt{2}}\left(\bar{u}_{r'}(\mathbf{p}')\gamma^\mu\left(1-\gamma^5\right)v_{r_1}(\mathbf{q}_1)\right)\left(\bar{u}_{r_2}(\mathbf{q}_2)\gamma_\mu\left(1-\gamma^5\right)u_r(\mathbf{p})\right) \quad (8\text{-}16)$$

where

$$\frac{G}{\sqrt{2}} = \frac{g^2}{8m_W^2} \qquad \left(\text{experimental value, } G = (1.16637 \pm 0.00002)\times 10^{-5}\ \text{GeV}^{-2}\right). \quad (8\text{-}17)$$

Observe that (8-16) could be represented as a Feynman diagram with an appropriate vertex factor as in Fig. 8-3, which has no propagator, but simply four external particles at a vertex. This is known as a <u>contact interaction</u>, presumably because the external particles are in direct contact at the same vertex, without a mediating propagator in between.

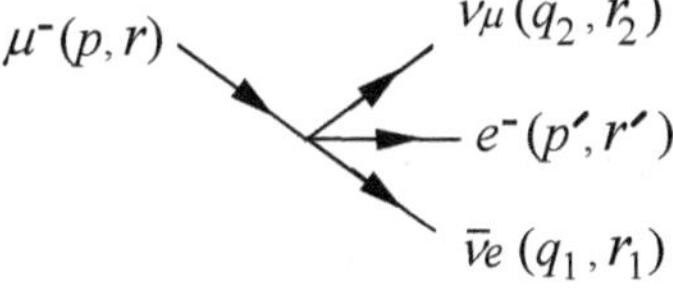

Contact interactions (an approximation) have no propagators with external particles all meeting at vertices

Figure 8-3. Approximate Feynman Diagram for Muon Decay (Contact Interaction)

Weak interactions are commonly evaluated in the simplified form of contact interactions, because the W and Z gauge bosons are so heavy. The great simplification of (8-16) comes at little cost in accuracy, in essence, by simply replacing the weak gauge boson propagator in the Feynman amplitude with $ig_{\mu\nu} / m_W^2$.

Historical Note

Weak interactions were originally modeled, by Fermi in 1934, as contact interactions that could describe nuclear beta decay. The constant <u>G</u> of (8-17) is thus known as the <u>Fermi weak-interaction coupling constant</u> (sometimes written as G_F), and the entire contact method as the <u>Fermi theory of weak interactions</u>. It was THE theory prior to GSW theory. But, it is not renormalizable, so was not

Contact interaction theory = Fermi theory

considered the final word on weak interactions. However, it is still useful today, as an excellent approximation for many processes.

End of Historical Note

Applying our general expression (8-8) to muon decay with the amplitude (8-16), results in

$$d\Gamma_{muon} \approx d\Gamma_{m=1} \approx (2\pi)^4 \delta^{(4)}(p' + q_1 + q_2 - p) \frac{m_\mu m_e m_{\nu_e} m_{\nu_\mu}}{E} \frac{1}{(2\pi)^9} \frac{d\mathbf{p}'}{E'} \frac{d\mathbf{q}_1}{E_1} \frac{d\mathbf{q}_2}{E_2} \left|\mathcal{M}_{m=1}\right|^2 . \quad (8\text{-}18)$$

Muon differential decay rate, particular 4-momenta and spin

We now need to average (8-18) over initial spin states, sum over all final spin states, and integrate over all final momenta. For the spin summing and averaging concept and methodology, see Vol. 1, pgs. 454, 459-460. (Note the symbol Γ used there is for a different entity than the symbol Γ we use in this chapter for decay rate.)

Spin Averaging and Summing

Incoming spin average and outgoing spin sum

Ignore 4-momenta, for the moment, and focus on spins, which are represented in $\mathcal{M}_{m=1}$. In doing spin sums and averages, it will shorten our work, if we derive (8-20) below (using the following relations).

Recalling some relations

$$\left(\gamma^\mu\right)^\dagger \gamma^0 = \gamma^0 \gamma^\mu \quad \text{Vol. 1, (4-148), pg. 122} \qquad \gamma^5 \gamma^\mu = -\gamma^\mu \gamma^5 \quad \text{Chap. 5 (5-51)} \qquad (8\text{-}19)$$

A relation that will help later

$$\gamma^5 \left(\gamma^\mu\right)^\dagger \gamma^0 = \gamma^5 \gamma^0 \gamma^\mu = -\gamma^0 \gamma^5 \gamma^\mu = \gamma^0 \gamma^\mu \gamma^5 \qquad (8\text{-}20)$$

$$\rightarrow \left(1-\gamma^5\right)\left(\gamma^\mu\right)^\dagger \gamma^0 = \gamma^0 \gamma^\mu \left(1-\gamma^5\right) \qquad (8\text{-}21)$$

Recall, also that $\left(\gamma^0\right)^\dagger = \gamma^0$, $\left(\gamma^5\right)^\dagger = \gamma^5$ and $\left(\gamma^5\right)^2 = 1$, which you should commit to memory, if you haven't already.

Then, averaging over incoming spins and summing over outgoing yields equations (8-22) and (8-23) (where with an eye to what comes next, we write out the spinor space indices in the last row of (8-23))

Carrying out the spin sums and averages

$$\underbrace{\sum_{r'=1}^{2} \sum_{r_1=1}^{2} \sum_{r_2=1}^{2}}_{\text{sum outgoing}} \underbrace{\frac{1}{2}\sum_{r=1}^{2}}_{\substack{\text{average}\\ \text{incoming}}} \left|\mathcal{M}_{m=1}\right|^2$$

$$= \sum_{r'=1}^{2} \sum_{r_1=1}^{2} \sum_{r_2=1}^{2} \frac{1}{2}\sum_{r=1}^{2} \left(\frac{-iG}{\sqrt{2}} \left(u^\dagger_{r'}(\mathbf{p}')\gamma^0\gamma^\mu\left(1-\gamma^5\right)v_{r_1}(\mathbf{q}_1)\right)\left(u^\dagger_{r_2}(\mathbf{q}_2)\gamma^0\gamma_\mu\left(1-\gamma^5\right)u_r(\mathbf{p})\right)\right)$$

$$\times\left(\frac{iG}{\sqrt{2}} \left(u^\dagger_{r}(\mathbf{p})\left(1-\gamma^5\right)\gamma^{\nu\dagger}\gamma^0 u_{r_2}(\mathbf{q}_2)\right)\left(v^\dagger_{r_1}(\mathbf{q}_1)\left(1-\gamma^5\right)\gamma^\dagger_\nu\gamma^0 u_{r'}(\mathbf{p}')\right)\right) \qquad (8\text{-}22)$$

$$= \frac{G^2}{4} \sum_{r'=1}^{2} \sum_{r_1=1}^{2} \sum_{r_2=1}^{2} \sum_{r=1}^{2} \left(\bar{u}_{r'}(\mathbf{p}')\gamma^\mu\left(1-\gamma^5\right)v_{r_1}(\mathbf{q}_1)\right)\left(\bar{u}_{r_2}(\mathbf{q}_2)\gamma_\mu\left(1-\gamma^5\right)u_r(\mathbf{p})\right)$$

$$\times\left(\bar{u}_{r}(\mathbf{p})\gamma^\nu\left(1-\gamma^5\right)u_{r_2}(\mathbf{q}_2)\right)\left(\bar{v}_{r_1}(\mathbf{q}_1)\gamma_\nu\left(1-\gamma^5\right)u_{r'}(\mathbf{p}')\right).$$

It will help if we re-write (8-22) with the spinor indices shown explicitly.

$$= \frac{G^2}{4} \sum_{r'=1}^{2} \sum_{r_1=1}^{2} \sum_{r_2=1}^{2} \sum_{r=1}^{2} \left(\bar{u}_{r'\alpha}(\mathbf{p}')\gamma^\mu_{\alpha\beta}\left(1-\gamma^5\right)_{\beta\gamma}v_{r_1,\gamma}(\mathbf{q}_1)\right)\left(\bar{u}_{r_2,\delta}(\mathbf{q}_2)\gamma_{\mu,\delta\varepsilon}\left(1-\gamma^5\right)_{\varepsilon\eta}u_{r,\eta}(\mathbf{p})\right)$$

$$\times\left(\bar{u}_{r,\alpha'}(\mathbf{p})\gamma^\nu_{\alpha'\beta'}\left(1-\gamma^5\right)_{\beta'\gamma'}u_{r_2,\gamma'}(\mathbf{q}_2)\right)\left(\bar{v}_{r_1,\delta'}(\mathbf{q}_1)\gamma_{\nu,\delta'\varepsilon'}\left(1-\gamma^5\right)_{\varepsilon'\eta'}u_{r',\eta'}(\mathbf{p}')\right). \qquad (8\text{-}23)$$

Now, re-arrange (8-23) to have spinors with the same 4-momenta juxtaposed, where we don't have to worry about order of spinor entities, as the spinor index labeling takes care of that for us.

$$=\frac{G^2}{4}\sum_{r'=1}^{2}\sum_{r_1=1}^{2}\sum_{r_2=1}^{2}\sum_{r=1}^{2} u_{r',\eta'}(\mathbf{p}')\bar{u}_{r',\alpha}(\mathbf{p}')v_{r_1,\gamma}(\mathbf{q}_1)\bar{v}_{r_1,\delta'}(\mathbf{q}_1)u_{r_2,\gamma'}(\mathbf{q}_2)\bar{u}_{r_2,\delta}(\mathbf{q}_2)$$

$$\times u_{r,\eta}(\mathbf{p})\bar{u}_{r,\alpha'}(\mathbf{p})\gamma^{\mu}_{\alpha\beta}\left(1-\gamma^5\right)_{\beta\gamma}\gamma_{\mu,\delta\varepsilon}\left(1-\gamma^5\right)_{\varepsilon\eta}\gamma^{\nu}_{\alpha'\beta'}\left(1-\gamma^5\right)_{\beta'\gamma'}\gamma_{\nu,\delta'\varepsilon'}\left(1-\gamma^5\right)_{\varepsilon'\eta'}. \tag{8-24}$$

Now, recall from Vol. 1, pg. 460, (17-75) and (17-76), the outer product relations (where we have shown indices explicitly in some parts, but imply them by virtue of order in others),

$$u_s(\mathbf{p})\bar{u}_s(\mathbf{p})=u_{s,\alpha}(\mathbf{p})\bar{u}_{s,\beta}(\mathbf{p})=\frac{\gamma^{\mu}_{\alpha\beta}p_{\mu}+mI_{\alpha\beta}}{2m}=\frac{\gamma^{\mu}p_{\mu}+m}{2m}=\frac{\not{p}+m}{2m}\quad\text{sum on } s \tag{8-25}$$

$$v_s(\mathbf{p})\bar{v}_s(\mathbf{p})=v_{s,\alpha}(\mathbf{p})\bar{v}_{s,\beta}(\mathbf{p})=\frac{\gamma^{\mu}_{\alpha\beta}p_{\mu}-mI_{\alpha\beta}}{2m}=\frac{\gamma^{\mu}p_{\mu}-m}{2m}=\frac{\not{p}-m}{2m}\quad\text{sum on } s\,. \tag{8-26}$$

Using those relations in (8-24) gives us

$$\sum_{r'=1}^{2}\sum_{r_1=1}^{2}\sum_{r_2=1}^{2}\frac{1}{2}\sum_{r=1}^{2}\left|\mathcal{M}_{m=1}\right|^2=\frac{G^2}{4}\left(\frac{\not{p}'+m_e}{2m_e}\right)_{\eta'\alpha}\left(\frac{\not{q}_1-m_{\nu e}}{2m_{\nu e}}\right)_{\gamma\delta'}\left(\frac{\not{q}_2+m_{\nu\mu}}{2m_{\nu\mu}}\right)_{\gamma'\delta}\left(\frac{\not{p}+m_{\mu}}{2m_{\mu}}\right)_{\eta\alpha'}$$

$$\times\gamma^{\mu}_{\alpha\beta}\left(1-\gamma^5\right)_{\beta\gamma}\gamma_{\mu,\delta\varepsilon}\left(1-\gamma^5\right)_{\varepsilon\eta}\gamma^{\nu}_{\alpha'\beta'}\left(1-\gamma^5\right)_{\beta'\gamma'}\gamma_{\nu,\delta'\varepsilon'}\left(1-\gamma^5\right)_{\varepsilon'\eta'}. \tag{8-27}$$

$$=\frac{G^2}{4}\gamma^{\mu}_{\alpha\beta}\left(1-\gamma^5\right)_{\beta\gamma}\left(\frac{\not{q}_1-m_{\nu e}}{2m_{\nu e}}\right)_{\gamma\delta'}\gamma_{\nu,\delta'\varepsilon'}\left(1-\gamma^5\right)_{\varepsilon'\eta'}\left(\frac{\not{p}'+m_e}{2m_e}\right)_{\eta'\alpha}$$

$$\times\gamma_{\mu,\delta\varepsilon}\left(1-\gamma^5\right)_{\varepsilon\eta}\left(\frac{\not{p}+m_{\mu}}{2m_{\mu}}\right)_{\eta\alpha'}\gamma^{\nu}_{\alpha'\beta'}\left(1-\gamma^5\right)_{\beta'\gamma'}\left(\frac{\not{q}_2+m_{\nu\mu}}{2m_{\nu\mu}}\right)_{\gamma'\delta} \tag{8-28}$$

Due to the smallness of the neutrino masses, we can ignore them in terms where they appear added to, or subtracted from, their own 4-momenta. We can thus write

Simplifying, helpful approximation by ignoring neutrino mass

$$=\frac{G^2}{4}\gamma^{\mu}_{\alpha\beta}\left(1-\gamma^5\right)_{\beta\gamma}\left(\frac{q_{1\zeta}\gamma^{\zeta}}{2m_{\nu e}}\right)_{\gamma\delta'}\gamma_{\nu,\delta'\varepsilon'}\left(1-\gamma^5\right)_{\varepsilon'\eta'}\left(\frac{\not{p}'+m_e}{2m_e}\right)_{\eta'\alpha}$$

$$\times\gamma_{\mu,\delta\varepsilon}\left(1-\gamma^5\right)_{\varepsilon\eta}\left(\frac{\not{p}+m_{\mu}}{2m_{\mu}}\right)_{\eta\alpha'}\gamma^{\nu}_{\alpha'\beta'}\left(1-\gamma^5\right)_{\beta'\gamma'}\left(\frac{q_{2\xi}\gamma^{\xi}}{2m_{\nu\mu}}\right)_{\gamma'\delta}. \tag{8-29}$$

In the first row, the term with m_e in the numerator is the trace (sum over α) of (8-30).

$$\text{Top row (8-29) mass term factor} = \gamma^{\mu}_{\alpha\beta}\left(1-\gamma^5\right)_{\beta\gamma}\left(\frac{q_{1\zeta}\gamma^{\zeta}}{2m_{\nu e}}\right)_{\gamma\delta'}\gamma_{\nu,\delta'\varepsilon'}\left(1-\gamma^5\right)_{\varepsilon'\eta'}\left(\frac{m_e}{2m_e}\right)_{\eta'\alpha}, \tag{8-30}$$

which given the second part of (8-19) and the fact (Vol. 1, (4-164), pg. 124) that the trace of an odd number of gamma matrices is zero plus (5-87) of Chap. 5, i.e.,

$$=\text{Tr}\left\{\gamma^{\mu}\left(1-\gamma^5\right)\left(\frac{q_{1\zeta}\gamma^{\zeta}}{2m_{\nu_e}}\right)\gamma_{\nu}\left(1-\gamma^5\right)\left(\tfrac{1}{2}\right)\right\}=\text{Tr}\left\{\gamma^{\mu}\left(\frac{q_{1\zeta}\gamma^{\zeta}}{4m_{\nu_e}}\right)\left(1+\gamma^5\right)\gamma_{\nu}\left(1-\gamma^5\right)\right\}$$

$$=\text{Tr}\left\{\gamma^{\mu}\left(\frac{q_{1\zeta}\gamma^{\zeta}}{4m_{\nu_e}}\right)\gamma_{\nu}\underbrace{\left(1-\gamma^5\right)\left(1-\gamma^5\right)}_{1-2\gamma^5+\left(\gamma^5\right)^2=2\left(1-\gamma^5\right)}\right\}=\text{Tr}\left\{\gamma^{\mu}\left(\frac{q_{1\zeta}\gamma^{\zeta}}{2m_{\nu_e}}\right)\gamma_{\nu}\left(1-\gamma^5\right)\right\} \tag{8-31}$$

$$=\left(\frac{q_{1\zeta}}{2m_{\nu_e}}\right)\left(\text{Tr}\left\{\gamma^{\mu}\gamma^{\zeta}\gamma_{\nu}\right\}-\text{Tr}\left\{\gamma^{\mu}\gamma^{\zeta}\gamma_{\nu}\gamma^5\right\}\right)=\left(\frac{q_{1\zeta}}{2m_{\nu_e}}\right)g_{\nu\rho'}\left(\text{Tr}\left\{\gamma^{\mu}\gamma^{\zeta}\gamma^{\rho'}\right\}-\text{Tr}\left\{\gamma^5\gamma^{\mu}\gamma^{\zeta}\gamma^{\rho'}\right\}\right)=0.$$

Similar logic applies to the muon mass term in the second row of (8-29), so we can drop all masses in the numerator. That leaves us with the following, where in (8-33) we raise and lower indices on spacetime inner products to give us factors in the form we seek,

$$m_\mu m_e m_{\nu_e} m_{\nu_\mu} \sum_{r'=1}^{2} \sum_{r_1=1}^{2} \sum_{r_2=1}^{2} \frac{1}{2}\sum_{r=1}^{2} \left|\mathcal{M}_{m=1}\right|^2 = \frac{G^2}{64}\gamma^\mu_{\alpha\beta}\left(1-\gamma^5\right)_{\beta\gamma} q_{1\zeta}\gamma^\zeta{}_{\gamma\delta'}\gamma_{\nu,\delta'\varepsilon'}\left(1-\gamma^5\right)_{\varepsilon'\eta'} p'_\rho \gamma^\rho_{\eta'\alpha}$$
$$\times \gamma_{\mu,\delta\varepsilon}\left(1-\gamma^5\right)_{\varepsilon\eta} p_\chi \gamma^\chi_{\eta\alpha'}\gamma^\nu_{\alpha'\beta'}\left(1-\gamma^5\right)_{\beta'\gamma'} q_{2\xi}\gamma^\xi_{\gamma'\delta}. \tag{8-32}$$

$$= \frac{G^2}{64}\left(p'_\rho \gamma^\rho_{\eta'\alpha}\gamma^\mu_{\alpha\beta}\left(1-\gamma^5\right)_{\beta\gamma} q_{1\zeta}\gamma^\zeta_{\gamma\delta'}\gamma^\nu_{\delta'\varepsilon'}\left(1-\gamma^5\right)_{\varepsilon'\eta'}\right)$$
$$\times\left(q_2^\xi \gamma_{\xi,\gamma'\delta}\gamma_{\mu,\delta\varepsilon}\left(1-\gamma^5\right)_{\varepsilon\eta} p^\chi \gamma_{\chi,\eta\alpha'}\gamma_{\nu,\alpha'\beta'}\left(1-\gamma^5\right)_{\beta'\gamma'}\right) \tag{8-33}$$

$$= \frac{G^2}{64}\underbrace{p'_\rho q_{1\zeta}\operatorname{Tr}\left\{\gamma^\rho\gamma^\mu\left(1-\gamma^5\right)\gamma^\zeta\gamma^\nu\left(1-\gamma^5\right)\right\}}_{D^{\mu\nu}}\underbrace{q_2^\xi p^\chi \operatorname{Tr}\left\{\gamma_\xi\gamma_\mu\left(1-\gamma^5\right)\gamma_\chi\gamma_\nu\left(1-\gamma^5\right)\right\}}_{E_{\mu\nu}}. \tag{8-34}$$

Let's look at the part of (8-34) we labeled $D^{\mu\nu}$, and use the second part of (8-19) again. *Evaluating traces*

$$\begin{aligned} D^{\mu\nu} &= p'_\rho q_{1\xi}\operatorname{Tr}\left\{\gamma^\rho\gamma^\mu\left(1-\gamma^5\right)\gamma^\xi\gamma^\nu\left(1-\gamma^5\right)\right\} = p'_\rho q_{1\xi}\operatorname{Tr}\left\{\gamma^\rho\gamma^\mu\gamma^\xi\left(1+\gamma^5\right)\gamma^\nu\left(1-\gamma^5\right)\right\} \\ &= p'_\rho q_{1\xi}\operatorname{Tr}\left\{\gamma^\rho\gamma^\mu\gamma^\xi\gamma^\nu\left(1-\gamma^5\right)\left(1-\gamma^5\right)\right\} = 2p'_\rho q_{1\xi}\operatorname{Tr}\left\{\gamma^\rho\gamma^\mu\gamma^\xi\gamma^\nu\left(1-\gamma^5\right)\right\} \\ &= 2p'_\rho q_{1\xi}\left(\operatorname{Tr}\left\{\gamma^\rho\gamma^\mu\gamma^\xi\gamma^\nu\right\} - \operatorname{Tr}\left\{\gamma^\rho\gamma^\mu\gamma^\xi\gamma^\nu\gamma^5\right\}\right). \end{aligned} \tag{8-35}$$

To evaluate (8-35), we'll need the gamma matrix trace relation of Vol. 1, (4-165), pg. 124,

$$\operatorname{Tr}\left(\gamma^\rho\gamma^\mu\gamma^\xi\gamma^\nu\right) = 4\left(g^{\rho\mu}g^{\xi\nu} - g^{\rho\xi}g^{\mu\nu} + g^{\rho\nu}g^{\mu\xi}\right) \tag{8-36}$$

along with a similar relation when one gamma matrix is γ^5, given in Appendix B of Chap. 5, where $\varepsilon^{\rho\mu\xi\nu}$ is the Levi-Civita tensor of rank 4,

$$\operatorname{Tr}\left\{\gamma^\rho\gamma^\mu\gamma^\xi\gamma^\nu\gamma^5\right\} = \operatorname{Tr}\left\{\gamma^5\gamma^\rho\gamma^\mu\gamma^\xi\gamma^\nu\right\} = -4i\varepsilon^{\rho\mu\xi\nu}. \tag{8-37}$$

With these, (8-35) becomes

$$D^{\mu\nu} = 8p'_\rho q_{1\xi}\left(g^{\rho\mu}g^{\xi\nu} - g^{\rho\xi}g^{\mu\nu} + g^{\rho\nu}g^{\mu\xi} + i\varepsilon^{\rho\mu\xi\nu}\right). \tag{8-38}$$

Result after traces evaluated

In parallel fashion, the second part of (8-34) labeled $E_{\mu\nu}$ becomes

$$E_{\mu\nu} = 8q_2^\gamma p^\delta\left(g_{\gamma\mu}g_{\delta\nu} - g_{\gamma\delta}g_{\mu\nu} + g_{\gamma\nu}g_{\mu\delta} + i\varepsilon_{\gamma\mu\delta\nu}\right). \tag{8-39}$$

With $D^{\mu\nu}$ and $E_{\mu\nu}$ given by (8-38) and (8-39), (8-34) is

$$m_\mu m_e m_{\nu_e} m_{\nu_\mu} \sum_{r'=1}^{2} \sum_{r_1=1}^{2} \sum_{r_2=1}^{2} \frac{1}{2}\sum_{r=1}^{2} \left|\mathcal{M}_{m=1}\right|^2 = \frac{G^2}{64}D^{\mu\nu}E_{\mu\nu}. \tag{8-40}$$

Evaluating $D^{\mu\nu}E_{\mu\nu}$ is an algebraic workload not involving any real learning, so you can either take my word for it, or check it out done for you in Appendix B. As mentioned in the preface of this book, you are better off using your time to digest additional physics than wasting it trying to crank out the algebra in a low-level math exercise. So, I do pedagogically meaningless, time-consuming things like this for you. At any rate, the result is

$$D^{\mu\nu}E_{\mu\nu} = \left(64p'_\rho q_{1\xi}q_2^\gamma p^\delta\right)\left(4\delta^\rho_\gamma\delta^\xi_\delta\right) = \left(64p'_\gamma q_{1\delta}q_2^\gamma p^\delta\right)(4) = 64\left(4\left(pq_1\right)\left(p'q_2\right)\right), \tag{8-41}$$

Final form of key part of analysis

where two four-vectors juxtaposed implies inner product. So, (8-40) becomes

$$m_\mu m_e m_{\nu_e} m_{\nu_\mu} \sum_{r'=1}^{2} \sum_{r_1=1}^{2} \sum_{r_2=1}^{2} \frac{1}{2}\sum_{r=1}^{2} \left|\mathcal{M}_{m=1}\right|^2 = 4G^2\left(pq_1\right)\left(p'q_2\right), \tag{8-42}$$

where neutrino masses have been taken as negligible compared to their energies. Thus, (8-18), becomes, after spin summing,

Muon differential decay rate ignoring ν masses and other branches, all spins, specific momenta

$$d\Gamma_{muon} \approx d\Gamma_{m=1} \approx \frac{4G^2}{(2\pi)^5 E}\delta^{(4)}\left(p' + q_1 + q_2 - p\right)\left(pq_1\right)\left(p'q_2\right)\frac{d\mathbf{p}'}{E'}\frac{d\mathbf{q}_1}{E_1}\frac{d\mathbf{q}_2}{E_2}. \tag{8-43}$$

Integration Over 3-Momenta

We next need to evaluate (8-43) for all possible final momenta. As I often do, I will write out all steps, including those that are very elementary, to save you time and tedium (like being off by a sign or factor of 2 and wasting considerable time trying to find out where).

Need to integrate over all possible momenta

Step 1: Assumptions

We have already assumed negligible neutrino masses.

1) That means

Step 1: Our assumptions

$$q_1^2 = \left(E_1\right)^2 - \left|\mathbf{q}_1\right|^2 = m_\nu^2 \approx 0 \;\rightarrow\; E_1 \approx \left|\mathbf{q}_1\right| \text{ and similarly, } q_2^2 \approx 0, \text{ so } E_2 \approx \left|\mathbf{q}_2\right| \tag{8-44}$$

We will also assume the following.

2) Electron mass is negligible compared to muon mass.

3) Electron mass is negligible compared to its energy (not always true but almost always), since almost all of the muon mass must be converted to the energy of the three final particles, one of which is the electron. Thus,

$$(E')^2 - \left|\mathbf{p}'\right|^2 = m_e^2 \ll (E')^2 \qquad \rightarrow \qquad E' \approx \left|\mathbf{p}'\right|. \tag{8-45}$$

Further, in the rest frame of the muon, 3-momentum is zero, so in that frame the sum of the 3-momenta of all final particles must equal zero. Thus, the greatest 3-momentum the electron could have is the case where it has opposite 3-momentum from the two neutrinos, and all 3 have colinear velocities.

Muon rest frame, electron velocity in opposite direction of neutrinos $\left|\mathbf{q}_1\right| + \left|\mathbf{q}_2\right| = \left|\mathbf{p}'\right|$

$$\text{From } E_1 + E_2 + E' = m_\mu \xrightarrow[\text{assumptions}]{\text{above}} \left|\mathbf{q}_1\right| + \left|\mathbf{q}_2\right| + \left|\mathbf{p}'\right| \approx m_\mu \tag{8-46}$$

$$\rightarrow \left|\mathbf{p}'\right| + \left|\mathbf{p}'\right| \approx m_\mu \;\rightarrow\; \left|\mathbf{p}'\right| \approx \tfrac{1}{2}m_\mu \approx E'.$$

In other words,

4) To high accuracy, maximum electron energy is ½m_μ when working in the muon rest frame.

Step 2: Deducing some helpful relations

Step 2: Additional Relations We'll Need

It will help if we define a new variable

$$q \equiv p - p' \qquad \left(q^\mu \equiv p^\mu - p'^\mu\right), \tag{8-47}$$

where

$$q^2 = q_0^2 - \mathbf{q}^2 \qquad \text{(invariant)}. \tag{8-48}$$

In the muon rest frame (where $\mathbf{p} = 0$), we will use key relations involving q, p, and p' as follows.

$$\begin{gathered} q^2 = \left(m_\mu - E'\right)^2 - \mathbf{p}'^2 = \left(m_\mu^2 - 2m_\mu E' + E'^2\right) - \left(E'^2 - m_e^2\right) = m_\mu^2 - 2m_\mu E' + m_e^2 \\ q^2 \approx m_\mu^2 - 2m_\mu E' = m_\mu\left(m_\mu - 2E'\right) \end{gathered} \tag{8-49}$$

$$pq = m_\mu q_0 = m_\mu\left(m_\mu - E'\right) \tag{8-50}$$

$$p'q = E'q_0 - \mathbf{p}'\cdot\mathbf{q} = E'\left(m_\mu - E'\right) - \mathbf{p}'\cdot\left(-\mathbf{p}'\right) = m_\mu E' - E'^2 + \mathbf{p}'^2 = m_\mu E' - m_e^2 \approx m_\mu E' \tag{8-51}$$

$$pp' = EE' - \mathbf{p}\cdot\mathbf{p}' = m_\mu E' - (0)\cdot\mathbf{p}' = m_\mu E' \tag{8-52}$$

Step 3: Integrate $d\Gamma_{m=1}$ Over Neutrino 3-momenta

Step 3: Integrate differential decay rate over neutrino $\mathbf{q}_1$ and $\mathbf{q}_2$

We can, for the moment, assume we have fixed muon 4-momenta and fixed electron 4-momenta, i.e., we integrate (8-43) over the neutrino momenta only. This means q of (8-47) is fixed, too. Then

(8-43) leads, where we change notation slightly on the differentials to clearly indicate they are 3-dimensional, to

$$\underbrace{\int d\Gamma_{m=1}}_{\text{over } \mathbf{q}_1,\mathbf{q}_2} \approx \frac{4G^2}{(2\pi)^5 E} p_\mu p'_\nu \frac{d^3\mathbf{p}'}{E'} \underbrace{\int \delta^{(4)}(q_1+q_2-q) q_1^\mu q_2^\nu \frac{d^3\mathbf{q}_1}{E_1}\frac{d^3\mathbf{q}_2}{E_2}}_{I^{\mu\nu}} \tag{8-53}$$

Isolating $I^{\mu\nu}$, the part of the integral over $\mathbf{q}_1$ and $\mathbf{q}_2$

If we find the integral $I^{\mu\nu}$, then we have found what we seek in this step. We will employ a trick, using Lorentz covariance, to help evaluate that integral.

A trick to evaluate $I^{\mu\nu}$

Finding $I^{\mu\nu}$

As can be seen from its expression in (8-53), the integral $I^{\mu\nu}$, after integration, only involves q, both in the form of $q^\mu q^\nu$ (where q^μ is a covariant 4-vector, as seen from its definition in (8-47)) and the invariant q^2, from (8-48). Thus, $I^{\mu\nu}$ is necessarily a covariant 4D tensor, the most general relation for which is

$$I^{\mu\nu} = g^{\mu\nu} A(q^2) + q^\mu q^\nu B(q^2). \tag{8-54}$$

Express $I^{\mu\nu}$ in most general Lorentz covariant form

If we can find A and B, we can determine $I^{\mu\nu}$ from (8-54). To that end, we will find the following relations, which are obtained from (8-54), useful.

$$g_{\mu\nu} I^{\mu\nu} = 4A(q^2) + q^2 B(q^2) \tag{8-55}$$

$$q_\mu q_\nu I^{\mu\nu} = q^2 A(q^2) + (q^2)^2 B(q^2). \tag{8-56}$$

Auxiliary relations with $I^{\mu\nu}$ that will help find $I^{\mu\nu}$, i.e., will help find A and B

If we can evaluate the left hand sides of (8-55) and (8-56), we will have two equations in two unknowns, A and B, and can then solve them for A and B. As these relations use q^2 explicitly, it will help eventually if we express that in a more useful form, as follows.

From the delta function in (8-53), we see q, in addition to (8-47), can be expressed as

$$q = q_1 + q_2. \tag{8-57}$$

Alternative way to express q that will help

So, from (8-44), for negligible neutrino masses,

$$q^2 = (q_1+q_2)(q_1+q_2) = q_1^2 + 2q_1 q_2 + q_2^2 \approx 2q_1 q_2. \tag{8-58}$$

Approximate form of q^2 that will help

Now, turning to the LHS of (8-55), using $I^{\mu\nu}$ from (8-53) along with (8-58), we have

$$\begin{aligned} g_{\mu\nu} I^{\mu\nu} &= \int \delta^{(4)}(q_1+q_2-q) g_{\mu\nu} q_1^\mu q_2^\nu \frac{d^3\mathbf{q}_1}{E_1}\frac{d^3\mathbf{q}_2}{E_2} \\ &= \int \delta^{(4)}(q_1+q_2-q)(q_1 q_2)\frac{d^3\mathbf{q}_1}{E_1}\frac{d^3\mathbf{q}_2}{E_2} = \int \delta^{(4)}(q_1+q_2-q)\frac{q^2}{2}\frac{d^3\mathbf{q}_1}{E_1}\frac{d^3\mathbf{q}_2}{E_2}. \end{aligned} \tag{8-59}$$

Evaluating first $I^{\mu\nu}$ auxiliary relation

From (8-47), and the fact that we are fixing p and p' during our integration over $\mathbf{q}_1$ and $\mathbf{q}_2$, we have

$$g_{\mu\nu} I^{\mu\nu} = \frac{q^2}{2} \underbrace{\int \delta^{(4)}(q_1+q_2-q)\frac{d^3\mathbf{q}_1}{E_1}\frac{d^3\mathbf{q}_2}{E_2}}_{I} = \frac{q^2}{2} I. \tag{8-60}$$

That reduces to a function of q^2 and an integral I, both invariant

The LHS of (8-60) and q^2 on the RHS are invariant, so the integral I must be invariant, as well. Hence, it can be evaluated in any reference frame (not just the muon rest frame which this analysis focuses on). The easiest frame to evaluate I turns out to be the center-of-momentum (COM) frame of the two neutrinos, where $\mathbf{q}_1 = -\mathbf{q}_2$, so from (8-57), $\mathbf{q} = 0$, and from (8-44), the neutrino energies are equal. We shall label that energy ω.

Because I invariant, can evaluate in any frame

$$\text{In neutrinos' COM frame} \quad \mathbf{q} = 0 \quad \text{and} \quad \begin{aligned} &\omega = E_1 = |\mathbf{q}_1| = E_2 = |\mathbf{q}_2| = \frac{E_1+E_2}{2} = \frac{q_0}{2} \\ &q^2 = (E_1+E_2)(E_1+E_2) = (2\omega)(2\omega) = 4\omega^2 \end{aligned}. \tag{8-61}$$

Pick most helpful frame, COM of the 2 neutrinos

Then, I of (8-60) becomes

$$I=\int\delta\left(E_1+E_2-q_0\right)\delta^{(3)}\left(\mathbf{q}_1+\mathbf{q}_2-\mathbf{q}\right)\frac{d^3\mathbf{q}_1}{E_1}\frac{d^3\mathbf{q}_2}{E_2}=\int\frac{\delta\left(2\omega-q_0\right)}{\omega^2}d^3\mathbf{q}_1$$
$$=\int\frac{\delta\left(2\omega-q_0\right)}{\omega^2}4\pi\left|\mathbf{q}_1\right|^2d\left|\mathbf{q}_1\right|\approx4\pi\int\frac{\delta\left(2\omega-q_0\right)}{\omega^2}\omega^2\frac{dq_0}{2}=\frac{4\pi}{2}\int\delta\left(2\omega-q_0\right)dq_0=2\pi. \tag{8-62}$$

In that frame, and thus all frames, $I=2\pi$

As this result is invariant, we can now return to our analysis in the muon rest frame, where the same value of I holds.

Return to muon rest frame, use $I=2\pi$ there

From (8-60), we see (8-55) is equal to

$$g_{\mu\nu}I^{\mu\nu}=\pi q^2=4A\left(q^2\right)+q^2B\left(q^2\right). \tag{8-63}$$

1st auxiliary relation equal to πq^2

In similar fashion, we will work out the LHS of (8-56).

$$q_\mu q_\nu I^{\mu\nu}=\int\delta^{(4)}\left(q_1+q_2-q\right)q_\mu q_\nu q_1^\mu q_2^\nu\frac{d^3\mathbf{q}_1}{E_1}\frac{d^3\mathbf{q}_2}{E_2}=\int\delta^{(4)}\left(q_1+q_2-q\right)\left(qq_1\right)\left(qq_2\right)\frac{d^3\mathbf{q}_1}{E_1}\frac{d^3\mathbf{q}_2}{E_2}. \tag{8-64}$$

Evaluating 2nd auxiliary relation in $I^{\mu\nu}$

(8-64) is invariant, so we will use the COM of the two neutrinos frame. Then, with (8-61), we express the inner products in (8-64) as

$$qq_1=\left(q_1+q_2\right)q_1=\left(E_1+E_2\right)E_1-\underbrace{\left(\mathbf{q}_1+\mathbf{q}_2\right)}_{=0}\cdot\mathbf{q}_1=\left(E_1+E_2\right)E_1=2\omega^2=\frac{q^2}{2}$$
$$qq_2=\left(E_1+E_2\right)E_2=2\omega^2=\frac{q^2}{2}. \tag{8-65}$$

Evaluating factors in that relation

Then, (8-64) becomes

$$q_\mu q_\nu I^{\mu\nu}=\frac{q^4}{4}\int\delta^{(4)}\left(q_1+q_2-q\right)\frac{d^3\mathbf{q}_1}{E_1}\frac{d^3\mathbf{q}_2}{E_2}=\frac{q^4}{4}I=\frac{q^4}{2}\pi, \tag{8-66}$$

so, with (8-56)

$$q_\mu q_\nu I^{\mu\nu}=\frac{\pi}{2}\left(q^2\right)^2=q^2A\left(q^2\right)+\left(q^2\right)^2B\left(q^2\right). \tag{8-67}$$

2nd auxiliary relation equal to $\frac{1}{2}\pi(q^2)^2$

We can solve (8-63) and (8-67) for A and B. Rather than taking the time to carry out the steps of this simple algebra problem, I will just give you the answer.

Do **Problem 2** to check the validity of (8-68).

$$A=\frac{\pi}{6}q^2\qquad\qquad B=\frac{\pi}{3}. \tag{8-68}$$

Solve two auxiliary relations in two unknowns A and B

Using (8-68) in (8-54), we find

$$I^{\mu\nu}=g^{\mu\nu}A\left(q^2\right)+q^\mu q^\nu B\left(q^2\right)=\frac{\pi}{6}\left(g^{\mu\nu}q^2+2q^\mu q^\nu\right). \tag{8-69}$$

And thus, find $I^{\mu\nu}$

End of Finding $I^{\mu\nu}$

With (8-69) in our original relation for the integral of $d\Gamma_{\mu=1}$ (8-53), we have

$$\underbrace{\int d\Gamma_{m=1}}_{\text{over }\mathbf{q}_1,\mathbf{q}_2}\approx\frac{4G^2}{(2\pi)^5E}p_\mu p'_\nu\frac{\pi}{6}\left(g^{\mu\nu}q^2+2q^\mu q^\nu\right)\frac{d^3\mathbf{p}'}{E'}$$
$$=\frac{2\pi}{3}\frac{G^2}{(2\pi)^5E}p_\mu p'_\nu\left(g^{\mu\nu}q^2+2q^\mu q^\nu\right)\frac{d^3\mathbf{p}'}{E'}=\frac{G^2}{3(2\pi)^4E}\left(\left(pp'\right)q^2+2\left(pq\right)\left(p'q\right)\right)\frac{d^3\mathbf{p}'}{E'}. \tag{8-70}$$

With that, we get final spin sums result, but still need momenta integration

This is the integral we sought over the neutrino momenta (for all spins). As a final step we need to integrate that over all final electron momenta.

Step 4: Integrate $d\Gamma_{m=1}$ Over Final Electron 3-momentum

To begin our integration over $\mathbf{p}'$ in the muon rest frame, first substitute the values in (8-49) to (8-52) into (8-70).

$$\underbrace{\int d\Gamma_{m=1}}_{\text{over } \mathbf{q}_1,\mathbf{q}_2} \approx \frac{G^2}{3(2\pi)^4 E}\Big(\left(m_\mu E'\right)m_\mu\left(m_\mu - 2E'\right) + 2m_\mu\left(m_\mu - E'\right)\left(m_\mu E'\right)\Big)\frac{d^3\mathbf{p}'}{E'}$$
$$= \frac{G^2 m_\mu^2}{3(2\pi)^4 E}E'\Big(\left(m_\mu - 2E'\right) + 2\left(m_\mu - E'\right)\Big)\frac{d^3\mathbf{p}'}{E'} = \frac{G^2 m_\mu^2}{3(2\pi)^4 E}\left(3m_\mu - 4E'\right)d^3\mathbf{p}'. \quad (8\text{-}71)$$

Step 4: Integrating $d\Gamma_{m=1}$ over momenta using spin sum results

In (8-71), we will want to use a relation we have used often, along with (8-45) for negligible electron mass,

$$d^3\mathbf{p}' = 4\pi\left|\mathbf{p}'\right|^2 d\left|\mathbf{p}'\right| \approx 4\pi\left(E'\right)^2 dE'. \quad (8\text{-}72)$$

Converting orthonormal coordinates to spherical for easier integration

Thus, with (8-71) and (8-72), we find the complete integral of (8-43) is

$$\Gamma_{m=1} = \underbrace{\int\Bigg(\underbrace{\int d\Gamma_{m=1}}_{\text{over } \mathbf{q}_1,\mathbf{q}_2}\Bigg)}_{\text{over } \mathbf{p}'} = \int \frac{G^2 m_\mu^2}{3(2\pi)^4 E}\left(3m_\mu - 4E'\right)4\pi(E')^2\,dE'$$
$$= \frac{2G^2 m_\mu^2}{3(2\pi)^3 E}\int\left(3m_\mu (E')^2 - 4(E')^3\right)dE'. \quad (8\text{-}73)$$

As noted in (8-46), the maximum energy the electron could have is ½m_μ (and the minimum is zero) in the muon's rest frame. Integrating (8-73) over those limits, we have, where $E = m_\mu$ in the muon's rest frame,

$$\Gamma_{m=1} = \frac{2G^2 m_\mu^2}{3(2\pi)^3 m_\mu}\int_0^{m_\mu/2}\left(3m_\mu (E')^2 - 4(E')^3\right)dE' = \frac{G^2 m_\mu}{3\bullet 4\pi^3}\left(\frac{3m_\mu\left(\frac{m_\mu}{2}\right)^3}{3} - \frac{4\left(\frac{m_\mu}{2}\right)^4}{4}\right)$$
$$= \frac{G^2 m_\mu}{12\pi^3}\left(\frac{m_\mu^4}{8} - \frac{m_\mu^4}{16}\right) = \frac{G^2 m_\mu}{12\pi^3}\left(\frac{m_\mu^4}{16}\right) = \frac{G^2 m_\mu^5}{192\pi^3}.. \quad (8\text{-}74)$$

Final result for $\Gamma_{m=1}$ after spin sums and momenta integration

Muon Decay Rate and Lifetime

Thus, we obtain a total decay rate for the muon, for our given approximations, of

$$\Gamma_{muon} \approx \Gamma_{m=1} \approx \frac{G^2 m_\mu^5}{192\pi^3}, \quad (8\text{-}75)$$

Muon total decay rate, approximate

and hence, a muon lifetime, to good approximation, of

$$\tau_\mu = \frac{1}{\Gamma_{muon}} \approx \frac{192\pi^3}{G^2 m_\mu^5}. \quad (8\text{-}76)$$

Muon lifetime, approximate

Note the inverse dependence on muon mass to the 5th power. Heavy particles decay faster, much faster.

Do **Problem 3** to show (8-76) gives a muon lifetime of about $2.15\text{x}10^{-6}$ seconds.

When one includes the other muon decay mode contributions (which are small) plus QED radiative corrections, a slightly different result predicts the muon lifetime to be very close to the experimental value, which is

$$\tau_\mu = (2.19703 \pm 0.00004)\times 10^{-6}\text{ sec}. \quad (8\text{-}77)$$

Muon lifetime, measured value

Actually, in practice, the more precise version of (8-76) is used with the experimentally measured muon mass and muon decay time (8-77) to determine G, the Fermi constant, to high accuracy. Given the definition of G (8-17), we find

Muon lifetime, mass measurement used to determine G

$$\frac{2G}{\pi\sqrt{2}}m_W^2 = \frac{g^2}{4\pi} \approx 32\times10^{-3}, \tag{8-78}$$

which is the same order of magnitude as the QED fine structure constant $e^2/4\pi \approx 1/137 \approx 7.3\text{x}10^{-3}$, and implies lowest perturbation order calculations should describe weak interactions well. (Of course, we already knew this a different way from the relation $e = g\sin\theta_W$ that we derived in Chap. 7.) Given that, in calculating cross sections, (8-78) is effectively further reduced by a factor of $1/m_W^2$, weak interactions are actually much weaker than QED interactions.

Weak coupling on order of QED, so perturbation OK

Could a Muon Decay into Quarks?

One could ask why we wouldn't have a muon decay mode like

$$\mu^- \to d + \bar{u} + \nu_\mu, \tag{8-79}$$

which would look like Fig. 8-2, pg. 269, with the electron and electron antineutrino replaced with the down and anti-up quarks. If the quarks were of the same color and anticolor, they could form a bound state and all charges would be conserved.

The answer is that the bound state that would be formed would have a total mass-energy greater than that of the original muon. The muon has mass 0.106 GeV, but a typical meson (quark-antiquark pair) is on the order of 0.135 GeV or greater. From the constraint of conservation of mass-energy, such decay is not possible.

Quarks too heavy for muon to decay into

8.2.4 Tau Decay

The story for tau particle decay is different, and more complex. For it, we could have decay into a quark-antiquark pair, since the tau, at 1.777 GeV, has more than enough mass to do so. It could also decay into a muon (plus neutrinos). Allowable decays for the tau, all of which conserve mass-energy, include

$$\begin{array}{ccc} \tau^- \to e^- + \bar{\nu}_e + \nu_\tau & \tau^- \to \mu^- + \bar{\nu}_\mu + \nu_\tau & \\ \tau^- \to d_r + \bar{u}_{\bar{r}} + \nu_\tau & \tau^- \to d_g + \bar{u}_{\bar{g}} + \nu_\tau & \tau^- \to d_b + \bar{u}_{\bar{b}} + \nu_\tau \\ \tau^- \to s_r + \bar{u}_{\bar{r}} + \nu_\tau & \tau^- \to s_g + \bar{u}_{\bar{g}} + \nu_\tau & \tau^- \to s_b + \bar{u}_{\bar{b}} + \nu_\tau , \end{array} \tag{8-80}$$

Tau heavy enough to decay into quarks

plus even rarer decays, like the latter two in (8-9).

Given the greater number of decay modes for the tau over the muon, one would expect it to decay faster, for that reason alone. However, the greater mass makes for an even stronger effect, as decay rate varies with mass to the fifth power. (See (8-75) for the muon case.) When all these effects are taken into account (which, though we now have the skills to do it, is well beyond what we have space for in this text), one finds the tau lifetime to be

$$\tau_\tau = 3.23\times10^{-13} \text{ seconds}, \tag{8-81}$$

a whole lot faster than a blink of an eye.

Tau heavier than muon, much shorter lifetime

8.2.5 Neutron Decay

We will not carry out the calculation for neutron beta decay, but simply lay out the fundamental approach to doing so.

Fig. 8-4 shows the primary mode of decay, which can be written several ways, i.e.,

$$\begin{gathered} n^0 \to p^+ + e^- + \bar{\nu}_e \quad \text{or} \quad n^0 \to p^+ + W^- \to p^+ + e^- + \bar{\nu}_e \\ \text{or} \quad udd \to uud + W^- \to uud + e^- + \bar{\nu}_e . \end{gathered} \tag{8-82}$$

Ways to write primary neutron decay mode

There is another decay mode, which occurs about 0.1% of the time and is similar to one of those in (8-9),

$$n^0 \rightarrow p^+ + e^- + \bar{\nu}_e + \gamma . \qquad (8\text{-}83)$$

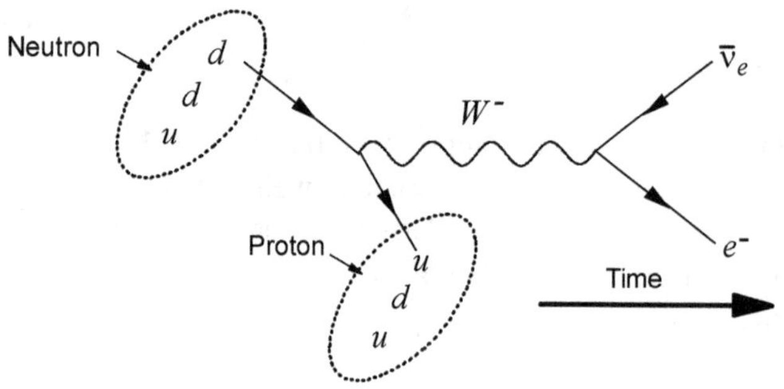

Neutron beta decay

Neutron
d
d
u
W^-
$\bar{\nu}_e$
Proton
u
d
u
e^-
Time

Figure 8-4. Neutron Beta Decay

Decay of a neutron is different when it is free from that when it is part of a nucleus. In the latter case, the binding energy of the nucleon affects the decay rate, and makes decay much slower. The different characteristics of each type of nucleus (different elements and isotopes) dictate how rapidly such a neutron decay can take place.

Neutrons in nuclei much longer lived than free neutron

For a free neutron, the lifetime is

$$\tau_n = 14 \text{ minutes}, 39.6 \text{ seconds} . \qquad (8\text{-}84)$$

Free neutron lifetime about 15 mins

Note that the down quark is heavier than the up quark, so we would not expect to see the reverse process, where a proton decays into a neutron (via a W^+ mediator). And we don't. Protons are extraordinarily stable against decay[1].

Proton doesn't decay to neutron because q_u lighter than q_d

8.2.6 W, Z, H Decays

In similar fashion to what we have done above, one can calculate the lifetime of any particle. For the electroweak bosons, in particular, one finds lifetimes of

W bosons	$3.2 \text{X} 10^{-25}$ seconds
Z boson	$2.6 \text{X} 10^{-25}$ seconds
Higgs boson	$1.6 \text{X} 10^{-22}$ seconds.

Decay times of bosons

The Higgs, though close in mass to the *W* and *Z*, has a much longer decay time due to different coupling constants (Yukawa couplings in the transition amplitude instead of *g*) and different decay branches.

8.3 Electroweak Interaction Scattering

We will investigate one case of electroweak scattering, electron-positron annihilation into a heavier lepton-antilepton pair. Other types of scattering follow a similar pattern.

Electroweak scattering: our focus $e^- + e^+ \rightarrow \mu^- + \mu^+$

8.3.1 Key Cross-section Relation

In Vol. 1, Chap. 17, we derived the differential cross-section in the center of momentum (COM) frame for an interaction involving two initial and two final particles, displayed therein as (17-68) on pg. 458. That relation is

$$\boxed{\left(\frac{d\sigma}{d\Omega_1'}\right)_{COM} = \frac{1}{64\pi^2 (E_1 + E_2)^2} \frac{|\mathbf{p}_1'|}{|\mathbf{p}_1|} \left(\prod_l^{\substack{\text{extern} \\ \text{ferms}}} 2m_l \right) |\mathcal{M}|^2 \quad \begin{cases} \text{COM, 2 initial \&} \\ \text{2 final particles,} \\ \text{elastic or inelastic.} \end{cases}} \qquad (8\text{-}85)$$

General relation for differential cross section, 2 particles in, 2 out (derived in Vol. 1)

We highlight (8-85) in a box, due to its importance, but you don't have to memorize it. Unprimed quantities refer to incoming particles; primed, to outgoing. Although we used (8-85) in Vol. 1 exclusively for QED interactions, its derivation was not so limited, and it is valid for all types of SM

[1] In the SM, protons simply don't decay. In certain more advanced, unproven theories, particularly grand unified theories, they may decay with lifetimes on the order of 10^{35} years. Large scale experiments to detect such decay have, so far, come up empty.

interactions. The factor of the square of the absolute value of $\mathcal{M}$, the Feynman amplitude, in (8-85) is a general characteristic of differential cross-sections for any interaction involving any number of particles.

Fig. 8-5 is a reproduction of Fig. 17-16 from Vol. 1 and shows the relationships between 3-momenta, energy, and scattering angles of the incoming and outgoing particles in the COM frame. The symbol l represents any charged lepton, i.e., electron, muon, or tau.

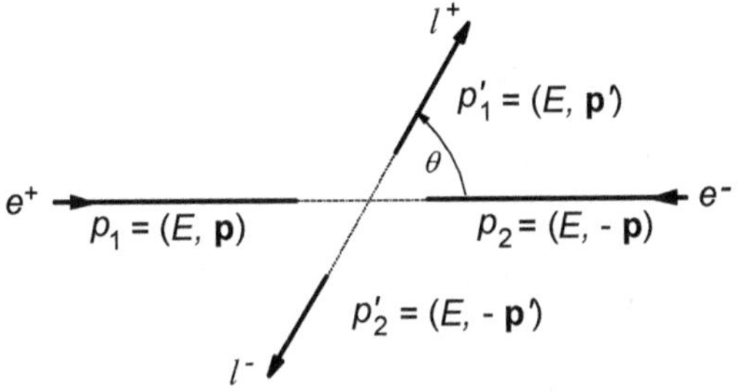

Kinematics and dynamics for $e^- + e^+ \rightarrow l^- + l^+$

Figure 8-5. Kinematics and Dynamics for the Process with Cross-Section (8-85)

Note that in the COM frame, $\mathbf{p}_1 = -\mathbf{p}_2$, and $\mathbf{p}'_1 = -\mathbf{p}'_2$, so if you know the 3-momentum of either incoming particle, you know it for the other, with the same thing being true for the outgoing particles.

8.3.2 Electron-Positron Scattering

In Vol. 1, Sects. 17.5.2 and 17.5.3, pgs. 463-471, we discussed electron-positron scattering due solely to QED interactions. In what follows, we will often be referencing the first of these sections, having a final lepton pair of either muon-antimuon or tau-antitau. We will focus on the reaction with final muons, as our result can be extrapolated to taus, simply by changing the muon mass m_μ to the tau mass m_τ in our final cross-section relation.

Thus, the interaction we wish to investigate is

The interaction we'll investigate closely

$$e^- + e^+ \rightarrow \mu^- + \mu^+, \tag{8-86}$$

which is represented diagrammatically, in lowest order, in Fig. 8-6.

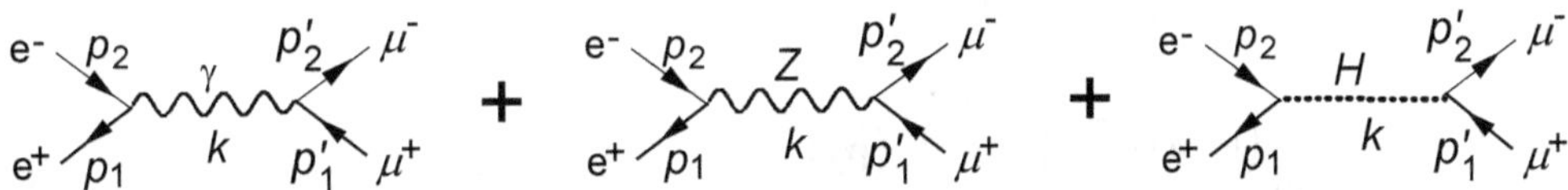

Feynman diagrams for that interaction at lowest order

Figure 8-6. Leading Order Contributions to $e^- + e^+ \rightarrow \mu^- + \mu^+$

At lowest order (no loops), there are three different bosons that can act as mediators, the photon, the Z gauge boson, and the Higgs, and in each case, the boson 4-momentum is $k = p_1 + p_2$. The corresponding Feynman amplitude is

Total amplitude = sum of sub-amplitudes

$$\mathcal{M} = \mathcal{M}_\gamma + \mathcal{M}_Z + \mathcal{M}_H, \tag{8-87}$$

where, from Feynman rules, we have (where (8-88) is the same as Vol. 1, (17-98), pg. 463), and the second part of (8-89), we explain below,

$$\mathcal{M}_\gamma = \bar{u}(\mathbf{p}'_2) ie\gamma^\alpha v(\mathbf{p}'_1) i\frac{-g_{\alpha\beta}}{k^2 + i\varepsilon} \bar{v}(\mathbf{p}_1) ie\gamma^\beta u(\mathbf{p}_2) = ie^2 \bar{u}(\mathbf{p}'_2)\gamma_\alpha v(\mathbf{p}'_1) \frac{1}{k^2 + i\varepsilon} \bar{v}(\mathbf{p}_1)\gamma^\alpha u(\mathbf{p}_2). \tag{8-88}$$

Photon mediator sub-amplitude

$$\mathcal{M}_Z = \bar{u}(\mathbf{p}_2')\left\{\frac{ig}{4\cos\theta_W}\right\}\gamma^\alpha\left(1-4\sin^2\theta_W-\gamma^5\right)v(\mathbf{p}_1')\left(i\frac{-g_{\alpha\beta}+k_\alpha k_\beta/m_Z^2}{k^2-m_Z^2+i\varepsilon}\right)$$
$$\times\bar{v}(\mathbf{p}_1)\left\{\frac{ig}{4\cos\theta_W}\right\}\gamma^\beta\left(1-4\sin^2\theta_W-\gamma^5\right)u(\mathbf{p}_2)$$
$$= i\frac{g^2}{16\cos^2\theta_W}\bar{u}(\mathbf{p}_2')\gamma_\alpha\left(1-4\sin^2\theta_W-\gamma^5\right)v(\mathbf{p}_1')\frac{1}{k^2-m_Z^2+i\varepsilon}$$
$$\times\bar{v}(\mathbf{p}_1)\gamma^\alpha\left(1-4\sin^2\theta_W-\gamma^5\right)u(\mathbf{p}_2) \tag{8-89}$$

Z boson mediator sub-amplitude

$$\mathcal{M}_H = \bar{u}(\mathbf{p}_2')\left(\frac{-i}{v}m_\mu\right)v(\mathbf{p}_1')\left(\frac{i}{k^2-m_H^2+i\varepsilon}\right)\bar{v}(\mathbf{p}_1)\left(\frac{-i}{v}m_e\right)u(\mathbf{p}_2)$$
$$= -\frac{i}{v^2}m_\mu m_e\,\bar{u}(\mathbf{p}_2')v(\mathbf{p}_1')\frac{1}{k^2-m_H^2+i\varepsilon}\bar{v}(\mathbf{p}_1)u(\mathbf{p}_2). \tag{8-90}$$

Higgs mediator sub-amplitude

Reduction of the Z Propagator

For this interaction, we can drop the term $k_\alpha k_\beta/m_Z^2$ from the Z boson propagator in (8-89) for the following reason.

Reducing the Z propagator

In Vol. 1, pg. 122, we showed how, using the Dirac equation, one derives relations (4-150) therein, which we repeat below as (8-91).

$$p_\mu\gamma^\mu u_r(\mathbf{p}) = m\,u_r(\mathbf{p}) \qquad p_\mu\gamma^\mu v_r(\mathbf{p}) = -m\,v_r(\mathbf{p})$$
$$\bar{u}_r(\mathbf{p})p_\mu\gamma^\mu = \bar{u}_r(\mathbf{p})m \qquad \bar{v}_r(\mathbf{p})p_\mu\gamma^\mu = -m\bar{v}_r(\mathbf{p}) \tag{8-91}$$

Since $k = p_1 + p_2 = p'_1 + p'_2$, we will find a term in (8-89) with the factor

$$\frac{k_\alpha k_\beta}{m_Z^2}\bar{v}(\mathbf{p}_1)\left\{\frac{ig}{4\cos\theta_W}\right\}\gamma^\beta\left(1-4\sin^2\theta_W-\gamma^5\right)u(\mathbf{p}_2)$$
$$= \frac{k_\alpha}{m_Z^2}\bar{v}(\mathbf{p}_1)k_\beta\gamma^\beta\left\{\frac{ig}{4\cos\theta_W}\right\}\left(1-4\sin^2\theta_W-\gamma^5\right)u(\mathbf{p}_2)$$
$$= \frac{k_\alpha}{m_Z^2}\underbrace{\bar{v}(\mathbf{p}_1)\left(p_{1\beta}+p_{2\beta}\right)\gamma^\beta}_{-m_e\bar{v}(\mathbf{p}_1)+\bar{v}(\mathbf{p}_1)p_{2\beta}\gamma^\beta}\left\{\frac{ig}{4\cos\theta_W}\right\}\left(1-4\sin^2\theta_W-\gamma^5\right)u(\mathbf{p}_2) \tag{8-92}$$
$$= \frac{k_\alpha}{m_Z^2}\left(-m_e\bar{v}(\mathbf{p}_1)\right)\left\{\frac{ig}{4\cos\theta_W}\right\}\left(1-4\sin^2\theta_W-\gamma^5\right)u(\mathbf{p}_2)$$
$$+\frac{k_\alpha}{m_Z^2}\bar{v}(\mathbf{p}_1)\left\{\frac{ig}{4\cos\theta_W}\right\}\left(1-4\sin^2\theta_W-\gamma^5\right)\underbrace{p_{2\beta}\gamma^\beta u(\mathbf{p}_2)}_{m_e u(\mathbf{p}_2)}.$$

The two terms in (8-92) cancel one another and we can therefore drop the $k_\alpha k_\beta/m_Z^2$ part of the propagator. This cancellation will not occur in all interactions with a massive vector field propagator, but it does in the present case.

So, in this case, we can drop the $k_\alpha k_\beta$ part of the propagator

Low and Intermediate Energy Scattering

The ratio of (8-90) to (8-89) is, given the relationship for $m_W = vg/2$ of (7-144) in Chap. 7, together with $m_H \approx m_Z$,

$$\frac{\mathcal{M}_H}{\mathcal{M}_Z}\ \text{of order}\ \frac{1}{g^2}\frac{m_\mu m_e}{v^2}\frac{k^2-m_Z^2}{k^2-m_H^2} = \frac{1}{g^2}\frac{m_\mu m_e}{(2m_W/g)^2}\frac{k^2-m_Z^2}{k^2-m_H^2} \approx \frac{m_\mu m_e}{4m_W^2} \ll 1. \tag{8-93}$$

Higgs contribution negligible at similar energy levels

So, to high accuracy, as long as k^2 is not close to m_H^2 (that is, $k^2 \leq m_Z^2$), we can neglect the contribution from the Higgs mediator sub-amplitude and take

Approximate amplitude at low and intermediate energies

$$\mathcal{M} \approx \mathcal{M}_\gamma + \mathcal{M}_Z \,. \tag{8-94}$$

In the referenced sections of Vol. 1, we calculated $\mathcal{M}_\gamma$. With that, we found $|\mathcal{M}_\gamma|^2$, used (8-85) ((17-68) in Vol. 1), did spin summing and averaging (to get (17-109) in Vol. 1), and found the differential cross-section in the COM frame ((17-113) in Vol. 1), which we then integrated to find the total cross-section ((17-115) in Vol. 1). We repeat these results below, where $\mathbf{p}' = \mathbf{p}'_1 = -\mathbf{p}'_2$, and $E = E_1 = E_2 = E'_1 = E'_2$ is the energy of each particle (which are all equal in the COM frame).

Photon mediator cross-section (from Vol. 1)

For $\mathcal{M}_\gamma$ alone
$$\left(\frac{d\sigma}{d\Omega'_1}\right)_{\substack{COM \\ e^-e^+\to\mu^-\mu^+}} = \frac{\alpha^2}{16E^4}\frac{|\mathbf{p}'|}{E}\left(E^2 + m_\mu^2 + |\mathbf{p}'|^2 \cos^2\theta\right) \qquad \begin{cases}\text{unknown spins,} \\ |\mathbf{p}'| = \sqrt{E^2 - m_\mu^2}\end{cases}, \tag{8-95}$$

and
$$\sigma_{\substack{COM \\ e^-e^+\to\mu^-\mu^+}} = \frac{\pi\alpha^2}{4E^4}\frac{|\mathbf{p}'|}{E}\left(E^2 + m_\mu^2 + \tfrac{1}{3}|\mathbf{p}'|^2\right) \qquad \begin{cases}\text{unknown spins,} \\ |\mathbf{p}'| = \sqrt{E^2 - m_\mu^2}\end{cases}. \tag{8-96}$$

From (8-85), we know we need to use the total amplitude (8-94) to find

Total amplitude includes interference terms

$$|\mathcal{M}|^2 \approx |\mathcal{M}_\gamma + \mathcal{M}_Z|^2 = (\mathcal{M}_\gamma + \mathcal{M}_Z)(\mathcal{M}_\gamma + \mathcal{M}_Z)^* = |\mathcal{M}_\gamma|^2 + \underbrace{\mathcal{M}_\gamma\mathcal{M}_Z^* + \mathcal{M}_Z\mathcal{M}_\gamma^*}_{\text{interference contribution}} + |\mathcal{M}_Z|^2, \tag{8-97}$$

which, as we see, is not simply the addition of the squares of the absolute values of the photon and Z amplitudes. The additional term comes from products of the two amplitudes, and their complex conjugates, and is known as the interference term, so named because a similar effect arises when two different waves interfere. In a sense, the photon and Z fields are overlapping and so interfere.

Differential cross section at low to intermediate energy, presented, not derived

Although you have the skills to deduce the cross-section, given (8-85), (8-88), (8-89), and (8-97), we will cut through that mountain of algebraic manipulation, of many pages including even more extensive spins sums and averages than we have seen before, to present the final result. I believe it will benefit you more if you spend the hours it would take going through that derivation to learn new material instead. (And if it makes you feel better, I have never actually done this myself.)

We do restrict that final result to the case where the incoming energy far exceeds the combined mass of the final muons. That is,

Dynamics relations

$$\text{for} \quad (p_1 + p_2)(p_1 + p_2) = (E_1 + E_2)^2 - (\mathbf{p}_1 + \mathbf{p}_2)^2 = (2E)^2 = E_{tot}^2 = k^2, \tag{8-98}$$

when

Simplifying assumption: total energy much more than muon mass

$$E_{tot}^2 = k^2 \gg m_\mu^2, \tag{8-99}$$

then for $\mathcal{M} = \mathcal{M}_\gamma + \mathcal{M}_Z$, where the scattering angle θ is the angle by which the final antimuon trajectory ($\mathbf{p}'_1$ direction) departs from the incoming positron trajectory ($\mathbf{p}_1$ direction),

$$\left(\frac{d\sigma}{d\Omega'_1}\right)_{\substack{COM \\ e^-e^+\to\mu^-\mu^+}} = F(k^2)(1 + \cos^2\theta) + G(k^2)\cos\theta, \tag{8-100}$$

and where, with g_V and g_A representing factors for the vector (no γ^5) and axial (γ^5) parts, respectively,

$$g_V = 2\sin^2\theta_W - \tfrac{1}{2} \qquad g_A = -\tfrac{1}{2}, \tag{8-101}$$

$$F(k^2) = \frac{\alpha^2}{4k^2}\left(\underbrace{1}_{\gamma\text{ term}} + \underbrace{\frac{g_V^2}{\sqrt{2}\pi}\frac{m_Z^2}{k^2 - m_Z^2}\left(\frac{k^2 G}{\alpha}\right)}_{\text{interference term}} + \underbrace{\frac{(g_V^2 + g_A^2)^2}{8\pi^2}\left(\frac{m_Z^2}{k^2 - m_Z^2}\right)^2\left(\frac{k^2 G}{\alpha}\right)^2}_{Z\text{ term}}\right) \tag{8-102}$$

$$G\left(k^2\right)=\frac{\alpha^2}{4k^2}\left(\underbrace{\frac{\sqrt{2}g_A^2}{\pi}\frac{m_Z^2}{k^2-m_Z^2}\left(\frac{k^2G}{\alpha}\right)}_{\text{interference term}}+\underbrace{\frac{g_V^2g_A^2}{\pi^2}\left(\frac{m_Z^2}{k^2-m_Z^2}\right)^2\left(\frac{k^2G}{\alpha}\right)^2}_{Z\text{ term}}\right). \qquad (8\text{-}103)$$

We will now examine $e^-+e^+\to\mu^-+\mu^+$ scattering over four energy regimes, low, intermediate, resonance (which we will explain), and above resonance.

We'll look at four energy regimes

Low Energy Regime

We define the low energy regime by (8-93) and (8-99). That is,

$$E_{tot}^2=k^2\ll m_Z^2\left(\approx 8{,}000\,\text{GeV}^2\right)<m_H^2 \quad \text{but} \quad E_{tot}^2=k^2\gg m_\mu^2. \qquad (8\text{-}104)$$

Low energy = relativistic, but << m_Z

The process is relativistic (kinetic energy is much greater than rest mass energies involved), but still low relative to the Higgs mass (so we can ignore the Higgs mediator sub-amplitude).

The largest non-photon term in (8-102) and (8-103) is the interference term in (8-103), which for $k \ll m_Z$ has magnitude

$$\frac{\sqrt{2}}{\pi}g_A^2(\sim 1)\frac{k^2G}{\alpha}\ll\frac{\sqrt{2}}{\pi}g_A^2(\sim 1)\frac{m_Z^2G}{\alpha}=\frac{\sqrt{2}}{\pi}(.5)^2(\sim 137)(8000)\left(1.2\times 10^{-5}\right)\approx 1.48\,. \qquad (8\text{-}105)$$

For $k \ll m_Z$ the interference term and the other non-photon terms have magnitude much less than 1, so all the non-photon terms in this regime can be ignored. This gives us an effectively QED-only differential cross-section

In this regime, only photon part significant

$$\left(\frac{d\sigma}{d\Omega_1'}\right)_{\substack{COM\\ e^-e^+\to\mu^-\mu^+}}=\frac{\alpha^2}{4k^2}\left(1+\cos^2\theta\right)=\frac{\alpha^2}{4E_{tot}^2}\left(1+\cos^2\theta\right) \quad (\text{low, yet relativistic, energy}), \qquad (8\text{-}106)$$

which turns out to be what we found in QED

which is just what we found in (17-117) of Vol. 1, for the QED case. We plotted that case in Vol. 1, Fig. 17-17, which we reproduce below as the LHS of Fig. 8-7. Note the front-back symmetry (symmetry about the plane orthogonal to the positron incoming 3-momentum). The antimuon direction is as likely to depart from the incoming positron direction in the +30° direction as the +150° direction.

As shown in equation (17-117) of Vol. 1, the total cross-section is

$$\sigma_{\substack{COM\\ e^-e^+\to\mu^-\mu^+}}=\frac{4\pi\alpha^2}{3E_{tot}^2} \quad (\text{low, yet relativistic, energy}), \qquad (8\text{-}107)$$

whose dependence on energy is shown on the RHS of Fig. 8-7.

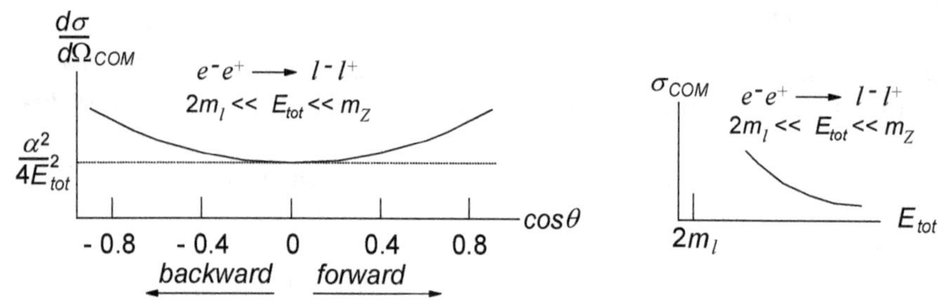

Front-back symmetry of differential cross-section, in this regime

Total cross-section decreases with energy, in this regime

Figure 8-7. Plot of Relation (8-106) and Its Integral over Solid Angle

Intermediate Energy Regime

As $E_{tot}^2=k^2$ increases, the photon contribution in (8-102) and (8-103) falls off inversely with k^2, the interference terms increase as $k^2-m_Z^2$ in the denominator gets smaller, and the Z terms increase with $k^2/\left(k^2-m_Z^2\right)^2$. However, for a particular range, the small G^2 factor in the Z terms overrides the increase due to k^2, so only the photon and interference terms are relevant. That is, the very small factor

As energy increases, interference terms become significant before Z term

$$\frac{m_Z^2}{\left|k^2 - m_Z^2\right|}\frac{k^2 G}{\alpha} \tag{8-108}$$

occurs to the first power in the interference terms and to the second power in the Z terms, so as k^2 grows, the interference terms will become significant well before the Z terms. In particular $G(k^2)$ is no longer negligible, so (8-100) contains a term in $cos\,\theta$.

That gives rise to a term in $cos\,\theta$, not just $cos^2\theta$

$$\begin{aligned}\left(\frac{d\sigma}{d\Omega_1'}\right)_{\substack{COM \\ e^-e^+\to\mu^-\mu^+}} &= \frac{\alpha^2}{4k^2}\left(1+\frac{g_v^2}{\sqrt{2}\pi}\frac{m_Z^2}{k^2-m_Z^2}\left(\frac{k^2G}{\alpha}\right)\right)\left(1+cos^2\theta\right)+\frac{\alpha^2}{4k^2}\frac{\sqrt{2}g_A^2}{\pi}\frac{m_Z^2}{k^2-m_Z^2}\left(\frac{k^2G}{\alpha}\right)cos\,\theta \\ &= \left(\frac{\alpha^2}{4E_{tot}^2}+\frac{\alpha}{4}\frac{g_v^2}{\sqrt{2}\pi}\frac{m_Z^2}{E_{tot}^2-m_Z^2}G\right)\left(1+cos^2\theta\right)+\frac{\alpha}{4}\frac{\sqrt{2}g_A^2}{\pi}\frac{m_Z^2}{E_{tot}^2-m_Z^2}G\,cos\,\theta.\end{aligned} \tag{8-109}$$

The differential cross-section is no longer symmetric in θ. There is a forward-backward asymmetry, as depicted for a typical energy level in the LHS of Fig. 8-8. Compare it to the LHS of Fig. 8-7.

Do **Problem 4** to find the total cross-section for (8-109) is (8-110).

$$\sigma_{\substack{COM \\ e^-e^+\to\mu^-\mu^+}} = \left(\frac{\alpha^2}{4E_{tot}^2}+\frac{\alpha}{4}\frac{g_v^2}{\sqrt{2}\pi}\frac{m_Z^2}{E_{tot}^2-m_Z^2}G\right)\frac{16\pi}{3} \qquad \text{(intermediate energy)}. \tag{8-110}$$

As long as E_{tot} is not approaching m_Z (which it doesn't in the intermediate energy range), the total cross-section will still tend to fall off with E_{tot}, though not as rapidly as in Fig. 8-7. As E_{tot} approaches m_Z, the total cross-section will start rising.

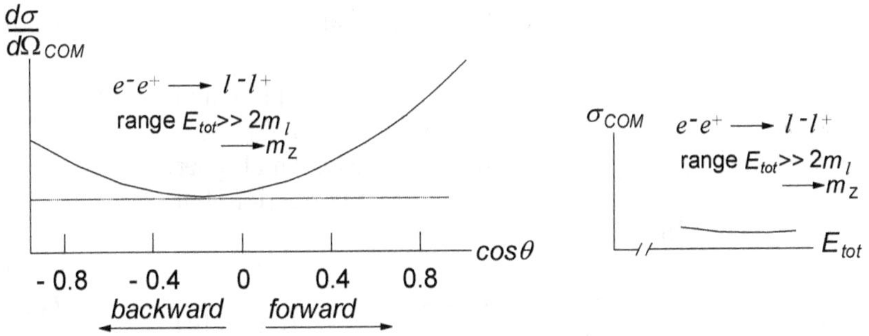

The $cos\,\theta$ term makes the differential cross-section front-back asymmetric

Figure 8-8. Plot of Relation (8-109) and Its Integral over Solid Angle

Resonance Energy Regime

As we transition from $E_{tot} << m_Z$ to $E_{tot} \approx m_Z$, the denominators of the Z terms in (8-102) and (8-103) will approach zero, and do so more rapidly than the denominators of the interference terms, so we can ignore the latter terms. The photon term will become negligible, as it has the square of E_{tot} in the denominator. We call this energy range the resonance regime for the reason we will give shortly. From (8-100) to (8-103), we find

For $E_{tot} \approx m_Z$, Z term dominates over photon and interference terms

This is the resonance regime

$$\left(\frac{d\sigma}{d\Omega'_1}\right)_{\substack{COM \\ e^-e^+\to\mu^-\mu^+}} \approx \frac{\alpha^2}{4k^2}\frac{\left(g_V^2+g_A^2\right)^2}{8\pi^2}\left(\frac{m_Z^2}{k^2-m_Z^2}\right)^2\left(\frac{k^2G}{\alpha}\right)^2\left(1+\cos^2\theta\right)$$

$$+\frac{\alpha^2}{4k^2}\frac{g_V^2g_A^2}{\pi^2}\left(\frac{m_Z^2}{k^2-m_Z^2}\right)^2\left(\frac{k^2G}{\alpha}\right)^2\cos\theta \qquad \left(\text{resonance energy } E_{tot}\approx m_z\right) \quad (8\text{-}111)$$

$$=\frac{\left(g_V^2+g_A^2\right)^2}{32\pi^2}\left(\frac{m_Z^2}{E_{tot}^2-m_Z^2}\right)^2E_{tot}^2G^2\left(1+\cos^2\theta\right)+\frac{g_V^2g_A^2}{4\pi^2}\left(\frac{m_Z^2}{E_{tot}^2-m_Z^2}\right)^2E_{tot}^2G^2\cos\theta .$$

Or, finally,

$$\left(\frac{d\sigma}{d\Omega'_1}\right)_{\substack{COM \\ e^-e^+\to\mu^-\mu^+}} \approx \left(\frac{m_Z^2}{E_{tot}^2-m_Z^2}\right)^2E_{tot}^2G^2\left(\frac{\left(g_V^2+g_A^2\right)^2}{32\pi^2}\left(1+\cos^2\theta\right)+\frac{g_V^2g_A^2}{4\pi^2}\cos\theta\right). \quad (8\text{-}112)$$

Resonance differential cross-section asymmetric

Integrating the last term from $\theta = 0^\circ$ to 180° yields zero, so from prior similar integrations, we know that

$$\sigma_{\substack{COM \\ e^-e^+\to\mu^-\mu^+}} \approx \frac{\left(g_V^2+g_A^2\right)^2}{32\pi^2}\left(\left(\frac{m_Z^2}{E_{tot}^2-m_Z^2}\right)^2E_{tot}^2G^2\right)\frac{16\pi}{3} \qquad \left(\text{resonance energy } E_{tot}\approx m_z\right)$$

$$\approx \frac{\left(g_V^2+g_A^2\right)^2}{6\pi}\frac{G^2m_Z^6}{\left(E_{tot}^2-m_Z^2\right)^2}. \quad (8\text{-}113)$$

Lowest order total cross-section near resonance (near $E_{tot} = m_Z$)

From (8-112), we can see that, like we had for the intermediate range, the differential cross-section will have front-back asymmetry in the resonance regime. From it and (8-113), we see that the cross-section (differential and total), as calculated here, will go to infinity when $E_{tot} = m_Z$. That, as you may guess, doesn't actually happen in the real world.

Lowest order resonance total cross-section infinite at $E_{tot} = m_Z$

The reason is that we have ignored higher order corrections to the denominator in (8-112), and thus, also in (8-113). For example, in the Z propagator of Fig. 8-6, we would have a fermion-antifermion self-energy loop, like we had for the photon propagator self-energy in QED. And higher order loops, as well. We won't go through the lengthy procedure of calculating that correction, but just cite the final result, where ${}_Z\Gamma$ is the total decay rate of the Z boson.

$$\sigma_{\substack{COM \\ e^-e^+\to\mu^-\mu^+}} \approx \frac{\left(g_V^2+g_A^2\right)^2}{6\pi}\frac{G^2m_Z^6}{\left(E_{tot}^2-m_Z^2\right)^2+m_Z^2\left({}_Z\Gamma\right)^2} \qquad \left(\begin{array}{c}\text{resonance regime with}\\ \text{higher order corrections}\end{array}\right). \quad (8\text{-}114)$$

Higher order corrections yield finite peak at $E_{tot} = m_Z$

The total decay rate of the Z boson comes into play because that boson decays into fermion-antifermion pairs, and when an internal loop begins to form in the Z propagator, that is effectively what is happening. And the reverse process occurs when the pair coalesces back into a Z.

The plot of (8-112) looks much like the LHS of Fig. 8-8, so we won't repeat that general shape here. However, the plot (8-114) looks much different from the RHS of Fig. 8-8, and it is depicted in the RHS of Fig. 8-9. The LHS of the figure shows a typical response of a dynamic system (be it mechanical such as a damped spring and mass or electrical such as an RLC circuit) to a variation in the externally driven ("driver") frequency (oscillating force mechanically, alternating voltage electrically). In such dynamic systems, the peak corresponds to a system resonance (at the resonant frequency ω_n). Hence, the adoption of similar nomenclature for peaks in cross-section response to input energy (which quantum mechanically is related to the wave frequency of the incoming particle/waves).

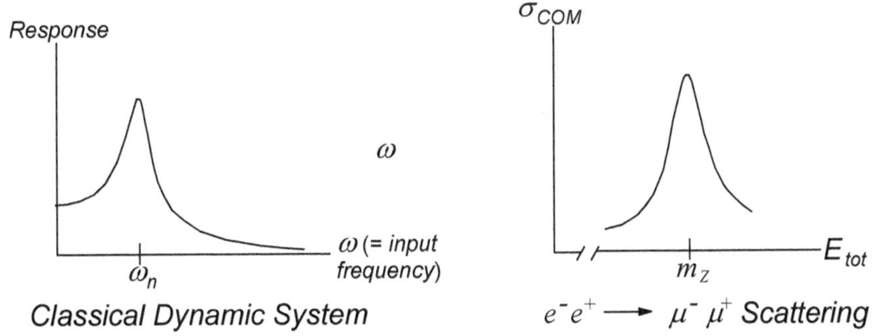

Figure 8-9. Resonance Plots of Classical System and QFT Scattering of (8-114)

The peak of the total cross-section occurs for (8-114) when $E_{tot} = m_Z$, and equals (where we don't cancel the mass factors, because we will reference the form of the denominator as shown shortly)

$$\text{Peak } \sigma_{COM \atop e^-e^+\to\mu^-\mu^+} \approx \frac{\left(g_V^2+g_A^2\right)^2}{6\pi}\frac{G^2 m_Z^6}{m_Z^2\left({}_Z\Gamma\right)^2}. \quad (8\text{-}115)$$

Total cross-section value at $E_{tot} = m_Z$

The width of the peak in Fig. 8-9 is considered to be the difference in energy levels between the two points on either side of the peak that have half the value of the peak. For those points, the denominator of (8-114) needs to be twice the denominator of (8-115). I claim the width, to good approximation, equals ${}_Z\Gamma$, which we can readily prove, by finding the denominator of (8-114) at the energy levels $m_Z - {}_Z\Gamma/2$ and $m_Z + {}_Z\Gamma/2$. For the first point, where near the end of the relation below, we use $m_Z >> {}_Z\Gamma$ to drop a term, and the denominator of the last factor in (8-114) is

Peak width is ${}_Z\Gamma$

$$\left(\left(m_Z-\frac{{}_Z\Gamma}{2}\right)^2-m_Z^2\right)^2+m_Z^2\left({}_Z\Gamma\right)^2=\left(m_Z^2-m_z\left({}_Z\Gamma\right)+\tfrac{1}{4}\left({}_Z\Gamma\right)^2-m_Z^2\right)^2+m_Z^2\left({}_Z\Gamma\right)^2$$
$$=\left(-m_z\left({}_Z\Gamma\right)+\tfrac{1}{4}\left({}_Z\Gamma\right)^2\right)^2+m_Z^2\left({}_Z\Gamma\right)^2\approx\left(-m_z\left({}_Z\Gamma\right)\right)^2+m_Z^2\left({}_Z\Gamma\right)^2=2m_Z^2\left({}_Z\Gamma\right)^2, \quad (8\text{-}116)$$

Proving it

which is twice the denominator of the last factor in (8-115) at the first half amplitude point.

A similar relation holds for the other point, as only the sign on the $m_Z({}_Z\Gamma)$ term just after the first equal sign in the second row changes, and that is erased when it is squared. So, the width of the peak is indeed ${}_Z\Gamma$.

Note that the width, i.e., the decay rate, ${}_Z\Gamma$ in cgs or other unit systems has units of (seconds)$^{-1}$, but our horizontal axis in Fig. 8-9 has units of GeV. To convert from (seconds)$^{-1}$ to GeV, we multiply ${}_Z\Gamma$ by $\hbar$ (= 6.58X10^{-25} GeV-s), so the width in GeV is $\hbar_Z\Gamma$. In natural units, where $\hbar = 1$ and is unitless, these quantities are the same.

In non-natural units, peak width is $\hbar\,{}_Z\Gamma$

Hence, the decay width is formally defined as $\hbar_Z\Gamma$, where ${}_Z\Gamma$ is the decay rate, but working in natural units, one generally finds ${}_Z\Gamma$ referred to as both decay rate and decay width.

$\hbar\,{}_Z\Gamma$ called the "decay width"; ${}_Z\Gamma$, the decay rate

For the Z resonance, $\hbar_Z\Gamma = 2.495 \pm 0.002$ GeV at $m_Z = 91.1876 \pm 0.0021$ GeV. In a plot covering 100 GeV, the peak would appear quite narrow.

High Energy Regime

By "high energy" we mean that E_{tot} is significantly more than m_Z. In this regime, (8-100) to (8-103) become, effectively,

For $E_{tot} >> m_Z$, Z term still dominates

This is high energy regime

$$\left(\frac{d\sigma}{d\Omega_1'}\right)_{COM \atop e^-e^+ \atop \to\mu^-\mu^+}\approx\frac{\alpha^2}{4E_{tot}^2}\left(\begin{matrix}\left(1+\frac{g_V^2}{\sqrt{2}\pi}m_Z^2\left(\frac{G}{\alpha}\right)+\frac{\left(g_V^2+g_A^2\right)^2}{8\pi^2}m_Z^4\left(\frac{G}{\alpha}\right)^2\right)\left(1+\cos^2\theta\right)\\+\left(\frac{\sqrt{2}g_A^2}{\pi}m_Z^2\left(\frac{G}{\alpha}\right)+\frac{g_V^2g_A^2}{\pi^2}m_Z^4\left(\frac{G}{\alpha}\right)^2\right)\cos\theta\end{matrix}\right)\quad\left(E_{tot}\gg m_Z\right), \quad (8\text{-}117)$$

Differential cross-section in high energy regime still asymmetric

which, due to the $\cos\theta$ term, is once again asymmetric. That term, once again, disappears when we integrate over the solid angle, leaving us with a high energy dependence that drops off with the square of the total energy. This is much like what we had in the low energy regime, except that now we have the influence of the weak interaction, showing up in the terms with G in them.

$$\sigma_{\substack{COM \\ e^-e^+\to\mu^-\mu^+}} \approx \frac{4\pi\alpha^2}{3E_{tot}^2}\left(1+\frac{g_V^2}{\sqrt{2}\pi}m_Z^2\left(\frac{G}{\alpha}\right)+\frac{\left(g_V^2+g_A^2\right)^2}{8\pi^2}m_Z^4\left(\frac{G}{\alpha}\right)^2\right) \quad (E_{tot}\gg m_Z). \tag{8-118}$$

Total cross-section in high energy regime

High Energy Higgs Resonance

As $|k| = E_{tot}$ approaches the Higgs mass, m_H = 125.1 GeV, the denominator of the Higgs sub-amplitude (8-90) becomes small and one might consider that sub-amplitude could no longer be ignored. However, the Higgs decays into many different particles and the branching ratio to a muon-antimuon pair is quite small. This is due, in large part, to the factor of $m_\mu m_e$ in (8-90), which for another final particle type, such as a tau-antitau pair, would be $m_\tau m_e$, and much larger than $m_\mu m_e$. For other particles, like bottom quarks, it would be even much greater.

Another resonance at Higgs mass energy

These mass factors result from the fact that the Higgs coupling (see Feynman rules 15-13 to 15-16) is proportional to the mass of the particles with which it shares a vertex. Greater Higgs coupling means greater mass of the particles to which it is coupled.

As a result, the Higgs resonance at m_H for the interaction we are dealing with, is not a very high peak, unlike the Z boson resonance.

Higgs resonance not pronounced

In practice, searches for Higgs resonances are not made by colliding leptons, but by colliding protons and antiprotons, as they do with the Large Hadron Collider (LHC) at CERN.

All Regimes in One Figure

Figure 8-10 shows the entire energy regime for $e^-+e^+\to\mu^-+\mu^+$ scattering. Note that no such interaction can occur unless the total incoming energy is greater than the combined masses of the two muons. We ignore the Higgs resonance because it is so small. The figure is illustrative only and not to scale. For example, the resonance peak should be much narrower, but then it would not be possible to fit in the label for the width in the space that would be available.

Total cross-section in all regimes graphically

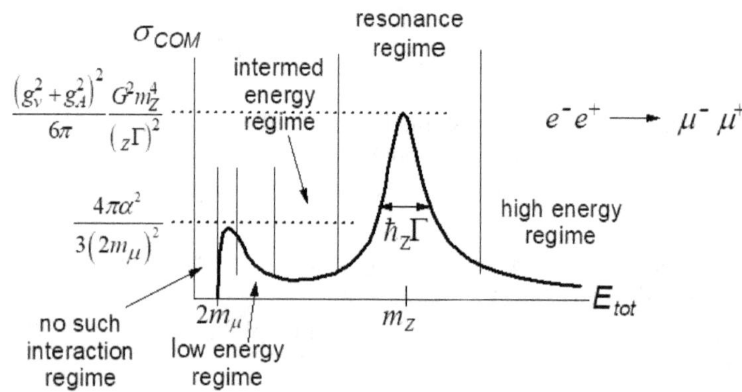

Figure 8-10. Total Cross-Section vs Total Energy for $e^-+e^+\to\mu^-+\mu^+$
(Qualitative only. Not to scale)

Additional Resonances, If Any

One could ask what it means if we were to have other resonances, at even higher energies. To answer that, first note that the cause of the Z boson resonance is the denominator in (8-113) (radiatively corrected in (8-114)), and that denominator came from the Z boson propagator in the amplitude $\mathcal{M}$. That propagator blows up when $k^2 = m_Z^2$ $\left(= E_{tot}^2 \text{ in our present case}\right)$.

The bottom line is that any virtual particle mediating any interaction will cause the amplitude (and thus the cross-section) to go to infinity at first order (without radiative corrections) when k^2 $\left(E_{tot}^2\right.$ in our case) of the propagator equals its mass squared. With radiative corrections, the amplitude and cross-section do not go to infinity, but they peak.

Additional resonance would imply another, higher mass, particle

So, if another particle existed that interacted via the electroweak (or any other) force and were much heavier than the Z and the Higgs, in $e^- + e^+$ collisions, we would see another peak when the interaction energy level reached that of the mass of the heavier particle.

This is the sort of thing experimentalists look for. An unexpected resonance in an interaction would imply the existence of a particle, hitherto unknown, and thus, concomitant new physics.

Experimental Determination of Physical Values

Can determine m_Z and $_Z\Gamma$ from experiment using resonance relation

Note that the shape, peak, and width of the cross-section around resonance depend on m_Z and $_Z\Gamma$. In fact, the peak location is just $E_{tot} = m_Z$ and the width, $\hbar\,_Z\Gamma$. With experimentally determined values for cross-section at various points at and near resonance, using (8-114), one can determine the mass of the Z boson and its decay rate. Generally, however, finding the width experimentally is not trivial, but researchers have been able to find extremely accurate values for it by fitting experimental values for cross-section and E_{tot} at a number of points with (8-114).

Breit-Wigner Formula

(8-114) is a special case of a more general resonance relation, good for many different interactions between two initial particles, known as the Breit-Wigner formula, after its discovers, Gregory Breit and Eugene Wigner.

We will not delve into deriving this relationship here, but merely state it for completeness. It is, where i indicates initial particles, X indicates final particles, m is the mass of the mediator, Γ_i is the decay rate of the mediator into the initial particles, Γ_X is the decay rate of the mediator into the final particles, Γ is the total decay rate of the mediator, J is the total spin of the mediator, s_1 is the spin on the first initial particle, and s_2 is the spin on the other initial particle.

Breit-Wigner general formula for resonance of two particle interaction

$$\sigma_{\substack{COM \\ i\rightarrow X}} = 4\pi\frac{m^2}{p^2}\frac{2J+1}{(2s_1+1)(2s_2+1)}\frac{\Gamma_i\Gamma_X}{\left(k^2-m^2\right)^2+m^2(\Gamma)^2} \qquad \left(k^2 \text{ near } m^2\right). \qquad \text{(8-119)}$$

The factors that may seem a little odd to us, the spins, decay rate Γ_i, and decay rate Γ_X, come from the amplitude in a way that would be more time consuming than it is worth for us right now to deduce.

8.4 Bhabha Scattering

In Vol. 1, Sec. 17.5.2, beginning on pg. 467, we extended the electron-positron collision to include final particles of an electron and a positron. In QED, this necessitated the introduction of a second lowest order amplitude (with a concomitant second Feynman diagram) to include the other way an electron and positron could interact to leave a final electron-positron pair. See Fig. 17-18, pg. 467, therein.

In the present case of higher energy levels, we need to include similar diagrams, not just for the photon mediator, but also for the Z and the Higgs mediators. That is, at lowest order, we would need sub-amplitudes for each of the Feynman diagrams in Fig. 8-11.

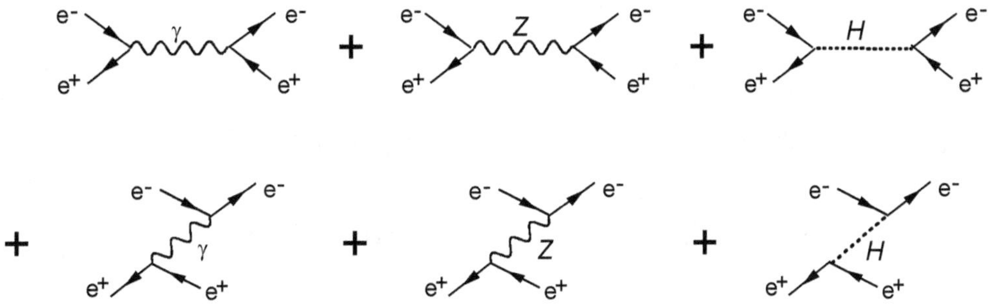

Feynman diagrams for electroweak Bhabha scattering

Figure 8-11. Feynman Diagrams (at Lowest Order) for $e^- + e^+ \rightarrow e^- + e^+$

We will not be analyzing the $e^- + e^+ \rightarrow e^- + e^+$ interaction with meditators beyond the photon here. One can see it would involve a lot more work than what we did for $e^- + e^+ \rightarrow \mu^- + \mu^+$, as there would be six sub-amplitudes (reduced to four without the negligible effects of the Higgs sub-amplitudes), yielding 36 (16 after eliminating the Higgs contributions) different terms in the total amplitude. You have the tools to do this, if you had to, but, thankfully, you don't have to.

8.5 Things to be Aware of

Better accuracy with higher order corrections

Just as a reminder, in the calculations in this chapter, we employ tree level amplitudes, where we take constants, such as particle masses and weak coupling, to be those that are measured in experiment at contemporary energy levels. Doing this yields very accurate, though approximate, theoretical results.

Higher order → running coupling constants

But calculations too complex for this book

However, to be precise, we would need to include higher order radiative corrections, including all sorts of loops and vertex corrections that arise due to the many interaction terms in electroweak theory. This would result in running coupling constants for *g* and *e*, similar in effect to what we found in QED for *e*, when we carried out renormalization. Given the complexity we found in doing this in Vol. 1, we would expect a parallel procedure in electroweak theory to be enormously, almost overwhelmingly, long and extensive. And we would be right. So, we will not be doing such things in the present volume.

8.6 Chapter Summary

8.6.1 Contact (Fermi) Theory

Propagators for massive vector fields in electroweak theory can generally be approximated, with little loss in accuracy, via (where *MV* stands for "massive vector" field, such as the *W* or *Z* field)

$$iD_{F\mu\nu}(k) = i\frac{-g_{\mu\nu} + k_\mu k_\nu / m_{MV}^2}{k^2 - m_{MV}^2} \rightarrow i\frac{g_{\mu\nu}}{m_{MV}^2} \quad \text{where} \quad \frac{G}{\sqrt{2}} = \frac{g_{\mu\nu}}{8m_{MV}^2}. \qquad (8\text{-}120)$$

This is called contact theory and *G*, the Fermi constant, after the original discoverer of contact theory.

8.6.2 Decay Summary

General Relation

From the results of Vol. 1, we deduced the decay rate for a given process, for a momentum range of $\mathbf{p}'_f$ to $\mathbf{p}'_f + d\,\mathbf{p}'_f$, as

$$d\Gamma_d = (2\pi)^4 \delta^{(4)}\left(\sum p_f - \sum p_i\right)\frac{1}{2E}\left(\prod_l 2m_l\right)\left(\prod_f \frac{d\mathbf{p}'_f}{(2\pi)^3 2E_f}\right)\left|\mathcal{M}_d\right|^2 \qquad \text{repeat of (8-8)}$$

Branches, Decay Rates, and Lifetimes

Particle decay can have different branches (or modes or channels), each having its own decay rate Γ_m, where the

$$\text{total decay rate (including all modes) for given particle is } \Gamma = \sum_m \Gamma_m. \qquad \text{repeat of (8-10)}$$

From that, we know the

$$\text{particle lifetime} \quad \tau = \frac{1}{\Gamma} = \frac{1}{\sum_m \Gamma_m} \quad \text{with particle half-life} = .693\,\tau. \qquad \text{repeat of (8-11)}$$

For a given branch, the branching ratio (the percentage of time we would see decay via that branch) is

$$B_m = \frac{\Gamma_m}{\Gamma} = \frac{\Gamma_m}{\sum_{m'} \Gamma_{m'}}. \qquad \text{repeat of (8-12)}$$

An Example: Muon Decay

Muon decay has three branches, but one, with final particle of muon neutrino, electron, and electron antineutrino, has a branching ratio of 0.986, so we can approximate decay rate by examining only that branch.

To evaluate muon decay, we

1. determine, using Feynman rules, the amplitude for the main branch,
2. approximate the W propagator via (8-120), (i.e., use contact theory),
3. employ the general decay rate relation (8-8) with this amplitude,
4. make reasonable simplifying assumptions about different masses and energy levels involved,
5. carry out spin averaging and summing, and
6. integrate over 3-momentum

The result is

$$\tau_\mu = \frac{1}{\Gamma_{muon}} \approx \frac{192\pi^3}{G^2 m_\mu^5}. \qquad \text{repeat of (8-76)}$$

This is on the order of 10^{-6} seconds. When one includes radiative corrections and other muon decay modes, the muon lifetime is calculated to be about $2.2 \text{X} 10^{-6}$ seconds, very close to the experimental value.

General Points

A particle cannot decay into particles whose masses sum to more than the original particle mass. Hence, a muon cannot decay into quarks or taus. A tau, on the other hand, is heavier than a muon and can decay via more branches into other particles, such as a muon and neutrinos and quarks and a neutrino.

Neutron decay occurs when a down quark decays into a (lighter) up quark along with an electron and an antineutrino. A free neutron decays, on average, in a little less than 15 minutes.

W, Z, and Higgs particles decay in about 10^{-22} seconds or less.

8.6.3 Scattering Summary

General Scattering Relation

From Vol. 1 results we have the general relation for differential cross-section of

$$\left(\frac{d\sigma}{d\Omega_1'}\right)_{COM} = \frac{1}{64\pi^2\left(E_1+E_2\right)^2}\frac{|\mathbf{p}_1'|}{|\mathbf{p}_1|}\left(\prod_l^{\substack{\text{extern}\\\text{ferms}}} 2m_l\right)|\mathcal{M}|^2 \quad \begin{cases}\text{COM, 2 initial \&}\\ \text{2 final particles,}\\ \text{elastic or inelastic.}\end{cases} \qquad \text{repeat of (8-85)}$$

An Example: Electron-positron Scattering to Muon-antimuon

To evaluate $e^- + e^+ \rightarrow \mu^- + \mu^+$ scattering, we

1. determine, using Feynman rules, the amplitude, which in this case has 3 sub-amplitudes, due to photon, Z, and Higgs mediators (propagators),
2. reduce the Z propagator (eliminate the term with $k_\mu k_\nu$ in it), but do not use contact theory,
3. approximate the amplitude by dropping the Higgs contribution, which is very small,
4. use the general COM frame scattering relation (8-85) with this amplitude,
5. make reasonable simplifying assumptions about different masses and energy levels involved,
6. carry out spin averaging and summing, and
7. convert 3-momentum vector to magnitude and direction depending on θ and ϕ.

The result is

$$\left(\frac{d\sigma}{d\Omega_1'}\right)_{\substack{COM \\ e^-e^+\to\mu^-\mu^+}} = F\left(k^2\right)\left(1+\cos^2\theta\right)+G\left(k^2\right)\cos\theta\,, \qquad \text{repeat of (8-100)}$$

where $F(k^2)$ and $G(k^2)$ are defined in (8-102) and (8-103). Note that we did not actually carry out steps 6 and 7, which are extensive, to say the least, but accepted results from others.

Differential cross-section (relation (8-100)) for low and intermediate total incoming energy levels is shown in Figs. 8-7 and 8-8, pgs. 282 and 283, respectively. The qualitative results for total cross-section over energy regimes is plotted in Fig. 8-10, pg. 286.

Resonances (peaks in total cross-section) occur at total energy levels equal to the propagator mass levels, due to the factor of $k^2 - m^2$ in the propagator denominator, where m is the propagator mass.

General Formula around Resonance

The Breit-Wigner relation for total cross-section in the region around resonance in any two-particle electroweak interaction is shown in (8-119).

Another Example: Bhabha Scattering

Fig. 8-11, pg. 287, shows Feynman diagrams for the $e^- + e^+ \rightarrow e^- + e^+$ (Bhabha) scattering, where full electroweak effects, rather than purely QED as in Vol. 1, are taken into account. There are twice as many sub-amplitudes as for $e^- + e^+ \rightarrow \mu^- + \mu^+$ scattering, and the problem involves a great deal of algebraic manipulation over many pages. We did not do this here.

8.6.4 Feynman Rules for Electroweak Theory

At low energy (true vacuum), Feynman's rules for electroweak interactions are shown below. Note that since QED is part of e/w theory, Feynman rules for QED are included in the list via rules 1 to 11. See Vol. 1, pg. 514 for Feynman's Rules for QED. The remaining rules are added for weak interactions. Parts A and B are general rules, and are the same as what we had for QED alone.

Feynman's Rules

A. The S matrix element (the transition amplitude) for a given interaction is

$$S_{fi} = \delta_{fi} + \left((2\pi)^4 \delta^{(4)}\left(\sum p_f - \sum p_i\right) \left(\prod_{\substack{\text{all external} \\ \text{bosons}}} \sqrt{\frac{1}{2V\omega_{\mathbf{k}}}} \right) \left(\prod_{\substack{\text{all external} \\ \text{fermions}}} \sqrt{\frac{m}{VE_{\mathbf{p}}}} \right) \right) \mathcal{M} \qquad \mathcal{M} = \sum_{n=1}^{\infty} \mathcal{M}^{(n)} \qquad (8\text{-}121)$$

where Σp_f is the total 4-momentum of all final particles, Σp_i is the total 4-momentum of all initial particles, and the contribution $\mathcal{M}^{(n)}$ comes from the nth order perturbation term of the S operator, $S^{(n)}$.

B. The Feynman amplitude $\mathcal{M}^{(n)}$ is obtained from all of the topologically distinct, connected (i.e., all lines connected to one another in a given diagram) Feynman diagrams which contain n vertices. The contribution to each $\mathcal{M}^{(n)}$ is obtained by the following.

Particles Interacting with Particles

1. For each vertex having a photon and two leptons, include a factor $ie\gamma^{\mu}$.
2. For each internal photon line, labeled by 4-momentum k, include a factor $iD_{F\mu\nu}(k) = i\dfrac{-g_{\mu\nu}}{k^2 + i\varepsilon}$
3. For each internal fermion line, labeled by 4-momentum p, include a factor $iS_F(p) = i\dfrac{\not p + m}{p^2 - m^2 + i\varepsilon}$
4. For each external line, include one of the following spinor factors, where **p** and **k** indicate basis states of corresponding 3-momenta, r represents spin state for fermions and polarization state for photons, [see Vol. 1, (4-93) and (4-94), pg. 111]

   a) for each initial fermion: $u_r(\mathbf{p})$ b) for each final fermion: $\bar{u}_r(\mathbf{p})$
   c) for each initial antifermion: $\bar{v}_r(\mathbf{p})$ d) for each final antifermion: $v_r(\mathbf{p})$
   e) for each initial photon: $\varepsilon_{r,\mu}(\mathbf{k})$ f) for each final photon: $\varepsilon_{r,\mu}(\mathbf{k})$

5. The spinor factors (γ matrices, S_F functions, spinors) for each fermion line are ordered so that, reading from right to left, they occur in the same sequence as following the fermion line in the direction of its arrows through the vertex. (Order is important as it conveys spinor matrix multiplication order when we do not show spinor indices.)

6. The four-momenta at each vertex are conserved (same total after as before).

   6a. For each anti-particle fermion propagator label the Feynman diagram 4-momentum with opposite sign of what it has physically and use this negative of the physical value in the propagator.

7. For each closed loop of internal fermions only (without photons or intermediate vector bosons, i.e., IVBs, inside the loop itself, like what we call a "photon loop" which internally has an electron and a positron), take the trace (in spinor space) of the resulting matrix and multiply by a factor of (–1).

8. For each 4-momentum q which is not fixed by 4-momentum conservation, carry out the integration $\left(1/(2\pi)^4\right)\int d^4q$ One such integration for each closed loop.

9. Multiply the expression by (–1) for each interchange of neighboring fermion operators (each associated with a particular spinor factor from rule #4) which would be required to place the expression in appropriate normal order. "Appropriate", when we are adding sub amplitudes, means each sub amplitude must be in the same, not just any, normal order of destruction and creation operators.

10. (Only needed when doing renormalization. Can ignore otherwise.)
    Add a mass counter term to any fermion line in a Feynman diagram →
    For each mass counter term diagram, add a term to the amplitude with a factor equal to $i\delta m$.

QED Particles Interacting with an External QED Potential Field

11. For each interaction of a charged particle with a static external photon field (a static potential field),
    a) include a factor $A_\mu^e(\mathbf{k}) = \int e^{-i\mathbf{k}\cdot\mathbf{x}} A_\mu^e(\mathbf{x}) d^3x$, and
    b) instead of $(2\pi)^4\,\delta^{(4)}(\Sigma p_f - \Sigma p_i)$ in (8-121), use $2\pi\,\delta(\Sigma E_f - \Sigma E_i)$.

Additional Electroweak: Particles Interacting with Particles

The following rules are only for calculations of the lowest non-vanishing order of perturbation theory.

12. For each internal massive vector boson (W or Z particle) line, labeled by 4-momentum k, include a factor
$$iD_{F\mu\nu}(k,m) = i\frac{\left(-g_{\mu\nu} + k_\mu k_\nu / m^2\right)}{k^2 - m^2 + i\varepsilon} \qquad m = m_W \text{ or } m_Z$$

13. For each external (initial or final) massive vector boson (W or Z), include a factor $\varepsilon_{r,\mu}(\mathbf{k})$. (If complex polarization vectors are used, instead include a factor $\varepsilon^*_{r,\mu}(\mathbf{k})$ for a final state vector boson.)

14. For each internal Higgs line, labeled by 4-momentum k, include a factor $i\Delta_F(k,m_H) = \dfrac{i}{k^2 - m_H^2 + i\varepsilon}$.

15. The GSW theory has 28 different vertexes, which arise from the interaction terms in the total electroweak Lagrangian of Chap. 7, pg. 251, i.e., from terms (7-144) to (7-151). We label each as 15-X, where X is a different number for each vertex. For each vertex having the fields shown in the "Fields" column, include the factor shown in the same row in the "Vertex Factor" column.

    The symbol l means lepton; $\bar{l}$, anti-lepton; q, quark; $\bar{q}$, anti-quark, H, Higgs, ν_l, neutrino; $\bar{\nu}_l$, anti-neutrino; A, photon; W and $W^\dagger$, W^+ and W^- (both notations are found in the literature, so we need to get comfortable using either interchangeably); and Z, the Z vector boson. We do not label the fermion fields with L or R superscripts because the projection of the LC and RC parts of the more general field is taken care of by P_L and P_R parts of the vertex factors.

    Note that Rule 15-1 is the same as Rule 1 above for QED. Although redundant, we include it below, so the list correlates with all the terms in the e/w Lagrangian of Chap. 7 referenced above.

Rule #	$\mathcal{L}$ Part	Fields	Eq	Term in Eq	Vertex Factor
15-1	$\mathcal{L}^{LB}$	$\bar{l}\,l\,A$	(7-145)	1	$ie\gamma^{\mu}$ (Rule #1 above repeated here.)
15-2		$\bar{\nu}_l\,l\,W$	"	2	$\frac{-ig}{2\sqrt{2}}\gamma^{\mu}\left(1-\gamma^5\right) \quad \left(=\frac{-ig}{\sqrt{2}}\gamma^{\mu}P_L\right)$
15-3		$\bar{l}\,\nu_l\,W^{\dagger}$	"	3	$\frac{-ig}{2\sqrt{2}}\gamma^{\mu}\left(1-\gamma^5\right)$
15-4		$\bar{\nu}_l\nu_l\,Z$	"	4	$\frac{-ig}{4\cos\theta_W}\gamma^{\mu}\left(1-\gamma^5\right)$
15-6		$\bar{l}\,l\,Z$	"	5 + 6	$\frac{ig}{4\cos\theta_w}\gamma^{\mu}\left(1-4\sin^2\theta_W-\gamma^5\right)$
15-7	$\mathcal{L}^{QB}$	$\bar{q}_u\,q_u\,A$	(7-146)	1	$-\frac{2}{3}ie\gamma^{\mu}$
15-8		$\bar{q}_d\,q_d\,A$	"	2	$\frac{1}{3}ie\gamma^{\mu}$
15-9		$\bar{q}_u\,q_d\,W$	"	3	$\frac{-ig}{2\sqrt{2}}\gamma^{\mu}\left(1-\gamma^5\right)V_{q_uq_d}$
15-10		$\bar{q}_d\,q_u\,W^{\dagger}$	"	4	$\frac{-ig}{2\sqrt{2}}\gamma^{\mu}\left(1-\gamma^5\right)V^{\dagger}_{q_dq_u}$
15-11		$\bar{q}_u\,q_u\,Z$	"	5 + 7	$\frac{-ig}{4\cos\theta_w}\gamma^{\mu}\left(1-\frac{8}{3}\sin^2\theta_W-\gamma^5\right)$
15-12		$\bar{q}_d\,q_d\,Z$	"	6 + 8	$\frac{ig}{4\cos\theta_w}\gamma^{\mu}\left(1-\frac{4}{3}\sin^2\theta_W-\gamma^5\right)$
15-13	$\mathcal{L}^{LH}$	$\bar{\nu}_l\nu_l\,H$	(7-147)	1 + 2	$\frac{-ig_{\nu_l}}{2\sqrt{2}}\left({}_{\nu}M_{\nu_l i}P^{\dagger}_{i\nu_{l'}}\left(1-\gamma^5\right)+P_{\nu_l i}\,{}_{\nu}M_{i\nu_{l'}}\left(1+\gamma^5\right)\right)$
15-14		$\bar{l}\,l\,H$	"	3	$\frac{-i}{\sqrt{2}}g_l=\frac{-i}{v}m_l$
15-15	$\mathcal{L}^{QH}$	$\bar{q}_u\,q_u\,H$	(7-148)	1	$\frac{-i}{\sqrt{2}}g_{q_u}=\frac{-i}{v}m_{q_u}$
15-16		$\bar{q}_d\,q_d\,H$	"	2	$\frac{-i}{\sqrt{2}}g_{q_d}=\frac{-i}{v}m_{q_d}$
15-17	$\mathcal{L}^{HH}$	$HHHH$	(7-149)	1	$-i6\lambda$
15-18		HHH	"	2	$-i6\lambda v$
15-19	$\mathcal{L}^{HB}$	$HHW^{\dagger}W$	(7-150)	1	$\frac{i}{2}g^2g^{\mu\nu}$ (for W^+_{μ}, W^-_{ν})
15-20		$HW^{\dagger}W$	"	2	$\frac{i}{2}g^2vg^{\mu\nu}$ (")
15-21		$HHZZ$	"	3	$\frac{ig^2}{2\cos^2\theta_W}g^{\mu\nu}$ (for Z_{μ}, Z_{ν})
15-22		HZZ	"	4	$\frac{ig^2v}{2\cos^2\theta_W}g^{\mu\nu}$ (")
15-23	$\mathcal{L}^{BB}$	$W^{\dagger}WZ$	(7-151)	Row 1	$ig\cos\theta_W\left\{g^{\alpha\beta}\left(k_1-k_2\right)+g^{\beta\gamma}\left(k_2-k_3\right)+g^{\gamma\alpha}\left(k_3-k_1\right)\right\}$ for $Z_{\alpha}(k_1), W^+_{\beta}(k_2), W^-_{\gamma}(k_3)$, all k_i inward
15-24		$W^{\dagger}WA$	"	Row 2	$ie\left\{g^{\alpha\beta}\left(k_1-k_2\right)+g^{\beta\gamma}\left(k_2-k_3\right)+g^{\gamma\alpha}\left(k_3-k_1\right)\right\}$ For $A_{\alpha}(k_1), W^+_{\beta}(k_2), W^-_{\gamma}(k_3)$, all k_i inward

15-25		$W^\dagger WZZ$	"	Row 3, 1st part	$ig^2 \cos^2\theta_W \{g^{\alpha\delta}g^{\beta\gamma} + g^{\alpha\gamma}g^{\beta\delta} + g^{\alpha\beta}g^{\gamma\delta}\}$ for $Z_\alpha, Z_\beta, W^+_\gamma, W^-_\delta$
15-26		$W^\dagger WAA$	"	Row 3, 2nd part	$ie^2 \{g^{\alpha\delta}g^{\beta\gamma} + g^{\alpha\gamma}g^{\beta\delta} + g^{\alpha\beta}g^{\gamma\delta}\}$ for $A_\alpha, A_\beta, W^+_\gamma, W^-_\delta$
15-27		$W^\dagger WZA$	"	Row 4	$ieg \cos\theta_W \{g^{\alpha\delta}g^{\beta\gamma} + g^{\alpha\gamma}g^{\beta\delta} - 2g^{\alpha\beta}g^{\gamma\delta}\}$ for $Z_\alpha, A_\beta, W^+_\gamma, W^-_\delta$
15-28		$(W^\dagger W)^2$	"	Row 5	$ig^2 \{-g^{\alpha\delta}g^{\beta\gamma} + 2g^{\alpha\gamma}g^{\beta\delta} - g^{\alpha\beta}g^{\gamma\delta}\}$ for $W^+_\alpha, W^-_\beta, W^+_\gamma, W^-_\delta$

8.7 Appendices

8.7.1 Appendix A. The Propagator for a Massive Vector Field

Derivation of the Massive Propagator

As we found in Sect. 5.4.3, pg. 157, a massive vector field has three independent components, because of the constraint nature imposes (which is not a mere gauge)

$$\partial_\mu Z^\mu = 0 \qquad \rightarrow \qquad k_\mu Z^\mu = 0 , \qquad (8\text{-}122)$$

where, to keep things simple (and without loss of generality), we align our x^3 axis in the direction of **k**. That is (where the "+ number" notation indicates new equations added in this revised version of the text that were not in the original version of the text),

$$k^\mu = \left(\omega_{\mathbf{k}}, 0, 0, |\mathbf{k}|\right) \qquad k_\mu = \left(\omega_{\mathbf{k}}, 0, 0, -|\mathbf{k}|\right) \qquad \text{where } \omega_{\mathbf{k}}^2 - \mathbf{k}^2 = m_Z^2 . \qquad (8\text{-}122)+1$$

We are free to choose any polarization basis to express the Z field in, and it is virtually always easiest if the basis vectors are unit vectors that are all orthogonal to one another. Thinking it is probably the most convenient, we might be tempted to choose the following, which are the ones we chose in Vol. 1, pgs. 142-143, (5-34) for (massless) photons.

$$\varepsilon_0^\mu = (1,0,0,0) \qquad \varepsilon_1^\mu = (0,1,0,0) \qquad \varepsilon_2^\mu = (0,0,1,0) \qquad \varepsilon_3^\mu = (0,0,0,1) \qquad (8\text{-}122)+2$$

As in Vol. 1, we label the polarization vectors as ε_r^μ where r labels the vectors and runs from 1 to 3 for the spatial polarization vectors, and $r = 0$ for the timelike polarization vector. The Greek μ labels the 4D spacetime components of each of the polarization vectors.

However, it turns out that (8-122)+2 actually is not the most convenient basis for massive fields. The reason for this is a bit subtle and involved, and is treated in an addendum to this appendix, which was added in the revision to this text and can be found at the end of this chapter.

The most convenient basis for massive fields (where primes distinguish basis vectors from those in (8-122)+2) is

$$\varepsilon_0'^\mu = \frac{1}{m_z}\left(\omega_{\mathbf{k}}, 0, 0, |\mathbf{k}|\right) \qquad \varepsilon_1'^\mu = (0,1,0,0) \qquad \varepsilon_2'^\mu = (0,0,1,0) \qquad \varepsilon_3'^\mu = \frac{1}{m_z}\left(|\mathbf{k}|, 0, 0, \omega_{\mathbf{k}}\right) . \qquad (8\text{-}123)$$

Hopefully, you can prove to yourself that these basis vectors are unit vectors and are orthogonal. Just remember that lowering spatial indices changes the spatial component sign, that $\omega_{\mathbf{k}}^2 - \mathbf{k}^2 = m^2$, and that length squared of spatial vectors is negative.

As shown in the cited addendum, the zeroth polarization component of the Z field in the (8-123) basis is zero. So, in (8-122), the $Z'_{r\text{-}0}$ polarization component is zero. Since the other three components in that basis are independent, (8-122) becomes

$$k_\mu \varepsilon_r'^\mu = 0 \ \text{ for } r = 1,2,3. \qquad (8\text{-}124)$$

(8-124) is then the valid constraint relation to use in the primed basis. The basis choice simplified things by making $Z'_{r\text{-}0} = 0$, so we can get all physics involved by simply working with the three spatial relations (8-124).

Note that (with repeated indices indicating summation)

$$\varepsilon_r'^0\varepsilon_r'^0 = 0+0+\frac{|\mathbf{k}||\mathbf{k}|}{m_Z^2} = \frac{\omega_\mathbf{k}^2 - m_Z^2}{m_Z^2} = -1 + \frac{k^0k^0}{m_Z^2} \qquad \varepsilon_r'^0\varepsilon_r'^1 = 0+0+0=0 \qquad \varepsilon_r'^1\varepsilon_r'^1 = 1+0+0=1$$

$$\varepsilon_r'^3\varepsilon_r'^3 = 0+0+\frac{\omega_\mathbf{k}^2}{m_Z^2} = \frac{m_Z^2 + |\mathbf{k}||\mathbf{k}|}{m_Z^2} = 1 + \frac{k^3k^3}{m_Z^2} \qquad \varepsilon_r'^3\varepsilon_r'^0 = 0+0+\frac{|\mathbf{k}|\omega_\mathbf{k}}{m_Z^2} = \frac{k^3k^0}{m_Z^2}. \tag{8-125}$$

The pattern in (8-125) can be generalized to

$$\sum_{r=1}^{3} \varepsilon_r'^\mu \varepsilon_r'^\nu = -g^{\mu\nu} + \frac{k^\mu k^\nu}{m_Z^2}, \tag{8-126}$$

which is the completeness relation for massive vector boson fields. This result is covariant and thus, independent of coordinate system and independent of polarization basis.

Now recall (Vol. 1, pg. 159), the Feynman propagator is defined as the vacuum expectation of time ordered operators, in this case,

$$iD_F^{\mu\nu}(x-y) = \langle 0|T\{Z^\mu(x)Z^\nu(y)\}|0\rangle . \tag{8-127}$$

In the process of doing this for photons, we needed to sum over polarization states and used the completeness relation of Vol. 1, pg. 161, (5-82), repeated below.

$$\text{For massless photons} \quad \sum_{r=0}^{3}\zeta_r\varepsilon_r^\mu\varepsilon_r^\nu = -g^{\mu\nu} \quad \text{where} \quad \zeta_0 = -1 \quad \zeta_1=\zeta_2=\zeta_3=1, \tag{8-128}$$

$$\text{results in} \quad iD_F^{\mu\nu}(k) = i\frac{-g^{\mu\nu}}{k^2+i\varepsilon} \quad \text{for photons.} \tag{8-129}$$

However, for massive vectors, we need to use (8-126), instead of (8-128), where here we have no ζ_0 factor . Hence, wherever we had the RHS of (8-128) in our derivation and final result for photons, we need to substitute the RHS of (8-126) for massive vectors. We also need to insert the mass squared term in the denominator, which for photons is zero but which arises in any boson propagator derivation, to get our final result,

$$iD_{F\mu\nu}(k) = i\frac{-g_{\mu\nu} + k_\mu k_\nu / m_Z^2}{k^2 - m_Z^2 + i\varepsilon}. \tag{8-130}$$

We get this result because massive vectors must obey (8-122), so their polarization vectors must obey (8-124). Doing that forces the polarization sum relation of (8-126) on us.

Comparing Massive and Massless Vector Propagators

Note that for massless vectors, $\omega_\mathbf{k} = |\mathbf{k}|$, so (8-123) would become

$$\varepsilon_0'^\mu = \frac{1}{m_z}(\omega_\mathbf{k},0,0,\omega_\mathbf{k}) \qquad \varepsilon_1'^\mu = (0,1,0,0) \qquad \varepsilon_2'^\mu = (0,0,1,0) \qquad \varepsilon_3'^\mu = \frac{1}{m_z}(\omega_\mathbf{k},0,0,\omega_\mathbf{k}), \tag{8-131}$$

except the mass in the denominator going to zero means the r = 0 and 3 polarization vectors are infinite, or perhaps better put, undefined. That is readily fixed by normalizing them, so we have the three polarization vectors for r = 1,2,3

$$\varepsilon_1^\mu = (0,1,0,0) \qquad \varepsilon_2^\mu = (0,0,1,0) \qquad \varepsilon_3^\mu = (1,0,0,1). \tag{8-132}$$

These are all orthogonal to one another and to the massless 4-momentum vector (where, again, the x^3 axis is aligned with **k**). Note the r = 3 polarization vector here is light-like, as it has zero length in 4D.

$$k^\mu = (\omega_\mathbf{k},0,0,\omega_\mathbf{k}) \qquad k_\mu = (\omega_\mathbf{k},0,0,-\omega_\mathbf{k}) \qquad k_\mu\varepsilon_3^\mu = \omega_\mathbf{k} - \omega_\mathbf{k} = 0. \tag{8-133}$$

Three independent polarization vectors (8-132), which are all orthogonal to k^μ, are forced on us by (8-122). But, for massless vectors that is not imposed on us by nature, but a choice of gauge (Lorenz) we make.

If we chose not to restrict the massless vector with a gauge (which would make calculations harder), we could then have four independent, orthogonal unit polarization vectors, which are not all orthogonal to k^μ,

$$\varepsilon_0^\mu = (1,0,0,0) \qquad \varepsilon_1^\mu = (0,1,0,0) \qquad \varepsilon_2^\mu = (0,0,1,0) \qquad \varepsilon_3^\mu = (0,0,0,1) \qquad (\text{no gauge}). \quad (8\text{-}134)$$

For this, for example,

$$\begin{aligned} \varepsilon_r^0 \varepsilon_r^0 &= 1+0+0+0=1 & \varepsilon_r^0 \varepsilon_r^1 &= 0+0+0+0=0 \\ \varepsilon_r^1 \varepsilon_r^1 &= 0+1+0+0=1 & \varepsilon_r^3 \varepsilon_r^3 &= 0+0+0+1=1, \end{aligned} \qquad (8\text{-}135)$$

generalized as (8-128), the completeness relations we used to deduce the photon propagator.

If we used (8-132) instead of (8-134), we get the same results,

$$\begin{aligned} \varepsilon_r^0 \varepsilon_r^0 &= 0+0+1=1 & \varepsilon_r^0 \varepsilon_r^1 &= 0+0+0=0 \\ \varepsilon_r^1 \varepsilon_r^1 &= 1+0+0=1 & \varepsilon_r^3 \varepsilon_r^3 &= 0+0+1=1, \end{aligned} \qquad (8\text{-}136)$$

and thus, the same completeness relations (8-128), and the same propagator (8-129).

Bottom line: The second term in the numerator of the vector propagator (8-130) arises for massive vector fields, but not for massless ones, regardless of whether we employ a gauge constraint on the massless fields, or not. For the massive case, or if we employ the Lorenz gauge in the massless case, three polarization vectors are orthogonal (in 4D) to the four-momentum k^μ.

<u>Assuming Massive Vector Field Not Subject to Constraint</u>

What we describe in this sub-section may seem strange right now, but it will play a role in our discussion in a later chapter of electroweak renormalization.

Suppose nature did not restrict us to (8-122). In that case, we could use (8-134) for our polarization vectors. They would not all have to be orthogonal to k^μ; we could choose them to take the form of (8-134), and thus we would have a propagator of form (8-129), without the second term in the numerator of (8-130).

Bottom line: If nature did not constrain massive vector fields to having 3 independent polarization vectors all orthogonal to k^μ, we could have a simple form for the propagator, the same form as the photon propagator.

All of the results of this section are summarized in Wholeness Chart 8-1.

Wholeness Chart 8-1. Vector Fields Polarization Vectors and Propagators

	<u>$m = 0$ Typical</u>	**<u>$m \neq 0$ Actual</u>**	**<u>$m = 0$ Alternative</u>**	**<u>$m \neq 0$ Hypothetical</u>**
Constraint	$\partial_\mu A^\mu = 0$ (Lorenz gauge)	$\partial_\mu Z^\mu = 0$ (Nature imposes)	No constraint	As at immediate left
Impact	$k_\mu\, \varepsilon_r{}^\mu = 0$ (Polariz vectors orthogonal to k_μ)	As at left	No orthogonality constraint with k^μ	As at immediate left
Orthogonality	$\varepsilon_r^\mu \varepsilon_{\mu s} = -\zeta_r \delta_{rs}$ 3 indep r	As at left	Can take as at left, but 4 indep r	Can take as at immediate left
Completeness	$\zeta_r \varepsilon_r^\mu \varepsilon_r^\nu = -g^{\mu\nu}$	As at left	Can take as at left, but 4 indep r	Can take as at immediate left
Above in definition of Feynman propagator yields below				
Propagator $iD_{F\mu\nu}$	$i\dfrac{-g_{\mu\nu}}{k^2 + i\varepsilon}$	$i\dfrac{-g_{\mu\nu} + k_\mu k_\nu / m_Z^2}{k^2 - m_Z^2 + i\varepsilon}$	$i\dfrac{-g_{\mu\nu}}{k^2 + i\varepsilon}$	$i\dfrac{-g_{\mu\nu}}{k^2 - m_Z^2 + i\varepsilon}$

8.7.2 Appendix B. Evaluating a Factor in the Muon Decay Relation

We first note contraction identities for the Levi-Civita tensor of rank 4, of which we will only use the first.

$$
\begin{gathered}
\varepsilon^{\rho\mu\xi\nu}\varepsilon_{\gamma\mu\delta\nu}=\varepsilon^{\mu\nu\rho\xi}\varepsilon_{\mu\nu\gamma\delta}=-2\left(\delta^{\rho}_{\gamma}\delta^{\xi}_{\delta}-\delta^{\rho}_{\delta}\delta^{\xi}_{\gamma}\right)\\
\varepsilon^{\rho\mu\xi\nu}\varepsilon_{\rho\mu\delta\nu}=\varepsilon^{\mu\nu\rho\xi}\varepsilon_{\mu\nu\rho\delta}=-6\delta^{\xi}_{\delta}\qquad\varepsilon^{\rho\mu\delta\nu}\varepsilon_{\rho\mu\delta\nu}=\varepsilon^{\mu\nu\rho\delta}\varepsilon_{\mu\nu\rho\delta}=-24.
\end{gathered}\tag{8-137}
$$

Also, recall
$$g^{\mu\nu}g_{\mu\nu}=4.\tag{8-138}$$

Then, repeating (8-38) and (8-39), for convenience

$$D^{\mu\nu}=8p'_{\rho}q_{1\xi}\left(\left(g^{\rho\mu}g^{\xi\nu}-g^{\rho\xi}g^{\mu\nu}+g^{\rho\nu}g^{\mu\xi}\right)+i\varepsilon^{\rho\mu\xi\nu}\right)\qquad\text{repeat of (8-38)}$$

$$E_{\mu\nu}=8q_2^{\gamma}p^{\delta}\left(\left(g_{\gamma\mu}g_{\delta\nu}-g_{\gamma\delta}g_{\mu\nu}+g_{\gamma\nu}g_{\mu\delta}\right)+i\varepsilon_{\gamma\mu\delta\nu}\right)\qquad\text{repeat of (8-39)}$$

we have
$$
\begin{aligned}
D^{\mu\nu}E_{\mu\nu}=64p'_{\rho}q_{1\xi}q_2^{\gamma}p^{\delta}&\underbrace{\left(g^{\rho\mu}g^{\xi\nu}-g^{\rho\xi}g^{\mu\nu}+g^{\rho\nu}g^{\mu\xi}+i\varepsilon^{\rho\mu\xi\nu}\right)}_{X^{\rho\mu\xi\nu}}\\
&\times\underbrace{\left(g_{\gamma\mu}g_{\delta\nu}-g_{\gamma\delta}g_{\mu\nu}+g_{\gamma\nu}g_{\mu\delta}+i\varepsilon_{\gamma\mu\delta\nu}\right)}_{X_{\gamma\mu\delta\nu}}.
\end{aligned}\tag{8-139}
$$

$$
\begin{aligned}
X^{\rho\mu\xi\nu}X_{\gamma\mu\delta\nu}=&\,g^{\rho\mu}g^{\xi\nu}g_{\gamma\mu}g_{\delta\nu}-g^{\rho\mu}g^{\xi\nu}g_{\gamma\delta}g_{\mu\nu}+g^{\rho\mu}g^{\xi\nu}g_{\gamma\nu}g_{\mu\delta}+ig^{\rho\mu}g^{\xi\nu}\varepsilon_{\gamma\mu\delta\nu}\\
&-g^{\rho\xi}g^{\mu\nu}g_{\gamma\mu}g_{\delta\nu}+g^{\rho\xi}g^{\mu\nu}g_{\gamma\delta}g_{\mu\nu}-g^{\rho\xi}g^{\mu\nu}g_{\gamma\nu}g_{\mu\delta}-ig^{\rho\xi}g^{\mu\nu}\varepsilon_{\gamma\mu\delta\nu}\\
&+g^{\rho\nu}g^{\mu\xi}g_{\gamma\mu}g_{\delta\nu}-g^{\rho\nu}g^{\mu\xi}g_{\gamma\delta}g_{\mu\nu}+g^{\rho\nu}g^{\mu\xi}g_{\gamma\nu}g_{\mu\delta}+ig^{\rho\nu}g^{\mu\xi}\varepsilon_{\gamma\mu\delta\nu}\\
&+i\varepsilon^{\rho\mu\xi\nu}g_{\gamma\mu}g_{\delta\nu}-i\varepsilon^{\rho\mu\xi\nu}g_{\gamma\delta}g_{\mu\nu}+i\varepsilon^{\rho\mu\xi\nu}g_{\gamma\nu}g_{\mu\delta}-\varepsilon^{\rho\mu\xi\nu}\varepsilon_{\gamma\mu\delta\nu}.
\end{aligned}\tag{8-140}
$$

$$
\begin{aligned}
X^{\rho\mu\xi\nu}X_{\gamma\mu\delta\nu}=&\,\overset{\boxed{1a}}{\delta^{\rho}_{\gamma}\delta^{\xi}_{\delta}}-\overset{\boxed{2a}}{g^{\xi\rho}g_{\gamma\delta}}+\overset{\boxed{3a}}{\delta^{\rho}_{\delta}\delta^{\xi}_{\gamma}}+\overset{\boxed{4a}}{ig^{\rho\mu}g^{\xi\nu}\varepsilon_{\gamma\mu\delta\nu}}\\
&-\overset{\boxed{2b}}{g^{\rho\xi}g_{\gamma\delta}}+\overset{\boxed{2c}}{4g^{\rho\xi}g_{\gamma\delta}}-\overset{\boxed{2d}}{g^{\rho\xi}g_{\gamma\delta}}-\overset{\boxed{5}}{ig^{\rho\xi}g^{\mu\nu}\varepsilon_{\gamma\mu\delta\nu}}\\
&+\overset{\boxed{3b}}{\delta^{\rho}_{\delta}\delta^{\xi}_{\gamma}}-\overset{\boxed{2e}}{g^{\rho\xi}g_{\gamma\delta}}+\overset{\boxed{1b}}{\delta^{\rho}_{\gamma}\delta^{\xi}_{\delta}}+\overset{\boxed{4b}}{ig^{\rho\nu}g^{\mu\xi}\varepsilon_{\gamma\mu\delta\nu}}\\
&+\overset{\boxed{6a}}{i\varepsilon^{\rho\mu\xi\nu}g_{\gamma\mu}g_{\delta\nu}}-\overset{\boxed{7}}{i\varepsilon^{\rho\mu\xi\nu}g_{\gamma\delta}g_{\mu\nu}}+\overset{\boxed{6b}}{i\varepsilon^{\rho\mu\xi\nu}g_{\gamma\nu}g_{\mu\delta}}-\overset{\boxed{8}}{\varepsilon^{\rho\mu\xi\nu}\varepsilon_{\gamma\mu\delta\nu}}.
\end{aligned}\tag{8-141}
$$

$$\boxed{1}=\overset{\boxed{1a}}{\delta^{\rho}_{\gamma}\delta^{\xi}_{\delta}}+\overset{\boxed{1b}}{\delta^{\rho}_{\gamma}\delta^{\xi}_{\delta}}=2\delta^{\rho}_{\gamma}\delta^{\xi}_{\delta}\tag{8-142}$$

$$\boxed{2}=-\overset{\boxed{2a}}{g^{\xi\rho}g_{\gamma\delta}}-\overset{\boxed{2b}}{g^{\rho\xi}g_{\gamma\delta}}+\overset{\boxed{2c}}{4g^{\rho\xi}g_{\gamma\delta}}-\overset{\boxed{2d}}{g^{\rho\xi}g_{\gamma\delta}}-\overset{\boxed{2e}}{g^{\rho\xi}g_{\gamma\delta}}=0\tag{8-143}$$

$$\boxed{3}=\overset{\boxed{3a}}{\delta^{\rho}_{\delta}\delta^{\xi}_{\gamma}}+\overset{\boxed{3b}}{\delta^{\rho}_{\delta}\delta^{\xi}_{\gamma}}=2\delta^{\rho}_{\delta}\delta^{\xi}_{\gamma}\tag{8-144}$$

$$
\begin{aligned}
\boxed{4}&=\overset{\boxed{4a}}{ig^{\rho\mu}g^{\xi\nu}\varepsilon_{\gamma\mu\delta\nu}}+\underbrace{\overset{\boxed{4b}}{ig^{\rho\nu}g^{\mu\xi}\varepsilon_{\gamma\mu\delta\nu}}}_{ig^{\rho\nu}g^{\xi\mu}\varepsilon_{\gamma\mu\delta\nu}}=ig^{\rho\mu}g^{\xi\nu}\varepsilon_{\gamma\mu\delta\nu}+ig^{\rho\mu}g^{\xi\nu}\varepsilon_{\gamma\nu\delta\mu}\\
&=ig^{\rho\mu}g^{\xi\nu}\left(\varepsilon_{\gamma\mu\delta\nu}+\varepsilon_{\gamma\nu\delta\mu}\right)=ig^{\rho\mu}g^{\xi\nu}\left(\varepsilon_{\gamma\mu\delta\nu}+\varepsilon_{\gamma\delta\mu\nu}\right)=ig^{\rho\mu}g^{\xi\nu}\left(\varepsilon_{\gamma\mu\delta\nu}-\varepsilon_{\gamma\mu\delta\nu}\right)=0.
\end{aligned}\tag{8-145}
$$

$\boxed{5}=-ig^{\rho\xi}g^{\mu\nu}\varepsilon_{\gamma\mu\delta\nu}=0$, since if $\mu=\nu$, $\varepsilon_{\gamma\mu\delta\nu}=0$, and if $\mu\neq\nu$, $g^{\mu\nu}=0$. (8-146)

Aside for general relativity (for future use after this text), where we can have $g^{\mu\nu}\neq 0$ for $\mu\neq\nu$:

For example, $g^{12}\varepsilon_{\gamma 1\delta 2}+g^{21}\varepsilon_{\gamma 2\delta 1}=g^{12}\varepsilon_{\gamma 1\delta 2}+g^{12}\varepsilon_{\gamma\delta 12}=g^{12}\left(\varepsilon_{\gamma 1\delta 2}-\varepsilon_{\gamma 1\delta 2}\right)=0.$

$$\boxed{6}=i\varepsilon^{\rho\mu\xi\nu}\overset{\boxed{6a}}{g_{\gamma\mu}g_{\delta\nu}}+i\varepsilon^{\rho\mu\xi\nu}\overset{\boxed{6b}}{g_{\gamma\nu}g_{\mu\delta}}=0 \quad \text{for same reason as } \boxed{4}$$

$$\boxed{7}=-i\varepsilon^{\rho\mu\xi\nu}g_{\gamma\delta}g_{\mu\nu}=0 \quad \text{for same reason as } \boxed{5} \tag{8-147}$$

$$\boxed{8}=-\varepsilon^{\rho\mu\xi\nu}\varepsilon_{\gamma\mu\delta\nu}=\overset{\boxed{8a}}{2\delta^{\rho}_{\gamma}\delta^{\xi}_{\delta}}-\overset{\boxed{8b}}{2\delta^{\rho}_{\delta}\delta^{\xi}_{\gamma}} \quad \text{from first row of (8-137)} \tag{8-148}$$

$$X^{\rho\mu\xi\nu}X_{\gamma\mu\delta\nu}=\overset{\boxed{1}}{2\delta^{\rho}_{\gamma}\delta^{\xi}_{\delta}}+\overset{\boxed{3}}{2\delta^{\rho}_{\delta}\delta^{\xi}_{\gamma}}+\overset{\boxed{8a}}{2\delta^{\rho}_{\gamma}\delta^{\xi}_{\delta}}-\overset{\boxed{8b}}{2\delta^{\rho}_{\delta}\delta^{\xi}_{\gamma}}=4\delta^{\rho}_{\gamma}\delta^{\xi}_{\delta}. \tag{8-149}$$

Then, from (8-139),

$$D^{\mu\nu}E_{\mu\nu}=\left(64p'_{\rho}q_{1\xi}q_2^{\gamma}p^{\delta}\right)\left(4\delta^{\rho}_{\gamma}\delta^{\xi}_{\delta}\right)=\left(64p'_{\gamma}q_{1\delta}q_2^{\gamma}p^{\delta}\right)(4)=64\left(4\left(pq_1\right)\left(p'q_2\right)\right). \quad \text{repeat of (8-41)}$$

8.8 Problems

1. Show that particle half-life is $t_{1/2}$ = .693 τ, where τ is the particle lifetime, i.e., the expected (mean) life. Hint: Assume decay is exponential, such that for a large number of particles the amount left undecayed after time t is $A(t)$ = $A(0)e^{-t/\tau}$. The probability density that any given particle will decay at time t is $\rho(t)=\dfrac{A(t)}{\int_0^{\infty}A(t)\,dt}$, where the probability that a particle will decay by the time t' has elapsed is $\int_0^{t'}\rho(t)\,dt$. Note that, from integral tables, $\int_{t_1}^{t_2}t\,e^{-t/\tau}dt=\left[\tau^2e^{-t/\tau}\left(\dfrac{t}{\tau}-1\right)\right]_{t_1}^{t_2}$. With that, show that for the exponential relation assumed, the mean lifetime is τ, and thus, the assumed relation is valid. Then, show the half-life is .693 τ.
2. Prove that (8-68) solves (8-63) and (8-67).
3. Show that (8-76) yields a muon lifetime of about 2.15 x10^{-6} seconds. Hint: You can use Vol. 1, Wholeness Chart 2-1, pg. 14 therein, to convert from GeV to seconds.
4. Show the total cross-section for (8-109) is (8-110).

8.9 Addendum to Appendix A (added in June 2022 revision)

8.9.1 Background

There are two bases we want to keep straight in our minds.

<u>The Usual Minkowski Basis</u>

The first of these is the usual spacetime basis for vector entities in flat spacetime. This is the Minkowski metric basis, where each orthogonal direction in space has a unit basis vector and the time direction also has a unit basis vector. These vectors can be represented as

$$\mathbf{i}_0,\mathbf{i}_1,\mathbf{i}_2,\mathbf{i}_3, \tag{8-150}$$

where we have 4D coordinate axes of x^0-x^1-x^2-x^3.

If we want to express each of these in terms of the entire basis, we have

$$\mathbf{i}_0=i_0^{\mu}=(1,0,0,0) \quad \mathbf{i}_1=i_1^{\mu}=(0,1,0,0) \quad \mathbf{i}_2=i_2^{\mu}=(0,0,1,0) \quad \mathbf{i}_3=i_3^{\mu}=(0,0,0,1), \tag{8-151}$$

which is actually rather trivial (though I hate to use that word.) Any vector field, such as the Z field, can be expressed in terms of its four components in that basis (and that is what we usually do).

$$Z^{\mu}=Z^0\mathbf{i}_0+Z^1\mathbf{i}_1+Z^2\mathbf{i}_2+Z^3\mathbf{i}_3=Z^0\begin{bmatrix}1\\0\\0\\0\end{bmatrix}+Z^1\begin{bmatrix}0\\1\\0\\0\end{bmatrix}+Z^2\begin{bmatrix}0\\0\\1\\0\end{bmatrix}+Z^3\begin{bmatrix}0\\0\\0\\1\end{bmatrix}=\begin{bmatrix}Z^0\\Z^1\\Z^2\\Z^3\end{bmatrix}. \tag{8-152}$$

Polarization Vector Basis

But, we could choose any set of independent vectors as basis vectors, not just those along the x^0-x^1-x^2-x^3 axes. They don't have to be unit vectors, but we will restrict our discussion to unit vectors only. That is, we could have other basis vectors, in fact there are an infinite number of choices, other than (8-151). We can express these as

$$\boldsymbol{\varepsilon}_0,\boldsymbol{\varepsilon}_1,\boldsymbol{\varepsilon}_2,\boldsymbol{\varepsilon}_3 \quad \text{generally, as } \boldsymbol{\varepsilon}_r \quad r=0,1,2,3\,. \tag{8-153}$$

In terms of our coordinate axes basis (the Minkowski basis), we can express these as

$$\boldsymbol{\varepsilon}_0=\varepsilon_0^\mu=\left(a_0,b_0,c_0,d_0\right) \quad \boldsymbol{\varepsilon}_1=\varepsilon_1^\mu=\left(a_1,b_1,c_1,d_1\right) \quad \boldsymbol{\varepsilon}_2=\varepsilon_2^\mu=\left(a_2,b_2,c_2,d_2\right) \quad \boldsymbol{\varepsilon}_3=\varepsilon_3^\mu=\left(a_3,b_3,c_3,d_3\right), \tag{8-154}$$

which is parallel to (8-151).

Note that r represents the particular basis vector, and μ the components of that basis vector in the Minkowski system.

We can represent the same Z vector field of (8-152) in this different basis as

$$Z^\mu=Z_r\boldsymbol{\varepsilon}_r=Z_{r=0}\boldsymbol{\varepsilon}_0+Z_{r=1}\boldsymbol{\varepsilon}_1+Z_{r=2}\boldsymbol{\varepsilon}_2+Z_{r=3}\boldsymbol{\varepsilon}_3=Z_{r=0}\varepsilon_0^\mu+Z_{r=1}\varepsilon_1^\mu+Z_{r=2}\varepsilon_2^\mu+Z_{r=3}\varepsilon_3^\mu$$

$$=Z_{r=0}\begin{bmatrix}a_0\\b_0\\c_0\\d_0\end{bmatrix}+Z_{r=1}\begin{bmatrix}a_1\\b_1\\c_1\\d_1\end{bmatrix}+Z_{r=2}\begin{bmatrix}a_2\\b_2\\c_2\\d_2\end{bmatrix}+Z_{r=3}\begin{bmatrix}a_3\\b_3\\c_3\\d_3\end{bmatrix}=\begin{bmatrix}Z_{r=0}a_0+Z_{r=1}a_1+Z_{r=2}a_2+Z_{r=3}a_3\\Z_{r=0}b_0+Z_{r=1}b_1+Z_{r=2}b_2+Z_{r=3}b_3\\Z_{r=0}c_0+Z_{r=1}c_1+Z_{r=2}c_2+Z_{r=3}c_3\\Z_{r=0}d_0+Z_{r=1}d_1+Z_{r=2}d_2+Z_{r=3}d_3\end{bmatrix}=\begin{bmatrix}Z^0\\Z^1\\Z^2\\Z^3\end{bmatrix}. \tag{8-155}$$

(8-155) and (8-152) are the exact same vector, just expressed in terms of different bases. For example, Z^1 is the amount of the Z field along the x^1 axis, i.e., aligned with the $\mathbf{i}_1$ basis vector. $Z_{r=1}$ is the amount of the Z field in the $\boldsymbol{\varepsilon}_1$ direction (which is generally not the same as the $\mathbf{i}_1$ vector).

As examples, Z^1 has generally four different components in the $\boldsymbol{\varepsilon}_r$ basis, each component aligned along one of the four $\boldsymbol{\varepsilon}_r$ basis vectors, though it only has one component (Z^1) in the Minkowski $\mathbf{i}_s$ basis ($s = 0,1,2,3$). $Z_{r=1}$ is the amount of the Z vector aligned with the $\boldsymbol{\varepsilon}_1$ vector, so only has one component ($Z_{r=1}$) in that basis, but it generally has four components in the Minkowski basis ($Z_{r=1}a_1$, $Z_{r=1}b_1$, $Z_{r=1}c_1$, $Z_{r=1}d_1$).

Typically, in classical theory and QFT, the $\boldsymbol{\varepsilon}_r$ vectors are taken to align with (linear) polarization directions of the field being studied. In general, the polarization directions do not align with the $\mathbf{i}_s$ directions of the x^0-x^1-x^2-x^3 coordinate system, though they can.

8.9.2 Traditional Polarization Vector Basis

In analyzing the photon field in Vol. 1, pgs. 142-143, it was found convenient to align our x^0-x^1-x^2-x^3 coordinate system such that the x^3 axis was aligned with field 3-momentum 3-vector $\mathbf{k}$. Since the polarizations were then in the x^1 and x^2 directions, things became very simplified for our polarization vectors $\boldsymbol{\varepsilon}_r$. That is,

$$\varepsilon_0^\mu=\left(1,0,0,0\right) \qquad \varepsilon_1^\mu=\left(0,1,0,0\right) \qquad \varepsilon_2^\mu=\left(0,0,1,0\right) \qquad \varepsilon_3^\mu=\left(0,0,0,1\right). \tag{8-156}$$

Note how this plays out for the case of a massive field, as in Appendix A of this chapter, pg. 293, where

$$Z^\mu=Z_{r=0}\boldsymbol{\varepsilon}_0+Z_{r=1}\boldsymbol{\varepsilon}_1+Z_{r=2}\boldsymbol{\varepsilon}_2+Z_{r=3}\boldsymbol{\varepsilon}_3=Z_{r=0}\varepsilon_0^\mu+Z_{r=1}\varepsilon_1^\mu+Z_{r=2}\varepsilon_2^\mu+Z_{r=3}\varepsilon_3^\mu=\begin{bmatrix}Z_{r=0}\\Z_{r=1}\\Z_{r=2}\\Z_{r=3}\end{bmatrix}=\begin{bmatrix}Z^0\\Z^1\\Z^2\\Z^3\end{bmatrix}. \tag{8-157}$$

We align our coordinate system such that $\mathbf{k}$ of the field Z is directed along the positive x^3 axis. So,

$$k^\mu=\left(\omega_{\mathbf{k}},0,0,|\mathbf{k}|\right) \qquad k_\mu=\left(\omega_{\mathbf{k}},0,0,-|\mathbf{k}|\right) \qquad \text{where} \quad \omega_{\mathbf{k}}^2-\mathbf{k}^2=m_Z^2 \tag{8-158}$$

For massive fields, we have the constraint

$$k_\mu Z^\mu=0 \qquad \text{repeat of (8-122) in Vol. 2.} \tag{8-159}$$

Using (8-157) and (8-158), we find (8-159) gives us

$$k_\mu Z^\mu=\left(\omega_{\mathbf{k}},0,0,-|\mathbf{k}|\right)\begin{pmatrix}Z_{r=0}\\Z_{r=1}\\Z_{r=2}\\Z_{r=3}\end{pmatrix}=\omega_{\mathbf{k}}Z_{r=0}-|\mathbf{k}|Z_{r=3}=0 \;\rightarrow\; Z_{r=0}=\frac{|\mathbf{k}|}{\omega_{\mathbf{k}}}Z_{r=3}\neq 0 \;\rightarrow\; Z^0=\frac{|\mathbf{k}|}{\omega_{\mathbf{k}}}Z^3\neq 0\,. \tag{8-160}$$

For massless fields like the photon, $|\mathbf{k}| = \omega_{\mathbf{k}}$, and as we found for photons, $Z_{r=0} = Z_{r=3}$. That simplified a lot of things, which we won't get into here. (See Vol. 1, Chap. 5). However, for massive fields $|\mathbf{k}| \neq \omega_{\mathbf{k}}$ and we lose those simplifications.

8.9.3 A Different Polarization Basis

We would like to employ a polarization vector basis that does simplify things for massive fields. In particular, it will help significantly if we can use such a basis where $Z'_{r=0} = 0$. It turns out that the following choice will do the job (as we will see). We use primes to distinguish from the traditional choice (8-156).

$$\begin{array}{llll} \varepsilon_0'^{\mu} = \frac{1}{m_z}\left(\omega_{\mathbf{k}},0,0,|\mathbf{k}|\right) & \varepsilon_1'^{\mu} = (0,1,0,0) & \varepsilon_2'^{\mu} = (0,0,1,0) & \varepsilon_3'^{\mu} = \frac{1}{m_z}\left(|\mathbf{k}|,0,0,\omega_{\mathbf{k}}\right) \\ \varepsilon'_{0\mu} = \frac{1}{m_z}\left(\omega_{\mathbf{k}},0,0,-|\mathbf{k}|\right) & \varepsilon'_{1\mu} = (0,-1,0,0) & \varepsilon'_{2\mu} = (0,0,-1,0) & \varepsilon'_{3\mu} = \frac{1}{m_z}\left(|\mathbf{k}|,0,0,-\omega_{\mathbf{k}}\right). \end{array} \tag{8-161}$$

Problem: Show that each vector in (8-161) has unit length and that it is orthogonal to all the other vectors in (8-161).

With (8-161), we get

$$Z^{\mu} = Z'_{r=0}\boldsymbol{\varepsilon}'_0 + Z'_{r=1}\boldsymbol{\varepsilon}'_1 + Z'_{r=2}\boldsymbol{\varepsilon}'_2 + Z'_{r=3}\boldsymbol{\varepsilon}'_3 = Z'_{r=0}\varepsilon_0'^{\mu} + Z'_{r=1}\varepsilon_1'^{\mu} + Z'_{r=2}\varepsilon_2'^{\mu} + Z'_{r=3}\varepsilon_3'^{\mu} = \begin{bmatrix} Z'_{r=0}\frac{\omega_{\mathbf{k}}}{m_z} + Z'_{r=3}\frac{|\mathbf{k}|}{m_z} \\ Z'_{r=1} \\ Z'_{r=2} \\ Z'_{r=0}\frac{|\mathbf{k}|}{m_z} + Z'_{r=3}\frac{\omega_{\mathbf{k}}}{m_z} \end{bmatrix} = \begin{bmatrix} Z^0 \\ Z^1 \\ Z^2 \\ Z^3 \end{bmatrix}, \tag{8-162}$$

which is the same Z vector as in (8-157), just expressed (everywhere except after the last equal sign) in terms of different polarization basis vectors.

Now, align our coordinate axis such that the x^3 axis is in the **k** direction, so we get (8-158). Then use (8-159) with (8-158) and (8-162).

$$0 = k_{\mu}Z^{\mu} = \left(\omega_{\mathbf{k}},0,0,-|\mathbf{k}|\right)\begin{bmatrix} Z'_{r=0}\frac{\omega_{\mathbf{k}}}{m_z} + Z'_{r=3}\frac{|\mathbf{k}|}{m_z} \\ Z'_{r=1} \\ Z'_{r=2} \\ Z'_{r=0}\frac{|\mathbf{k}|}{m_z} + Z'_{r=3}\frac{\omega_{\mathbf{k}}}{m_z} \end{bmatrix} = Z'_{r=0}\frac{\omega_{\mathbf{k}}^2}{m_z} + Z'_{r=3}\frac{\omega_{\mathbf{k}}|\mathbf{k}|}{m_z} - Z'_{r=0}\frac{|\mathbf{k}|^2}{m_z} - Z'_{r=3}\frac{\omega_{\mathbf{k}}|\mathbf{k}|}{m_z}$$
$$= Z'_{r=0}\frac{\omega_{\mathbf{k}}^2 - |\mathbf{k}|^2}{m_z} = Z'_{r=0}m_Z \qquad \rightarrow \qquad Z'_{r=0} = 0\,. \tag{8-163}$$

Hence, if $Z'_{r=0} = 0$, then (8-162) becomes

$$Z^{\mu} = Z'_{r=1}\boldsymbol{\varepsilon}'_1 + Z'_{r=2}\boldsymbol{\varepsilon}'_2 + Z'_{r=3}\boldsymbol{\varepsilon}'_3 = Z'_{r=1}\varepsilon_1'^{\mu} + Z'_{r=2}\varepsilon_2'^{\mu} + Z'_{r=3}\varepsilon_3'^{\mu} = \begin{bmatrix} Z'_{r=3}\frac{|\mathbf{k}|}{m_z} \\ Z'_{r=1} \\ Z'_{r=2} \\ Z'_{r=3}\frac{\omega_{\mathbf{k}}}{m_z} \end{bmatrix} = \begin{bmatrix} Z^0 \\ Z^1 \\ Z^2 \\ Z^3 \end{bmatrix}, \tag{8-164}$$

and any factors with $\boldsymbol{\varepsilon}'_0$ drop out of any calculation with Z^{μ}. The constraint equation (8-159) in this basis forces the zeroth polarization component in this basis to be zero.

Thus, the constraint (8-159), with the part of (8-164) after the second equal sign, becomes

$$0 = k_{\mu}Z^{\mu} = Z'_{r=1}k_{\mu}\varepsilon_1'^{\mu} + Z'_{r=2}k_{\mu}\varepsilon_2'^{\mu} + Z'_{r=3}k_{\mu}\varepsilon_3'^{\mu}\,. \tag{8-165}$$

Since the Z'_r (r = 1,2,3) in (8-165) are all independent, each term in (8-165) must independently go to zero. Thus, we have

$$k_{\mu}\varepsilon_r'^{\mu} = 0 \quad \text{for each } r = 1,2,3 \qquad \text{this is (8-124).}$$

Chapter 9

Electroweak Symmetries: Continuous & Discrete

"I have a much easier time imagining how we would understand the big bang, even though we can't do it yet, than I can imagine understanding consciousness."
Edward Witten

9.0 Chapter Overview

In prior chapters, we have noted, and shown by examples, that weak isospin charge, hypercharge, and electric charge are conserved in electroweak interactions. We also showed that the presence of mass terms in the Lagrangian led to non-conservation of weak charges. (See Chap. 7, Sect. 7.4.6.) But we have not formally proven those things. In this chapter, we will use Noether's theorem (see Vol. 1, (6-31), pg. 174) to do so.

We will then explore Lorentz invariance (symmetry) in greater depth, including the effect of the Lorentz transformation on chirality and helicity, as well as the implications of changing (transforming) charge, parity (spatial inversion), and temporal order in particle interactions. The lattermost processes are known as C, P, and T transformations, respectively, and when all three are carried out in succession, by CPT. Finally, we will investigate how and why lepton number and baryon number are conserved in interactions.

So, in this chapter, we will explore

Four areas of study of symmetry in this chapter

- the application of Noether's theorem to electroweak theory and resulting implications for conservation laws,
- the Lorentz transformation in QFT revisited,
- C, P, and T (charge, parity, and time) transformations, and
- lepton and baryon number conservation.

9.1 Noether's Theorem and Electroweak Theory

A summary of the results of this section can be found in Wholeness Chart 9-1, pg. 341. You may wish to follow along with that chart as you progress through the section.

9.1.1 Review of Noether's Theorem

From the above noted reference in Vol. 1, Noether's theorem can be stated as follows.

Noether's theorem

If the Lagrangian density $\mathcal{L}(\phi^r, \phi^r{}_{,\mu})$ is symmetric in form with respect to a transformation in ϕ^r which is a function of a parameter α, i.e., $\phi^r(x^\mu) \rightarrow \phi'^r(x^\mu,\alpha)$, then the four current

$$j^\mu\left(\phi^r,\phi^r{}_{,\nu}\right)=\frac{\partial\mathcal{L}}{\partial\phi^r{}_{,\mu}}\frac{\partial\phi'^r}{\partial\alpha}\qquad\left(\text{sum on } r \text{ and } \frac{\partial\phi'^r}{\partial\alpha}=\frac{\partial}{\partial\alpha}\phi'^r\left(x^\mu,\alpha\right)\right)\tag{9-1}$$

has zero four-divergence, i.e., $\partial_\mu j^\mu = 0$. Thus, its zeroth component j^0 integrated over all space is conserved. Of course, so then is qj^0 integrated over all space conserved, where q is a constant that turns (9-1) into a charge density.

This is true for global transformations, where the parameter α is not a function of x^μ, as well as for local transformations, where $\alpha = \alpha(x^\mu)$.

9.1.2 Application in Principle of Noether's Theorem to Electroweak Theory

In QED interaction theory, we apply (9-1), where there is a single parameter $\alpha = \alpha(x^\mu)$. (See Vol. 1, Chap. 11.) That theory is a $U(1)$ theory, under which symmetry transformations are carried out by variations of the single parameter α, and wherein the Lie algebra generator is comprised of the 1X1 identity "matrix".

QED review of Noether's theorem

In electroweak theory, which is an $SU(2)XU(1)$ theory, the $SU(2)$ symmetry part entails three parameters, $\omega_i\,(x^\mu)$ (i = 1,2,3); and the $U(1)$ symmetry part, one, $f(x^\mu)$. (See Chap. 6 herein.) The $SU(2)$ generators are the three 2X2 Pauli matrices σ_i. The $U(1)$ generator is 1, i.e., the 1X1 identity.

Electroweak application of Noether's theorem

We will look first at the high energy (false vacuum) theory, which is simplest because, for it, the Lagrangian is symmetric. We will then extrapolate our findings to the low energy (true vacuum) case.

First, at false vacuum

Note that we can transform using any single one of the four parameters without employing any of the others, and the Lagrangian will still be symmetric (at least in the high energy case) under that one-parameter transformation alone. In other words, as the parameters are independent, we can take three of them equal to zero, and symmetry will still hold as we vary the fourth.

9.1.3 High Energy Theory and Conserved Currents

In this section, we will need to refer to the electroweak Lagrangian (6-48) of Chap. 6 for the false vacuum. We showed, in that chapter, how it is symmetric under the transformation set of Wholeness Chart 6-5, pg. 181. We now use that transformation set with Noether's theorem (9-1).

Use interaction $\mathcal{L}$ of Chap. 6

The $U(1)$ Four-current

Finding the Four-current for Leptons and Higgs

Consider (9-1) for the $U(1)$ transformation of Wholeness Chart 6-5, where the parameter α is f. We ignore quarks for the time being, and the Lagrangian is that of (6-48), with the covariant derivative terms expressed explicitly in (6-52). Our four-current is then, where we use the subscript h for hypercharge, the charge associated with the false vacuum $U(1)$ Lie algebra, and we keep the ordering of operator fields intact in applying Noether's theorem,

Finding four-current for hypercharge (without quarks, yet)

$$
\begin{aligned}
{}_h j^\mu_{no\,qrk} = & \underbrace{\frac{\partial\mathcal{L}}{\partial\Psi^L_{l,\mu}}}_{i\bar{\Psi}^L_l\gamma^\mu}\frac{\partial\Psi'^L_l}{\partial f} + \frac{\partial\bar{\Psi}'^L_l}{\partial f}\underbrace{\frac{\partial\mathcal{L}}{\partial\bar{\Psi}^L_{l,\mu}}}_{=0} + \underbrace{\frac{\partial\mathcal{L}}{\partial\psi^R_{l,\mu}}}_{i\bar{\psi}^R_l\gamma^\mu}\frac{\partial\psi'^R_l}{\partial f} + \frac{\partial\bar{\psi}'^R_l}{\partial f}\underbrace{\frac{\partial\mathcal{L}}{\partial\bar{\psi}^R_{l\mu}}}_{=0} + \underbrace{\frac{\partial\mathcal{L}}{\partial\psi^R_{\nu_l,\mu}}}_{i\bar{\psi}^R_{\nu_l}\gamma^\mu}\frac{\partial\psi'^R_{\nu_l}}{\partial f} \\
& + \frac{\partial\bar{\psi}'^R_{\nu_l}}{\partial f}\underbrace{\frac{\partial\mathcal{L}}{\partial\bar{\psi}^R_{\nu_l\mu}}}_{=0} + \underbrace{\frac{\partial\mathcal{L}}{\partial\Phi_{,\mu}}}_{(D^\mu\Phi)^\dagger}\frac{\partial\Phi'}{\partial f} + \frac{\partial\Phi'^\dagger}{\partial f}\underbrace{\frac{\partial\mathcal{L}}{\partial\Phi^\dagger_{,\mu}}}_{D^\mu\Phi} + \underbrace{\frac{\partial\mathcal{L}}{\partial\tilde{\Phi}_{,\mu}}}_{=0}\frac{\partial\tilde{\Phi}'}{\partial f} \\
& + \frac{\partial\mathcal{L}}{\partial W^\nu_{1,\mu}}\frac{\partial W'^\nu_1}{\partial f} + \frac{\partial\mathcal{L}}{\partial W^\nu_{2,\mu}}\frac{\partial W'^\nu_2}{\partial f} + \frac{\partial\mathcal{L}}{\partial W^\nu_{3,\mu}}\frac{\partial W'^\nu_3}{\partial f} + \frac{\partial\mathcal{L}}{\partial B^\nu_{,\mu}}\frac{\partial B'^\nu}{\partial f}.
\end{aligned}
\qquad (9\text{-}2)
$$

Since

$$
\begin{aligned}
\frac{\partial\Phi'}{\partial f} &= \frac{\partial}{\partial f}\left(1 + \tfrac{i}{2}\omega_i(x)\sigma_i + \tfrac{i}{2}f(x)\right)\Phi = \tfrac{i}{2}\Phi \\
\frac{\partial\Phi'^\dagger}{\partial f} &= \frac{\partial}{\partial f}\Phi^\dagger\left(1 - \tfrac{i}{2}\omega_i(x)\sigma_i - \tfrac{i}{2}f(x)\right) = -\tfrac{i}{2}\Phi^\dagger,
\end{aligned}
\qquad (9\text{-}3)
$$

the middle two terms in the second row of (9-2) become

$$
\frac{\partial\mathcal{L}}{\partial\Phi_{,\mu}}\frac{\partial\Phi'}{\partial f} + \frac{\partial\Phi'^\dagger}{\partial f}\frac{\partial\mathcal{L}}{\partial\Phi^\dagger_{,\mu}} = \tfrac{i}{2}\left(D^\mu\Phi\right)^\dagger\Phi - \tfrac{i}{2}\Phi^\dagger\left(D^\mu\Phi\right) = \tfrac{i}{2}\left(\left(D^\mu\Phi\right)^\dagger\Phi - \Phi^\dagger\left(D^\mu\Phi\right)\right). \qquad (9\text{-}4)
$$

In the last row of (9-2) (where we note the derivative with respect to f is different from the derivative with respect to $\partial^\nu f$)

$$
W'^\nu_i = W^\nu_i - \tfrac{1}{g}\partial^\nu\omega_i(x) - \varepsilon_{ijk}\omega_j(x)W^\nu_k \rightarrow \frac{\partial W'^\nu_i}{\partial f} = 0 \;\text{ and }\; B'^\nu = B^\nu - \tfrac{1}{g'}\partial^\nu f(x) \rightarrow \frac{\partial B'^\nu}{\partial f} = 0, \qquad (9\text{-}5)
$$

so, we can drop the last row of (9-2), leaving us with

$$ {}_h j^{\mu}_{no\,qrk} = i\bar{\Psi}^L_l \gamma^{\mu} \frac{\partial \Psi'^L_l}{\partial f} + i\bar{\psi}^R_l \gamma^{\mu} \frac{\partial \psi'^R_l}{\partial f} + i\bar{\psi}^R_{\nu_l} \gamma^{\mu} \frac{\partial \psi'^R_{\nu_l}}{\partial f} + \frac{i}{2}\left(\left(D^{\mu}\Phi\right)^{\dagger} \Phi - \Phi^{\dagger}\left(D^{\mu}\Phi\right)\right). \quad (9\text{-}6) $$

With

$$ \frac{\partial \Psi'^L_l}{\partial f} = \frac{\partial}{\partial f}\underbrace{\left(1 + \tfrac{i}{2}\omega_i\sigma_i + iYf\right)}_{Y=-1/2}\Psi^L_l = iY\Psi^L_l = -\frac{i}{2}\Psi^L_l $$

$$ \frac{\partial \psi'^R_l}{\partial f} = \frac{\partial}{\partial f}\underbrace{\left(1 + iYf\right)}_{Y=-1}\psi^R_l = iY\psi^R_l = -i\psi^R_l \qquad \frac{\partial \psi'^R_{\nu_l}}{\partial f} = \frac{\partial}{\partial f}\psi^R_{\nu_l} = 0, \quad (9\text{-}7) $$

in (9-6), we end up with

$$ \begin{aligned} {}_h j^{\mu}_{no\,qrk} &= \tfrac{1}{2}\bar{\Psi}^L_l \gamma^{\mu}\Psi^L_l + \bar{\psi}^R_l \gamma^{\mu}\psi^R_l + \tfrac{i}{2}\left(\left(D^{\mu}\Phi\right)^{\dagger}\Phi - \Phi^{\dagger}\left(D^{\mu}\Phi\right)\right) \\ &= \tfrac{1}{2}\begin{pmatrix}\bar{\psi}^L_{\nu_l} & \bar{\psi}^L_l\end{pmatrix}\gamma^{\mu}\begin{pmatrix}\psi^L_{\nu_l} \\ \psi^L_l\end{pmatrix} + \bar{\psi}^R_l \gamma^{\mu}\psi^R_l + \tfrac{i}{2}\left(\left(D^{\mu}\Phi\right)^{\dagger}\Phi - \Phi^{\dagger}\left(D^{\mu}\Phi\right)\right) \quad (9\text{-}8) \\ &= \underbrace{\tfrac{1}{2}\bar{\psi}^L_{\nu_l}\gamma^{\mu}\psi^L_{\nu_l} + \tfrac{1}{2}\bar{\psi}^L_l\gamma^{\mu}\psi^L_l + \bar{\psi}^R_l\gamma^{\mu}\psi^R_l}_{{}_h j^{\mu}_{lep}} + \underbrace{\tfrac{i}{2}\left(\left(D^{\mu}\Phi\right)^{\dagger}\Phi - \Phi^{\dagger}\left(D^{\mu}\Phi\right)\right)}_{{}_h j^{\mu}_{Higgs}}, \end{aligned} $$

Result: four-current for hypercharge (without quarks, yet)

where we define the lepton and Higgs sub-parts of the "no quark" hypercharge current as shown.

We can express the Higgs sub-part of (9-8), with the aid of (6-124), where the left arrow over the derivative in the second row means the derivative applies to the field on its left, as

$$ \begin{aligned} {}_h j^{\mu}_{Higgs} &= \tfrac{i}{2}\left(\left(D^{\mu}\Phi\right)^{\dagger}\Phi - \Phi^{\dagger}\left(D^{\mu}\Phi\right)\right) \\ &= \tfrac{i}{2}\left(\Phi^{\dagger}\left(\overleftarrow{\partial}^{\mu} - \tfrac{i}{2}g\sigma_k W^{\mu}_k - i\overbrace{\tfrac{g'}{2}}^{Y=1/2}B^{\mu}\right)\Phi - \Phi^{\dagger}\left(\partial^{\mu} + \tfrac{i}{2}g\sigma_k W^{\mu}_k + i\tfrac{g'}{2}B^{\mu}\right)\Phi\right) \\ &= \tfrac{i}{2}\left(\Phi^{\dagger,\mu}\Phi - \tfrac{i}{2}g\Phi^{\dagger}\sigma_k\Phi W^{\mu}_k - \tfrac{i}{2}g'\Phi^{\dagger}\Phi B^{\mu} - \Phi^{\dagger}\Phi^{,\mu} - \tfrac{i}{2}g\Phi^{\dagger}\sigma_k\Phi W^{\mu}_k - \tfrac{i}{2}g'\Phi^{\dagger}\Phi B^{\mu}\right) \quad (9\text{-}9) \\ &= \underbrace{\tfrac{i}{2}\left(\Phi^{,\mu\dagger}\Phi - \Phi^{\dagger}\Phi^{,\mu}\right)}_{\text{free field part of Higgs current}} + \underbrace{\tfrac{g}{2}\Phi^{\dagger}\sigma_k\Phi W^{\mu}_k + \tfrac{g'}{2}\Phi^{\dagger}\Phi B^{\mu}}_{\text{interaction part of Higgs current}}. \end{aligned} $$

Expanding Higgs sub-part of hypercharge 4-current

The free field terms in (9-9), apart from a factor of –½ (which will be relevant here), correspond to what we found in Vol.1, (3-88), pg. 62, for the four-current of a free scalar,

$$ \text{free field} \quad {}_{scalar}j^{\mu} = -i\left(\Phi^{\dagger,\mu}\Phi - \Phi^{\dagger}\Phi^{,\mu}\right) \quad \text{of Vol. 1}. \quad (9\text{-}10) $$

Free Higgs part like scalar 4-current we have seen before

By incorporating (9-9) into (9-8), we end up with

$$ \begin{aligned} {}_h j^{\mu}_{no\,qrk} = &\underbrace{\tfrac{1}{2}\bar{\psi}^L_{\nu_l}\gamma^{\mu}\psi^L_{\nu_l} + \tfrac{1}{2}\bar{\psi}^L_l\gamma^{\mu}\psi^L_l + \bar{\psi}^R_l\gamma^{\mu}\psi^R_l}_{{}_h j^{\mu}_{lep}} \\ &+ \underbrace{\underbrace{\tfrac{i}{2}\left(\Phi^{,\mu\dagger}\Phi - \Phi^{\dagger}\Phi^{,\mu}\right)}_{\text{free field part of Higgs current}} + \underbrace{\tfrac{g}{2}\Phi^{\dagger}\sigma_k\Phi W^{\mu}_k + \tfrac{g'}{2}\Phi^{\dagger}\Phi B^{\mu}}_{\text{interaction part of Higgs current}}}_{{}_h j^{\mu}_{Higgs}}. \quad (9\text{-}11) \end{aligned} $$

Expanded form of hypercharge 4-current (no quarks, yet)

The three lepton terms in the first row of (9-11), again apart from multiplicative constants, have a lot of similarities with the QED (also $U(1)$ symmetry) four-current of Vol. 1, (11-26),

$$ {}_{QED}j^{\mu} = \bar{\psi}_l \gamma^{\mu}\psi_l \quad \text{of Vol. 1}. \quad (9\text{-}12) $$

Lepton part like QED four-current

Note from (6-52), where we had substituted the covariant derivative terms into the Lagrangian, we had the lepton part of that Lagrangian, $\mathcal{L}^L$, equal to

$$\mathcal{L}^L = \underbrace{i\bar{\Psi}_l^L \partial\!\!\!/ \Psi_l^L + i\bar{\psi}_l^R \partial\!\!\!/ \psi_l^R + i\bar{\psi}_{\nu_l}^R \partial\!\!\!/ \psi_{\nu_l}^R}_{\text{free lepton } \mathcal{L}_0^L} \underbrace{+\overbrace{\frac{g'}{2}\bar{\Psi}_l^L B\!\!\!/ \Psi_l^L}^{Y=-1/2} + \overbrace{g'\bar{\psi}_l^R B\!\!\!/ \psi_l^R}^{Y=-1}}_{\text{lepton } B \text{ interaction } {}_h\mathcal{L}^L} \underbrace{-\frac{g}{2}\bar{\Psi}_l^L W\!\!\!\!/_j \sigma_j \Psi_l^L}_{\text{lepton } W \text{ interaction } {}_W\mathcal{L}^L} . \qquad (9\text{-}13)$$

Note that we can write the B interaction part, using (9-8), and realizing that the neutrino does not couple to the B field as it has $Y = 0$, as the

B-lepton interaction part of $\mathcal{L}$ in terms of hypercharge current

$$\text{lepton hypercharge interaction Lagrangian} = {}_h\mathcal{L}^L = g'\,{}_h j^\mu_{lep} B_\mu , \qquad (9\text{-}14)$$

the $U(1)$ (hypercharge) coupling constant times the $U(1)$ (hypercharge) lepton four-current times the $U(1)$ (hypercharge) field, reminiscent of the QED interaction term $e\bar{\psi}_l \gamma^\mu \psi_l A_\mu = e\left({}_{QED}\, j^\mu\right) A_\mu$.

Conservation of Hypercharge

From Noether's theorem, we know, as long as we are not including quarks at this point,

Zero 4-divergence of 4-current here means conserved quantity = hypercharge

$$\partial_\mu \,{}_h j^\mu_{no\,qrk} = 0 \qquad \rightarrow \qquad {}_h Q_{no\,qrk} = -g' \int {}_h j^0_{no\,qrk} dV \quad \text{is conserved} . \qquad (9\text{-}15)$$

The Hypercharge Operator (without quarks, for now)

You should now read in Vol. 1, Sect. 3.71, pg 62; Sect. 3.8.1, pg. 64; Sect. 4.7.2, pg. 112; and from the bottom of pg. 291 through to the middle of pg. 292. From that, with some thought, you can realize that ${}_hQ_{no\,qrk}$ is an operator, which, when all the integration of (9-15) over 3D volume is carried out (via same steps as shown in the Vol. 1 sections cited), is (9-16) below (parallel to (3-98), (4-104) and (11-27) in Vol. 1). Terms in (9-11) that are bilinear in fields give rise to number operators in (9-16). Note that the antiparticle of a LC lepton is a RC antilepton, N represents number operator, $\bar{N}$ represents anti-particle number operator, and sub/superscript symbol meaning should be obvious,

Hypercharge operator (no quarks yet) in terms of number operators

$${}_hQ_{noqrk} = -g' \sum_{l,\mathbf{p},r} \left(\begin{matrix} \frac{1}{2}\left(N^L_{\nu_l,r}(\mathbf{p}) - \bar{N}^R_{\nu_l,r}(\mathbf{p})\right) + \frac{1}{2}\left(N^L_{l,r}(\mathbf{p}) - \bar{N}^R_{l,r}(\mathbf{p})\right) \\ +\left(N^R_{l,r}(\mathbf{p}) - \bar{N}^L_{l,r}(\mathbf{p})\right) \end{matrix} \right) + \frac{g'}{2}\sum_{\mathbf{p}}\left(N_H(\mathbf{p}) - \bar{N}_{\bar{H}}(\mathbf{p})\right)$$
$$+ \underbrace{\begin{matrix}\text{interaction terms mixing construction/destuction} \\ \text{operators of two } \Phi \text{ with either a } W^\mu_k \text{ or } B^\mu .\end{matrix}}_{T_{inter}} \qquad (9\text{-}16)$$

What about the interaction terms, comprising three operators, each of which could be a creation or a destruction operator for one of the two Higgs and either a W or B field? Let's look at an example of a single LC electron state and determine the expectation (measured) value of hypercharge, as in (9-17) below. All the number operators, except the one for LC electrons, will yield zero. We show only one of the possible T_{inter} terms (that creates a W and two Higgs particles), as the others will have the same result.

T_{inter} terms in hypercharge operator yield zero expectation value for every ket

$$\begin{aligned} \langle e^{-L} | {}_hQ_{no\,qrk} | e^{-L} \rangle &= \langle e^{-L} | \left(0 + 0 - \frac{g'}{2} + 0 + \ldots + T_{inter}\right) | e^{-L} \rangle \\ &= \langle e^{-L} | -\frac{g'}{2} | e^{-L} \rangle + \langle e^{-L} | -\frac{g}{2} | W_1, H, H, e^{-L} \rangle + \ldots \\ &= -\frac{g'}{2}\underbrace{\langle e^{-L} || e^{-L} \rangle}_{1} - \frac{g}{2}\underbrace{\langle e^{-L} || W_1, H, H, e^{-L} \rangle}_{0} + 0 + \ldots = -\frac{g'}{2} . \end{aligned} \qquad (9\text{-}17)$$

So, we can just ignore those terms

All of the terms in T_{inter} will create a ket with different particles from the bra (or alternatively, simply destroy the ket). This means that all of the T_{inter} terms contribute zero to the expectation value and hence, can be ignored in the operator. We can thus work with an effective hypercharge operator, which we can simply call the <u>hypercharge operator</u>, from now on,

And thus, have an effective operator without those terms,

$$\begin{aligned} {}_hQ_{noqrk} = -g' \sum_{l,\mathbf{p},r} &\left(\frac{1}{2}\left(N^L_{\nu_l,r}(\mathbf{p}) - \bar{N}^R_{\nu_l,r}(\mathbf{p})\right) + \frac{1}{2}\left(N^L_{l,r}(\mathbf{p}) - \bar{N}^R_{l,r}(\mathbf{p})\right) + \left(N^R_{l,r}(\mathbf{p}) - \bar{N}^L_{l,r}(\mathbf{p})\right) \right) \\ &+ \frac{g'}{2}\sum_{\mathbf{p}}\left(N_H(\mathbf{p}) - \bar{N}_{\bar{H}}(\mathbf{p})\right), \end{aligned} \qquad (9\text{-}18)$$

which we will treat as the hypercharge operator

Lepton, Higgs, B, and W_i Hypercharges

(9-18) operating on a multiparticle ket will give us a hypercharge eigenvalue equal to the total hypercharge of all the particles (not counting quarks quite yet) in the ket. For examples, we can find the expectation (measured) value of the hypercharge on states, as in

Total hypercharge for some multiparticle states

$$\left\langle \nu_\mu^L \right| {}_hQ_{noqrk} \left| \nu_\mu^L \right\rangle = \left\langle \nu_\mu^L \right| -\tfrac{g'}{2} \left| \nu_\mu^L \right\rangle = -\tfrac{g'}{2}$$

$$\left\langle 3e^{-L}, 2e^{+L}, \nu_e^R \right| {}_hQ_{noqrk} \left| 3e^{-L}, 2e^{+L}, \nu_e^R \right\rangle = \left\langle 3e^{-L}, 2e^{+L}, \nu_e^R \right| -g'\left(\tfrac{3}{2} - 2 + 0\right) \left| 3e^{-L}, 2e^{+L}, \nu_e^R \right\rangle = \tfrac{g'}{2}$$

$$\left\langle 4e^{-L}, 2e^{-R}, \bar{\nu}_e^L \right| {}_hQ_{noqrk} \left| 4e^{-L}, 2e^{-R}, \bar{\nu}_e^L \right\rangle = \left\langle 4e^{-L}, 2e^{-R}, \bar{\nu}_e^L \right| -g'\left(\tfrac{4}{2} + 2 + 0\right) \left| 4e^{-L}, 2e^{-R}, \bar{\nu}_e^L \right\rangle = -4g' \quad (9\text{-}19)$$

$$\left\langle H \right| {}_hQ_{noqrk} \left| H \right\rangle = \left\langle H \right| \tfrac{g'}{2} \left| H \right\rangle = \tfrac{g'}{2}$$

$$\left\langle B \right| {}_hQ_{noqrk} \left| B \right\rangle = \left\langle B \right| 0 \left| B \right\rangle = 0 \qquad \left\langle W_i \right| {}_hQ_{noqrk} \left| W_i \right\rangle = \left\langle W_i \right| 0 \left| W_i \right\rangle = 0.$$

Of course, in cases where we show multiple electrons in a state, they cannot, by Pauli's exclusion principle, have the same **p** and r values.

Particle hypercharges Yg' what we found before by less sophisticated means

From (9-16), and the examples following it, we should be able to readily glean that LC leptons have hypercharge of $-g'/2$; RC charged leptons, $-g'$; RC neutrinos, zero; the Higgs, $+g'/2$; the B boson, zero; and all three W_i, zero. These agree with what we listed in Wholeness Chart 7-9, pg. 243, and symbolized there by Yg'. Antiparticles have the same magnitude as their particle siblings, but opposite signs. Note the Y value we use for each field in the covariant derivative equals the number of g′ charge units the associated particle carries.

Examples of Hypercharge Conservation

We consider the interaction of Fig. 6-3, pg. 185, $e^{-L} + e^{+R} \rightarrow W_3 \rightarrow \nu_e^L + \bar{\nu}_e^R$, and determine the total hypercharge before the first vertex, immediately after that vertex, and then after the final vertex.

Showing hypercharge conserved in a particular interaction

$$\left\langle e^{-L}, e^{+R} \right| {}_hQ_{noqrk} \left| e^{-L}, e^{+R} \right\rangle = \left\langle e^{-L}, e^{+R} \right| -g'\left(\tfrac{1}{2} - \tfrac{1}{2}\right) \left| e^{-L}, e^{+R} \right\rangle = \left\langle e^{-L}, e^{+R} \right| 0 \left| e^{-L}, e^{+R} \right\rangle = 0$$

$$\left\langle W_3 \right| {}_hQ_{noqrk} \left| W_3 \right\rangle = \left\langle W_3 \right| 0 \left| W_3 \right\rangle = 0 \qquad (9\text{-}20)$$

$$\left\langle \nu_e^L, \bar{\nu}_e^R \right| {}_hQ_{noqrk} \left| \nu_e^L, \bar{\nu}_e^R \right\rangle = \left\langle \nu_e^L, \bar{\nu}_e^R \right| -g'\left(\tfrac{1}{2} - \tfrac{1}{2}\right) \left| \nu_e^L, \bar{\nu}_e^R \right\rangle = \left\langle \nu_e^L, \bar{\nu}_e^R \right| 0 \left| \nu_e^L, \bar{\nu}_e^R \right\rangle = 0.$$

The initial total hypercharge is zero, and it remains so throughout the interaction. Hypercharge is conserved.

Do **Problem 1** to show hypercharge is conserved for the interaction of Fig. 6-5, pg. 187.

Hypercharge Four-current for Quarks

Same procedure for quarks

Quarks behave just as leptons do, except that they have different hypercharges (reflected in their individual Y values – see (6-111)). Applying Noether's theorem to the quark part of the high energy interaction Lagrangian (6-112), we would find

Quark hypercharge 4-current

$${}_hj^\mu_{qrk} = -\tfrac{1}{6}\bar{\Psi}^L_q \gamma^\mu \Psi^L_q - \tfrac{2}{3}\bar{\psi}^R_{qu} \gamma^\mu \psi^R_{qu} + \tfrac{1}{3}\bar{\psi}^R_{qd} \gamma^\mu \psi^R_{qd}, \text{ and} \qquad (9\text{-}21)$$

$$\text{the quark hypercharge interaction Lagrangian} = {}_h\mathcal{L}^Q = g'\, {}_hj^\mu_{qrk} B_\mu . \qquad (9\text{-}22)$$

Do **Problem 2** to find (9-21).

So, as long as we are only dealing with quarks,

Zero 4-divergence of 4-current here means conserved hypercharge

$$\partial_\mu\, {}_hj^\mu_{qrk} = 0 \quad \rightarrow \quad {}_hQ_{qrk} = -g' \int {}_hj^0_{qrk}\, dV \;\text{ is conserved}. \qquad (9\text{-}23)$$

Hypercharge Operator for Quarks

In parallel with what we found for leptons,

$$
{}_hQ_{qrk} = g'\sum_{u,d,\mathbf{p},r} \tfrac{1}{6}\left(N^L_{q_u,r}(\mathbf{p}) - \bar{N}^R_{q_u,r}(\mathbf{p})\right) + \tfrac{1}{6}\left(N^L_{q_d,r}(\mathbf{p}) - \bar{N}^R_{q_d,r}(\mathbf{p})\right) + \tfrac{2}{3}\left(N^R_{q_u,r}(\mathbf{p}) - \bar{N}^L_{q_u,r}(\mathbf{p})\right) - \tfrac{1}{3}\left(N^R_{q_d,r}(\mathbf{p}) - \bar{N}^L_{q_d,r}(\mathbf{p})\right). \tag{9-24}
$$

Quark hypercharge operator in terms of number operators

Both up-type and down-type LC quark have hypercharges of +1/6; an up-type RC quark, +2/3; a down-type RC quark, –1/3. Just as with leptons, these charges match those of Wholeness Chart 7-9.

Quark hypercharges match what we found before

Following the same logic as with leptons, quark hypercharge is conserved in high-energy interactions, as well, provided we are only dealing with quarks.

Total Hypercharge Currents, Lagrangian, and Operator

So, we have the

$$
\begin{matrix}\text{total hypercharge} \\ \text{interaction Lagrangian}\end{matrix} = {}_h\mathcal{L} = {}_h\mathcal{L}^L + {}_h\mathcal{L}^H + {}_h\mathcal{L}^Q = g'\underbrace{\left({}_hj^\mu_{lep} + {}_hj^\mu_{Higgs} + {}_hj^\mu_{qrk}\right)}_{\text{total hypercharge current } {}_hj^\mu} B_\mu . \tag{9-25}
$$

Total hypercharge current in $\mathcal{L}$ terms

The hypercharge operator, including leptons, Higgs, and quarks, is therefore the

$$
\text{total hypercharge operator} = {}_hQ = {}_hQ_{lep} + {}_hQ_{Higgs} + {}_hQ_{qrk}
$$

$$
= -g'\sum_{l,\mathbf{p},r}\left(\tfrac{1}{2}\left(N^L_{\nu_l,r}(\mathbf{p}) - \bar{N}^R_{\nu_l,r}(\mathbf{p})\right) + \tfrac{1}{2}\left(N^L_{l,r}(\mathbf{p}) - \bar{N}^R_{l,r}(\mathbf{p})\right) + \left(N^R_{l,r}(\mathbf{p}) - \bar{N}^L_{l,r}(\mathbf{p})\right)\right) \tag{9-26}
$$

$$
+\frac{g'}{2}\sum_{\mathbf{p}}\left(N_H(\mathbf{p}) - \bar{N}_{\bar{H}}(\mathbf{p})\right) + g'\sum_{\substack{u,d,\\ \mathbf{p},r}}\left(\begin{matrix}\tfrac{1}{6}\left(N^L_{q_u,r}(\mathbf{p}) - \bar{N}^R_{q_u,r}(\mathbf{p})\right) + \tfrac{1}{6}\left(N^L_{q_d,r}(\mathbf{p}) - \bar{N}^R_{q_d,r}(\mathbf{p})\right) \\ + \tfrac{2}{3}\left(N^R_{q_u,r}(\mathbf{p}) - \bar{N}^L_{q_u,r}(\mathbf{p})\right) - \tfrac{1}{3}\left(N^R_{q_d,r}(\mathbf{p}) - \bar{N}^L_{q_d,r}(\mathbf{p})\right)\end{matrix}\right).
$$

Total hypercharge operator including leptons, Higgs, and quarks

Bottom Line for Hypercharge Charges Yg′

Different particles have different hypercharges Yg', but total hypercharge is conserved in any interaction at high energy (false vacuum level), including interactions that may mix leptons and quarks, such as $e^{-L} + e^{+R} \rightarrow W_3 \rightarrow u^L + \bar{u}^R$.

Total hypercharge conserved at high energy

<u>SU(2) Four-currents</u>

As one may guess, four-current analysis for the weak isospin charges $I_3^W g$ associated with $SU(2)$ theory follows directly for each of the ω_i ($i = 1,2,3$) parameters in the symmetric $SU(2)$ transformation. We will first examine the symmetry under the variation of ω_3.

Finding four-current for weak isospin charge (without quarks)

Finding the Four-current for W_3 Interactions (no quarks for now)

Referring again to the electroweak Lagrangian (without quarks for now) (6-48), (6-52), and (6-60), and the symmetry transformations of Wholeness Chart 6-5, we can find the four-current associated with the parameter ω_3.

$$
\begin{aligned}
{}_3j^\mu_{noqrk} = {} & \underbrace{\frac{\partial\mathcal{L}}{\partial\Psi^L_{l,\mu}}}_{i\bar{\Psi}^L_l\gamma^\mu}\frac{\partial\Psi'^L_l}{\partial\omega_3} + \frac{\partial\bar{\Psi}'^L_l}{\partial\omega_3}\underbrace{\frac{\partial\mathcal{L}}{\partial\bar{\Psi}^L_{l,\mu}}}_{=0} + \underbrace{\frac{\partial\mathcal{L}}{\partial\psi^R_{l,\mu}}}_{i\bar{\psi}^R_l\gamma^\mu}\frac{\partial\psi'^R_l}{\partial\omega_3} + \frac{\partial\bar{\psi}'^R_l}{\partial\omega_3}\underbrace{\frac{\partial\mathcal{L}}{\partial\bar{\psi}^R_{l\mu}}}_{=0} + \underbrace{\frac{\partial\mathcal{L}}{\partial\psi^R_{\nu_l,\mu}}}_{i\bar{\psi}^R_{\nu_l}\gamma^\mu}\frac{\partial\psi'^R_{\nu_l}}{\partial\omega_3} \\
& + \frac{\partial\bar{\psi}'^R_{\nu_l}}{\partial\omega_3}\underbrace{\frac{\partial\mathcal{L}}{\partial\bar{\psi}^R_{\nu_l,\mu}}}_{=0} + \underbrace{\frac{\partial\mathcal{L}}{\partial\Phi_{,\mu}}}_{(D^\mu\Phi)^\dagger}\frac{\partial\Phi'}{\partial\omega_3} + \frac{\partial\Phi'^\dagger}{\partial\omega_3}\underbrace{\frac{\partial\mathcal{L}}{\partial\Phi^\dagger_{,\mu}}}_{D^\mu\Phi} + \underbrace{\frac{\partial\mathcal{L}}{\partial\tilde{\Phi}_{,\mu}}}_{=0}\frac{\partial\tilde{\Phi}'}{\partial\omega_3} \\
& \underbrace{+\frac{\partial\mathcal{L}}{\partial W^\nu_{1,\mu}}\frac{\partial W'^\nu_1}{\partial\omega_3} + \frac{\partial\mathcal{L}}{\partial W^\nu_{2,\mu}}\frac{\partial W'^\nu_2}{\partial\omega_3} + \frac{\partial\mathcal{L}}{\partial W^\nu_{3,\mu}}\frac{\partial W'^\nu_3}{\partial\omega_3}}_{{}_3j^\mu_W} + \frac{\partial\mathcal{L}}{\partial B^\nu_{,\mu}}\frac{\partial B'^\nu}{\partial\omega_3}.
\end{aligned} \tag{9-27}
$$

Various lepton factors in (9-27) are

$$\frac{\partial \Psi_l'^L}{\partial \omega_3} = \frac{\partial}{\partial \omega_3}\left(\Psi_l^L + \tfrac{i}{2}\omega_i(x)\sigma_i\Psi_l^L - \tfrac{i}{2}f(x)\Psi_l^L\right) = \tfrac{i}{2}\sigma_3\Psi_l^L$$
$$\frac{\partial \psi_l'^R}{\partial \omega_3} = \frac{\partial}{\partial \omega_3}\left(\psi_l^R - if(x)\psi_l^R\right) = 0 \qquad \frac{\partial \psi_{\nu_l}'^R}{\partial \omega_3} = \frac{\partial \psi_{\nu_l}^R}{\partial \omega_3} = 0, \tag{9-28}$$

meaning only the first term in the first row of (9-27) survives.

Higgs factors in (9-27) are

$$\frac{\partial \Phi'}{\partial \omega_3} = \frac{\partial}{\partial \omega_3}\left(1 + \tfrac{i}{2}\omega_i(x)\sigma_i + \tfrac{i}{2}f(x)\right)\Phi = \tfrac{i}{2}\sigma_3\Phi$$
$$\frac{\partial \Phi'^\dagger}{\partial \omega_3} = \frac{\partial}{\partial \omega_3}\Phi^\dagger\left(1 - \tfrac{i}{2}\omega_i(x)\sigma_i - \tfrac{i}{2}f(x)\right) = -\tfrac{i}{2}\Phi^\dagger\sigma_3 \tag{9-29}$$

We'll do the easier factors for the W boson parts first.

$$\frac{\partial W_1'^\nu}{\partial \omega_3} = \frac{\partial}{\partial \omega_3}\left(W_1^\nu - \frac{1}{g}\partial^\nu\omega_1(x) - \varepsilon_{1jk}\omega_j(x)W_k^\nu\right) = -\varepsilon_{13k}W_k^\nu = -\varepsilon_{132}W_2^\nu = W_2^\nu$$
$$\frac{\partial W_2'^\nu}{\partial \omega_3} = -\varepsilon_{231}W_1^\nu = -W_1^\nu \qquad \frac{\partial W_3'^\nu}{\partial \omega_3} = -\varepsilon_{33k}W_k^\nu = 0 \qquad \frac{\partial B'^\nu}{\partial \omega_3} = \frac{\partial}{\partial \omega_3}\left(B^\nu - \frac{1}{g'}\partial^\nu f(x)\right) = 0. \tag{9-30}$$

So, in the last row of (9-27), only the first and second terms are non-zero. Evaluating the remaining factors in that row, we have, using (6-60),

$$\begin{aligned}
\frac{\partial \mathcal{L}}{\partial W^\nu_{1\;,\mu}} &= \frac{\partial}{\partial W^\nu_{1\;,\mu}}\left(-\tfrac{1}{4}G_i^{\alpha\beta}G_{i\,\alpha\beta}\right)\\
&= \frac{\partial}{\partial W^\nu_{1\;,\mu}}\left(-\tfrac{1}{4}W_i^{\alpha\beta}W_{i\alpha\beta} + g\varepsilon_{ijk}W_{i\alpha}W_{j\beta}W_k^{\beta,\alpha} - \tfrac{1}{4}g^2\varepsilon_{ijk}\varepsilon_{ilm}W_j^\alpha W_k^\beta W_{l\alpha}W_{m\beta}\right)\\
&= \underbrace{\frac{\partial}{\partial W^\nu_{1\;,\mu}}\left(-\tfrac{1}{4}\left(W_i^{\alpha,\beta} - W_i^{\beta,\alpha}\right)\left(W_{i\,\alpha,\beta} - W_{i\,\beta,\alpha}\right)\right)}_{\text{call this } X_1} + \underbrace{\frac{\partial}{\partial W^\nu_{1\;,\mu}}\left(g\varepsilon_{ijk}W_{i\alpha}W_{j\beta}\partial^\alpha W_k^\beta\right)}_{\text{call this } Y_1}
\end{aligned} \tag{9-31}$$

$$\begin{aligned}
X_1 &= -\frac{1}{4}\left(\frac{\partial\left(W_i^{\alpha,\beta} - W_i^{\beta,\alpha}\right)}{\partial W^\nu_{1\;,\mu}}\right)\left(W_{i\,\alpha,\beta} - W_{i\,\beta,\alpha}\right) - \tfrac{1}{4}\left(W_i^{\alpha,\beta} - W_i^{\beta,\alpha}\right)\frac{\partial\left(W_{i\,\alpha,\beta} - W_{i\,\beta,\alpha}\right)}{\partial W^\nu_{1\;,\mu}}\\
&= -\frac{1}{4}\left(\frac{\partial W_i^{\alpha,\beta}}{\partial W^\nu_{1\;,\mu}} - \frac{\partial W_i^{\beta,\alpha}}{\partial W^\nu_{1\;,\mu}}\right)\left(W_{i\,\alpha,\beta} - W_{i\,\beta,\alpha}\right) - \tfrac{1}{4}\left(W_i^{\alpha,\beta} - W_i^{\beta,\alpha}\right)\left(\frac{\partial W_{i\,\alpha,\beta}}{\partial W^\nu_{1\;,\mu}} - \frac{\partial W_{i\,\beta,\alpha}}{\partial W^\nu_{1\;,\mu}}\right)\\
&= -\tfrac{1}{4}\delta_{i1}\left(\delta_\nu^\alpha g^{\beta\mu} - \delta_\nu^\beta g^{\alpha\mu}\right)\left(W_{i\,\alpha,\beta} - W_{i\,\beta,\alpha}\right) - \tfrac{1}{4}\delta_{i1}\left(W_i^{\alpha,\beta} - W_i^{\beta,\alpha}\right)\left(g_{\alpha\nu}\delta_\beta^\mu - g_{\beta\nu}\delta_\alpha^\mu\right)\\
&= -\tfrac{1}{4}\left(W_{1\nu}{}^{,\mu} - W^\mu_{1\;,\nu}\right) + \tfrac{1}{4}\left(W^\mu_{1\;,\nu} - W_{1\nu}{}^{,\mu}\right) - \tfrac{1}{4}\left(W_{1\nu}{}^{,\mu} - W^\mu_{1\;,\nu}\right) + \tfrac{1}{4}\left(W^\mu_{1\;,\nu} - W_{1\nu}{}^{,\mu}\right)\\
&= -W_{1\nu}{}^{,\mu} + W^\mu_{1\;,\nu}
\end{aligned} \tag{9-32}$$

Also,

$$\begin{aligned}
Y_1 &= \frac{\partial}{\partial W^\nu_{1\;,\mu}}\left(g\varepsilon_{ijk}W_i^\alpha W_{j\beta}\partial_\alpha W_k^\beta\right) = g\varepsilon_{ij1}W_i^\alpha W_{j\beta}\frac{\partial W^\beta_{1\;,\alpha}}{\partial W^\nu_{1\;,\mu}} = g\varepsilon_{ij1}W_i^\alpha W_{j\beta}\delta_\nu^\beta\delta_\alpha^\mu\\
&= g\varepsilon_{ij1}W_i^\mu W_{j\nu} = g\left(W_2^\mu W_{3\nu} - W_3^\mu W_{2\nu}\right)
\end{aligned} \tag{9-33}$$

With (9-30), (9-31), (9-32) and (9-33), the first term in the last row of (9-27) then becomes

$$\frac{\partial \mathcal{L}}{\partial W^{\nu}_{1\,,\mu}}\frac{\partial W_1^{\prime\nu}}{\partial \omega_3}=\left(X_1+Y_1\right)W_2^{\nu}=\left(W^{\mu}_{1\,,\nu}-W_{1\nu}{}^{,\mu}\right)W_2^{\nu}+g\left(W_2^{\mu}W_{3\nu}-W_3^{\mu}W_{2\nu}\right)W_2^{\nu}. \tag{9-34}$$

Extrapolating that result to the next term in that row, we have

$$\frac{\partial \mathcal{L}}{\partial W^{\nu}_{2\,,\mu}}\frac{\partial W_2^{\prime\nu}}{\partial \omega_3}=-\left(W^{\mu}_{2\,,\nu}-W_{2\nu}{}^{,\mu}\right)W_1^{\nu}-g\left(W_3^{\mu}W_{1\nu}-W_1^{\mu}W_{3\nu}\right)W_1^{\nu}. \tag{9-35}$$

Thus, (9-27), with the aid of (9-28), (9-29), (9-34), and (9-35), becomes

$$\begin{aligned}{}_3 j^{\mu}_{no\,qrk}=&\underbrace{-\tfrac{1}{2}\bar{\Psi}^L_l\gamma^{\mu}\sigma_3\Psi^L_l}_{{}_3 j^{\mu}_{lep}}+\underbrace{\tfrac{i}{2}\left(\left(D^{\mu}\Phi\right)^{\dagger}\sigma_3\Phi-\Phi^{\dagger}\sigma_3 D^{\mu}\Phi\right)}_{{}_3 j^{\mu}_{Higgs}}+\\ &\underbrace{\left(W^{\mu}_{1\,,\nu}-W_{1\nu}{}^{,\mu}\right)W_2^{\nu}-\left(W^{\mu}_{2\,,\nu}-W_{2\nu}{}^{,\mu}\right)W_1^{\nu}+g\left(W_2^{\mu}W_{3\nu}-W_3^{\mu}W_{2\nu}\right)W_2^{\nu}-g\left(W_3^{\mu}W_{1\nu}-W_1^{\mu}W_{3\nu}\right)W_1^{\nu}}_{{}_3 j^{\mu}_{W}}.\end{aligned} \tag{9-36}$$

Result: four-current for weak isospin charge (without quarks)

From (9-13), we can see that the lepton part of (9-36) is the four-current part of the Lagrangian for the lepton-W_3 interaction,

$$\text{lepton } W_3 \text{ inteaction Lagrangian}\quad {}_{W_3}\mathcal{L}^L=-\frac{g}{2}\bar{\Psi}^L_l\gamma^{\mu}W_{3\mu}\sigma_3\Psi^L_l=-\frac{g}{2}\bar{\Psi}^L_l\gamma^{\mu}\sigma_3\Psi^L_l W_{3\,\mu}=g\,{}_3 j^{\mu}_{lep}W_{3\,\mu}. \tag{9-37}$$

More generally, (9-36) is the four-current for the W_3 interaction for all but quarks, so,

$$\text{all but quarks } W_3 \text{ interaction Lagrangian}\quad {}_{W_3}\mathcal{L}_{no\,qrk}=g\,{}_3 j^{\mu}_{no\,qrk}W_{3\,\mu}. \tag{9-38}$$

W_3-lepton part of L in terms of weak isospin current

We won't go through all the algebra to show this explicitly, but you can see fairly readily that the last row of (9-36) into (9-38) gives us terms in g multiplied by three factors of W fields plus terms in g^2 multiplied by four factors of W fields, just as we find in the (9-31) part of $\mathcal{L}^B$ for the W fields, $-\frac{1}{4}G_i^{\alpha\beta}G_{i\,\alpha\beta}$. Similarly, the Higgs part of (9-36) in (9-38) gives us the part of $\mathcal{L}^H$ in (6-48) that prescribes the Higgs interaction with the W_3 field.

Conservation of Isospin Charge (without quarks for now)

From Noether's theorem, when no quarks are involved,

$$\partial_{\mu}\,{}_3 j^{\mu}_{no\,qrk}=0 \quad\rightarrow\quad {}_3\mathcal{Q}_{no\,qrk}=-g\int {}_3 j^{\,0}_{no\,qrk}dV \text{ is conserved}, \tag{9-39}$$

Zero 4-divergence of 4-current here means conserved quantity = weak isospin charge

where ${}_3\mathcal{Q}_{no\,qrk}$ is called the weak isospin charge operator (without quarks).

Weak Isospin Charge Operator (without quarks)

As we saw with the hypercharge operator, only terms bilinear in the same type field in the 4-current survive both the integration of (9-39) and the evaluation of the expectation value of isospin charge. Thus, from (9-36),

$$\begin{aligned}{}_3\mathcal{Q}_{no\,qrk}&=\frac{g}{2}\int\bar{\Psi}^L_l\gamma^0\sigma_3\Psi^L_l\,dV-i\frac{g}{2}\int\Phi^{\dagger,0}\sigma_3\Phi-\Phi^{\dagger}\sigma_3\Phi^{,0}dV\\ &=\frac{g}{2}\int\begin{bmatrix}\psi^{\dagger L}_{\nu_l} & \psi^{\dagger L}_{l}\end{bmatrix}\gamma^0\gamma^0\begin{bmatrix}1 & \\ & -1\end{bmatrix}\begin{bmatrix}\psi^L_{\nu_l}\\ \psi^L_l\end{bmatrix}dV\\ &\quad -i\frac{g}{2}\int\left(\frac{1}{\sqrt{2}}\begin{bmatrix}\phi_a^{\dagger,0} & \phi_b^{\dagger,0}\end{bmatrix}\begin{bmatrix}1 & \\ & -1\end{bmatrix}\frac{1}{\sqrt{2}}\begin{bmatrix}\phi_a\\ \phi_b\end{bmatrix}-\frac{1}{\sqrt{2}}\begin{bmatrix}\phi_a^{\dagger} & \phi_b^{\dagger}\end{bmatrix}\begin{bmatrix}1 & \\ & -1\end{bmatrix}\frac{1}{\sqrt{2}}\begin{bmatrix}\phi_a^{,0}\\ \phi_b^{,0}\end{bmatrix}\right)dV\\ &=\frac{g}{2}\int\left(\psi^{\dagger L}_{\nu_l}\psi^L_{\nu_l}-\psi^{\dagger L}_{l}\psi^L_l\right)dV-i\frac{g}{4}\int\left(\phi_a^{\dagger,0}\phi_a-\phi_b^{\dagger,0}\phi_b-\phi_a^{\dagger}\phi_a^{,0}+\phi_b^{\dagger}\phi_b^{,0}\right)dV.\end{aligned} \tag{9-40}$$

With all our experience in finding number operators (see Vol. 1, Wholeness Chart 5-4, pg. 158) using spinor multiplications and integration over 3D volume, we should be comfortable accepting that bilinear terms in (9-40) give rise to number operators. And so, (9-40) becomes

$$
{}_3Q_{no\,qrk} = \frac{g}{2}\sum_{l,\mathbf{p},r}\left(\left(N^L_{\nu_l,r}(\mathbf{p}) - \bar{N}^R_{\nu_l,r}(\mathbf{p})\right) - \left(N^L_{l,r}(\mathbf{p}) - \bar{N}^R_{l,r}(\mathbf{p})\right)\right)
+\frac{g}{2}\left(\sum_{\mathbf{p}}\left(N_a(\mathbf{p}) - \bar{N}_a(\mathbf{p})\right) - \sum_{\mathbf{p}}\left(N_b(\mathbf{p}) - \bar{N}_b(\mathbf{p})\right)\right). \tag{9-41}
$$

Weak isospin charge operator (no quarks yet) in terms of number operators

Lepton, Higgs, B, and W Weak Isospin Charges $I_3^W g$

Since (9-41) lacks number operators for the B and W bosons, they must have zero isospin charge I_3^W, as we showed in Wholeness Chart 7-9, pg. 243. Note that while the W_1 and W_2 particles are not in isospin charge eigenstates, their expectation values will equal zero. We would measure +1 charge as often as we would measure –1, with the average being zero. (9-41) is the operator we use to find expectation values, so this holds together. More on this a little later.

Do **Problem 3** to show the isospin charges $I_3^W g$ on particular particles equal what we showed in Wholeness Chart 7-9, and to find the total isospin charge on a multiparticle state.

Examples of Isospin Charge Conservation

Let's look at the same interaction of (9-20), and follow the isospin charges this time, instead of the hypercharges, as the interaction proceeds.

$$
\begin{gathered}
\langle e^{-L},e^{+R}|\,{}_3Q_{no\,qrk}\,|e^{-L},e^{+R}\rangle = \langle e^{-L},e^{+R}|\,g\left(-\tfrac{1}{2}+\tfrac{1}{2}\right)|e^{-L},e^{+R}\rangle = \langle e^{-L},e^{+R}|\,0\,|e^{-L},e^{+R}\rangle = 0 \\
\langle W_3|\,{}_3Q_{no\,qrk}\,|W_3\rangle = \langle W_3|\,0\,|W_3\rangle = 0 \\
\langle \nu_e^L,\bar{\nu}_e^R|\,{}_3Q_{no\,qrk}\,|\nu_e^L,\bar{\nu}_e^R\rangle = \langle \nu_e^L,\bar{\nu}_e^R|\,g\left(\tfrac{1}{2}-\tfrac{1}{2}\right)|\nu_e^L,\bar{\nu}_e^R\rangle = \langle \nu_e^L,\bar{\nu}_e^R|\,0\,|\nu_e^L,\bar{\nu}_e^R\rangle = 0.
\end{gathered} \tag{9-42}
$$

Showing weak isospin charge conserved in a particular interaction

And we see that isospin charge is indeed conserved throughout the interaction.

Isospin Four-current for Quarks

With regard to isospin, the quark parts of the Lagrangian parallel their leptonic siblings directly, having identical derivative terms in the Lagrangian. They also have identical $SU(2)$ transformations under variations in ω_3. So, we can extrapolate the results of (9-36) directly to get

$$
{}_3j^\mu_{qrk} = -\frac{1}{2}\bar{\Psi}^L_q\gamma^\mu\sigma_3\Psi^L_q \tag{9-43}
$$

Quark part of weak isospin 4-current

Isospin Charge for Quarks

We can thus also extrapolate the results of (9-41) directly to get

$$
{}_3Q_{qrk} = -g\int {}_3j^0_{qrk}\,dV = \frac{g}{2}\sum_{u,d,\mathbf{p},r}\left(\left(N^L_{q_u,r}(\mathbf{p}) - \bar{N}^R_{q_u,r}(\mathbf{p})\right) - \left(N^L_{q_d,r}(\mathbf{p}) - \bar{N}^R_{q_d,r}(\mathbf{p})\right)\right). \tag{9-44}
$$

Quark weak isospin charge operator in terms of number operators

Total Isospin Current, Lagrangian, and Operator

Parallel with (9-25), we have

$$
\begin{matrix}\text{total isopsin charge}\\ \text{interaction Lagrangian}\end{matrix} = {}_3\mathcal{L} = {}_3\mathcal{L}^L + {}_3\mathcal{L}^H + {}_3\mathcal{L}^Q = g\underbrace{\left({}_3j^\mu_{lep} + {}_3j^\mu_{Higgs} + {}_3j^\mu_{qrk}\right)}_{\text{total isospin current } {}_3j^\mu}W_{3\mu}\,. \tag{9-45}
$$

Total weak isospin current in $\mathcal{L}$ terms

The isospin charge operator, including leptons, Higgs, and quarks, is therefore the

$$
\text{total isospin charge operator } {}_3Q = {}_3Q_{no\,qrk} + {}_3Q_{qrk}
$$

$$
\begin{aligned}
&= \frac{g}{2}\sum_{l,\mathbf{p},r}\left(\left(N^L_{\nu_l,r}(\mathbf{p}) - \bar{N}^R_{\nu_l,r}(\mathbf{p})\right) - \left(N^L_{l,r}(\mathbf{p}) - \bar{N}^R_{l,r}(\mathbf{p})\right)\right) \\
&\quad + \frac{g}{2}\left(\sum_{\mathbf{p}}\left(N_a(\mathbf{p}) - \bar{N}_a(\mathbf{p})\right) - \sum_{\mathbf{p}}\left(N_b(\mathbf{p}) - \bar{N}_b(\mathbf{p})\right)\right) \\
&\quad + \frac{g}{2}\sum_{u,d,\mathbf{p},r}\left(\left(N^L_{q_u,r}(\mathbf{p}) - \bar{N}^R_{q_u,r}(\mathbf{p})\right) - \left(N^L_{q_d,r}(\mathbf{p}) - \bar{N}^R_{q_d,r}(\mathbf{p})\right)\right).
\end{aligned} \tag{9-46}
$$

Total weak isospin charge operator including leptons, Higgs, and quarks

Bottom Line for Isospin Charges $I_3^W g$

Total weak isospin charge conserved at high energy

Different particles have different weak isospin charges $I_3^W g$, but total weak isospin charge is conserved in any interaction at high energy (false vacuum level), including interactions that may mix leptons and quarks, such as $e^{-L} + e^{+R} \rightarrow W_3 \rightarrow u^L + \bar{u}^R$.

Four-current and Operator for Variations in ω_1

4-current associated with variations in ω_1 (without quarks yet)

Consider the four-current we obtain for variations in ω_1, similar to (9-27).

$$
\begin{aligned}
{}_1j^{\mu}_{no\,qrk} = {} & \underbrace{\frac{\partial\mathcal{L}}{\partial\Psi^L_{l,\mu}}}_{i\bar{\Psi}^L_l\gamma^\mu}\frac{\partial\Psi'^L_l}{\partial\omega_1} + \frac{\partial\bar{\Psi}'^L_l}{\partial\omega_1}\underbrace{\frac{\partial\mathcal{L}}{\partial\bar{\Psi}^L_{l,\mu}}}_{=0} + \underbrace{\frac{\partial\mathcal{L}}{\partial\psi^R_{l,\mu}}}_{i\bar{\psi}^R_l\gamma^\mu}\frac{\partial\psi'^R_l}{\partial\omega_1} + \frac{\partial\bar{\psi}'^R_l}{\partial\omega_1}\underbrace{\frac{\partial\mathcal{L}}{\partial\bar{\psi}^R_{l\mu}}}_{=0} + \underbrace{\frac{\partial\mathcal{L}}{\partial\psi^R_{\nu_l,\mu}}}_{i\bar{\psi}^R_{\nu_l}\gamma^\mu}\frac{\partial\psi'^R_{\nu_l}}{\partial\omega_1} \\
& + \frac{\partial\bar{\psi}'^R_{\nu_l}}{\partial\omega_1}\underbrace{\frac{\partial\mathcal{L}}{\partial\bar{\psi}^R_{\nu_l,\mu}}}_{=0} + \underbrace{\frac{\partial\mathcal{L}}{\partial\Phi_{,\mu}}}_{(D^\mu\Phi)^\dagger}\frac{\partial\Phi'}{\partial\omega_1} + \frac{\partial\Phi'^\dagger}{\partial\omega_1}\underbrace{\frac{\partial\mathcal{L}}{\partial\Phi^\dagger_{,\mu}}}_{D^\mu\Phi} + \underbrace{\frac{\partial\mathcal{L}}{\partial\tilde{\Phi}_{,\mu}}}_{=0}\frac{\partial\tilde{\Phi}'}{\partial\omega_1} \\
& + \underbrace{\frac{\partial\mathcal{L}}{\partial W^\nu_{1\,,\mu}}\frac{\partial W'^\nu_1}{\partial\omega_1} + \frac{\partial\mathcal{L}}{\partial W^\nu_{2\,,\mu}}\frac{\partial W'^\nu_2}{\partial\omega_1} + \frac{\partial\mathcal{L}}{\partial W^\nu_{3\,,\mu}}\frac{\partial W'^\nu_3}{\partial\omega_1}}_{{}_1j^\mu_W} + \frac{\partial\mathcal{L}}{\partial B^\nu_{,\mu}}\underbrace{\frac{\partial B'^\nu}{\partial\omega_1}}_{=0} .
\end{aligned}
\qquad (9\text{-}47)
$$

Following the same steps we took to go from (9-27) to (9-36), we end up with

$$
\begin{aligned}
{}_1j^{\mu}_{no\,qrk} = {} & \underbrace{-\tfrac{1}{2}\bar{\Psi}^L_l\gamma^\mu\sigma_1\Psi^L_l}_{{}_1j^\mu_{lep}} + \underbrace{\tfrac{i}{2}\Big(\left(D^\mu\Phi\right)^\dagger\sigma_1\Phi - \Phi^\dagger\sigma_1 D^\mu\Phi\Big)}_{{}_1j^\mu_{Higgs}} + \\
& \underbrace{\left(W^{\mu}_{2\,,\nu} - W_{2\nu}{}^{,\mu}\right)W^\nu_3 - \left(W^{\mu}_{3\,,\nu} - W_{3\nu}{}^{,\mu}\right)W^\nu_2 + g\left(W^\mu_3 W_{1\nu} - W^\mu_1 W_{3\nu}\right)W^\nu_3 - g\left(W^\mu_1 W_{2\nu} - W^\mu_2 W_{1\nu}\right)W^\nu_2}_{{}_1j^\mu_W} .
\end{aligned}
\qquad (9\text{-}48)
$$

Charge Operator for ω_1 Variation Symmetry

Zero 4-divergence of 4-current means conserved quantity

With Noether's theorem, we have

$$
\partial_\mu\, {}_1j^\mu_{no\,qrk} = 0 \qquad \rightarrow \qquad {}_1Q_{no\,qrk} = -g\int {}_1j^0_{no\,qrk}\,dV \quad \text{is conserved}. \qquad (9\text{-}49)
$$

As with the ω_3 symmetry, all terms in (9-48) with three fields will drop out when we take the expectation value of the charge operator in (9-49). The other terms will be of the form

Associated charge operator (without quarks)

$$
\begin{aligned}
{}_1Q_{no\,qrk} &= \frac{g}{2}\int \bar{\Psi}^L_l\gamma^0\sigma_1\Psi^L_l\,dV - i\frac{g}{2}\int \Phi^{\dagger,0}\sigma_1\Phi - \Phi^\dagger\sigma_1\Phi^{,0}\,dV \\
&= \frac{g}{2}\int\begin{bmatrix}\psi^{\dagger L}_{\nu_l} & \psi^{\dagger L}_l\end{bmatrix}\gamma^0\gamma^0\begin{bmatrix} & 1\\ 1 & \end{bmatrix}\begin{bmatrix}\psi^L_{\nu_l}\\ \psi^L_l\end{bmatrix}dV \\
&\quad - i\frac{g}{2}\int\left(\frac{1}{\sqrt{2}}\begin{bmatrix}\phi^{\dagger,0}_a & \phi^{\dagger,0}_b\end{bmatrix}\begin{bmatrix} & 1\\ 1 & \end{bmatrix}\frac{1}{\sqrt{2}}\begin{bmatrix}\phi_a\\ \phi_b\end{bmatrix} - \frac{1}{\sqrt{2}}\begin{bmatrix}\phi^\dagger_a & \phi^\dagger_b\end{bmatrix}\begin{bmatrix} & 1\\ 1 & \end{bmatrix}\frac{1}{\sqrt{2}}\begin{bmatrix}\phi^{,0}_a\\ \phi^{,0}_b\end{bmatrix}\right)dV \\
&= \frac{g}{2}\int\left(\psi^{\dagger L}_{\nu_l}\psi^L_l + \psi^{\dagger L}_l\psi^L_{\nu_l}\right)dV - i\frac{g}{4}\int\left(\phi^{\dagger,0}_a\phi_b + \phi^{\dagger,0}_b\phi_a - \phi^\dagger_a\phi^{,0}_b - \phi^\dagger_b\phi^{,0}_a\right)dV.
\end{aligned}
\qquad (9\text{-}50)
$$

When we carry out the integration in (9-50) over volume and do the spinor multiplications, we won't get number operators, since there are no terms therein bilinear in the same field type. Every term will contain a creation or destruction operator of one kind of particle times a creation or destruction operator of a different kind of particle. And thus, in an expectation value, the ket, after operation on it by ${}_1Q_{no\,qrk}$, will be different from the bra, with the result of zero expectation value. That is,

Expectation value of this charge for any state is zero

$$
\left\langle state\,\#1\right|{}_1Q_{no\,qrk}\left|state\,\#1\right\rangle = \left\langle state\,\#1\right|\frac{g}{2}\left|state\,\#2\right\rangle = \frac{g}{2}\left\langle state\,\#1\middle|state\,\#2\right\rangle = 0 . \qquad (9\text{-}51)
$$

The expectation value of ${}_1Q_{no\ qrk}$ is zero for any and all (multi)particle states.

Charge Conservation for ${}_1Q_{no\ qrk}$

Hence, in any interaction, at each stage, the expectation value for this operator is zero, and therefore, charge associated with this charge operator is conserved.

So, charge of zero is always conserved

Quark four-current and Charge Operator for Variations in ω_1

All of the above results for leptons are the same for quarks. The expectation value of ${}_1Q_{qrk}$ is zero for any and all (multi)particle states, and as this never changes during an interaction, the associated charge is conserved.

Same result for quarks: this charge zero for all

Effective Operator ${}_1Q_{no\ qrk}$

Since the expectation value of ${}_1Q$ is always zero, it is itself effectively zero, and with no loss in generality, we can simply take it to be (which we will henceforth)

Effective total charge operator ${}_1Q = 0$

$$ {}_1Q = {}_1Q_{no\,qrk} + {}_1Q_{qrk} = 0\,. \tag{9-52}$$

Four-current and Operator for Variations in ω_2

All of the above results for the symmetry variation in ω_1 are the same for the symmetry variation in ω_2. The expectation value of the corresponding charge operator ${}_2Q = {}_2Q_{no\ qrk} + {}_2Q_{qrk}$ is always zero, and thus, such charge is always conserved in interactions. Hence, we will also take

Same result for ω_2

Effective total charge operator ${}_2Q = 0$

$$ {}_2Q = {}_2Q_{no\,qrk} + {}_2Q_{qrk} = 0\,. \tag{9-53}$$

Bottom Line: Electroweak Four-currents, Charges, and Conservation at False Vacuum

We have shown, using Noether's theorem, that hypercharge Yg' and weak isospin charge $I_3{}^W g$ have the values shown in Wholeness Chart 7-9, and that those charges are conserved in all electroweak interactions at the false vacuum energy level. The respective four-currents and charge operators are summarized in Wholeness Chart 9-1, pg. 341.

False vacuum: charges agree with what we found earlier and total charge conserved

Electric Charge Conservation

As noted in Chap. 7, and encapsulated in the last row of Wholeness Chart 9-1, for a given particle, number of e units of electric charge N_Q equals number of g units of weak isospin charge plus number of g' units of hypercharge.

Electric charge conserved since hypercharge and weak isospin charge are

$$ N_Q = I_3^W + Y \tag{9-54}$$

So, if the number of units of weak isospin charge and number of units of hypercharge are both conserved, then so is the number of units of electric charge. Electroweak charge conservation implies electric charge conservation.

9.1.4 Low Energy Theory and Conserved Currents

Low Energy Mass Terms Destroy Symmetry

Recall, as we have noted (pgs. 191 and 196), mass terms in the Lagrangian spoil its symmetry under $SU(2)\times U(1)$ transformations. After symmetry breaking, at contemporary energy levels, we do, of course, have mass terms, whereas at the false vacuum, we did not.

True vacuum: Mass terms → no symmetry → no conservation

If we no longer have a symmetric Lagrangian, then Noether's theorem does not apply, and charge need not be conserved. We showed how this works using a Feynman diagram on pg. 243, where a LC electron can morph into a RC electron due to the mass term in the Lagrangian, and in the process, weak isospin charge and hypercharge are not conserved.

But Weak Charges Still Conserved at Vertices

However, if, in a given interaction, we ignore the mass terms (in essence, ignoring the fact that particles of certain electroweak charges can morph into particles with different electroweak charges), then the electroweak charges can still be considered conserved. And thus, our charge operators and conservation laws we developed at high energy can still be considered valid.

But charge conserved at vertices, can use same charge operators

This means that charge is still always conserved at interaction vertices, such as those in many vertex relations for Feynman rules shown at the end of Chap. 8.

Electric Charge Still Conserved at Low Energy

Electric charge is always conserved, even taking into account the mass terms, since the QED Lagrangian is symmetric under the $U(1)$ transformation, even with non-zero fermion mass terms, (See Vol. 1, pgs. 293-296, where this is shown. A massive photon would, however, destroy the QED symmetry.)

Electric charge always conserved

Photon and Z Boson Charges

We know the photon and Z intermediate vector boson are linear combinations of the B and W_3 bosons, and since the latter have zero hypercharge and zero weak isospin charge, then so do the former. This, of course, from (9-54), means that electric charge on the photon and Z are zero, as well.

Photon and Z charge neutral, all charge types

W Boson Charges

From Wholeness Chart 7-9, we see that the W^+ and W^- fields have definite weak isospin charge (i.e., they are in charge eigenstates), though the W_1 and W_2 fields, of which the W^+ and W^- fields are composed, do not (i.e., not in charge eigenstates). Specifically, we found (in this chapter) no number operators in the charge operators associated with W_1 and W_2.

W bosons weak isospin charge complicated; hypercharge always zero

Resolving this fully is a little tricky, since we don't really have symmetry at low energy, so we can't (without much convolution) deduce a divergence-less four-current and from that, conserved charge operators for the W^+ and W^- bosons. However, we can posit a four-current of the usual form and from it, deduce a charge operator, even though the associated charge may not be conserved in all interactions.

We'll simplify with an intuitive look

We start by recalling

$$W_{1\alpha} = \frac{W_\alpha^+ + W_\alpha^-}{\sqrt{2}} \quad W_{2\alpha} = i\frac{W_\alpha^+ - W_\alpha^-}{\sqrt{2}} \qquad W_\alpha^+ = \frac{W_{1\alpha} - iW_{2\alpha}}{\sqrt{2}} \quad W_\alpha^- = \frac{W_{1\alpha} + iW_{2\alpha}}{\sqrt{2}} . \tag{9-55}$$

Once again tapping our experience with number operators, we should be comfortable accepting (9-56) below.

$$\begin{aligned} {}_3Q_{W^+} &= -g\int {}_3j_{W^+}^0 dV = ig\int\left(\left(W_\alpha^{+,0}\right)^\dagger W^{+\alpha} - \left(W^{+\alpha}\right)^\dagger W_\alpha^{+,0}\right)dV \\ &= -g\sum_{\mathbf{k}} \left(N_{W^+}(\mathbf{k}) - \bar{N}_{W^+}(\mathbf{k})\right) = -g\sum_{\mathbf{k}} \left(N_{W^+}(\mathbf{k}) - N_{W^-}(\mathbf{k})\right), \end{aligned} \tag{9-56}$$

where N symbolizes number operator and the antiparticle of the W^+ is the W^-. From (9-56), we can conclude the total particle number operator for the W^+ is

$${}_{tot\,partic}N_{W^+} = \sum_{\mathbf{k}}\left(N_{W^+}(\mathbf{k}) - N_{W^-}(\mathbf{k})\right). \tag{9-57}$$

Also,

$$\begin{aligned} {}_3Q_{W^-} &= -g\int {}_3j_{W^-}^0 dV = ig\int\left(\left(W_\alpha^{-,0}\right)^\dagger W^{-\alpha} - \left(W^{-\alpha}\right)^\dagger W_\alpha^{-,0}\right)dV \\ &= -g\sum_{\mathbf{k}} \left(N_{W^-}(\mathbf{k}) - \bar{N}_{W^-}(\mathbf{k})\right) = -g\sum_{\mathbf{k}} \left(N_{W^-}(\mathbf{k}) - N_{W^+}(\mathbf{k})\right), \end{aligned} \tag{9-58}$$

Finding weak isospin charge operators for true vacuum W bosons

with the total particle number operator

$${}_{tot\,partic}N_{W^-} = \sum_{\mathbf{k}}\left(N_{W^-}(\mathbf{k}) - N_{W^+}(\mathbf{k})\right) \tag{9-59}$$

Note that (9-57) and (9-59) have opposite signs, as we would expect on physical grounds for particles and antiparticles.

Using (9-55), (9-57), and (9-59), we can find the weak isospin charge operator for W_1 as

$$\begin{aligned} {}_3Q_{W_1} &= -g\int {}_3j_{W_1}^0 dV = ig\int\left(\left(W_{1\alpha}^{,0}\right)^\dagger W_1^{\alpha} - \left(W_1^{\alpha}\right)^\dagger W_{1\alpha}^{,0}\right)dV \\ &= ig\int\left(\left(\frac{W_\alpha^{+,0} + W_\alpha^{-,0}}{\sqrt{2}}\right)^\dagger\left(\frac{W^{+\alpha} + W^{-\alpha}}{\sqrt{2}}\right) - \left(\frac{W^{+\alpha} + W^{-\alpha}}{\sqrt{2}}\right)^\dagger\left(\frac{W_\alpha^{+,0} + W_\alpha^{-,0}}{\sqrt{2}}\right)\right)dV \end{aligned}$$

$$
{}_3Q_{W_1} = i\frac{g}{2}\int \begin{pmatrix} \left(W_\alpha^{+,0}\right)^\dagger W^{+\alpha} \underbrace{+\left(W_\alpha^{+,0}\right)^\dagger W^{-\alpha} + \left(W_\alpha^{-,0}\right)^\dagger W^{+\alpha}}_{\substack{\text{drops out of expectation value as} \\ \text{not bilinear in same field, so ignore}}} + \left(W_\alpha^{-,0}\right)^\dagger W^{-\alpha} \\ -\left(W^{+\alpha}\right)^\dagger W_\alpha^{+,0} \underbrace{-\left(W^{+\alpha}\right)^\dagger W_\alpha^{-,0} - \left(W^{-\alpha}\right)^\dagger W_\alpha^{+,0}}_{\text{drops out of expectation value}} - \left(W^{-\alpha}\right)^\dagger W_\alpha^{-,0} \end{pmatrix} dV
$$

$$
\begin{aligned}
&= i\frac{g}{2}\int\left(\left(W_\alpha^{+,0}\right)^\dagger W^{+\alpha} - \left(W^{+\alpha}\right)^\dagger W_\alpha^{+,0} + \left(W_\alpha^{-,0}\right)^\dagger W^{-\alpha} - \left(W^{-\alpha}\right)^\dagger W_\alpha^{-,0}\right)dV \\
&= -\frac{g}{2}\sum_{\mathbf{k}}\left(N_{W^+}(\mathbf{k}) - N_{W^-}(\mathbf{k})\right) - \frac{g}{2}\sum_{\mathbf{k}}\left(N_{W^-}(\mathbf{k}) - N_{W^+}(\mathbf{k})\right) = 0.
\end{aligned} \tag{9-60}
$$

W_1 isospin charge operator effectively zero

With (9-55), (9-57), and (9-59), and proceeding in parallel fashion to (9-60), one finds the weak isospin charge operator for W_2 is

$$
\begin{aligned}
{}_3Q_{W_2} &= -g\int {}_3j_{W_2}^{,0}\,dV = ig\int\left(\left(W_{2\alpha}^{,0}\right)^\dagger W_2^\alpha - \left(W_2^\alpha\right)^\dagger W_{2\alpha}^{,0}\right)dV \\
&= -\frac{g}{2}\sum_{\mathbf{k}}\left(N_{W^+}(\mathbf{k}) - N_{W^-}(\mathbf{k})\right) - \frac{g}{2}\sum_{\mathbf{k}}\left(N_{W^-}(\mathbf{k}) - N_{W^+}(\mathbf{k})\right) = 0.
\end{aligned} \tag{9-61}
$$

W_2 isospin charge operator effectively zero

The weak isospin charge expectation values for either the W_1 or the W_2 will be (9-60) and (9-61), respectively, i.e., zero. So, we would expect a zero number operator in our isospin charge operator for those two fields. And that is what we found in (9-46).

Weak isospin W_1 & W_2 charge operators have zero expectation value

Note from (9-55), and from what we mentioned earlier, that the fields W_1 and W_2 are not in isospin eigenstates. They are each comprised of a W^+ plus a W^- field. The W^+ field is in a weak isospin charge state with eigenvalue +1 (or $+g$ if we include the coupling constant). The W^- field is in a weak isospin charge state with eigenvalue –1. On average (which is the expectation), we would find zero charge on either a W_1 or W_2 particle. Half the time we would measure +1 from the collapse to the W^+, and half the time, –1 from the collapse to the W^-.

W_1 & W_2 not in isospin eigenstates; W^+ & W^- are

Bottom line: For calculations below the electroweak symmetry breaking scale, where we focus on the W^+ and W^- fields, the W^+ has $+g$ weak isospin charge and the W^- has $-g$. They both have hypercharge of zero. They then, via (9-54), have electric charges of $+e$ and $-e$, respectively. All charges are conserved at all vertices.

W^+ & W^- are eigenstates of all charges: electric $+e$, $-e$; hypercharge 0, weak isospin $+g$, $-g$

Charge Operators for Low Energy

At low energy, we can use as our isospin charge operator, the high energy operator (9-46) plus

$$
{}_3Q_W = g\sum_{\mathbf{k}}\left(N_{W^+}(\mathbf{k}) - N_{W^-}(\mathbf{k})\right). \tag{9-62}
$$

W^+ & W^- weak isospin charge operator

The hypercharge operators at low energy are the same as at high energy, and in both cases, all gauge bosons have zero such charge.

Higgs Charge at Low Energy

Recall that the unitary gauge employed at low energy means ϕ_a of (9-40) is zero, and so, at low energy, we can delete the terms in (9-46) having subscript a. We can then replace b by H to represent the Higgs of our present-day universe.

True vacuum Higgs: top part of doublet zero, do delete from weak isospin operator

Summary Relations

As noted earlier, our final results, for hypercharge and weak isospin charge operators at high and low energy, are summarized in Wholeness Chart 9-1, pg. 341, at the end of this chapter.

Do **Problem 4** to show the isospin charge is conserved for one particular interaction.

9.1.5 Comparison of Methods to Determine Charge

Only Diagonal Generators Yield Charge Operators

Note that for $SU(2)$, only terms in the Lagrangian with the diagonal matrix generator σ_3 operating on the doublet Ψ^L (for leptons or quarks) gave rise to a weak isospin charge operator ($_3Q$) whose expectation value was not zero. Terms with non-diagonal matrix generators σ_1 and σ_2 gave rise to charge operators $_1Q$ and $_2Q$, whose expectation values for every state are zero, so we can, without loss of generality, take those operators themselves as zero.

Only diagonal SU(2) generators yield charge operators with non-zero expectation values

This is a general principle for any $SU(n)$ group. Only terms with diagonal matrix generators, i.e., those of the Cartan sub-algebra, result in non-zero charge operators. Off diagonal components lead to terms in the charge operator that are bilinear in two different fields, as we saw going from (9-50) to (9-52), and thus yield zero for charge expectation value of any state.

SU(n): Only Cartan subalgebra generators as charge operators

Compare to Group Theory of Chapter 2

Note further, as an example, from Chap. 2, (2-127), the following, where in that chapter, for $SU(2)$, we used the symbol X_3 for the Pauli matrix σ_3, since we were generalizing therein.

$$\frac{g}{2}X_3\Psi_e^L = \frac{g}{2}\sigma_3\Psi_e^L = \frac{g}{2}\begin{bmatrix}1 & 0\\ 0 & -1\end{bmatrix}\begin{bmatrix}0\\ \psi_e^L\end{bmatrix} = -\frac{g}{2}\begin{bmatrix}0\\ \psi_e^L\end{bmatrix} = -\frac{g}{2}\Psi_e^L$$
$$\frac{g}{2}X_3\Psi_{\nu_e}^L = \frac{g}{2}\sigma_3\Psi_{\nu_e}^L = \frac{g}{2}\begin{bmatrix}1 & 0\\ 0 & -1\end{bmatrix}\begin{bmatrix}\psi_{\nu_e}^L\\ 0\end{bmatrix} = \frac{g}{2}\begin{bmatrix}\psi_{\nu_e}^L\\ 0\end{bmatrix} = \frac{g}{2}\Psi_{\nu_e}^L . \qquad (9\text{-}63)$$

SU(2) example with Cartan (diagonal) generator

The σ_3 matrix times $g/2$ operating on a doublet comprising only the LC electron yields an eigenvalue of $-\,g/2$, which is the weak isospin charge (called simply "weak charge" in Chap. 2) on the LC electron. The same operation on a doublet comprising only an LC neutrino yields an eigenvalue of $+g/2$, the weak isospin charge on the LC neutrino. These are the same values we found in the present chapter using the weak isospin charge operator. The two approaches are equivalent.

For $SU(2)$ singlets, such as the RC electron or the RC electron neutrino, in Chap. 2, we found, where the X_3 matrix in the 1D (singlet) rep is zero,

$$\frac{g}{2}X_3\psi_e^R = \frac{g}{2}[0]\psi_e^R = 0 \qquad \frac{g}{2}X_3\psi_{\nu_e}^R = \frac{g}{2}[0]\psi_{\nu_e}^R = 0 . \qquad (9\text{-}64)$$

For singlets, Cartan generator = 0 → charge = 0

The eigenvalues for RC leptons are zero, which equals their weak isospin charge, as found in this chapter. So, the two approaches are equivalent for both singlets and doublets.

Note that $g/2$ is the factor we multiply $i\sigma_i W_i^\mu$ by in the covariant derivative D^μ (see (6-47)) to determine the $SU(2)$ interaction terms in the Lagrangian. That factor ultimately arises in the weak isospin charge operator (9-46) to give us the charge values LC particles have.

Basic charge unit = factor in covariant derivative

Three Ways We Have Determined Charges

Additionally, in Chap. 6, Sect. 6.5, beginning on pg. 182, we found weak isospin charges on particular particles by looking directly at relevant terms in the Lagrangian, and extrapolating from what we knew about similar terms in the QED Lagrangian.

So, we have deduced weak isospin charge and hypercharge values for the SM particles via three different, but compatible, ways.

1. Group theory Cartan subalgebra operating on doublets and singlets (Chap. 2),
2. Examining terms in the electroweak Lagrangian and comparing to similar terms in QED (Chap. 5), and
3. Integrating the $\mu = 0$ component of the Noether four-current over all space to get each charge operator, and then finding its expectation value for single particle states (this chapter).

Three ways to determine particle charge

9.2 Lorentz Transformation Revisited

In Vol. 1, Wholeness Chart 6-3, pg. 172, we show the Lorentz transformation of scalars, vectors, tensors, and spinors; and note that the Lagrangian density is invariant, a world scalar. Here, we want to examine how spinor products, γ^5, helicity, and chirality transform.

9.2.1 Lorentz Transformation of Spinors

In Vol. 1, (6-22), pg. 171, and also in Chap. 5 herein, (5-64) and (5-65), we noted that the Lorentz transformation of a spinor field (taking rotations θ_i therein as zero) is carried out by the operator

$$D = e^{-i\mathbf{M}\cdot\mathbf{Q}} \qquad M^k = -\tfrac{i}{2}\gamma^0\gamma^k,\; Q^k = \left(v^1, v^2, v^3\right) \qquad v \ll c \tag{9-65}$$

Lorentz transformation for spinors, symbol D

such that

$$\psi'\left(x'^{\mu}\right) = D\psi\left(x^{\mu}\right) \xrightarrow[\text{written out}]{\text{with spinor indices}} \psi'_{\alpha}\left(x'^{\mu}\right) = D_{\alpha\beta}\psi_{\beta}\left(\Lambda^{\mu}_{\ \nu}x^{\nu}\right). \tag{9-66}$$

Note that some authors use the symbol *S*, instead of *D*. We can rewrite *D* as

$$D = e^{\frac{1}{2}\gamma^0\gamma^k v_k}, \tag{9-67}$$

where we need to keep in mind, as we progress, that the velocity v_k in *D* is the relative velocity of the two frames, not that of any particular particle (unless it is at rest in the second frame, of course).

Do **Problem 5** to show that *D* is not unitary.

That *D* is not unitary should not be too surprising, since, as we noted in Chap. 2, pg. 14, the Lorentz transformation $\Lambda_{\mu}^{\ \nu}$ itself is not unitary (or, as we would say for a real matrix, not orthogonal). That is $\Lambda^{\dagger\ \nu}_{\ \mu} = \Lambda^{T\ \nu}_{\ \mu} \neq \Lambda^{-1\ \nu}_{\ \ \mu}$. Since *D* is not unitary, we cannot do what we normally do with $\bar{\psi}$ for operators on ψ. That is, we cannot use its complex conjugate transpose as an operation on $\bar{\psi}$, i.e.,

D not unitary

$$\bar{\psi}' \neq \bar{\psi}D^{\dagger}. \tag{9-68}$$

The question then arises as to what operator, temporarily call it $\bar{D}$, does transform $\bar{\psi}$ to a second Lorentz system. To answer this, consider that $\bar{\psi}\psi$ is an inner product (in spinor space), i.e., it is a scalar, so, as our theory is a 4D spacetime theory, it must be a world scalar, and thus, invariant. Hence,

$$\bar{\psi}\psi = \bar{\psi}'\psi' = \left(\bar{\psi}\bar{D}\right)\left(D\psi\right) = \bar{\psi}\bar{D}D\psi\,. \tag{9-69}$$

From (9-69), we see that $\bar{D}$ is the inverse of *D*, i.e.,

$$\bar{D} = D^{-1} = e^{-\frac{1}{2}\gamma^0\gamma^m v_m}. \tag{9-70}$$

So, need to use D^{-1}, not $D^{\dagger}$

Thus,

$$\bar{\psi}' = \bar{\psi}\bar{D} = \bar{\psi}D^{-1} = \bar{\psi}e^{-\frac{1}{2}\gamma^0\gamma^m v_m} \qquad \psi' = D\psi = e^{\frac{1}{2}\gamma^0\gamma^k v_k}\psi\,. \tag{9-71}$$

There is related material on this topic in Chap. 5, Sect. 5.3, beginning on pg. 153.

9.2.2 Lorentz Transformation of Spinor Products and Gamma Matrices

Vector Four-current

One could ask how the four-current relation $\bar{\psi}\gamma^{\mu}\psi$, which we see repeatedly in QED, transforms. We stated this without proof on the afore-referenced page in Vol. 1.

We can deduce this rather quickly by noting that, as stated in Vol. 1, pg. 25, the Lagrangian density is Lorentz invariant, so the QED $\mathcal{L}$ is Lorentz invariant, and so its interaction term is invariant, i.e.,

A term in $\mathcal{L}$, Lorentz invariant like all terms

$$e\bar{\psi}\gamma^{\mu}\psi A_{\mu} = e\left(\bar{\psi}\gamma^{\mu}\psi\right)' A'_{\mu}\,. \tag{9-72}$$

Since the 4D inner product of two four-vectors is invariant, and A_{μ} is a four-vector, $\bar{\psi}\gamma^{\mu}\psi$ must behave, under the transformation, like a four-vector, as well. That is, where $\Lambda^{\mu}_{\ \nu}$ represents the Lorentz transformation in spacetime,

Since A_{μ} is 4-vector, other factor must transform as 4-vector

$$\Lambda^{\mu}_{\ \nu}\left(\bar{\psi}\gamma^{\nu}\psi\right) = \left(\bar{\psi}\gamma^{\mu}\psi\right)' = \bar{\psi}'\gamma'^{\mu}\psi'\,. \tag{9-73}$$

Lorentz transformation of other factor

This means, with (9-71),

$$\Lambda^{\mu}_{\ \nu}\left(\bar{\psi}D^{-1}D\gamma^{\nu}D^{-1}D\psi\right) = \Lambda^{\mu}_{\ \nu}\bar{\psi}'\left(D\gamma^{\nu}D^{-1}\right)\psi' = \bar{\psi}'\gamma'^{\mu}\psi'\,, \tag{9-74}$$

Lorentz transformation of gamma matrices

$$\gamma'^{\mu} = \Lambda^{\mu}_{\ \nu} D \gamma^{\nu} D^{-1} . \qquad (9\text{-}75)$$

(9-75) implies gamma matrices, like spinors, change under Lorentz transformation. However, that gets complicated, and we can simplify by just thinking in terms of what (9-72) is telling us. Whatever frame we are working in, if we simply take our fields and gamma matrices as we always have, terms in the Lagrangian keep their same form under Lorentz transformations, and all will be OK.

But whatever frame we are in, we can use ψ, γ^{μ} as we have learned them

The bottom line: $\bar{\psi}\gamma^{\mu}\psi V_{\mu}$, where V_{μ} is any four-vector, is invariant, meaning we can use the relations we have developed all along for fields and gamma matrices in any frame in $\mathcal{L}$. $\bar{\psi}\gamma^{\mu}\psi$ is called a vector four-current, or simply a vector current.

The other factor is a vector 4-current

Axial Four-current

How about the four-current we have seen repeatedly in electroweak theory $\bar{\psi}\gamma^{\mu}\gamma^{5}\psi$? Similar logic applies. Consider the term in the Lagrangian, which must be Lorentz invariant,

4-current-like factors with γ^5 in them

$$\frac{g}{\cos\theta_W}\bar{\psi}_e\gamma^{\mu}\tfrac{1}{2}\left(1-\gamma^5\right)\psi_e Z_{\mu} = \frac{g}{2\cos\theta_W}\left(\bar{\psi}_e\gamma^{\mu}\left(1-\gamma^5\right)\psi_e Z_{\mu}\right)'$$
$$= \frac{g}{2\cos\theta_W}\left(\bar{\psi}_e\gamma^{\mu}\left(1-\gamma^5\right)\psi_e\right)' Z'_{\mu} = \frac{g}{2\cos\theta_W}\left(\bar{\psi}_e\gamma^{\mu}\psi_e\right)' Z'_{\mu} - \frac{g}{2\cos\theta_W}\left(\bar{\psi}_e\gamma^{\mu}\gamma^5\psi_e\right)' Z'_{\mu} . \qquad (9\text{-}76)$$

Since Z_{μ} is a four-vector, $\bar{\psi}_e\gamma^{\mu}\gamma^5\psi_e$ must also transform like a four-vector, under Lorentz transformation. This is true for proper Lorentz transformations, which do not entail reflection (spatial inversion) of spatial coordinate axes, i.e., $x^i \rightarrow -x^i$ or time (time reversal), i.e., $t \rightarrow -t$. There is a little wrinkle for improper Lorentz transformations, discussed shortly, which does entail such reflection.

Transform like vector 4-currents under proper Lorentz transformation

The bottom line: For reasons to be discussed, this wrinkle causes us to deem $\bar{\psi}_e\gamma^{\mu}\gamma^5\psi_e$ an axial four-current, or simply, an axial current. For proper Lorentz transformations, it transforms like a four-vector. Like the vector current, in the Lagrangian, we can use relations we have developed all along for fields and gamma matrices in any frame, since $\bar{\psi}\gamma^{\mu}\gamma^5\psi V_{\mu}$ is Lorentz invariant.

Called axial currents

γ^5 Matrix

Do **Problems 6 and 7** to show (9-77) and (9-78).

If you did the suggested problems, you showed that the proper Lorentz transformation of γ^5 is

$$\gamma'^5 = D\gamma^5 D^{-1}, \qquad (9\text{-}77)$$

Transformation of γ^5

which leads to the world scalar (under proper Lorentz transformation) of

$$\bar{\psi}'_e\gamma'^5\psi'_e = \bar{\psi}_e\gamma^5\psi_e . \qquad (9\text{-}78)$$

A world scalar under proper Lorentz transformation

Note on Gamma Matrices and Transformations

To be clear, the transformation on gamma matrices (9-75) is valid for evaluating interactions when changing to a frame which has relative velocity with respect to the original frame, i.e., for a frame boost. We do not use transformed gamma matrices when we observe a particle (or work with its field) in the original frame and then observe a second particle (or work with its field) of the same type that has a different velocity from the first particle (or field). The second particle (field) might be considered "boosted" relative to the first, but we use the same gamma matrices, and the same mathematical form for spinors for the second particle (field), provided we are observing both from the same frame.

Transformations between frames, not for different particles in same frame

And again, as long as we remain in the same frame, we can use the form for gamma matrices and spinors we have been working with all along. The (proper) Lorentz transformed entities can be used by the observer (at rest) in the original frame to predict what the observer (at rest) in the 'boosted' frame would measure for different physical observables.

In given frame, we use ψ, γ^{μ} relations we know

Summary Relations

Although we have focused on the electron spinor field, with subscript *e*, the transformation relationships we have found above are valid for any fermion field.

The Lorentz transformation properties of the entities we examined above are summarized in Wholeness Chart 9-2, on pg. 342, at the end of the chapter.

9.2.3 Lorentz Transformations of Helicity and Chirality

Review

We looked at chirality and helicity for velocities $v \rightarrow c$ in Chap. 5, pgs. 137-141, which is summarized in the wholeness chart at the end of that section. As we showed there in the Weyl rep, *L* and *R* Weyl spinor fields, which we now call LC and RC fields, as speed approaches that of light, approach LH (left helicity) and RH (right helicity) fields, respectively. For massless particles, traveling at the speed of light, helicity equals chirality, i.e., the particle is in both a helicity eigenstate and a chirality eigenstate.

For v=c, helicity eigenstate = chiral eigenstate

We found those results using the Weyl rep, but as we showed in Chap. 5, general relations between spinors and matrices in one rep are true in any rep. A spinor in an eigenstate in one rep, is a spinor in that same eigenstate in any rep. The components of the spinors and the matrices can change from rep to rep, but relationships between them (equations) do not. The relationships are spinor space covariant with respect to transformations between reps.

Gamma matrix relations and eigenstates same in any rep

So, at the speed of light, a particle/field is simultaneously in both a chirality and a helicity eigenstate, regardless of the rep it is represented in.

Helicity

An Intuitive, Non-mathematical Look

For massive particles, helicity changes under Lorentz transformation, as the following well-known example illustrates. Consider a massive particle with its spin axis and its 3-momentum vector aligned having RH (right hand helicity) as seen from our frame. Then consider a second frame with velocity relative to ours aligned with the particle 3-velocity, but with higher speed. From the point of view of the second frame, the spin part of the helicity will continue to spin in the same direction, but the 3-momentum will be in the opposite direction from that seen in our frame. Hence, the helicity has been reversed to LH by a boost. So, helicity is not Lorentz invariant.

Physically; helicity changes with velocity, changes under Lorentz transformation

This effect is not limited to ultra-relativistic velocities, since the speeds involved in the prior example can be small.

Relativistically, there is yet another effect from boosts (via proper Lorentz transformations), since spin direction of a massive particle, as seen from the boosted frame, looks different than in the original frame (unlike the non-relativistic case, where spin direction of a massive particle looks the same in all frames). See Vol. 1, Box 4-2, pg. 95, and Fig. 4-1, pg. 96.

A Mathematical Look

In Vol. 1, (4-39), (4-49), (4-120), and (4-121) we showed the helicity operators for fermions in RQM and QFT to be

$$\text{RQM}\ \Sigma_{\mathbf{p}} = \mathbf{\Sigma}\cdot i_{\mathbf{p}} = \mathbf{\Sigma}\cdot\frac{\mathbf{p}}{|\mathbf{p}|} = \Sigma_1\frac{p^1}{|\mathbf{p}|} + \Sigma_2\frac{p^2}{|\mathbf{p}|} + \Sigma_3\frac{p^3}{|\mathbf{p}|} \qquad \Sigma_i = \frac{1}{2}\begin{bmatrix}\sigma_i & \\ & \sigma_i\end{bmatrix} \tag{9-79}$$

$$\begin{aligned}\text{QFT}\ \Sigma_{\mathbf{p}} &= \int \psi^\dagger\left(\mathbf{\Sigma}\cdot\frac{\mathbf{p}}{|\mathbf{p}|}\right)\psi d^3x = \int \psi^\dagger\left(\Sigma_i\frac{p^i}{p}\right)\psi d^3x \\ &= \sum_{r,\mathbf{p}}\frac{m}{E_{\mathbf{p}}}\left(u_r^\dagger(\mathbf{p})\Sigma_i\frac{p^i}{p}u_r(\mathbf{p})N_r(\mathbf{p}) + v_r^\dagger(\mathbf{p})\Sigma_i\frac{p^i}{p}v_r(\mathbf{p})\bar{N}_r(\mathbf{p})\right).\end{aligned} \tag{9-80}$$

Mathematically, helicity varies with velocity

We will not go any deeper into the mathematics of helicity transformation here, but simply call attention to the fact that 3-momentum **p** and energy $E_{\mathbf{p}}$ are not Lorentz invariant, and since (9-79) and (9-80) contain these quantities, helicity expectation values must be different in different Lorentz frames.

Bottom Line

Generally, helicity (its expectation value) is not invariant under Lorentz transformation, though for massless particles, it is invariant.

Helicity invariant for massless particles, not so for massive ones

Chirality

Chirality, in contrast to helicity, is Lorentz invariant.

But, chirality is invariant

An Intuitive, Non-mathematical Look

Consider a weak interaction involving a W mediator, as seen from two different frames having non-zero relative velocity. One would expect that the LC parts of the particles interact, as seen from both frames, and the RC parts would not. If this were not so, at least one of the observers would witness weak interactions mediated by a W, in which the RC spinor parts played a role. But RC spinor parts play no role in such an interaction, regardless of who is observing it. That is, the LC contribution to an interaction is the same, regardless of who views it. And there is no RC contribution, regardless of who views it.

Intuitively, expect chirality to be invariant

Seen a bit more concretely, an RC neutrino is not going to suddenly interact with an RC electron when observed from a different reference frame. Only LC neutrinos and electrons interact and that must be true for all observers.

By this logic, ψ^L must remain LC under Lorentz transformation.

A Mathematical Look

Note from (9-71) and (9-77), for any ψ^L,

$$\begin{aligned}\psi^L = \tfrac{1}{2}\left(1-\gamma^5\right)\psi \quad \rightarrow \quad \psi^{L'} &= D\psi^L = D\tfrac{1}{2}\left(1-\gamma^5\right)\psi = D\tfrac{1}{2}\left(1-\gamma^5 D^{-1}D\right)\psi \\ &= \tfrac{1}{2}\left(1-D\gamma^5 D^{-1}\right)D\psi = \tfrac{1}{2}\left(1-\gamma'^5\right)\psi' = \psi^{L'}.\end{aligned} \tag{9-81}$$

Mathematically, chirality is invariant

ψ^L may have different form in different Lorentz frames, but it is still LC in those frames.

Bottom Line

Though fermion fields change form under Lorentz transformation, the property of chirality does not. An LC field in one Lorentz frame is an LC field in all Lorentz frames.

9.3 Charge, Parity, and Time Transformations

Charge, parity, and time transformations, designated as C, P, and T, play a major role in QFT. We will look at them first from a "words only" point of view to describe them simply, in their essence, then from a mathematical point of view to see the details of what they entail and how they are employed in QFT.

C, P, and T transformations

9.3.1 A Simplified, Non-mathematical Overview

Charge Transformation

Antiparticles have opposite charge from particles. Charge transformation of a field or particle has a simple meaning. It turns a particle/field into its antiparticle/field, and an antiparticle/field into its particle/field. It switches the signs of the charges.

C changes charges, i.e., particle ↔ antiparticle

We symbolize the operator that performs such an operation, not surprisingly, by the letter C.

Parity Transformation

Parity transformations are a little trickier. Whereas a charge transformation switches signs of charges, a parity transformation switches signs of the spatial coordinates. x^1 goes to $-x^1$; x^2 to $-x^2$; and x^3 to $-x^3$. It is essentially a mirror reflection, or spatial inversion, of each axis. We symbolize the parity operator, again unsurprisingly, by P.

Vectors switch signs of their components, too. For example, a velocity vector pointing in space along the original x^1 axis in a positive direction, after the parity transformation, points along the new x^1 axis in the negative direction. Under the transformation, $\mathbf{v} \rightarrow \mathbf{v}' = -\mathbf{v}$, and this is true for $\mathbf{v}$ pointing in any direction. All of its three scalar components switch signs, $v^i \rightarrow v'^i = -v^i$.

P changes sign of 3D true vectors

Much as we had active and passive interpretations of rotations and Lorentz boosts, we can think of a parity transformation either as switching coordinate axis directions while the physical vector stays the same, or as coordinates staying the same while the vector switches such that it points in the opposite direction. In both cases, vector component values all take the opposite sign. The two interpretations are equally valid, and be aware that discussions of parity, here or in other literature, can often switch suddenly from one to the other.

Can think of P as passive or active transformation, x^i axes switch directions or vector switches

Note, however, what happens to classical angular momentum, as illustrated in Fig. 9-1.

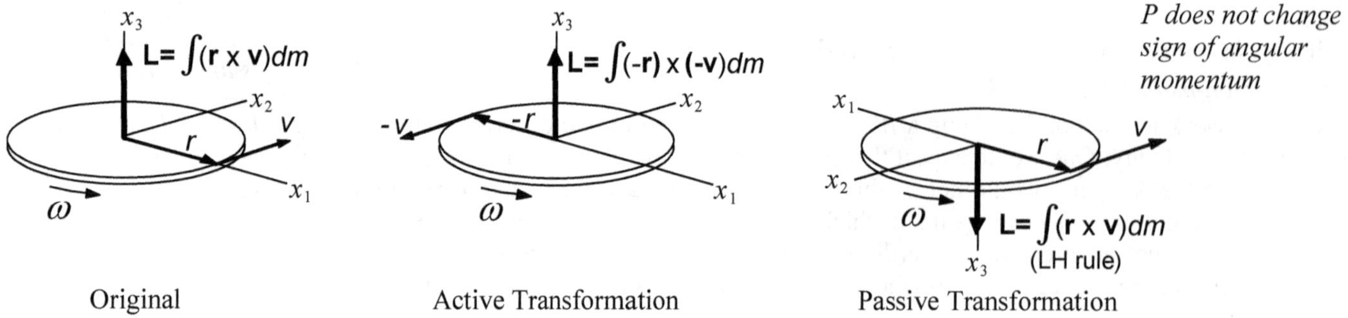

P does not change sign of angular momentum

Figure 9-1. Parity Change Leaves Angular Momentum Unchanged

An active parity transformation takes $\mathbf{v} \rightarrow -\mathbf{v}$ and $\mathbf{r} \rightarrow -\mathbf{r}$, and that leaves the sign of **L** unchanged. But we said that vectors change sign under parity transformations, so what is up with **L**? The answer is that **L** is what is called a pseudo-vector or an axial vector. It does not change sign under parity transformations and differs in that regard from what normal vectors, which we call true vectors, or simply just vectors, do. The nomenclature "axial" is related to angular momentum being about an axis (specifically, here, an axis perpendicular to two true vectors).

P does not change sign of 3D axial (pseudo) vectors

A passive parity transformation reverses the direction of the coordinate axes. That is, we now have a left-hand coordinate system, in which, for cross products, by their very definition, we need to use the left-hand, not the right-hand, rule. This causes **L** (angular momentum, a pseudo-vector) to point in the opposite direction in physical space. However, the values (signs) of its vector components remain unchanged, as the vector gets reversed in direction in physical space by the passive parity transformation, but the spatial axes get reversed too.

And this is true for both active and passive parity transformations

In both the active and passive cases, expressed in terms of vector components, the cross product $L_i = \varepsilon_{ijk} r_j v_k$ goes to $L_i = \varepsilon_{ijk}(-r_j)(-v_k)$, and the components L_i are unchanged.

Spin is just quantized angular momentum, so we can extrapolate our results to QFT. Spin doesn't change under parity transformations, i.e., it is parity invariant.

This means that a particle, classical or quantum, with given helicity, will have its 3-momentum components reversed under parity, but not its spin components. So, its helicity will flip sign. Parity changes the sign of helicity.

P reverses helicity

The effects on fields and chirality are more subtle, and we will need to hold off discussion of those until we look at the actual mathematics underlying parity transformation. But, we will find that chirality, like helicity, is reversed under *P*.

Time Transformations

Time transformations, symbolized by $\hat{T}$, or *T*, are like parity transformations, except they are for the time coordinate instead of the space coordinates. Under time transformation, $t \rightarrow -t$, i.e., $x^0 \rightarrow -x^0$.

Time transformations have subtleties we will examine later

You may have heard that Feynman and others considered that particles moving backward in time act as antiparticles moving forward in time. With this interpretation, we would expect a time transformation to change a particle/field into an antiparticle/field (with time in that case considered going forward) with its direction of travel reversed (and thus, its 3-momentum reversed).

They can be thought of as switching particles and antiparticles

This is true in some degree, but there are subtleties involved, and we will need to look more closely at the mathematics behind it all to understand why. We do that a bit later in this chapter.

Interactions Under *C, P, T*

We usually consider *C, P,* and *T* transformations as applied to an entire interaction and ask if the probability of the interaction changes when we apply one or more of *C, P, T* to every particle in the interaction. We know probability equals the square of the absolute value of the transition (scattering) amplitude and that depends explicitly on the Lagrangian. So, if the Lagrangian is symmetric (unchanged) under *C, P,* or *T*, then we would expect the probability to be unchanged, as well. And, as we will show shortly with examples, it is.

So, symmetry under a *C, P,* or *T* transformation means invariance of probability under that transformation.

Symmetry under C,P, or T means probability unchanged

C Symmetry

For charge transformations, we would ask if switching all particles to antiparticles changes the probability of interaction. For the $e^+ + e^- \rightarrow \gamma \rightarrow \mu^+ + \mu^-$ interaction just discussed, obviously there is no change, but that is a trivial example, as nothing really changes. Consider instead, the interaction of the first row of (9-82) below. Carry out a charge transformation on this, and we have an RC particle and an LC antiparticle incoming. But they don't interact via the weak force, as only their LC particle and RC antiparticle counterparts do.

$$\begin{aligned} &e^{-L} + \bar{\nu}_e^R \rightarrow W^- \rightarrow \mu^{-L} + \bar{\nu}_\mu^R \\ &\quad \overset{C}{\Rightarrow} \; e^{+L} + \nu_e^R \xrightarrow[\text{happen}]{\text{can't}} W^+ \xrightarrow[\text{happen}]{\text{can't}} \mu^{+L} + \nu_\mu^R . \end{aligned} \tag{9-82}$$

Example of lack of C symmetry in e/w theory

So, C transformations in electroweak theory are not always invariant. In general, C is not a symmetry operation for electroweak interactions. Probability in the (9-82) case goes from a non-zero value to zero under C.

In QED, all interactions are C symmetric, as the positron (every antiparticle) regardless of chirality has the same e/m coupling to photons as the electron (every particle).

QED has full C symmetry

P Symmetry

For parity transformations, we would expect helicity to change, but the more important question is what does the P operation do to chirality. We will see how this works mathematically shortly, but for now, just accept that P changes LC to RC, and vice versa. Observe what happens when we apply it to the interaction in the top row of (9-82).

$$\begin{aligned} &e^{-L} + \bar{\nu}_e^R \rightarrow W^- \rightarrow \mu^{-L} + \bar{\nu}_\mu^R \\ &\quad \overset{P}{\Rightarrow} \; e^{-R} + \bar{\nu}_e^L \xrightarrow[\text{happen}]{\text{can't}} W^- \xrightarrow[\text{happen}]{\text{can't}} \mu^{-R} + \bar{\nu}_\mu^L . \end{aligned} \tag{9-83}$$

Example of lack of P symmetry in e/w theory

This, too, is prohibited, i.e., the probability is zero, because neither RC particles nor LC antiparticles interact with the electroweak mediator W^-.

So, P transformations in electroweak theory are not always invariant. In general, P is not a symmetry operation for electroweak interactions. Probability in this case goes from a finite value to zero under P.

Combined Operation CP

But look what happens when we do both C and P operations on this interaction.

$$\begin{aligned} &e^{-L} + \bar{\nu}_e^R \rightarrow W^- \rightarrow \mu^{-L} + \bar{\nu}_\mu^R \\ &\quad \overset{CP}{\Rightarrow} \; e^{+R} + \nu_e^L \rightarrow W^+ \rightarrow \mu^{+R} + \nu_\mu^L . \end{aligned} \tag{9-84}$$

Example of CP transformation

The transformed interaction indeed does happen. The question now is whether it happens with the same probability. As it turns out, it does, and we will shortly see how to determine this by examining the Lagrangian.

For now, we note that many interactions obey CP symmetry, though, importantly, some do not. More on this later.

E/w theory has some CP violation (lack of CP symmetry)

Time $\hat{T}$ Symmetry

The subtleties noted earlier with regard to the time transformation have to do with the fact that there are actually two different ways in QFT to do it, which we label $\hat{T}$ and T. The first of these is more intuitive, but the second is found most often in the literature. We will consider the first of these now, and save the second for later.

Two ways to model time transformations

We consider the easier one here, labeled $\hat{T}$

Under $\hat{T}$, particles become antiparticles (and vice versa), 3-momenta change signs (since particles are traveling in opposite directions), and chirality switches (LC $\leftrightarrow$ RC).

Regarding the change in direction of travel, we can simply think that instead of an incoming particle coming into the interaction zone from the left, it is coming in from the right. The momentum direction is changed, but incoming particles stay incoming, and as well, outgoing particles stay outgoing.

So, under a $\hat{T}$ transformation, as an example, we would have

$$e^{-L} + \bar{\nu}_e^R \to W^- \to \mu^{-L} + \bar{\nu}_\mu^R$$
$$\overset{\hat{T}}{\Rightarrow} \; e^{+R} + \nu_e^L \;\to\; W^+ \;\to\; \mu^{+R} + \nu_\mu^L , \tag{9-85}$$

which is actually the same result as (9-84). You may hear or read that the time transformation is the same as the CP transformation, and in this context, it is true.

$\hat{T}$ acts just like CP

Combined Operation $CP\hat{T}$

Now, look what happens if we first do CP, as in (9-84), and then follow that with $\hat{T}$.

$$e^{-L} + \bar{\nu}_e^R \to W^- \to \mu^{-L} + \bar{\nu}_\mu^R$$
$$\overset{CP}{\Rightarrow} \; e^{+R} + \nu_e^L \;\to\; W^+ \to \mu^{+R} + \nu_\mu^L \;\; \overset{\hat{T}}{\Rightarrow} \;\; e^{-L} + \bar{\nu}_e^R \to W^- \to \mu^{-L} + \bar{\nu}_\mu^R . \tag{9-86}$$

$CP\hat{T}$ gives back original interaction; symmetric under $CP\hat{T}$

A full $CP\hat{T}$ operation gives us back what we started with. $\hat{T} = (CP)^{-1}$, probability of interaction is unchanged under $CP\hat{T}$, and the theory is symmetric under it. And this is universally true. There is no $CP\hat{T}$ violation in QFT. Note, $(CP)(CP)$ also gives back the original interaction, as we just flip charges and axes a second time. So, we can also consider $\hat{T} = CP$.

Discrete vs Continuous Symmetries

Note that C, P, and $\hat{T}$ (or T) are discrete transformations. For example, P switches each point on the x^1 axis to a single point on the $-x^1$ axis. And so, for all points. They are not rotated continuously, as we would have in a 3D rotation transformation.

C, P, T are discrete transformations

Similarly, C changes a charge q to $-q$, in one shot. It does not gradually change the charge from one value to other ones. In the charge plot of Wholeness Chart 6-6, pg. 183, particles and antiparticles have discrete locations, and C exchanges one for the other in one action. They do not slide along the charge axis to their new position after C operation.

And, similarly, T switches one point in time on the time axis to its mirror image position on the negative time axis. It doesn't continuously move from one value of time to another.

Hence, C, P, and T operations are called discrete transformations. Rotation is a continuous transformation, as is a Lorentz transformation (which varies with continuous variation of **v**, the relative velocity between two frames). Similarly, $SU(n)$ transformations are continuous, as they vary with continuous parameters, such as ω_i for e/w $SU(2)$.

Note that Noether's theorem is only applicable to continuous symmetries, so we will need to take a different avenue to investigate the impact of C, P, and T.

So, C, P, and T not amenable to Noether's theorem

9.3.2 The Mathematics of Discrete Symmetries

In this section we will determine the actual mathematical form for the C, P, and $\hat{T}$ (and T) transformations for scalars, spinors, and vectors. For transformations on spinors, we will generally work in the standard rep, except when a different rep makes analysis easiest. We will assume all gamma matrices and spinors are in the standard rep unless labeled otherwise. We recognize that any result we get for one rep will be valid in any rep since *relations* between gamma matrices and spinors are the same in any rep, even though the matrices and spinors themselves are different.

The math behind C, P, and T

Wholeness Chart 9-3, pg. 342, summarizes the results of all of this, and you may wish to follow along with that as you progress through the following steps of development.

Relations We'll Need

General Relations, Good in Any Rep

We will be using the following relations (9-87), (9-88), and (9-89) from Vol. 1, Chap. 4, Appendix A, pg. 122, and this volume, Chap. 5, Appendix B, pg. 160.

$$\gamma^0\gamma^\mu = \gamma^{\mu\dagger}\gamma^0 \quad \xrightarrow{\text{equivalently}} \quad \gamma^\mu\gamma^0 = \gamma^0\gamma^{\mu\dagger} \tag{9-87}$$
$$\gamma^\mu\gamma^\nu = -\gamma^\nu\gamma^\mu \text{ for } \mu \neq \nu \qquad \gamma^1\gamma^1 = \gamma^2\gamma^2 = \gamma^3\gamma^3 = -1 \qquad \gamma^0\gamma^0 = 1 \tag{9-88}$$
$$\gamma^5\gamma^\mu = -\gamma^\mu\gamma^5 \qquad \gamma^5\gamma^5 = 1 \qquad \gamma^5 = \gamma^{5\dagger} \tag{9-89}$$

Key general gamma matrix relations we'll use

We will also need the following relation, which we prove in Appendix A of this chapter.

$$\gamma^{\mu\dagger} = \gamma_{\mu}. \tag{9-90}$$

Relations Good in the Standard Rep

When we work in the standard rep, which is our default rep unless otherwise stated, the gamma matrices obey the following relations, which you can check yourself, if you wish, from (5-2)a of Chap. 5.

Standard Rep

$$\begin{array}{lll}
\left(\gamma^0\right)^* = \gamma^0 & \left(\gamma^0\right)^T = \gamma^0 & \left(\gamma^0\right)^\dagger = \gamma^0 \\
\left(\gamma^1\right)^* = \gamma^1 & \left(\gamma^1\right)^T = -\gamma^1 & \left(\gamma^1\right)^\dagger = -\gamma^1 \\
\left(\gamma^2\right)^* = -\gamma^2 & \left(\gamma^2\right)^T = \gamma^2 & \left(\gamma^2\right)^\dagger = -\gamma^2 \\
\left(\gamma^3\right)^* = \gamma^3 & \left(\gamma^3\right)^T = -\gamma^3 & \left(\gamma^3\right)^\dagger = -\gamma^3 \\
\left(\gamma^5\right)^* = \gamma^5 & \left(\gamma^5\right)^T = \gamma^5 & \left(\gamma^5\right)^\dagger = \gamma^5 .
\end{array} \tag{9-91}$$

Key standard rep gamma matrix relations we'll use

Charge Transformations

We seek transformations that will switch particles and antiparticles for each field type.

First, look at C

Scalars

Note what happens when we complex conjugate a scalar field.

$$\begin{aligned}
\phi(x) &= \sum_{\mathbf{k}} \frac{1}{\sqrt{2V\omega_{\mathbf{k}}}} a(\mathbf{k}) e^{-ikx} + \sum_{\mathbf{k}} \frac{1}{\sqrt{2V\omega_{\mathbf{k}}}} b^\dagger(\mathbf{k}) e^{ikx} \\
\phi^*(x) &= \sum_{\mathbf{k}} \frac{1}{\sqrt{2V\omega_{\mathbf{k}}}} b(\mathbf{k}) e^{-ikx} + \sum_{\mathbf{k}} \frac{1}{\sqrt{2V\omega_{\mathbf{k}}}} a^\dagger(\mathbf{k}) e^{ikx} .
\end{aligned} \tag{9-92}$$

In the top row we destroy particles and create antiparticles. In the bottom row, we destroy antiparticles and create particles. We have switched particles with antiparticles, which is our goal for the C operator. So, from (9-92), we can conclude that, for scalars, the charge transformation is simply complex conjugation, where repeated operation gives us our original field back, and we use superscript c to indicate charge conjugated field,

$$\phi \overset{C}{\Rightarrow} \phi^c \equiv C\phi = \phi^* \qquad \phi^* \overset{C}{\Rightarrow} \phi^{*c} \equiv C\phi^* = \phi \qquad C^2\phi = \phi \qquad C^2\phi^* = \phi^* . \tag{9-93}$$

Scalar under C transformation

Spinors

The charge conjugation operator for spinors in the standard rep (Chap. 5, (5-2) and (5-3)), labeled herein as C and proven below, is defined by

Spinor under C, proof follows

$$\psi \overset{C}{\Rightarrow} \psi^c = C\psi = -i\gamma^2\psi^* . \tag{9-94}$$

That is, it is defined by first taking the complex conjugate of the spinor field then pre-multiplying by $-i\gamma^2$.

Let's prove it by starting with the spinor field in the standard rep and $-i\gamma^2$ written out explicitly.

Use standard rep

$$\begin{aligned}
\psi &= \sum_{\mathbf{p}} \sqrt{\frac{m}{VE_{\mathbf{p}}}} \left(c_1(\mathbf{p})\psi_{\mathbf{p}}^{(1)} + c_2(\mathbf{p})\psi_{\mathbf{p}}^{(2)} + d_2^\dagger(\mathbf{p})\psi_{\mathbf{p}}^{(3)} + d_1^\dagger(\mathbf{p})\psi_{\mathbf{p}}^{(4)} \right) \\
&= \sum_{\mathbf{p}} \sqrt{\frac{m}{VE_{\mathbf{p}}}} \left(c_1(\mathbf{p})u_1(\mathbf{p})e^{-ipx} + c_2(\mathbf{p})u_2(\mathbf{p})e^{-ipx} + d_1^\dagger(\mathbf{p})v_1(\mathbf{p})e^{ipx} + d_2^\dagger(\mathbf{p})v_2(\mathbf{p})e^{ipx} \right).
\end{aligned} \tag{9-95}$$

$$\text{In standard rep} \quad -i\gamma^2 = -i\begin{bmatrix} & & & -i \\ & & i & \\ & i & & \\ -i & & & \end{bmatrix} = \begin{bmatrix} & & & -1 \\ & & 1 & \\ & 1 & & \\ -1 & & & \end{bmatrix}. \tag{9-96}$$

So, to find ψ^c, we use (9-96) to operate on the complex conjugate of (9-95), and we will look at each term one at a time in (9-95), for given $\mathbf{p}$. First, the first term,

$$C\left(c_1(\mathbf{p})\psi_{\mathbf{p}}^{(1)}\right)=-i\gamma^2\left(c_1(\mathbf{p})u_1(\mathbf{p})e^{-ipx}\right)^*=\begin{bmatrix} & & & -1\\ & & 1 & \\ & 1 & & \\ -1 & & & \end{bmatrix} c_1^\dagger(\mathbf{p})\sqrt{\frac{E+m}{2m}}\begin{pmatrix}1\\0\\ \frac{p^3}{E+m}\\ \frac{p^1+ip^2}{E+m}\end{pmatrix}^* e^{ipx}$$

$$=\begin{bmatrix} & & & -1\\ & & 1 & \\ & 1 & & \\ -1 & & & \end{bmatrix} c_1^\dagger(\mathbf{p})\sqrt{\frac{E+m}{2m}}\begin{pmatrix}1\\0\\ \frac{p^3}{E+m}\\ \frac{p^1-ip^2}{E+m}\end{pmatrix} e^{ipx}=c_1^\dagger(\mathbf{p})\underbrace{\sqrt{\frac{E+m}{2m}}\begin{pmatrix}-\frac{p^1-ip^2}{E+m}\\ \frac{p^3}{E+m}\\ 0\\ -1\end{pmatrix}}_{-v_1(\mathbf{p})} e^{ipx}=-c_1^\dagger(\mathbf{p})\underbrace{v_1(\mathbf{p})e^{ipx}}_{\psi_{\mathbf{p}}^{(4)}}. \tag{9-97}$$

Do **Problems 8** to show (9-98) below, the charge conjugation of the second term in (9-95).

$$C\left(c_2(\mathbf{p})\psi_{\mathbf{p}}^{(2)}\right)=-i\gamma^2\left(c_2(\mathbf{p})u_2(\mathbf{p})e^{-ipx}\right)^*=c_2^\dagger(\mathbf{p})v_2(\mathbf{p})e^{ipx}=c_2^\dagger(\mathbf{p})\psi_{\mathbf{p}}^{(3)}. \tag{9-98}$$

Additionally, which you can take my word for or work out yourself,

$$C\left(d_2^\dagger(\mathbf{p})\psi_{\mathbf{p}}^{(3)}\right)=-i\gamma^2\left(d_2^\dagger(\mathbf{p})v_2(\mathbf{p})e^{ipx}\right)^*=d_2(\mathbf{p})u_2(\mathbf{p})e^{-ipx}=d_2(\mathbf{p})\,\psi_{\mathbf{p}}^{(2)}, \text{ and} \tag{9-99}$$

$$C\left(d_1^\dagger(\mathbf{p})\psi_{\mathbf{p}}^{(4)}\right)=-i\gamma^2\left(d_1^\dagger(\mathbf{p})v_1(\mathbf{p})e^{ipx}\right)^*=-d_1(\mathbf{p})u_1(\mathbf{p})e^{-ipx}=-d_1(\mathbf{p})\psi_{\mathbf{p}}^{(1)}. \tag{9-100}$$

(9-97) to (9-100), with (9-94), give us

$$\psi^c=C\psi$$
$$=\sum_{\mathbf{p}}\sqrt{\frac{m}{VE_{\mathbf{p}}}}\left(-c_1^\dagger(\mathbf{p})v_1(\mathbf{p})e^{ipx}+c_2^\dagger(\mathbf{p})v_2(\mathbf{p})e^{ipx}+d_2(\mathbf{p})u_2(\mathbf{p})e^{-ipx}-d_1(\mathbf{p})u_1(\mathbf{p})e^{-ipx}\right). \tag{9-101}$$

C transformation of ψ in standard rep

Comparing (9-101) to (9-95), and recognizing that the minus signs are insignificant phase factors, we see that the charge conjugate field ψ^c creates particles that have the same spin as the antiparticles that ψ created. And ψ^c destroys antiparticles with the same spin as the particles that ψ destroyed. And so, ψ^c, acting as an operator, will indeed switch particles with antiparticles from what ψ does, acting as an operator. Thus, we have proven (9-94).

We will find it easier to work with $\psi^\dagger$, rather than $\bar{\psi}$, and we will eventually need some other relation for spinor field configurations, as follows.

By taking the complex conjugate of both sides of (9-94), we have

$$\psi^* \overset{C}{\Rightarrow} (-i)^*\left(\gamma^2\right)^*\psi=i\left(-\gamma^2\right)\psi=-i\gamma^2\psi=-i\gamma^2\left(\psi^*\right)^*. \tag{9-102}$$

Mathematical purists may have issues with deducing (9-102) from (9-94), for reasons we won't get into, but we could simply take (9-102) as a definition (that eventually yields a viable theory).
The transpose of (9-102), along with (9-91) gives us

$$\psi^\dagger \overset{C}{\Rightarrow} -i\psi^T\left(\gamma^2\right)^T=-i\psi^T\gamma^2=-i\left(\psi^\dagger\right)^*\gamma^2. \tag{9-103}$$

C transformation of $\psi^\dagger$ in standard rep

For future reference, we will also want the relation derived by taking the complex conjugate of (9-103),

$$\psi^T \overset{C}{\Rightarrow} \left(-i\psi^T\gamma^2\right)^*=i\psi^\dagger\left(\gamma^2\right)^*=i\psi^\dagger\left(-\gamma^2\right)=-i\psi^\dagger\gamma^2=-i\left(\psi^T\right)^*\gamma^2. \tag{9-104}$$

With an eye to future needs, we have expressed the final form of each of (9-102), (9-103), and (9-104) with the original field configuration in parentheses.

Spinor Inner Products

Hence, from (9-94), (9-103), and (9-88), where in the second line below, we take the transpose,

C and inner products

$$\bar{\psi}\psi = \psi^\dagger\gamma^0\psi \overset{C}{\Rightarrow} \left(-i\psi^T\gamma^2\right)\gamma^0\left(-i\gamma^2\psi^*\right) = -\psi^T\gamma^2\gamma^0\gamma^2\psi^* = \psi^T\gamma^2\gamma^2\gamma^0\psi^*$$
$$= \underbrace{-\psi^T\gamma^0\psi^* = \psi^\dagger\gamma^0\psi}_{\substack{\text{switching spinor fields}\\ \text{changes sign}}} = \bar{\psi}\psi. \qquad (9\text{-}105)$$

$\bar{\psi}\psi$ invariant under C

(9-105) is an inner product, a scalar in spinor space, which for numbers means it equals its own transpose. However, here we have two fermion fields, and in taking the transpose, we switch their positions, meaning we get an extra factor of −1 from the reversed order of the creation and destruction operators that are contained (implied) in the expressions for the fields.

Transformations in spinor space do not affect spinor space scalar quantities, just as transformations in 3D space do not affect 3D inner product scalars. So, (9-105) holds, in general, for any spinor space rep, not just the standard rep, which we used to prove it.

Note that different authors employ different symbols and/or operations in *C, P,* and *T* transformations, so, it is not always easy comparing one text to another, although the final results (further on herein) in all cases are the same. For example, some authors have a plus sign in front of the RHS of (9-94), and some have even more complicated expressions for it, so best to stay with one text until you fathom the essence of the discrete transformations. Herein, we are going with the most commonly used relations.

Caution in comparing texts, as there are different approaches

As a further example of the differences you may see, it turns out (we won't take the time to show it, as we don't need it) that in the Majorana rep, both $-i\gamma^2\psi^*$ and $i\psi^*$ change particles to antiparticles. It is strictly speaking, only correct to use the former of these to find the charge transformation, as only that form is correct in all reps. But you may see sources opting for the second expression, because it is simpler. Yes, it is simpler. But also, yes, it is less consistent and more confusing.

Do **Problem 9** to show (9-106).

Note that

$\bar{\psi}\gamma^5\psi$ invariant under C

$$\bar{\psi}\gamma^5\psi \overset{C}{\Rightarrow} \bar{\psi}^c\gamma^5\psi^c = \bar{\psi}\gamma^5\psi. \qquad (9\text{-}106)$$

Vector Four-currents

C and vector 4-currents

To deduce the effect of charge conjugation on vector four-currents, we will need (9-107) below, which can be found using relations (9-87), (9-88), and (9-91).

$$\text{In standard rep } \left(\gamma^0\right)^*\gamma^2 = \gamma^0\gamma^2 = -\gamma^2\gamma^0 \quad \left(\gamma^1\right)^*\gamma^2 = \gamma^1\gamma^2 = -\gamma^2\gamma^1 \quad \left(\gamma^2\right)^*\gamma^2 = -\gamma^2\gamma^2$$
$$\left(\gamma^3\right)^*\gamma^2 = \gamma^3\gamma^2 = -\gamma^2\gamma^3 \quad \xrightarrow[\text{general}]{\text{in}} \quad \left(\gamma^\mu\right)^*\gamma^2 = -\gamma^2\gamma^\mu. \qquad (9\text{-}107)$$

Again, wherever we derive product relationships between gamma matrices in one rep, such relationships will hold in all reps.

Then, where subscripts 1 and 2 stand for any fields in a four-current, with (9-94), (9-103), and the gamma matrix relations (9-87), (9-88), (9-91), and (9-107),

$$j^\mu_{vec} = \bar{\psi}_1\gamma^\mu\psi_2 \overset{C}{\Rightarrow} j^{c\,\mu}_{vec} = \bar{\psi}^c_1\gamma^\mu\psi^c_2 = \left(\psi^\dagger_1\right)^c\gamma^0\gamma^\mu\psi^c_2 = \left(-i\left(\psi^\dagger_1\right)^*\gamma^2\right)\gamma^0\gamma^\mu\left(-i\gamma^2\psi^*_2\right)$$
$$= -\left(\psi^T_1\gamma^2\right)\gamma^0\gamma^\mu\left(\gamma^2\psi^*_2\right) = \psi^T_1\gamma^0\gamma^2\gamma^\mu\gamma^2\psi^*_2 = \psi^T_1\left(\gamma^2\right)^\dagger\gamma^0\gamma^\mu\gamma^2\psi^*_2$$
$$= \psi^T_1\left(\gamma^2\right)^\dagger\left(\gamma^\mu\right)^\dagger\gamma^0\gamma^2\psi^*_2 = \psi^T_1\left(\gamma^2\right)^\dagger\left(\gamma^\mu\right)^\dagger\left(\gamma^2\right)^\dagger\gamma^0\psi^*_2 \quad \text{transpose,}$$
$$\left(\begin{array}{l}\text{include } (-1) \text{ factor for fermion switch, transpose does not}\\ \text{change component values; only switches column to row vector}\end{array}\right)$$

$$= -\psi_2^\dagger \gamma^0 (\gamma^2)^* (\gamma^\mu)^* (\gamma^2)^* \psi_1 = -\psi_2^\dagger \gamma^0 (-\gamma^2)(\gamma^\mu)^* (-\gamma^2)\psi_1$$
$$= -\psi_2^\dagger \gamma^0 \gamma^2 (\gamma^\mu)^* \gamma^2 \psi_1 = \psi_2^\dagger \gamma^0 \gamma^2 \gamma^2 \gamma^\mu \psi_1 = -\psi_2^\dagger \gamma^0 \gamma^\mu \psi_1 = -\bar{\psi}_2 \gamma^\mu \psi_1 . \qquad (9\text{-}108)$$

Charge conjugation of a vector four-current switches the positions of the fields, plus changes the sign, and since this is true in one rep, it is true in all.

Vector 4-current C transformation, sign change and field switch

Axial Four-current

Do **Problem 10** to find (9-109).

Similarly, the charge conjugation of the axial current is

$$j^\mu_{axi} = \bar{\psi}_1 \gamma^\mu \gamma^5 \psi_2 \quad \overset{C}{\Rightarrow} \quad j^{\mu\, c}_{axi} = \bar{\psi}_2 \gamma^\mu \gamma^5 \psi_1 . \qquad (9\text{-}109)$$

Axial 4-current C transformation, no sign change, field switch only

Charge conjugation of an axial four-current switches the positions of the fields, but does not change the sign.

Vector Bosons

In QED, the interaction term in the Lagrangian has the form

$$e\left({}_{QED}\, j^\mu A_\mu\right) = e\bar{\psi}\gamma^\mu \psi A_\mu . \qquad (9\text{-}110)$$

QED interaction term in $\mathcal{L}$

Since, the photon is chargeless, and its own antiparticle, we might expect it would be invariant under C transformations. However, we also know QED is charge conjugation invariant, since positrons interact in the same way, to the same degree, with photons as electrons do. Thus, we need to have the QED Lagrangian invariant under C.

But, the four-current in (9-110) changes sign under C, as we saw in (9-108), so if A_μ were to remain A_μ, the interaction part of the Lagrangian would not be invariant, since it would change sign. This would mean that in the interaction Dirac equation (Vol. 1, (7-16)), the interaction part would have a different sign, i.e., the governing equation for QED would not be invariant.

Photon 4-vector needs to keep QED term in $\mathcal{L}$ invariant

Therefore, experiment dictates that we define the effect of charge conjugation of the photon field as introducing a minus sign, i.e.,

$$A_\mu \quad \overset{C}{\Rightarrow} \quad -A_\mu . \qquad (9\text{-}111)$$

Photon 4-vector under C transformation

All terms in the free photon Lagrangian are bilinear in the field A_μ (its derivatives, actually), so introduction of the minus sign of (9-111) leaves that invariant, as well.

Related, though a bit more complicated, logic for the massive intermediate vector bosons leads one to the charge transformation on them as (where we recall that there are two commonly used notations with $W_\mu^+ = W_\mu$ and $W_\mu^- = W_\mu^\dagger$)

$$Z_\mu \quad \overset{C}{\Rightarrow} \quad -Z_\mu \qquad W_\mu \quad \overset{C}{\Rightarrow} \quad -W_\mu^\dagger \qquad W_\mu^\dagger \quad \overset{C}{\Rightarrow} \quad -W_\mu . \qquad (9\text{-}112)$$

Massive boson 4-vectors under C transformation

Partial Derivatives

As we have derivatives in our Lagrangian, we need to consider C, P, and $\hat{T}$ (and T) transformations on them. For C, there is no change, as the derivatives are with respect to real, not complex, numbers, and they involve spacetime coordinates, not charges of the particles.

$$\partial_\mu \quad \overset{C}{\Rightarrow} \quad \partial_\mu . \qquad (9\text{-}113)$$

Derivatives under C transformation

<u>Parity Transformations</u>

We now examine the effects of parity change ($x^i \rightarrow -x^i$, i= 1,2,3) on different fields and their derivatives.

Second. look at P

Scalars

We recall from earlier discussion, how a (true) 3D vector changes sign under parity. A scalar has no direction, so it doesn't change.

$$\phi\left(t,x^i\right) \overset{P}{\Rightarrow} \phi\left(t,-x^i\right). \tag{9-114}$$

Scalar under P transformation, invariant

Pseudo-scalars

A pseudo-scalar is a scalar-type entity that transforms like (9-114), but with a minus sign. An example of which is $V = \mathbf{a}\cdot(\mathbf{b}\times\mathbf{c})$, that yields the volume of a parallelepiped in 3D-space, spanned by three non-coplanar vectors; **a**, **b** and **c**. All three true vectors change directions (signs).

Pseudo-scalar under P, sign change

$$\text{pseudo-scalar } \pi\left(t,x^i\right) \overset{P}{\Rightarrow} -\pi\left(t,-x^i\right) \tag{9-115}$$

Spinors

Chirality under parity change is easier to understand using the Weyl rep, and we will use that in the following, labeled with a subscript W to distinguish it from the (unlabeled) standard rep matrices used elsewhere.

Recall from earlier that helicity reverses under P transformation. Then recall that for particles traveling at the speed of light, helicity and chirality eigenstates ("states" here in the eigen sense, as they are "fields" in the QFT field vs state sense) are one and the same. And, as we saw in Chap. 5, (5-18) (repeated below as (9-116)), in the Weyl rep for this case, the top two spinor components of ψ are in LH and LC eigenstates. The bottom two components are in RH and RC states.

Recall Weyl rep

$$\underbrace{{}_W\psi = \begin{bmatrix} {}_W\psi^L \\ {}_W\psi^R \end{bmatrix} = \begin{bmatrix} {}_W\psi_1 \\ {}_W\psi_2 \\ {}_W\psi_3 \\ {}_W\psi_4 \end{bmatrix}}_{\text{Weyl rep spinor field}} \qquad \underbrace{\begin{bmatrix} {}_W\psi^L \\ 0 \end{bmatrix} = \begin{bmatrix} {}_W\psi_1 \\ {}_W\psi_2 \\ 0 \\ 0 \end{bmatrix}}_{L\text{ Weyl spinor}} = \underbrace{{}_W\psi^L}_{\substack{\text{often}\\\text{used}\\\text{symbol}}} \qquad \underbrace{\begin{bmatrix} 0 \\ {}_W\psi^R \end{bmatrix} = \begin{bmatrix} 0 \\ 0 \\ {}_W\psi_3 \\ {}_W\psi_4 \end{bmatrix}}_{R\text{ Weyl spinor}} = \underbrace{{}_W\psi^R}_{\substack{\text{often}\\\text{used}\\\text{symbol}}} \tag{9-116}$$

Under P, the LH (LC) state should turn into the RH (RC) state; and the RH (RC) state should turn into the LH (LC) state. Note from the form of γ^0 in the Weyl rep, what that matrix operation does for us. With I the 2X2 identity matrix,

$$ {}_W\gamma^0\,{}_W\psi = \begin{bmatrix} & & 1 & \\ & & & 1 \\ 1 & & & \\ & 1 & & \end{bmatrix} \begin{bmatrix} {}_W\psi_1 \\ {}_W\psi_2 \\ {}_W\psi_3 \\ {}_W\psi_4 \end{bmatrix} = \begin{bmatrix} 0 & I \\ I & 0 \end{bmatrix} \begin{bmatrix} {}_W\psi^L \\ {}_W\psi^R \end{bmatrix} = \begin{bmatrix} {}_W\psi^R \\ {}_W\psi^L \end{bmatrix} = \begin{bmatrix} {}_W\psi_3 \\ {}_W\psi_4 \\ {}_W\psi_1 \\ {}_W\psi_2 \end{bmatrix}. \tag{9-117}$$

γ^0 in Weyl rep for v=c switches helicity and chirality

The γ^0 matrix operating on a spinor field, when $v = c$, reverses helicity (and chirality).

From the discussion following (5-18), we know that at speeds less than c, helicity is generally not the same as chirality, and the R and L portions of ψ are no longer (in general) in helicity eigenstates. However, they remain in RC and LC eigenstates, regardless of velocity.

While the helicity operator (Vol. 1, (4-121)) varies with **p** (and thus, **v**) of the field, γ^0 does not. Hence, if ${}_W\psi^L$ and ${}_W\psi^R$ remain LC and RC for any field velocity, and γ^0 remains unchanged for any field velocity, then, with regard to chirality (but not helicity), (9-117) holds for any field velocity. Further, that result holds in any spinor rep, as all relations between spinor fields do. As a result, we can conclude that, in general,

γ^0 in Weyl rep for any v switches chirality

$$\psi\left(t,x^i\right) \overset{P}{\Rightarrow} \gamma^0\psi\left(t,-x^i\right). \tag{9-118}$$

From the complex conjugate of (9-118), or simply by definition,

$$\psi^* \overset{P}{\Rightarrow} \left(\gamma^0\right)^*\psi^* = \gamma^0\psi^*, \tag{9-119}$$

and from the transpose of (9-119),

$$\psi^\dagger \overset{P}{\Rightarrow} \psi^\dagger\left(\gamma^0\right)^\dagger = \psi^\dagger\gamma^0. \tag{9-120}$$

And from the transpose of (9-118),

$$\psi^T \overset{P}{\Rightarrow} \psi^T\left(\gamma^0\right)^T = \psi^T\gamma^0. \tag{9-121}$$

Spinor Inner Products

The inner product of two spinors is a scalar, and it transforms as a scalar.

$$\bar{\psi}_1\left(t,x^i\right)\psi_2\left(t,x^i\right)=\psi_1^\dagger\left(t,x^i\right)\gamma^0\psi_2\left(t,x^i\right)\overset{P}{\Rightarrow}\left(\psi_1^\dagger\left(t,-x^i\right)\gamma^0\right)\gamma^0\left(\gamma^0\psi_2\left(t,-x^i\right)\right)$$
$$=\psi_1^\dagger\left(t,-x^i\right)\gamma^0\psi_2\left(t,-x^i\right)=\bar{\psi}_1\left(t,-x^i\right)\psi_2\left(t,-x^i\right). \quad (9\text{-}122)$$

$\bar{\psi}_1\psi_2$ *invariant under P*

Note, however, that

$$\bar{\psi}_1\left(t,x^i\right)\gamma^5\psi_2\left(t,x^i\right)=\psi_1^\dagger\left(t,x^i\right)\gamma^0\gamma^5\psi_2\left(t,x^i\right)\overset{P}{\Rightarrow}\psi_1^\dagger\left(t,-x^i\right)\gamma^0\gamma^0\gamma^5\gamma^0\psi_2\left(t,-x^i\right)$$
$$=-\psi_1^\dagger\left(t,-x^i\right)\gamma^0\gamma^5\psi_2\left(t,-x^i\right)=-\bar{\psi}_1\left(t,-x^i\right)\gamma^5\psi_2\left(t,-x^i\right). \quad (9\text{-}123)$$

$\bar{\psi}_1\gamma^5\psi_2$ *changes sign under P*

transforms like a scalar except for the switch in sign. It is a pseudo-scalar.

Vector Four-currents

Using (9-118), (9-87), and (9-90), we have

$$\bar{\psi}_1\left(t,x^i\right)\gamma^\mu\psi_2\left(t,x^i\right)\overset{P}{\Rightarrow}\left(\psi_1^\dagger\left(t,-x^i\right)\gamma^0\right)\gamma^0\gamma^\mu\left(\gamma^0\psi_2\left(t,-x^i\right)\right)=\psi_1^\dagger\left(t,-x^i\right)\gamma^\mu\gamma^0\psi_2\left(t,-x^i\right)$$
$$=\psi_1^\dagger\left(t,-x^i\right)\gamma^0\gamma^{\mu\dagger}\psi_2\left(t,-x^i\right)=\psi_1^\dagger\left(t,-x^i\right)\gamma^0\gamma_\mu\psi_2\left(t,-x^i\right)=\bar{\psi}_1\left(t,-x^i\right)\gamma_\mu\psi_2\left(t,-x^i\right). \quad (9\text{-}124)$$

Vector 4-currents lower 4D index under P

Parity turns a contravariant vector four-current into a covariant one. The lowered space indices, μ = 1,2,3 have different sign from the raised ones, as we expect in a parity operation. (This is for the choice of Minkowski metric with signature (+,–,–,–).)

Axial Four-currents

Similarly, but different by a sign change, for axial currents,

$$\bar{\psi}_1\left(t,x^i\right)\gamma^\mu\gamma^5\psi_2\left(t,x^i\right)\overset{P}{\Rightarrow}-\bar{\psi}_1\left(t,-x^i\right)\gamma_\mu\gamma^5\psi_2\left(t,-x^i\right). \quad (9\text{-}125)$$

Axial 4-currents lower 4D index under P and change

(9-125) is also known as a pseudo-vector (four-current), which is a more general term for a vector-type entity in any field of study (not just QFT) that changes sign upon parity change, compared with a true vector.

Axial 4-currents behave like pseudo-vectors

Do **Problem 11** to prove (9-125).

Chirality

Recall (which you should be committing to memory, by now) that

$$\bar{\psi}_1^L\gamma^\mu\psi_2^L=\bar{\psi}_1\tfrac{1}{2}\left(1+\gamma^5\right)\gamma^\mu\tfrac{1}{2}\left(1-\gamma^5\right)\psi_2=\bar{\psi}_1\gamma^\mu\tfrac{1}{2}\left(1-\gamma^5\right)\tfrac{1}{2}\left(1-\gamma^5\right)\psi_2$$
$$=\bar{\psi}_1\gamma^\mu\tfrac{1}{4}\left(1-2\gamma^5+\left(\gamma^5\right)^2\right)\psi_2=\bar{\psi}_1\gamma^\mu\tfrac{1}{2}\left(1-\gamma^5\right)\psi_2. \quad (9\text{-}126)$$

From (9-124), (9-125), and (9-126), we see that

$$\bar{\psi}_1^L\gamma^\mu\psi_2^L=\bar{\psi}_1\gamma^\mu\tfrac{1}{2}\left(1-\gamma^5\right)\psi_2=\tfrac{1}{2}\bar{\psi}_1\gamma^\mu\psi_2-\tfrac{1}{2}\bar{\psi}_1\gamma^\mu\gamma^5\psi_2$$
$$\overset{P}{\Rightarrow}\tfrac{1}{2}\bar{\psi}_1\gamma_\mu\psi_2+\tfrac{1}{2}\bar{\psi}_1\gamma_\mu\gamma^5\psi_2=\bar{\psi}_1\gamma_\mu\tfrac{1}{2}\left(1+\gamma^5\right)\psi_2=\bar{\psi}_1^R\gamma_\mu\psi_2^R. \quad (9\text{-}127)$$

P reverses chirality of LC and RC 4-currents

The parity transformation flips LC and RC. It reverses chirality, as we promised it would.

Vector Bosons

As we saw, true 3D vectors change sign under parity. However, in QFT our vectors are 4D, and the 0th coordinate, the time-like coordinate, does not change sign. Our coordinate axes alignment for any vector is the same as that for the position vector, so if $(t, x^i) \rightarrow (t, -x^i)$, then for any vector V_μ, we have $(V_0, V_i) \rightarrow (V_0, -V_i)$. But note that $x^i = -x_i$, and hence, $V^i = -V_i$. This means, since the μ = 0 component does not change, but the μ = 1,2,3 components change sign, that

$$A_\mu(t,x^i) \overset{P}{\Rightarrow} A^\mu(t,-x^i) \qquad Z_\mu(t,x^i) \overset{P}{\Rightarrow} Z^\mu(t,-x^i) \qquad W_\mu^\pm(t,x^i) \overset{P}{\Rightarrow} W^{\pm\mu}(t,-x^i). \tag{9-128}$$

4-vector under P raises/lowers 4D index

If you have any trouble understanding why the argument x^i also changes sign, see Appendix B.

The inverse of (9-128), where the original vector is contravariant is, of course, also true.

Note our interaction terms with vector four-currents and spacetime inner products, do not change. For example,

Vector 4-current interaction terms unchanged under P, axial 4-current changes sign

$$\bar{\psi}_1\gamma^\mu\psi_2 A_\mu \overset{P}{\Rightarrow} \bar{\psi}_1\gamma_\mu\psi_2 A^\mu . \tag{9-129}$$

Interaction terms with axial four-currents will, via (9-125), change sign.

Partial Derivatives

Since, under parity, $(t, x^i) \to (t, -x^i)$, and partial derivatives ∂_μ are with respect to the 4D coordinates, we have

$$\partial_0 = \frac{\partial}{\partial t} \overset{P}{\Rightarrow} \partial_0 = \frac{\partial}{\partial t} = \partial^0 \qquad \partial_i = \frac{\partial}{\partial x^i} \overset{P}{\Rightarrow} -\partial_i = -\frac{\partial}{\partial x^i} = \frac{\partial}{\partial x_i} = \partial^i , \tag{9-130}$$

or, generally,

$$\partial_\mu \overset{P}{\Rightarrow} \partial^\mu \qquad \partial^\mu \overset{P}{\Rightarrow} \partial_\mu . \tag{9-131}$$

Derivatives under P raise/lower 4D index

However, note that terms in the Lagrangian with derivatives in them also flip the sign of x^i in ψ. For example, examining just one term in (9-95),

$$\begin{aligned}&\bar{\psi}(t,x^i)\gamma^\mu\partial_\mu\psi(t,x^i)\\ &\xrightarrow[\text{term}]{\text{one}} \bar{\psi}_{\mathbf{p}}^{(1)}(t,x^i)\gamma^\mu\partial_\mu\psi_{\mathbf{p}}^{(1)}(t,x^i) = \bar{u}_1(\mathbf{p})e^{ipx}(t,x^i)\gamma^\mu\partial_\mu u_1(\mathbf{p})e^{-ipx} \\ &= \bar{u}_1(\mathbf{p})e^{ipx}(t,x^i)\gamma^0\partial_0\, u_1(\mathbf{p})e^{-ipx} + \bar{u}_1(\mathbf{p})e^{ipx}(t,x^i)\gamma^i\partial_i\, u_1(\mathbf{p})e^{-i\left(E_{\mathbf{p}}t-p^ix^i\right)}.\end{aligned} \tag{9-132}$$

When we do a parity transformation, we have $\partial_i \to -\partial_i$, but also in the exponent $x^i \to -x^i$. As a result, (9-132) is unchanged. The general rule is that terms in $\mathcal{L}$ with derivatives in them will be unchanged under parity transformation.

Time Transformations, Part 1

For time transformations, $(t, x^i) \to (-t, x^i)$, but as noted above, there are two versions of time transformation, which we designate as $\hat{T}$ and T. We'll examine the simpler, $\hat{T}$, first.

Two types of time transformation, here labeled $\hat{T}$ and T

First, $\hat{T}$, the Simpler

Scalars

The simpler is $\hat{T}$ and we examine it first

As noted, we would expect the $\hat{T}$ transformation to change a particle/field into an antiparticle/field. So,

$\hat{T}$ on scalar fields

$$\phi(t,x^i) \overset{\hat{T}}{\Rightarrow} \phi^*(-t,x^i) = \phi^\dagger(-t,x^i) \qquad \phi^*(t,x^i) = \phi^\dagger(t,x^i) \overset{\hat{T}}{\Rightarrow} \phi(-t,x^i) \tag{9-133}$$

Spinors

As noted with (9-85), $\hat{T}$ switches particles/fields and antiparticles/fields, just like C. And it switches chirality, just like P. We saw that for spinors, these processes were accomplished by pre-multiplying the complex conjugate field by $-i\gamma^2$ as in (9-94) and by γ^0, as in (9-118). Since $\hat{T}$ does the same as CP, at least to within an arbitrary phase factor, we can take our time transformation as

$$\psi(t,x^i) \overset{\hat{T}}{\Rightarrow} \gamma^0\gamma^2\psi^*(-t,x^i). \tag{9-134}$$

$\hat{T}$ on spinor fields

Taking the complex conjugate of both sides of (9-134), or again, by definition, we have

$$\psi^* \overset{\hat{T}}{\Rightarrow} (\gamma^0)^*(\gamma^2)^*\psi = \gamma^0(-\gamma^2)\psi = -\gamma^0\gamma^2\psi = -\gamma^0\gamma^2(\psi^*)^*, \tag{9-135}$$

and from the transpose of (9-135)

$$\psi^{\dagger} \overset{\hat{T}}{\Rightarrow} \psi^{T}\left(\gamma^{2}\right)^{\dagger}\left(\gamma^{0}\right)^{\dagger} = \psi^{T}\left(-\gamma^{2}\right)\gamma^{0} = -\psi^{T}\gamma^{2}\gamma^{0} = -\left(\psi^{\dagger}\right)^{*}\gamma^{2}\gamma^{0}. \tag{9-136}$$

The transpose of (9-134) gives us

$$\psi^{T} \overset{\hat{T}}{\Rightarrow} \psi^{\dagger}\left(\gamma^{2}\right)^{T}\left(\gamma^{0}\right)^{T} = \psi^{\dagger}\gamma^{2}\gamma^{0} = \left(\psi^{T}\right)^{*}\gamma^{2}\gamma^{0}. \tag{9-137}$$

Spinor Inner Products

The scalar inner product of spinor fields is invariant under $\hat{T}$.

$$\begin{aligned}\bar{\psi}(t,x^i)\psi(t,x^i) &= \psi^{\dagger}(t,x^i)\gamma^0\psi(t,x^i) \overset{\hat{T}}{\Rightarrow} \left(-\psi^T\gamma^2\gamma^0\right)\gamma^0\left(\gamma^0\gamma^2\psi^*\right) = -\psi^T\gamma^2\gamma^0\gamma^2\psi^* \\ &= \psi^T\gamma^0\gamma^2\gamma^2\psi^* = -\psi^T\gamma^0\psi^* = \underbrace{\psi^{\dagger}\left(\gamma^0\right)^T\psi}_{\leftarrow\text{transposed}} = \psi^{\dagger}\gamma^0\psi = \bar{\psi}(-t,x^i)\psi(-t,x^i).\end{aligned} \tag{9-138}$$

$\hat{T}$ leaves scalar inner product unchanged

The pseudo-scalar inner product changes sign.

$$\begin{aligned}\bar{\psi}(t,x^i)\gamma^5\psi(t,x^i) &= \psi^{\dagger}(t,x^i)\gamma^0\gamma^5\psi(t,x^i) \overset{\hat{T}}{\Rightarrow} \left(-\psi^T\gamma^2\gamma^0\right)\gamma^0\gamma^5\left(\gamma^0\gamma^2\psi^*\right) \\ &= -\psi^T\gamma^2\gamma^5\gamma^0\gamma^2\psi^* = -\psi^T\gamma^2\gamma^2\gamma^5\gamma^0\psi^* = \psi^T\gamma^5\gamma^0\psi^* \\ &= \underbrace{-\psi^{\dagger}\left(\gamma^0\right)^T\left(\gamma^5\right)^T\psi}_{\text{prior relation transposed}} = -\psi^{\dagger}\gamma^0\gamma^5\psi = -\bar{\psi}(-t,x^i)\gamma^5\psi(-t,x^i).\end{aligned} \tag{9-139}$$

$\hat{T}$ changes sign on pseudo-scalar inner product

Vector Four-current

$$\begin{aligned}\bar{\psi}_1(t,x^i)\gamma^{\mu}\psi_2(t,x^i) = \psi_1^{\dagger}(t,x^i)\gamma^0\gamma^{\mu}\psi_2(t,x^i) &\overset{\hat{T}}{\Rightarrow} \left(-\psi_1^T\gamma^2\gamma^0\right)\gamma^0\gamma^{\mu}\left(\gamma^0\gamma^2\psi_2^*\right) \\ &= -\psi_1^T\gamma^2\gamma^{\mu}\gamma^0\gamma^2\psi_2^*.\end{aligned} \tag{9-140}$$

for $\mu = 1,3$

$$\begin{aligned}&= \psi_1^T\gamma^{\mu}\gamma^2\gamma^0\gamma^2\psi_2^* = -\psi_1^T\gamma^{\mu}\gamma^0\gamma^2\gamma^2\psi_2^* = \psi_1^T\gamma^{\mu}\gamma^0\psi_2^* \\ &= \underbrace{-\psi_2^{\dagger}\left(\gamma^0\right)^T\left(\gamma^{\mu}\right)^T\psi_1}_{\text{prior relation transposed}} = \underbrace{-\psi_2^{\dagger}\gamma^0\left(-\gamma^{\mu}\right)\psi_1}_{\text{stand rep}} = \underbrace{\bar{\psi}_2\gamma^{\mu}\psi_1}_{\text{all reps}}\end{aligned} \tag{9-140)a$$

for $\mu = 2$,

$$\begin{aligned}&= -\psi_1^T\gamma^2\gamma^2\gamma^0\gamma^2\psi_2^* = \psi_1^T\gamma^0\gamma^2\psi_2^* = -\psi_1^T\gamma^2\gamma^0\psi_2^* \\ &= \psi_2^{\dagger}\left(\gamma^0\right)^T\left(\gamma^2\right)^T\psi_1 = \psi_2^{\dagger}\gamma^0\gamma^2\psi_1 = \bar{\psi}_2\gamma^2\psi_1\end{aligned} \tag{9-140)b$$

for $\mu = 0$,

$$\begin{aligned}&= -\psi_1^T\gamma^2\gamma^0\gamma^0\gamma^2\psi_2^* = -\psi_1^T\gamma^0\gamma^2\gamma^2\gamma^0\psi_2^* = \psi_1^T\gamma^0\gamma^0\psi_2^* \\ &= -\psi_2^{\dagger}\left(\gamma^0\right)^T\left(\gamma^0\right)^T\psi_1 = -\psi_2^{\dagger}\gamma^0\gamma^0\psi_1 = -\bar{\psi}_2\gamma^0\psi_1.\end{aligned} \tag{9-140)c$$

For μ = 1,2,3, the vector four-current keeps the same sign. For μ = 0, it changes. We can coalesce all this into

$$\bar{\psi}_1(t,x^i)\gamma^{\mu}\psi_2(t,x^i) \overset{\hat{T}}{\Rightarrow} -\bar{\psi}_2(-t,x^i)\gamma_{\mu}\psi_1(-t,x^i). \tag{9-141}$$

$\hat{T}$ on four-currents

Axial Four-Current

We won't take the time to derive the $\hat{T}$ transformations for terms with γ^5 in them, as $\hat{T}$ is not prevalent in the literature, so it is generally not worth our while. I simply state it for the axial four-current, and note it is the same as the vector four-current.

$$\bar{\psi}_1(t,x^i)\gamma^{\mu}\gamma^5\psi_2(t,x^i) \overset{\hat{T}}{\Rightarrow} -\bar{\psi}_2(-t,x^i)\gamma_{\mu}\gamma^5\psi_1(-t,x^i). \tag{9-142}$$

Vector Bosons

In order to keep QED interaction terms like (9-110) invariant under $\hat{T}$, we will need to have four-vectors change sign for $\mu = 0$, but not for $\mu = 1,2,3$. This is accomplished via

$\hat{T}$ on four-vectors

$$A^{\mu}\left(t,x^i\right) \overset{\hat{T}}{\Rightarrow} -A_{\mu}\left(-t,x^i\right) \qquad Z^{\mu}\left(t,x^i\right) \overset{\hat{T}}{\Rightarrow} -Z_{\mu}\left(-t,x^i\right) \qquad W^{\pm\mu}\left(t,x^i\right) \overset{\hat{T}}{\Rightarrow} -W_{\mu}^{\mp}\left(-t,x^i\right). \tag{9-143}$$

Partial Derivatives

$$\partial_0 = \frac{\partial}{\partial t} \overset{\hat{T}}{\Rightarrow} -\partial_0 = -\frac{\partial}{\partial t} = -\partial^0 \qquad \partial_i = \frac{\partial}{\partial x^i} \overset{\hat{T}}{\Rightarrow} \partial_i = \frac{\partial}{\partial x^i} = -\frac{\partial}{\partial x_i} = -\partial^i, \tag{9-144}$$

so,

$\hat{T}$ on derivatives

$$\partial_{\mu} \overset{\hat{T}}{\Rightarrow} -\partial^{\mu} \qquad \partial^{\mu} \overset{\hat{T}}{\Rightarrow} -\partial_{\mu}. \tag{9-145}$$

<u>$CP\hat{T}$ Transformations</u>

$CP\hat{T}$ transformations

The summaries in Wholeness Chart 9-3, pg. 342, can help in the following. Note that for entities such as (9-142), where we lower an originally raised spacetime index, if the original relation had a lower index, we raise it, just as we show for the partial derivatives of (9-145).

Spinor Fields

So, $CP\hat{T}$ on ψ gives us

$$\begin{aligned}
\psi\left(t,x^i\right) &\overset{C}{\Rightarrow} -i\gamma^2\psi^*\left(t,x^i\right) \overset{P}{\Rightarrow} \gamma^0\left(-i\gamma^2\psi^*\left(t,-x^i\right)\right) = -i\gamma^0\gamma^2\psi^*\left(t,-x^i\right) \\
&\overset{\hat{T}}{\Rightarrow} \gamma^0\gamma^2\left(-i\gamma^0\gamma^2\psi^*\left(-t,-x^i\right)\right)^* = \gamma^0\gamma^2(-i)^*\left(\gamma^0\right)^*\left(\gamma^2\right)^*\psi\left(-t,-x^i\right) = i\gamma^0\gamma^2\gamma^0\left(-\gamma^2\right)\psi\left(-t,-x^i\right) \\
&= -i\gamma^0\gamma^2\gamma^0\gamma^2\psi\left(-t,-x^i\right) = i\gamma^0\gamma^0\gamma^2\gamma^2\psi\left(-t,-x^i\right) = -i\psi\left(-t,-x^i\right),
\end{aligned} \tag{9-146}$$

which apart from an arbitrary phase factor of $-i$, brings us back to the original spinor. We could have eliminated the phase change by defining (9-134) with an extra factor of i, but at least one other popular text takes it as stated above. So, in hopes of avoiding confusion on your part when, and if, you consult other texts, I have defined it as (9-134).

Vector Four-currents

Under $CP\hat{T}$, vector four-currents are invariant. Using (9-108), (9-124), and (9-141), we have

Vector and axial four-currents invariant under $CP\hat{T}$

$$\begin{gathered}
\bar{\psi}_1\gamma^{\mu}\psi_2 \overset{C}{\Rightarrow} -\bar{\psi}_2\gamma^{\mu}\psi_1 \overset{P}{\Rightarrow} -\bar{\psi}_2\gamma_{\mu}\psi_1 \overset{\hat{T}}{\Rightarrow} \bar{\psi}_1\gamma^{\mu}\psi_2 \\
\bar{\psi}_1\left(t,x^i\right)\gamma^{\mu}\psi_2\left(t,x^i\right) \overset{CP\hat{T}}{\Rightarrow} \bar{\psi}_1\left(-t,-x^i\right)\gamma^{\mu}\psi_2\left(-t,-x^i\right).
\end{gathered} \tag{9-147}$$

Axial Four-currents

Under $CP\hat{T}$, axial vector four-currents are invariant, as well. With (9-109), (9-125), and (9-142),

$$\begin{gathered}
\bar{\psi}_1\gamma^{\mu}\gamma^5\psi_2 \overset{C}{\Rightarrow} \bar{\psi}_2\gamma^{\mu}\gamma^5\psi_1 \overset{P}{\Rightarrow} -\bar{\psi}_2\gamma_{\mu}\gamma^5\psi_1 \overset{\hat{T}}{\Rightarrow} \bar{\psi}_1\gamma^{\mu}\gamma^5\psi_2 \\
\bar{\psi}_1\left(t,x^i\right)\gamma^{\mu}\gamma^5\psi_2\left(t,x^i\right) \overset{CP\hat{T}}{\Rightarrow} \bar{\psi}_1\left(-t,-x^i\right)\gamma^{\mu}\gamma^5\psi_2\left(-t,-x^i\right).
\end{gathered} \tag{9-148}$$

Four-vectors

Four-vectors are invariant, also. For example, with (9-112), (9-128), and (9-143),

Four-vectors invariant under $CP\hat{T}$

$$W_{\mu}^{+} \overset{C}{\Rightarrow} -W_{\mu}^{-} \overset{P}{\Rightarrow} -W^{-\mu} \overset{\hat{T}}{\Rightarrow} W_{\mu}^{+} \qquad W_{\mu}^{+}\left(t,x^i\right) \overset{CP\hat{T}}{\Rightarrow} W_{\mu}^{+}\left(-t,-x^i\right). \tag{9-149}$$

Interaction Terms

Given (9-147) to (9-149), interaction terms under $CP\hat{T}$ are invariant,. For example,

$$\underbrace{-\frac{g}{\sqrt{2}}\bar{\psi}^L_{\nu_e}\gamma^\mu\psi^L_e W^+_\mu}_{\text{all arguments }(t,x^i)} \overset{CP\hat{T}}{\Rightarrow} \underbrace{-\frac{g}{\sqrt{2}}\bar{\psi}^L_{\nu_e}\gamma^\mu\psi^L_e W^+_\mu}_{\text{all arguments }(-t,-x^i)} \tag{9-150}$$

Interaction terms invariant under $CP\hat{T}$

Free Field Terms

From (9-105), (9-122), and (9-138), we see that mass terms, i.e., terms of form $m\bar{\psi}\psi$ are invariant under any one of C, P, or $\hat{T}$. So, they are invariant under $CP\hat{T}$.

Mass terms invariant under $CP\hat{T}$

For terms with derivatives, from (9-113), (9-131), and (9-145), we see that the partial derivative changes sign under $CP\hat{T}$.

$$\partial_\mu \overset{C}{\Rightarrow} \partial_\mu \overset{P}{\Rightarrow} \partial^\mu \overset{\hat{T}}{\Rightarrow} -\partial_\mu \qquad \partial_\mu \overset{CP\hat{T}}{\Rightarrow} -\partial_\mu\,. \tag{9-151}$$

But since x^μ becomes $-x^\mu$ under $CP\hat{T}$ in the argument of ψ, the derivative of ψ is invariant. That is,

$$\partial_\mu\psi\left(x^\nu\right) \overset{CP\hat{T}}{\Rightarrow} \left(-\partial_\mu\right)\psi\left(-x^\nu\right) = \partial_\mu\psi\left(x^\nu\right) \tag{9-152}$$

Kinetic terms invariant under $CP\hat{T}$ if account made for $x^\mu \rightarrow -x^\mu$ in ψ

So, in this context, free field derivative terms can be considered invariant under $CP\hat{T}$.

$$\bar{\psi}\gamma^\mu\partial_\mu\psi \overset{CP\hat{T}}{\Rightarrow} \bar{\psi}\gamma^\mu\partial_\mu\psi\,. \tag{9-153}$$

The same effect occurs for derivatives of boson fields. So, terms with such fields and their derivatives are invariant, too. Beyond that, free boson terms are generally of bilinear form with derivatives in each factor, for example, $\partial_\mu A_\nu \partial^\mu A^\nu$, so we also get sign cancellations that way.

Bottom line for $CP\hat{T}$: The Lagrangian is invariant under $CP\hat{T}$ transformation.

$\mathcal{L}$ invariant under $CP\hat{T}$

Inverses of C, P, and $\hat{T}$

Note that since C, P, and $\hat{T}$ are discrete symmetries, two successive operations of any one of them leaves us what we started with. For example, x^i under P becomes $-x^i$, and $-x^i$ under P becomes x^i. Each of C, P, and $\hat{T}$ is its own inverse. The same will be true of T, the second type of time transformation, when we get to it, as well as any combination of these operations such as CP.

Closer Look at Interactions

An Example

In order to keep things as simple and brief as possible, consider the transition amplitude S_{fi} for only the first vertex of (9-84) under a CP transformation.

Example of CP transformation on states invariance

$$e^{-L} + \bar{\nu}^R_e \rightarrow W^- \overset{CP}{\Rightarrow} e^{+R} + \nu^L_e \rightarrow W^+, \tag{9-154}$$

The transition amplitude, at lowest order, is found in the manner of Vol. 1 (8-12), where here we use the electroweak Lagrangian instead of the QED one we used there, and N means normal ordering. The relevant Lagrangian terms are shown herein in Chap. 7, (7-64). So, we have

$$\begin{aligned} S_{fi} &= \langle W^-|S_{oper}|e^{-L},\bar{\nu}^R_e\rangle \xrightarrow[\text{order}]{\text{lowest}} \langle W^-|S^{(1)}|e^{-L},\bar{\nu}^R_e\rangle = \langle W^-|(-i)\int d^4x_1 N\{-\mathcal{L}_I\}_{x_1}|e^{-L},\bar{\nu}^R_e\rangle \\ &= S^{(1)}_{fi} = i\langle W^-|\int d^4x_1 N\{\ldots \underbrace{-\frac{g}{\sqrt{2}}\bar{\psi}^L_{\nu_e}W\!\!\!/^{+}\psi^L_e}_{\boxed{1}} \underbrace{-\frac{g}{\sqrt{2}}\bar{\psi}^L_e W\!\!\!/^{-}\psi^L_{\nu_e}}_{\boxed{2}} + \ldots\}_{x_1}|e^{-L},\bar{\nu}^R_e\rangle. \end{aligned} \tag{9-155}$$

Only term $\boxed{1}$ survives, and it gives us

$$S^{(1)}_{fi} = i\langle W^-|\int d^4x_1 N\left\{-\frac{g}{\sqrt{2}}\bar{\psi}^L_{\nu_e}W\!\!\!/^{+}\psi^L_e\right\}_{x_1}|e^{-L},\bar{\nu}^R_e\rangle = -i\frac{g}{\sqrt{2}}(const)\langle W^-\|W^-\rangle = -i\frac{g}{\sqrt{2}}(const)\,. \tag{9-156}$$

Now, let's consider transforming the bra and ket states under CP, as in (9-154).

$$CP \text{ of } S_{fi} \xrightarrow[\text{order}]{\text{lowest}} i\langle W^+|\int d^4x_1 N\{\ldots \underbrace{-\frac{g}{\sqrt{2}}\bar{\psi}^L_{\nu_e}W\!\!\!/^{+}\psi^L_e}_{\boxed{1}} \underbrace{-\frac{g}{\sqrt{2}}\bar{\psi}^L_e W\!\!\!/^{-}\psi^L_{\nu_e}}_{\boxed{2}} + \ldots\}_{x_1}|e^{+R},\nu^L_e\rangle. \tag{9-157}$$

Now, only term [2] survives, and gives us, with the same (*const*) value as before,

$$CP \text{ of } S_{fi} \xrightarrow[\text{order}]{\text{lowest}} -i\langle W^+ | \int d^4x_1 N\left\{\frac{g}{\sqrt{2}}\bar{\psi}_e^L W^- \psi_{\nu_e}^L\right\}_{x_1} |e^{+R}, \nu_e^L\rangle$$

$$S_{fi}^{(1)} \overset{CP}{\Rightarrow} -i\frac{g}{\sqrt{2}}(const)\langle W^+ || W^+\rangle = -i\frac{g}{\sqrt{2}}(const). \tag{9-158}$$

Hence, the transition amplitude, and thus, the probability of interaction, remains unchanged when we transform the states.

Now, instead of transforming the states, consider transforming the Lagrangian.

Same example, but CP transformation on fields

Do **Problem 12** to find the CP transformation of term [1] in (9-155).

With the result of Problem 12, and with that, intuiting the CP transform of term [2], or by simply using the last two rows of the CP column in Wholeness Chart 9-3, we find

$$CP \text{ of } S_{fi} = \langle W^- | CP \text{ of } S_{oper} | e^{-L}, \bar{\nu}_e^R\rangle \xrightarrow[\text{order}]{\text{lowest}} \langle W^- | CP \text{ of } \{S^{(1)}\} | e^{-L}, \bar{\nu}_e^R\rangle$$

$$S_{fi}^{(1)} \overset{CP}{\Rightarrow} i\langle W^- | \int d^4x_1 N\{\ldots \underbrace{-\frac{g}{\sqrt{2}}\bar{\psi}_e^L W^- \psi_{\nu_e}^L}_{CP \text{ of } [1] = [1']} \underbrace{-\frac{g}{\sqrt{2}}\bar{\psi}_{\nu_e}^L W^+ \psi_e^L}_{CP \text{ of } [2] = [2']} + \ldots\}_{x_1} |e^{-L}, \bar{\nu}_e^R\rangle. \tag{9-159}$$

Note that the original [1] CP transformed equals the original [2]; and vice versa. Each of [1] and [2] becomes the other. And that means this pair of terms in the Lagrangian is invariant under CP.

So, now consider (9-159), for which only the CP of [2] (i.e., [2']) survives, and which has become the same as (9-156),

$$CP \text{ of } S_{fi} \xrightarrow[\text{order}]{\text{lowest}} i\langle W^- | \int d^4x_1 N\left\{-\frac{g}{\sqrt{2}}\bar{\psi}_{\nu_e}^L W^+ \psi_e^L\right\}_{x_1} |e^{-L}, \bar{\nu}_e^R\rangle$$

$$S_{fi}^{(1)} \overset{CP}{\Rightarrow} -i\frac{g}{\sqrt{2}}(const)\langle W^- || W^-\rangle = -i\frac{g}{\sqrt{2}}(const). \tag{9-160}$$

CP transformation on fields has same result as CP on states

The transition amplitude has not changed under the transformation because the Lagrangian is invariant under the transformation. Although we have only looked at one vertex interaction term, the result holds to all orders, as higher orders simply contain more vertex interaction terms multiplied together and linked via contractions.

Bottom line for this example:

Two key results

1) We obtain the same transformed transition amplitude, whether we transform the states or the Lagrangian. Either method works, though the latter is easier.

$$CP \text{ of } S_{fi} = \langle CP \text{ of} f | S_{oper} | CP \text{ of } i\rangle = \langle f | CP \text{ of } S_{oper} | i\rangle \tag{9-161}$$

2) If the Lagrangian is invariant under a CP transformation, then the transformed transition amplitude equals the original transition amplitude.

$$CP \text{ of } S_{fi} = \text{ original } S_{fi} \text{ if } \mathcal{L} \text{ is } CP \text{ invariant}. \tag{9-162}$$

Theory

We can generalize the above results.

Generalizing results to Θ, any combination of C, P, and $\hat{T}$ (or T)

C, P, and $\hat{T}$ (and T, which we haven't examined yet) can be considered as operations on fields. They can also be considered, as we did above, as operations on states, i.e., on kets and bras, in the context of states as vectors in a Hilbert space[1]. Those states, as we know, are operated on by fields

---

[1] There can be some confusion as to whether we take our fields (such as ψ) or our states (such as $|e^-\rangle$) as vectors in Hilbert space. It actually can be either, so one needs to be careful about context. In one context, the fields are considered Hilbert space vectors so operators (transformations), symbolized here by $\mathcal{O}$, operate on vectors in that space as in $\mathcal{O}\psi$. The ψ, in turn, operate on separate entities, the quantum states (to create and destroy particles in those states). In the second context, the quantum states are taken as vectors in Hilbert space, and the operators (transformations), symbolized by Θ, act directly on those states, as in $\Theta|e^-\rangle$. In that second context, there are still the operators ψ, but if we want to transform them under Θ, then we need to do so via $\Theta\psi\Theta^{-1}$. It is that second

such as ψ and A_μ. In our present context, they can also be operated on by charge, parity, and time operations. We will use the generic symbol Θ to symbolize any one, or a combination of any two or three, of those three operators on states. Thus, for example,

Θ operating on a state

$$\text{for }\Theta\text{ as charge operator}\qquad \Theta\left|e^{-L}\right\rangle=\left|e^{+L}\right\rangle. \tag{9-163}$$

In the context of Hilbert space, a charge operation on the field ψ, which operates on a state in that space, is

Θ operating on an operator that operates on states

$$\text{for }\Theta\text{ as charge operator on states,}\quad\text{field }\psi^c=\Theta\psi\Theta^{-1}\qquad \bar{\psi}^c\psi^c=\Theta\bar{\psi}\Theta^{-1}\Theta\psi\Theta^{-1}. \tag{9-164}$$

In our earlier context, we had this represented by (9-94) and (9-105).

Now, from Vol. 1, (7-62) and (7-66), we know the transition amplitude for a given interaction is

$$S_{fi}=\langle f|S|i\rangle=\left\langle f\left|e^{-i\int\mathcal{H}_I^I d^4x}\right|i\right\rangle=\left\langle f\left|e^{i\int\mathcal{L}_I d^4x}\right|i\right\rangle \tag{9-165}$$

We ask what happens to the transition amplitude if we change the incoming and outgoing particles by charge conjugation, parity switch, and/or time transformation, represented by the Θ operation. Let's use the example of (9-84) for CP transformation.

$$S_{fi}=\left\langle \mu^{-L},\bar{\nu}_\mu^R\left|e^{i\int\mathcal{L}_I d^4x}\right|e^{-L},\bar{\nu}_e^R\right\rangle$$

$$\underset{\Rightarrow}{CP\text{ on states}}\quad S'_{fi}=\left\langle \mu^{-L},\bar{\nu}_\mu^R\left|\Theta^{-1}S\,\Theta\right|e^{-L},\bar{\nu}_e^R\right\rangle=\left\langle \mu^{+R},\nu_\mu^L\left|e^{i\int\mathcal{L}_I d^4x}\right|e^{+R},\nu_e^L\right\rangle. \tag{9-166}$$

Θ on states in a transition amplitude

If $S'_{fi}=S_{fi}$, then the CP transformed state interaction has the same probability as the original interaction.

Conversely, consider what happens if we perform the CP transformation on S instead of the state.

Θ on fields in S in a transition amplitude

$$\underset{\Rightarrow}{CP\text{ on }S}\quad S'_{fi}=\left\langle \mu^{-L},\bar{\nu}_\mu^R\left|\Theta S\Theta^{-1}\right|e^{-L},\bar{\nu}_e^R\right\rangle. \tag{9-167}$$

Same transition amplitude either way

Now, Θ here represents CP, though more generically, it is a symbol that stands for any one, or a combination of any two or three of the C, P and $\hat{T}$ (or T) operators, each of which, if performed twice, gives back what we had originally. Thus, as noted two pages earlier, each such operator is its own inverse, so, $\Theta\,\Theta=I$ and $\Theta^{-1}\,\Theta^{-1}=I$. Inserting these identities, respectively, right before the ket and right after the bra in (9-167), we get (9-166), i.e., the two equations are the same thing. So, we conclude that CP transformation of the operator S, which is comprised of fields, gives us the same transition amplitude (and thus, probability) as CP transformation of the states. We can do either one, whichever is easier.

Now note that

$$S=e^{i\int\mathcal{L}_I d^4x}=I+i\int\mathcal{L}_I d^4x_1-\tfrac{1}{2!}T\left\{\left(\int\mathcal{L}_I d^4x_1\right)\left(\int\mathcal{L}_I d^4x_2\right)\right\}-\dots \tag{9-168}$$

$$\Theta S\Theta^{-1}=\Theta I\Theta^{-1}+i\int\Theta\mathcal{L}_I\Theta^{-1}d^4x_1-\tfrac{1}{2!}T\left\{\left(\int\Theta\mathcal{L}_I\Theta^{-1}d^4x_1\right)\left(\int\Theta\mathcal{L}_I\Theta^{-1}d^4x_2\right)\right\}-\dots \tag{9-169}$$

(9-169) equals (9-168) if $\mathcal{L}$ is symmetric under the Θ transformation. For the case of the (9-84) interaction under CP, this is true, so the transition amplitude for the original and transformed interactions is the same, and they occur with equal probability. The conclusion holds for any Θ and any interaction.

If $\mathcal{L}$ symmetric in Θ, transition amplitude is also

Bottom line in general: We have the same general conclusions as we did for the "Bottom line for this example" above, (9-161) and (9-162). Further, it is easier to examine the Lagrangian for symmetry than the states, so that is what we will do and what is invariably done in the literature.

General conclusions same as earlier special case example

1) Θ transformation of $S_{fi}=\langle\Theta\text{ of }f|S|\Theta\text{ of }i\rangle=\langle f|\Theta\text{ of }S|i\rangle$, (9-170)

2) Θ transformation of $S_{fi}=$ original S_{fi} if $\mathcal{L}$ symmetric under Θ. (9-171)

Easiest to check symmetry of $\mathcal{L}$ under Θ

---

context that we employ in this section. In the rest of this chapter, we use the first context for C, P, and $\hat{T}$ (and T) operations. One needs to learn to be somewhat agile in flipping back and forth between these contexts, as both are used commonly in the literature. (And as one last, hopefully not too confusing point, to be precise, the states are actually in Fock space, a generalization of Hilbert space.) For more on this, see Sect. 2.8.15, pg. 58a and Wholeness Chart 2-13, pg. 58b, which were added in the July 2025 revision of this text.

Example of *CP* Violation

CP violation example, related to CKM matrix

As an example where *CP* is not invariant, consider the CKM matrix V terms in the true vacuum electroweak Lagrangian, pg. 251, (7-146).

$$\mathcal{L} = \ldots \underbrace{-\frac{g}{\sqrt{2}}\bar{\psi}^L_{q_u}\gamma^\alpha V_{q_u q_d}\psi^L_{q_d}W^+_\alpha}_{\boxed{3}} \underbrace{-\frac{g}{\sqrt{2}}\bar{\psi}^L_{q_d}\gamma^\alpha V^\dagger_{q_d q_u}\psi^L_{q_u}W^-_\alpha}_{\boxed{4}} + \ldots \qquad (9\text{-}172)$$

Now, transform terms $\boxed{3}$ and $\boxed{4}$ above under *CP*. Wholeness Chart 9-3 can help, particularly the last two rows.

$$\boxed{3} = -\frac{g}{\sqrt{2}}\bar{\psi}^L_{q_u}\gamma^\alpha V_{q_u q_d}\psi^L_{q_d}W^+_\alpha \overset{CP}{\Rightarrow} -\frac{g}{\sqrt{2}}\bar{\psi}^L_{q_d}\gamma^\alpha V_{q_u q_d}\psi^L_{q_u}W^-_\alpha = -\frac{g}{\sqrt{2}}\bar{\psi}^L_{q_d}\gamma^\alpha V^T_{q_d q_u}\psi^L_{q_u}W^-_\alpha = \boxed{3'}$$

$$\boxed{4} = -\frac{g}{\sqrt{2}}\bar{\psi}^L_{q_d}\gamma^\alpha V^\dagger_{q_d q_u}\psi^L_{q_u}W^-_\alpha \overset{CP}{\Rightarrow} -\frac{g}{\sqrt{2}}\bar{\psi}^L_{q_u}\gamma^\alpha V^\dagger_{q_d q_u}\psi^L_{q_d}W^+_\alpha = -\frac{g}{\sqrt{2}}\bar{\psi}^L_{q_u}\gamma^\alpha V^*_{q_u q_d}\psi^L_{q_d}W^+_\alpha = \boxed{4'}. \qquad (9\text{-}173)$$

Imaginary part of CKM matrix leads to CP violation

$\boxed{3'}$ only equals $\boxed{4}$ if the *CKM* matrix V is real. Ditto for $\boxed{4'}$ and $\boxed{3}$. Recall from Chap. 7, (7-119), that the CKM matrix has four real variables, but one of those, α, occurs in the complex exponential $e^{i\alpha}$, a factor in several components of the matrix. So, we would expect some *CP* violation to occur in quark electroweak interactions, because the Lagrangian is not *CP* invariant in these quark terms.

And this does turn up in experiments, perhaps most notably in *K* meson interactions, where it was first discovered in 1964, and for which a Nobel Prize was awarded in 1980 to James Cronin and Val Fitch. *K* mesons, also called Kaons, comprise a bound state of a strange quark (or antiquark) and an up or down antiquark (or quark).

CP violation occurs, but rare, and result of generation changing

Bottom line: *CP* violation does occur in nature, but it is rare. It is evidenced by generation changing at a vertex, as governed for quarks by the CKM matrix. If that matrix were real, there would be no *CP* violation, at least via this route. But the CKM matrix is not real.

Time Transformations, Part 2

The other, more relevant time transformation T

Now, *T*, the More Common Time Transformation

We now, finally, examine the other time transformation *T*, the more prevalent one in the literature.

Non-mathematical, Physical View of T and CPT

1st, physical look at T

T does not switch particles and antis, but switches initial and final

The $\hat{T}$ transformation switched particles and antiparticles, as well as *LC* and *RC*, but kept the time order of an interaction the same, i.e., incoming particles remain incoming, and outgoing remain outgoing. See top right part of Fig. 9-2. The *T* transformation, on the other hand, keeps particles as particles (and antiparticles as antiparticles) and maintains the same chirality, but reverses the time order of the interaction. For the interaction (9-85), under *T*, instead of $\hat{T}$ (See Fig. 9-2, bottom right),

Example of T on an interaction

$$e^{-L} + \bar{\nu}^R_e \rightarrow W^- \rightarrow \mu^{-L} + \bar{\nu}^R_\mu \overset{T}{\Rightarrow} \mu^{-L} + \bar{\nu}^R_\mu \rightarrow W^- \rightarrow e^{-L} + \bar{\nu}^R_e. \qquad (9\text{-}174)$$

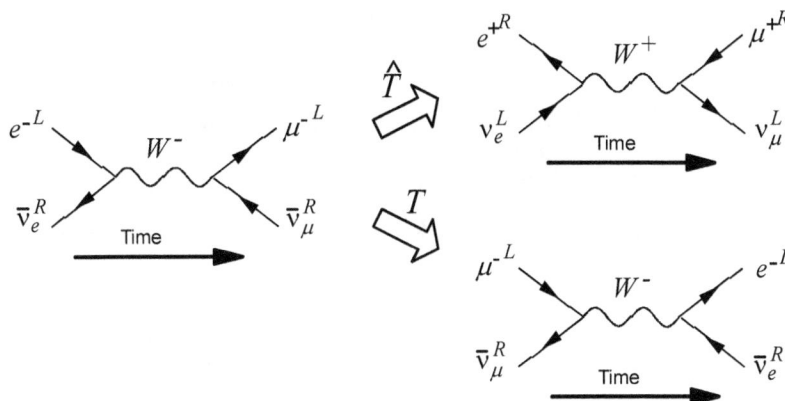

Diagram comparison of T and $\hat{T}$

Figure 9-2. Transforming an Interaction under *T* and $\hat{T}$

Under T, incoming particles switch with outgoing particles, concomitant with 3-momenta changing sign for all particles, as they are headed in opposite directions. Spin is also reversed, which can be visualized physically as ω of Fig. 9-1, pg. 341, having opposite sign, as time is going backward. Hence, helicity is unchanged, which implies chirality is, as well. The whole process is like watching a movie of the original interaction played in reverse.

T: $\mathbf{p}_i \rightarrow -\mathbf{p}_i$, spin reverses, helicity unchanged; chirality unchanged

Given T, under CPT, instead of (9-86), we have

Same example, now under CPT

$$\begin{aligned} & e^{-L} + \bar{\nu}_e^R \rightarrow W^- \rightarrow \mu^{-L} + \bar{\nu}_\mu^R \\ & \overset{CP}{\Rightarrow} \; e^{+R} + \nu_e^L \rightarrow W^+ \rightarrow \mu^{+R} + \nu_\mu^L \;\overset{T}{\Rightarrow}\; \mu^{+R} + \nu_\mu^L \rightarrow W^+ \rightarrow e^{+R} + \nu_e^L, \end{aligned} \tag{9-175}$$

as depicted in Fig. 9-3.

Diagram comparison of CPT and CP$\hat{T}$

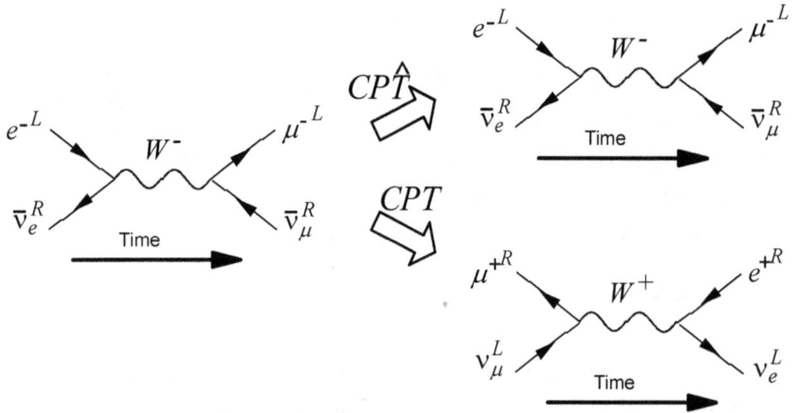

Figure 9-3. Transforming an Interaction under *CPT* and *CP$\hat{T}$*

Invariance under $CP\hat{T}$ is actually trivial, as we just get the original interaction back. Invariance under CPT is quite a different matter. It was first posited by Eugene Wigner in 1932. Rigorous proofs can be found in Streater and Wightman[1], and Greaves and Thomas[2], though they are not for the mathematically faint-hearted. Extensive and valuable discussions, for those wishing to pursue the matter more deeply than we do herein, can be found dispersed throughout several sections of both Ticciati[3] and Weinberg[4].

$CP\hat{T}$ gives original interaction back; CPT does not, and is more profound

In essence, what CPT invariance says is that if we exchange particles with antiparticles, left chirality with right, and reverse the time order (of the external particles) of an interaction, we will get the same transition amplitude, and thus, the same probability of interaction. CPT invariance has never been seen to be violated in any experiments. And theory, as noted above, predicts it to be true of any quantum field theory.

CPT symmetry never seen violated in experiment

Mathematical Look at *T* Transformations

Now, math look at T

A key part of the T transformation is $i \rightarrow -i$ (i.e., complex conjugation), in all factors of all terms in the Lagrangian. This, as it will turn out, makes things like spin work out correctly. We also need to remember to take $\mathbf{p} \rightarrow -\mathbf{p}$.

Scalars

Scalars are unchanged under T.

$$\phi(t,x^i) \overset{T}{\Rightarrow} \phi(-t,x^i) \qquad \phi^*(t,x^i) \overset{T}{\Rightarrow} \phi^*(-t,x^i)$$
$$\text{plus } i \rightarrow -i,\ \mathbf{k} \rightarrow -\mathbf{k} \qquad \text{plus } i \rightarrow -i,\ \mathbf{k} \rightarrow -\mathbf{k} \tag{9-176}$$

Scalars under T: note $t \rightarrow -t$; $i \rightarrow -i$, $\mathbf{k} \rightarrow -\mathbf{k}$

[1] R. F. Streater and A. S. Wightman, *PCT, Spin and Statistics, and All That*. Princeton University Press (1989).
[2] H. Greaves and T. Thomas, The CPT Theorem, *Studies in* History *and Philosophy of Modern Physics,* 45 (2014), 46-66, https://arxiv.org/abs/1204.4674.
[3] R. Ticciati, *Quantum Field Theory for Mathematicians*, Cambridge University Press (1999).
[4] S. Weinberg, *The Quantum Theory of Fields*, *Volume 1*, Cambridge University Press (1995).

Spinors

We first simply present the T transformation for spinors, and after that, justify it.

$$\underbrace{\psi\left(t,x^i\right)\overset{T}{\Rightarrow}\gamma^1\gamma^3\psi\left(-t,x^i\right)}_{\text{plus } i\to -i,\ \mathbf{p}\to-\mathbf{p}}. \tag{9-177}$$

Spinors under T

Note the change in sign for all factors containing an i. To justify the above, first determine the form of the Dirac matrices of (9-177) in the standard rep,

$$\gamma^1\gamma^3=\begin{bmatrix} & & & 1\\ & & 1 & \\ & -1 & & \\ -1 & & & \end{bmatrix}\begin{bmatrix} & & 1 & \\ & & & -1\\ -1 & & & \\ & 1 & & \end{bmatrix}=\begin{bmatrix} & 1 & & \\ -1 & & & \\ & & & 1\\ & & -1 & \end{bmatrix}. \tag{9-178}$$

Now, consider the action of (9-178) on ψ, as in (9-177), where

Showing action of T on spinors

$$\begin{aligned}\psi&=\sum_{\mathbf{p}}\sqrt{\frac{m}{VE_{\mathbf{p}}}}\left(c_1(\mathbf{p})\psi_{\mathbf{p}}^{(1)}+c_2(\mathbf{p})\psi_{\mathbf{p}}^{(2)}+d_2^\dagger(\mathbf{p})\psi_{\mathbf{p}}^{(3)}+d_1^\dagger(\mathbf{p})\psi_{\mathbf{p}}^{(4)}\right)\\ &=\sum_{\mathbf{p}}\sqrt{\frac{m}{VE_{\mathbf{p}}}}\left(c_1(\mathbf{p})u_1(\mathbf{p})e^{-ipx}+c_2(\mathbf{p})u_2(\mathbf{p})e^{-ipx}+d_1^\dagger(\mathbf{p})v_1(\mathbf{p})e^{ipx}+d_2^\dagger(\mathbf{p})v_2(\mathbf{p})e^{ipx}\right).\end{aligned} \tag{9-179}$$

For given $\mathbf{p}$, the first term in (9-179) under T, in the standard rep, is

$$\begin{aligned}T\left(c_1(\mathbf{p})\psi_{\mathbf{p}}^{(1)}\right)&=\underbrace{\gamma^1\gamma^3\left(c_1(\mathbf{p})u_1(\mathbf{p})e^{-i(Et-\mathbf{p}\cdot\mathbf{x})}\right)}_{\text{plus } t\to-t,\ i\to-i,\ \mathbf{p}\to-\mathbf{p}}=\begin{bmatrix} & 1 & & \\ -1 & & & \\ & & & 1\\ & & -1 & \end{bmatrix}c_1(-\mathbf{p})\sqrt{\frac{E+m}{2m}}\begin{pmatrix}1\\0\\-\dfrac{p^3}{E+m}\\-\dfrac{p^1-ip^2}{E+m}\end{pmatrix}e^{i(-Et+\mathbf{p}\cdot\mathbf{x})}\\ &=c_1(-\mathbf{p})\underbrace{\sqrt{\frac{E+m}{2m}}\begin{pmatrix}0\\-1\\-\dfrac{p^1-ip^2}{E+m}\\ \dfrac{p^3}{E+m}\end{pmatrix}}_{-u_2(\mathbf{p})}e^{-i(Et-\mathbf{p}\cdot\mathbf{x})}=-c_1(-\mathbf{p})u_2(\mathbf{p})e^{-ipx}=-c_1(-\mathbf{p})\psi_{\mathbf{p}}^{(2)}.\end{aligned} \tag{9-180}$$

The T transformation (9-177) flips the spin from u_1 to u_2, where again, the minus sign in front is an insignificant phase factor. It also reverses 3-momentum. Note that since $i\to -i$, $t\to -t$, and $\mathbf{p}\to-\mathbf{p}$, the exponent remains unchanged.

T reverses spin and 3-momentum, as noted physically before

So, as stated earlier, the T transformation, at least for the first term in (9-179), reverses both 3-momentum and spin. We will not go through the tedium of showing it for the other three terms, but just accept that the same conclusion would be drawn, were we to do so.

It remains to be seen if T, as defined in (9-177), also exchanges incoming particles with outgoing. We will deduce this shortly.

Similar to what we have done before for C, P, and $\hat{T}$ transformations, from (9-177), we find from complexification of (9-177), followed by transposing the prior relations,

$$\psi^*\overset{T}{\Rightarrow}\left(\gamma^1\right)^*\left(\gamma^3\right)^*\psi^*=\gamma^1\gamma^3\psi^* \tag{9-181}$$

$$\psi^\dagger\overset{T}{\Rightarrow}\psi^\dagger\left(\gamma^3\right)^T\left(\gamma^1\right)^T=\psi^\dagger\left(-\gamma^3\right)\left(-\gamma^1\right)=\psi^\dagger\gamma^3\gamma^1 \tag{9-182}$$

$$\psi^T\overset{T}{\Rightarrow}\psi^T\left(\gamma^3\right)^*\left(\gamma^1\right)^*=\psi^T\gamma^3\gamma^1, \tag{9-183}$$

where we need to remember here and whenever we use the T transformation, that $i\to -i$ and $\mathbf{p}\to-\mathbf{p}$ are also included in (9-181) to (9-183).

Gamma Matrices

Changing $i \rightarrow -i$ in T applies to each entire term in the Lagrangian, not just the fields. This is unlike the other discrete transformations we have studied, as they have no effect on constants, whereas T does. This means that

$$\gamma^{\mu} \overset{T}{\Rightarrow} \left(\gamma^{\mu}\right)^{*} \qquad (9\text{-}184)$$

Gamma matrices under T change, since i → −i

Spinor Inner Products

Again, we find the (scalar) inner product to be invariant under T.

$$\begin{aligned}\bar{\psi}\psi = \psi^{\dagger}\gamma^{0}\psi &\overset{T}{\Rightarrow} \left(\psi^{\dagger}\gamma^{3}\gamma^{1}\right)\left(\gamma^{0}\right)^{*}\left(\gamma^{1}\gamma^{3}\psi\right) = \psi^{\dagger}\gamma^{3}\gamma^{1}\gamma^{0}\gamma^{1}\gamma^{3}\psi \\ &= \psi^{\dagger}\gamma^{0}\gamma^{3}\gamma^{1}\gamma^{1}\gamma^{3}\psi = -\psi^{\dagger}\gamma^{0}\gamma^{3}\gamma^{3}\psi = \psi^{\dagger}\gamma^{0}\psi = \bar{\psi}\psi.\end{aligned} \qquad (9\text{-}185)$$

(Spinor) inner product under T

Do **Problem 13** to show (9-186).

The pseudo-scalar transformation is

$$\bar{\psi}_1\gamma^{5}\psi_2 \overset{T}{\Rightarrow} \bar{\psi}_1\gamma^{5}\psi_2 . \qquad (9\text{-}186)$$

Pseudo-scalar inner product under T

Vector Four-currents

The γ^{μ} therein are all real in the standard rep, except when μ=2, and then, $(\gamma^2)^* = -\gamma^2$.

$$\begin{aligned}\bar{\psi}_1\gamma^{\mu}\psi_2 = \psi_1^{\dagger}\gamma^{0}\gamma^{\mu}\psi_2 &\overset{T}{\Rightarrow} \left(\psi_1^{\dagger}\gamma^{3}\gamma^{1}\right)\left(\gamma^{0}\right)^{*}\left(\gamma^{\mu}\right)^{*}\left(\gamma^{1}\gamma^{3}\psi_2\right) = \psi_1^{\dagger}\gamma^{3}\gamma^{1}\gamma^{0}\left(\gamma^{\mu}\right)^{*}\gamma^{1}\gamma^{3}\psi_2 \\ &= \psi_1^{\dagger}\gamma^{0}\gamma^{3}\gamma^{1}\left(\gamma^{\mu}\right)^{*}\gamma^{1}\gamma^{3}\psi_2 = \bar{\psi}_1\gamma^{3}\gamma^{1}\left(\gamma^{\mu}\right)^{*}\gamma^{1}\gamma^{3}\psi_2 .\end{aligned} \qquad (9\text{-}187)$$

For $\mu = 0$

$$(9\text{-}187) = \bar{\psi}_1\gamma^{3}\gamma^{1}\gamma^{0}\gamma^{1}\gamma^{3}\psi_2 = -\bar{\psi}_1\gamma^{3}\gamma^{0}\gamma^{1}\gamma^{1}\gamma^{3}\psi_2 = \bar{\psi}_1\gamma^{3}\gamma^{0}\gamma^{3}\psi_2 = -\bar{\psi}_1\gamma^{0}\gamma^{3}\gamma^{3}\psi_2 = \bar{\psi}_1\gamma^{0}\psi_2 . \quad (9\text{-}188)$$

For $\mu = 1$ $\quad (9\text{-}187) = \bar{\psi}_1\gamma^{3}\gamma^{1}\gamma^{1}\gamma^{1}\gamma^{3}\psi_2 = -\bar{\psi}_1\gamma^{3}\gamma^{1}\gamma^{3}\psi_2 = \bar{\psi}_1\gamma^{1}\gamma^{3}\gamma^{3}\psi_2 = -\bar{\psi}_1\gamma^{1}\psi_2 \qquad (9\text{-}189)$

For $\mu = 2$,

$$\begin{aligned}(9\text{-}187) &= \bar{\psi}_1\gamma^{3}\gamma^{1}\left(-\gamma^{2}\right)\gamma^{1}\gamma^{3}\psi_2 = \bar{\psi}_1\gamma^{3}\gamma^{2}\gamma^{1}\gamma^{1}\gamma^{3}\psi_2 = -\bar{\psi}_1\gamma^{3}\gamma^{2}\gamma^{3}\psi_2 . \\ &= \bar{\psi}_1\gamma^{2}\gamma^{3}\gamma^{3}\psi_2 = -\bar{\psi}_1\gamma^{2}\psi_2 .\end{aligned} \qquad (9\text{-}190)$$

For $\mu = 3$, similar to $\mu = 1$, $\quad (9\text{-}187) = -\bar{\psi}_1\gamma^{3}\psi_2 . \qquad (9\text{-}191)$

For $\mu = 0$, there is no sign change, but for $\mu = 1,2,3$, there is. Because of the relations

$$\gamma_0 = g_{00}\gamma^{0} = \gamma^{0} \qquad \gamma_1 = g_{11}\gamma^{1} = -\gamma^{1} \qquad \gamma_2 = g_{22}\gamma^{2} = -\gamma^{2} \qquad \gamma_3 = g_{33}\gamma^{3} = -\gamma^{3}, \qquad (9\text{-}192)$$

we can express (9-187), with a lowered spacetime index, as

$$\bar{\psi}_1\gamma^{\mu}\psi_2 \overset{T}{\Rightarrow} \bar{\psi}_1\gamma_{\mu}\psi_2 . \qquad (9\text{-}193)$$

Vector four-currents under T

Axial Four-currents

Do **Problem 14** to show (9-194).

For axial four-currents, we have

$$\bar{\psi}_1\gamma^{\mu}\gamma^{5}\psi_2 \overset{T}{\Rightarrow} \bar{\psi}_1\gamma_{\mu}\gamma^{5}\psi_2 . \qquad (9\text{-}194)$$

Axial four-currents under T

Vector Bosons

As 3-momentum changes sign under T, we expect other 3-vectors will as well. That is, they switch their spacetime index from covariant to contravariant (or vice versa). So, for four-vectors, where the time component does not change sign, we can get this effect by raising and lowering spacetime indices, as in

$$V_\alpha \overset{T}{\Rightarrow} V^\alpha \qquad\qquad V^\alpha \overset{T}{\Rightarrow} V_\alpha . \tag{9-195}$$

Four-vectors under T

Sample Interaction under T Transformation

Let's consider (9-193) and (9-194), in the context of Fig. 9-2. First, those two relations change the same way, which tells us there is no chirality change. If (9-194) had a minus sign on the RHS, then we would have such a change[1].

Axial, vector 4-currents under T determine chirality under T

On the LHS of Fig. 9-2, we employ Lagrangian vertex terms of (7-145) from Chap. 7, and for the first vertex on the left, take the subscript 1 in (9-193) and (9-194) to be ν_e, and subscript 2 to be e. This is because, at the first vertex, one must destroy both the particle (electron) and antiparticle (anti electron neutrino) in the state ket; and these destruction operators are contained in the fields having the stated subscripts. For the second vertex, we take the subscripts, respectively, to be μ and ν_μ, to find the second order transition amplitude,

$$\begin{aligned} S_{fi} &\xrightarrow[\text{order}]{\text{second}} = \left\langle \mu^{-L}_{\mathbf{p}'_1}, \bar{\nu}^{R}_{\mu\,\mathbf{p}'_2} \left| S^{(2)} \right| e^{-L}_{\mathbf{p}_1}, \bar{\nu}^{R}_{e\,\mathbf{p}_2} \right\rangle \\ &= -\frac{1}{2!}\frac{g^2}{2}\left\langle \mu^{-L}_{\mathbf{p}'_1}, \bar{\nu}^{R}_{\mu\,\mathbf{p}'_2} \left| \iint d^4x_1 d^4x_2 N\left\{ (\bar{\psi}^L_\mu \gamma^\alpha \psi^L_{\nu_\mu} W^-_\alpha)_{x_1} (\bar{\psi}^L_{\nu_e} \gamma^\beta \psi^L_e W^+_\beta)_{x_2} + \dots \right\} \right| e^{-L}_{\mathbf{p}_1}, \bar{\nu}^{R}_{e\,\mathbf{p}_2} \right\rangle, \end{aligned} \tag{9-196}$$

An example interaction

where the terms not shown yield zero for the given bra and ket. Now, under T acting on states, according to Fig. 9-2, the term shown in (9-196) becomes

$$\begin{aligned} &\overset{T}{\Rightarrow} -\frac{1}{2!}\frac{g^2}{2}\left\langle e^{-L}_{-\mathbf{p}_1}, \bar{\nu}^{R}_{e\,-\mathbf{p}_2} \left| \iint d^4x_1 d^4x_2 N\left\{ (\bar{\psi}^L_\mu \gamma^\alpha \psi^L_{\nu_\mu} W^-_\alpha)_{x_1} (\bar{\psi}^L_{\nu_e} \gamma^\beta \psi^L_e W^+_\beta)_{x_2} \right\} \right| \mu^{-L}_{-\mathbf{p}'_1}, \bar{\nu}^{R}_{\mu\,-\mathbf{p}'_2} \right\rangle \\ &= 0. \end{aligned} \tag{9-197}$$

This example under T on states seems to yield different $S_{fi} = 0$

We get zero for the probability of interaction, but we know that is not correct, as the T transformed interaction of Fig. 9-2 is not prohibited theoretically or experimentally. So, what is going on?

The answer is this. The Lagrangian, in addition to the terms shown in (9-196), has other terms (see (7-145) again) of the form

$$-\frac{g}{\sqrt{2}}\bar{\psi}^L_e \gamma^\alpha \psi^L_{\nu_e} W^-_\alpha \qquad\qquad -\frac{g}{\sqrt{2}}\bar{\psi}^L_{\nu_\mu} \gamma^\beta \psi^L_\mu W^+_\beta , \tag{9-198}$$

which, when incorporated into the transition amplitude calculation, give us

$$\begin{aligned} S_{fi} &\overset{T}{\Rightarrow} S'_{fi} \xrightarrow[\text{order}]{\text{second}} = \left\langle e^{-L}_{-\mathbf{p}_1}, \bar{\nu}^{R}_{e\,-\mathbf{p}_2} \left| S^{(2)} \right| \mu^{-L}_{-\mathbf{p}'_1}, \bar{\nu}^{R}_{\mu\,-\mathbf{p}'_2} \right\rangle \\ &= -\frac{1}{2!}\frac{g^2}{2}\left\langle e^{-L}_{-\mathbf{p}_1}, \bar{\nu}^{R}_{e\,-\mathbf{p}_2} \left| \iint d^4x_1 d^4x_2 N\left\{ (\bar{\psi}^L_e \gamma^\alpha \psi^L_{\nu_e} W^-_\alpha)_{x_1} (\bar{\psi}^L_{\nu_\mu} \gamma^\beta \psi^L_\mu W^+_\beta)_{x_2} + \dots \right\} \right| \mu^{-L}_{-\mathbf{p}'_1}, \bar{\nu}^{R}_{\mu\,-\mathbf{p}'_2} \right\rangle, \end{aligned} \tag{9-199}$$

But using other terms in $\mathcal{L}$, yields non-zero S'_{fi}

where the terms not shown yield zero. We know we will get the same probability as in the original interaction if the complex conjugate transpose of (9-199) equals (9-196), i.e., if $(S'_{fi})^\dagger = S_{fi}$. That is what we shall now prove.

First, we note that any reaction with all three momenta in (9-199) in opposite directions must have the same probability of occurrence, as an interaction cannot depend on arbitrary assignment of signs on direction. The relative relationships of 3-momenta are relevant, and those are preserved with flipping directions of travel on all particles. We can also flip dummy indices x_1 and x_2, as well as α and β. Thus, (9-199) can be re-expressed as

$$S'_{fi} = -\frac{1}{2!}\frac{g^2}{2}\left\langle e^{-L}_{\mathbf{p}_1}, \bar{\nu}^{R}_{e\,\mathbf{p}_2} \left| \iint d^4x_1 d^4x_2 N\left\{ (\bar{\psi}^L_e \gamma^\beta \psi^L_{\nu_e} W^-_\beta)_{x_2} (\bar{\psi}^L_{\nu_\mu} \gamma^\alpha \psi^L_\mu W^+_\alpha)_{x_1} + \dots \right\} \right| \mu^{-L}_{\mathbf{p}'_1}, \bar{\nu}^{R}_{\mu\,\mathbf{p}'_2} \right\rangle . \tag{9-199)+1$$

The complex conjugate transpose of (9-199)+1 equals (9-196). So, probability is unchanged under T transformation on states. QED.

[1] There is a subtlety here, as, strictly speaking, this is only true when there is no exchange of particles with antiparticles. When there is such an exchange, as with C and $\hat{T}$, for mathematically convoluted reasons we won't get into, one would need the vector and axial four-currents to change sign to keep the same chirality. This has to do with the fact that antiparticles with the same helicity as particles have opposite chirality.

Recall, we said we could either transform the states to determine if an interaction occurs with the same probability, or the Lagrangian (and not the states). So, let's transform the Lagrangian under T, but leave the states unchanged. Doing that to the Lagrangian vertex terms in (9-196), via (9-193) and (9-194), we get

$$S_{fi} \overset{T}{\Rightarrow} S'_{fi} \xrightarrow[\text{order}]{\text{second}} = \left\langle \mu^-_{\mathbf{p}'_1}, \bar{\nu}^R_{\mu\,\mathbf{p}'_2} \middle| S'^{(2)} \middle| e^{-L}_{\mathbf{p}_1}, \bar{\nu}^R_{e\,\mathbf{p}_2} \right\rangle$$
$$= -\frac{1}{2!}\frac{g}{2}\left\langle \mu^-_{\mathbf{p}'_1}, \bar{\nu}^R_{\mu\,\mathbf{p}'_2} \middle| \iint d^4x_1 d^4x_2 N\left\{ (\underbrace{\bar{\psi}^L_{\mu}\gamma_\alpha \psi^L_{\nu_\mu} W^{-\alpha})_{x_1} (\bar{\psi}^L_{\nu_e}}\gamma_\beta \psi^L_e W^{+\beta})_{x_2} \right\} \middle| e^{-L}_{\mathbf{p}_1}, \bar{\nu}^R_{e\,\mathbf{p}_2} \right\rangle, \tag{9-200}$$

Same interaction under T on $\mathcal{L}$: invariant S_{fi} = invariant probability

which is the same as (9-196).

The probability for the particular interaction of Fig. 9-2 is invariant under T, and we showed this two different ways, by transforming the states and by transforming the Lagrangian. This invariance of probability under T is not the case for all interactions, as we will discuss a bit later.

Most, not all, interactions invariant under T

Partial Derivatives

Under T, $t \to -t$, so the time derivative will change signs, but the spatial derivatives will not. So,

$$\partial_\mu \overset{T}{\Rightarrow} -\partial^\mu . \tag{9-201}$$

But in our free kinetic spinor terms, we also have a factor of i. And under T, all i factors in the constants change sign. So, we can instead write

$$i\partial_\mu \overset{T}{\Rightarrow} i\partial^\mu . \tag{9-202}$$

Derivatives under T

T as an Anti-unitary Operation

Mathematically, T is known as an <u>antiunitary operator</u>. For a unitary operator U, we have the inner product and addition relations

$$\langle U\psi_1 | U\psi_2 \rangle = \langle \psi_1 | \psi_2 \rangle \qquad U(a\psi_1 + b\psi_2) = aU\psi_1 + bU\psi_2 . \tag{9-203}$$

But for T, because it includes the process $i \to -i$, which acts on constants as well as fields,

$$\langle T\psi_1 | T\psi_2 \rangle = \langle \psi_1 | \psi_2 \rangle^* \qquad T(a\psi_1 + b\psi_2) = a^*T\psi_1 + b^*T\psi_2 , \tag{9-204}$$

T is antiunitary, due to action $i \to -i$

and inner products are not invariant under T. They turn into their complex conjugates. Further, T is not linear, as the RHS of (9-204) does not conform to the linearity rule of the RHS of (9-203). Thus, T is part of a class of operators known as antiunitary.

<u>*CPT* Transformations</u>

It will help if we refer to Wholeness Chart 9-3, pg. 342. Note that, in the chart, all transformations refer back to the first column. For example, under CP, the derivative index is raised, and under T, it is raised as well, but the raising is with respect to the original (first column) lowered derivative. If the original term is lowered, then both CP and T raise it. They actually reverse the upper/lower index location. So, in a CPT transformation, T would actually lower the index after CP has raised it.

CPT transformations

CPT on Fields and Four-currents

We deduce some CPT transformations on fields below.

For scalars, from (9-93), (9-114), and (9-176),

$$\phi \overset{C}{\Rightarrow} \phi^* \overset{P}{\Rightarrow} \phi^* \overset{T}{\Rightarrow} \phi^* \qquad \phi^* \overset{C}{\Rightarrow} \phi \overset{P}{\Rightarrow} \phi \overset{T}{\Rightarrow} \phi \tag{9-205}$$

CPT of scalars

For spinors, we use (9-94), then (9-119), then (9-181), to get

$$\psi \overset{C}{\Rightarrow} -i\gamma^2\psi^* \overset{P}{\Rightarrow} \gamma^0\left(-i\gamma^2\psi^*\right) \overset{T}{\Rightarrow} \gamma^1\gamma^3\left(+i\gamma^0\left(\gamma^2\right)^*\psi^*\right) = \gamma^1\gamma^3\left(-i\gamma^0\gamma^2\psi^*\right)$$
$$= -i\gamma^0\gamma^1\gamma^3\gamma^2\psi^* = i\gamma^0\gamma^1\gamma^2\gamma^3\psi^* = \gamma^5\psi^* . \tag{9-206}$$

CPT of spinors

For the Hermitian conjugate spinor, we use (9-103), then (9-121), then (9-183), to find

$$\psi^{\dagger} \overset{C}{\Rightarrow} -i\psi^T\gamma^2 \overset{P}{\Rightarrow} \left(-i\psi^T\gamma^2\right)\gamma^0 \overset{T}{\Rightarrow} \left(-i\psi^T\gamma^2\gamma^0\right)\gamma^3\gamma^1 = i\psi^T\gamma^0\gamma^2\gamma^3\gamma^1$$
$$= i\psi^T\gamma^0\gamma^1\gamma^2\gamma^3 = \psi^T\gamma^5 = \left(\psi^{\dagger}\right)^*\gamma^5. \tag{9-207}$$

Note that the entire transformed field after one transformation is used as the field to be transformed in the subsequent transformation. For example in (9-207), after *C*, the new field is $-i\psi^T\gamma^2$, and that is *P* transformed as a spinor transpose, such as ψ^T in (9-121), not like ψ or $\psi^{\dagger}$. That is, the *P* transformation of $-i\psi^T\gamma^2$ in (9-207) is $(-i\psi^T\gamma^2)\gamma^0$, not $-i(\psi^T\gamma^0)\gamma^2$ or $\gamma^0(-i\psi^T\gamma^2)$.

At each step in CPT, use entire transformation of prior step, and form of transformed field from prior step

For vectors, from (9-112), (9-128), and (9-195),

$$V_\mu \overset{C}{\Rightarrow} -V_\mu^{\dagger} \overset{P}{\Rightarrow} -V^{\mu\dagger} \overset{T}{\Rightarrow} -V_\mu^{\dagger}\,. \tag{9-208}$$

CPT of vectors

For vector four-currents, from (9-108), (9-124), and (9-193),

$$\bar{\psi}_1\gamma^\mu\psi_2 \overset{C}{\Rightarrow} -\bar{\psi}_2\gamma^\mu\psi_1 \overset{P}{\Rightarrow} -\bar{\psi}_2\gamma_\mu\psi_1 \overset{T}{\Rightarrow} -\bar{\psi}_2\gamma^\mu\psi_1\,. \tag{9-209}$$

CPT of four-currents

For axial four-currents, from (9-109), (9-125), and (9-194),

$$\bar{\psi}_1\gamma^\mu\gamma^5\psi_2 \overset{C}{\Rightarrow} \bar{\psi}_2\gamma^\mu\gamma^5\psi_1 \overset{P}{\Rightarrow} -\bar{\psi}_2\gamma_\mu\gamma^5\psi_1 \overset{T}{\Rightarrow} -\bar{\psi}_2\gamma^\mu\gamma^5\psi_1\,. \tag{9-210}$$

CPT of axial currents

Note that in taking the *T* transformation in (9-209) and (9-210), we have already carried out the $\gamma^\mu \overset{T}{\Rightarrow} \left(\gamma^\mu\right)^*$ operation when we deduced the transformations on the four-currents, so we don't do it again here. Similarly, taking of *i* to –*i* is done under *T* for all terms and constants in the Lagrangian, but when we deduced the transformation of quantities shown in Wholeness Chart 9-3 with *i* factors in them (like the free kinetic terms), we took this into account. So, we don't have to do it again when we transform terms in the Lagrangian under *CPT*. Just use the terms shown in the final column, last four rows, of that chart, and all should be well.

$i \rightarrow -i$ under T already included in derivation of individual terms in $\mathcal{L}$

CPT and Terms in $\mathcal{L}$

Mass type terms in the Lagrangian, under *CPT*, are unchanged. From (9-105), (9-122), and (9-185)

$$m\bar{\psi}\psi \overset{C}{\Rightarrow} m\bar{\psi}\psi \overset{P}{\Rightarrow} m\bar{\psi}\psi \overset{T}{\Rightarrow} m\bar{\psi}\psi \tag{9-211}$$

CPT of mass terms

Note from Wholeness Chart 9-3 what *CPT* does to original interaction terms.

$$\bar{\psi}_1\gamma^\mu\psi_2V_\mu \overset{CPT}{\Rightarrow} \bar{\psi}_2\gamma^\mu\psi_1V_\mu^{\dagger} \qquad \bar{\psi}_1\gamma^\mu\gamma^5\psi_2V_\mu \overset{CPT}{\Rightarrow} \bar{\psi}_2\gamma^\mu\gamma^5\psi_1V_\mu^{\dagger}\,. \tag{9-212}$$

CPT on general case four-currents

Now, consider an actual term in $\mathcal{L}$, represented in the first vertex shown in Fig. 9-2 and expressed mathematically in (9-196), which transforms like (9-212), i.e., (with the last part below derived in Appendix C),

$$-\frac{g}{\sqrt{2}}\bar{\psi}^L_{\nu_e}\gamma^\beta\psi^L_eW^+_\beta \overset{CPT}{\Rightarrow} -\frac{g}{\sqrt{2}}\bar{\psi}^L_e\gamma^\beta\psi^L_{\nu_e}W^-_\beta = \left(-\frac{g}{\sqrt{2}}\bar{\psi}^L_{\nu_e}\gamma^\beta\psi^L_eW^+_\beta\right)^{\dagger} \tag{9-213}$$

CPT on example vertex term

CPT is equivalent to taking the Hermitian conjugate. And in this example, the transformed term is identical to the LHS of (9-198), i.e., to a term that was in the original Lagrangian.

CPT same as complex conjugate transpose

And that original term gets transformed as

$$-\frac{g}{\sqrt{2}}\bar{\psi}^L_e\gamma^\beta\psi^L_{\nu_e}W^-_\beta \overset{CPT}{\Rightarrow} -\frac{g}{\sqrt{2}}\bar{\psi}^L_{\nu_e}\gamma^\beta\psi^L_eW^+_\beta = \left(-\frac{g}{\sqrt{2}}\bar{\psi}^L_e\gamma^\beta\psi^L_{\nu_e}W^-_\beta\right)^{\dagger}. \tag{9-214}$$

Two terms transform into each other

Each of these two terms on the LHS of (9-213) and (9-214) gets transformed into the other. This leaves these two terms in the Lagrangian invariant under *CPT*. It also means the Hermitian conjugate of these two terms together is invariant. This should not be too surprising, since we know $\mathcal{L}$ is real, it must be unchanged under complex conjugation.

So, CPT on those two terms leaves $\mathcal{L}$ invariant

We can generalize. $\mathcal{L}$ is real, so it equals its own complex conjugate. If all terms in $\mathcal{L}$ undergo *CPT*, the result is $\mathcal{L}^{\dagger}$, but that is simply $\mathcal{L}$. This is *CPT* symmetry. $\mathcal{L}$ is unchanged under *CPT*.

In general, $\mathcal{L} = \mathcal{L}^{\dagger}$ and $\mathcal{L}^{\dagger}$ is CPT of $\mathcal{L}$

$$\mathcal{L} = \mathcal{L}^{\dagger} \quad \mathcal{L} \overset{CPT}{\Rightarrow} = \mathcal{L}^{\dagger} \qquad \text{so,} \qquad \mathcal{L} \overset{CPT}{\Rightarrow} = \mathcal{L}\,. \tag{9-215}$$

$\mathcal{L}$ is CPT invariant

This is what theory tells us. The characteristic of $\mathcal{L} = \mathcal{L}^{\dagger}$ is known as the hermiticity of the Lagrangian.

Bottom line: $\mathcal{L}$ is symmetric under *CPT*, and the *CPT* transformation of $\mathcal{L}$ is the same as taking the Hermitian conjugate of $\mathcal{L}$.

Looking for CPT Violation in Experiments

CPT invariance never seen violated in experiment

CPT violation can be investigated experimentally by determining the probability of a particular interaction, and then determining the probability of the *CPT* transform of that interaction. It has been looked for in experiments and never found. If it ever is found, it would revolutionize QFT, as we know it, since the terms in our current version of $\mathcal{L}$ would need revision.

CPT and the CKM Matrix

CP violation due to CKM matrix disappears under CPT

Note that under *CPT*, we do not get a violation from the CKM matrix. This is because, after the *CP* transformation of (9-173), we apply *T*, which takes *i* to –*i* and flips the CKM matrix back to what it originally was. Specifically, where the boxed numbers below reference (9-173), we have (where *V* here represents the CKM matrix, not a general four-vector)

Because T takes $i \rightarrow -i$ in CKM matrix

$$\begin{aligned} \boxed{3} &= -\frac{g}{\sqrt{2}} \bar{\psi}^{L}_{q_u} \gamma^{a} V_{q_u q_d} \psi^{L}_{q_d} W^{+}_{\alpha} \overset{CPT}{\Rightarrow} -\frac{g}{\sqrt{2}} \bar{\psi}^{L}_{q_d} \gamma^{\alpha} V^{\dagger}_{q_d q_u} \psi^{L}_{q_u} W^{-}_{\alpha} = \boxed{4} \\ \boxed{4} &= -\frac{g}{\sqrt{2}} \bar{\psi}^{L}_{q_d} \gamma^{\alpha} V^{\dagger}_{q_d q_u} \psi^{L}_{q_u} W^{-}_{\alpha} \overset{CPT}{\Rightarrow} -\frac{g}{\sqrt{2}} \bar{\psi}^{L}_{q_u} \gamma^{a} V_{q_u q_d} \psi^{L}_{q_d} W^{+}_{\alpha} = \boxed{3}. \end{aligned} \tag{9-216}$$

Decay of a Particle and Its Antiparticle

Particle decay and CPT

Consider the *CPT* transformed interaction depicted on the lower right of Fig. 9-3, pg. 342,

First examine a typical interaction

$$\mu^{+R} + \nu^{L}_{\mu} \rightleftharpoons e^{+R} + \nu^{L}_{e}\,, \tag{9-217}$$

where the reaction can go in either direction, as indicated by the bidirectional arrows. To understand this, first consider the transition amplitude for the forward arrow of (9-217), where we ignore the terms in $\mathcal{L}$ that will yield zero,

$$S_{fi} = -\frac{1}{2!}\frac{g^2}{2} \left\langle e^{+R}, \nu^{L}_{e} \right| \iint d^4x_1 d^4x_2 \, N\left\{ (\bar{\psi}^{L}_{\nu_e} \gamma^{\alpha} \psi^{L}_{e} W^{+}_{\alpha})_{x_1} (\bar{\psi}^{L}_{\mu} \gamma^{\beta} \psi^{L}_{\nu_\mu} W^{-}_{\beta})_{x_2} \right\} \left| \mu^{+R}, \nu^{L}_{\mu} \right\rangle \tag{9-218}$$

Now, consider that of the interaction in the opposite direction, using only the vertex factors from $\mathcal{L}$ that yield a non-zero result,

$$\underset{oppos}{S_{fi}} = -\frac{1}{2!}\frac{g^2}{2} \left\langle \mu^{+R}, \nu^{L}_{\mu} \right| \iint d^4x_1 d^4x_2 \, N\left\{ (\bar{\psi}^{L}_{\nu_\mu} \gamma^{\beta} \psi^{L}_{\mu} W^{+}_{\beta})_{x_2} (\bar{\psi}^{L}_{e} \gamma^{\alpha} \psi^{L}_{\nu_e} W^{-}_{\alpha})_{x_1} \right\} \left| e^{+R}, \nu^{L}_{e} \right\rangle, \tag{9-219}$$

which is the complex conjugate transpose of (9-218). Since probability is the product of the transition amplitude and its complex conjugate, the probability for (9-218) is the same as that for (9-219). So, either way the reaction proceeds, we get the same result, the same probability.

An interaction has same probability forward and backward

1st bottom line: A given interaction occurs in either direction with the same probability.

This is a little different from the *T* transformation (which also reverses incoming and outgoing particles), as that includes taking *i* to –*i*, whereas here we are not doing that.

Apply that rule to decay

Same rate for particle decay and creation

Now, for decays and their reverse processes, this means a given interaction occurs either via a particle breaking up into less massive ones, or less massive particles coming together to form the more massive one. And these processes occur at the same rate[1], since the probability of interaction is the same in either direction.

---

[1] You may have heard that unstable particles commonly decay, but their constituents don't commonly come together to form the "parent" particle. This is because the energy levels of requisite particles in the environment are not (collectively) at the level of the energy of the parent. If they were, in a dense environment of constituent particles, the interaction would go on in both directions. The interaction needs to be at an energy level at least equal to the mass-energy of the parent particle, or it cannot proceed. With a parent particle simply sitting there, we are guaranteed to have this much energy. For constituent particles, this is only true if they have enough kinetic plus mass energy to equal that of the parent mass-energy. If the interaction is at the requisite energy level, it can proceed, in either direction. The decay always has the requisite energy. The reverse does not, but depends on other conditions. A plot of probability vs energy level would show a cutoff below the rest mass-energy of the parent particle, and this is true for the

Now consider the decay of a muon, as we examined in Chap. 8, but now under *CPT*,

$$\mu^- \to e^- + \nu_\mu + \bar{\nu}_e \quad \overset{CPT}{\Rightarrow} \quad e^+ + \bar{\nu}_\mu + \nu_e \to \mu^+ \text{ same probability as } \mu^+ \to e^+ + \bar{\nu}_\mu + \nu_e . \qquad (9\text{-}220)$$

So, CPT on a decay plus reversal of decay means particle and antiparticle decay at same rate

Under *CPT*, the Lagrangian is symmetric, so the transition amplitude for the transformed interaction is the same as that of the original one. And the reversed direction of the transformed interaction has the same transition amplitude as the non-reversed one. Thus, the transition amplitude for the antimuon decay is the same as that of the muon.

2nd bottom line: *CPT* invariance dictates that a particle and its antiparticle decay at the same rate.

Often said: CPT means particle and its antiparticle decay at same rate

Order of Operation of C,P,T

QFT texts often state that the order of operation of *C, P,* and *T* (or $\hat{T}$) does not matter, and one gets the same result, for example, of doing *C* first before *P* (*CP* transformation) as one does by doing *P* before *C* (*PC* transformation). Similarly, *CPT* and *TCP* are said to yield the same result.

These statements are true for terms in the Lagrangian, but not strictly so for the fields themselves, as we are about to see.

The Order of C and P

In (9-146) we showed that by operating on ψ first with *C*, then with *P*, we obtain

$$\psi \overset{C}{\Rightarrow} -i\gamma^2\psi^* \overset{P}{\Rightarrow} \gamma^0\left(-i\gamma^2\psi^*\right) = -i\gamma^0\gamma^2\psi^* . \qquad (9\text{-}221)$$

However, consider the operation in reverse

$$\text{in reverse} \qquad \psi \overset{P}{\Rightarrow} \gamma^0\psi \overset{C}{\Rightarrow} -i\gamma^2\left(\gamma^0\psi\right)^* = -i\gamma^2\gamma^0\psi^* = i\gamma^0\gamma^2\psi^* . \qquad (9\text{-}222)$$

Changing order of C, P, and/or T can change sign of transformed field

We get the opposite sign. In this sense, order matters, though again, the difference is only in phase, and thus, not really consequential. Similarly, from (9-120) and (9-103)

$$\psi^\dagger \overset{C}{\Rightarrow} -i\psi^T\gamma^2 \overset{P}{\Rightarrow} \left(-i\psi^T\gamma^2\right)\gamma^0 = -i\psi^T\gamma^2\gamma^0 \qquad (9\text{-}223)$$

$$\psi^\dagger \overset{P}{\Rightarrow} \psi^\dagger\gamma^0 \overset{C}{\Rightarrow} -i\left(\psi^\dagger\gamma^0\right)^*\gamma^2 = -i\psi^T\gamma^0\gamma^2 = i\psi^T\gamma^2\gamma^0 , \qquad (9\text{-}224)$$

and again, we see a sign switch by switching the order of *C* and *P*. However, for Lagrangian entities, such as $\bar{\psi}\psi = \psi^\dagger\gamma^0\psi$ and $\bar{\psi}\gamma^\mu\psi = \psi^\dagger\gamma^0\gamma^\mu\psi$ the sign changes cancel, provided we operate on both ψ and $\psi^\dagger$ in the same order, i.e., both via *CP* or both via *PC*.

But, $\mathcal{L}$ terms unchanged if we use the same order on all fields

The Order of C, P, and T (or $\hat{T}$)

We can generalize. The sign of the transformed fields may vary with order of operation of *C, P,* and *T* (or $\hat{T}$), but the signs of terms in the Lagrangian remain unchanged, provided all fields are transformed in the same order of the discrete symmetries.

Different Signs in Different Texts

Different texts can show different signs for ψ and $\psi^\dagger$ under *CP* or *CPT* because they take the order of the operations differently. This can be a source of great confusion if one doesn't understand the point made in this subsection.

9.4 Other Conserved Quantities

9.4.1 Lepton and Baryon Number Conservation

Recall from Chap. 6, that we first showed the symmetry of $\mathcal{L}$ separately under *SU*(2) and *U*(1) solely for leptons, with no quarks involved. In particular, we employed the parameters ω_i (i = 1,2,3) and *f* under the variation of which, the Lagrangian was shown to be invariant. Later, we showed $\mathcal{L}$ was also symmetric under the same symmetry for quarks, and we used the same parameters ω_i and *f*. From this, using Noether's theorem in this chapter, we deduced weak isospin and hypercharges for both leptons and quarks. As they are conserved in interactions (ignoring the mass terms), the electric

Find lepton and baryon (quark) number conservation from Noether's theorem

---

interaction going in either direction. When conditions in a large collection of particles are such that the interaction can occur in either direction, the system is said to be in equilibrium.

charge (see last row of Wholeness Chart 9-1, pg. 341) is also conserved. In fact, the electric charge is conserved even including the lepton mass terms, as in QED.

Note, however, that we could have symmetry just for leptons using parameters ω_i and f, and then also have symmetry just for quarks, using different parameters, say ξ_i and α. Then, we would get a conserved lepton charge, and a separate conserved quark charge. Each of these can be multiplied by any constant we find suitable.

More relevant, at present, is $U(1)$ symmetry alone for leptons, where we can take this constant as 1, and we use (see Wholeness Chart 6-5, pg. 181), $Y=1$ for all leptons. Then, note what happens to a typical interaction term under this symmetry transformation.

Just do U(1) symmetry for leptons and no quarks

$$-\frac{g}{\sqrt{2}}\bar{\psi}_{\nu_e}^{L}W^{+}\psi_e^{L} \overset{New\,U(1)}{\Rightarrow} -\frac{g}{\sqrt{2}}\left(\bar{\psi}_{\nu_e}^{L}-i\alpha\bar{\psi}_{\nu_e}^{L}\right)W^{+}\left(\psi_e^{L}+i\alpha\psi_e^{L}\right)$$
$$=-\frac{g}{\sqrt{2}}\bar{\psi}_{\nu_e}^{L}W^{+}\psi_e^{L}+i\alpha\frac{g}{\sqrt{2}}\bar{\psi}_{\nu_e}^{L}W^{+}\psi_e^{L}-i\alpha\frac{g}{\sqrt{2}}\bar{\psi}_{\nu_e}^{L}W^{+}\psi_e^{L}+O\left(\alpha^2\right)=-\frac{g}{\sqrt{2}}\bar{\psi}_{\nu_e}^{L}W^{+}\psi_e^{L}. \tag{9-225}$$

The term is invariant under this $U(1)$ symmetry, and, as it turns out, so are all interaction terms. Thus, we end up with a conserved quantity

And we find lepton number conservation

$$N_{lep}=\sum_{l,\mathbf{p},r}\left(N_{\nu_l,r}(\mathbf{p})-\bar{N}_{\nu_l,r}(\mathbf{p})+N_{l,r}(\mathbf{p})-\bar{N}_{l,r}(\mathbf{p})\right), \tag{9-226}$$

which we define as the lepton number. All leptons, electrically charged or not, have a lepton number of 1. All antileptons have a lepton number of –1. You should be able to convince yourself, rather quickly, by looking at Feynman diagrams, such as that of Figs. 9-2 and 9-3, as well as others we have dealt with, that lepton number is indeed conserved.

Parallel logic for quarks with 1/3 baryon number

Parallel logic holds for quarks, though, for historical reasons, focusing on baryon number, rather than quark number, is conventional. So, quarks have baryon number of 1/3, and antiquarks, of –1/3, with the total baryon number conserved in interactions, as well as for free quarks/baryons[1].

Lepton & baryon number conservation arise from global U(1) symmetry

Note that lepton and baryon number conservation each arise from a global symmetry (α is constant in (9-225)), whereas charge conservation arises from a local symmetry.

9.4.2 Helicity vs Chirality: Conservation and Invariance

Helicity is an inner product of angular momentum **L** and linear momentum **p**. For a collection of particles before interaction, the sum total of all $\mathbf{L}_i$ is the same after interaction. Angular momentum is conserved. The same is true of the sum of all particle $\mathbf{p}_i$. As a result, total helicity is also conserved.

Helicity conserved, but not invariant

On the other hand, helicity is not invariant, as we have discussed before. An observer boosted to a velocity greater than a given particle's velocity will see the direction of **p** reversed from that of an unboosted observer, and thus, see a different helicity.

Chirality, in contrast, is not conserved, as we saw in Fig. 7-9, pg. 243. Due to the mass term, an LC particle can turn into an RC particle, and vice versa.

Chirality invariant, but not conserved

But chirality is invariant, in that a different observer will not see RC particles interacting in some frame of reference, that did not so interact in the original frame.

These results are summarized in Wholeness Chart 9-4 below.

9.5 Chapter Summary

The core results of this chapter are summarized in the following wholeness charts, which should need no further elaboration, plus a short overview of *C, P,* and *T* transformations with diagrams and a brief summary of lepton and baryon number conservation.

[1] One exception, shown by Gerardus 't Hooft, is the hypothesized Adler–Bell–Jackiw anomaly, yet to be actually observed and beyond the scope of this text. See G. 't Hooft, "Symmetry breaking through Bell-Jackiw anomalies", *Phys. Rev. Lett.* 37 (1976) 8.

Wholeness Chart 9-1. Electroweak Four-Currents and Charges

False Vacuum

Four-current and Charge Operator	
General principles	$\mathcal{L}$ symmetric under α variation, $j^{\mu} = \dfrac{\partial \mathcal{L}}{\partial \phi^{r}{}_{,\mu}} \dfrac{\partial \phi'^{r}}{\partial \alpha} \rightarrow \partial_{\mu} j^{\mu} = 0 \rightarrow Q = -(\text{coup const}) \int j^{0} dV$ conserved
Hypercharge	${}_{h}\mathcal{L} = {}_{h}\mathcal{L}^{L} + {}_{h}\mathcal{L}^{H} + {}_{h}\mathcal{L}^{Q} = g' {}_{h}j^{\mu} B_{\mu}$ ${}_{h}Q = -g' \int {}_{h}j^{0} dV$ $\langle \phi_i \vert {}_{h}Q \vert \phi_i \rangle = Y\, g'$ (hypercharge of ϕ_i particle)
Weak isospin	${}_{3}\mathcal{L} = {}_{3}\mathcal{L}^{L} + {}_{3}\mathcal{L}^{H} + {}_{3}\mathcal{L}^{Q} = g\, {}_{3}j^{\mu} W_{3\,\mu}$ ${}_{3}Q = -g \int {}_{3}j^{0} dV$ $\langle \phi_i \vert {}_{3}Q \vert \phi_i \rangle = I_3^W g$ (weak isospin charge of ϕ_i)
Other $SU(2)$	${}_{1}\mathcal{L} = {}_{1}\mathcal{L}^{L} + {}_{1}\mathcal{L}^{H} + {}_{1}\mathcal{L}^{Q} = g\, {}_{1}j^{\mu} W_{1\,\mu}$ ${}_{1}Q = -g \int {}_{1}j^{0} dV$ $\langle s \vert {}_{1}Q \vert s \rangle = 0$ (s is any state of particles)
	${}_{2}\mathcal{L} = {}_{2}\mathcal{L}^{L} + {}_{2}\mathcal{L}^{H} + {}_{2}\mathcal{L}^{Q} = g\, {}_{2}j^{\mu} W_{2\,\mu}$ ${}_{2}Q = -g \int {}_{2}j^{0} dV$ $\langle s \vert {}_{2}Q \vert s \rangle = 0$ (s is any state of particles)
Conclusion	Can take operators ${}_{1}Q = {}_{2}Q = 0$
Charge Operator in Terms of Number Operators	
Hypercharge	${}_{h}Q = {}_{h}Q_{lep} + {}_{h}Q_{Higgs} + {}_{h}Q_{qrk}$ $= -g' \sum_{l,\mathbf{p},r} \left(\frac{1}{2}\left(N^{L}_{\nu_l,r}(\mathbf{p}) - \overline{N}^{R}_{\nu_l,r}(\mathbf{p}) \right) + \frac{1}{2}\left(N^{L}_{l,r}(\mathbf{p}) - \overline{N}^{R}_{l,r}(\mathbf{p}) \right) + \left(N^{R}_{l,r}(\mathbf{p}) - \overline{N}^{L}_{l,r}(\mathbf{p}) \right) \right) + \underbrace{\frac{g'}{2} \sum_{\mathbf{p}} \left(N_H(\mathbf{p}) - \overline{N}_{\overline{H}}(\mathbf{p}) \right)}_{N_H(\mathbf{p}) = N_a(\mathbf{p}) + N_b(\mathbf{p})}$ $+ g' \sum_{u,d,\mathbf{p},r} \left(\frac{1}{6}\left(N^{L}_{q_u,r}(\mathbf{p}) - \overline{N}^{R}_{q_u,r}(\mathbf{p}) \right) + \frac{1}{6}\left(N^{L}_{q_d,r}(\mathbf{p}) - \overline{N}^{R}_{q_d,r}(\mathbf{p}) \right) + \frac{2}{3}\left(N^{R}_{q_u,r}(\mathbf{p}) - \overline{N}^{L}_{q_u,r}(\mathbf{p}) \right) - \frac{1}{3}\left(N^{R}_{q_d,r}(\mathbf{p}) - \overline{N}^{L}_{q_d,r}(\mathbf{p}) \right) \right)$
Weak isospin	${}_{3}Q = {}_{3}Q_{lep} + {}_{3}Q_{Higgs} + {}_{3}Q_{qrk} + {}_{3}Q_{W}$ $= \frac{g}{2} \sum_{l,\mathbf{p},r} \left(\left(N^{L}_{\nu_l,r}(\mathbf{p}) - \overline{N}^{R}_{\nu_l,r}(\mathbf{p}) \right) - \left(N^{L}_{l,r}(\mathbf{p}) - \overline{N}^{R}_{l,r}(\mathbf{p}) \right) \right) + \frac{g}{2} \left(\sum_{\mathbf{p}} \left(N_a(\mathbf{p}) - \overline{N}_a(\mathbf{p}) \right) - \sum_{\mathbf{p}} \left(N_b(\mathbf{p}) - \overline{N}_b(\mathbf{p}) \right) \right)$ $+ \frac{g}{2} \sum_{u,d,\mathbf{p},r} \left(\left(N^{L}_{q_u,r}(\mathbf{p}) - \overline{N}^{R}_{q_u,r}(\mathbf{p}) \right) - \left(N^{L}_{q_d,r}(\mathbf{p}) - \overline{N}^{R}_{q_d,r}(\mathbf{p}) \right) \right) + g \sum_{\mathbf{p}} \left(N_{W^+}(\mathbf{p}) - N_{W^-}(\mathbf{p}) \right)$

True Vacuum

Four-current and Charge Operator	
General	$\mathcal{L}$ not symmetric due to mass terms $\rightarrow$ yet still $j^{\mu} = \dfrac{\partial \mathcal{L}}{\partial \phi^{r}{}_{,\mu}} \dfrac{\partial \phi'^{r}}{\partial \alpha}$, but $Q = -(\text{coup const}) \int j^{0} dV$ not conserved Q still conserved at vertices
Hypercharge	Interaction terms same as at false vacuum, but unitary gauge used for Higgs; A_{μ}, Z_{μ}, $W^{\pm}{}_{\mu}$ instead of B_{μ}, $W_{i\,\mu}$
Weak isospin	Interaction terms same as at false vacuum, but unitary gauge used for Higgs; A_{μ}, Z_{μ}, $W^{\pm}{}_{\mu}$ instead of B_{μ}, $W_{i\,\mu}$
Charge operator in Terms of Number Operators	
Hypercharge	Same as false vacuum except unitary gauge used for Higgs where $N_a = 0$, so $N_H = N_b$
Weak isospin	Same as false vacuum except unitary gauge used for Higgs where $N_a = 0$, so $N_H = N_b$
Electric Charge Relation to Hypercharge and Weak Isospin Charge	
	$N_Q = I_3^W + Y$

Wholeness Chart 9-2. Lorentz Transformation Properties

Entity	**Transforms as**
$\bar{\psi}\psi$	scalar
$\bar{\psi}\gamma^{\mu}\psi$	vector
$\bar{\psi}T^{\mu\nu}\psi$	2nd rank tensor
$\bar{\psi}\gamma^{\mu}\gamma^{5}\psi$	pseudo-vector
$\bar{\psi}\gamma^{5}\psi$	pseudo-scalar

Wholeness Chart 9-3. C, P, and T Transformations

	C	**P**	**$\hat{\text{T}}$**	**CP**	**CP$\hat{\text{T}}$**	**T**	**CPT**
Physical effect		$\mathbf{x} \to -\mathbf{x}$	$t \to -t$		$x^{\mu} \to -x^{\mu}$	$t \to -t$	$x^{\mu} \to -x^{\mu}$
Other effect						$i \to -i$	
Particles n	$n \leftrightarrow n^{c}$	n	$n \leftrightarrow n^{c}$	$n \leftrightarrow n^{c}$	n	n	$n \leftrightarrow n^{c}$
3-momenta $\mathbf{p}$	$\mathbf{p}$	$\mathbf{p} \to -\mathbf{p}$	$-\mathbf{p}$	$\mathbf{p} \to -\mathbf{p}$	$\mathbf{p}$	$\mathbf{p} \to -\mathbf{p}$	$\mathbf{p}$
Spin σ	σ	σ	σ	σ	σ	$\sigma \to -\sigma$	$\sigma \to -\sigma$
Incoming/outgoing	same	same	same	same	same	in $\leftrightarrow$ out	in $\leftrightarrow$ out
Fields							
Scalar, true ϕ	ϕ^{*}	ϕ	ϕ^{*}	ϕ^{*}	Invar	ϕ	ϕ^{*}
pseudo π	π^{*}	$-\pi$	$-\pi^{*}$	$-\pi^{*}$	Invar	$-\pi$	π^{*}
Spinor ψ	$-i\gamma^{2}\psi^{*}$	$\gamma^{0}\psi$	$\gamma^{0}\gamma^{2}\psi^{*}$	$-i\,\gamma^{0}\gamma^{2}\psi^{*}$	$-i\psi$	$\gamma^{1}\gamma^{3}\psi$	$\gamma^{5}\psi^{*}$
$\psi^{\dagger}$	$-i(\psi^{\dagger})^{*}\gamma^{2}$	$\psi^{\dagger}\gamma^{0}$	$-(\psi^{\dagger})^{*}\gamma^{2}\gamma^{0}$	$-i(\psi^{\dagger})^{*}\gamma^{2}\gamma^{0}$	$i\psi^{\dagger}$	$\psi^{\dagger}\gamma^{3}\gamma^{1}$	$(\psi^{\dagger})^{*}\gamma^{5}$
Vector V_{μ}	$-V_{\mu}{}^{\dagger}$	V^{μ}	$-V^{\mu\dagger}$	$-V^{\mu\dagger}$	Invar	V^{μ}	$-V_{\mu}^{\dagger}$
Inner products							
True scalar $\bar{\psi}_1\psi_2$	$\bar{\psi}_2\psi_1$	$\bar{\psi}_1\psi_2$	$\bar{\psi}_2\psi_1$	$\bar{\psi}_2\psi_1$	Invar	$\bar{\psi}_1\psi_2$	$\bar{\psi}_2\psi_1$
Pseudo-scalar $\bar{\psi}_1\gamma^{5}\psi_2$	$\bar{\psi}_2\gamma^{5}\psi_1$	$-\bar{\psi}_1\gamma^{5}\psi_2$	$-\bar{\psi}_2\gamma^{5}\psi_1$	$-\bar{\psi}_2\gamma^{5}\psi_1$	Invar	$\bar{\psi}_1\gamma^{5}\psi_2$	$-\bar{\psi}_2\gamma^{5}\psi_1$
Gamma matrices							
γ^{μ}	γ^{μ}	γ^{μ}	γ^{μ}	γ^{μ}	Invar	$\gamma^{\mu\,*}$	$\gamma^{\mu\,*}$
Four currents							
Vector $\bar{\psi}_1\gamma^{\mu}\psi_2$	$-\bar{\psi}_2\gamma^{\mu}\psi_1$	$\bar{\psi}_1\gamma_{\mu}\psi_2$	$-\bar{\psi}_2\gamma_{\mu}\psi_1$	$-\bar{\psi}_2\gamma_{\mu}\psi_1$	Invar	$\bar{\psi}_1\gamma_{\mu}\psi_2$	$-\bar{\psi}_2\gamma^{\mu}\psi_1$
Axial $\bar{\psi}_1\gamma^{\mu}\gamma^{5}\psi_2$	$\bar{\psi}_2\gamma^{\mu}\gamma^{5}\psi_1$	$-\bar{\psi}_1\gamma_{\mu}\gamma^{5}\psi_2$	$-\bar{\psi}_2\gamma_{\mu}\gamma^{5}\psi_1$	$-\bar{\psi}_2\gamma_{\mu}\gamma^{5}\psi_1$	Invar	$\bar{\psi}_1\gamma_{\mu}\gamma^{5}\psi_2$	$-\bar{\psi}_2\gamma^{\mu}\gamma^{5}\psi_1$
Partial derivatives							
$i\partial_{\mu}$	$i\partial_{\mu}$	$i\partial^{\mu}$	$-i\partial^{\mu}$	$i\partial^{\mu}$	$-i\partial_{\mu}$	$i\partial^{\mu}$	$i\partial_{\mu}$
Lagrangian terms							
Free mass $m\bar{\psi}\psi$	$m\bar{\psi}\psi$	$m\bar{\psi}\psi$	$m\bar{\psi}\psi$	$m\bar{\psi}\psi$	Invar	$m\bar{\psi}\psi$	Invar
Free kinetic $i\bar{\psi}\gamma^{\mu}\partial_{\mu}\psi$	$-i\bar{\psi}\gamma^{\mu}\partial_{\mu}\psi$	$i\bar{\psi}\gamma_{\mu}\partial^{\mu}\psi$	$i\bar{\psi}\gamma_{\mu}\partial^{\mu}\psi$	$-i\bar{\psi}\gamma_{\mu}\partial^{\mu}\psi$	Invar	$i\bar{\psi}\gamma_{\mu}\partial^{\mu}\psi$	Invar
Interactn $\bar{\psi}_1\gamma^{\mu}\psi_2V_{\mu}$	$\bar{\psi}_2\gamma^{\mu}\psi_1V_{\mu}^{\dagger}$	$\bar{\psi}_1\gamma^{\mu}\psi_2V_{\mu}$	$\bar{\psi}_2\gamma_{\mu}\psi_1V^{\mu\dagger}$	$\bar{\psi}_2\gamma^{\mu}\psi_1V_{\mu}^{\dagger}$	Invar	$\bar{\psi}_1\gamma_{\mu}\psi_2V^{\mu}$	$\bar{\psi}_2\gamma^{\mu}\psi_1V_{\mu}^{\dagger}$
Interactn $\bar{\psi}_1\gamma^{\mu}\gamma^{5}\psi_2V_{\mu}$	$-\bar{\psi}_2\gamma^{\mu}\gamma^{5}\psi_1V_{\mu}^{\dagger}$	$-\bar{\psi}_1\gamma^{\mu}\gamma^{5}\psi_2V_{\mu}$	$\bar{\psi}_2\gamma_{\mu}\gamma^{5}\psi_1V^{\mu\dagger}$	$\bar{\psi}_2\gamma^{\mu}\gamma^{5}\psi_1V_{\mu}^{\dagger}$	Invar	$\bar{\psi}_1\gamma_{\mu}\gamma^{5}\psi_2V^{\mu}$	$\bar{\psi}_2\gamma^{\mu}\gamma^{5}\psi_1V_{\mu}^{\dagger}$

CP and CPT Transformations

Figure 9-4 shows Feynman diagrams for a typical interaction transformed under CP and CPT.

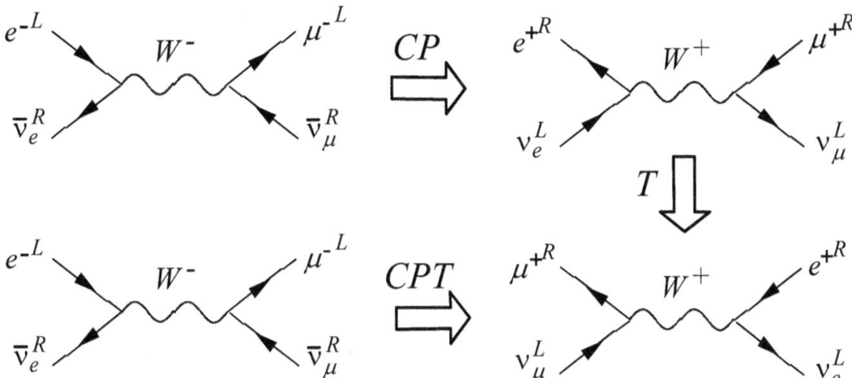

Figure 9-4. Diagrams of CP and CPT Transformations for Example Interaction

According to current theory, validated to date by experiment, transforming the Lagrangian under *CPT* is the same as taking its Hermitian conjugate, and gives the original Lagrangian back (it is symmetric under the operation).

$$\mathcal{L} \overset{CPT}{\Rightarrow} \mathcal{L}^{\dagger} \; (= \mathcal{L}) . \qquad \text{(9-227)}$$

This means each term in $\mathcal{L}$ is either its own complex conjugate transpose or that there is another term in $\mathcal{L}$ that is the first term's complex conjugate transpose. This fact is often used to write a given term, or terms, out in full, followed by "+ *h.c.*", an abbreviation for "plus the Hermitian conjugate".

Lepton and Baryon Number Conservation

From the independent $U(1)$ symmetries of leptons and quarks in the SM, Noether's theorem tells us that lepton number, which for leptons is 1 and for antileptons is –1, is conserved. So is baryon number, the value of which for quarks is 1/3 and for antiquarks, –1/3.

Wholeness Chart 9-4. Comparison of Helicity and Chirality

	Chirality	**Helicity**
Physical description	Related to weak charge	Related to handedness
Projection operator	$P^{L,R} = \frac{1}{2}\left(1 \mp \gamma^5\right)$	$\Pi^{L,R}(\mathbf{p}) = \frac{1}{2}\left(1 \mp \frac{\boldsymbol{\Sigma}\cdot\mathbf{p}}{\lvert\mathbf{p}\rvert}\right)$
In limit $v \to c$ (or $m = 0$)	Chirality eigenstate = helicity eigenstate	$\Pi^{L,R}(\mathbf{p}) \to P^{L,R}$
Lorentz transform properties	Invariant	Not invariant (for $v \neq c$)
Parity reversal	Changes sign, i.e., RC $\leftrightarrow$ LC	Changes sign, i.e., RH $\leftrightarrow$ LH
Conservation properties	Not conserved (due to mass terms)	Conserved
Weak charges non-conservation	"Vacuum eats weak charge."	
Weak charges	Invariant	

9.6 Appendices

9.6.1 Appendix A. A Relation We'll Need

By examining the gamma matrices expressions for the standard, Weyl, and Majorana reps you can readily see (9-228). Actually, you only have to look at one of the reps (for example, the standard rep version (5-2)a, as relations between gamma matrices are identical in different reps (even though the matrices themselves change).

$$\gamma^{0\dagger}=\gamma^{0} \qquad \gamma^{i\dagger}=-\gamma^{i} \tag{9-228}$$

Then,

$$\gamma_0=g_{0\mu}\gamma^{\mu}=\gamma^{0} \qquad \gamma_i=g_{i\mu}\gamma^{\mu}=-\gamma^{i} \quad\rightarrow\quad \gamma_0^{\dagger}=\gamma^{0\dagger} \qquad \gamma_i^{\dagger}=-\gamma^{i\dagger} \tag{9-229}$$

Comparing the last two relations of (9-229) with (9-228), we have

$$\gamma^{0\dagger}=\gamma_0 \qquad \gamma^{i\dagger}=\gamma_i \quad\rightarrow\quad \gamma^{\mu\dagger}=\gamma_{\mu}, \tag{9-230}$$

which is (9-90).

9.6.2 Appendix B. Parity Transformation of a Vector Visualized

Figure 9-5 should be almost self-explanatory. (A picture is worth a thousand words.) Under a parity transformation, a 3D vector will have the sign of its components change, and also the sign of the 3D position in its argument. The transformation shown is a passive transformation, i.e., we switch directions of the coordinate axes, but not the vector itself. After the transformation, the vector points in the direction of the $-x$ axis, so its component value is the negative of what it was before. And it is located at the negative of its x coordinate before.

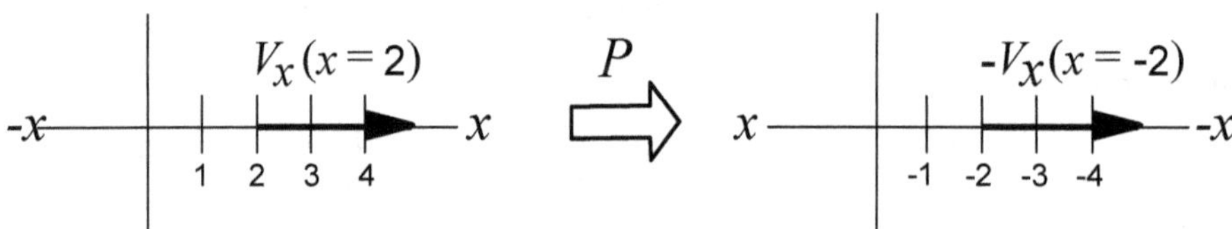

Figure 9-5. Parity Transformation of a 3D Vector

9.6.3 Appendix C. Hermitian Conjugates of Terms in the Lagrangian

Note first that

$$\text{real scalar quantity } \left(\psi^{\dagger}\psi\right)^{\dagger}=\psi^{\dagger}\psi \qquad \text{not} \qquad -\psi^{\dagger}\psi, \tag{9-231}$$

even though in taking the complex conjugate transpose (Hermitian conjugate), we switched the order of the fermion operators. Had we assumed the switching would entail an extra minus sign entering in, we would not have gotten the correct answer. A real scalar remains unchanged under complexification. It does not change sign.

The lesson here is that under Hermitian conjugation, fermion fields effectively commute. This is unlike what happens during C, P, and T transformations, and in so many other QFT situations, where we have to assume Fermi fields anti-commute.

In a sense, one can think of the complex conjugate transpose as simply a different way to express the same multiplication order of the factors involved. It doesn't change the physics, just the way we express the same physics. In any event, that is the way it all works.

Now, consider the Hermitian conjugate on the RHS of (9-213), where we recall (which again, you should be committing to memory by now)

$$\begin{aligned}&\bar{\psi}_1^{L}\gamma^{\alpha}\psi_2^{L}=\bar{\psi}_1\tfrac{1}{2}\left(1+\gamma^5\right)\gamma^{\alpha}\tfrac{1}{2}\left(1-\gamma^5\right)\psi_2=\bar{\psi}_1\gamma^{\alpha}\tfrac{1}{2}\left(1-\gamma^5\right)\tfrac{1}{2}\left(1-\gamma^5\right)\psi_2=\bar{\psi}_1\gamma^{\alpha}\tfrac{1}{2}\left(1-\gamma^5\right)\psi_2\\&\bar{\psi}_1^{R}\gamma^{\alpha}\psi_2^{R}=\bar{\psi}_1\gamma^{\alpha}\tfrac{1}{2}\left(1+\gamma^5\right)\psi_2.\end{aligned} \tag{9-232}$$

Then,

$$
\begin{aligned}
\left(\bar{\psi}^L_{\nu_e}\gamma^\beta\psi^L_e W^+_\beta\right)^\dagger &= \left(\psi^\dagger_{\nu_e}\gamma^0\gamma^\beta\tfrac{1}{2}\left(1-\gamma^5\right)\psi_e W^+_\beta\right)^\dagger \\
&= \psi^\dagger_e\tfrac{1}{2}\left(1-\left(\gamma^5\right)^\dagger\right)\left(\gamma^\beta\right)^\dagger\left(\gamma^0\right)^\dagger\psi_{\nu_e}W^-_\beta = \psi^\dagger_e\tfrac{1}{2}\left(1-\gamma^5\right)\left(\gamma^\beta\right)^\dagger\gamma^0\psi_{\nu_e}W^-_\beta \\
&= \psi^\dagger_e\tfrac{1}{2}\left(1-\gamma^5\right)\gamma^0\gamma^\beta\psi_{\nu_e}W^-_\beta = \psi^\dagger_e\gamma^0\tfrac{1}{2}\left(1+\gamma^5\right)\gamma^\beta\psi_{\nu_e}W^-_\beta \\
&= \bar{\psi}_e\gamma^\beta\tfrac{1}{2}\left(1-\gamma^5\right)\psi_{\nu_e}W^-_\beta = \bar{\psi}^L_e\gamma^\beta\psi^L_{\nu_e}W^-_\beta ,
\end{aligned}
\tag{9-233}
$$

which is what we stated without proof in (9-213). But note, had we presumed fermion fields in Hermitian conjugation anti-commute, we would have introduced a minus sign in the third line of (9-233) and gotten the wrong answer.

9.7 Problems

1. Show that hypercharge is conserved in the interaction $\mu^{-L} \rightarrow W^- + \nu^L_\mu \rightarrow \bar{\nu}^R_e + e^{-L} + \nu^L_\mu$.
2. Show that ${}_h j^\mu_{qrk} = -\frac{1}{6}\bar{\Psi}^L_q\gamma^\mu\Psi^L_q - \frac{2}{3}\bar{\psi}^R_{q_u}\gamma^\mu\psi^R_{q_u} + \frac{1}{3}\bar{\psi}^R_{q_d}\gamma^\mu\psi^R_{q_d}$.
3. Show the isospin charge I^W_3 on an LC electron, in number of increments of g, is –½; on an RC electron, zero; on an RC positron +½; on an LC neutrino (of any generation), +½; and on a multiparticle state with an LC electron, an RC electron, and an RC antineutrino, is – 1.
4. Show the isospin charge is conserved for the interaction of Problem 1.
5. Prove that $D = e^{\frac{1}{2}\gamma^0\gamma^k v_k}$ is not unitary.
6. Deduce the Lorentz group transformation relation of γ^5. Hint: Use the invariance of the last term in (9-76) in the manner we followed to find (9-75), which is the transformation relation for γ^μ.
7. Show that $\bar{\psi}_e\gamma^5\psi_e$ transforms like a scalar under proper Lorentz transformations.
8. Prove that $C\left(c_2(\mathbf{p})\psi^{(2)}_\mathbf{p}\right) = -i\gamma^2\left(c_2(\mathbf{p})u_2(\mathbf{p})e^{-ipx}\right)^* = c^\dagger_2(\mathbf{p})v_2(\mathbf{p})e^{ipx} = c^\dagger_2(\mathbf{p})\psi^{(3)}_\mathbf{p}$. Use the standard rep in your proof.
9. Show $\bar{\psi}\gamma^5\psi \overset{C}{\Rightarrow} \bar{\psi}^c\gamma^5\psi^c = \bar{\psi}\gamma^5\psi$.
10. Find $j^\mu_{axi} = \bar{\psi}_1\gamma^\mu\gamma^5\psi_2 \overset{C}{\Rightarrow} j^{c\,\mu}_{axi} = \bar{\psi}_2\gamma^\mu\gamma^5\psi_1$. Hint: Follow the same steps as for the vector four-current, but note when we transpose, we have to flip positions of γ^5 and γ^μ, which we didn't have to do for the vector four-current. Note that $(\gamma^5)^T = \gamma^5$ in the standard basis, as $(\gamma^5)^\dagger = \gamma^5$, in general, and in the standard basis, γ^5 is real.
11. Prove $\bar{\psi}_1\left(t,x^i\right)\gamma^\mu\gamma^5\psi_2\left(t,x^i\right) \overset{P}{\Rightarrow} -\bar{\psi}_1\left(t,-x^i\right)\gamma_\mu\gamma^5\psi_2\left(t,-x^i\right)$.
12. Show that the *CP* transform of $-\frac{g}{\sqrt{2}}\bar{\psi}^L_{\nu_e}W^+\psi^L_e$ is $-\frac{g}{\sqrt{2}}\bar{\psi}^L_e W^-\psi^L_{\nu_e}$.
13. Prove $\bar{\psi}_1\gamma^5\psi_2 \overset{T}{\Rightarrow} \bar{\psi}_1\gamma^5\psi_2$. Hint: Use the standard rep where γ^5 is real.
14. Show $\bar{\psi}_1\gamma^\mu\gamma^5\psi_2 \overset{T}{\Rightarrow} \bar{\psi}_1\gamma_\mu\gamma^5\psi_2$. Hint: Use the standard rep where γ^5 is real.

Chapter 10

Neutrinos

"I started writing about neutrinos because I love them. They are quite magical really; the Universe is completely swarming with them (they are the second most abundant particle after photons) but we know practically nothing about them. And what we do know is really weird."
Lily Asquith, particle physicist

10.0 Preliminaries

10.0.1 History of the Neutrino

If you are reading this book, you have no doubt heard a lot about the neutrino, its history, and its distinctive properties. Nevertheless, we will review some of these briefly, before engaging in some of the underlying mathematics governing the particle's behavior.

The fascinating story of the neutrino

The neutrino was first suggested by Wolfgang Pauli in 1930 in an effort to explain the seeming disappearance of energy, momentum, and angular momentum during beta decay. Others, like Bohr, tried to explain the phenomenon in other ways, such as energy in quantum mechanics only being conserved statistically. Pauli felt it should be carried away by a non-detectable particle, which he originally called a "neutron". But in 1932, when James Chadwick discovered what we today call a neutron, Enrico Fermi, one of very few who originally embraced Pauli's idea, dubbed it a neutrino ("little neutral one" in Italian).

In 1934, Fermi wrote a paper describing his then new theory of beta decay, which included neutrinos and unified them with the positron proposed by Dirac, but it was not well received. *Nature* rejected it, saying it was "too remote from reality". It was later accepted by an Italian journal, but basically ignored by the greater physics community.

However, in time, the accumulation of experimental measurements of the energy spectra of beta particles showed there is a strict limit on the energy of electrons from beta decay, that limit being the difference between rest mass-energies of the original neutron and the final proton. Such a limit should not exist if the conservation of energy were invalid. The most natural explanation was that a yet unidentified particle was carrying away the missing energy, and the Pauli-Fermi position gained traction.

In 1956, Clyde Cowan, Frederick Reines, F. B. Harrison, Herald W. Kruse, and A. D. McGuire detected the neutrino, a result that was rewarded almost forty years later with the 1995 Nobel Prize to Reines.

In 1962, Leon M. Lederman, Melvin Schwartz and Jack Steinberger detected a second family of neutrino (muon neutrino) and later won the 1998 Nobel Prize for their work. The third family neutrino (tau neutrino) was discovered experimentally in 1975.

For decades, neutrinos were thought to be massless, and the SM was conceived and evolved under that presumption.

In the late 1960s, the Homestake experiment (sometimes called the Davis experiment), headed by experimentalist Raymond Davis, Jr. and theorist John N. Bahcall, found only about one third of the neutrinos coming from the sun that theory predicted. According to theory, sun production via fusion is virtually 100% electron neutrinos, and detection methods, for neutrinos interacting with electrons in Earth-based detectors, measured only electron neutrinos. In time, other experiments measured the same neutrino deficiency, and this became known as the solar neutrino problem, which remained unsolved for three decades. Davis ended up sharing the 2002 Nobel Prize for the discovery, along with Masatoshi Koshiba, who did related work in Japan.

The resolution of the problem was actually first put forth in 1957 by Bruno Pontecorvo, who proposed that neutrinos (and antineutrinos) could change families (oscillate) as they propagated. Five years later Maki, Nakagawa, and Sakata formed the conceptual foundation for the quantitative theory of neutrino flavor oscillation, which was further elaborated on by Pontecorvo in 1967. The theory is grounded in the PMNS matrix, named after these four theorists, that we studied in Chap. 7. According to that theory, and as we are about to study, neutrinos, if (and only if) they have mass, morph continually from one flavor to another as they propagate. So, after a time, any neutrino flux would be comprised of roughly one third of each flavor (electron, muon, and tau neutrinos). And thus, a device such as those used in this time period, that reacted solely to electron neutrinos, would only measure a fraction of all the neutrinos in the flux.

Near the turn of the century, experiments began, primarily at the Sudbury Neutrino Observatory and the Super-Kamiokande reactor, that could detect different neutrino flavors, the former from the sun and the latter, from nuclear reactors. These confirmed the neutrino oscillation model, and led to the awarding of the 2015 Nobel Prize to Takaaki Kajita of Japan and Arthur B. McDonald of Canada. And this confirmed that neutrinos do, indeed, have mass.

10.0.2 Neutrino Facts

Neutrino odds and ends

The main source of neutrinos in the universe is the Big Bang, as relic neutrinos from it have been traveling the universe ever since. They number about 330 per cm^3, which makes them more numerous than all the atoms that make up all the stars, planets, black holes, dust, and rocks in the cosmos, all of which combined have a baryon number density of 2.5X10^{-7} per cm^3. In the standard model particle inventory, neutrinos are only outnumbered by photons, which have a density of about 410 per cm^3.

Active galactic nuclei produce neutrinos in staggering numbers, as do supernova explosions, which blast out on the order of 10^{58} of them in a few seconds (and which constitute 99% or so of the supernova energy).

The main source of neutrinos on Earth, however, is the sun. Starting with the fusion of two protons, nuclear reactions in the core of the Sun produce about 2X10^{38} electron neutrinos per second. Though only one neutrino is emitted for every million photons, approximately 3% of the total energy radiated by the sun is in the form of neutrinos.

The sun sends neutrinos towards us at such a rate that hundreds of billions of them pass through your eyeballs every second. You don't see or feel them, of course, because, as we learned, they only interact via the weak force, which is feeble indeed. Neutrinos are truly ghostlike in nature.

Other sources on Earth include the interactions of high energy cosmic rays in the upper atmosphere, radioactive elements within the planet, human-made nuclear reactors, and particle accelerators like those at CERN.

10.0.3 Chapter Overview

Topics in this chapter

In this chapter we will investigate

- neutrino oscillations,
- neutrinos as possible Majorana particles, and
- the see-saw mechanism, a conjectured explanation for the miniscule neutrino masses.

10.1 Neutrino Oscillations

10.1.1 Physical Overview

In the conventional basis leptons do not change generations at a vertex

Recall from Chap. 7 (particularly, pgs. 229-231) that under the conventional lepton basis, which we call Basis I, lepton families do not mix at vertices. As defined in that basis, an electron can share a vertex with an electron neutrino, but not with a muon or tau neutrino. So, interactions with electrons (as opposed to muons and taus), such as predominate in the sun, would produce electron neutrinos, and some of those would head towards the Earth.

However, in that basis, the electron neutrino is not in a mass eigenstate, but is a superposition of three mass eigenstates. Upon measurement of its mass, the electron neutrino state would collapse, randomly, to one of the three mass values, corresponding to one of the three mass eigenstates. So, electron neutrino propagation actually consists of propagation of three individual component states. In RQM, we would think of this as three mass eigenstate wave functions added together, with appropriate amplitudes, to comprise a general (electron flavor) wave function. The magnitude of each amplitude would govern the probability of measurement of the particular mass value of that mass eigenstate.

But in that basis neutrinos not in mass eigenstates

The wrinkle here is that each mass eigenstate propagates at a different speed, since the speed (phase velocity) of a quantum mechanical wave (of form as in (10-1)) is a function of its mass. For example, for the mass eigenstate labeled 1,

Mass eigenstates propagate at different speeds

$$\text{in natural units,}\quad E_1=\sqrt{\mathbf{p}_1^2+m_1^2}\qquad \omega_1=E_1\qquad \text{with wave form }\; e^{-i\left(E_1 t-\mathbf{p}_1\cdot\mathbf{x}\right)}=e^{-i\left(\omega_1 t-\mathbf{k}_1\cdot\mathbf{x}\right)}$$
$$\text{phase velocity }\; v_1=\frac{\omega_1}{k_1}=\frac{E_1\left(\mathbf{p}_1,m_1\right)}{\left|\mathbf{p}_1\right|}. \tag{10-1}$$

The phase velocity of each such state varies with its mass. Hence, in our superposition, the mass eigenstates all propagate with different speeds and interfere with one another. In so doing, they change the shape of the wave, and this changing goes on continually as the state propagates.

The interference means some parts of one mass eigenwave may constructively interfere with one of the other mass eigenwaves at some points in spacetime and destructively interfere with it at other points. But, each mass eigenstate is, in turn, itself composed of flavor eigenstates, so we are also getting interference between the flavor eigenstates of each mass eigenstate. Where those flavor eigenstates of one particular flavor constructively interfere, we would see increased probability of measuring that flavor; where they destructively interfere, a decreased probability.

Different mass eigenstate waves interfere

Hence, as the initial state propagates, we see a continual shifting in probabilities of measurement of different flavors. When $t = 0$, the state is 100% one particular flavor. But, over time, the other two flavors become more and more likely of being measured, though the probabilities themselves oscillate continually.

So, probability of measuring a given flavor oscillates

Note that if neutrinos all had the same mass, then there would be no difference in component wave speed and no flavor oscillation. This includes the case where all neutrinos are massless, so detection of neutrino flavor oscillation is proof that neutrinos have mass.

Of course, if all neutrinos had the same mass, then flavor eigenstates would also be mass eigenstates, and we would not have any of this extensive analysis. But Nature is not so simple. It is what it is.

If all neutrinos had same mass (including zero), then all waves at same speed → no oscillation

Bottom line: A neutrino is produced with a particular flavor, but propagation is governed by mass, so mass eigenstate components propagate at different speeds and interfere with each other. The mass eigenstates are themselves composed of flavor eigenstates, and as these mass eigenstates interfere, the probability of measuring a given flavor neutrino changes (oscillates) over time.

10.1.2 The Mathematics of Neutrino Oscillation

We will think in terms of neutrinos coming from the sun, but our conclusions will be valid for any source. In the sun, the main source of neutrinos is the fusion of two hydrogen nuclei (protons), one stage in the process of forming deuterium. (This subsequently interacts with another proton to yield helium and a gamma ray.) There are several ways to write this.

$$ {}^1_1H+{}^1_1H\;\rightarrow\;{}^2_1H+e^++\nu_e\quad\text{OR}\quad p^++p^+\rightarrow p^+n+e^++\nu_e$$
$$\text{OR}\quad uud+uud\rightarrow uud+udd+e^++\nu_e\quad\text{OR}\quad u\rightarrow d+W^+\rightarrow d+e^++\nu_e\,. \tag{10-2}$$

Most common neutrino reaction in sun

Now, we know the electron neutrino is a superposition of three mass eigenstates related by the PMNS matrix we studied in Chap. 7. From (7-93), where, as noted later in that section, we drop the primes and wavy underlines that helped us there pedagogically ($\alpha = e,\ \mu,\ \tau$ signifies flavor, and i = 1,2,3, mass eigenstates of the field),

$$\begin{pmatrix}\psi^L_{\nu_e}\\ \psi^L_{\nu_\mu}\\ \psi^L_{\nu_\tau}\end{pmatrix}=\underbrace{\begin{bmatrix}P_{e1} & P_{e2} & P_{e3}\\ P_{\mu 1} & P_{\mu 2} & P_{\mu 3}\\ P_{\tau 1} & P_{\tau 2} & P_{\tau 3}\end{bmatrix}}_{\text{PMNS matrix } P}\begin{pmatrix}\psi^L_{\nu_1}\\ \psi^L_{\nu_2}\\ \psi^L_{\nu_3}\end{pmatrix} \quad \text{or} \quad \psi^L_{\nu_\alpha}=P_{\nu_\alpha \nu_i}\psi^L_{\nu_i}\,. \qquad (10\text{-}3)$$

PMNS matrix, relating flavor to mass eigenstates

Neutrino oscillation is most easily and simply treated by analyzing states in RQM. As the fields create and destroy states having the same mixture of flavor and mass eigenstates, we also have (10-4) below (see Appendix A) for neutrino states (where we henceforth streamline by dropping the "L" superscript and take subscript ν_α as simply α),

Oscillation easiest to handle in RQM

$$|\nu_\alpha\rangle = P^*_{\alpha i}|\nu_i\rangle\,. \qquad (10\text{-}4)$$

In RQM, we can describe the mass eigenstates as plane waves of form

$$|\nu_i(x,t)\rangle = e^{-i(E_i t - p_i x)}|\nu_i(0)\rangle \quad \text{(no sum on } i\text{)}, \qquad (10\text{-}5)$$

Consider mass eigenstates as plane weaves

where we will consider the x axis aligned with $\mathbf{p}$, so p_i here will mean 3-momentum in the direction of travel for the ith mass eigenstate, and the zero argument on the RHS means we are taking the initial time $t = 0$. (See Appendix B for more detail on the form of the initial state.)

Neutrinos are almost invariably ultra-relativistic, as they are so light-weight almost any amount of energy they would pick up in an interaction would have to be primarily in the form of kinetic energy (and thus, very high speed, given the low mass). In that limit, $E_i \approx p_i >> m_i$, and we would have,

$$E_i = \sqrt{p_i^2 + m_i^2} = p_i\sqrt{1+\frac{m_i^2}{p_i^2}} \approx p_i + \frac{m_i^2}{2p_i} \qquad (10\text{-}6)$$

Energy a function of mass

(10-6) into (10-5) gives us

$$|\nu_i(x,t)\rangle = e^{-i\left(p_i t + \frac{m_i^2}{2p_i}t - p_i x\right)}|\nu_i(0)\rangle \qquad \text{(no sum on } i\text{)}. \qquad (10\text{-}7)$$

So, propagation frequency a function of mass

Since ultra-relativistic neutrinos travel close to the speed of light (in practice, within at least one part in 10^6), in natural units $t = L$, where L is the distance the neutrino has traveled from time $t = 0$, i.e., $x = L$, also. Hence, (10-7) can be expressed, where E is the expectation energy of the flavor eigenstate and is approximately equal to E_i ($\approx p_i$), as

$$|\nu_i(L)\rangle = e^{-i\left(p_i L + \frac{m_i^2}{2p_i}L - p_i L\right)}|\nu_i(0)\rangle = e^{-i\frac{m_i^2}{2p_i}L}|\nu_i(0)\rangle \approx e^{-i\frac{m_i^2}{2E}L}|\nu_i(0)\rangle \quad \text{(no sum on } i\text{)}. \qquad (10\text{-}8)$$

Ultra-relativistic wave function

For flavor eigenstates, with (10-4) in (10-8), we have (where we assign the indices i and j, and α and β, in a manner that will help later and now sum on i and j)

$$\begin{aligned}|\nu_\beta(L)\rangle &= P^*_{\beta j}|\nu_j(L)\rangle \approx P^*_{\beta j}e^{-i\frac{m_j^2}{2E}L}|\nu_j(0)\rangle \qquad & \langle\nu_\beta(L)| &\approx \langle\nu_j(0)|P_{\beta j}e^{i\frac{m_j^2}{2E}L}\\ |\nu_\alpha(0)\rangle &\approx P^*_{\alpha i}|\nu_i(0)\rangle & \langle\nu_\alpha(0)| &\approx \langle\nu_i(0)|P_{\alpha i}.\end{aligned} \qquad (10\text{-}9)$$

Ultra-relativistic neutrino flavor state in terms of mass eigenstates

In what follows, we won't actually need the bra for $t = 0$, but we provide it for completeness.

The transition amplitude for measuring a flavor state β, when the initial flavor state was α is

$$\langle\nu_\beta(L)|\nu_\alpha(0)\rangle = \langle\nu_j(0)|e^{i\frac{m_j^2}{2E}L}P_{\beta j}P^*_{\alpha i}|\nu_i(0)\rangle = e^{i\frac{m_j^2}{2E}L}P_{\beta j}P^*_{\alpha i}\,\delta_{ij} = e^{i\frac{m_i^2}{2E}L}P_{\beta i}P^*_{\alpha i}\,. \qquad (10\text{-}10)$$

Transition amplitude from one flavor state to another

You can do **Problem 1** if you would like to deduce relation (10-10) via QFT. If you do, you should read Appendix B of this chapter first.

The probability is the square of the absolute value of (10-10), (where underlining means no sum on α and β)

$$P_{\alpha\to\beta}=\left|\langle\nu_\beta(L)|\nu_\alpha(0)\rangle\right|^2=\left(e^{-i\frac{m_j^2}{2E}L}P_{\underline{\alpha}j}P^*_{\underline{\beta}j}\right)\left(e^{i\frac{m_i^2}{2E}L}P_{\underline{\beta}i}P^*_{\underline{\alpha}i}\right)=e^{i\frac{L}{2E}\left(m_i^2-m_j^2\right)}P_{\underline{\alpha}j}P^*_{\underline{\beta}j}P_{\underline{\beta}i}P^*_{\underline{\alpha}i}$$
$$=\left(\cos\left(\frac{L}{2E}\left(m_i^2-m_j^2\right)\right)+i\sin\left(\frac{L}{2E}\left(m_i^2-m_j^2\right)\right)\right)\left(Re\left\{P_{\underline{\alpha}j}P^*_{\underline{\beta}j}P_{\underline{\beta}i}P^*_{\underline{\alpha}i}\right\}+i\,Im\left\{P_{\underline{\alpha}j}P^*_{\underline{\beta}j}P_{\underline{\beta}i}P^*_{\underline{\alpha}i}\right\}\right). \qquad (10\text{-}11)$$

Probability from transition amplitude

The square of the absolute value must be real, so after multiplying factors in (10-11), any imaginary quantities must sum to zero. That leaves us with

$$P_{\alpha\to\beta}=\cos\left(\frac{L}{2E}\left(m_i^2-m_j^2\right)\right)Re\left\{P_{\underline{\alpha}j}P^*_{\underline{\beta}j}P_{\underline{\beta}i}P^*_{\underline{\alpha}i}\right\}-\sin\left(\frac{L}{2E}\left(m_i^2-m_j^2\right)\right)Im\left\{P_{\underline{\alpha}j}P^*_{\underline{\beta}j}P_{\underline{\beta}i}P^*_{\underline{\alpha}i}\right\}. \qquad (10\text{-}12)$$

Now, if the PMNS matrix were complex, it would lead to CP violation, just as the CKM matrix being complex does. But, there has been no evidence of CP violation with neutrinos, so we can consider the factor $Im\left\{P_{\underline{\alpha}j}P^*_{\underline{\beta}j}P_{\underline{\beta}i}P^*_{\underline{\alpha}i}\right\}$ equals zero, or at least is immeasurably close to it. This is supported (or at least not at present contradicted) by (7-106), which shows the only complex factor $e^{-i\delta}$ in any PMNS matrix components has δ in the neighborhood of 180^0. Thus, we will ignore the last term in (10-12).

No evidence PMNS matrix is complex, so drop imaginary part

Using the trig relation

$$\cos 2\theta=1-2\sin^2\theta \quad\rightarrow\quad \cos 2\left(\theta_i-\theta_j\right)=1-2\sin^2\left(\theta_i-\theta_j\right), \qquad (10\text{-}13)$$

(10-12) becomes, where in the underbrackets, we use (7-88),

$$P_{\alpha\to\beta}=\left(1-2\sin^2\left(\frac{L}{4E}\left(m_i^2-m_j^2\right)\right)\right)\left(P_{\underline{\alpha}j}P^*_{\underline{\beta}j}P_{\underline{\beta}i}P^*_{\underline{\alpha}i}\right)$$
$$=\underbrace{\left(P_{\underline{\alpha}j}P^*_{\underline{\beta}j}P_{\underline{\beta}i}P^*_{\underline{\alpha}i}\right)}_{\substack{P_{\underline{\alpha}j}\left(P^*_{j\underline{\beta}}\right)^T P_{\underline{\beta}i}\left(P^*_{i\underline{\alpha}}\right)^T=P_{\underline{\alpha}j}P^\dagger_{j\underline{\beta}}P_{\underline{\beta}i}P^\dagger_{i\underline{\alpha}}\\ =\left(PP^\dagger\right)_{\underline{\alpha\beta}}\left(PP^\dagger\right)_{\underline{\beta\alpha}}=\delta_{\underline{\alpha\beta}}\delta_{\underline{\beta\alpha}}}}-2\sin^2\left(\frac{L}{4E}\left(m_i^2-m_j^2\right)\right)\left(P_{\underline{\alpha}j}P^*_{\underline{\beta}j}P_{\underline{\beta}i}P^*_{\underline{\alpha}i}\right) \qquad (10\text{-}14)$$
$$=\delta_{\alpha\beta}-2\sin^2\left(\frac{L}{4E}\left(m_i^2-m_j^2\right)\right)\left(P_{\underline{\alpha}j}P^*_{\underline{\beta}j}P_{\underline{\beta}i}P^*_{\underline{\alpha}i}\right).$$

Master relation: probability from one flavor state to another

The only thing that varies in (10-14) is the distance L traveled. At different values for L, the probability of measuring the β flavor state changes. In fact, it oscillates sinusoidally with distance, just as one commonly hears. At $L = 0$, probability of finding the α flavor state is 1, which, of course, is what we would expect and lends credibility to the derived relation.

For given masses and energy level, it only depends on distance

Note that if neutrinos were massless, or all had the same mass, we would see no oscillation, and the original flavor would not change. So, again we note that detection of neutrino oscillations proves they have mass.

The mass factor in the argument of the $\sin^2$ factor is commonly written

$$\Delta m_{ij}^2=m_i^2-m_j^2, \qquad (10\text{-}15)$$

and the argument itself, in units other than natural, is

$$\frac{L}{4E}\Delta m_{ij}^2 \xrightarrow[\text{units}]{\text{non-natural}} \frac{\Delta m_{ij}^2c^3L}{4\hbar E} \xrightarrow[E\text{ in GeV}]{m^2\text{ in eV}^2,\ L\text{ in km}} 1.27\Delta m_{ij}^2\frac{L}{E}. \qquad (10\text{-}16)$$

Probability $\sin^2$ factor argument, non-natural units

You may have heard that neutrino oscillation detection can only pin down the difference in the squares of the masses, and not the actual masses. Now, from (10-14), you can see why.

In using (10-14), we keep the α and β flavor states fixed, but sum over i and j, which gives us a number of sine squared terms added for each pair of α and β. As P is a 3X3 matrix, the relation is very cumbersome. However, we can illustrate its use by presuming there are only two generations of neutrinos and thus, work with a simpler 2X2 matrix.

10.1.3 A Simplified Example: Only Two Neutrino Generations

Two-flavor example

While it can be illustrative to work with (10-14) using only two neutrino generations, there actually are practical cases where it is appropriate. In cosmic ray interactions with atmospheric nucleons, for example, the resulting neutrino mixing is primarily between muon and tau neutrinos, with the electron neutrino playing little role. And for solar neutrinos, one can, for computational purposes, consider the superposition of ν_μ and ν_τ to be a single generation.

So, let's consider a 2X2 version of the PMNS matrix, which parallels the Cabibbo matrix (7-122), a 2X2 approximation of the CKM matrix for quarks. From (7-105), where $\theta_{12} = \theta$, $\theta_{13} \approx 0$, $\theta_{23} \approx 0$,

2X2 approximation of PMNS matrix

$$P \approx \begin{bmatrix} \cos\theta & \sin\theta & \\ -\sin\theta & \cos\theta & \\ & & 1 \end{bmatrix} \xrightarrow[\text{to 2D}]{\text{3D}} P \approx \begin{bmatrix} \cos\theta & \sin\theta \\ -\sin\theta & \cos\theta \end{bmatrix}. \tag{10-17}$$

So, with $\Delta m_{12}^2 = \Delta m^2$, (10-14) becomes

$$P_{\substack{\alpha\to\beta,\alpha\neq\beta \\ i,j\to 1,2}} = -2\left(\sin^2\left(\frac{\Delta m^2 L}{4E}\right) P_{\beta 2} P_{\underline{\alpha} 2} P_{\beta 1} P_{\underline{\alpha} 1} + \sin^2\left(\frac{-\Delta m^2 L}{4E}\right) P_{\beta 1} P_{\underline{\alpha} 1} P_{\beta 2} P_{\underline{\alpha} 2} \right). \tag{10-18}$$

In the evaluation of (10-18), in the matrix P, we take $\alpha = 1$ (for the first fixed flavor α) and $\beta = 2$ for the second fixed flavor), and note that $\sin^2\theta = \sin^2(-\theta)$ to get

$$\begin{aligned} P_{\substack{\alpha\to\beta,\alpha\neq\beta \\ i,j\to 1,2}} &= -2\sin^2\left(\frac{\Delta m^2 L}{4E}\right)\left(P_{22}P_{12}P_{21}P_{11} + P_{21}P_{11}P_{22}P_{12}\right) \\ &= -2\sin^2\left(\frac{\Delta m^2 L}{4E}\right)\left(\cos\theta\sin\theta(-\sin\theta)\cos\theta + (-\sin\theta)\cos\theta\cos\theta\sin\theta\right) \\ &= 4\sin^2\left(\frac{\Delta m^2 L}{4E}\right)\sin^2\theta\cos^2\theta. \end{aligned} \tag{10-19}$$

Using the trig relation

$$\sin(2\theta) = 2\sin\theta\cos\theta \quad \rightarrow \quad \sin^2\theta\cos^2\theta = \tfrac{1}{4}\sin^2(2\theta) \tag{10-20}$$

in (10-19), along with (10-16) and (10-14), gives us

Probabilities for measuring either of two flavors

$$\begin{aligned} P_{\substack{\alpha\to\beta,\alpha\neq\beta \\ i,j\to 1,2}} &= \sin^2(2\theta)\sin^2\left(\frac{\Delta m^2 L}{4E}\right) \xrightarrow[E \text{ in GeV}]{m^2 \text{ in eV}^2,\ L \text{ in km}} = \sin^2(2\theta)\sin^2\left(1.27\frac{\Delta m^2 L}{E}\right) \\ P_{\substack{\alpha\to\beta,\alpha=\beta \\ i,j\to 1,2}} &= 1 - \sin^2(2\theta)\sin^2\left(1.27\frac{\Delta m^2 L}{E}\right), \end{aligned} \tag{10-21}$$

where the second line in (10-21) is found by requiring total probability of both the $\alpha = \beta$ and $\alpha \neq \beta$ cases to be unity.

These probabilities are plotted in Fig. 10-1.

Plot of those probabilities

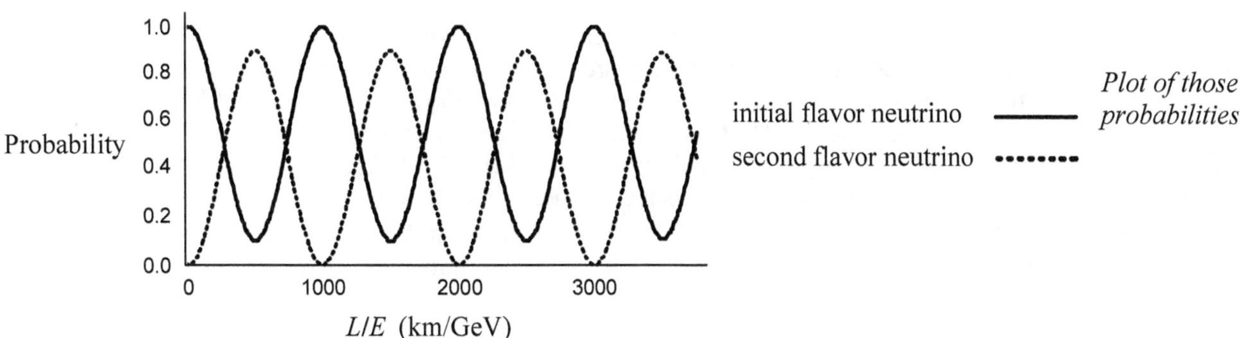

Figure 10-1. Oscillations between Two Neutrino Generations

We start with 100% chance of measuring the first flavor and once per cycle, again have 100% chance of doing so. The maximum probability of measuring the second flavor neutrino is $sin^2(2\theta)$, and the minimum is zero. If Δm^2 is larger, we get more tightly squeezed oscillation in space (shorter wavelengths).

10.1.4 Three Neutrino Generations

We will not do the calculation for the three-generation case, but merely show a typical graph from those who actually do such things[1].

Plot of probabilities for three flavors

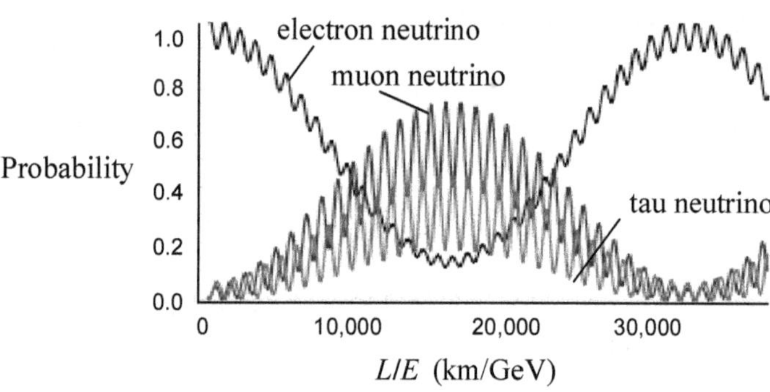

Figure 10-2. Oscillation of Initial Electron Neutrino for Three Generations

The figure uses 2015 vintage data for the mixing angle parameters θ_{ij} and assumes $\delta = 0^\circ$ (or equivalently 180°), i.e., that the PMNS matrix is real. If the PMNS matrix is not real, not only will the probabilities change, but those for antineutrinos will be different than for neutrinos (due to CP violation).

Fig. 10-2 also assumes what is called the normal mass hierarchy, i.e.,

Normal mass order

$$m_1 \leq m_2 \leq m_3 \,. \tag{10-22}$$

Since oscillations only measure differences between the squares of the masses, there is currently (2025) no way to tell the actual value of the individual masses, although certain other experiments have been able to establish limits[2]. Another possible mass ordering is the inverted mass hierarchy

Inverted mass order

$$m_3 \leq m_1 \leq m_2 \,. \tag{10-23}$$

The question of which is correct is known as the neutrino mass hierarchy problem.

Since only the squared mass differences are measured, it is possible that the lowest neutrino mass is precisely zero, without contradicting experiment, though few theorists think that likely.

Do **Problem 2** to see how only knowing the differences in the squares of the masses does not allow us to determine the actual masses of each neutrino.

10.2 Neutrinos as Possible Majorana Particles

Antineutrinos and neutrinos are electrically neutral particles, so they could possibly be the same particle. As we learned in Chap. 5, particles that have this property are hypothetical at this point and go by the name Majorana particles. Neutrinos that might be Majorana in nature, though electrically neutral, would still have different chirality. Differences between neutrinos and antineutrinos observed in experiment could be merely the result of them having two different chiralities. As of 2025, it is not known whether neutrinos are Majorana, or the type we have been studying, called Dirac fermions.

Neutrinos could be Majorana fermions, but no one knows for sure

In principle, the true nature of neutrinos can be determined experimentally, but doing so is far from trivial. For example, with Dirac neutrinos, we would need a neutrino and an antineutrino in order

---

[1] M. Balázs, https://demonstrations.wolfram.com/NeutrinoOscillations.

[2] See, for example, a summary of the KATRIN experiment, which, as of late 2019, had established a maximum upper limit for the electron neutrino of about 1 eV, at https://physics.aps.org/articles/v12/129.

to see neutrino annihilation. But with Majorana neutrinos, two of them, produced from the identical interactions (so we know they are the same particle), could annihilate, since each, being its own antiparticle, is also the antiparticle of the other. However, as we know, observing neutrinos is notoriously difficult, and they interact with one another so rarely that detecting two of them annihilating in the same interaction is a challenge of the highest order. Yet, such experiments have been going on for some time, and as of 2025, none have yielded anything close to conclusive results.

Several of these are looking for neutrinoless double beta decay, where two neighboring protons, in the same or different nuclei, would decay producing antineutrinos, and these would annihilate, leaving a neutrinoless result.

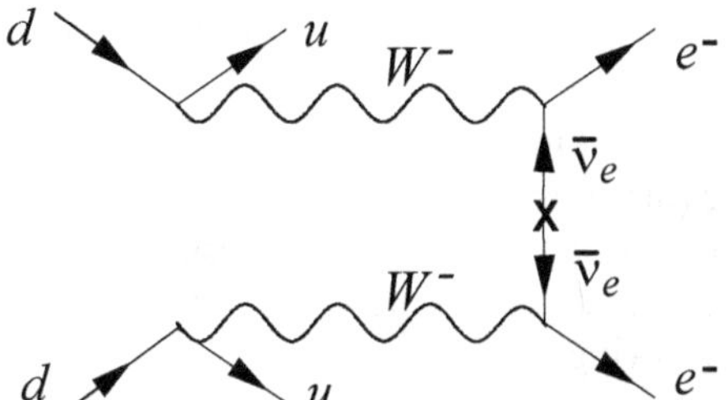

Neutrinoless double beta decay experiment could one day prove they are Majorana

Figure 10-3. Neutrinoless Double Beta Decay Feynman Diagram with $\bar{\nu}_e = \nu_e$

This would obviously be a lepton-number violating process, as we started with zero total lepton number, but end up, counting the two electrons, with a lepton number of two. So, Majorana fermions would violate lepton number conservation, whereas Dirac fermions obey it.

Probes of the cosmic neutrino background are also seeking to determine whether the neutrinos are Dirac or Majorana.

In general, though, in almost all cases, Majorana neutrinos behave virtually indistinguishably from Dirac neutrinos.

10.3 The Seesaw Mechanism

10.3.1 Notation Variation

It is common in the literature to designate spinor fields with a shorthand notation, such as follows, where superscript "*c*" means charge conjugation (*C* transformation),

Alternative (common) notation

$$\psi_e \to e \qquad \psi_\nu \to \nu \qquad \psi_\nu^L \to \nu_L \qquad \bar{\psi}_\nu^L \to \bar{\nu}_L \qquad \psi_\nu^{Lc} \to \nu_L^c \qquad \bar{\psi}_\nu^{Lc} \to \bar{\nu}_L^c . \qquad (10\text{-}24)$$

To help us get used to the alternate notation, we will use it exclusively in this section.

In summary, the spinor fields in the new shorthand notation (for neutrinos, as that will be our immediate focus) have the properties listed in Wholeness Chart 10-1.

Quantum field operator properties in alternate notation

Wholeness Chart 10-1. Properties of Fields in Alternative Notation

ν_L	destroys an LC neutrino and creates an RC antineutrino,
$\bar{\nu}_L$	creates " " " and destroys " " "
ν_L^c	creates " " " and destroys " " " (does same as $\bar{\nu}_L$)
$\bar{\nu}_L^c$	destroys " " " and creates " " " (does same as ν_L)

and for *R* subscript, interchange $L \leftrightarrow R$ everywhere above

Note that the subscript *L* or *R* always refers to particles. For a non-conjugated field, no overbar means destroys particles, overbar means creates particles, and antiparticle actions for the same field are just reversed from particle actions (particle ↔ antiparticle, LC ↔ RC, destroy ↔ create).

Charge conjugation has same effect as overbar

Charge conjugating a field has the same effect on particle/antiparticle and creation/destruction as an overbar (overbar is effectively a complex conjugate transpose (plus a γ^0 multiplication)). That is, the overbar and the superscript "c" have the same effect. The charge conjugation merely lets us have the overbar (row) operator effect in a non-overbar (column) vector.

10.3.2 Background

It may seem unusual to have such low values for masses of neutrinos, when all other particles like electrons, quarks, etc. are much heavier, with their masses relatively closely grouped. (See Fig. 7-6, pg. 239.) Given that particles get mass via the Higgs mechanism, why, for example, should neutrinos be 10^6 times less massive than the electron and all their other sibling elementary particles. That is, why would the coupling to the Higgs field be so many orders of magnitude less? This seems unnatural, at best.

Extremely low neutrino mass seems unnatural

One might not be too surprised if the Higgs-neutrino coupling were zero, giving rise to zero mass particles. One might likewise not be too surprised if the coupling resulted in masses on the order of the Higgs, or even grand unified theory (GUT), symmetry breaking scale.

Consider the quite reasonable possibility that after symmetry breaking, two types of neutrino exist, one having zero mass (no Higgs coupling) and the other having (large) mass of the symmetry breaking scale. As we will see, it turns out that reasonable superpositions of these fields can result in light neutrinos (like those observed) and a very heavy neutrino (of symmetry breaking scale, and therefore too heavy to be observed). We detail this mechanism in the following sections.

Zero mass and heavy mass neutrinos coupled to Higgs, could still yield what we see

10.3.3 Fundamental Math Concept Underlying the Seesaw Mechanism

Consider a real, two-dimensional space with a matrix (tensor) expressed in one set of orthonormal basis vectors (wavy overbar symbol) for that space as

Math background that will help us

$$\tilde{\mathcal{M}} = \begin{bmatrix} 0 & 0 \\ 0 & 100 \end{bmatrix}. \qquad (10\text{-}25)$$

Start with this matrix

Now, if we consider a new set of basis vectors (no wavy overbar symbol), rotated by an angle ϕ from the original basis, then the matrix components change, of course, and can be found by

$$\mathcal{M} = \begin{bmatrix} \cos\phi & \sin\phi \\ -\sin\phi & \cos\phi \end{bmatrix}\begin{bmatrix} 0 & 0 \\ 0 & 100 \end{bmatrix}\begin{bmatrix} \cos\phi & -\sin\phi \\ \sin\phi & \cos\phi \end{bmatrix} = \begin{bmatrix} 100\sin^2\phi & 100\cos\phi\sin\phi \\ 100\cos\phi\sin\phi & 100\cos^2\phi \end{bmatrix}. \qquad (10\text{-}26)$$

Rotate basis by a small angle

Note how this matrix looks if ϕ is small, say $\phi = 2°$, with $\cos\phi = .99939$ and $\sin\phi = .03490$.

$$\mathcal{M} = \begin{bmatrix} .122 & 3.488 \\ 3.488 & 99.878 \end{bmatrix} \qquad (10\text{-}27)$$

Get a matrix like this

We get an upper left diagonal component almost 3 orders of magnitude smaller than the lower right component, which is approximately the same as the original lower right component. The off-diagonal components equal the geometric mean of the diagonal components, i.e., $\sqrt{\left(100\cos^2\phi\right)\left(100\sin^2\phi\right)}$, and are not as small as the upper left component, but significantly smaller that the lower right one.

The fundamental point is that by starting with a matrix of form like (10-25), and transforming to another basis, rotated by a small angle from the original, we get a matrix of form like (10-27).

10.3.4 Dirac vs Majorana Mass Terms in the Lagrangian

Background

We first recall from Chap. 9 that, mathematically, charge conjugation of the neutrino field can be expressed as

$$\nu^c = C\nu = -i\gamma^2\nu^* \qquad \bar{\nu}^c = \bar{\nu}C = i\nu^T\gamma^0\gamma^2. \qquad (10\text{-}28)$$

Recalling charge conjugation for spinors

Possible Mass Term in the Lagrangian

We don't know a great deal, experimentally, about neutrino mass, but on general theoretical grounds, two distinct classes of neutrino mass terms are allowed in the Lagrangian. These are called Dirac and Majorana mass terms.

Two forms of mass terms allowable, called Dirac and Majorana

But note, before we get into it, that Majorana mass terms have nothing to do with the Majorana representation in spinor space. One can use any representation for the fields of which Majorana and

Dirac mass terms are composed. And at first blush, neither do Majorana mass terms imply the associated particles/fields are Majorana fermions, though we will shortly have more to say about this. For now, we will begin by assuming that both Dirac and Majorana mass terms could be either Dirac type particles or Majorana type particles (in any representation we like.)

The Dirac mass terms, which are the usual terms dealt with in introductory QFT, have form, where we are considering mass eigenstate fields, not flavor eigenstate fields, and where only a single mass eigenstate is shown,

Dirac mass terms

$$-m_D\left(\bar{\nu}_L\nu_R+\bar{\nu}_R\nu_L\right) \qquad \left(=-m_D\bar{\nu}\nu\right). \tag{10-29}$$

On the other hand, Majorana mass terms, which may look unfamiliar to the uninitiated, have form

Majorana mass terms

$$-\frac{1}{2}m_M^L\left(\bar{\nu}_L\nu_L^c+\bar{\nu}_L^c\nu_L\right)-\frac{1}{2}m_M^R\left(\bar{\nu}_R\nu_R^c+\bar{\nu}_R^c\nu_R\right), \tag{10-30}$$

Take care that the symbol $\nu_L\nu_L$ is used by some authors for the $\bar{\nu}_L^c\nu_L$ term of (10-30), with similar changes for other terms, where one must keep in mind for such notation that inner product in spinor space is implied, even though there is no obvious transpose term (row vector on left) in $\nu_L\nu_L$.

As shown in Appendix C, half of the terms in (10-29) and (10-30), are Hermitian conjugates (which are complex conjugate transposes and different from the charge conjugates associated with the C transformation) of the other mass terms. That is

Hermitian conjugation in mass type terms

$$\begin{gathered}-m_D\left(\bar{\nu}_L\nu_R+\bar{\nu}_R\nu_L\right)=-m_D\bar{\nu}_L\nu_R+h.c.\\ -\frac{1}{2}m_M^L\left(\bar{\nu}_L\nu_L^c+\bar{\nu}_L^c\nu_L\right)-\frac{1}{2}m_M^R\left(\bar{\nu}_R\nu_R^c+\bar{\nu}_R^c\nu_R\right)=-\frac{1}{2}m_M^L\bar{\nu}_L\nu_L^c-\frac{1}{2}m_M^R\bar{\nu}_R\nu_R^c\ +h.c.\end{gathered} \tag{10-31}$$

Impact of the Majorana Mass Terms on Theory

Note that the first term in (10-29) destroys an RC particle and creates an LC one. This is similar to what we saw for electrons in Feynman diagram Fig. 7-9, pg. 243, which illustrates one way weak isospin charge is not conserved. In the present case, an LC neutrino has +½ weak isospin charge and an RC neutrino has zero weak isospin charge, so again, weak isospin charge is not conserved. Lepton number, however, is conserved, as we started with a neutrino (not an anti-neutrino) and ended up with a neutrino.

Somewhat similarly, the first term in (10-30) can destroy an RC antineutrino (weak isospin charge = 0) and create an LC neutrino and thus, also does not conserve weak charge. But, importantly, in addition, it does not conserve lepton number (which the Dirac mass terms do.) We started with an antineutrino (lepton number –1) and ended up with a neutrino (lepton number +1).

Majorana mass terms mean lepton number non-conservation, particle turns to antiparticle

Wholeness Chart 10-2. Weak Charge and Lepton Number Conservation

	Dirac mass terms	Majorana mass terms
Conserves weak isospin charge I_3^W?	No	No
Conserves hypercharge Y?	No	No
Conserves electric charge number $I_3^W + Y$?	Yes	Yes
Conserves chirality?	No	No
Conserves lepton number?	Yes	No

Majorana Mass Terms Lead to Presuming Neutrinos = Antineutrinos

Because of the Majorana mass terms, free neutrinos can change as they propagate from neutrino to antineutrino (and vice versa), so there is no way to distinguish their particle form from their antiparticle form. Thus, it is generally acknowledged that neutrinos with mass terms such as (10-30) are their own antiparticles, i.e., they are Majorana fermions.

Because particle turns to antiparticle, no difference between them

If neutrinos are their own antiparticles, in order to maintain chirality conservation at an interaction vertex, they would have to be able to assume either left or right chirality. That is, Majorana neutrinos, can be either LC or RC. They would take on different chiralities in interactions (much as the same particle in any interaction can take on different spins and helicities).

So, Majorana mass terms mean neutrinos are Majorana types

Further, since the neutrino fields of (10-29) are the same fields as those of (10-30), the Dirac mass fields must also be Majorana fields. As long as we have Majorana mass terms for neutrinos, whether or not we have Dirac mass terms, the neutrinos in the theory are Majorana type. We can only have Dirac neutrinos if we have Dirac mass terms with no Majorana mass terms.

So, all neutrino fields, even those in Dirac mass terms, are Majorana

Lepton numbers for Majorana neutrinos are essentially meaningless. Interactions with Majorana neutrinos, such as those of Fig. 10-3, can violate lepton number conservation. In practice, any lepton number violation we might discover would have to come from Majorana type neutrinos.

Interactions with lepton number violation must have Majorana fermions

Bottom line: If there are Majorana mass terms in the Lagrangian, all neutrino fields in all terms in the Lagrangian are Majorana. Majorana neutrinos can take on either L or R chirality and interactions with them can violate lepton number conservation.

The Impact on Theory of Neutrinos Being Majorana Fermions

Then, for example, the field ν_L (and $\overline{\nu}_L^c$) in (10-30) would destroy a LC Majorana neutrino of mass m_M^L and create a RC Majorana neutrino of the same mass. It would be the same neutrino with different chirality. $\overline{\nu}_L$ and ν_L^c would do the reverse, create the LC Majorana neutrino and destroy the RC one.

Conversely, the field ν_R (and $\overline{\nu}_R^c$) would destroy a RC Majorana neutrino of mass m_M^R and create a LC Majorana neutrino of the same mass. $\overline{\nu}_R$ (and ν_R^c) would do the reverse.

The intriguing aspect of this is that there are now two different kinds of neutrinos, those with mass m_M^L and those with mass m_M^R.[1] If the latter were extremely heavy, say on the weak symmetry breaking scale, it would never be seen. As a real particle it would decay very rapidly and have long since disappeared from the universe. As a propagator particle, it would be so heavy that the denominator of the propagator would drive the entire propagator to effectively zero.

Two kinds of neutrinos with different masses in Majorana mass terms

If one very heavy, it would never be seen

The Dirac type mass terms (10-29) would then become irrelevant, as the ν_R factor would never find an m_M^R type neutrino to destroy, since none would be around, either as real or virtual particles. Similarly, the $\overline{\nu}_R$ factor could never create such a neutrino, as the mass-energy required to do so would simply not be available in our present universe.

All interaction terms would remain the same, except the neutrinos involved would be considered to be their own antiparticles, and these could take on either left or right chirality. LC neutrinos would interact in the usual way via the weak force, with RC ones being restricted in the usual way.

Bottom line: By adding neutrino Majorana mass terms to the Lagrangian and considering neutrinos to be Majorana fermions, we get neutrinos of two different masses. If one is light and the other very heavy, we get a consistent weak interaction theory.

Expressing in Terms of Non-Diagonal Mass Matrix

It will help if we express (10-29) and (10-30) together in terms of a mass matrix

$$\mathcal{M} = \begin{bmatrix} m_M^L & m_D \\ m_D & m_M^R \end{bmatrix}, \tag{10-32}$$

so then, with (10-31),

$$\begin{aligned}
\mathcal{L}_{\substack{mass\\terms}} &= -\tfrac{1}{2}\begin{pmatrix}\overline{\nu}_L & \overline{\nu}_R^c\end{pmatrix}\mathcal{M}\begin{pmatrix}\nu_L^c \\ \nu_R\end{pmatrix} + h.c. = -\tfrac{1}{2}\begin{pmatrix}\overline{\nu}_L & \overline{\nu}_R^c\end{pmatrix}\begin{pmatrix}m_M^L\nu_L^c + m_D\nu_R \\ m_D\nu_L^c + m_M^R\nu_R\end{pmatrix} + h.c. \\
&= -\tfrac{1}{2}\left(m_M^L\overline{\nu}_L\nu_L^c + m_D\overline{\nu}_L\nu_R + m_D\overline{\nu}_R^c\nu_L^c + m_M^R\overline{\nu}_R^c\nu_R\right) \\
&\qquad -\tfrac{1}{2}\left(m_M^L\overline{\nu}_L^c\nu_L + m_D\overline{\nu}_R\nu_L + m_D\overline{\nu}_L^c\nu_R^c + m_M^R\overline{\nu}_R\nu_R^c\right) \\
&= -\tfrac{1}{2}\left(m_D\overline{\nu}_L\nu_R + m_D\overline{\nu}_R^c\nu_L^c + m_D\overline{\nu}_R\nu_L + m_D\overline{\nu}_L^c\nu_R^c\right) \\
&\qquad -\tfrac{1}{2}m_M^L\left(\overline{\nu}_L\nu_L^c + \overline{\nu}_L^c\nu_L\right) - \tfrac{1}{2}m_M^R\left(\overline{\nu}_R\nu_R^c + \overline{\nu}_R^c\nu_R\right).
\end{aligned} \tag{10-33}$$

Re-expressing all mass terms with a non-diagonal matrix

[1] To hopefully avoid some confusion, we are talking about a completely different thing here when we say different neutrino masses than the different mass eigenstates ν_1, ν_2, and ν_3 of Sect. 10.1. Here, we are talking about a scenario where m_M^L will eventually be considered the mass of ν_1. And parallel, but essentially separate, scenarios would lead to ν_2, and ν_3.

For the next to last line of (10-33), where from (9-105) we know the charge conjugate of the mass term is the original mass term,

$$
\begin{aligned}
&-\tfrac{1}{2}\left(m_D\bar{\nu}_L\nu_R + m_D\bar{\nu}_R\nu_L + m_D\bar{\nu}_R^c\nu_L^c + m_D\bar{\nu}_L^c\nu_R^c\right) \\
&\qquad = -\tfrac{1}{2}m_D\bar{\nu}\nu - \tfrac{1}{2}m_D\bar{\nu}^c\nu^c = -m_D\bar{\nu}\nu = -m_D\left(\bar{\nu}_L\nu_R + \bar{\nu}_R\nu_L\right),
\end{aligned}
\qquad (10\text{-}34)
$$

which shows that the first line of (10-33), in terms of a mass matrix, equals (10-29) plus (10-30).

As we will see, the matrix in (10-32) is the neutrino space analog of the matrix in (10-27) of Section 10.3.3.

This mass matrix like prior math example

Thus, we can write (10-33), with the Hermitian conjugate expressed explicitly, as

$$
\mathcal{L}_{\substack{mass\\terms}} = -\frac{1}{2}\left(\bar{\nu}_L\;\bar{\nu}_R^c\right)\mathcal{M}\begin{pmatrix}\nu_L^c\\ \nu_R\end{pmatrix} + h.c. = -\frac{1}{2}\left(\bar{\nu}_L\;\bar{\nu}_R^c\right)\mathcal{M}\begin{pmatrix}\nu_L^c\\ \nu_R\end{pmatrix} - \frac{1}{2}\left(\bar{\nu}_L^c\;\bar{\nu}_R\right)\mathcal{M}\begin{pmatrix}\nu_L\\ \nu_R^c\end{pmatrix}. \qquad (10\text{-}35)
$$

Diagonal components are Majorana masses; non-diagonal, Dirac

Note the diagonal components of $\mathcal{M}$ give us the Majorana mass terms of (10-30), while the off-diagonal components give us the Dirac mass terms of (10-29). Keep this in mind as we progress.

10.3.5 See-sawing

Suppose, as suggested earlier, that Higgs or GUT symmetry breaking gave Majorana mass, but not Dirac mass, to neutrinos. That is, coupling to the Higgs field (or fields in GUTs) was not done in a way that led to Dirac mass terms. So, the mass matrix would be diagonal, unlike (10-32), of form

Finally, the see-saw mechanism itself

$$
\tilde{\mathcal{M}} = \begin{bmatrix} m_\nu & 0 \\ 0 & M \end{bmatrix}. \qquad (10\text{-}36)
$$

Consider a diagonal neutrino mass matrix

As noted earlier, all neutrinos then must be Majorana, and since for them, as can be seen from Wholeness Chart 10-1, pg. 353, if particles are the same as antiparticles,

$$
\nu_L^c = \nu_R \qquad \bar{\nu}_L^c = \bar{\nu}_R \qquad \nu_R^c = \nu_L \qquad \bar{\nu}_R^c = \bar{\nu}_L \qquad \text{(for Majorana particles)}. \qquad (10\text{-}36)\text{+}1
$$

Then, (10-33) becomes (with wavy underlines to distinguish the more massive neutrino from the less massive one and a notation change in the second line to that commonly found elsewhere)

$$
\begin{aligned}
\mathcal{L}_{\substack{mass\\terms}} &= -\frac{m_\nu}{2}\left(\bar{\nu}_L\nu_L^c + \bar{\nu}_L^c\nu_L\right) - \frac{M}{2}\left(\bar{\utilde{\nu}}_R\utilde{\nu}_R^c + \bar{\utilde{\nu}}_R^c\utilde{\nu}_R\right) = -\frac{m_\nu}{2}\left(\bar{\nu}_L\nu_R + \bar{\nu}_R\nu_L\right) - \frac{M}{2}\left(\bar{\utilde{\nu}}_R\utilde{\nu}_L + \bar{\utilde{\nu}}_L\utilde{\nu}_R\right) \\
&= -\frac{m_\nu}{2}\bar{\nu}\nu - \frac{M}{2}\bar{\utilde{\nu}}\utilde{\nu} = -\frac{m_\nu}{2}\bar{\nu}\nu - \frac{M}{2}\bar{N}N = -\tfrac{1}{2}\left(\bar{\nu}\;\bar{N}\right)\tilde{\mathcal{M}}\begin{pmatrix}\nu\\ N\end{pmatrix}.
\end{aligned}
\qquad (10\text{-}37)
$$

yields Lagrangian mass terms expressed via two new kinds of neutrinos, ν and N

ν and N are the fields directly coupled to the Higgs. The eigenstates are then

$$
\tilde{V}_1 = \begin{pmatrix}\nu\\ 0\end{pmatrix} \qquad \tilde{V}_2 = \begin{pmatrix}0\\ N\end{pmatrix}. \qquad (10\text{-}38)
$$

Eigenvectors of this diagonal mass matrix

Now consider transforming the wavy overbar vectors and matrix in the same way we transformed (10-25) to (10-27), by some rotation through an angle ϕ. This would give us a non-diagonal matrix like (10-32).

Now rotate basis by a small ϕ, as with math example earlier, to get non-diagonal matrix

Now consider the reverse process, where in our world, we measure quantities like those of (10-35) (with no wavy overbars and the non-diagonal matrix (10-32)), and we want to determine what the relation (10-37) (with diagonal matrix (10-36)) looks like.

This is just an eigenvalue problem, where we find (10-36) from (10-32), with m_ν and M the eigenvalues, and $\tilde{V}_1$ and $\tilde{V}_2$, the eigenvectors. That is, we could think of our fields in two different, but essentially equivalent, ways: 1) a mix of Majorana and Dirac mass terms with the column vector of fields in (10-35), or 2) pure Majorana mass terms associated with the mass matrix of (10-36), whose associated fields are represented by the different column vector $(\nu\; N)^T$.

Heuristically, finding $(\nu\; N)^T$ from $(\nu_L{}^c\; \nu_R)^T$ can be thought of as "rotating" our basis vectors in an abstract space until we find an alignment giving the fields vector the components $(\nu\; N)^T$.

Since we have an eigenvalue problem, the state ("state" in a vector space sense, not in a QFT sense) vectors in (10-35) have components ν_L^c and ν_R (and their conjugates), and these are linear superpositions of ν and N.

Assuming that all of this is the case in the real world (we have no way of knowing via experiments to date), what would the mass matrix (10-36) look like in order to give us the kind of masses (either m_D or m_M^L) that we see? Remember we are looking for a reason why neutrino mass is so much lower than that of other particles.

That reason posits that the field components of the vector in (10-37) are the ones directly coupled to the Higgs field. It works best if the mass $m_\nu = 0$, as that means there is no Higgs coupling for the ν field, but there is such coupling for the N. (And (10-36) then becomes the exact analog of (10-25) in Section 10.3.3.) Note that if we take $m_\nu \neq 0$, but $m_\nu << M$, we would still be left with our original problem, which is "why is one mass so much smaller than the others?". Having zero mass is easier to explain (no coupling) than extremely low mass (extremely small coupling.)

Theory works best for $m_\nu = 0$

Given the treatment of Section 10.3.3, we can immediately draw conclusions about the magnitudes of the four components of (10-32), given (10-36) with the upper left component equal to zero and the $(\nu\ N)^T$ basis being close to the $(\nu_L{}^c\ \nu_R)^T$ basis. That is, we have the mass hierarchy we need,

Like prior math example, get components of transformed matrix

$$M \approx m_M^R >> m_D > m_M^L \approx 0 \ , \qquad (10\text{-}39)$$

These have natural hierarchy of masses

where the Dirac mass m_D is the geometric mean of the left and right Majorana masses, the diagonal components of (10-32). That is,

$$m_M^R m_M^L = m_D{}^2 \ . \qquad (10\text{-}40)$$

Relation between diagonal and off-diagonal components

Note that for given value of m_D, a higher value for m_M^R means a lower value for m_M^L and vice versa. This is the reason for the name "see-saw mechanism".

Heaviest and lightest have see-saw relation

From (10-30), we see that the LC Majorana mass neutrino is very light, whereas the RC Majorana mass neutrino is very heavy. It is so heavy, in fact, assuming M is on the order of the weak symmetry breaking scale, that it never manifests itself in the world as we see it. In other words, in this scenario, the RC Majorana mass neutrino is essentially sterile (at current energy levels in the universe).

RC mass very heavy, thus, essentially sterile

We conclude this section by noting that, in more advanced theories, there are different varieties of see-saw mechanism, but these are beyond our present scope.

10.3.6 Distinction between Majorana Representation, Mass Terms, and Particles

As a review, the adjective "Majorana" is applied to three distinctly different things, which we need to distinguish between.

Distinguishing between three different things labeled "Majorana"

The first use most people see of this term is for one of three representations of gamma matrices and spinors. As noted, this use of "Majorana" has nothing to do with the Majorana mass terms of this chapter. Everything in this chapter can be done in any one of the three representations.

With (10-29) and (10-30), we began dealing with the second use of the term regarding Majorana vs. Dirac type mass terms in the Lagrangian. The neutrino spinors and gamma matrices in these terms can be represented by any one of the three representations above.

The third use of the term refers to type of neutrino, Dirac vs Majorana. (Neutrinos are the only fermions that can be either of Dirac type or Majorana type. All other fermions have electric charge, so they cannot be their own antiparticles, which have opposite charge.)

However, importantly, we found herein that if we include Majorana mass terms (2nd use of "Majorana") for neutrinos in the Lagrangian, then to maintain consistency with extant theory and experiment, neutrinos need to be Majorana fermions (3rd use of "Majorana").

In summary, we can have

- Majorana representation in spinor space (it or one of other 2 reps can be used for any of below)
- Majorana vs Dirac mass terms in Lagrangian
- Majorana vs Dirac type particles (with both type mass terms, have to have Majorana particles).

10.3.7 Possible Physical Scenarios

There are three possible scenarios for the two neutrino types.

3 possibilities for Dirac and Majorana neutrinos

1) Dirac and Majorana neutrinos both interact weakly, and what we see in experiments is a blend of both. (Not considered likely by most.)
2) Only Dirac neutrinos interact weakly, and we don't ever observe Majorana neutrinos in any experiments. (No Majorana mass terms.)
3) Only Majorana neutrinos interact weakly, and we don't ever observe Dirac neutrinos in any experiments. (Majorana mass terms with or without Dirac mass terms.)

10.3.8 Summary of the See-saw Mechanism

Bottom line: see-saw mechanism

If the see-saw mechanism exists, then we have both types of mass terms (10-29) and (10-30), with

$$M \approx m_M^R >> m_D > m_M^L \approx 0\,, \tag{10-41}$$

and for which the fields represent Majorana neutrinos. The Higgs couples to the very heavy N neutrino, but due to a slight mixing, this is transformed into the fields which interact weakly, the light ν_L and the very heavy ν_R. The ν_R is so heavy it is effectively sterile in our contemporary universe. The ν_L is what we measure in experiments. What we now call an antineutrino would just be a Majorana neutrino with R chirality.

All this is natural, as the N is coupled to the Higgs, but its sibling ν is not, with the net result being the very light neutrinos we observe.

10.4 Neutrino Odds and Ends

10.4.1 Sterile Neutrinos

Fourth neutrino flavor could exit

There is speculation that a fourth type of neutrino may exist that does not interact with matter like the three known neutrino flavors do and thus, is effectively sterile. This could be the superheavy neutrino of the see-saw mechanism or a fourth flavor neutrino.

Experiments to date inconclusive

As of 2025, the existence of such particles has been hinted at in various experiments, and data from the Wilkinson Microwave Anisotropy Probe of the cosmic background radiation is compatible with either three or four types of neutrinos.

We will not discuss this further, as it is a research project all unto itself.

10.4.2 Neutrinos and the Standard Model

Massive neutrinos "broke" old SM, but new SM doing fine

One may often hear it said that "massive neutrinos break the standard model". Such statements are a bit historically oriented, as the early SM was formulated with massless neutrinos. Today's SM seems alive and well, and has neutrinos with mass.

10.5 Appendices

10.5.1 Appendix A: PMNS Matrix and States

We can deduce (10-4) from (10-3), as follows.

$$\psi_{\nu_\alpha}^L = P_{\nu_\alpha \nu_i}\psi_{\nu_i}^L \xrightarrow[\text{notation}]{\text{streamlined}} \psi_\alpha = P_{\alpha i}\psi_i \text{ and } \psi_\alpha^c = \underbrace{-i\gamma^2\left(P_{\alpha i}\psi_i\right)^*}_{P_{\alpha i}^*\left(-i\gamma^2\psi_i^*\right)} = P_{\alpha i}^*\psi_i^c\,. \tag{10-42}$$

So, we can create an antineutrino ket via

$$\psi_{\mathbf{p}\,\alpha}|0\rangle = \left|\bar{\nu}_{\mathbf{p}\,\alpha}\right\rangle = P_{\alpha i}\psi_{\mathbf{p}i}|0\rangle = P_{\alpha i}\left|\bar{\nu}_{\mathbf{p}\,i}\right\rangle \qquad \to \left|\bar{\nu}_{\mathbf{p}\,\alpha}\right\rangle = P_{\alpha i}\left|\bar{\nu}_{\mathbf{p}\,i}\right\rangle. \tag{10-43}$$

We want to do a similar thing to create a neutrino ket, rather than an antineutrino ket. In order to get the same form for the constants (column spinor, not row spinor, etc.), we will use the charge conjugate of $\psi_{\mathbf{p}\,\alpha}$ rather than the adjoint $\bar{\psi}_{\mathbf{p}\,\alpha}$ to create the neutrino ket.

$$\psi_{\mathbf{p}\,\alpha}^c|0\rangle = \left|\nu_{\mathbf{p}\,\alpha}\right\rangle = P_{\alpha i}^*\psi_{\mathbf{p}i}^c|0\rangle = P_{\alpha i}^*\left|\nu_{\mathbf{p}\,i}\right\rangle \qquad \to \left|\nu_{\mathbf{p}\,\alpha}\right\rangle = P_{\alpha i}^*\left|\nu_{\mathbf{p}\,i}\right\rangle. \tag{10-44}$$

This is (10-4), and we suppress the **p** subscript in going from there to (10-10).

10.5.2 Appendix B: A Closer Look at the Neutrino Oscillation Derivation

There are some subtleties in the derivation of relation (10-10). First note that in (10-5),

$$\left|\nu_i(0)\right\rangle \xrightarrow[\text{space}]{\text{in coordinate}} \frac{1}{\sqrt{V}}e^{i\left(E_i(0)-p_i x'\right)} = \frac{1}{\sqrt{V}}e^{-ip_i x'}, \tag{10-45}$$

where we are assuming a wave packet can be well approximated by a complex sinusoid with finite length $\Delta x'$ in the x' direction, such that $V = \Delta x'\Delta y'\Delta z'$. In order to keep things simple, we are also ignoring some relativistic effects in the coefficient of (10-45) that arise in RQM, but would cancel out in the final analysis.

Then note that in (10-8), the arguments are actually time values, not space values, where $c = 1$. Hence, the last term in the exponent of the first part of that relation is really $p_i x$ not $p_i t$, and we actually have a static wave form that varies over x,

$$\left|\nu_i(t=L)\right\rangle = e^{-i\left(p_i L + \frac{m_i^2}{2p_i}L - p_i x\right)}\left|\nu_i(t=0)\right\rangle \tag{10-46}$$

We will be integrating over a relatively short distance along the x axis, however, and throughout that distance, x will be, to very good approximation, equal to L. So, what we are really saying in (10-8) is that at time $t = L/c$ ($c = 1$), we have a static wave of form (10-46), over a region where x doesn't change much from the value L. This gives us (10-8).

$$\left|\nu_i(t=L)\right\rangle \approx e^{-i\left(p_i L + \frac{m_i^2}{2p_i}L - p_i L\right)}\left|\nu_i(t=0)\right\rangle = e^{-i\frac{m_i^2}{2p_i}L}\left|\nu_i(t=0)\right\rangle \approx e^{-i\frac{m_i^2}{2E}L}\left|\nu_i(t=0)\right\rangle \quad \text{repeat of (10-8)}$$

(10-9) follows without change. Now, note the bracket on the LHS of (10-10) entails an integration over the entire wave. But, we are assuming that wave only has volume $\Delta x'\Delta y'\Delta z'$. So, in detail, (10-10) is

$$\begin{aligned}\left\langle \nu_\beta(t=L)\middle|\nu_\alpha(t=0)\right\rangle &= \left\langle \nu_j(t=0)\right| e^{i\frac{m_j^2}{2E}L} P_{\beta j}P_{\alpha i}^*\left|\nu_i(t=0)\right\rangle \\ &= e^{i\frac{m_j^2}{2E}L} P_{\beta j}P_{\alpha i}^* \int \frac{1}{\sqrt{V}}e^{ip_j x'}\frac{1}{\sqrt{V}}e^{-ip_i x'}dx'dy'dz' = e^{i\frac{m_j^2}{2E}L} P_{\beta j}P_{\alpha i}^*\delta_{ij} = e^{i\frac{m_i^2}{2E}L} P_{\beta i}P_{\alpha i}^*.\end{aligned} \tag{10-47}$$

10.5.3 Appendix C. Hermitian Conjugates of Some Terms

We have gotten used to mass terms in the Lagrangian of form (where we drop the minus sign)

$$\begin{aligned} m\bar{\nu}\nu &= m\bar{\nu}\tfrac{1}{2}\left(1-\gamma^5\right)\nu + m\bar{\nu}\tfrac{1}{2}\left(1+\gamma^5\right)\nu \\ &= m\bar{\nu}\tfrac{1}{2}\left(1-\gamma^5\right)\tfrac{1}{2}\left(1-\gamma^5\right)\nu + m\bar{\nu}\tfrac{1}{2}\left(1+\gamma^5\right)\tfrac{1}{2}\left(1+\gamma^5\right)\nu = m\left(\bar{\nu}_R\nu_L + \bar{\nu}_L\nu_R\right).\end{aligned} \tag{10-48}$$

We want to find the Hermitian conjugate of the next to the last term.

$$\begin{aligned}\left(\bar{\nu}_R\nu_L\right)^\dagger &= \left(\bar{\nu}\tfrac{1}{2}\left(1-\gamma^5\right)\nu\right)^\dagger = \left(\nu^\dagger\gamma^0\tfrac{1}{2}\left(1-\gamma^5\right)\nu\right)^\dagger = \nu^\dagger\tfrac{1}{2}\left(1-\left(\gamma^5\right)^\dagger\right)\left(\gamma^0\right)^\dagger\nu \\ &= \nu^\dagger\tfrac{1}{2}\left(1-\gamma^5\right)\gamma^0\nu = \nu^\dagger\gamma^0\tfrac{1}{2}\left(1+\gamma^5\right)\nu = \bar{\nu}\tfrac{1}{2}\left(1+\gamma^5\right)\nu = \bar{\nu}_L\nu_R.\end{aligned} \tag{10-49}$$

So, we can write the mass term (10-48) as

$$m\bar{\nu}\nu = m\bar{\nu}_R\nu_L + h.c. \tag{10-50}$$

Do **Problem 3** to prove the second line of (10-31) is valid.

10.6 Problems

1. Find the probability of neutrino flavor change using QFT. That is, find (10-10), as the rest of the analysis is the same as we did in Sect. 10.1.2.

   Background:

   This problem is a bit unlike anything we have developed our QFT tools for, as we have previously worked virtually exclusively with interactions, and here we have a free field/particle case. We have, to date, employed the interaction picture (IP), where the state evolution is governed by the interaction part of the Hamiltonian (or Lagrangian), the field evolution is governed by the free part, and we have found interaction transition amplitudes between initial and final states. (See Vol. 1, Wholeness Chart 8-4, pg. 248.) By contrast, in the present case, there is no interaction. The neutrino is propagating freely.

   We can use a similar procedure, but take the interaction Hamiltonian as zero. In other words, the initial and final states are static, unchanging, just as in the Heisenberg picture, with all evolution being carried out by the fields. Physically and intuitively, this may seem strange, but the expectation value (the transition amplitude) will be the same in any picture.

   In formulating this approach, we need to be mindful of the form of the fermion four-current, and that the expectation value of its zeroth component is the expectation value of the probability density (or equivalently with an "*e*" in front, charge density). That is

   $$\text{expectation value of probability density of fermion} = \langle \psi_{\mathbf{p},r} | \bar{\psi}\gamma^0\psi | \psi_{\mathbf{p},r} \rangle = \langle \psi_{\mathbf{p},r} | \psi^\dagger\psi | \psi_{\mathbf{p},r} \rangle .$$

   The point is that the operator $\psi^\dagger$ creates the fermion ket (after reduction to $|0\rangle$ by ψ) that will match the bra. It is not done by the full operator $\bar{\psi}$. We will adopt this principle in our solution.

   Hints:

   1) The spin and 3-momentum throughout will be the same, so assume i) a spin of $r = 1$ (any spin would do) ii) three-momentum in the x^3 direction. and iii) ultra-relativistic speeds, so u_1 for any mass eigenstate i goes to $\sqrt{\dfrac{E_i}{2m_i}}(1,0,1,0)^T$ (Vol. 1, (4-20), pg. 89 for $p^3 \to E$). Then, $u_{1i}^\dagger u_{1j} = \sqrt{\dfrac{E_i}{m_i}\dfrac{E_j}{m_j}}$.

   2) $|\nu_\alpha(t=0)\rangle = \psi^\dagger_{\underset{t=0}{\alpha}}|0\rangle = P^*_{\alpha i}\psi^\dagger_{\underset{t=0}{i}}|0\rangle$, $|\nu_\beta(t=L)\rangle = \psi^\dagger_{\underset{t=L}{\beta}}|0\rangle = P^*_{\beta j}\psi^\dagger_{\underset{t=L}{j}}|0\rangle$, and $\langle\nu_\beta(t=L)| = \langle 0|\psi_{\underset{t=L}{\beta}} = \langle 0|P_{\beta j}\psi_{\underset{t=L}{j}}$.

   3) We do not invoke normal ordering, as that arose from the Wick's theorem applied for interactions, when converting from time ordering to normal ordering. Here, we simply keep the time order.

   4) Then, find $\langle\nu_\beta(t=L)|\nu_\alpha(t=0)\rangle$, which will be a probability density over space.

   5) Then integrate the probability density over the small volume where the measurement will be taken to get probability.

2. Show how we cannot determine neutrino masses from experimental results giving us the difference between the squares of the masses. Hint: Write down the three equations involving the masses squared and assume the value found in experiment for each equation equals the constants *A,B,* and *C*. Then, show that *C* is always equal to *B* – *A*, and thus, we get three equations in three unknowns, but only two of the equations are independent, so the three unknowns are indeterminate.

3. Show that taking Hermitian conjugate of $\bar{\nu}_L\nu_L^c$ gives us $\bar{\nu}_L^c\nu_L$. By parallel logic, we can then recognize that the Hermitian conjugate of $\bar{\nu}_R\nu_R^c$ is $\bar{\nu}_R^c\nu_R$.

Chapter 11

Additional Topics in Electroweak Theory

"The more we discover, the less we seem to know. That's physics in a nutshell."
Leonard Susskind

11.0 Chapter Overview

There are a number of loose ends in electroweak theory we need to delve into. These fit into two general categories: 1) topics that are not overarching, but restricted to particular areas, and 2) broad-brush concepts that more or less envelop the entire theory. Combined, they number well over a dozen different topics, so we will not overview the substance of each of them in this introduction. Rather, we will simply name and categorize them in the following list, and then look at them more closely, one at a time, in the remainder of the chapter.

Topics covered in this chapter

- Specific, not all-encompassing, topics
  Mandelstam variables, equal times contractions and tadpole diagrams, mean free path, Fierz identities, anomalies, Higgs as a fifth force.
- General concepts that pervade the theory as a whole
  Graph of particle charges, renormalization, three coupling constants (their energy dependency), effective theories, electroweak theory via path integrals, uncertainty principle in QFT, entangled particles, spins greater than 2, the reality (or lack thereof) of virtual particles.

Note that many of these topics are not restricted to electroweak theory, but extend to other areas of QFT, as well.

11.1 Specific Topics

11.1.1 Mandelstam Variables

Consider the various ways (channels) two incoming particles can interact to produce two outgoing particles via a propagator. These are represented in the three diagrams of Fig. 11-1, where the incoming particles are labeled by 1 and 2; and the outgoing, by 3 and 4. The three different channels are labeled *s, t,* and *u*.

3 channels for 2 in, 2 out scattering

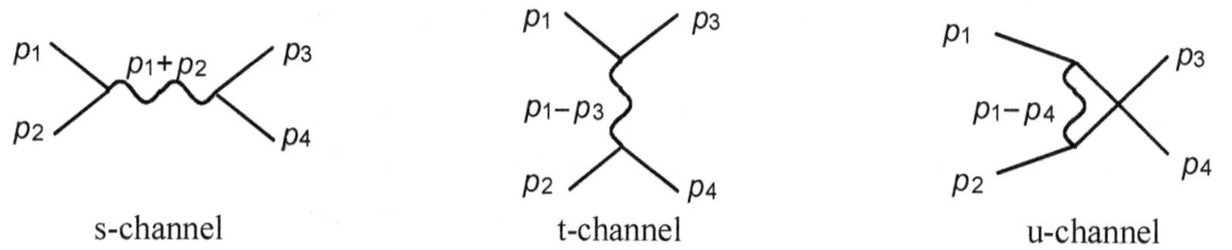

p_1 p_1+p_2 p_3 p_2 p_4
s-channel

p_1 p_3 p_1-p_3 p_2 p_4
t-channel

p_1 p_3 p_1-p_4 p_2 p_4
u-channel

Figure 11-1. *s, t, u* Channels for 2 → 2 Scattering

Since we have the same initial and final particles in each channel, the Feynman amplitudes for the three channels are added (rather than adding probabilities when there are different final particles). Also, since the initial particles are the same in each case, and the final ones are as well, the only

s, t, u called Mandelstam variables

difference in the amplitudes for the different channels will be the propagators. These will have forms, for massive propagators and the definitions of the Mandelstam variables *s, t,* and *u*, of

$$
\begin{aligned}
&s-\text{channel} \quad \frac{g_{\mu\nu}+k_\mu k_\nu/m^2}{(p_1+p_2)^2-m^2+i\varepsilon}=\frac{g_{\mu\nu}+k_\mu k_\nu/m^2}{s-m^2+i\varepsilon} \qquad s\equiv k^\alpha k_\alpha=(p_1+p_2)^2=(p_3+p_4)^2 \\
&t-\text{channel} \quad \frac{g_{\mu\nu}+k_\mu k_\nu/m^2}{(p_1-p_3)^2-m^2+i\varepsilon}=\frac{g_{\mu\nu}+k_\mu k_\nu/m^2}{t-m^2+i\varepsilon} \qquad t\equiv k^\alpha k_\alpha=(p_1-p_3)^2=(p_2-p_4)^2 \\
&u-\text{channel} \quad \frac{g_{\mu\nu}+k_\mu k_\nu/m^2}{(p_1-p_4)^2-m^2+i\varepsilon}=\frac{g_{\mu\nu}+k_\mu k_\nu/m^2}{u-m^2+i\varepsilon} \qquad u\equiv k^\alpha k_\alpha=(p_1-p_4)^2=(p_2-p_3)^2,
\end{aligned}
\tag{11-1}
$$

Propagators for the 3 channels and s, t, u definitions

s, t, and *u* are a great aid in analysis, since they are Lorentz invariant, so they can be used in any frame (COM, lab, or any other). Pick the easiest frame to calculate them in, then use them in any other frame you like. Their use is very common.

s, t, u all Lorentz invariant, and so aid in calculations

Do **Problem 1** to show (11-2).

The Mandelstam variables satisfy

$$s+t+u=\sum m_i^2 \qquad i=\text{external particles} \tag{11-2}$$

Summation law for s, t, u

Mandelstam variables are generally used in 2 particles in, 2 out, and 1 particle in, 3 out scattering, although generalizations exist for other interactions. In addition, they also work when the propagator is a fermion, instead of a boson, and when at least some of the external particles are bosons.

11.1.2 Equal Times Contractions and Tadpole Diagrams

There is a reason for why we do not include equal-times-contractions in our application of the Wick theorem in QED. When included in Feynman amplitudes, they yield zero. The following shows how.

Consider the following QED *S* matrix element with an equal time contraction at x_1, which would arise via Wick's theorem.

$$S_X^{(2)}=-e^2\int d^4x_1d^4x_2N\left\{\left(\bar{\psi}A\!\!\!/\psi\right)_{x_1}\left(\bar{\psi}A\!\!\!/\psi\right)_{x_2}\right\} \tag{11-3}$$

Equal times contraction for fermions in QED

(11-3) represents certain interactions, two of which are shown by the diagrams of Fig. 11-2.

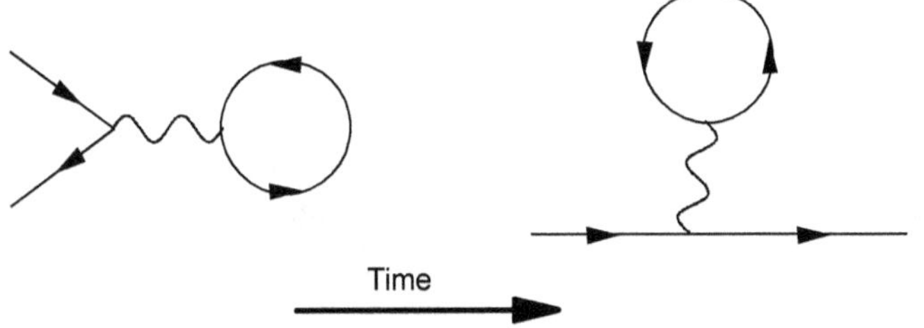

Feynman diagrams for this contraction include a tadpole diagram

Figure 11-2. Feynman Diagrams for Equal Times Contractions of Fermions

The LHS figure is expressed mathematically by

$$S_{EqTim}=\left\langle 0\left|S_X^{(2)}\right|e^-_{\mathbf{p}_1,r_1},e^+_{\mathbf{p}_2,r_2}\right\rangle=\sqrt{\frac{m}{VE_{\mathbf{p}_1}}}\sqrt{\frac{m}{VE_{\mathbf{p}_2}}}(2\pi)^4\delta^{(4)}(p_1+p_2)\mathcal{M}_{EqTim} \tag{11-4}$$

Because the outgoing energy must be zero (as we only have the vacuum outgoing), the incoming energy must also be zero. But this is only true if each of the incoming particles has zero energy, i.e., if they are real, they don't exist. The delta function in (11-4) is only non-zero for zero energy in, which means no particles in. The math tells us we can't have such an interaction.

Conservation laws mean 1st diagram makes no contribution

The RHS figure is a little more problematic. It entails a contraction (fermion propagator) whose energy is integrated from $-\infty$ to $+\infty$, even if the virtual photon has zero energy (since the electron energy out must equal the electron energy in). Mathematically, the loop propagator doesn't have to equal zero, at least at first blush.

But conservation obeyed in 2nd diagram

For each diagram of Fig. 11-2, we actually have a second diagram where the fermion loop has arrows in the opposite direction. That is, one diagram has a fermion for a loop and the other, an anti-fermion. The two amplitudes (really sub-amplitudes) must be added to get the total amplitude.

But (see Vol. 1, (7-75), pg. 202), the antifermion propagator has the same magnitude but opposite sign of the fermion propagator. So, the two sub-amplitudes cancel, leaving no contribution from the loop. This is one application of Furry's theorem (Vol. 1, Sect. 14.1.2, pg. 341), which states that any diagram having a fermion loop with an odd number of indices (one index in Fig. 11-2) makes zero contribution to the transition amplitude.

Fermion and antifermion loops cancel

Note that equations of earlier text versions numbered (11-5) and (11-6), as well as associated text, have been deleted, as of the July 2025 revision, and replaced with the prior two paragraphs.

The diagrams of Fig. 11-2 are called tadpole diagrams, for an obvious reason: part of each looks like a tadpole. The transition amplitudes for these diagrams equal zero in QED, and since we can only have tadpole diagrams in QED with fermion propagator loops, we don't have to worry about it in that theory.

Called tadpole diagram

= 0 in QED

But how about other theories, like GSW electroweak theory? It turns out that, in those, we do have contributing tadpole diagrams with boson propagator loops. These do not have antiparticle propagators with opposite signs, so we don't get the cancellation we get with fermions. Consider, for example, the second Higgs self-interacting term $-\lambda v\sigma^3$ of (7-149), which leads to a vertex having 3 Higgs particles. Two of these can lead to a Feynman diagram like Fig. 11-3.

But, boson loops of W,Z, gluons, Higgs not subject to Furry's theorem

Tadpole diagram for Higgs boson arises in e/w theory

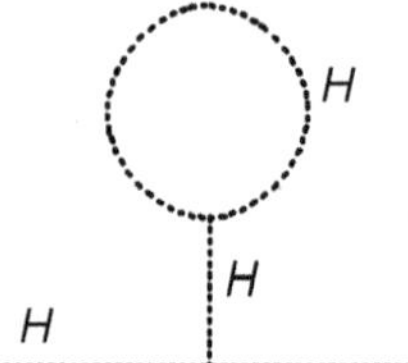

Figure 11-3. Feynman Diagram for Equal Times Contraction of Higgs Boson

This yields a non-zero contribution to a Feynman amplitude and so, must be part of the evaluation of any such amplitude involving Higgs fields. Physically, it seems weird, as the loop starts at one point in spacetime, travels through spacetime, and returns to the same point in spacetime (which could be a closed time-like loop in relativity language). Yet, mathematically, it arises in the theory, so must be taken into account. And recall that virtual particles are not constrained to obey certain physical laws in the way real particles are. (They aren't on-shell, for one thing.)

Non-zero contribution of tadpoles from bosons

Seems weird, but the math demands it

You may want to do **Problem 2** to see another type of Feynman diagram for Higgs particles, which we haven't seen before.

We will not delve into this topic at any greater length, as our purpose here is to make the reader aware of this, in the event she/he one day becomes involved in calculating transition amplitudes involving Higgs or other bosons forming equal times propagators[1].

Fortunately, most of the more fundamental calculations in electroweak theory, and all that we do in this book, do not involve such things, and for that reason, we will do little more with them now.

[1] For more on this topic, see M.E. Peskin and D.V. Schroeder, *An Introduction to Quantum Field Theory* (Perseus 1995), pgs.354-373.

11.1.3 Mean Free Path

Examples and definition in words of mean free path

You have probably heard statements such as a neutrino can pass through a light-year of lead, or a cosmic ray proton can pass through 10 km of the atmosphere (at sea level). These numbers represent what is called the mean free path of a given particle through a given medium. That is, a neutrino (of a particular energy) would travel, on average, through a light-year of lead before interacting with an elementary particle in the lead. Protons travel, on average, 10 km through the atmosphere (modified at different altitudes). Some will get stopped sooner, some later.

In this section we will deduce the relation to calculate the mean free path of a given particle through a given medium. We might expect it would depend on the interaction cross-section and also on the density of particles in the medium, and we would be correct. An increase in either one would increase the number of interactions, and thereby decrease the mean free path.

Deducing mean free path formula

Consider Fig. 11-4, with a target medium and an incoming particle beam, collimated so that any given beam particle trajectory is inside the cylinder swept out by the area A shown and parallel to the cylinder wall. We chose to model it graphically as a classical interaction, for illustrative reasons, but the conclusion will be valid quantum mechanically, as well. (See Vol. 1, the first few pages of Chap. 7 plus Wholeness Chart 17-1, pg. 444, for a review of scattering and cross-sections.) The cross-section for these two particles interacting is σ, which can be interpreted classically as the cross-sectional area of the target particle, where the beam particle is effectively point-like. All possible interactions will have to take place inside the target medium in the region defined by the cylinder whose walls align axially with those of the cylinder enclosing the incoming collimated beam.

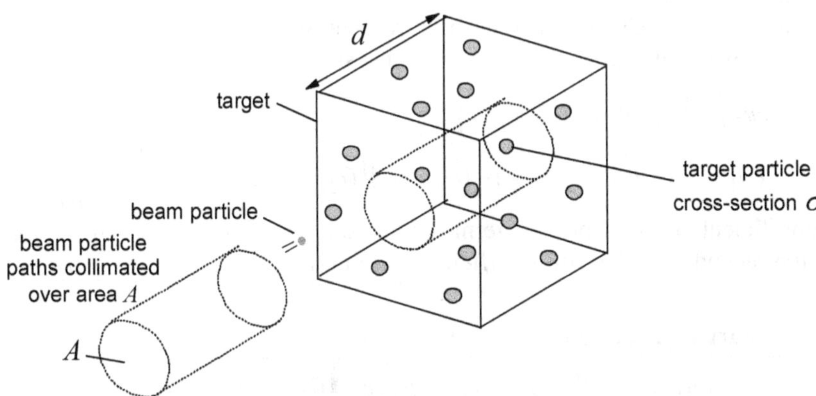

Figure 11-4. Collisions of a Particle Passing Through a Medium

Consider a single incoming beam particle. The chance of it hitting a given single target particle is the ratio of the target particle cross-section to the total area over which the beam particle trajectory is confined.

$$\text{odds of a single beam particle hitting a single given target particle } = \frac{\sigma}{A}. \tag{11-7}$$

The odds of any interaction at all for the beam particle is simply (11-7) times the number of target particles N_t inside the tube of length d and cross-sectional area A shown with dashed outline in Fig. 11-4. With n_t the target particle density, we have the

$$\text{odds of a single beam particle hitting any of the target particles } = \frac{\sigma}{A}N_t = \frac{\sigma}{A}n_t A d = \sigma n_t d\,^{1}. \tag{11-8}$$

The average number of collisions per unit distance is (11-8) divided by d times 1, the number of incoming beam particles,

---

[1] We are assuming none of the target particles is hidden, wholly or in part, behind another target particle. This is reasonable, since the target particle cross-section in QFT is typically many, many orders of magnitude less than the area A, and the distance between target particles is immense compared to the equivalent diameter of the cross-section.

$$\text{average number of collisions per unit length } = \frac{\sigma n_t d}{d} = \sigma n_t . \tag{11-9}$$

The average distance traveled by the beam particle until there is a collision is the inverse of (11-9),

$$\boxed{\text{average distance traveled per collision } = \text{ mean free path} = \frac{1}{\sigma n_t}} . \tag{11-10}$$

Mean free path formula

Our final formula for mean free path is (11-10). It is simple and makes sense - just the inverse of the product of the cross-section and the target particle number density, so, as we surmised earlier, it decreases with an increase in either of those parameters.

I have put the formula in a box, indicating it could be good to commit to memory, because it is relatively easy to memorize, and having it at one's fingertips may come in handy in the course of one's career.

11.1.4 Fierz Transformations

Sometimes one can have expressions for transition amplitudes in one form, which can be more simply handled in another form, with the fields in different sequence. For but one example,

$$\left(\bar{\psi}_e^L \gamma^\alpha \psi_{\nu_e}^L\right)\left(\bar{\psi}_{\nu_e}^L \gamma_\alpha \psi_e^L\right) \xrightarrow[\text{form}]{\text{preferable}} \propto \left(\bar{\psi}_e^L \gamma^\alpha \psi_e^L\right)\left(\bar{\psi}_{\nu_e}^L \gamma_\alpha \psi_{\nu_e}^L\right). \tag{11-11}$$

An entire set of relations to transform spinor entities from one form to another has been developed, initially by Markus Fierz, called Fierz identities, or sometimes Fierz-Pauli-Kofink identities. We aren't going to derive them here, nor spend time applying them. We simply want to note them, and apprise you, the reader, of their existence, in case they may someday prove to be of value to you.

Fierz transformations change order of fields in transition amplitudes

The five types of spinor entities (scalar, vector current, tensor current, axial vector current, pseudo-scalar) listed in Wholeness Chart 9-2, at the end of Chap. 9, can each be re-expressed in terms of those five entities re-arranged. For example, where we don't show how, but just state it,

$$\begin{aligned}\left(\bar{\psi}_1\gamma^\mu\psi_2\right)\left(\bar{\psi}_2\gamma_\mu\psi_1\right) &= \left(\bar{\psi}_1\psi_1\right)\left(\bar{\psi}_2\psi_2\right) - \tfrac{1}{2}\left(\bar{\psi}_1\gamma^\mu\psi_1\right)\left(\bar{\psi}_2\gamma_\mu\psi_2\right) \\ &\quad - \tfrac{1}{2}\left(\bar{\psi}_1\gamma^\mu\gamma^5\psi_1\right)\left(\bar{\psi}_2\gamma_\mu\gamma^5\psi_2\right) - \left(\bar{\psi}_1\gamma^5\psi_1\right)\left(\bar{\psi}_2\gamma^5\psi_2\right).\end{aligned} \tag{11-12}$$

Wholeness Chart 11-1 shows the coefficients in front of each term for the re-expression of any term in the first column. For example, the second row gives the coefficients for (11-12).

We don't derive them, but just show chart to determine certain ones

Wholeness Chart 11-1. Fierz Identities Coefficients*

Spinor Entity	$(\bar{\psi}_1\psi_1)(\bar{\psi}_2\psi_2)$	$(\bar{\psi}_1\gamma^\mu\psi_1)(\bar{\psi}_2\gamma_\mu\psi_2)$	$(\bar{\psi}_1\sigma^{\mu\nu}\psi_1)(\bar{\psi}_2\sigma_{\mu\nu}\psi_2)$	$(\bar{\psi}_1\gamma^\mu\gamma^5\psi_1)(\bar{\psi}_2\gamma_\mu\gamma^5\psi_2)$	$(\bar{\psi}_1\gamma^5\psi_1)(\bar{\psi}_2\gamma^5\psi_2)$
$(\bar{\psi}_1\psi_2)(\bar{\psi}_1\psi_2)$	1/4	1/4	–1/4	–1/4	1/4
$(\bar{\psi}_1\gamma^\mu\psi_2)(\bar{\psi}_2\gamma_\mu\psi_1)$	1	–1/2	0	–1/2	–1
$(\bar{\psi}_1\sigma^{\mu\nu}\psi_2)(\bar{\psi}_2\sigma_{\mu\nu}\psi_1)$	–3/2	0	–1/2	0	–3/2
$(\bar{\psi}_1\gamma^\mu\gamma^5\psi_2)(\bar{\psi}_2\gamma_\mu\gamma^5\psi_1)$	–1	–1/2	0	–1/2	1
$(\bar{\psi}_1\gamma^5\psi_2)(\bar{\psi}_2\gamma^5\psi_1)$	1/4	–1/4	–1/4	1/4	1/4

* $\sigma^{\mu\nu} = \frac{i}{2}\left[\gamma^\mu, \gamma^\nu\right]$

An often-used Fierz transformation not shown in the chart is

$$\left(\bar{\psi}_1^L\gamma^\alpha\psi_2^L\right)\left(\bar{\psi}_3^L\gamma_\alpha\psi_4^L\right) = -\left(\bar{\psi}_1^L\gamma^\alpha\psi_4^L\right)\left(\bar{\psi}_3^L\gamma_\alpha\psi_2^L\right), \tag{11-13}$$

A commonly used Fierz transformation

which, for 1=4=e and 2=3= ν_e, is just what we were looking for in (11-11).

Note that some authors use different sign conventions, wherein all the signs in Wholeness Chart 11-1 are reversed, as is the sign after the equal sign in (11-13). The signs come out in the wash in probability calculations, as there we take the square of the absolute value of the transition amplitude.

We will be doing little more with Fierz transformations in this book. If we do use them at some point, we will simply quote the one(s) needed for the case at hand[1].

11.1.5 Anomalies

Consider a classical symmetry that upon quantization, does not remain a symmetry in quantum theory. One example is scaling symmetry, in which the classical interaction strength (the fine structure constant α in electrodynamics) is the same at any energy level, but which is energy level dependent in QFT.

Classical e/m independent of energy scale; QED is dependent

A symmetry that exists classically, but is broken in QFT is called an anomaly. Anomalies in QFT comprise a fairly extensive field of study[2],which we will not delve into deeply here.

Anomaly: Classical sym broken in QFT

There are two main classes of anomaly, local (gauge) and global (non-gauge). Scaling symmetry, known more commonly as conformal symmetry[3], is a global symmetry. It is broken in the quantum theory, yet the quantum theory is still consistent.

2 types of anomaly: non-gauge (global) and gauge (local)

More serious are gauge anomalies, for which the underlying theory, as it turns out, must be inconsistent. In general, a quantum theory can remain consistent if it contains non-gauge anomalies, but not if it contains gauge anomalies.

QFT can be consistent with non-gauge, but not gauge

The GSW theory has no gauge anomalies. In formulating models beyond the SM, one constraint that must be met is the absence of gauge anomalies. Any model with them will be internally inconsistent.

GSW theory free of gauge anomalies

11.1.6 Higgs as a Fifth Force

The Higgs field is bosonic and can act as a virtual particle mediator between fermions, just as photons, intermediate vector bosons, gluons, and the hypothetical graviton do. Further, it has coupling constants such as g_e, g_μ, etc. linking its Lagrangian terms with fermions, just as the boson fields mediating other interactions do. So, can it be considered a fifth force?

The answer depends on your definition of "fundamental force". If we define it as any interaction mediated by any type of boson, the Higgs is a fifth such force.

Whether Higgs is a 5th force or not depends on definition

However, the more common use of the term among field theorists restricts it to gauge interactions, which arise from $SU(n)$ or $U(1)$ symmetries and are mediated by gauge fields of spin 1 or 2. In this context, the Higgs field is not a fifth force.

Physics professor Barak Shoshany summed it up aptly online.

> *In the end it doesn't really matter. It's like the question of whether Pluto should count as a planet or not. In the end, whether it's considered a planet or a dwarf planet is a matter of definition and defining it one way or the other doesn't enhance our understanding in any way. The same with the question of whether the Higgs interaction counts as a "fundamental force" or not.*

---

[1] Derivations of virtually all Fierz identities can be found in L. B. Okun, *Leptons and Quarks.* (North-Holland 1980), Section 29.3.4.

[2] See, for examples, R. Ticciati, *Quantum Field Theory for Mathematicians* (Cambridge 1999), Chapter 13, and M.D. Schwartz, *Quantum Field Theory and the Standard Model* (Cambridge 2014), Chapter 30.

[3] For reasons which are explained at the link "Conformal and Scale Invariant Transformations" at the website for this book (URL on page opposite pg. 1)

11.2 General Topics Pervading Overall Theory

11.2.1 Graph of Various Particle Charges

Fig. 11-5 is a slightly modified version of a neat graphic one can find in many places on the internet[1] that displays the weak hypercharge, weak isospin charge, and electric charge of elementary particles in the SM. For simplicity, only the first generation is shown, as the second and third generations mirror the first in charges.

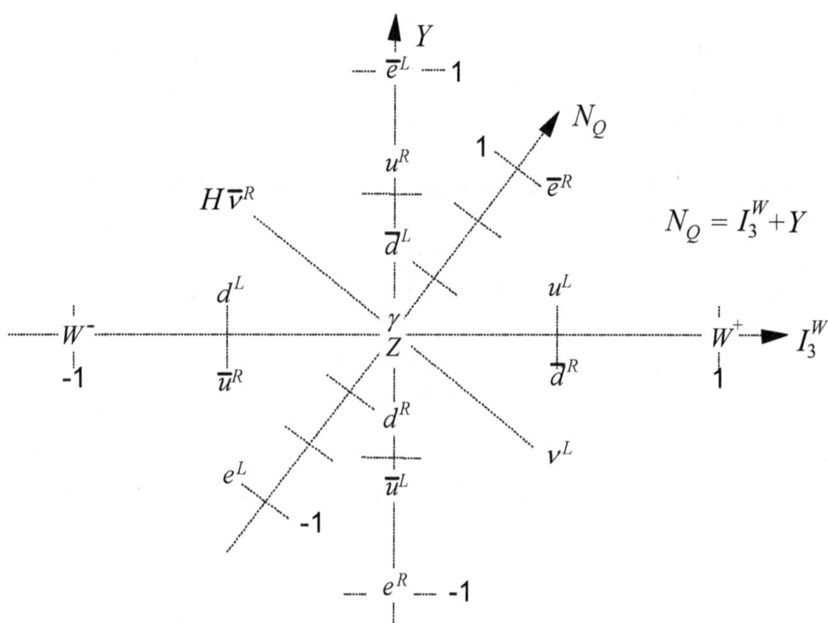

Particle charges graphically

Figure 11-5. Hypercharge, Weak Isospin, and Electric Charges Graphically

11.2.2 Renormalization of Electroweak Theory

In Chap. 7, Sect. 7.4.12, pg. 246, we briefly discussed renormalization of the GSW electroweak theory and noted that an in-depth treatment of it is very complicated and beyond the scope of this book. We will, however, in this section, look a little more closely at it to gain some rudimentary understanding of how it is carried out.

GSW theory renormalization uses non-unitary gauge

As noted in the earlier chapter, 't Hooft and Veltman took a different route to renormalization that makes use of a non-unitary gauge, instead of the unitary one we have employed herein, and that concomitantly, modifies the Lagrangian in a manner we are to discuss shortly. This modification comprises insertion into the Lagrangian of one or more terms, which collectively equal zero, and which end up subtracting out unwanted terms from the field equations. This is called the minimal subtraction (quite different from "minimal substitution") method of renormalization. The method we used in Vol. 1 is called the on-shell renormalization method.

Called minimal subtraction method of renormalization

A Big Part of the Problem

One big obstacle to electroweak renormalization is the form of the propagator for massive intermediate vector bosons, the form of which we deduced in Appendix A of Chap. 8. For a W boson, this is

Massive vector propagator we found in prior studies

$$iD_{F\mu\nu}(k) = i\frac{-g_{\mu\nu} + k_\mu k_\nu / m_W^2}{k^2 - m_W^2 + i\varepsilon}. \qquad (11\text{-}14)$$

Recall that Feynman diagram loops are integrated over k from $-\infty$ to $+\infty$ (in four dimensions), and that, prior to renormalization, such integrations lead to divergences in the transition amplitude. In

1 For one, https://en.wikipedia.org/wiki/Electroweak_interaction#/media/File:Electroweak.svg. Creative Commons license CC BY-SA 3.0 from https://commons.wikimedia.org/wiki/User:Cjean42.

QED, we managed to tame those divergences, as delineated in great detail in Vol. 1, Part 3. In doing that, though, for photons, we did not have the second term in the numerator of (11-14).

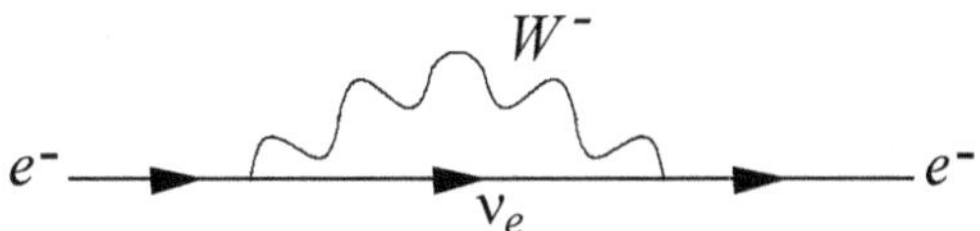

Figure 11-6. Fermion Self-Energy - One Loop in Electroweak Theory

For large k in (11-14), the propagator reduces to a constant i / m_W^2. Thus, our integration in a loop such as that shown in Fig. 11-6, where Λ is a high-energy cut-off parameter, and where the lepton propagator is proportional to $1/k$, becomes proportional to

Naïve power counting for loop integral of Fig. 11-6

$$\int_{-\Lambda}^{+\Lambda} \frac{i}{m_W^2}\frac{1}{k}d^4k \propto \int_0^{+\Lambda} \frac{i}{m_W^2}\frac{1}{k}k^3 dk \propto \frac{\Lambda^3}{m_W^2}. \tag{11-15}$$

(11-15) is still naïve power counting, which as we learned, may be greater than the actual power when all things are considered, but which is also a far cry from our naïve maximum power dependence in QED, which is of order $k^2 = \Lambda^2$, our naïve power dependence for the QED equivalent of Fig. 11-6, which is of order $k = \Lambda$ (Vol. 1, Wholeness Chart 9-1, pg. 265), and our full evaluation of the loop, which is of order $ln\ \Lambda$ (Vol. 1, (15-131), pg. 397).

2nd term in vector propagator → higher power count than QED

Each loop within a loop would introduce an additional factor of Λ^3 / m_W^2, so at higher orders of perturbation theory, progressively more severe divergences arise. It turns out (don't worry about how for now, it is convoluted) that to cancel these, we would need additional renormalization constants (like e and m in QED) at each order, and thus, end up needing an infinite number of such constants (which are simply not available in the theory).

This makes GSW theory non-renormalizable (at least in unitary gauge)

Bottom line: With the on-shell renormalization approach, the theory is non-renormalizable.

Contrast this with our QED approach which was renormalizable because it could absorb the infinities into two quantities, the mass and charge of the electron. The degree of divergence in QED was not dependent on the number of vertices n, whereas in the GSW theory, as posed, it is. (Don't worry about exactly how. It is a bit convoluted.) In that latter theory, the problem with divergences arises, in large part, from the $k_\mu k_\nu / m_W^2$ term in the propagator. In QED this term was absent.

2nd term in vector propagator is big block to renormalization

QED Revisited

It will help if we look at electrodynamics theory from a slightly different point of view.

Review of QED

Traditionally, putting the free photon part of the QED Lagrangian of Vol. 1, (11-36),

$$\mathcal{L}_0^1 = -\tfrac{1}{4}F_{\alpha\beta}F^{\alpha\beta} = \tfrac{1}{2}\left(A_{\beta,\alpha}A^{\alpha,\beta} - A_{\alpha,\beta}A^{\alpha,\beta}\right), \tag{11-16}$$

Free photon Lagrangian

into the Euler-Lagrange equation,

$$\frac{\partial}{\partial x^\nu}\left(\frac{\partial \mathcal{L}}{\partial A_{\mu,\nu}}\right) - \frac{\partial \mathcal{L}}{\partial A_\mu} = 0 , \tag{11-17}$$

into stationary action equation

yields, as we showed in Vol. 1, Chap. 11, Prob. 2,

$$\partial^\mu\left(\partial_\nu A^\nu(x)\right) - \partial^\nu\partial_\nu A^\mu(x) = 0 . \tag{11-18}$$

yields free field equation

We then employ the Lorenz gauge condition

$$\partial_\nu A^\nu(x) = 0 \tag{11-19}$$

Lorenz gauge gives us Maxwell equation

(or in QFT, the Gupta-Bleuler weak Lorenz condition), whereby the first term drops out to give us the familiar Maxwell equation.

However, we could have reformulated the Lagrangian by adding a term equal to zero, which essentially doesn't really change the Lagrangian, such that

$$\mathcal{L}_0^1 = \tfrac{1}{2}\left(\partial_\alpha A_\beta \partial^\beta A^\alpha - \partial_\beta A_\alpha \partial^\beta A^\alpha\right) - \tfrac{1}{2}\left(\partial_\alpha A^\alpha\right)^2 . \qquad (11\text{-}20)$$

Could instead, add term to $\mathcal{L}$ that equals zero due to chosen gauge

Including the last term in (11-20) in (11-17) yields an extra term in (11-18), and we have

$$\partial^\mu\left(\partial_\nu A^\nu(x)\right) - \partial^\nu\partial_\nu A^\mu(x) - \partial^\mu\left(\partial_\nu A^\nu(x)\right) = 0 \quad \rightarrow \quad \partial^\nu\partial_\nu A^\mu(x) = 0 , \qquad (11\text{-}21)$$

This also gives us Maxwell equation

which is Maxwell's equation.

If you aren't immediately comfortable with (11-21) do **Problem 3** to show it.

As noted above, the process of including a zero term in the Lagrangian, which subtracts out a particular term in the field equation, is known as minimal subtraction. In this case, the zero term was the Lorenz gauge. It doesn't change anything we can measure physically, but it can change the underlying math to make it more amenable to solutions.

Inserting zero term in $\mathcal{L}$ that subtracts out term in field equation called minimal subtraction

The added term in (11-20) is called the gauge-fixing term. By adding it to the Lagrangian, we are effectively applying the associated gauge (Lorenz here) to the field equation, so we get (11-21).

Zero term called gauge fixing term

Note that we can define a general QED Lagrangian with a constant λ as part of the gauge fixing term,

$$\mathcal{L}_0^1 = \tfrac{1}{2}\left(\partial_\alpha A_\beta \partial^\beta A^\alpha - \partial_\beta A_\alpha \partial^\beta A^\alpha\right) - \tfrac{\lambda}{2}\left(\partial_\alpha A^\alpha\right)^2 , \qquad (11\text{-}22)$$

Generalized form of gauge fixing term with constant λ in $\mathcal{L}$

which yields the field equation

$$\partial^\alpha\partial_\alpha A^\mu(x) - (1-\lambda)\partial^\mu\left(\partial_\nu A^\nu(x)\right) = 0 . \qquad (11\text{-}23)$$

Field equation with λ

For $\lambda = 1$, we have the results we have just shown above. In this context, this is known as the Feynman gauge or the Feynman-'t Hooft gauge (even though it is essentially the same as the Lorenz gauge in classical electromagnetism). $\lambda = 0$ is known as the Landau gauge and yields the field equation (11-18) that is not so simple.

Feynman gauge $\lambda = 1$; Landau gauge $\lambda = 0$

As we learned in Chap. 5, we cannot use the Lorenz gauge for quantum fields and instead employed the Gupta-Bleuler weak Lorenz condition. (Vol. 1, pgs. 151-154.) In essence, we would apply that in the above, rather than (11-19), but that becomes a bit unwieldy, so we will proceed in this section by simply assuming the Lorenz gauge of classical electrodynamics. For our purposes in this section, the results will be the same.

As long as we are consistent throughout, we can use whatever gauge we like. The physically measured quantities will remain the same. So, we want to choose a gauge (and hence a gauge fixing term) that suits our purpose in any given situation. One gauge may make it easier to calculate transition amplitudes; another, to renormalize a theory.

Instead of a gauge to aid calculations, choose one to aid renormalization

So far, we have used gauges that help us do calculations. 't Hooft chose to use a gauge that helped in renormalization.

Gauges, Massive Vector Fields, and Renormalization

Because the photon is massless, we obtained a propagator that lacks the second term in the numerator of (11-14), and this made a huge difference in the renormalizability of QED, compared to electroweak theory. But what if we could do something in the latter theory similar to (11-22), i.e., add some gauge fixing term(s) that would give us propagators without the troublesome term in the numerator? That is the approach 't Hooft took.

The goal: a gauge that eliminates the 2nd term in the W,Z propagators

Before proceeding you should review Appendix A of Chap. 8 (or read it for the first time if you haven't already). We are going to use a trick devised by 't Hooft that parallels the last column of Wholeness Chart 8-1 therein.

The GSW theory is, as we know, pretty complicated, so we will, instead, use the Higgs model to demonstrate the element of the 't Hooft approach. That model employs the scalar field ϕ for the Higgs field, rather than the $SU(2)$ doublet Φ of GSW, and as well, only has a single vector field, rather than the four such fields in GSW theory.

Our example: simpler Higgs model, not GSW

So, from Chap. 7 (see Wholeness Chart 7-3, pg. 219), the Higgs field in the Higgs model, before any gauge is applied, is

$$\phi - \frac{v}{\sqrt{2}} = \frac{1}{\sqrt{2}}(\sigma + i\eta), \tag{11-24}$$

where the Lagrangian contains the vector field A^μ (not a photon here) and is

Higgs model Lagrangian

$$\mathcal{L} = \tfrac{1}{2}\left|\partial^\mu \sigma\right|^2 - \tfrac{1}{2}(2\lambda v^2)\sigma^2 + \tfrac{1}{2}\left|\partial^\mu \eta\right|^2 - \tfrac{1}{4}F_{\mu\nu}F^{\mu\nu} + \tfrac{1}{2}m^2 A_\mu A^\mu \\ + mA_\mu\left(\partial^\mu \eta\right) + \text{cubic \& quartic interaction terms.} \tag{11-25}$$

The kinetic part, with $F_{\mu\nu}$, has the form (11-16), and is in terms of the particular vector field A^μ in use here. $m = qv$, which becomes the mass of A^μ at contemporary energy levels. All but the cubic and quartic terms, which are not shown explicitly, are bilinear in the fields.

Problem: A^μ and η fields coupled

There are issues with the first term in the second row, as it is bilinear in the A^μ and η fields, so they will be coupled in the lowest order (free) field equations. That means we cannot consider the two fields as independent, normal fields.

Solved in unitary gauge by taking $\eta = 0$

This problem was solved in Chap. 7 via the unitary gauge, whereby we took $\eta = 0$. That left us with terms in A^μ that give rise to the massive vector free field equation

$$\partial^\mu\left(\partial_\nu A^\nu\right) - \partial^\nu \partial_\nu A^\mu + m^2 A^\mu = 0\ . \tag{11-26}$$

Instead, here, we take 't Hooft gauge

However, we are free to choose a different gauge, and 't Hooft chose one whereby η satisfies

$$\partial_\mu A^\mu(x) - m\eta(x) = 0\ . \tag{11-27}$$

Then, to the Lagrangian, he added the zero-value gauge fixing term

Use it as gauge fixing term

$$-\tfrac{1}{2}\left(\partial_\mu A^\mu - m\eta\right)^2 = 0, \tag{11-28}$$

to yield

$\mathcal{L}$ in this gauge

$$\mathcal{L} = \tfrac{1}{2}\left|\partial^\mu \sigma\right|^2 - \tfrac{1}{2}(2\lambda v^2)\sigma^2 + \tfrac{1}{2}\left|\partial^\mu \eta\right|^2 - \tfrac{1}{4}F_{\mu\nu}F^{\mu\nu} + \tfrac{1}{2}m^2 A_\mu A^\mu \\ - \tfrac{1}{2}\left(\partial_\mu A^\mu\right)^2 + m\left(\partial_\mu A^\mu\right)\eta - \tfrac{1}{2}(m\eta)^2 + mA_\mu\left(\partial^\mu \eta\right) + \text{cubic/quartic terms.} \tag{11-29}$$

The terms

$$m\left(\partial_\mu A^\mu\right)\eta + mA_\mu\left(\partial^\mu \eta\right) = m\,\partial_\mu\left(A^\mu \eta\right), \tag{11-30}$$

Total derivative term makes no contribution, so can be dropped

represent a total derivative. Recall that all we have learned of QFT started in Chap. 2 with the action $S = \int \mathcal{L} d^4x$ being stationary between fixed initial and final field configurations. Varying the path in between led to the Euler-Lagrange equation, and from that, the entire theory. But a total derivative integrated over spacetime between fixed initial and final spacetime configurations is independent of path. It doesn't vary as the path does. Thus, terms in $\mathcal{L}$ that are total derivatives will disappear from the theory.

Said another way, our transition amplitude is a function of the S operator (not the action "S"), i.e., $S = T\left\{e^{-i\int \mathcal{H}_I d^4x}\right\} = T\left\{e^{i\int \mathcal{L}_I d^4x}\right\}$. Any total derivative term in $\mathcal{L}_I$ when integrated in the exponent will result in a constant value in the exponent, i.e., merely a phase change in S. And as we know, a phase change affects nothing measurable.

Hence, we can ignore the terms in (11-29) that sum to a total derivative, leaving us with

Leaving effective $\mathcal{L}$ in this gauge

$$\mathcal{L} = \tfrac{1}{2}\left|\partial^\mu \sigma\right|^2 - \tfrac{1}{2}(2\lambda v^2)\sigma^2 + \tfrac{1}{2}\left|\partial^\mu \eta\right|^2 - \tfrac{1}{4}F_{\mu\nu}F^{\mu\nu} + \tfrac{1}{2}m^2 A_\mu A^\mu \\ - \tfrac{1}{2}\left(\partial_\mu A^\mu\right)^2 - \tfrac{1}{2}(m\eta)^2 + \text{cubic/quartic terms.} \tag{11-31}$$

The gauge fixing term ends up canceling a term in field equation

With (11-31) into the Euler-Lagrange equation for A^μ, one obtains the free field equation, where the first two terms come from the fourth term in (11-31), as we have seen before,

$$\partial^\alpha \partial_\alpha A^\mu - \partial^\mu \partial_\alpha A^\alpha + \partial^\mu \partial_\alpha A^\alpha + m^2 A^\mu = 0\ , \tag{11-32}$$

or

$$\left(\partial^\alpha \partial_\alpha + m^2\right) A^\mu = 0 . \tag{11-33}$$

leaving simplified (Proca) field equation for massive field

A similar scalar equation, like those we have seen before, results for η. The two fields are not only uncoupled under the gauge (11-27), but we obtain simplified field equations like (11-33), as well.

But, there is an even greater benefit from the gauge choice (11-27). We are no longer constrained by $\partial_\mu A^\mu = 0$, so our polarization vectors, as we learned in Appendix A of Chap. 8, do not have to be orthogonal nor have only three of them independent. This is the situation of the last column in Wholeness Chart 8-1 of that appendix.

Bigger benefit: polariz vecs no longer have to be what we had before

In other words, under this gauge, we can have a propagator like that of the last row, last column of that wholeness chart. We can have a propagator like the photon propagator (but with a mass term in the denominator) without that pesky second term in the numerator.

And that means a propagator without troublesome 2nd term

And as we saw before, a theory with such a propagator can be renormalized.

And that means theory is renormalizable

This in effect, is what 't Hooft and Veltman did for GSW theory. They used more than one gauge, all known now as 't Hooft gauges, as there are more than a single vector field in GSW theory (unlike the Higgs model.)

Same approach used in GSW model

Returning to the Higgs model, note that with the new gauge (11-27), we have reintroduced the η field, which had been eliminated in the unitary gauge and for which there are no such physical η particles in our world. For this reason, the η particle is called a ghost particle, and it is much like the longitudinal and scalar photons we learned about in QED (Vol. 1, Chap 5). They and η do not exist as real particles, but they contribute as virtual particles associated with propagators. Similar ghost particles arise for other fields in the GSW theory.

Here, gauge $\eta \neq 0$; GSW similar with other fields

But, no such real particles exist; called "ghost particles"

Going further with this subject is beyond the scope of this book, but hopefully, you have some feeling now for the way the renormalization problem was solved for electroweak theory[1].

11.2.3 Running Coupling Constants

In Vol. 1, pgs. 315-317, we derived the relationship for the dependence of the electromagnetic coupling "constant" (the fine structure constant α) as a function of energy that resulted from Feynman amplitude contributions beyond tree level. In the process, we also derived the renormalization group equation (RGE) for QED, where p represents energy level,

$$p\frac{\partial}{\partial p}\alpha(p) = \alpha^2(p) 8\pi b_n = \beta(p,b_n), \tag{11-34}$$

RGE with beta function

where b_n is a function of the number of possible particle-antiparticle pair types in a loop and the charges on those particles. (See Vol. 1, (12-29), pg. 314.) The quantity on the RHS of (11-34) is known as the beta function.

The same thing can be done for weak and strong interactions. That is, in the process of renormalization, we find the weak and strong coupling constants depend on energy as well. As you may guess, this is not trivial work, so we merely cite general results herein from doing so.

In short, we can find RGEs for weak and strong couplings, with their own beta functions depending on weak and strong charges, and the particle-antiparticle pairs interacting thereby. When we do that, we find the beta functions for them are negative, rather than positive, as in QED. This means the $SU(2)$ and $SU(3)$ coupling constants decrease with energy level (opposite from $U(1)$ coupling).

Beta functions positive for e/m, negative for weak and strong forces

The net result is displayed in Fig. 11-7. Note that because the weak and strong coupling constants, at high energy, are effectively inversely proportional to the log of the energy, it is common practice to plot their inverses, as this produces straight line graphs.

Note that in the SM, the lines do not meet at a common point. If they did, it would imply a single high energy coupling constant for all three forces that, due to the differences in radiative corrections at different energies, breaks into three as the energy level decreases. This would suggest a grand unified theory underpinning the SM that would become apparent when (certain conjectured) symmetries are restored at high energy. It turns out that when supersymmetry (SUSY) is merged with

In SUSY, they do converge

[1] More extensive treatments can be found in E.S. Abers and B.W. Lee, *Gauge Theories, Physics Reports*, 9C, No. 1 (1973) and J.C. Taylor, *Gauge Theories and Weak Interactions* (Cambridge 1976).

the SM, the lines actually do meet. This has been one of the main motivators for the extensive work done in SUSY by so many over so many years.

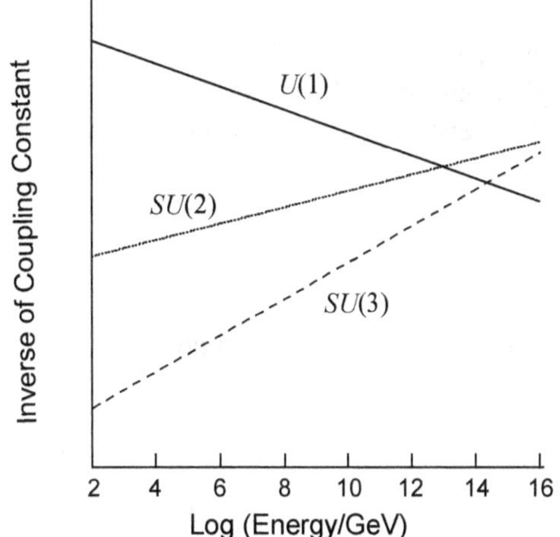

Figure 11-7. Running Coupling Constants

Coupling constants variation with energy

In SM, they don't converge

11.2.4 Effective Theories

In physics, an effective theory is a type of approximation for an underlying physical theory, which could be a quantum field theory, a statistical mechanics model, or models of other phenomena, when we don't know the underling theory itself. Newton's mechanics was, in fact, an effective theory, long before we knew the more precise special theory of relativity. At the usual scales humans work at (speeds much slower than light) Newton's mechanics is very effective, but it is only an approximation of a deeper, more all-encompassing theory good at any speed. In this context, many suspect general relativity may itself be an effective theory, the lower gravity approximation of some much broader theory (which may be superstring or loop quantum gravity based).

Effective theory: valid over particular scale, approximation of truer theory

QED and the SM are also only effective theories. They work pretty well at low energies, but we know (from problems such as gauge hierarchy and more) that it cannot be the final answer. At high energies, the physics must be different from QFT as we understand it.

SM is an effective theory

It is key to realize that renormalization entails an implicit assumption that current theory governs up to the highest energy levels. In fact, historically, that assumption led to the infinities arising in the transition amplitudes when one naively calculates the effect of integrating loops over all energy levels. But, perhaps (even likely), the physics is so different at high energies that these infinities, naturally and simply, do not arise. If that is indeed the case, it may seem that the entire renormalization program is a chimera, a delusional exercise that filled up a good part of many good physicists' careers, but is just not related to the ultimate reality.

Renormalization assumes SM valid at all energy scales

And yet, one result of the renormalization process is the scaling of coupling constants with energy, which has been verified in many experiments, at least up to energy levels those experiments can reach. This is a vindication, of sorts, for renormalization, though again, only for a limited energy scale. There must be something right with it, if it correctly predicts those dependencies.

Yet renormalization predicts correct coupling scaling

So, where are we today on this issue?

For many years it was widely believed that any valid quantum theory had to be renormalizable, and in fact, that belief was vital in the development of the modern SM. However, that position is no longer considered a sacred cow. Renormalizable theories still do hold a special status in physics, though for different reasons (such as the running coupling constants predictions, for example) than those originally motivating their development. But, the prevailing sentiment today is that non-renormalizable models can, in fact, be valid as effective theories, because the divergence problems may simply not exist in truer theories at higher energy. This has, as one might suspect, spawned the growth of an entire area of research that keeps many theorists busy every day.

Now, renormalization not considered essential for a viable theory

Further, there may be a hierarchy of energy scales, with different theories having validity at different scales, each theory being an effective one for the more accurate theory above it in the hierarchy. There is no guarantee that a truer theory is, in fact, a field theory.

Probably a hierarchy of scales, suitable effective theory at each level

There is also no guarantee that a completely true theory (read "theory of everything") actually exists. The universe may be such that no intellectually graspable theoretical framework can ever encompass it all. In that case, the term "effective theory" would only have meaning in a relative context. It would be "effective" relative to some deeper theory, even though that deeper theory would itself have to be merely "effective", as well.

Almost anything goes, as the boundaries on the possibilities are simply not known[1].

Bottom line: Renormalizability can be a good guide in developing quantum models, but unlike what was believed for a long time, it is not essential. Effective theories that are not renormalizable, but make good predictions at lower energy levels, may have validity at those levels and be derivable as a low energy limit of a more complete theory.

11.2.5 Electroweak Theory via Path Integrals

The big advantage of path integrals is their suitability for interactions, such as the strong interaction in hadron formation, which cannot be handled via perturbation within the canonical approach that has been employed extensively by us so far. Electroweak interactions entail small coupling constants, so expansions to higher orders entail subsequently smaller and smaller contributions to transition amplitudes. Strong interactions for quarks forming hadrons have significantly larger coupling constants, whereby higher order terms do not diminish readily to inconsequential levels, and so, are not handled well by perturbative techniques and canonical quantization.

Overview of how PI approach might be used in electroweak theory

However, since one can use path integrals with large coupling constants, one can also use it with small ones, so it can be used in electroweak and QED theories, as well, if one wished. In fact, we did apply the PI approach to electrodynamics in Chap. 4, which you may wish to review by looking over Wholeness Chart 4-4 at the end of that chapter.

We can use the same approach for electroweak theory, as we used for QED earlier. That is, we find the free generating function Z_0, from it find the interacting generating function Z, from that find the Green function $G^{\mu \dots}$, and from that the transition amplitude S_{fi}.

Recall that relations for the free generating functions for a general vector source field J_k and a general fermion source field σ are

$$Z_0[J_\kappa]=e^{-\frac{i}{2}\left[J_\mu D_F^{\mu\nu}J_\nu\right]} \qquad Z_0[\sigma,\bar{\sigma}]=e^{-i[\bar{\sigma}S_F\sigma]}$$
$$[AKB]=\int A(\underline{x})K(\underline{x},\underline{y})B(\underline{y})d^4\underline{x}d^4\underline{y} \tag{11-35}$$

For electroweak interactions at high energy (false vacuum), where sub-subscripts 1,2,3 represent the three W_i and other subscript meanings should be obvious, we have

$$Z_0[\text{all fields}]=Z_0[J_{1\kappa}]Z_0[J_{2\kappa}]Z_0[J_{3\kappa}]Z_0[J_{B\kappa}]Z_0[\sigma_e,\bar{\sigma}_e]Z_0[\sigma_\nu,\bar{\sigma}_\nu]. \tag{11-36}$$

We only work with one generation in (11-36), and for simplicity, ignore quarks. Further, care needs to be taken with LC and RC fermions, but we ignore that distinction as well, again, to keep it simple. The following is then more symbolic than precise, but hopefully, still makes the relevant point.

We then have (11-37), where in the second line we have to sum over all combinations of fermion and boson source fields,

$$Z[\text{all fields}]=e^{ie\int I_\delta(x)d^4x}Z_0[\text{all fields}]$$
$$\text{where } I_\delta(x)=\left(-\frac{1}{i}\frac{\delta}{\delta\sigma_i(x)}\right)\gamma_{k\mu}\left(\frac{1}{i}\frac{\delta}{\delta\bar{\sigma}_j(x)}\right)\left(\frac{1}{i}\frac{\delta}{\delta J_{k\mu}(x)}\right) \qquad i,j=e,\nu \quad k=1,2,3,B. \tag{11-37}$$

[1] For more on this topic, see the extensive discussion by S. Weinberg in Chapter 12 of *The Quantum Theory of Fields Vol. 1* (Cambridge 1995).

With (11-37) into

$$G^{\mu ..} = (-1)^{\bar{n}} \left(\frac{1}{i}\right)^n \left. \frac{\delta^n Z[\text{all fields}]}{\delta J_\mu(x_1)..\delta\bar{\sigma}(y_1)..\delta\sigma(z_1)} \right|_0 \qquad (11\text{-}38)$$

where the fields in the denominator are the legs of the Green function, which will end up being the initial and final fields in a transition amplitude.

(11-38) is a Green function composed of all propagators. When we substitute fields for the propagators composing the legs, the legs become the fields that create and destroy the external (incoming and outgoing) particles in the interaction. Changing the leg fields to particular particle spinors and vectors, we get the transition amplitude for those particles.

We won't go through the tedium of doing an actual example, as, for electroweak theory, this approach really buys us nothing over the canonical approach. But, we do want to be aware that there are two possible routes to finding interaction probabilities (and thus, cross sections and decay rates), and with the steps above, we hopefully can see how this would be done in an alternative manner to what we have been doing for the past 200 or more pages.

We could use PI in e/w theory, but not needed for our work

11.2.6 Uncertainties at Vertices

Recall that

$$\int_{V=\infty}\int_{T=\infty} e^{-i(P_f - P_i)x}\, dVdT = (2\pi)^4 \delta^{(4)}(P_f - P_i). \qquad (11\text{-}39)$$

Our amplitude calcs with $V,T = \infty \rightarrow$ delta functions, 4-momentum conservation

In finding transition amplitudes (the long way, not the Feynman rules way) in Vol. 1, Chap. 8, we integrated over effectively infinite volume and time to get delta functions with arguments like those of (11-39), i.e., the difference between total incoming 4-momentum and total outgoing 4-momentum at a vertex. (See Vol. 1, (8-30), pg. 222, for example.) This forced complete conservation of 4-momentum on us at every vertex, and thus, for the entire interaction overall. (See also, Vol. 1, pg. 454.)

This worked because we can calculate cross-sections and decay rates for this hypothetical case (of infinite volume and time) easily, thanks to the delta functions that arise. And cross-sections and decay rates, as we learned, are the same (for the same particles) regardless of experimental setup (regardless of the volume and time elapsed in a given experiment).

Same σ and decay rates as with finite V,T, so it works

However, had we used more realistic V and T values, we would not have gotten the delta function of (11-39), but a more cumbersome relation. And without that delta function, we do not have 100% conservation of energy and 3-momentum at vertices. In other words, for finite volume, we have uncertainty in 3-momentum; and for finite time, we have uncertainty in energy.

But with finite V,T, 4-momentum uncertainty

Sound familiar? This is the QFT version of the uncertainty principle. The smaller the spatial region we confine our measurement to, the less certain we are of 3-momentum. The shorter the time interval, the less certain we are of energy.

This is uncertainty principle at work in QFT

Measurements of things like cross-section and decay half-lives are statistical in nature. So, had we used finite V and T in our analyses (and created an inordinate amount of extra work for ourselves), we would still have gotten cross-sections and half-life determinations, but they would have been statistical in nature. Their averages would be what we found as precise numbers taking an easier mathematical route.

Measurements and predictions all statistical, so it works

In the real world, interactions are not spread out over the whole space and history of the universe, but the spacetime regions are typically large enough in space, and long enough in time compared to the particles' spacetime interaction region, that the infinite V and T work pretty well. However, you may have heard hand-waving arguments that the propagator "borrows" energy and 3-momentum "from the vacuum" or whatever, in order to carry out its mediator functions. In the context of this section, this has some truth to it (except maybe the "vacuum borrowing" part). For finite V and T, the 4-momentum associated with a propagator at a vertex is not fixed (not precisely conserved). There is some variation, some uncertainty.

But note uncertainty in propagator 4-momenta in finite V,T case

This issue of 4-momentum uncertainty generally gets lost in QFT courses, because the infinite V and T are assumed early on, and never addressed again. Hopefully, this section helps a little in clarifying the matter.

11.2.7 Entangled Particles

Note that particles become entangled when they interact with one another. Free particles can have any 3-momenta, spins, and energies independent of what other free particles may have. However, consider two particles interacting (like an electron and positron) to produce two other particles (like a muon and anti-muon). The original particles are free and can have any values for physical quantities. However, the particular values for these particles in any given case can get distributed in any of a continuous number of ways among the final two particles.

Particles leaving an interaction entangled, until measured

For example, one of these final particles could have any value of 3-momentum from zero to the sum total of 3-momenta from the two initial particles. Either final particle could have spin aligned in any direction. But, once we measure the 3-momentum of one of the particles, the 3-momentum of the other is fixed. It must "collapse" to the value which, when added to that of the first particle, equals the original total 3-momentum (ignoring, for the moment, the uncertainty discussed in the prior section). Similarly, with spin. Measure it for one of the final particles, and the other is instantaneously equal to the value for which the sum of the final spins equals the sum of the initial ones.

Measuring spin, etc. for one, collapses spin, etc. on other

And this is what happens in scattering experiments. If a detector picks up a readily detectable particle (like an electron) 4-momentum, then, for a two final particle event, we know the 4-momentum of the other particle (which may be undetectable, like a neutrino). The two final particles are entangled, each with indefinite physical characteristics, until one of them is measured. Then, the other collapses into the state it must be in to obey conservation laws.

This, in a nutshell, is entanglement in QFT terms.

This is QFT version of entanglement

11.2.8 Spins Greater than Two

As we've seen (and will see in QCD), spins of elementary particles in the SM are limited to 0, ½, and 1. The conjectured graviton has spin 2, and SUSY theories include a spin 3/2 gravitino. No known elementary particles have spin greater than 2, though there are composite particles (combinations of quarks) that do.

The question arises as to whether there could be elementary particles with spins higher than 2. This has spawned a significant amount of research, of which a detailed account is well beyond the scope of this text.

Can we have elementary particles with spin > 2?

As a brief summary, a number of "no-go" theorems have shown that interactions of massless higher spin particles with themselves and with lower spin particles cannot occur and be consonant with basic principles of QFT such as Lorentz invariance[1]. In other words, for the SM, as it stands, such particles are prohibited.

No-go theorems say not in the SM

This comes with caveats, however, as certain advanced theories, such as some versions of superstring theory, can accommodate elementary particles with spin greater than two. As with many things in physics, the final verdict is still out, and awaits a more complete model of nature.

But possible in other models

11.2.9 How Real are Virtual Particles?

We talk a lot about virtual particles, but from whence they come?

In QFT, we start with a collection of nonlinear, coupled partial differential equations. For QED, these numbered two (Vol. 1, (7-18) and (7-19)), the Maxwell equation with its interaction term including fermions, and the Dirac equation with its interaction term including the photon. In principle, we simply solve them for ψ and A^{μ}, as functions of space and time, and we have a completely solved theory of QED. In this, there are no virtual particles, i.e., no propagators. There are simply the final real fields, for which we have explicit mathematical expressions describing their evolution through time, before, while, and after they interact.

Ideal QFT solutions are closed form, with no virtual particles

However, as we know, finding explicit, closed form solutions is not possible for any meaningful interactions. It is not possible in QED, the simplest of quantum field theories with only two coupled equations. We may lose sight of this in electroweak theory, where we focus on the Lagrangian, rather than the field equations. But, that Lagrangian ultimately gives us a host of coupled, nonlinear, partial differential equations in a bunch of fields. These include four vector boson fields, two LC and two

But closed solutions not possible

---

[1] For a little bit more on this, at a student level, see M. Schwartz, *Quantum Field Theory and the Standard Model* (Cambridge 2014), pgs. 135-139.

RC lepton fields (for each generation), two LC and two RC quark fields (per generation), and the Higgs field. This means thirteen coupled, nonlinear, PDEs in thirteen unknowns. If we include all generations, we are talking 29 PDEs in 29 unknowns. In short, our extant branches of the SM can simply not be solved in closed form.

So, creative people like Feynman applied perturbation theory and got approximate (though in many cases very, very accurate) solutions. In the course of this, Feynman developed the amplitude and diagrams that bear his name, and the various orders of approximation in perturbation theory are displayed graphically in the latter as collections of lines and vertices. These lines and vertices represent different terms and factors in the transition amplitude.

So, we use perturbation, and math entities assume virtual particle character or status

Thus, virtual particles in Feynman diagrams stand for mathematical factors that arise in the development of perturbation theory. We give them life, when we talk about them as if they are actual entities transiting between real particles and maintaining certain conservation laws as they do. But, they may be just a fiction, symbols, and nothing more, for parts of a scheme we have developed as a substitute for the complete, closed form solution. And the latter, as we said, contains no virtual particles.

But they may only be symbols, with no actual reality

Bottom line: Whether or not virtual particles actually exist, or are just a convenient concept helping us grapple with something deeper and more complete, is a philosophical question. As we discussed early on in Vol. 1, there is no way to measure them. They may, or may not, be a part of our physical world.

11.2.10 An Issue with the Spin 1 Lagrangian

The Issue

Many authors begin the development of quantization from classical field theory by postulating the relationship

A common way (not our way) to quantize leads to a problem

$$\left[\phi^r(x),\pi_s(y)\right]=i\delta^r_s\delta(\mathbf{x}-\mathbf{y}), \tag{11-40}$$

where ϕ^r is any field, π_s is the conjugate momentum for the field ϕ^s, and all other commutators are zero. From this, one can derive the coefficient commutation relations (Vol. 1, pgs. 51-53), which underlie, and are the very foundation of, all of QFT.

Starting with (11-40) is a little different from our approach in Vol. 1, where we started with the Poisson bracket relations of classical field theory and then took those brackets over into commutators (with an added factor of i). The difference may seem virtually negligible, since with our approach, we ended up with (11-40), anyway (with one exception to be explained).

However, it is then often noted[1] that this leads to an issue for photon field quantization, and by extension, to any spin 1 field (such as the Z, W_i, and gluon fields). To gain insight into this issue, note first that for a photon, (11-40) becomes

Starting with spin 1 field-momentum commutator

$$\left[A^\mu(x),\pi^\nu(y)\right]=ig^{\mu\nu}\delta(\mathbf{x}-\mathbf{y}). \tag{11-41}$$

Then, consider the free part of the symmetric Lagrangian (for a derivation of the last part of (11-42), see Vol. 1, pg. 288, (11-7) and (11-8))

with a symmetric $\mathcal{L}$

$$\mathcal{L}=-\tfrac{1}{4}F_{\alpha\beta}F^{\alpha\beta}=-\tfrac{1}{2}\left(\partial_\alpha A_\beta\partial^\alpha A^\beta-\partial_\beta A_\alpha\partial^\alpha A^\beta\right)\qquad\left(\text{symmetric Lagrangian}\right), \tag{11-42}$$

from which we determine the conjugate momentum

$$\begin{aligned}\pi^\mu&=\frac{\partial\mathcal{L}}{\partial\dot{A}_\mu}=\frac{\partial}{\partial A_{\mu,0}}\left(-\tfrac{1}{2}\left(A_{\beta,\alpha}A^{\beta,\alpha}-A_{\alpha,\beta}A^{\beta,\alpha}\right)\right)\\&=-\tfrac{1}{2}\left(\delta^\mu_\beta\delta^0_\alpha A^{\beta,\alpha}+A_{\beta,\alpha}g^{\mu\beta}g^{0\alpha}-\delta^\mu_\alpha\delta^0_\beta A^{\beta,\alpha}-A_{\alpha,\beta}g^{\mu\beta}g^{0\alpha}\right).\\&=-\tfrac{1}{2}\left(A^{\mu,0}+A^{\mu,0}-A^{0,\mu}-A^{0,\mu}\right)=-A^{\mu,0}+A^{0,\mu}.\end{aligned} \tag{11-43}$$

Note, however, that for the zeroth component of (11-43), we get

[1] See F. Mandl and G. Shaw, *Quantum Field Theory*, 2nd ed, (Wiley 2010), pgs.74-75.

$$\pi^0 = -\dot{A}_0 + \dot{A}_0 = 0\,, \tag{11-44}$$

$\pi^0 = 0$

which is quite incompatible with the quantization postulate (11-41). That is,

$$\left[A^0(x),\pi^0(y)\right] = 0 \neq ig^{00}\delta(\mathbf{x}-\mathbf{y})\,, \tag{11-45}$$

and that is inconsistent with the commutation postulate

and we cannot derive the associated coefficient commutation relations for the photon.

<u>The Commonly Proposed Solution</u>

The oft cited solution is to use the non-symmetric Lagrangian proposed by Fermi, and which we used in Vol. 1 prior to our chapter on symmetry (Chap. 11, therein, where we showed we really needed to use the symmetric version of $\mathcal{L}$). Fermi's Lagrangian is

$$\mathcal{L} = -\tfrac{1}{2}\left(\partial_\alpha A_\beta\right)\left(\partial^\alpha A^\beta\right) \qquad \left(\text{non-symmetric Fermi Lagrangian}\right), \tag{11-46}$$

where

$$\pi^\mu = \frac{\partial\mathcal{L}}{\partial\dot{A}_\mu} \qquad \Rightarrow\; \pi^0 = \frac{\partial\mathcal{L}}{\partial\dot{A}_0} = \frac{\partial}{\partial\dot{A}_0}\left(-\tfrac{1}{2}\left(\partial_\alpha A_\beta\right)\left(\partial^\alpha A^\beta\right)\right) = -\dot{A}^0\,, \tag{11-47}$$

Using a non-symmetric $\mathcal{L} \rightarrow \pi^0 \neq 0$, consistent commutator, and correct coefficient commutators

the LHS of (11-45) is no longer identically zero, we can equate it to the RHS, and we end up with the appropriate coefficient commutation relations for QED.

It is argued (see the cited reference for details) that we should quantize with the non-symmetric Lagrangian (11-46) and after quantization, impose the Lorenz condition, as this will make the non-symmetric development of the theory equivalent to the symmetric one.

<u>The Problem with This "Resolution" of the Issue</u>

But then, no matter how one slices it, the fundamental Lagrangian from which we deduce our theory is not symmetric, and the ramifications of this are manifold (and undesirable).

But then, we have a non-symmetric QFT

<u>An Alternative Resolution</u>

The appendix of this chapter shows that by starting with classical Poisson brackets and mapping them to quantum commutators, we can avoid the issue. In particular, (11-45) does not arise.

See appendix for alternative solution with symmetric $\mathcal{L}$

11.3 Summary of Topics in this Chapter

This chapter does not comprise a single overarching theme, like all the other chapters in this book, but rather contains short presentations of a number of disparate topics. So, our summary will be simply a sentence or few for each topic.

Mandelstam Variables

For 1 and 2 designating incoming particles, 3 and 4, outgoing, the Lorentz invariant Mandelstam variables are

$$s = \left(p_1 + p_2\right)^2 \qquad t = \left(p_1 - p_3\right)^2 \qquad u = \left(p_1 - p_4\right)^2\,, \tag{11-48}$$

$$\text{where} \quad s + t + u = \sum m_i^2 \qquad i = \text{external particles}\,. \tag{11-49}$$

Equal Times Contractions and Tadpoles

Fermion equal times contractions (propagators) equal zero because the positron propagator has the same magnitude as, but opposite sign from, the electron propagator, and both must be added for the loop. This is Furry's theorem for a single vertex fermion loop.

Boson equal times contractions (propagators) are not identically zero, lead to tadpole Feynman diagram parts, and must be evaluated mathematically.

Mean Free Path

The average path length before a particle entering a target medium (of particle number density n_t) interacts with a target particle, where σ is the cross-section for the two-particle interaction, is the

$$\text{mean free path} = \frac{1}{\sigma n_t}\,. \tag{11-50}$$

Fierz Transformations

Fierz transformations allow us to re-order fields in a transition amplitude. Wholeness Chart 11-1, pg. 366, comprises a table of some of them. One commonly used transformation is

$$\left(\bar{\psi}_1^L \gamma^\alpha \psi_2^L\right)\left(\bar{\psi}_3^L \gamma_\alpha \psi_4^L\right) = -\left(\bar{\psi}_1^L \gamma^\alpha \psi_4^L\right)\left(\bar{\psi}_3^L \gamma_\alpha \psi_2^L\right). \qquad (11\text{-}51)$$

Anomalies

An anomaly in QFT comprises a symmetry that exists classically, but is broken in QFT. Scaling (or conformal) symmetry (invariance in energy level) is one example, as coupling constants in the SM vary with energy level. There are non-gauge (global) anomalies, such as scaling, and gauge (local) anomalies, the latter leading to inconsistencies in a quantum theory. The GSW theory of electroweak interactions is free of gauge anomalies.

Higgs Field as a Fifth Force

The Higgs field is bosonic and is coupled to fermions, but is not a gauge field, as it is not one of the spin 1 vector fields of $SU(2)\text{X}U(1)$. Whether one considers it a 5th force or not is a matter of definition, though theorists generally consider it is not.

Particle Charges

A graph of weak isospin charge, hypercharge, and electric charge of various particles can be found on pg. 368.

Renormalization of Electroweak Theory

't Hooft and Veltman renormalized electroweak theory by using gauges other than the unitary gauge to obtain massive vector propagators without the troublesome $k_\mu k_\nu / m^2$ term in the numerator. Their method is called the minimal subtraction method and differs from the on-shell method we used in QED.

Running Coupling Constants

As in QED, renormalization of electroweak and QCD theories leads to coupling constants that vary with interaction energy level. The renormalization group equations (RGE) for the weak ($SU(2)$) and strong ($SU(3)$) coupling constants have negative beta functions, whereas the beta function for $U(1)$ theory is positive. So, the former two decrease with energy level, but the latter increases.

The three coupling constants do not meet at a common point at very high energy in the SM, but they do in supersymmetric theories. In a unified theory, they would be expected to converge at some energy level.

Effective Theories

An effective theory is one that makes good predictions over certain ranges, but is not valid over all ranges. Newton's mechanics compared to special relativity is one example. Newton's gravity theory compared to general relativity is another. NRQM compared to QFT is yet another. And so is the SM, as we know it is not the final answer at high energy.

Renormalization assumes a current theory (QED or GSW electroweak, for examples) is valid at all energy levels, so renormalizability is not really a necessary criterion for a viable effective theory in our contemporary universe. Yet, historically, it played a key role in the development of the SM, and it makes accurate predictions for running coupling constants. So, renormalizable theories still hold a special place in physics, even though they are not as sacrosanct as they once were.

Electroweak Theory via Path Integrals

One can develop electroweak theory using path integrals, though it is an unneeded, additional complication, and virtually never employed in QFT texts. The canonical quantization approach works extremely well for electroweak theory.

Uncertainty in QFT

In the development of QFT via the canonical approach, we take the volume and time of an interaction as infinite, and this simplifies the mathematics by leading to delta functions forcing the incoming and outgoing total momenta to be equal. The final results for cross-sections and decay rates are the same in any experimental setup, i.e., the same for experiments with finite volume and finite time intervals.

However, in more realistic cases, with finite volume and time, we do not get delta functions in the transition amplitudes, so there is uncertainty in outgoing vs incoming 4-momentum at a vertex. This is the QFT version of the uncertainty principle. The smaller the space and time scales, the more uncertain the 3-momentum and energy are.

Entangled Particles

When particles interact, the properties (momentum, spin, ...) of the final particles are not precisely known. They are not in eigenstates, until the given property of all but one of the final particles has been measured. The last such particle then collapses into an eigenstate that conserves the particular property.

Spins Greater than Two

SM particles all have spins of 0, ½, or 1. Theories of gravity entail spins of 2, and some other theories, spins of 3/2, as well. Certain no-go theorems, based on fundamental principles of QFT, such as Lorentz and gauge invariances, state that it is not possible to have spins greater than 2. However, the SM is an effective theory, and some more advanced models, which their founders hope might be truer theories, include higher spin particles.

The Reality of Virtual Particles

Virtual particles arise in perturbation theory solutions in QED and electroweak theory, but would not be a part of complete, closed solutions, if we could find such solutions. So, it is an open (and probably unanswerable) question whether virtual particles represent actual entities in the physical world, or are merely useful metaphors for purely mathematical factors arising in transition amplitudes.

Issue with the Spin 1 Lagrangian

If one quantizes by starting with field and conjugate momentum commutators and a symmetric part of the free vector Lagrangian, the zeroth component of the conjugate momentum is zero, and thus inconsistent with the postulated commutators (which are non-zero). The "resolution" to this issue is commonly taken to be a reversion to the non-symmetric spin 1 Lagrangian, which seems less than satisfactory.

By instead beginning quantization with the classical Poisson bracket relations and mapping them to quantum commutators, one can keep a symmetric Lagrangian, obtain the proper coefficient commutation relations, and resolve the issue in a wholly satisfactory manner.

Closing the Curtain on Electroweak Theory

This pretty much wraps up Part 3 of this book on electroweak theory. Next up is Part 4 on strong interactions.

11.4 Appendix. Resolving the Spin 1 Quantization Issue

This appendix presents a resolution of the quantization issue discussed in Sect. 11.2.10, pg. 377.

11.4.1 Classical Background for Symmetric $\mathcal{L}$

Instead of starting quantization with (11-40), consider mapping the Poisson brackets of classical field theory (Vol. 1, Wholeness Chart 2-2, pg. 21, last column, near the end) to the commutators of QFT (Vol. 1, Wholeness Chart 2-5, pg. 31, last column, near the end).

$$\underbrace{\left\{A^{\mu}(x),\pi^{\nu}(y)\right\}}_{\text{Poisson brackets}}=g^{\mu\nu}\delta(\mathbf{x}-\mathbf{y})\xrightarrow[\text{QFT}]{\text{to}}\underbrace{\left[A^{\mu}(x),\pi^{\nu}(y)\right]}_{\text{Commutators}}=ig^{\mu\nu}\delta(\mathbf{x}-\mathbf{y}) \qquad (11\text{-}52)$$

Poisson brackets for fields are defined, in general (where the δ/δ derivative is defined in the cited reference), as

$$\{u,v\} \equiv \left(\frac{\delta u}{\delta A_\alpha} \frac{\delta v}{\delta \pi^\alpha} - \frac{\delta u}{\delta \pi^\alpha} \frac{\delta v}{\delta A_\alpha} \right) \delta(\mathbf{x}-\mathbf{y}). \tag{11-53}$$

For $u = A^\mu$, $v = \pi^\nu$, this becomes

$$\{A^\mu,\pi^\nu\} = \left(\frac{\delta A^\mu}{\delta A_\alpha} \frac{\delta \pi^\nu}{\delta \pi^\alpha} - \frac{\delta A^\mu}{\delta \pi^\alpha} \frac{\delta \pi^\nu}{\delta A_\alpha} \right) \delta(\mathbf{x}-\mathbf{y}) = \left(g^{\mu\alpha}\delta^\nu_\alpha - 0\right)\delta(\mathbf{x}-\mathbf{y}) = g^{\mu\nu}\delta(\mathbf{x}-\mathbf{y}). \tag{11-54}$$

Note, however, that if either $u = 0$ or $v = 0$ in (11-53), then the Poisson bracket equals zero, *not* (11-54). Importantly, in our case, we have $v = \pi^0 = 0$, via (11-44), even in the classical theory with the classical Lagrangian (11-42). So, if $u = A^0$, and $v = \pi^0 = 0$, instead of (11-54), we get

$$\{A^0,\pi^0\} = \{A^0,0\} = 0 \tag{11-55}$$

Some, more mathematically inclined, might consider (11-55) ill-defined, since it would entail a derivative in (11-54) of zero with respect to zero, but we can circumvent that issue, if we wish, by simply defining Poisson brackets as zero whenever either u or v is zero.

Summary of Poisson Brackets for Symmetric e/m $\mathcal{L}$:

$$\begin{aligned} &\{A^\mu,\pi^\nu\} = g^{\mu\nu}\delta(\mathbf{x}-\mathbf{y}) && \nu \neq 0 \\ &\{A^0,\pi^0\} = 0 && \text{since } \pi^0 = 0. \end{aligned} \tag{11-56}$$

The key point: In classical theory with a symmetric electrodynamic Lagrangian, Poisson brackets equal zero when $v = \pi^0$, because $\pi^0 = 0$. They do not obey (11-54), just as in QFT, the LHS of (11-45) equals zero and does not obey (11-41), the quantum correlate of (11-54). We have a similar situation in classical theory as what has been attributed to the quantized theory.

Resolution: We should not quantize by assuming (11-41), but by assuming that the Poisson brackets of (11-56) correlate with commutators.

11.4.2 Quantizing the Symmetric $\mathcal{L}$

Spatial Components of A^μ and Its Conjugate Momentum

First, consider (11-52) only for μ and ν not equal to 0, i.e., take $\mu = i$ and $\nu = j$. The only non-zero terms are for $i = j$, and we work only with the $i = 1$ case as an example, since the other two cases will be directly parallel. Then, (11-52) with (11-43) becomes

$$\begin{aligned} ig^{11}\delta(\mathbf{x}-\mathbf{y}) &= \left[A^1,\pi^1\right] = \left[A^1,\left(-\left(A^{1,0}-A^{0,1}\right)\right)\right] = -A^1\left(A^{1,0}-A^{0,1}\right) + \left(A^{1,0}-A^{0,1}\right)A^1 \\ &= -A^1A^{1,0} + A^1A^{0,1} + A^{1,0}A^1 - A^{0,1}A^1 = -\left(A^1A^{1,0}-A^{1,0}A^1\right) + \left(A^1A^{0,1}-A^{0,1}A^1\right) \\ &= -\left[A^1,A^{1,0}\right] + \left[A^1,A^{0,1}\right]. \end{aligned} \tag{11-57}$$

If we follow the procedure of Vol. 1, pgs.52-53, we find the second commutator in the last line of (11-57) equals zero and gives us

$$\left[A^1,A^{0,1}\right] = 0 \quad \rightarrow \quad \left[a_0(\mathbf{k}),a_1^\dagger(\mathbf{k}')\right] = \left[a_1(\mathbf{k}),a_0^\dagger(\mathbf{k}')\right] = 0, \tag{11-58}$$

and the first commutator in the same line gives us

$$-\left[A^1,A^{1,0}\right] = -\left[A^1,\dot{A}^1\right] = ig^{11}\delta(\mathbf{x}-\mathbf{y}) = -i\delta(\mathbf{x}-\mathbf{y}) \quad \rightarrow \quad \left[a_1(\mathbf{k}),a_1^\dagger(\mathbf{k}')\right] = \delta(\mathbf{k}-\mathbf{k}'). \tag{11-59}$$

These results extend immediately to $i = j = 2$ or 3.

Conclusion: For i,j = 1,2 or 3, we can quantize the symmetric $\mathcal{L}$ theory and get the standard coefficient commutation relations of QFT.

Time Components of A^μ and π^ν

Consider now $\mu = \nu = 0$ in the commutator part of (11-52), where due to (11-56), instead of g^{00}, we should have zero.

$$\left[A^0,\pi^0\right] = \left[A^0,0\right] = 0. \tag{11-60}$$

This parallels what we found from the classical theory for Poisson brackets, via (11-55), and in that case, there is no conflict in quantizing by going from Poisson brackets to commutators.

The important point is that from (11-60), we can easily still have $\left[A^0, A^{0,0}\right] \neq 0$. So, we only need one little wrinkle in our quantization procedure. That is, we can take, as part of that procedure (a postulate), that the commutator of the time component of the photon field with its time derivative takes a form parallel to what we deduced for the other such commutators in (11-59), but with $\mu = \nu = 0$, i.e.,

$$-\left[A^0, A^{0,0}\right] = -\left[A^0, \dot{A}^0\right] = ig^{00}\delta(\mathbf{x}-\mathbf{y}). \qquad (11\text{-}61)$$

(11-61) is the starting point from which the accepted coefficient commutation relations are derived, i.e.

$$-\left[A^0, A^{0,0}\right] = ig^{00}\delta(\mathbf{x}-\mathbf{y}) \quad \rightarrow \quad \left[a_0(\mathbf{k}), a_0^\dagger(\mathbf{k}')\right] = -\delta(\mathbf{k}-\mathbf{k}'). \qquad (11\text{-}62)$$

Conclusion: For quantization beginning with Poisson brackets and a symmetric Lagrangian, one can derive all of the appropriate spin 1 coefficient commutation relations, and the issue is resolved.

11.4.3 Summary of Resolution of the Quantization Issue

For μ and $\nu = 1,2,3$,

$$\left\{A^i, \pi^j\right\} = g^{ij}\delta(\mathbf{x}-\mathbf{y}) \xrightarrow[\text{QFT}]{\text{to}} \left[A^i, \pi^j\right] = ig^{ij}\delta(\mathbf{x}-\mathbf{y})$$
$$\rightarrow \left[A^i, \dot{A}^j\right] = ig^{ij}\delta(\mathbf{x}-\mathbf{y}) \quad \rightarrow \quad \left[a_i(\mathbf{k}), a_j^\dagger(\mathbf{k}')\right] = \delta_{ij}\delta(\mathbf{k}-\mathbf{k}'). \qquad (11\text{-}63)$$

For $\mu = \nu = 0$,

$$\left\{A^0, \pi^0\right\} = 0 \qquad \xrightarrow[\text{QFT}]{\text{to}} \left[A^0, \pi^0\right] = 0$$
$$\xrightarrow{\text{postulate}} \left[A^0, \dot{A}^0\right] = ig^{00}\delta(\mathbf{x}-\mathbf{y}) \rightarrow \left[a_0(\mathbf{k}), a_0^\dagger(\mathbf{k}')\right] = -\delta(\mathbf{k}-\mathbf{k}').$$

Note that for $\mu = 0$ and $\nu = 1,2$, or 3, $g^{\mu\nu} = 0$, so (11-52) is zero for both the classical and QFT cases, so $a_0(\mathbf{k})$ commutes with $a_j^\dagger(\mathbf{k})$ for $j = 1,2$, or 3.

11.5 Problems

1. Show that $s + t + u = \sum m_i^2$ where i labels the external particles, and s, t, and u are the Mandelstam variables. Hint: The conservation of energy and 3-momentum will help.
2. Show part of a Feynman diagram with an equal times propagator that arises from the Lagrangian term $-\frac{\lambda}{4}\sigma^4$.
3. From $\mathcal{L}_0^1 = \frac{1}{2}\left(\partial_\alpha A_\beta \partial^\beta A^\alpha - \partial_\beta A_\alpha \partial^\beta A^\alpha\right) - \frac{1}{2}\left(\partial_\alpha A^\alpha\right)^2$, use the Euler-Lagrange equation to derive the field equation $\partial^\nu \partial_\nu A^\mu(x) = 0$.

Part Four

Strong Interactions

"[1966 to 1979] was a great time to be a high-energy theorist, the period of the famous triumph of quantum field theory. And what a triumph it was, in the old sense of the word: a glorious victory parade, full of wonderful things brought back from far places to make the spectator gasp with awe and laugh with joy."

Sidney Coleman

Chapter 12 Quantum Chromodynamics: The Basics

Chapter 13 QCD Symmetries: Continuous and Discrete

Chapter 14 Hadron Composition

Chapter 15 QCD Interaction Theory

Chapter 16 QCD Renormalization and Coupling Constant

Chapter 17 QCD Experiments

Chapter 12

Quantum Chromodynamics: The Basics

"Don't believe any calculation in meson theory that uses a Feynman diagram."
Richard Feynman

12.0 Preliminaries

What did Feynman mean by this?

He was referring to the unsuitability of perturbation theory for the strong interactions underlying hadron formation (of which meson theory is a part). Feynman diagrams were designed as pictorial symbols for transition sub-amplitudes, which proliferate in number for higher orders. For each sub-amplitude, we need to add a term to our transition amplitude calculation. In a perturbation theory, such as QED or electroweak theory, the higher order terms, with more vertices in higher order diagrams, make successively smaller contributions to the amplitude.

In general, QCD hadron formation not amenable to perturbation approach

In a non-perturbative theory, this is not true. Higher order terms make significant contributions, so one cannot simply cut off one's calculation at a particular order to get a desired degree of accuracy in the result. One cannot take a certain number of Feynman diagrams, containing up to a certain number of vertices, and be done with it. In effect, the whole idea of the diagrams doesn't work when you may need an extraordinarily large number of them to represent reality. Picking some subset of this huge number of diagrams doesn't work. As Feynman notes, do not trust any calculation that does.

So, Feynman diagrams can't be used in QCD analyses of quarks bound in hadrons

Nevertheless, we will use Feynman diagrams to represent quark/gluon interactions, and there are two reasons why. First, from Fig. 11-7, pg. 373, we see that, due to radiative corrections, the running strong coupling constant is strong at lower energy (greater distance) levels, but weaker at high energy (shorter distance). So, bound quarks at close distances to one another are weakly bound, and in fact, approach free quarks in the high energy asymptotic limit. (This, as you may be aware, is known as asymptotic freedom.) So, for high energy interactions, the strong force becomes so weak that it can actually be analyzed using perturbation theory. Historically, as we will discuss later, the $SU(3)$ color theory of quarks was confirmed by probing nucleons with high energy electrons and obtaining results that matched the predictions from perturbation analyses of that theory. More on this in later chapters.

But, they can be used for high energy probes of nucleons, as perturbation there OK

Second, even in the low energy case, with quarks further apart and bound together tightly, where perturbation theory does not work, Feynman diagrams can serve as aids to visualization. They can be illustrative, but do not offer much help analyzing quarks bound together inside hadrons or mesons.

And they can be used as visualization aids for hadron

But for now, first things first – the basics underlying strong interactions.

12.0.1 Chapter Overview

In this chapter, we will use methods that parallel what we did for $U(1)$ QED and for $SU(2)\text{X}U(1)$ electroweak theory. General methodologies used there also work for $SU(3)$ theory. That is, we will

- model free quark fields (which don't actually exist in nature) and free gluon fields as we did free lepton and vector boson fields earlier,
- employ minimal substitution to find interaction terms in the QCD Lagrangian,
- show that the QCD Lagrangian is symmetric under $SU(3)$ transformations, and
- take a simplified look at typical QCD interactions.

Topics in this chapter

You may want to review sections of Chap. 2 having to do with QCD before proceeding on with this chapter.

12.0.2 Degree of Difficulty of QCD

QCD is harder than QED or electroweak theory for two reasons.

First, it is grounded in an $SU(3)$ symmetry, rather than the simpler $SU(2)$ and $U(1)$ symmetries of those theories. We have (as we learned in Chap. 2) eight Lie algebra generators, rather than three or one, so we will have eight vector bosons (gluons) mediating the strong force, rather than three or one. Additionally, the Cartan subalgebra has two 3X3 matrices, rather than a single 2X2 or 1X1, so each particle will have two QCD charges (labeled ε_3 and ε_8 in Chap. 2) rather than just one charge, as in the other theories. (Though, for convenience, we ultimately meld them into a single type of charge (called color), as described in Chap. 2.)

QCD hard because:
1) SU(3) theory and two, not one, kind of charge
2) perturbation approach not generally valid

Second, the strong interaction is, as its name conveys, strong, so, for determining bound states, perturbation theory, the saving theoretical grace of our other theories, cannot be used.

But, in other senses, QCD is easier than the electroweak theory for several reasons.

First, the $SU(3)$ symmetry, the foundation of QCD, does not break at low energy. It held just after the Big Bang, and it holds today. So, in studying it, we won't have all the complications symmetry breaking entails. Indeed, it took us 200 or so pages to cover all of the ramifications to electroweak theory of that breaking. Not a factor in strong interactions.

Second, the color force has no analog to the chirality of electroweak interactions. We don't have to concern ourselves with LC and RC fields, or some similar differentiation between color fields. All quarks interact via the strong force in the same way. We don't have to worry about singlet quarks vs triplet quarks behaving differently under strong $SU(3)$ interactions. For quarks, there are only QCD triplets, no QCD singlets. Quark composites, such as hadrons or mesons, could be considered singlets, as they have zero total color charge, but quarks themselves are never singlets.

Third, due to the reason mentioned in the prior paragraph, the strength of interaction is the same between any two quarks of *different* color; and it is the same between any two quarks of the *same* color. This is far different from leptons, whose electroweak interaction strengths vary widely from particle type to particle type.

QCD easy because:
1) unbroken sym
2) color, not chirality
3) all particles same force strength
4) gluons massless
5) no leptons

Fourth, as experiment has confirmed, gluons are massless, so we don't get the problematic second term in the numerator of their propagators, and we can use the more straightforward polarization vectors of the same kind as those for the QED photon.

Finally, leptons don't interact via the strong force, as they are QCD singlets (with zero charge), so we can just ignore them. In contrast, quarks interact via both the weak and strong force, so we had to deal with both leptons and quarks when we studied electroweak theory.

12.1 The "Free" Quark and "Free" Gluon Lagrangian

In dealing with strong interactions, we will ignore the (irrelevant for QCD) RC and LC type casting from electroweak theory, and unless stated otherwise, restrict the flavor symbol f, which in prior chapters has stood for any quark or lepton, to just quarks, i.e., to $f = u, d, c, s, t, b$.

12.1.1 Idealizations of Our Model

As you have no doubt heard, free (non-interacting) quarks have never been found in nature, as they are constrained (see Chap. 2) to color neutral configurations, such as mesons and hadrons. If one tries to separate two quarks bound in a meson, they will move farther apart until, at some distance, the binding between them will break, and instead of two separate quarks, two new mesons (each of a bound quark and anti-quark) will spring into existence. The increase in binding energy as the quarks are pulled apart gets converted into mass for two new quarks. This is known as <u>quark confinement</u>. It is the other side of the asymptotic freedom coin.

Even though quarks confined, we will model free quarks to start

Nevertheless, we will start with a model of free quarks and free vector bosons (gluons) that will, in the course of our development, ultimately end up in an unbreakable embrace. This, of course, is a purely theoretical starting point.

12.1.2 "Free" Quarks

<u>We Can Simply Consider Massive Quarks</u>

Particles gain mass when the electroweak symmetry breaks via the Higgs mechanism, which we studied extensively in Chap. 7. In dealing with strong interactions, since the $SU(3)$ symmetry never

Can consider quark mass inherent when modeling strong force

breaks, we can start by assuming quarks already have mass. We don't have to deal with the issue of them acquiring that mass, but just take mass as an inherent quality. This is a reasonable assumption for any time after the electroweak epoch (later than 10^{-12} seconds after the Big Bang).

The Free Quark Lagrangian

As we have learned from decades of experiment, quarks are fermions, like leptons, so in a hypothetical free state, they obey the Dirac equation, and have free Lagrangian terms that parallel those for leptons. (See Chap. 7, (7-144).) That is,

$$\mathcal{L}_0 = \mathcal{L}_0^L + \mathcal{L}_0^Q + \mathcal{L}_0^B + \mathcal{L}_0^H$$
$$\mathcal{L}_0^Q = i\bar{\psi}_f \partial\!\!\!/ \psi_f - m_f \bar{\psi}_f \psi_f . \qquad (12\text{-}1)$$

Free Lagrangian and free quark part

Quark Triplets

Since we are working in $SU(3)$, it will help if we re-express the second row of (12-1) in terms of color triplets. Designating color (r, g, b) by the subscript a, we have (where we sum on f and a)

$$\mathcal{L}_0^Q = i\bar{\psi}_{fa} \partial\!\!\!/ \psi_{fa} - m_{fa} \bar{\psi}_{fa} \psi_{fa} = i\bar{\Psi}_f \partial\!\!\!/ \Psi_f - m_f \bar{\Psi}_f \Psi_f \qquad \Psi_f(x) = \begin{bmatrix} \psi_{fr} \\ \psi_{fg} \\ \psi_{fb} \end{bmatrix} . \qquad (12\text{-}2)$$

Free quark part in triplet form

Recall that we use upper case Ψ for doublets in electroweak theory, and we will also use them for triplets in QCD. $\bar{\Psi}_f \Psi_f$ (no sum on f for this example) will represent the inner product in color space ($SU(3)$ space) of a triplet field of flavor f with its own adjoint.

Review of Group Theory Parts of a Field

In Chap. 2, pg. 41, we saw how a field can have quite a number of different indices. In general

$$\Psi^h_{fa} \qquad h = L, R \qquad f = u, d, c, s, t, b, e, \nu_e, \mu, \nu_\mu, \tau, \nu_\tau \qquad a = r, g, b . \qquad (12\text{-}3)$$

Spinor field and its many indices

For a given flavor f, the h index represents a part of (factor in) the field that is operated on by $SU(2)$ matrix operators. The a index represents the part of the field that is operated on by $SU(3)$ matrix operators. For certain combinations of values for these indices, we get fields that form a component in a doublet or triplet. For example, for $h = L$, $f = u$, and $a = g$, we have (as shown on the cited page)

$$\text{LC green up quark} \quad \Psi^L_{ug} = \underbrace{\sum_{r,\mathbf{p}} \sqrt{\frac{m}{VE_\mathbf{p}}} \left(c_{ur}(\mathbf{p}) u_r(\mathbf{p}) e^{-ipx} + d^\dagger_{ur}(\mathbf{p}) v_r(\mathbf{p}) e^{ipx} \right)}_{\text{general solution to Dirac equation for up quark, } \psi_u} \begin{bmatrix} 1 \\ 0 \end{bmatrix}_W \begin{bmatrix} 0 \\ 1 \\ 0 \end{bmatrix}_S = \psi_u \begin{bmatrix} 1 \\ 0 \end{bmatrix}_W \begin{bmatrix} 0 \\ 1 \\ 0 \end{bmatrix}_S , \qquad (12\text{-}4)$$

One example of a spinor field and its parts

a field that is both the upper component in a weak doublet and the middle component of a color triplet. This field would destroy an LC, green up quark and create an RC anti-green, anti-up quark. In this context, the creation and destruction operators in (12-4) "sense" the non-zero components of the doublet and triplet.

For $h = L$, $f = e$, and $a = 0$, we get

$$\text{LC electron} \quad \Psi^L_e = \underbrace{\sum_{r,\mathbf{p}} \sqrt{\frac{m}{VE_\mathbf{p}}} \left(c_{er}(\mathbf{p}) u_r(\mathbf{p}) e^{-ipx} + d^\dagger_{er}(\mathbf{p}) v_r(\mathbf{p}) e^{ipx} \right)}_{\text{general solution for electron field, } \psi_e} \begin{bmatrix} 0 \\ 1 \end{bmatrix}_W = \psi_e \begin{bmatrix} 0 \\ 1 \end{bmatrix}_W , \qquad (12\text{-}5)$$

Another example

which is a color singlet and the lower component of a weak doublet.

In (12-4) and (12-5) there are also spinor indices, which, as is common practice, are hidden and not designated with a subscript. In addition to the action of $SU(2)$ and $SU(3)$ operators, the fields in (12-4) and (12-5) can, as well, be acted on by $U(1)$ operators, like the Dirac operator in the Dirac equation (Vol. 1, (4-54).) Information on energy, three-momentum, and charge is also contained in (12-5). The quantum field is indeed a rich mosaic of parts.

There are also spinor indices (hidden)

End of Review

12.1.3 "Free" Gluons

The mediators of the strong force, the gluons, are vector bosons, like their counterparts in QED and electroweak theory. So, they obey the same Proca equation (though they are massless, so $m = 0$, therein), and thus, they have free Lagrangian terms of similar form. If we use A_i^μ as our symbol for gluons (distinguished from the photon symbol by the subscript i), where i designates the type of gluon, then similar to what we had for photons, W, and B bosons (see (6-45), pg. 175),

Gluon field tensor like photon and IVBs

$$F_i^{\mu\nu}(x) = \partial^\nu A_i^\mu(x) - \partial^\mu A_i^\nu(x) \qquad i = 1,2,\ldots,8. \tag{12-6}$$

$F_i^{\mu\nu}$ is distinguished from the electromagnetic tensor $F^{\mu\nu}$ of QED by the subscript i. The free gluon term in the Lagrangian is then

Free gluon $\mathcal{L}$ term like photon and IVBs

$$\mathcal{L}_0^G = -\tfrac{1}{4} F_i^{\mu\nu} F_{i\,\mu\nu}. \tag{12-7}$$

12.2 The QCD Interaction Lagrangian

12.2.1 Strongly Interacting Quarks

To find interactions, we assume our method of minimal substitution (see Chap. 6, Sect. 6.3.7, pg. 175) works in QCD, just as it did in QED and electroweak theory. However, instead of using a covariant derivative with $SU(2)$ and $U(1)$ Lie algebra generator matrices, we use $SU(3)$ generator matrices λ_i (Gell-Mann matrices). (See Chap. 2, (2-80).) We also use an $SU(3)$ (strong) coupling constant g_S.

Use minimal substitution in QCD, like QED & e/w

Thus, our QCD covariant derivative is

QCD covariant derivative

$$D^\mu = \partial^\mu + i\frac{g_S}{2}\lambda_j A_j^\mu, \tag{12-8}$$

where, for convenience, we repeat the Gell-Mann matrix definitions from Chap. 2,

The eight SU(3) generators (the conventional basis)

$$\lambda_1 = \begin{bmatrix} 0 & 1 & 0 \\ 1 & 0 & 0 \\ 0 & 0 & 0 \end{bmatrix} \quad \lambda_2 = \begin{bmatrix} 0 & -i & 0 \\ i & 0 & 0 \\ 0 & 0 & 0 \end{bmatrix} \quad \lambda_3 = \begin{bmatrix} 1 & 0 & 0 \\ 0 & -1 & 0 \\ 0 & 0 & 0 \end{bmatrix} \quad \lambda_4 = \begin{bmatrix} 0 & 0 & 1 \\ 0 & 0 & 0 \\ 1 & 0 & 0 \end{bmatrix}$$
$$\lambda_5 = \begin{bmatrix} 0 & 0 & -i \\ 0 & 0 & 0 \\ i & 0 & 0 \end{bmatrix} \quad \lambda_6 = \begin{bmatrix} 0 & 0 & 0 \\ 0 & 0 & 1 \\ 0 & 1 & 0 \end{bmatrix} \quad \lambda_7 = \begin{bmatrix} 0 & 0 & 0 \\ 0 & 0 & -i \\ 0 & i & 0 \end{bmatrix} \quad \lambda_8 = \frac{1}{\sqrt{3}}\begin{bmatrix} 1 & 0 & 0 \\ 0 & 1 & 0 \\ 0 & 0 & -2 \end{bmatrix}, \tag{12-9}$$

with anti-symmetric structure constants f_{ijk} satisfying

$$\left[\lambda_i, \lambda_j\right] = i2 f_{ijk}\lambda_k. \tag{12-10}$$

Substituting (12-8) for ∂^μ in (12-2) gives us, where we ignore the electroweak contribution here and throughout this chapter,

Full QCD Lagrangian; free plus interaction (ignoring e/w part)

$$\mathcal{L}^Q = \underbrace{i\bar{\Psi}_f \gamma_\mu \partial^\mu \Psi_f - m_f \bar{\Psi}_f \Psi_f}_{\mathcal{L}_0^Q} \underbrace{- \frac{g_S}{2}\bar{\Psi}_f \gamma_\mu \lambda_j \Psi_f A_j^\mu}_{\mathcal{L}^{QG}}. \tag{12-11}$$

The γ^μ acts on the 4D spinor space part of the field, and λ_j acts on the 3D $SU(3)$ triplet part of the field.

$\mathcal{L}^{QG}$ in (12-11) is the part of the Lagrangian governing strong interactions. If we were to use perturbation theory for QCD, this term would give us vertices with one gluon and two quarks. Since the λ_i are not all diagonal, we can have different quark colors at a vertex with a single gluon.

Two quark fields and one gluon in interaction term

Do **Problem 1** to show this. You really should do this problem, as it illustrates a key point.

If we have different colored quarks at a vertex with a gluon, and if color charge is conserved (which we will later show it is), then the gluons themselves must carry color charge. If a red quark changes to a blue one at a vertex with an existing gluon, the gluon must carry away red and anti-blue color.

We later show color conserved, so gluons must carry color, too

12.2.2 Interacting Gluons

In Chap. 6, we substituted the electroweak covariant derivative (6-47) into the W_i^μ part of the free Lagrangian $-\frac{1}{4}W_i^{\mu\nu}W_{i\,\mu\nu}$ and found (6-60), i.e., $-\frac{1}{4}G_i^{\mu\nu}G_{i\,\mu\nu}$, where ε_{ijk} therein are the structure constants for the $SU(2)$ Lie algebra. We can do the same thing for the $SU(3)$ theory. We won't repeat all the steps we went through in Chap. 6, but simply deduce, by inspection of that result, the parallel relation for QCD gluons, where we employ the $SU(3)$ structure constants f_{ijk}, instead of ε_{ijk} and use a caret to distinguish the gluon tensor $\hat{G}_i^{\mu\nu}$ from the electroweak tensor $G_i^{\mu\nu}$. Thus, using (12-8) in (12-6), we find, similar to (6-58),

Minimal substitution in gluon tensor

$$F_i^{\mu\nu} \xrightarrow[\text{derivative}]{\text{with covariant}} D^\nu A_i^\mu - D^\mu A_i^\nu = \hat{G}_i^{\mu\nu} = F_i^{\mu\nu} + g_s f_{ijk} A_j^\mu A_k^\nu \qquad (12\text{-}12)$$

and similar to (6-60), from (12-7), the purely gluonic contribution to the Lagrangian,

Purely gluonic parts of the Lagrangian

$$\mathcal{L}^G = -\tfrac{1}{4}\hat{G}_i^{\mu\nu}\hat{G}_{i\,\mu\nu} = \underbrace{-\tfrac{1}{4}F_i^{\mu\nu}F_{i\,\mu\nu}}_{\mathcal{L}_0^G} \; \underbrace{+g_s f_{ijk}A_{i\mu}A_{j\nu}\partial^\mu A_k^\nu - \tfrac{1}{4}g_s^2 f_{ijk}f_{ilm}A_j^\mu A_k^\nu A_{l\mu}A_{m\nu}}_{\mathcal{L}^{GG}} . \qquad (12\text{-}13)$$

Note that, like electroweak theory, the strong force vector bosons (gluons) have terms in the Lagrangian with three and four factors of their fields. This means that if we were to apply perturbation theory to QCD, from the terms in (12-11) and (12-13), we can see we would have vertices in Feynman diagrams looking like those of Fig. 12-1, where the helical lines represent gluons.

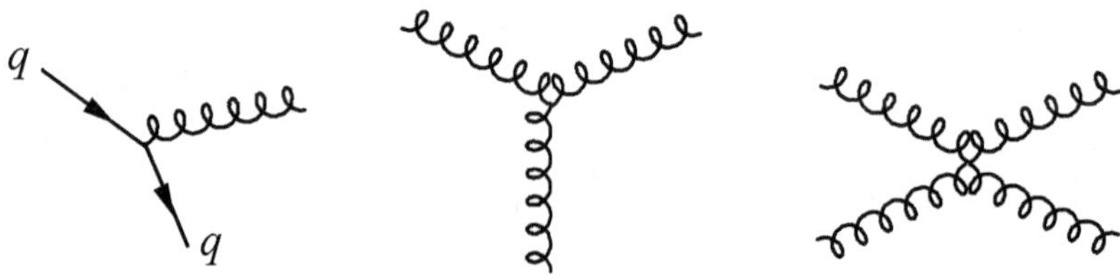

Figure 12-1. Types of QCD Vertices

Each purely gluonic field equation is nonlinear in gluon fields, like SU(2) IVBs

The terms with three and four factors of gluon fields also mean the field equation for a given gluon not interacting with quarks has terms with two and three factors of those fields. And that means that field equation is nonlinear in the gluon fields. As we have mentioned before, this nonlinearity occurs in $SU(n)$ theories ($n > 1$), but not in $U(1)$ theories. Gluons, like the W fields, are self-interacting.

Of course, when interactions with fermions are also included, the Lagrangian gives rise in QCD, electroweak, and QED theories to nonlinear partial differential field equations (which are coupled). The nonlinearity one hears about for $n > 1$ special unitary theories, but not for $U(1)$ theories, refers only to the purely vector boson field equations, without fermion interactions. In truth, it is almost a misnomer, but it is common practice.

But, all fully interacting QFT field equations are nonlinear

12.2.3 The Full QCD Lagrangian

The QCD Lagrangian is much simpler than the GSW electroweak one. It is simply (12-11) plus (12-13),

Total, any energy, QCD Lagrangian $\mathcal{L} = \mathcal{L}^Q + \mathcal{L}^G = \mathcal{L}_0^Q + \mathcal{L}^{QG} + \mathcal{L}_0^G + \mathcal{L}^{GG}$

$$= \underbrace{i\bar{\Psi}_f \partial\!\!\!/ \Psi_f - m_f \bar{\Psi}_f \Psi_f}_{\mathcal{L}_0^Q} \underbrace{-\frac{g_s}{2}\bar{\Psi}_f A\!\!\!/_i \lambda_i \Psi_f}_{\mathcal{L}^{QG}}$$

$$\underbrace{\underbrace{-\frac{1}{4}F_i^{\mu\nu}F_{i\,\mu\nu}}_{\mathcal{L}_0^G} \underbrace{+g_s f_{ijk} A_{i\mu} A_{j\nu} \partial^\mu A_k^\nu - \frac{1}{4} g_s^2 f_{ijk} f_{ilm} A_j^\mu A_k^\nu A_{l\mu} A_{m\nu}}_{\mathcal{L}^{GG}}}_{-\frac{1}{4}\hat{G}_i^{\mu\nu}\hat{G}_{i\mu\nu}} \tag{12-14}$$

where $F_i^{\mu\nu}(x) = \partial^\nu A_i^\mu(x) - \partial^\mu A_i^\nu(x) \quad i = 1,2,\ldots,8,$ and $A\!\!\!/_j = \gamma_\mu A_j^\mu$.

Total QCD $\mathcal{L}$ at all energy levels

Note there are no *QQ* terms in (12-14), which means we have no vertices with three or four quarks. As we have seen before in electroweak theory, the non-linear terms come from bosons, not fermions.

12.2.4 Symmetry of the QCD Lagrangian

Objective: Show QCD $\mathcal{L}$ symmetric

Most (perhaps all) texts do not actually show, step-by-step, that the quark Lagrangian is gauge invariant. We do that in the following.

The Gauge Transformations

Extrapolating from the transformations of *SU*(2) electroweak theory in Wholeness Chart 6-3, pg. 179, to *SU*(3) QCD theory, we anticipate (and will later prove) that the QCD Lagrangian is invariant under finite transformations of the following form, where the group parameters ω_i are now eight in number. (In Chap. 2, we used α_i for these parameters but the parallels with *SU*(2) will be more apparent if we use the symbols ω_i.)

Parallel procedure to SU(2) theory

Finite SU(3) Transformation

Finite transformations, as we should be aware of by now, are very complicated. For the record, we show them for the quark triplet, but not, due to the complexity, for the gluons.

$$\Psi_f(x) \to \Psi'_f(x) = e^{\frac{i}{2}\lambda_n\omega_n(x)}\Psi_f(x)$$
$$\bar{\Psi}_f(x) \to \bar{\Psi}'_f(x) = \bar{\Psi}_f(x) e^{-\frac{i}{2}\lambda_m\omega_m(x)} \tag{12-15}$$

The QCD finite transformations

Note that some authors define what we call ω_i as $g_s\omega_i$. It doesn't really matter, as long as one stays consistent throughout, since the ω_i are arbitrary parameters, anyway.

It is easier to deal with infinitesimal transformations, i.e., very small ω_i. Thus, we extrapolate from Wholeness Chart 6-4, pg. 180, to find the following.

Infinitesimal SU(3) Transformation (Small ω_i)

$$\Psi_f(x) \to \Psi'_f(x) = \left(1 + \frac{i}{2}\lambda_n\omega_n(x)\right)\Psi_f(x)$$
$$\bar{\Psi}_f(x) \to \bar{\Psi}'_f(x) = \bar{\Psi}_f(x)\left(1 - \frac{i}{2}\lambda_m\omega_m(x)\right) \tag{12-16}$$

The QCD infinitesimal transformations

$$A_i^\mu(x) \to A_i'^\mu(x) = A_i^\mu(x) - \frac{1}{g_s}\partial^\mu\omega_i(x) - f_{ijk}\omega_j(x)A_k^\mu(x) \tag{12-17}$$

Gauge Invariance of the QCD Lagrangian

We promised (12-14) is invariant under (12-16) and (12-17), so, let's check it.

Invariance of the Kinetic Plus Quark-Gluon Interaction Terms

Let's first transform the terms in the first row of (12-14), except for the mass term, under (12-16) and (12-17). The proof parallels, step-by-step, what we did for the high energy *W* bosons of *SU*(2) theory in Chap. 6, (6-128) to (6-137), where we take $\sigma_i \to \lambda_i$, $W_i \to A_i^\mu$, structure constants $\varepsilon_{ijk} \to f_{ijk}$, and $i = 1,..,8$ instead of $i = 1,2,3$. However that proof also included *B* bosons relations, so it is not so easy to follow by making the above substitutions. Hence, we will do the *SU*(3) proof separately below.

Showing kinetic and $\mathcal{L}^{QG}$ terms invariant

As the transformations are for infinitesimal values of the parameters ω_i, we can ignore all terms higher than first order in ω_i. Note that the ∂^μ in the transformation (12-17) acts only on $\omega(x)$, and not on any Ψ that may follow.

$$\text{First row of (12-14) except mass terms} = i\bar{\Psi}_f\gamma_\mu\partial^\mu\Psi_f - \frac{g_s}{2}\bar{\Psi}_f\gamma_\mu A_i^\mu\lambda_i\Psi_f$$

$$\xrightarrow[A_i^\mu \to A_i^\mu - \frac{1}{g_s}\partial^\mu\omega_i - f_{ijk}\omega_j A_k^\mu]{\Psi_f \to \left(1+\frac{i}{2}\lambda_n\omega_n\right)\Psi_f} i\bar{\Psi}_f\left(1-\frac{i}{2}\lambda_m\omega_m\right)\gamma_\mu\partial^\mu\left(\left(1+\frac{i}{2}\lambda_n\omega_n\right)\Psi_f\right)$$
$$-\frac{g_s}{2}\bar{\Psi}_f\left(1-\frac{i}{2}\lambda_m\omega_m\right)\gamma_\mu\lambda_i\left(A_i^\mu - \frac{1}{g_s}\partial^\mu\omega_i - f_{ijk}\omega_j A_k^\mu\right)\left(1+\frac{i}{2}\lambda_n\omega_n\right)\Psi_f . \tag{12-18}$$

1st line of (12-18) =

$$i\bar{\Psi}_f\gamma_\mu\partial^\mu\Psi_f + \underbrace{i\bar{\Psi}_f\gamma_\mu\partial^\mu\left(\left(\frac{i}{2}\lambda_n\omega_n\right)\Psi_f\right)}_{\underbrace{i\bar{\Psi}_f\gamma_\mu\left(\frac{i}{2}\lambda_n\partial^\mu\omega_n\right)\Psi_f}_{\text{Call this } X} + \underbrace{i\bar{\Psi}_f\gamma_\mu\left(\frac{i}{2}\lambda_n\omega_n\right)\partial^\mu\Psi_f}_{\text{Call this } Y}}$$
$$\underbrace{+i\bar{\Psi}_f\left(-\frac{i}{2}\lambda_m\omega_m\right)\gamma_\mu\partial^\mu\Psi_f}_{-Y} + \text{terms of order } \omega_i^2, \tag{12-19}$$

where two of the terms in (12-19) cancel.

$$2^{\text{nd}}\text{ line of (12-18)} = -\frac{g_s}{2}\bar{\Psi}_f\gamma_\mu\lambda_i\left(A_i^\mu\right)\Psi_f \underbrace{-\frac{g_s}{2}\bar{\Psi}_f\gamma_\mu\lambda_i\left(-\frac{1}{g_S}\partial^\mu\omega_i\right)\Psi_f}_{-X} - \frac{g_s}{2}\bar{\Psi}_f\gamma_\mu\lambda_i\left(-f_{ijk}\omega_j A_k^\mu\right)\Psi_f$$
$$+ \text{terms of order } \omega_i^2 - \frac{g_s}{2}\bar{\Psi}_f\gamma_\mu\lambda_i\left(A_i^\mu\right)\left(\frac{i}{2}\lambda_n\omega_n\right)\Psi_f \tag{12-20}$$
$$-\frac{g_s}{2}\bar{\Psi}_f\left(-\frac{i}{2}\lambda_m\omega_m\right)\gamma_\mu\lambda_i\left(A_i^\mu\right)\Psi_f + \text{terms of order } \omega_i^2, \omega_i^3 .$$

Dropping all higher order terms and adding (12-19) plus (12-20), we have

$$(12\text{-}18) = \begin{array}{l} i\bar{\Psi}_f\gamma_\mu\partial^\mu\Psi_f - \frac{g_s}{2}\bar{\Psi}_f\gamma_\mu\lambda_i A_i^\mu\Psi_f + \overbrace{\frac{g_s}{2}\bar{\Psi}_f\lambda_i f_{ijk}\omega_j A_k^\mu\gamma_\mu\Psi_f}^{\boxed{1}} \\ \overbrace{-i\frac{g_s}{4}\bar{\Psi}_f\lambda_i\lambda_n\omega_n A_i^\mu\gamma_\mu\Psi_f}^{\boxed{2}} \overbrace{+i\frac{g_s}{4}\bar{\Psi}_f\lambda_m\lambda_i\omega_m A_i^\mu\gamma_\mu\Psi_f}^{\boxed{3}} . \end{array} \tag{12-21}$$

The last two terms can be re-written as

$$\boxed{2}+\boxed{3} = -i\frac{g_s}{4}\bar{\Psi}_f\lambda_k\lambda_j\omega_j A_k^\mu\gamma_\mu\Psi_f + i\frac{g_s}{4}\bar{\Psi}_f\lambda_j\lambda_k\omega_j A_k^\mu\gamma_\mu\Psi_f$$
$$= i\frac{g_s}{4}\bar{\Psi}_f\left(\lambda_j\lambda_k - \lambda_k\lambda_j\right)\omega_j A_k^\mu\gamma_\mu\Psi_f = i\frac{g_s}{4}\bar{\Psi}_f\left[\lambda_j,\lambda_k\right]\omega_j A_k^\mu\gamma_\mu\Psi_f . \tag{12-22}$$

Using (12-10) in the above, and the anti-symmetry of f_{jki} after the first equal sign below, we have

$$\boxed{2}+\boxed{3} = i\frac{g_s}{4}\bar{\Psi}_f\left(i2f_{jki}\lambda_i\right)\omega_j A_k^\mu\gamma_\mu\Psi_f = -\frac{g_s}{2}\bar{\Psi}_f\lambda_i f_{ijk}\omega_j A_k^\mu\gamma_\mu\Psi_f , \tag{12-23}$$

which cancels term $\boxed{1}$ in (12-21). Finally, we have the transformed (12-18) equal to the only terms left in (12-21). That is,

$$(12\text{-}18) = i\bar{\Psi}_f\gamma_\mu\partial^\mu\Psi_f - \frac{g_s}{2}\bar{\Psi}_f\gamma_\mu\lambda_i A_i^\mu\Psi_f , \tag{12-24}$$

which is the same as the original (12-18).

$$i\bar{\Psi}_f\gamma_\mu\left(\partial^\mu + i\frac{g_s}{2}A_i^\mu\lambda_i\right)\Psi_f \xrightarrow[A_i^\mu \to A_i^\mu - \frac{1}{g_s}\partial^\mu\omega_i - f_{ijk}\omega_j A_k^\mu]{\Psi_f \to \left(1+\frac{i}{2}\lambda_n\omega_n\right)\Psi_f} i\bar{\Psi}_f\gamma_\mu\left(\partial^\mu + i\frac{g_s}{2}A_i^\mu\lambda_i\right)\Psi_f . \tag{12-25}$$

These terms invariant

We conclude that the quark kinetic plus quark-gluon interaction parts of the QCD Lagrangian are symmetric under the coupled transformation (12-16) & (12-17).

Invariance of the Mass Terms

Do **Problem 2** to show the mass terms in (12-14) are symmetric under the $SU(3)$ transformation.

Mass terms invariant

Invariance of Purely Gluonic Terms

The invariance of $-\frac{1}{4}\hat{G}_{i\,\mu\nu}\hat{G}_i^{\mu\nu}$ in (12-14) parallels, step-by-step, what we did for $-\frac{1}{4}G_{i\,\mu\nu}G_i^{\mu\nu}$ in electroweak theory in Appendix A of Chap. 6, (6-138) to (6-150), with eight gluons A_i^{μ} instead of three W_i^{μ} bosons, eight matrix generators instead of three, and structure constants f_{ijk} instead of ε_{ijk}.

Purely gluonic terms invariant

Bottom line: The QCD Lagrangian is symmetric under the $SU(3)$ gauge transformation.

So, QCD $\mathcal{L}$ invariant

You can show the $SU(3)$ covariant derivative of the Ψ_f transforms just like the spinor field itself by doing **Problem 3**.

12.3 Typical QCD Interactions Simplified

Diagrams

From the example of Problem 1 we can see that gluons can carry color charges. Take the case where, at a spacetime event (vertex), we have an incoming red quark, an outgoing green quark, and an outgoing gluon. In that case, the gluon must carry both red color (to conserve the incoming red charge) and anti-green color (to conserve zero green charge).

At least some gluons carry color charge

So, gluons carry two kinds of color charge. Every combination of color and anti-color charges are carried by the eight gluons. And gluons have zero electric charge. Examples of some interactions one can run into with quarks and gluons are shown in Fig. 12-2. Note that, since gluons have no electric charge, they cannot change a quark of one electric charge into a quark of a different electric charge at a vertex.

They carry color and anti-color

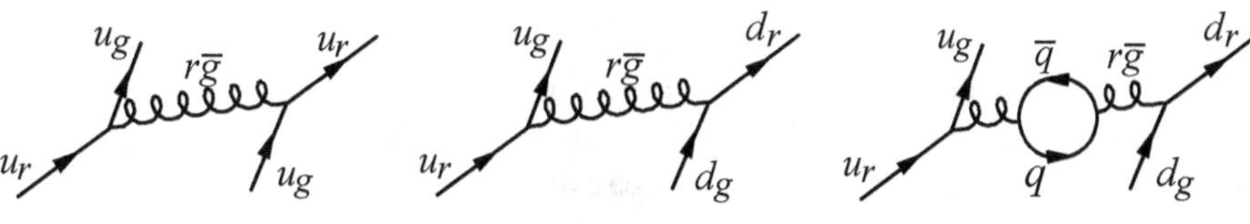

Some ways quarks and gluons interact and one way they don't

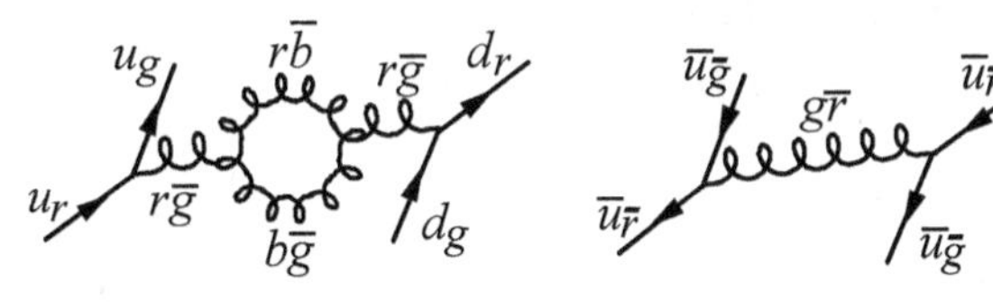

Not possible

d_g u_r u_r $r\bar{g}$ d_g

Gluons carry no electric charge

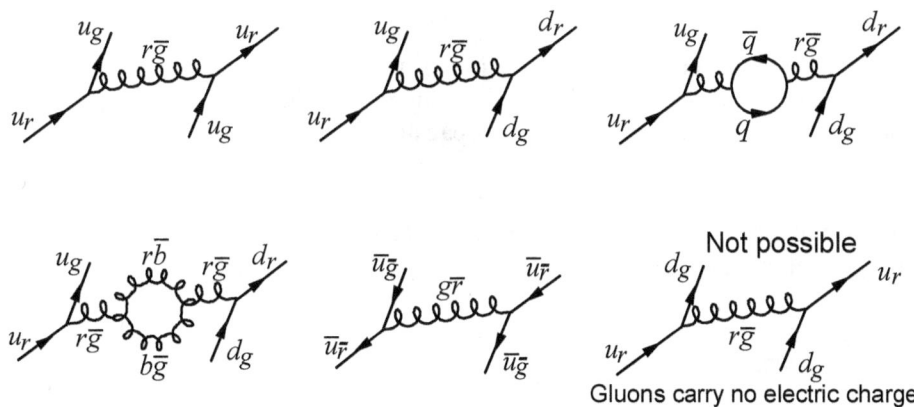

Figure 12-2. Possible and Impossible Interactions with Gluons

Math Behind the Diagrams

Before reading this section, I strongly suggest you look over the part of Chap. 7 from (7-153) to (7-157) with a focus on what happens in converting the $W_{1\mu}$ and $W_{2\mu}$ fields in the $SU(2)$ covariant derivative to the W_{μ}^{+} and W_{μ}^{-} fields that we use to find transition amplitudes. We will follow a parallel procedure in $SU(3)$ theory.

To keep things simple at the start, we will look first at only the Lagrangian terms in $\mathcal{L}^{QG}$ of (12-14) with $i = 1,2$ and $f = u$.

Similar to what we found in Chap. 7 for GSW electroweak theory, where

$$W_\mu^+ = \frac{W_{1\mu} - iW_{2\mu}}{\sqrt{2}} \quad W_\mu^- = \frac{W_{1\mu} + iW_{2\mu}}{\sqrt{2}} \qquad W_{1\mu} = \frac{W_\mu^+ + W_\mu^-}{\sqrt{2}} \quad W_{2\mu} = i\frac{W_\mu^+ - W_\mu^-}{\sqrt{2}}, \tag{12-26}$$

we define (where the plus and minus superscripts in (12-27) do *not* denote electric charge, as gluons are electrically chargeless, but are used in parallel with the notation of (12-26), where they do denote electric charge)

Defining new fields similar to what we did in electroweak theory

$$T_\mu^+ = \frac{A_{1\mu} - iA_{2\mu}}{\sqrt{2}} \quad T_\mu^- = \frac{A_{1\mu} + iA_{2\mu}}{\sqrt{2}} \qquad A_{1\mu} = \frac{T_\mu^+ + T_\mu^-}{\sqrt{2}} \quad A_{2\mu} = i\frac{T_\mu^+ - T_\mu^-}{\sqrt{2}}. \tag{12-27}$$

Then consider the particular terms in $\mathcal{L}^{QG}$ of (12-14)

Two $\mathcal{L}$ terms we want to express in the new fields

$$-\frac{g_s}{2}\bar{\Psi}_u A_1 \lambda_1 \Psi_u - \frac{g_s}{2}\bar{\Psi}_u A_2 \lambda_2 \Psi_u, \tag{12-28}$$

and make the substitutions of (12-27) and (12-9), to obtain

$$-\frac{g_s}{2}\begin{bmatrix}\bar{u}_r & \bar{u}_g & \bar{u}_b\end{bmatrix} A_1 \begin{bmatrix}0 & 1 & 0\\ 1 & 0 & 0\\ 0 & 0 & 0\end{bmatrix}\begin{bmatrix}u_r\\ u_g\\ u_b\end{bmatrix} - \frac{g_s}{2}\begin{bmatrix}\bar{u}_r & \bar{u}_g & \bar{u}_b\end{bmatrix} A_2 \begin{bmatrix}0 & -i & 0\\ i & 0 & 0\\ 0 & 0 & 0\end{bmatrix}\begin{bmatrix}u_r\\ u_g\\ u_b\end{bmatrix}$$

$$= -\frac{g_s}{2}\begin{bmatrix}\bar{u}_r & \bar{u}_g & \bar{u}_b\end{bmatrix}\begin{bmatrix}0 & A_1 - iA_2 & 0\\ A_1 + iA_2 & 0 & 0\\ 0 & 0 & 0\end{bmatrix}\begin{bmatrix}u_r\\ u_g\\ u_b\end{bmatrix}$$

$$= -\frac{g_s}{2}\begin{bmatrix}\bar{u}_r & \bar{u}_g & \bar{u}_b\end{bmatrix}\underbrace{\left(A_1 - iA_2\right)}_{\sqrt{2}T^+}\begin{bmatrix}0 & 1 & 0\\ 0 & 0 & 0\\ 0 & 0 & 0\end{bmatrix}\begin{bmatrix}u_r\\ u_g\\ u_b\end{bmatrix} \tag{12-29}$$

$$-\frac{g_s}{2}\begin{bmatrix}\bar{u}_r & \bar{u}_g & \bar{u}_b\end{bmatrix}\underbrace{\left(A_1 + iA_2\right)}_{\sqrt{2}T^-}\begin{bmatrix}0 & 0 & 0\\ 1 & 0 & 0\\ 0 & 0 & 0\end{bmatrix}\begin{bmatrix}u_r\\ u_g\\ u_b\end{bmatrix}$$

Two $\mathcal{L}$ terms expressed via newly defined fields

$$= -\frac{g_s}{\sqrt{2}}\bar{u}_r T^+ u_g - \frac{g_s}{\sqrt{2}}\bar{u}_g T^- u_r.$$

The last term in (12-29) represents the LHS vertex in the first diagram in Fig. 12-2. An incoming red up quark is destroyed, an outgoing green up quark is created, as is an outgoing $r\bar{g}$ gluon. T_μ^- creates that gluon, i.e., it is the first part, mathematically, of the $r\bar{g}$ gluon propagator.

Each term can represent a vertex for red and green up quarks interacting

Similarly, the next to last term in (12-29) represents the RHS vertex in the first diagram in Fig. 12-2. An incoming green up quark is destroyed, an outgoing red up quark is created, and a $r\bar{g}$ gluon is destroyed. T_μ^+ represents that second part, mathematically, of the gluon propagator.

With Wick's theorem, we would find, at first order, where x_2 represents the LHS vertex and x_1, the RHS vertex,

Lowest order part of S operator with new fields can represent $rg \rightarrow rg$

$$\text{a term in the } S \text{ operator} \quad -\frac{1}{2!}\frac{g_s^2}{2}\int N\left\{(\bar{u}_r\gamma^\mu u_g T_\mu^+)_{x_1}(\bar{u}_g\gamma^\nu u_r T_\nu^-)_{x_2}\right\}d^4x_1 d^4x_2. \tag{12-30}$$

Note that the last two terms in (12-29), as reflected in (12-30) can also represent the middle diagram in the second row of Fig. 12-2. On the LHS vertex, now at x_1, an incoming anti-red/up quark is destroyed, an outgoing anti-green/up is created, and an outgoing $g\bar{r}$ gluon propagator is emitted. On the RHS vertex, at x_2, the gluon propagator is absorbed, an incoming anti-green/up quark is destroyed, and an outgoing anti-red/up quark is created.

It can also represent $\bar{r}\bar{g} \rightarrow \bar{r}\bar{g}$

As we have seen before, and is a general rule, the propagator of a particle going from x_2 to x_1 is the propagator of an antiparticle going from x_1 to x_2.

Note that we do not have to use T_μ^- and T_μ^+ in place of $A_{1\mu}$ and $A_{2\mu}$, just as we didn't have to use W_μ^- and W_μ^+ in place of $W_{1\mu}$ and $W_{2\mu}$ in electroweak theory. It just makes it easier if we do so. Otherwise, in e/w theory, if we had used $W_{1\mu}$ and $W_{2\mu}$, they would not be in weak isospin or electric charge eigenstates, i.e., they would not have definite charge, but would have to "collapse" to a given

Theory valid with either original or new fields

Interaction conceptually easier with new ones

charge eigenstate in a particular interaction – complicated! On the other hand, W_μ^- and W_μ^+ are in charge eigenstates. They have definite charge and so, are far easier to use in calculations.

Similarly, our gluon $A_{1\mu}$ is not in an eigenstate of color/anti-color charge, and we would have to use both $A_{1\mu}$ and $A_{2\mu}$ as part of the first diagram in Fig. 12-2, with them "collapsing" to the case where the virtual gluon carried $r\bar{g}$ charge. This would be an unneeded and unwelcome complication that is eliminated by using T_μ^- and T_μ^+ instead. Said another way, with T_μ^- and T_μ^+ we have associated $SU(3)$ matrices with only a single non-zero component, and this obviously simplifies things, as they then link only one type of color to another.

Mediator fields in charge eigenstates easier, as with newly defined fields

---

You can show the same thing for red and blue quarks as we showed for red and green in the above by doing **Problem 4**.

From Prob. 4,we have

$$U_\mu^+ = \frac{A_{4\mu} - iA_{5\mu}}{\sqrt{2}} \quad U_\mu^- = \frac{A_{4\mu} + iA_{5\mu}}{\sqrt{2}} \qquad A_{4\mu} = \frac{U_\mu^+ + U_\mu^-}{\sqrt{2}} \quad A_{5\mu} = i\frac{U_\mu^+ - U_\mu^-}{\sqrt{2}} \tag{12-31}$$

We can extend what we have done for rg quark combinations above and rb combinations in Problem 4 to gb combinations. That is,

$$V_\mu^+ = \frac{A_{6\mu} - iA_{7\mu}}{\sqrt{2}} \quad V_\mu^- = \frac{A_{6\mu} + iA_{7\mu}}{\sqrt{2}} \qquad A_{6\mu} = \frac{V_\mu^+ + V_\mu^-}{\sqrt{2}} \quad A_{7\mu} = i\frac{V_\mu^+ - V_\mu^-}{\sqrt{2}}, \tag{12-32}$$

with parallel results to what we have seen previously for those other combinations.

These results generalize to all terms with off-diagonal λ_i

As an aside, you may wonder why we don't simply define the $SU(3)$ generators such that the off diagonal matrices have a single non-zero component, as in (12-29), rather than the Gell-Mann matrices, which are more complicated. The answer is that, as we showed in Chap. 2, Prob. 26, a unitary Lie group must have Lie algebra generators that are Hermitian (and traceless).

The S operator terms, and thus the transition amplitudes, are the same in each case above except for the fields. The transition amplitudes for interactions between rg, rb, gb and their antiparticles will be the same, with the interactions being the same strength, for all possible color combinations.

Be aware that virtually no other texts present this material in the way I am doing herein. I submit it is inherent in those other texts, but hidden away behind the scenes. Hopefully, what I explicate above and below will help in getting some mental handle on all of this strong interaction stuff.

But now, how about the diagonal matrix generators λ_3 and λ_8 and their associated fields $A_3{}^\mu$ and $A_8{}^\mu$? Because I know what is coming, I will define new fields P_1, P_2, and P_3 as satisfying (where we temporarily drop the spacetime indices to streamline notation and make it easier for the "mind's eye")

$$\begin{aligned} A_3 + \frac{1}{\sqrt{3}}A_8 &= \sqrt{2}\left(\tfrac{2}{3}P_1 - \tfrac{1}{3}P_2 - \tfrac{1}{3}P_3\right) \\ -A_3 + \frac{1}{\sqrt{3}}A_8 &= \sqrt{2}\left(-\tfrac{1}{3}P_1 + \tfrac{2}{3}P_2 - \tfrac{1}{3}P_3\right) \\ -\frac{2}{\sqrt{3}}A_8 &= \sqrt{2}\left(-\tfrac{1}{3}P_1 - \tfrac{1}{3}P_2 + \tfrac{2}{3}P_3\right). \end{aligned} \tag{12-33}$$

Newly defined fields for the original fields (associated with the diagonal λ_i)

As we have done before, we can use new fields, defined in terms of original fields, as suits our need, since they are gauge fields. They themselves are never measured, and measurable quantities remain unchanged.

As an aside, both sides of (12-33) sum to zero, which means the sum of the first two equations equals the negative of the last, so there are only two independent equations. Since there are three unknown fields P_1, P_2, and P_3, given that the fields $A_3{}^\mu$ and $A_8{}^\mu$ are considered known and independent, only one of the P_1, P_2, and P_3 is independent. Specify one of them, and the other two are fixed (because they must be the solutions to two independent equations). The relationships between the three P fields are thus fixed.

Only one of newly defined fields P_1, P_2, and P_3 is independent

However, we don't need to actually solve for P_1, P_2, and P_3 . We only have to know that (12-33) holds.

Returning to (12-14) and the only parts of $\mathcal{L}^{QG}$ we have yet to examine, where we again show just the $f = u$ term for simplicity, are

$$-\frac{g_s}{2}\bar{\Psi}_u \not{A}_3 \lambda_3 \Psi_u - \frac{g_s}{2}\bar{\Psi}_u \not{A}_8 \lambda_8 \Psi_u$$

$$= -\frac{g_s}{2}\begin{bmatrix}\bar{u}_r & \bar{u}_g & \bar{u}_b\end{bmatrix}\not{A}_3\begin{bmatrix}1 & 0 & 0\\ 0 & -1 & 0\\ 0 & 0 & 0\end{bmatrix}\begin{bmatrix}u_r\\ u_g\\ u_b\end{bmatrix} - \frac{g_s}{2}\begin{bmatrix}\bar{u}_r & \bar{u}_g & \bar{u}_b\end{bmatrix}\not{A}_8\frac{1}{\sqrt{3}}\begin{bmatrix}1 & 0 & 0\\ 0 & 1 & 0\\ 0 & 0 & -2\end{bmatrix}\begin{bmatrix}u_r\\ u_g\\ u_b\end{bmatrix} \qquad (12\text{-}34)$$

Expressing diagonal $\mathcal{L}$ terms via original $A_3{}^{\mu}$, $A_8{}^{\mu}$

$$= -\frac{g_s}{2}\bar{u}_r\not{A}_3 u_r - \frac{g_s}{2}\bar{u}_g\left(-\not{A}_3\right)u_g - \frac{g_s}{2}\bar{u}_r\frac{1}{\sqrt{3}}\not{A}_8 u_r - \frac{g_s}{2}\bar{u}_g\frac{1}{\sqrt{3}}\not{A}_8 u_g - \frac{g_s}{2}\bar{u}_b\left(-\frac{2}{\sqrt{3}}\not{A}_8\right)u_b$$

$$= -\frac{g_s}{2}\bar{u}_r\left(\not{A}_3 + \frac{1}{\sqrt{3}}\not{A}_8\right)u_r - \frac{g_s}{2}\bar{u}_g\left(-\not{A}_3 + \frac{1}{\sqrt{3}}\not{A}_8\right)u_g - \frac{g_s}{2}\bar{u}_b\left(-\frac{2}{\sqrt{3}}\not{A}_8\right)u_b .$$

Re-expressing those terms via new fields P_1, P_2, and P_3

Substituting (12-33) into (12-34), we have

$$-\frac{g_s}{2}\bar{\Psi}_u \not{A}_3 \lambda_3 \Psi_u - \frac{g_s}{2}\bar{\Psi}_u \not{A}_8 \lambda_8 \Psi_u$$

$$= \underbrace{-\frac{g_s}{\sqrt{2}}\bar{u}_r\left(\tfrac{2}{3}\not{P}_1 - \tfrac{1}{3}\not{P}_2 - \tfrac{1}{3}\not{P}_3\right)u_r}_{rr \text{ vertex}} \underbrace{-\frac{g_s}{\sqrt{2}}\bar{u}_g\left(-\tfrac{1}{3}\not{P}_1 + \tfrac{2}{3}\not{P}_2 - \tfrac{1}{3}\not{P}_3\right)u_g}_{gg \text{ vertex}} \underbrace{-\frac{g_s}{\sqrt{2}}\bar{u}_b\left(-\tfrac{1}{3}\not{P}_1 - \tfrac{1}{3}\not{P}_2 + \tfrac{2}{3}\not{P}_3\right)u_b}_{bb \text{ vertex}} . \qquad (12\text{-}35)$$

So, what does (12-35) mean in terms of interactions similar to what we've shown in Fig. 12-2? Well, each diagram in Fig. 12-3 below shows one of the ways quarks can interact strongly with colorless gluons as mediating virtual particles, as dictated by (12-35). The symbol $c\bar{c}$ is used to indicate a gluon with no net color charge. Such gluons can be thought of much like the gluons we looked at earlier in Fig. 12-2, except that here the color and anti-color are of the same type, so the net color is zero.

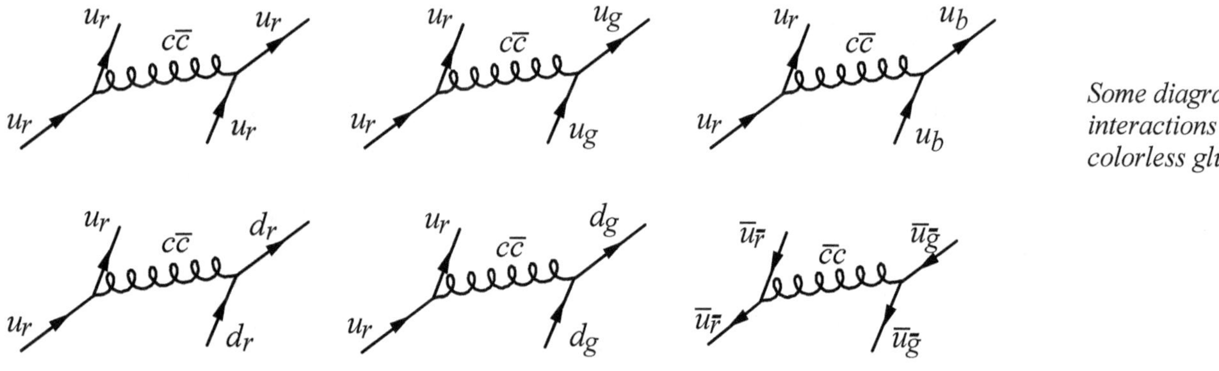

Some diagrams of interactions with colorless gluons

Figure 12-3. Colorless Gluon Interactions

In (12-35), as may become clearer shortly, the P_1 field carries colors $r\bar{r}$; the P_2 field, colors $g\bar{g}$; and the P_3 field, colors $b\bar{b}$. So, the total propagator in each case in (12-35) is colorless, but for each vertex type, the mix of $r\bar{r}$; $g\bar{g}$; and $b\bar{b}$ making up $c\bar{c}$ is different.

Now, consider the first diagram in Fig. 12-3. Each of its vertices comes from the terms we labeled as rr vertex in (12-35). Each vertex has an incoming and an outgoing red (up) quark. In the transition amplitude, we would get three propagator terms, since there is no propagator formed from different fields, such as P_1 and P_2. Thus, the transition amplitude, at lowest order, would be proportional to

Transition amplitude for $rr \to rr$ using those terms

for $i = rr, f = rr$, lowest order,

$$S_{fi} \propto \int \left\{ \left(\tfrac{2}{3}P_1^{\mu}\right)\left(\tfrac{2}{3}P_1^{\nu}\right) + \left(-\tfrac{1}{3}P_2^{\mu}\right)\left(-\tfrac{1}{3}P_2^{\nu}\right) + \left(-\tfrac{1}{3}P_3^{\mu}\right)\left(-\tfrac{1}{3}P_3^{\nu}\right) \right\} dx_1 dx_2$$

$$\propto \int \underbrace{\left\{ \underbrace{\tfrac{4}{9} i D_1^{\mu\nu}\left(x_1 - x_2\right)}_{r\bar{r} \text{ propagator}} + \underbrace{\tfrac{1}{9} i D_2^{\mu\nu}\left(x_1 - x_2\right)}_{g\bar{g} \text{ propagator}} + \underbrace{\tfrac{1}{9} i D_3^{\mu\nu}\left(x_1 - x_2\right)}_{b\bar{b} \text{ propagator}} \right\}}_{\text{total } c\bar{c} \text{ propagator}} dx_1 dx_2 . \qquad (12\text{-}36)$$

Fig. 12-4 represents (12-36) in a Feynman diagram.

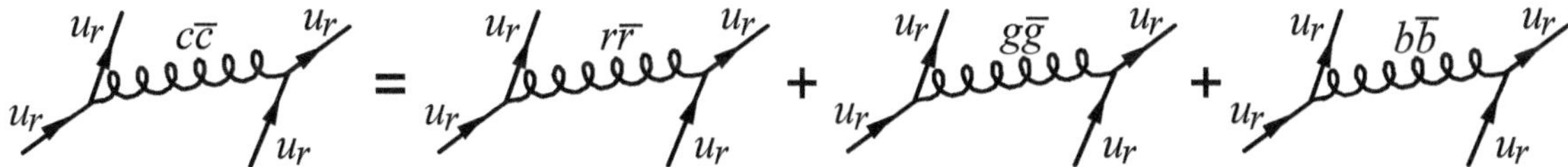

Figure 12-4. Parts of the Colorless Gluon Virtual Particle

Recall the propagators $iD_1^{\mu\nu}$, $iD_2^{\mu\nu}$, and $iD_3^{\mu\nu}$ in (12-36) are just numbers, all having the same form, and they are sandwiched between the same bra and ket (outgoing and incoming particles). So, even though they represent different color/anti-color combinations, we can simply take each of them as the same numerical quantity and label it $iD_{c\bar{c}}^{\mu\nu}(x_1 - x_2)$. Then, we find (12-36) can be written as

Re-expressing $rr \rightarrow rr$ transition amplitude

for $i = rr, f = rr$, lowest order, $$S_{fi} \propto \int \tfrac{2}{3} iD_{c\bar{c}}^{\mu\nu}(x_1 - x_2)\, dx_1 dx_2 \,. \qquad (12\text{-}37)$$

Now, do **Problem 5** to repeat the above process for the second diagram in Fig. 12-3, with incoming red and green (up) quarks and outgoing red and green (up) quarks.

In doing Problem 5, we obtain

Colorless gluon $rg \rightarrow rg$ transition amplitude

for $i = rg, f = rg$, lowest order, $$S_{fi} \propto -\int \tfrac{1}{3} iD_{c\bar{c}}^{\mu\nu}(x_1 - x_2)\, dx_1 dx_2 \,. \qquad (12\text{-}38)$$

Recall from Vol. 1, Sects. 16.2.5 and 16.2.6, pgs. 409-410, that the sign of the transition amplitude tells us whether an interaction is attractive or repulsive. Attraction has a negative sign; repulsion, a positive one. The interaction between the same two colored (red) quarks has a transition amplitude (12-37) with positive sign, and that jibes with what we have been saying since Chap. 2. The conclusion generalizes readily: the same color quarks repel.

Signs on transition amplitudes → like colors repel, unlike attract

In contrast, the amplitude for interaction between two different colored (red and green) quarks of (12-38) has a negative sign, and is attractive. And this, again, is what we have been saying from the beginning. And this too, generalizes. Different color quarks attract.

Repulsion stronger than attraction

Further, the repulsion for same color quarks is twice as strong as the attraction between different colored ones. That leads us to deduce that in Fig. 2-2, pg. 55, the inner product of any two color vectors is proportional to the (lowest order) transition amplitude with colorless gluon mediator. The inner product of the red vector with itself, in terms of the ε_3-ε_8 coordinates is 1/3. The inner product of the red vector with the green is –1/6. In general, two vectors in the figure will be attractive if they are separated by more than 90°; and repulsive, if separated by less than 90°. Further, the greater the angular separation for attraction, the greater the attraction. The closer the angular separation for repulsion, the greater the repulsion. All this is true only for cases where the incoming quarks do not change color at a vertex.

For cases where quark colors do change at a vertex, as depicted in Fig. 12-2, with different color quarks incoming, transition amplitudes, such as (12-30), are negative (provided we keep the same normal order as in (12-37) and (12-38)). Different color incoming quarks attract in those cases, too.

Bottom line: Different color quarks attract. Same color quarks repel. For quarks not changing color at a vertex (where the propagator has no color charge), same color quarks repel with twice the strength that different color quarks attract.

P fields perspective makes it easier to see equivalence of rr, gg, bb

To see what we have gained by introducing the P_1, P_2, and P_3 fields, compare the last line of (12-34) with the last line of (12-35). If you had never seen (12-35), or the treatment herein after that relation, you might wonder, as I once did for quite a while, how the interaction vertices of *rr*, *gg*, and *bb* in (12-34) could be essentially the same. Surely, the different colors interact with themselves at a vertex in the same way. Yet, how this happens may be obscure when the fields are expressed as A_3 and A_8, like we do in (12-34). It is transparent when the same relation is expressed in terms of P_1, P_2, and P_3, as we do in (12-35).

An Alternative Route to the Same Result

Note that instead of redefining our fields as in (12-27), (12-31), (12-32), and (12-33), we could have converted our Lie algebra generators to different matrices, i.e.,

Can define different matrices instead of different fields

$$\tilde{\lambda}_1 = \sqrt{2}\begin{bmatrix} 0 & 1 & 0 \\ 0 & 0 & 0 \\ 0 & 0 & 0 \end{bmatrix} \quad \tilde{\lambda}_2 = \sqrt{2}\begin{bmatrix} 0 & 0 & 0 \\ 1 & 0 & 0 \\ 0 & 0 & 0 \end{bmatrix} \quad \tilde{\lambda}_3 = \sqrt{2}\begin{bmatrix} \frac{2}{3} & 0 & 0 \\ 0 & -\frac{1}{3} & 0 \\ 0 & 0 & -\frac{1}{3} \end{bmatrix} \quad \tilde{\lambda}_4 = \sqrt{2}\begin{bmatrix} 0 & 0 & 1 \\ 0 & 0 & 0 \\ 0 & 0 & 0 \end{bmatrix}$$
$$\tilde{\lambda}_5 = \sqrt{2}\begin{bmatrix} 0 & 0 & 0 \\ 0 & 0 & 0 \\ 1 & 0 & 0 \end{bmatrix} \quad \tilde{\lambda}_6 = \sqrt{2}\begin{bmatrix} 0 & 0 & 0 \\ 0 & 0 & 1 \\ 0 & 0 & 0 \end{bmatrix} \quad \tilde{\lambda}_7 = \sqrt{2}\begin{bmatrix} 0 & 0 & 0 \\ 0 & 0 & 0 \\ 0 & 1 & 0 \end{bmatrix} \quad \tilde{\lambda}_8 = \sqrt{2}\begin{bmatrix} -\frac{1}{3} & 0 & 0 \\ 0 & \frac{2}{3} & 0 \\ 0 & 0 & -\frac{1}{3} \end{bmatrix}, \tag{12-39}$$

Alternative matrices (not Gell-Mann)

with another matrix that is not independent of the (12-39) defined by

$$\tilde{\lambda}_d = -\left(\tilde{\lambda}_3 + \tilde{\lambda}_8\right) = -\sqrt{2}\begin{bmatrix} \frac{2}{3} & 0 & 0 \\ 0 & -\frac{1}{3} & 0 \\ 0 & 0 & -\frac{1}{3} \end{bmatrix} - \sqrt{2}\begin{bmatrix} -\frac{1}{3} & 0 & 0 \\ 0 & \frac{2}{3} & 0 \\ 0 & 0 & -\frac{1}{3} \end{bmatrix} = \sqrt{2}\begin{bmatrix} -\frac{1}{3} & 0 & 0 \\ 0 & -\frac{1}{3} & 0 \\ 0 & 0 & \frac{2}{3} \end{bmatrix}. \tag{12-40}$$

Then, the quark-gluon terms in (12-14) would become

$$\mathcal{L}^{QG} = -\frac{g_s}{2}\bar{\Psi}_f \not{A}_i \lambda_i \Psi_f \xrightarrow[\text{basis}]{\text{alternative}} -\frac{g_s}{2}\bar{\Psi}_f \Big(\not{T}^+ \tilde{\lambda}_1 + \not{T}^- \tilde{\lambda}_2 + \not{U}^+ \tilde{\lambda}_4 + \not{U}^- \tilde{\lambda}_5$$
$$+ \not{V}^+ \tilde{\lambda}_6 + \not{V}^- \tilde{\lambda}_7 + \not{P}_1 \tilde{\lambda}_3 + \not{P}_2 \tilde{\lambda}_8 + \not{P}_3 \tilde{\lambda}_d \Big) \Psi_f, \tag{12-41}$$

Same result as we had with field redefinition

with the same results we found in (12-29) to (12-38), represented in Figs. 12-2, 12-3, and 12-4.

Either by redefining our fields or redefining our basis matrices, we can bring the theory into a form that is easier for our minds to grasp, a form where gluons carry distinct color/anti-color combinations between quarks of distinct colors. However, the different definitions are but changes in gauge, and do not affect what we observe. One set of definitions does not represent reality any more closely than any other. Each is just a different model of reality, and we should feel comfortable using whichever particular model helps us attack a particular problem best.

Different definitions like gauges, same observations

Note carefully, however, that the newly defined matrices (12-39) do not form a Lie algebra generator basis because they are not Hermitian (a basic requirement for $SU(n)$ Lie algebras). They are more of a trick to help us mentally, rather than a mathematically viable part of the theory.

Actual Calculations

All of this Sect. 12.3 is intended to provide some intuitive feeling for QCD interactions … how quarks change, what the propagators in a convenient form might look like, etc. However, when one gets down to actually calculating transition amplitudes, the gluon fields in the form of A_i^μ, each associated with a particular Gell-Mann matrix, are the ones one typically uses, not the more conceptually viable ones of this section. In using the A_i^μ fields, it is not generally possible to grasp what all of the various terms and factors are doing, in the sense of what we can visualize as particular particle exchanges and transmutations. But, with the T, U, V, and P fields, it is.

New gluon fields easier conceptually; originals used for computation

The T, U, V, and P fields are similar to the W^+ and W^- fields of electroweak theory. We could have calculated all electroweak interactions in terms of the W_1 and W_2 fields, which are not in charge eigenstates. But it was easier to visualize what was happening (and calculate transition amplitudes) using the W^+ and W^- fields, which are in charge eigenstates. The W^+ and W^- are linear combinations of the W_1 and W_2, just as the T, U, V, and P fields are linear combinations of the A_i fields. The A_1, A_2, A_4, A_5, A_6, and A_7 fields are not in color eigenstates, but the T, U, and V fields are, so we can visualize the latter more readily in Feynman diagrams. Similarly, even though the A_3 and A_8 fields are in color charge eigenstates (of zero net color charge), they are not in the simple form of one of $r\bar{r}$, $g\bar{g}$, or $b\bar{b}$. However, the fields P_1, P_2, and P_3 are each in such simple form.

This is like e/w, with W^+ and W^- fields vs the original W_1 and W_2

And again, all that we have done in this section is based on lowest order relations, and only in certain specialized cases do they give accurate results for a theory like QCD, which, more generally, is not amenable to perturbation methods.

This section is a simple view, only accurate in high energy case when QCD perturbation OK

We will soon begin the formal calculations.

12.4 Chapter Summary

We started with the (hypothetical, as quarks are not free) free QCD Lagrangian,

$$\text{QCD } \mathcal{L}_0 = \underbrace{i\bar{\Psi}_f \partial\!\!\!/ \Psi_f - m_f \bar{\Psi}_f \Psi_f}_{\mathcal{L}_0^Q} \underbrace{- \tfrac{1}{4} F_i^{\mu\nu} F_{i\mu\nu}}_{\mathcal{L}_0^G} \qquad i = 1,2,\ldots,8. \quad f = u,d,c,s,t,b \tag{12-42}$$

where $-\frac{1}{4}F_i^{\mu\nu}F_{i\mu\nu} = -\frac{1}{4}\left(\partial^\nu A_i^\mu - \partial^\mu A_i^\nu\right)\left(\partial_\nu A_{i\mu} - \partial_\mu A_{i\nu}\right) = -\frac{1}{2}\left(\left(\partial^\nu A_i^\mu\right)\partial_\nu A_{i\mu} - \left(\partial^\nu A_i^\mu\right)\partial_\mu A_{i\nu}\right)$,

and substituted the $SU(3)$ covariant derivative,

$$\partial^\mu \;\Rightarrow\; D^\mu = \partial^\mu + i\frac{g_s}{2}\lambda_j A_j^\mu , \tag{12-43}$$

to end up with the

Total, any energy, QCD Lagrangian

$$\text{QCD } \mathcal{L} = \underbrace{i\bar{\Psi}_f \partial\!\!\!/ \Psi_f - m_f \bar{\Psi}_f \Psi_f}_{\mathcal{L}_0^Q} \; \underbrace{- \frac{g_s}{2}\bar{\Psi}_f A\!\!\!/_j \lambda_j \Psi_f}_{\mathcal{L}^{QG}} \tag{12-44}$$

$$\underbrace{\underbrace{-\frac{1}{4}F_i^{\mu\nu}F_{i\mu\nu}}_{\mathcal{L}_0^G} \; \underbrace{+ g_s f_{ijk} A_{i\mu} A_{j\nu} \partial^\mu A_k^\nu - \frac{1}{4} g_s^2 f_{ijk} f_{ilm} A_j^\mu A_k^\nu A_{l\mu} A_{m\nu}}_{\mathcal{L}^{GG}}}_{\mathcal{L}^G = -\frac{1}{4}\hat{G}_i^{\mu\nu}\hat{G}_{i\mu\nu}} .$$

We then showed the total QCD $\mathcal{L}$ (12-44) is symmetric under the local (gauge) $SU(3)$ transformation set

$$\Psi_f \rightarrow \Psi'_f = \left(1 + \frac{i}{2}\lambda_n \omega_n\right)\Psi_f \qquad \bar{\Psi}_f \rightarrow \bar{\Psi}'_f = \bar{\Psi}_f\left(1 - \frac{i}{2}\lambda_m \omega_m\right)$$
$$A_i^\mu \rightarrow A_i'^\mu = A_i^\mu - \frac{1}{g_s}\partial^\mu \omega_i - f_{ijk}\omega_j A_k^\mu . \tag{12-45}$$

We followed this by using lowest order diagrams and amplitudes (which don't really do the job at low energy, but work at high energy), to show that the fields A_i^μ can be transformed to other fields, which are in eigenstates of color/anti-color (including zero color boson fields) and which simplify visualization of QCD interactions. Using this approach, we justified the claim that like colors repel and unlike colors attract.

12.5 Problems

1. Show, by example, that the quark-gluon interaction terms $\mathcal{L}^{QG} = -\frac{g_s}{2}\bar{\Psi}_f \gamma_\mu \lambda_j \Psi_f A_j^\mu$ in the Lagrangian mean quarks can change color at a spacetime event. Pick the Gell-Mann matrix λ_1 of (12-9) and consider an up quark in and an up quark out.
2. Show that the Lagrangian term $-m_f \bar{\Psi}_f \Psi_f$ is symmetric under the transformation set (12-16) and (12-17).
3. Show that $D^\mu \Psi_f$ transforms under the transformation set (12-16) and (12-17) in the same manner Ψ does. You will need the structure constants relation (12-10). This is a problem with a lot of tedium, so you might just want to look at the solution in the solution booklet. Your call.

4. Take $i = 4$ and 5 and $f = d$ in the quark-gluon interaction terms in (12-14), define

$$U_\mu^+ = \frac{A_{4\mu} - iA_{5\mu}}{\sqrt{2}} \quad U_\mu^- = \frac{A_{4\mu} + iA_{5\mu}}{\sqrt{2}} \qquad A_{4\mu} = \frac{U_\mu^+ + U_\mu^-}{\sqrt{2}} \quad A_{5\mu} = i\frac{U_\mu^+ - U_\mu^-}{\sqrt{2}}$$

, and find the S operator term with the U gluon propagator parallel to (12-30). Draw Feynman type diagrams for the lowest order interaction of an incoming red down quark with an incoming down blue quark, and also for an incoming anti-blue down with an incoming anti-red down quark.

5. For an incoming red quark and outgoing red quark at one vertex and an incoming green and outgoing green quark at a second vertex, show that the transition amplitude is –½ that of two red incoming and two red outgoing quarks.

Chapter 13

QCD Symmetries: Continuous and Discrete

"Above all, don't get wiser as you get older. If you do, you will become too inhibited to try the impossible, and one can achieve the limits of the possible only by occasionally venturing beyond them. The famous proverb should really be transposed: Angels rush in where fools fear to tread."

Harry Suhl, esteemed condensed-matter physicist.

13.0 Chapter Overview

In this chapter we will explore

- general principles for $SU(n)$ symmetry of the Lagrangian for any n,
- QCD four-currents from $SU(3)$ symmetry and associated color charge conservation,
- flavor symmetries,
- discrete symmetries in QCD, particularly, what is known as the strong CP problem, and its proposed resolution, the Peccei-Quinn theory with the conjectured axion field.

Topics in this chapter

13.1 Symmetry of Any SU(n) Quantum Field Theory

In Chaps. 6 and 12, we showed, by brute force examination of each term, how the high energy electroweak and the strong interaction Lagrangians are symmetric under $SU(2)$ and $SU(3)$ group transformations, respectively. But, as Richard Feynman famously said, a good physicist can find a given result in more than one way. So, here we look at another way of determining special unitary symmetry, a more general way with more far-ranging usefulness.

We'll look at general principles for symmetry of SU(n) theories

13.1.1 Generalizing Symmetry of Covariant Derivative Term in $\mathcal{L}$

QED $U(1)$ Covariant Derivative Transformation

In Vol. 1, Chap 11, Prob. 9, we showed that in QED, the covariant derivative of ψ transforms under local $U(1)$ in the same way as ψ. That is, where x represents the spacetime coordinate of a 4D point,

$$\left(D^{\mu}\psi\right)' = e^{-i\alpha(x)}D^{\mu}\psi\,, \qquad (13\text{-}1)$$

provided

$$D^{\mu}\psi = \left(\partial^{\mu} - ieA^{\mu}\right)\psi\,, \qquad (13\text{-}2)$$

$$\psi' = e^{-i\alpha}\psi\text{ , and} \qquad (13\text{-}3)$$

$$A'^{\mu} = A^{\mu} - \frac{1}{e}\partial^{\mu}\alpha\,. \qquad (13\text{-}4)$$

For an infinitesimal transformation, (13-3) and (13-1) become

$$\psi' = \left(1 - i\alpha\right)\psi \qquad (13\text{-}5)$$

$$\left(D^{\mu}\psi\right)' = \left(1 - i\alpha\right)D^{\mu}\psi\,. \qquad (13\text{-}6)$$

QED $D^{\mu}\psi$ transforms under U(1) like ψ

QCD $SU(3)$ Covariant Derivative Transformation

In this volume, Chap. 12, Prob. 3, we showed the covariant derivative of the color triplet Ψ_f transforms under local $SU(3)$ in the same way Ψ_f does. That is, for an infinitesimal transformation,

$$\left(D^{\mu}\Psi_f\right)' = \left(1 + i\frac{\lambda_m}{2}\omega_m\right)D^{\mu}\Psi_f, \tag{13-7}$$

QCD $D^{\mu}\Psi_f$ transforms under SU(3) like Ψ_f

provided
$$D^{\mu}\Psi_f = \left(\partial^{\mu} + ig_S\frac{\lambda_i}{2}A_i^{\mu}\right)\Psi_f, \tag{13-8}$$

$$\Psi'_f = \left(1 + i\frac{\lambda_m}{2}\omega_m\right)\Psi_f, \text{ and} \tag{13-9}$$

$$A_i'^{\mu} = A_i^{\mu} - \frac{1}{g_s}\partial^{\mu}\omega_i - f_{ijk}\omega_j A_k^{\mu}. \tag{13-10}$$

For (13-7) to hold, we also need the following commutation relations to hold.

$$\left[\frac{\lambda_i}{2}, \frac{\lambda_j}{2}\right] = if_{ijk}\frac{\lambda_k}{2} \qquad \text{or equivalently} \quad \left[\lambda_i, \lambda_j\right] = i2f_{ijk}\lambda_k. \tag{13-11}$$

Weak Interaction $SU(2)$ Covariant Derivative Transformation

We didn't do it when we studied electroweak theory, but if we had, we would have found, for the weak doublet there under a local, infinitesimal $SU(2)$ transformation, that

$$\left(D^{\mu}\Psi_l^L\right)' = \left(1 + i\frac{\sigma_m}{2}\omega_m\right)D^{\mu}\Psi_l^L, \tag{13-12}$$

Weak $D^{\mu}\Psi_l^L$ transforms under SU(2) like Ψ_l^L

provided
$$D^{\mu}\Psi_l^L = \left(\partial^{\mu} + ig\frac{\sigma_i}{2}W_i^{\mu}\right)\Psi_l^L, \tag{13-13}$$

$$\Psi_l'^L = \left(1 + i\frac{\sigma_m}{2}\omega_m\right)\Psi_l^L, \text{ and} \tag{13-14}$$

$$W_i'^{\mu} = W_i^{\mu} - \frac{1}{g}\partial^{\mu}\omega_i - \varepsilon_{ijk}\omega_j W_k^{\mu}. \tag{13-15}$$

For (13-12) to hold, we also need the following commutation relations to hold.

$$\left[\frac{\sigma_i}{2}, \frac{\sigma_j}{2}\right] = i\varepsilon_{ijk}\frac{\sigma_k}{2} \qquad \text{or equivalently} \quad \left[\sigma_i, \sigma_j\right] = i2\varepsilon_{ijk}\sigma_k. \tag{13-16}$$

Relation (13-12) can be found using the solution to Prob. 3 of Chap. 12 for $SU(3)$ and (13-16), then following the same steps as in that problem, but substituting the doublet Ψ_l^L for the triplet Ψ_f, bosons W_i for A_i^{μ}, structure constants ε_{ijk} for f_{ijk}, g for g_S, and matrix generators σ_i for λ_i, along with taking $i = 1,2,3$ instead of 1 to 8.

General Covariant Derivative Transformation and Symmetry in $SU(n)$ Theories

Comparing (13-2) through (13-6) with (13-8) through (13-11) and with (13-13) through (13-16), you may see a pattern arising, which we can generalize to any $SU(n)$, as follows. That is, where a tilde under a symbol represents entities in an $SU(n)$ theory (i.e., an n-tuplet, an nXn matrix generator, one of $n^2 - 1$ boson gauge fields, or the structure constants of the $SU(n)$ group), and g_n represents the associated coupling constant, we have

$$\left(D^{\mu}\underset{\sim}{\Psi}\right)' = \left(1 + i\frac{\underset{\sim}{\lambda}_m}{2}\omega_m\right)D^{\mu}\underset{\sim}{\Psi}, \tag{13-17}$$

In general, $D^{\mu}\underset{\sim}{\Psi}$ transforms under SU(n) like $\underset{\sim}{\Psi}$

provided
$$D^{\mu}\underset{\sim}{\Psi} = \left(\partial^{\mu} + ig_n\frac{\underset{\sim}{\lambda}_i}{2}\underset{\sim}{A}_i^{\mu}\right)\underset{\sim}{\Psi} \tag{13-18}$$

$$\underset{\sim}{\Psi}' = \left(1 + i\frac{\underset{\sim}{\lambda}_m}{2}\omega_m\right)\underset{\sim}{\Psi} \tag{13-19}$$

$$\underset{\sim}{A}_i'^{\mu} = \underset{\sim}{A}_i^{\mu} - \frac{1}{g_n}\partial^{\mu}\omega_i - \underset{\sim}{f}_{ijk}\omega_j\underset{\sim}{A}_k^{\mu}, \tag{13-20}$$

and
$$\left[\frac{\underset{\sim}{\lambda}_i}{2}, \frac{\underset{\sim}{\lambda}_j}{2}\right] = i\underset{\sim}{f}_{ijk}\frac{\underset{\sim}{\lambda}_k}{2} \qquad \text{or equivalently} \quad \left[\underset{\sim}{\lambda}_i, \underset{\sim}{\lambda}_j\right] = i2\underset{\sim}{f}_{ijk}\underset{\sim}{\lambda}_k. \tag{13-21}$$

Note that (13-21) is true for any $SU(n)$ theory, so stating it explicitly is actually redundant. Also, I used the symbol Y_i in Chap. 2 for what we use $\underset{\sim}{\lambda}_i$ for above and hope that will not cause confusion.

So, as a general rule, in any $SU(n)$ quantum field theory, where relations (13-18) and (13-20) hold, the covariant derivative will change under the transformation just like the n-tuplet changes.

Thus, the Lagrangian term with the covariant derivative is invariant, i.e., (with higher order terms in the infinitesimal parameter ω_m effectively zero, and as we showed in Chap. 2, Prob. 26, Lie algebra generator basis matrices $\underset{\sim}{\lambda}_i$ for any unitary group are Hermitian)

$$\bar{\underset{\sim}{\Psi}} D^\mu \underset{\sim}{\Psi} \xrightarrow[\text{transformation}]{SU(n)} \left(\bar{\underset{\sim}{\Psi}} D^\mu \underset{\sim}{\Psi}\right)' = \bar{\underset{\sim}{\Psi}} \underbrace{\left(1 - i\frac{\underset{\sim}{\lambda}_i^\dagger}{2}\omega_i\right)\left(1 + i\frac{\underset{\sim}{\lambda}_m}{2}\omega_m\right)}_{\underset{\sim}{\lambda}_i^\dagger = \underset{\sim}{\lambda}_i} D^\mu \underset{\sim}{\Psi} = \bar{\underset{\sim}{\Psi}} D^\mu \underset{\sim}{\Psi} . \quad (13\text{-}22)$$

Hence, in general, $\bar{\underset{\sim}{\Psi}} D^\mu \underset{\sim}{\Psi}$ is invariant under SU(n)

Bottom line: This means all free lepton terms and all lepton-gauge boson interaction terms in $\mathcal{L}$ are symmetric under an $SU(n)$ transformation.

This is true as long as (13-18) and (13-20) hold, where the transformation of the n-tuplets (13-19) and the commutation relations (13-21) are generally inherent in any $SU(n)$ group.

These general principles provide us with a huge shortcut in ascertaining the symmetry of a wide range of quantum field theories. We don't have to examine each lepton-gauge boson interaction term in $\mathcal{L}$ separately. (13-22) means they are all automatically symmetric.

13.1.2 Generalizing Symmetry of Gauge Boson Terms in $\mathcal{L}$

If, as in the prior section, there are general symmetry principles for Lagrangian fermion terms, both free and interacting with gauge bosons via the covariant derivative, in $SU(n)$ theories, we can ask if a similar thing happens with gauge bosons. In fact, it does, as we explore below.

<u>QED *U*(1) Photon Terms Symmetry</u>

$$\text{Free photon term in } \mathcal{L}: -\tfrac{1}{4}F_{\nu\beta}F^{\nu\beta} = -\tfrac{1}{4}\left(\partial_\beta A_\nu - \partial_\nu A_\beta\right)\left(\partial^\beta A^\nu - \partial^\nu A^\beta\right) = -\tfrac{1}{2}\left(\partial_\beta A_\nu \partial^\beta A^\nu - \partial_\nu A_\beta \partial^\beta A^\nu\right) \quad (13\text{-}23)$$

$$\text{Free plus self-interacting photon terms in } \mathcal{L}: \; -\tfrac{1}{2}\left(D_\beta A_\nu D^\beta A^\nu - D_\nu A_\beta D^\beta A^\nu\right). \quad (13\text{-}24)$$

With

$$D_\beta = \partial_\beta - ieA_\beta , \quad (13\text{-}25)$$

we showed in Vol. 1, (11-41), pg. 297, that (13-24) = (13-23), i.e.,

$$-\tfrac{1}{2}\left(D_\beta A_\nu D^\beta A^\nu - D_\nu A_\beta D^\beta A^\nu\right) = -\tfrac{1}{2}\left(\partial_\beta A_\nu \partial^\beta A^\nu - \partial_\nu A_\beta \partial^\beta A^\nu\right) = -\tfrac{1}{4}F_{\nu\beta}F^{\nu\beta} , \quad (13\text{-}26)$$

and there are no photon self-interacting terms. (QED is a linear theory, in this sense.)

In Vol. 1, (11-36), pg. 294, we showed that under a local $U(1)$ group transformation on ψ, i.e.,

$$\psi' = e^{-i\alpha}\psi \quad (13\text{-}27)$$

(13-26) is symmetric, i.e.,

$$\left(-\tfrac{1}{4}F_{\nu\beta}F^{\nu\beta}\right)' = -\tfrac{1}{4}F_{\nu\beta}F^{\nu\beta} , \quad (13\text{-}28)$$

provided

$$A'_\nu = A_\nu - \tfrac{1}{e}\partial_\nu \alpha . \quad (13\text{-}29)$$

QED: photon only terms invariant under U(1)

<u>Weak *SU*(2) *W* Boson Terms Symmetry</u>

In Chap. 6 of this text, we showed the free plus self-interacting W boson terms in $\mathcal{L}$ at high energy can be found by minimal substitution of the weak covariant derivative, parallel to what we did in QED.

$$\text{Free } W \text{ boson term in } \mathcal{L}: \; -\tfrac{1}{4}W_i^{\mu\nu}W_{i\mu\nu} = -\tfrac{1}{4}\left(\partial^\nu W_i^\mu - \partial^\mu W_i^\nu\right)\left(\partial_\nu W_{i\mu} - \partial_\mu W_{i\nu}\right) \quad (13\text{-}30)$$

With D_μ of (13-13) inserted in place of the partial derivatives in (13-30), we have

Free plus self-interacting W boson terms in $\mathcal{L}$:

$$-\tfrac{1}{4}\left(D^\nu W_i^\mu - D^\mu W_i^\nu\right)\left(D_\nu W_{i\,\mu} - D_\mu W_{i\nu}\right) = -\tfrac{1}{4} G_i^{\mu\nu} G_{i\,\mu\nu} \tag{13-31}$$

where $G_i^{\mu\nu}$ can be expressed as

$$G_i^{\mu\nu} = \partial^\nu W_i^\mu - \partial^\mu W_i^\nu - g\varepsilon_{ijk} W_j^\nu W_k^\mu . \tag{13-32}$$

On pgs. 199-200, we showed (13-32) is symmetric under the W boson part of the $SU(2)$ transformation (13-15), i.e.,

$$\left(G_i^{\mu\nu}\right)' = G_i^{\mu\nu}, \tag{13-33}$$

Weak: W boson only terms invariant under SU(2)

and so, therefore, is the term (13-31) in the Lagrangian.

Strong $SU(3)$ Gluon Terms Symmetry

In (12-6) and (12-7) of Chap. 12 in this text, we saw the free gluon Lagrangian term is (13-34) below.

$$\text{Free gluon term in } \mathcal{L}: \quad -\tfrac{1}{4} F_i^{\mu\nu} F_{i\,\mu\nu} = -\tfrac{1}{4}\left(\partial^\nu A_i^\mu - \partial^\mu A_i^\nu\right)\left(\partial_\nu A_{i\,\mu} - \partial_\mu A_{i\nu}\right) \tag{13-34}$$

With minimal substitution of (13-8) in (13-34), we get

Free plus self-interacting A boson terms in $\mathcal{L}$:

$$-\tfrac{1}{4}\left(D^\nu A_i^\mu - D^\mu A_i^\nu\right)\left(D_\nu A_{i\,\mu} - D_\mu A_{i\nu}\right) = -\tfrac{1}{4} \hat{G}_i^{\mu\nu} \hat{G}_{i\,\mu\nu}, \tag{13-35}$$

where, in parallel with (13-32),

$$\hat{G}_i^{\mu\nu} = \partial^\nu A_i^\mu - \partial^\mu A_i^\nu - g_s f_{ijk} A_j^\nu A_k^\mu . \tag{13-36}$$

On pg. 391, we noted that showing (13-36) is symmetric under the gluon $SU(3)$ transformation (13-10) parallels, step-by-step, what we did to show the W boson symmetry in (13-33). So,

$$\left(\hat{G}_i^{\mu\nu}\right)' = \hat{G}_i^{\mu\nu} . \tag{13-37}$$

QCD: gluon only terms invariant under SU(3)

Generalizing Gauge Boson Terms Symmetry in $SU(n)$ Theories

So, in general, for $n > 3$, if

$$-\tfrac{1}{4} \underset{\sim}{F}_i^{\mu\nu} \underset{\sim}{F}_{i\,\mu\nu} = -\tfrac{1}{4}\left(\partial^\nu \underset{\sim}{A}_i^\mu - \partial^\mu \underset{\sim}{A}_i^\nu\right)\left(\partial_\nu \underset{\sim}{A}_{i\,\mu} - \partial_\mu \underset{\sim}{A}_{i\nu}\right), \tag{13-38}$$

$$D^\mu \underset{\sim}{A}_j^\nu = \left(\partial^\mu + i g_n \frac{\underset{\sim}{\lambda}_i}{2} \underset{\sim}{A}_i^\mu\right) \underset{\sim}{A}_j^\nu, \text{ and} \tag{13-39}$$

Provided SU(n) theory conforms to these criteria,

$$\underset{\sim}{A}_i'^\mu = \underset{\sim}{A}_i^\mu - \frac{1}{g_n}\partial^\mu \omega_i - \underset{\sim}{f}_{ijk}\omega_j \underset{\sim}{A}_k^\mu \qquad \text{(repeat of (13-20),} \tag{13-40}$$

then

$$-\tfrac{1}{4}\left(D^\nu \underset{\sim}{A}_i^\mu - D^\mu \underset{\sim}{A}_i^\nu\right)\left(D_\nu \underset{\sim}{A}_{i\,\mu} - D_\mu \underset{\sim}{A}_{i\nu}\right) = -\tfrac{1}{4} \underset{\sim}{G}_i^{\mu\nu} \underset{\sim}{G}_{i\,\mu\nu}, \tag{13-41}$$

$$\text{where } \underset{\sim}{G}_i^{\mu\nu} = \partial^\nu \underset{\sim}{A}_i^\mu - \partial^\mu \underset{\sim}{A}_i^\nu - g_n \underset{\sim}{f}_{ijk} \underset{\sim}{A}_j^\nu \underset{\sim}{A}_k^\mu, \tag{13-42}$$

and so,

$$\left(\underset{\sim}{G}_i^{\mu\nu}\right)' = \underset{\sim}{G}_i^{\mu\nu}, \tag{13-43}$$

gauge boson only terms invariant under SU(n)

i.e., the gauge boson only terms in the Lagrangian are symmetric.

13.1.3 Mass Term Symmetry

Mass Terms in QED, Weak, and Strong Theories

As we saw in Vol. 1, Sect. 11.5.4, pg. 296, in order to have a symmetric Lagrangian under $U(1)$ transformations, the gauge boson in $U(1)$ theory (QED photon there) must be massless. We also saw, in that chapter, that the fermion mass term was symmetric.

QED: fermion mass term invariant, photon mass term not

Weak: neither fermion nor gauge boson mass terms invariant

We saw this behavior for gauge bosons again in this volume in Chap. 6 (Sect. 6.8.1, pg. 190), summarized in Wholeness Chart 6-9, pg. 196, except that in electroweak theory, the fermion mass terms are not symmetric. (See Appendix B of that chapter.)

QCD: fermion mass terms invariant, gauge boson mass terms not

In QCD, the gauge boson (gluon) behavior under $SU(3)$ for massive gluons parallels that for W bosons under $SU(2)$, i.e., with massive gluons, $\mathcal{L}$ is not symmetric. However, as we showed in Prob. 2 of Chap. 12, in QCD, fermion mass terms in $\mathcal{L}$ are symmetric.

Weak case not invariant as doublet components are not same mass

The difference between the fermion mass terms in QCD and electroweak theory (one symmetric, the other not) occurs because in electroweak theory, the doublet contains fields with different masses (e.g., electron neutrino and electron), whereas in QCD, each component of the triplet has the same mass. (See the following section for details.)

Generalizing Mass Term Symmetries in $SU(n)$ Theories

For an n-tuplet in an $SU(n)$ theory with all components of the n-tuplet having the same mass,

$$m\bar{\Psi}\Psi \xrightarrow[\text{transformation}]{SU(n)} \left(m\bar{\Psi}\Psi\right)' = m\bar{\Psi}\left(1-i\frac{\lambda_i}{2}\omega_i\right)\left(1+i\frac{\lambda_m}{2}\omega_m\right)\Psi = m\bar{\Psi}\Psi \; . \tag{13-44}$$

For an n-tuplet in an $SU(n)$ theory with components of the n-tuplet having different masses, we need to employ a mass matrix M, which can be diagonal, but without all diagonal components being equal. That is,

$$\begin{aligned}\bar{\Psi}M\Psi \xrightarrow[\text{transformation}]{SU(n)} \left(\bar{\Psi}M\Psi\right)' &= \bar{\Psi}\left(1-i\frac{\lambda_i}{2}\omega_i\right)M\left(1+i\frac{\lambda_m}{2}\omega_m\right)\Psi \\ &\approx \bar{\Psi}M\Psi - i\bar{\Psi}\frac{\lambda_i}{2}\omega_i M\Psi + i\bar{\Psi}M\frac{\lambda_m}{2}\omega_m\Psi = \bar{\Psi}M\Psi - i\bar{\Psi}\left[\frac{\lambda_i}{2}\omega_i M - M\frac{\lambda_m}{2}\omega_m\right]\Psi \\ &= \bar{\Psi}M\Psi - i\frac{\omega_i}{2}\bar{\Psi}\left[\lambda_i, M\right]\Psi \neq \bar{\Psi}M\Psi .\end{aligned} \tag{13-45}$$

SU(n): fermion mass terms invariant if all n-tuplet fields have same mass

In (13-45), we only get symmetry if the generator matrices commute with the mass matrix. If all components of Ψ have the same mass, then the mass matrix can be expressed as a single mass value m times the identity matrix, and in that case, the commutator in the last line of (13-45) is zero (we have symmetry for the mass terms).

SU(n): gauge boson terms invariant only if they are massless

Bottom line: In general, for any $SU(n)$ theory to be symmetric, the gauge bosons must be massless. The fermions can have mass provided all components of any n-tuplet have the same mass, in which case the fermion mass terms in $\mathcal{L}$ are symmetric.

13.1.4 Higgs Type Terms and Symmetry

SU(n): other terms, like Higgs in SU(2), need to be evaluated case-by-case

The Higgs terms in high energy electroweak theory (See Wholeness Chart 6-8, pg. 195) couple directly with fermions and, via the covariant derivative, with gauge bosons. These terms are added, somewhat *ad hoc*, in order to give us all of the behavior we studied as energy level drops below that of the electroweak epoch, i.e., as symmetry breaks and particles gain mass. Terms with Higgs fields are not, *a priori*, a part of a typical $SU(n)$ theory.

So, such atypical terms must be investigated separately, on their own, to determine whether they are symmetric or not. For GSW theory, we did this and showed that they are, indeed, symmetric.

13.1.5 Overview of SU(n) Symmetry Principles

Summary

SU(n) general symmetry principles summary at end of chapter

General principles for any $SU(n)$ quantum field theory symmetry are summarized in the Chapter Summary, Sect. 13.5.1, Wholeness Chart 13-1, pg. 414.

Impact

Many advanced theories have $n > 3$, such as most grand unified theories, which commonly employ the $SU(5)$ group. All such theories can be considered symmetric, without further ado, if they conform to the general principles listed in Sect. 13.5.1.

Historical Note

For pedagogic reasons, in this book we have started with minimal substitution and the covariant derivative, then showed these led to symmetric theories. In practice, however, as QFT was being developed, it is more likely that the basic principles of symmetry for $SU(n)$ theories were the guiding tenets, and these led, in turn, to the covariant derivative and the method of minimal substitution. When teaching, however, it can sometimes make for a clearer presentation, if the cart is put before the horse.

13.2 Noether's Theorem and QCD

Noether's theorem and QCD charges

13.2.1 Review of Noether's Theorem

In Chap. 9, Sect. 9.1.1, we reviewed Noether's theorem, but, for convenience, we restate its expression for four-current

Noether's theorem

$$j^{\mu}\left(\phi^{r},\phi^{r}{}_{,\nu}\right)=\frac{\partial\mathcal{L}}{\partial\phi^{r}{}_{,\mu}}\frac{\partial\phi'^{r}}{\partial\alpha}\qquad\left(\text{sum on } r \text{ and } \frac{\partial\phi'^{r}}{\partial\alpha}=\frac{\partial}{\partial\alpha}\phi'^{r}\left(x^{\mu},\alpha\right)\right),\tag{13-46}$$

and note that if $\mathcal{L}$ is symmetric with respect to a transformation in ϕ^r which is a function of a parameter α, i.e., $\phi^r(x^\mu)\rightarrow\phi'^r(x^\mu,\alpha)$, then $\partial_\mu j^\mu = 0$. Thus, Q (= the spatial integral of qj^0) is conserved, where q is a constant that turns (13-46) into a charge density.

13.2.2 The SU(3) Four-current

For convenience, we also re-express the QCD Lagrangian and the $SU(3)$ transformation set of Chap. 12 (pg. 389).

QCD Lagrangian

$$\begin{aligned}\mathcal{L}= {}& i\bar{\Psi}_f\gamma^{\mu}\Psi_{f,\mu}-m_f\bar{\Psi}_f\Psi_f-\tfrac{g_s}{2}\bar{\Psi}_f\gamma^{\mu}A_{i\mu}\lambda_i\Psi_f+\tfrac{1}{2}\left(\partial_{\nu}A_{i\mu}\partial^{\mu}A_i^{\nu}-\partial_{\mu}A_{i\nu}\partial^{\mu}A_i^{\nu}\right)\\ &+g_sf_{ijk}A_{i\mu}A_{j\nu}\partial^{\mu}A_k^{\nu}-\tfrac{1}{4}g_s^2f_{ijk}f_{ilm}A_j^{\mu}A_k^{\nu}A_{l\mu}A_{m\nu}\end{aligned}\tag{13-47}$$

QCD SU(3) transformation set

$$\Psi'_f=\left(1+i\frac{\lambda_n}{2}\omega_n\right)\Psi_f\tag{13-48}$$

$$A'_{i\mu}=A_{i\mu}-\frac{1}{g_s}\partial_{\mu}\omega_i-f_{ijk}\omega_jA_{k\mu}\tag{13-49}$$

From (13-47) to (13-49), we can find the relevant factors of (13-46), with the fields in (13-47) replacing ϕ^r, and ω_i replacing the parameter α. We will also use the Greek letter β in place of μ in (13-46), since μ is used in (13-47) and (13-49). The Greek letter α will now represent a 4D component and *not* the parameter of the symmetry transformation.

From (13-47)

Stuff we'll need to apply Noether's theorem to QCD

$$\frac{\partial\mathcal{L}}{\partial\Psi_{f,\beta}}=\frac{\partial\left(i\bar{\Psi}_f\gamma^{\mu}\Psi_{f,\mu}\right)}{\partial\Psi_{f,\beta}}=i\bar{\Psi}_f\gamma^{\mu}\delta_{\mu}^{\beta}=i\bar{\Psi}_f\gamma^{\beta}\qquad\qquad\frac{\partial\mathcal{L}}{\partial\bar{\Psi}_{f,\beta}}=0\tag{13-50}$$

$$\begin{aligned}\frac{\partial\mathcal{L}}{\partial A_{l\alpha,\beta}}&=\frac{\frac{1}{2}\partial\left(A_{i\mu,\nu}A_i^{\nu,\mu}-A_{i\nu,\mu}A_i^{\nu,\mu}\right)}{\partial A_{l\alpha,\beta}}+\frac{\partial\left(g_sf_{ijk}A_{i\mu}A_{j\nu}A_k^{\nu,\mu}\right)}{\partial A_{l\alpha,\beta}}\\&=\tfrac{1}{2}\left(\delta_{il}\delta_{\mu}^{\alpha}\delta_{\nu}^{\beta}A_i^{\nu,\mu}+\delta_{il}g^{\alpha\nu}g^{\beta\mu}A_{i\mu,\nu}-\delta_{il}\delta_{\nu}^{\alpha}\delta_{\mu}^{\beta}A_i^{\nu,\mu}-\delta_{il}g^{\alpha\nu}g^{\beta\mu}A_{i\nu,\mu}\right)\\&\qquad+g_sf_{ijk}A_{i\mu}A_{j\nu}\left(\delta_{lk}g^{\alpha\nu}g^{\beta\mu}\right)\\&=\tfrac{1}{2}\left(A_l^{\beta,\alpha}+A_l^{\beta,\alpha}-A_l^{\alpha,\beta}-A_l^{\alpha,\beta}\right)+g_sf_{ijl}A_i^{\beta}A_j^{\alpha}\\&=A_l^{\beta,\alpha}-A_l^{\alpha,\beta}+g_sf_{lij}A_i^{\beta}A_j^{\alpha}.\end{aligned}\tag{13-51}$$

From (13-48) and (13-49),

$$\frac{\partial\Psi'_f}{\partial\omega_i}=i\frac{\lambda_n}{2}\delta_{in}\Psi_f=i\frac{\lambda_i}{2}\Psi_f\tag{13-52}$$

$$\frac{\partial A_i'^{\mu}}{\partial\omega_l}=-\frac{1}{g_s}\partial_{\mu}\left(\frac{\partial\omega_i}{\partial\omega_l}\right)-f_{ijk}\left(\frac{\partial\omega_j}{\partial\omega_l}\right)A_k^{\mu}=-\frac{1}{g_s}\partial_{\mu}\left(\delta_{il}\right)-f_{ijk}\delta_{jl}A_k^{\mu}=-f_{ilk}A_k^{\mu}=f_{ikl}A_k^{\mu}.\tag{13-53}$$

Using the above results in Noether's theorem, we have

$$
\begin{aligned}
{}_{QCD}j_i^{\beta} &= \frac{\partial \mathcal{L}}{\partial \Psi_{f,\beta}} \frac{\partial \Psi'_f}{\partial \omega_i} + \frac{\partial \mathcal{L}}{\partial A_{l\alpha,\beta}} \frac{\partial A'_{i\alpha}}{\partial \omega_l} = i\bar{\Psi}_f \gamma^\beta i \frac{\lambda_i}{2} \Psi_f + \left(A_l^{\beta,\alpha} - A_l^{\alpha,\beta} + g_s f_{lmj} A_m^\beta A_j^\alpha \right) f_{ikl} A_{k\alpha} \\
&= -\bar{\Psi}_f \gamma^\beta \frac{\lambda_i}{2} \Psi_f + f_{ikl} \left(A_l^{\beta,\alpha} - A_l^{\alpha,\beta} \right) A_{k\alpha} + g_s f_{lmj} f_{ikl} A_m^\beta A_j^\alpha A_{k\alpha}.
\end{aligned} \tag{13-54}
$$

The eight Noether QCD four-currents

13.2.3 The SU(3) Charge Operators

By integrating the $\beta = 0$ component of (13-54) over all space, we obtain the eight QCD charge operators (for $i = 1, \ldots, 8$), where we introduce the coupling constant in the manner we have done before (see Wholeness Chart 9-1 at the end of Chap. 9, pg. 341),

The eight Noether QCD charge operators

$$
{}_{QCD}Q_i = -g_s \int {}_{QCD}j_i^0(x)\, d^3x. \tag{13-55}
$$

Now, recall from Chap. 9 that our interest is with the expectation values for (13-55), i.e.,

$$
\text{expectation value of } {}_{QCD}Q_i = {}_{QCD}\bar{Q}_i = \langle state |\, {}_{QCD}Q_i \,| state \rangle \tag{13-56}
$$

The eight Noether QCD charge expectation values

because they represent what we actually measure. However, for any state, the last term in (13-54) with three creation and/or destruction operators will turn the original ket into a state that must be different from the original state, i.e., orthogonal to the original bra, and give us a zero contribution to the expectation value. The next to last term, due to f_{ikl}, is zero unless $k \neq l$, and for that case, the bra and ket must be different, i.e., orthogonal and leading to zero expectation value.

Thus, the only effective term in the last line of (13-54) is the first one, which is bilinear in fields. And so, we can work with an effective charge operator that includes only that term. It will yield the same expectation value for any state as the full expression (13-54), and thus, in comparing theory to experiment, will be indistinguishable from the full expression.

Given that, we adopt a QCD charge operator for each i of

The eight charge operators in terms of fields

$$
{}_{QCD}Q_i = -g_s \int {}_{QCD}j_i^0(x)\, d^3x = g_s \int \bar{\Psi}_f \gamma^0 \frac{\lambda_i}{2} \Psi_f\, d^3x = g_s \int \Psi_f^\dagger \frac{\lambda_i}{2} \Psi_f\, d^3x. \tag{13-57}
$$

In (13-57), the λ_i matrices sandwiched between the color triplets will give us terms bilinear in fields of various colors. But only for the diagonal matrices, will we have creation color operators on the RHS matched by the same color destruction operators on the LHS. Terms without such a color matching result in a zero contribution in (13-56), so once again, we can streamline our QCD charge operators by considering only the matrices in (13-57) that do not yield zero expectation value (the diagonal ones) to comprise an effective charge operator.

Thus, what we can, with no loss in generality or prediction capability, consider only two of our eight QCD charge operators to be non-zero, those containing the Cartan subalgebra matrices for $i = 3$ and 8.

Taking charge operators with zero expectation value as zero doesn't affect predictions

$$
{}_{QCD}Q_3 = \frac{g_s}{2} \int \Psi_f^\dagger \lambda_3 \Psi_f\, d^3x \qquad {}_{QCD}Q_8 = \frac{g_s}{2} \int \Psi_f^\dagger \lambda_8 \Psi_f\, d^3x \qquad \text{effectively, } {}_{QCD}Q_i = 0 \text{ for all other } i. \tag{13-58}
$$

Note the repeated index f means we sum over all flavors. Note also that we now have two charge operators, as opposed to only one in the $SU(2)$ electroweak theory.

Our only two QCD charge operators that are effectively non-zero

$$
{}_{QCD}Q_3 = \frac{g_s}{2} \int \begin{bmatrix} \psi_{fr}^\dagger & \psi_{fg}^\dagger & \psi_{fb}^\dagger \end{bmatrix} \begin{bmatrix} 1 & 0 & 0 \\ 0 & -1 & 0 \\ 0 & 0 & 0 \end{bmatrix} \begin{bmatrix} \psi_{fr} \\ \psi_{fg} \\ \psi_{fb} \end{bmatrix} d^3x \tag{13-59}
$$

$$
{}_{QCD}Q_8 = \frac{g_s}{2} \int \begin{bmatrix} \psi_{fr}^\dagger & \psi_{fg}^\dagger & \psi_{fb}^\dagger \end{bmatrix} \frac{1}{\sqrt{3}} \begin{bmatrix} 1 & 0 & 0 \\ 0 & 1 & 0 \\ 0 & 0 & -2 \end{bmatrix} \begin{bmatrix} \psi_{fr} \\ \psi_{fg} \\ \psi_{fb} \end{bmatrix} d^3x. \tag{13-60}
$$

In a manner we have seen repeatedly before, we can carry out the integration in (13-59) and hopefully recognize that it will give us charge operators in terms of number operators. Thus, where the second subscript on the number operators designates color (the three components in the triplet Ψ_f), s represents spin, and an overbar on the number operator indicates antiparticles, we have

$$ {}_{QCD}Q_3 = \frac{g_s}{2}\int\left(\psi^\dagger_{fr}\psi_{fr} - \psi^\dagger_{fg}\psi_{fg}\right)d^3x $$
$$ = g_s \sum_{f,\mathbf{p},s}\left\{\left(\tfrac{1}{2}N_{fr,s}(\mathbf{p}) - \tfrac{1}{2}N_{fg,s}(\mathbf{p})\right) - \left(\tfrac{1}{2}\bar{N}_{fr,s}(\mathbf{p}) - \tfrac{1}{2}\bar{N}_{fg,s}(\mathbf{p})\right)\right\}, \tag{13-61} $$

First non-zero charge operator in terms of number operators

$$ {}_{QCD}Q_8 = \frac{g_s}{2}\int\frac{1}{\sqrt{3}}\left(\psi^\dagger_{fr}\psi_{fr} + \psi^\dagger_{fg}\psi_{fg} - 2\psi^\dagger_{fb}\psi_{fb}\right)d^3x $$
$$ = g_s \sum_{f,\mathbf{p},s}\left\{\begin{matrix}\left(\frac{1}{2\sqrt{3}}N_{fr,s}(\mathbf{p}) + \frac{1}{2\sqrt{3}}N_{fg,s}(\mathbf{p}) - \frac{1}{\sqrt{3}}N_{fb,s}(\mathbf{p})\right) \\ -\left(\frac{1}{2\sqrt{3}}\bar{N}_{fr,s}(\mathbf{p}) + \frac{1}{2\sqrt{3}}\bar{N}_{fg,s}(\mathbf{p}) - \frac{1}{\sqrt{3}}\bar{N}_{fb,s}(\mathbf{p})\right)\end{matrix}\right\}. \tag{13-62} $$

Second non-zero charge operator in terms of number operators

Do **Problem 1** to derive (13-61) from (13-59). Can you assume (13-62) follows in similar fashion from (13-60)?

13.2.4 Examples of Color Expectation for States

We know that quarks cannot exist in isolation, but let's consider a red quark state purely as a hypothetical example, and determine its SU(3) charges, which we designate (as we did in Chap. 2) by ε_3 and ε_8. The result will be valid for any red quark with any spin and three-momentum, so we won't bother to specify those values explicitly when we write down the ket.

$$ {}_{QCD}Q_3|u_r\rangle = g_s \sum_{f,\mathbf{p},s}\left\{\left(\tfrac{1}{2}N_{fr,s}(\mathbf{p}) - \tfrac{1}{2}N_{fg,s}(\mathbf{p})\right) - \left(\tfrac{1}{2}\bar{N}_{fr,s}(\mathbf{p}) - \tfrac{1}{2}\bar{N}_{fg,s}(\mathbf{p})\right)\right\}|u_r\rangle $$
$$ = g_s\left(\tfrac{1}{2} - 0 - 0 + 0\right)|u_r\rangle = \frac{g_s}{2}|u_r\rangle \quad \rightarrow \quad \varepsilon_3 = \frac{g_s}{2} \tag{13-63} $$

i=3 charge operator eigenvalue for red quark

$$ {}_{QCD}Q_8|u_r\rangle = g_s \sum_{f,\mathbf{p},s}\left\{\begin{matrix}\left(\frac{1}{2\sqrt{3}}N_{fr,s}(\mathbf{p}) + \frac{1}{2\sqrt{3}}N_{fg,s}(\mathbf{p}) - \frac{1}{\sqrt{3}}N_{fb,s}(\mathbf{p})\right) \\ -\left(\frac{1}{2\sqrt{3}}\bar{N}_{fr,s}(\mathbf{p}) + \frac{1}{2\sqrt{3}}\bar{N}_{fg,s}(\mathbf{p}) - \frac{1}{\sqrt{3}}\bar{N}_{fb,s}(\mathbf{p})\right)\end{matrix}\right\}|u_r\rangle \tag{13-64} $$
$$ = g_s\left(\frac{1}{2\sqrt{3}} + 0 - 0 - 0 - 0 + 0\right)|u_r\rangle = \frac{g_s}{2\sqrt{3}}|u_r\rangle \quad \rightarrow \quad \varepsilon_8 = \frac{g_s}{2\sqrt{3}}. $$

i=8 charge operator eigenvalue for red quark

As we noted in Chap. 2, the strong charges, just like the weak charges, are often simply designated without the associated coupling constant. In other words, the charge in (13-63) is often considered simply ½ instead of $g_S/2$, with the g_S being implicit.

We would, of course, get the same values for any other flavor quark (*d, c, s, t, b*) of red color.

Now, look at what we found for the *SU*(3) eigenvalues of the Cartan subalgebra in Chap. 2, Wholeness Chart 2-7b, pg. 49. They match what we just found. Also, look at our eigenvalue plot Fig. 2-2, pg. 55 (repeated on pg. 415 at end of this chapter), where we plotted the red quark eigenvalues ε_3 and ε_8. Same thing. It all hangs together.

Red quark eigenvalues match those of Chap. 2 for Cartan subalgebra

We can, and of course we do, designate the two values found using the *SU*(3) charge operators in (13-63) and (13-64) by what we call the red color charge. Similar results apply for green and blue quarks, as illustrated in the Chap. 2 referenced sections, and as you can find by working it out yourself.

Similar results for green and blue quarks

Do **Problem 2** to find the *SU*(3) charge operator eigenvalues for a green and a blue quark.

How about the charge on a positively charged (electrically) pi-meson, which can be considered as the normalized state

Positively charged pion state

$$\left|\pi^{+}\right\rangle=\frac{1}{\sqrt{3}}\left(\left|u_{r}\bar{d}_{r}\right\rangle+\left|u_{g}\bar{d}_{g}\right\rangle+\left|u_{b}\bar{d}_{b}\right\rangle\right)? \tag{13-65}$$

Applying (13-63) and (13-64) to (13-65), we have

$$\begin{aligned}{}_{QCD}Q_{3}\left|\pi^{+}\right\rangle&=g_{s}\sum_{f,\mathbf{p},s}\left\{\begin{matrix}\left(\frac{1}{2}N_{f\,r,s}(\mathbf{p})-\frac{1}{2}N_{f\,g,s}(\mathbf{p})\right)\\ -\left(\frac{1}{2}\bar{N}_{f\,r,s}(\mathbf{p})-\frac{1}{2}\bar{N}_{f\,g,s}(\mathbf{p})\right)\end{matrix}\right\}\frac{1}{\sqrt{3}}\left(\left|u_{r}\bar{d}_{r}\right\rangle+\left|u_{g}\bar{d}_{g}\right\rangle+\left|u_{b}\bar{d}_{b}\right\rangle\right)\\ &=g_{s}\frac{1}{\sqrt{3}}\left(\left(\frac{1}{2}-\frac{1}{2}\right)\left|u_{r}\bar{d}_{r}\right\rangle+\left(-\frac{1}{2}+\frac{1}{2}\right)\left|u_{g}\bar{d}_{g}\right\rangle+(0)\left|u_{b}\bar{d}_{b}\right\rangle\right)=0\quad\rightarrow\quad\varepsilon_{3}=0\end{aligned} \tag{13-66}$$

π^{+} has zero i = 3 charge eigenvalue

$${}_{QCD}Q_{8}\left|\pi^{+}\right\rangle=$$

$$g_{s}\sum_{f,\mathbf{p},s}\left\{\begin{matrix}\left(\frac{1}{2\sqrt{3}}N_{f\,r,s}(\mathbf{p})+\frac{1}{2\sqrt{3}}N_{f\,g,s}(\mathbf{p})-\frac{1}{\sqrt{3}}N_{f\,b,s}(\mathbf{p})\right)\\ -\left(\frac{1}{2\sqrt{3}}\bar{N}_{f\,r,s}(\mathbf{p})+\frac{1}{2\sqrt{3}}\bar{N}_{f\,g,s}(\mathbf{p})-\frac{1}{\sqrt{3}}\bar{N}_{f\,b,s}(\mathbf{p})\right)\end{matrix}\right\}\frac{1}{\sqrt{3}}\left(\left|u_{r}\bar{d}_{r}\right\rangle+\left|u_{g}\bar{d}_{g}\right\rangle+\left|u_{b}\bar{d}_{b}\right\rangle\right) \tag{13-67}$$

$$=g_{s}\frac{1}{\sqrt{3}}\left(\left(\frac{1}{2\sqrt{3}}-\frac{1}{2\sqrt{3}}\right)\left|u_{r}\bar{d}_{r}\right\rangle+\left(\frac{1}{2\sqrt{3}}-\frac{1}{2\sqrt{3}}\right)\left|u_{g}\bar{d}_{g}\right\rangle+\left(-\frac{1}{\sqrt{3}}+\frac{1}{\sqrt{3}}\right)\left|u_{b}\bar{d}_{b}\right\rangle\right)=0\rightarrow\varepsilon_{8}=0.$$

π^{+} has zero i = 8 charge eigenvalue

The pion has zero $SU(3)$ charge operator eigenvalues and thus, zero color charge. (See Fig. 2-2 again, where the pion is located at the origin.)

So, π^{+} has zero net color

All of the above is independent of the flavors of quark in the meson, so any meson can be shown to have zero color in the same manner as we did above for the π^{+}.

Similar results for any meson

Do **Problem 3** to find the $SU(3)$ charge operator eigenvalues for a nucleon.

From Prob. 3 and the prior work in this section, we can recognize that all mesons and baryons have zero net color, which shouldn't be news, but now we have shown it formally starting from Noether's theorem.

Noether tells us all mesons and baryons are colorless

13.2.5 Summary of QCD Charges

Wholeness Chart 13-2, at the end of this chapter, summarizes the results of this section.

13.3 Noether's Theorem and Quark Flavor Symmetry

In this section, we are going to forget about color, i.e., ignore $SU(3)$ theory for strong force interactions. Instead, we are going to focus on quark flavors. For N_f quark flavors, we will discuss $SU(N_f)$ theory with an internal flavor space, which may seem a bit strange at this point, but let's plow ahead and see where it takes us.

We will investigate flavor space for quarks

13.3.1 Vector Flavor Currents

We will ignore leptons altogether, so numeric values of the f subscript in this Sect. 13.3 will represent, in order from 1 to 6, the u, d, c, s, t, b quarks where N_f equals the total number of quarks of type f in the model. The part of the QCD Lagrangian containing quarks (see (12-14), pg. 389), where a and b are color indices, and repeated indices are summed, is

$$\mathcal{L}_{q}=\mathcal{L}_{0}^{Q}+\mathcal{L}^{QG}=\bar{\psi}_{f}\left(i\gamma_{\mu}D^{\mu}-m_{f}\right)\psi_{f}=\bar{\psi}_{fa}\left(i\gamma_{\mu}\left(\partial^{\mu}+i\frac{g_{s}}{2}\underbrace{A_{j}^{\mu}\lambda_{j\,ab}}_{\underline{A}^{\mu}}\right)-m_{f}I_{ab}\right)\psi_{fb}. \tag{13-68}$$

The quark part of the QCD Lagrangian

To keep notation simple, we will drop the color indices (a, b) on quark fields and compact the gluon term $A_{j}^{\mu}\lambda_{j\,ab}$ into the symbol $\underline{A}^{\mu}$. Thus, for our present purposes, (13-68) can be written as

$$\mathcal{L}_q = \bar{\psi}_f \left(i\gamma^\mu \left(\partial_\mu + i\frac{g_s}{2} A_\mu \right) - m_f \right) \psi_f \qquad f = 1,\ldots,N_f , \tag{13-69}$$

That same Lagrangian dropping color indices

or in terms of the flavor multiplet Ψ and the mass matrix M,

$$\mathcal{L}_q = \bar{\Psi}\left(i\not\partial - \frac{g_s}{2}\not A - M \right)\Psi \qquad \Psi = \begin{bmatrix} u \\ d \\ \vdots \end{bmatrix} \qquad M = \begin{bmatrix} m_u & & \\ & m_d & \\ & & \ddots \end{bmatrix}. \tag{13-70}$$

That same Lagrangian in multiplet form

Note that in flavor space, $\not A$ equals the identity matrix times $A_j \lambda_{j\,ab}$, i.e., it is a diagonal matrix with the same term in every component on the diagonal. Each of those terms is 3X3 in color space.

Now, *if* the masses of the quarks were all the same, the Lagrangian (13-70) would be symmetric under the *SU(N_f) global* transformation set

$$\left(\text{flavor where } \omega_i \neq \omega_i(x)\text{; that is, } \omega_i \text{ is independent of the spacetime point}\right)$$
$$\Psi' = \Psi e^{-i\frac{\lambda_i}{2}\omega_i} \xrightarrow{\text{infinitesimal}} \Psi' = \Psi + \delta\Psi = \Psi - i\frac{\lambda_i}{2}\omega_i\Psi \qquad \delta A_{\mu i} = 0, \tag{13-71}$$

$\mathcal{L}$ symmetric for global flavor transformation if all quark masses equal

where the λ_i are the *SU(N_f)* Lie algebra generators, of total number $N_f^2 - 1$.

Do **Problem 4** to prove the last statement.

In reality, of course, the masses are not all equal, so we are only talking about an *approximate* symmetry here that could have some validity at energies much above the masses involved.

Only approximate symmetry, since quark masses not all equal

But assuming that symmetry is valid, from Noether's theorem, we have $N_f^2 - 1$ conserved currents, where the V subscript refers to vector type current, and we are careful to distinguish these currents from others we have studied,

$$j^\mu_{Vi}(x) = \frac{\partial \mathcal{L}_q}{\partial \Psi_{,\mu}}\frac{\partial \Psi'}{\partial \omega_i} = \frac{1}{2}\bar{\Psi}\gamma^\mu \lambda_i \Psi \qquad (\text{flavor currents}), \tag{13-72}$$

Flavor currents

$$\text{where} \qquad \partial_\mu j^\mu_{Vi} = 0 \qquad Q_{Vi} = \int j^0_{Vi}\, d^3x = \text{constant} \qquad (\text{flavor "charges"}). \tag{13-73}$$

Flavor "charges" approximately conserved

As with electroweak and color symmetries, the flavor symmetry charges Q_{Vi} associated with the diagonal matrices of the Cartan subalgebra give us eigenvalues we can associate with each field (and thus, each single particle state) and use to label, and distinguish between, different fields/states.

Charge eigenvalues can be used to label particles

One may wonder, given all our prior work with symmetries, currents, and charges for QED, electroweak, and strong interactions if all of the above implies some new force, mediated by new gauge bosons between different flavor quarks, i.e., a kind of "flavor interaction" dependent on the particular flavor charges of different flavor quarks. It does not. Strong interactions are mediated by gluons, not "flavorons". There is no strong interaction between *u* and *d* quarks contingent upon their flavors, but only on their color charges.

No flavor forces, though focus on flavor symmetry

Nevertheless, the (approximate) symmetries in flavor space will serve us well in work yet to come. As we do that work, keep in mind that the flavor "charges" are not really charges in the sense of being involved in interactions, but they are useful (at least for charges associated with the diagonal flavor space generators) because they provide eigenvalues, noted above for designating, and distinguishing between, various flavors of quark.

Note that it is common to consider only the *u* and *d* quarks ($N_f = 2$), or only the *u, d,* and *s* quarks $N_f = 3$), because their masses are quite low relative to the *c, t,* and *b* quarks, and so can be considered approximately equal (relatively) for certain purposes. Doing so, can, for example, help in deducing the makeup of composite quark states (baryons and mesons), which we will explore in a later chapter.

Approximation of equal masses for u, d or u, d, s quarks often suitable

This symmetry, where all quarks under consideration have equal masses, is known as the Wigner-Weyl mode, named after two physicists we are well familiar with.

13.3.2 Axial Flavor Currents

We can repeat the above analysis, but instead of the transformation (13-71), consider the chiral transformation incorporating the γ^5 matrix,

Prior was vector symmetry; now axial

$$\Psi' = \Psi + \delta\Psi = \Psi - \frac{i}{2}\underset{\sim}{\lambda}_i \omega_i \gamma^5 \Psi \qquad \delta \underset{\sim}{A}_\mu = 0 \qquad \text{where} \quad \omega_i \neq \omega_i(x)$$
$$\Psi^{\dagger\prime} = \Psi^\dagger + \delta\Psi^\dagger = \Psi^\dagger + \frac{i}{2}\Psi^\dagger \omega_i \underset{\sim}{\lambda}_i \gamma^5 \qquad (13\text{-}74)$$
$$\bar{\Psi}' = \Psi^{\dagger\prime}\gamma^0 = \Psi^\dagger\gamma^0 + \delta\Psi^\dagger\gamma^0 = \bar{\Psi} + \frac{i}{2}\Psi^\dagger \omega_i \underset{\sim}{\lambda}_i \gamma^5\gamma^0 = \bar{\Psi} - \frac{i}{2}\omega_i \Psi^\dagger \gamma^0 \underset{\sim}{\lambda}_i \gamma^5 = \bar{\Psi} - \frac{i}{2}\omega_i \bar{\Psi}\underset{\sim}{\lambda}_i \gamma^5 .$$

The axial transformation

Now, consider the Lagrangian of (13-70), where we assume the standard rep for the γ^μ, and equal quark masses m,

$$\delta \mathcal{L}_q = \delta\bar{\Psi}\left(i\partial\!\!\!/ - \frac{g_s}{2}\underset{\sim}{A\!\!\!/} - m\right)\Psi + \bar{\Psi}\left(-\frac{g_s}{2}(\delta \underset{\sim}{A\!\!\!/})\right)\Psi + \bar{\Psi}\left(i\partial\!\!\!/ - \frac{g_s}{2}\underset{\sim}{A\!\!\!/} - m\right)\delta\Psi$$
$$= \left(-\frac{i}{2}\bar{\Psi}\gamma^5 \underset{\sim}{\lambda}_j \omega_j\right)\left(i\gamma^\mu\partial_\mu - \frac{g_s}{2}\gamma^\mu \underset{\sim}{A}_\mu - m\right)\Psi \; + \; 0$$
$$+ \bar{\Psi}\left(i\gamma^\mu\partial_\mu - \frac{g_s}{2}\gamma^\mu \underset{\sim}{A}_\mu - m\right)\left(-\frac{i}{2}\underset{\sim}{\lambda}_j \omega_j \gamma^5 \Psi\right)$$
$$= \frac{1}{2}\omega_j \bar{\Psi}\underset{\sim}{\lambda}_j \gamma^5\gamma^\mu\partial_\mu\Psi + i\frac{g_s}{4}\omega_j \underset{\sim}{A}_\mu \bar{\Psi}\gamma^5\gamma^\mu \underset{\sim}{\lambda}_j \Psi + m\frac{i}{2}\omega_j \bar{\Psi}\underset{\sim}{\lambda}_j \gamma^5\Psi \qquad (13\text{-}75)$$
$$+ \frac{1}{2}\omega_j \bar{\Psi}\underset{\sim}{\lambda}_j \gamma^\mu\gamma^5\partial_\mu\Psi + i\frac{g_s}{4}\omega_j \underset{\sim}{A}_\mu \bar{\Psi}\gamma^\mu\gamma^5 \underset{\sim}{\lambda}_j \Psi + m\frac{i}{2}\omega_j \bar{\Psi}\underset{\sim}{\lambda}_j \gamma^5\Psi$$
$$= -\frac{1}{2}\omega_j \bar{\Psi}\underset{\sim}{\lambda}_j \gamma^\mu\gamma^5\partial_\mu\Psi - i\frac{g_s}{4}\omega_j \underset{\sim}{A}_\mu \bar{\Psi}\gamma^\mu\gamma^5 \underset{\sim}{\lambda}_j \Psi + m\frac{i}{2}\omega_j \bar{\Psi}\underset{\sim}{\lambda}_j \gamma^5\Psi$$
$$+ \frac{1}{2}\omega_j \bar{\Psi}\underset{\sim}{\lambda}_j \gamma^\mu\gamma^5\partial_\mu\Psi + i\frac{g_s}{4}\omega_j \underset{\sim}{A}_\mu \bar{\Psi}\gamma^\mu\gamma^5 \underset{\sim}{\lambda}_j \Psi + m\frac{i}{2}\omega_j \bar{\Psi}\underset{\sim}{\lambda}_j \gamma^5\Psi = im\omega_j \bar{\Psi}\underset{\sim}{\lambda}_j \gamma^5\Psi .$$

Lagrangian under axial transformation witjh equal quark masses

Not symmetric

Contrary to the ordinary (vector) flavor symmetry, where equal masses yielded a symmetric Lagrangian, axial flavor symmetry only exists if all quark masses are zero.

For the case where quarks are all massless, we have, again via Noether's theorem, conserved axial currents

But for all $m_q = 0$, $\mathcal{L}$ is symmetric

$$j^\mu_{Ai}(x) = \frac{\partial \mathcal{L}_q}{\partial \Psi_{,\mu}}\frac{\partial \Psi'}{\partial \omega_i} = \frac{1}{2}\bar{\Psi}\gamma^\mu\gamma^5 \underset{\sim}{\lambda}_i \Psi \qquad \text{(flavor)} \qquad (13\text{-}76)$$

where $\qquad \partial_\mu j^\mu_{Ai} = 0 \qquad Q_{Ai} = \int j^0_{Ai}\, d^3x = \text{constant} \qquad (\text{flavor, massless quarks}). \qquad (13\text{-}77)$

So, for all $m_q = 0$, have conserved flavor charges

13.3.3 Massless Quarks and Chiral Flavor Symmetry

Assuming massless quarks may, at this point, seem a bit outlandish, but consider that the rest mass of a proton or neutron (≈ 935 MeV) is on the order of a hundred times the sum total of the rest masses of the constituent quarks (≈ 9 to 11.5 MeV). The remainder comes from the mass-energy of the interactions involved in binding the three quarks together, a large part of which comes from the cloud of virtual gluons. Approximating quark masses as vanishingly small, relative to the mass-energy of the interactions involved, can then be quite justifiable.

m_q=0 OK approximation for hadrons, as little of their mass from m_q

Consider, also, that in the early universe, before particles acquired mass, flavor symmetry truly existed, so incorporating such symmetry into analysis in that case is more than appropriate.

Also OK before e/w epoch, as m_q=0 then

So, we take a brief look at the implications of zero mass quarks, where both the flavor (vector flavor, to be precise, though the use of just "flavor" is common) transformation and axial flavor transformation leave the Lagrangian invariant. Then, we can construct (define) chiral currents from (13-72) and (13-76),

$$\text{(flavor, massless multiplets)}$$
$$j_i^{\mu L} \equiv \frac{1}{2}\left(j^\mu_{Vi} - j^\mu_{Ai}\right) = \left(\frac{1}{2}\bar{\Psi}\gamma^\mu \underset{\sim}{\lambda}_i \Psi - \frac{1}{2}\bar{\Psi}\gamma^\mu\gamma^5 \underset{\sim}{\lambda}_i \Psi\right) = \bar{\Psi}\gamma^\mu \frac{1}{2}\left(1-\gamma^5\right)\underset{\sim}{\lambda}_i \Psi = \frac{1}{2}\bar{\Psi}^L \gamma^\mu \underset{\sim}{\lambda}_i \Psi^L \qquad (13\text{-}78)$$
$$j_i^{\mu R} \equiv \frac{1}{2}\left(j^\mu_{Vi} + j^\mu_{Ai}\right) = \left(\frac{1}{2}\bar{\Psi}\gamma^\mu \underset{\sim}{\lambda}_i \Psi + \frac{1}{2}\bar{\Psi}\gamma^\mu\gamma^5 \underset{\sim}{\lambda}_i \Psi\right) = \bar{\Psi}\gamma^\mu \frac{1}{2}\left(1+\gamma^5\right)\underset{\sim}{\lambda}_i \Psi = \frac{1}{2}\bar{\Psi}^R \gamma^\mu \underset{\sim}{\lambda}_i \Psi^R ,$$

LC and RC (chiral) flavor currents

with associated charges

$$Q_{Li} \equiv \frac{1}{2}\left(Q_{Vi} - Q_{Ai}\right) \quad Q_{Ri} \equiv \frac{1}{2}\left(Q_{Vi} + Q_{Ai}\right) \quad (\text{chiral flavor "charges" [massless quarks]}). \; (13\text{-}79)$$

Chiral (flavor) charges

Each charge operator in (13-79) contains the ith Lie algebra matrix generator $\underset{\sim}{\lambda}_i$, so those operators obey commutation relations associated with the respective Lie algebra generators.

Do **Problem 5** to show (13-80).

From (13-78) and (13-79), we have, where f_{ijk} are the Lie algebra structure constants,

$$\left[Q_{Li},Q_{Lj}\right]=i\,f_{ijk}\,Q_{Lk} \qquad \left[Q_{Ri},Q_{Rj}\right]=i\,f_{ijk}\,Q_{Rk} \qquad \left[Q_{Li},Q_{Rj}\right]=0. \quad \text{(flavor, massless quarks)} \tag{13-80}$$

(13-80) tells us the right and left chiral groups are independent, i.e., decoupled and operate separately as their own subgroups. This means the whole chiral group is actually a direct product of the LC and RC flavor subgroups, i.e.,

LC and RC flavor groups are independent

$$\text{Chiral flavor group} = SU\left(N_f\right)_L \otimes SU\left(N_f\right)_R . \tag{13-81}$$

There are two separately conserved types of currents. For given i, those for LC, and RC, quark fields, respectively. As we have learned, only the diagonal matrix generators lead to non-zero charge expectation values (charge values we might measure), so, for example, in flavor $SU(3)$, only the $i = 3$ and 8 flavor charge operators will provide us with eigenvalues to enable differentiation between different chiral charged fields/particles.

LC and RC charges separately conserved (for massless quarks)

13.3.4 Chiral Symmetry Breaking

There are two kinds of breaking of chiral symmetry, explicit and spontaneous.

2 ways to break chiral flavor symmetry

Explicit Chiral Symmetry Breaking

Real world quarks, in our present era, have masses, so chiral symmetry cannot be an exact symmetry of the QCD Lagrangian. When quarks attain mass, the symmetry is broken, axial in any case, and vector for non-equal masses. This is known as explicit chiral symmetry breaking.

In e/w symmetry breaking, quarks get mass (explicit chiral symmetry breaking)

In some cases, we can work with an approximate chiral symmetry. For example, as noted earlier, the *u, d, s* quarks are relatively light, and the approximate symmetry can be useful for them. It can generally not be used for the heavy quarks *c, t, b*.

Spontaneous Chiral Symmetry Breaking

We have learned that quarks get mass via spontaneous electroweak symmetry breaking, so one might expect the explicit symmetry breaking of the prior subsection to be considered spontaneous symmetry breaking.

However, electroweak symmetry breaking is independent of strong interactions, so we could have a separate spontaneous symmetry breaking within the strong sector. In the present context, this is different from explicit chiral symmetry breaking, and goes by the name of spontaneous chiral symmetry breaking. The model for this is known as the Nambu-Goldstone mode of realization of chiral symmetry.

If separate symmetry breaking inside strong sector → spontaneous breaking

13.3.5 And More

There is much more entailed in chiral symmetry and its breaking, but it is beyond the scope of this book. Good further treatments of it can be found in Sazdijian[1,2] and the references cited therein.

We've just scratched the surface of chiral symmetry

13.3.6 Summary of Quark Flavor Symmetry

Wholeness Chart 13-3, pg. 416, summarizes the main results of this section.

13.4 The Strong CP Problem

13.4.1 Discrete Symmetries and Strong Interactions

Terms in the QCD Lagrangian are generally symmetric under *C, P,* and *T* transformations, with the notable exception of *P* violation arising due to certain factors we are about to discuss. Since that

P only discrete symmetry of note in QCD

---

[1] H. Sazdjian, Introduction to chiral symmetry in QCD (slide show), Student Lectures, Confinement 12 (2016) . https://indico.cern.ch/event/353906/contributions/2261788/

[2] H. Sazdjian, Introduction to chiral symmetry in QCD, EPJ Web of Conferences 137, 0200 (2017), arXiv:1612,04078 [hep-ph] (13 Dec 2016).

Lagrangian is invariant under charge conjugation, this means *CP* is violated, and because *CPT* is (we believe) never violated, *T* is also violated (in such a way as to cancel the *P* violation).

So, from theory, we would expect to have CP violation, but none is observed experimentally; hence the famous issue in field theory circles, called the strong *CP* problem.

No QCD P symmetry → strong CP problem

13.4.2 Our Approach to the Strong CP Problem

As readers of this text know, I strive to give each of you a solid conceptual foundation in all things presented herein by providing, sometimes in great detail, virtually every facet of derivations, analyses, and logic forming that foundation. I loathe invoking the escape clause "it can be shown".

However, for the strong CP problem, there is a wealth of background material one needs in order to truly understand it and its various possible solutions, and covering that material would take us far astray from the primary objective of this book – to give you a sound footing in the basic principles of the standard model. So, we will have to be satisfied, for now, with an encapsulation of the basic tenants of the strong CP issue, without full comprehension of how they were arrived at, plus a list of references for deeper study, when and if it is appropriate for your career path.

We will short-cut and condense our treatment of it here

13.4.3 Two Causes of CP Non-invariance in the QCD Lagrangian

There are two sources for *CP* violation in QCD. One has to do with our good old friend the mass matrix of electroweak theory flavor space, which is generally not diagonal and generally complex. The other arises from an anomaly, which we discuss briefly below.

Two sources for strong CP violation

The Mass Matrix Contribution to *CP* Violation

The mass matrix terms in a flavor basis have form

1) the mass matrix

$$\mathcal{L}_{mass} = \bar{\psi}^R_{f'} M_{ff'} \psi^L_f + h.c. \qquad (13\text{-}82)$$

where we used a unitary transformation to diagonalize $M_{f'f}$ and get real number masses on the diagonal with the transformed fields then expressed in the mass basis. Prior to carrying out the diagonalization, the mass matrix was complex, so its determinant could be expressed as

$$Det\, M = |M| e^{i\beta} \qquad \beta = Arg\{Det\, M\}. \qquad (13\text{-}83)$$

A lengthy analysis we won't get into leads to a *CP* violating term in the Lagrangian of form (where we employ the 4th rank Levi-Civita tensor, use (13-36) for gluon fields, and drop the caret [hat] notation of (13-36), as is common in the literature)

$$\mathcal{L}_\beta = -\beta \frac{g_s^2}{32\pi^2} \varepsilon_{\mu\nu\delta\rho} G_i^{\delta\rho} G_i^{\mu\nu} = -\beta \frac{g_s^2}{32\pi^2} \tilde{G}_{i\,\mu\nu} G_i^{\mu\nu} \qquad (13\text{-}84)$$

CP violating contribution to $\mathcal{L}$ from mass matrix

Note the commonly used shorthand notation employing the tilde over the field tensor.

You can show the CP violation of (13-84) by doing **Problem 6**, or save time by just taking my word for it, as there is a lot of other material here we are just accepting without deducing, as well.

The Anomaly Contribution

Consider the axial flavor symmetry, in the limit of massless quarks, of (13-76) and (13-77), where the axial current has zero divergence, at least from a direct consideration of the Lagrangian. However, when one includes quantum corrections from a particular triangle diagram, we get an anomaly. (See Sect. 11.1.5, pg. 367.) Known as the chiral anomaly, it contributes to our axial current, making its divergence non-zero, i.e.,

2) chiral anomaly

$$\partial_\mu j^\mu_{Ai} \neq 0. \qquad (13\text{-}85)$$

This ends up contributing a term to the Lagrangian, where θ is a constant between 0 and 2π, of form

$$\mathcal{L}_\theta = \theta \frac{g_s^2}{32\pi^2} \tilde{G}_{i\,\mu\nu} G_i^{\mu\nu}. \qquad (13\text{-}86)$$

CP violating contribution to $\mathcal{L}$ from chiral anomaly

That is, by adding (13-86) to the Lagrangian, it takes the place of the anomaly contribution we got from quantum effects, so we can use it in lieu of going through all the steps to calculate that anomaly.

Like (13-84), (13-86) is not *CP* invariant.

The Two Causes Added Together

So, our two *CP* violating terms in the QCD Lagrangian combined are (13-84) plus (13-86),

$$\mathcal{L}_{CP\,viol} = \theta \frac{g_s^2}{32\pi^2} \tilde{G}_{i\,\mu\nu} G_i^{\mu\nu} - \beta \frac{g_s^2}{32\pi^2} \tilde{G}_{i\,\mu\nu} G_i^{\mu\nu} = \overbrace{(\theta - \beta)}^{\bar{\theta}} \frac{g_s^2}{32\pi^2} \tilde{G}_{i\,\mu\nu} G_i^{\mu\nu}. \tag{13-87}$$

Add 1) and 2) to get one term in $\mathcal{L}$ with $\bar{\theta}$ factor

θ and β are both between 0 and 2π, so, for any reasonable choice of the two constants, we should have rather significant *CP* violation due to their difference $\bar{\theta}$.

13.4.4 The Problem

From experiment, it has been determined that $\bar{\theta}$ must be incredibly small. That is,

Experimental limit on $\bar{\theta}$

$$\bar{\theta} \le 10^{-10}, \tag{13-88}$$

which raises the question, "why is it so small"? It is extraordinarily unnatural to expect to subtract two constants and get a number 10 billion times smaller than either. This is the strong *CP* problem.

$\bar{\theta} = \theta - \beta$ so small is strong CP problem

We do not carry out the analysis behind the experimental result here, as again, it is extensive and not so beneficial at this point in our development of QCD and the standard model. We do note that strong *CP* violation should produce an electric dipole moment of the neutron. However, increasingly accurate measurements over many years have found no such dipole moment. The maximum possible value for it, whereby it would still be too small to be detected, gives rise to the constraint (13-88).

$\bar{\theta} \ne 0$ leads to neutron electric dipole, but none is measured

13.4.5 The Peccei-Quinn Solution

Various possible solutions to the strong *CP* problem have been put forth, but the one generally acknowledged as the best was proposed by Roberto Peccei and Helen Quinn in 1977.

The Peccei-Quinn solution

They suggested that there may be another quantum field with additional terms in the Lagrangian, which we can designate as *a*, and which is a massless scalar (pseudo-scalar, actually, but we won't concern ourselves with that detail, here). The field *a* acts much like the Higgs field in electroweak theory. In the beginning, the universe had part of its effective potential dependent on *a*, which we will call $\mathcal{V}_{effec}$, at a non-stable location on a $\mathcal{V}_{effec}$ vs *a* plot. That is, the universe tends to seek a lower potential, just as it did with our Higgs potential, which we illustrated with the Mexican hat diagram.

Postulate another field a, giving rise to a potential, much like Higgs field

The Peccei-Quinn Lagrangian is (13-89) below, where superscripts designate free and interaction terms; subscripts *a*, ψ, and *A* respectively, the *a*, quark, and gluon fields; f_a is an unknown constant, and $\mathcal{V}_{effec}$ is an effective potential depending on $\bar{\theta}$ and *a*.

$$\begin{aligned}\mathcal{L} &= \mathcal{L}_{SM} + \mathcal{L}_a^0 + \mathcal{L}_a^I + \mathcal{L}_{CP\,viol} - \mathcal{V}_{effec}(\bar{\theta}, a) = \mathcal{L}_{SM} + \mathcal{L}_a^0 + \mathcal{L}_{a\psi} + \mathcal{L}_{aA} + \mathcal{L}_{CP\,viol} - \mathcal{V}_{effec}(\bar{\theta}, a) \\ &= \mathcal{L}_{SM} + \tfrac{1}{2}(\partial_\mu a)(\partial^\mu a) + \mathcal{L}_{a\psi} + \frac{g_s^2}{32\pi^2} \frac{a}{f_a} \tilde{G}_{i\,\mu\nu} G_i^{\mu\nu} + \bar{\theta} \frac{g_s^2}{32\pi^2} \tilde{G}_{i\,\mu\nu} G_i^{\mu\nu} - \mathcal{V}_{effec}(\bar{\theta}, a).\end{aligned} \tag{13-89}$$

Peccei-Quinn $\mathcal{L}$ incorporating the field a

From analysis we again won't get into, involving chiral transformations, the effective potential can be deduced to be of form

$$\mathcal{V}_{effec}(\bar{\theta}, a) \propto -\cos\left(\bar{\theta} + \frac{a(x)}{f_a}\right). \tag{13-90}$$

Form of the a field potential (taken from other sources)

After the universe slides off its non-stable state, the potential reaches a true vacuum minimum, so

$$\frac{\partial \mathcal{V}_{effec}}{\partial a} \propto \sin\left(\bar{\theta} + \frac{a}{f_a}\right)\left(\frac{1}{f_a}\right) = 0 \qquad a_{\mathcal{V}\,min} = -f_a \bar{\theta} \quad (\text{VEV of } a). \tag{13-91}$$

Finding the minimum of that potential

In (13-89), at the minimum, $\mathcal{V}_{effec}$ takes a constant value (and thus, as we have learned, can be ignored since it affects nothing we can measure). Importantly, the *CP* violating term in $\bar{\theta}$ drops out, i.e., two terms in the lst line of (13-89) become, after the true vacuum minimum (13-91) has been reached,

$$\text{At min } \mathcal{V}_{effec} \rightarrow \frac{g_s^2}{32\pi^2} \frac{a}{f_a} \tilde{G}_{i\,\mu\nu} G_i^{\mu\nu} + \bar{\theta} \frac{g_s^2}{32\pi^2} \tilde{G}_{i\,\mu\nu} G_i^{\mu\nu} = (-\bar{\theta} + \bar{\theta}) \frac{g_s^2}{32\pi^2} \tilde{G}_{i\,\mu\nu} G_i^{\mu\nu} = 0 \tag{13-92}$$

At that minimum, QCD $\mathcal{L}$ has no CP violation

We then have no *CP* violation, consistent with experiment. The only unknown is f_a, different values for which will shift the VEV value of $a_{\mathcal{V}min}$ according to (13-91), but not affect the final result (13-92). Any value of f_a works to minimize the potential and cancel the $\bar{\theta}$ term, and thus leave us

f_a unknown, can have any value and still solves strong CP problem

with a *CP* invariant Lagrangian. This contrasts with our earlier situation, where $\overline{\theta}$, which could range between 0 and 2π, had to be, in seemingly random fashion, less than 10^{-10}.

The Peccei-Quinn solution is natural. It is simply the effective potential seeking its lowest state.

A natural result: no fine-tuning required

13.4.6 Axions

The a field acts a lot like the Higgs field, and Steven Weinberg called it a "Higglet". But Frank Wilczek dubbed it the axion, a name he saw on a laundry detergent, which may be appropriate because it "cleaned up" the problem. Weinberg later consented to that appellation.

Particles created and destroyed by field a, called axion

Recall that the Higgs field could be thought of in two ways. One, as a high energy field, before symmetry breaking, which we designated by ϕ. But after symmetry breaking, when the field took on the VEV in the trough of the Mexican hat potential, it was more appropriate to deem $\sigma = \phi - \phi_{\mathcal{V}\,min}$ as the Higgs field, where the term $\frac{1}{2}m_\sigma^2\sigma^2$ represented small oscillations about the potential minimum, at which $\sigma = 0$.

In a similar fashion, when the axion field a gets a VEV, at the $\mathcal{V}_{effec}$ minimum, we can consider a new version of the axion field $a' = a - a_{\mathcal{V}min}$, where $a' = 0$ at the minimum, and oscillations about the minimum are reflected in a mass term for a'.

Call the new axion field, centered on $a_{\mathcal{V}min}$, a'

Related analysis can lead to an expression for that mass as a function of f_a. Astrophysical and cosmological observations have led to both an upper and lower bound on that mass. From those, it has been deduced that the mass lies between approximately 10^{-5} and 1 eV, extremely light by any measure. Those observations also yield bounds on f_a, which is known as the axion decay constant, as it is related to the mass, and we know that more massive particles decay faster than less massive ones.

Axion decay constant f_a related to axion mass (which is extremely light)

Axions have to date not been detected and are a leading candidate for dark matter.

In 2025, axions still undetected and prime candidate for dark matter

13.4.7 Further Reading

There is a great deal of ground, which we have either skipped over or never addressed, related to axions and the Peccei-Quinn solution to the strong CP problem. More extensive treatments are listed in this footnote[1] and the references cited therein. Be forewarned, however, that the material is dense and complex, and needs a good deal of time and effort to fully fathom.

Much to this topic we have skimmed over or not covered

13.4.8 Summary of the Strong CP Problem

CP violation arises in the QCD Lagrangian via two paths: 1) the mass matrix diagonalization yielding a term with a factor β, and, 2) a chiral anomaly resulting in effectively the same term but with a factor θ instead of β. The first of these is already in our original Lagrangian, and the latter term can be incorporated by including that term directly into the original Lagrangian and then ignoring the anomaly calculation.

Synopsis of the strong CP problem

These two terms can be combined into one by employing $\overline{\theta} = \theta - \beta$. This term should be of order 1 and give rise to an electric dipole moment for the neutron, but no such moment has ever been found. The experimental limit is $\overline{\theta} \leq 10^{-10}$. This is known as the strong *CP* problem.

The Peccei-Quinn solution entails the introduction of a new pseudo-scalar field a, known as the axion, in the Lagrangian. The potential due to this field is of a particular form, and the early universe is presumed to be at a non-stable level of that potential. The universe then spontaneously moves to a lower, stable potential, where a, behaving in similar fashion to the Higgs, obtains a VEV. Physics at that potential is more readily modeled by a new field a', centered on the potential minimum.

At that minimum, the *CP* violating terms in the Lagrangian are canceled, naturally, so there is no such violation and no prediction of a neutron electric dipole moment. Cosmologic and astrophysical observations imply the mass of the hypothesized axion a' is low, even lower than that of neutrinos.

---

[1] R.D. Peccei and H.R. Quinn, *CP* Conservation in the Presence of Pseudoparticles, *Phys Rev Lett.* **38**(25), 1440-1443 (June 1977); R.D. Peccei, The Strong CP Problem and Axions, [hep-ph]0607268 (2006); J. E. Kim and G. Carosi, Axions and the Strong CP Problem, *Rev.Mod.Phys.*82:557-602,2010 [hep-ph] 0807.3125 (2009); Pierre Sikivie, Dark Matter Axions [hep-ph] 0909.0949; Pierre Sikivie, Pool Table Analogy with Axion Physics, *Phys. Tod.* (Dec 1996), 22-27; R. Ticciati, *Quantum Field Theory for Mathematicians,* Cambridge (1999), 475-479; M.D. Schwartz, *Quantum Field Theory and the Standard Model,* Cambridge (2014), 609-613.

13.5 Chapter Summary

13.5.1 SU(n) Symmetry

Symmetry of an *SU*(*n*) quantum field theory is assured, without extensive analysis, if it embraces the criteria below.

Wholeness Chart 13-1. General Principles for *SU*(*n*) Symmetry

For free fermion and interacting fermion-gauge boson terms

Criteria

Lagrangian term	$\bar{\underset{\sim}{\Psi}} D^\mu \underset{\sim}{\Psi}$	(13-93)
Covariant derivative	$D^\mu \underset{\sim}{\Psi} = \left(\partial^\mu + ig_n \frac{\underset{\sim}{\lambda_i}}{2} \underset{\sim}{A}_i^\mu\right)\underset{\sim}{\Psi}$	repeat of (13-18)
Transformations	$\underset{\sim}{\Psi}' = \left(1 + i\frac{\lambda_m}{2}\omega_m\right)\underset{\sim}{\Psi}$ $\underset{\sim}{A}_i'^\mu = \underset{\sim}{A}_i^\mu - \frac{1}{g_n}\partial^\mu\omega_i - f_{ijk}\omega_j \underset{\sim}{A}_k^\mu$	repeat of (13-19) repeat of (13-20)

Result

The above relations mean		
	$\left(D^\mu \underset{\sim}{\Psi}\right)' = \left(1 + i\frac{\lambda_m}{2}\omega_m\right) D^\mu \underset{\sim}{\Psi}$	repeat of (13-17)
and that means covariant derivative terms are symmetric, i.e.,		
	$\left(\bar{\underset{\sim}{\Psi}} D^\mu \underset{\sim}{\Psi}\right)' = \bar{\underset{\sim}{\Psi}}\left(1 - i\frac{\lambda_i}{2}\omega_i\right)\left(1 + i\frac{\lambda_m}{2}\omega_m\right)\underset{\sim}{\Psi} = \bar{\underset{\sim}{\Psi}} D^\mu \underset{\sim}{\Psi}$	repeat of (13-22)

For gauge boson only terms

Criteria

Lagrangian term	$-\frac{1}{4}\underset{\sim}{F}_i^{\mu\nu}\underset{\sim}{F}_{i\,\mu\nu} = \left(\partial^\nu \underset{\sim}{A}_i^\mu - \partial^\mu \underset{\sim}{A}_i^\nu\right)\left(\partial_\nu \underset{\sim}{A}_{i\,\mu} - \partial_\mu \underset{\sim}{A}_{i\,\nu}\right)$	repeat of (13-38)
Covariant derivative (like (13-18))	$D^\mu \underset{\sim}{A}_j^\nu = \left(\partial^\mu + ig_n \frac{\underset{\sim}{\lambda_i}}{2} \underset{\sim}{A}_i^\mu\right)\underset{\sim}{A}_j^\nu$	repeat of (13-39)
Transformation	repeat of (13-20) above	

Result

The above relations mean		
	$-\frac{1}{4}\left(D^\nu \underset{\sim}{A}_i^\mu - D^\mu \underset{\sim}{A}_i^\nu\right)\left(D_\nu \underset{\sim}{A}_{i\,\mu} - D_\mu \underset{\sim}{A}_{i\,\nu}\right) = -\frac{1}{4}\underset{\sim}{G}_i^{\mu\nu}\underset{\sim}{G}_{i\,\mu\nu}$	repeat of (13-41)
	$\underset{\sim}{G}_i^{\mu\nu} = \partial^\nu \underset{\sim}{A}_i^\mu - \partial^\mu \underset{\sim}{A}_i^\nu - g_n \underset{\sim}{f}_{ijk} \underset{\sim}{A}_j^\nu \underset{\sim}{A}_k^\mu$	repeat of (13-42)
and the boson only terms are symmetric, i.e.,		
	$\left(\underset{\sim}{G}_i^{\mu\nu}\right)' = \underset{\sim}{G}_i^{\mu\nu}$, so $\left(-\frac{1}{4}\underset{\sim}{G}_i^{\mu\nu}\underset{\sim}{G}_{i\,\mu\nu}\right)' = -\frac{1}{4}\underset{\sim}{G}_i^{\mu\nu}\underset{\sim}{G}_{i\,\mu\nu}$	repeat of (13-43)

For mass terms

Criteria

Lagrangian terms where $\underset{\sim}{m}$ is a scalar	$\underset{\sim}{m}\bar{\underset{\sim}{\Psi}}\underset{\sim}{\Psi}$	(13-94)
Transformation	repeat of (13-19) above	

Result

Mass terms are symmetric	$\left(\underset{\sim}{m}\bar{\underset{\sim}{\Psi}}\underset{\sim}{\Psi}\right)' = \underset{\sim}{m}\bar{\underset{\sim}{\Psi}}\underset{\sim}{\Psi}$	repeat of (13-44)

13.5.2 QCD Four-currents and Charge Operators

Wholeness Chart 13-2. QCD Four-Currents and Charges

Four-current and Charge Operator	
General principles	$\mathcal{L}$ symmetric under α variation, $j^{\mu} = \dfrac{\partial \mathcal{L}}{\partial \phi^{r}{}_{,\mu}} \dfrac{\partial \phi'^{r}}{\partial \alpha} \rightarrow \partial_{\mu} j^{\mu} = 0 \rightarrow Q = -(coup\ const)\int j^{0} dV$ conserved
QCD charge operators	${}_{QCD}\mathcal{L} \rightarrow {}_{QCD}Q_i = -g_s \int {}_{QCD}j_i^{0} dV$ $\rightarrow \langle \phi_j \vert {}_{QCD}Q_i \vert \phi_j \rangle = \varepsilon_i g_s$ (ith QCD charge expectation value of ϕ_j particle)
	Can take operators ${}_{QCD}Q_i = 0$ $i = 1,2,4,5,6,7$ since associated λ_i not diagonal $\rightarrow$ expectation ${}_{QCD}\bar{Q}_i = 0$
Charge Operator in Terms of Number Operators	
QCD charge operators via number operators	${}_{QCD}Q_3 = g_s \sum_{f,\mathbf{p},s} \left\{ \left(\tfrac{1}{2} N_{f\,r,s}(\mathbf{p}) - \tfrac{1}{2} N_{f\,g,s}(\mathbf{p}) \right) - \left(\tfrac{1}{2} \bar{N}_{f\,r,s}(\mathbf{p}) - \tfrac{1}{2} \bar{N}_{f\,g,s}(\mathbf{p}) \right) \right\}$ ${}_{QCD}Q_8 = g_s \sum_{f,\mathbf{p},s} \left\{ \begin{matrix} \left(\frac{1}{2\sqrt{3}} N_{f\,r,s}(\mathbf{p}) + \frac{1}{2\sqrt{3}} N_{f\,g,s}(\mathbf{p}) - \frac{1}{\sqrt{3}} N_{f\,b,s}(\mathbf{p}) \right) \\ - \left(\frac{1}{2\sqrt{3}} \bar{N}_{f\,r,s}(\mathbf{p}) + \frac{1}{2\sqrt{3}} \bar{N}_{f\,g,s}(\mathbf{p}) - \frac{1}{\sqrt{3}} \bar{N}_{f\,b,s}(\mathbf{p}) \right) \end{matrix} \right\}$
Quark charges	${}_{QCD}Q_3 \vert q_r \rangle = \frac{g_s}{2} \vert q_r \rangle$ $\quad \varepsilon_3 = \frac{1}{2}$ $\quad {}_{QCD}Q_8 \vert q_r \rangle = \frac{g_s}{2\sqrt{3}} \vert q_r \rangle$ $\quad \varepsilon_8 = \frac{1}{2\sqrt{3}}$ ${}_{QCD}Q_3 \vert q_g \rangle = -\frac{g_s}{2} \vert q_g \rangle$ $\quad \varepsilon_3 = -\frac{1}{2}$ $\quad {}_{QCD}Q_8 \vert q_g \rangle = \frac{g_s}{2\sqrt{3}} \vert q_g \rangle$ $\quad \varepsilon_8 = \frac{1}{2\sqrt{3}}$ ${}_{QCD}Q_3 \vert q_b \rangle = (0) \vert q_b \rangle$ $\quad \varepsilon_3 = 0$ $\quad {}_{QCD}Q_8 \vert q_b \rangle = -\frac{g_s}{\sqrt{3}} \vert q_b \rangle$ $\quad \varepsilon_8 = -\frac{1}{\sqrt{3}}$
Converted to color charges	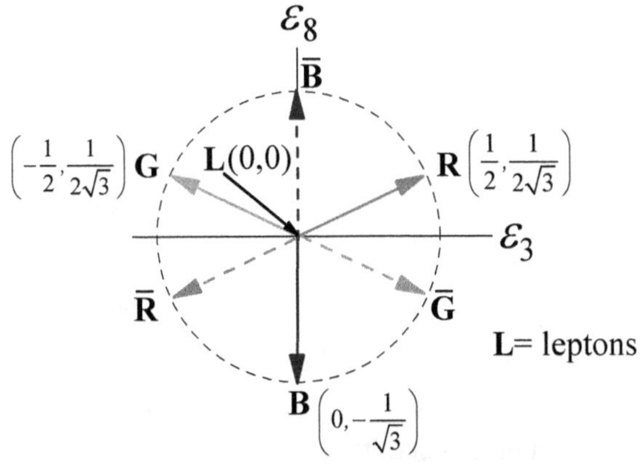 ε_8 $\bar{\mathbf{B}}$ $\left(-\frac{1}{2}, \frac{1}{2\sqrt{3}}\right)$ **G** $\quad$ L(0,0) $\quad$ **R** $\left(\frac{1}{2}, \frac{1}{2\sqrt{3}}\right)$ ε_3 $\bar{\mathbf{R}}$ $\quad$ $\bar{\mathbf{G}}$ L= leptons **B** $\left(0, -\frac{1}{\sqrt{3}}\right)$

13.5.3 Flavor Symmetries

Wholeness Chart 13-3 summarizes Sect. 13.3 on quark flavor symmetry.

Wholeness Chart 13-3. Quark Flavor Symmetries

Masses	$SU(N_f)$ Type of Transformation	Form of Global Transformation	Form of 4-Current	$\mathcal{L}$ Symmetric?	Flavor "Charge" Conserved?
m_q all same	Vector	$\delta\Psi_V = -\frac{i}{2}\lambda_i\,\omega_i\Psi$	$j^\mu_{Vi} = \frac{1}{2}\bar{\Psi}\gamma^\mu\lambda_i\Psi$	Y	$\partial_\mu j^\mu_{Vi} = 0 \quad Q_{Vi} = const$
	Axial	$\delta\Psi_A = -\frac{i}{2}\lambda_i\,\omega_i\gamma^5\Psi$	$j^\mu_{Ai} = \frac{1}{2}\bar{\Psi}\gamma^\mu\gamma^5\lambda_i\Psi$	N	$\partial_\mu j^\mu_{Ai} \neq 0 \quad Q_{Ai} \neq const$
	Left Chiral	$\delta\Psi_L = \delta\Psi_V - \delta\Psi_A$	$j^\mu_{Li} = \frac{1}{2}\left(j^\mu_{Vi} - j^\mu_{Ai}\right)$	N	$\partial_\mu j^\mu_{Li} \neq 0 \quad Q_{Li} \neq const$
	Right Chiral	$\delta\Psi_R = \delta\Psi_V + \delta\Psi_A$	$j^\mu_{Ri} = \frac{1}{2}\left(j^\mu_{Vi} + j^\mu_{Ai}\right)$	N	$\partial_\mu j^\mu_{Ri} \neq 0 \quad Q_{Ri} \neq const$
m_q all = 0	Vector	As vector above	As vector above	Y	$\partial_\mu j^\mu_{Vi} = 0 \quad Q_{Vi} = const$
	Axial	As axial above	As axial above	Y	$\partial_\mu j^\mu_{Ai} = 0 \quad Q_{Ai} = const$
	Left Chiral	As LC above	As LC above	Y	$\partial_\mu j^\mu_{Li} = 0 \quad Q_{Li} = const$
	Right Chiral	As RC above	As RC above	Y	$\partial_\mu j^\mu_{Ri} = 0 \quad Q_{Ri} = const$
m_q different	Any of above	As each above	As each above	N	$\partial_\mu j^\mu_{any\,i} \neq 0 \quad Q_{any\,i} \neq const$

13.5.4 The Strong CP Problem

See the summary in Sect. 13.4.8 on pg. 413.

13.6 Problems

1. Derive (13-61) from (13-59).
2. Find the $SU(3)$ charge operator eigenvalues ε_3 and ε_8 for *i*) a green quark and *ii*) a blue quark, and *iii*) an anti-green quark, then *iv*) compare the results with Fig. 2-2 of Chap. 2 (which is same figure as in Wholeness Chart 13-2 at the end of Chap. 13). Then, *v*) find ε_3 and ε_8 for an electron.
3. Find the $SU(3)$ charge operator eigenvalues for a proton. What is the color charge of the proton? Do you expect the same results for a neutron? Do you expect the same results for any baryon, regardless of the flavors of its component quarks and its electric charge?
4. Show that, if N_f quark masses are all equal, the QCD Lagrangian is symmetric under the flavor transformation (13-71) for the quark flavor multiplet with the N_f component fields.

5. Prove (13-80). Hint: The charge $Q_{V\,i}$ includes the generator λ_i, so it has the same commutation relations as those generators do. The same is true of $Q_{A\,i}$. However, an axial charge has a spinor space factor of γ^5 in it, so a product of $Q_{V\,i}$ and $Q_{A\,i}$ will have that same factor. A product of $Q_{A\,i}$ and $Q_{A\,j}$, on the other hand, means a product $\gamma^5\,\gamma^5 = I$, so it will have no factor of γ^5 in it. So then, we have

$$\left[Q_{V\,i},Q_{V\,j}\right]=i\,f_{ijk}\,Q_{V\,k} \qquad \left[Q_{V\,i},Q_{A\,j}\right]=i\,f_{ijk}\,Q_{A\,k} \qquad \left[Q_{A\,i},Q_{A\,j}\right]=i\,f_{ijk}\,Q_{V\,k},$$

which we can use with the definitions of LC and RC flavor charges to prove (13-80).

6. Substitute the changes under *CP* of Wholeness Chart 9-3 at the end of Chapter 9 for the gluon fields and the spacetime derivatives in (13-84) and show that the term is not invariant under *CP*.

Chapter 14

Hadron Composition

"[Science is] about overcoming our own ideas and continually going beyond common sense."

"Every hope of saying "Well, we are almost there, we've solved all the problems" is nonsense."

"The beauty of physics is the revolution, the continuous change in our way of grasping the world. ... this is a great adventure, and it's an adventure that is far from over."
Carlo Rovelli

14.0 Chapter Overview

We know quarks are the building blocks of baryons and mesons, three each for the former, two (quark/antiquark) for the latter. We also know that there exists a huge proliferation of different kinds of hadrons, a "particle zoo", as it has been called.

What we may not know is that group representation theory can play a significant role in analyzing which quarks form which hadrons in which ways, and how we distinguish between them all. This chapter lays a foundation for understanding how that all plays out.

Before getting into that, however, we need to double back and pick up one of the loose ends in group theory that we did not cover in Chap. 2. This involves a new (for us in this book) classification of group matrix representations into two categories, either "reducible" or "irreducible", the definitions of which we are about to delineate.

So, the topics in this chapter are

- reducible and irreducible representations of groups,
- an example of same from non-relativistic quantum mechanics, and
- group theory with flavor multiplets applied to hadron formation.

Topics in this chapter

14.1 More Group Theory

14.1.1 Direct Sums

Consider a group **C** represented by the 5-dimensional matrices C of the form exhibited in (14-1) (where the matrix components could be real or complex and generally are functions of one or more parameters). Components left blank signify zero values. Note the meaning of the $\oplus$ sign, which implies what is called a direct sum (of matrix representations).

$$C=\begin{bmatrix} a_{11} & a_{12} & a_{13} & & \\ a_{21} & a_{22} & a_{23} & & \\ a_{31} & a_{32} & a_{33} & & \\ & & & b_{11} & b_{12} \\ & & & b_{21} & b_{22} \end{bmatrix}=A\oplus B \quad \text{where } A=\begin{bmatrix} a_{11} & a_{12} & a_{13} \\ a_{21} & a_{22} & a_{23} \\ a_{31} & a_{32} & a_{33} \end{bmatrix} \quad B=\begin{bmatrix} b_{11} & b_{12} \\ b_{21} & b_{22} \end{bmatrix} \qquad (14\text{-}1)$$

Direct sum of matrices nXn and mXm yields (n+m)X(n+m) matrix

When the matrix C operates on a five-component vector v, as in (14-2) below, the A matrix only acts on the top three components and has no effect on the bottom two. Similarly, the B matrix only acts on the bottom two components and does nothing to the top three.

$$\mathbf{Cv}=\mathbf{v}' \xrightarrow[\text{coordinate system}]{\text{representation in some specific}} Cv=\begin{bmatrix} a_{11} & a_{12} & a_{13} & & \\ a_{21} & a_{22} & a_{23} & & \\ a_{31} & a_{32} & a_{33} & & \\ & & & b_{11} & b_{12} \\ & & & b_{21} & b_{22} \end{bmatrix}\begin{bmatrix} v_1 \\ v_2 \\ v_3 \\ v_4 \\ v_5 \end{bmatrix}=\begin{bmatrix} v'_1 \\ v'_2 \\ v'_3 \\ v'_4 \\ v'_5 \end{bmatrix}=v'. \quad (14\text{-}2)$$

nXn & mXm submatrices act independently on vector comps in indep subspaces

The A and B matrices act independently on different components of a vector and can be considered independent matrices acting on independent vector spaces (of dimensions 3 and 2, respectively.) A and B are submatrices of C. The form of the set of matrices C in (14-1) and (14-2) is said to be block diagonal, for what hopefully is a fairly obvious reason, i.e., submatrix blocks along the diagonal and zeroes everywhere else.

Direct sum matrix dimen = sum of submatrix dimens

The dimension of the set of matrices C is 5 which equals the dimension of A (=3) plus the dimension of B (=2). More generally, in direct summing, the resulting matrix dimension equals the sum of the dimensions of the matrices being direct summed. A operates on a subspace of the 5-dimensional space. That subspace has 3 dimensions. B operates on a subspace of 2 dimensions.

Now consider a similarity transformation T, which acts on the matrix group representation C, that fills up at least some of the original zero value matrix components of C.

$$TCT^{-1}=\begin{bmatrix} \underset{\sim}{c}_{11} & \underset{\sim}{c}_{12} & \underset{\sim}{c}_{13} & \underset{\sim}{c}_{14} & \underset{\sim}{c}_{15} \\ \underset{\sim}{c}_{21} & \underset{\sim}{c}_{22} & \underset{\sim}{c}_{23} & \underset{\sim}{c}_{24} & \underset{\sim}{c}_{25} \\ \underset{\sim}{c}_{31} & \underset{\sim}{c}_{32} & \underset{\sim}{c}_{33} & \underset{\sim}{c}_{34} & \underset{\sim}{c}_{35} \\ \underset{\sim}{c}_{41} & \underset{\sim}{c}_{42} & \underset{\sim}{c}_{43} & \underset{\sim}{c}_{44} & \underset{\sim}{c}_{45} \\ \underset{\sim}{c}_{51} & \underset{\sim}{c}_{52} & \underset{\sim}{c}_{53} & \underset{\sim}{c}_{54} & \underset{\sim}{c}_{55} \end{bmatrix}=\underset{\sim}{C} \qquad \begin{gathered}\text{Same abstract operation } \mathbf{C}\text{, expressed}\\ \text{in different coordinate system.}\end{gathered} \quad (14\text{-}3)$$

In other coord systems, subspace independence may not be obvious

We can think of the set of $\underset{\sim}{C}$ matrices as representing the same group **C**, just expressed in different form. That is, T has essentially changed our coordinate system (a passive transformation). Matrices and vectors in the new coordinate system are denoted with tildes or "squiggles" underneath; those in the old system, as plain letters. (Note the tilde underneath has a different meaning here than in Chap. 13. Unfortunately, there are only so many different symbols one can use in developing a theory.) So, for vectors,

$$\underset{\sim}{v}=Tv \qquad \text{(Same abstract vector } \mathbf{v}\text{, expressed in different coordinate system.)} \quad (14\text{-}4)$$

$$\text{Still, in an abstract sense,} \quad \mathbf{Cv}=\mathbf{v}' \xrightarrow[\text{new coordinate system}]{\text{but expressed in this}} \underset{\sim}{C}\,\underset{\sim}{v}=\underset{\sim}{v}' \quad \left(\text{where } \underset{\sim}{v}'=Tv'\right). \quad (14\text{-}5)$$

The T transformation gives us different components for the matrices and the column vectors, even though the abstract group operation **C** carries out the same operation on the same abstract vector **v**.

Subspace independence still there in the new coordinate system

For example, the operation carried out by C could be the rotation of a 5D vector in the 5D space. The vector would represent a (coordinate-system-independent) physical quantity like the position vector, which is rotated through a particular angle about a particular axis. That angle, that axis, and the vector length remain the same physically regardless of which coordinate system we prefer to view them in. But the components of the matrix representing the rotation, and the components of the 5D vector, will be different in different coordinate systems. The right-hand sides of (14-2) and (14-5) represent that same rotated vector, as observed in different coordinate systems. The T transformation changes the coordinate system.

Of course, the group operation represented by the set of matrices C could be any number of things, not just rotation. But similar logic applies, regardless of the particular type of operation **C** carries out.

14.1.2 Reducible vs Irreducible Representations

Subspace independence doesn't change with coord system we may choose, but not obvious in many systems

Note that if we had started with (14-3), instead of (14-1), we would not be immediately aware that there were two independent subspaces on which **C** operates. They would still be there, but it would not be obvious. However, by transforming $\underset{\sim}{C}$ with T^{-1}, i.e., $T^{-1}\underset{\sim}{C}T$, we would get C, and then it would be obvious.

In going from $\underset{\sim}{C}$ to C, we have *reduced* the matrix to block diagonal form, i.e., reduced the group representation to two independent submatrices along the diagonal. One could then imagine a scenario where either A or B matrices could be further reduced to block diagonal form with submatrices within them, such as

$$\underset{\smile}{A} = \hat{T}A\hat{T}^{-1} = \begin{bmatrix} \underset{\smile}{a}_{11} & \underset{\smile}{a}_{12} & \\ \underset{\smile}{a}_{21} & \underset{\smile}{a}_{22} & \\ 0 & 0 & \underset{\smile}{a}_{33} \end{bmatrix}, \tag{14-6}$$

where the $\cup$ marking underneath quantities has a different meaning than that of Chap. 13.

However, if we cannot further reduce A or B, in a manner such as that of (14-6), then we say they are irreducible. An irreducible matrix representation cannot be transformed to block diagonal form. A reducible matrix representation can be. $\underset{\sim}{C}$ of (14-3) is reducible because a similarity transformation exists that can bring it into block diagonal form, like C of (14-1). Note that a direct sum of matrices, as in (14-1), always gives us a reduced matrix.

A reducible matrix rep is transformable to block diagonal form

Two representations of a given group, such as C in (14-2) and $\underset{\sim}{C}$ in (14-3), are said to be equivalent representations if they are related by a similarity transformation, such as T in (14-3). Note the matrix representation may or may not be reducible, yet still have different, equivalent representations. To be equivalent means one can be transformed into the other. Equivalence of D and D' means there is some transformation T such that $D' = TDT^{-1}$, nothing more (i.e., no block diagonal necessarily implied.)

D and TDT^{-1} are equivalent reps

14.1.3 Invariant Subspaces

We first cite the formal mathematical definition of what is known as an invariant subspace, then examine the previous example, in light of it.

An invariant subspace W of an operator $\mathbf{C}$ has the property that all vectors $\mathbf{w}$ in W are transformed by $\mathbf{C}$ into vectors that are also in W.

$$\mathbf{w} \in W \quad \Rightarrow \quad \mathbf{Cw} \in W \tag{14-7}$$

The subset of vectors $\mathbf{v}_A = (v_1, v_2, v_3, 0, 0)^T$ in (14-2) comprises an invariant subspace (among all the vectors in the 5D vector space) under C because the action of C on any and all of these vectors $\mathbf{v}_A$ yields vectors that are still within the same subset of vectors. That is, the transformed vector is still of form $\mathbf{v}_A$; its last two components are still zero. Similarly for the subset of vectors $\mathbf{v}_B = (0, 0, 0, v_4, v_5)^T$ in (14-2). The two invariant subspaces behave independently under C.

Invariant subspace = vector space outside of which group action does not take a vector

By way of our prior example of 5D rotation, any vector initially in a 2D plane spanned by the basis vectors along the 4th and 5th axes in (14-2) would be rotated by $\mathbf{C}$ inside that plane. $\mathbf{C}$ could never rotate such a vector outside the plane. Even if we change coordinate axes via the transformation T of (14-3), that same physical vector would not be rotated by $\mathbf{C}$ outside of that original physical 2D plane. It may be rotated out of a coordinate plane defined by the x_4', x_5' axes in the new primed coordinate system, and have non-zero values in any or all of the five vector components in the new primed coordinates, but it would not be rotated out of the original 2D plane formed by the x_4, x_5 axes. That lack of ability of $\mathbf{C}$ to move original vectors out of the original plane indicates $\mathbf{C}$ has a subgroup that has its own independent action on vectors in that vector subspace.

Parallel logic applies, of course, to the other invariant subspace, the 3D volume spanned by the x_1, x_2, x_3 axes outside of which $\mathbf{C}$ would not rotate any vector originally inside that volume.

A trivial example of an invariant subspace is the vector $\mathbf{0}$, since any group operation on it will yield $\mathbf{0}$. Another trivial example is the 5D vector space above for real $\mathbf{v}$, $\mathbb{R}^5$, as any action of any (real) operator will also be in $\mathbb{R}^5$. We will generally be concerned with non-trivial invariant subspaces.

14.1.4 Definition of Reducible Representation

Bottom line, simple language: In cases of practical relevance to us, for groups represented as matrices, we can define a reducible matrix representation of a group as one for which a similarity transformation (a transformation such as T^{-1} of (14-3) or $\hat{T}$ in (14-6)) can result in block diagonal form (such as in (14-2) or (14-6)). An irreducible representation could not be transformed to such form.

Note that the same transformation must block diagonalize *all* members of the representation.

14.1.5 Things to Note

The 3D rotation group *SO*(3) representation of (2-11) in Chap. 2 is irreducible because any vector in the 3D space can be rotated into any other vector. Nothing in (2-11) of that chapter constricts any 3D vector to any particular 2D plane, i.e., any particular subspace. That group representation cannot be block diagonalized into submatrices, and the 3D vector space it acts on is an invariant subspace.

However, we could imagine another rotation matrix group that constricts rotation to a 2D plane, and such a matrix would be reducible to a 2D submatrix and a 1D submatrix along the diagonal. The associated 2D vector space would be an invariant subspace under the group within the 3D space. This matrix group would be a subgroup of the full *SO*(3) group discussed in the prior paragraph.

In a similar vein, the group **C** represented by (14-1) (with real components) is a subgroup of the full *SO*(5) group (which includes all possible rotations in 5D real space, as characterized herein).

Symbol ⊕ means two subgroup reps in resulting group rep

When we see the symbol ⊕ for direct summing two group representations together to form a larger group representation (as in (14-1)), we should recognize that this new larger group representation has two subgroup representations, the ones direct summed together via the ⊕ symbol. This is *not* a binary operation, as defined in Chap. 2 in Wholeness Chart 2-1, pg. 9. A binary operation occurs between members of a single set. The symbol ⊕ means we are adding (direct summing) two different sets (subgroup representations, really). It is an operation that combines group representations, not an operation between elements of a given group.

This is not a group binary operation, but a combination of group representations

14.1.6 Direct Products and Eigenvalues

<u>Outer Product of Vectors from Different Vector Spaces</u>

You should now review Sects. 2.2.8 and 2.2.9 of Chap. 2 and then consider the following update of what we mean by (2-34) and (2-35).

Reviewing part of Chap. 2 on outer products and operators

In those relations, **w** represents the color triplet part of a quark field and **y**, the LC weak doublet. In the vector space of fields, **w**⊗**y** is an outer, or tensor, product of the two fields (each of which is a vector in its own vector space).

The operator we called **B** could be any operator acting in the weak isospin vector space. It operates on the **y** doublet, but not the **w** triplet. If we like, we can take **B** to be the *SU*(2) subalgebra matrix generator σ_3, or more conventionally, $\frac{1}{2}\sigma_3$.

The operator **A** acts on the **w** color triplet, but not the **y** doublet. At one point in the referenced section, I suggested we could consider **A** as a color operator, whose eigenvalue could be the color of the particular field **w**. That was actually a liberty I took to make concepts easier to digest, but we now know there is no such operator, but instead the two *SU*(3) Cartan subalgebra operators λ_3 and λ_8, or more conventionally, $\frac{1}{2}\lambda_3$ and $\frac{1}{2}\lambda_8$. Combining the eigenvalues for those two operators allowed us to use the shortcut designation of color for each of the three components of the color triplet. So, if we like, we can consider **A** in this discussion as one of $\frac{1}{2}\lambda_3$ or $\frac{1}{2}\lambda_8$. We could also consider a second operator **A**′ as the other of those operators.

So, a given quark field has an LC doublet part (e.g., component LC *u* or LC *d*) and a color triplet part (*r, g,* or *b* component). The point we want to make here is that, more generally, for **w** and **y** in different vector spaces (which they are in this example), we will have separate eigenvalues from the *SU*(2) operator $\frac{1}{2}\sigma_3$ and from the *SU*(3) operators $\frac{1}{2}\lambda_3$ or $\frac{1}{2}\lambda_8$. More generally, **A** and **B** will (when vectors **w** and **y** are in eigenstates) give separate eigenvalues (with which we can label and designate the outer product field **v** = **w**⊗**y**.) That is, for **w** and **y** in eigenstates, with $\varepsilon_{\mathbf{A}}$ and $\varepsilon_{\mathbf{B}}$ representing their respective eigenvalues,

$$\begin{gathered}\mathbf{Aw}=\varepsilon_{\mathbf{A}}\mathbf{w} \qquad\qquad \mathbf{By}=\varepsilon_{\mathbf{B}}\mathbf{y}\\ \mathbf{Av}=\mathbf{A}(\mathbf{w}\otimes\mathbf{y})=(\mathbf{Aw})\otimes\mathbf{y}=\varepsilon_{\mathbf{A}}(\mathbf{w}\otimes\mathbf{y})=\varepsilon_{\mathbf{A}}\mathbf{v}\\ \mathbf{Bv}=\mathbf{B}(\mathbf{w}\otimes\mathbf{y})=\ \mathbf{w}\otimes\mathbf{By}\ =\varepsilon_{\mathbf{B}}(\mathbf{w}\otimes\mathbf{y})=\varepsilon_{\mathbf{B}}\mathbf{v}.\end{gathered} \tag{14-8}$$

For different vec spaces, an operator acts only on vec from its vec space

Eigenvalues do not add

Do **Problem 1** to express (14-8) in terms of the color and weak isospin Cartan subalgebra operators and their eigenvalues, where **v** is a green LC down quark.

Outer Product of Vectors from the Same Vector Space

The point we want to make in this Sect. 14.1.6 is that things are different when our outer product is of vectors from the same vector space, rather than different vector spaces, as in the case above. Consider two vectors $\mathbf{z}_1$ and $\mathbf{z}_2$ in the same space, upon which a Lie group operator $\mathbf{C}$ operates. Also, consider the two vectors are each in an eigenstate of $\mathbf{C}$, and we form an outer product from the two.

$$\mathbf{C}\mathbf{z}_1 = \varepsilon_{c1}\mathbf{z}_1 \qquad \mathbf{C}\mathbf{z}_2 = \varepsilon_{c2}\mathbf{z}_2 \qquad \mathbf{u} = \mathbf{z}_1 \otimes \mathbf{z}_2 \tag{14-9}$$

Then, we define, in similar fashion to the product rule of elementary calculus,

For same vec space, an operator acts on every outer product component

$$\begin{aligned}\mathbf{C}\mathbf{u} = \mathbf{C}(\mathbf{z}_1 \otimes \mathbf{z}_2) &\equiv (\mathbf{C}\mathbf{z}_1) \otimes \mathbf{z}_2 + \mathbf{z}_1 \otimes (\mathbf{C}\mathbf{z}_2) = \varepsilon_{c1}\mathbf{z}_1 \otimes \mathbf{z}_2 + \varepsilon_{c2}\mathbf{z}_1 \otimes \mathbf{z}_2 \\ &= (\varepsilon_{c1} + \varepsilon_{c2})\mathbf{z}_1 \otimes \mathbf{z}_2 = (\varepsilon_{c1} + \varepsilon_{c2})\mathbf{u}.\end{aligned} \tag{14-10}$$

For outer products of more than two vectors, the same procedure holds. For example,

$$\mathbf{C}(\mathbf{z}_1 \otimes \mathbf{z}_2 \otimes \mathbf{z}_3) = (\varepsilon_{c1} + \varepsilon_{c2} + \varepsilon_{c3})(\mathbf{z}_1 \otimes \mathbf{z}_2 \otimes \mathbf{z}_3). \tag{14-11}$$

and eigenvalues add

Bottom line: If our outer product is of vectors from the same vector space, and these vectors are in eigenstates of a particular Lie group operator, then the eigenvalue of the outer product equals the sum of the eigenvalues of the vectors making up the outer product.

14.1.7 Eigenstates of QFT Fields vs States

We have noted before (Wholeness Chart 2-13, pg. 58b, and footnote on pg. 329) that in QFT, we generally take our vector space as the space of quantum fields, which is a Hilbert space (for the fields) with operators such as the σ_i of $SU(2)$ and the λ_i of $SU(3)$, operating on those fields. Each such field is an outer product of an $SU(3)$ field, an $SU(2)$, and a $U(1)$ field, since the SM is $SU(3)$X$SU(2)$X$U(1)$.

However, we can, in another context, consider our vector space, instead, to be the Fock space of states. Then, we have operators such as the charge operators ${}_iQ$ of weak isospin (Wholeness Chart 9-1 at end of Chap. 9, pg. 341) and the ${}_{QCD}Q_i$ of color (Wholeness Chart 13-2 at end of Chap. 13, pg. 415) operating on those states, Note, for example, that the weak isospin charge operator ${}_3Q$ has the same eigenvalue when acting on a single LC up quark state $|u^L\rangle$, as the operator $\frac{1}{2}\sigma_3$ has when operating on an LC up quark field Ψ_u^L. The same thing holds for all operators on fields and their associated charge operators on states. As another example, ${}_{QCD}Q_8$ operating on a single blue quark state $|q_b\rangle$, has the same eigenvalue as $\frac{1}{2}\lambda_8$ operating on a blue quark field Ψ_{q_b}.

In QFT, vector space can be Hilbert space of fields or Fock space of states

Now, in all our work so far, our quantum fields have populated a Hilbert space, which has a single field for each independent "axis". Every field in this Hilbert space is an outer product, since the SM is $SU(3)$X$SU(2)$X$U(1)$, but we only have, for example, a single RC green down quark field. We haven't had a field with two or more quark fields in it, e.g., we haven't had one RC green down quark field and one RC anti-red, anti-down quark field together forming an outer product to make a composite field. That is, we haven't had outer products of an $SU(3)$ field with another $SU(3)$ field; or an $SU(2)$ field with another $SU(2)$ field. (We have worked with inner products of quantum fields of the same vector space dimension n, such as $\bar{\psi}\psi$, in our Lagrangian, Hamiltonian, and transition amplitudes, but that gives us a scalar, not a vector in a vector space.)

The Fock space of states is different from the Hilbert space of fields. There, we commonly have two or more particles in the same multiparticle state, e.g., a meson comprised of a quark and an antiquark. If we thought of the states as wavefunctions, the component particle wave functions would be multiplied together. (Recall NRQM wave function for the hydrogen atom where the electron and proton wave functions are multiplied together. The two particles are bound together, just as the quark and anti-quark are in a meson.) Eigenvalues, such as e/m or weak isospin charge, for such composite (multiparticle) states, are additive. For the e/m charge on a meson made of a u quark and a $\bar{d}$ quark we add the charges, 2/3 + 1/3 = 1 (where we drop the factor of e, as is common). This is effectively how we arrive at total eigenvalue for an outer product field, as in (14-10) and (14-11), so we can think of a multiparticle state as an outer product of single particle states.

In QFT, states are multiparticle, so use Fock space for states

In QFT, fields are single (not multiple), so use Hilbert space for fields

We summarize this section in Wholeness Chart 14-1 below.

Wholeness Chart 14-1. Our Use So Far of Operators and Eigenvalues for Fields and States

	Fields			**States**		
	Vec	**Oper**	**Eigenvalue**	**Vec**	**Oper**	**Eigenvalue**
General	As at right, but only for $\mathbf{z}_1$, not direct product vectors			$\mathbf{z}_1$	**C**	$\mathbf{C}\mathbf{z}_1 = \varepsilon_{c1}\mathbf{z}_1 \qquad \varepsilon_{c1}$ = eigenvalue
				$\mathbf{z}_1 \otimes \mathbf{z}_2$		$\mathbf{C}(\mathbf{z}_1 \otimes \mathbf{z}_2) = (\varepsilon_{c1}+\varepsilon_{c2})(\mathbf{z}_1 \otimes \mathbf{z}_2) \quad \varepsilon_{c1}+\varepsilon_{c2}$= eigval
$U(1)$	$\psi_{e\,up}$	Spin Σ_i	$\Sigma_i \psi_{e\,up} = \frac{1}{2}\psi_{e\,up}$ spin = ½	$\lvert e^-_{up}\rangle$	${}_{QED}\Sigma_i$	${}_{QED}\Sigma_i \lvert e^-_{up}\rangle = \frac{1}{2}\lvert e^-_{up}\rangle$ spin = ½
			(See Vol. 1, pgs. 93-100, 113)	$\lvert 3e^-_{up}\rangle$		${}_{QED}\Sigma_i \lvert 3e^-_{up}\rangle = 3/2\, \lvert 3e^-_{up}\rangle$ spin =3/2
		Charge	$q = -e\,I$ (we didn't study)	$\lvert 2e^-\rangle$	Q	$Q\lvert 2e^-\rangle = -2e\lvert 2e^-\rangle \quad q = -2e$
$SU(2)$	Ψ^L_e	$\frac{g}{2}\sigma_3$	$g\frac{\sigma_3}{2}\Psi^L_e = -g\frac{1}{2}\Psi^L_e \qquad I^W_3 = -\frac{1}{2}g$	$\lvert e^{-L}\rangle$	${}_3Q$	${}_3Q\lvert e^{-L}\rangle = -g\frac{1}{2}\lvert e^{-L}\rangle \quad I^W_3 = -\frac{1}{2}g$
$SU(3)$	Ψ_{ug}	$\frac{g_s}{2}\lambda_3$	$g_s\frac{\lambda_3}{2}\Psi_{ug} = -g_s\frac{1}{2}\Psi_{ug} \quad \varepsilon_3 = -\frac{1}{2}g_s$	$\lvert u_g\rangle$	${}_{QCD}Q_3$	${}_{QCD}Q_3\lvert u_g\rangle = -g_s\frac{1}{2}\lvert u_g\rangle \quad \varepsilon_3 = -\frac{1}{2}g_s$
		$\frac{g_s}{2}\lambda_8$	$g_s\frac{\lambda_8}{2}\Psi_{ug} = g_s\frac{1}{2\sqrt{3}}\Psi_{ug} \quad \varepsilon_8 = \frac{1}{2\sqrt{3}}g_s$		${}_{QCD}Q_8$	${}_{QCD}Q_8\lvert u_g\rangle = g_s\frac{1}{2\sqrt{3}}\lvert u_g\rangle \quad \varepsilon_8 = \frac{1}{2\sqrt{3}}g_s$
				$\lvert u_r u_g d_b\rangle$		${}_{QCD}Q_3\lvert u_r u_g d_b\rangle = 0\lvert u_r u_g d_b\rangle \quad \varepsilon_3 = 0$
						${}_{QCD}Q_8\lvert u_r u_g d_b\rangle = 0\lvert u_r u_g d_b\rangle \quad \varepsilon_8 = 0$
$SU(N_f)$ $N_f = 3$	Ψ_d	$\frac{1}{2}\lambda_3$	$\frac{1}{2}\lambda_3\Psi_d = -\frac{1}{2}\Psi_d \quad \varepsilon_3 = -\frac{1}{2}$	$\lvert d\rangle$	Q_{V3}	$Q_{V3}\lvert d\rangle = -\frac{1}{2}\lvert d\rangle \quad \varepsilon_3 = -\frac{1}{2}$
		$\frac{1}{2}\lambda_8$	$\frac{1}{2}\lambda_8\Psi_d = \frac{1}{2\sqrt{3}}\Psi_d \quad \varepsilon_8 = \frac{1}{2\sqrt{3}}$		Q_{V8}	$Q_{V8}\lvert d\rangle = \frac{1}{2\sqrt{3}}\lvert d\rangle \quad \varepsilon_8 = \frac{1}{2\sqrt{3}}$
				$\lvert uud\rangle$		$Q_{V3}\lvert uud\rangle = \left(\frac{1}{2}+\frac{1}{2}-\frac{1}{2}\right)\lvert uud\rangle \quad \varepsilon_3 = \frac{1}{2}$
						$Q_{V8}\lvert uud\rangle = \frac{1}{\sqrt{3}}\left(\frac{1}{2}+\frac{1}{2}+\frac{1}{2}\right)\lvert uud\rangle \quad \varepsilon_8 = \frac{3}{2\sqrt{3}}$

Note that, as covered in the last chapter, we can also work in a completely different vector space than the three of the SM, $SU(3)\text{X}SU(2)\text{X}U(1)$, and that is the $SU(N_f)$ flavor space. All of the above logic holds for that space, as well. The "flavor charge" eigenstates will have eigenvalues for each particle "flavor charge" and those will be additive for multiparticle states.

Where we are going with this: In this chapter, we will be examining composite states, i.e., hadrons, which comprise more than one elementary particle, and we want to be able to identify and distinguish between such states using appropriate eigenvalues, such as color, spin, electric charge, and as we will see, importantly, flavor.

14.2 Example from NRQM: Spin of Composites

14.2.1 A Review of Spin Addition

Overviewing spin sums for composite particles

In NRQM you almost certainly studied the addition of angular momentum, in particular, spin angular momentum. This involved things like Clebsch-Gordan coefficients, and it went something like this.

Take two or more particles of given spin, consider them bound together, and determine the possible spin states of the resultant composite particle. For example, the particles could be a quark and antiquark bound together as a meson. Or a neutron and a proton bound together to form a deuterium nucleus. Or three quarks forming a nucleon. Or many other combinations.

Our example: proton and neutron forming deuteron

By way of example, consider the second case above, the common textbook problem of a spin ½ neutron and a spin ½ proton forming a deuteron. As the dominant deuteron state has zero orbital angular momentum, we ignore the orbital contribution to angular momentum in what follows. Note that j typically symbolizes total angular momentum, which equals total spin s_T plus orbital angular momentum l, but here $l = 0$, so j for us will represent spin.

The combination of the neutron and proton can lead to a nucleus with total spin 1 (½ + ½) or zero (½ – ½). The spin 1 state could be aligned in the positive *z* direction (spin up), the negative *z* direction (spin down), or in the *x-y* plane (neither up nor down, but not zero).

We label these states with quantum numbers *m* (total spin in *z* direction) and *j* (representing absolute value of the total spin), and use the ket symbol $|j, m\rangle$, where for $j = 0$, $m = 0$, and for $j = 1$, $m = 1$, 0, or –1. This gives rise to four possible spin states, where we, here and throughout, ignore the spacetime dependent part of the wave function and focus only on the spin part,

$$|\,j,m\rangle \xrightarrow[\text{spin states}]{\text{possible}} |1,1\rangle \quad |1,0\rangle \quad |1,-1\rangle \quad |0,0\rangle . \tag{14-12}$$

Possible deuteron spin eigenstates

(14-12) represents specific spin states and are eigenstates of the spin operator in the *z* direction and the spin total absolute value operator. The eigenvalues are represented by *j* and *m*, so, we can use these numbers as quantum numbers to label each specific spin state. We note in passing that the $|0,0\rangle$ state is unstable and only exists in nature in a virtual state, but that is irrelevant to our analysis here.

Prior to measurement, the general state of the nucleus is not an eigenstate of spin, as the spin is unknown (and before measurement, unknowable). Upon measurement, it collapses into a particular eigenstate. We will not concern ourselves more with this, however, as we will only be focusing here on deducing the various possible spin eigenstates (14-12).

Note that we started with two particles, the spin of each being represented via a two-component spinor (vector in a 2D vector space) having two basis spinor vectors. So, for each particle, we are working in a 2D vector space (upon which *SU*(2) spin operators operate). When we combine the two particles into a composite, we end up with four basis vectors, as in (14-12), and thus, we are working in a 4D vector space (upon which certain operators operate). This raises the question of whether group theory might be helpful in all of this. It is, as we begin to show in the next section.

2D spin spaces of neutron and proton yield 4D spin space for deuteron

14.2.2 Group Representation Theory and Spin Addition

As we know from NRQM, spin for a fermion can be described via a two-component vector, known as a spinor (NRQM type), and the operators for the spin vector components in real 3D physical space can be represented via the three Pauli matrices operating on that 2D, generally complex, spinor space. (Recall from Chap. 2, the homomorphism between *SO*(3) and *SU*(2). Each can characterize the same physical phenomenon.)

We start by representing the spin eigenstates of the neutron and proton in 2D spinor space (RHS of each relation in (14-13) below), where here $j = ½$ and $m = +½$, $–½$ connect us with the notation of the prior section.

$$|\,j,m\rangle \;\rightarrow\; \left|\tfrac{1}{2},\tfrac{1}{2}\right\rangle_n = \begin{bmatrix}1\\0\end{bmatrix}_n \quad \left|\tfrac{1}{2},-\tfrac{1}{2}\right\rangle_n = \begin{bmatrix}0\\1\end{bmatrix}_n \quad \left|\tfrac{1}{2},\tfrac{1}{2}\right\rangle_p = \begin{bmatrix}1\\0\end{bmatrix}_p \quad \left|\tfrac{1}{2},-\tfrac{1}{2}\right\rangle_p = \begin{bmatrix}0\\1\end{bmatrix}_p . \tag{14-13}$$

Spin eigenstates of neutron and proton separately

A general state for either is then

$$\begin{bmatrix}C_{+n}\\C_{-n}\end{bmatrix} = C_{+n}\begin{bmatrix}1\\0\end{bmatrix}_n + C_{-n}\begin{bmatrix}0\\1\end{bmatrix}_n \qquad \begin{bmatrix}C_{+p}\\C_{-p}\end{bmatrix} = C_{+p}\begin{bmatrix}1\\0\end{bmatrix}_p + C_{-p}\begin{bmatrix}0\\1\end{bmatrix}_p . \tag{14-14}$$

General spin state for either particle is a sum of eigenstates

Now, as we learned in NRQM, a composite state wave function is obtained by outer multiplying the individual wave functions. That gives us

$$\begin{aligned}\begin{bmatrix}C_{+n}\\C_{-n}\end{bmatrix}\begin{bmatrix}C_{+p}\\C_{-p}\end{bmatrix} &= \underbrace{C_{+n}C_{+p}}_{C_{++}}\begin{bmatrix}1\\0\end{bmatrix}_n\begin{bmatrix}1\\0\end{bmatrix}_p + \underbrace{C_{+n}C_{-p}}_{C_{+-}}\begin{bmatrix}1\\0\end{bmatrix}_n\begin{bmatrix}0\\1\end{bmatrix}_p \\ &\quad + \underbrace{C_{-n}C_{+p}}_{C_{-+}}\begin{bmatrix}0\\1\end{bmatrix}_n\begin{bmatrix}1\\0\end{bmatrix}_p + \underbrace{C_{-n}C_{-p}}_{C_{--}}\begin{bmatrix}0\\1\end{bmatrix}_n\begin{bmatrix}0\\1\end{bmatrix}_p ,\end{aligned} \tag{14-15}$$

Combining general spin state parts of the two wave functions

where we streamline by introducing new notation for the constants, and note that we will have to normalize (14-15) to give a total probability of one.

The point is that we now have four basis vectors

$$\begin{bmatrix}1\\0\end{bmatrix}_n\begin{bmatrix}1\\0\end{bmatrix}_p \quad \begin{bmatrix}1\\0\end{bmatrix}_n\begin{bmatrix}0\\1\end{bmatrix}_p \quad \begin{bmatrix}0\\1\end{bmatrix}_n\begin{bmatrix}1\\0\end{bmatrix}_p \quad \begin{bmatrix}0\\1\end{bmatrix}_n\begin{bmatrix}0\\1\end{bmatrix}_p, \qquad (14\text{-}16)$$

Basis vectors for the combined state

and thus, with the composite particle, we are working in a 4D complex vector space of states.

This may be starting to look a little familiar. It is, what, in group theory terminology, we call an outer (or tensor) product. We have outer multiplied a 2D complex vector space by another 2D complex vector space and gotten a 4D complex vector space. And here the group operator on the vector space is the *SU*(2) spin operator (with three spin operator generators corresponding to the three directions in physical space). In representation theory notation, we write this as

Above combination is outer product of 2D spinors

$$SU(2)\otimes SU(2),\text{ or via shorthand notation as } 2\otimes 2. \qquad (14\text{-}17)$$

We can write the vectors as two component tensor products with two indices, each having values of + or – (or alternatively, 1 or 2).

$$\begin{aligned}&\begin{bmatrix}1\\0\end{bmatrix}_n\otimes\begin{bmatrix}1\\0\end{bmatrix}_p=\begin{bmatrix}1\\0\end{bmatrix}_n\begin{bmatrix}1\\0\end{bmatrix}_p=v_{++}=v_{11} \qquad \begin{bmatrix}1\\0\end{bmatrix}_n\otimes\begin{bmatrix}0\\1\end{bmatrix}_p=\begin{bmatrix}1\\0\end{bmatrix}_n\begin{bmatrix}0\\1\end{bmatrix}_p=v_{+-}=v_{12}\\ &\begin{bmatrix}0\\1\end{bmatrix}_n\otimes\begin{bmatrix}1\\0\end{bmatrix}_p=\begin{bmatrix}0\\1\end{bmatrix}_n\begin{bmatrix}1\\0\end{bmatrix}_p=v_{-+}=v_{21} \qquad \begin{bmatrix}0\\1\end{bmatrix}_n\otimes\begin{bmatrix}0\\1\end{bmatrix}_p=\begin{bmatrix}0\\1\end{bmatrix}_n\begin{bmatrix}0\\1\end{bmatrix}_p=v_{--}=v_{22},\end{aligned} \qquad (14\text{-}18)$$

New symbols for the four basis vectors

where *r* is the index for the neutron and *s* is the index for the proton,

$$v_{rs}=v_r\otimes v_s \qquad r=1,2 \qquad s=1,2. \qquad (14\text{-}19)$$

Alternatively, as will be important in what is to follow, we can designate these four basis vectors with one index, call it *t*, having values 1,2,3,4. Either way, we have four basis vectors and are working in a 4D complex vector space (upon which certain operators can act). Thus, where we use the symbol w_t to distinguish our four-component, single index notation from our prior notation with two indices,

$$w_t=\begin{bmatrix}w_1\\w_2\\w_3\\w_4\end{bmatrix}=\begin{bmatrix}v_{++}\\v_{+-}\\v_{-+}\\v_{--}\end{bmatrix}=\begin{bmatrix}v_{11}\\v_{12}\\v_{21}\\v_{22}\end{bmatrix}. \qquad (14\text{-}20)$$

Can represent four basis vectors as components in 4D vector in 4D space

Under (14-17), our spin operators $\frac{1}{2}\sigma_i$ for each of the neutron and proton act on the product space vector. In accord with (14-8), the total spin in the *z* direction can be found via

$$\begin{aligned}\left(\tfrac{1}{2}{}_n\sigma_3+\tfrac{1}{2}{}_p\sigma_3\right)v_{11}&=\tfrac{1}{2}\left(\begin{bmatrix}1&\\&-1\end{bmatrix}_n\begin{bmatrix}1\\0\end{bmatrix}_n\right)\begin{bmatrix}1\\0\end{bmatrix}_p+\tfrac{1}{2}\begin{bmatrix}1\\0\end{bmatrix}_n\left(\begin{bmatrix}1&\\&-1\end{bmatrix}_p\begin{bmatrix}1\\0\end{bmatrix}_p\right)\\&=\tfrac{1}{2}\begin{bmatrix}1\\0\end{bmatrix}_n\begin{bmatrix}1\\0\end{bmatrix}_p+\tfrac{1}{2}\begin{bmatrix}1\\0\end{bmatrix}_n\begin{bmatrix}1\\0\end{bmatrix}_p=\left(\tfrac{1}{2}+\tfrac{1}{2}\right)\begin{bmatrix}1\\0\end{bmatrix}_n\begin{bmatrix}1\\0\end{bmatrix}_p=\begin{bmatrix}1\\0\end{bmatrix}_n\begin{bmatrix}1\\0\end{bmatrix}_p.\end{aligned} \qquad (14\text{-}21)$$

Summing neutron and proton spins in z direction for first basis vector

The *z* direction spin eigenvalue for the v_{11} composite state is +1, as we had for the *m* value in the first composite state of (14-12).

Similarly,

$$\begin{aligned}\left(\tfrac{1}{2}{}_n\sigma_3+\tfrac{1}{2}{}_p\sigma_3\right)v_{12}&=\tfrac{1}{2}\left(\begin{bmatrix}1&\\&-1\end{bmatrix}_n\begin{bmatrix}1\\0\end{bmatrix}_n\right)\begin{bmatrix}0\\1\end{bmatrix}_p+\tfrac{1}{2}\begin{bmatrix}1\\0\end{bmatrix}_n\left(\begin{bmatrix}1&\\&-1\end{bmatrix}_p\begin{bmatrix}0\\1\end{bmatrix}_p\right)\\&=\tfrac{1}{2}\begin{bmatrix}1\\0\end{bmatrix}_n\begin{bmatrix}0\\1\end{bmatrix}_p-\tfrac{1}{2}\begin{bmatrix}1\\0\end{bmatrix}_n\begin{bmatrix}0\\1\end{bmatrix}_p=\left(\tfrac{1}{2}-\tfrac{1}{2}\right)\begin{bmatrix}1\\0\end{bmatrix}_n\begin{bmatrix}0\\1\end{bmatrix}_p=0,\end{aligned} \qquad (14\text{-}22)$$

And for the second basis vector

and we have zero spin in the *z* direction, as in the second and last states of (14-12), where $m = 0$. Similarly, the *z* direction spin values for v_{21} and v_{22} are $m = 0$ and -1, respectively.

Results for third and fourth basis vectors

Do **Problem 2** to prove the last statement.

It is then shown in most standard quantum mechanics texts, in a fairly lengthy analysis (which we can't spare the time to go into), that for the operator S_T^2, which is the square of the total spin, i.e.,

Can do same for total spin squared S_T^2, but two of basis vectors not eigenstates

$$S_{iT} = \frac{1}{2}{}_n\sigma_i + \frac{1}{2}{}_p\sigma_i \qquad S_T^2 = \left(\frac{1}{2}{}_n\sigma_i + \frac{1}{2}{}_p\sigma_i\right)\left(\frac{1}{2}{}_n\sigma_i + \frac{1}{2}{}_p\sigma_i\right), \tag{14-23}$$

the v_{11} and v_{22} are eigenstates, but the v_{12} and v_{21} are not. We can, however, obtain four product space basis states that are each eigenstates of both S_T^2 and ½σ_3, by transforming (14-16) (and concomitantly, (14-20)), as follows.

Transform to new basis, where each basis vector an eigenstate of both operators

$$\begin{gathered} \hat{v}_{11} = v_{11} = \begin{bmatrix}1\\0\end{bmatrix}_n\begin{bmatrix}1\\0\end{bmatrix}_p \qquad \hat{v}_{12} = \frac{1}{\sqrt{2}}\left(v_{12}+v_{21}\right) = \frac{1}{\sqrt{2}}\begin{bmatrix}1\\0\end{bmatrix}_n\begin{bmatrix}0\\1\end{bmatrix}_p + \frac{1}{\sqrt{2}}\begin{bmatrix}0\\1\end{bmatrix}_n\begin{bmatrix}1\\0\end{bmatrix}_p \\ \hat{v}_{21} = \frac{1}{\sqrt{2}}\left(v_{12}-v_{21}\right) = \frac{1}{\sqrt{2}}\begin{bmatrix}1\\0\end{bmatrix}_n\begin{bmatrix}0\\1\end{bmatrix}_p - \frac{1}{\sqrt{2}}\begin{bmatrix}0\\1\end{bmatrix}_n\begin{bmatrix}1\\0\end{bmatrix}_p \qquad \hat{v}_{22} = v_{22} = \begin{bmatrix}0\\1\end{bmatrix}_n\begin{bmatrix}0\\1\end{bmatrix}_p \end{gathered} \tag{14-24}$$

Do **Problem 3** to show (14-24) as a 4D matrix transformation like T in (14-4) is for 5D.

The new basis vectors (with "hats" over them) of (14-24) correspond to (14-12), where

$$|\,j,m\rangle \xrightarrow[\text{spin states}]{\text{possible}} \hat{v}_{11} = |1,1\rangle \qquad \hat{v}_{12} = |1,0\rangle \qquad \hat{v}_{21} = |0,0\rangle \qquad \hat{v}_{22} = |1,-1\rangle, \tag{14-25}$$

j,m eigenvalues of the two operators for the new basis

each having definite z direction spin m and definite total spin j (i.e., are eigenstates of S_T^2 and ½σ_3, though we haven't shown that directly).

The new basis vectors (with "hats") (14-24) are linear combinations of the old ones (no "hats"). For examples, from the first line of (14-24),

New basis vectors are linear combinations of old basis vectors

$$\begin{gathered} |1,1\rangle = \hat{v}_{11} = C_{\hat{1}\hat{1}11}v_{11} + C_{\hat{1}\hat{1}12}v_{12} + C_{\hat{1}\hat{1}21}v_{21} + C_{\hat{1}\hat{1}22}v_{22} = v_{11} = \begin{bmatrix}1\\0\end{bmatrix}_n\begin{bmatrix}1\\0\end{bmatrix}_p \\ C_{\hat{1}\hat{1}11} = 1 \qquad C_{\hat{1}\hat{1}12} = C_{\hat{1}\hat{1}21} = C_{\hat{1}\hat{1}22} = 0 \end{gathered} \tag{14-26}$$

$$\begin{gathered} |1,0\rangle = \hat{v}_{12} = C_{\hat{1}\hat{2}11}v_{11} + C_{\hat{1}\hat{2}12}v_{12} + C_{\hat{1}\hat{2}21}v_{21} + C_{\hat{1}\hat{2}22}v_{22} = \frac{1}{\sqrt{2}}\underbrace{\begin{bmatrix}1\\0\end{bmatrix}_n\begin{bmatrix}0\\1\end{bmatrix}_p}_{v_{12}} + \frac{1}{\sqrt{2}}\underbrace{\begin{bmatrix}0\\1\end{bmatrix}_n\begin{bmatrix}1\\0\end{bmatrix}_p}_{v_{21}} \\ C_{\hat{1}\hat{2}11} = C_{\hat{1}\hat{2}22} = 0 \qquad C_{\hat{1}\hat{2}12} = C_{\hat{1}\hat{2}21} = \frac{1}{\sqrt{2}}, \end{gathered} \tag{14-27}$$

Coefficients in linear combinations are called Clebsch-Gordan coefficients

where the $C_{\hat{i}\hat{j}ij}$ called Clebsch-Gordan coefficients. They essentially relate the new basis vectors to the original ones. Commonly, most of them are zero, as in (14-26) and (14-27).

Do **Problem 4** to find the Clebsch-Gordan coefficients for the second row of (14-24).

14.2.3 Subspaces and Irreducible Representations for Spin Addition

Now consider what happens when we operate on our new basis states for our composite particle (deuteron) with our $SU(2)\otimes SU(2)$ matrix operators other than ½ ${}_n\sigma_3$ and ½ ${}_p\sigma_3$. From (14-23) and (14-24),

Spin operators ½σ_i acting on the new basis vectors

$$\begin{aligned} S_{1T}\hat{v}_{11} &= \left(\frac{1}{2}{}_n\sigma_1 + \frac{1}{2}{}_p\sigma_1\right)\hat{v}_{11} = \frac{1}{2}\left(\begin{bmatrix} & 1\\1 & \end{bmatrix}_n\begin{bmatrix}1\\0\end{bmatrix}_n\right)\begin{bmatrix}1\\0\end{bmatrix}_p + \frac{1}{2}\begin{bmatrix}1\\0\end{bmatrix}_n\left(\begin{bmatrix} & 1\\1 & \end{bmatrix}_p\begin{bmatrix}1\\0\end{bmatrix}_p\right) \\ &= \frac{1}{2}\begin{bmatrix}0\\1\end{bmatrix}_n\begin{bmatrix}1\\0\end{bmatrix}_p + \frac{1}{2}\begin{bmatrix}1\\0\end{bmatrix}_n\begin{bmatrix}0\\1\end{bmatrix}_p = \frac{1}{2}v_{21} + \frac{1}{2}v_{12} = \frac{1}{\sqrt{2}}\hat{v}_{12}. \end{aligned} \tag{14-28}$$

The x direction spin operator S_{1T} changes $\hat{v}_{11}$ into $\hat{v}_{12}$ (times $\frac{1}{\sqrt{2}}$).

$$
\begin{aligned}
\left(\tfrac{1}{2}{}_n\sigma_1+\tfrac{1}{2}{}_p\sigma_1\right)\hat{v}_{12} &= \left(\tfrac{1}{2}{}_n\sigma_1+\tfrac{1}{2}{}_p\sigma_1\right)\tfrac{1}{\sqrt{2}}\left(v_{12}+v_{21}\right) \\
&= \tfrac{1}{2}{}_n\sigma_1\left(\tfrac{1}{\sqrt{2}}\begin{bmatrix}1\\0\end{bmatrix}_n\begin{bmatrix}0\\1\end{bmatrix}_p+\tfrac{1}{\sqrt{2}}\begin{bmatrix}0\\1\end{bmatrix}_n\begin{bmatrix}1\\0\end{bmatrix}_p\right)+\tfrac{1}{2}{}_p\sigma_1\left(\tfrac{1}{\sqrt{2}}\begin{bmatrix}1\\0\end{bmatrix}_n\begin{bmatrix}0\\1\end{bmatrix}_p+\tfrac{1}{\sqrt{2}}\begin{bmatrix}0\\1\end{bmatrix}_n\begin{bmatrix}1\\0\end{bmatrix}_p\right) \\
&= \tfrac{1}{2}\tfrac{1}{\sqrt{2}}\left(\begin{bmatrix} & 1\\1 & \end{bmatrix}_n\begin{bmatrix}1\\0\end{bmatrix}_n\right)\begin{bmatrix}0\\1\end{bmatrix}_p+\tfrac{1}{2}\tfrac{1}{\sqrt{2}}\left(\begin{bmatrix} & 1\\1 & \end{bmatrix}_n\begin{bmatrix}0\\1\end{bmatrix}_n\right)\begin{bmatrix}1\\0\end{bmatrix}_p+\tfrac{1}{2}\tfrac{1}{\sqrt{2}}\begin{bmatrix}1\\0\end{bmatrix}_n\left(\begin{bmatrix} & 1\\1 & \end{bmatrix}_p\begin{bmatrix}0\\1\end{bmatrix}_p\right) \\
&\qquad +\tfrac{1}{2}\tfrac{1}{\sqrt{2}}\begin{bmatrix}0\\1\end{bmatrix}_n\left(\begin{bmatrix} & 1\\1 & \end{bmatrix}_p\begin{bmatrix}1\\0\end{bmatrix}_p\right) \\
&= \tfrac{1}{2}\tfrac{1}{\sqrt{2}}\begin{bmatrix}0\\1\end{bmatrix}_n\begin{bmatrix}0\\1\end{bmatrix}_p+\tfrac{1}{2}\tfrac{1}{\sqrt{2}}\begin{bmatrix}1\\0\end{bmatrix}_n\begin{bmatrix}1\\0\end{bmatrix}_p+\tfrac{1}{2}\tfrac{1}{\sqrt{2}}\begin{bmatrix}1\\0\end{bmatrix}_n\begin{bmatrix}1\\0\end{bmatrix}_p+\tfrac{1}{2}\tfrac{1}{\sqrt{2}}\begin{bmatrix}0\\1\end{bmatrix}_n\begin{bmatrix}0\\1\end{bmatrix}_p \\
&= \tfrac{1}{\sqrt{2}}\begin{bmatrix}0\\1\end{bmatrix}_n\begin{bmatrix}0\\1\end{bmatrix}_p+\tfrac{1}{\sqrt{2}}\begin{bmatrix}1\\0\end{bmatrix}_n\begin{bmatrix}1\\0\end{bmatrix}_p=\tfrac{1}{\sqrt{2}}v_{11}+\tfrac{1}{\sqrt{2}}v_{22}=\tfrac{1}{\sqrt{2}}\hat{v}_{11}+\tfrac{1}{\sqrt{2}}\hat{v}_{22}.
\end{aligned}
\tag{14-29}
$$

The x direction spin operators change $\hat{v}_{12}$ into a linear combination of $\hat{v}_{11}$ and $\hat{v}_{22}$.

$$
\begin{aligned}
\left(\tfrac{1}{2}{}_n\sigma_1+\tfrac{1}{2}{}_p\sigma_1\right)\hat{v}_{21} &= \left(\tfrac{1}{2}{}_n\sigma_1+\tfrac{1}{2}{}_p\sigma_1\right)\tfrac{1}{\sqrt{2}}\left(v_{12}-v_{21}\right) \\
&= \tfrac{1}{2}{}_n\sigma_1\left(\tfrac{1}{\sqrt{2}}\begin{bmatrix}1\\0\end{bmatrix}_n\begin{bmatrix}0\\1\end{bmatrix}_p-\tfrac{1}{\sqrt{2}}\begin{bmatrix}0\\1\end{bmatrix}_n\begin{bmatrix}1\\0\end{bmatrix}_p\right)+\tfrac{1}{2}{}_p\sigma_1\left(\tfrac{1}{\sqrt{2}}\begin{bmatrix}1\\0\end{bmatrix}_n\begin{bmatrix}0\\1\end{bmatrix}_p-\tfrac{1}{\sqrt{2}}\begin{bmatrix}0\\1\end{bmatrix}_n\begin{bmatrix}1\\0\end{bmatrix}_p\right) \\
&= \tfrac{1}{2}\tfrac{1}{\sqrt{2}}\left(\begin{bmatrix} & 1\\1 & \end{bmatrix}_n\begin{bmatrix}1\\0\end{bmatrix}_n\right)\begin{bmatrix}0\\1\end{bmatrix}_p-\tfrac{1}{2}\tfrac{1}{\sqrt{2}}\left(\begin{bmatrix} & 1\\1 & \end{bmatrix}_n\begin{bmatrix}0\\1\end{bmatrix}_n\right)\begin{bmatrix}1\\0\end{bmatrix}_p+\tfrac{1}{2}\tfrac{1}{\sqrt{2}}\begin{bmatrix}1\\0\end{bmatrix}_n\left(\begin{bmatrix} & 1\\1 & \end{bmatrix}_p\begin{bmatrix}0\\1\end{bmatrix}_p\right) \\
&\qquad -\tfrac{1}{2}\tfrac{1}{\sqrt{2}}\begin{bmatrix}0\\1\end{bmatrix}_n\left(\begin{bmatrix} & 1\\1 & \end{bmatrix}_p\begin{bmatrix}1\\0\end{bmatrix}_p\right)
\end{aligned}
\tag{14-30a}
$$

$$
=\tfrac{1}{2}\tfrac{1}{\sqrt{2}}\begin{bmatrix}0\\1\end{bmatrix}_n\begin{bmatrix}0\\1\end{bmatrix}_p-\tfrac{1}{2}\tfrac{1}{\sqrt{2}}\begin{bmatrix}1\\0\end{bmatrix}_n\begin{bmatrix}1\\0\end{bmatrix}_p+\tfrac{1}{2}\tfrac{1}{\sqrt{2}}\begin{bmatrix}1\\0\end{bmatrix}_n\begin{bmatrix}1\\0\end{bmatrix}_p-\tfrac{1}{2}\tfrac{1}{\sqrt{2}}\begin{bmatrix}0\\1\end{bmatrix}_n\begin{bmatrix}0\\1\end{bmatrix}_p=0. \tag{14-30b}
$$

The x direction spin operators change $\hat{v}_{21}$ into 0.

Do **Problem 5** to show the x direction spin operators change $\hat{v}_{22}$ into $\hat{v}_{12}$ (times a constant).

Operating on any of 3 basis vectors yields a linear combo of those same 3 basis vectors

From (14-28) to (14-30), along with Problem 5, we see that the x direction spin operators in our $SU(2)\otimes SU(2)$ outer product space "rotate" $\hat{v}_{11}$, $\hat{v}_{12}$, and $\hat{v}_{22}$ into vectors that are linear combinations of those three vectors, but do not do that with $\hat{v}_{21}$. In fact, they destroy that vector.

In similar fashion, which we won't go to the trouble to show, the y direction spin operators $\frac{1}{2}\,{}_n\sigma_2$ and $\frac{1}{2}\,{}_p\sigma_2$ will "rotate" $\hat{v}_{11}$, $\hat{v}_{12}$, and $\hat{v}_{22}$ into vectors that are linear combinations of those three vectors, but do not do that with $\hat{v}_{21}$. They turn $\hat{v}_{21}$ to zero.

Further, the z direction spin operators $\frac{1}{2}\,{}_n\sigma_3$ and $\frac{1}{2}\,{}_p\sigma_3$ acting on $\hat{v}_{11}$ yield $\hat{v}_{11}$; acting on $\hat{v}_{22}$ yield $-\hat{v}_{22}$; acting on $\hat{v}_{12}$ yield zero; and acting on $\hat{v}_{21}$ yield zero.

These 3 basis vectors span an invariant subspace

What is this telling us? Well, since those operators "rotate" complex vectors within a vector space, and they are a complete set of such "rotation" operators for our case, this tells us that the three basis vectors, namely $\hat{v}_{11}$, $\hat{v}_{12}$, and $\hat{v}_{22}$, when acted on by any combination of these spin operators, will leave us with a vector that lies within the space spanned by those same three vectors. That is, those three vectors form an invariant subspace (complex) of dimension 3. The group operations on that subspace comprise a 3D subgroup.

4th basis vector cannot turn into a combo of the other 3, and vice versa

And the vector $\hat{v}_{21}$, when acted on by any combination of the three spin operators, will yield zero. In other words, there is no way we can "rotate" $\hat{v}_{21}$ into the 3D subspace. And there is no way we can "rotate" any combination of $\hat{v}_{11}$, $\hat{v}_{12}$, and $\hat{v}_{22}$ into $\hat{v}_{21}$. $\hat{v}_{21}$ alone forms its own 1D invariant subspace. $\hat{v}_{11}$, $\hat{v}_{12}$, and $\hat{v}_{22}$ form a triplet and $\hat{v}_{21}$ forms a singlet.

Explicitly, where the $\underset{\sim}{c}_{ij}$ represent a 4D transformation,

Showing the 4D matrix rep reduced to irreducible matrices

$$\begin{bmatrix} v'_{11} \\ v'_{12} \\ v'_{21} \\ v'_{22} \end{bmatrix} = \begin{bmatrix} \underset{\sim}{c}_{11} & \underset{\sim}{c}_{12} & \underset{\sim}{c}_{13} & \underset{\sim}{c}_{14} \\ \underset{\sim}{c}_{21} & \underset{\sim}{c}_{22} & \underset{\sim}{c}_{23} & \underset{\sim}{c}_{24} \\ \underset{\sim}{c}_{31} & \underset{\sim}{c}_{32} & \underset{\sim}{c}_{33} & \underset{\sim}{c}_{34} \\ \underset{\sim}{c}_{41} & \underset{\sim}{c}_{42} & \underset{\sim}{c}_{43} & \underset{\sim}{c}_{44} \end{bmatrix} \begin{bmatrix} v_{11} \\ v_{12} \\ v_{21} \\ v_{22} \end{bmatrix} \xrightarrow[\text{other basis}]{\text{transform to}} \begin{bmatrix} \hat{v}'_{11} \\ \hat{v}'_{12} \\ \hat{v}'_{22} \\ \hat{v}'_{21} \end{bmatrix} = \begin{bmatrix} a_{11} & a_{12} & a_{13} & \\ a_{21} & a_{22} & a_{23} & \\ a_{31} & a_{32} & a_{33} & \\ & & & b_{11} \end{bmatrix} \begin{bmatrix} \hat{v}_{11} \\ \hat{v}_{12} \\ \hat{v}_{22} \\ \hat{v}_{21} \end{bmatrix}. \tag{14-31}$$

b_{11} is a complex number, so the transformation of $\hat{v}_{21}$ to $\hat{v}'_{21}$ is a 1D complex transformation. The a_{ij} represent a 3D matrix group transformation.

SU(2)⊗SU(2)direct product yields direct sum SU(3)⊕ U(1)

The invariant subspaces on the RHS of (14-31) were not evident before we transformed to the new basis (14-24). What we have shown, in fact, is that a direct product $SU(2)\otimes SU(2)$ of matrix representations is comprised of a direct sum of a 3D group rep and a 1D group rep. In (14-31), we reduced a reducible complex 4D group rep to irreducible 3D and 1D reps. This is generally expressed as

Symbolized like this

$$2\otimes 2 = 3\oplus 1\,. \tag{14-32}$$

This result good for any SU(2) group, not just spin operators

Although (14-32) was derived for the case of physical spins on physical particles, it is generally valid in a pure mathematical sense and is applicable to any case where we have a direct product of an *SU*(2) group representation with another *SU*(2) group representation.

Getting back to the physical interpretation, the new basis (with "hats") is typically expressed with different symbols,

Converting back to prior notation

$$\begin{bmatrix} \hat{v}_{11} \\ \hat{v}_{12} \\ \hat{v}_{22} \\ \hat{v}_{21} \end{bmatrix} = \begin{bmatrix} \zeta_{11} \\ \zeta_{10} \\ \zeta_{1-1} \\ \zeta_{00} \end{bmatrix} \qquad \zeta_{jm} = |\,j,m\rangle \qquad \zeta_{11} = |1,1\rangle \quad \zeta_{10} = |1,0\rangle \quad \zeta_{1-1} = |1,-1\rangle \quad \zeta_{00} = |0,0\rangle, \tag{14-33}$$

where j represents total spin and m the spin in the z direction.

For the 3D group rep, we could consider, instead of the total spin sum operators above, the particular *SU*(3) Cartan subalgebra generators $\frac{1}{2}\lambda_3$ and $\frac{1}{2}\lambda_8$, and assign the eigenvalues of those generators to the three parts of the triplet. By doing so, we could label ζ_{11}, ζ_{10}, and ζ_{1-1} with the ε_3 and ε_8 eigenvalues instead of the j and m values. And then we would have a graph much like the figure in Wholeness Chart 13-2, pg. 415, where ζ_{11}, ζ_{10}, and ζ_{1-1} take the places, respectively, of the R, G, and B quarks.

14.2.4 *Summary of SU*(2)⊗*SU*(2)

Summarizing SU(2)⊗SU(2)

When we outer product two 2D complex vector spaces, which are each acted upon by an *SU*(2) group matrix representation, we end up with a 4D complex space having two invariant subspaces, one of 3D and one of 1D, upon which two irreducible representations act. Each basis vector, in each of these subspaces, is a linear combination of the original four basis vectors that resulted from the outer product. In each such linear combination, the coefficient in front of each such original basis vector is known as a Clebsch-Gordan coefficient. These coefficients can be calculated somewhat laboriously, and we did not do that here. Tables[1] exist for them, and for all possible *SU*(*n*) direct product spaces you will run into in physics.

In the present case, we can label the 3D and 1D subspace basis vectors in either of two ways, each of which employs two quantum numbers. In one of these, they are m, which, physically, represents the total z direction spin, and j, which, physically, represents the total overall spin magnitude. In the other of these, we can use the *SU*(3) operator quantum numbers ε_3 and ε_8, which are the eigenvalues of the Cartan subalgebra generators $\frac{1}{2}\lambda_3$ and $\frac{1}{2}\lambda_8$. In that case, we have a triplet (3D space) and a singlet (1 space), where the singlet has $\varepsilon_3 = \varepsilon_8 = 0$.

Wholeness Chart 14-2 summarizes this.

---

[1] Try the Particle Data Group website, https://pdg.lbl.gov/, or Wikipedia.

Wholeness Chart 14-2. Summary of Matrix Representations of $SU(2) \otimes SU(2)$

SU(2) Bases	Original Basis $SU(2) \otimes SU(2)$		New Basis $SU(2) \otimes SU(2)$		Irreducible Representations and Independent Subspaces
$\left\vert\frac{1}{2},\frac{1}{2}\right\rangle_n = \begin{bmatrix}1\\0\end{bmatrix}_n$	$\begin{bmatrix}1\\0\end{bmatrix}_n\begin{bmatrix}1\\0\end{bmatrix}_p = v_{11}$		$\hat{v}_{11} = v_{11} \qquad = \zeta_{11} = \vert 1,1\rangle$	eigenstates of total spin and *z* direc spin	$\begin{bmatrix}\begin{bmatrix} & & \\ & 3D & \\ & & \end{bmatrix} & \\ & [1D]\end{bmatrix}\begin{bmatrix}\zeta_{11}\\ \zeta_{10}\\ \zeta_{1-1}\\ \zeta_{00}\end{bmatrix}$
$\left\vert\frac{1}{2},-\frac{1}{2}\right\rangle_n = \begin{bmatrix}0\\1\end{bmatrix}_n$	$\begin{bmatrix}1\\0\end{bmatrix}_n\begin{bmatrix}0\\1\end{bmatrix}_p = v_{12}$	Not eigen states of total spin *j*	$\hat{v}_{12} = \frac{1}{\sqrt{2}}v_{12} + \frac{1}{\sqrt{2}}v_{21} = \zeta_{10} = \vert 1,0\rangle$		
$\left\vert\frac{1}{2},\frac{1}{2}\right\rangle_p = \begin{bmatrix}1\\0\end{bmatrix}_p$	$\begin{bmatrix}0\\1\end{bmatrix}_n\begin{bmatrix}1\\0\end{bmatrix}_p = v_{21}$		$\hat{v}_{21} = \frac{1}{\sqrt{2}}v_{12} - \frac{1}{\sqrt{2}}v_{21} = \zeta_{00} = \vert 0,0\rangle$		$3 \oplus 1$
$\left\vert\frac{1}{2},-\frac{1}{2}\right\rangle_p = \begin{bmatrix}0\\1\end{bmatrix}_p$	$\begin{bmatrix}0\\1\end{bmatrix}_n\begin{bmatrix}0\\1\end{bmatrix}_p = v_{22}$		$\hat{v}_{22} = v_{22} \qquad = \zeta_{1-1} = \vert 1,-1\rangle$ Clebsch-Gordan coefficients = $1, \frac{1}{\sqrt{2}}, -\frac{1}{\sqrt{2}}, 0$		Can label ζ_{ij} with *SU*(3) Cartan eigenvalues ε_3 and ε_8, or with *j* and *m*.

14.2.5 Things to Note about Direct Products and Direct Sums

Vector Space vs Matrix Representation

We can consider use of the symbol ⊕, in relations like (14-32), in two equivalent ways. One, as a direct sum of vector spaces, such that a 3D vector direct summed with a 2D vector yields a 5D vector. That is, we can focus on the vector space aspect.

Alternatively, we can consider it a direct sum of matrix representations as *A* and *B* in (14-1). That is, we can focus on the matrix operator aspect.

The use of ⊕ generally encompasses both meanings, though a particular author may put primary attention on one or the other.

Similarly, use of the direct product symbol ⊗ can focus on the tensor (outer) product of vector spaces or on the direct product of matrix representations.

⊕ and ⊗ refer to both vector spaces and matrix reps

Group Representations in Direct Product/Direct Sum Relations

For relations like (14-32), the LHS generally refers to *SU*(*n*) groups (*n* =2, in this example) in the fundamental matrix representation of dimension *n*. The RHS, on the other hand, is a little trickier, and a full group theory explanation of it would take us too far afield from our present purpose. Hopefully, it will suffice, at this point, to note simply that the numbers on the RHS of (14-32) stand for different dimension representations, and different subgroups, of *SU*(2)X*SU*(2). We are focused on matrix representations, so those numbers, for us, designate matrix dimensions of irreducible submatrices, which would occupy, as blocks, the diagonal of a parent matrix.

Meaning of numbers in direct product and direct sum relations

Direct Product and Direct Sum of Groups vs Matrices

In this text, the expressions "direct product" and "direct sum" will refer specifically to matrix representations of groups, as discussed in Chap. 2 and this present chapter. Take care, when reading other literature, that the same terms, applied to abstract groups, have somewhat different meanings, which we will not delve into herein.

Direct product/sum defs different in abstract group theory and representation theory

We employ the latter

14.3 Generalizing Direct Products and Direct Sums

The relation (14-32) is only the beginning of the theory of direct products and direct sums for matrix representations. There are many other such relations for different direct products with different numbers of groups and different degree *n*, resulting in different-dimension independent (invariant) subspaces in each case.

It would take a good deal of space, and much work, to derive all these relations, and we won't do that here. We do list the most relevant of these, in Wholeness Chart 14-3 below, where an overbar, for us, will mean an antiparticle multiplet (though in formal group theory overbar implies anti-symmetry).

Many different SU(n) representation direct products exist

Wholeness Chart 14-3. Some *SU*(*n*) Direct Product, Direct Sum Relations

A summary of some of them

$$1\otimes 2=2 \qquad 2\otimes 2=3\oplus 1 \qquad 2\otimes \bar{2}=3\oplus 1 \tag{14-34}$$

$$3\otimes \bar{3}=8\oplus 1 \qquad 3\otimes 3=6\oplus 3 \qquad 6\otimes 3=10\oplus 8 \tag{14-35}$$

$$3\otimes 2=4\oplus 2 \qquad 2\otimes 2\otimes 2=4\oplus 2\oplus 2 \qquad 3\otimes 3\otimes 3=10\oplus 8\oplus 8\oplus 1 \tag{14-36}$$

Matrix dimensions add and multiply like ordinary numbers

Note how the products and sums multiply and add up much like ordinary numbers do.

There is a method for deducing relations like those in the chart for any size multiplet and any number of direct products, called Young tableau. We will not delve into that topic here, as it would divert us from more immediate goals, though I may someday cover it on the book website[1].

14.4 Hadrons: Combining Flavors

Hadrons combine quark flavors like deuteron combines neutron and proton

In hadrons, we have different flavor quarks forming composites, similar in nature to our prior example of different nucleons (a neutron and a proton) forming a composite (deuteron). The interaction holding hadrons together is, of course, the strong, or color, interaction, independent of the flavors involved (more or less, as different masses for different flavors can affect the interaction).

But, when, in the middle of the last century, experimentalists started finding an inordinate number of different hadrons, the focus soon grew on the flavor of the quarks composing them. And, in fact, that is the only real way to distinguish between hadrons. A neutron and a proton are each composed of three quarks, employing the same color interaction to hold them together. The difference between them lies in the flavors of their components. The neutron is *udd* and the proton *uud*. The positively charged pi meson, where overbar symbolizes antiparticle, is $u\bar{d}$; the negatively charged one, $\bar{u}d$; and the neutral one, a superposition of $u\bar{u}$ and $d\bar{d}$.

Use group theory with flavor for hadrons like spin group for deuteron

As usual when particles are bound together, we multiply their wave functions (in NRQM and RQM) to get the composite wave function. This corresponds, in QFT, to outer multiplying individual particle states. Hence, we can use the group theory results deduced earlier in this chapter. We'll start with pi mesons, but we need to clear up one point, first.

14.4.1 New Wrinkle for Analyzing Composite Particles

For composite particles: need to modify view of fields in Hilbert space

The conclusion of Sect. 14.1.6 and Wholeness Chart 14-1, pg. 423, contrasting multiparticle states populating a Fock space (where outer products of single particle states exist) with quantum fields populating a Hilbert space (with no outer products of single fields), needs to be modified for analyzing composite particle states. Here is why.

Multiparticle states such as the pi meson $|\pi^+\rangle = |u\bar{d}\rangle$ have eigenvalues like (in the ideal massless quark case for flavor *SU*(3)) ε_3 and ε_8, which are related to flavor "charge", and which we can find by operating on them with the operators Q_{V3} and Q_{V8}, where the charges of the component particles add. (See last row and column of the referenced wholeness chart.) For example, where antiparticles have opposite values from particles,

$$Q_{V3}\left|u\bar{d}\right\rangle = \left(\tfrac{1}{2}+\tfrac{1}{2}\right)\left|u\bar{d}\right\rangle = \left|u\bar{d}\right\rangle \qquad \varepsilon_3 = 1. \tag{14-37}$$

Kets are scalars, with no $SU(N_f)$ vector structure

But, note that, in practice, our multiparticle states, the kets, have no flavor *SU*(3) vector structure. For them, the u and $\bar{d}$ are not of forms $[u,0,0]^T$ and $[0,\bar{d},0]^T$, but are simply scalars. Nor does Q_{V3} have a matrix structure. It is simply a collection of number operators with each having coefficient equal to the ε_3 eigenvalue of its corresponding quark (*u, d*, or *s*).

The point is, if we want to use group theory to analyze composite particles, we need to use some structure whereby the vector space comprises three-component vectors and the operators are *SU*(3) matrices.

To do so, first recall that a quantum field is, in essence, an outer product of vector spaces, as in

$$\text{LC green up quark} \quad \Psi^L_{ug} = \sum_{r,\mathbf{p}}\sqrt{\frac{m}{VE_{\mathbf{p}}}}\left(c_{ur}(\mathbf{p})u_r(\mathbf{p})e^{-ipx} + d^\dagger_{ur}(\mathbf{p})v_r(\mathbf{p})e^{ipx}\right)\begin{bmatrix}1\\0\end{bmatrix}_W\begin{bmatrix}0\\1\\0\end{bmatrix}_S = \psi_u\begin{bmatrix}1\\0\end{bmatrix}_W\begin{bmatrix}0\\1\\0\end{bmatrix}_S . \tag{14-38}$$

[1] For more on Young tableau, see H. Georgi, *Lie Algebras in Particle Physics*, 2nd ed. (CRC Press 2018); H.J. Lipkin, *Lie Groups for Pedestrians*, 2nd ed. (Dover 1966); A. Das and S. Okubo, *Lie Algebras for Physicists*, (Hindustan Book Agency 2014)

And the operators on that field, such as $\frac{1}{2}\lambda_i$ and $\frac{1}{2}\sigma_i$, are matrices. Further, in flavor space, we have multiplets (i.e., vectors in a vector space) and operators like

$$\Psi = \begin{bmatrix} u \\ d \\ s \end{bmatrix} \qquad \tfrac{1}{2}\underset{\sim}{\lambda}_3 = \tfrac{1}{2}\begin{bmatrix} 1 & & \\ & -1 & \\ & & 0 \end{bmatrix} \quad \text{etc.} \tag{14-39}$$

Recall further, that the eigenvalues, such as ε_3 for the quantum field multiplet Ψ under the operation of $\frac{1}{2}\underset{\sim}{\lambda}_3$, are the same as those of the corresponding state operated on by the corresponding operator, such as Q_{V3}. For example,

Fields and associated states have same eigenvalues

$$\tfrac{1}{2}\underset{\sim}{\lambda}_3\Psi_d = \tfrac{1}{2}\begin{bmatrix} 1 & & \\ & -1 & \\ & & 0 \end{bmatrix}\begin{bmatrix} 0 \\ d \\ 0 \end{bmatrix} = -\tfrac{1}{2}\begin{bmatrix} 0 \\ d \\ 0 \end{bmatrix} = -\tfrac{1}{2}\Psi_d \qquad Q_{V3}\,|d\rangle = -\tfrac{1}{2}|d\rangle, \quad \text{for both, } \varepsilon_3 = -\tfrac{1}{2}. \tag{14-40}$$

What this means, is that if we wish to use group theory to determine eigenvalues for states, we can use fields, in the form of multiplets (vectors in a vector space) Ψ, as a surrogate for particles (states). Eigenvalues for multiparticle states are additive, and so are eigenvalues for outer products of field multiplets (vectors). So, for the sake of analysis, we can model the multiparticle composite states as outer products of their corresponding fields. This, in fact, is how we will proceed.

Composite particle analysis: field multiplet outer products in lieu of multiparticle kets

Bottom line: Even though we will talk of composites like mesons and baryons, as particles such as $|uud\rangle$, we will analyze them as outer products of flavor multiplet fields, such as $\Psi_u \otimes \Psi_u \otimes \Psi_d$. All eigenvalues are the same in both cases.

In other words, for the purposes of analyzing composite particles, we will employ Fock space for fields, instead of the more limited Hilbert space.

For composite particles: expand Hilbert space of fields to Fock space

14.4.2 An Example: the Pion

Let's consider the family of pi mesons (or pions) composed of various combinations of up or down quarks with anti-down and anti-up quarks, generally symbolized by π^0, π^+, and π^-, where the superscripts denote the electric charge. As noted in the prior chapter, if the quarks are massless, we get a conserved flavor "charge", and so can assign quantum numbers to the quarks corresponding to those charges. Thus, the resulting pion would have quantum numbers as well, associated with, and deducible from, the flavor quantum numbers of the constituent quarks.

Simplest example: pions from u,d quarks

Well, as noted in the prior chapter, for many hadrons (neutrons, protons, and mesons included) total hadron mass exceeds, by significant measure, the sum of the masses of the constituent quarks. Most of the hadron mass is in the form of kinetic and binding mass-energy of the quarks. In such cases, it is not unreasonable to approximate the masses of the quarks involved as zero. Then, we can employ the flavor symmetries of a given $SU(N_f)$ group acting on a vector space of N_f different flavor quarks.

Pion mass >> m_u+m_d, so can assume $m_u = m_d = 0$ and use flavor symmetry

In that spirit, consider the flavor space of u and d quarks (upon which the $SU(2)$ group acts), and also the flavor space of $\bar{u}$ and $\bar{d}$ antiquarks (upon which $SU(2)$ also acts). By forming an outer product of them, we should get mesons, and our results will parallel those for the deuteron.

Similar to (14-13) through (14-20), we have

$$\underbrace{\begin{bmatrix} u \\ d \end{bmatrix} \otimes \begin{bmatrix} \bar{u} \\ \bar{d} \end{bmatrix}}_{\text{symbolically}} = \underbrace{\left(C_+\begin{bmatrix} 1 \\ 0 \end{bmatrix} + C_-\begin{bmatrix} 0 \\ 1 \end{bmatrix}\right) \otimes \left(\bar{C}_+\begin{bmatrix} \bar{1} \\ 0 \end{bmatrix} + \bar{C}_-\begin{bmatrix} 0 \\ \bar{1} \end{bmatrix}\right)}_{\text{as vectors in 2D}}$$

$$= \underbrace{C_+\bar{C}_+}_{C_{++}}\underbrace{\begin{bmatrix} 1 \\ 0 \end{bmatrix}\begin{bmatrix} \bar{1} \\ 0 \end{bmatrix}}_{v_{11}} + \underbrace{C_+\bar{C}_-}_{C_{+-}}\underbrace{\begin{bmatrix} 1 \\ 0 \end{bmatrix}\begin{bmatrix} 0 \\ \bar{1} \end{bmatrix}}_{v_{12}} + \underbrace{C_-\bar{C}_+}_{C_{-+}}\underbrace{\begin{bmatrix} 0 \\ 1 \end{bmatrix}\begin{bmatrix} \bar{1} \\ 0 \end{bmatrix}}_{v_{21}} + \underbrace{C_-\bar{C}_-}_{C_{--}}\underbrace{\begin{bmatrix} 0 \\ 1 \end{bmatrix}\begin{bmatrix} 0 \\ \bar{1} \end{bmatrix}}_{v_{22}} \tag{14-41}$$

$$= \underbrace{C_{++}u\bar{u} + C_{+-}u\bar{d} + C_{-+}\bar{u}d + C_{--}d\bar{d}}_{\text{symbolically}}.$$

Outer product of u,d isospin particle and antiparticle states

with direct product basis vectors

The v_{ij} basis vectors here do not readily seem to span invariant subspaces and the associated form of the resulting 4D group representation does not readily appear reducible to irreducible sub-reps.

But, as with the deuteron 4D spin space, that does become apparent with the appropriate transformation to a different basis.

We note now that the transformations to invariant subspaces for particles and antiparticles vary somewhat, not in principle but in detail, from the deuteron spin example. We are not going to derive these, but just take what the mathematicians give us. That is, we will just accept, from tables, the Clebsch-Gordan coefficients that others before us have worked out for the last relation of (14-34),

SU(2)⊗SU(2) here for flavor like with spin

$$2\otimes\bar{2}=3\oplus 1. \qquad (14\text{-}42)$$

So, in similar fashion as (14-24) through (14-33), we have, with somewhat different Clebsch-Gordan coefficients due to the antiparticle factor in the direct product,

Transform to new basis vectors as did with spin

$$\begin{aligned}\hat{v}_{11}&=-v_{12}=-u\bar{d} & \hat{v}_{12}&=\frac{1}{\sqrt{2}}\left(v_{11}-v_{22}\right)=\frac{1}{\sqrt{2}}u\bar{u}-\frac{1}{\sqrt{2}}d\bar{d}\\ \hat{v}_{21}&=\frac{1}{\sqrt{2}}\left(v_{11}+v_{22}\right)=\frac{1}{\sqrt{2}}u\bar{u}+\frac{1}{\sqrt{2}}d\bar{d} & \hat{v}_{22}&=v_{21}=d\bar{u}\,,\end{aligned} \qquad (14\text{-}43)$$

The $3\oplus 1$ invariant subspaces are

As with spin, get invariant subspace for 3D triplet and 1D singlet

$$\text{triplet}=\begin{bmatrix}\hat{v}_{11}\\ \hat{v}_{12}\\ \hat{v}_{22}\end{bmatrix}=\begin{bmatrix}\breve{v}_{1}\\ \breve{v}_{2}\\ \breve{v}_{3}\end{bmatrix}=\begin{bmatrix}-u\bar{d}\\ \frac{1}{\sqrt{2}}\left(u\bar{u}-d\bar{d}\right)\\ d\bar{u}\end{bmatrix} \qquad \text{singlet}=\hat{v}_{21}=\breve{v}=\frac{1}{\sqrt{2}}\left(u\bar{u}+d\bar{d}\right), \qquad (14\text{-}44)$$

Several choices for quantum numbers summarized in chart

where each composite state has the same $SU(3)$ Cartan subalgebra eigenvalues as the corresponding R, G, and B quarks of color $SU(3)$. In Wholeness Chart 14-4, the <u>total flavor isospin I^F</u> is analogous to our value j for the deuteron. <u>I_3^F isospin</u> is analogous to m. Much like weak hypercharge, but here for flavor, we define a <u>flavor hypercharge Y^F</u>, which has little value in this example, but will in the next.

We have introduced the F superscripts above to emphasize that we are working in flavor space and should not confuse isospin therein with physical spin I_3 or weak isospin I_3^W, or flavor hypercharge with weak hypercharge Y. However, in the literature, flavor isospin and hypercharge are generally written without such superscript, so we will now switch to that symbolism. However, keep it clearly in mind for the rest of this chapter that these quantities are purely for flavor space, and do not confuse them with the quantities the same symbols were used for earlier in this book.

Wholeness Chart 14-4. Quantum Numbers of Pions

State	Particle	Total Flavor Isospin $I=I^F$	Flavor Isospin $I_3=I_3^F$	Flavor ε_3	Flavor ε_8	Charge Q	Flavor Hypercharge $Y=Q-I_3=Y^F$
$-u\bar{d}$	π^+	1	1	$\frac{1}{2}$	$\frac{1}{2\sqrt{3}}$	+1	0
$\frac{1}{\sqrt{2}}\left(u\bar{u}-d\bar{d}\right)$	π^0	1	0	$-\frac{1}{2}$	$\frac{1}{2\sqrt{3}}$	0	0
$d\bar{u}$	π^-	1	−1	0	$-\frac{1}{\sqrt{3}}$	−1	0
$\frac{1}{\sqrt{2}}\left(u\bar{u}+d\bar{d}\right)$	π^0	0	0	0	0	0	0

Graphs for various choices of q numbers

We can plot these on the LHS of Fig. 14-1 as we did for quark colors, in the middle as total isospin I vs I_3 isospin, or on the RHS as flavor hypercharge Y vs I_3.

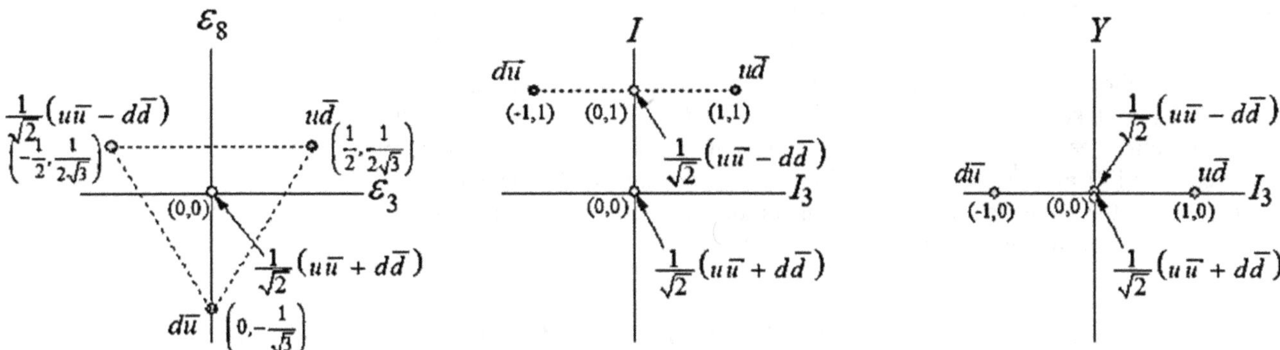

Figure 14-1. Pi Meson Triplet and Singlet: Three Ways to Plot

Y vs I_3 choice is most common in literature

The plots on the LHS and in the middle of the figure are generally not used in the literature, but we show them for comparison. Also, quantum numbers other than total (flavor) isospin I, like (flavor) hypercharge, for example, are more commonly used for such plots, as we are about to see.

Note that ε_3 and ε_8 are flavor eigenvalues, having nothing to do with any kind of actual charge (not electric, weak, hyper, or color), so they do not represent something we can measure physically with instruments. They are a theoretical tool and have nothing to do with properties (charges, spins, momenta, energies) we would detect in the real world. So, they aren't really helpful in labeling and identifying particles we might run into in experiments. The same can be said for (flavor) isospins I_3 and I, as they are related to flavor space multiplets.

However, charge Q does distinguish between different types of mesons in the real world, so it is a meaningful quantum number. Combining it with the theoretical (but unmeasurable directly) isospin I_3 gives us Y, so Y has meaning with regard to measurements, and has become the quantum number of choice for hadron plots like those of Fig. 14-1.

14.4.3 Example #2: Pions and More from u, d, and s

Mesons from u, d, s quarks

The third lightest quark, after the u and d, is the strange quark s, whose bare mass is just under 100 MeV, a lot more that the u and d, which are both under 5 MeV, but still much lighter than the next heaviest, the charm quark, at almost 1.3 GeV. The s quark mass does, however, get close to the typical QCD interaction energy (effective mass of a quark in a baryon) of somewhat over 300 MeV, so assuming it is massless, in order to have conserved flavor "charge" may be a bit more of a stretch. Nevertheless, even with massive quarks, we can start with the u, d, and s quarks together in an $SU(3)$ triplet, and combine those quarks, similar to what we did above for the u and d quarks, and still get a meaningful method for keeping track of the hadrons created by combining quarks.

The relevant SU(3) flavor triplet

To this end, consider the $SU(3)$ quark flavor triplet

$$\begin{bmatrix} u \\ d \\ s \end{bmatrix}, \tag{14-45}$$

which we need to outer multiply by its antiparticle equivalent to get two quark composites. From the first part of (14-35),

Two 3-flavor group reps in a direct product yield sum of irreducible sub-reps

$$3 \otimes \bar{3} = 8 \oplus 1, \tag{14-46}$$

we will obtain an octet and a singlet, representing the composite states. But note, that if we wish to graph the results, in a manner comparable to the LHS of Fig. 14-1, the plot for the octet would require seven dimensions, not two, since there are seven Cartan subalgebra generators for $SU(8)$. Not feasible for three dimensional beings like us using two dimensional sheets of paper. We could, as a more practical alternative, plot the values of total isospin I and directional isospin I_3, as in the middle of Fig. 14-1, but historically, and for practical reasons previously stated, this is generally not done.

A new quantum number, strangeness

What is done has evolved from experimental discoveries, in the mid 1900s, of unexpected, and thereby "strange", composite particles (though at the time, they were not known to be composite). A characteristic known as strangeness was attributed to such particles, and with the advent of quark theory, one particular quark was deemed to carry this property, and imbue hadrons, of which it is a part, with it. That is, the strange quark was assigned an empirically motivated quantum number, called strangeness, of value $S = -1$. It turns out that by applying (14-46) to (14-45), one can keep track of the mesons formed rather well, and use a 2D plot, by employing, and keeping track of, S and I_3.

In this picture, the three quarks have the quantum numbers shown in Wholeness Chart 14-5. Note the isospin values I_3 are taken from prior values of the u and d in $SU(2)$, and the s is assigned zero. Personally, this practice seems strange to me, as we are now working in $SU(3)$, where one would expect the values to be 1, 0, and −1, respectively. However, the advantage of this method, the standard practice, is that we get the same final quantum numbers for u and d combinations here in $SU(3)$, as we did in $SU(2)$ above. So, we are going with the standard practice.

Wholeness Chart 14-5. Quantum Numbers for *u*, *d*, and *s* Quarks

Quarks	Spin j	Baryon Num B	Flavor Charge Q	Flavor Isospin I_3	Strangeness S	Flavor Hypercharge $Y = Q - I_3$
u	1/2	1/3	2/3	1/2	0	1/6
d	1/2	1/3	−1/3	−1/2	0	1/6
s	1/2	1/3	−1/3	0	−1	−1/3

Flavor charge = electric charge = Q above

Relevant q numbers for flavor $SU(3)\otimes SU(3)$

Note that Y can also be expressed, at least in the present case, in terms of the strangeness S as

$$Y = Q - I_3 = \frac{B+S}{2}, \tag{14-47}$$

Two ways to express q number Y

and since B is the same for all quarks, Y is a direct indication of S. And, as we mentioned earlier, in combination with I_3, one can also deduce charge Q from it. So, Y (in combination with I_3) can tell us a lot about the particles involved, and play a role in distinguishing between them.

Returning to (14-46), we note that, as before, the quantum numbers of combinations of the three quarks are additive. From the LHS of (14-46), we have

$$\begin{array}{lllll} v_{11} = u\bar{u} & v_{12} = u\bar{d} & v_{13} = u\bar{s} & v_{21} = d\bar{u} & v_{22} = d\bar{d} \\ v_{23} = d\bar{s} & v_{31} = s\bar{u} & v_{32} = s\bar{d} & v_{33} = s\bar{s}. & \end{array} \tag{14-48}$$

Original basis vectors from direct product

Again, we will take the Clebsch-Gordan coefficients from tables to express our combinations of the three quarks as components in an invariant 8D subspace and a 1D subspace.

$$\text{octet} = \begin{bmatrix} \hat{v}_{11} \\ \hat{v}_{12} \\ \hat{v}_{13} \\ \hat{v}_{21} \\ \hat{v}_{22} \\ \hat{v}_{23} \\ \hat{v}_{31} \\ \hat{v}_{32} \end{bmatrix} = \begin{bmatrix} \breve{v}_1 \\ \breve{v}_2 \\ \breve{v}_3 \\ \breve{v}_4 \\ \breve{v}_5 \\ \breve{v}_6 \\ \breve{v}_7 \\ \breve{v}_8 \end{bmatrix} = \begin{bmatrix} u\bar{d} \\ u\bar{s} \\ d\bar{s} \\ d\bar{u} \\ s\bar{u} \\ s\bar{d} \\ \frac{1}{\sqrt{2}}\left(u\bar{u} - d\bar{d}\right) \\ \frac{1}{\sqrt{6}}\left(u\bar{u} + d\bar{d} - 2s\bar{s}\right) \end{bmatrix} \qquad \text{singlet} = \hat{v}_{33} = \breve{v} = \frac{1}{\sqrt{3}}\left(u\bar{u} + d\bar{d} + s\bar{s}\right) \tag{14-49}$$

New, transformed basis vectors as 8D and 1D invariant subspaces

These are eigenstates of Y (and S), as well as I_3, so can use eigenvalues as quantum numbers

---

Do **Problem 6** to find relevant quantum numbers for the singlet and several components of the octet.

Each component in (14-49) is an eigenstate of charge Q and I_3, and thus of Y and S as well, since the B eigenvalue for all of them is zero. The associated eigenvalues can be used as quantum numbers to label them. Each then represents a composite particle (meson) with specific strangeness, charge, etc. We will not list all of their eigenvalues/quantum numbers, as we did for our earlier example in Wholeness Chart 14-4. We do, however, plot the relevant quantum numbers Y vs I_3 in Fig. 14-2, and S vs I_3 in Fig. 14-3.

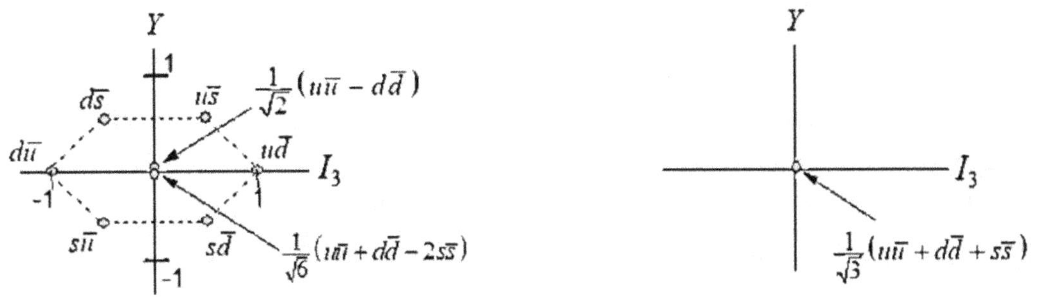

Graphing each new basis vector (composite particle) via Y vs I_3

Figure 14-2. Meson Octet and Singlet for *u, d, s* Quark Combination *Y* vs I_3 Quantum Numbers

Note that many authors define Y as twice what we do, so the dotted lines on the LHS of Fig. 14-2 look like a regular hexagon, not a squashed one as it does here. Plotting S vs I_3, where $S = 2Y$ since $B = 0$ here, does give us a regular hexagon, as we see in Fig. 14-3. We defined flavor hypercharge Y the way we did so that it would parallel our definition of weak hypercharge. (Many authors define weak hypercharge as twice what we do, as well, but we took it as we did in earlier chapters because I felt it was more transparent and pedagogically helpful defined that way. You will find both definitions in the literature.)

Y is often defined as twice our definition

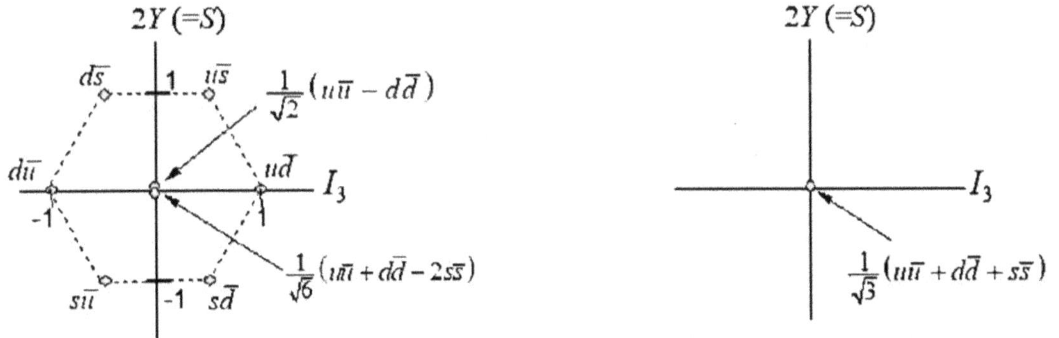

Graphing with 2Y (=strangeness S) yields regular hexagon

Figure 14-3. Meson Octet and Singlet for *u, d, s* Quark Combination *2Y* vs I_3 Quantum Number

These states correspond to observed composite particles known as pions, K-mesons, the η meson, and the η' meson, where the seemingly random symbolism and nomenclature is a relic from the days before quarks were known.

New basis states with eigenvalue, measurable q numbers in terms of original basis states

$$K^0(d\bar{s}) \qquad K^+(u\bar{s}) \qquad K^0(s\bar{d}) \qquad K^-(s\bar{u}) \qquad \pi^-(d\bar{u}) \qquad \pi^+(u\bar{d})$$
$$\pi^0\left(\frac{1}{\sqrt{2}}(u\bar{u}-d\bar{d})\right) \qquad \eta\left(\frac{1}{\sqrt{6}}(u\bar{u}+d\bar{d}-2s\bar{s})\right) \qquad \eta'\left(\frac{1}{\sqrt{3}}(u\bar{u}+d\bar{d}+s\bar{s})\right) \tag{14-50}$$

In actuality, the observed state known as the η, is an admixture (superposition) of what we have shown as η and η' in (14-50), which is to be expected since the two have the same eigenvalues (quantum numbers) and can therefore form a state that has the same eigenvalues. However, the η' component of that observed state is quite small, so η in (14-50) approximates the observed state η.

As a historical note, the octet of Fig. 14-3, discovered first experimentally, was deemed in 1961 by Murray Gell-Mann and Israeli physicist Yuval Ne'eman as the eightfold way, the name coming from the noble eightfold path of Buddhism.

There is a wealth of information associated with meson formation we have not presented. For example, we have only considered the zero orbital momentum states, and have ignored determination of the spin states of (14-50). More on this can be found in Halzen and Martin[1].

Much more to this subject, like $l \neq 0$ states, we won't do here

Further, we have ignored the heavy quarks, *c, t,* and *b*, which, in principle, would lead to $SU(6) \otimes SU(6)$ and spawn a plethora (36, to be exact) of composite states. Heavier states decay more readily, so they are generally not observed in nature, nor at lower energy levels of particle accelerators.

If included all six flavors would have 36 mesons

[1] F. Halzen and A. D. Martin, *Quarks and Leptons: An Introductory Course in Modern Particle Physics* (Wiley 1984) Chapter 2.

14.4.4 Example #3: Baryons from u, d, and s Quarks

We could look at baryons formed solely from up and down quark doublets using the direct product relation in the middle of (14-36),

We'll skip doing baryons from u, d quarks

$$2 \otimes 2 \otimes 2 = 4 \oplus 2 \oplus 2, \tag{14-51}$$

but instead, will skip to the more general case of up, down, and strange quarks.

The flavor triplet (14-45) can form baryons with direct products between three such triplets, as in the last part of (14-36),

Baryons from u, d, s

The relevant direct product relation

$$3 \otimes 3 \otimes 3 = 10 \oplus 8 \oplus 8 \oplus 1. \tag{14-52}$$

We will not go through the details of finding the particular *u, d, s* combinations in each of the decuplet, the two octets, and the singlet states, but cut to the chase with Fig. 14-4. We will, however, plot $2Y$ on the vertical axis to get regular polygon shapes.

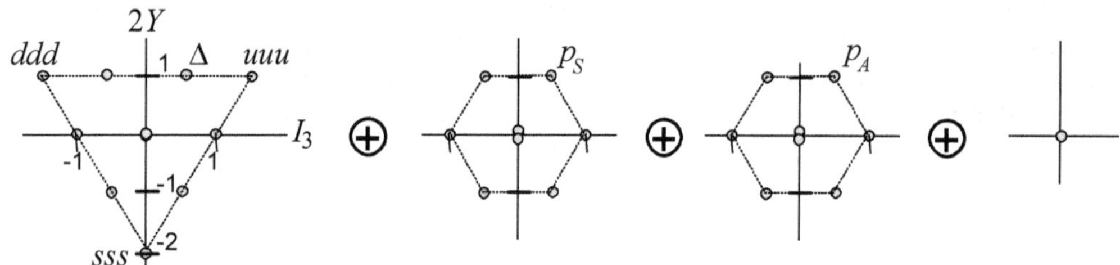

Figure 14-4. Baryon Multiplets for *u, d, s* Quark Combinations

Three of the labeled states in Fig. 14-4 are

Example of some states in different multiplets with same eigenvalues

$$\Delta = \frac{1}{\sqrt{3}}\left(uud + \left(ud + du\right)u\right) \tag{14-53}$$

$$p_S = \frac{1}{\sqrt{6}}\left(\left(ud + du\right)u - 2uud\right) \tag{14-54}$$

$$p_A = \frac{1}{\sqrt{2}}\left(ud - du\right)u\,, \tag{14-55}$$

where the subscripts S and A denote symmetry and antisymmetry under exchange of the first two quarks. Note that, otherwise, the p_S and p_A states have mixed symmetry.

To deduce the ground state proton *uud*, however, we need to also consider the spins of our three quarks, which give rise to the spin states of all the composite particles. This is done in parallel with the above analysis using the middle relation of (14-36),

We ignore here the spin parts of the states

$$2 \otimes 2 \otimes 2 = 4 \oplus 2 \oplus 2 \quad \text{for spin (not flavor) parts of states,} \tag{14-56}$$

to account for the spins. We will not do this, but refer the reader to Halzen and Martin again[1]. The bottom line is that we end up with ground state (lowest mass) baryons in an octet and a decuplet, as shown in Fig. 14-5.

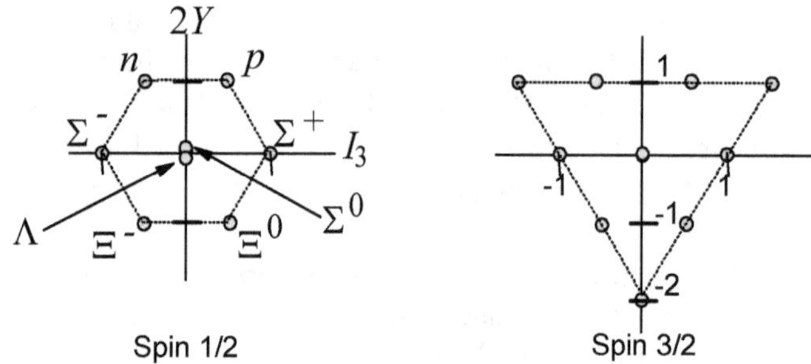

Figure 14-5. Ground State Baryon Octet and Decuplet for *u, d, s* Combinations

The symbols in the octet of Fig. 14-5 are composed of quarks in the following combinations.

---

[1] *ibid*, pgs. 51-55.

Ground state baryons from u,d,s quarks with distinct q numbers

$$\begin{array}{llll} p = uud & n = udd & \Sigma^- = dds & \Sigma^+ = uus \\ \Xi^- = ssd & \Xi^0 = ssu & \Sigma^0 = uds & \Lambda = uds \end{array} \qquad (14\text{-}57)$$

where Σ^0 has total (flavor) isospin $I = 1$ and Λ has $I = 0$.

One can see how the famous/ infamous particle zoo arose from the quark structure of matter, and how, before that structure was known, the experimental results were so perplexing.

14.4.5 Heavy Quarks and Their Composites

Heavy c, b, t quarks form hadrons too, and can use group theory for them

Even though, in all but the pre-electroweak epoch, we cannot justify approximating the heavy quarks *c, b, t* as either massless or having all equal masses, we can use group representation theory to help us analyze and keep track of composite particles they can form. Flavor "charges" cannot be conserved in such case, but conceptually, we can continue to think in terms of quantum numbers like S, Y, and I_3, as well as other such numbers, appropriate to the heavy quarks.

Being heavy, they decay rapidly and need high energy to be produced

Composite particles made of heavy quarks decay readily, due to their large masses, so they are not something typically observed in nature, such as in cosmic rays. And when they are formed, for example, in particle accelerator collisions, they are so short lived that they are never measured directly. Evidence for them comes from detection of the other particles they decay into.

The Charm Quark

History and background of the charm quark and its composites

In November 1974, a group of experimentalists colliding electrons and positrons found a narrow resonance near a COM energy of 3.1 GeV, and shortly thereafter another such resonance at 3.7 GeV. As we learned, a resonance implies a virtual particle with a mass at the resonance mass-energy level. And when there is a virtual particle, there is also a real particle. The researchers called the first particle ψ and the second ψ'. Independently, another group colliding protons found the same first resonance as well and labeled it J.

The ψ and ψ' particles were immediately recognized as bound quark/antiquark states (one ground, one excited) of a new particle, dubbed the charm quark. The charm-anticharm composite was christened charmonium, and its ground state, the J/ψ particle. Given the then known strange quark with a –1/3 charge, its partner, with a +2/3 charge had been anticipated for some time. Here was the proof, and both the experimentalists involved and theorists everywhere celebrated what is known as the "November revolution".

To incorporate the charm quark into our quark composite analyses, one needs to add another quantum number, called, not unexpectedly, "charm", and labeled C. The charm quark had C =1; the *u, d, s* quarks, C = 0. We then need to update our flavor hypercharge to

$$Y = \frac{B+S+C}{2}, \qquad (14\text{-}58)$$

and use the appropriate direct product relations, which for baryons is

$$4 \otimes 4 \otimes 4 = 20 \oplus 20 \oplus 20 \oplus \bar{4}. \qquad (14\text{-}59)$$

You might guess that this gets pretty complicated in a hurry, and we won't be going further with it. If you are interested, see Halzen and Martin[1] for more.

The Bottom Quark

History and background of the bottom quark and its composites

In 1973, physicists Makoto Kobayashi and Toshihide Maskawa first suggested the existence of a new quark, having characteristics of what we now know as the bottom quark, as a means of explaining certain observed CP violation. Two years later, Haim Harari dubbed the new quark the "bottom", though it is also sometimes called the "beauty" quark.

It was first observed experimentally in 1977 at Fermilab by a group led by Leon Lederman when particle collisions yielded bottomonium (also called the upsilon meson Υ), a meson composed of bottom and anti-bottom quarks.

The bare mass of the bottom quark is approximately 4.2 GeV, and its transitions are almost exclusively into or from a top quark. However, due to the CKM matrix, which we studied in Chap. 7, pg. 238, it can decay into charm and up quarks, rarely in terms of transition probabilities, but still very

[1] *ibid*, pgs. 57-67.

rapidly in terms of human time. In particular, the elements V_{ub} and V_{cb} in the CKM matrix cause decays into *u* and *c* quarks to be suppressed, so much so, that most composite particles with *b* quarks actually decay more slowly than their siblings with *c* quarks. Given that heavier elementary particles typically decay more rapidly than lighter ones, this is counterintuitive.

The term B meson is used to denote any meson containing an anti-bottom quark and a *u, d, s,* or *c* (but not a *b*, as that is bottomonium) quark. Because B mesons typically decay a bit more quickly than mesons containing a charm quark ($\sim10^{-13}$ s compared to $\sim10^{-12}$ s), and because they are so heavy, they are relatively easy to identify in collision experiments via a technique called "B-tagging". Due to this, they are the easiest particles to use in CP violation investigations.

Bottom baryons have been observed, as well.

The Top Quark

History and background of the top quark and its composites

The top quark (also known as the truth quark) was discovered in 1995 in experiments at Fermilab. It has a mass of approximately 173 GeV, far greater than any other quark.

In theory, a top/anti-top meson, named toponium (also, the theta meson), should form. But, due to the large top quark mass, the top quark lifetime is so short ($\sim5X10^{-25}$ s) that toponium is not expected to be, and has not been, observed in nature. Thus, it is, in actuality, only a hypothetical particle.

The top quark lifetime is so brief, at only about a twentieth of the strong interaction timescale, it does not hang around long enough to form hadrons. In other words, when it is created (as in particle colliders), it decays before it forms part of a meson or baryon. And thus, it is, in effect, the one quark that is considered to be "directly observed" (even though it is only its decay products that are actually detected). But, in essence, it is considered to provide experimentalists a means to study a "bare" quark.

Since the masses of the elementary particles are determined, according to the SM, via coupling to the Higgs, and the strongest such coupling is for the most massive particles, experiments with the top quarks provide a unique means to detect flaws in the extant theory. Because of this, top quarks are widely studied to discriminate between various theories that could transcend the SM.

14.5 Things to be Aware of

14.5.1 Nuclei

The strong force holds baryons together through the exchange of gluons between quarks. It turns out that nucleons are held together inside a nucleus via the strong force, with such exchanges, as well.

Protons and neutrons hold together via color residual effects

In somewhat similar fashion to commercial cling wrap using static electricity to cling to a glass, via exchange of virtual photons, nucleons in close proximity of one another can exchange some of the gluons in the cloud of gluons produced by the quarks inside of, and fluttering around the edges of, each nucleon. This exchange acts in such a way as to attract nucleons to one another, in a robust enough way to overcome the repulsive force from multiple protons clustered inside a nucleus.

Analysis of this is complex, to say the least, and well beyond the scope of this book, so we will leave it at that for now.

14.5.2 Nucleon Spin

For a number of years, proton and neutron spin of ½ was considered to be the result of simple addition of the ½ spins of the composite three quarks. The ground state would have two +½ spins and one –½ spin, yielding a total spin value of ½. However, in the late 1980s, experiments began to reveal something far different.

Proton spin of ½ only 30% due to three valence quark spins

It turns out that the spin of the three non-virtual quarks (called valence quarks) averages out to be only about 30% or so of the total spin of the proton. The remaining 70% comes from the spins of the cloud of virtual gluons plus the orbital momenta of the quarks and gluons about and among themselves. Experimentally, about 40% (± 10%) of the total spin was found to come from the gluons.

Intuitively, this is very perplexing. The immediate question is how, amidst that confusing mess of additions of various angular momenta, do we end up with a proton spin of exactly ½? No one has any kind of simple visualization or logic for this.

Proton spin composed of quark, gluon spins plus orbital angular momenta, virtual and real

But, a calculational technique known as Lattice QCD (related to gauge lattice regularization of Vol.1, pg. 374), with the aid of supercomputers, has predicted the gluon contribution to be 50%, with a margin of error of a few percent. The remaining 20^{+}% must come from orbital angular momentum, as the sea quarks (virtual quark/antiquark loops in gluon trajectories) make negligible contribution.

This was all unexpected, is quite amazing, and shows us once again that we have plenty of reason to be humble in the face of Nature's majesty and depth, despite all our advances in knowledge over the past century or two.

14.5.3 Anti-screening

Anti-screening model often used as mental guide to strong coupling dependence on energy

The decrease in the strong force coupling constant with increasing energy, contrasted with an increase for the e/m coupling constant, is often explained heuristically as "anti-screening" (as opposed to "screening", discussed in Vol. 1, pg. 317.) The gluons carry color charge, whereas the virtual photons do not carry electric charge, so at closer distances, the former adds to the effective color charge, whereas the latter does the reverse.

But don't accept the idea uncritically

As I mention in Vol. 1, I am not a fan of the screening concept explanation. In reasoning through it, I have applied certain concepts, such as the field inside a spherical shell of uniform distribution of charge is zero, and the field outside acts as if the whole charge were located at the center. In doing so, I have not been able to make the idea work to my satisfaction. And so, I feel the coupling constant dependence on energy level is simply a result of the mathematics, and it is difficult to paint a mental picture that we, as limited humans, can grasp. I am open, however to having my position modified, by those who know this concept more intimately than I.

I mention screening because you will see it elsewhere. But draw your own conclusions as to its validity.

14.6 Chapter Summary

14.6.1 Direct Sums, Direct Products, and Irreducible Representations

Direct sums of matrix group representations, are best understood by example, as in (14-1).

$$C=\begin{bmatrix} a_{11} & a_{12} & a_{13} & & \\ a_{21} & a_{22} & a_{23} & & \\ a_{31} & a_{32} & a_{33} & & \\ & & & b_{11} & b_{12} \\ & & & b_{21} & b_{22} \end{bmatrix}=A\oplus B \qquad A=\begin{bmatrix} a_{11} & a_{12} & a_{13} \\ a_{21} & a_{22} & a_{23} \\ a_{31} & a_{32} & a_{33} \end{bmatrix} \quad B=\begin{bmatrix} b_{11} & b_{12} \\ b_{21} & b_{22} \end{bmatrix} \qquad \text{repeat of (14-1)}$$

C as shown is in a basis where it is reduced. In another basis, many (or all) of the zero components in (14-1) are non-zero, and we can label the matrix therein as $\underset{\sim}{C}$, which is reducible (to block diagonal form) under suitable transformation. C of (14-1) acting on a 5D vector will not mix the top three components of the vector with the bottom two. The respective 3D and 2D vector spaces are thus invariant (independent) subspaces under the action of C. Those subspaces remain invariant if we transform to another basis where the matrix has form $\underset{\sim}{C}$, although it will not be obvious in that basis. A matrix representation that cannot be reduced to block diagonal form under any possible transformation is known as irreducible.

Direct products of irreducible matrix representations lead to a direct sum of irreducible matrix representations. For examples, where the numbers on the LHS indicate the degree n of each irreducible $SU(n)$ group fundamental representation and overbar, for us, means antiparticles,

$$2\otimes\overline{2}=3\oplus 1 \qquad 3\otimes\overline{3}=8\oplus 1 \qquad 3\otimes 3\otimes 3=10\oplus 8\oplus 8\oplus 1\,. \tag{14-60}$$

Each number on the left side of an equal sign in (14-60) represents the number of fundamental particles considered. The number of multiplications represents the number of particles joined together. In the second case, for example, we have three kinds of particles that are joined in pairs to form composites.

The composite particles in each invariant subspace are found by transforming from the original outer product basis to a basis where the irreducible subgroup representation structure is evident, i.e., where the direct product matrix group representation has block diagonal form. Tables exist for the coefficients relating the two bases, and such coefficients are known as Clebsch-Gordan coefficients.

The eigenvalue of an $SU(n)$ operator acting on an outer product of vectors from the same vector space (or its associated antiparticle vector space) equals the sum of eigenvalues from that operator acting independently on each of the vectors composing the outer product.

14.6.2 Hadron Composition via Group Theory

Relations like (14-60) help us construct the composite particles formed from more fundamental particles (which for quarks are categorized by flavor).

The first direct product there results in a 4D complex matrix group, having 3D and 1D irreducible sub-representations, which acts on a 4D vector space (representing composite particles) composed of a 3D invariant subspace and a 1D invariant subspace. Basis vectors of each invariant subspace have eigenvalues for certain $SU(2)$ operators, whose eigenvalues can be used to plot them in graphs. In practice, the I_3 flavor isospin and the $Y = Q - I_3$ (or $2Y$) flavor hypercharge eigenvalues are used, where Q is formulated for this purpose as an $SU(2)$ matrix operator on field multiplets.

$$\text{Matrix operators on field multiplets } I_3 = \begin{bmatrix} \frac{1}{2} & \\ & -\frac{1}{2} \end{bmatrix} \quad Q = \begin{bmatrix} \frac{2}{3} & \\ & -\frac{1}{3} \end{bmatrix} \quad Y = Q - I_3 = \begin{bmatrix} \frac{1}{6} & \\ & \frac{1}{6} \end{bmatrix}. \quad (14\text{-}61)$$

See the RHS of Fig. 14-1, pg. 433, for a graph showing this case for the flavor eigenvalues I_3 and Y of the composite particles (pi mesons there), where each eigenvalue equals the sum of the eigenvalues of the more fundamental particles in that composite.

Finally, some such composite particles, from different invariant subspaces, with the same eigenvalues (quantum numbers), may form linear superpositions equal to the composite particle one would detect in experiments.

The same general process applies to more complicated direct products, such as in the RHS of (14-60).

14.7 Problems

1. Take v in (14-8) as a green LC down quark and find its color and weak isospin Cartan subalgebra eigenvalues.
2. Show that the z direction spin values for v_{21} and v_{22} of (14-18) and (14-20) are $m = 0$ and -1, respectively.
3. Show (14-24) as a 4D matrix transformation of the form of T in (14-4).
4. Find the Clebsch-Gordan coefficients for the second row of (14-24).
5. Show the x direction spin operators change $\hat{v}_{22}$ of (14-24) into $\hat{v}_{12}$ (times a constant).
6. Find I_3, S, and Y for the singlet and the first, third, and eighth components in the octet of (14-49).

Chapter 15

QCD Interaction Theory

*"A child said, 'What is the grass?' fetching it to me with full hands;
How could I answer the child?. . . I do not know what it is any more than he."*
Walt Whitman

"To sense that behind anything that can be experienced there is a something that our mind cannot grasp and whose beauty and sublimity reaches us only indirectly and as a feeble reflection, this is religiousness. In this sense I am religious. To me it suffices to wonder at these secrets and to attempt humbly to grasp with my mind a mere image of the lofty structure of all that there is."
Albert Einstein

15.0 Preliminaries

15.0.1 Background

We have noted the strong interaction is inherently unsuitable for perturbation analysis, since its coupling constant is large. This is true for its role in binding hadrons and nuclei together. However, the strong coupling "constant", like the weak and QED coupling constants, varies with energy level, due to radiative corrections, as we saw in Fig. 11-7, pg. 373. In opposite fashion from the QED coupling constant, it decreases with the interaction energy/momentum level (and thus, decreases at shorter time and distance scales).

QCD g_s varies oppositely from QED e with energy/momentum

At very close distance (high energy/momentum), in fact, quarks are so loosely bound to one another that they are, effectively, free. As we have noted before, the force holding quarks together approaches zero, as the energy/momentum level approaches infinity, and this is called asymptotic freedom.

High energy (close distance) g_s small and perturbation possible

On the flip side, low energy/momentum (greater length scales and further apart) quarks are held together very tightly. Much like a spring connecting two particles behaves, the strong force gets stronger at greater distances. This prevents them from ever being found in isolation and is known as quark confinement, or <u>infrared slavery</u>. Asymptotic freedom and quark confinement go hand-in-hand.

Low energy (long distance) g_s large, and perturbation not possible

Asymptotic freedom means we actually can use perturbation theory for the strong interaction, provided the quarks being analyzed are close enough together. In fact, this was the means by which QCD became the acknowledged theory of strong interactions, as we discuss briefly below.

So, in analyzing the strong interaction, the two regimes, high energy and low energy, are handled in qualitatively different ways. For high energy (short distance) experiments, we can use perturbation theory. For low energy (greater distance) experiments, we need a non-perturbative approach.

High energy QCD scattering perturbative; hadron QCD not

<u>Non-perturbative QCD</u>

In the early 1970s, Kenneth Wilson developed a method for handling nonperturbative theories of physics known as <u>lattice gauge theory.</u> He initially applied it in a number of different areas, including ferromagnetism and critical behavior of fluids, but then, importantly for QFT, turned to strong interactions.

Lattice gauge theory does non-perturbative QCD, numerically

In QCD, lattice gauge theory breaks spacetime into a lattice of small 4D volumes, where quarks occupy the sites of the lattice. The spacetime continuum is thereby discretized, and the associated QCD functional integral can be estimated numerically, in essence, by adding values from all the lattice

Functional integral evaluated as sum over tiny 4D volumes

points, rather than integrating. In principle, with enough computing power, the 4D volumes can be taken smaller and smaller, until a meaningful result is obtained.

By the early 1980s, the approach had yielded encouraging, but far from perfect, results for hadron masses, and was limited in that it could not include effects from sea quarks. By 2004, however, computation power had increased to the point where this could be done, and hadron mass calculations came into fairly good alignment with experiment. By 2008, they were accurate to within a few percent.

With today's computer power, very accurate results

Note that lattice gauge theory is the basis of the gauge lattice regularization technique described in Vol. 1, Chap. 15. (See Wholeness Chart 15-2, pg. 397, therein.)

Unfortunately, lattice gauge theory is a study unto itself, and one that is beyond the scope of this book. So, we will restrict our work herein to perturbative QCD. It will be limited, by nature, but will be a good foundation for further study into non-perturbative QCD, should your career take you in that direction[1].

Perturbative QCD

It turns out that high energy scattering of an electron with a quark in a proton, or electron-positron annihilation producing hadrons, can be analyzed via perturbative QCD. So, in this chapter, we will lay the foundation for such analysis. Even though we will be employing perturbation, where the canonical quantization approach works well, we will use functional integral methods instead, as they are essential for non-perturbative QCD, and thus, a valuable tool we should have in our QFT toolkit.

In this chapter, we stick to perturbative QCD

15.0.2 Some QCD History

In the 1950s and early 1960s, physicists had become a bit overwhelmed by the proliferation of so many different kinds of hadrons discovered in experiments. The prevailing view was one of nuclear democracy, where every hadron was considered just as fundamental, or possibly just as composite, as any other. But there were so many of them, and many hoped for some way to make better sense of it all.

Mid 1900s, many hadrons discovered without a good theory

Enter Murray Gell-Mann and George Zweig, who in 1964 independently introduced the idea that mesons and baryons could be thought of as comprised of more primary elements. Gell-Mann called them quarks, Zweig called them "aces", and the former name prevailed. Both of them and others considered the quarks to be merely an organizing structure, however, and not actual particles.

Until promising quark theory came along

That all changed in 1968, when SLAC (Stanford Linear Accelerator Center originally, now SLAC National Accelerator Laboratory) experiments indicated that hadrons were definitely composites of smaller particles. Feynman dubbed these particles partons, and that name has come to be used for both of what we now call quarks and gluons.

Scattering experiments indicated quarks were real entities

The SLAC experiments, involving deep-inelastic scattering of high energy electrons off protons, displayed a cross section dependence on only the ratio $|\mathbf{q}|^2/\nu$, where $\mathbf{q}$ is the 3-momentum transfer from the electron and ν is the electron energy transfer. James Bjorken showed that a proton model based on individual point-like partons yielded just such a dependence, and the relation describing it is known today as Bjorken scaling. A key aspect of the analysis was Bjorken's presumption that the partons (quarks) were perfectly free inside the nucleon. This, of course, aligns with the asymptotic freedom we now know to be the case.

Theory indicated quarks appeared to be free inside hadrons

But even with Bjorken's result, an underlying theory of it all was still lacking. That was resolved in 1973 by David Gross, David Politzer, and Franck Wilczek.

Gross had originally set out, in 1972, to prove there was, as many believed at the time, no quantum field theory that could explain deep inelastic scaling, i.e., there were no asymptotically free field theories. He and his student, Wilczek, used renormalization group theory (Vol. 1, pgs. 317, 320 and this volume, pg. 372) to determine the sign of the β function therein. That sign, if positive, means the coupling constant of the associated theory ($U(1)$, $SU(2)$, or $SU(3)$) increases with energy (shorter distance). QED has a positive β function. If the β function is negative, the coupling constant decreases with energy (shorter distance). A positive β function for the strong interaction field theory would disqualify it as an explanation for the experiments.

Three theorists showed SU(3) interaction could explain it via negative QCD beta function

To Gross's surprise, after laborious calculations, he and Wilczek found the QCD β function is negative. And thus, $SU(3)$ color field theory does describe what was being found at SLAC.

---

[1] For starters, try Lattice Quantum Chromodynamics Comes of Age, *Physics Today*, Feb 2004, 45-51.

Independently, Politzer was trying to generalize dynamical symmetry breaking to non-abelian theories and also found the β function to be negative.

Gross and Wilczek's paper appeared alongside Politzer's in the June 25, 1973 issue of *Physical Review Letters*. In 2004, the three of them shared the Nobel Prize in physics for their discovery.

15.0.3 Chapter Overview

In this chapter, we will cover

Topics in this chapter

- a mathematical example that will help in this chapter,
- the Faddeev-Popov approach to QCD,
- the generating functional for that approach, and
- the Feynman rules for perturbative QCD derived from that approach.

15.1 The Faddeev-Popov Method

15.1.1 An Insightful Mathematical Example

A useful mathematical prelude

Sidney Coleman[2] originated the following example for ordinary integration, which we will find helpful, as it parallels the more complex integration required to find the QCD generating functional.

As background, we need to recall (4-73) of Chap. 4, where $\bar{\theta}_i$ and θ_j are Grassmann variables, and we use the symbol ζ_i here in place of η_i there, because we will use the latter symbol for something different later in this chapter,

$$\int e^{-\bar{\theta}_i A_{ij}\theta_j+\bar{\theta}_i\zeta_i+\bar{\zeta}_i\theta_i}\,d\bar{\theta}_1 d\theta_1\ldots d\bar{\theta}_n d\theta_n=\left(Det\,A_n\right)e^{\bar{\zeta}_i A_{ij}^{-1}\zeta_j}\,. \tag{15-1}$$

For $\zeta = 0$, this becomes

$$Det\,A_n=\int e^{-\bar{\theta}_i A_{ij}\theta_j}\,d\bar{\theta}_1 d\theta_1\ldots d\bar{\theta}_n d\theta_n\,. \tag{15-2}$$

We then start our example with a well-defined integral of n dimensions, where integral signs without limits imply integration from $-\infty$ to $+\infty$, and all z_j throughout this section are real.

A well-defined integral over n variables

$$Z=\int\ldots\int e^{-X}\,dz_{k+1}\ldots dz_{k+n}\qquad X=X\left(z_{k+1},\ldots,z_{k+n}\right). \tag{15-3}$$

Given (15-3), the integral of $k+n$ dimensions,

Made ill-defined with k more integration variables

$$Y=\int\ldots\int e^{-X}\,dz_1\ldots dz_k dz_{k+1}\ldots dz_{k+n}\,, \tag{15-4}$$

is divergent (ill-defined), due to the $z_1, \ldots, z_k$ parts of the integration.

Can be made well defined over k + n variables

Our objective is to determine a well-defined integral in $k+n$ dimensions that is the equivalent of (15-3). To begin, we constrain the values of $z_1, \ldots, z_k$ via k suitable relations

$$f_i\left(z_1,\ldots,z_k,z_{k+1},\ldots,z_{k+n}\right)=f_i\left(z_j\right)=0\qquad i=1,\ldots k\quad j=1,\ldots,k+n\,. \tag{15-5}$$

Using constraint equations and Dirac delta functions

Because X does not depend on any of the $z_1, \ldots, z_k$, it does not depend on any of the f_i, so (15-3) can be re-expressed with Dirac delta functions and the f_i as

$$\begin{aligned}Z&=\int e^{-X}\delta\left(f_1\right)\delta\left(f_2\right)\ldots\delta\left(f_k\right)df_1 df_2\ldots df_k dz_{k+1}\ldots dz_{k+n}\qquad X=X\left(z_{k+1},\ldots,z_{k+n}\right)\\&=\int e^{-X}\prod_{i=1}^{k}\delta\left(f_i\right)df_1 df_2\ldots df_k dz_{k+1}\ldots dz_{k+n}\,.\end{aligned} \tag{15-6}$$

As a reminder, when we change variables of integration inside an integration, the differential part needs to be modified by the determinant of the matrix made up of the partial derivatives of the original variables with respect to the new variables, as in (15-7) below, where the matrix is known formally as the Jacobian. We don't prove this well-known mathematical relation here, but intuitively this makes some sense as $df_i=\dfrac{\partial f_i}{\partial z_{i'}}dz_{i'}$. So, switching from the f_i of (15-6) to the $z_1, \ldots, z_k$, we have

---

[2] S. Coleman, *Aspects of Symmetry*, Cambridge (1985), Chap. 5, Sect. 5.2.

$$Z=\int e^{-X}\,Det\left(\frac{\partial f_i}{\partial z_{i'}}\right)\prod_{i=1}^{k}\delta(f_i)dz_1\ldots dz_k dz_{k+1}\ldots dz_{k+n} \qquad i'=1,\ldots,k. \tag{15-7}$$

Changing integration variables involves a determinant

With (15-2), this becomes the relation that will guide us in discussions of the generating functional, where the θ_i and $\bar{\theta}_i$ find their way into the relationship because they are part of the means we use to re-express the determinant,

$$\begin{aligned}Z&=\int e^{-X}\left(\int e^{-\sum_{i,i'=1}^{k}\bar{\theta}_i\frac{\partial f_i}{\partial z_{i'}}\theta_{i'}}d\bar{\theta}_1 d\theta_1\ldots d\bar{\theta}_k d\theta_k\right)\prod_{i=1}^{k}\delta(f_i)dz_1\ldots dz_k dz_{k+1}\ldots dz_{k+n}\\&=\int e^{-X-\sum_{i,i'=1}^{k}\bar{\theta}_i\frac{\partial f_i}{\partial z_{i'}}\theta_{i'}}\left(\prod_{i=1}^{k}\delta(f_i)\right)\left(\prod_{i=1}^{k}d\bar{\theta}_i d\theta_i\right)\left(\prod_{i=1}^{k+n}dz_i\right)\\&=\int e^{-\tilde{X}}\left(\prod_{i=1}^{k}\delta(f_i)\right)\left(\prod_{i=1}^{k}d\bar{\theta}_i d\theta_i\right)\left(\prod_{i=1}^{k+n}dz_i\right) \qquad \text{where } \tilde{X}\equiv X+\sum_{i,i'=1}^{k}\bar{\theta}_i\frac{\partial f_i}{\partial z_{i'}}\theta_{i'}.\end{aligned} \tag{15-8}$$

New, well-defined form of integral over k+n variables z_i (plus over k pairs of Grassmann numbers)

15.1.2 Issues with the Gluon Generating Functional

Before starting this section, you may wish to review Chaps. 3 and 4, or at least the Wholeness Charts 3-7 and 4-4 at the ends of them.

<u>Gluons Only Lagrangian</u>

Consider the gluons-only part of the QCD Lagrangian from (12-12) and (12-13) of Chap. 12, pg. 388,

$$\mathcal{L}^G=-\tfrac{1}{4}\hat{G}_i^{\mu\nu}\hat{G}_{i\,\mu\nu}, \tag{15-9}$$

Gluons only Lagrangian

where

$$\hat{G}_i^{\mu\nu}=D^\nu A_i^\mu-D^\mu A_i^\nu=\underbrace{\partial^\nu A_i^\mu-\partial^\mu A_i^\nu}_{F_i^{\mu\nu}}+g_s f_{ijk}A_j^\mu A_k^\nu. \tag{15-10}$$

<u>The Associated Generating Functional</u>

The generating functional, for gluons-only, where we have eight classical sources $J_{i\kappa}(x)$, $i=1,\ldots,8$, each one associated with a gluon field $A_i^\kappa(x)$, is

$$Z\left[J_{i\kappa}\right]=\frac{\int e^{iX\left[J_{i\kappa}\right]}\mathcal{D}A}{\int e^{iX\left[J_{i\kappa}=0\right]}\mathcal{D}A}=\frac{\int e^{iX\left[J_{i\kappa}\right]}\mathcal{D}A}{N}, \tag{15-11}$$

Generating functional for gluons only

with

$$\mathcal{D}A=\prod_{i=1}^{8}\prod_{\mu=0}^{3}\mathcal{D}A_i^\mu(x). \tag{15-12}$$

Note that, given the denominator N of (15-11), $Z[J_{i\kappa}=0]=1$. The augmented action is

$$X\left[J_{i\kappa}\right]=\int\left(\mathcal{L}^G+J_{i\kappa}(x)A_i^\kappa(x)\right)d^4x=\int\left(-\tfrac{1}{4}\hat{G}_i^{\mu\nu}\hat{G}_{i\,\mu\nu}+J_{i\kappa}(x)A_i^\kappa(x)\right)d^4x. \tag{15-13}$$

For the free gluon case, where there are no gluon-gluon interactions, we have

$$\hat{G}_i^{\mu\nu}\xrightarrow{\text{free}}=F_i^{\mu\nu}=\partial^\nu A_i^\mu-\partial^\mu A_i^\nu, \tag{15-14}$$

which looks a lot like the photon field tensor $F^{\mu\nu}$ of QED. So, we might expect we could use Lorenz conditions on all eight gluons, like we did for the one photon, i.e.,

$$\partial_\mu A_i^\mu=0 \qquad i=1,\ldots,8 \tag{15-15}$$

Might expect to use Lorenz gauge condition with gluons

But it doesn't work

However, the justification for this in electrodynamics was based on the fact that for a given set of classical fields $A^\mu(x)$, we could always find a gauge transformation for which the transformed four-potential $A^\mu(x)$ obeyed the Lorenz condition. But the proof there relied on the simple form of the transformation (see Vol. 1, Wholeness Chart 5-1, pg. 141, where the sign on f was arbitrary, and here we change f to g (which is not a coupling constant in this case), so as not to confuse it with other entities in this chapter we use the symbol f for)

$$A^{\mu} \rightarrow A'^{\mu} = A^{\mu} - \partial^{\mu} g = A^{\mu} - \frac{1}{e}\partial^{\mu}\alpha \,. \tag{15-16}$$

In the present case, we have

$$A_i'^{\mu} = A_i^{\mu} - \frac{1}{g_s}\partial^{\mu}\omega_i - f_{ijk}\omega_j A_k^{\mu} , \tag{15-17}$$

the more complicated form of which means the Lorenz condition approach cannot be generalized to $SU(3)$, or, more generally, other $SU(n)$ theories. Trying it leads to inconsistencies in the theory. (Take my word for it.) We can, however, do something somewhat similar, as we shall shortly see.

Returning to (15-11) and (15-13), we note that for a massless spin 1 field, there are only two independent dynamical variables, the two transverse components of the field. In electrodynamics, the Lorenz condition is one constraint on the four components.

Gluons, massless like photons, have only two independent (transverse) comps

In the gluon generating functional, however, we carry out a path integral over four components μ of each gluon field, as if each such component were independent of the others, when in fact, only two are independent. This, in turn, means that the generating functional (15-11) is ill-defined, much like (15-4), which integrates over more variables than just the independent ones. That is, we are effectively integrating over additional variables that play no role in the actual integral we are interested in. We need to eliminate the integrals over these extra degrees of freedom.

Generating functional integrates over all four, so ill-defined

The Approach to Get a Well-defined Generating Functional

The Russian physicists Ludvig Faddeev and Victor Popov discovered a way to make the QCD functional integral well defined. They did this by introducing constraints, similar in concept to those of (15-5), on the degrees of freedom (field components) of the gluon fields.

Faddeev and Popov showed how to make it well-defined, similar to our earlier example

The full procedure is quite complex, so we will not go through a complete, rigorous presentation of it. We will, however, outline and justify its main features, and thereby, hopefully, make the final result seem reasonable.

15.1.3 The Generating Functional via Faddeev-Popov

The Faddeev-Popov procedure starts with a set of gauge conditions, constraints on the gluon field components, using expressions of form

$$f_i\left(A_i^{\mu}(x)\right) \equiv \partial_{\mu} A_i^{\mu}(x) - h_i(x) \qquad i = 1,2,\ldots,8 , \tag{15-18}$$

where the eight h_i are arbitrary functions. It imposes the gauge conditions, or constraints

Start with certain constraint equations

$$f_i\left(A_i^{\mu}(x)\right) = 0 \qquad \rightarrow \qquad \partial_{\mu} A_i^{\mu}(x) = h_i(x) , \tag{15-19}$$

which look a little like the Lorenz gauge, except here we have h_i, where there we had zero.

Faddeev-Popov then introduced the functional integrals

$$\int \mathcal{D}\eta \mathcal{D}\tilde{\eta} \equiv \prod_{i=1}^{8} \int \mathcal{D}\eta_i \mathcal{D}\tilde{\eta}_i \tag{15-20}$$

and functional integrals over Grassmann fields

where the $\eta_i(x)$ and $\tilde{\eta}_i(x)$ are Grassmann fields rather than discrete Grassmann variables such as θ_i and $\bar{\theta}_i$, as in (15-2). Furthermore, they are scalars, not spinor or vector fields.

They then showed explicitly, whereas we will only justify by comparison with (15-7), that, for f_i defined by (15-18), we can re-write (15-11) as

$$Z\left[J_{i\mu}\right] \propto \int e^{iX}\, Det\left(g_s \frac{\delta f_i}{\delta \omega_{i'}}\right) \prod_{i=1}^{8} \delta[f_i] \mathcal{D}A , \tag{15-21}$$

to get a relation similar to that of our prior math example

with the generalization of (15-2), where to streamline notation, we now write dx for d^4x,

$$Det\left(g_s \frac{\delta f_i}{\delta \omega_{i'}}\right) \propto \int e^{\,i g_s \int \eta_i(x) \frac{\delta f_i}{\delta \omega_{i'}} \tilde{\eta}_{i'}(x') dx dx'} \mathcal{D}\eta \mathcal{D}\tilde{\eta} , \tag{15-22}$$

where we make use of the determinant relation

and where $\delta[f_i]$ are functional delta functions defined via

$$\int F[f_i]\delta[f_i]\mathcal{D}f_i = F[f_i = 0]. \tag{15-23}$$

At first sight, (15-21) may not look quite like what we would expect from the analogous relation (15-7). That is, we would expect derivatives in (15-21) to be with respect to the longitudinal and scalar components of A_i^μ. However, since in (15-7), $\frac{\partial f_i}{\partial z_{i'}} = \frac{\partial f_i}{\partial z_k}\frac{\partial z_k}{\partial z_{i'}}$, by analogy, $\frac{\delta f_i}{\delta \omega_{i'}} = \frac{\delta f_i}{\delta A_k^\nu}\frac{\delta A_k^\nu}{\delta \omega_{i'}}$ in (15-21). The extra factor of g_s also comes in because of the way we have defined our gluon field transformation (15-17). (Some authors use a transformation set where the second and third terms of (15-17) are multiplied by g_s.)

Evaluating the determinant with the given constraints

Due to the functional delta function in (15-21), we only need to evaluate the determinant therein when all $f_i = 0$. To this end, note that

$$\frac{\delta f_i}{\delta \omega_{i'}} = \frac{\delta f_i}{\delta A_k^\nu}\frac{\delta A_k^\nu}{\delta \omega_{i'}} \xrightarrow[\text{all } f_i=0]{\text{for our case}} \left.\frac{\delta f_i}{\delta \omega_{i'}}\right|_{\text{all } f_i=0} = \left.\frac{\delta f_i}{\delta A_k^\nu}\frac{\delta A_k^\nu}{\delta \omega_{i'}}\right|_{\text{all } f_i=0}, \tag{15-24}$$

so, we need to evaluate each of the two factors to the right of the equal sign in (15-24).

To begin with, note that neither A_μ nor A'_μ in (15-17) need to solve (15-19) for a valid theory. Unlike our use of the Lorenz gauge condition for electromagnetism, we have not imposed those constraints on our QCD theory, so far.

But, let's consider the transformation of (15-17), where the original field does solve (15-19). And let's change our notation such that we label that original field $A_i^{(0)\mu}$, and the transformed field, which generally does not solve (15-19), as A_i^μ (no prime). Then, we have

$$A_i^\mu = A_i^{(0)\mu} - \frac{1}{g_s}\partial^\mu \omega_i - f_{ijk}\omega_j A_k^{(0)\mu}. \tag{15-25}$$

Now if we impose the condition on A_i^μ in (15-25), and in (15-24) too, that it must also satisfy (15-19), for the given h_i, then we have

$$\left.A_i^\mu\right|_{\text{all } f_i=0} = A_i^{(0)\mu} = \left.A_i^\mu\right|_{\text{all } \omega_i=0}, \tag{15-26}$$

and so, (15-24) becomes

$$\frac{\delta f_i}{\delta \omega_{i'}} = \frac{\delta f_i}{\delta A_k^\nu}\frac{\delta A_k^\nu}{\delta \omega_{i'}} \xrightarrow[f_i=0]{\text{for our case}} \left.\frac{\delta f_i}{\delta \omega_{i'}}\right|_{\text{all } \omega_i=0} = \left.\frac{\delta f_i}{\delta A_k^\nu}\frac{\delta A_k^\nu}{\delta \omega_{i'}}\right|_{\text{all } \omega_i=0}. \tag{15-27}$$

Now, let's proceed to evaluate the two factors on the RHS of (15-27). First,

$$\begin{aligned}\frac{\delta A_k^\nu(x)}{\delta \omega_{i'}(x')} &= \frac{\delta}{\delta \omega_{i'}(x')}\left(A_k^{(0)\nu}(x) - \frac{1}{g_s}\partial^\nu \omega_k(x) - f_{kjl}\omega_j A_l^{(0)\nu}(x)\right) \\ &= -\frac{1}{g_s}\partial^\nu \frac{\delta \omega_k(x)}{\delta \omega_{i'}(x')} - f_{kjl}\frac{\delta \omega_j(x)}{\delta \omega_{i'}(x')}A_l^{(0)\nu}(x) = -\frac{1}{g_s}\partial^\nu \delta(x-x')\delta_{ki'} - f_{kjl}\delta(x-x')\delta_{ji'}A_l^{(0)\nu}(x).\end{aligned} \tag{15-28}$$

So,

$$\begin{aligned}\left.\frac{\delta A_k^\nu(x)}{\delta \omega_{i'}(x')}\right|_{\text{all } \omega_i=0} &= \left.-\left(\frac{1}{g_s}\delta_{ki'}\partial_{x'}^\nu + f_{ki'l}A_l^{(0)\nu}(x')\right)\delta(x-x')\right|_{\text{all } \omega_i=0} \\ &= -\left(\frac{1}{g_s}\delta_{ki'}\partial_{x'}^\nu + f_{ki'l}A_l^\nu(x')\right)\delta(x-x').\end{aligned} \tag{15-29}$$

Second,

$$\frac{\delta f_i}{\delta A_k^\nu} = \frac{\delta}{\delta A_k^\nu}\left(\partial_\mu A_i^\mu - h_i\right) = \partial_\mu \frac{\delta A_i^\mu}{\delta A_k^\nu} = \partial_\mu \delta_{ik}\delta_\nu^\mu. \tag{15-30}$$

Putting (15-29) and (15-30) into (15-27), we have

$$\xrightarrow[\text{all } f_i=0]{\text{for our case}} \quad \left.\frac{\delta f_i}{\delta \omega_{i'}}\right|_{\text{all } \omega_i=0} = \partial_\mu \delta_{ik} \delta^\mu_\nu \left(-\left(\frac{1}{g_s} \delta_{ki'} \partial^\nu_{x'} + f_{ki'l} A^\nu_l (x') \right) \delta(x-x') \right)$$
$$= -\partial_\nu \left(\frac{1}{g_s} \delta_{ii'} \partial^\nu_{x'} + f_{ii'l} A^\nu_l (x') \right) \delta(x-x'). \qquad (15\text{-}31)$$

Putting (15-31) into (15-22), we have

$$Det\left(g_s \frac{\delta f_i}{\delta \omega_{i'}} \right) \propto \int e^{i\int \eta_i(x)\left(-\partial_\nu \left(\delta_{ii'}\partial^\nu_{x'} + g_s f_{ii'l} A^\nu_l (x') \right) \delta(x-x') \right) \tilde{\eta}_{i'}(x') dx dx'} \mathcal{D}\eta \mathcal{D}\tilde{\eta} \qquad (15\text{-}32)$$

Intermediate result for the determinant

Do **Problem 1** to show (15-33).

$$i\int \eta_i(x)\left(-\partial_\nu \left(\delta_{ii'}\partial^\nu_{x'} \right) \delta(x-x') \right) \tilde{\eta}_{i'}(x') dx dx' = i\int \left(\partial_\mu \eta_i(x) \right)\left(\partial^\mu \tilde{\eta}_i(x) \right) dx \qquad (15\text{-}33)$$

With (15-33) into (15-32), we have

$$Det\left(g_s \frac{\delta f_i}{\delta \omega_{i'}} \right) \propto \int e^{i\int \overbrace{\left(\partial_\mu \eta_i(x) \right)\left(\partial^\mu \tilde{\eta}_i(x) + g_s f_{ijk} \tilde{\eta}_j(x) A^\mu_k(x) \right)}^{\mathcal{L}^g} dx} \mathcal{D}\eta \mathcal{D}\tilde{\eta}\,, \qquad (15\text{-}34)$$

Final result for the determinant

$$\mathcal{L}^g = \overbrace{\left(\partial_\mu \eta_i(x) \right) \partial^\mu \tilde{\eta}_i(x)}^{\mathcal{L}^g_0} + \overbrace{g_s f_{ijk} \left(\partial_\mu \eta_i(x) \right) \tilde{\eta}_j(x) A^\mu_k(x)}^{\mathcal{L}^{gG}}, \qquad (15\text{-}35)$$

where, for reasons we are about to see, the η_i and $\tilde{\eta}_i$ are called ghost fields, and $\mathcal{L}^g$ is called the ghost term Lagrangian. Thus, we can express (15-21) as

Grassmann fields we introduced = ghosts

$$Z[J_{i\mu}] \propto \int e^{i\left(X + \int \mathcal{L}^g dx \right)} \prod_{i=1}^{8} \delta[f_i] \mathcal{D}A \mathcal{D}\eta \mathcal{D}\tilde{\eta} = \int e^{i\int \left(\mathcal{L}^G + \mathcal{L}^g + J_{i\kappa} A^\kappa_i \right) dx} \prod_{i=1}^{8} \delta[f_i] \mathcal{D}A \mathcal{D}\eta \mathcal{D}\tilde{\eta}. \qquad (15\text{-}36)$$

Original generating functional with ghosts

Finally, we need to remove the functional delta functions in (15-36). For this, we will multiply (15-36) by the constant (recall a functional integral for a given function is a number)

Need to eliminate the functional deltas

$$C \equiv \int e^{-\frac{i}{2}\sum_i (h_i(x))^2} \prod_{j=1}^{8} \mathcal{D}h_j\,. \qquad (15\text{-}37)$$

Use a constant equal to functional integral in h_i

Note that since $Z[J_{i\mu}]$ is gauge invariant, it is independent of the choice of $h_i(x)$. Thus, we have

$$Z[J_{i\mu}] \propto C Z[J_{i\mu}] \propto \int \left(Z[J_{i\mu}] \right) e^{-\frac{i}{2}\sum_i (h_i(x))^2} \prod_{j=1}^{8} \mathcal{D}h_j$$
$$\propto \int\int e^{i\int \left(\mathcal{L}^G + \mathcal{L}^g + J_{i\kappa} A^\kappa_i \right) dx} \prod_{i=1}^{8} \delta[f_i] e^{-\frac{i}{2}\sum_i (h_i(x))^2} \prod_{j=1}^{8} \mathcal{D}h_j \mathcal{D}A \mathcal{D}\eta \mathcal{D}\tilde{\eta}. \qquad (15\text{-}38)$$

Note that the sub-integral in (15-38) over h_i can be re-written, using (15-18) as

$$\int \prod_{i=1}^{8} \delta\left[\partial_\mu A^\mu_i(x) - h_i(x) \right] e^{-\frac{i}{2}\sum_i (h_i(x))^2} \prod_{j=1}^{8} \mathcal{D}h_j = e^{-\frac{i}{2}\sum_i \left(\partial_\mu A^\mu_i \right)^2}. \qquad (15\text{-}39)$$

The constant with the delta function = a factor in generating functional

Hence, we end up with our final result for the generating functional for gluons in the absence of quarks, where again, $N = Z[J_{i\,\mu} = 0]$, which, again as we learned in Chap. 4, we can ignore for connected Green functions,

$$Z\left[J_{i\mu}\right]=\frac{1}{N}\int e^{i\int\left(\mathcal{L}^{G}-\frac{1}{2}\sum_{i}\left(\partial_{\mu}A_{i}^{\mu}\right)^{2}+\mathcal{L}^{g}+J_{i\kappa}A_{i}^{\kappa}\right)dx}\mathcal{D}A\mathcal{D}\eta\mathcal{D}\tilde{\eta}=\frac{1}{N}\int e^{i\int\left(\mathcal{L}^{G'}+\mathcal{L}^{g}+J_{i\kappa}A_{i}^{\kappa}\right)dx}\mathcal{D}A\mathcal{D}\eta\mathcal{D}\tilde{\eta}$$
$$\mathcal{L}^{G'}=\mathcal{L}^{G}-\frac{1}{2}\sum_{i}\left(\partial_{\mu}A_{i}^{\mu}\right)^{2}\qquad\mathcal{L}^{g}=\mathcal{L}_{0}^{g}+\mathcal{L}^{gG}. \tag{15-40}$$

Final result: a well-defined functional integral for gluons

Note that the Faddeev-Popov method leads to additional terms for our original exponent $iX[J_{i\kappa}]$ of (15-11) and (15-13), the ghost Lagrangian $\mathcal{L}^{g}$ of (15-35) (made of two subparts) and a sum of terms $-\frac{1}{2}\sum_{i}\left(\partial_{\mu}A_{i}^{\mu}\right)^{2}$, which are added to the gluons-only Lagrangian $\mathcal{L}^{G}$. It also includes additional integrations over η_i and $\tilde{\eta}_i$. But, whereas (15-11) is ill-defined, (15-40) is not.

But some things added: ghosts and a summation of terms in $\left(\partial_{\mu}A_{i}^{\mu}\right)^{2}$

15.1.4 Ghosts

Ghosts Aren't Real

The ghost Lagrangian $\mathcal{L}^{g}$ of (15-35) has terms with factors of the ghost fields, η_i and $\tilde{\eta}_i$ (actually, derivatives thereof) and gluon fields A_i^{μ}, and this means they will interact with one another. (For the path integral approach (PIA), this will require a little modification of what we have seen before, but we will get to that a bit later.)

The ghost fields are *scalar* Grassmann fields, so the particles they create and destroy are spin-zero fermions. But, as we learned in Vol. 1 (Sect. 4.6.4, pg. 110), such particles cannot be real (physical). Fermions must have spin ½ (or hypothetically, spin 3/2). Yet, the ghosts can play a role as virtual particles, i.e., they can have propagators. This is similar to the photon longitudinal and scalar components, which aren't seen in the real world, but need to be included in the photon propagator.

Ghosts are scalars with anti-commutation relations, so aren't real

Bottom line: The Faddeev-Popov ghost fields are a convenient mathematical trick, but are not real particles.

Ghosts are a math trick that helps analysis

Ghosts Can be Exorcised

By choosing a different gauge, other than (15-18), the ghost fields can be eliminated (exorcised, if we wish to maintain the supernatural nomenclature). A gauge that does this is the axial gauge, where $n_\mu = (0,0,0,1)$ or any other fixed spacelike four-vector,

It can be shown that in a different gauge, ghosts disappear

$$f_i\left(A_i^{\mu}(x)\right)=n_{\mu}A_i^{\mu}(x)-h_i(x)=0\qquad i=1,2,\ldots,8, \tag{15-41}$$

though we won't be showing how.

The axial gauge is used to prove the equivalence of the canonical and PI approaches, for non-perturbative QCD theory, and is mentioned often in the literature on that topic. However, it does not work so well for perturbation theory, our focus in the remainder of this part of the book, so, as noted, we will not consider it further herein.

That the ghost quanta can, indeed, be eliminated from the theory demonstrates that they are not physical particles (which could not be eliminated and still yield a viable theory).

Real (physical) particles cannot disappear from a valid theory, so ghosts are not real

15.1.5 Gauge Fixing

We can rearrange the gluons-only part of the Lagrangian in a way that will be helpful. We first show how this is done with a QED example.

Re-arranging the gluon $\mathcal{L}$ to help analysis

Gauge Fixing in QED

Let's apply the Faddeev-Popov method to the electromagnetic field, which is similar to what we have just seen for quarks, except that there is only one gauge boson field (the photon), and we have no photon-photon interactions, so we have no terms like the last one in (15-10). We only have one pair of ghost fields, $\eta(x)$ and $\tilde{\eta}(x)$, so, $\mathcal{L}^{gG}$ of (15-35) is zero, and our generating functional is

Illustrative example: Faddeev-Popov for e/m

$$Z\left[J_{\mu}\right]=\frac{1}{N}\int e^{i\int\left(\mathcal{L}_{0}^{e/m}+\mathcal{L}_{0}^{g}+J_{\kappa}A^{\kappa}\right)dx}\mathcal{D}A\mathcal{D}\eta\mathcal{D}\tilde{\eta}, \tag{15-42}$$

where

$$\mathcal{L}_0^{e/m} = -\frac{1}{4}F_{\mu\nu}F^{\mu\nu} - \frac{1}{2}\left(\partial_\mu A^\mu\right)^2 \qquad F^{\mu\nu} = \partial^\nu A^\mu - \partial^\mu A^\nu \,, \tag{15-43}$$

and the free ghost term (see (15-35)) is

$$\mathcal{L}_0^g = \left(\partial_\mu \eta\right)\partial^\mu \tilde{\eta} \,. \tag{15-44}$$

Do **Problem 2** to show (15-45).

If you did Problem 2, you saw that, for the PIA, we can express the Faddeev-Popov photon Lagrangian of (15-43) as

Faddeev-Popov PIA Lagrangian for photons

$$\mathcal{L}_0^{e/m} = -\frac{1}{4}F_{\mu\nu}F^{\mu\nu} - \frac{1}{2}\left(\partial_\mu A^\mu\right)^2 = -\frac{1}{2}\left(\partial_\mu A_\nu\right)\left(\partial^\mu A^\nu\right), \tag{15-45}$$

where the last part is the first photon Lagrangian, proposed by Fermi, that we studied in Chap. 5 of Vol. 1. When the Lorenz gauge condition is imposed, the symmetric photon Lagrangian and the Fermi Lagrangian are equal, effectively at least, in the PIA.

So, we can express (15-42) as

Faddeev-Popov photon generating functional

$$Z\left[J_\mu\right] = \frac{1}{N}\int e^{i\int\left(\mathcal{L}_0^{e/m} + \mathcal{L}_0^g + J_\kappa A^\kappa\right)dx}\mathcal{D}A\mathcal{D}\eta\mathcal{D}\tilde{\eta} = \frac{1}{N}\int e^{i\int\left(-\frac{1}{2}\left(\partial_\mu A_\nu \partial^\mu A^\nu\right) + \left(\partial_\mu \eta\right)\partial^\mu \tilde{\eta} + J_\kappa A^\kappa\right)dx}\mathcal{D}A\mathcal{D}\eta\mathcal{D}\tilde{\eta} \,. \tag{15-46}$$

An aside:

Since $\mathcal{L}_0^g$ is the only function of the ghost fields $\eta(x)$ and $\tilde{\eta}(x)$ in (15-46), we can factor out

In this case, if we wanted, we could absorb ghost part into N

$$\int e^{i\int \mathcal{L}_0^g dx}\mathcal{D}\eta\mathcal{D}\tilde{\eta} \,, \tag{15-47}$$

and absorb it into the normalization constant N. This leaves us with

$$Z\left[J_\mu\right] = \frac{1}{N}\int e^{i\int\left(\mathcal{L}_0^{e/m} + J_\kappa A^\kappa\right)dx}\mathcal{D}A = \frac{1}{N}\int e^{i\int\left(-\frac{1}{2}\left(\partial_\mu A_\nu \partial^\mu A^\nu\right) + J_\kappa A^\kappa\right)dx}\mathcal{D}A \,, \tag{15-48}$$

where $Z[J_\mu = 0] = 1$.

End of aside.

The term

e/m gauge-fixing term

$$-\frac{1}{2}\left(\partial_\mu A^\mu\right)^2 \tag{15-49}$$

in (15-45) and its sibling, the last term in the bottom row of (15-40), are known as gauge-fixing terms (see pg. 370). They arose via (15-39), because we fixed our gauge as $f_i = 0$, for all i.

Of course, in electrodynamics, we typically employ the Lorenz condition, where (15-49) is zero, i.e., $h_i = 0$, in (15-19), but here, for the sake of illustration, we are considering the more general case, where the specific Lorenz condition may not be true.

Gauge Fixing in QCD

QCD analogous to prior QED case

We can extrapolate the results from QED above directly to QCD simply by expanding our single photon field to eight gluon fields.

The Lagrangian

For *free* (not self-interacting) gluons along with the ghost fields, parallel to (15-45) and (15-44),

Faddeev-Popov $\mathcal{L}$ for free gluons and ghosts only

$$\text{Faddeev-Popov, no quarks} \quad \mathcal{L}_{0,\text{nq}} = -\frac{1}{4}F_{i\mu\nu}F_i^{\mu\nu} - \frac{1}{2}\left(\partial_\mu A_i^\mu\right)^2 + \left(\partial_\mu \eta_i\right)\partial^\mu \tilde{\eta}_i = -\frac{1}{2}\left(\partial_\mu A_{i\nu}\right)\left(\partial^\mu A_i^\nu\right) + \left(\partial_\mu \eta_i\right)\partial^\mu \tilde{\eta}_i \,. \tag{15-50}$$

Including quarks, we have

Faddeev-Popov $\mathcal{L}$ for free gluons, ghosts, and quarks

$$\text{Faddeev-Popov, with quarks} \quad \mathcal{L}_0 = -\frac{1}{2}\left(\partial_\mu A_{i\nu}\right)\left(\partial^\mu A_i^\nu\right) + \left(\partial_\mu \eta_i\right)\partial^\mu \tilde{\eta}_i + \bar{\psi}_a\left(i\partial\!\!\!/ - m\right)\psi_a \,, \tag{15-51}$$

where letters $a,b,\ldots$ will designate color and to keep things simplest, we will only work with a single quark flavor, for now, so there are no flavor labels.

With minimal substitution in (15-51) for the quark and gluon terms, plus $\mathcal{L}^{gG}$, the last term of (15-35), which we have already found represents the ghost interactions, we obtain the full Lagrangian

$$\mathcal{L} = \mathcal{L}_0 + \mathcal{L}_I , \qquad (15\text{-}52)$$

where (see the end of Chap. 12 for the first three terms below)

$$\text{Faddeev-Popov } \mathcal{L}_I = \underbrace{-\tfrac{g_s}{2}\bar{\psi}_a\left(\lambda_j\right)_{ab}\gamma_\mu\psi_b A_j^\mu}_{\mathcal{L}^{QG}} \underbrace{+ g_s f_{ijk}A_{i\mu}A_{j\nu}\partial^\mu A_k^\nu - \tfrac{1}{4}g_s^2 f_{ijk}f_{ilm}A_j^\mu A_k^\nu A_{l\mu}A_{m\nu}}_{\mathcal{L}^{GG}}$$
$$\underbrace{+ g_s f_{ijk}\left(\partial_\mu\eta_i\right)\tilde{\eta}_j A_k^\mu}_{\mathcal{L}^{gG}} . \qquad (15\text{-}53)$$

Faddeev-Popov $\mathcal{L}$ for interacting gluons, ghosts, and quarks

These terms, when displayed in a Feynman diagram, represent, respectively, vertices with i) two quarks and one gluon, ii) three gluons, iii) four gluons, and iv) two ghosts and a gluon. Note from Fig. 15-1 that we depict ghosts with dashed lines.

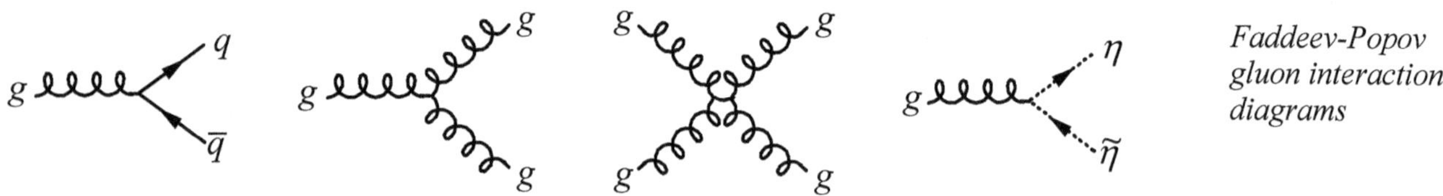

Faddeev-Popov gluon interaction diagrams

Figure 15-1. Gluon Vertices with Quarks, Other Gluons, and Ghosts

15.1.6 The Faddeev-Popov Generating Functional with Quarks

The full generating functional in the Faddeev-Popov approach including gluons, quarks, and ghosts is, where we define X', S_i, $\tilde{S}_i$, and $\mathcal{L}_S$ below,

$$Z\left[J_{i\kappa},\sigma_a,\bar{\sigma}_a,S_i,\tilde{S}_i\right] = \frac{1}{N}\int e^{iX'}\mathcal{D}A\mathcal{D}\bar{\psi}\mathcal{D}\psi\mathcal{D}\eta\mathcal{D}\tilde{\eta} , \qquad (15\text{-}54)$$

Faddeev-Popov generating functional

$$X' \equiv \int\left(\mathcal{L}_0 + \mathcal{L}_I + \mathcal{L}_S\right)dx \qquad (15\text{-}55)$$

The source Lagrangian terms in the action X' are

$$\mathcal{L}_S(x) = J_{i\kappa}(x)A_i^\kappa(x) + \bar{\sigma}_a(x)\psi_a(x) + \bar{\psi}_a(x)\sigma_a(x) + S_i(x)\eta_i(x) + \tilde{S}_i(x)\tilde{\eta}_i(x) , \qquad (15\text{-}56)$$

Faddeev-Popov source Lagrangian

where S_i, $\tilde{S}_i$, are scalar Grassmann source fields associated with the ghost fields having $i = 1, 2,\dots,8$, and σ_a, $\bar{\sigma}_a$ are spinor Grassmann source fields associated with quark fields having $a = r, g, b$. Integrals include all components of the fields. See (15-12), (15-20), and, where ζ below is a spinor index,

$$\mathcal{D}\bar{\psi}\mathcal{D}\psi = \prod_a \prod_\zeta \mathcal{D}\bar{\psi}_{a\zeta}\mathcal{D}\psi_{a\zeta} . \qquad (15\text{-}57)$$

As usual, if we have connected Green function diagrams, we can ignore N in (15-54), but, to be complete, we recall it is

$$N = \int e^{iX}\mathcal{D}A\mathcal{D}\bar{\psi}\mathcal{D}\psi\mathcal{D}\eta\mathcal{D}\tilde{\eta} \qquad X = \int \mathcal{L}dx = \int\left(\mathcal{L}_0 + \mathcal{L}_I\right)dx \qquad \left(\mathcal{L}_S = 0\right) , \qquad (15\text{-}58)$$

Normalization factor, which can be ignored for connected Green function

so,

$$Z[0,0,0,0,0] = 1 , \qquad (15\text{-}59)$$

Green functions are found for QCD in the same manner we did for QED in Chap. 4, which is shown in Wholeness Chart 4-4 at the end of that chapter. That is, where X in (15-60) can be either X of (15-58) or X' of (15-55),

$$G^{\mu\cdots}(x_1,.y_1,.\,z_1,.) = \left\langle A_i^\mu(x_1)\ldots\psi_a(y_1)\ldots\bar\psi_b(z_1)\ldots\right\rangle$$
$$= \frac{1}{N}\int e^{iX}\left\{A_1(x_1)\ldots\psi(y_1)\ldots\bar\psi(z_1)\ldots\right\}\mathcal{D}A\mathcal{D}\bar\psi\mathcal{D}\psi\mathcal{D}\eta\mathcal{D}\tilde\eta, \qquad (15\text{-}60)$$

Faddeev-Popov Green function with real (non-ghost) legs

where fields inside the {} brackets are the legs (which eventually are external particles in a Feynman diagram). As we learned in Chap. 4, the Green functions can be found by taking derivatives of the generating functional, then taking the source fields equal to zero in the result, i.e., for quark and gluon fields,

$$\left\langle A_i^\mu(x_1)\ldots\psi_a(y_1)\ldots\bar\psi_b(z_1)\ldots\right\rangle = (-1)^{\bar n}\left(\frac{1}{i}\right)^n \frac{\delta^n Z\left[J_k,\sigma,\bar\sigma\right]}{\delta J_{i\mu}(x_1)..\delta\bar\sigma_a(y_1)..\delta\sigma_b(z_1)}\Bigg|_0, \qquad (15\text{-}61)$$

Green function from generating functional, quark and gluon legs

where n is the total number of leg fields and $\bar n$ is the total number of adjoint quark leg fields.

Note that (15-61) represents the only Green functions one needs to calculate transition amplitudes for physical processes, since ghost particles cannot occur in physical states. However, it turns out to be helpful if we also work with Green functions having ghost field legs. For example, where $\bar n = 0$, since ghosts are bosons, not fermions,

$$\left\langle A_i^\mu(x_1)\ldots\tilde\eta_j(y_1)\ldots\eta_k(z_1)\ldots\right\rangle = \left(\frac{1}{i}\right)^n \frac{\delta^n Z\left[J_k,S,\tilde S\right]}{\delta J_{i\mu}(x_1)..\delta\tilde S_j(y_1)..\delta S_k(z_1)}\Bigg|_0. \qquad (15\text{-}62)$$

Green function from generating functional, gluon and ghost legs

We find Green functions in momentum space in the same way as in QED. That is, where B_i is any source field, and the accepted convention has all momenta q_i directed inwards,

$$(2\pi)^4\delta^{(4)}(q_1+q_2+\ldots+q_n)\left\langle B_1(q_1)B_2(q_2)\ldots B_n(q_n)\right\rangle$$
$$= \int\prod_{i=1}^n e^{-iq_ix_i}\left\langle B_1(x_1)B_2(x_2)\ldots B_n(x_n)\right\rangle dx_1dx_2\ldots dx_n. \qquad (15\text{-}63)$$

Finding Green function in momentum space from position space

15.1.7 Free Gluon, Quark, and Ghost Fields

The Free Generating Functionals: Fundamental Form

For non-interacting fields, the interaction Lagrangian is zero, so (15-54) becomes

$$Z_0\left[J_{i\kappa},\sigma_a,\bar\sigma_a,S_i,\tilde S_i\right] = \frac{1}{N_0}\int e^{iX_0'}\mathcal{D}A\mathcal{D}\bar\psi\mathcal{D}\psi\mathcal{D}\eta\mathcal{D}\tilde\eta \qquad X_0' = \int\left(\mathcal{L}_0+\mathcal{L}_S\right)dx \qquad (15\text{-}64)$$

Faddeev-Popov generating functional for free fields only

with

$$N_0 = \int e^{i\int\mathcal{L}_0 dx}\mathcal{D}A\mathcal{D}\bar\psi\mathcal{D}\psi\mathcal{D}\eta\mathcal{D}\tilde\eta\,, \qquad (15\text{-}65)$$

so that

$$Z_0\left[0,0,0,0,0\right] = 1\,. \qquad (15\text{-}66)$$

Note that, for the individual free fields, from (15-51) and (15-56),

For free gluon fields
$$\begin{cases} Z_0\left[J_{i\kappa}\right] = \dfrac{1}{N_1}\int e^{iX\left[J_{i\kappa}\right]}\mathcal{D}A & \text{with } \sigma_a = \bar\sigma_a = S_i = \tilde S_i = 0 \\ X\left[J_{i\kappa}\right] = \int\left(-\tfrac{1}{2}\left(\partial_\mu A_{i\nu}\right)\left(\partial^\mu A_i^\nu\right) + J_{i\kappa}A_i^\kappa\right)dx & \end{cases} \qquad (15\text{-}67)$$

Faddeev-Popov free field generating functionals for individual separate fields

For free quark fields
$$\begin{cases} Z_0\left[\sigma_a,\bar\sigma_a\right] = \dfrac{1}{N_2}\int e^{iX\left[\sigma_a,\bar\sigma_a\right]}\mathcal{D}\bar\psi\mathcal{D}\psi & \text{with } J_{i\kappa} = S_i = \tilde S_i = 0 \\ X\left[\sigma_a,\bar\sigma_a\right] = \int\left(\bar\psi_a\left(i\partial\!\!\!/ - m\right)\psi_a + \bar\sigma_a(x)\psi_a(x) + \bar\psi_a(x)\sigma_a(x)\right)dx & \end{cases} \qquad (15\text{-}68)$$

For free ghost fields
$$\begin{cases} Z_0\left[S_i,\tilde{S}_i\right]=\dfrac{1}{N_3}\int e^{iX\left[S_i,\tilde{S}_i\right]}\mathcal{D}\eta\,\mathcal{D}\tilde{\eta} \qquad \text{with } J_{i\kappa}=\sigma_a=\bar{\sigma}_a=0 \\ X\left[S_i,\tilde{S}_i\right]=\int\left(\left(\partial_\mu\eta_i\right)\partial^\mu\tilde{\eta}_i+S_i(x)\eta_i(x)+\tilde{S}_i(x)\tilde{\eta}_i(x)\right)dx \end{cases} \tag{15-69}$$

where the normalizations N_i in (15-67) to (15-69) are all taken such that $Z_0[0]=1$ in each case.

Recall from Chap. 4 for QED, that we formed the total free generating functional for all fields from the product of the free generating functionals for the individual fields. This is a general relation, applicable beyond QED, and we can apply it here for QCD, as well. Thus, (15-64) becomes

Free fields generating functional = product of separate free field generating functionals

For all QCD fields free $\quad Z_0\left[J_{i\kappa},\sigma_a,\bar{\sigma}_a,S_i,\tilde{S}_i\right]=Z_0\left[J_{i\kappa}\right]Z_0\left[\sigma_a,\bar{\sigma}_a\right]Z_0\left[S_i,\tilde{S}_i\right],$ (15-70)

Do **Problem 4** to deduce (15-70) explicitly for QCD.

Propagators

The propagators for the three types of fields parallel what we already know for scalar, spinor, and vector propagators. In (15-71) to (15-73) below, scripts i and j indicate gluon field or ghost field and range from 1 to 8. Scripts a and b represent color and range over r, g, b.

Faddeev-Popov QCD propagators

gluon
$$\begin{cases} iD_{F\,ij}^{\mu\nu}(x-y)=\underbrace{A_i^\mu(x)A_j^\nu(y)}=\underbrace{A_j^\nu(y)A_i^\mu(x)}=\delta_{ij}iD_F^{\mu\nu}(x-y) \\ =\delta_{ij}\dfrac{1}{(2\pi)^4}\int e^{-ik(x-y)}iD_F^{\mu\nu}(k)\,d^4k \quad \text{where } D_F^{\mu\nu}(k)=\dfrac{-g^{\mu\nu}}{k^2+i\varepsilon} \end{cases} \tag{15-71}$$

quark
$$\begin{cases} iS_{F\,ab}(x-y)=\underbrace{\psi_a(x)\bar{\psi}_b(y)}=-\underbrace{\bar{\psi}_b(y)\psi_a(x)}=\delta_{ab}iS_F(x-y) \\ =\delta_{ab}\dfrac{1}{(2\pi)^4}\int e^{-ip(x-y)}iS_F(p)\,d^4p \quad \text{where } S_F(p)=\dfrac{1}{\not p-m+i\varepsilon} \end{cases} \tag{15-72}$$

ghost
$$\begin{cases} i\Delta_{F\,ij}(x-y)=\underbrace{\tilde{\eta}_i(x)\eta_j(y)}=-\underbrace{\eta_j(y)\tilde{\eta}_i(x)}=\delta_{ij}i\Delta_F(x-y) \\ =\delta_{ij}\dfrac{1}{(2\pi)^4}\int e^{-ik(x-y)}i\Delta_F(k)\,d^4k \quad \text{where } \Delta_F(k)=\dfrac{1}{k^2+i\varepsilon} \end{cases} \tag{15-73}$$

The Free Generating Functionals: More Useful Form

There is a more convenient way of expressing the free generating functionals, using the relation near the end of Wholeness Chart 4-4, at the end of Chap. 4. First, recall the definition of the square brackets with three fields,

Math relation useful in PIA

$$[AKB]\equiv\int A(\underline{x})K(\underline{x},\underline{y})B(\underline{y})\,d^4\underline{x}\,d^4\underline{y}. \tag{15-74}$$

With (15-74), and with what we deduced in Chap. 4, we have

Separate free field generating functionals in more useful form

gluons $\quad Z_0\left[J_{i\kappa}\right]=e^{-\frac{i}{2}\left[J_{i\kappa}D_{F\,ij}^{\kappa\lambda}J_{j\lambda}\right]}$ (15-75)

quarks $\quad Z_0\left[\sigma_a,\bar{\sigma}_a\right]=e^{-i\left[\bar{\sigma}_aS_{F\,ab}\sigma_b\right]}$ (15-76)

ghosts $\quad Z_0\left[S_i,\tilde{S}_i\right]=e^{-i\left[S_i\Delta_{F\,ij}\tilde{S}_j\right]}.$ (15-77)

15.2 QCD Perturbation Theory

Faddeev-Popov QCD perturbation theory using PIA

In the present section, we will develop QCD perturbation theory along a parallel track to what we did in QED in Vol. 1, but do so, instead, via the PIA (path integral approach) of Chaps. 3 and 4 of this text. The steps we follow, including those of the prior chapters and the prior section of this chapter, are outlined in Wholeness Chart 15-1 below.

Note that in Chaps. 3 and 4, we used the generating functional to help prove the equivalence of the CA (canonical approach) and the PIA, but it is easier and more straightforward to do perturbation using Green functions directly, without reference to the generating functional.

Generating functional good for proofs; Green functions for calculation in perturbation theory

Generating functional	→	useful for proving equivalence of CA and PIA, and the validity of the Faddeev-Popov method
Green functions	→	useful for perturbation calculations

Wholeness Chart 15-1. Steps of QCD Perturbation Theory

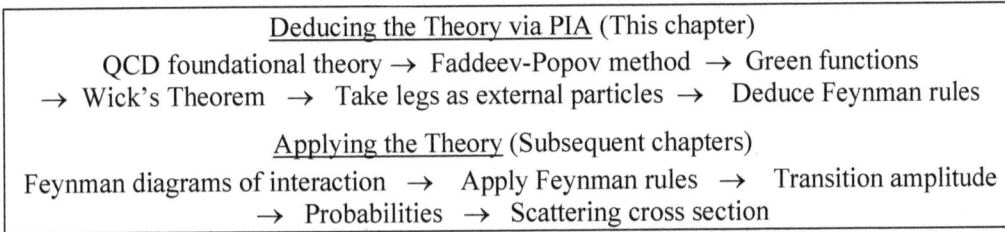

Deducing the Theory via PIA (This chapter)
QCD foundational theory → Faddeev-Popov method → Green functions
→ Wick's Theorem → Take legs as external particles → Deduce Feynman rules

Applying the Theory (Subsequent chapters)
Feynman diagrams of interaction → Apply Feynman rules → Transition amplitude
→ Probabilities → Scattering cross section

Steps for developing QCD perturbation theory via PIA

Steps for applying that theory to problems

In many cases, where the QCD and QED procedures are similar, we will merely draw conclusions from what we showed more rigorously in QED, and not go through the tedium of proving every step.

We will draw on QED parallels to shortcut full QCD analysis

15.2.1 The Perturbation Expansion

Extrapolating from QED

You probably should go back and review parts of Chaps. 3 and 4, particularly Sects. 3.1.4 and 3.1.5, Wholeness Chart 3-3, pg. 86, and Wholeness Chart 4-4, pg. 126. We will be particularly concerned with the QCD sibling of (4-23), which is also found in Wholeness Chart 4-4, and which we repeat below for convenience.

$$\text{QED}\quad G^{\mu..}(x_{1,.}y_{1,.}z_{1,.}) = \left\langle A^{\mu}(x_1)\ldots\psi(y_1)\ldots\bar{\psi}(z_1)\ldots\right\rangle = \frac{\int\left\{e^{i\int \mathcal{L}d^4x}A^{\mu}(x_1)..\psi(y_1)..\bar{\psi}(z_1)\right\}\mathcal{D}A\mathcal{D}\bar{\psi}\mathcal{D}\psi}{\int e^{i\int \mathcal{L}d^4x}\mathcal{D}A\mathcal{D}\bar{\psi}\mathcal{D}\psi}. \tag{15-78}$$

Recalling QED Green function

We need to extrapolate (15-78) from QED to QCD, which means turning photons into gluons, taking fermions as colored quarks, and including ghosts. We will also need to focus on the interaction part of the Lagrangian, as we will be determining interaction amplitudes. When we do all that, we have our Green function for perturbation QCD, with c and d representing color,

$$\text{QCD}\quad G^{\mu..}(x_1,\ldots y_1,\ldots z_1,\ldots u_1\ldots w_1) = \left\langle A_i^{\mu}(x_1)\ldots\psi_c(y_1)\ldots\bar{\psi}_d(z_1)\ldots\tilde{\eta}_j(u_1)\ldots\eta_k(w_1)\ldots\right\rangle$$
$$= \frac{\int\left\{e^{i\int \mathcal{L}d^4x}A_i^{\mu}(x_1)\ldots\psi_c(y_1)\ldots\bar{\psi}_d(z_1)\ldots\tilde{\eta}_j(u_1)\ldots\eta_k(w_1)\ldots\right\}\mathcal{D}A\mathcal{D}\bar{\psi}\mathcal{D}\psi\mathcal{D}\eta\mathcal{D}\tilde{\eta}}{\int e^{i\int \mathcal{L}d^4x}\mathcal{D}A\mathcal{D}\bar{\psi}\mathcal{D}\psi\,\mathcal{D}\eta\mathcal{D}\tilde{\eta}}, \tag{15-79}$$

Extrapolating to QCD Green function of Faddeev-Popov

where, as always, for connected Green functions, we can ignore the denominator.

Can ignore denominator for connected diagrams

Helpful Insight: Green Functions with $\mathcal{L}_0$

Employing Wick's Theorem:

In (15-79) for the free field case ($\mathcal{L}_I = 0$), we have, where the last line is in the CA form,

$$\left\langle A_i^{\mu}(x_1)\ldots\psi_c(y_1)\ldots\bar{\psi}_d(z_1)\ldots\tilde{\eta}_j(u_1)\ldots\eta_k(w_1)\ldots\right\rangle_{\mathcal{L}_0}$$
$$= \int e^{i\int \mathcal{L}_0 dx}\left\{A_i^{\mu}(x_1)\ldots\psi_c(y_1)\ldots\bar{\psi}_d(z_1)\ldots\tilde{\eta}_j(u_1)\ldots\eta_k(w_1)\ldots\right\}\mathcal{D}A\mathcal{D}\bar{\psi}\mathcal{D}\psi\mathcal{D}\eta\mathcal{D}\tilde{\eta} \tag{15-80}$$
$$= \left\langle 0\left|T\left\{SA_i^{\mu}(x_1)\ldots\psi_c(y_1)\ldots\bar{\psi}_d(z_1)\ldots\tilde{\eta}_j(u_1)\ldots\eta_k(w_1)\ldots\right\}\right|0\right\rangle,\qquad S = e^{i\int \mathcal{L}_I(x)dx} = e^0 = I.$$

Can use Wick's theorem in Green function when $\mathcal{L}_I = 0$

Note that, due to the CA form, we can conclude that we can use Wick's theorem with (15-80). We can turn the relation, in either the CA or PIA form, into terms with propagators.

Expressing the PIA Green Function:

It will save space, and actually make things a little easier to understand, if we demonstrate the main point of this subsection using QED, as it has fewer fields than (15-80). When we finish, we can readily extrapolate the results to QCD.

The $\mathcal{L}_0$ QED Green function equivalent of (15-80) is

$$\begin{aligned}\left\langle A^{\mu}(x_1)\ldots\psi(y_1)\ldots\bar{\psi}(z_1)\right\rangle_{\mathcal{L}_0} &= \int e^{i\int\mathcal{L}_0 dx}\left\{A^{\mu}(x_1)\ldots\psi(y_1)\ldots\bar{\psi}(z_1)\right\}\mathcal{D}A\mathcal{D}\bar{\psi}\mathcal{D}\psi \\ &= \left\langle 0\left|T\left\{SA^{\mu}(x_1)\ldots\psi(y_1)\ldots\bar{\psi}(z_1)\right\}\right|0\right\rangle, \quad \text{where here } S = Te^{i\int\mathcal{L}_I(x)dx} = e^0 = I.\end{aligned} \tag{15-81}$$

The QED Green function for interacting fields in the PIA is

$$\begin{aligned}\left\langle A^{\mu}(x_1)\ldots\psi(y_1)\ldots\bar{\psi}(z_1)\right\rangle &= \int e^{i\int\mathcal{L}dx}\left\{A^{\mu}(x_1)\ldots\psi(y_1)\ldots\bar{\psi}(z_1)\right\}\mathcal{D}A\mathcal{D}\bar{\psi}\mathcal{D}\psi \\ &= \int e^{i\int\mathcal{L}_0 dx}e^{i\int\mathcal{L}_I dx}\left\{A^{\mu}(x_1)\ldots\psi(y_1)\ldots\bar{\psi}(z_1)\right\}\mathcal{D}A\mathcal{D}\bar{\psi}\mathcal{D}\psi \\ &= \int e^{i\int\mathcal{L}_0 dx}\left(1+i\int\mathcal{L}_I dx+\ldots\right)\left\{A^{\mu}(x_1)\ldots\psi(y_1)\ldots\bar{\psi}(z_1)\right\}\mathcal{D}A\mathcal{D}\bar{\psi}\mathcal{D}\psi \\ &= \int e^{i\int\mathcal{L}_0 dx}\left\{A^{\mu}(x_1)\ldots\psi(y_1)\ldots\bar{\psi}(z_1)\right\}\mathcal{D}A\mathcal{D}\bar{\psi}\mathcal{D}\psi \\ &\quad +\int e^{i\int\mathcal{L}_0 dx}\left(i\int\mathcal{L}_I dx+\ldots\right)\left\{A^{\mu}(x_1)\ldots\psi(y_1)\ldots\bar{\psi}(z_1)\right\}\mathcal{D}A\mathcal{D}\bar{\psi}\mathcal{D}\psi \\ &= \int e^{i\int\mathcal{L}_0 dx}\left\{A^{\mu}(x_1)\ldots\psi(y_1)\ldots\bar{\psi}(z_1)\right\}\mathcal{D}A\mathcal{D}\bar{\psi}\mathcal{D}\psi \\ &\quad +\int e^{i\int\mathcal{L}_0 dx}\left(i\int e\bar{\psi}(x)\not{A}(x)\psi(x)dx+\ldots\right)\left\{A^{\mu}(x_1)\ldots\psi(y_1)\ldots\bar{\psi}(z_1)\right\}\mathcal{D}A\mathcal{D}\bar{\psi}\mathcal{D}\psi,\end{aligned} \tag{15-82}$$

where we can express the last part as

$$\begin{aligned}\left\langle A^{\mu}(x_1)\ldots\psi(y_1)\ldots\bar{\psi}(z_1)\right\rangle &= \left\langle A^{\mu}(x_1)\ldots\psi(y_1)\ldots\bar{\psi}(z_1)\right\rangle_{\mathcal{L}_0} \\ &\quad +ie\int\left\langle\left(\bar{\psi}(x)\not{A}(x)\psi(x)\right)A^{\mu}(x_1)\ldots\psi(y_1)\ldots\bar{\psi}(z_1)\right\rangle_{\mathcal{L}_0}dx+\ldots\end{aligned} \tag{15-83}$$

Using Wick's theorem on the first term after the equal sign in (15-83) yields only disconnected propagators. No photon or fermions will share a vertex, which makes sense, since that is a free-fields-only Green function. As we are disregarding disconnected Green functions, we can disregard that term. Thus, at first order, (15-83) becomes

$$\left\langle A^{\mu}(x_1)\ldots\psi(y_1)\ldots\bar{\psi}(z_1)\right\rangle \approx i\int\left\langle\left(e\bar{\psi}(x)\not{A}(x)\psi(x)\right)A^{\mu}(x_1)\ldots\psi(y_1)\ldots\bar{\psi}(z_1)\right\rangle_{\mathcal{L}_0}dx\,. \tag{15-84}$$

The exact same logic, with more fields involved, can be used for QCD, so the equivalent form of (15-84) in that theory is, where $\mathcal{L}_I$ is for QCD, rather than QED,

$$\begin{aligned}&\left\langle A_i^{\mu}(x_1)\ldots\psi_c(y_1)\ldots\bar{\psi}_d(z_1)\ldots\tilde{\eta}_j(u_1)\ldots\eta_k(w_1)\ldots\right\rangle \\ &\quad\approx i\int\left\langle\mathcal{L}_I(x)A_i^{\mu}(x_1)\ldots\psi_c(y_1)\ldots\bar{\psi}_d(z_1)\ldots\tilde{\eta}_j(u_1)\ldots\eta_k(w_1)\ldots\right\rangle_{\mathcal{L}_0}dx.\end{aligned} \tag{15-85}$$

Faddeev-Popov Green function in terms of a free field Green function

We are about to make good use of (15-85).

15.2.2 Vertex Factors

Formal, lengthy QCD analysis → factors arise for vertices

Need to include in Feynman's rules

We will look in this section at first-order perturbation theory using (15-79), where we ignore the denominator, expressed in the form of (15-85), to determine factors in the transition amplitude corresponding to QCD vertices in Feynman diagrams. We will find that each different vertex type has what is called a "vertex factor" in the transition amplitude, and that is what we will determine herein for each of the vertices of Fig. 15-1. These vertex factors will become part of the Feynman rules for perturbative QCD.

Similar to what we did in QED and e/w

Note that in QED, there was only one type of vertex (two fermions and a photon) and the vertex factor for Feynman rules there was $ie\gamma^{\mu}$. In GSW electroweak theory, we had many such vertex factors, as there were many types of vertices. See Feynman's rules for electroweak theory, pgs. 290-293.

The Quark-Gluon Vertex

Applying to 1st specific case: quark-quark-gluon vertex

Consider one gluon and two quarks. The appropriate part of $\mathcal{L}_I$ in (15-85) is $\mathcal{L}^{QG}$ of (15-53),

$$\mathcal{L}_I = \mathcal{L}^{QG} = -\frac{g_S}{2}\bar{\psi}_a\left(\lambda_j\right)_{ab}\gamma_\mu\psi_b A_j^\mu . \qquad (15\text{-}86)$$

Then, for first order, we use (15-86) in (15-85) (with ν in place of μ, since we have μ in (15-86)),

1st order qqG legs Green function in terms of the $\mathcal{L}_0$ Green function

$$\begin{aligned}\left\langle A_i^\nu(x_1)\psi_c(y_1)\bar{\psi}_d(z_1)\right\rangle &\approx -i\frac{g_S}{2}\int\left\langle\left\{\bar{\psi}_a(x)\left(\lambda_j\right)_{ab}\gamma_\mu\psi_b(x)\right\}A_j^\mu(x)A_i^\nu(x_1)\psi_c(y_1)\bar{\psi}_d(z_1)\right\rangle_{\mathcal{L}_0}dx\\ &= -i\frac{g_S}{2}\int\underbrace{\left\langle A_j^\mu(x)A_i^\nu(x_1)\psi_c(y_1)\left\{\bar{\psi}_a(x)\left(\lambda_j\right)_{ab}\gamma_\mu\psi_b(x)\right\}\bar{\psi}_d(z_1)\right\rangle_{\mathcal{L}_0}}_{F_{icd}^\nu}dx. \end{aligned} \qquad (15\text{-}87)$$

where we define F_{icd}^ν as shown, and the second line follows from the first by switching fermions an even number of times. F_{icd}^ν, like (15-80), can be expanded via Wick's theorem to get

Expanding via Wick's theorem

$$\begin{aligned}F_{icd}^\nu = {} & A_j^\mu(x)A_i^\nu(x_1)\psi_c(y_1)\bar{\psi}_a(x)\left(\lambda_j\right)_{ab}\gamma_\mu\psi_b(x)\bar{\psi}_d(z_1)\\ &+A_j^\mu(x)A_i^\nu(x_1)\psi_c(y_1)\bar{\psi}_a(x)\left(\lambda_j\right)_{ab}\gamma_\mu\psi_b(x)\bar{\psi}_d(z_1)\end{aligned} . \qquad (15\text{-}88)$$

where we can discard the disconnected part

The second row represents two disconnected diagrams, as shown in the LHS of Fig. 15-2. The first is simply a single diagram with three propagators connected at spacetime event x, i.e., a vertex, as shown in the RHS.

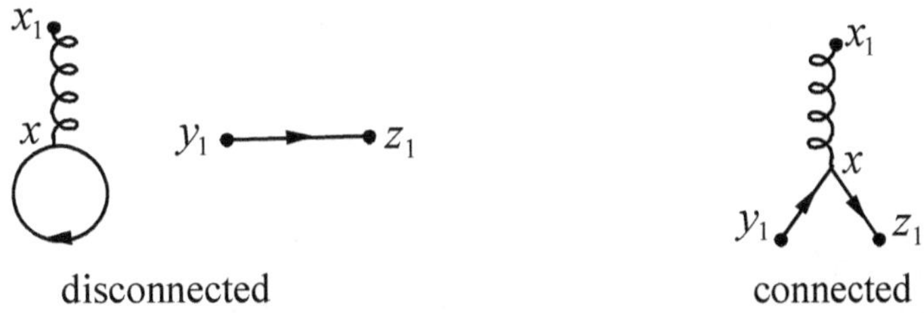

Figure 15-2. Connected and Disconnected Parts of (15-88)

Since we are only concerned with connected Green functions, and since the disconnected one cancels with a factor in the denominator in (15-79), anyway, we ignore the second row of (15-88). With (15-71) and (15-72), this gives us (where the Kronecker deltas change ab to cd and j to i),

$$F_{icd}^\nu = iD_F^{\mu\nu}(x_1 - x)iS_F(y_1 - x)\gamma_\mu\left(\lambda_i\right)_{cd}iS_F(x - z_1), \qquad (15\text{-}89)$$

which, into (15-87), gives us

1st order qqG Green function, in position space

$$\left\langle A_i^\nu(x_1)\psi_c(y_1)\bar{\psi}_d(z_1)\right\rangle = -i\frac{g_S}{2}\int iD_F^{\mu\nu}(x_1 - x)iS_F(y_1 - x)\gamma_\mu\left(\lambda_i\right)_{cd}iS_F(x - z_1)dx . \qquad (15\text{-}90)$$

With (15-63), we find the corresponding momentum space Green function

$$\begin{aligned}&(2\pi)^4\delta^{(4)}(k + p' + p)\left\langle A_i^\nu(k)\psi_c(p')\bar{\psi}_d(p)\right\rangle\\ &= \int e^{-ikx_1}e^{-ip'y_1}e^{-ipz_1}\left\langle A_i^\nu(x_1)\psi_c(y_1)\bar{\psi}_d(z_1)\right\rangle dx_1dy_1dz_1 \qquad (15\text{-}91)\\ &= -i\frac{g_S}{2}\int e^{-ikx_1}e^{-ip'y_1}e^{-ipz_1}iD_F^{\mu\nu}(x_1 - x)iS_F(y_1 - x)\gamma_\mu\left(\lambda_i\right)_{cd}iS_F(x - z_1)dx\,dx_1dy_1dz_1,\end{aligned}$$

where, we recall, all momenta are directed inward.

Do **Problem 5** to show (15-92).

From the result of Problem 5, we find the momentum space Green function in (15-91) to be

$$\left\langle A_i^{\gamma}(k)\psi_c(p')\bar{\psi}_d(p)\right\rangle = iD_F^{\mu\nu}(k)iS_F(p+k)\left(-i\tfrac{g_s}{2}\gamma_\mu(\lambda_i)_{cd}\right)iS_F(p), \tag{15-92}$$

and in momentum space

where we have a

$$\underline{\text{gluon-quark-quark vertex factor}} \;\; -ig_s\gamma_\mu\left(\frac{\lambda_i}{2}\right)_{cd} \qquad \left(\text{some authors use } T_i = \frac{\lambda_i}{2}\right). \tag{15-93}$$

From which we read off the vertex factor

This is represented graphically in the Feynman diagram of Fig. 15-3(a). That figure also shows vertex factors for the other types of QCD vertices, which we will be deducing in the next few pages.

This vertex diagram and its factor, plus all others in QCD

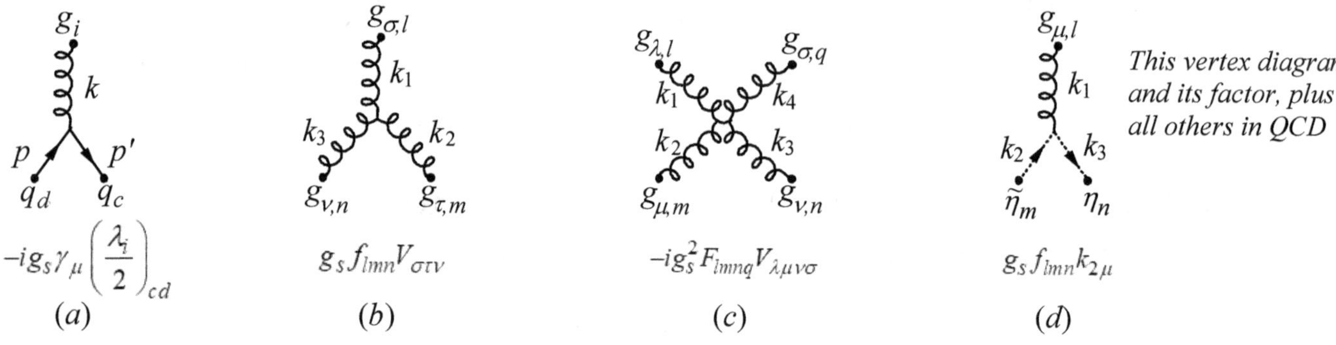

Figure 15-3. QCD Vertex Factors via Faddeev-Popov Approach

The vertex factor (15-93) will pop up in Feynman rules every time we have a vertex with two quarks and a gluon. Whereas in QED, we just had $ie\gamma^\mu$ (for a photon gauge boson, instead of a gluon), in QCD, we have (15-93).

Can use PIA or CA to find vertex factors in QED, e/w, or QCD

Note that we could have used the PIA and Green functions in QED to determine the factor $ie\gamma^\mu$, instead of the CA that we did use in Vol. 1. This would parallel what we did above, except there would be no color indices such as a and b, only one gauge boson with $\lambda_i/2$ replaced by 1, and $-e$ instead of g_s.

<u>The Three-Gluon Vertex</u>

2nd case: GGG vertex

For the three-gluon vertex, shown in Fig. 15-3(b), we will follow a similar procedure, but use an interaction Lagrangian equal to the second term in (15-53), and start with the 1st order Green function, where the symbol $g_{\mu\nu}$ represents the Minkowski (flat spacetime) metric, as it typically does in QFT,

$$\begin{aligned}\left\langle A_l^{\alpha}(x_1)A_m^{\beta}(x_2)A_n^{\gamma}(x_3)\right\rangle &\approx i\int \left\langle \mathcal{L}_I(x)A_l^{\alpha}(x_1)A_m^{\beta}(x_2)A_n^{\gamma}(x_3)\right\rangle_{\mathcal{L}_0} dx \\ &= i\int \left\langle g_s f_{ijk}A_{i\mu}(x)A_{j\nu}(x)\partial^\mu A_k^{\nu}(x)A_l^{\alpha}(x_1)A_m^{\beta}(x_2)A_n^{\gamma}(x_3)\right\rangle_{\mathcal{L}_0} dx \\ &= i\,g_s g_{\mu\sigma}g_{\nu\tau}f_{ijk}\int \underbrace{\left\langle A_i^{\sigma}(x)A_j^{\tau}(x)\partial^\mu A_k^{\nu}(x)A_l^{\alpha}(x_1)A_m^{\beta}(x_2)A_n^{\gamma}(x_3)\right\rangle_{\mathcal{L}_0}}_{F} dx.\end{aligned} \tag{15-94}$$

1st order GGG legs Green function with appropriate $\mathcal{L}_I$ term

Our symbol F should have subscript and superscripts attached to it, but we will streamline by not using them. Employing Wick's theorem to find all combinations of contractions in F of (15-94), we know that any with contractions between two gluon fields that are both functions of x will give us disconnected Green functions (like we found in the second row of (15-88) and the LHS of Fig. 15-2). Since we can ignore those, we are left with only contractions between pairs of gluon fields, one of which is a function of x, and the other a function of either x_1, x_2, or x_3. Let's look at one of these terms and then deduce the remaining ones via symmetry. For

Expand via Wick's theorem → several terms

$$F_1 = \underbrace{A_i^{\sigma}(x)A_l^{\alpha}(x_1)}\underbrace{A_j^{\tau}(x)A_m^{\beta}(x_2)}\partial^\mu \underbrace{A_k^{\nu}(x)A_n^{\gamma}(x_3)}, \tag{15-95}$$

Evaluating one of those terms

we have one part of (15-94), given (15-71), as

$$\left\langle A_l^{\alpha}(x_1)A_m^{\beta}(x_2)A_n^{\gamma}(x_3)\right\rangle_1 \approx ig_s g_{\mu\sigma}g_{\nu\tau}f_{lmn}\int iD_F^{\sigma\alpha}(x-x_1)iD_F^{\tau\beta}(x-x_2)\frac{\partial}{\partial x_\mu}iD_F^{\nu\gamma}(x-x_3)dx. \tag{15-96}$$

Green function in position space for that term

We again want to convert to momentum space, as that is what Feynman rules are in terms of.

$$(2\pi)^4 \delta^{(4)}(k_1+k_2+k_3)\left\langle A_l^{\alpha}(k_1) A_m^{\beta}(k_2) A_n^{\gamma}(k_3)\right\rangle_1$$
$$= \int e^{-ik_1x_1} e^{-ik_2x_2} e^{-ik_3x_3} \left\langle A_l^{\alpha}(x_1) A_m^{\beta}(x_2) A_n^{\gamma}(x_3)\right\rangle_1 dx_1 dx_2 dx_3 \qquad (15\text{-}97)$$
$$= ig_s g_{\mu\sigma} g_{\nu\tau} f_{lmn} \int e^{-ik_1x_1} e^{-ik_2x_2} e^{-ik_3x_3} iD_F^{\sigma\alpha}(x-x_1) iD_F^{\tau\beta}(x-x_2) \frac{\partial}{\partial x_\mu} iD_F^{\nu\gamma}(x-x_3) dx\, dx_1 dy_1 dz_1 .$$

Substituting explicit relations for the position space propagators of (15-71) into (15-97), we get, after a lot of algebra (though you should, by now, have a sense for how these transformations to momentum space come out)

$$\left\langle A_l^{\alpha}(k_1) A_m^{\beta}(k_2) A_n^{\gamma}(k_3)\right\rangle_1 = g_s g_{\mu\sigma} g_{\nu\tau} f_{lmn} k_3^{\mu} iD_F^{\sigma\alpha}(k_1) iD_F^{\tau\beta}(k_2) iD_F^{\nu\gamma}(k_3), \qquad (15\text{-}98)$$

Green function in momentum space for that term

where $k_1 + k_2 + k_3 = 0$ and as usual, four-momenta are considered directed inward in Fig. 15-3(b).

The remaining contributions come from interchanging the fields in x_1, x_2, or x_3 in (15-95) with one another. For example, from interchanging the last two of these in (15-95), we would have

$$F_2 = \underbrace{A_i^{\sigma}(x) A_l^{\alpha}(x_1)} \underbrace{A_j^{\tau}(x) A_n^{\gamma}(x_3)} \underbrace{\partial^{\mu} A_k^{\nu}(x) A_m^{\beta}(x_2)}, \qquad (15\text{-}99)$$

Evaluating another one of the terms

and this ultimately leads a sibling relation of (15-98),

$$\left\langle A_l^{\alpha}(k_1) A_m^{\beta}(k_2) A_n^{\gamma}(k_3)\right\rangle_2 = g_s g_{\mu\sigma} g_{\nu\tau} f_{lnm} k_2^{\mu} iD_F^{\sigma\alpha}(k_1) iD_F^{\tau\gamma}(k_3) iD_F^{\nu\beta}(k_2). \qquad (15\text{-}100)$$

Given the anti-symmetry of the structure constants, (15-100) can be re-written as

$$\left\langle A_l^{\alpha}(k_1) A_m^{\beta}(k_2) A_n^{\gamma}(k_3)\right\rangle_2 = -g_s g_{\mu\sigma} g_{\nu\tau} f_{lmn} k_2^{\mu} iD_F^{\sigma\alpha}(k_1) iD_F^{\tau\gamma}(k_3) iD_F^{\nu\beta}(k_2) \qquad (15\text{-}101)$$

or, by further exchanging the dummy variables ν and τ, as

$$\left\langle A_l^{\alpha}(k_1) A_m^{\beta}(k_2) A_n^{\gamma}(k_3)\right\rangle_2 = -g_s g_{\mu\sigma} g_{\nu\tau} f_{lmn} k_2^{\mu} iD_F^{\sigma\alpha}(k_1) iD_F^{\tau\beta}(k_2) iD_F^{\nu\gamma}(k_3). \qquad (15\text{-}102)$$

Green function in momentum space for this other term

Adding (15-98) and (15-102), we have

$$\left\langle A_l^{\alpha}(k_1) A_m^{\beta}(k_2) A_n^{\gamma}(k_3)\right\rangle_1 + \left\langle A_l^{\alpha}(k_1) A_m^{\beta}(k_2) A_n^{\gamma}(k_3)\right\rangle_2$$
$$= g_s g_{\mu\sigma} f_{lmn} g_{\nu\tau} \left(k_3^{\mu} - k_2^{\mu}\right) iD_F^{\sigma\alpha}(k_1) iD_F^{\tau\beta}(k_2) iD_F^{\nu\gamma}(k_3) \qquad (15\text{-}103)$$
$$= g_s f_{lmn} g_{\nu\tau} \left(k_{3\sigma} - k_{2\sigma}\right) iD_F^{\sigma\alpha}(k_1) iD_F^{\tau\beta}(k_2) iD_F^{\nu\gamma}(k_3).$$

Green function in momentum space for these two terms

There are 3! ways to re-arrange the factors in (15-95) with arguments of x_i, so we have four more non-zero contributions to (15-94). In total, we have

$$\begin{aligned}
F_1 &= \underbrace{A_i^{\sigma}(x) A_l^{\alpha}(x_1)} \underbrace{A_j^{\tau}(x) A_m^{\beta}(x_2)} \left(\underbrace{\partial^{\mu} A_k^{\nu}(x)) A_n^{\gamma}(x_3)}\right) \\
F_2 &= \underbrace{A_i^{\sigma}(x) A_l^{\alpha}(x_1)} \underbrace{A_j^{\tau}(x) A_n^{\gamma}(x_3)} \left(\underbrace{\partial^{\mu} A_k^{\nu}(x)) A_m^{\beta}(x_2)}\right) \\
F_3 &= \underbrace{A_i^{\sigma}(x) A_m^{\beta}(x_2)} \underbrace{A_j^{\tau}(x) A_l^{\alpha}(x_1)} \left(\underbrace{\partial^{\mu} A_k^{\nu}(x)) A_n^{\gamma}(x_3)}\right) \\
F_4 &= \underbrace{A_i^{\sigma}(x) A_m^{\beta}(x_2)} \underbrace{A_j^{\tau}(x) A_n^{\gamma}(x_3)} \left(\underbrace{\partial^{\mu} A_k^{\nu}(x)) A_l^{\alpha}(x_1)}\right) \\
F_5 &= \underbrace{A_i^{\sigma}(x) A_n^{\gamma}(x_3)} \underbrace{A_j^{\tau}(x) A_m^{\beta}(x_2)} \left(\underbrace{\partial^{\mu} A_k^{\nu}(x)) A_l^{\alpha}(x_1)}\right) \\
F_6 &= \underbrace{A_i^{\sigma}(x) A_n^{\gamma}(x_3)} \underbrace{A_j^{\tau}(x) A_l^{\alpha}(x_1)} \left(\underbrace{\partial^{\mu} A_k^{\nu}(x)) A_m^{\beta}(x_2)}\right).
\end{aligned} \qquad (15\text{-}104)$$

The six parts of the GGG 1st order Green function

The last four will pair up, much like (15-95) and (15-99) did to give us (15-103), and when we add all six of them together, we get

$$\left\langle A_l^{\alpha}(k_1) A_m^{\beta}(k_2) A_n^{\gamma}(k_3)\right\rangle = \overbrace{g_s f_{lmn} V_{\sigma\tau\nu}}^{\text{vertex factor}} iD_F^{\sigma\alpha}(k_1) iD_F^{\tau\beta}(k_2) iD_F^{\nu\gamma}(k_3) \qquad (15\text{-}105)$$

Combining all six, we get the GGG vertex factor

where the *l, m, n* are gluon indices, and

3-gluon vertex factor is

$$g_s f_{lmn} V_{\sigma\tau\nu} = g_s f_{lmn}\left\{g_{\nu\tau}\left(k_{3\sigma}-k_{2\sigma}\right)+g_{\tau\sigma}\left(k_{2\nu}-k_{1\nu}\right)+g_{\sigma\nu}\left(k_{1\tau}-k_{3\tau}\right)\right\}. \qquad (15\text{-}106)$$

In QED, we don't have a three-photon vertex, so we can't compare this result to the corresponding vertex factor in that theory, as we did for the two quark/one gluon vertex above. The vertex factor (15-106) is shown in Fig. 15-3(b).

The Four-Gluon Vertex Factor

For the four-gluon vertex, we use interaction Lagrangian of the third term in (15-53), and follow similar steps to what we did for the three-gluon vertex. We will opt to forgo the algebraic tedium and simply accept what others before us, who have gone through that tedium, have found, which is

3rd case: GGGG vertex

$$\left\langle A_l^\alpha(k_1)A_m^\beta(k_2)A_n^\gamma(k_3)A_q^\delta(k_4)\right\rangle = \overbrace{-ig_s^2F_{lmnq}V_{\lambda\mu\nu\sigma}}^{\text{vertex factor}}\, iD_F^{\lambda\alpha}(k_1)\,iD_F^{\mu\beta}(k_2)\,iD_F^{\nu\gamma}(k_3)\,iD_F^{\sigma\delta}(k_4) \qquad (15\text{-}107)$$

1st order GGGG legs Green function in momentum space

where the 4-gluon vertex factor is

$$-ig_s^2F_{lmnq}V_{\lambda\mu\nu\sigma} = -ig_s^2\left\{\begin{array}{l} f_{ilm}f_{inq}\left(g_{\lambda\nu}g_{\mu\sigma}-g_{\mu\nu}g_{\lambda\sigma}\right) \\ \quad + f_{inm}f_{ilq}\left(g_{\nu\lambda}g_{\mu\sigma}-g_{\mu\lambda}g_{\nu\sigma}\right)+f_{iln}f_{imq}\left(g_{\lambda\mu}g_{\nu\sigma}-g_{\nu\mu}g_{\lambda\sigma}\right)\end{array}\right\}. \qquad (15\text{-}108)$$

GGGG vertex function

This is shown in Fig. 15-3(c).

The Gluon-Ghost Vertex

4th case: ghost-ghost-gluon vertex

In similar fashion to what we did for the two-quark, one gluon vertex, we can find the

gluon-ghost-ghost vertex factor $\qquad g_s f_{lmn}k_{2\mu}, \qquad$ (15-109)

$\eta\eta G$ vertex factor

which is shown in Fig. 15-3(d).

Do **Problem 6** to show (15-109).

15.3 Feynman Rules for Strong Interactions

Feynman rules for QCD can be found in similar fashion to how they were found in QED. One simply analyzes a slew of transition amplitudes (found in the PIA by taking Green function legs as external particles), notices patterns that arise, and then, formulates rules abiding to those patterns. This, of course, is a long and tedious process, and once again, we will avoid such a process and be content with simply citing what those who have actually carried it out have found.

QCD Feynman rules, incorporating vertex factors

15.3.1 The QCD Rules

The first several rules closely parallel what we have already seen in the QED and electroweak Feynman rules, with certain modifications unique to QCD. There are two additional rules, however, unlike any we have seen previously, and which are specific to QCD.

Feynman's Rules for QCD

Symbols *a, b, c, ...* represent color; *i, j, k,...* represent associated gluon field (eight in all).

A. The *S* matrix element (the transition amplitude) for a given interaction is

$$S_{fi} = \delta_{fi} + \left((2\pi)^4\delta^{(4)}\left(\sum p_f - \sum p_i\right)\left(\prod_{\text{ext bosons}}\sqrt{\frac{1}{2V\omega_{\mathbf{k}}}}\right)\left(\prod_{\text{ext fermions}}\sqrt{\frac{m}{VE_{\mathbf{p}}}}\right)\right)\mathcal{M} \qquad \mathcal{M}=\sum_{n=1}^{\infty}\mathcal{M}^{(n)} \qquad (8\text{-}110)$$

where Σp_f is the total 4-momentum of all final particles, Σp_i is the total 4-momentum of all initial particles, and the contribution $\mathcal{M}^{(n)}$ comes from the *n*th order perturbation term of the *S* operator, $S^{(n)}$.

B. The Feynman amplitude $\mathcal{M}^{(n)}$ is obtained from all of the topologically distinct, connected (i.e., all lines connected to one another in a given diagram) Feynman diagrams which contain *n* vertices. The contribution to each $\mathcal{M}^{(n)}$ is obtained by the following.

Rules Similar in Concept to QED and Electroweak Theories

1. For each vertex of given type, include the factor associated with that type in the following.

Rule #	$\mathcal{L}$ Part	Fields	Eq	Vertex Factor
1(a)	$\mathcal{L}^{QG}$	qqG	(15-93)	$-ig_s\gamma_\mu\left(\frac{\lambda_i}{2}\right)_{ab}$
1(b)	$\mathcal{L}^{GG}$	GGG	(15-106)	$g_s f_{lmn} V_{\sigma\tau\nu} = g_s f_{lmn}\left\{g_{\nu\tau}\left(k_{3\sigma}-k_{2\sigma}\right)+g_{\tau\sigma}\left(k_{2\nu}-k_{1\nu}\right)+g_{\sigma\nu}\left(k_{1\tau}-k_{3\tau}\right)\right\}$
1(c)	$\mathcal{L}^{GG}$	$GGGG$	(15-108)	$-ig_s^2 F_{lmnq} V_{\lambda\mu\nu\sigma} = -ig_s^2\left\{\begin{matrix} f_{ilm}f_{inq}\left(g_{\lambda\nu}g_{\mu\sigma}-g_{\mu\nu}g_{\lambda\sigma}\right)+f_{inm}f_{ilq}\left(g_{\nu\lambda}g_{\mu\sigma}-g_{\mu\lambda}g_{\nu\sigma}\right) \\ +f_{iln}f_{imq}\left(g_{\lambda\mu}g_{\nu\sigma}-g_{\nu\mu}g_{\lambda\sigma}\right)\end{matrix}\right\}$
1(d)	$\mathcal{L}^{gG}$	$\eta\eta G$	(15-109)	$g_s f_{lmn} k_{2\mu}$

2. For each internal line type, include the propagator factor associated with that type below.

Rule #	Field	Propagator
2(a)	gluon line, labeled by 4-momentum k	$iD_{F\,ij}^{\mu\nu}(k)=i\delta_{ij}\frac{-g^{\mu\nu}}{k^2+i\varepsilon}$
2(b)	quark line, labeled by 4-momentum p	$iS_{F\,ab}(p)=i\delta_{ab}\frac{1}{\not{p}-m+i\varepsilon}$
2(c)	ghost line, labeled by 4-momentum k	$i\Delta_{F\,ij}(k)=\delta_{ij}\frac{1}{k^2+i\varepsilon}$

3. For each external line, include one of the following spinor factors, where **p** and **k** indicate basis states of corresponding 3-momenta, r represents spin state for quarks and polarization state for gluons. Only transverse states for gluons are allowed, i.e., $r=1,2$.

   a) for each initial quark: $u_{ar}(\mathbf{p})$ b) for each final quark: $\bar{u}_{ar}(\mathbf{p})$

   c) for each initial antiquark: $\bar{v}_{ar}(\mathbf{p})$ d) for each final antiquark: $v_{ar}(\mathbf{p})$

   e) for each initial gluon: $\varepsilon_{ir\mu}(\mathbf{k})$ f) for each final gluon: $\varepsilon^*_{ir\mu}(\mathbf{k})$

4. The spinor factors (γ matrices, S_F functions, spinors) for each quark line are ordered so that, reading from right to left, they occur in the same sequence as following the quark line in the direction of its arrow through the vertex. (Order is important as it conveys spinor matrix multiplication order when we do not show spinor indices.)

5. The four-momenta at each vertex are conserved (same total after as before).

   5a. For each anti-particle quark propagator, label the Feynman diagram 4-momentum with opposite sign of what it represents physically and use this negative of the physical value in the propagator.

6. For each closed loop of internal quarks only (without gluons inside the loop itself, a "gluon loop" which internally has a quark and anti-quark), take the trace (in spinor space) of the resulting matrix and multiply by a factor of (–1).

7. For each 4-momentum q which is not fixed by 4-momentum conservation, carry out the integration $(1/2\pi)^4\int d^4q$. One such integration for each closed loop.

8. Multiply the expression by (–1) for each interchange of neighboring Grassmann fields (each associated with a particular spinor factor from rule #3) which would be required to place the expression in appropriate normal order. "Appropriate", when we are adding sub amplitudes, means each sub amplitude must be in the same, not just any, normal order of destruction and creation operators.

Additional Types of Rules for QCD

9. For each closed loop of ghosts, multiply by a factor of (–1).

10. For each closed loop of gluons, multiply by the relevant "symmetry factor" S, described below.

15.3.2 Clarification of the Additional Types of Rules

Rules unique to QCD

Rule # 9

Note that Rule #9 arises in the same way Rule # 6 does. It is a result of the ghost fields being Grassmann fields, which anti-commute, like fermion fields do.

Ghosts as Grassmann fields give us Rule #9

Rule # 10

For a diagram like Fig. 15-5(a), we can employ rule #1(b) for each of the two three-gluon vertices, i.e., we would have to include a factor of $(g_s f_{lmn} V_{\sigma\tau\nu})^2$ in the transition amplitude. This is a higher order diagram without loops (a tree level diagram).

Rule #10 needs elaboration

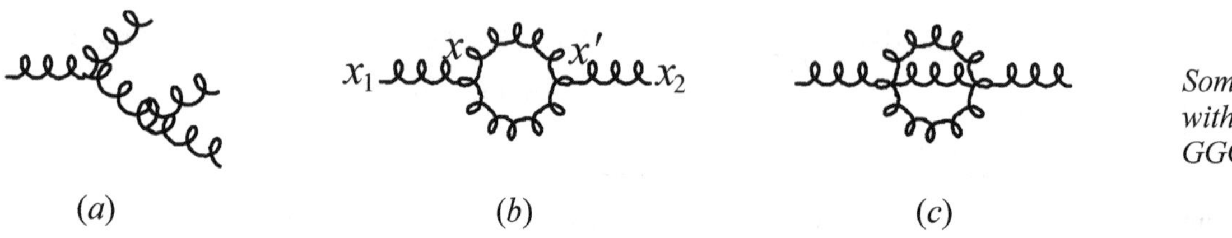

Some diagrams with GGG and GGGG vertices

Figure 15-5. Gluon Vertices in Different Diagrams

However, that rule does not do the job when we have higher order diagrams with loops as in Figs. 15-5(b) and (c). To see why, first consider how we got the vertex factor in Rule #1(b).

No gluon loops, don't need Rule #10; with gluon loops, need it

We obtained that factor by starting with (15-94) for three gluons,

Review of how we got GGG vertex factor

$$\left\langle A_l^\alpha(x_1) A_m^\beta(x_2) A_n^\gamma(x_3)\right\rangle \approx i\int \underbrace{\Big\langle \overbrace{g_s f_{ijk} g_{\mu\sigma} g_{\nu\tau} A_i^\sigma(x) A_j^\tau(x)\partial^\mu A_k^\nu(x)}^{\mathcal{L}_3(x)} A_l^\alpha(x_1) A_m^\beta(x_2) A_n^\gamma(x_3)\Big\rangle_{\mathcal{L}_0}}_{F} dx, \qquad (15\text{-}111)$$

where F is comprised of 3! separate terms (15-104), not 6!, because any contraction between fields at x will lead to disconnected diagrams, which we are ignoring as they are not relevant, since they cancel the denominator in our Green function, and we are ignoring that, too.

where we had 3! = 6 contributing terms

But, diagrams with gluon loops can have two (or more) vertices of the same type in the same loop, as in Fig. 15-5(b) (two three-gluon vertices) or Fig. 15-5(c) (two four-gluon vertices). Note what happens, then, for example, with our Green function for the first of these,

Consider two gluon loop case of Fig. 15-5(b)

$$\left\langle A_l^\alpha(x_1) A_m^\beta(x_2)\right\rangle \xrightarrow[\text{order}]{\text{2nd}} \propto -\frac{1}{2}\int \underbrace{\left\langle \mathcal{L}_3(x)\mathcal{L}_3(x') A_l^\alpha(x_1) A_m^\beta(x_2)\right\rangle_{\mathcal{L}_0}}_{F^{2nd}} dxdx',$$

$$\text{where } \mathcal{L}_3(x) = g_s g_{\mu\sigma} g_{\nu\tau} f_{ijk} A_i^\sigma(x) A_j^\tau(x)\partial^\mu A_k^\nu(x), \qquad (15\text{-}112)$$

$$\mathcal{L}_3(x') = g_s g_{\mu'\sigma'} g_{\nu'\tau'} f_{i'j'k'} A_{i'}^{\sigma'}(x') A_{j'}^{\tau'}(x')\partial^{\mu'} A_{k'}^{\nu'}(x').$$

Green function for that case

In Fig. 15-5 (b), the interchange of x and x' to make a separate diagram is already taken care of by the Feynman rules. So, we only have to look at the contractions between the fields in x_1, x, x', and x_2 shown there.

The field in x_1 can contract with any one of the three fields in x (3 ways). Then each of the remaining two fields in x can then contract with any one of the three fields in x' (2X3 = 6 ways). That leaves one field in x' and one field in x_2, which must contract with one another, as they are the only fields left. 3 times 6 is 18, which means there are a total of 18 possible ways to form the contractions, i.e., 18 separate terms in F^{2nd} of (15-112).

Analysis shows half as many terms as assumed to get Rule #1(b)

But in arriving at rule #1(b), for a single three-gluon vertex, we were considering 3! = 6 different terms in F. Going by that rule for Fig. 15-5(b), would mean we should be considering 6^2 = 36 terms. 36/18 = 2, so we would get an amplitude that is twice what the correct one is.

The answer lies in Rule #10, where our symmetry factor S in this case is ½. Using that and all the other rules, we cancel the extra (erroneous) impact of Rule #1(b) and get the correct amplitude.

So, need to multiply by symmetry factor S=1/2, as Rule #10 prescribes

Similar analyses can be made to determine symmetry factors for different combinations of gluon loops. For example, for the loops of Fig. 15-5(*c*), the symmetry factor works out to be 1/6.

Other symmetry factors for gluon loop diagrams

15.4 Chapter Summary

15.4.1 Background

QCD is not amenable to perturbation for analysis of hadron formation because the strong coupling constant g_s "runs" with energy, and has a large value at low energy (large distance) levels. As quarks separate from one another, the force holding them gets stronger, such that they can never be found in isolation, and this is known as quark confinement or infrared slavery. A method of QCD analysis in this range, called lattice gauge theory, employs numerical integration of QCD path integrals, using powerful computers, and therein lies the power of the PIA over the CA for non-perturbative QFT.

However, at large energy (short distance), g_s becomes so small that perturbation is possible. As quarks get close together, the attraction between them gets weak, so much so, that one can employ the usual QFT perturbation techniques. That is, one could use the CA, but in this text, we have chosen to use the PIA instead, to provide practice with using the PIA for strong interactions.

As energy levels approach infinity (distance approaches zero), the attraction between quarks approaches zero, and this is known as asymptotic freedom.

Experiments in the late 1960s and early 1970s indicated hadrons had substructure, i.e., they were composed of particles called partons, or, alternatively, quarks and gluons. The evidence strongly suggested that these substructure particles, when bombarded with high energy electrons, interacted as if they were free, and not confined.

In 1973, Gross, Politzer, and Wilczek showed that $SU(3)$ color field theory explained the experimental results. Specifically, they showed that, within that theory, the β function for the strong coupling constant was negative, implying both asymptotic freedom and quark confinement.

15.4.2 The Faddeev-Popov Approach to QCD

The fact that massless vector bosons are constricted by nature to have only two independent components, the transverse components, means that in the PIA, we would be integrating over more field components than simply the independent ones. This causes the integral to be ill-defined.

Faddeev and Popov came up with a method to resolve this issue, but it resulted in an extra gluon term in the Lagrangian plus the introduction of new, fictitious fields called ghosts. These ghost fields do not generate real particles (that are seen in the physical universe), but they must be included in the theory as propagators (giving rise to virtual ghost particles). They are a mathematical trick. Thinking of ghost particles as actually existing can help in visualization, but they are really only a trick.

With the Faddeev-Popov Lagrangian, one can then find the associated generating functional, free generating functional, Green functions, and propagators for gluons, quarks, and ghosts.

15.4.3 QCD Perturbation and Feynman Rules

Generating functionals are quite useful in proving certain theoretical aspects of QFT, but Green functions are the weapon of choice for attacking perturbative QCD.

In doing so, one begins with the Green function for a given interaction, where the legs will eventually represent the external particles. From these, and lengthy, detailed analysis, one can come up with the short-cut analysis method of Feynman rules for QCD, pg. 458.

These rules parallel, in many respects, the Feynman rules we learned for QED and electroweak theory. However, they are augmented by rules that include i) ghost loops, and ii) symmetry factors S that arise in gluon loops (when one carries out the full, detailed analysis), which would otherwise be incorrect by the factor $1/S$.

These Feynman rules can then be used directly, without further reference to the PIA or Green functions, to analyze QCD processes that involve only feeble interactions (short distance scales or high energy, where the interaction is nearly free).

15.5 Problems

1. Integrate by parts to show (15-33). Hint: In $d(uv) = udv + vdu$, take $du = \partial_\nu u\, dx$ (which is really $\frac{\partial u}{\partial x^\nu} dx^\nu$), $u = \eta_i$, and $v = \partial^\nu \tilde{\eta}_{i'}$.

2. Show (15-45). Hint: Exact differentials will drop out of $\mathcal{L}$ in $\int \mathcal{L} dx$, if we assume fields vanish at infinity. Express quantities so they have terms that are exact differentials. Also note that

$$-\tfrac{1}{4} F_{\mu\nu} F^{\mu\nu} = -\tfrac{1}{4}\left(\partial_\nu A_\mu - \partial_\mu A_\nu\right)\left(\partial^\nu A^\mu - \partial^\mu A^\nu\right)$$
$$= -\frac{1}{4}\left(\underbrace{\partial_\nu\left(A_\mu\left(\partial^\nu A^\mu - \partial^\mu A^\nu\right)\right)}_{X1} \underbrace{-\partial_\mu\left(A_\nu\left(\partial^\nu A^\mu - \partial^\mu A^\nu\right)\right)}_{X2} \underbrace{-2A_\mu\left(\partial_\nu \partial^\nu A^\mu - \partial_\nu \partial^\mu A^\nu\right)}_{X3}\right).$$

3. This problem was deleted in the July 2025 revision (the second revision).

4. Deduce (15-70) explicitly for the Faddeev-Popov form of QCD.

5. Derive (15-92).

6. Derive (15-109), the two ghost, one gluon vertex factor $g_s f_{ijk} k_{2\nu}$.

Chapter 16

QCD Renormalization and Coupling Constant

"Nature does not always shave with Occam's razor"
Old saying

16.0 Preliminaries

From the discussion of effective theories in Chap. 11, Sect. 11.2.4, pg. 373, we recognize that QCD is really only an effective theory, as we do not know the behavior of elementary particles at extremely high energy levels, so the type of integrals we use in renormalization, extending from $-\infty$ to $+\infty$, are almost certainly not valid over their entire range. Yet, we can have useful effective theories that are valid over the range of energies we can access in experiment (including astrophysical observations), even though we don't know what the more correct, more all-encompassing theory looks like.

Renormalization not critical for effective theories

Such effective theories need not be renormalizable, in order to be useful. However, as pointed out in the referenced section, in the effective theories $U(1)$, $SU(2)$, and $SU(3)$ of the standard model, renormalization nevertheless provides us with valid relationships for the dependence of coupling constants on energy level. And so, renormalization is useful. And so, also, we need to understand it, to some reasonable degree, for QCD, in order to evaluate strong interaction experiments at varying energy levels, where the coupling constant differs. This difference, it turns out, changes a non-perturbative theory when such coupling constant is large, to a perturbative theory when it is small. And to quote Robert Frost, "And that has made all the difference."

Yet valid for obtaining coupling dependence on attainable energy levels

In Vol. 1, we took almost a hundred pages to cover renormalization of QED, and it was anything but trivial. In fact, I personally consider it to be the most difficult to learn of all the material covered in that volume. QCD is a more complicated theory, so one would expect that a complete coverage of renormalization for it would be even more extensive and complex. And such expectations would be correct.

So, in this chapter, we will streamline the presentation of QCD renormalization and provide an overview of the process, rather than a full detail rendition, in an attempt to give you the reader a conceptual foundation for, and general broad-brush understanding of, the fundamental tenets of the subject. As we do so, I will have to resort, with some regularity, to a phrase I detest using while teaching, "it can be shown". We can consider such leaps of faith in similar light to how we view integral tables, the many relations between spinor gamma matrices, or even Feynman rules. None of us has ever traced the steps others have used to derive all of those things, but we accept that those others knew what they were doing, and that so many before us have used their results successfully.

We will streamline the very complex topic of QCD renormalization

When and if your career path takes you more deeply into renormalization theory, you can take the time to explore, and digest more fully, the particular results that we herein simply accept without fully justifying with complete derivations.

16.0.1 Chapter Overview

We begin this chapter by investigating primitive divergences, the factors that make transition amplitudes go to infinity (before renormalization) which, in various combinations, are responsible for all divergences, at all orders, of a given theory. In QED, our primitive divergences were the one loop photon propagator, the one loop fermion propagator, and the vertex. In that theory, all divergences, at any order (more vertices and more lines in a Feynman diagram) could be built up from just those three types of primitive divergence. In QCD, there are more than three primitive divergence types, and we will deduce all of them.

Subjects we will cover

With that as a basis, we will then derive a simple criterion that any divergent transition amplitude must satisfy. And from there, we will demonstrate how any theory with a coupling constant having (in natural units) a negative dimension (for units of $(\text{MeV})^M$ where $M < 0$) is non-renormalizable.

Following that, we will explore renormalization *per se* in QCD, where we will use a different renormalization method (or scheme, as it is commonly called) than we did in Vol. 1 for QED. What we used there is called the on-shell scheme. What we will use in this chapter is known as the modified minimal subtraction scheme, or $\overline{\text{MS}}$ scheme, for short. For practice, and to acclimate ourselves to this different renormalization procedure, we will first apply it to QED, and compare it to the on-shell scheme of Vol. 1. We will then apply the $\overline{\text{MS}}$ scheme to QCD.

From there, we will then conclude the chapter by deducing the dependence of the strong coupling constant on mass-energy and show that the beta function (see Vol. 1, pg. 317) for QCD is negative. This, in turn, leads to the conclusion that quarks obey both asymptotic freedom and infrared slavery.

16.0.2 Topics in This Chapter

In this chapter, we will introduce and develop

- primitive divergences,
- a necessary criterion for renormalization using external particles only,
- non-renormalizability of negative dimension coupling constant theories,
- modified minimal subtraction ($\overline{\text{MS}}$) scheme renormalization for QED,
- $\overline{\text{MS}}$ scheme renormalization applied to QCD,
- the running QCD coupling constant, and
- asymptotic freedom due to the negative QCD beta function.

16.1 Primitive Divergences

There are a number of ways one can categorize Feynman diagrams, and we summarized those in Vol. 1, pgs. 353-354, Wholeness Charts 14-2 and 14-3, which you may wish to review before proceeding further in this section.

One of these categorizations is what are called primitive divergences, where the word "primitive" implies that all other divergences are not primitive, i.e., are composed of the primitive ones.

One way to categorize Feynman diagrams: primitive divergences

16.1.1 Primitive Divergences in QED

In QED, we have three types of these, depicted in Fig. 16-1, which is a reproduction of Fig. 13-1 of Vol. 1. With each are the symbols for the respective divergent integrals, which are regularized in the process of renormalization.

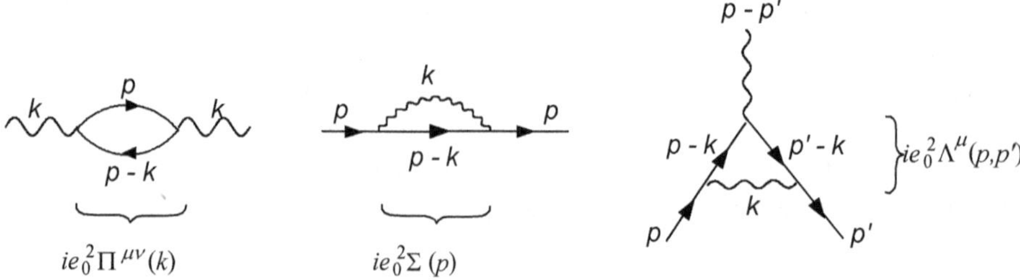

Figure 16-1. Primitive Divergences in QED

In QED, we know a primitive divergence diagram when we see one, i.e., those in Fig. 16-1 are. But, we need a general definition to help us distinguish them, in more complicated theories, from other divergences and from non-divergent parts of transition amplitudes. We also need to be clear, in forming such a definition, how it differs from the definition of proper vs. improper diagrams, which may seem similar. In doing so, it can prove clearer and easier to understand, if we start with a lot of examples, as in Wholeness Charts 16-1 and 16-2.

Types of Feynman Diagrams

Examples of proper and improper Feynman graphs are shown in Wholeness Chart 16-1.

Review by examples of proper vs improper diagrams

Wholeness Chart 16-1. Examples of Proper and Improper QED Feynman Diagrams

Proper	Improper
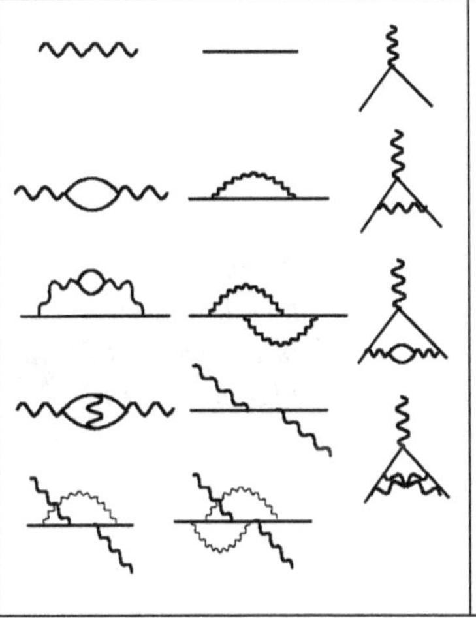	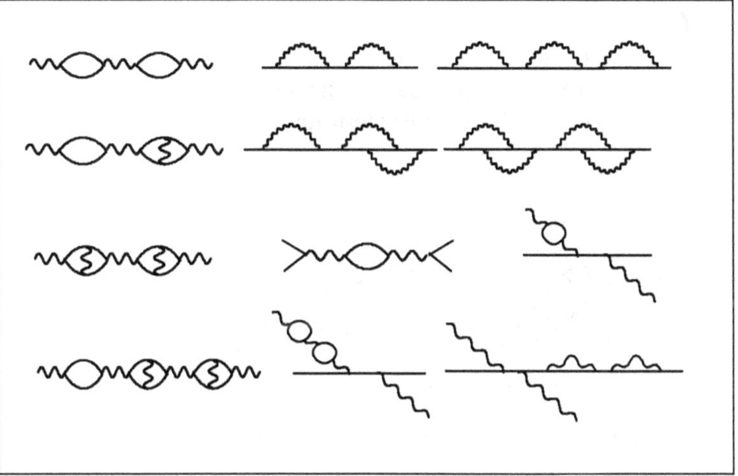

In many cases, an improper diagram has side-by-side loops, whereas proper diagrams do not, though that is not a general principle. More precisely, correctly, and generally, an improper Feynman diagram is a non-tree level (reducible) diagram having at least one internal line which, if cut, would leave two disconnected diagrams. A proper Feynman diagram is any i) tree-level (irreducible, skeleton) diagram or ii) a non-tree level (reducible) diagram having no such internal line.

Definition of proper and improper diagrams

Proper and improper diagrams will play some role in this chapter, so to avoid confusion we have included the above discussion of them to distinguish them clearly from primitively divergent diagrams, which we are about to explore in greater depth. We begin our look at the latter diagrams with the examples in Wholeness Chart 16-2, where we will explain the higher order vertex primitive divergences shortly.

Do **Problem 1** to gain practice in distinguishing between proper and improper Feynman diagrams.

Wholeness Chart 16-2. Primitively Divergent vs Not Primitively-divergent QED Feynman Diagrams

Examples: primitive divergence vs not primitive-divergence

Convergent	Divergent	
	Primitively-divergent	Divergent but Not Primitively
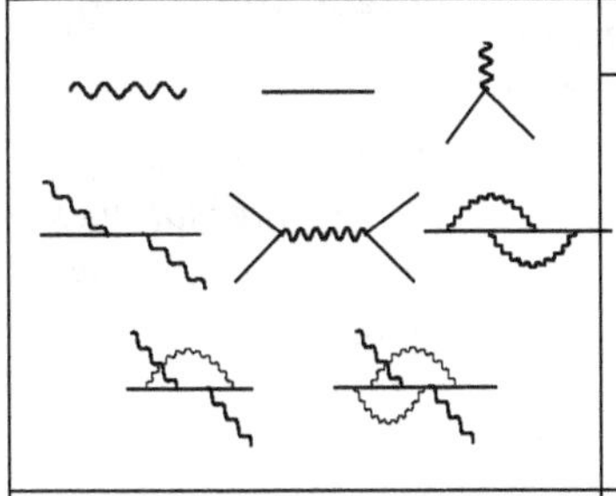	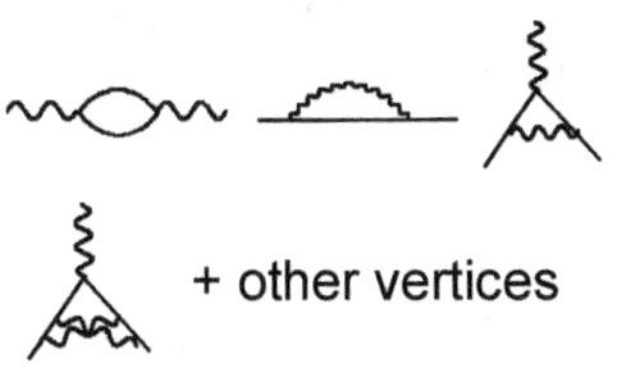 + other vertices	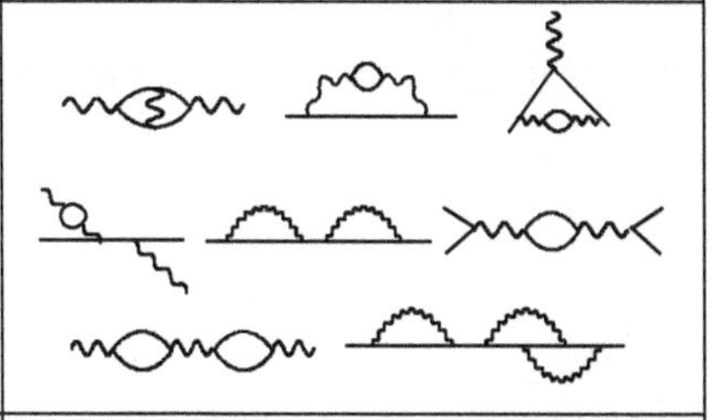
Not primitively-divergent		Not primitively-divergent

In QED, we can recognize the primitively divergent graphs (and concomitant divergent amplitudes), from which all divergent graphs (amplitudes) can be built. But, we need a precise, general case definition, which is as follows.

Definitions: primitive vs not primitive-divergence

A primitively divergent graph is a divergent graph for which, for every possible internal line that could be cut, we would get a non-divergent graph. A not primitively-divergent graph is either i) a non-divergent graph, or ii) a divergent graph having at least one internal line, which, if cut, would leave at least one divergent graph. Some examples of the latter are shown in the last column of Wholeness Chart 16-2.

Do **Problem 2** to gain some practice in distinguishing between primitively divergent, non-divergent, and not primitively-divergent Feynman diagrams.

Primitively divergent graph is a proper graph

Note that a primitively divergent graph is a proper graph, but a proper graph is not necessarily a primitively divergent one.

Three types of primitive divergence in QED

2nd order photon and fermion divergences plus vertex at 2nd and higher orders

Note further that higher order vertex corrections can be primitively divergent, as in the example shown in Fig. 16-2. This is why we were careful to say earlier that there are three "types" of QED primitive divergences, the 2nd order photon, the 2nd order fermion, and the vertex (to all orders).

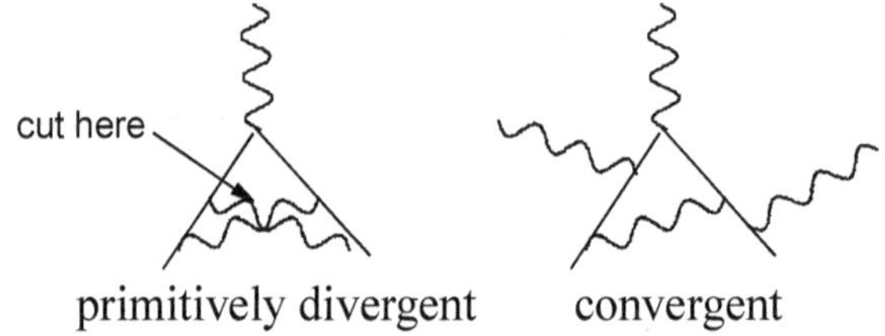

Figure 16-2. A Higher Order Vertex Primitive Divergence

Necessary Criterion for Divergence of Proper Diagrams

Power counting of integrals → superficial degree of divergence K

You should now read (or re-read) Sects. 9.1.3 to 9.1.5 in Vol. 1, pgs. 259-263, and look over the wholeness chart at the end of that chapter titled "Loop Corrections". This covers what we refer to as naïve power counting, or superficial power counting, or the superficial degree of divergence.

As shown therein, if p (or k) is the indeterminate four-momentum in a loop or vertex correction in a proper diagram over which we integrate from $-\infty$ to $+\infty$, then we can determine the maximum degree of divergence K via naïve power counting, as p^K (or k^K). If K=2, the diagram diverges, at most, quadratically with energy. If K=1, at most, linearly; and if K=0, at most, logarithmically. For K<0, the graph is convergent. K is the superficial (maximum) degree of divergence. "Superficial" because the actual divergence, when one does the full calculation, may be less than K. But, it can never be more.

Any divergent proper diagram must have $K \geq 0$

But some convergent proper diagrams can have $K \geq 0$

So, any proper graph with $K< 0$ must converge. Any such graph that diverges must have $K\geq 0$. Some convergent proper graphs can still have $K\geq 0$, but any divergent such graph must have $K\geq 0$.

Thus, the necessary, but not sufficient, condition for divergence of a proper diagram is

$$K \geq 0, \text{ where the superficial divergence is proportional to } p^K \left(\text{or } k^K\right). \tag{16-1}$$

Determining K

Relation for finding K from internal lines with d 4-momentum integrations from $-\infty$ to $+\infty$

We first state a relation for K, for proper diagrams, then discuss why it is valid. That expression is, in terms of d, the number of internal momenta not fixed by four-momentum conservation at a vertex; f_i, the number of internal fermion lines; and b_i, the number of internal photon lines,

$$K = 4d - f_i - 2b_i \quad \text{for proper diagrams.} \tag{16-2}$$

Note, that the integral

$$\int_{-\infty}^{\infty} d^4 p \propto \int_0^{\infty} p^3 dp \propto p^4 \Big|_0^{\infty} \propto \lim_{p\to\infty} p^4 \qquad \text{so}, K = 4\,, \tag{16-3}$$

where we are trying to make a point, so ignore the fact that p^3 is odd and the result is actually zero in this case. For the fermion and boson propagators over which a loop integral would take place, we have (where the fermion mass term becomes negligible in the limits)

4-momentum order for fermion and photon propagator integrations

$$\int_{-\infty}^{\infty} \frac{1}{\not p - m} d^4 p \propto \int_0^{\infty} \frac{1}{p} p^3 dp \propto p^3 \Big|_0^{\infty} \qquad \int_{-\infty}^{\infty} \frac{1}{k^2} d^4 k \propto \int_0^{\infty} \frac{1}{k^2} k^3 dk \propto k^2 \Big|_0^{\infty} . \tag{16-4}$$

The point is that if we had no propagators in a single loop (as in (16-3)), which is never really the case), we would get divergence as p^4, or $K= 4$. For each fermion propagator in that loop (as in the LHS of (16-4)), this is reduced by one; for each photon propagator (as in the RHS of (16-4)), reduced by two. Thus, for a single loop, (16-2) is valid. The divergence of two such loops aligned in succession, as in the first diagram in the second column of Wholeness Chart 16-1, is not determined via (16-2), as that relationship is valid only for proper diagrams. In improper diagrams, there are often independent loop integrals, each having a different indeterminate four-momentum over which integration takes place. A separate relation like (16-2) would be valid for each such loop.

Now consider the proper diagrams in Wholeness Chart 16-3, which are different contributions to the fermion in/fermion out diagram, where we calculate K for each. We will here consider the fermion line to be a propagator, i.e., to be a single internal fermion line. Thus, from (16-2), it has $K = -1$ (not ≥ 0), so it must converge.

Note for the examples in the chart that, for the same external particles, all divergent proper diagrams have the same superficial K value. It doesn't matter what the internal part of the diagram looks like. No matter of order the divergent proper diagram is, K is the same. We will find that this pattern holds in general.

Wholeness Chart 16-3. Superficial Divergences for the Fermion In/Fermion Out Proper Diagrams

Fermion divergences above zeroth order all have same K, independent of internal structure

Proper Diagram	d	f_i	b_i	$K = 4d - f_i - 2b_i$
	0	1	0	$0 - 1 - 0 = -1$
	1	1	1	$4(1) - 1 - 2(1) = 1$
	2	3	2	$4(2) - 3 - 2(2) = 1$
	3	5	3	$4(3) - 5 - 2(3) = 1$

Do **Problem 3** for more practice calculating the superficial divergence K, and in particular, to show it is not valid, in general, for improper diagrams.

Using examples, we will now deduce the following relationship between the total number d of indeterminate internal four-momenta, the number of internal lines (propagators), and the number of vertices (number of Dirac delta functions) n for any diagram, whether proper or improper, divergent or convergent, provided it has at least one vertex,

Finding d, number of indeterminate 4-momenta, from f_i, b_i, and n

$$\begin{aligned} d &= (\text{number of internal lines}) - (\text{number of constraints on indeterminate four-momenta}) \\ &= f_i + b_i - (n-1) \qquad \text{for any diagram with at least one vertex.} \end{aligned} \tag{16-5}$$

For the second diagram in Wholeness Chart 16-3, we have two internal lines ($f_i = 1$ and $b_i = 1$) and two vertices ($n = 2$). From (16-5), we should have one indeterminate four-momentum d, which, in fact, is what we do have.

Justify via examples

For the third diagram in the chart, (16-5) gives us $3 + 2 - (4 - 1) = 2$, which is what we knew already. For the fourth diagram, we have $5 + 3 - (6 - 1) = 3$, which again matches the chart value.

The $(n-1)$ number of constraints follows from n total Dirac delta functions, but the external momenta are fixed by $p_{initial} = p_{final}$, so we have $(n-1)$ constraints left on the internal momenta.

n δ-function constraints, but one is external, leaving (n–1) internal

Note that (16-5) holds for all diagrams, except for the bare (no loops) fermion diagram and the bare (no loops) photon diagram. We can see this for the bare fermion in the first row of Wholeness Chart 16-3, where $d = 0$, but taking $f_i = 1$ again, (16-5) would give us $d = 2$. A similar result holds for the bare photon. But, for all other diagrams, it is true. So, (16-5) holds for all diagrams with at least one vertex in them.

This relation for d good for all diagrams with at least one vertex

---

Do **Problem 4** for more practice finding d from f_i, b_i, and n in other diagrams.

Now, again using examples at first, we will deduce the following relationships for total vertices n, provided $n \neq 0$, from just fermions or just bosons, where f_e is the number of external fermions, and b_e is the number of external bosons.

Finding number of vertices n from just fermions or just bosons

$$n = \frac{f_e}{2} + f_i \qquad n = b_e + 2b_i \quad \text{for any diagram with at least one vertex.} \tag{16-6}$$

Justify via examples

Consider the three proper diagrams in the second row of diagrams in Wholeness Chart 16-1. For the photon single loop, we have $n = 0 + 2 = 2$ using the first relationship of (16-6) and $n = 2 + 0 = 2$ for the second.

For the fermion single loop, we have $n = 2/2 + 1 = 2$ using the first relationship of (16-6) and $n = 0 + 2(1) = 2$ for the second. For the vertex single loop, we have $n = 2/2 + 2 = 3$, and $n = 1 + 2(1) = 3$.

Do **Problem 5** for more practice finding n from f_i, b_i, f_e, and b_e in other diagrams.

(16-6) holds for all diagrams, proper or not, except for the bare fermion and bare photon diagrams, i.e., it holds for all diagrams with at least one vertex in them.

These relations for n hold for all diagrams with one or more vertices

Now, we combine (16-2), (16-5), and (16-6), as follows.

$$\begin{aligned} K &= 4d - f_i - 2b_i = 4\left(f_i + b_i - (n-1)\right) - f_i - 2b_i = 3f_i + 2b_i - 4n + 4 \\ &= 3\left(n - \frac{f_e}{2}\right) + \left(n - b_e\right) - 4n + 4 = 4 - \tfrac{3}{2} f_e - b_e . \end{aligned} \tag{16-7}$$

Recalling (16-1), from (16-7), we have

Combining above yields K relation solely in terms of external particles

$$K = 4 - \tfrac{3}{2} f_e - b_e \geq 0 \text{, a necessary condition for any divergent proper diagram with a vertex.} \tag{16-8}$$

This is a somewhat amazing and significant result, as K, the superficial divergence, depends only on the number of external bosons and fermions and is independent of the internal structure of the graph, provided it has at least one vertex. This is what we hinted at in our earlier comment about Wholeness Chart 16-3.

Same K for same external particles regardless of internal structure

This means, for example, that any diagram with vertices in Wholeness Chart 16-3, or any other such proper diagram with exactly two external fermion lines (and no external boson lines), will have the same superficial divergence. Hence, the primitive divergence (second figure in Wholeness Chart 16-3) for two external fermion lines (and no external boson lines) has the same superficial divergence as any proper diagram with vertices and two external fermion lines. And the same conclusion holds for any combination involving the same number of external fermions and external bosons.

So, with (16-8), we can determine the superficial divergence of any primitively divergent diagram, since any such diagram is a proper diagram and has vertices. And the various combinations of external particles for which $K \geq 0$ will represent all possible primitively divergent diagrams. That is, (16-8) can be used to eliminate a vast number of external particle combinations that cannot result in primitive divergences. It can tell us which ones we need to consider as primitive. For example, for $f_e = 2$ and $b_e = 2$, we get $K = -1$, so any such proper diagram is convergent.

Using only external particles, K can identify all possible primitively divergent diagrams

Wholeness Chart 16-4 shows all such possible combinations of external particles for QED. In the chart, we employ the shaded circle symbol, as we did in Chap. 3, to imply different possible internal structures, although primitive divergences need to be proper diagrams. We also note the actual divergences, which we derived in Vol. 1.

Only a few combinations yield $K \geq 0$

For the triangle graph in the penultimate row of Wholeness Chart 16-4, we saw in Vol. 1 (pg. 341) that due to Furry's theorem, every fermion triangle graph has a counterpart with the same magnitude, but opposite sign, that contributes to the amplitude, so all such graphs make zero total contribution.

Every triangle graph always cancelled by other triangle graph

The fermion rectangle diagram in the last row, representing light-by-light scattering, has superficial logarithmic divergence. However, as we showed in Vol. 1, pg. 329, when calculated, it actually turns out to be strongly convergent.

Fermion rectangle graph actually converges

As we said earlier, $K \geq 0$ is only a necessary condition that divergent proper graphs must satisfy. We can have convergent graphs that also satisfy it, and the last two in Wholeness Chart 16-4 are examples of such graphs in QED.

Do **Problem 6** to show that other combinations of external fermions and bosons have $K < 0$, so all associated diagrams must be convergent.

From Prob. 6, along with working through other combinations, one can see we are left with only three types of primitive divergence: the one-loop photon, the one-loop fermion, and the vertex.

So, we are left with 3 types of primitive divergence in QED

Wholeness Chart 16-4. All Possible Primitively Divergent Graphs for QED

Diagram	*fe*	*be*	$K = 4 - \frac{3}{2} f_e - b_e$	**Superficial Divergence**	**Actual Divergence**
	0	2	$4-0-2=2$	quadratic	logarithmic
	2	0	$4-3-0=1$	linear	logarithmic
	2	1	$4-3-1=0$	logarithmic	logarithmic
Certain higher order	2	1	$4-3-1=0$	logarithmic	logarithmic
	0	3	$4-0-3=1$	linear	none
	0	4	$4-0-4=0$	logarithmic	none

Summary of Primitive Divergences for QED

So, we have shown what we already knew, that there are three types of primitive divergences in QED, the second order self-energy divergences for the photon and fermion, and the proper diagram vertex divergence to all orders. All possible divergences in QED are built up from these.

All QED divergences built from these 3 types of primitive divergence

Any irreducible (skeleton, tree-level) diagram converges. When we insert any of these three primitive divergences, in any number, into such a diagram we get transition amplitude modifications in terms of the $\Pi^{\mu\nu}$, Σ, and Λ^{μ} of Figure 16-1, pg. 464 (where the Λ^{μ} is to all orders), and those modifications are divergent. When we renormalize $\Pi^{\mu\nu}$, Σ, and Λ^{μ}, we redefine the coupling constant (electron charge) and fermion mass in such a way as to eliminate the infinities, and the theory becomes finite at all orders. The redefined mass and charge of the theory are then the physical mass and charge measured in experiment.

16.1.2 Primitive Divergences in QCD

Determining *K*

QCD primitive divergences

We can extrapolate the above analysis of QED primitive divergences to QCD. In QCD, the initial relationship for *K*, parallel to (16-2), where we explain each term further below, is

K for QCD in terms of internal lines

$$K = 4d - f_i - 2b_i - 2\eta_i + n_{3G} + \left(n_{gG} - \eta_e / 2\right) \text{ for proper diagrams.} \qquad (16\text{-}9)$$

Explaining where each term in K comes from ↓

The b_i represent bosons, so for QCD, these are the number of gluon propagators. η_i is the number of internal ghost lines, which, since they are scalars with boson-type propagators, have the same effect as bosons (photons in QED or gluons in QCD), represented by b_i. That is, the k^2 in the denominator reduces the (superficial) divergence by 2, for each such propagator.

From Feynman rule 1(b) for QCD, pg. 459, we see that the three-gluon vertex (Fig. 15-3(b) pg. 456) includes a factor in the numerator with terms linear in the gluon four-momentum *k*. This will increase the (superficial) divergence, rather than decreasing it, and the increase is by one power of four-momentum. Hence the positive sign with the coefficient of 1 on the n_{3G} term in (16-9).

From Feynman rule 1(c), we see that there are no four-momentum factors in the four-gluon vertex (Fig. 15-3(c)), so such vertices have no effect on *K*, and (16-9) has no term representing them.

The term in parentheses arises from ghost-gluon vertices (Fig. 15-3(d)) and needs a bit of explanation, which we now get to. The symbol n_{gG} represents the number of ghost-gluon vertices and η_e, the number of external ghost lines.

From Feynman rule 1(d), we see that such a vertex has one factor of four-momentum of one of the ghosts. However, this is only for the ghost line *entering* the vertex (arrow inward), and only for propagators, not external lines, which do not have indeterminate four-momenta, so don't contribute to *K*. An example of the effect of this on *K* can be gleaned from Fig. 16-3.

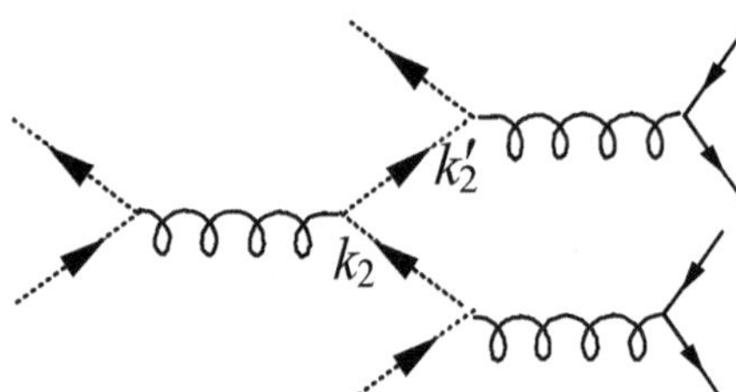

Figure 16-3. An Example of the Ghost-Gluon Vertex Contributions to *K*

The external ghost lines in the figure (where we assume external ghosts are possible at this point) make no contribution to *K*, as they are not propagators. The only two four-momenta that come into play are at the second vertex from the left and the one just above it to the right, each of which has an incoming (arrow inward) propagator four-momentum that contributes to the superficial divergence. The total number of contributions to *K* is two. The total number of external ghosts is 4. From the parentheses term in (16-9), we have

$$n_{gG} - \frac{\eta_e}{2} = 4 - \frac{4}{2} = 2, \qquad (16\text{-}10)$$

which is what it should be from looking at Fig. 16-3. Of course, the transition amplitude for this interaction converges (since it is tree level), but we are only using it to illustrate the purpose of relation (16-10).

This is only one example, but hopefully can give you some confidence, as we generalize. The term in parentheses of (16-9) will always give us the correct contribution to *K* for ghost-gluon vertices.

The number of vertices n_{QCD} in any QCD graph is

Vertices in QCD

$$n_{QCD} = n_{qG} + n_{3G} + n_{4G} + n_{gG}, \qquad (16\text{-}11)$$

where the first symbol after the equal sign is for quark-gluon vertices, and the third symbol is for four-gluon vertices. Each such vertex has an associated Kronecker delta function constraining four-momenta, and, as in QED, one of them is eliminated by external momentum conservation, as a constraint on internal momentum. That leaves, as in QED, $(n-1)$ constraints on the $f_i + b_i + \eta_i$ internal four-momenta. So, the total number of independent QCD unfixed four-momenta d (over which integration takes place from $-\infty$ to $+\infty$) is

Number of unfixed 4-momenta

$$d = f_i + b_i + \eta_i - \left(n_{QCD} - 1\right) \text{ for diagrams with at least one vertex.} \tag{16-12}$$

Directly parallel to the first part of (16-6), via the same logic, we have

Relations parallel to QED for quarks and ghosts

$$n_{qG} = \frac{f_e}{2} + f_i \qquad n_{gG} = \frac{\eta_e}{2} + \eta_i \text{ for diagrams with at least one vertex.} \tag{16-13}$$

And parallel to the second part of (16-6), via similar logic that we are about to explain, we have

Relations different in QCD for bosons (gluons)

$$b_e + 2b_i = n_{qG} + 3n_{3G} + 4n_{4G} + n_{gG} \text{ for diagrams with at least one vertex.} \tag{16-14}$$

Explaining where terms in boson (gluon) relation come from ↓

The first and fourth terms after the equal sign in (16-14) arise in the same way n in the second part of (16-6) did in QED, which we explored in Problem 5. In QED, each two fermion, one boson vertex contributed to the total vertices number n and that equaled the LHS of (16-14), as we found with a number or examples there. Here the two quark, one gluon vertex makes such a contribution, as does the two ghost, one gluon vertex.

But now, we have additional vertices composed of three gluons and four gluons, so let's look at some examples, in order to gain some comfort with (16-14). Note carefully, that for determining primitive divergence diagrams, one commonly assumes gluons (as well as ghosts) can be external. That assumption lets us conceptualize the primitive divergences and their associated graphs, but such diverging graphs always appear inside diagrams where ghosts and gluons are internal, not external.

In Fig. 16-4(a), we have $b_i = 0$, $b_e = 3$. and $n_{3G} = 1$. So, (16-14) gives us 3 + 2(0) = 3(1), which is indeed correct. In Fig. 16-4(b), we have $b_i = 0$, $b_e = 4$. and $n_{4G} = 1$, and (16-14) gives us 4 + 2(0) = 4(1), which is also correct. In Fig. 16-4(c), we have $b_i = 2$, $b_e = 2$. and $n_{3G} = 2$. So, (16-14) gives us 2 + 2(2) = 3(2), again correct. Hopefully, we can then accept that (16-14) is valid generally.

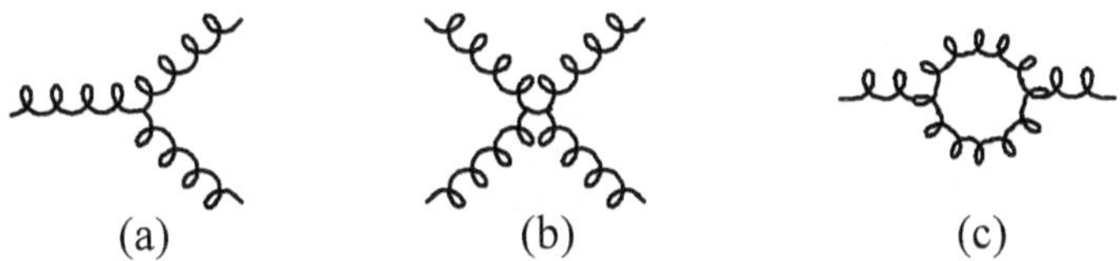

Figure 16-4. Examples of Relation (16-14)

Now, as with QED, we want to find K solely in terms of external particles. To that end, start with a repeat of (16-9) below, then successively use (16-12) and (16-11), then solve (16-13) and (16-14) for internal particle numbers and substitute the results. That is,

Combining above yields QCD K relation in terms of external particles

$$\begin{aligned} K &= 4d - f_i - 2b_i - 2\eta_i + n_{3G} + \left(n_{gG} - \eta_e / 2\right) \\ &= 4\left(f_i + b_i + \eta_i - \left(n_{QCD} - 1\right)\right) - f_i - 2b_i - 2\eta_i + n_{3G} + \left(n_{gG} - \eta_e / 2\right) \\ &= 3f_i + 2b_i + 2\eta_i - 4\left(n_{qG} + n_{3G} + n_{4G} + n_{gG} - 1\right) + n_{3G} + \left(n_{gG} - \eta_e / 2\right) \\ &= 3\left(n_{qG} - \frac{f_e}{2}\right) + \left(n_{qG} + 3n_{3G} + 4n_{4G} + n_{gG} - b_e\right) + \left(2n_{gG} - \eta_e\right) \\ &\qquad - 4\left(n_{qG} + n_{3G} + n_{4G} + n_{gG} - 1\right) + n_{3G} + \left(n_{gG} - \eta_e / 2\right) \\ &= (0)n_{qG} - \tfrac{3}{2}f_e + (0)n_{3G} + (0)n_{4G} + (0)n_{gG} - b_e - \eta_e + 4 - \tfrac{1}{2}\eta_e, \end{aligned} \tag{16-15}$$

or finally,

Final result: QCD K solely in terms of external particles

$$K = 4 - \tfrac{3}{2}\left(f_e + \eta_e\right) - b_e, \tag{16-16}$$

where $K \geq 0$ is necessary for any divergent QCD proper graph with at least one vertex.

Note that but for the external ghost number, this is the same relation as we had for QED, (16-8). And as with QED, for the same external particles, any internal structure, at any order, of any proper diagram with at least one vertex, has the same K.

As with QED, since all primitively divergent diagrams are proper diagrams with vertices, we can use (16-16) to tell us which diagrams, with which particular combinations of external particles, could possibly be primitively divergent. For example, any diagrams with more than 4 external bosons must converge (since $K < 0$) and cannot be primitively divergent. The same is true of any diagram with three or more external quarks. Or three or more external ghosts. In fact, the only QCD diagrams that could possibly be primitively divergent (have $K \geq 0$) are those in Wholeness Chart 16-5.

Eliminating QCD diagrams that cannot be primitively divergent

In the chart, we again employ the shaded circle symbol, as we did in Wholeness Chart 16-4, to imply different possible internal structures, where we keep in mind that primitive divergences need to be proper diagrams. For example, in the first diagram for gluon self-energy, the shaded circle can be either a quark/antiquark loop, a ghost/antighost loop, or a gluon/gluon loop. The shaded circle in the last diagram could be a quark rectangle, similar to that in Wholeness Chart 16-4, a ghost rectangle, or a gluon rectangle.

The ghost self-energy (third) diagram, unlike its fermion self-energy sibling, converges. One can understand this by noting that the internal ghost propagator has a k^2 in the denominator, while the fermion propagator (second diagram) has p. Since the actual fermion self-energy is logarithmically divergent, the lowest possible divergence, the actual ghost self-energy would be less divergent than that, i.e., not divergent.

The same logic applies to the ghost-ghost-gluon vertex (5th) diagram. The two internal ghost lines in the second order vertex loop now each have factors of k^2 in the denominator, whereas in the fermion-fermion-gluon vertex, the two internal fermion lines each only have p in the denominator. Hence, the ghost-ghost-gluon vertex diverges less than the fermion-fermion-gluon vertex. Since the latter is actually only logarithmically divergent, the ghost-ghost-gluon vertex must converge.

The last two diagrams in the chart parallel similar ones in QED, i.e., they are convergent, for similar reasons. The remaining three diagrams all diverge, but showing so is a major task, the basic steps of which (but not the details) we will shortly enumerate.

Wholeness Chart 16-5. All Possible Primitively Divergent Graph Types for QCD

7 possibly primitively divergent QCD graphs, but only 3 actually are

Diagram	fe	be	ηe	$K = 4 - \frac{3}{2}(f_e + \eta_e) - b_e$	Superficial Divergence	Actual Divergence
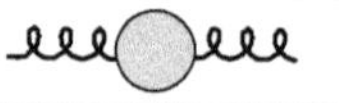	0	2	0	$4 - 2 = 2$	quadratic	logarithmic
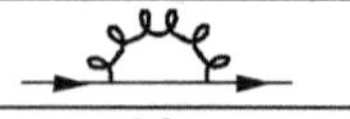	2	0	0	$4 - 3 = 1$	linear	logarithmic
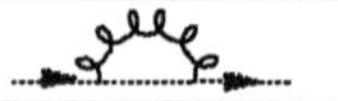	0	0	2	$4 - 3 = 1$	linear	none
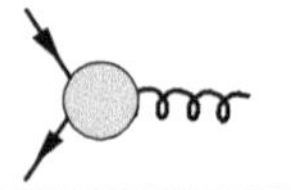	2	1	0	$4 - 3 - 1 = 0$	logarithmic	logarithmic
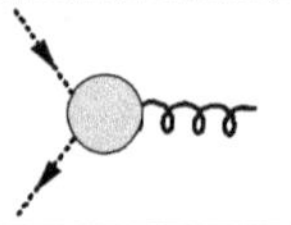	0	1	2	$4 - 3 - 1 = 0$	logarithmic	none
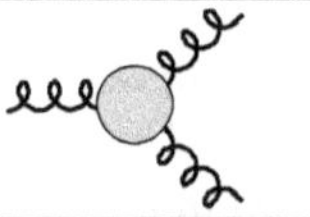	0	3	0	$4 - 3 = 1$	linear	none
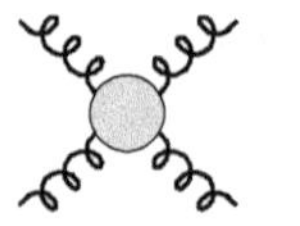	0	4	0	$4 - 4 = 0$	logarithmic	none

Summary of Primitive Divergences for QCD

So, we have shown that there are seven types of possible primitive divergences in QCD, though only three of them actually are. Those three, like their QED brethren, diverge logarithmically.

All QCD divergences are built from the actual QCD primitive divergences.

All QCD divergences built from 3 types of primitive divergence

16.2 Renormalization in Different Theories

16.2.1 QED vs Weak vs Strong Theories

QED renormalization → dependence of e (and thus, α) on energy level

A key result of the QED renormalization process is the dependence of the coupling constant e on the interaction energy level. In Vol.1, Chap. 12, we deduced this dependence and summarized the result in (12-41) therein in terms of α, the fine structure constant, which, as we know, is proportional to e^2. We repeated this result in (11-34) of Chap. 11 herein, and repeat it again as (16-17) below. In that relation, p is energy level and b_n depends on the type of interactions in QED, as shown in Vol. 1, (12-29) and (12-30).

Renormalization group equation

$$p\frac{\partial}{\partial p}\alpha(p)=\alpha^2(p)8\pi b_n=\beta(p,b_n) \qquad (16\text{-}17)$$

(16-17) is the renormalization group equation, or RGE, and β is known as the beta function. In QED, b_n is positive, so β is positive, and thus, the coupling constant α (or equivalently, e) increases with increase in energy.

QED β function > 0 → α (and e) increases with energy level

In the references cited above, we noted without proof (which is an extensive task) that the beta function for weak interactions is negative. Thus, the weak coupling constant decreases with increase in energy. See Fig. 11-7, pg. 373.

From the referenced figure, one can see that the strong coupling constant also decreases with increase in energy, to greater degree than the weak coupling constant, and this, of course, is the basis for asymptotic freedom and infrared slavery. At high energy (short distance), the QCD coupling becomes weaker than it is at low energy (greater distance). We will soon deduce the negative beta function for QCD.

We will find QCD $\beta < 0 \rightarrow g_s$ decreases with energy

16.2.2 Renormalization Criteria

A field theory is deemed renormalizable if a finite number of parameters (such as masses and coupling constants) can be redefined so that transition amplitudes do not diverge when the energy level approaches infinity. That is, the predictions are insensitive to the energy cutoff, typically labeled Λ, provided such cutoff is much higher than the energy level of the interaction being evaluated.

Theory renormalizable if finite number of parameters can absorb all infinities

A non-renormalizable theory, on the other hand, can be still be made finite, well-defined, and correctly predictive of experiment, but only below a certain high energy cut-off. Predictions from such a theory would diverge at energies above that cut-off. Such a theory would be an effective theory, good over a limited range of energy levels, and applicable therein, but not universally valid.

Non-renormalizable theory can make good predictions below certain energy

As we have noted before, all of our theories are really effective theories, as we do not know how nature behaves at arbitrarily high energies. Renormalization assumes the theory we are dealing with is correct to all energy levels, and we are just re-arranging it to accord with what we measure. Yet, as we have also noted before, renormalization gives us predictions for things like coupling constant dependence on energy, so it still has great utility, even if it is not the final answer.

All our theories really effective as very high energy physics unknown

16.3 Other Interaction Terms

In QED, electroweak, and QCD theories, we have particular interaction terms in the Lagrangian. These theories, as formulated, are known as minimal gauge theories, since they possess the minimal number and types of interaction terms to correctly describe nature as we see it in experiments.

Minimal QFT has the interaction terms we've worked with so far

But one could ask if other terms in the Lagrangian might exist that could lead to viable theories, i.e., theories that are internally consistent, if not a 100% match with empirical results. Such theories are known as non-minimal gauge theories.

Non-minimal theories have other interaction terms

In this section, we determine criteria for such additional interaction terms in the Lagrangian.

16.3.1 Non-minimal Interaction Terms

Non-minimal interaction terms $\mathcal{L}'_I$ are terms which could be added to the Lagrangian of a given theory without destroying gauge invariance. There are many possible such terms. However, if we require the theory to be renormalizable, these terms generally will not fit the bill, as we are about to show.

Non-minimal interaction terms must satisfy gauge invariance

QED

As we showed in Vol. 1, Chap. 11, our formulation of QED is symmetric under the $U(1)$ gauge transformation set

An example of non-minimal QED

$$\psi(x) \rightarrow \psi'(x) = e^{-i\alpha(x)}\psi(x) \qquad A_\mu(x) \rightarrow A'_\mu(x) = A_\mu(x) - \tfrac{1}{e}\partial_\mu \alpha(x). \quad (16\text{-}18)$$

The electromagnetic field tensor

$$F^{\mu\nu}(x) = \partial^\nu A^\mu(x) - \partial^\mu A^\nu(x), \quad (16\text{-}19)$$

is independently symmetric under (16-18), since

$$\begin{aligned} F^{\mu\nu}(x) &\rightarrow F'^{\mu\nu}(x) = \partial^\nu A'^\mu(x) - \partial^\mu A'^\nu(x) \\ &= \partial^\nu A^\mu(x) - \partial^\mu A^\nu(x) - \partial^\nu \tfrac{1}{e}\partial^\mu \alpha(x) + \partial^\mu \tfrac{1}{e}\partial^\nu \alpha(x) = \partial^\nu A^\mu(x) - \partial^\mu A^\nu(x) = F^{\mu\nu}(x). \end{aligned} \quad (16\text{-}20)$$

Similarly, the term

$$\bar{\psi}(x)\tfrac{i}{2}\left(\gamma^\mu\gamma^\nu - \gamma^\nu\gamma^\mu\right)\psi(x) = \bar{\psi}(x)\sigma^{\mu\nu}\psi(x), \quad (16\text{-}21)$$

where the definition of $\sigma^{\mu\nu}$ can be gleaned, is independently symmetric under (16-18), as well, which, hopefully, is fairly obvious, since the gamma matrices are invariant.

Hence, any term of form

One possible non-minimal QED interaction term in $\mathcal{L}$

$$g_1\bar{\psi}\sigma^{\mu\nu}\psi F_{\mu\nu} \quad (16\text{-}22)$$

will be invariant under (16-18), and so is a viable candidate for an interaction term in the Lagrangian. It turns out, however, that (16-22) leads to a predicted additional component in the magnetic moment of the electron (and its sibling charged fermions), which does not show up in experiment.

So, even though (16-22) satisfies the gauge invariance requirements of a good quantum field theory, it does not agree with experiment, and so, is not included in the theory.

Note that from (16-19), (16-22) leads to terms of form

It contains a derivative not found in the minimal QED interaction term

$$g_1\bar{\psi}\sigma^{\mu\nu}\psi\partial_\nu A_\mu . \quad (16\text{-}23)$$

Such terms, as all terms do in the Lagrangian, must have total natural unit dimension of four, as we discussed in Sect. 1.3.4 at the end of Chap. 1. The fermions each have dimension 3/2, the photon 1, and the derivative 1, which means $\bar{\psi}\sigma^{\mu\nu}\psi\partial_\nu A_\mu$ (no coupling constant) must have dimension 5. So, the only way for (16-23) to have total dimension 4 is if the coupling constant g_1 has dimension –1. This is unlike any coupling constants we have so far seen in QED, electroweak, or QCD theories, which were all dimensionless[1].

That derivative means the coupling constant has dimension –1

Note that *any* additional field, instead of a derivative, would have the same effect, as such field would have a positive dimension (1 for a vector, 3/2 for a spinor, and 1 for a scalar), and thus, constrain the coupling constant to negative dimension. However, such terms are generally not gauge invariant, so would not qualify as non-minimal terms.

An extra field, instead of an extra derivative, has the same effect on g_1 (i.e., negative dimension)

QCD and Other Theories

Precisely the same logic holds for other interactions, such as the strong interaction, where the gluon has the same dimension as the photon. Any non-minimal QCD terms (terms other than those we have been working with, such as (16-23)) will have coupling constants with negative dimension.

More generally, in any QFT, a non-minimal term means coupling constant dimension < 0

[1] We note in passing that, in the interaction picture, the fields satisfy the same equations of motion, and the same commutation relations, as the free fields, subject to the condition that no terms in $\mathcal{L}_I$ contain derivatives of the fields. However, it turns out that, regardless of the presence of derivatives in terms or not, the S matrix expansion remains valid, when the free-field commutation relations are used. So, Feynman rules for such terms continue to be valid (subject to some fairly straightforward modifications to vertex factors we will not get into). For details, see N.N. Bogoliubov and D.V. Shirkov, *Introduction to the Theory of Quantized Fields*, 3rd ed., Wiley (NY, 1979), Sect. 21.

The Value of this Finding

We are about to show that any theory with negative dimension coupling constants cannot be renormalized, and this is an important finding, as it aids us in ruling out non-minimal terms in the Lagrangian for any theory we require to be renormalizable.

As we will see, negative dimension coupling constant means theory is non-renormalizable

16.3.2 Negative Dimension Coupling Constants and Renormalizability

QED

A theory with a term like (16-23) will, due to the derivative, yield a vertex term in the Feynman amplitude like

Our non-minimal QED example Feynman amplitude contribution

$$\mathcal{M}_{non-min} = i\,g_1\,\bar{u}(\mathbf{p}')\sigma^{\mu\nu}k_\nu\,u(\mathbf{p})\varepsilon_\mu(\mathbf{k}), \tag{16-24}$$

where the critical difference from minimal QED is the additional factor of photon four-momentum. Since each photon propagator is attached to two vertices, we get an extra factor of k^2 arising in the amplitude for each photon propagator. And that cancels the factor of k^2 in the denominator of that propagator, which otherwise reduces the superficial divergence and makes K of (16-2) smaller. The contribution from internal bosons to K is then zero, rather than $2b_i$. This means our non-minimal theory (with non-minimal terms in $\mathcal{L}_I$) has

The partial derivative yields a 4-momentm

and that cancels the photon propagator effect

$$K_{non-min} = 4d - f_i\,. \tag{16-25}$$

Every vertex still has one photon line and two fermion lines, so (16-5) and (16-6) remain the same.

Do **Problem 7** to derive (16-26) from (16-25).

From Prob.7, we find, instead of (16-8), where the subscript meaning should be apparent,

That results in $K_{non-min}$, the superficial divergence relation for non-minimal theories

$$K_{non-min} = 4 - \tfrac{3}{2}f_e - 2b_e + n \geq 0\,, \tag{16-26}$$

as a necessary condition for any divergent proper diagram with at least one vertex. We re-write (16-26) as (16-27) below, in terms of a function of only external fermions and bosons, which we label K_0.

which can be expressed in terms of a function of f_e, b_e, along with n

$$K_{non-min} = \underbrace{4 - \tfrac{3}{2}f_e - 2b_e}_{K_0(f_e,b_e)} + n = \underbrace{K - b_e}_{K_0(f_e,b_e)} + n \geq 0 \qquad (K = K_{min})\,. \tag{16-27}$$

The part of (16-27) after the second equal sign is incidental, presented only for comparison purposes, and not needed in what follows.

From (16-27), the superficial divergence increases with increasing n (number of vertices), whereas in the comparable relation (16-8) of the minimal theory, there is no such n dependence. Hence, higher-order diagrams in the non-minimal theory mean higher K values and, as the order (number of vertices) goes to infinity, we will get K approaching infinity. Although K values only indicate superficial divergence, the actual divergences can generally only be somewhat smaller than K, so the actual divergences go to infinity, as well. Such an infinite number of different divergences cannot be absorbed into a finite number of redefined physical parameters. Conclusion: for non-minimal interaction terms (such as, for example, (16-23)), the non-minimal theory is non-renormalizable.

This means we need infinite parameters to renormalize

We can re-express (16-27), in terms of the coupling constant from our earlier example, as

Re-expressing $K_{non-min}$ in terms of the coupling constant dimension

$$K_{non-min} = K_0\left(f_e, b_e\right) - nD_{g_1} \geq 0\,, \tag{16-28}$$

where D_{g_1} is the dimension of the coupling constant g_1, which is –1 in this case.

This result can be generalized to other non-minimal terms having coupling constants with other dimensions, and to Lagrangians with more than one type of non-minimal term. Where n_i is the order of the graph in the particular coupling g_i, we have

Most general form of QED $K_{non-min}$, for any number of non-minimal terms

$$K_{non-min} = K_0\left(f_e, b_e\right) - \sum_i n_i D_{g_i} \geq 0\,. \tag{16-29}$$

(16-29) makes sense, as we know a more negative dimension coupling constant implies more factors of four-momentum in the numerator, and thus, greater divergence in integrals containing four-momenta.

But again, for any given non-minimal term, with given i, we need a theory that is good as the number of vertices goes to infinity. In such case, each set of diagrams with a particular number of vertices would need its own physical parameter to be redefined, in order to be renormalizable. In other words, for any non-minimal term, we need an infinite number of such parameters available, which there never are. Beyond that, renormalizable theories need such parameters to be finite in number.

In any case, negative dimension coupling $\rightarrow \infty$ number of parameters needed $\rightarrow$ non-renormalizable

Any theory of quantum electrodynamics with a coupling constant having negative dimension is non-renormalizable.

QCD and Other Theories

All of the stated results for QED, non-minimal terms, and negative dimension coupling constants are extendible directly to other theories, including QCD. The final result, which is very useful in formulating and investigating possible alternatives to the SM, is the following general principle.

Same true for other field theories

Bottom line: Any quantum field theory with a coupling constant having negative dimension is non-renormalizable.

For $g_i < 0$, non-renormalizable

16.4 The Renormalization Procedure

16.4.1 Review of QED Renormalization

In this subsection we offer a brief summary of the renormalization process, but you may find it beneficial to review parts of Vol. 1 where we cover renormalization. Particularly relevant parts are Sect. 13.1, pg. 322 therein, Wholeness Chart 14-4, beginning on pg. 368, Sect. 13.6, beginning on pg. 332, Chap. 13 summary on pg. 336, Chap. 14 summary on pg. 366, and the final results of Chap. 15 Sects. 15.6, 15.7, and 15.8, pgs. 388-399.

The basic steps, with the QED photon propagator renormalization as an example, are as follows. The example is only for 2nd order, to keep things simple, but the same procedure applies to nth order.

Summary of steps in QED renormalization

For each of the photon propagator, the fermion propagator, the vertex, and external lines, do the following.

1. Draw all relevant Feynman diagrams to n^{th} order (2nd order shown).

2. Add all of the associated Feynman amplitudes to get the total contribution to the total amplitude, which will have an integral evaluated over an internal 4-momentum from $-\infty$ to $+\infty$, where that integral is divergent.

$$iD_{F\alpha\beta}(k) \Rightarrow iD^{2nd}_{F\alpha\beta} = iD_{F\alpha\beta}(k) + iD_{F\alpha\mu}(k) ie_0^2 \Pi^{\mu\nu}(k) iD_{F\nu\beta}(k)$$

(In the example, $\Pi^{\mu\nu}$ contains the divergent integral from the loop.)

3. Employ a regularization procedure temporarily employing a parameter Λ. In this procedure, the problematic integral converges for finite Λ, though for $\Lambda \rightarrow \infty$, it becomes the original integral and diverges.

$$iD^{2nd}_{F\alpha\beta}(k) = iD_{F\alpha\beta}(k)\left(1 + e_0^2\left(2b_n \ln\frac{k}{\Lambda}\right) - (\text{finite part})\right) \quad \text{(higher order terms dropped)}$$

(b_n is a constant that depends on the number of fermion flavors the loop could be composed of. See Vol. 1, (12-29) and (12-30), pg. 314 and below.)

4. Redefine parameters of the theory, such as mass and coupling constant, in such a way as to absorb the problematic integral, so that when $\Lambda \rightarrow \infty$, the amplitude still converges.

$$e^2 = e_0^2 Z_\gamma^{2nd} \qquad Z_\gamma^{2nd} = 1 + e_0^2\left(2b_n \ln\frac{k}{\Lambda}\right) = 1 + e_0^2\left(b_n \ln\frac{k^2}{\Lambda^2}\right)$$

(For $\Lambda \rightarrow \infty$, $e_0 \rightarrow 0$, but e stays finite and equal to the electric charge we measure in experiment. m is defined in terms of m_0 and Λ, such that as $\Lambda \rightarrow \infty$, m is constant and equal to the mass we measure in experiment.)

5. The result is a finite theory, where one can use tree diagrams with e and m, instead of e_0 and m_0, plus modified versions of the propagators and vertex. The coupling constant e becomes a function of energy level k, i.e., $e = e(k^2)$, which can be gleaned from step 4.

For reference,

$$b_n = \sum_{a=1}^{n} \frac{1}{12\pi^2} \lambda_a Q_a^2 \qquad \lambda_a = \text{number of possible pair types}, \;\; Q_a = \text{charge in units of } e_0 \,. \qquad (16\text{-}30)$$

Each contribution to (16-30) kicks in when the energy level of the interaction exceeds twice the mass of a given particle (since there are two particles in the loop). If energy were high enough such that all particles played a role, we would have

$$b_n = \frac{1}{12\pi^2}\{ \underbrace{3(1)^2}_{\text{leptons}} + \underbrace{3\cdot 3\left(\tfrac{2}{3}\right)^2}_{\substack{\text{charge 2/3}\\ \text{quarks}}} + \underbrace{3\cdot 3\left(\tfrac{1}{3}\right)^2}_{\substack{\text{charge 1/3}\\ \text{quarks}}} + \underbrace{1(1)^2}_{W^+,W^-} \} = \frac{1}{12\pi^2} 9 = \frac{3}{4\pi^2} \,. \qquad (16\text{-}31)$$

16.4.2 Different Renormalization Schemes

There are different renormalization schemes

As noted in Vol. 1 and in this volume (Chap. 11, Sect. 11.2.2, pg. 368 and the chapter overview at the beginning of this chapter), there are different ways to renormalize, the two most common of which are the on-shell scheme and the minimal subtraction ($\overline{\text{MS}}$) scheme. Wholeness Chart 16-6 illustrates the similarities and differences between the two, which we thereafter explore in greater depth. Either scheme can be used for QED, though the on-shell scheme is typically used for it, at least in introductory courses. The $\overline{\text{MS}}$ scheme, on the other hand, is more commonly applied to QCD.

Vol. 1: on-shell scheme

This volume: $\overline{MS}$ scheme

In the on-shell scheme, we renormalize the bare mass m_0 to employ the physical (rest or inertial) mass m, which is independent of mass-energy scale μ. In the $\overline{\text{MS}}$ scheme, we still have a constant bare mass m_0, but the renormalized mass is not constant, and is a function of the mass-energy level μ. The $\overline{\text{MS}}$ coupling constant, similar to that of the on-shell scheme, is also a function of μ. We denote the $\overline{\text{MS}}$ values by $m_r = m_r(\mu)$ and $e_r = e_r(\mu)$.

On-shell: m constant

$\overline{MS}$ scheme: m function of energy level μ

Because we do not have a fixed fermion physical mass in the $\overline{\text{MS}}$ scheme, our particles and fields therein are off-shell, at virtually any energy level. In contrast, in the on-shell scheme, all real particles have a fixed physical mass m, and so, are on-shell. Hence, the name for that scheme.

Both: same bare mass, bare coupling, bare propagators

In both schemes, we start with the same bare propagators and vertex functions from which the normalization problem originally arose, as they lead to infinite transition amplitudes. That is, steps 1 and 2 above are the same in both approaches. The difference comes in step 3 and thereafter, so the resultant modified propagators and modified vertices are different in the two schemes. In both, dimensional regularization is the preferred method of regularizing.

2 schemes compared

Wholeness Chart 16-6. Comparison of On-Shell and $\overline{\text{MS}}$ Renormalization Schemes

			Vertices and Propagators			
Scheme	**Mass**	**Naming**	**Bare**	**Modified**	**Regularization**	**Mass vs Energy Scale**
On-shell	$m_0 \rightarrow$ m – constant	"On-shell" since use physical mass m	Note comment in box below	Note comment in box below	Dimensional regularization or other methods	Physical vs bare, $m \neq m_0$ m, m_0 constant
$\overline{\text{MS}}$	$m_0 \rightarrow m_r(\mu)$ μ= energy scale	Not on shell, as use $m_r \neq m$	Same as on-shell	Same form as on-shell but mass, coupling, and Σ, Π, Λ different	Dimensional regularization	Physical vs $\overline{\text{MS}}$, $m \neq m_r$ m same at any energy μ, but m_r depends on μ

16.4.3 Different Parameters Λ and η

From Vol. 1, Chap. 15, (15-78), via dimensional regularization, we have obtained, where m is the constant value for the fermion physical (inertial) mass measured in experiment and γ is the Euler-Mascheroni constant,

Dimensional regularization example

$$\text{For } D=4-\eta, \quad \frac{1}{(2\pi)^D}\int\frac{1}{\left(p^2-m^2\right)^2}d^Dp \underset{\substack{\eta\to 0\\ D\to 4}}{\to} \frac{i}{(4\pi)^2}\left(\frac{2}{\eta}-\gamma+\mathcal{O}(\eta)\right)\left(1-\tfrac{\eta}{2}\ln m^2\right)=\frac{i}{(4\pi)^2}\left(\frac{2}{\eta}-\gamma-2\ln m\right), \tag{16-32}$$

where the point we want to make is that throughout our regularization process, we get a factor of $\left(\frac{2}{\eta}-\gamma-2\ln m\right)$ coming into play. In (16-32), $\eta\to 0$ makes the integral divergent. For $\eta\neq 0$, it converges.

Shortly later in Vol. 1, in (15-80), we adopted Λ instead of η, where the integral diverges with $\ln\Lambda$ instead of $1/\eta$. That is,

Relation between Λ and η in on-shell scheme

$$\text{for finite } \Lambda \text{ and } \eta \text{ small, } \ln\Lambda=\frac{1}{\eta}-\frac{\gamma}{2}. \qquad \text{For } \Lambda\to\infty \text{ and } \eta\to 0,\quad \ln\Lambda=\frac{1}{\eta}. \tag{16-33}$$

Sometimes we used Λ and sometimes we used η in showing a particular integral is divergent with either $\Lambda\to\infty$ or, alternatively, with $\eta\to 0$.

In the following sections, with $\overline{\text{MS}}$ renormalization scheme, we will also employ dimensional regularization, and use the similar, though not quite the same, relationship between Λ and η

Also use dimensional regularization in $\overline{MS}$

Relation between Λ and η in $\overline{MS}$ scheme

$$\ln\Lambda=\frac{1}{\eta}-\frac{\gamma}{2}+\frac{\ln 4\pi}{2}. \tag{16-34}$$

16.5 $\overline{MS}$ Renormalization of QED

We start by illustrating, in the present section, the $\overline{\text{MS}}$ scheme applied to the simplest part of the standard model, QED. Then, in the next section, we apply the same $\overline{\text{MS}}$ scheme to QCD.

16.5.1 The Fermion Propagator

In Vol. 1, in Chap. 13, (13-53) and (13-54), we showed the 2nd order fermion propagator is

First, background on QED renormalization of fermion propagator

$$iS_F^{2nd}(p)=iS_F(p)+iS_F(p)ie_0^2\Sigma(p)iS_F(p), \tag{16-35}$$

where, for m_0 as the (constant) bare mass of the fermion,

QED propagator in terms of bare mass, bare charge, and infinite Σ

$$iS_F^{2nd}(p)=\frac{i}{\not p-m_0+e_0^2\Sigma(p)+i\varepsilon}+H.O. \tag{16-36}$$

and, as shown in Wholeness Chart 14-4, pg. 369, column (III), second row,

$$e_0^2\Sigma(p)=i\frac{e_0^2}{(2\pi)^4}\int iD_{F\alpha\beta}(k)\gamma^\alpha iS_F(p-k)\gamma^\beta d^4k=i\frac{e_0^2}{(2\pi)^4}\int i\frac{-g_{\alpha\beta}}{k^2+i\varepsilon}\gamma^\alpha i\frac{1}{\not p-m+i\varepsilon}\gamma^\beta d^4k. \tag{16-37}$$

We used dimensional regularization, where

Use dimensional regularization to evaluate Σ

$$e_0^2\Sigma(p)=i\frac{e_0^2}{(2\pi)^D}\int iD_{F\alpha\beta}(k)\gamma^\alpha iS_F(p-k)\gamma^\beta d^Dk \qquad D=4-\eta, \tag{16-38}$$

and when we were all finished, we set $\eta=0$.

<u>In the On-Shell Renormalization Scheme</u>

In the on-shell scheme that we used in Vol. 1, we found (16-37) (Vol. 1, (13-8) and (15-131)) to be (where we note the symbols A and B used in Vol. 1)

What we found for QED Σ in on-shell scheme

$$e_0^2\Sigma(p) = \underbrace{-\frac{3m}{8\pi^2}\ln\frac{\Lambda}{m}}_{e_0^2 A(\Lambda,m)} + e_0^2\left(\not p - m\right)\underbrace{\frac{1}{8\pi^2}\ln\Lambda}_{B(\Lambda)} + \underbrace{e_0^2\left(\not p - m\right)\Sigma_c\left(\not p - m\right)}_{\text{complicated, finite}}. \tag{16-39}$$

We can re-write this as

$$e_0^2\Sigma(p) = \frac{e_0^2}{8\pi^2}\left(\not p - 4m\right)\ln\Lambda + e_0^2\frac{3m}{8\pi^2}\ln m + e_0^2\left(\not p - m\right)\Sigma_c\left(\not p - m\right), \tag{16-40}$$

and if we use the LHS of (16-33), we can express this in terms of η, instead of Λ,

$$e_0^2\Sigma(p) = \frac{e_0^2}{16\pi^2}\left(\not p - 4m\right)\left(\frac{2}{\eta} - \gamma\right) + e_0^2\frac{3m}{8\pi^2}\ln m + e_0^2\left(\not p - m\right)\Sigma_c\left(\not p - m\right). \tag{16-41}$$

With this Σ, we redefined mass and coupling constant to renormalize the theory

With the bare coupling constant e_0 and mass m_0 renormalized (to absorb the infinities that arise), and incorporation of parallel results for the photon propagator and QED vertex, the theory became well-defined. That is, transition amplitudes no longer diverged.

In the $\overline{\text{MS}}$ Renormalization Scheme

QED fermion propagator in $\overline{\text{MS}}$ scheme

In the $\overline{\text{MS}}$ scheme, we still have a constant bare mass m_0, but the renormalized mass is not constant, and is a function of the mass-energy scale μ. The coupling constant, as in the on-shell scheme, is also a function of μ. We denote these by $m_r = m_r(\mu)$ and $e_r = e_r(\mu)$. Because we do not have a fixed physical mass of the fermion, i.e., m, our particles and fields will be off-shell, at virtually any energy level. The $\overline{\text{MS}}$ scheme is not an on-shell scheme.

In $\overline{\text{MS}}$ scheme, employ a modified bare coupling

In this scheme, we take the dimensionless bare coupling constant to be

$$\tilde{e}_0 = \mu^{-\eta/2}e_0, \tag{16-42}$$

and

Finding Σ in $\overline{\text{MS}}$ scheme via dimensional regularization

$$e_0^2\Sigma(p) = i\frac{\tilde{e}_0^2\mu^\eta}{(2\pi)^D}\int iD_{F\alpha\beta}(k)\gamma^\alpha iS_F(p-k)\gamma^\beta d^Dk \qquad D = 4-\eta, \tag{16-43}$$

instead of (16-38).

Recall from Chap. 15, that evaluating relations like (16-43) is an exercise in tedium, taking many pages. So, we will cut to the chase and avoid that tedium by simply presenting the final result, which (it can be shown) with (16-34), is

Result of that regularization

$$e_0^2\Sigma(p) = \frac{\tilde{e}_0^2}{16\pi^2}\left(\not p - 4m_0\right)\left(\frac{2}{\eta} - \gamma + \ln 4\pi\right) + \tilde{e}_0^2\underbrace{\Sigma_r(p)}_{\text{finite}} = \frac{\tilde{e}_0^2}{16\pi^2}\left(\not p - 4m_0\right)2\ln\Lambda + \tilde{e}_0^2\underbrace{\Sigma_r(p)}_{\text{finite}}, \tag{16-44}$$

where $\Sigma_r(p)$, as noted, is finite and equals (you don't have to memorize this ☺)

$$\Sigma_r(p) = \frac{1}{16\pi^2}\left(2m_0 - \not p\right) - \frac{2}{16\pi^2}\int_0^1\left(\not p\left(1-z\right) - 2m_0\right)\ln\left(\frac{m_0^2 z - p^2 z(1-z)}{\mu^2}\right)dz. \tag{16-45}$$

Terms have been dropped in (16-44) and (16-45) that vanish as $\eta \to 0$ (when the original theory is recovered).

Note that, in general, different renormalization schemes employ the same relation (16-36), but separate $e_0^2\Sigma(p)$ into divergent and finite parts differently. Compare (16-44) to (16-41) (which is the same as(16-40)).

Same result in streamlined form

(16-44) can be rearranged as

$$e_0^2\Sigma(p) = -\tilde{e}_0^2\left(\frac{4m_0}{16\pi^2}2\ln\Lambda\right) + \tilde{e}_0^2\not p\frac{1}{16\pi^2}2\ln\Lambda + \tilde{e}_0^2\Sigma_r(p). \tag{16-46}$$

When we take $\eta \to 0$ $(\Lambda \to \infty)$, from (16-42), we will have $\tilde{e}_0^2 = e_0^2$, so we can see some parallels between (16-46) and (16-40).

From here on, we do the same thing we did in the on-shell scheme, but with a different expression for $e_0^2\Sigma(p)$. That is, we plug (16-46) into (16-36), instead of (16-40) into (16-36). Again, we avoid the tedium and cite the end result,

$$iS_F^{2nd}(p)=\frac{iZ_{MS\,f}^{2nd}}{\not p-m_r+e_{r\,f}^2\Sigma_r(p)+i\varepsilon}+O\left(\tilde{e}_0^4\right), \qquad (16\text{-}47)$$

Resulting $\overline{MS}$ fermion propagator at 2nd order

where

$$Z_{MS\,f}^{2nd}=1-e_{r\,f}^2\frac{1}{16\pi^2}2\ln\Lambda \qquad m_r=Z_{MS\,f}^{2nd}\left(m_0+e_{r\,f}^2\frac{4m_0}{16\pi^2}2\ln\Lambda\right) \qquad e_{r\,f}^2\equiv\tilde{e}_0^2Z_{MS\,f}^{2nd}. \qquad (16\text{-}48)$$

Fermion Z, mass, and coupling in $\overline{MS}$ QED renormalization

The renormalized charge due to just the fermion propagator is $e_{r\,f}$. We will find the total renormalized charge will include this, plus factors from the photon propagator and the vertex modification. Keep in mind that $Z_{MS\,f}^{2nd}$ varies with energy level, since it depends on $\tilde{e}_0$ (via $e_{r\,f}^2$), which itself depends on μ via (16-42).

From (16-48),

$$e_{r\,f}^2\equiv\tilde{e}_0^2\left(1-e_{r\,f}^2\frac{1}{16\pi^2}2\ln\Lambda\right) \quad\rightarrow\quad e_{r\,f}^2=\tilde{e}_0^2+O\left(\tilde{e}_0^4\right). \qquad (16\text{-}49)$$

Re-expressed QED coupling due to fermion $\overline{MS}$ renormalization

where the $\equiv$ symbol means the expression is good to all orders.

Note that (16-47) is the *bare* propagator, i.e., it derives from (16-36). So, we can define a *renormalized* fermion propagator, for reasons which will (hopefully) become more evident, as

$$iS_{F\,r}^{2nd}(p)=\left(Z_{MS\,f}^{2nd}\right)^{-1}iS_F^{2nd}(p)=\frac{i}{\not p-m_r+e_{r\,f}^2\Sigma_r(p)+i\varepsilon}+O\left(\tilde{e}_0^4\right). \qquad (16\text{-}50)$$

Renormalized QED fermion propagator

The form of (16-50) holds at nth order, where $Z_{MS\,f}^{nth}$ is different from $Z_{MS\,f}^{2nd}$, for $n\neq 2$. That is,

$$iS_{F\,r}^{nth}(p)=\left(Z_{MS\,f}^{nth}\right)^{-1}iS_F^{nth}(p). \qquad (16\text{-}51)$$

Even though the bare propagator $iS_F^{2nd}(p)$ in (16-50) is independent of the mass-energy scale μ (from (16-36)), the renormalized propagator $iS_{F\,r}^{2nd}(p)$ is μ dependent, since, at the least, $\Sigma_r(p)$ is (from (16-45)), even when $\eta\rightarrow 0$.

Whereas in the on-shell scheme, the renormalized mass m is equal to the physical (inertial) mass and constant, in the $\overline{MS}$ the renormalized mass m_r is not necessarily so. We can find an expression for it in terms of the physical mass by noting that the propagator (16-50) must have a pole when $\not p=m$, i.e., the denominator must go to zero then. Hence, the denominator becomes

$$m-m_r+e_r^2\Sigma_r(p=m)=0 \quad\rightarrow\quad m_r=m+e_r^2\Sigma_r(p=m). \qquad (16\text{-}52)$$

Renormalized mass in terms of physical mass for $\overline{MS}$ scheme

Since both e_r and Σ_r are finite and μ dependent, so is m_r, i.e., $m_r=m_r(\mu)$.

The dependence of the coupling constant e_r on μ is determined, not just by the fermion propagator modifications, but also by the photon propagator and vertex modifications. To find that dependence, we will have to deduce those additional relations, which we do in the subsections subsequent to this one.

Bottom line: Our final result of this subsection, for second order is (16-50), and at nth order, (16-51), the respective renormalized QED fermion propagators under the $\overline{MS}$ renormalization scheme, with 2nd order renormalized charge from (16-48) and a comparable expression at n^{th} order,

$$e_{r\,f}\equiv\tilde{e}_0\left(Z_{MS\,f}^{2nd}\right)^{1/2} \qquad e_{r\,f}\equiv\tilde{e}_0\left(Z_{MS\,f}^{nth}\right)^{1/2}. \qquad (16\text{-}53)$$

Key result: effect on coupling from fermion $\overline{MS}$ renormalization

16.5.2 The Photon Propagator

In Vol. 1, Chap. 13, (13-51) and (13-52), or more succinctly, the first row (below the title row) of Wholeness Chart 14-4, pg. 368 therein, we showed the 2nd order photon propagator is

Photon propagator at 2nd order

$$iD_{F\alpha\beta}^{2nd} = iD_{F\,\alpha\beta}(k) + iD_{F\,\alpha\mu}(k)ie_0^2\Pi^{\mu\nu}(k)iD_{F\,\nu\beta}(k)\,, \tag{16-54}$$

which we repeated in step 2 on pg. 476. In (16-54),

$$e_0^2\Pi^{\mu\nu}(k) = \frac{-ie_0^2}{(2\pi)^4}\text{Tr}\int iS_F(p)\gamma^\mu iS_F(p-k)\gamma^\nu d^4p\,. \tag{16-55}$$

In the On-shell Scheme

For the on-shell renormalization, we used dimensional regularization to find the top row of (16-56) below, and (16-33) to get the bottom row.

Find $\Pi^{\mu\nu}$ via dimensional regularization

$$\begin{aligned} e_0^2\Pi^{\mu\nu}(k) &= -e_0^2 b_n g^{\mu\nu}k^2 2\left(\frac{1}{\eta}-\frac{\gamma}{2}\right) + e_0^2 b_n g^{\mu\nu}k^2 2\,ln\,k - e_0^2\underbrace{g^{\mu\nu}k^2\Pi_c(k^2)}_{\text{finite}} \\ &= -e_0^2 b_n g^{\mu\nu}k^2 2(ln\,\Lambda - ln\,k) - e_0^2 g^{\mu\nu}k^2\Pi_c(k^2). \end{aligned} \tag{16-56}$$

What we found for QED $\Pi^{\mu\nu}$ in on-shell scheme

In the $\overline{\text{MS}}$ Scheme

In the $\overline{\text{MS}}$ scheme, instead of (16-56), and with (16-34), we get (it can be shown)

$$\begin{aligned} e_0^2\Pi^{\mu\nu}(k) &= -\tilde{e}_0^2 b_n g^{\mu\nu}k^2\left(\frac{2}{\eta}-\gamma+ln\,4\pi\right) + \tilde{e}_0^2 b_n g^{\mu\nu}k^2 2\,ln\,k - \tilde{e}_0^2 g^{\mu\nu}k^2\Pi_r(k^2) \\ &= -\tilde{e}_0^2 b_n g^{\mu\nu}k^2 2(ln\,\Lambda - ln\,k) - \tilde{e}_0^2 g^{\mu\nu}k^2\Pi_r(k^2), \end{aligned} \tag{16-57}$$

QED $\Pi^{\mu\nu}$ in $\overline{\text{MS}}$ scheme

with the last term, finite and μ dependent, having

$$\Pi_r(k^2) = -\frac{1}{2\pi^2}\int_0^1 z(1-z)\,ln\left(\frac{m^2-k^2z(1-z)}{\mu^2}\right)dz\,. \tag{16-58}$$

The photon propagator at 2nd order is

$$iD_{F\alpha\beta}^{2nd}(k) = \frac{-iZ_{MS\,\gamma}^{2nd}g_{\alpha\beta}}{k^2+e_{r\gamma}^2k^2\Pi_r(k^2)+i\varepsilon}, \tag{16-59}$$

where

$$\begin{aligned} Z_{MS\,\gamma}^{2nd} &= 1-\tilde{e}_0^2 b_n\left(\frac{2}{\eta}-\gamma+ln\,4\pi\right)+\tilde{e}_0^2 b_n 2\,ln\,k = \underbrace{1-e_{r\gamma}^2 b_n\left(\frac{2}{\eta}-\gamma+ln\,4\pi\right)}_{\text{true at 2nd order}}+\tilde{e}_0^2 b_n 2\,ln\,k \\ &= 1-e_{r\gamma}^2 b_n 2\,ln\,\Lambda+\tilde{e}_0^2 b_n 2\,ln\,k = 1-e_{r\gamma}^2 b_n 2\,ln\frac{\Lambda}{k}, \end{aligned} \tag{16-60}$$

Photon Z factor in $\overline{\text{MS}}$ scheme

and the renormalized charge due solely to the photon propagator is

$$e_{r\gamma} = \tilde{e}_0\left(Z_{MS\,\gamma}^{2nd}\right)^{1/2}. \tag{16-61}$$

Thus, we can define a renormalized photon propagator as

$$iD_{F\,r\alpha\beta}^{2nd}(k) = \frac{-i\,g_{\alpha\beta}}{k^2+e_{r\gamma}^2k^2\Pi_r(k^2)+i\varepsilon} = \left(Z_{MS\,\gamma}^{2nd}\right)^{-1} iD_{F\alpha\beta}^{2nd}(k)\,. \tag{16-62}$$

Re-expressed QED photon propagator in $\overline{\text{MS}}$ scheme

The result extends to nth order,

$$iD_{F\,r\alpha\beta}^{nth}(k) = \left(Z_{MS\,\gamma}^{nth}\right)^{-1} iD_{F\alpha\beta}^{nth}(k)\,, \tag{16-63}$$

though the expression for Π_r, and thus $Z_{MS\,\gamma}^{nth}$ is different from that at 2nd order.

Bottom line: Our final result of this subsection, for second order, is (16-62), the renormalized QED photon propagator under the $\overline{\text{MS}}$ renormalization scheme with (16-60), along with renormalized charge from (16-61). Comparable relations exist for n^{th} order,

$$e_{r\gamma} \equiv \tilde{e}_0\left(Z_{MS\,\gamma}^{2nd}\right)^{1/2} \qquad e_{r\gamma} \equiv \tilde{e}_0\left(Z_{MS\,\gamma}^{nth}\right)^{1/2}. \tag{16-64}$$

Key result: effect on QED coupling from photon $\overline{\text{MS}}$ renormalization

16.5.3 The QED Vertex

In the On-Shell Scheme

In the fourth row of Vol 1 Wholeness Chart 14-4, pg. 368, we showed the 2nd order vertex correction, with (13-9) of Chap. 9 therein and (16-33) herein, as (where we show the $L(\Lambda)$ symbol used in Vol. 1)

$$\gamma^{\mu}_{2nd}(p,p') = \gamma^{\mu} + e_0^2 \Lambda^{\mu}(p,p') = \gamma^{\mu} + e_0^2\left(\frac{1}{16\pi^2}\left(\frac{2}{\eta} - \gamma\right)\gamma^{\mu} + \Lambda^{\mu}_c(p,p')\right)$$
$$= \gamma^{\mu} + e_0^2\left(\underbrace{\frac{1}{16\pi^2} 2\ln\Lambda}_{L(\Lambda)}\gamma^{\mu} + \underbrace{\Lambda^{\mu}_c(p,p')}_{\text{finite}}\right), \qquad (16\text{-}65)$$

2nd order vertex factor in on-shell scheme

where the penultimate term in each line $\to \infty$, as $\Lambda \to \infty$, or alternatively, as $\eta \to 0$.

In the $\overline{\text{MS}}$ Scheme

In the $\overline{\text{MS}}$ scheme, we get (it can be shown)

$$\gamma^{\mu}_{2nd}(p,p') = \gamma^{\mu} + \underbrace{\tilde{e}_0^2\left(\frac{1}{16\pi^2}\left(\frac{2}{\eta} - \gamma + \ln 4\pi\right)\gamma^{\mu} + \underbrace{\Lambda^{\mu}_r(p,p')}_{\text{finite}}\right)}_{e_0^2\Lambda^{\mu}(p,p')}$$
$$= \gamma^{\mu} + \tilde{e}_0^2\left(\frac{1}{16\pi^2} 2\ln\Lambda\,\gamma^{\mu} + \underbrace{\Lambda^{\mu}_r(p,p')}_{\text{finite}}\right), \qquad (16\text{-}66)$$

2nd order vertex factor in $\overline{MS}$ scheme

where Λ^{μ}_r is a function of mass-energy level μ and differs from Λ^{μ}_c. With

$$Z^{2nd}_{MSV} = 1 + \frac{\tilde{e}_0^2}{16\pi^2} 2\ln\Lambda, \qquad (16\text{-}67)$$

2nd order vertex Z in $\overline{MS}$ scheme

(16-66) becomes

$$\gamma^{\mu}_{2nd}(p,p') = \gamma^{\mu} Z^{2nd}_{MSV} + \tilde{e}_0^2 \Lambda^{\mu}_r(p,p'). \qquad (16\text{-}68)$$

2nd order $\overline{MS}$ vertex factor in terms of Z

We define an $\overline{\text{MS}}$ *renormalized* vertex from (16-68),

$$\gamma^{\mu}_r \equiv \frac{\gamma^{\mu}_{2nd}(p,p')}{Z^{2nd}_{MSV}} \qquad (16\text{-}69)$$

with renormalized charge,

$$e_{rV} \equiv \tilde{e}_0 Z^{2nd}_{MSV} \qquad Z^{2nd}_{MSV} = 1 + \frac{\tilde{e}_0^2}{16\pi^2} 2\ln\Lambda = \underbrace{1 + \frac{e_{rV}^2}{16\pi^2} 2\ln\Lambda}_{\text{true at 2nd order}} \qquad (16\text{-}70)$$

Key result: effect on QED $\overline{MS}$ coupling from vertex

such that

$$e_{rV}\gamma^{\mu}_r = \tilde{e}_0 Z^{2nd}_{MSV} \frac{\gamma^{\mu}_{2nd}(p,p')}{Z^{2nd}_{MSV}} = \tilde{e}_0 \gamma^{\mu}_{2nd}(p,p'). \qquad (16\text{-}71)$$

Note that some authors define the Z factor for the vertex (16-67) as the inverse of what we do here and in Vol. 1. It is conventional as to whether one defines a Z factor as one way or as the inverse of that way, as long as one stays consistent throughout a given analysis.

Bottom line: Our final result of this subsection, the renormalized QED vertex under the $\overline{\text{MS}}$ renormalization scheme for second order is (16-69) with comparable results at n^{th} order (where Z^{nth}_{MSV} differs from Z^{2nd}_{MSV}). The renormalized charge, due solely to vertex corrections, is (16-70) at second order, with a comparable result at n^{th} order.

16.5.4 Combining These Results

In Vol. 1, Wholeness Chart 14-4, pg. 371, Columns (X) and (XI), we summarized the result of subsection (X) Amplitude Determination of pgs. 344 to 348. And that is this.

We combine the fermion, photon, and vertex $\overline{MS}$ effects on QED coupling

The renormalized vertex has a factor of e_r, which is modified from e_0 (or $\tilde{e}_0$ in the $\overline{\text{MS}}$ scheme) by the photon, fermion, and vertex renormalization factors. Each vertex has two fermion lines and one photon line, each of which bring in a correction factor, along with the vertex correction factor. We showed two examples of how this works in Vol. 1, pgs. 360-362 and then generalized.

Thus, we get a renormalized vertex factor as

$$\overline{MS}\text{ scheme } i\tilde{e}_0\gamma^{\mu}_{2nd}$$
$$\rightarrow\; i\tilde{e}_0\left(\left(\gamma^{\mu}Z^{2nd}_{MS\,V}+\tilde{e}_0^2\Lambda^{\mu}_r(p,p')\right)+O\left(e_0^5\right)\right)(2\text{ fermion propagators})(1\ \gamma\text{ propagator}). \tag{16-72}$$

From the above, along with (16-53), (16-64), and (16-70), we have[1]

$\overline{MS}$ relation combining those effects

$$e_r=\tilde{e}_0 Z^{2nd}_{MS\,V}\left(Z^{2nd}_{MS\,f}\right)^{1/2}\left(Z^{2nd}_{MS\,f}\right)^{1/2}\left(Z^{2nd}_{MS\,\gamma}\right)^{1/2}=\tilde{e}_0 Z^{2nd}_{MS\,V}\left(Z^{2nd}_{MS\,f}\right)\left(Z^{2nd}_{MS\,\gamma}\right)^{1/2}. \tag{16-73}$$

It is shown in the appendix that

A key, helpful Ward identity simplifies things

$$Z^{2nd}_{MS\,V}=\frac{1}{Z^{2nd}_{MS\,f}}, \tag{16-74}$$

which is known as a Ward identity. That means (16-73) becomes

Resulting total Z factor effect on coupling is just square root of photon Z factor

$$e_r=\tilde{e}_0\left(Z^{2nd}_{MS\,\gamma}\right)^{1/2}=e_0\mu^{-\eta/2}\left(Z^{2nd}_{MS\,\gamma}\right)^{1/2}, \tag{16-75}$$

and this relation turns out to be good to all orders, i.e.,

$$e_r=\tilde{e}_0\left(Z^{nth}_{MS\,\gamma}\right)^{1/2}=e_0\mu^{-\eta/2}\left(Z^{nth}_{MS\,\gamma}\right)^{1/2}. \tag{16-76}$$

Focusing on 2nd order relations, from (16-60) and (16-34), we have

Square root of photon Z factor

$$\left(Z^{2nd}_{MS\,\gamma}\right)^{1/2}=\left(1-\tilde{e}_0^2 b_n\left(\frac{2}{\eta}-\gamma+ln4\pi\right)+\tilde{e}_0^2 b_n 2\,ln\,k\right)^{1/2}$$
$$=1-\tilde{e}_0^2 b_n\left(\frac{1}{\eta}-\frac{\gamma}{2}+\frac{ln4\pi}{2}\right)+\tilde{e}_0^2 b_n\,ln\,k=1-\tilde{e}_0^2 b_n\,ln\Lambda+\tilde{e}_0^2 b_n\,ln\,k=1+\tilde{e}_0^2 b_n\,ln\frac{k}{\Lambda} \tag{16-77}$$

In (16-75), this gives us

That leaves us with the relationship between bare $\overline{MS}$ coupling and $\overline{MS}$ renormalized coupling

$$e_r(k,\Lambda)=\tilde{e}_0\left(1+\tilde{e}_0^2 b_n\,ln\frac{k}{\Lambda}\right). \tag{16-78}$$

Since we will want to compare *different* energy levels, commonly labeled p and μ, we can express (16-78) as

Energy level symbols commonly used instead of k are p and μ

$$e_r(p,\Lambda)=\tilde{e}_0\left(1+\tilde{e}_0^2 b_n\,ln\frac{p}{\Lambda}\right),\quad\text{and} \tag{16-79}$$

$$e_r(\mu,\Lambda)=\tilde{e}_0\left(1+\tilde{e}_0^2 b_n\,ln\frac{\mu}{\Lambda}\right). \tag{16-80}$$

Taking the ratio of (16-79) to (16-80), we have, again to second order,

Using e_r relation to deduce a key result

$$\frac{e_r(p,\Lambda)}{e_r(\mu,\Lambda)}=\frac{\tilde{e}_0\left(1+\tilde{e}_0^2 b_n\,ln\frac{p}{\Lambda}\right)}{\tilde{e}_0\left(1+\tilde{e}_0^2 b_n\,ln\frac{\mu}{\Lambda}\right)}=1+\tilde{e}_0^2 b_n\,ln\frac{p}{\Lambda}-\tilde{e}_0^2 b_n\,ln\frac{\mu}{\Lambda}$$
$$=1+\tilde{e}_0^2 b_n\,ln\,p-\tilde{e}_0^2 b_n\,ln\Lambda-\tilde{e}_0^2 b_n\,ln\,\mu+\tilde{e}_0^2 b_n\,ln\Lambda=1+\tilde{e}_0^2 b_n\,ln\frac{p}{\mu}. \tag{16-81}$$

---

[1] As noted, some authors employ symbolism where the vertex Z factor is the inverse of what we have here, so it appears in the denominator of the relation comparable to (16-73), rather than the numerator.

Or,

$$e_r(p) = e_r(\mu)\left(1 + \tilde{e}_0^2 b_n \ln\frac{p}{\mu}\right). \tag{16-82}$$

Now, take the partial derivative of (16-82) with respect to p, and evaluate at $p = \mu$,

$$\begin{aligned}\frac{\partial e_r(\mu)}{\partial \mu} = \left.\frac{\partial e_r(p)}{\partial p}\right|_{p=\mu} &= e_r(\mu)\left.\left(\tilde{e}_0^2 b_n \frac{\partial \ln p}{\partial p} - \tilde{e}_0^2 b_n \frac{\partial \ln \mu}{\partial p}\right)\right|_{p=\mu} \\ &= \left(e_r(\mu)\tilde{e}_0^2 b_n \frac{1}{p}\right)_{p=\mu} = e_r(\mu)\tilde{e}_0^2 b_n \frac{1}{\mu}.\end{aligned} \tag{16-83}$$

Note, from (16-78), that

$$e_r^2(\mu,\Lambda) = \tilde{e}_0^2\left(1 + \tilde{e}_0^2 b_n \ln\frac{\mu}{\Lambda}\right)^2 = \tilde{e}_0^2 + H.O., \tag{16-84}$$

where we are still considering Λ as finite, so that at second order, we can ignore the higher order terms. Re-arranging (16-83) with (16-84), we have, with Λ considered fixed, the renormalization group equation, or RGE, for QED of

$$\boxed{\mu\frac{\partial e_r(\mu)}{\partial \mu} = e_r^3(\mu) b_n = \beta_e(\mu, b_n),} \tag{16-85}$$

That result: RGE for QED

where, as we noted in Vol. 1, pgs. 314-317, as energy increases, b_n takes on different values at thresholds for μ equal to twice the mass of the fermion flavor making up each half of the second order loop.

β_e is the beta function for the RGE in the coupling constant e_r, and since b_n is always positive, the QED coupling e_r increases with increasing energy, as we know it does.

Do **Problem 8** to derive (16-86).

(16-85) can be re-expressed in terms of the fine structure "constant" α as

$$\boxed{\mu\frac{\partial \alpha(\mu)}{\partial \mu} = \alpha^2(\mu) 8\pi b_n = \beta_\alpha(\mu, b_n)}, \tag{16-86}$$

Alternative form: QED RGE for fine structure constant

where β_α is the beta function for the RGE, and is closely related to β_e. A positive value for one means a positive value for the other, and an increase in both e and α with energy μ. In Vol. 1, we used β for what we here call β_α. Different authors may call either β_α or β_e the beta function β.

These RGEs are the same as we obtained from the on-shell renormalization scheme. See Vol. 1, (12-41) and (12-42), pg. 317. This must be so, if they are to describe the coupling (or fine structure) constant we measure in experiment, which indeed they do.

Note that we obtained these RGEs using only one loop diagrams. That is, we did not need higher order diagrams, which represent higher order terms in the transition amplitude. The RGE effectively encodes the higher order terms without having to specify exactly what those terms are, and that is a very big advantage in many cases.

Only need lowest order to get RGE

16.5.5 The Renormalization Part

We have used the $\overline{\text{MS}}$ renormalization scheme to show how the QED coupling constant dependence on energy is found. We have not shown explicitly that it actually renormalizes the theory and eliminates the infinities that arise from the primitive divergences. And we will not do that explicitly, but merely state that it does indeed do so. As $\Lambda \to \infty$ (or equivalently as $\eta \to 0$), the bare coupling e_0 and bare mass m_0 are taken such as to make the renormalized coupling e_r and mass m_r finite. As a byproduct of that, e_r is found to depend on energy μ.

We focused on effect on e_r, but this approach also makes theory finite (not shown)

That is, the $\overline{\text{MS}}$ scheme really does renormalize QED, but our interest here has been determining the result of that renormalization on the QED coupling constant.

16.6 $\overline{MS}$ Renormalization of QCD

We now extrapolate what we have learned of the $\overline{\text{MS}}$ renormalization scheme to QCD. We continue doing what we have been doing in this chapter, in many steps of the process, of presenting just the results of quite lengthy calculations, in the spirit of the adage "it can be shown". This may not be fully satisfying, but it gets us to the goal without being bogged down for exceptionally long periods of time[1].

We use same $\overline{MS}$ scheme to renormalize QCD

Before looking at the renormalization process for the QCD primitive divergences, we need to understand two things: color factors and null diagrams.

16.6.1 Color Factors

What Are Color Factors

The vertex factors of Feynman QCD rule #1 (pg. 459), associated with the diagrams of Fig. 15-3 (pg. 456), contain Gell-Mann matrices λ_i ((2-80), pg. 35) and structure constants f_{ijk} (Table 2-2, pg. 36). In QCD calculations of transition amplitudes and in renormalization procedures, one needs to evaluate the various combinations of these factors that appear in Feynman amplitudes, and those quantities are known as color factors.

Color factors appear in amplitudes due to factors of λ_i & f_{ijk}

Useful Relations for Finding Color Factors

In this subsection, we present tools, relations between the structure constants and the matrices, that will later help us determine particular color factors we will need.

The following relations are useful in finding color factors

We know the Gell-Mann matrices are traceless

$$Tr\lambda_i = 0 \qquad i = 1,2,\ldots 8 \tag{16-87}$$

and obey the commutation relations

$$\left[\frac{\lambda_i}{2},\frac{\lambda_j}{2}\right] = if_{ijk}\frac{\lambda_k}{2} \quad \text{or} \quad \left[\lambda_i,\lambda_j\right] = i2f_{ijk}\lambda_k\,, \tag{16-88}$$

where the structure constants are completely anti-symmetric. Exchanging any two indices, flips the sign.

$$f_{ijk} = f_{jki} = f_{kij} = -f_{jik} = -f_{ikj} = -f_{kji}\,. \tag{16-89}$$

Structure constants, when any two of i, j, and k are the same, are zero, e.g., $f_{227} = 0$.

Do **Problem 9** to gain some comfort with (16-90).

With Table 2-2, one can, if one takes the time, verify

$$f_{ijk}f_{ljk} = 3\delta_{il}\,, \tag{16-90}$$

but, hopefully, by doing the problem suggested, you can just accept that it is true for all values of i.

Do **Problem 10** to gain some comfort with (16-91).

Another relation that the suggested problem helps us get comfortable with, and that can be proven using every Gell-Mann matrix, is

$$Tr\left(\frac{\lambda_i}{2}\frac{\lambda_j}{2}\right) = \frac{1}{2}\delta_{ij}\,. \tag{16-91}$$

Do **Problem 11** to prove (16-92).

---

[1] More detailed, with lengthier, but still not 100% complete, analysis of much of the material found herein on renormalization of QCD can be found in F. Mandl and G. Shaw, *Quantum Field Theory*, 2nd ed (Wiley 2010), pgs. 330-352.

Yet another useful relation is (where, as usual, repeated indices mean summation)

$$\frac{\lambda_i}{2}\frac{\lambda_i}{2}=\tfrac{4}{3}I \quad \left(\text{or commonly expressed as just } =\tfrac{4}{3}\right). \tag{16-92}$$

(16-87) to (16-92) can be utilized to derive further identities. Two useful ones, which we prove after stating them, are

$$if_{ijk}\frac{\lambda_j}{2}\frac{\lambda_k}{2}=-\tfrac{3}{2}\left(\frac{\lambda_i}{2}\right), \text{ and} \tag{16-93}$$

$$\frac{\lambda_j}{2}\frac{\lambda_i}{2}\frac{\lambda_j}{2}=-\tfrac{1}{6}\left(\frac{\lambda_i}{2}\right). \tag{16-94}$$

Proof of (16-93):

With (16-88) and (16-89),

$$if_{ijk}\frac{\lambda_j}{2}\frac{\lambda_k}{2}=if_{ijk}\left[\frac{\lambda_j}{2},\frac{\lambda_k}{2}\right]+if_{ijk}\frac{\lambda_k}{2}\frac{\lambda_j}{2}=-f_{ijk}f_{jkl}\frac{\lambda_l}{2}-if_{ijk}\frac{\lambda_j}{2}\frac{\lambda_k}{2}=-f_{ijk}f_{ljk}\frac{\lambda_l}{2}-if_{ijk}\frac{\lambda_j}{2}\frac{\lambda_k}{2}. \tag{16-95}$$

Then, rearranging and using (16-90), we have

$$2if_{ijk}\frac{\lambda_j}{2}\frac{\lambda_k}{2}=-3\delta_{il}\frac{\lambda_l}{2} \quad\rightarrow\quad if_{ijk}\frac{\lambda_j}{2}\frac{\lambda_k}{2}=-\frac{3}{2}\frac{\lambda_i}{2}. \tag{16-96}$$

Proof of (16-94):

To get the results after the second equal sign below, use (16-88) and (16-92), then, after that, (16-93), to obtain

$$\begin{aligned}\frac{\lambda_j}{2}\frac{\lambda_i}{2}\frac{\lambda_j}{2}&=\frac{\lambda_j}{2}\left[\frac{\lambda_i}{2},\frac{\lambda_j}{2}\right]+\frac{\lambda_j}{2}\frac{\lambda_j}{2}\frac{\lambda_i}{2}=if_{ijk}\frac{\lambda_j}{2}\frac{\lambda_k}{2}+\frac{4}{3}\frac{\lambda_i}{2}\\&=-\frac{3}{2}\frac{\lambda_i}{2}+\frac{4}{3}\frac{\lambda_i}{2}=\left(-\frac{9}{6}+\frac{8}{6}\right)\left(\frac{\lambda_i}{2}\right)=-\frac{1}{6}\left(\frac{\lambda_i}{2}\right).\end{aligned} \tag{16-97}$$

End of proofs

Color Factors and Null Diagrams

Certain Feynman diagrams in QCD make zero contribution to transition amplitudes and primitive divergences because the associated color factor vanishes. Several of these are shown in Fig. 16-5, and to evaluate the color factors, we will use Feynman Rules (pg. 459).

Some diagrams one might expect to contribute make zero contribution to amplitudes

In Fig. 16-5(a), we have a 3-gluon vertex and a gluon propagator leading to a Feynman amplitude

$$\mathcal{M}\propto S_{sym}\bar{u}_b(\mathbf{p}')i\delta_{jm}D_F^{\nu\alpha}(k)\left(g_s f_{ijm}V_{\mu\nu\alpha}\right)i\delta_{ij}D_F^{\mu\nu}(k')\left(-ig_s\gamma_\mu\left(\frac{\lambda_i}{2}\right)_{ba}\right)u_a(\mathbf{p}) \tag{16-98}$$

where S_{sym} is the symmetry factor, and the color factor in the amplitude is found from

$$\underbrace{g_s f_{ijm}V_{\mu\nu\alpha}}_{GGG\text{ vertex}}\;\underbrace{i\delta_{ij}\frac{-g^{\mu\nu}}{k'^2+i\varepsilon}}_{\text{gluon propagator}}=\underbrace{0}_{\text{see below}}. \tag{16-99}$$

Since the structure constant f_{ijk} is antisymmetric in i and j, and δ_{ij} is symmetric, this vanishes. Alternatively, if $i=j$, the structure constant is zero. If $i\neq j$, the Kronecker delta is zero.

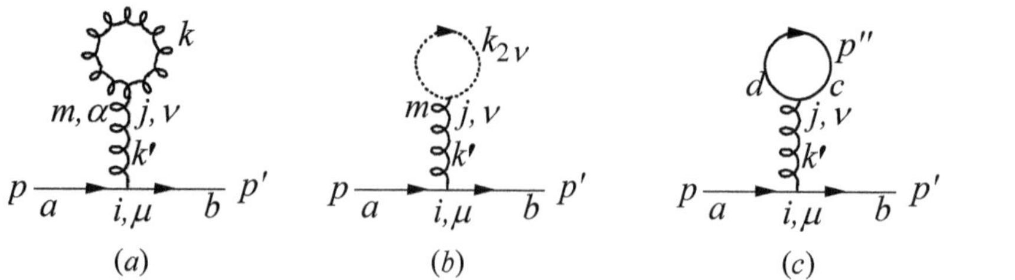

In particular, these three shown

Figure 16-5. Some Null Diagrams in QCD

Do **Problem 12** to show the color factor in Fig. 16-5(b) is zero, so it makes no contribution to Feynman amplitudes.

For Fig. 16-5(c), the quark loop starts and ends at the same color index. So, we get a color factor, where we use (16-87) at the end, of

$$\underbrace{-ig_s\gamma_\nu\left(\frac{\lambda_j}{2}\right)_{cd}}_{qqG\text{ vertex}}\underbrace{i\delta_{cd}\frac{1}{\not p-m+i\varepsilon}}_{\text{quark loop propagator}}\propto\left(\lambda_j\right)_{cd}\delta_{cd}=Tr\lambda_j=0\,. \tag{16-100}$$

So, that diagram also makes no contribution to any Feynman diagram.

We showed (16-100) to help in learning to evaluate color factors, but as we saw on pg. 364, any diagram like Fig. 16-5(c) with a closed fermion loop is null, anyway, for the reasons delineated there.

In the remainder of the chapter, we will simply consider the contribution to the amplitude of all diagrams with loops like those in Fig. 16-5 is zero.

16.6.2 The QCD Renormalized Quark Propagator

As we saw in Wholeness Chart 16-5, pg. 472, similar to QED, there are only three types of primitive divergence in QCD, the gluon self-energy, the quark self-energy, and the quark-quark-gluon vertex. We will look at each of these and present the form, with appropriate Z factors, they take when renormalized according to the $\overline{MS}$ scheme. We will then combine these results, as we did in QED for e, to determine the dependence of g_S, the QCD coupling constant, on energy. As the reader has been forewarned, for each primitive divergence type, we will skip an extensive amount of developmental material and just present the final result.

Evaluate QCD quark propagator higher order as we did QED fermion propagator

We start in this sub-section with the self-energy of the quark propagator, the second order contribution of which is illustrated in Fig. 16-6.

Figure 16-6. The Quark Self-Energy at Second Order

Similar to (16-37), where here we take g_{S0} to represent the bare QCD coupling constant, which we have previously taken as g_S,

$$g_{S0}^2\Sigma_{ac}(p)=i\frac{g_{S0}^2}{(2\pi)^4}C_{ac}\int iD_{F\alpha\beta}(k)\gamma^\alpha iS_F(p-k)\gamma^\beta d^4k\,, \tag{16-101}$$

QCD Σ_{ab} found like QED Σ

with C_{ac} being a color factor arising from the two vertex factors and the delta function in the gluon propagator. It has the value (see (16-92)),

but need to include color factor

$$C_{ac}=\left(\frac{\lambda_i}{2}\right)_{ab}\left(\frac{\lambda_j}{2}\right)_{bc}\delta_{ij}=\left(\frac{\lambda_i}{2}\right)_{ab}\left(\frac{\lambda_i}{2}\right)_{bc}=\tfrac{4}{3}\delta_{ac}=\tfrac{4}{3}\left(\text{since } a \text{ must}=c\right). \tag{16-102}$$

a is the color of the incoming quark, but the outgoing quark must be the same color, so we need to take $a=c$.

Hence, (16-101) becomes

$$g_{S0}^2\Sigma_{ac}(p)=\tfrac{4}{3}g_{S0}^2\frac{i}{(2\pi)^4}\int iD_{F\alpha\beta}(k)\gamma^\alpha iS_F(p-k)\gamma^\beta d^4k\,, \tag{16-103}$$

which gives us Σ_{ab} in terms of bare QCD coupling

which is the same as the QED fermion propagator (16-37) except for the factor of 4/3 and g_{s0} instead of e_0. Hence, we can employ all of the $\overline{\text{MS}}$ scheme analysis we did there directly to here. Thus, we employ a modified coupling constant

$$\tilde g_{S0}=g_{S0}\mu^{-\eta/2}\,, \tag{16-104}$$

$\overline{MS}$ modified bare QCD coupling

and after much calculation, find (16-103) is

$$g_{S0}^2 \Sigma_{ac}(p) = \frac{4}{3} g_{Srq}^2 \frac{1}{16\pi^2}\left(\frac{2}{\eta} - \gamma + \ln 4\pi\right) + \ldots$$
$$= \frac{4}{3} g_{Srq}^2 \frac{1}{16\pi^2} 2\ln\Lambda + \ldots = g_{Srq}^2 \frac{1}{6\pi^2}\ln\Lambda + \ldots \qquad (16\text{-}105)$$

Σ_{ab} in more suitable form

The renormalized charge due only to the quark self-energy, comparable to (16-53), is

$$g_{rq} \equiv \tilde{g}_{S0}\left(Z_{\overline{MS}q}^{2nd}\right)^{1/2} \qquad (16\text{-}106)$$

Key result: effect on QCD coupling from quark $\overline{MS}$ renormalization

where

$$Z_{\overline{MS}q}^{2nd} = 1 - \frac{4}{3} g_{Srq}^2 \frac{1}{16\pi^2} 2\left(\frac{1}{\eta} - \frac{\gamma}{2} + \frac{\ln 4\pi}{2}\right) = 1 - g_{Srq}^2 \frac{1}{6\pi^2}\ln\Lambda. \qquad (16\text{-}107)$$

Quark QCD Z factor

16.6.3 The Renormalized Gluon Propagator

Fig. 16-7 shows the three non-zero contributions to the modified gluon propagator at second order.

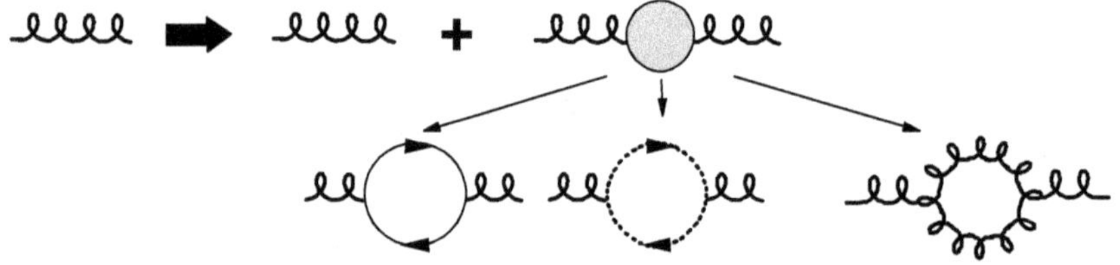

Figure 16-7. The Gluon Self-Energy Contributions at Second Order

The contributions from the quark, ghost, and gluon loops are calculated in the usual way, i.e., by adding the contributions to the propagator from each diagram. The calculation is lengthy, however, so we will just quote the final results, which parallel the photon propagator results (16-54), (16-60), and (16-64), and where n_q is the number of quark flavors (which is 6 in the SM).

$$g_{S0}^2 \Pi_{ij}^{\mu\nu}(k) = \delta_{ij}\left(k^\mu k^\nu - k^2 g^{\mu\nu}\right)\frac{\tilde{g}_0^2}{16\pi^2}\left(\frac{2n_q}{3} - 5\right)\left(\frac{2}{\eta} - \gamma + \ln 4\pi\right) + \ldots$$
$$= \delta_{ij}\left(k^\mu k^\nu - k^2 g^{\mu\nu}\right)\frac{\tilde{g}_0^2}{16\pi^2}\left(\frac{2n_q}{3} - 5\right) 2\ln\Lambda + \ldots, \qquad (16\text{-}108)$$

QCD $\Pi_{ij}^{\mu\nu}$ found like QED $\Pi^{\mu\nu}$

which is comparable to (16-57). This ultimately gives us

$$g_{rG} \equiv \tilde{g}_{S0}\left(Z_{\overline{MS}G}^{2nd}\right)^{1/2}, \qquad (16\text{-}109)$$

Key result: effect on QCD coupling from gluon $\overline{MS}$ renormalization

with,

$$Z_{\overline{MS}G}^{2nd} = 1 - \frac{g_{rG}^2}{16\pi^2}\left(2n_q - 5\right)\left(\left(\frac{2}{\eta} - \gamma + \ln 4\pi\right) - 2\ln k\right) = 1 - \frac{g_{rG}^2}{8\pi^2}\left(2n_q - 5\right)\left(\ln\Lambda - \ln k\right). \qquad (16\text{-}110)$$

Gluon QCD Z factor

16.6.4 The Renormalized QCD Vertex

The second order QCD vertex correction diagrams are shown in Fig. 16-8.

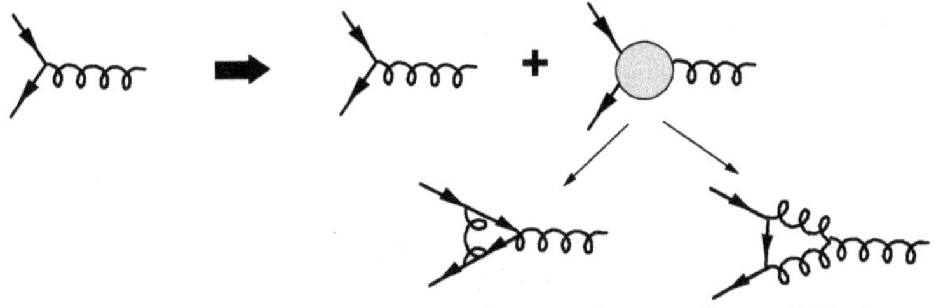

Figure 16-8. The QCD Vertex Corrections at Second Order

Comparable to the QED vertex relation (16-65), after much calculation, we have

$$\tilde{g}_{S0}^2 \Lambda_i^{2nd\,\mu}(p,p') = \frac{13\tilde{g}_{S0}^2}{48\pi^2}\left(\frac{2}{\eta} - \gamma + \ln 4\pi\right)\left(\frac{\lambda_i}{2}\right)\gamma^\mu + \ldots = \frac{13\tilde{g}_{S0}^2}{48\pi^2} 2\ln\Lambda\left(\frac{\lambda_i}{2}\right)\gamma^\mu + \ldots, \quad (16\text{-}111)$$

QCD Λ_i^μ found like QED Λ^μ

and comparable to (16-70) and (16-67)

$$g_{rV} \equiv \tilde{g}_{S0} Z_{\overline{MS}\,V\,QCD}^{2nd}, \quad (16\text{-}112)$$

Key result: effect on QCD coupling from vertex $\overline{MS}$ renormalization

$$Z_{\overline{MS}\,V\,QCD}^{2nd} = 1 + \frac{13g_r^2}{48\pi^2}\left(\frac{2}{\eta} - \gamma + \ln 4\pi\right) = 1 + \frac{13g_r^2}{48\pi^2} 2\ln\Lambda . \quad (16\text{-}113)$$

Vertex QCD Z factor

16.6.5 Combining These Results

In the same manner as we found for QED with (16-73), the full correction to the strong coupling constant comes from the collective effect from the quark self-energy correction, the gluon self-energy correction, and the vertex correction.

$$g_r = \tilde{g}_{S0}\left(Z_{\overline{MS}\,q}^{2nd}\right)\left(Z_{\overline{MS}\,G}^{2nd}\right)^{1/2} Z_{\overline{MS}\,V\,QCD}^{2nd} = g_{S0}\mu^{-\eta/2}\left(Z_{\overline{MS}\,q}^{2nd}\right)\left(Z_{\overline{MS}\,G}^{2nd}\right)^{1/2} Z_{\overline{MS}\,V\,QCD}^{2nd}, \quad (16\text{-}114)$$

Total effect on QCD coupling from $\overline{MS}$ renormalization

where in QCD, the quark and vertex factors do not cancel like the fermion and vertex factors did in QED. It can be shown that (16-114) ultimately equals

$$g_r = \tilde{g}_{S0}\left(1 - \frac{\tilde{g}_{S0}^2}{16\pi^2}\underbrace{\left(11 - \frac{2n_q}{3}\right)}_{\beta_0}\ln\frac{k}{\Lambda} + H.O.\right). \quad (16\text{-}115)$$

QCD coupling in terms of energy level k and Λ

Note the different sign after the 1 in (16-115) from the comparable QED result (16-78).

We can find the dependence of coupling on energy level, given its value at another level, in the same manner as we did in QED with (16-81). That is, at second order, with (16-115)

$$\begin{aligned}\frac{g_r(p,\Lambda)}{g_r(\mu,\Lambda)} &= \frac{\tilde{g}_{S0}\left(1 - \frac{\tilde{g}_{S0}^2}{16\pi^2}\beta_0\ln\frac{p}{\Lambda}\right)}{\tilde{g}_{S0}\left(1 - \frac{\tilde{g}_{S0}^2}{16\pi^2}\beta_0\ln\frac{\mu}{\Lambda}\right)} = 1 - \frac{\tilde{g}_{S0}^2}{16\pi^2}\beta_0\ln\frac{p}{\Lambda} + \frac{\tilde{g}_{S0}^2}{16\pi^2}\beta_0\ln\frac{\mu}{\Lambda} \\ &= 1 - \frac{\tilde{g}_{S0}^2}{16\pi^2}\beta_0\left(\ln p - \ln\Lambda - \ln\mu + \ln\Lambda\right) = 1 - \frac{\tilde{g}_{S0}^2}{16\pi^2}\beta_0\left(\ln p - \ln\mu\right) \quad (16\text{-}116) \\ &= 1 - \frac{\tilde{g}_{S0}^2}{16\pi^2}\beta_0\ln\frac{p}{\mu},\end{aligned}$$

or

$$g_r(p) = g_r(\mu)\left(1 - \frac{\tilde{g}_{S0}^2}{16\pi^2}\beta_0 \ln\frac{p}{\mu}\right) = \frac{g_r(\mu)}{1 + \frac{\tilde{g}_{S0}^2}{16\pi^2}\beta_0 \ln\frac{p}{\mu}} \quad \text{(at second order).} \qquad (16\text{-}117)$$

QCD coupling in terms of energy level p and reference energy level μ

To get the RGE for QCD, we follow the same steps we did for QED in (16-83).

$$\frac{\partial g_r(\mu)}{\partial \mu} = \left.\frac{\partial g_r(p)}{\partial p}\right|_{p=\mu} = -\frac{g_r(\mu)}{16\pi^2}\beta_0\left.\left(\tilde{g}_{S0}^2\frac{\partial \ln p}{\partial p} - \tilde{g}_{S0}^2\frac{\partial \ln \mu}{\partial p}\right)\right|_{p=\mu}$$
$$= -\left.\left(\frac{1}{16\pi^2}g_r(\mu)\tilde{g}_{S0}^2\beta_0\frac{1}{p}\right)\right|_{p=\mu} = -\frac{1}{16\pi^2}g_r(\mu)\tilde{g}_{S0}^2\beta_0\frac{1}{\mu}, \qquad (16\text{-}118)$$

where in QCD, comparable to (16-84) in QED, we have $g_r^2(\mu) = \tilde{g}_{S0}^2 + H.O.$, so (16-118) becomes

$$\boxed{\mu\frac{\partial g_r(\mu)}{\partial \mu} = -\frac{\beta_0}{16\pi^2}g_r^3(\mu) = -\frac{1}{16\pi^2}\left(11 - \frac{2n_q}{3}\right)g_r^3(\mu) = \beta_{g_r}}, \qquad (16\text{-}119)$$

QCD RGE for coupling constant

where the beta function β_{g_r}, unlike in QED, is negative. The strong coupling constant decreases with increase in energy level, and as we have noted earlier, this leads to both asymptotic freedom and infrared slavery.

Note that if the number of quark flavors n_q were to exceed 16, the beta function would be positive. We would then have neither asymptotic freedom nor quark confinement, and the universe as we know it would not exist. In fact, due to the fine tuning of nature's constants needed to produce our universe, even a minor change in the number of quark flavors would drastically alter many things needed for our existence, such as, for one example, nuclear fusion, and thus, star formation.

QCD beta function magnitude and sign depend on number of quark flavors

In QED, by contrast, the beta function value depended on the number of electrically charged fundamental particles (via b_n), but its sign did not.

In QED, the RGE can be expressed in terms of the coupling constant or the fine structure constant. We can do a similar thing in QCD by defining, via analogy with α,

$$\alpha_s \equiv \frac{g_r^2}{4\pi}. \qquad (16\text{-}120)$$

QCD analog of fine structure constant

Thus, (16-117) becomes

$$\alpha_s(p) = \frac{\alpha_s(\mu)}{\left(1 + \frac{\beta_0}{4\pi}\alpha_s(\mu)\ln\frac{p}{\mu}\right)^2} = \frac{\alpha_s(\mu)}{1 + 2\frac{\beta_0}{4\pi}\alpha_s(\mu)\ln\frac{p}{\mu}} = \frac{\alpha_s(\mu)}{1 + \frac{\beta_0}{4\pi}\alpha_s(\mu)\ln\frac{p^2}{\mu^2}}, \qquad (16\text{-}121)$$

α_s in terms of energy level p and reference energy level μ

If the value of $\alpha_s(\mu)$ is determined at energy level μ, then one can find $\alpha_s(p)$ at another energy level p. We will plot (16-121) and compare it with experiment in the next chapter.

Also, (16-119) can be converted to the RGE for α_s,

$$\boxed{\mu\frac{\partial \alpha_s(\mu)}{\partial \mu} = -\frac{1}{2\pi}\left(11 - \frac{2n_q}{3}\right)\alpha_s^2(\mu) = -\frac{\beta_0}{2\pi}\alpha_s^2(\mu) = \beta_{\alpha_s}}, \qquad (16\text{-}122)$$

QCD RGE in terms of α_s

where again, the negative sign of the beta function means decreasing "QCD fine structure constant", and thus, the QCD coupling constant, with increasing energy.

16.7 Chapter Summary

16.7.1 Primitive Divergences

QED

In this chapter we showed the superficial divergence K for a QED proper diagram can be found by simply counting the external photon and fermion lines according to (16-7)

$$K = 4 - \frac{3}{2}f_e - b_e. \qquad \text{QED, repeat of (16-7)}$$

Every divergent proper graph must have $K \geq 0$, but some graphs with $K \geq 0$ actually do converge. Since every primitively-divergent graph is a proper graph, we can find the primitively-divergent graphs in QED by examining every possible combination of external and internal lines. In doing so, we eliminate all possible combinations of external particles whose graphs must converge, and we are left with the five types of diagrams in Wholeness Chart 16-4, pg. 469. By further examination, two of these are eliminated, and we find three primitive divergences in QED, the photon self-energy, the fermion self-energy, and the vertex.

QCD

The superficial divergences in QCD are found in parallel fashion with K of (16-16)

$$K = 4 - \tfrac{3}{2}\left(f_e + \eta_e\right) - b_e . \qquad \text{QCD, repeat of (16-16)}$$

After considering all possible combinations of external lines, we found all possible QCD primitively divergent graphs to be those of Wholeness Chart 16-5, pg. 472. With further consideration, we eliminated all but three of those types, the gluon self-energy, the quark self-energy, and the QCD vertex.

16.7.2 Renormalization in Different Theories

Renormalizability

A field theory is renormalizable if a finite number of parameters (like mass and coupling constant) can be redefined such that transition amplitudes do not diverge as energy level approaches infinity.

Non-Minimal Interaction Theories

All of the interaction terms we have studied in QED, electroweak, and QCD field theories are needed to match theory with experiment, so they are the minimal terms we need for a good theory, and such terms are called minimal interaction terms.

Additional terms that keep the Lagrangian symmetric are possible, however, and such terms are called non-minimal. Theories with non-minimal terms must have coupling constant with negative dimension and have superficial divergences K according to (16-28),

$$K_{non-min} = 4 - \tfrac{3}{2} f_e - 2b_e - nD_{g_1} . \qquad \text{repeat of (16-28)}$$

D_{g_1} is the dimension of the coupling constant for the non-minimal interaction term, and n is the number of vertices representing that term in a given diagram.

Since the coupling constant dimension is negative, each higher order (more vertices) diagram would have a greater K than those of lesser orders. And since every interaction entails an addition of diagrams of all orders (to infinity, in principle) we would have contributions from an infinite number of diagrams with $K \to \infty$ as $n \to \infty$. For a given set of external lines, the superficial divergence would approach infinity. While this is only a superficial divergence measure, an infinite superficial divergence does imply an actual infinite divergence. We would need an infinite number of parameters to renormalize, so non-minimal theories, theories with negative dimension coupling constant(s), are non-renormalizable.

16.7.3 QCD Renormalization

QCD renormalization via the $\overline{\text{MS}}$ scheme parallels that of QED. In the process of such renormalization, both the mass and coupling constant become functions of energy level μ, i.e. $m_r = m_r(\mu)$ and $g_r = g_r(\mu)$.

The resulting RGE, parallel to that for the QED fine structure constant, is (16-122) ,

$$\mu \frac{\partial \alpha_s(\mu)}{\partial \mu} = -\frac{1}{2\pi}\left(11 - \frac{2n_q}{3}\right)\alpha_s^2(\mu) = \beta_{\alpha_s} \qquad \alpha_s(\mu) \equiv \frac{g_r^2}{4\pi} . \qquad \text{repeat of (16-122)}$$

For the six quark flavors in the SM, $n_q = 6$, so the beta function is negative, i.e., the coupling decreases with increasing energy (closer distance). This leads to asymptotic freedom and quark confinement, the hallmarks of the QCD theory.

16.8 Appendix: Deriving (16-74) (Added in June 2022 revision)

To derive (16-74), start with the first part of (16-48), the 2nd order part of (16-53), and the latter part of (16-70), each given a new equation number below.

$$Z_{MS\,f}^{2nd} = 1 - e_{r\,f}^{2} \frac{1}{16\pi^2} 2\ln\Lambda \tag{16-123}$$

$$e_{r\,f} \equiv \tilde{e}_0 \left(Z_{MS\,f}^{2nd} \right)^{1/2} \tag{16-124}$$

$$Z_{MS\,V}^{2nd} = 1 + \frac{\tilde{e}_0^{2}}{16\pi^2} 2\ln\Lambda \tag{16-125}$$

(16-124) into (16-123) gives us

$$Z_{MS\,f}^{2nd} = 1 - Z_{MS\,f}^{2nd} \frac{\tilde{e}_0^{2}}{16\pi^2} 2\ln\Lambda \quad \rightarrow \quad Z_{MS\,f}^{2nd} \left(1 + \frac{\tilde{e}_0^{2}}{16\pi^2} 2\ln\Lambda \right) = 1 . \tag{16-126}$$

(16-125) into (16-126) yields

$$Z_{MS\,f}^{2nd} Z_{MS\,V}^{2nd} = 1 , \tag{16-127}$$

which is (16-74).

16.9 Problems

1. For the six diagrams in the second row of diagrams in each column of Wholeness Chart 16-1, show that each is either proper or improper.
2. Show that the first diagram in the second column of Wholeness Chart 16-2 is primitively divergent. Then, show that the first and second diagrams in the second row of the third column are not primitively divergent.
3. Repeat what we did in Wholeness Chart 16-3 for photon in/photon out proper diagrams.
4. Use (16-5) to find the number of indeterminate four-momenta d for the diagrams of Problem 3.
5. Use (16-6) to find the number of vertices n, two different ways, for the first diagram in Wholeness Chart 16-1, for each diagram in the third row of proper diagrams, and each diagram in the second row of improper diagrams.
6. Use (16-8) to show that, in QED, the following combinations of external lines in a proper Feynman diagram converge: i) $f_e = 2$, $b_e = 3$; ii) $f_e = 4$, $b_e = 0$; iii) $f_e = 0$, $b_e = 5$.
7. Prove (16-26).
8. Prove (16-86).
9. Show that $f_{1jk} f_{1jk} = 3$ and $f_{8jk} f_{8jk} = 3$.
10. Show that $Tr\left(\frac{\lambda_1}{2} \frac{\lambda_1}{2} \right) = \frac{1}{2}$ and $Tr\left(\frac{\lambda_1}{2} \frac{\lambda_2}{2} \right) = 0$.
11. Prove $\frac{\lambda_i}{2} \frac{\lambda_i}{2} = \frac{4}{3}$.
12. Show that Fig. 16-5(b) makes no contribution to any Feynman amplitude.

Chapter 17

QCD Experiments

"The real measure of success is the number of experiments that can be crowded into 24 hours."
Thomas Edison

17.0 Preliminaries

17.0.1 Background and Scope of this Chapter

As noted earlier in the text, because the strong coupling constant varies with energy level, QCD interactions can be analyzed with perturbation theory when interaction energy level is high (short distances), but cannot be when energy level is low (larger distances). This means quark interactions to form hadrons must be evaluated via non-perturbative means. And it means high energy electron scattering off nucleons and high energy electron-positron scattering to form hadrons can be handled perturbatively.

Progress in hadron formation analysis has been made via lattice gauge theory, which is a topic beyond the scope of this book. In a nutshell, as discussed at the beginning of Chap. 15, it breaks spacetime into small discrete 4D volumes and uses computers to numerically integrate the associated path integrals. In recent years, calculated results have come quite close to experimental measurements.

Non-perturbative hadron analysis via lattice gauge theory not treated in this book

We will not discuss anything more in this chapter about non-perturbative QCD, but will focus exclusively on experimental verification of perturbative solution predictions.

In Chaps. 16 and 17 of Vol. 1, we did an in-depth, step-by-small-step treatment of QED scattering and other predictions (like the Coulomb potential and the anomalous magnetic moment), which took over 80 pages. In this chapter, we will not be carrying out such extensive in-depth analysis for QCD, but rather focus on a grand overview of the fundamental concepts behind the comparison of theory with experiment. In doing so, we will 1) extrapolate QED scattering results to QCD, and 2) in many cases, simply state results of long, involved analyses, the details of which can be found in references to be cited.

This chapter focuses on comparing perturbative QCD predictions with experiment

So, the purpose of this chapter will be to give you, the reader, a foundation in the basic principles of deducing experimental predictions from more basic QCD theory, and then comparing those predictions with empirical results.

17.0.2 Chapter Overview

Topics in this chapter

In this chapter, we will compare experiment with theory for

- measurements of QCD α_s as a function of energy μ,
- elastic scattering off the proton,
- deep inelastic scattering off the proton, and
- high energy electron-positron scattering producing hadrons.

17.1 QCD Coupling vs Energy

In Chap. 16, (16-121), we found the QCD analog of the fine structure constant, for different energy levels p and μ, where n_q is the number of quark flavors that readily form loops at energy p,

$$\alpha_s(p)=\frac{\alpha_s(\mu)}{1+\frac{1}{4\pi}\left(11-\frac{2n_q}{3}\right)\alpha_s(\mu)\ln\frac{p^2}{\mu^2}}. \qquad (17\text{-}1)$$

The reference level μ is typically taken as the mass of the Z boson (91.3 GeV), at which experiment has found

$$\alpha_s(m_Z)=0.118\pm.002. \qquad (17\text{-}2)$$

With this, the comparison of (17-1) with experiment is shown in Fig. 17-1, and little more comment need be made.

Strong coupling dependence on energy: theory vs. experiment

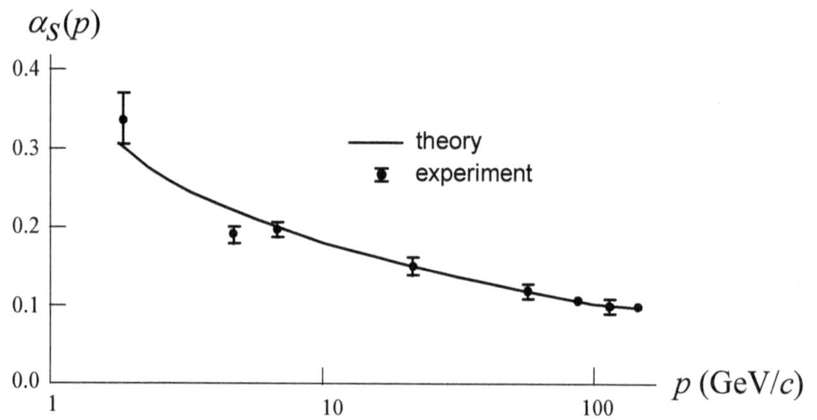

Figure 17-1. QCD Coupling Constant α_s vs. Energy
Adapted from W. M. Yao et al, *Jour. of Phys.*, Vol. 33, 2006, Sect. 9, pgs 112-116.

17.2 High Energy Scattering off Nucleons

Two types of electron scattering off protons

Electron beam scattering off nucleons has been carried out since the 1960s and played a key role in deducing the quark composite structure of protons and neutrons. In the elastic case, where p represents the proton and we get no change in type of the incident particles[1],

elastic

$$e^- + p \rightarrow e^- + p. \qquad (17\text{-}3)$$

Alternatively, we could get inelastic scattering, where a new particle is created. We will designate one of more of these particles by the symbol X, so

inelastic

$$e^- + p \rightarrow e^- + X, \qquad (17\text{-}4)$$

and the two possibilities are shown, with labels we will be using, in Fig. 17-2.

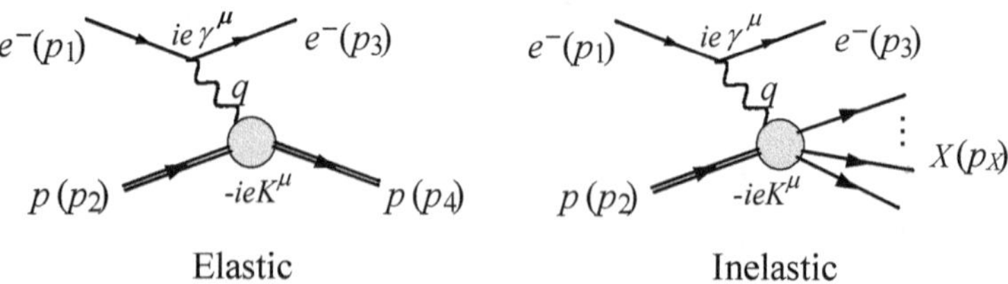

Figure 17-2. Electron Scattering off a Proton

[1] Recall from Vol. 1 that in elastic scattering, all energy changes are kinetic and particles do not change their identity (say from up quark to down quark). Inelastic scattering, by comparison, exchanges kinetic energy with the mass-energy of particles, so particles can change identity (since different flavor particles have different masses.)

The results of such experiments depend strongly on the energy level of the electron, so in Fig. 17-2, we represent such different possible results with a shaded circle. The K^{μ} symbolizes the various possible factors for different internal structures of the shaded circle.

Results depend strongly on energy level

17.2.1 Different Energy Regimes

There are four distinct energy regimes we need to consider, as follows.

4 energy regimes with different results

1. Very low electron energies, such that the photon in Fig. 17-2 has wavelength significantly greater than the radius r_p of the proton.

$$\lambda \gg r_p \tag{17-5}$$

In this case, the scattering behaves effectively as if the proton were point-like.

2. Low, but not as low, electron energies where the photon wavelength is comparable to the proton radius.

$$\lambda \approx r_p \tag{17-6}$$

Here, the proton looks like an extended charged object, rather than a point. Such experiments are useful for determining the size of the proton.

3. High electron energies where the wavelength is short enough to detect sub-structure, i.e., valence quarks, in the proton.

$$\lambda < r_p \tag{17-7}$$

4. Very high electron energies at which scattering also occurs off sea quarks and the proton appears to be a conglomeration of gluons, valence quarks, and sea quarks.

$$\lambda \ll r_p \tag{17-8}$$

17.2.2 Elastic vs Inelastic Scattering

With regard to elastic vs inelastic scattering of protons, things get a little ambiguous, because the incoming photon can boost the energy of a composite quark (without changing the quark flavor since it is a QED interaction). Thus, the *uud* composite is no longer in the lowest energy level, which corresponds to a proton. For example, the hadron could then be a Δ^+ particle, made of *uud* quarks, but having a greater mass than the proton, because it is not in the ground state and has more internal energy. As an aside, the Δ^+ decays, commonly, via the strong interaction to a proton and a neutral pion.

Quarks may not change, but proton does (quarks in elastic interaction)

The point is, if we think in terms of the proton and Δ^+, the interaction is inelastic. If we think in terms of the composite quarks (which only change energy level, not flavor), then it is elastic.

This could be very confusing, but keep in mind that in this chapter, unless we specify otherwise, when we say "inelastic", we mean the electron-proton interaction leaves us with a final electron and something other than just a proton. The final products can include many other types of hadrons.

Inelastic here means proton changes to other particle(s)

As examples, the first two types of interactions below (where "G" means gluon) are inelastic; the last, elastic.

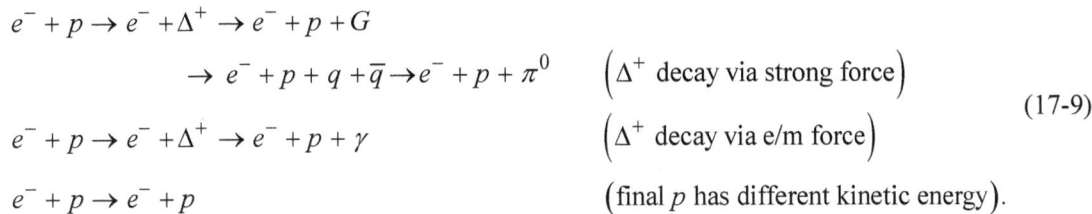

$$\begin{aligned}
&e^- + p \rightarrow e^- + \Delta^+ \rightarrow e^- + p + G \\
&\qquad\qquad \rightarrow e^- + p + q + \bar{q} \rightarrow e^- + p + \pi^0 && \left(\Delta^+ \text{ decay via strong force}\right) \\
&e^- + p \rightarrow e^- + \Delta^+ \rightarrow e^- + p + \gamma && \left(\Delta^+ \text{ decay via e/m force}\right) \\
&e^- + p \rightarrow e^- + p && \left(\text{final } p \text{ has different kinetic energy}\right).
\end{aligned} \tag{17-9}$$

Examples of electron-proton interaction

The last of (17-9) is the most common result.

Aside

Inelastic interactions are most commonly thought of in terms of weak interactions, whereby flavor changes are mediated by W vector bosons. However, we should keep in mind that inelastic interactions can also occur via strong interactions, as shown, for example, in Fig. 17-3, where u symbolizes up quark and s, strange quark.

One example of inelastic QCD interaction

Figure 17-3. Inelastic Interaction Mediated by Gluon

17.2.3 Form Factors and Structure Functions

Scattering amplitudes are different for point-like targets vs extended targets, and also for stationary (immovable, no recoil, during collisions) targets vs non-stationary ones. The former in each case (point-like and stationary) are, of course, idealizations that don't really occur in nature. But, by using such approximations, in many instances, we can get very accurate results.

Scattering different for point particles than for extended ones

Elastic Scattering and Form Factors

Immovable Target

In Vol. 1, we derived the QED (elastic) differential cross-section for a point-like immovable target, the result of which is shown there in (17-172), pg. 479, which is known as Mott scattering, and which we repeat below. The subscript label "1" corresponds to the incoming electron, as we used it in Fig. 17-2 (though there the target proton is not necessarily considered immovable). Z is the total number of positive charges for the target, which for a proton is +1.

$$\left(\frac{d\sigma}{d\Omega'}\right)_{Mott}=\frac{(\alpha Z)^2}{4E_1^2 v_1^4 \sin^4(\theta/2)}\left(1-v_1^2\sin^2(\theta/2)\right)\quad\begin{cases}\text{elastic, relativistic or not,}\\ \text{lab frame, immovable,}\\ \text{point-like target, } v_1=|\mathbf{p}_1|/E_1\end{cases}\qquad(17\text{-}10)$$

Elastic, immovable point target differential cross-section

Now, when we have an extended (non-point-like) target, we have to modify (17-10) by what is known as a form factor (actually the square of its absolute value), designated here as F, that is generally a function of the virtual photon 3-momentum $\mathbf{q}$.

$$\left(\frac{d\sigma}{d\Omega'}\right)_{\substack{non-point-like,\\ immovable}}=\left(\frac{d\sigma}{d\Omega'}\right)_{Mott}\left|F(\mathbf{q})\right|^2 .\qquad(17\text{-}11)$$

Elastic, immovable extended target cross-section with form factor

For an immovable spherical target, it is *relatively* straightforward to evaluate F. We won't do that, but you can find it elsewhere[1]. The more complicated the shape of the target, the more difficult it is to calculate F, though it can be determined by comparing the experimental cross-section to (17-11).

We will overview form factors and not study them extensively

The transition amplitude resulting from Fig. 17-2, and leading to (17-11), would have form

$$\mathcal{M}\left(e^-p\rightarrow e^-p\right)\ \alpha\ \alpha^2\left(\bar{u}(\mathbf{p}_3)\gamma^{\mu}(\mathbf{p}_1)\right)\frac{g_{\mu\nu}}{q^2+i\varepsilon}\left(\bar{u}(\mathbf{p}_4)K^{\nu}u(\mathbf{p}_2)\right),\qquad(17\text{-}12)$$

where

$$K^{\nu}=\gamma^{\nu}F(\mathbf{q})\quad\text{(immovable, extended target).}\qquad(17\text{-}13)$$

Form factor in associated amplitude

Note that (17-10) depends only on the incoming electron energy (speed), but that (17-11) also depends on the momentum transferred by the virtual photon to the target.

Recoiling Target

For a recoiling point-like target, we adopt a relation that others have found for electron-muon QED scattering (we never did this in Vol. 1), and just substitute the mass of the target particle for the muon mass. We simply cite that appropriate relation[2], where the electron mass is taken as negligible,

$$\left(\frac{d\sigma}{d\Omega_1'}\right)_{\substack{lab\\ e^-t\rightarrow e^-t}}=\frac{(\alpha Z)^2}{4E_1^2\sin^4\frac{\theta}{2}}\frac{E_3}{E_1}\left(\cos^2\frac{\theta}{2}-\frac{q^2}{2M_t^2}\sin^2\frac{\theta}{2}\right)\quad\begin{cases}\text{elastic, relativistic or not,}\\ \text{lab frame, initially stationary,}\\ \text{recoiling, point-like target,}\\ \mu\rightarrow t\text{ (target particle)}\end{cases}.\qquad(17\text{-}14)$$

Elastic, recoiling point target cross-section

The factor $\frac{E_3}{E_1}=\frac{1}{1+2\left(E_1/M_t\right)\sin^2\frac{\theta}{2}}$ arises from the recoil of the target.

[1] F. Halzen and A. Martin, *Quarks & Leptons: An Introductory Course in Modern Particle Physics,* Wiley (1984), pgs. 173-175.

[2] *ibid.*, (8.10), pg. 175.

For a non-point-like target, an analysis[1] we won't get into yields the Rosenbluth formula, where κ is the anomalous magnetic moment of the target,

$$\left(\frac{d\sigma}{d\Omega_1'}\right)_{\substack{lab \\ e^-p\to e^-p}} = \frac{(\alpha Z)^2}{4E_1^2 \sin^4\frac{\theta}{2}}\frac{E_3}{E_1}\left(A\cos^2\frac{\theta}{2} - B\frac{q^2}{2M_p^2}\sin^2\frac{\theta}{2}\right) \begin{cases}\text{elastic, relativistic or not,} \\ \text{lab frame, initially stationary,} \\ \text{recoiling, non-point-like target,} \\ t\to p \text{ (target = proton)}\end{cases} \quad (17\text{-}15)$$

$$A = F_1^2\left(q^2\right) - \frac{\kappa^2 q^2}{4M_p^2}F_2^2\left(q^2\right) \qquad B = \left(F_1\left(q^2\right) + \kappa F_2\left(q^2\right)\right)^2 .$$

Elastic, recoiling extended target cross-section with form factors

F_1 and F_2 are our form factors, which in this case are similar to what we saw earlier for the immovable target in that they are only functions of the virtual photon momentum. If the target object in (17-15) were point-like, then it would reduce to (17-14), i.e., A and $B \to 1$ for any value of q^2. In this case, the form factors would be constants. They would not change with q^2.

Bottom line: Form factors convert amplitudes, and thus, cross-sections, from representing point-like targets to representing extended targets, and they do this for immovable and recoiling target particles, though they are generally applied to elastic interactions. For point-like targets, the form factors are independent of photon 4-momentum q.

Form factors covert point-like target amplitudes to non-point-like target amplitudes

In what follows, we will be concerned solely with recoiling targets, both point-like and extended.

Relevant Form of Cross-Sections

We will need to define a commonly used variable.

$$\nu \equiv \frac{p_2 q}{M_p} = \frac{p_2^\mu q_\mu}{M_p} \xrightarrow[\text{initial rest frame}]{\text{in proton target}} \nu = E_1 - E_3 = E_\gamma = q_0 . \quad (17\text{-}16)$$

Definition of a commonly used Lorentz invariant variable

Do **Problem 1** to prove the RHS of (17-16).

Note that ν is a scalar invariant, though we can calculate it easily in the initial proton rest frame.

For what we will be doing, it is more relevant to express cross-sections as differentials, not just with respect to solid angle Ω, but with respect to the energy of the outgoing electron E_3, as well. Again, we will just cut to the chase and present the final result, whereby (17-14) and (17-15) are converted to

We now need to use differential cross-section with respect to energy

$$\left(\frac{d\sigma}{dE_3 d\Omega_1'}\right)_{\substack{lab \\ e^-t\to e^-t}} = \frac{4(\alpha Z)^2 (E_3)^2}{q^4}\times$$
$$\left(\cos^2\frac{\theta}{2} - \frac{q^2}{2M_t^2}\sin^2\frac{\theta}{2}\right)\delta\left(\nu + \frac{q^2}{2M_t}\right) \begin{cases}\text{elastic, relativistic or not,} \\ \text{lab frame, initially stationary,} \\ \text{recoiling, point-like target,} \\ \mu\to t \text{ (target particle)}\end{cases} \quad (17\text{-}17)$$

That cross-section for elastic, recoiling point-like target

and

$$\left(\frac{d\sigma}{dE_3 d\Omega_1'}\right)_{\substack{lab \\ e^-p\to e^-p}} = \frac{4(\alpha Z)^2 (E_3)^2}{q^4}\times$$
$$\left(A\left(q^2\right)\cos^2\frac{\theta}{2} - B\left(q^2\right)\frac{q^2}{2M_p^2}\sin^2\frac{\theta}{2}\right)\delta\left(\nu + \frac{q^2}{2M_p}\right) \begin{cases}\text{elastic, relativistic or not,} \\ \text{lab frame, initially stationary,} \\ \text{recoiling, non-point-like target,} \\ t\to p \text{ (target = proton)}\end{cases} \quad (17\text{-}18)$$

That cross-section for elastic, recoiling, extended target

Note from the Dirac delta functions in the above, since ν is positive (from (17-16)), q^2 must be negative for any non-zero cross-section, i.e., it must be spacelike.

---

[1] *ibid.*, pgs. 175-177.

Inelastic Scattering and Structure Functions

Turning to inelastic case

For inelastic scattering, we also need to consider cross-sections and amplitudes for non-point-like targets with recoil, but additionally, we have to take into account the transmutation of the target particle into other particles. Known relationships for inelastic collisions between point-like particles (which include recoil herein) have to be modified to include extended target particle structure. These entities that enfold all of this are called structure functions.

Structure functions account for internal structure and inelastic interactions

The term "form factor" tends to be used for an extended particle interacting elastically, whereas the more complex nature of inelastic interactions, with particle sub-structure playing a key role, engenders use of the term "structure function". Form factors incorporate particular forms (shapes and sizes) of the extended particle. Structure functions take form factors into account plus internal structure that leads to final particles which differ from the initial particles.

The study of structure functions is not trivial, and we will do little more with them than cite general conclusions and relevant characteristics for them.

We will overview structure functions and not study them deeply

With (17-16), and where we again cite Halzen and Martin[1] for those interested in the details, for inelastic scattering $e^- p \rightarrow e^- X$, we have, somewhat similar to (17-17) and (17-18),

Differential cross-section with respect to energy for inelastic, recoiling, extended target

$$\left(\frac{d\sigma}{dE_3 d\Omega_1'}\right)_{\substack{lab \\ e^- p \rightarrow e^- X}} = \frac{4(\alpha Z)^2 (E_3)^2}{q^4} \times \left(W_2\left(\nu,q^2\right)\cos^2\frac{\theta}{2} + 2W_1\left(\nu,q^2\right)\sin^2\frac{\theta}{2}\right) \quad \begin{cases} \text{inelastic, relativistic or not,} \\ \text{lab frame, initially stationary,} \\ \text{recoiling, non-point-like target,} \\ \text{(target = proton here).} \end{cases} \qquad (17\text{-}19)$$

W_1 and W_2 here are structure functions

W_1 and W_2 are structure functions for electron-proton inelastic scattering, whose precise form would be quite unknown before experiment. They are functions of two invariant scalars, ν and q^2, whereas in the elastic case, the comparable form factors were functions only of q^2. Note that the delta functions in (17-17) and (17-18) are embedded in W_1 and W_2, and not shown explicitly in (17-19).

17.2.4 Scattering with Point-like Substructure of Proton

Imagine yourself as one of the early researchers into the then unknown structure of the proton. You reason that if it is an extended object that is broken up by the incoming virtual photon, you will need a relationship like (17-19), with unknown structure functions W_1 and W_2 that you hope to determine by experiment.

However, if the proton is made of effectively point-like components (think quarks) off of which the photon scatters elastically, then the cross-section would look like (17-17), where M_t equals the mass of the point-like particle (for us, m_q) and not the total mass of the proton. In this case, if we were trying to find W_1 and W_2 of (17-19), we would find that they would have to be of form to make (17-19) identical to (17-17). That is, they would need to be

Form of W_1 and W_2 for point-like constituents of proton

$$\begin{aligned} 2W_1^{point} &= -\frac{q^2}{2M_t^2}\delta\left(\nu + \frac{q^2}{2M_t}\right) = -\frac{q^2}{2m_q^2}\delta\left(\nu + \frac{q^2}{2m_q}\right) \\ W_2^{point} &= \delta\left(\nu + \frac{q^2}{2M_t}\right) = \delta\left(\nu + \frac{q^2}{2m_q}\right). \end{aligned} \qquad (17\text{-}20)$$

In fact, (17-20) is what they found, thereby proving that the proton was made of smaller entities we have grown to call quarks. However, we need to massage (17-20) to put it in the form commonly found in the literature.

Researchers want to reformulate W_1 and W_2 into dimensionless quantities

Researchers prefer to work with the arguments of the structure functions that are dimensionless. To do that, we utilize a well-known relation for Dirac delta functions, and find

[1] *ibid.*, Chaps. 8 and 9.

$$\delta(ax) = \frac{1}{a}\delta(x) \quad \rightarrow \quad \delta\left(\nu + \frac{q^2}{2m_q}\right) = \delta\left(\nu\left(1 + \frac{q^2}{2m_q\nu}\right)\right) = \frac{1}{\nu}\delta\left(1 + \frac{q^2}{2m_q\nu}\right). \qquad (17\text{-}21)$$

Thus, (17-21) turns (17-20) into

Form of preferred dimensionless quantities

$$2m_q W_1^{point} = -\frac{q^2}{2m_q\nu}\delta\left(1 + \frac{q^2}{2m_q\nu}\right) \qquad \nu W_2^{point} = \delta\left(1 + \frac{q^2}{2m_q\nu}\right), \qquad (17\text{-}22)$$

which are themselves dimensionless. (17-22) corresponds to what one commonly finds in the literature, rather than (17-20).

These independent of q^2, for $q^2 = -2m_q\nu$

The important thing to notice about (17-22) is that, provided we keep $q^2 = -2m_q\nu$, the two structure functions are constant. So, if we plotted the structure functions vs q^2, but in the process kept $2m_q\nu = -q^2$, there would be no variation in (17-22).

Experiment found this, so point-like free parton model is vindicated

When researchers did this, they indeed found a horizontal plot as q^2 was varied. Keep in mind that all of this is applicable for very high q^2, where inelastic electron-proton scattering occurs. It goes by the name deep inelastic scattering.

Called Bjorken scaling

James Bjorken predicted this behavior for protons composed of free, point-like objects (which he called partons), and this invariance with q^2 scaling is called Bjorken scaling, which was mentioned at the beginning of Chap. 15. As you can imagine, this was all a very big deal at the time. It heralded the discovery of quarks.

17.3 Precise Makeup of the Proton

Further analysis of structure functions can tell us a lot more about the makeup of the proton, and again, we will simply summarize the results of a fair amount of theoretical analysis. Fig. 17-4 below comprises the essence of that summary.

Structure functions with deeper scattering can tell us more about the structure of nucleons

In the figure, x represents the fraction of the proton mass-energy for the particular parton the photon is interacting with. The vertical axis is a form of a structure function that represents content density. That is, it effectively represents the number (if viewed as a histogram) of photon-parton collisions at a particular mass-energy of the parton. As a continuous curve, it represents the collisions per unit mass-energy, i.e., a collision density. Each plot is discussed separately below.

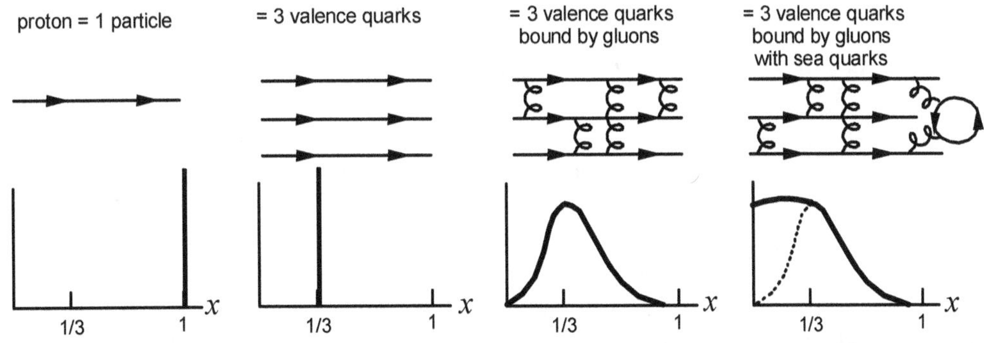

Figure 17-4. Proton Substructure Revealed by Structure Function Form
x = fraction of proton mass-energy (of the parton that the photon collides with);
vertical axis = a particular structure function reflecting number of scattering events
(Adapted from Halzen and Martin[1], Fig. 9.7)

The first plot is what one would find if the proton were a single object. $x = 1$ means the photon interacted with a particle that possessed all of the mass-energy of the proton.

The second plot is what one would find if there were three quarks in the proton and nothing else. No gluons, no sea quarks, just valence quarks. Every interaction would be with a quark having 1/3 the total mass-energy of the proton, i.e., $x = 1/3$.

[1] *ibid.*, pg. 301.

The third plot shows what we would find if the proton had not only the three quarks, but a lot of gluons continually transferring 4-momentum between those three quarks. The gluons could carry any amount of mass-energy between that of the entire proton (in the limit) and none (in the opposite limit). That is, we would detect a continuous range of mass-energy for the target particle, whose density peak would center on the average quark value.

The fourth plot includes sea quarks, formed in the quark-antiquark loops of the transiting gluons. In the model, as shown, the sea quarks contain a significant portion of the total mass-energy of the proton.

Testing strongly supports valence quark, gluon, and sea quark model

Testing has validated the model of the fourth plot, and that is how we know the proton is composed of valence quarks, gluons, and sea quarks. With time, experiments have measured the percentages of each more accurately, and the theoretical (lattice gauge) models have improved to the point where agreement between theory and experiment is fairly good.

All that we have said above is applicable to the study of the neutron as well, and of course, this has been done, though we won't be studying it herein.

17.4 Hadron Jets from Electron-Positron Collisions

17.4.1 Two-Jet Interactions

High energy electron-positron scattering produces quark-antiquark pairs, in addition to lepton-antilepton pairs

Colliding electrons and positrons at high energy, as we've seen (Vol 1, Fig. 1-1, pg. 2 and Fig. 17-16, pg. 465), can create muon/antimuon or tau/antitau pairs, but it can also create quark/antiquark pairs, as in (17-23) and Fig. 17-5 below (where we will be sticking to the COM frame).

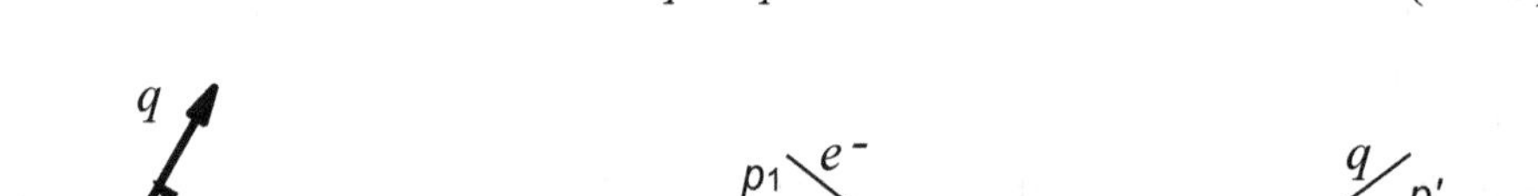

$$e^- + e^+ \rightarrow q + \bar{q} \qquad (17\text{-}23)$$

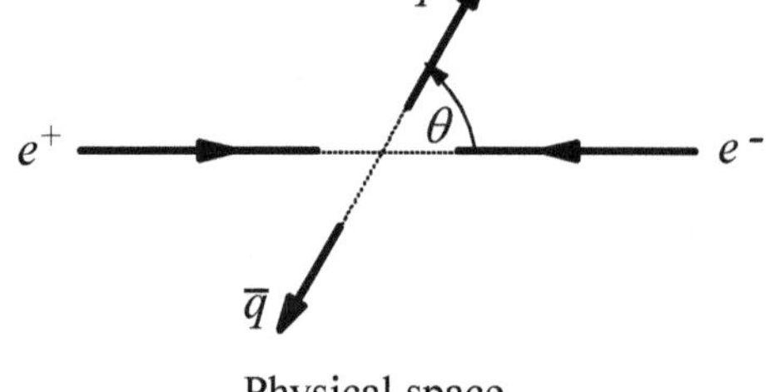

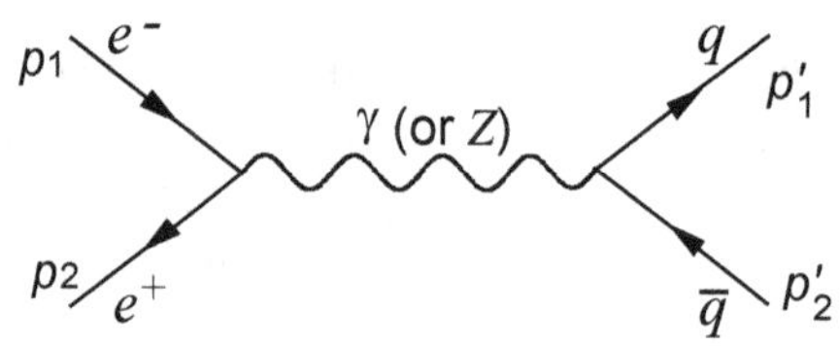

Figure 17-5. Electron-positron Collision Forming Quark-antiquark Pair

Colored quark-antiquark must fragment into colorless products (hadrons)

However, quarks are colored, and as we know, they cannot exist in isolation, so must quickly form (colorless) hadrons. Since the strong interaction is weak at close distances, the quark of given color and its antiquark of the associated anti-color can initially head off in different (opposite to conserve 3-momentum) directions. But eventually, as they get further away and the strong force gets stronger, they will interact via QCD interactions to form colorless products.

There are many ways this interaction can take place, and they are quite complicated. Just one such way, where we label quarks by their color rather than *q*, is illustrated in Fig. 17-6. The process of the two quarks transforming into jets of hadrons is called fragmentation.

One of many ways fragmentation occurs

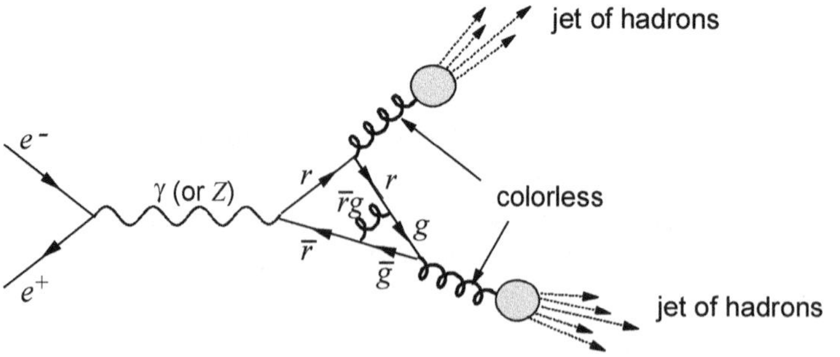

Figure 17-6. One Way Quarks and Antiquarks Form Colorless Products

The energy involved in these collisions is high, typically on the order of 15 to 40 GeV, far greater than the masses of the leptons and quarks involved. When the two quarks produced initially get far enough apart, they must interact via the QCD interaction to neutralize color, and even though that interaction is strong, the amount of 4-momentum transferred is low. This is because the strength of interaction affects the probability that it will occur, but the momentum transferred shows up in the denominators of the gluon and quark propagators involved. And if that momentum is smaller, the probability of interaction is greater. For the gluon propagators involved, 4-momentum approaching zero maximizes the propagator value. For the quarks involved, 4-momentm approaching the quark mass maximizes the propagator value. But this is still far less that the 4-momentum of the interaction.

Due to nature of QCD, fragmentation entails little momentum transfer

Another commonly used explanation for this cites the inverse relationship between distance and momentum. When the two quarks are further apart the momentum (and thus, energy) transfer between them is less, much less than the total momentum (energy) of the interaction.

In other words, very little momentum (compared to that of each of the initial quark and antiquark) is transferred between the flow of particles in one direction with the flow of particles in the opposite direction. Hence, we end up with two jets of hadrons moving in directions that are very close to opposite.

So, jets directions align with original quark and antiquark directions

Note that if the strong interaction were stronger at closer distances, we would get lots of quarks and gluons manifesting from the original two quarks almost right away. We would not have just two particles streaming away from the collision site, but many. So, the 3-momentum would be distributed among all of those particles, instead of just two particles. And this means we would not get two collimated jets of hadrons, but a spray of them in all directions.

If QCD strong at short distance, would not have two distinct jets

Bottom line: The presence of hadron jets in high energy electron-positron collisions indicates that the strong interaction is weak at close distances, but strong at further distances. Early experiments showing this were of great historical importance, as they demonstrated both asymptotic freedom and infrared slavery. Random sprays of hadrons would have been counter-indicative.

Jets support assumptions of asymptotic freedom and infrared slavery

17.4.2 Three and Four-Jet Interactions

The two-jet emissions of the prior section are the most common hadron jet outcome, but not the only such outcome. For example, three-jet emissions can occur in the manner depicted in Fig. 17-7.

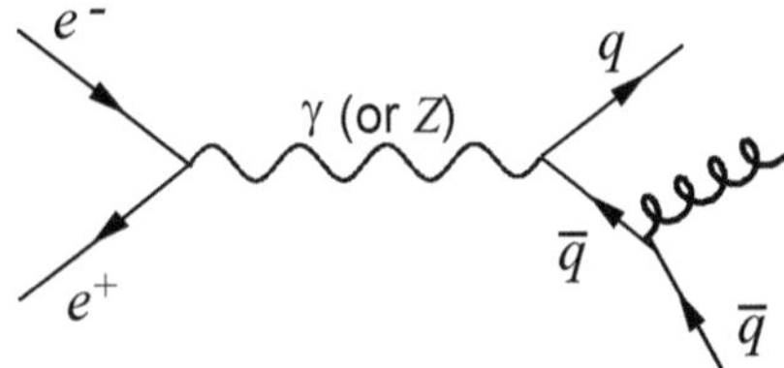

Figure 17-7. Three-Jet Scattering

Two-jet events comprise about 90% of all such hadron-producing interactions. Three-jet events comprise about 9%; four-jet events, roughly 1%; and higher-number jet events, a vanishingly low percentage.

Three and more jets events occur but more rarely

Note that high energy gluons, like the one shown in Fig. 17-7 and those leading to the jets in Fig. 17-6, are called hard gluons and their emission, hard radiation. Due to their high energy, they have low QCD coupling. Low energy gluons, like the $r\bar{g}$ gluon in Fig. 17-6, are called soft gluons, and their emission, soft radiation. Soft gluons have high QCD coupling, but do not transfer much energy or momentum, whereas hard gluons do.

17.4.3 Quantitative Look at Hadron Jet Events

Two-Jet Events

We can make use of the relation we deduced in Vol. 1, Chap. 17, for the differential cross-section for high energy electron-positron collisions that produce muons and antimuons, or taus and antitaus.

Quantitatively, can model 2-jet events like lepton-antilepton production

$$e^- + e^+ \rightarrow \mu^- + \mu^+ \quad \text{or} \quad \tau^- + \tau^+ . \tag{17-24}$$

That relation is (17-117) of Vol. 1, which is

$$\left(\frac{d\sigma}{d\Omega'}\right)_{\substack{COM \\ e^-e^+\to l^-l^+}} = \frac{\alpha^2}{4E_{tot}^2}\left(1+\cos^2\theta\right) \qquad \begin{cases}\text{unknown spins} \\ E_{tot} \gg 2m_l,\ l \neq e.\end{cases} \tag{17-25}$$

with same scattering cross-section

(17-25) can be used for the interaction (17-23), if we simply modify the fine structure constant to take into account the fractional electric charge of the quarks, i.e.,

but need to take into account different charge on quarks

$$\alpha^2 \to \left(\tfrac{2}{3}\right)^2 \alpha^2 \text{ for up type quarks,} \quad \left(\tfrac{1}{3}\right)^2 \alpha^2 \text{ for down type quarks.} \tag{17-26}$$

Doing this, one has a differential cross-section for the jets with dependence on θ just as in (17-25), i.e., the same form as Fig. 17-17, Vol. 1, pg. 467. Experimentally, just such a dependence is found, corroborating the two-jet model of QCD.

Just what is found experimentally, further supporting QCD model

Three-Jet Events

One can deduce the differential cross-section for three-jet events, where the scattering angle θ is determined as follows. Consider the energies of the three jets $E_1 > E_2 > E_3$, and transform to the center of mass frame of jets 2 and 3. In that frame, jets 2 and 3 are colinear and head in opposite directions. The angle θ is then defined as the angle between jet 1 and the common line of flight of jets 2 and 3.

It turns out that the cross-section in three-jet events depends on the spin of the gluon. A spin 0 boson assumption leads to a cross-section that is a different function of the scattering angle θ than a spin 1 boson assumption. This difference is depicted in Fig. 17-8, where it can be seen that vector (spin 1), rather than scalar (spin 0), QCD gauge bosons are strongly supported by experiment.

3-jet event cross-section different for spin 0 gluons and spin 1 gluons

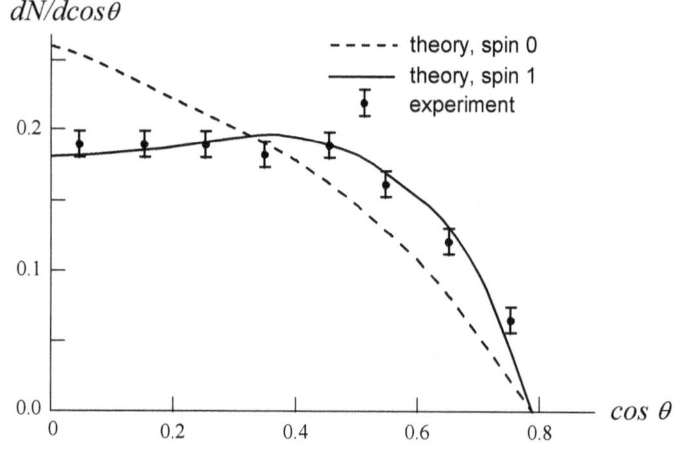

Experiment supports spin 1 gluon of QCD theory

Figure 17-8. Three-Jet Theory vs Experiment for Spin 0 vs Spin 1 Gluons
N = number of events per unit solid angle
(Adapted from R. Brandelik et al, *Phys. Lett* **B27** (1980) pg. 453
and Sl. L. Wu, *Phys. Rep.* **107** (1984) pg. 59)

17.5 Chapter Summary

17.5.1 QCD Coupling Constant

As predicted by renormalization analysis, the QCD coupling constant (and its sibling correlate to the fine structure constant, α_S) varies with the interaction mass-energy level. Unlike the QED coupling constant, but similar to the weak coupling constant, it decreases with increasing mass-energy (decreasing distance). This gives rise to asymptotic freedom and quark confinement.

17.5.2 Deep Inelastic Scattering

Electron scattering off a proton can be elastic (with electron and proton as final particles) or inelastic (with different final particles). The electron sheds a photon via the QED interaction and that photon interacts with the proton, also via the QED interaction (which, of course, is the only way the photon interacts).

There are four distinct energy regimes.

1. Very low energy with long photon wavelength at which the proton looks like a point object and interaction is generally elastic.

2. A somewhat higher energy level where the wavelength is on the order of the proton radius, and the proton looks like an extended object. The interaction is typically elastic and the experiment can provide a measure of the radius of the proton.

3. High electron energy where the wavelength is shorter still, such that it can interact with, and detect, valence quarks in the proton. This can be elastic or inelastic and provides information on the sub-structure of the proton.

4. Very high electron energy, where the wavelength is short enough to also scatter off sea quarks and provide information about them, as well as about the gluons from which they arise. This is generally inelastic.

Regime 4 and regime 3, when it is inelastic, go by the name deep inelastic scattering. Valence quarks, gluons, and sea quarks are known collectively as partons.

Form factors are modifications to the amplitudes (and thus, cross-sections) which were originally formulated assuming particles behave like points, and those factors take into account the extended nature of particles. Structure functions are further modifications to the amplitudes (and cross-sections) that account for internal structure of composite particles.

By comparing theoretical work using structure functions to results from experiments, researchers can determine, to various degrees, the internal structure of composite particles. One particular such analysis assumed point-like constituents of nucleons, where those constituents interacted with the photon as if they were free. It predicted Bjorken scaling, wherein structure constants, provided a particular variable is held constant, are invariant with q^2, where q^μ is the photon four-momentum. Subsequent experimental observation of this particular scaling ushered in the quark paradigm.

Bjorken scaling was evidenced by experiments in regime #3 above and indicated the presence of valence quarks. Gluons do not interact electromagnetically, so photon scattering, at any energy level, can tell us nothing about them in a direct way. But in the deeper (higher energy) inelastic scattering experiments of regime #4, photons can interact with sea quarks, and since they arise within gluons, such experiments can provide information on both sea quarks and gluons. The behavior of structure functions versus the fraction of proton mass-energy for partons led to our present model of nucleons as composed of valence quarks, gluons, and sea quarks.

17.5.3 Hadron Jets

High energy annihilation of electrons with positrons can result in quark-antiquark pair production and subsequent hadron formation from the pair.

Theoretical assumptions of essentially free quark and antiquark upon creation (when they are close together) and strong QCD interaction later (when they are further apart) lead to predictions of final product collimated hadron jets. These predictions are born out in experiment.

Two-jet events are predicted to have differential cross-section dependence on scattering angle that is the same as that observed for final muon-antimuon products, and this is precisely what is seen.

Three-jet events have a different theoretical prediction for gluons having spin 0 than for gluons having spin 1. Experiment clearly favors the latter, and this aligns well with QCD theory as we know it.

17.5.4 Non-perturbative QCD

The discoveries we've discussed were made via investigations, theoretical and experimental, into the strong force at energy levels where perturbation theory can be applied. Further research, both theoretical and experimental, in non-perturbative QCD, has been carried out for decades already and continues to be carried out.

This research is primarily focused on the nature of the binding that holds quarks together to form hadrons, which occurs at large (relatively) quark separation distances, and therefore large strong coupling. Theoretically, this involves lattice gauge theory in which path integrals are evaluated numerically. The ever-increasing power of supercomputers has enabled theorists to come closer and closer into agreement with experiment.

We do not delve, in any depth, into non-perturbative QCD in this text.

The Curtain Falls on Quantum Chromodynamics

This ends Part 4 of this book on the theory and experimental results of QCD. Part 5 summarizes the entirety of Vol. 1 and Vol. 2.

17.6 Problem

1. Show the RHS of (17-16).

Part Five
Summary of the Standard Model and Beyond

"Nothing makes me more hopeful [than symmetry] that our generation of human beings may actually hold the key to the universe in our hands - that perhaps in our lifetimes we may be able to tell why all of what we see in this immense universe of galaxies and particles is logically inevitable."
Steven Weinberg

"We know a lot of things, but what we don't know is a lot more."
Edward Witten

Chapter 18 Looking Forward and Looking Backward: QFT Summary and What's Next

Chapter 18

Looking Backward and Looking Forward: QFT Summary and What's Next

"Surely someday, we can believe, we will grasp the central idea of it all as so simple, so beautiful, so compelling that we will all say to each other, 'Oh, how could it have been otherwise! How could we all have been so blind so long!"
John Wheeler, at the end of his *It from Bit* essay

18.0 Preliminaries

Wheeler's is one perspective on the ultimate fate of science. I will close this chapter, this book, and this two-volume series with another, which is closer to my own feelings on the matter.

18.0.1 Where We've Been

As we noted in Chap 1 of Vol. 1, that chapter was a map of QFT. Well, it was a treasure map, and after following it over extensive, rugged terrain, you have now gained the treasure. For most of us who do this kind of work, understanding QFT is worth far more than gold doubloons and pieces of eight. Savor what you have done. It is priceless.

As Feynman said in his Nobel lecture, "I find myself … before one new corner of nature's pattern of beauty and true majesty revealed. That was my reward."

18.0.2 Chapter Overview

In this chapter, we will overview what we have covered in these two volumes, i.e.,

- steps to developing QFT,
- the fundamental particles lineup,
- parameters of the theory,
- the complete Lagrangian of the entire SM,
- Feynman rules for each branch of QFT,
- renormalization,
- the vacuum, and
- experimental confirmation of the theory.

In the process, we will recognize that all theories beyond general relativity and what we have studied are, as of this writing, speculative. QFT reaches the pinnacle of humankind's current understanding of the physical universe. In other words, you have reached the vanguard. You know the structure of creation as well as anyone.

Following the overview of QFT, we will examine

- the shortcomings of the SM, i.e., the problems it does not resolve.

We will follow that with a brief look at

- lines of thinking for possible resolutions of QFT's unsolved problems, and
- resources that may present good introductions to those lines of thinking.

18.1 Big Picture Overview of QFT

18.1.1 Steps to Developing QFT and the SM

Wholeness Chart 7-12, The Standard Model in a Nutshell (pg. 252b) by Boyle is an excellent one page overview of the entire standard model of particle physics. I submit that captures and summarizes the essence of this book.

The Canonical Approach

Wholeness Chart 18-1 shows an overview of the most fundamental steps in the canonical quantization development of the $SU(3)XSU(2)XU(1)$ standard model of QFT. If you have progressed through Vol. 1 and this volume, this should all be familiar, so I will make little comment on it.

Wholeness Chart 18-1. Milestones in the Canonical Development of Quantum Theories

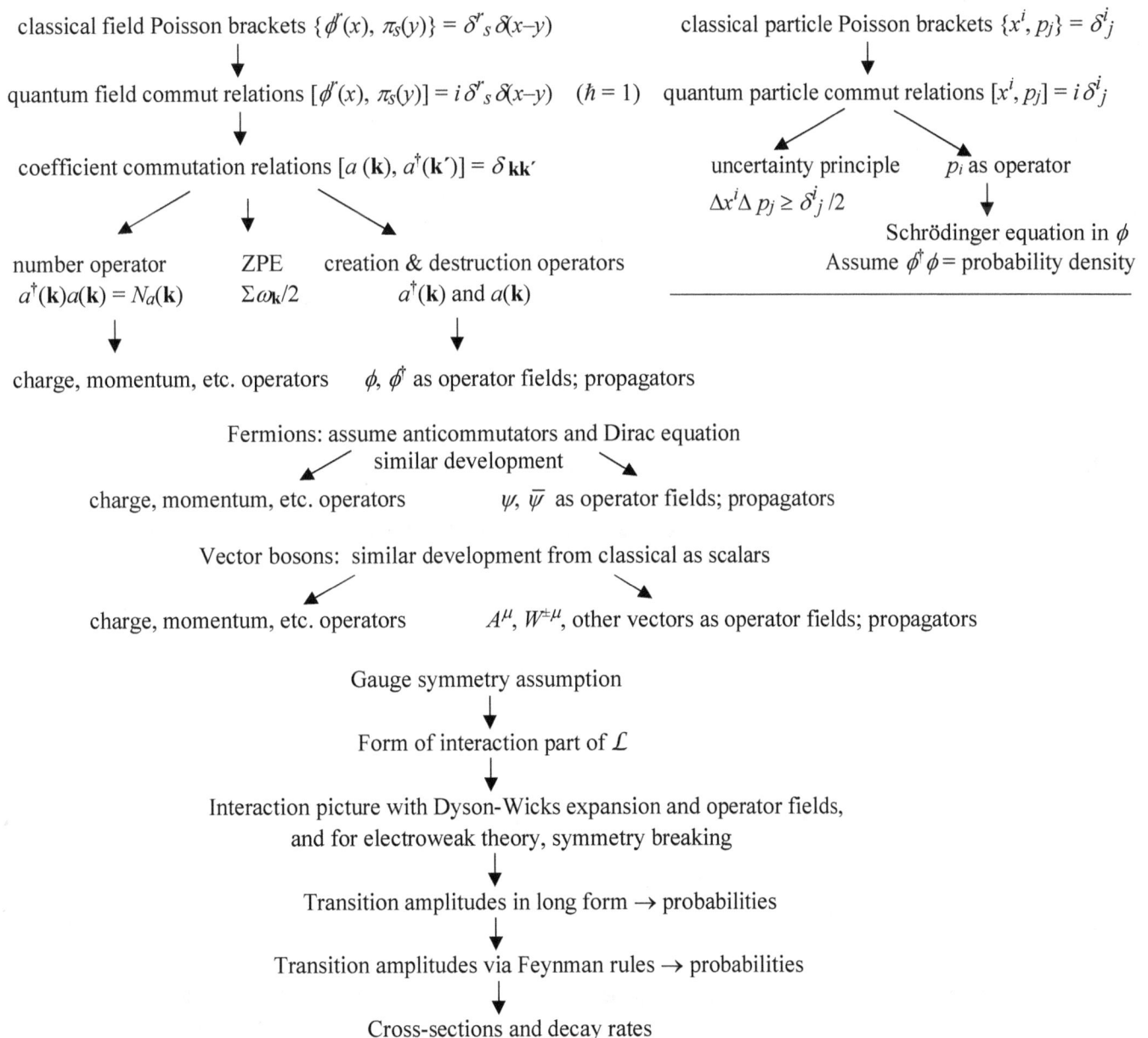

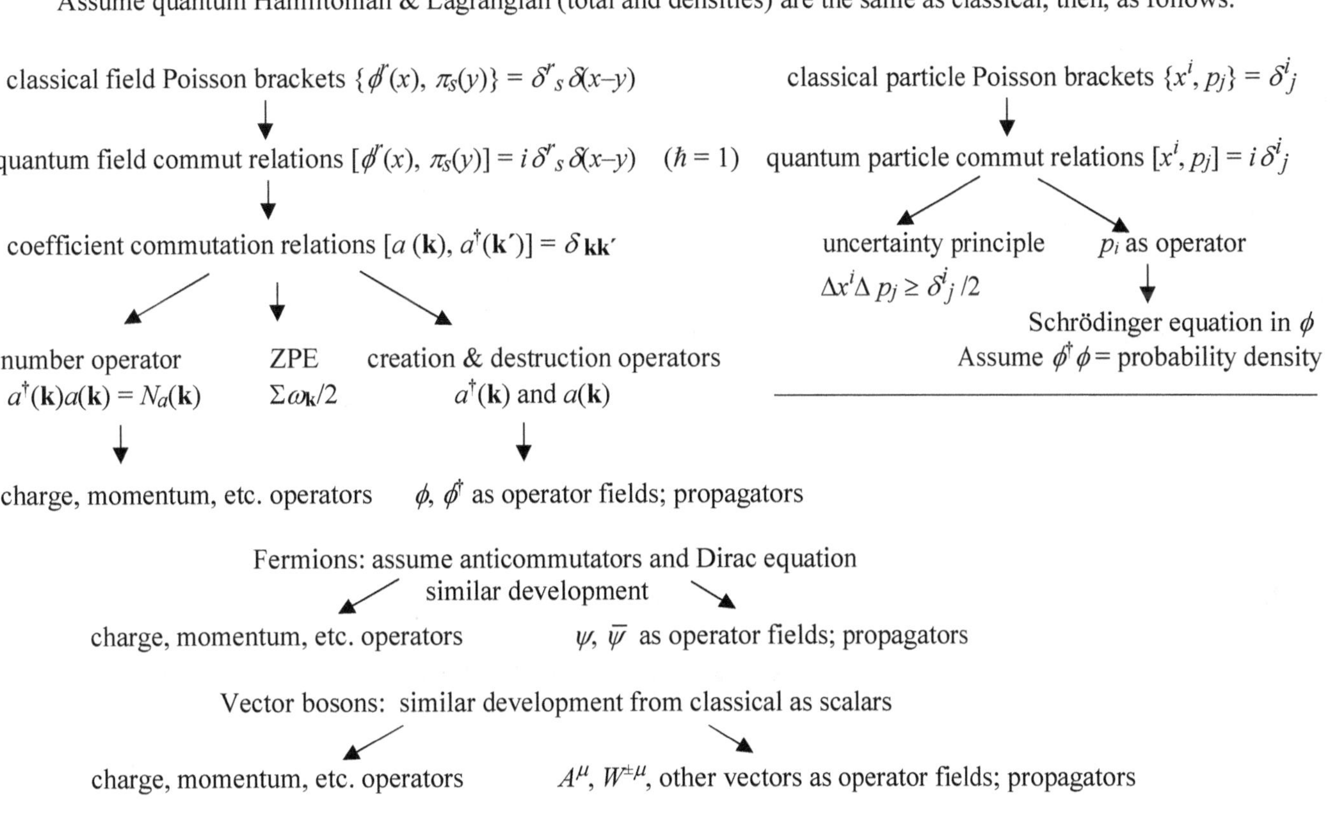

Gauge symmetry assumption

↓

Form of interaction part of $\mathcal{L}$

↓

Interaction picture with Dyson-Wicks expansion and operator fields, and for electroweak theory, symmetry breaking

↓

Transition amplitudes in long form → probabilities

↓

Transition amplitudes via Feynman rules → probabilities

↓

Cross-sections and decay rates

The Path Integral Approach

Wholeness Chart 18-2 overviews the fundamental steps in the PI approach to the SM.

Wholeness Chart 18-2. Milestones in the Path Integral Development of Quantum Theories

Assume quantum Lagrangian (total and density) is same as classical, then, as follows.

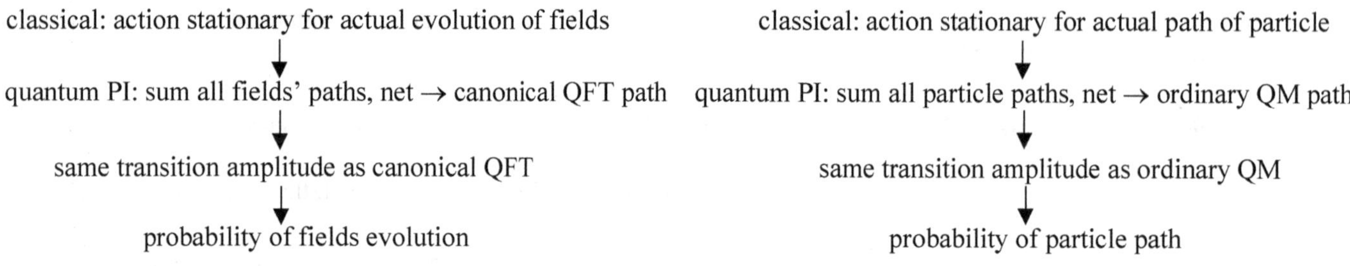

QFT PI approach employs Green functions expressed as path integrals, where legs eventually taken as external particles. Generating functional is used to prove transition amplitude is the same as that found in canonical approach.

18.1.2 The Building Blocks

"Matter is a lush tapestry, woven from a complex assortment of threads. Diverse subatomic particles weave together to fabricate the universe we inhabit... [It is] an ornate brocade."

Emily Conover, *Science News*, April 10, 2021

The Threads in the Tapestry

The interweaving of the tapestry is well displayed in Wholeness Chart 18-3. The top row darkened ellipses/circle represent what we usually consider to be particles. Below them, the darkened circles represent the various force mediators of the SM. Lines symbolize interactions (terms in the Lagrangian) that link the various particles and force mediators. Note the W, Higgs, and gluons are self-interacting, but the photon and Z are not.

Wholeness Chart 18-3. The Interaction Threads of the SM Tapestry

(Adapted from English Wikipedia)

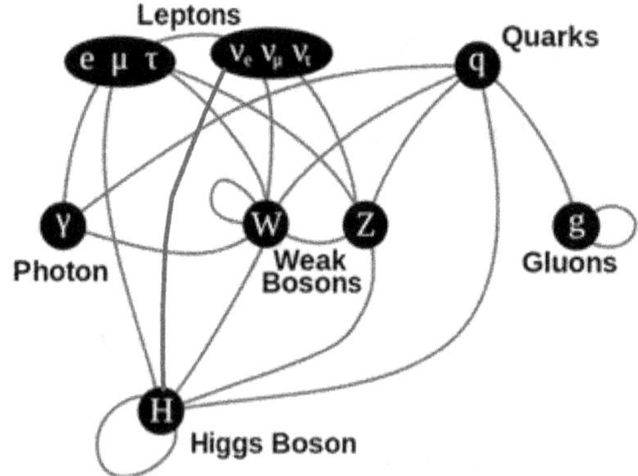

The Particles of the SM

The question "How many different particles are there in the SM?" has different answers, depending on how you define "particles". Are LC and RC electrons different particles? How about different colored up quarks? And antiparticles? Are they classified as separate?

Wholeness Chart 18-4 summarizes the number of SM particles via different categorization schemes. Equal signs in the chart heading do not literally mean equality, but are shorthand for "assuming they count as one particle".

You may read in places that the total number of elementary particles is 25. By this counting there are 6 quarks, 3 charged leptons, 3 neutrinos, 3 intermediate vector bosons, 8 gluons, 1 photon, and 1 Higgs. Quark and lepton antiparticles are not counted. For example, the positron is not included, as it is the antiparticle of the electron. But for the bosons, antiparticles are included. The W^- is the antiparticle of the W^+, but both are counted. So, this method of counting seems internally inconsistent.

However, in the high energy description, we have three W particles, the W_1, W_2, and W_3, which are all different, none being an antiparticle of any other. So, with the high energy description, the elementary particle count of 25 is correct.

Wholeness Chart 18-4 is for the low energy description

Wholeness Chart 18-4. Counting the Particles in the SM

Particles	**Anti = partic LC = RC** $q_r = q_g = q_b$	**Anti ≠ partic LC = RC** $q_r = q_g = q_b$	**Anti = partic LC ≠ RC** $q_r = q_g = q_b$	**Anti = partic LC = RC** $q_r \neq q_g \neq q_b$	**Anti = partic LC ≠ RC** $q_r \neq q_g \neq q_b$	**All diff**
leptons	6	12	12	6	12	24
quarks	6	12	12	18	36	72
Fermion total	12	24	24	24	48	96
photon	1	1	1	1	1	1
W	1	2	1	1	1	2
Z	1	1	1	1	1	1
Higgs	1	2	1	1	1	2
gluons	8	8	8	8	8	8
Boson total	12	14	12	12	12	14
Grand total	**24**	**38**	**36**	**36**	**60**	**110**

The Parameters of the SM

Wholeness Chart 18-5 counts the parameters of the SM, where we note there are only two possible assumptions about what constitutes a parameter. If the strong CP violating parameter is truly zero, then there are 26 such parameters. These are dictated by Nature. As we know, if any of several of these were much different, the universe we know could not exist.

Further parameters from general relativity, which are not part of the SM, include the gravitational constant G and the cosmological constant Λ responsible for the accelerating expansion of the universe. If dark matter is really not dark matter, but a modification to our classical theory of gravity, then there would be at least one more GR-related parameter.

Wholeness Chart 18-5. Counting the Parameters in the SM

Parameter	**Number for Zero Strong CP Violation**	**Number for Non-zero Strong CP Violation**
Lepton masses (Yukawa couplings g_l)	6	6
Quark masses (Yukawa couplings g_q)	6	6
$U(1)$ coupling g'	1	1
$SU(2)$ coupling g	1	1
$SU(3)$ coupling g_S	1	1
Weinberg angle θ_W	1	1
Electroweak symmetry breaking μ, λ	2	2
CKM mixing angles	3	3
CKM phase angle	1	1
PMNS mixing angles	3	3
PMNS phase angle	1	1
Strong CP violation $\bar{\theta}$	0	1
Total	**26**	**27**

18.1.3 The Complete Lagrangian of the SM

We can express the entire SM Lagrangian density (in the high energy electroweak epoch) in short form using minimal substitution with the covariant derivative that includes all SM gauge boson fields. That is, substitute

$$\partial^\mu \to D^\mu = \partial^\mu + i\frac{g}{2}\sigma_j W_j^\mu + ig'YB^\mu + i\frac{g_S}{2}\lambda_j A_j^\mu \tag{18-1}$$

into the (false vacuum)

$$\textbf{high energy, free Lagrangian} \quad \mathcal{L}_0 = \mathcal{L}_0^L + \mathcal{L}_0^Q + \mathcal{L}_0^B + \mathcal{L}_0^G + \mathcal{L}_0^H + \mathcal{L}^{LH} + \mathcal{L}^{QH}$$

$$\begin{aligned}
&\mathcal{L}_0^L = i\left(\bar{\Psi}_l^L \partial\!\!\!/ \Psi_l^L + \bar{\psi}_l^R \partial\!\!\!/ \psi_l^R + \bar{\psi}_{\nu_l}^R \partial\!\!\!/ \psi_{\nu_l}^R\right) && \text{(leptons, massless)}\\
&\mathcal{L}_0^Q = i\left(\bar{\Psi}_{qa}^L \partial\!\!\!/ \Psi_{qa}^L + \bar{\psi}_{qa}^R \partial\!\!\!/ \psi_{qa}^R\right) && \text{(quarks, massless)}\\
&\mathcal{L}_0^B = -\frac{1}{4}W_i^{\mu\nu}W_{i\,\mu\nu} - \frac{1}{4}B^{\mu\nu}B_{\mu\nu} && \text{(electroweak gauge bosons)}\\
&\mathcal{L}_0^G = -\frac{1}{4}F_i^{\mu\nu}F_{i\,\mu\nu} && \text{(gluons, QCD gauge bosons)} \qquad (18\text{-}2)\\
&\mathcal{L}_0^H = \left(\partial^\mu\Phi\right)^\dagger\left(\partial_\mu\Phi\right) - \mu^2\Phi^\dagger\Phi - \lambda\left(\Phi^\dagger\Phi\right)^2 && \text{(Higgs boson)}\\
&\mathcal{L}^{LH} = -{}_AY_{ll'}\bar{\Psi}_l^L\Phi\psi_{l'}^R - {}_BY_{l\nu_{l'}}\bar{\Psi}_l^L\tilde{\Phi}\psi_{\nu_{l'}}^R + h.c. && \text{(lepton-Higgs coupled)}\\
&\mathcal{L}^{QH} = -{}_CY_{qq_d}\bar{\Psi}_q^L\Phi\psi_{q_d}^R - {}_DY_{qq_u}\bar{\Psi}_q^L\tilde{\Phi}\psi_{q_u}^R + h.c. && \text{(quark-Higgs coupled)}
\end{aligned}$$

$$W_i^{\mu\nu} = \partial^\nu W_i^\mu - \partial^\mu W_i^\nu \qquad B^{\mu\nu} = \partial^\nu B^\mu - \partial^\mu B^\nu \qquad F_i^{\mu\nu} = \partial^\nu A_i^\mu - \partial^\mu A_i^\nu,$$

to give us the (false vacuum)

high energy, free plus interacting Lagrangian

$$\begin{aligned}
&i\left(\bar{\Psi}_l^L D\!\!\!\!/\, \Psi_l^L + \bar{\psi}_l^R D\!\!\!\!/\, \psi_l^R + \bar{\psi}_{\nu_l}^R D\!\!\!\!/\, \psi_{\nu_l}^R\right) + i\left(\bar{\Psi}_{qa}^L D\!\!\!\!/\, \Psi_{qa}^L + \bar{\psi}_{qa}^R D\!\!\!\!/\, \psi_{qa}^R\right)\\
&-\frac{1}{4}G_i^{\mu\nu}G_{i\,\mu\nu} - \frac{1}{4}B^{\mu\nu}B_{\mu\nu} - \frac{1}{4}\hat{G}_i^{\mu\nu}\hat{G}_{i\,\mu\nu} + \left(D^\mu\Phi\right)^\dagger\left(D_\mu\Phi\right) - \mu^2\Phi^\dagger\Phi - \lambda\left(\Phi^\dagger\Phi\right)^2\\
&\qquad - {}_AY_{ll'}\bar{\Psi}_l^L\Phi\psi_{l'}^R - {}_BY_{l\nu_{l'}}\bar{\Psi}_l^L\tilde{\Phi}\psi_{\nu_{l'}}^R + h.c.\\
&\qquad - {}_CY_{qq_d}\bar{\Psi}_q^L\Phi\psi_{q_d}^R - {}_DY_{qq_u}\bar{\Psi}_q^L\tilde{\Phi}\psi_{q_u}^R + h.c.
\end{aligned} \tag{18-3}$$

$$G_i^{\mu\nu} = D^\nu W_i^\mu - D^\mu W_i^\nu \qquad B^{\mu\nu} = \partial^\nu B^\mu - \partial^\mu B^\nu \qquad \hat{G}_i^{\mu\nu} = D^\nu A_i^\mu - D^\mu A_i^\nu.$$

(18-3) is the most compact way to write out the full SM Lagrangian. To get the low energy (true vacuum) equivalent we need to substitute

$$\begin{aligned}
&\Phi(x) = \frac{1}{\sqrt{2}}\begin{bmatrix}\phi_1 + i\phi_2\\ \phi_3 + i\phi_4\end{bmatrix} \qquad \tilde{\Phi} = \frac{1}{\sqrt{2}}\begin{pmatrix}\phi_3 - i\phi_4\\ -\phi_1 + i\phi_2\end{pmatrix} \qquad \phi_1, \phi_2, \phi_3, \phi_4 \text{ real}\\
&\Phi(x) = \frac{1}{\sqrt{2}}\begin{bmatrix}0\\ \phi_3\end{bmatrix} = \frac{1}{\sqrt{2}}\begin{bmatrix}0\\ \sigma + v\end{bmatrix} \qquad \tilde{\Phi} = \frac{1}{\sqrt{2}}\begin{pmatrix}\sigma + v\\ 0\end{pmatrix} \qquad \text{in unitary gauge,}
\end{aligned} \tag{18-4}$$

the Weinberg relations,

$$\begin{bmatrix}B_\mu\\ W_{3\mu}\end{bmatrix} = \begin{bmatrix}\cos\theta_W & -\sin\theta_W\\ \sin\theta_W & \cos\theta_W\end{bmatrix}\begin{bmatrix}A_\mu\\ Z_\mu\end{bmatrix}, \tag{18-5}$$

and

$$W_{1\mu} = \frac{W_\mu^+ + W_\mu^-}{\sqrt{2}} \qquad W_{2\mu} = i\frac{W_\mu^+ - W_\mu^-}{\sqrt{2}}. \tag{18-6}$$

We have carried out those substitutions and shown the complete, true vacuum electroweak part of $\mathcal{L}$ in Chap. 7 and the complete QCD part of $\mathcal{L}$ in Chap. 12. At low energy, the QED $U(1)$ theory, with coupling constant e, becomes distilled out of the electroweak theory.

18.1.4 Nutshell Summary for How Mass Arises in E/W Symmetry Breaking

The following is a streamlined overview of how mass arises from Higgs symmetry breaking that may be easier to keep in memory than the many steps we went through doing this in Chap. 7.

At high energy, before Higgs symmetry breaking, we have Yukawa coupling of the Higgs to fermions.

$$\text{Some terms in } \mathcal{L} = -\,{}_AY_{ll'}\,\overline{\Psi}_l^L\,\Phi\,\psi_{l'}^R \xrightarrow[\text{only electron term}]{\text{Basis I, for leptons,}} -g_e\overline{\Psi}_e^L\,\Phi\,\psi_e^R \qquad (18\text{-}7)$$

After symmetry breaking, we have

$$\Phi \rightarrow \frac{1}{\sqrt{2}}\phi \qquad\qquad \phi = \sigma + v\,, \qquad (18\text{-}8)$$

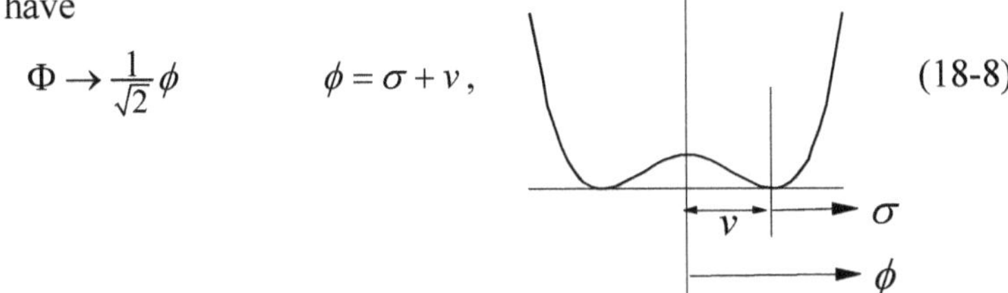

and (18-7) becomes

$$-\frac{1}{\sqrt{2}}g_e\overline{\psi}_e^L\,\phi\,\psi_e^R = -\frac{g_e}{\sqrt{2}}\overline{\psi}_e^L\,(\sigma+v)\,\psi_e^R = \underbrace{-\frac{g_e}{\sqrt{2}}\sigma\overline{\psi}_e^L\,\psi_e^R}_{\text{interaction term}}\;\underbrace{-\overbrace{\frac{vg_e}{\sqrt{2}}}^{m_e}\overline{\psi}_e^L\,\psi_e^R}_{\text{mass term}} \qquad (18\text{-}9)$$

(18-7) is a pure interaction at high energy, but due to the Higgs defined as σ at low energy via (18-8), we get a mass term arising (plus we still have an interaction between the fermions and the new Higgs σ). (18-8) arose from symmetry breaking on the Mexican hat, so the symmetry breaking gave rise to massive fields/particles.

Recall that

$$\overline{\psi}_e^R\psi_e^L + \overline{\psi}_e^L\psi_e^R = \overline{\psi}_e\psi_e\,. \qquad (18\text{-}10)$$

In addition to (18-9), from symmetry breaking in similar fashion, we get a term like $-\frac{vg_e}{\sqrt{2}}\overline{\psi}_e^R\,\psi_e^L$. So, adding this to the last term in (18-9), and using (18-10), we get mass terms of the familiar form $-\frac{vg_e}{\sqrt{2}}\overline{\psi}_e\psi_e = -m_e\overline{\psi}_e\psi_e$.

18.1.5 Feynman Rules for Each Branch of the SM

Applying the interaction picture, the commutation/anti-commutation relations, and the Dyson-Wicks expansion with the Lagrangian (18-3) (in high or low energy form) one can calculate transition amplitudes for any given SM interaction. This process is extensive and tedious, however, and Feynman devised a (relatively) simple set of rules one can use instead.

The same rules can be found via the path integral approach, as we showed, in part, in Chap. 15.

Feynman rules for QED can be found in Vol. 1, pg. 514.

Feynman rules for electroweak interactions can be found in this volume starting on pg. 290.

Feynman rules for QCD interactions can be found in this volume starting on pg. 458.

18.1.6 Renormalization

All of the above is hardly trivial, but the situation gets worse when one tries to calculate transition amplitudes in a naïve way. As we saw, divergent integrals emerge from higher order contributions needing integration over 4-momenta ranging from $-\infty$ to $+\infty$. Expunging these from QFT is a major undertaking, but thanks to the efforts of many, it has been achieved.

A renormalizable theory must have a finite number of primitive divergences, from which all divergences arise. And it must have enough adjustable parameters (like mass and charge in QED) that can be massaged to absorb the finite number of infinities from the primitive divergences. QED, electroweak, and QCD theories turn out to be renormalizable.

Effective theories, ones that work over a particular energy range (that is typically our contemporary energy regime), can describe Nature as we know it, and we don't know Nature's behavior at the energy levels where such a theory probably fails. QED, electroweak, and QCD theories are, indeed, only effective theories.

Effective theories do not have to be renormalizable, since we accept that they will not be valid at very high energy. However, renormalizing each of the different branches of the SM leads to a coupling constant dependence on energy for each branch that is observed. So, renormalization, even for the effective theories of the SM, reflects, in some way, what is happening in the real world, and thus has definite value.

18.1.7 General Procedures in QFT for Any Model

For any quantum field theory, QED, electroweak, QCD, supersymmetry, grand unified, superstring, etc., we follow the same procedure to find the relevant charges (which are conserved) carried by the particles in the theory. The steps in this procedure are as follows.

1. Find a transformation under which $\mathcal{L}$ is symmetric.
2. Apply Noether's theorem, using $\mathcal{L}$ and the given transformation, to find one or more 4-currents j^μ. (If more than one, we label them, such as j_i^μ.)
3. From Noether's theorem, $\partial_\mu j^\mu = 0$. So, $\int j^0 dV$ is conserved. We multiply by "q", a convenient constant taken as the charge on one particle, so $q\int j^0 dV = Q$ is conserved.
4. Q is a charge operator, composed of number operators, whose eigenvalue when operating on a (multiparticle) state equals the total charge on the state. E.g.,

$$Q\left|3\phi_{\mathbf{p}1},2\phi_{\mathbf{p}2}\right\rangle = q\sum_{\mathbf{p}}\left(N(\mathbf{p})-\bar{N}(\mathbf{p})\right)\left|3\phi_{\mathbf{p}1},2\phi_{\mathbf{p}2}\right\rangle = q(3+2)\left|3\phi_{\mathbf{p}1},2\phi_{\mathbf{p}2}\right\rangle = 5q\left|3\phi_{\mathbf{p}1},2\phi_{\mathbf{p}2}\right\rangle \quad (18\text{-}11)$$

5. In $SU(n)$ theories, we will get other operators from the procedure of the above steps, which do not have eigenvalues for states. Although these arise naturally from Noether's theorem, they are, in the SM, not of great use and commonly just ignored.

Particles are discerned and defined by the charges they carry. We know a neutrino is a neutrino because the only charge it carries is weak charge. We know a quark from a lepton because the former feels all three SM forces, while the latter only feels two. What charges a given particle has is determined by its eigenvalues under the charge operators, as, for example, the operator of (18-11) applied to a single particle state.

And what the charge operators are is determined by the symmetries of the Lagrangian. The symmetries determine what type particles we find in our world, and so, those symmetries define the universe. An amazing concept, actually.

This is an overriding principle that can be considered to capture the essence of QFT in a nutshell.

18.1.8 What is the Vacuum?

I have elaborated my perspective on the vacuum in Chap. 10 of Vol. 1 and will not repeat any of that here.

We do want to note that there are several, not always compatible, definitions of what, exactly, the vacuum is. These are as follows, where I borrow from Nikolić[1].

The vacuum can be considered

1. a state having no particles at all,
2. a state that is annihilated by a destruction operator,
3. the state having the lowest possible energy, or
4. a local minimum of a potential.

These are closely related, but not always exactly the same. For example, the false vacuum of electroweak theory has zero scalar particles, but is not in the lowest energy state.

This listing may help to alleviate confusion when the term "vacuum" is used in different contexts.

18.1.9 Experimental Confirmation of the SM

In Vol. 1, Chaps. 16 and 17 covered the striking experimental confirmations of QED, including, but not limited to, the predictions of the coulomb potential, the anomalous magnetic moment, the Lamb shift, and many diverse scattering results.

In this volume, parts of Chap. 7 and elsewhere discussed the experimental detection of the Higgs boson, along with the measurement of its mass, as well as the CKM matrix impact on scattering experiments. Chap. 8 discussed certain other corroborated predictions of electroweak theory, including scattering, resonances, and particle decays. Chap. 10 extended that to include neutrino oscillations.

Chap. 17 detailed non-perturbative QCD experimental results and how they led early researchers to adopt the QFT model of the strong interaction. It briefly discussed, but did not cover in any depth, the remarkable correlations with experiments on hadron composition engendered by the non-perturbative approach to QCD, involving lattice gauge theory and numerical analysis.

---

[1] H. Nikolić, Is the Zero Point Energy Physical?, *Annals of Physics* 383 (2017) 181-195. arxiv.org/abs/1702.03291.

Conclusion of This Presentation of QFT and the SM as We Know It

This ends our study of quantum field theory and the standard model as it is understood on the publication date of this volume. Directions the theory may take in subsequent years and decades are discussed in the next section.

18.2 What's Next

Where we go from here in unraveling Nature's mysteries is pretty much dictated by experimental observation, and comparison of that to extant theory. In that regards, consider the following questions, which that theory and present experiment cannot answer.

18.2.1 Unsolved Problems in the Standard Model

The SM cannot be a complete theory, or it would be able to answer the following questions.

1. Why are there exactly three families of elementary fermions?
2. Why are the masses of particles what they are, and in particular, why the huge discrepancy between neutrino masses and those of other particles?
3. Why are the coupling constants what they are, and are they related (unified) in any way?
4. Why is electric charge quantized?
5. Why is the positive charge of the proton exactly equal in magnitude (or at least to one part in 10^{20}) to the negative charge of the electron?
6. What is spin exactly and why is it quantized in ½ integer amounts?
7. Why do hadrons, despite being composed of a conglomeration of valence quarks, gluons, and sea quarks have exact multiples of ½ integer spin?
8. How is the gauge hierarchy issue resolved by nature?
9. Why is strong CP violation zero, or at least as close to zero as we can measure?
10. Can a free proton decay?
11. What happens well above the electroweak symmetry breaking scale?
12. Do fundamental particles have substructure?
13. Why is there more matter than antimatter?
14. How does dark matter fit into QFT?
15. Can the observed small cosmological constant be predicted from QFT?
16. Can gravity be included in QFT?
17. Can all the interaction types be unified?
18. Is zero-point-energy a physical reality?

This is quite a list, and none of the above touch on the basic, still unanswered question of how to interpret quantum mechanics, itself. It is all quite humbling, actually.

Further questions arising from experimental discrepancies from theory as of the publication date of this volume involve, for one, differences from theory in decay patterns of certain exotic pions. Are these indicative of new physics (new elementary fields)? Time, hopefully, will tell.

18.2.2 Potential Solutions to Unsolved Problems

We have discussed possible solutions to part of #2 (the see-saw mechanism) and to #9 (Peccei-Quinn theory) in the above list.

Supersymmetry (SUSY) posits a symmetry between fermions and bosons and in so doing, offers fairly elegant solutions to #8 (gauge hierarchy) and #14 (dark matter candidates). But it has fallen on rather hard times of late, experimentally. The window in parameter space where it remains viable is small and continues to shrink.

A number of grand unified theories (GUTs) have been proposed to bring the $U(1)$, $SU(2)$, and $SU(3)$ interactions under one tent, commonly $SU(5)$, and thereby resolving #3 (disparate coupling constants), and possibly #10 (proton decay). A number, though not all (particularly "flipped $SU(5)$"), of such theories have been disproven.

Superstrings, loop quantum gravity, and some other models hold promise for turning gravitation into a quantum field, but while a great deal has been done in these areas, none seems close to a clear resolution of #15 (cosmological constant), #16 (quantum gravity), or #17 (unification).

Of course, there are many theorists worldwide working on each of the listed unresolved issues, with a concomitant large number of speculative models, any one of which could one day prove to be a big part of the resolution of one or many of those issues.

18.2.3 Resources for Study Beyond the SM

For the next stages in your study of physics, I suggest the texts listed below.

Physics Pervaded by Symmetry

First, I recommend the superb, highly pedagogic book on the all-pervasive nature of symmetry in physics by Schwichtenberg[1]. It ties all of physics up into one nice, holistic package.

Supersymmetry

Aitchison's text[2] is, in my opinion, a good, reasonably transparent, entry-level introduction to SUSY. See the pedagogic notes on SUSY I have posted on the website of this book, which, I submit, can help considerably in working one's way through Aitchison, or other texts.

GUTs

The time since I've done much with grand unified theories is measured in decades, so I am not the one to recommend resources for learning about them. I used Ross[3], but don't remember it as being particularly transparent.

General Relativity

I may be an old-timer, but I found Misner, Thorne, and Wheeler[4], when it came out, to be, pedagogically, far above average. In my day, it was my favorite physics textbook.

A more recent book, which, in my opinion, contains exceptionally lucid explanations is Moore[5]. Another, which is better than most at explanations, is Carroll[6]. On this book's website, I have posted pedagogic aids to special relativity, general relativity, and cosmology, which could help newcomers to those subjects.

Quantum Fields in Classical Curved Spacetime

Certain gravitational effects can be analyzed effectively using QFT in a classical spacetime. This is not quantum gravity, but a model using quantum fields interacting on a non-Minkowski (curved) spacetime background. Mukhanov and Winitzki[7] provide an exceptional introduction to this topic, and I have written pedagogic notes, posted on the website of this book, as an additional aid to help readers over some of the rougher terrain therein.

Superstrings and Other Quantum Gravity Theories

Zwiebach[8] has written an outstanding, very pedagogic introduction to string theory with a smattering of its more advanced form, M theory. It was written for undergraduates, so contains little supersymmetry, but in my opinion, is a must read for anyone beginning the study of superstrings. And again, I have posted an aid on this book's website that can help those studying Zwiebach or other texts.

I have never studied loop quantum gravity or other quantum gravity theories, so, I cannot make recommendations on where to go to learn those subjects.

Regarding your future, whatever your chosen path, I wish you the best and a host of deeply satisfying moments therein.

18.2.4 The Future of Physics

Where is our beloved subject headed? What yet undreamed-of paradigm shifts await us? Will John Wheeler's perspective, as quoted at the beginning of this chapter, come to pass? Or not?

My own sense of the matter is summed up in the following quote by John Horgan in *Scientific American* on September 20, 2021, shortly before the publication of this volume. I leave you with that.

"My hope, and expectation, is that the world will keep us guessing forever.

---

[1] J. Schwichtenberg, *Physics from Symmetry* (2nd ed., Springer, 2018).
[2] I. Aitchison, *Supersymmetry in Particle Physics: An Elementary Introduction* (Cambridge, 2007).
[3] G. G. Ross, *Grand Unified Theories* (Benjamin/Cummings 1985)
[4] C. W. Misner, K. S. Thorne, and J. A. Wheeler, *Gravitation* (Freeman 1973).
[5] T. A. Moore, *A General Relativity Workbook*, University Science Books (2012).
[6] S.M. Carroll, *Spacetime and Geometry*, Cambridge (2019).
[7] V. F. Mukhanov, and S. Winitzki, *Introduction to Quantum Effects in Gravity* (Cambridge 2007)
[8] B. Zwiebach, *A First Course in String Theory*, 2nd ed., (Cambridge 2009)

Index

⊕, ⊗. *See* direct sum and direct product

[*AKB*]. *See* functional, short-hand notation

't Hooft, Gerardus, 178

't Hooft gauge, 372

2D rotation, 22

2D rotation group, 12, 23, *See also* *SO*(2)

3D rotation group, 13, 25, *See also* *SO*(3)

Abelian group, 21, 55
- definition of, 10
- linearity vs nonlinearity, 191, 388

aces, 442

algebra
- overview, 8, 9
- representation, 23

anomalous magnetic moment, 493, 497
- muon, 512

anomaly, 367, 379
- chiral, 411
- gauge, 367, 379
- global, 367, 379

antiparticle
- color eigenvaues, 56
- state eigenvalues, 50

antiquark, 57

anti-screening, 439

antiunitary operator, 336

associative, 9

associative algebra, 8

asymptotic freedom, 384, 385, 441, 461, 464

axial current, 152

axial gauge, 448

axial vector, 316

axial vector current. *See* axial current

axion, 413
- decay constant, 413

B meson, 438

B vector boson, 171

Baker-Campbell-Hausdorf, 28

baryon
- colorless, 57
- from *u,d*, 436
- from *u,d,s*, 436
- number, 340, 343
- octet from *u,d,s*, 436
- preview, 2
- q number plots from *u,d,s*, 436

basis, 14
- choice of, 228, 236
- flavor. *See* flavor basis
- hybrid Basis I, 229
- hybrid Basis II, 231
- hybrid basis IIq, 236
- hybrid basis Iq, 236
- leptons vs quarks, 229
- mass. *See* mass basis
- summary of hybrid I vs II, 232
- summary of hybrid Iq vs IIq, 237
- transformation, 228

basis vector
- in Lie algebra vector space, 25

beta decay, 277, 346

beta function, 372
- QCD, 442, 464
- QED, 473

Bhabha scattering. *See* scattering, Bhabha
- from Green function, 72

binary operation, 9, 21

Bjorken
- James, 442
- scaling, 442, 499, 503

block diagonal. *See* matrix, block diagonal

boosts
- fields, 154
- spinor. *See* spinor, boosts

boson
- e/w not actually change with energy, 244
- type interaction mediates, 2

bottom quark, 437

bottomonium, 437

branching ratio, 268, 288

Breit-Wigner formula, 287, 290

B-tagging, 438

C transformations, 315, 319
- derivatives, 322
- four-currents, 321
- scalars, 319
- spinors, 319
- symmetry, 317
- vectors, 322

C,P, and *T* transformations. *See* *CPT* transformations
- discrete symmetries, 318

CA, 114

Cabibbo
- angle, 238
- matrix, 238
- Nicola, 238

Cabibbo-Kobayashi-Maskawa. *See* CKM matrix

canonical quantization, 1, 2
- comparison to path integral, 126

Cartan subalgebra, 48–51, 311
- Abelian nature, 55
- exponentiated, 55
- observables, 51

Cartesian product, 18

Casimir operator, 55

characterization
- of physical phenomenon, 22
- summary, 34

charge
- color. *See* color charge
- e/w summary chart, 183, 243, 368
- eigenstates, 227
- electric. *See* electric charge
- finding from Cartan subalgebra, 184
- flavor. *See* flavor charge
- hypercharge. *See* hypercharge
- low energy conserved at vertices, 308
- operators, e/w summary, 341
- operators, QCD, 405
- operators, QCD summary, 415
- reading off from Lagrangian terms, 182
- relation between electric, weak, and hypercharge, 183, 308, 341
- renormalization, 480, 481, 482, 483

SU(2) non weak isospin, 308
summary of fermion e/w charges, 196
three ways to determine, 311
types, comparison chart, 2
weak isopsin. *See* weak isospin charge

charge transformation. *See C* transformations

charge, parity, time transformations. *See CPT* and *C,P,T* transformations

charged current, 245

charm quark, 437

charmonium, 437

chiral anomaly. *See* anomaly, chiral

chiral field. *See* also, Weyl field
same as Weyl field, 149

chiral symmetry. *See* symmetry, chiral

chirality, 40, 132, 148, 157, 298
compared to helicity, 340
compared to heliclity, summary, 343
equals helicity for *v*=*c*, 139
gamma matrix, 160
invariant, 340
left, 40, 151
low energy, RC and LC don't exist, 248
non-conservation, 340
operator, 149
right, 40, 151
right, e/w interactions at low energy, 244
summary, 152
symbols for, in this book, 151

circle, shaded, 69

CKM matrix, 236, *See CP* transformations, CKM matrix
component values, 237
CPT transformations. *See CPT* transformations, CKM matrix

Clebsch-Gordan coefficients, 423, 426, 428, 434, 439

Clifford algebra, 135

closure, 9, 22

Coleman
Sidney, 443

color charge, 399, 405
compared to other charges, 2
examples, 406
summary, 415
table for fermions, 50

color factors, 485
useful relations for finding, 485

color singlet, 45, 50

color triplet, 46

colorless composite state. *See* QCD, colorless composite state

COM frame, 274, 278
kinematics & dynamics, 279

commutation
clarifying note, 191
issue with vector bosons, 377

commutative, 9

completeness relation, 294

complex groups, 15

composite particle. *See* hadron

Compton scattering
from Green function, 83

condensate
Higgs. *See* Higgs, condensate

confinement
quark. *See* quark, confinement

conformal symmetry. *See* symmetry, scaling

connected
Green function diagram, 80, 82

contact interaction, 269

contact theory, 288

correlation function, 100, 103, *See* Green function, called correlation function

Coulomb gauge, 156

Coulomb potential, 493

coupling constants
e/w relationship between, 224
e/w running, 246
minimal substitution, 479
MS running, 477
negative dimension, 464, 474, 476, 491
QCD running, 464, 494, 502
QED running, 477
running, 372, 379, 493
strong, 441

covariant derivative
electroweak, 175, 400
entire SM, 510
QCD, 387, 400
QED, 175, 399
SU(*n*), 400

CP
PMNS matrix, 350
transformation, 350
violation, 350, 411, 438
violation example, 330

CP transformations, 317
CKM matrix, 330
example, 328, 343
Lagrangian, 329
transition amplitude, 329

CPT transformations, 298, 315, 316, 336
CKM matrix, 338
decay of particle vs antiparticle, 338
example, 343
Lagrangian, 337
no symmetry violations, 338
particle decay, 338
summary, 342
symmetry, 318
vs *CPT̂*, 332

CPT̂ transformations
four-currents, 327
four-vectors, 327
interaction terms, 327
Lagrangian, 328
scalars, 327

crossed reactions, 84

crossing relations. *See* crossed reactions

cross-section
differential, 497, 501
differential, general relation, 278, 289
electron-positron scattering, 289
Mott, 496
plot vs energy, 286
total, 283
total, peak, 285

current. *See* four-current, neutral current, or charged current

curved spacetime
QFT in. *See* QFT, in curved spacetime

d.o.f.. *See* degrees of freedom

decay, 265, 266
beta. *See* beta decay
branch, 268, 288

channel, 268, 288
Higgs, 278, 289
mode, 268, 288
muon, 268, 276, 289, 339
muon branches, 268
neutron, 277
particle vs antiparticle, 338
rate, 265, 288
rate, differential, 267
rate, particular process, 267
rate, total, 267, 268, 287
summary, e/w, 288
tau, 277
W and Z, 278, 284, 289
width, 285
deep inelastic scattering. *See* scattering, deep inelastic
degree
of a group, 12, 54
degrees of freedom, 155
GSW model, 222
Higgs model, 218
massive vector field. *See* massive vector field, degrees of freedom
massless vector field, 217
scalar field, 217
delta particle, 495
destroy the vacuum, 213
deuteron, 423, 430
differential cross-section. *See* cross-section, differential
differential decay rate. *See* decay, rate, differential
dimension
for natural units, 4
of a Lie algebra, 53
of a Lie group, 53
of a representation, 53, 61
of different quantum fields, 4
of vector space, 12
dimensional regularization, 477
Dirac algebra, 135
Dirac equation
solutions, 133
Dirac-Pauli rep. *See* standard rep
Dirac-Pauli representation. *See* standard rep
direct product, 18–21, 59, 425, 429, 439
to direct sum relations, 430
direct sum, 418, 421, 429, 439
to direct product relations, 430
disconnected
Green function diagram, 73, 78, 80, 82
discrete transformations. *See* transformation, discrete
distributive, 9
divergence
maximum degree. See superficial power counting
dots, small
in diagrams, 67
double cover, 33
doublet, 44, 48
down quark, 422, 431, 436
dual vector, 53
D_μ. *See* covariant derivative
effective theories, 373, 379, 463, 511
eigenstate
field vs state, 422
eigenvalue, 55, 440
additive for multiparticle states, 431
eigenvector, 55
eightfold way, 435
elastic scattering. *See* scattering, elastic
electric charge, 183
compared to other charges, 2
electron doublet, 173
electron-positron collisions, 289, 500, 503
electroweak
addiitional topics, 362
comparison of Goldstone, Higgs, GSW, 221
covariant derivative. *See* covariant derivative
epoch, 162
GSW model postulates, 182
high energy symmetry. *See* Lagrangian, e/w free, high energy
interaction, part of book found in, 3
interaction, preview, 1
path integral approach. *See* path integral, electroweak theory
renormalization. *See* renormalization, e/w theory
scattering. *See* scattering, electroweak
symmetry, 162
symmetry breaking, 162
theory history, 194
entanglement, 376, 380
equal times contractions, 363, 378
equilibrium, 338-339 footnote
equivalent representation. *See* representation, equivalent
Euler-Mascheroni constant,, 477
explicit symmetry breaking. *See* symmetry, explicit breaking
exponentiation of Lie algebra. *See* $SO(2)$, $SO(3)$, $SU(2)$, and $SU(3)$
Faddeev
Ludvig, 445
Faddeev-Popov
generating functional. *See* generating functional, Faddeev-Popov
Faddeev-Popov method, 443, 445, 461
false vacuum, 208
Lagrangian, 221
Fermi
constant, 288
Lagrangian. *See* Lagrangian, Fermi's
Fermi theory, 269, 288
coupling constant, 269
Fermi, Enrico, 346
fermion
flavor, 41
mass terms, 208, 220
switching field order, 103
type inteaction felt, 2
Feynman amplitude
comparing approaches, 2
electron-positron scattering, 279
Feynman diagram
examples for e/w theory, 266
from Green function diagram, 67, 73, 76, 82
proper, 466
proper vs improper, 465
ways to categorize, 464
Feynman gauge. *See* Feynman-'t Hooft gauge

Feynman rules, 461
  each branch of SM, 511
  electroweak theory, 188, 252, 266, 290
  QCD theory, 458
Feynman-'t Hooft gauge, 370
field
  as a vector in Hilbert space, 329
  outer product form, 41
  overview, 8, 9
  parts of in group theory, 386
  source. *See* source, field
Fierz identities, 366
Fierz transformations. *See* Fierz identities
Fierz-Pauli-Kofink identities. *See* Fierz identities
fine structure constant, 484
  analogy for QCD, 490, 494, 502
flavor
  basis, 227
  charge, 408, 423
  eigenstates, 227
  eigenstates, measurement, 239
  hadron formation, 430
  hypercharge, 432, 434
  isospin, 432
  multiplet fields, 431
  space, 423, 431
  symmetry. *See* symmetry, flavor
Fock space, 39, 330, 422, 430
form factor, 496, 497, 503
  vs structure function, 498
four-current
  axial, 408, 411
  e/w summary, 341
  e/w, example, 245
  flavor, 407
  Higgs, 300
  leptons, 300, 305
  QCD, 399
  QCD summary, 415
  quarks, 302, 404
  SU(2) non weak isospin, 307
  SU(2) weak isospin, 303, 306
  U(1) hypercharge, 299
  vector, 407
Fourier transforms, 66
fragmentation, 500
functional, 101
  differentiation, 92, 101
  generating. *See* generating functional
  short-hand notation [AKB], 90
functional integral. *See* path integral
functional quantization. *See* path integral, *See* path integral quantization
fundamental representation, 45
  common names, 45
g. *See* weak coupling constant
g'. *See* hypercharge, coupling constant
G, 288, See Fermi theory, coupling constant
g_A, 281
gamma matrices
  chirality matrix, 148, 160
  Majorana rep, 142
  standard rep, 133
  Weyl rep, 136
gauge
  anomaly. *See* anomaly, gauge
  axial. *See* axial gauge
  covariant derivative. *See* covariant derivative
  field, 40, 172
  fixing, 370
  invariance, 191
  invariance, QCD, 389, 397
  symmetry, 39, 193
  transformation, 40, 172
gauge fixing
  QCD, 449
  QED, 448
gauge hierarchy, 247
Gell-Mann
  Murray, 442
Gell-Mann matrices, 387
  8 parameter composite, 17
  individual matrices, 35
Gell-Mann-Nishijima relation, 183
general relativity
  QFT in. *See* QFT, in curved spacetime
generating functional, 65, 92, 120
  different form for, 95
  differentiation, 93
  Faddeev-Popov, 450
  free field, 96, 99, 120, 121, 123
  free field yields whole theory, 101, 125
  free QCD, 452
  free, most useful form, 452
  ghost, 451
  gluon, 444, 447, 451
  interacting from free, 99, 122
  interactions, 99, 120
  most useful for, 453
  normalized, 100
  path integral form, 115, 120
  QCD, 445
  quark, 451
  relation to Green function, 95, 120
  summary, 102, 126
generator, 25, *See also SO(2)*, *SO*(3), *SU*(2) and *SU*(3)
  summary of. *See* Lie algebra, summary of generators
  U(1), 166, 214
geometric mean, 358
ghost, 372, 447, 448
  generating functional. *See* generating functional, ghost
  self-energy, 472
Glashow
  Sheldon, 168
Glashow/Salam/Weinberg. *See* GSW
gluon
  generating functional. *See* generating functional, gluon
  hypothetical "free", 387
  interacting, 388
  massless, 385
  preview, 2
  self-energy, 472
  spin determination, 502, 503
God particle. *See* Higgs, as God particle
Goldstone
  Jeffrey, 168, 211
  model, 168, 207, 208
  model, comparison to Higgs and GSW, 169
  model, summary, 214
  model, symmetry, 212
  model, wholeness chart, 214
  theorem, 214
Goldstone boson. *See* Nambu-Goldstone boson
Grand unified theories. *See* GUTs

Grassmann
algebra, 87, 127
anti-commutators, 90
differentiation, 88
field, 64, 87, 89, 92, 450
functional, 89
functional differentiation, 89
generator, 88
integration over Grassman variables, 127
number, 87
paired fields, 89
path integrals, 128
summary of properties, 91
variable, 64, 87, 443
gravitational charge
compared to other charges, 2
graviton, 2
gravity, 2
Green function, 64, 118
2-point, 79, 94
4-point, 81, 95
alternative notation, 85
as propagator, 66
called correlation function, 100
canonical overview, 86
converting to transition amplitude, 119
denominator, 71, 80
diagram, 67, 73, 76, 82
electroweak theory, 375
for Dirac equation, 66
for Klein-Gordon equation, 65
for Maxwell equation, 66
free photon field propagator, 97
free spinor field propagator, 99
Heisenberg picture, 101, 104
leg momenta, 75, 76
leg subscripts, 76
legs, 67, 118
momentum (time flow), 76
momentum space, 71
n-point, 67, 84
numerator, 80
ones that vanish, 80
order of term, 84
other pictures, 101, 104
path integral form, 118, 454
QCD, 450, 461
QFT methodology, 67, 70, 84
relation to generating functional, 95, 120
summary, 102, 126
useful for perturbation, 453
usual math, 65, 84
vacuum bubbles, 81
visualizing why it works, 78, 119
Gross
David, 442
group. *See also*, group theory
Abelian. *See* Abelian group
action on a vector space, 21
complex. See complex groups
finite, 21
infinite, 21
Lie. *See* Lie group
operation, 11, *Also see* binary operation
overview, 9
QFT, 10
representation, 10, 11, 21, 22, 31, 44–46, 51, 52, 61
types of, 21, 60
with tensor elements, 58
group theory, 418, *Also see* group
angular momentum, 8, 20
formal symbols, 22
hadron formation, 418
overview, 8, 59–60
part of book found in, 3
spin addition, 424
two approaches, 59
weights, 50
GSW
Higgs potential, 221
Lagrangian at true vacuum, 223, 250
model, 168, 207, 220
model and unification, 241
model, comparison to Goldstone and Higgs, 169
model, summary, 239
model, wholeness chart, 239
symmetry, 220
theory, overview, 195
Gupta-Bleuler weak Lorenz condition, 156
GUTs, 242, 513
g_V, 281
hadron, 418, 430, 440
mass, 442
preview, 2
hadron jets. *See* jets
half-life, 268
hard gluons, 501
hard radiation, 501
Heisenberg fields, 112
helicity, 132, 298
compared to chirality, 340
compared to chirality, summary, 343
conservation, 340
equals chirality for v=c, 139
left, 151
non-invariant, 340
right, 151
symbols for, in this book, 151
Hermitian
generator, 52, 64
matrix, 24
operator, 24
hermiticity
Lagrangian under *CPT*, 337
Higgs
as 5th force, 193, 367, 379
as God particle, 249
bestow mass, 162
condensate, 211, 263
coupling terms in GSW model, 175
decay. *See* decay, Higgs
detection, 247
doublet, 169, 220, 222
false vs true vacuum field, 211, 223
field at true vacuum, 222
field, in differnt models, 168
field, in Goldstone model, 208
field, in GSW model, 169, 221
field, in Higgs model, 216
field, must be neutral scalar, 246
field, unique form of potential, 246
GSW model coupling terms, 187
lepton coupling, 225
lifetime, 278
mass, 224, 239, 247
mass term, 211
mass, guage hierarchy. *See* gauge hierarchy
model, 168, 207, 216
model, comparison to Goldstone and GSW, 169
model, Lagrangian, 216
model, summary, 218
model, wholeness chart, 219
particle, 3, 4
particle creation from field, 241
Peter, 168
potential, 216, 221, 222

potential in Goldstone and Higgs models, 170
potential in GSW model, 169
propagator. *See* propagator, Higgs
quark couplings, 189, 235
spin, from experiment, 247
high energy interactions. *See* interactions, electroweak, high energy, *See* scattering, high energy
higher order
Green function conversion, 83
Hilbert space, 329, 422, 430
Homestake experiment, 346
HOT, 17
hypercharge, 171, 174, 183
bosons, 309
conservation, 301, 302, 303
coupling constant, 172
expectation value, 302
flavor. *See* flavor, hypercharge
leptons, 301
operator, 301, 303
quarks, 189
I_3^W. *See* weak isospin charge
identity, 9
improper Feynman diagram. *See* Feynman diagram, proper vs improper
inelastic scattering. *See* scattering, inelastic
infrared slavery, 441
interaction operator, 99
interactions
electron-positron. *See* electron-positron collisions
electroweak, high energy, 184
high energy e/w example #1, 184
high energy e/w example #2, 186
high energy e/w example #3, 187
high energy e/w other examples, 187
QCD, 387, 391
reversible, 339
typical quark, 189
interference term. *See* scattering, interference term
intermediate vector boson, 2, 132
invariant subspace, 420, 427
inverse, 9
irreducible representation, 54, 418, 419, 439
isomorphic, 22, 166
isospin
flavor. *See* flavor, isospin
isospin charge. *See* weak isospin charge
isospin doublet. *See* weak isospin doublet
IVB. *See* intermediate vector boson
J/psi particle. *See* charmonium
Jacobi identity, 55
jets, 500, 503
K, superficial degree of divergence. *See* superficial power counting
kernel, 90
Kobayashi
Makoto, 437
Lagrangian
color symmetry, 40
CPT invariance, 337
e/w interaction, bosons, 176
e/w interaction, Higgs, 178
e/w interaction, high energy, 171
e/w interaction, lepton-Higgs, 178
e/w interaction, leptons, 176
e/w quark, 188
e/w, free, high energy, 171, 174
e/w, interaction, high energy, 221
e/w, interaction, low energy, 223, 234, 250, 292
e/w, lepton-boson terms, 224
e/w, quark-boson terms, 236
Faddeev-Popov, 450
false vs true vacuum symmetry, 213
Fermi's, 378
gauge fixing term, 449
ghost term, 447
Higgs-lepton coupling, 225
Higgs-quark coupling, 235
high energy, 174, 176, 510
how to show symmetric, 181, 197
interaction terms transformations, 226
invariance, electroweak, 179
kinetic term transformation, 226
lepton weak interaction terms, 173
low energy, 171, 510
low energy e/w, deducing each term, 252
mass terms, 190
potential. *See* potential, Lagrangian
QCD, 387, 388, 397
QCD, "free", 386
QED, 369, 377
QED symmetry, 40
scalar boson, 162
showing high energy e/w symmetry, 196
source. *See* source, Lagrangian
spin 1 issue, 377
standard model, entire, 510
summary of e/w lepton terms, 176
summary of e/w quark terms, 189
symmetry and renormalizable, 178
weak interaction term, 152
weak symmetry, 40
Landau gauge, 370
lattice gauge theory, 441, 493, 503
lattice QCD, 438
LC. *See* chirality, left
Lederman
Leon, 437
left chirality. *See* chirality, left
legs. *See* Green function, legs
lepton
number, 339, 343
preview, 2
zero color eigenvalue, 56
Levi-Civita symbol, 26, 272
LH. *See* helicity, left
Lie algebra
2nd binary operation, 26
definition, 23
graphic analogy, 51
intro to, 10, 23
relation to manifolds, 59
representation, 23
summary of generators, 36
Lie bracket, 23
Lie group, 21
definition, 11

intro to, 10
multiparameter, 13
one parameter, 11
overview, 60
representation, 11, 34, 37, 44–46, 51, 52, 61

lifetime
muon, 289
particle, 268, 288

Lorentz transformation, 14
and invariance, 298
as a group, 14
axial four-current, 313
chirality, 315
entities in the Lagrangian, summary, 342
four-current, 312
gamma matrices, 313
helicity, 314
spinor representation, 55, 312

Lorenz condition. *See* Lorenz gauge

Lorenz gauge, 123, 369

lowering
operation, 34
operator, 56

LSZ reduction formula, 101, 106

Majorana
three uses of the term, 358

Majorana field
in standard rep, 145, 158
same in any rep, 147

Majorana particles, 142, 352, 355
do they exist?, 146

Majorana rep, 132, 142
basis solutions, 142, 143
Dirac equation in, 142
gamma matrices, 142
general solution, 143
real solution, 145
transformation from standard rep, 142

Majorana representation. *See* Majorana rep

Mandelstam variables, 362, 378

manifold, 59

Maskawa
Toshihide, 437

mass
basis, 227, 235
eigenstates, 227, 235
eigenstate measurement, 230, 239
hadron. *See* hadron, mass
particles, plot and list of, 239
QED, 476

mass hierarchy
neutrino, inverted, 352
neutrino, normal, 352

mass terms
and symmetry. *See* symmetry, and mass terms
charged lepton, 225
Dirac, 354
fermion. *See* fermion mass terms
Lagrangian. *See* Lagrangian, mass terms
Majorana, 354
neutrino. *See* neutrino mass terms
quark. *See* quark mass terms
scalar. *See* scalar mass terms
vector. *See* vector mass terms

massive spin 1 field. *See* massive vector field

massive vector field, 154
degrees of freedom, 155, 156, 158, 218
field equation. *See* Proca equation
propagator, 266, *See* propagator, massive vector field

matrix
block diagonal, 419
diagonal, 49, 51, 55
group, 11
reduced, 420
subgroup, 22

matter-antimatter asymmetry, 512

maximum degree of divergence. *See* superficial power counting

Maxwell's equation, 155

mean free path, 365, 378

meson
colorless, 57
preview, 2
q numbers plots for *u,d,s*, 434

Mexican hat
particle creation, 241
plot, 170, 208
possible changes over time, 241
possible different shapes, 241
potential, 208
slightly stable false vacuum, 241

minimal gauge theories, 473

minimal substitution
electroweak, 175
entire SM, 510
QCD, 384
QED, 175

minimal subtraction. *See* renormalization, minimal subtraction

Minkowski
metric, 14
norm, 14

modified minimal subtraction. *See* renormalization, minimal subtraction

Møller scattering,
from Green function, 79

MS renormalization. See renormalization, minimal subtraction

multiparticle state, 430

multiplet, 43–48, 55, 431

muon
decay. *See* decay, muon
lifetime, 276
rest frame, 273

naïve power counting. *See* superficial power counting

Nambu
Yoichiro, 211

Nambu-Goldstone
boson, 211, 214
mode of chiral symmetry, 410

natural units. *See* units

neutral current, 245
no flavor changing, 246

neutrino
breaking the SM, 359
facts about, 347
heavy, sterile, 356
its history, 346
mass, 347, 354
mass hierarchy, 352, 358
mass terms, 227, 229, 233, 354
oscillation, example, 351
oscillations, 238, 347
possible Majorana fermions, 347, 352, 355
primary solar interaction, 348
right chiral, 183, 194

sample interactions, 230, 231
solar neutrino problem, 347
sources, 347
sterile, 359
neutrinoless double beta decay, 353
neutron
decay. *See* decay, neutron
lifetime, 278
Noether's theorem, 41, 246, 298, 404
flavor, 407
in e/w theory, 299
no-go theorems
spin higher than 2, 376, 380
non-Abelian, 55
group definition, 10
nonlinearity
and non-Abelian theories, 191
clarification, 192, 388
QCD, 388
non-minimal gauge theories, 473, 491
non-relativistic quantum mechanics. *See* NRQM
notation, 4
alternate for fields, 353
November revolution, 437
NRQM, 2
path integral approach, 111
nuclear democracy, 442
null diagrams, 486
$O(n)$, 12
observable, 51
on-shell renormalization. *See* renormalization, on-shell
operators
inverse, 124
on states vs fields, 46, 58a, 58b
order
of a group, 54
of a matrix, 54
of a tensor, 54
orthogonal
group. *See* $O(n)$
outer product, 18–21, 41, 421, 425
and eigenvalues, 421
quark field, 430
P transformations, 315, 322
derivatives, 325
four-currents, 324
scalars, 322
spinors, 323
symmetry, 317
vectors, 324
parameter
infinitesimal, 14
number in SM. *See* standard model, parameters list
parametrization, 13, 22, 25, 33
shortcut form, 34
summary, 34
parity transformation. *See P* transformations
particle
number, 165
number density, 165
number in the SM. *See* standard model, number of particles
Particle data book. *See* Particle Data Group
Particle Data Group, 428
particle masses. *See* mass, particles
particle zoo, 418
parton, 442, 499
path integral, 101, 110
action, 113
approach validated for QFT, 125
comparison to canonical, 114, 126
electroweak theory, 374, 379
evaluation of fermionic integral, 127
fermionic, 118
fields not operators, 112
graphical comparison to canonical, 115
ones we can evaluate, 118
overview in QFT, 110
proving equivalent to canonical, 119
QCD, 453
QED, 113
QFT, 1, 111
QFT from NRQM, 113
QFT postulates, 112
quantization, 2
quantization for QFT, 112
same as functional integral, 101
SM, overview, 508
transition amplitude, 113
Pauli matrices, 30
definition, 173
key indentities, 173
Pauli, Wolfgang, 346
Peccei-Quinn theory. *See* strong CP problem
perturbation theory
QCD. *See* QCD, perturbation theory
suitability for QCD, 384, 441, 461
photon
preview, 2
photon propagator. *See* propagator, photon
free, 84, 96
from Green function, 79
physics
its future, 513
PI. *See* path integral
pi meson, 430, 431
plots of quantum numbers, 432
quantum numbers summary, 432
PIA, 110, 114
pion. *See* pi meson
PMNS matrix, 228, 348
component values, 233
CP violation if complex, 350
Poincaré group, 22, 39
polarization
bases for, 297
sums, 294
vectors, 293
Politzer
David, 442
Pontecorvo-Maki-Nakagawa-Sakata. *See* PMNS matrix
Popov
Victor, 445
potential, 163
effective operator, 165
expectation value, 165
Goldstone model, 208
GSW model, 169, 221
Higgs model, 216
in terms of true vacuum fields, 210
Lagrangian terms, 163
minimum, 209, 216, 222
plotted vs fields, 164, 170
power counting, 369

primitive divergence, 464
  QCD, 466, 469, 471, 472, 491
  QED, 465, 469, 490
Proca equation, 107, 132, 155, 157, 372
projection operators
  Weyl/chiral field, 150
  Weyl/chiral field, properties, 150
propagator. *See* Green function for different equations
  ghost, 452
  gluon, 452
  Higgs, 279
  massive vector field, 293, 368, 372, 379
  photon, 279, 294
  quark, 452
  renormalization. See renormalization, particular propagator
  summary of boson types, 295
  W vector boson, 269
  Z vector boson, 279, 280
proper Feynman diagram. *See* Feynman diagram, proper vs improper, proper
proton
  substructure. *See* scattering, and proton substructure
pseudo-vector, 316
QCD. *See also*, strong interaction
  alternative matrices, 396
  binding nucleons, 438
  color, 41
  colorless composite states, 57
  compared to QED, 385
  history, 442
  inelastic interactions, 495
  interactions, 384, 387
  Lagrangian. *See* Lagrangian, QCD
  non-perturbative, 493, 503
  part of book found in, 3
  path integral. *See* path integral, QCD
  perturbation theory, 442, 452, 461, 463, 493
  primitive divergences. *See* primitive divergence, QCD
  renormalization. *See* renormalization, QCD
  symmetry. *See* symmetry, local QCD
  symmetry factor. *See* symmetry factor
  transition amplitude example, 394
QED
  Lagrangian. *See* Lagrangian, QED
  part of book found in, 3
  preview, 1
  primitive divergences. *See* primitive divergence, QED
  renormalization. See renormalization, QED
  review, 369
  symmetry. *See* symmetry, local QED review
QFT, 1
  canonical approach overview, 507
  in curved spacetime, 513
  PI approach overview, 508
quantization
  issue with spin 1 fields, 377, 380
quantum chromodynamics. *See* QCD
quantum electrodynamics. *See* QED
quantum field theory. *See* QFT
quantum gravity, 2
quark
  beauty. *See* bottom quark
  bottom. *See* bottom quark
  charm. *See* charm quark
  color eigenvalue plot, 56
  color states, 41
  confinement, 385, 441
  down. *See* down quark
  e/w theory, 188, 234
  generating functional. *See* generating functional, quark
  heavy, 437
  hypercharge. *See* hypercharge, quarks
  hypothetical free, 384, 385, 397
  left chiral, 188
  mass terms, 235
  naming, 442
  preview, 2
  q numbers for *u,d,s*, 434
  sample interactions, 237
  sea. *See* sea quark
  self-energy, 472
  strange. *See* strange quark
  top. *See* top quark
  truth. *See* top quark
  valence. *See* valence quark
  weak isospin charge. *See* weak isospin charge, quarks
raising
  operation, 34
  operator, 56
rank
  of a matrix, 54
  of a tensor, 54, 59
  of Lie algebra, 54
RC. *See* chirality, right
reducible representation, 55, 418, 419, 439
regularization, 476
  dimensional. See dimensional regularization
renormalization, 476
  and effective theories, 373, 463
  basic steps, 476
  charge. See charge, renormalization
  comparing on-shell and MS, 477
  coupling constant. See renormalization, QED, QCD, and e/w theory
  criterion for, 473, 491
  e/w theory, 246, 368, 379
  fermion propagator, 480
  gluon propagator, 488
  group equation, 372, 473, 484, 489, 491
  in different theories, 473
  mass. See mass, QED, See renormalization, QED and QCD
  massive vector fields, 370
  minimal subtraction, 368, 379, 464, 477, 478, 484
  MS. *See* renormalization, minimal subtraction
  negative coupling dimension, 475, 491
  on-shell, 369, 477, 478
  photon propagator, 476, 481
  QCD, 463, 485, 489
  QED, 476, 484
  quark propagator, 487
  summary overview, 511
  vertex, 488
rep. *See* representation of a group or algebra

representation
equivalent, 420
irreducible. *See* irreducible representation
reducible. *See* reducible representation
spinor. *See* spinor reps
representation of a group. *See* group, representation and Lie group, representation
representation of an algebra. *See* algebra, representation and Lie algebra, representation
representation space, 11
resonance, 283
RGE. *See* renormalization, group equation
RH. *See* helcity, right
right chiral neutrinos. *See* neutrino, right chiral
right chirality. *See* chirality, right
Rosenbluth formula, 497
RQM, 2
running coupling constants. *See* coupling constants, running
S operator, 70
for QED, 70
s,t,u channels. *See* Mandelstam variables
Salam
Abdus, 168, 265
scalar
complex characterization, 166
field, high energy, 168
field, real and imaginary parts, 163
general conclusions for, 168
Lagrangian, 167, 168
mass terms, 207
real characterization, 166
scalar boson, 162
scalar operation, 23
scaling symmetry. *See* symmetry, scaling
scattering, 265
and proton substructure, 499, 503
Bhabha, 287, 290
deep inelastic, 442, 493, 495, 499, 502
elastic, 493, 494, 495, 496
electron off nucleon, 494
electron off proton, 498, 502
electron-positron, 279, 289
electroweak, 278
energy regimes, 495, 502
extended target, 496
form factor. *See* form factor
high energy, 285, 494
immovable target, 496
inelastic, 494, 495, 498
interference term, 281
intermediate energy, 280, 282
low energy, 280, 282
Mott, 496
point-like target, 496
QED, 493
recoiling target, 496, 498
resonance regime, 283, 286
resonance, Higgs, 286
structure function. *See* structure function
summary, e/w, 289
screening, 439
sea quark, 439, 500, 503
see-saw mechanism, 347, 353
sign changes
fermion fields in amplitudes. *See* fermion, switching field order
similarity transformation. *See* transformation, similarity
single integral sign
short hand for multiple, 72
singlet, 43–48, 44, 50, 56
SM. *See* standard model
SO(2), 12, 22, 23, 25
2D rep, 44
characterize 2D rotation, 12, 22
doublet, 44
exponentialtion of Lie algebra, 24
generator, 25
parametrization, 25
representation, 12
singlet, 44
transformation, 12
SO(3), 13
characterize 3D rotation, 22
exponentiation of Lie algebra, 28
generators, 25
Lie algebra summary, 28
parametrization, 28
representation, 13, 25
singlet, 44
small parameters, 29
structure constants, 28
transformation, 13
triplet, 44
SO(3,1), 15, *See also* Lorentz transformation as a group
SO(*n*), 12, *See also* special orthogonal group
summary, 30
soft gluons, 501
soft radiation, 501
solar neutrino problem. *See* neutrino, solar problem
source
field, 92, 99, 120
Lagrangian, 92, 120
special
constraint on matrix, 58
special orthogonal group. *See also* *SO*(*n*)
definition, 12
special unitary group. *See also* *SU*(*n*)
definition, 15
focus of this book, 59
spin, 20, 22, 33, 46, *See also* group theory, angular momentum
addition, 423
eigenstates, 424
higher than 2, 376, 380
nucleon, 438
proton, 438
sums and averages, 270
spin addition
group theory. *See* group theory, spin addition
spin operator
Weyl rep. *See* Weyl rep, spin operator
spinor
4D, 55
boosts, 153, 158
inner product, 147
multiplication, 146-147
spinor propagator
free field, 99
spinor reps, 133, 157
summary, 147
spontaneous symmetry breaking. *See* symmetry, breaking

standard model
as direct product of groups, 19, 22, 40
beyond it, 513
experimental confirmation, overview, 511
Feynman rules. *See* Feynman rules, each branch of SM
Lagrangian. *See* Lagrangian, standard model, entire
number of particles, 508
overview, 506
overview of interactions,schematic, 508
parameters list, 509
part of book summarizing, 3, 507
preview, 1
unsolved problems, 512
standard rep, 132
gamma matrices, 133
standard representation. *See* fundamental representation, *See* standard rep
state
as a vector in Fock space, 330
strange quark, 433
strangeness, 433
related to flavor hypercharge, 434
strong coupling constant, 50
strong CP problem, 399, 411
experimental results, 412
Peccei-Quinn solution, 412
summary, 413
strong force. *See* strong interaction
strong force charge. *See* color charge
conservation, 41
strong interaction. *See also*, QCD
color and eigenvalues, 50
preview, 1, 2, 19
strong singlet. *See* color singlet
structure constants. *See* $SO(3)$, $SU(2)$ and $SU(3)$, structure constants
structure function, 498, 503
$SU(1)$, 15
$SU(2)$, 15
Cartan subalgebra, 48, 49
characterize 3D rotation, 22
doublet, 45, 172
exponentiation of Lie algebra, 31
generators, 30, 56
parametrization, 32
representation, 16, 34
singlet, 45
small parameters, 32
structure constants, 30
transformations, 40, 41
$SU(3)$, 16
Cartan subalgebra, 50
exponentiation of Lie algebra, 36
generators, 35, 56, 57
Lie algebra, 34
parametrization, 37
QCD, 385
representation, 16
singlet, 45, 50
small parameters, 17, 34, 36
structure constants, 35
transformations, 40, 41, 389
triplet, 46
$SU(n)$, 15
Cartan generators, 51
generators, 51
representation, 51
summary, 414
subgroup, 22, 23, 427, 439, *Also see* group, reducible vs irreducible
submatrix, 22, 419
subspace, 426
invariant. *See* invariant subspace
Sudbury Neutrino Observatory, 347
superficial degree of divergence. *See* superficial power counting
superficial divergence. *See* superficial power counting
superficial power counting, 466
non-minimal theories, 475
QCD, 471, 491
QED, 467, 490
Super-Kamiokande, 347
superstrings, 513
supersymmetry, 248, 513
SUSY. *See* supersymmetry
symmetry
all pervasive in Nature, 513
and mass terms, 190, 196, 203, 243, 308, 391, 402, 409
breaking and particle creation, 241
breaking apparent, not real, 242
breaking, summary of 3 models, 249
chiral, 409
chiral breaking, explicit, 410
chiral breaking, spontaneous, 410
chiral, summary, 416
conformal. *See* symmetry, scaling
discrete. *See* *C,P,T* transformations and transformation, discrete
explicit breaking, 242, 410
external, 39
flavor, 399, 407
flavor, summary, 416
gauge bosons in $SU(n)$, 402
Higgs type terms in $SU(n)$, 403
in various spaces, 43
internal, 39
lacking with mass terms, 190
local, electroweak, 178, 400
local, QCD, 384, 389, 397, 400
local, QED review, 178, 399
local, $SU(n)$, 400, 403
QFT, 39
quantum mechanics, 39
scaling, 367
$SU(n)$, general principles, 399, 414
under $SU(n)$, 38
symmetry factor, 459, 486
T transformations, 316, 331, 332
antiunitary, 336
derivatives, 336
example, 335
four-currents, 334
gamma matrices, 334
inner products, 334
scalars, 332
spinors, 332
vectors, 334
$\hat{T}$ transformations, 316, 325
derivatives, 327
four-currents, 326
scalars, 325
spinors, 325
symmetry, 317
vectors, 327
tadpole diagram, 364, 378

tangent space, 24
tau
  decay. *See* decay, tau
  lifetime, 277
tensor
  as group element, 58
tensor product, 19–21, 41, 59
  of vectors, 19
theta meson. *See* toponium
time transformation. *See* T and $\hat{T}$ transformations
top quark, 438
toponium, 438
total cross-section. *See* cross-section, total
total decay rate. *See* decay, rate, total
transformation
  2D rotation, 12
  3D rotation, 10
  bosons not really $U(1)$, 193
  continuous, 318
  coupled, 178
  discrete, 318, 399, 410
  finite, electroweak, 179
  finite, QCD, 389
  gauge, 39, *See* gauge, transformation
  infinitesimal, 167, 197
  infinitesimal, electroweak, 180, 181
  infinitesimal, QCD, 389, 397
  local vs global, 168
  Lorentz. *See* Lorentz transformation
  similarity, 419, 420
  symmetric, 167
  unitary. *See* unitary, transformation
transition amplitude
  comparing approaches, 2
  long vs short way, 265
  QCD example, 394
triplet, 44
trivial representation, 45
true vacuum, 209
$U(1)$, 15
  generator. *See* generator, $U(1)$
  representation, 22
$U(n)$, 15
uncertainty principle
  QFT, 375, 380
unification
  e/w. *See* GSW, model, and unification
unital algebra, 8
unitarity
  Hermitian generators, 52
unitary
  constraint on matrix, 57
  matrix, 57
  transformation, 15, 39
unitary algebra. *See* unital algebra
unitary gauge
  GSW model, 222
  Higgs model, 218
unitary group. *See* $U(n)$
units, 4
unsolved problems
  SM. *See* standard model, unsolved problems
upsilon meson. *See* bottomonium
vacuum
  definition of, 166, 511
  degeneracy, 212
  destroy. *See* destroy the vacuum
  eats charge, 243
  energy, 211
  false. *See* false vacuum
  true. *See* true vacuum
vacuum expectation value. *See* VEV
valence quark, 438, 500, 503
vector, 316
  axial, 316
  mass terms, 208
  true, 316
vector bosons
  electroweak, high vs low energy, 171
  issue with commutation relations, 377
vector current, 152
vector space, 53
  of fields, 329
  of states, 329
  overview, 8, 9
vertex factor
  electroweak, 266, 269, 291
  QCD, 454
vertex function, 83
VEV
  of potential, 210
virtual particles
  do they exist, 376, 380
W^-, 132, 172, 220
W^+, 132, 172, 220
W vector boson, 171
  decay. *See* decay, W and Z
  lifetime, 278
  propagator. *See* propagator, W vector boson
weak charge. *See* weak isospin charge
  table of fermions, 49
weak charge operator, 48
weak coupling constant, 49, 173
weak doublet, 45, 46, 48, *See* weak isospin doublet
weak force. *See* weak interaction
weak force charge, 48
  compared to other charges, 2
  conservation, 41
weak interaction, 132, *See also* electroweak:interaction
  preview, 1, 19
weak isospin, 174
  charge. *See* weak isospin charge
  charge conservation, 306
  charge operator, 306
  doublet, 174
  Higgs, 305
  leptons, 305
  operator, 305
  quarks, 306
  singlet, 174
weak isospin charge, 182, 305
  bosons, 309, 310
  leptons, 182, 196
  quarks, 189
  summary, 196
weak mixing angle. *See* Weinberg mixing angle
weak singlet, 45
weights. *See* group theory, weights
Weinberg
  Steven, 168
Weinberg mixing angle, 172, 220
Weyl equations, 138

Weyl fermion, 141
  chirality, 148
Weyl field, 149, *See* also, chiral field
  creation and destruction effects, 151
  same in any rep, 147
Weyl rep, 132, 136
  basis solutions, 136
  creation and destruction operators, 137, 140
  Dirac equation in, 138
  gamma matrices, 136
  highly relativistic case, 137
  kinematics vs dynamics, 140
  L and R fields, 139, 140
  spin operator, 139
  summary, 141
  transformation from standard rep, 136
  value of, 137
Weyl representation. *See* Weyl rep
Weyl spinors, 138
Wick's theorem, 453
Wilczek
  Frank, 442
Wilson
  Kenneth, 441
$W_i^{\mu\nu}$, 174
X, 19, 21
Yang-Mills fields, 191
Young tableau, 430
Yukawa
  couplings, 178, 233
  matrices, 178
  matrices, leptons, 225
  matrices, quarks, 235
Z. *See* generating functional
Z particle. *See* Z vector boson
Z vector boson, 132, 171
  decay. *See* decay, W and Z
  lifetime, 278
  propagator. *See* propagator, Z vector boson
Zo. *See* generating functional, free field
Zweig
  George, 442
Γ_d. *See* decay, rate, particular process
Ψ:lepton doublet, 173
γ^5, 149, *See* also, chirality operator
  matrix, 148, 160
  properties, 149
  relations, 160
θ_W. *See* Weinberg mixing angle